Mineral Physics—In Memory of Orson Anderson

Mineral Physics—In Memory of Orson Anderson

Special Issue Editor

Robert Cooper Liebermann

MDPI • Basel • Beijing • Wuhan • Barcelona • Belgrade

Special Issue Editor
Robert Cooper Liebermann
Stony Brook University
USA

Editorial Office
MDPI
St. Alban-Anlage 66
4052 Basel, Switzerland

This is a reprint of articles from the Special Issue published online in the open access journal *Minerals* (ISSN 2075-163X) from 2019 to 2020 (available at: https://www.mdpi.com/journal/minerals/special_issues/Mineral_Physics_Memory_Orson_Anderson)

For citation purposes, cite each article independently as indicated on the article page online and as indicated below:

LastName, A.A.; LastName, B.B.; LastName, C.C. Article Title. *Journal Name* **Year**, *Article Number*, Page Range.

ISBN 978-3-03936-541-8 (Hbk)
ISBN 978-3-03936-542-5 (PDF)

Cover image courtesy of Robert Cooper Liebermann.

Contents

About the Special Issue Editor xi

Preface to "Mineral Physics—In Memory of Orson Anderson" xiii

Robert Cooper Liebermann
The Orson Anderson Era of Mineral Physics at Lamont in the 1960s
Reprinted from: *Minerals* **2019**, *9*, 342, doi:10.3390/min9060342 1

William A. Bassett
The Takahashi–Bassett Era of Mineral Physics at Rochester in the 1960s
Reprinted from: *Minerals* **2020**, *10*, 344, doi:10.3390/min10040344 19

Robert Cooper Liebermann
The Birth of Mineral Physics at the ANU in the 1970s
Reprinted from: *Minerals* **2020**, *10*, 163, doi:10.3390/min10020163 28

Robert Cooper Liebermann
My Career as a Mineral Physicist at Stony Brook: 1976–2019
Reprinted from: *Minerals* **2019**, *9*, 761, doi:10.3390/min9120761 36

Hans J. Mueller and Heiner Vollstädt
Complex Geophysical Investigations under Extreme P,T–Conditions at Zentralinstitut für Physik der Erde (ZIPE) (1970–1990)
Reprinted from: *Minerals* **2020**, *10*, 412, doi:10.3390/min10050412 56

William A. Bassett
Cornell's Role in Developing Synchrotron Radiation for Mineral Physics
Reprinted from: *Minerals* **2020**, *10*, 514, doi:10.3390/min10060514 78

Chaowen Xu and Toru Inoue
Phase Relations in MAFSH System up to 21 GPa: Implications for Water Cycles in Martian Interior
Reprinted from: *Minerals* **2019**, *9*, 559, doi:10.3390/min9090559 86

Masaki Akaogi, Taisuke Tajima, Masaki Okano and Hiroshi Kojitani
High-Pressure and High-Temperature Phase Transitions in Fe_2TiO_4 and Mg_2TiO_4 with Implications for Titanomagnetite Inclusions in Superdeep Diamonds
Reprinted from: *Minerals* **2019**, *9*, 614, doi:10.3390/min9100614 95

Rongrong Lin and Leon Thomsen
Validation of Digital Rock Physics Algorithms
Reprinted from: *Minerals* **2019**, *9*, 669, doi:10.3390/min9110669 107

Agnès Dewaele
Equations of State of Simple Solids (Including Pb, NaCl and LiF) Compressed in Helium or Neon in the Mbar Range
Reprinted from: *Minerals* **2019**, *9*, 684, doi:10.3390/min9110684 116

Wade Mans, Jin S. Zhang, Ming Hao, Joseph R. Smyth, Dongzhou Zhang, Gregory J. Finkelstein and Przemyslaw Dera
Hydrogen Effect on the Sound Velocities of Upper Mantle Omphacite
Reprinted from: *Minerals* **2019**, *9*, 690, doi:10.3390/min9110690 124

Jan V.M. Borgomano, Jérôme Fortin and Yves Guéguen
Cracked, Porous Rocks and Fluids: Moon and Earth Paradox
Reprinted from: *Minerals* **2019**, *9*, 693, doi:10.3390/min9110693 . . . . **133**

Jing Yang, Yingwei Fei, Xiaojun Hu, Eran Greenberg and Vitali B. Prakapenka
Effect of Carbon on the Volume of Solid Iron at High Pressure: Implications for Carbon Substitution in Iron Structures and Carbon Content in the Earth's Inner Core
Reprinted from: *Minerals* **2019**, *9*, 720, doi:10.3390/min9120720 . . . . **142**

Ian Jackson
Viscoelastic Behaviour from Complementary Forced-Oscillation and Microcreep Tests
Reprinted from: *Minerals* **2019**, *9*, 721, doi:10.3390/min9120721 . . . . **157**

Suguru Takahashi, Eiji Ohtani, Daijo Ikuta, Seiji Kamada, Tatsuya Sakamaki, Naohisa Hirao and Yasuo Ohishi
Thermal Equation of State of Fe_3C to 327 GPa and Carbon in the Core
Reprinted from: *Minerals* **2019**, *9*, 744, doi:10.3390/min9120744 . . . . **170**

Vasilije V. Dobrosavljevic, Wolfgang Sturhahn and Jennifer M. Jackson
Evaluating the Role of Iron-Rich (Mg,Fe)O in Ultralow Velocity Zones
Reprinted from: *Minerals* **2019**, *9*, 762, doi:10.3390/min9120762 . . . . **179**

Karine Gouriet, Philippe Carrez and Patrick Cordier
Ultimate Mechanical Properties of Forsterite
Reprinted from: *Minerals* **2019**, *9*, 787, doi:10.3390/min9120787 . . . . **191**

Hiroki Ichikawa and Taku Tsuchiya
Ab Initio Thermoelasticity of Liquid Iron-Nickel-Light Element Alloys
Reprinted from: *Minerals* **2020**, *10*, 59, doi:10.3390/min10010059 . . . . **201**

Britany L. Kulka, Jonathan D. Dolinschi, Kurt D. Leinenweber, Vitali B. Prakapenka, and Sang-Heon Shim
The Bridgmanite–Akimotoite–Majorite Triple Point Determined in Large Volume Press and Laser-Heated Diamond Anvil Cell
Reprinted from: *Minerals* **2020**, *10*, 67, doi:10.3390/min10010067 . . . . **212**

Florian Tian-Siang Hua, Przemyslaw Dera and Jennifer Kung
Compressional Behavior of Hydrous Orthoenstatite: Insight into the Nature of LVZ under Continental Plate
Reprinted from: *Minerals* **2020**, *10*, 71, doi:10.3390/min10010071 . . . . **224**

Oliver Tschauner, Chi Ma, Antonio Lanzirotti and Matthew G. Newville
Riesite, a New High Pressure Polymorph of TiO_2 from the Ries Impact Structure
Reprinted from: *Minerals* **2020**, *10*, 78, doi:10.3390/min10010078 . . . . **239**

Satoru Urakawa, Toru Inoue, Takanori Hattori, Asami Sano-Furukawa, Shinji Kohara, Daisuke Wakabayashi, Tomoko Sato, Nobumasa Funamori and Ken-ichi Funakoshi
X-ray and Neutron Study on the Structure of Hydrous SiO_2 Glass up to 10 GPa
Reprinted from: *Minerals* **2020**, *10*, 84, doi:10.3390/min10010084 . . . . **247**

Eric Edmund, Francesca Miozzi, Guillaume Morard, Eglantine Boulard, Alisha Clark, Frédéric Decremps, Gaston Garbarino, Volodymyr Svitlyk, Mohamed Mezouar and Daniele Antonangeli
Axial Compressibility and Thermal Equation of State of Hcp Fe–5wt% Ni–5wt% Si
Reprinted from: *Minerals* **2020**, *10*, 98, doi:10.3390/min10020098 . . . . **260**

Ruilian Tang, Jiuhua Chen, Qiaoshi Zeng, Yan Li, Xue Liang, Bin Yang and Yu Wang
Study on the High-Pressure Behavior of Goethite up to 32 GPa Using X-Ray Diffraction, Raman, and Electrical Impedance Spectroscopy
Reprinted from: *Minerals* **2020**, *10*, 99, doi:10.3390/min10020099 . . . . **272**

Francesca Miozzi, Jan Matas, Nicolas Guignot, James Badro, Julien Siebert and Guillaume Fiquet
A New Reference for the Thermal Equation of State of Iron
Reprinted from: *Minerals* **2020**, *10*, 100, doi:10.3390/min10020100 . . . . **285**

Tyler Perez, Gregory J. Finkelstein, Olivia Pardo, Natalia V. Solomatova and Jennifer M. Jackson
A Synchrotron Mössbauer Spectroscopy Study of a Hydrated Iron-Sulfate at High Pressures
Reprinted from: *Minerals* **2020**, *10*, 146, doi:10.3390/min10020146 . . . . **298**

Jennifer Girard, Reynold E. Silber, Anwar Mohiuddin, Haiyan Chen and Shun-ichiro Karato
Development of a Stress Sensor for In-Situ High-Pressure Deformation Experiments Using Radial X-Ray Diffraction
Reprinted from: *Minerals* **2020**, *10*, 166, doi:10.3390/min10020166 . . . . **309**

Camelia Veronica Stan, Earl Francis O'Bannon III, Pavel Mukhin, Nobumichi Tamura and Larissa Dobrzhinetskaya
X-ray Laue Microdiffraction and Raman Spectroscopic Investigation of Natural Silicon and Moissanite
Reprinted from: *Minerals* **2020**, *10*, 204, doi:10.3390/min10030204 . . . . **320**

Gabriel D. Gwanmesia, Matthew L. Whitaker, Lidong Dai, Alwin James, Haiyan Chen, Richard S. Triplett and Nao Cai
The Elastic Properties of β-Mg_2SiO_4 Containing 0.73 wt.% of H_2O to 10 GPa and 600 K by Ultrasonic Interferometry with Synchrotron X-Radiation
Reprinted from: *Minerals* **2020**, *10*, 209, doi:10.3390/min10030209 . . . . **336**

Richard L. Rowland II, Barbara Lavina, Kathleen E. Vander Kaaden, Lisa R. Danielson and Pamela C. Burnley
Thermal Analysis, Compressibility, and Decomposition of Synthetic Bastnäsite-(La) to Lanthanum Oxyfluoride
Reprinted from: *Minerals* **2020**, *10*, 212, doi:10.3390/min10030212 . . . . **357**

Eric Edmund, Michel Gauthier, Daniele Antonangeli, Simon Ayrinhac, Silvia Boccato, Thibault Deletang, Marc Morand, Yiuri Garino, Paraskevas Parisiades and Frédéric Decremps
Picosecond Acoustics Technique to Measure the Sound Velocities of Fe-Si Alloys and Si Single-Crystals at High Pressure
Reprinted from: *Minerals* **2020**, *10*, 214, doi:10.3390/min10030214 . . . . **373**

Dongyuan Zhou, Junjie Dong, Yanhan Si, Feng Zhu and Jie Li
Melting Curve of Potassium Chloride from in situ Ionic Conduction Measurements
Reprinted from: *Minerals* **2020**, *10*, 250, doi:10.3390/min10030250 . . . . **387**

Steeve Gréaux, Youmo Zhou, Yoshio Kono, Akihiro Yamada, Yuji Higo and Tetsuo Irifune
Thermoelastic Properties of $K_{0.7}Na_{0.3}AlSi_3O_8$ Hollandite and $NaAlSi_2O_6$ Jadeite: Implication for the Fate of the Subducted Continental Crust in the Deep Mantle
Reprinted from: *Minerals* **2020**, *10*, 261, doi:10.3390/min10030261 . . . . **400**

Ayaka Saito, Hiroyuki Kagi, Shiho Marugata, Kazuki Komatsu, Daisuke Enomoto, Koji Maruyama and Jun Kawano
Incorporation of Incompatible Strontium and Barium Ions into Calcite ($CaCO_3$) through Amorphous Calcium Carbonate
Reprinted from: *Minerals* **2020**, *10*, 270, doi:10.3390/min10030270 . . . . **414**

Sasithorn Chornkrathok, Dongzhou Zhang and Przemyslaw Dera
Structure and Behavior of the Ni End-Member Schreibersite Ni_3P under Compression to 50 GPa
Reprinted from: *Minerals* **2020**, *10*, 306, doi:10.3390/min10040306 . . . . **427**

Nancy L. Ross, Theresa A. Detrie and Zhenxian Liu
High-Pressure Raman and Infrared Spectroscopic Study of Prehnite
Reprinted from: *Minerals* **2020**, *10*, 312, doi:10.3390/min10040312 . . . . **441**

Wendy R. Panero and Razvan Caracas
Stability and Solid Solutions of Hydrous Alumino-Silicates in the Earth's Mantle
Reprinted from: *Minerals* **2020**, *10*, 330, doi:10.3390/min10040330 . . . . **448**

Brent Delbridge and Miaki Ishii
Seismic Wave Speeds Derived from Nuclear Resonant Inelastic X-Ray Scattering for Comparison with Seismological Observations
Reprinted from: *Minerals* **2020**, *10*, 331, doi:10.3390/min10040331 . . . . **460**

Yingzhe Li and Jay D. Bass
Single Crystal Elastic Properties of Hemimorphite, a Novel Hydrous Silicate
Reprinted from: *Minerals* **2020**, *10*, 425, doi:10.3390/min10050425 . . . . **479**

Murli H. Manghnani, Quentin Williams, Teruyuki Matsui, Peter C. Schultz and Charles R. Kurkjian
Effects of Composition, Pressure, and Temperature on the Elastic Properties of SiO_2–TiO_2 Glasses: An Integrated Ultrasonic and Brillouin Study
Reprinted from: *Minerals* **2020**, *10*, 481, doi:10.3390/min10050481 . . . . **488**

Ross J. Angel, Francesca Miozzi and Matteo Alvaro
Limits to the Validity of Thermal-Pressure Equations of State
Reprinted from: *Minerals* **2019**, *9*, 562, doi:10.3390/min9090562 . . . . **500**

Frank D. Stacey
Equations of State for the Deep Earth: Some Fundamental Considerations
Reprinted from: *Minerals* **2019**, *9*, 636, doi:10.3390/min9100636 . . . . **511**

Tomoo Katsura and Yoshinori Tange
A Simple Derivation of the Birch–Murnaghan Equations of State (EOSs) and Comparison with EOSs Derived from Other Definitions of Finite Strain
Reprinted from: *Minerals* **2019**, *9*, 745, doi:10.3390/min9120745 . . . . **520**

Jun Tsuchiya, Risa Nishida and Taku Tsuchiya
First Principles Calculation of the Stability of Iron Bearing Carbonates at High Pressure Conditions
Reprinted from: *Minerals* **2020**, *10*, 54, doi:10.3390/min10010054 . . . . **539**

J. Michael Brown and Baptiste Journaux
Local-Basis-Function Equation of State for Ice VII–X to 450 GPa at 300 K
Reprinted from: *Minerals* **2020**, *10*, 92, doi:10.3390/min10020092 . . . . **551**

Tony Yu, Clemens Prescher, Young Jay Ryu, Feng Shi, Eran Greenberg, Vitali Prakapenka, Peter Eng, Joanne Stubbs, Yoshio Kono, Guoyin Shen, Heather Watson, Mark L. Rivers, Stephen R. Sutton and Yanbin Wang
A Paris-Edinburgh Cell for High-Pressure and High-Temperature Structure Studies on Silicate Liquids Using Monochromatic Synchrotron Radiation
Reprinted from: *Minerals* **2019**, *9*, 715, doi:10.3390/min9110715 . . . . **575**

Yingxin Liu, Haijian Li, Xiaojing Lai, Feng Zhu, Robert P. Rapp and Bin Chen
Casting Octahedra for Reproducible Multi-Anvil Experiments by 3D-Printed Molds
Reprinted from: *Minerals* **2020**, *10*, 4, doi:10.3390/min10010004 . . . . **589**

Zhicheng Jing, Tony Yu, Man Xu, Julien Chantel and Yanbin Wang
High-Pressure Sound Velocity Measurements of Liquids Using In Situ Ultrasonic Techniques in a Multianvil Apparatus
Reprinted from: *Minerals* **2020**, *10*, 126, doi:10.3390/min10020126 . . . . **599**

About the Special Issue Editor

Robert Cooper Liebermann As an undergraduate student at Caltech in the early 1960s, I spent most of my time playing intercollegiate football and indulging in student body politics. However, in my senior year, I was fortunate to take a seminar reading course in geophysics led by Don Anderson, who introduced me to studies of the deep interior of the Earth.

In my early days as a seismology graduate student at Columbia University, I conducted research at the Lamont Geological Observatory on techniques to study the relative excitation of surface waves by earthquakes and underground nuclear explosions. In 1965, Orson Anderson joined the geology faculty at Columbia and was looking for prospective graduate students. I began to work with Orson and his research colleague Ed Schreiber, using the techniques of ultrasonic interferometry to measure sound velocities in minerals at elevated pressures and temperatures (in those days, to 0.7 GPa and 200 °C); the research activities of their laboratory are described in "The Orson Anderson Era of Mineral Physics at Lamont in the 1960s" in this volume. I have continued to use such techniques for the remainder of my career at the Australian National University (see "The Birth of Mineral Physics at the ANU in the 1970s") and at Stony Brook University (see "My Career as a Mineral Physicist at Stony Brook: 1976–2019" in this volume); see also Liebermann, R. C., "The Role of Serendipity in My Career in Mineral Physics: 1968 to 2013", *Phys. Earth Planet. Interiors* **2014**, *228*, 307–323.

In 1976, I took up a faculty position at the Department of Geosciences at Stony Brook University. Over the next half a century, in collaboration with graduate students from the U. S., China and Russia, and postdoctoral colleagues from Australia, France and Japan, I pursued studies related to the elastic properties of minerals (and their structural analogues) at high pressures and temperatures. In the 1980s, together with Donald Weidner (Director of the Mineral Physics Institute), I established the Stony Brook High Pressure Laboratory, and established the first modern multi-anvil, high-pressure laboratory in North America, using apparatus imported from Japan. With these facilities, we were able to achieve pressure in excess of 20 GPa and simultaneous temperatures above 2000 °C (see also Liebermann, R. C., "Multi-anvil, high-pressure apparatus: A half century of development and progress", *High Pressure Research* **2011**, *31*, 493–532). In 1991, in collaboration with Alexandra Navrotsky at Princeton University and Charles Prewitt at the Geophysical Laboratory, I founded the NSF Science and Technology Center for High Pressure Research. In 2003, I served as President of COMPRES: Consortium for Mineral Physics Research in Earth Sciences. I formally retired in 2014 and have spent the ensuing years watching over the research projects of my Stony Brook colleague Professor Baosheng Li and writing articles on the history of mineral physics.

In collaboration with Professor Lars Ehm at Stony Brook and Gabriel Gwanmesia from Delaware State University, I have created a diversity program called "A Career Path for African-American Students from Historically Black Colleges and Universities to National Laboratories". This initiative teaches marketable skills, places students in internships, and fosters the professional career tracks of underrepresented minorities. It has been supported by a new program called "Opportunities for Enhancing Diversity in the Geosciences of the NSF" (EOS, *Earth and Space Sciences News* **2016**, *97*, 9–11).

Preface to "Mineral Physics—In Memory of Orson Anderson"

ORSON LAMAR ANDERSON

December 3, 1924–June 19, 2019

Orson Anderson was born in December 1924 in Price, Utah. He completed all of his college education at the University of Utah. His PhD advisors were Henry Eyring and Walter Elsasser. Once he graduated, he worked at Bell Telephone Laboratories and was introduced to physical acoustics by Warren Mason and Harold McSkimin. After a few years at American Standard Company, he joined the Lamont Geological Observatory of Columbia University in 1963 and later became a Professor at the Department of Geology. In 1971, he moved to the University of California, Los Angeles (UCLA), where he remained until his retirement in 2002. During his years at UCLA, he served as Director of the Institute of Geophysics and Planetary Physics (IGPP).

In early 2019, I wrote a paper entitled "The Orson Anderson Era of Mineral Physics at Lamont in the 1960s", and began to explore options for its publication. When the Assistant Editor for *Minerals*, Ms. Jingjing Yang, agreed to consider my paper, she also inquired as to whether I would like to the be the Guest Editor for a Special Issue in honor of Orson Anderson. After asking prospective authors about the viability of such a Special Issue, I accepted her invitation, with the hope and expectation that it would be a wonderful present for his 95th birthday. This Special Issue is the result. It contains original scientific papers, as well as historical reviews of the field of mineral physics (and also rock physics). Mineral physics is the study of mineralogical problems through the application of condensed matter physics. In reality, mineral physicists use not only physics, but also solid-state chemistry; they study not only minerals, but all materials related to natural minerals (e.g., structural analogs, but also glasses, melts and fluids). Mineral and rock physics are intimately connected to many other geoscience disciplines, including seismology, planetary science, petrology, geochemistry, geomagnetism, and geodynamics, and even materials and climate science (see the Williams–Lattimore diagram below; copyright R. C. Liebermann).

The papers in this Special Issue are grouped into four categories: Reviews, Experimental Science, Theoretical Science and Technological Developments. These papers include those from; first, authors covering five generations of mineral physicists, including contemporaries of Orson (e.g., William Bassett, Frank Stacey), the next generation of leaders in mineral physics throughout the world (e.g., Michael Brown, Eiji Ohtani), current leaders in the field (e.g., Agnes Dewaele, Jun Tsuchiya), senior graduate students (e.g., Jan Borgonmano, Vasilije Dobrosavlijevic, Francesca Miozzi), and an undergraduate student (i.e., Tyler Perez). Note that Tyler, a student of Jennifer Jackson at Caltech, is an academic great-great grandson of Orson Anderson (Anderson > Liebermann > Bass > Jackson > Perez).

Regrettably, Orson died in June 2019, before we could complete and publish this Special Issue, and thus, it will be published "In Memory of Orson Anderson." However, his daughter Bonnie reported that she was able to read my paper on the Orson Anderson era at Lamont to him before he died.

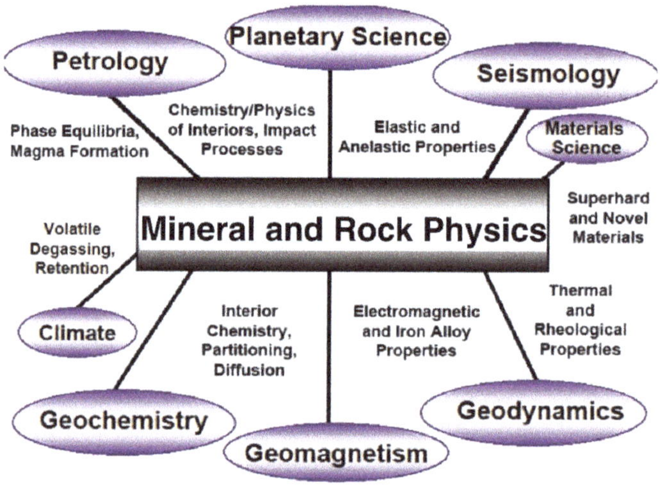

Corrigendum: The woman in the photo in Figure 1 of my paper on "The Birth of Mineral Physics at the ANU in the 1970s" was not Mrs. Akimoto, the wife of Professor Syun-iti Akimoto of the University of Tokyo. The woman is, in fact, Nobuko Yagi, the wife of Professor Kenzo Yagi of Hokkaido University and the mother of Takehiko Yagi of the University of Tokyo. I apologize to the Yagi family for this oversight.

Robert Cooper Liebermann
Special Issue Editor

minerals

MDPI

Review

The Orson Anderson Era of Mineral Physics at Lamont in the 1960s

Robert Cooper Liebermann

Mineral Physics Institute and Department of Geosciences, Stony Brook University, Stony Brook, NY 11794-2100, USA; Robert.Liebermann@stonybrook.edu; Tel.: 1-(631)-632-1968

Received: 30 April 2019; Accepted: 28 May 2019; Published: 4 June 2019

Abstract: From 1964 to the early 1970s, Orson Anderson led a research program at the Lamont Geological Observatory in the newly-emerging field of "mineral physics". In collaboration with colleagues Edward Schreiber and Naohiro Soga, Orson exploited the techniques of physical acoustics to study the behavior of the sound velocities of minerals at elevated pressures and temperatures. This research program also included semi-empirical studies of the relationships between the bulk modulus and the molar volume of solids, the use of lattice dynamics to calculate the elastic moduli of cubic structures as a function of pressure to predict instabilities, and theoretical investigations of the Lagrangian and Eulerian formulations of finite strain equations of state.

Keywords: mineral physics; ultrasonic interferometry; resonant ultrasound spectroscopy; law of corresponding states; equations-of-state; Columbia University; American Geophysical Union

1. Background on Mineral Physics

According to Robert Hazen [1], "mineral physics is the study of mineralogical problems through the application of condensed matter physics". In reality, mineral physicists use not only physics but also solid-state chemistry. In addition, they study not only minerals but all materials related to natural minerals (e.g., structural analogs, but also glasses, melts, and fluids). The father of mineral physics is widely accepted to be Francis Birch, whose seminal paper in 1952 summarized the state of knowledge of the elasticity and constitution of the Earth's interior at that time [2]; however, Birch did not use the name "mineral physics". Orson often cited the German physicist Eduard Grüneisen as the person who first introduced the term mineral physics [3]. Over the subsequent two decades, many laboratories began to conduct experiments on the physical properties of minerals at high pressures and temperatures. Among these were Orson Anderson at Lamont (and later at the University of California Los Angeles/UCLA), John Jamieson at the University of Chicago, Alvin Van Valkenburg at the National Bureau of Standards, William Bassett and Taro Takahashi at the University of Rochester, Ho-kwang Mao and Peter Bell at the Carnegie Institution of Washington, Gene Simmons at the Massachusetts Institute of Technology, Mineo Kumazawa at Nagoya University, Thomas Ahrens at the California Institute of Technology, Gerhard Barsch (and later Earl Graham) at the Pennsylvania State University, Hartmut Spetzler at the University of Colorado, Francis Birch (and later Richard O'Connell) at Harvard University, Murli Manghnani at the University of Hawaii, Syun-iti Akimoto at the University of Tokyo, Jean-Paul Poirier at the Institut de Physique du Globe in Paris, and Robert Liebermann at the Australian National University (and later at Stony Brook University).

2. The Mineral Physics Laboratory at Lamont

The Mineral Physics Laboratory at the Lamont Geological Observatory was established in the early 1960s by Orson Anderson. From 1965 to 1970, I was a graduate student, later postdoc, in that laboratory (Figure 1). The story of how the lab was founded is best told in Orson's own words.

Figure 1. In 1976, Orson gave one of the invited talks at a workshop on seismic anisotropy in the Castle of Liblice in Czechoslovakia, convened by Vladislav Babuska, Jaroslava Plomerova, and I.

In the following paragraphs, I shall quote from a correspondence written in May 2008 [4] by Anderson to Michael Purdy (Director of Lamont) and to the author in 2008 [5] and 2014 [6], with the permission of Orson (whose words will be highlighted in italics throughout this paper).

I [Anderson] was a physicist at Bell Labs for 10 years. I received my PhD in physics at the University of Utah under the supervision of Henry Eyring and Walter Elsasser in the field of physical acoustics in 1951 and was snapped up by Bell Telephone Laboratories, who needed physicists trained in acoustics but found them sparse in the graduate lists of the country. Transistor physics was blooming at the time, but the telephone industries still ran on electrical mechanical devices, and physical acoustics is the mainstay of the research needed for such devices. I prospered in my science career at Bell Labs under the tutelage of Warren Mason and Hal McSkimin.

It was a custom at Bell Labs that an important visitor be accompanied for his entire visit by a host who is a young scientist. A Bell Vice President, William Baker, asked me to be the host of Lamont Director Maurice Ewing during his one-day visit in 1962. At the luncheon, I was seated between Baker and Ewing. Ewing said that the most significant lab to him that he visited was the lab of his host. Ewing said that one of Lamont's perplexing problems was to determine the thickness of sediments over the basement rocks. The analysis of the sonar signal determined the lapsed time, not the thickness. The primary difficulty is that the velocity of sound through the sediment layer is often not known. Ewing thought that they needed the kind of lab that Anderson had and the technical support that went with it. So, he boldly asked my VP if Lamont could borrow me to establish a physical acoustics lab at Lamont. Vice President Baker said yes you could have him two or three days a week. So, it was settled that I was to build a lab at Lamont equipped to handle the sediment problem and to find and hire the scientists necessary to run the lab. I felt a little bit like I was in the slave market, so I spoke up and said I will agree to do that if Lamont will take me on some of their ocean research expeditions.

I hired two postdocs on behalf of Lamont to run the new lab: Edward Schreiber from Alfred University in 1963 and Naohiro Soga from the Research Division of American Standard in 1964 (where he had been working for two years after finishing graduate school at Kyoto University). Both had PhDs in ceramics and industrial chemistry, and I trained them in physical acoustics at my lab at Bell Labs. Schreiber took charge of the high-pressure experiments and Soga the high-temperature experiments. By 1965, Soga and Schreiber, as postdocs, were producing good relevant papers at Lamont, and I had been seduced away from Bell Labs and was now an adjunct professor of geology at Columbia and a full-time employee at Lamont. My research grants paid my salary.

In 1967, Professor Charles Drake of Columbia, the Chairman of the Department of Geology, had me appointed a professor of Geology at Columbia University [an appointment which Orson's graduate students had been urging for sometime]. I was asked to teach mineralogy. Much later on, Drake told me that Professor Rhodes Fairbridge of Columbia said that, in view of my record in teaching minerals, the biggest mistake that the Department ever made was to appoint me a professor in it. I hasten to add that Drake did not agree with Fairbridges' opinion. In addition, Orson cited Max Born, Lawrence Bragg, Peter Debye, Peter Paul Ewald, Andre Guinier, and Max von Laue as physicists who had made major contributions to mineralogy.

To learn geophysics, Orson became editor of the Journal of Geophysical Research. It was during that period in which all of the basic plate tectonic papers were published under his editorship.

Even though laboratories in this field of experimental mineralogy arose simultaneously in the two decades after Birch's 1952 paper, the field did not have a name. The name "mineral physics" first appeared at Lamont. Director Ewing [known affectionately as "Doc" Ewing by all staff] observed to me that my lab was the only one at Lamont without a name; so, I pondered what the name should be and thought it should have physics in it because we were using the tools of physical acoustics, and it ought to be related to geological sciences, so the word mineral seemed an appropriate adjective. Thus, a sign labeled "Mineral Physics" appeared on the door of our lab in an old shed, which was located just north of the new seismology building. As far as I can determine, the words "mineral physics" were not used before the sign was printed and posted. A photograph of the sign, which was made by Ed Schreiber, is shown in Figure 2 [7].

Figure 2. "Mineral Physics" sign from Lamont in 1965. The name proposed by Orson Anderson. The sign and photo were made by Edward Schreiber [7].

This is why Orson maintains that mineral physics was born at Lamont in the year 1964 [5]. The name "mineral physics" was subsequently adopted by the All-Union Committee of the American Geophysical Union (AGU), for which Orson served at the founding chair. The name has become part of the vocabulary of researchers around the world; by 2007, there were more than 30 mineral physics laboratories in academic and governmental institutions in the U.S.

The research program in Orson's laboratory at Lamont included experimental, semi-empirical, and theoretical studies. The principal experimental technique was the determination of sound velocities using ultrasonic interferometry (including both pulse–echo overlap and pulse superposition versions) to a pressure of 2 kbar (0.2 GPa) and a temperature of 200 °C. Minerals investigated included polycrystalline aggregates or single crystals of MgO-periclase [8], Al_2O_3-corundum [9], Mg_2SiO_4-forsterite [10], $MgAl_2O_4$-spinel [11], zincite-ZnO [12], garnet [13], alpha-SiO_2 [14], CaO [15], Fe_2O_3-hematite [16], $NiFe_2O_4$-ferrite [17], and $(Mg,Fe)_2SiO_4$-olivine [18]. Advances in ceramic technology made available high-quality polycrystalline specimens suitable for high-frequency ultrasonic experiments [i.e., low porosity, small grain size, free of cracks, and densities close to theoretical values derived from X-ray diffraction]. Most of these studies were led by Ed Schreiber and Naohiro Soga, along with visiting scientists (Mineo Kumazawa and Murli Manghnani) and graduate student Robert Liebermann. For a summary of these ultrasonic studies, see Figure 3 and the review paper in Reviews of Geophysics [19]; see also [20]. *In Anderson's 1995 book "Equations of State of Solids for Geophysics and Ceramic Science" [21], he states: Ed and I showed the community that precision ultrasonic measurements of sound velocity done under pressure [modest indeed by today's standards] could yield valid equations of state in the shock wave pressure regime. Naohiro and I demonstrated that high-temperature elasticity obeyed certain fundamental laws in solid-state physics first presented by Max Born and Eduard Grüneisen in the 1920s and 1930s.* Their paper [22] was selected as the outstanding contribution to ceramic literature in 1966 and brought them the Ross Coffin Purdy Award of the American Ceramic Society in 1968.

TABLE 1. Physical Properties

Property (at P=1 bar, T=298 °K)		Units	Corundum α-Al_2O_3 Poly'xl	Periclase MgO Poly'xl	Periclase MgO Single xl	Spinel $MgAl_2O_4$ Single xl	Forsterite Mg_2SiO_4 Poly'xl	Garnet Al-Py Single xl	Hematite α-Fe_2O_3 Poly'xl	Quartz α-SiO_2 Single xl	Lime CaO Poly'xl	Zincite ZnO Poly'xl	Bromellite BeO Poly'xl
Density theoretical	ρ	g/cc	3.988[10]	3.584[1]	3.584[1]		3.214[1]		5.274[1]	2.648[1]	3.345[1]	5.675[1]	3.008[1]
bulk		g/cc	3.972[2]	3.5797[5]	3.5833[9]	3.6193[10]	3.021[12]	4.160[16]	5.2539[17]	2.6485[20]	3.285[24]	5.624[27]	3.000[30]
Mean atomic weight	$\bar{M}$	g	20.39	20.16	20.16	20.36[10]	20.10	23.79[16]	31.94	20.03	28.04	40.68	12.51
Thermal expansivity	α	/°K x10⁶	16.3[3]	31.5[6]	31.5[6]	16.2[11]	24.0[13]	21.6[16]	32.9[18]	36.6[20,21]	28.1[25]	15.0[28]	17.7[31]
Specific heat	C_p	erg/g/°K x10⁶	7.83[4]	9.25[7]	9.25[7]	8.03[4]	8.38[4]	7.61[16]	6.50[19]	7.41[22]	7.64[26]	4.57[29]	10.22[32]
Compressional velocity	v_p	km/sec	10.845[2]	9.661[5]	9.692	9.914	7.586[12]	8.531[16]	7.901[17]	6.047[23]	7.945[24]	5.939[27]	12.05[30]
Shear velocity	v_s	km/sec	6.373[2]	5.997[5]	6.041	5.645	4.359[12]	4.762[16]	4.163[17]	4.092[23]	4.812[24]	2.803[27]	7.345[30]
Bulk modulus	K_S	kb	2521[2]	1624[5]	1622[9]	2020[10]	973[12]	1770[16]	2066[17]	377[23]	1059[24]	1394[27]	2201[30]
	K_T	kb	2505[2]	1601	1599[9]	2009	967[12]	1757[16]	2027[17]	374	1049[24]	1389[27]	2186[30]
Shear modulus	μ	kb	1613[2]	1288[5]	1308	1153[10]	574	943[16]	910[17]	444	761[24]	442[27]	1618[30]
Poisson's ratio	σ_S		0.236[2]	0.186[5]	0.182	0.260[10]	0.254[12]	0.274[16]	0.308[17]	0.077	0.210[24]	0.357[27]	0.205[30]
Pressure derivatives													
$(\partial v_p/\partial P)_T$		km/sec/kb x10³	5.18[2]	8.66[5]	8.35	4.9	10.3[12]	7.84[16]	4.67[17]	13.7[23]	10.4[24]	3.64[27]	6.48[30]
$(\partial v_s/\partial P)_T$		km/sec/kb x10³	2.21[2]	4.23[5]	4.02	0.43	2.45[14]	2.17[16]	0.63[17]	-3.4[23]	2.9[24]	-3.19[27]	0.33[30]
$(\partial K_S/\partial P)_T$			3.98[2]	4.58[5]	4.50[9]	4.18[10]	4.87	5.43[16]	4.53[17]	6.4[23]	5.23[24]	4.78[27]	5.52[30]
$(\partial K_T/\partial P)_T$			3.99[2]	(4.6)	4.52[9]	4.19[10]	(4.9)	5.45[16]	(4.5)	(6.4)	5.26[24]	(4.8)	5.50[30]
$(\partial \mu/\partial P)_T$			1.76[2]	2.62	2.56[9]	0.75	1.3	1.40[16]	0.73[17]	0.45	1.64	-0.69	0.88[30]
$(\partial \sigma_S/\partial P)_T$		/kb x10⁴	1.06	1.95[5]	2.04	3.0	5.8	3.03[16]	2.3[17]	48.	6.5[24]	6.5[27]	4.6[30]
Temperature derivatives													
$(\partial v_p/\partial T)_P$		km/sec/°K x10⁴	-3.6[2]	-5.0[8]	-5.2	-3.1[11]	-4.1[15]	-3.9[16]	-1.5[33]	-2.7[23]	-5.1[24]	-1.87[27]	-2.8[30]
$(\partial v_s/\partial T)_P$		km/sec/°K x10⁴	-3.1[2]	-4.8[8]	-4.4	-2.2[11]	-2.9[15]	-2.2[16]	-0.61[33]	+0.05[23]	-3.7[24]	-0.39[27]	-2.0[30]
$(\partial K_S/\partial T)_P$		kb/°K	-0.14	-0.12	-0.16[9]	-0.13[11]	-0.11	-0.20[16]	-0.17[33]	-0.10	-0.14[24]	-0.13[27]	-0.12[30]
$(\partial K_T/\partial T)_P$		kb/°K	-0.21	-0.27	-0.31	(-0.2)	(-0.15)	-0.27	-0.29[33]	(-0.1)	(-0.2)	(-0.2)	(-0.2)
$(\partial \mu/\partial T)_P$		kb/°K	-0.18	-0.25	-0.23[9]	-0.11[11]	-0.9	-0.11[16]	-0.06[33]	-0.015	-0.14	-0.02	-0.12
$(\partial \sigma_S/\partial T)_P$		/°K x10⁵	1.3	2.9	2.0	0.54	0.9	.023[16]	0.2[33]	-7.2	1.1[24]	-0.65[27]	0.41[30]
Critical thermal gradients													
$(\partial T/\partial P)_{v_p}$		°K/kb	14.	17.	16.	16.	25.	20.	31.	51.	20.	20.	23.
$(\partial T/\partial P)_{v_s}$		°K/kb	7.1	8.8	9.1	2.0	8.4	10.	10.	680.	7.9	-82.	1.6
$(\partial T/\partial P)_{\rho}$		°K/kb	24.	20.	20.	31.	43.	26.	15.	73.	34.	48.	26.
Grüneisen parameters													
γ_{th}			1.32[2]	1.54	1.54	1.13	0.92	1.21[16]	1.99[17]	0.703	1.19[24]	0.814[27]	1.27[30]
γ_p			1.53	1.77	1.71	1.33	1.65	1.95	1.53[17]	1.18	1.71	1.18[27]	1.52[30]
γ_s			1.20	1.46	1.40	0.49	0.88	1.13	0.64[17]	0.023	0.97	-1.25[27]	0.43[30]
γ_{LT}			1.23[2]	1.49	1.43	0.56	0.95	1.20[16]	0.70[17]	0.18	1.04[24]	-1.13[27]	0.54[30]
γ_{HT}			1.31[2]	1.56	1.50	0.77	1.14	1.40[16]	0.94[17]	0.41	1.22[24]	-0.44[27]	0.79[30]
δ_s			3.4	2.3	3.1	4.0	4.7	5.3[16]	2.5	7.3[23]	4.7[24]	6.2[27]	3.1[30]
Debye temperature		°K	1029	934	941	887	647	745[16]	641[17]	572	654	410	1270[30]

Figure 3. Summary of elastic properties of minerals from Anderson, Schreiber, Liebermann, and Soga [19].

Another experimental technique adopted and improved by Naohiro Soga and associates involved vibrating specimens of isotropic symmetry to their natural resonances and measuring the frequencies of their normal modes [23,24]. Initially, these studies used spherical specimens of tektites and lunar beads, but, later, these were extended to rectangular specimens (and the technique was termed rectangular parallelepiped resonance, or RPR) and were particularly useful in measuring the resonances at high temperatures (extension to high pressures was not feasible, as the spherical normal modes were damped by the applied pressure).

When Orson decided to extend this technique to anisotropic specimens, he engaged Columbia undergraduate Harold Demarest to pursue this objective. In the following paragraph, Orson describes the success of this initiative.

A Lamont graduate student, Harry Demarest, decided to see if he could measure the velocity of sound in a small cube of a mineral with cubic symmetry, but he had to do research to find the required equations that yield the velocity of sound. His experimental results showed a series of resonant bands (a spectrum), and Demarest developed the theory for analyzing this spectrum to yield the elastic constants and velocities of sound. His sample was a single crystal, with cubic symmetry, of MgO. This was very exciting because we had opened the door to measuring the elastic constant of a small body having any crystallography symmetry. Furthermore, we did not need glue to bond the crystal to the transducer but used gravity to hold the crystal between the transducers touching opposite corners of the cube. Using glue for the transducers limits temperature measurements to about 700 °C. Without glue, we could hope for measurements for temperatures up to a large fraction of the melting point (in the range 2000–2500 °C). Demarest graduated from Lamont with his PhD, and we replaced him with Dr. Ichiro Ohno as a Lamont postdoc. Ohno was an accomplished theorist from Nagoya who undertook the theoretical problem of analyzing the resonance spectrum of a crystal with any symmetry. At our lab at Lamont, we had invented a new field in physical acoustics, it is now called RUS (Resonant Ultrasound Spectroscopy). For pioneering the field of RUS, the Acoustical Society of America gave me a citation. This citation brought welcome new funding to Lamont and the Mineral Physics Lab from the Office of Naval Research. More importantly, our lab had pierced the barrier to find physical properties at Earth's mantle and core conditions because of the high temperature capability of RUS. This new work of mineral physics gave the high-temperature answers to help theorists develop the new theory of superplumes [25].

There were two, somewhat amusing episodes during these sound velocity studies:

In 1969, Mineo Kumazawa had arrived at Lamont to study the elasticity of single-crystal diamondz (following earlier work at Bell Labs by colleagues of Anderson). Instead, he was encouraged to study single-crystal olivines. Orson purchased a large single crystal of Burmese peridot from a gem dealer in New York City for a cost of more than $5000; he was so proud of that crystal that he published a photo and news item in the local newspaper in Palisades, NY. After Kumazawa cut the precious crystal into rectangular parallelepipeds for the acoustic studies, someone broke into the mineral physics lab at night and stole the largest pieces (which now had no commercial value). Kumazawa was devastated and embarrassed, and had no recourse but to try to perform his sound velocity measurements on the small pieces remaining, often termed "curf loss". He succeeded, and he and Orson published a seminal paper in the *Journal of Geophysical Research* [18].

The other "amusing" episode occurred in 1970 and involved laboratory measurements of the sound velocities in lunar rocks returned from the Moon by the Apollo missions. In Orson's own words:

This was the time of the Apollo landings, and we received five rocks from the Moon to measure their physical properties. We were amazed to discover that the surface of the Moon had a velocity of sound of about 1.8–2.1 km per second, approximately that of seawater. Schreiber and I published a paper in Science [26] showing that the average sound velocity of the Moon rocks was close to provolone cheese. This paper brought many request for reprints. Of course, our task was to find why the velocity was so low in the lunar crust. We searched and found small glass spheroids (1 mm or less) in the lunar soils. They were formed by the liquid splash from meteorites hitting the lunar surface. We found the sound velocity of the lunar glass spheroids to be about 6.5 km per second [23], which is what the rocks should have had by their composition. After several years, we found that the velocity of sound of lunar material was low because it had been extensively cracked in a complete vacuum. There was no liquid meniscus between the surface cracks and the sound would not cross over the cracks. More importantly to physical acoustics was our discovery that we could measure the velocity of sound in a tiny sphere. This led to a substantial innovation in physical acoustics because we were resonating a very tiny isotropic sphere and obtaining the velocity of sound. The equations used are found in Rayleigh's classical book. See discussion of sphere resonant techniques above.

In the introduction to of a memorial volume of PAGEOPH dedicated to Ed Schreiber (Figure 4) [27], William Bassett and I quoted from their 1970 Science paper "Properties and Composition of Lunar Materials: Earth Analogies".

To account for these very low velocities, we decided to consider materials other than those listed initially by Birch or the later more detailed compilations of Anderson and Liebermann [28]. [The search was aided by considerations of much earlier speculations concerning the nature of the Moon by Erasmus in 1542 [29]; "With this merry toy . . . he made his friends to believe the Moon to be made of green cheese"], and a significant group of materials was found which have velocities that cluster around those actually observed for lunar rocks (see Figure 5).

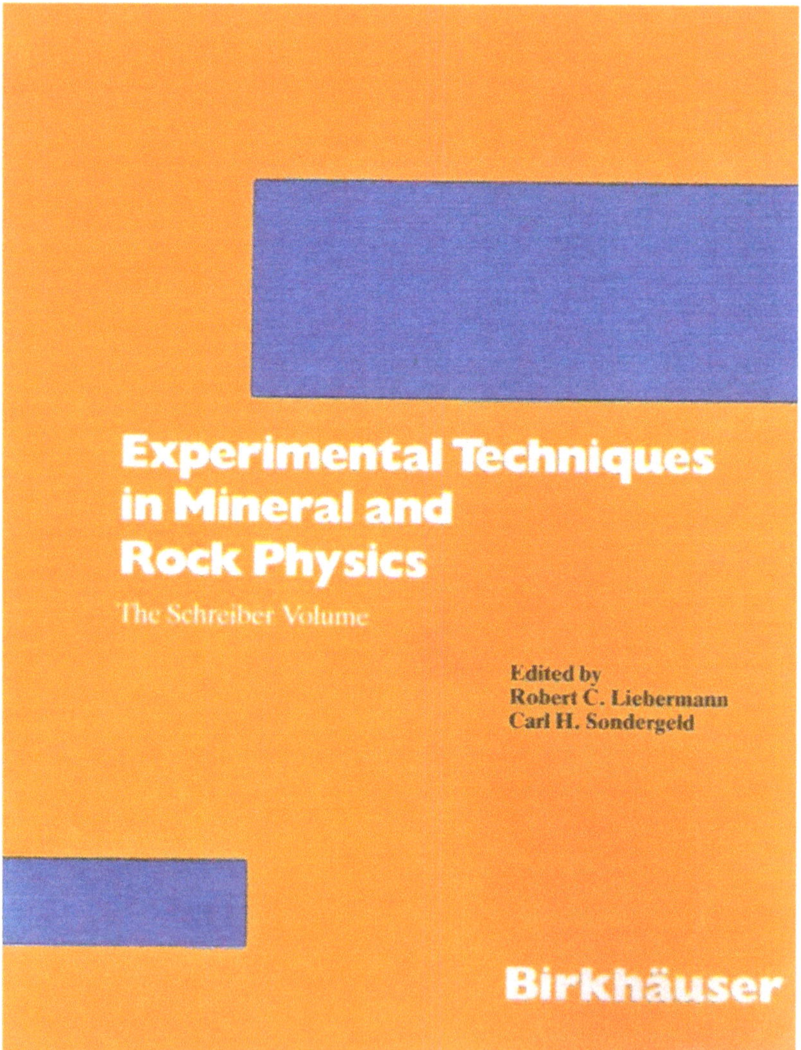

Figure 4. The Schreiber Volume edited by Liebermann and Sondergeld [27].

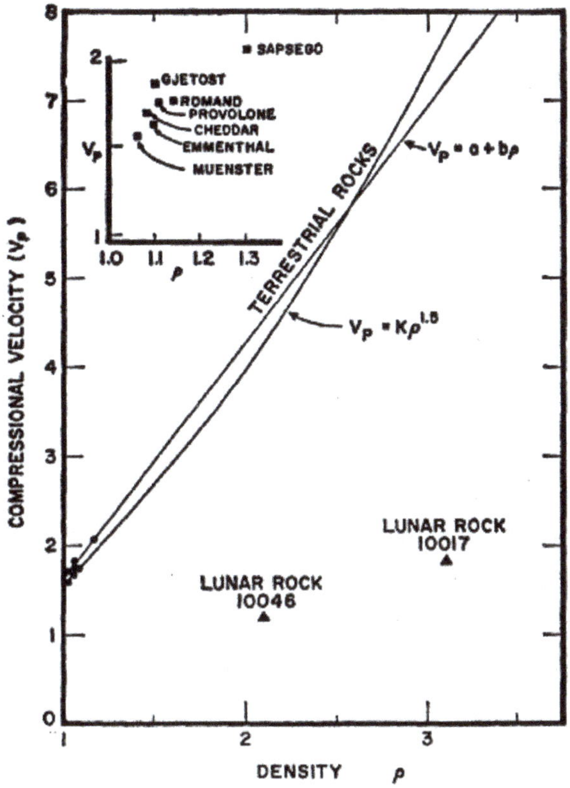

Figure 5. From Schreiber and Anderson [26] showing the comparison between sound velocities for the lunar rocks and for various earth materials.

The materials [cheeses] were chosen so as to represent a broad geographic distribution in order to preclude any bias that might be introduced by regional sampling. It is seen that these materials exhibit compressional velocities that are in consonance with those measured for lunar rocks—which leads us to suspect that perhaps old hypotheses are best after all and should not be lightly discarded.

*As seen in Figure 5, the $V_p = a + bp$ curve of Birch [30] for the rock types diabase, gabbros, and eclogites fit the cheeses surprisingly well. This apparent inconsistency, in that the cheeses do obey these relationships by having a velocity appropriate to their density, in contrast to the lunar rocks with which they compare so well, may readily accounted for when one considers how much better **aged** the lunar materials are."*

In 1970, Anderson and Schreiber used these velocity measurements in a Christmas card (Figure 6) for which they wrote, "It brightens the spirits in times like these to know the Moon is made of cheese".

It brightens the spirits
In times like these,
To know the Moon
Is made of cheese.

Seasons Greetings

Dear Bob and Barbara.
I'm moving to UCLA this spring.
Ed and Chris will stay. We'll split
the lab.
Best regards
Orson

Lunar rocks and cheeses	Sound Velocity, V_P (Kilometers/second)
Lunar Rock 10017	1.84
Gjetost (Norway)	1.83
Provolone (Italy)	1.75
Romano (Italy)	1.75
Cheddar (Vermont)	1.72
Emmenthal (Swiss)	1.65
Muenster (Wisconsin)	1.57
Lunar Rock 10046	1.25

(Science, 168, 1579, 1970)

MINERAL PHYSICS LABORATORY
LAMONT-DOHERTY GEOLOGICAL OBSERVATORY
OF COLUMBIA UNIVERSITY

Figure 6. From Christmas card sent by Schreiber and Anderson to colleagues in 1970. Based on Schreiber and Anderson [26] showing velocities of lunar rocks compared with terrestrial cheeses.

From 1964 to the early 1970s, Anderson and his colleagues published many papers in peer-reviewed scientific journals; after leaving Lamont for the University of California in 1971, Orson put these Collected Reprints into a special volume to share with his friends and colleagues (see Figure 7). In addition to his 1995 book (Figure 8), Orson also was a co-author, with Ed Schreiber and Naohiro Soga, of a book entitled "Elastic Constants and Their Measurement" [31] (see Figure 9), in which they discussed the theory of elasticity and various experimental techniques to measure elastic properties, indirect methods to estimate elastic constants, and thermodynamic approaches to estimate the pressure and temperature derivatives of elastic constants.

Figure 7. Collected Reprints from the mineral physics laboratory of Orson Anderson at Lamont from 1964 to 1970.

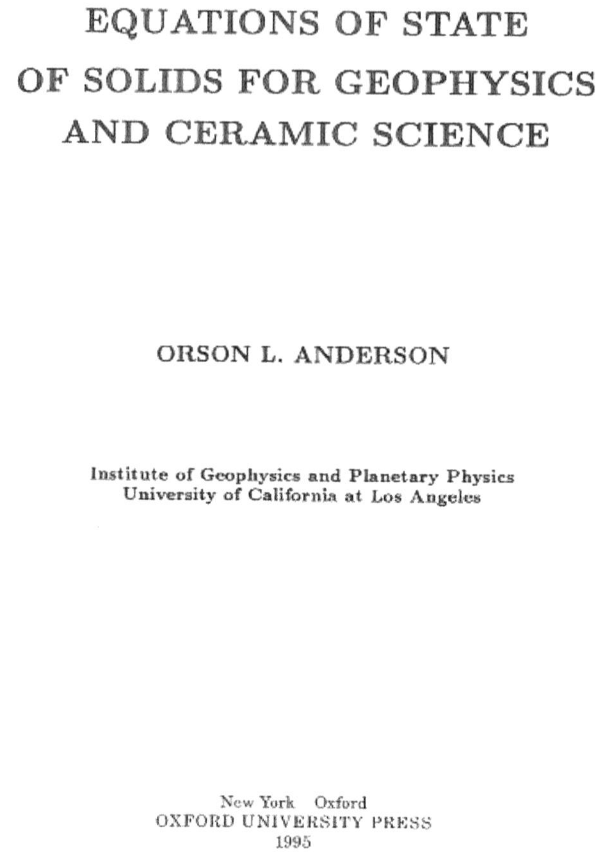

EQUATIONS OF STATE
OF SOLIDS FOR GEOPHYSICS
AND CERAMIC SCIENCE

ORSON L. ANDERSON

Institute of Geophysics and Planetary Physics
University of California at Los Angeles

New York Oxford
OXFORD UNIVERSITY PRESS
1995

Figure 8. 1995 book by Orson Anderson based on notes for his solid-state geophysics course at Lamont Geological Observatory in the 1960s.

Figure 9. Book by Schreiber, Anderson, and Soga [31].

During his eight years on the geology faculty at Columbia University, Orson supervised four graduate students: Bob Liebermann, whom he recruited from the Seismology Group at Lamont; Leon Thomsen, whom he recruited from the Goddard Institute for Space Science in New York; Harry Demarest, whom he recruited when he was an undergraduate at Columbia; and Nicholas Warren, who followed Orson to UCLA. In the acknowledgements for his 1995 book [21], Orson thanked us: *Leon for teaching me Lagrangian theory, Bob for teaching me persistence, and Harry for laying the computational foundation of resonant ultrasound spectroscopy.*

Orson initially co-opted me to do library research for a review paper he had promised to write for the Air Force Office of Scientific Research (AFOSR). In those days before electronic journals, it was necessary to access hard copies and take notes or, if lucky, make photocopies. Fortunately, Columbia has an excellent library containing journals going back to the 19th century, and I also visited the Library of Congress during an AGU meeting in Washington. This 1966 report for the AFOSR was later published in Volume IVB of Physical Acoustics, edited by Warren Mason [28].

Both Thomsen and I graduated in 1969 with the theses "On the Fourth-Order Anharmonic Equation of State of Solids" [32] and "Effect of Iron Content upon the Elastic Properties of Oxides and Some Applications to Geophysics" [33], respectively. In collaboration with Subir Banerjee (then at the Ampex Corporation and now at the University of Minnesota), we discovered anomalies in the compressional and shear properties of hematite in the region of the Morin transition, which we ascribed to magnetoelastic interactions [34,35].

The semi-empirical research in Orson's laboratory included investigations of the relationship of the bulk modulus K to the volume V of oxides [36], which later evolved to a proposed "law of corresponding states for oxides" [37]; for restrictions of this "law", see [38]. The latter showed that Anderson's law is valid only for oxides having their mean atomic weight to be near the value of 20 or 21, while other oxides follow the more common law of chemical physics prescribed by lattice dynamics.

During this same time period, Don Anderson, at Caltech, had formulated a "seismic equation of state" [39]. The two Andersons (Don and Orson) spent a few years dueling over which equations of state was the more appropriate; see [40]. As the two protagonists had almost identical initials, D.L. and O.L., many authors began to spell out their first names when referencing their papers.

In another semi-empirical research initiative, Anderson and Liebermann [41] used lattice dynamics and a Born repulsive potential between atoms to calculate the elastic constants of cubic lattices NaCl, CsCl, and ZnS as functions of pressure and to predict the vanishing of a shear constant (either c_{44} or $c' = 1/2(c_{11} - c_{12})$); see also [42].

The primary theoretical research in the mineral physics lab in the 1960s was conducted by graduate student Leon Thomsen, to whom I am indebted for the following history [43].

As described in detail above, Orson's group got the opportunity to measure acoustic velocities on lunar rocks, which had been recovered on the Apollo mission. Comparing with terrestrial analogs, they found that the lunar rock velocities were close to that of sapsago cheese, a hard cheese from Switzerland. As such, they actually published a scholarly paper likening the Moon to green cheese [26].

Perhaps because of this, I took Orson's class in mineral physics. During that semester, a paper by Anderson appeared in the *Journal of Geophysical Research*, of which he was the editor. I detected a logical error in that paper, and I brought it to Orson's attention. Orson said, "You are correct; write a short paper discussing my error. I will see that it gets accepted by the *Journal of Geophysical Research* under your name". This was a great lesson in academic ethics [44].

I switched my focus to Geophysics and moved my office from the Goddard Institute for Space Science to Lamont. Orson guided my research into a PhD thesis [32,45]. This was a challenge to the conventional (Eulerian) formulation of finite strain theory promoted by the de facto dean of mineral physics, Francis Birch of Harvard; see details in Section 1 of my 1970 paper [45]. My thesis turned out to be a theoretical dead end, but it provided me with the basic mathematical foundation and insights which eventually led me to make important contributions in the analysis of seismic anisotropy" (end of quote from Thomsen).

At Columbia, Orson taught a graduate course on "solid-state geophysics", which started with elementary lattice dynamics and continued to equations-of-state of solids. I adapted these notes and used them at the Australian National University from 1973 to 1976 and at Stony Brook from 1977 to 2013. These lecture notes later formed the basis for his 1995 book "Equations of State of Solids for Geophysics and Ceramic Science" (see Figure 8) [21].

As a native of Utah, Orson had a fascination with the Four Corners area of the western U.S. As he was now a professor of geology at Columbia and using ultrasonic techniques to measure sound velocities of rocks and minerals, Anderson organized field trips in 1966 and 1967 to the diatremes of northeastern Arizona, including those at Mule Ear, Garnet Ridge, Cane Valley, and Buell Park.

To reach Mule Ear, it was necessary to hire a rubber raft and a guide to transport us from Bluff, Utah, to the field site (Figure 10a). The Columbia team was a strange one for geological field work, as it included a solid-state physicist (Anderson), a ceramist (Schreiber), an industrial chemist (Soga), and two seismologists (Peter Molnar and the author), so-called "strangers in paradise;" fortunately, we had an excellent field guide, Eugene Shoemaker from the US Geological Survey in Flagstaff, Arizona (Figure 10b). Though we did not find any diamond-bearing rocks, we did obtain many interesting specimens and photos (see Figure 10c). In 1968, we drove to the Garnet Ridge diatreme, where Orson and Mineo Kumazawa posed for special photos (Figure 10d,e). Kumazawa, Helmstaedt, and Masaki [46] later published a paper on the elastic properties of eclogite xenoliths from the diatremes of the Eastern Colorado Plateau, including a lherzolite from the Mule Ear diatreme (see photos of field trips to diatremes at end of text Figure 10a–e).

In the late 1960s, the Lamont mineral physics team dispersed, and the Mineral Physics Laboratory closed. The sign was inherited by Ed Schreiber and later given to Orson, who still has it in Utah (see Figure 2 from [7]).

Naohiro Soga moved to Rice University and subsequently to Kyoto University, where he continued to apply lattice dynamics to inorganic network-forming materials such as silicates; he deduced their interatomic bond strengths, as well as other types of force constants arising from the network structural vibrations in order to clarify their physical properties.

Orson Anderson moved to the University of California Los Angeles (UCLA) in 1971 and served on the faculty until 2002.

Edward Schreiber moved to Queens College of the City University of New York. He died tragically in a car accident in 1991.

In 1998, a Topical Session of the American Geophysical Union was convened in honor of Orson Anderson, who gave the last talk. A collection papers was later published in *American Mineralogist* (Volume 85, 2000), which represented topics of interest to Orson, who was a co-author on four of the papers [47].

Figure 10. *Cont.*

Figure 10. *Cont.*

Figure 10. *Cont.*

Figure 10. (**a**) Orson Anderson and Naohiro Soga being transported to the Mule Ear diatreme (in the background) via a rubber raft piloted by guide Kenny Ross; (**b**) field geologist Eugene Shoemaker showing outcrop from the Mule Ear diatreme to Edward Schreiber (ceramist), Orson Anderson (solid-state physicist with back turned), Naohiro Soga (industrial chemist), and Peter Molnar (seismologist); (**c**) Ed Schreiber viewing Alhambra Rock; (**d**) Orson Anderson on the Garnet Ridge diatreme; (**e**) Mineo Kumazawa at the Garnet Ridge diatreme.

Funding: This paper was written with support from research grants to Baosheng Li and myself from the National Science Foundation (EAR-1524078) and the Department of Energy NNSA (De-NA0002907).

Acknowledgments: In 2008, to mark the 30th anniversary of the first scientific conference focused on mineral physics (at Airlie House in Virginia in 1977), Charles Prewitt and I proposed to write a paper on the subsequent evolution of this new research field for *EOS, Transactions of the American Geophysical Union*; our initial submission to *EOS* was returned for major revision. As time progressed and we procrastinated, it was not until 2012 at Joint COMPRES Annual Meeting and High-Pressure Mineral Physics Seminar-8 at the Granlibakken Conference Center on Lake Tahoe before we returned to this topic and published a paper in *Physics of the Earth and Planetary Interiors* entitled: "From Airlie House in 1977 to Granlibakken in 2012: 35 Years of Evolution of Mineral Physics" [7]. Also in 2008, I wrote to Sean Solomon (Director of the Lamont) offering to write a historical piece for the Lamont "Latest Earth Sciences News" website on the Orson Anderson era of mineral physics in the 1960s. Though Sean encouraged me to submit an historical news item, again I procrastinated. Recently, I decided to resurrect my paper on the developments in the Mineral Physics Laboratory of Orson Anderson in the 1960s at Lamont and submit it to *Minerals* (to provide access to a wider readership); my earlier paper [43], Naohiro Soga [48] contains some of the discussion presented here. I thank Orson Anderson for the permission to quote liberally from his personal communications with Lamont and me, and I thank Sean Solomon for encouraging me to write this historical piece. I have profited from correspondence from Charlotte Schreiber, Naohiro Soga, William Bassett, Leon Thomsen, and Peter Molnar. I especially thank Leon [43], Naohiro Soga [49], and Peter Molnar [3] for their insightful contributions to previous drafts of this paper. I thank David Walker and an anonymous reviewer for their comments and suggestions, which have improved the readability of this paper.

Conflicts of Interest: The author declares no conflict of interest.

References

1. Hazen, R.M. What is mineral physics? *EOS Trans. Am. Geophys. Union* **1994**, *65*, 746.
2. Birch, F. Elasticity and constitution of the Earth's interior. *J. Geophys. Res.* **1952**, *57*, 227–286. [CrossRef]
3. Molnar, P. Personal communication to Robert Liebermann, March 2019.
4. Anderson, O.L. Personal communication to Michael Purdy, May 2008.
5. Anderson, O.L. Personal communication to Robert Liebermann, March 2008.
6. Anderson, O.L. Personal communication to Robert Liebermann, January 2014.
7. Liebermann, R.C.; Prewitt, C.T. From Airlie House in 1977 to Granlibakken in 2012: 35 years of evolution of mineral physics. *Phys. Earth Planet. Inter.* **2014**, *228*, 36–45. [CrossRef]
8. Anderson, O.L.; Schreiber, E. Pressure derivatives of the sound velocities of polycrystalline magnesia. *J. Geophys. Res.* **1965**, *70*, 5241–5248. [CrossRef]
9. Schreiber, E.; Anderson, O.L. Pressure derivatives of the sound velocities of polycrystalline alumina. *J. Am. Ceram. Soc.* **1966**, *49*, 184–190. [CrossRef]
10. Schreiber, E.; Anderson, O.L. Pressure derivatives of the sound velocities of polycrystalline forsterite with 6% porosity. *J. Geophys. Res.* **1966**, *71*, 762–764. [CrossRef]
11. Schreiber, E. Elastic moduli of single-crystal spinel at 25 °C and to 2 Kbar. *J. Appl. Phys.* **1967**, *38*, 2508–2511. [CrossRef]
12. Soga, N.; Anderson, O.L. Anomalous behavior of the shear sound velocity under pressure for polycrystalline ZnO. *J. Appl. Phys.* **1967**, *38*, 2985–2988. [CrossRef]
13. Soga, N. Elastic constants of garnet under pressure and temperature. *J. Geophys. Res.* **1967**, *72*, 4227–4234. [CrossRef]
14. Soga, N. Temperature derivatives of isotropic sound velocities of alpha-quartz. *J. Geophys. Res.* **1968**, *73*, 827–829. [CrossRef]
15. Soga, N. New measurements of the sound velocity of calcium oxide and its relation to Birch's law. *J. Geophys. Res.* **1967**, *72*, 5157–5159. [CrossRef]
16. Liebermann, R.C.; Schreiber, E. Elastic constants of polycrystalline hematite as a function of pressure to 3 Kbars. *J. Geophys. Res.* **1968**, *73*, 6585–6590. [CrossRef]
17. Liebermann, R.C. Pressure and temperature dependence of the elastic properties of polycrystalline trevorite ($NiFe_2O_4$). *Phys. Earth Planet. Inter.* **1972**, *6*, 360–365. [CrossRef]
18. Kumazawa, M.; Anderson, O.L. Elastic moduli, pressure derivatives and temperature derivatives of single-crystal olivine and single-crystal forsterite. *J. Geophys. Res.* **1969**, *74*, 5961–5972. [CrossRef]
19. Anderson, O.L.; Schreiber, E.R.; Liebermann, C.; Soga, N. Some elastic constant data on minerals relevant to geophysics. *Rev. Geophys.* **1968**, *6*, 491–524. [CrossRef]
20. Anderson, O.L.; Liebermann, R.C. Elastic constants of oxide compounds used to estimate the properties of the earth's interior. In *Applications of Modern Physics to the Earth and Planetary Interiors*; Runcorn, S.K., Ed.; John Wiley: London, UK, 1969; pp. 425–449.
21. Anderson, O.L. Equations of State of Solids for Geophysics and Ceramic Science. Oxford University Press: Oxford, UK, 1995.
22. Soga, N.; Anderson, O.L. High-temperature elastic properties of polycrystalline MgO and Al_2O_3. *J. Am. Ceram. Soc.* **1966**, *49*, 355–359. [CrossRef]
23. Soga, N.; Anderson, O.L. Elastic properties of tektites measured by the resonance sphere technique. *J. Geophys. Res.* **1967**, *72*, 1733–1739. [CrossRef]
24. Soga, N.; Schreiber, E. Porosity dependence of sound velocity and Poisson's ratio for polycrystalline MgO determined by the resonance sphere technique. *J. Am. Ceram. Soc.* **1968**, *151*, 465–466. [CrossRef]

25. Yuen, D.A.; Maruyama, S.; Karato, S.I.; Windley, B.F. *Superplumes: Beyond Plate Tectonics*; Springer-Verlag: Berlin, Germany, 2007.
26. Schreiber, E.; Anderson, O.L. Properties and composition of lunar materials: Earth analogies. *Science* **1970**, *168*, 1579–1580. [CrossRef]
27. Liebermann, R.C.; Bassett, W.A. Reflections on the career of Edward Schreiber. In *Experimental Techniques in Mineral and Rock Physics*; Liebermann, R.C., Sondergeld, C.H., Eds.; The Schreiber Volume, Birkhauser Verlag: Berlin, Germany, 1993; Also published in *Pure Appl. Geophys.* **1993**, *141*, 209–657.
28. Anderson, O.L.; Liebermann, R.C. Sound velocities in rocks and minerals: Experimental methods, Extrapolations to very high pressures, and results. In *Physical Acoustics*; Mason, W.P., Ed.; Academic Press: New York, NY, USA, 1968; Volume IV, pp. 329–472.
29. Erasmus. *Adagia*; 1542.
30. Birch, F. The velocity of compressional waves in rocks to 10 kilobars. *J. Geophys. Res.* **1960**, *65*, 1083–1102. [CrossRef]
31. Schreiber, E.; Anderson, O.L.; Soga, N. *Elastic Constants and Their Measurement*; McGraw-Hill Book Company: New York, NY, USA, 1973.
32. Thomsen, L. On the fourth-order anharmonic equation of state of solids. Ph.D. Thesis, Columbia University, New York, NY, USA, 1969.
33. Liebermann, R.C. Effect of iron content upon the elastic properties of oxides and some applications to geophysics. Ph.D. Thesis, Columbia University, New York, NY, USA, 1969.
34. Liebermann, R.C.; Banerjee, S.K. Anomalies in the compressional and shear properties of hematite in the region of the Morin transitions. *J. Appl. Phys.* **1970**, *41*, 1414–1416. [CrossRef]
35. Liebermann, R.C.; Banerjee, S.K. Magnetoelastic interactions in hematite: Implications for geophysics. *J. Geophys. Res.* **1971**, *76*, 2735–2757. [CrossRef]
36. Anderson, O.L.; Nafe, J.E. Bulk modulus-volume relationship for oxide compounds and related geophysical problems. *J. Geophys. Res.* **1965**, *70*, 3951–3963. [CrossRef]
37. Anderson, O.L. A proposed law of corresponding states for oxide compounds. *J. Geophys. Res.* **1966**, *71*, 4963–4971. [CrossRef]
38. Anderson, O.L.; Soga, N. A restriction to the law of corresponding states. *J. Geophys. Res.* **1967**, *72*, 5754–5757. [CrossRef]
39. Anderson, D.L. A seismic equation of state. *Geophys. J. Roy. Astro. Soc.* **1967**, *13*, 9–30. [CrossRef]
40. Anderson, D.L.; Anderson, O.L. The bulk modulus-volume relationship for oxides. *J. Geophys. Res.* **1970**, *75*, 3494–3500. [CrossRef]
41. Anderson, O.L.; Liebermann, R.C. Equation for the elastic constants and their pressure derivatives for three cubic lattices and some geophysical applications. *Phys. Earth Planet. Inter.* **1970**, *5*, 61–85. [CrossRef]
42. Anderson, O.L. Elastic constants of the central force model for three cubic structures: Pressure derivatives and equations of state. *J. Geophys. Res.* **1970**, *75*, 2719–2740. [CrossRef]
43. Thomsen, L. Personal communication to Robert Liebermann, February 2019.
44. Thomsen, L.; Anderson, O.L. On the high temperature equation of state of solids. *J. Geophys. Res.* **1969**, *74*, 981–991. [CrossRef]
45. Thomsen, L. On the fourth-order anharmonic equation of state of solids. *J. Phys. Chem. Solids* **1970**, *31*, 2003–2016. [CrossRef]
46. Kumazawa, M.; Helmstaedt, H.; Masaki, K. Elastic properties of eclogite xenoliths from the diatremes of the Eastern Colorado plateau and their implication to upper mantle structure. *J. Geophys. Res.* **1971**, *76*, 1231–1247. [CrossRef]
47. Liebermann, R.C.; Isaak, D.G. Dedication to Orson L. Anderson. *Am. Mineral.* **2000**, *85*, 269.
48. Liebermann, R.C. The role of serendipity in my career in mineral physics: 1963–2013. *Phys. Earth Planet. Inter.* **2014**, *228*, 307–323. [CrossRef]
49. Soga, N. Personal communication to Robert Liebermann, March 2019.

minerals

MDPI

Editorial

The Takahashi–Bassett Era of Mineral Physics at Rochester in the 1960s

William A. Bassett

Department of Earth and Atmospheric Sciences, Cornell University, Ithaca, NY 14853, USA; wab7@cornell.edu

Received: 16 March 2020; Accepted: 9 April 2020; Published: 12 April 2020

Abstract: The late Taro Takahashi earned a particularly well-deserved reputation for his research at Lamont Geological Observatory on carbon dioxide and its transfer between the atmosphere and the oceans. However, his accomplishments in Mineral Physics, the field embracing the high-pressure–high-temperature properties of materials, has received less attention in spite of his major contributions to this emerging field focused on the interiors of Earth and other planets. In 1963, I was thrilled when he was offered a faculty position in the Geology Department at the University of Rochester, where I had recently joined the faculty. Taro and I worked together for the next 10 years with our talented students exploring the blossoming field just becoming known as Mineral Physics, the name introduced by Orson Anderson and Ed Schreiber, who were also engaged in measuring physical properties at high pressures and temperatures. While their specialty was ultrasonic velocities in minerals subjected to high pressures and temperatures, ours was the determination of crystal structures, compressibilities, and densities of such minerals as iron, its alloys, and silicate minerals, especially those synthesized at high-pressure, such as silicates with the spinel structure. These were materials expected to be found in the Earth's interior and could therefore provide background for the interpretation of geophysical observations.

Keywords: mineral physics; earth interior; diamond anvil cell; high pressure

---

1. Introduction

Geophysics, the study of the macro properties of Earth, was very active in the mid-twentieth century. Such properties included seismic signals (at all depths), gravity, magnetism, moments of inertia, free oscillations, and heat flow. The study of meteorites and volcanic rocks provided clues about planetary interiors and suggested the most likely compositions. At the same time, it was becoming clear that there was not enough understanding of the most fundamental properties of matter at the conditions existing in the interiors of Earth and other planets, properties that could possibly be measured in laboratories equipped with apparatus capable of achieving those pressures and temperatures.

Taro Takahashi and I were good friends while we were graduate students in geology at Columbia University in the latter part of the 1950s. When he and I had finished and defended our PhD theses, he was invited by Maurice Ewing, Director of Columbia University's Lamont Geological Observatory, to undertake a post-doctoral position to participate in excursions around the oceans on the Vema, Lamont's research vessel. I once visited him at Lamont and climbed to the crow's nest on the Vema, an adventure which gave me a taste of what it must have been like for him to go to sea and experience the excitement of making valuable contributions to oceanography. On the Vema, Taro collected samples of sea water and analyzed them in the Vema chemistry lab immediately after they were collected. Doc Ewing's primary research interest, and what he is most famous for, was seismic exploration of the oceans and the rocks in the floor under the oceans. Over meals, while both were pursuing their research interests on the Vema, Taro would ask Doc about what the seismic studies were revealing. One day, as he learned more about seismic research, he asked Doc what he thought caused the Mohorovičić discontinuity (Moho for short) that separates rocks conducting seismic signals at ~6 km per second from deeper rocks conducting seismic signals at ~8 km per second, and has a depth that varies dramatically under oceans and continents. It is a major feature of the earth's interior and was attracting a lot of attention which eventually led to the theory of plate tectonics, the great paradigm shift in geology. Doc replied that there were two possible explanations, either a change in composition or a change in physical properties. It was clear to both that, at that time, the compositions and physical properties were simply too poorly understood to answer the question adequately.

2. End of Post-Doc Years and Chance to Pursue Research on Earth's Interior

When Taro's two-year post-doc appointment was up, he decided it was time to find a position at a research university where he might be able to find a more definitive answer to the question he had asked Doc Ewing. He was convinced that finding an answer to that and numerous other questions about Earth's interior would depend on more accurate data regarding the properties of minerals at the pressures and temperatures in the Earth's interior, and that would need to be done in a lab with special equipment. He accepted a faculty appointment at Alfred University in Upstate New York in around 1959. One of the first things Taro did was write a letter to Percy Bridgman, Professor of Physics at Harvard University and considered the father of high-pressure research, to request some of his papers. Bridgman replied with an extensive collection of his reprints. As it turned out, Bridgman had sent that valuable collection to Taro shortly before he died. It was only a few years after that a research team at the General Electric Research Lab succeeded in synthesizing diamond using the belt apparatus invented by Tracy Hall. When General Electric executives told Tracy that the belt design belonged to GE, Tracy left GE and returned to his hometown, Provo, Utah. There, he joined the Brigham Young University faculty, where he invented the tetrahedral multi-anvil press which was also able to synthesize diamond. It was a derivation of the tetrahedral multi-anvil apparatus that Taro installed in his lab at Alfred University [1] (Figure 1). He and his students analyzed the physical properties of minerals at high pressures, especially minerals with a composition believed to be the same as, or similar to, the components of the earth's interior.

Soon after I completed my thesis, I took a post-doc position in the Chemistry Department at Brookhaven National Laboratory measuring the ages of minerals from young volcanic rocks, a job made possible because two of the most sensitive mass spectrometers in the world resided there in Oliver Schaeffer's lab. This was very different from going to sea on the Vema, but was very exciting in other ways. When my post-doc appointment ended in 1961, I took a position in the Geology Department at the University of Rochester, not far from Alfred.

Shortly after I had settled into my new job, the chairman of the Geology Department, Bob Sutton, asked me who I thought the department should add to its faculty next. Without hesitation, I said, "Taro Takahashi". I, too, had installed equipment, a GE X-ray machine so I could study minerals; Taro seemed a perfect choice. Much to my pleasure, Bob took me seriously, and we invited Taro to pay us a visit. Shortly after that, Taro was offered a position in the Geology Department at Rochester and generous space to carry on his research into the effects of high pressure on minerals. Outfitting his lab with equipment, his move from Alfred to Rochester went well, except for one unfortunate incident. He brought with him the large plumber's vise (near the foreground in Figure 1) that he needed for some of his experiments. When the campus plumbers left after installing a badly needed sink, he realized that his vise was missing and could not be found anywhere. It was a serious loss but, at the same time, it was a source of amusement when we tried to imagine the plumbers asking each other if the vise could possibly have belonged to the new guy just moving in, and if he could possibly need a plumber's vise for his research.

Figure 1. Alfred University, 1960: Taro Takahashi's high-pressure lab equipped with a tetrahedral multi-anvil apparatus. Taro is on the right and his graduate student at the time, Ed Schreiber, is in the center. It was shortly after this (about 1963) that Taro accepted an offer at the University of Rochester and continued his high-pressure research, while Ed moved to Lamont Geological Observatory as a member of the research faculty to pursue high-pressure experiments in the Mineral Physics laboratory of Orson Anderson. (Photo courtesy of Taro Takahashi and Bob Liebermann).

3. Married Life and Friends

I met Jane, my wife-to-be, soon after I moved to Rochester. She too was a geologist. We fell in love and married, and bought a farm in Churchville, a few miles west of Rochester and had our first child, Kari. Taro was a frequent visitor to the farm, where he became acquainted with Kari. They seemed to hit it off right away (Figure 2).

Taro met Elaine, his wife-to-be, at Rochester also. When he proposed to her, he gave her an engagement ring with a pearl and a diamond, suggesting that her passion for biology might join his passion for oceanography and high-pressure research. I had given Jane an engagement ring bearing two diamonds the same size. That must have been an omen as, in our case, I had not yet learned what kind of research I would be conducting for the rest of my life as a result of my collaboration with Taro. After they married, Taro and Elaine both became frequent visitors to our farm. Taro and I liked to write papers there, with me at the typewriter and Taro looking over my shoulders saying things like, "Isn't there a simpler way to say that?" or "How about inserting a reference here?". When he was not making comments like that, he was generating new ideas and new ways to convey them through print; I would keep typing, but then I would get to comment. There was no delete button on my typewriter, of course, just a rapid-firing X key. Our pages often had many Xed out words, sometimes nine tenths of a page. However, I loved writing papers that way and I think he did too. It was exciting and fast (it was, after all, an electric typewriter). Taro and I produced 21 peer-reviewed papers with our students (see Appendix A). Computer aids to writing would have to wait, but we did make use of the primitive computers then available to help interpret data.

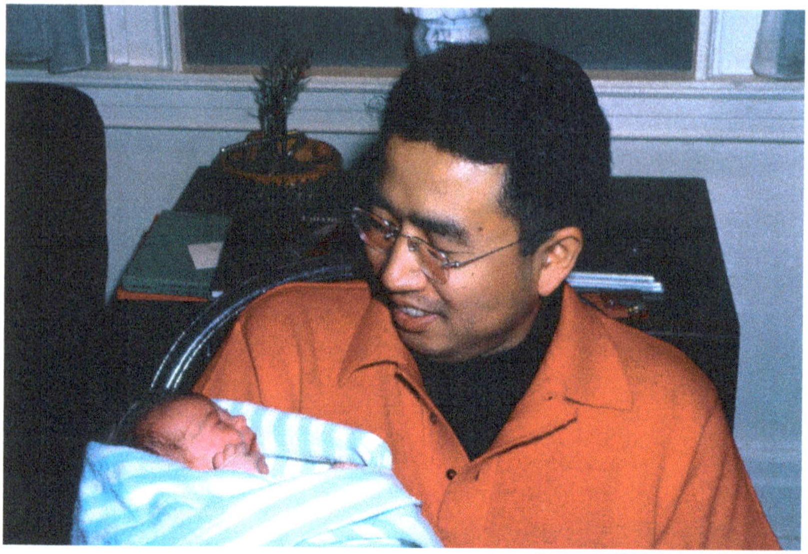

Figure 2. Taro was a frequent visitor to our farm a few miles west of Rochester. Here, he holds Kari, our first-born child. I'm not sure, of course, but he does look as if he might be telling her that someday she too may be a geologist. If so, he was right.

4. The Diamond Anvil Cell

Taro and I talked about the properties of minerals at very high pressures and temperatures and ways to extend the range of pressure and provide simultaneous temperature so we could use X-rays and other analytical techniques to examine their properties. One day, Taro came to my office, very excited about something, and said he had just seen a notice for a paper to be given in a few days at a conference in New York City, describing an instrument called a diamond anvil cell that would let a person look directly at a sample through the diamonds, using a microscope. That sounded too good to be true, and so we drove to New York City and listened to Alvin Van Valkenburg talk about the simple, beautiful, clever, little instrument and saw pictures of samples taken at high pressure [2]. If visible light could be used to observe samples at high pressures, why not X-rays? We introduced ourselves and made arrangements to pay a visit to Van's lab at the National Bureau of Standards in Washington, DC. One of the first things we learned there was that the idea of using X-rays had already been thought of and put to use by Charlie Weir, a member of the Bureau of Standard's high-pressure team [2]. Taro asked if they would object to having our machinist make a diamond anvil cell designed for use with X-rays, and they said that they did not object. I did not even know that we had a machinist. Fortunately, Taro knew a very able machinist, Phil Stook, on one of the upper floors of a large, drafty, old brick factory in Hornell, NY who used to make transmissions for army tanks but now did odd jobs for people.

Taro and I drove back from Washington to Rochester, marveling at the bright, new way that had just opened up before us. He was sure that Phil would be willing to tackle something as small as a diamond anvil cell, in spite of his earlier job. The first few cells that Phil made for us in that old brick factory worked perfectly [3]. However, there were many problems yet to be solved; my GE X-ray diffractometer was woefully inadequate for all the ideas we and our students were having, and so we purchased additional Norelco and Jarrell Ash X-ray machines, each with four ports and much better designed for use with X-ray film cameras. Needless to say, the samples were microscopic in size, as that was the only way to achieve the high pressures (up to 300 kilobars) that we hoped to go to. No manufacturers made collimators suitable for producing such a small (~0.1 mm) X-ray beam, and so we consulted nearby Corning Glass works. They sent us a boxful of empty thermometer stems of various sizes made of lead glass. They were perfect when we cut them to length. The best results were achieved after days (and nights) of exposure. Our students soon fell into a ritual of developing the films at the end of each week when they were finished with their classes and could measure the diffraction lines to solve for crystal structure and molar volume. The small X-ray films (35 × 35 mm) were difficult to read accurately, especially with their dark backgrounds, due to the tiny sample sizes. We decided that focusing monochromatic sources would greatly improve the quality of our films. Although I found bent-crystal focusing monochromators for sale, there was no product for mounting them on our X-ray machines. So, I had Phil make a prototype box with 1/8″ brass walls to hold and tune the monochromator and told him it must not leak any of the X-rays. When he had finished, he showed me that it could hold water without leaking. That was safer than I had expected, but I decided that a few of them for the most challenging samples would serve nicely. From then on, we had sharp lines that stood out from the lighter backgrounds.

Once we learned what a capable machinist Phil was, we got together and persuaded the Rochester Geology Department to hire Phil and find space so we could establish a machine shop to make more diamond anvil cells, but with ever changing designs as we came to realize just how versatile the instrument could be, as a result of its extraordinarily simple principle. Diagrams of just a few of the many different designs that evolved are shown in reference [4].

We made a great team because Taro had a strong background in solid-state physics, thermodynamics, and math, at which he was far more capable than I was. He befriended some of the faculty in the Physics Department, who provided help in interpreting our data. Taro was very good at guiding our students to squeeze as much information as possible from their data.

I, on the other hand, loved dreaming up new designs for the diamond anvil cell. There were so many possibilities and so much to be studied. I am very grateful for Taro's grasp of the need for lab measurements, and his ability to interpret our data and guide our students just as Orson Anderson at Lamont, at about the same time, contributed to our understanding of planetary interiors through ultrasonic measurements of elastic properties, and his guidance of such able colleagues as Bob Liebermann and Ed Schreiber.

5. Research Results

There are many interesting parameters to be measured as a function of pressure. Crystal structure is essential for determining molar volume which, in turn, is needed for determining densities. The effect of pressure on these very basic properties is needed for interpreting field observations resulting from geophysical research. Therefore, these properties seemed like the most important ones to start with, especially as we were bringing together X-ray diffraction (my specialty) and pressure (Taro's specialty).

Fortunately, the crystal structures of most of the samples we wanted to study were well known or simple enough that they were easily determined. Initially, we did not worry about the question of hydrostaticity, but we soon found that deviatoric stress could lead to inaccurate molar volume measurements. We adopted Alvin Van Valkenburg's gasketing method of immersing samples in a fluid pressure medium or a solid soft enough to introduce negligible deviatoric stress. We used X-ray diffraction patterns of NaCl, or sometimes Au, as the pressure marker. In fact, we, along with others, devoted much effort to making the NaCl pressure scale accurate. We wrote 22 papers, mostly on the isothermal compression of minerals important to the Earth's interior. Figure 3 shows the plots of molar volume versus pressure of four of the mineral phases that were of greatest interest at the time.

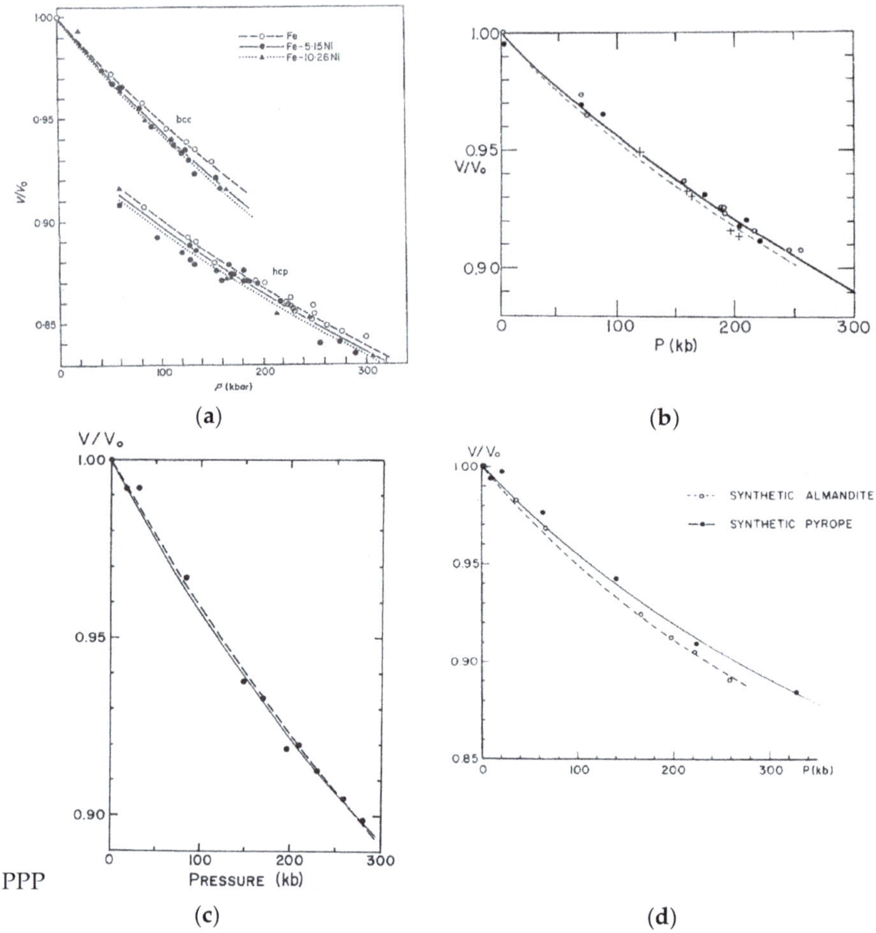

PPP

Figure 3. Four examples of early measurements of isothermal compression at 25 °C of phases important for interpreting data concerning the Earth's interior using lattice parameters based on X-ray diffraction (**a**) Fe, Fe-5.15 Ni , Fe-10.26 Ni [5]; (**b**) Three spinel phases of synthesized Fe_2SiO_4 (open circles), $Fe_{1.8}Mg_{0.2}SiO_4$ (pluses), $Fe_{1.6}Mg_{0.4}SiO_4$ (closed circles) [6]; (**c**) Spinel phase of synthesized Co_2SiO_4 dashed and solid lines by two slightly different least-square fits [7]; (**d**) Two garnet compositions [8].

6. High Temperatures Simultaneous with High Pressure

One day in the mid-1960s, while Taro and I were having lunch at a local restaurant, we were discussing the desirability of being able to raise the temperature of our samples simultaneously with high pressure. Resistance heating was certainly a possibility, but the achievable temperatures would not get us very far into the Earth's interior. Taro said, "Maybe we could use a laser beam to heat our samples to higher temperatures while they are under pressure". Lasers were very new at that point, so new that when we discussed our idea with a faculty member in the Department of Physics at the University of Rochester, he commented that "Maybe lasers will have a use after all". The idea immediately appealed to me, but lasers were not something you could place an order for. I thought maybe it would be good to get some experience with resistance heating and collecting data at temperatures up to perhaps ~800 °C. That was challenge enough, and it was only after Taro and I parted ways that I finally succeeded in coming up with a good design for resistance heating. A few years later, lasers became available and I visited the Howard Hughes Laser lab in California in 1968 with a diamond anvil cell loaded with graphite at high pressure. I shot the sample with their ruby laser and looked to see if the graphite had become transparent, indicating that it had transformed into diamond. It had not. Even though the test run was a failure, I was so convinced that Taro was right that I purchased the laser anyway. When I set it up in my lab and tried the experiment again, it failed again. The solution was easy, however. The diamond anvils were such good thermal conductors that the graphite simply was not hot enough for the transition. I mixed particles of graphite with NaCl and tried again. This time, the black particles of graphite not only became transparent, but they clearly had a very high refractive index that made them stand out from the salt. If a different sample was too transparent for the laser beam to heat it, graphite could be mixed with it to make it dark. Heating a sample by laser was an excellent way to promote phase transitions with kinetic barriers, like that of graphite to diamond. Most silicates of interest also had kinetic barriers and so the study of high-pressure phases of many silicates could benefit from laser heating. Measuring temperature, however, was a new challenge if pressure–temperature phase boundaries were to be determined or new phases were to be found. Blackbody emission spectra were the solution to that, as long as the sample was black or nearly black. Laser heating with more powerful YAG lasers became so intense that diamond particles in NaCl could be melted to form spherical droplets with a giant fullerene structure [9], and eventually anvils themselves could be heated to diamond melting temperatures without anything dark being present [10].

We accomplished a great deal of research on a variety of phases, nearly all believed to be components of planetary interiors, and attended many conferences during the ~ten years that Taro and I worked together. We had some very talented students: Ho-Kwang (Dave) Mao, who has become one of the leading researchers in mineral physics, is a member of the National Academy of Sciences, and founded HPSTAR, a new research institute in China patterned after Carnegie's Geophysical Laboratory. Lin-gun (John) Liu joined Ted Ringwood's Lab in Canberra, Australia. A few months later, Ted visited me in Rochester and told me that John had just successfully synthesized and identified the first silicate perovskite phase [11]. Ted was convinced that it was the so-called "post-spinel" phase, a phase that soon came to be considered the most abundant mineral in the Earth. It was not found as a naturally occurring phase until 2014, when it was given the mineral name bridgmanite in honor of Percy Bridgman [12].

After 2 years at Alfred and 7 years at Rochester, Taro took a sabbatical at Caltech working with Don Anderson and Tom Ahrens. After that, he felt he had accomplished what he had told Doc Ewing he wanted to do concerning the Earth's interior by finding ways to determine physical properties such as crystal structures, densities, and compressibilities that surely influence the velocities of seismic signals, as well as ways to solve compositions based on those physical properties at high pressures. He decided that the time had come to return to his geochemical research at Lamont Geological Observatory, just as he had told Doc he would do. Amazingly, while at Rochester, Taro had remotely continued his carbon dioxide research with Wally Broecker at Lamont.

Taro had, indeed, been instrumental in launching a new field which soon came to be known as mineral physics thanks to Orson Anderson and Ed Schreiber. I missed Taro and Elaine as friends and colleagues, but continued my interest in mineral physics, moving to Cornell University where there was the Cornell High Energy Synchrotron Source (CHESS), which soon proved to be one of the most important new technologies for conducting research in our new field of study.

Taro's suggestion of laser heating had opened a new realm of research into examination of the properties of minerals deep in the Earth's interior, eventually leading researchers all the way to the pressure-temperature conditions in the Earth's core, and even the interiors of larger planets. Diamond anvil cells and multi-anvil devices have evolved side by side over the decades following the 1960s, each filling increasingly diverse needs. Nothing can compare to the multi-anvil devices when sample size matters, or the synthesis of new phases is needed for further testing. However, it is hard to beat the diamond anvil cells if the experimenter wants to travel to a different lab or conference with a sample already under high-pressure and easily slipped into a suitcase or a pocket. In the decades that followed, some giant advances in technology have motivated users to find ways to use them. Synchrotron sources along with digital methods for recording X-ray patterns now provide such small, intense X-ray beams and rapid imaging that exposures which used to require a week can now be made in seconds and, in some cases, fractions of seconds, making it possible to record a movie of a phase transition or a chemical reaction. Synchrotron radiation also made it possible to introduce a range of spectroscopic techniques that could yield information never before possible. I hope that my descriptions, along with Bob Liebermann's descriptions of the early days of mineral physics at Lamont, will further motivate newcomers to explore ways to delve even more deeply as technology advances [1].

Taro, who passed away on 3 December 2019, will be sorely missed by all of us who knew him. It was an honor to have had the chance to work so closely with such a remarkable and versatile research scientist, with such a keen mind and such great ideas. If it had not been for Taro, I might never have joined such a fascinating branch of science at the time of its inception.

Acknowledgments: I especially want to thank Bob Liebermann for inviting me to tell this story about Taro Takahashi's lesser-known contributions to one of today's most active new branches of science. I appreciate the help given to me by my friend, Elise Skalwold, in helping me prepare this paper.

Conflicts of Interest: The authors declare no conflict of interest.

Appendix A

List of Peer Reviewed Papers by Takahashi, Bassett, and Students:

- Bassett, W.A.; Takahashi, T. Specific volume measurements of crystalline solids at pressures up to 200 kilobars by X-ray diffraction, In *1964 Symposium on High-pressure Technology*; Giardini, A.A., Lloyd, E.C., Eds.; American Society of Mechanical Engineers: New York, NY, USA, 1964; pp. 1–7.
- Bassett, W.A.; Takahashi, T. Silver iodide polymorphs. *Am. Miner.* 1965, *50*, 1576–1594.
- Bassett, W.A.; Takahashi, T.; Stook, P. X-ray diffraction and optical observations on crystalline solids up to 300 kilobars. *Rev. Sci. Instrum.* 1967, *38*, 37–42.
- Mao, H.K.; Bassett, W.A.; Takahashi, T. Effect of pressure on crystal structure and lattice parameters of iron up to 300 kilobars. *J. Appl. Phys.* 1967, *38*, 272–276.
- Takahashi, T.; Bassett, W.A. The effect of pressure on the density of possible mantle-core minerals up to 300 kilobars, In *U.S. Progress Report, International Upper Mantle Project*; National Academy of Sciences, 1967; pp. 134–136.
- Bassett, W.A.; Takahashi, T.; Mao, H.K.; Weaver, J.S. Pressure-induced phase transformation in NaCl. *J. Appl. Phys.* 1968, *39*, 319–325.
- Takahashi, T.; Bassett, W.A.; Mao, H.K. Isothermal compression of the alloys of iron up to 300 kilobars at room temperature: Iron-nickel alloys. *J. Geophys. Res.* 1968, *73*, 4717–4725.
- Mao, H.K.; Takahashi, T.; Bassett, W.A.; Weaver, J.S.; Akimoto, S. Effect of pressure and temperature on the molar volumes of wustite and three $(Fe,Mg)_2SiO_4$ spinel solid solutions, *J. Geophys. Res.* 1969, *74*, 1061–1069.
- Takahashi, T.; Mao, H.K.; Bassett, W.A. X-ray diffraction study of a high pressure polymorph of lead, *Science* 1969, *165*, 1352–1353.
- Bassett, W.A.; Takahashi, T.; Campbell, J.K. Volume changes for the B1-B2 phase transformations in three potassium halides at room temperature, *Trans. Am. Crystallographic Assoc.* 1969, *5*, 93–103.
- Mao, H.K.; Takahashi, T.; Bassett, W.A. Isothermal compression of the spinel phase of Ni_2SiO_4 up to 300 kilobars at room temperature, *Phys. Earth Planet. Interiors* 1970, *3*, 51–53.
- Takahashi, T.; Liu, L.G. Compression of ferromagnesian garnets and the effects of solid solutions on the bulk modulus. *J. Geophys. Res.* 1970, *75*, 5757–5766.
- Liu, L.G.; Takahashi, T.; Bassett, W.A. Effect of pressure and temperature on the lattice parameters of rhenium. *J. Phys. Chem. Solids* 1970, *31*, 1345–1351.

- Mao, H.K.; Bassett, W.A.; Takahashi, T. High pressure phase transformation in magnetite. *Annu. Rep. Dir. Geophys. Lab.* 1970, *68*, 249–250.
- Mao, H.K.; Takahashi, T.; Bassett, W.A. Study of lead up to 180 kilobars. *Annu. Rep. Dir. Geophys. Lab.* 250–253.
- Weaver, J.S.; Takahashi, T.; Bassett, W.A. Calculation of the P-V relation for sodium chloride up to 300 kilobars at 25 °C, In *Accurate Characterization of the High-Pressure Environment: Proceedings*; Lloyd, E.C., Ed.; US National Bureau of Standards: Gaithersburg, MD, USA, 1971; pp. 189–199.
- Bassett, W.A.; Takahashi, T. Effect of pressure on possible mantle minerals up to 300 kilobars; In *Upper Mantle Project*; National Research Council of the National Academies (NRC), 1971; pp. 181–183.
- Liu, L.G.; Bassett, W.A.; Takahashi, T. Effect of pressure on lattice parameters of stishovite. *J. Geophys. Res.* 1974, *79*, 1160–1164.
- Mao, H.K.; Takahashi, T.; Bassett, W.A.; Kinsland, G.L.; Merrill, L. Isothermal compression of magnetite to 320 kbar and pressure-induced phase transformation. *J. Geophys. Res.* 1974, 79, 1165–1170.
- Liu, L.G.; Bassett, W.A.; Takahashi, T. Isothermal compressions of a spinel phase of Co_2SiO_4 and magnesian ilmenite. *J. Geophys. Res.* 1974, *79*, 1171–1174.
- Bassett, W.A.; Takahashi, T. X-ray diffraction studies up to 300 kilobars. In *Advances in High Pressure Research*; Wentorf, R., Ed.; Academic Press London and New York, 1974; Volume 4, pp. 165–247.

References

1. Liebermann, R.C. Multi-anvil, high pressure apparatus: A half-century of development and progress. *High Press. Res.* **2011**, *31*, 493–532. [CrossRef]
2. Bassett, W.A. Diamond Anvil Cell, 50th Birthday. *High Press. Res.* **2009**, *29*, 163–186. [CrossRef]
3. Bassett, W.A.; Takahashi, T.; Stook, P. X-ray diffraction and optical observations on crystalline solids up to 300 kilobars. *Rev. Sci. Instrum.* **1967**, *38*, 37–42. [CrossRef]
4. Bassett, W.A. The diamond cell and the nature of the Earth's mantle. *Ann. Rev. Earth Planet. Sci.* **1979**, *7*, 357–384. [CrossRef]
5. Takahashi, T.; Bassett, W.A.; Mao, H.K. Isothermal compression of the alloys of iron up to 300 kilobars at room temperature: Iron-nickel alloys. *J. Geophys. Res.* **1968**, *73*, 4717–4725. [CrossRef]
6. Mao, H.K.; Takahashi, T.; Bassett, W.A.; Weaver, J.S.; Akimoto, S. Effect of pressure and temperature on the molar volumes of wustite and three $(Fe,Mg)_2SiO_4$ spinel solid solutions. *J. Geophys. Res.* **1969**, *74*, 1061–1069. [CrossRef]
7. Liu, L.G.; Bassett, W.A.; Takahashi, T. Isothermal compression of a spinel phase of Co_2SiO_4 and magnesian ilmenite. *J. Geophys. Res.* **1974**, *79*, 1171–1174. [CrossRef]
8. Takahashi, T.; Liu, L.G. Compression of ferromagnesian garnets and the effect of solid solutions on the bulk modulus. *J. Geophys. Res.* **1970**, *75*, 5757–5766. [CrossRef]
9. Bassett, W.A.; Weathers, M.S. Fullerene structures produced from melted diamond at high pressure by laser heating. In *Proceedings of the 1993 AIRAPT Conference in Colorado Springs*; American Institute of Physics: Melville, NY, USA, 1994; pp. 651–653.
10. Gold, J.S.; Bassett, W.A.; Weathers, M.S.; Bird, J.M. Melting of diamond. *Science* **1984**, *225*, 921–922. [CrossRef] [PubMed]
11. Liu, L. Silicate Perovskite from Phase Transformation of Pyrope-Garnet at High Pressure and Temperature. *Geophys. Res. Letters* **1974**, *1*, 277–280. [CrossRef]
12. Tschauner, O.; Ma, C.; Beckett, J.R.; Prescher, C.; Prakapenka, V.B.; Rossman, G.R. Discovery of bridgmanite, the most abundant mineral in Earth, in a shocked meteorite. *Science* **2014**, *346*, 1100–1102. [CrossRef] [PubMed]

minerals

MDPI

Editorial

The Birth of Mineral Physics at the ANU in the 1970s

Robert Cooper Liebermann

Mineral Physics Institute and Department of Geosciences, Stony Brook University, Stony Brook, NY 11794-2100, USA; Robert.Liebermann@stonybrook.edu; Tel.: +1-(631)-766-5711

Received: 20 January 2020; Accepted: 10 February 2020; Published: 12 February 2020

Abstract: In 1970, I established the first mineral physics laboratory in Australia at the Australian National University (ANU) under the auspices of A.E. (Ted) Ringwood. Over the next six years, we published 25 research papers in peer-reviewed journals, many of them in collaboration with graduate students, Ian Jackson and Leonie Jones. This research was focused on measurements of sound velocities in minerals (and their structural analogues) at high pressures and temperatures, as well as studies of melting and elastic shear instabilities in materials and applications of these experimental data to interpreting seismic models of the Earth's interior.

Keywords: mineral physic; ultrasonics; sound velocities; high-pressure phases; melting and elastic shear instabilities

1. Introduction

Introduction reproduced with permission from Liebermann [1], published by Elsevier B.V., 2014.

In 1969, when I was a graduate student at Columbia University's Lamont Geological Observatory, my advisor Orson Anderson attended an international conference at the Australian National University (ANU) in Canberra, Australia, to deliver our joint paper on lattice dynamic calculations of the elastic moduli of cubic lattices [2]. This was the meeting at which A.D. Wadsley collapsed on the stage and died while Frank Press was delivering his talk (T. Takahashi, personal communication, March 2013). During that meeting, A.E. (Ted) Ringwood informed Anderson that he was looking for a postdoc to set up an ultrasonics laboratory; subsequently, Richard O'Connell declined to apply for this job and I was offered the post. Enroute from Lamont to Canberra, I spent 10 months as a postdoc at the Seismological Laboratory at Caltech; during that period, I profited from advice and guidance on equipment for my new ultrasonics lab at the ANU from Hartmut Spetzler and Rick O'Connell.

At the Birch Symposium at Harvard in 1970, I met Ted Ringwood and we laid plans to purchase equipment so that it would be on campus when I arrived in late September. John Jaeger (then the Chair of the Department of Geophysics and Geochemistry at the ANU) visited the Anderson lab at Lamont later that year and told me that he hoped I knew how to operate and maintain all that electronic equipment as the ANU did not have suitable staff to do so; that turned out to be apocryphal as the Department not only had an excellent machine shop, but also several very talented electronic staff. In addition, Ringwood and David Green were blessed with two superb Senior Technical Officers, Alan Major and William Hibberson, who were tremendous assets to my new research program.

My six years at the ANU were a fantastic period of exposure to high-pressure geosciences. While I was still at Lamont, Frank Stacey from the University of Queensland was a visiting scientist. In early 1971, Stacey invited me to give a paper at the Australian–New Zealand Association of Science (ANZAS) in Brisbane. Later, he recommended to Ian Jackson and Leonie Jones, both graduates of the Physics Department of the University of Queensland, that they apply for a graduate fellowship to study at the ANU under my supervision. When they arrived to begin their graduate research programs, Ringwood was nervous about their lack of exposure to Earth sciences, so he insisted that they take "remedial" courses in the geology department. After both completed introductory geology, Jackson took the 3rd year mineralogy course and topped the class; Ringwood never raised the issue again.

In 1973, the ANU established the Research School of Earth Sciences (RSES) which provided the opportunity for the Department of Geophysics and Geochemistry to move out from under the control of the Research School of Physical Sciences. Anton Hales arrived from the University of Texas at Dallas to become the Founding Director of the RSES; over the next three years, I learned much from Hales about scientific leadership and administration.

2. Arrival and First Years at the ANU

In October 1970, my wife Barbara and I arrived in Canberra with our one-year old daughter, Karen in tow. As Ted Ringwood was overseas, David Green welcomed us and installed us in Apt. #1 of the Garran University flats in the Woden valley. John Cleary led the search for a car, ending in a 2-door Datsun coupe (a little difficult to install a baby carrier in the rear seat).

On one of my first days in the Department of Geophysics and Geochemistry, Prof. Jaeger introduced me to the faculty and students and offered me a yard of ale with instructions to down it in one gulp. The ultrasonic equipment ordered when I was still at Lamont had been delivered, and with the help of electronics specialist, Eddie Penikis, I began to set up the new ultrasonics laboratory.

With Alan Major, we began to learn how to hot-press polycrystalline specimens suitable for ultrasonic measurements of elastic wave velocities, using synthetic powders prepared by Alan and Bill Hibberson [3]. The first specimens were fabricated in a piston-cylinder apparatus at pressures less than 35 kbar (3.5 GPa) and temperatures up top 1200 °C; we initially focused on germanate analogues of silicate mineral phases. We tested the polycrystals in the ultrasonics lab using a polishing jig and specimen holder fabricated in the Department machine shop (both of which are still currently in use in our lab at Stony Brook).

The first velocity experiment at high pressure were conducted in a liquid-media apparatus to 7.5 kbar in laboratory of Mervyn Paterson with assistance of Bill McIntyre; Bill had been assigned by Jaeger to work with me while Jaeger was "wintering" in Northern Hemisphere. Ringwood was *very* surprised to learn about this assignment but had no basis to object.

The first results on rutile and spinel compounds were obtained by the pulse transmission ("time-of-flight") technique and reported at the IUGG meeting in Moscow, Russia, in August 1970, less than one year after establishing the new ultrasonics lab at the ANU, with technical assistance of new electronics technician Doug Mayson [4]. These "time-of-flight" measurements were collected to 7.5 kbar at room T and the data at highest P were taken to represent pore free specimen [4–11].

The next advance in ultrasonics was to employ the pulse superposition technique to obtain more precise measures of travel times to 7.5 kbar. Values reported were those "back-extrapolated" to 1 bar to obtain velocities representative of a pore-free specimen; these data were then compared with results from single crystals in other labs [5,11–14]. In later experiments, the measurements of velocities using the pulse superposition technique were sufficiently precise to allow determination of the pressure derivatives.

3. Research of Graduate Students Ian Jackson and Leonie Jones

Ian Jackson initially worked on melting and elastic shear instabilities in alkali halides [15] and engaged in some hard-sphere modeling of melting with Bob Watts of the Research School of Physical Sciences.

Ian later studied the disproportionation of spinels into mixed oxides and revealed the significance of cation configuration and implications for the mantle [16]. and the elastic properties of ($Mg_xFe_{1-x}O$) solid solutions [11].

His PhD dissertation (1976) was entitled "Phase equilibria and elastic properties in silicate analogue systems: studies of melting and polymorphic phase transformations".

Leonie Jones initially worked on the elastic and thermal properties of fluoride and oxide analogues with the rock salt, fluorite, rutile, perovskite structures [17] and later the elasticity of aluminate, titanate, stannate and germanate compounds with the perovskite structure [18].

Her Ph.D. dissertation (1976) was entitled "High-temperature elastic properties of fluoride and oxide analogues".

4. Related Research (1971–1976)

In 1976, I worked on phase equilibria and elastic properties of pyrolite model for the oceanic upper mantle with David Green, who presented our paper at the IUGG international congress in Grenoble, France [19]. Ringwood expressed surprise that I was collaborating with his colleague and competitor when I was working "under the supervision of Ringwood" (Figure 1).

Figure 1. Visit of Professor Syun-iti Akimoto of the University of Tokyo to Research School of Earth Sciences (RSES) in 1975: L to R: David Green, Ted Bence from Stony Brook, Don Anderson from Caltech, Ted Ringwood, Mrs. Akimoto and author.

We also worked on a number of other mineral physics projects:

Elasticity systematics and applications to interpretation of Earth models [5,20–22].

Hot-pressing of polycrystalline specimens of high-pressure phases [23]—presented at 4th International Conference of High Pressure—25–29 November 1974, Kyoto, Japan. See also [24].

Elasticity of anorthite and nature of the lunar crust [25].

Elasticity of stishovite—most challenging project to hot-press and measure sound velocities in polycrystalline stishovite [9,10]. No anisotropy observed. Consistent with later Brillouin scattering data from Weidner's laboratory [26].

Elasticity and phase equilibria of spinel disproportion reactions and thermomechanical regime of descending lithospheric plates [16,27].

5. Interactions with Other Research Groups in the RSES

I was welcomed into the rock mechanics research group of Mervyn Paterson and Bruce Hobbs. There I met and interacted with Jim Boland and John Fitzgerald, specialists in transmission electron microscopy. I also had extensive contact with John Cleary and others in their seismological group.

As there were no formal courses required for graduate students in the Institute of Advanced Studies at the ANU, I initiated a series of informal seminars on Monday evenings for Ian Jackson and Leonie Jones. In addition to reading and discussing current geophysical literature, we profited from participation of visiting scientists, including Don Anderson, Adam Dziewonski and Ralph

Lapwood. It was during that era that geophysicists were exploring use of inverse theory to analyze observational data to obtain velocity and density models of the Earth's interior. To help us understand this approach, Prof. Lapwood kindly offered a set of tutorial lectures on inverse theory, using the book by Cornelius Lanzcos.

After arriving in 1973 to become the first Director of RSES, Anton Hales became a regular attendee at our weekly seminars. In addition to bringing his decades of geophysical knowledge, he provided funds for me to buy cheese and crackers to sustain us through the evening (and to leaven the beer which we drank).

When the Department of Geophysics and Geochemistry evolved to become the Research School of Earth Sciences (RSES) in 1973, it became necessary to establish a Faculty Board and produce bylaws; I helped Ian MacDougall to draft these bylaws and we served together on the Faculty Board. During one of the board meetings, some members challenged Director Hales on the appropriateness of his contract with the U. S. Department of Defense (to install and operate seismic stations on the Australian continent); Hales abruptly adjourned the meeting and deputized me to talk to Faculty Board members and convince them that it was Hales' right to hold such a contract.

6. Post-1970 Research Collaboration between ANU and Stony Brook

In 1976, I was offered a faculty position as an Associate Professor in the Department of Earth and Space Sciences (later Department of Geosciences) at Stony Brook University and was encouraged by Orson Anderson (my Ph.D. advisor) to accept.

After a postdoctoral appointment in the laboratory of Thomas Ahrens at Caltech, Ian Jackson returned to the ANU in the late 1970s to take up a faculty position in the Petrophysics Group of Mervyn Paterson in the Research School of Earth Sciences.

Jackson and I continued to collaborate on research projects, capitalizing on ultrasonic techniques developed in Jackson's lab at the ANU and synthesis techniques to fabricate polycrystalline specimens of high-pressure phases of mantle minerals developed in my lab at Stony Brook. Gabriel Gwanmesia and Baosheng Li hot-pressed the polycrystals at Stony Brook, and Sally Rigden measured the sound velocities at the ANU during visits by Gwanmesia and the author (Figure 2a,b) [28–34].

(**a**)

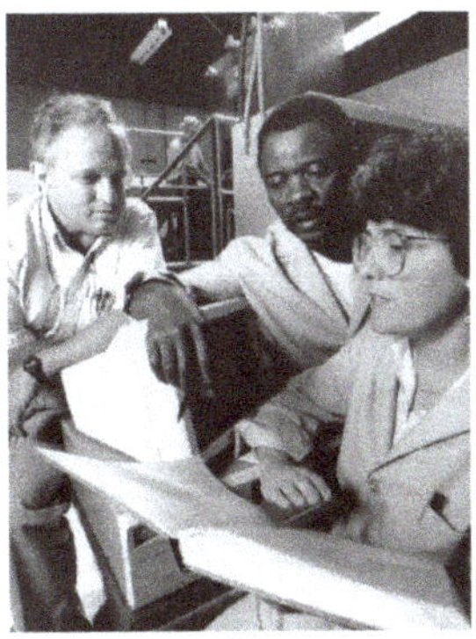

(**b**)

Figure 2. (**a**) Gabriel Gwanmesia and Sally Rigden with the ultrasonic interferometer in Ian Jackson's laboratory at the ANU. (**b**) Sally Rigden from the ANU with Gabriel Gwanmesia and the author in High Pressure Laboratory at Stony Brook in 1991. Reproduced with permission from Liebermann [1], published by Elsevier B.V., 2014.

In 1994, Baosheng Li, then a graduate student and now a Professor in the Department of Geosciences, developed techniques to incorporate ultrasonic interferometric measurements of sound velocities in situ in the Kennedy-Getting high-pressure apparatus in collaboration with Ian Jackson (my first graduate student and subsequently the Director of the Research School of Earth Sciences at the ANU; see Figure 3a,b). This pioneering development provided the opportunity to conduct sound velocity experiments to pressures of 10 GPa [35] and was exploited over the next few years by many graduate students and postdocs.

(a)

(b)

Figure 3. (**a**) May 1994. Baosheng Li and Ian Jackson in the Stony Brook High Pressure Laboratory. Ian spent 6 weeks in our laboratory working with Baosheng on interfacing the ANU ultrasonic interferometry system with the 1000-ton split-cylinder apparatus of the Kennedy–Getting type [35]. Reproduced with permission from Liebermann [1], published by Elsevier B.V., 2014. (**b**) Baosheng Li and Kennedy-Getting press with Walker-type module in High Pressure Lab in 1994. Reproduced with permission from Liebermann [1], published by Elsevier B.V., 2014.

As part of her doctoral research at the ANU, Jennifer Kung spent six months at Stony Brook on the synthesis of polycrystalline specimens of $ScAlO_3$-perovskite with the aid of Gabriel Gwanmesia [36,37]. When she was a postdoctoral research associate at Stony Brook from 2000–2005, Jennifer measured the high-temperature velocities of orthoenstatite in Jackson's lab (Figure 4) [38]; Sytle Antao (graduate student of John Parise) accompanied Jennifer to measure the sound velocities of magnesium ferrite spinel [39].

Figure 4. Jennifer Kung with high-temperature, ultrasonic apparatus in laboratory of Ian Jackson at the RSES at the ANU. Circa 2011.

In the most recent collaboration between the ANU and Stony Brook, Ting Chen and her colleagues measured the sound velocities of polycrystalline coesite at high pressure and temperature in Baosheng Li's laboratory [40] using a specimen hot-pressed by the author in a girdle—anvil apparatus in 1975, and so very "well-aged" [23].

In the author's most recent visit to Australia, he and his wife were hosted by Maggie and Ian Jackson at their cottage on Lilli Pilli beach, which we remember fondly from our many family visits there from 1970 to 1976 (Figure 5).

Figure 5. Ian Jackson and the author at Lilli Pilli, Australia, 2006.

Funding: This paper was written with support from a research grant to Baosheng Li and the author from the National Science Foundation (EAR-1524078).

Acknowledgments: I would like to dedicate this paper to the memory of A.E. (Ted) Ringwood who offered me a research position at the ANU in 1970 and provided technical support and scientific advice over the ensuing 6 years (and, only in exceptional cases, requested to be a co-author). I am grateful to Alan Major, Bill Hibberson and Doug Mayson for their technical contributions to our ultrasonic laboratory and to Ian Jackson and Leonie Jones for their doctoral studies under my supervision. My post-ANU collaboration with Jackson's Petrophysics group was energized by Sally Rigden and Jennifer Kung from the ANU and Gabriel Gwanmesia and Baosheng Li from Stony Brook. I also thank Ian Jackson and Jennifer Kung for their comments and input to early versions of this paper. During my 6 years at the ANU, I maintained an ongoing correspondence with Orson Anderson, my PhD advisor from Columbia University, who kept me up to date on faculty searches in U.S. universities and offered me counsel on which positions might be the best for me on my return to the U.S.

Conflicts of Interest: The author declares no conflict of interest.

References

1. Liebermann, R.C. The role of serendipity in my career in mineral physics: 1968 to 2013. *Phys. Earth Planet. Inter.* **2014**, *228*, 307–323. [CrossRef]
2. Anderson, O.L.; Liebermann, R.C. Equations for the pressure derivatives of the elastic constants for three cubic lattices, and some geophysical applications. *Phys. Earth Planet. Inter.* **1970**, *3*, 61–85. [CrossRef]
3. Liebermann, R.C. Elastic properties of minerals determined from ultrasonic or compression data. *Phys. Earth Planet. Inter.* **1972**, *5*, 213–217. [CrossRef]
4. Liebermann, R.C. Compressional velocities of polycrystalline olivine, spinel and rutile minerals. *Earth Planet. Sci. Lett.* **1972**, *17*, 263–268. [CrossRef]
5. Liebermann, R.C. Elastic properties of germanate analogues of olivine, spinel and beta polymorphs of (Mg, Fe)$_2$SiO$_4$. *Nat. Phys. Sci.* **1973**, *244*, 105–107. [CrossRef]
6. Liebermann, R.C. Elasticity of pyroxene-garnet and pyroxene-ilmenite phase transformations in germinates. *Phys. Earth Planet. Inter.* **1974**, *8*, 361–374. [CrossRef]
7. Liebermann, R.C.; Mayson, D.J. Elastic properties of polycrystalline diopside ($CaMgSi_2O_6$). *Phys. Earth Planet. Inter.* **1976**, *11*, 1–4. [CrossRef]
8. Liebermann, R.C. Elasticity of ilmenites. *Phys. Earth Planet. Inter.* **1976**, *12*, 5–10. [CrossRef]
9. Liebermann, R.C.; Ringwood, A.E.; Major, A. Elasticity of stishovite. *Earth Planet. Sci. Lett.* **1976**, *32*, 129–140. [CrossRef]
10. Liebermann, R.C.; Ringwood, A.E. Some comments on the elasticity of stishovite as determined by ultrasonic and static compression techniques. In *High-Pressure Research: Applications in Geophysics*; Manghnani, M.H., Akimoto, S., Eds.; Academic Press: New York, NY, USA, 1977; pp. 343–349.
11. Jackson, I.; Liebermann, R.C.; Ringwood, A.E. The elastic properties of ($Mg_xFe_{1-x}O$) solid solutions. *J. Phys. Chem. Miner.* **1978**, *3*, 11–31. [CrossRef]
12. Liebermann, R.C. Elastic properties of polycrystalline SnO_2 and GeO_2: Comparison with stishovite and rutile data. *Phys. Earth Planet. Inter.* **1973**, *7*, 461–465. [CrossRef]
13. Liebermann, R.C. Elasticity of olivine (a), beta (b) and spinel (g) polymorphs of germanates and silicates. *Geophys. J. R. Astron. Soc.* **1975**, *42*, 899–929. [CrossRef]
14. Liebermann, R.C. Elasticity of the ilmenite-perovskite phase transformation in $CdTiO_3$. *Earth Planet. Sci. Lett.* **1976**, *29*, 326–332. [CrossRef]
15. Jackson, I.N.S.; Liebermann, R.C. Melting and elastic shear instability of alkali halides. *J. Phys. Chem. Solids* **1974**, *35*, 1115–1119. [CrossRef]
16. Jackson, I.N.S.; Liebermann, R.C.; Ringwood, A.E. Disproportionation of spinels into mixed oxides: Significance of cation configuration and implications for the mantle. *Earth Planet. Sci. Lett.* **1974**, *24*, 203–208. [CrossRef]
17. Jones, L.E.A.; Liebermann, R.C. Elastic and thermal properties of fluoride and oxide analogues with the rocksalt, fluorite, rutile, perovskite structures. *Phys. Earth Planet. Inter.* **1974**, *9*, 101–107. [CrossRef]
18. Liebermann, R.C.; Jones, L.E.A.; Ringwood, A.E. Elasticity of aluminate, titanate, stannate and germanate compounds with the perovskite structure. *Phys. Earth Planet. Inter.* **1977**, *14*, 165–178. [CrossRef]
19. Green, D.H.; Liebermann, R.C. Phase equilibria and elastic properties of pyrolite model for the oceanic upper mantle. *Tectonophysics* **1976**, *32*, 61–92. [CrossRef]

20. Liebermann, R.C.; Ringwood, A.E. Birch's law and polymorphic phase transformations. *J. Geophys. Res.* **1973**, *78*, 6926–6932. [CrossRef]
21. Liebermann, R.C. On velocity density systematics, polymorphic phase transformations, and the transition zone of the earth's mantle. *Comments Earth Sci. Geophys.* **1973**, *3*, 127–133.
22. Liebermann, R.C. Elasticity of the olivine-spinel and olivine-beta phase transformations and the 400 km. discontinuity of the mantle. *J. Geophys. Res.* **1973**, *78*, 7015–7017. [CrossRef]
23. Liebermann, R.C.; Ringwood, A.E.; Mayson, D.J.; Major, A. Hot-pressing of polycrystalline aggregates at very high pressures for ultrasonic measurements. In Proceedings of the 4th International Conference on High Pressure, Kyoto, Japan, 25–29 November 1974; pp. 495–502.
24. Gwanmesia, G.D.; Li, B.; Liebermann, R.C. Hot pressing of polycrystals of high-pressure phases of mantle minerals in multi-anvil apparatus. *Pure Appl. Geophys.* **1993**, *141*, 467–484. [CrossRef]
25. Liebermann, R.C.; Ringwood, A.E. Elastic properties of anorthite and the nature of the lunar crust. *Earth Planet. Sci. Lett.* **1976**, *31*, 69–74. [CrossRef]
26. Weidner, D.J.; Bass, J.D.; Ringwood, A.E.; Sinclair, E. The single-crystal elastic moduli of stishovite. *J. Geophys. Res.* **1982**, *87*, 40–474. [CrossRef]
27. Liebermann, R.C.; Jackson, I.; Ringwood, A.E. Elasticity and phase equilibria of spinel disproportionation reactions. *Geophys. J. R. Astron. Soc.* **1977**, *50*, 553–586. [CrossRef]
28. Rigden, S.M.; Jackson, I.N.S.; Niesler, H.; Liebermann, R.C.; Ringwood, A.E. Pressure dependence of the elastic wave velocities for Mg_2GeO_4 spinel up to 3 GPa. *Geophys. Res. Lett.* **1988**, *15*, 605–608. [CrossRef]
29. Gwanmesia, G.; Liebermann, R.C.; Guyot, F. Hot-pressing and characterization of polycrystals of ß-Mg_2SiO_4 for acoustic velocity measurements. *Geophys. Res. Lett.* **1990**, *17*, 1331–1334. [CrossRef]
30. Gwanmesia, G.D.; Rigden, S.; Jackson, I.; Liebermann, R.C. Pressure dependence of elastic wave velocity for β-Mg_2SiO_4 and the composition of the Earth's mantle. *Science* **1990**, *250*, 794–797. [CrossRef]
31. Rigden, S.M.; Gwanmesia, G.D.; FitzGerald, J.D.; Jackson, I.; Liebermann, R.C. Spinel elasticity and seismic structure of the transition zone of the mantle. *Nature* **1991**, *354*, 143–145. [CrossRef]
32. Rigden, S.M.; Gwanmesia, G.; Jackson, I.; Liebermann, R.C. Progress in high-pressure ultrasonic interferometry: The pressure dependence of elasticity of high-pressure Mg_2SiO_4 polymorphs and constraints on the composition of the transition zone of the Earth's mantle. In *High Pressure Research: Applications to Earth and Planetary Sciences*; Syono, Y., Manghnani, M.H., Eds.; Terra Scientific Publishing Co.: Tokyo, Japan; American Geophysical Union: Washington, DC, USA, 1992; pp. 167–182.
33. Rigden, S.M.; Gwanmesia, G.D.; Liebermann, R.C. Elastic wave velocities of a pyrope-majorite garnet to 3 GPa. *Phys. Earth Planet. Inter.* **1994**, *84*, 35–44. [CrossRef]
34. Li, B.; Rigden, S.M.; Liebermann, R.C. Elasticity of stishovite at high pressure. *Phys. Earth Planet. Inter.* **1996**, *96*, 113–127. [CrossRef]
35. Li, B.; Jackson, I.; Gasparik, T.; Liebermann, R.C. Elastic wave velocity measurement in multi-anvil apparatus to 10 GPa using ultrasonic interferometry. *Phys. Earth Planet. Inter.* **1996**, *98*, 79–91. [CrossRef]
36. Kung, J.; Rigden, S.; Gwanmesia, G. Elasticity of $ScAlO_3$ at high pressure. *Phys. Earth Planet. Inter.* **2000**, *118*, 65–75. [CrossRef]
37. Kung, J.; Rigden, S.; Jackson, I. Silicate perovskite analogue $ScAlO_3$: Temperature dependence of elastic moduli. *Phys. Earth Planet. Inter.* **2000**, *120*, 299–314. [CrossRef]
38. Kung, J.; Jackson, I.; Liebermann, R.C. High-temperature elasticity of polycrystalline orthoenstatite ($MgSiO_3$). *Am. Mineral.* **2011**, *96*, 577–585. [CrossRef]
39. Antao, S.M.; Jackson, I.; Li, B.; Kung, J.; Chen, J.; Hassan, I.; Liebermann, R.C.; Parise, J.B. High-temperature elasticity of magnesioeferrite spinel. *Phys. Chem. Miner.* **2007**, *34*, 345–350. [CrossRef]
40. Chen, T.; Liebermann, R.C.; Zou, Y.; Li, Y.; Qi, X.; Li, B. Tracking silica in Earth's upper mantle using new sound velocity data for coesite to 5.8 GPa and 1073 K. *Geophys. Res. Lett.* **2017**, *44*, 7757–7765. [CrossRef]

minerals

MDPI

Editorial

My Career as a Mineral Physicist at Stony Brook: 1976–2019

Robert Cooper Liebermann

Mineral Physics Institute and Department of Geosciences, Stony Brook University, Stony Brook, NY 11794-2100, USA; Robert.Liebermann@stonybrook.edu; Tel.: +001-(631)-766-5711

Received: 14 November 2019; Accepted: 6 December 2019; Published: 7 December 2019

Abstract: In 1976, I took up a faculty position in the Department of Geosciences of Stony Brook University. Over the next half century, in collaboration with graduate students from the U.S., China and Russia and postdoctoral colleagues from Australia, France and Japan, we pursued studies of the elastic properties of minerals (and their structural analogues) at high pressures and temperatures. In the 1980s, together with Donald Weidner, we established the Stony Brook High Pressure Laboratory and the Mineral Physics Institute. In 1991, in collaboration with Alexandra Navrotsky at Princeton University and Charles Prewitt at the Geophysical Laboratory, we founded the NSF Science and Technology Center for High Pressure Research.

Keywords: mineral physics; ultrasonic interferometry; high-pressure multi-anvil apparatus; synchrotron X-radiation; CHiPR (Center for High Pressure Research); COMPRES (Consortium for Materials Properties Research in Earth Sciences); Stony Brook University

---

1. Introduction

In March 2013, I was invited by Tetsuo Irifune to give the keynote address at the Final Symposium of G-COE and TANDEM program at Ehime University in Matsuyama, Japan. I chose that opportunity to reflect on my career in mineral physics with the title: "Mineral Physics and Bob-san from 1963 to 2013: Role of Serendipity." Yanbin Wang was in the audience in Matsuyama and suggested that I submit a manuscript on this topic to Physics of the Earth and Planetary Interiors for a Special Issue for which he was the Guest Editor. When Jay Bass heard of this talk, he recommended that I be asked to reprise this talk after the banquet at the 2013 Annual Meeting of COMPRES at Lake Geneva, Wisconsin in June 2013; I modified the title for the U.S. audience: "Role of Serendipity in My Career in Mineral Physics: 1963 to 2013." In October 2013, Dan Davis (the Chair of the Department of Geosciences) invited me to give this talk at the weekly Geosciences Colloquium and later to adapt it for the Geosciences Newsletter. This paper is the result and focuses on my career in mineral physics at Stony Brook from 1976 to 2019 and it compliments my earlier paper in Physics of the Earth and Planetary Interiors [1].

2. Before Stony Brook

After being an undergraduate in geophysics at Caltech and obtaining a PhD at Columbia University under Orson Anderson, I spent 6 years at the Australian National University (ANU) in Canberra working in the high-pressure laboratory of Ted Ringwood. When I decided to explore the opportunities to return to the U.S. to a tenure-track faculty position, I was fortunate that the Department of Earth and Space Sciences at Stony Brook was conducting a faculty search led by Don Weidner, who was open-minded enough to consider hiring another mineral physicist. I was encouraged to negotiate an attractive start-up package by Charlie Prewitt and Ted Bence (who were on sabbaticals in Australia at Monash University in Melbourne and the ANU, respectively).

3. Start of my Career at Stony Brook (1976 to 1983)

I arrived in December 1976, shortly after the birth of our third child (two of whom are Australians) and was welcomed by the faculty, graduate students, and staff. Several of the senior graduate students of Prewitt and Weidner (Alan Kafka, Hubert King, Louise Levien, and Mike Vaughan) took me under their wing and "taught" me how to become a faculty member.

When the Provost Sei Sujishi learned that my first proposal to NSF Geophysics (which I wrote while still in Canberra) had been approved for funding, he immediately freed up the start-up funds which enable me to place an order for a 500-ton hydraulic press from Harwood Engineering and a girdle-anvil, high-pressure apparatus from Pressure Systems Research. This equipment was delivered in September 1977 and installed in Room 375 of the ESS Building with the help of Tony Vidmar (our new technician), Al Catalano (machinist) and Alan Major (consultant from Ringwood's lab at the ANU) (see Figure 1a,b). Vidmar and I had been flown in Fred Gwinner's plane to Boston to visit the Harwood factory to assess the needs for installing the new press (Fred was foreman of the departmental Machine Shop). On our return to campus, we had to educate the Facilities and Engineering unit at Stony Brook about the difference between weight/force and pressure, as they were nervous about installing such a heavy press on the third floor of the Earth and Space Sciences Building.

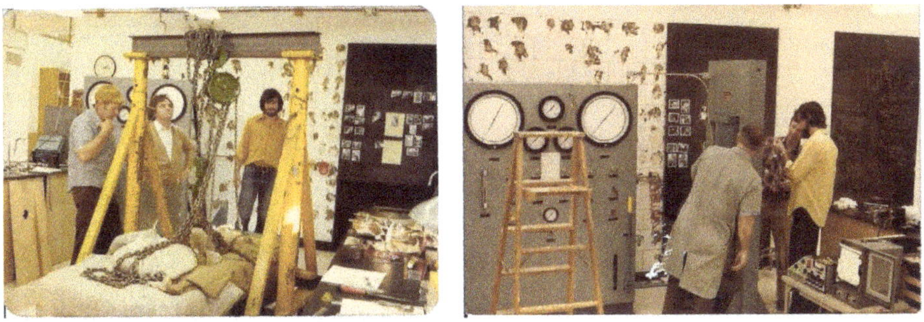

(a) (b)

Figure 1. Installation of Harwood 500-ton press and girdle-anvil apparatus in 1977. (**a**) Bob and technician Tony Vidmar contemplating how to bring the press to upright position; (**b**) Al Catalano (machinist extraordinaire) and Alan Major from ANU with Tony Vidmar checking hydraulic connections to press.

In Spring 1977, I served on the Graduate Committee and had a chance to get an early look at the applications for Fall 1977. We were successful in recruiting Paula Davidson from Brown University, with lobbying from Louise Levien and as the result of a return visit to campus by Paula. From my colleagues Charlie Sclar at Lehigh University and Taro Takahashi at Queens College of CUNY, I learned that Jay Bass was unhappy at Arizona State University and looking for a new institution to continue his Ph.D. studies. I contacted Jay and invited him to apply to our Department, which he did and arrived in August 1977 with Paula. Although both Paula and Jay ultimately decided to pursue their doctoral dissertations under the supervision of other Earth Science faculty (Don Lindsley for Paula and Don Weidner for Jay), I have always taken a special pride in helping to recruit them to Stony Brook.

My first teaching experience at Stony Brook was in the Spring 1977 semester, in which I offered ESS 607: Topics in Geophysics; I chose to adapt my informal lectures on "solid-state geophysics" from the ANU and enjoyed having Robin Reichlin and Doug Anderson in the class (along with Kafka, King, Levien and Vaughan who sat in for moral support). This course was modeled after one I took from Orson Anderson at Columbia in the 1960s (and which later evolved into his book "Equations of State of Solids for Geophysics and Ceramic Science" [2]). In Spring 1978, this course was approved as ESS 556: Solid State Geophysics; over the next 35 years, I have taught this course another 13 times and profited from having an outstanding series of geophysics and geochemistry graduate students enroll (see list of these students below Appendix A). I have also taught a shortened versions in France at the Université Paris XI in Orsay in 1983 and at the Université Paul Sabatier in Toulouse in 2002 (where is had been incorporated as a foundation course for their Master's students and it now taught by Frédéric Béjina).

The period of 1976–1983 was an active period of recruiting new faculty in geophysics by Don Weidner and me. We were very fortunate that the strong geochemistry group stepped back and allowed the new geophysics program to reach maturity. During this period, many new geophysics faculty were hired, some of whom stayed to pursue their careers (Teng-fong Wong and Dan Davis) while others moved on to other institutions (Brad Hager, Jay Melosh, Rob Comer, Cliff Thurber). The addition of Wong and Davis expanded the research programs in experimental geophysics while the later appointments of Bill Holt and Lianxing Wen opened up new opportunities in global geodynamics and seismology.

Most significantly, Ann Lattimore joined the group as an Administrative Assistant in the mineral physics group in the early 1980s and served with distinction for the next 25 years until her retirement in 2007 (see Figure 2 from her retirement party).

Figure 2. Ann Lattimore with Don Weidner and Bob on occasion of her retirement in 2007.

While my own research laboratory was in its infancy, I profited by collaboration with a series of visiting scientists who came to the Brillouin spectroscopy laboratory of Don Weidner to measure the elasticity of single crystal of mantle minerals which they had synthesized in their home laboratories, including Eiji Ito, Hiroshi Sawamoto, Nozumu Hamaya and Akira Fukizawa. When Don published his new elastic moduli of single-crystal SiO_2-coesite [3], he included ultrasonic sound velocity data which I had obtained on a polycrystalline specimen at the ANU. As his first research project at Stony Brook, Jay Bass engaged a junior faculty member in the Department of Applied Mathematics in an investigation to obtain the elastic properties of minerals from joint inversion of data from acoustic and static compression experiments [4].

My first graduate student at Stony Brook, Barbara Leitner, synthesized a single crystal of pyrope garnet in Don Lindsley's petrology lab and measured its elastic properties in Weidner's Brillouin lab for her M.S. thesis project [5]. In collaboration with Charlie Prewitt and his postdoc Satoshi Sasaki, we documented the crystal structure of $CaGeO_3$-perovskite using a specimen synthesized in my lab at the ANU [6].

During this same period, Susan Narbut and Ann Sirinides completed their M.S. theses in my laboratory on the elasticity of two-phase aggregates produced by phase transformations involving disproportionation reactions. A separate project on attenuation and dispersion in anelastic materials pursued by Ann Singer was not successful; however, during my sabbatical in France in 1983–1984 supported by a grant from the NSF, I published a theoretical paper with Jean-Paul Poirier on the activation volume for creep and its variation with depth in the Earth's lower mantle [7].

In addition to the undergraduate students cited below, two students obtained B.S. Honors thesis under my supervision: Peter Lellis worked on isostructural systematics of elasticity in mantle minerals and Kirk Maasch evaluated data from acoustic and static compression experiments on the elastic behavior of hematite [8].

In 1979, the National Research Council of the U.S. National Academy of Sciences commissioned a study of "The Impact of Technology on Geophysics." I was invited to prepare and publish a chapter in this study entitled "The impact of technology on high-pressure geophysics [9].

When I visited Stony Brook for an interview for a faculty position in 1976, the Chair Pete Palmer asked me what my plans were for conducting geophysical surveys of Long Island; I think that Pete was disappointed when I told him I had no such plans. However, a few years later, the U.S. Nuclear Regulatory Commission, which was funding a Northeast Seismic Network to monitor local earthquake activity, invited Weidner and me to acquire seismic monitoring equipment and install it on Long Island where the Shoreham nuclear plant was scheduled to open shortly. Over the next 5 years, we operated seismic stations in the Mashomack Preserve on Shelter Island and in the Caumsett State Park on Lloyd's Neck. For their M.S. thesis research, Ellyn Schlesinger conducted a seismic noise survey on Long Island and Richard Wilkinson conducted an intensity survey to confirm the location and focal mechanism of an earthquake which occurred beneath Long Island Sound near Greenport in October 1981. When Neville Carter left for Texas A&M, Wong and I inherited his liquid pressure system, which led to a M.S. thesis by Tom Ruubel measuring velocities in natural rocks from the Ramapo fault system. While the pursuit of this local seismology project diverted some resources from our other funded projects (including important contributions by Kafka, Leitner and Vaughan), it gave us an opportunity to engage a series of undergraduates in research; these included Emanuel Caiti, Claire Teuten, Linda Gunderson, and Kirk Maasch, as well as Noel Barstow (a structural geology graduate student), Thomas Caruso (a seismology graduate student), Schlesinger, and Wilkinson. We also occupied temporary seismic stations, including one in the Morton Wildlife Refuge on Jessup Neck (only accessible using Mike Vaughan's canoe). One of our helpers in this field work was a paleontology graduate student who insisted that we purchase deer tick collars to protect him from lyme disease.

As described in detail below, the Stony Brook mineral physics team (Liebermann, Prewitt and Weidner) submitted a major instrumentation proposal to the NSF Earth Sciences Division in August 1983 for a new generation of high-pressure apparatus. I then departed for France for a year-long sabbatical leave, my first at Stony Brook. I was lodged at the Université Paris XI in Orsay, outside Paris, with the research group headed by Olivier Jaoul; we commenced a study of atomic diffusion of Fe and Si in olivine at high pressure [10], which evolved to a 20-year collaboration driven primarily by a series of graduate students and postdocs (Yves Bertran-Alvarez [11,12], Paul Raterron [13], Frédéric Béjina [14,15]). As we were not able to transfer our family to Tokyo in January 1984 due to the lack of appropriate accommodation for our "large" family (us and three kids under age 15), we decided to remain in the Paris region and Jean-Paul Poirier generously provided some financial support; this led to a strong collaborative effort between the laboratories of Poirier in Paris and ours in Stony Brook focused using transmission electron microscopy to understand the mechanism of phase transformations in minerals at high pressures (driven by postdocs and visiting scientists Jannick Ingrin [16], James Boland [17], François Guyot [18], Isabelle Martinez [19], Laurence Galoisy (from the laboratory of Georges Calas)) who worked with our graduate students Anne Remsberg [20], Yanbin Wang [21] and Gabriel Gwanmesia [22].

4. Establishment Stony Brook High Pressure Laboratory and Evolution to CHiPR and COMPRES: 1984 to 2019

In 1971, as I was attending the International Union of Geodesy and Geophysics (IUGG) Congress in Moscow and making an academy exchange visit to Czechoslovakia, Ringwood encouraged me to visit the high-pressure laboratories in Japan enroute back to Canberra. Guided by Naohiro Soga and Mineo Kumazawa, I spent ten days visiting labs in Tokyo, Nagoya, Kyoto and Osaka.

In 1983, the NSF Division of Earth Sciences created a new program for Instrumentation and Facilities. We (Liebermann, Prewitt and Weidner) decided to submit a proposal for a modern, multi-anvil, high-pressure lab modeled on those in Japan. Our proposal was funded in late 1983, we began to consider the options for acquiring high-pressure apparatus for our laboratories at Stony Brook University.

By the mid-1980s, most of the new developments in multi-anvil, high-pressure apparatus had become concentrated in Japan. In summer 1984, we went on a "shopping trip" to companies and laboratories in Japan to search for high pressure equipment, including Tokyo, Tsukuba, Nagoya, and Misasa. We decided to import two different types of high-pressure apparatus to Stony Brook: (1) a DIA-type, cubic-anvil apparatus modeled after MAX-80 at the Photon Factory (PF) in Tsukuba and named SAM-85 (Figure 3a); and (2) a Kawai-type, 2000-ton uniaxial, split-sphere apparatus modeled on the 5000-ton version in the laboratory of Eiji Ito and named USSA-2000 (Figure 3b).

(a)

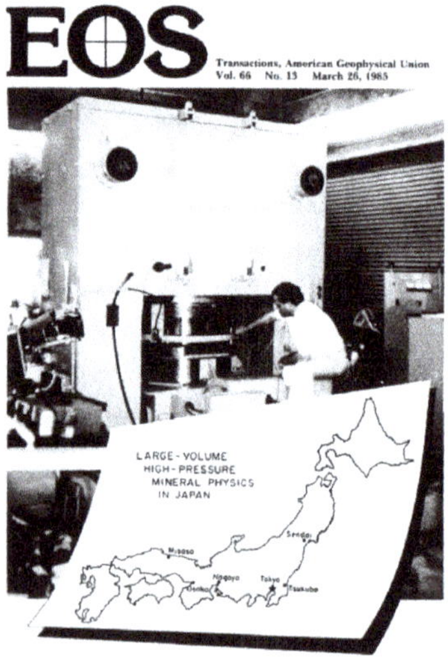

(b)

Figure 3. (**a**) MAX-80 at Photon Factory in Tsukuba in 1984. Don Weidner, author and Charlie Prewitt with Osamu Shimomura. This modified DIA apparatus was designed by a team led by Osamu Shimomura and Takehiko Yagi and installed on a X-ray beamline at the Photon Factory in Tsukuba, Japan in the early 1980s; it was called MAX-80 for "multi-anvil type X-ray system designed in 1980." (**b**) Author in laboratory of Eiji Ito in ISEI, Misasa, Japan with USSA-5000 apparatus. This is a Kawai-type, double-stage, multi-anvil apparatus. This USSA-5000 apparatus later served as the model for multi-anvil laboratories at Stony Brook University in the U.S. (Liebermann, Prewitt and Donald Weidner), at the University of Alberta in Canada (Christopher Scarfe), and the Bayerisches Geoinstitut in Germany (David Rubie).

In the summer of 1985, SAM-85 was installed initially in Room 375 with the guidance and assistance of Osamu Shimomura,Yosiko Sato and Hisao Kanda (Figure 4a,b). One of the first experiments in SAM-85 was the synthesis of artificial diamond by Hisao Kanda (see Figure 4c).

(**a**) (**b**)

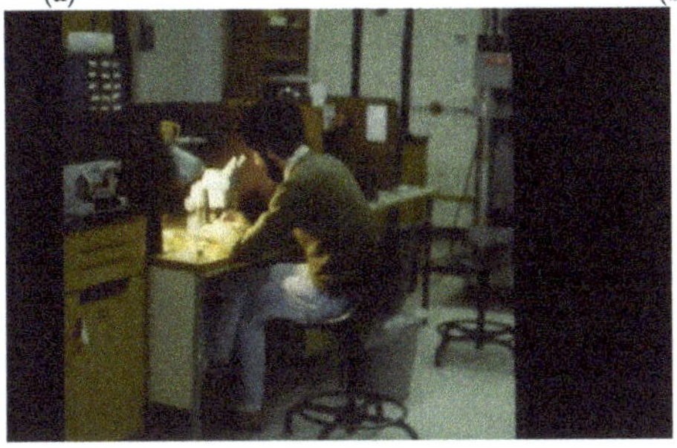

(**c**)

Figure 4. (**a**) Team of Gabriel Gwanmesia, Osamu Shimomura, Anne Remsberg, Yosiko Sato, Tibor Gasparik and Charlie Prewitt with author after installation of SAM-85 (Six Anvil Machine named for Osaamu Shimomura and Osamu Fukunaga, who designed the apparatus). Reproduced with permission from Liebermann [1], published by Elsevier B.V., 2014. (**b**) Consultation among first users of SAM-85: Jannick Ingrin, Anne Remsberg, Hisao Kanda, Tibor Gasparik and Gabriel Gwanmesia. (**c**) Hisao Kanda preparing to synthesize artificial diamonds for the first time at Stony Brook in SAM-85.

In December 1985, the USSA-2000 apparatus was installed in a new High Pressure Laboratory at Stony Brook (Figure 5a,b). During these installations, we profited from the guidance and advice of many Japanese colleagues (Figure 5c). From 1985 to 2002, Tibor Gasparik (Stony Brook Ph.D., 1981) served as the Manager of the High-Pressure Lab, with Bob as the Director (Figure 5d). The laboratory as it existed in 1985 following the dedication ceremony overseen by the author, Don Weidner and Charlie Prewitt (Figure 6).

During the 1985–1986 year, three Japanese scientists spent 9–12 months each as visitors to the High-Pressure Laboratory: Hisao Kanda from the National Institute for Research in Inorganic Materials in Tsukuba, Manabu Kato from Nagoya University and Hiroshi Watanabe from Osaka University (see photo in Figure 5c). Kanda discovered an unused high-pressure apparatus on the loading dock in the Department of Chemistry at Stony Brook. After some detective work, we learned that this press belonged to William LeNoble in Chemistry, who had inherited it from the U.S. Army Benet Weapons Laboratory in Watervliet, NY. He generously gave us the press and, after a difficult passage across campus when the forklift became stuck in the mud, it was installed in the High-Pressure Laboratory. Our talented machine shop staff refurbished the press and pumping system and Virginia Haniford, one of Don Weidner's graduate students, painted it a bright green (see Figure 7a). This press contained two rams, one of 700 ton capacity and one of 300 ton, and had originally been designed by George Kennedy and Ivan Getting at UCLA. We purchased a Walker-type multi-anvil module, inserted it in this press, and christened it the Uniaxial Split-cylinder Apparatus with 1000 ton rams (USCA-1000).

We also moved the 500-ton Harwood press and SAM-85 to the High-Pressure Lab and Gabriel Gwanmesia conducted pressure calibration experiments on both of them for his M.S. thesis (Figure 7b).

Thus, by mid-1986, there were four distinct high-pressure systems operational in the High-Pressure Laboratory: the USSA-2000, the Kennedy-Getting press, the Harwood press and SAM-85 and the lab was fully-equipped and nice and clean (Figure 7c).

(**a**) (**b**)

(**c**) (**d**)

Figure 5. (**a**) Installing USSA-2000 in the High-Pressure Laboratory at Stony Brook University in December 1985. This building had originally been a cooling tower and had fallen into disuse. Charlie Prewitt "discovered" it and we were able to convince the university to renovate it to house the new multi-anvil, high-pressure apparatus in 1985. (**b**) The installation on 4 December 1985 was recorded by Channel 55 News in New York City, who asked us to halt the installation until their cameraman could arrive; this required us to instruct the crane operator to hold the heavy press in mid-air for a half hour. (**c**) Al Catalano inserting guideblock into 2000-ton press, watched by Tibor Gasparik, Hisao Kanda, the author, and Manabu Kato. (**d**) Author with Kawai-type guideblock in USSA-2000 press with Japanese dolls as good-luck charms (gift of Sumitomo company).

Over the next five years, many graduate students and postdocs exploited the new opportunities for research using these high-pressure apparatus (Figure 7d).

One of the first experiments in the USSA-2000 apparatus was conducted by Anne Remsberg who studied the transformation in Co_2SiO_4 from the olivine to the spinel phase in collaboration with Jim Boland and Tibor Gasparik; the olivine polymorph is especially colorful (see Figure 7e) and Anne often wore her magenta blouse when presenting talks on her research [20] (see Figure 4b above).

For his Ph.D. research at Stony Brook, Gabriel Gwanmesia adapted the hot-pressing techniques we had developed at the ANU in the 1970s for multi-anvil, high-pressure apparatus [23], in particular the 2000-ton uniaxial split-sphere apparatus (USSA-2000). Subsequently, ultrasonic interferometric techniques were used to measure the sound wave velocities in polycrystalline specimens of the wadsleyite and ringwoodite phases of Mg_2SiO_4 [24,25] in the piston-cylinder apparatus at the ANU in collaboration with Sally Rigden and Ian Jackson (Figure 7f,g); later, this Australian collaboration was extended to pyrope-majorite garnets by Gwanmesia [26], stishovite by Baosheng Li [27], magnesium ferrite by Sytle Antao [28] and enstatite by Jennifer Kung [29].

Figure 6. Don Weidner, Bob Liebermann and Charlie Prewitt at dedication ceremony for the USSA-2000 in December 1985.

In 1986, we were awarded a grant from the NSF for "Acquisition and Maintenance of Multi-anvil, High Pressure Apparatus" at Stony Brook. With the financial support for the High-Pressure Lab secured on a continuing basis, the university decided to create a new research unit using funding from the Graduate Research Initiative of the State University of New York. This unit was named the Mineral Physics Institute, with funding to support the operation of the High-Pressure Lab and other research labs in mineral physics and chemistry, and Donald Weidner was appointed as the Founding Director.

In 1986, Charlie Prewitt moved to Washington, D.C. become the Director of the Geophysical Laboratory of the Carnegie Institution of Washington and Alex Navrotsky moved from Arizona State University to Princeton University. These transitions created new opportunities for collaboration. In 1991, a proposal for a NSF Science and Technology Center for High Pressure Research (CHiPR) was funded for a total of 11 years to 2002, with total funding of $36 million. Stony Brook served as the headquarters of CHiPR with Weidner as the Principal Investigator (PI) and Liebermann as the Co-PI, and branch campuses at Princeton (Co-PI Navrotsky) and Carnegie (Co-PI Prewitt) (Figure 8).

(a) (b)

Figure 7. *Cont.*

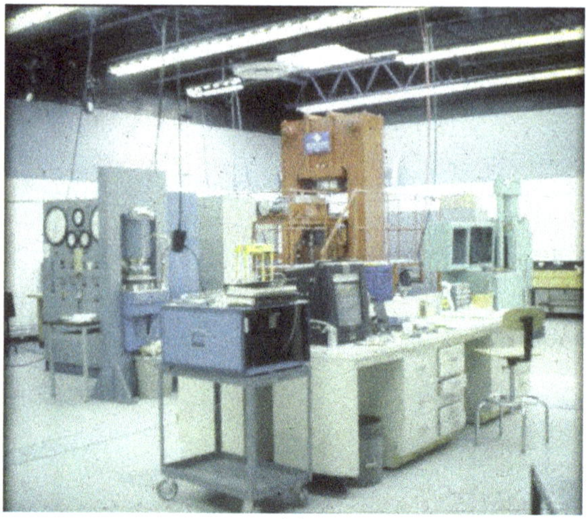

(**c**)

(**d**)

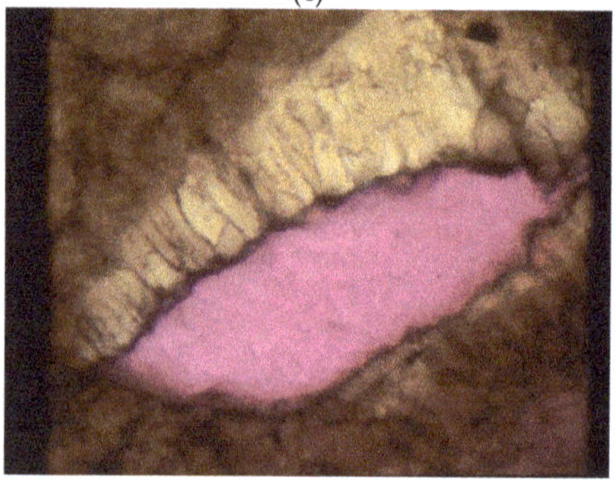

(**e**)

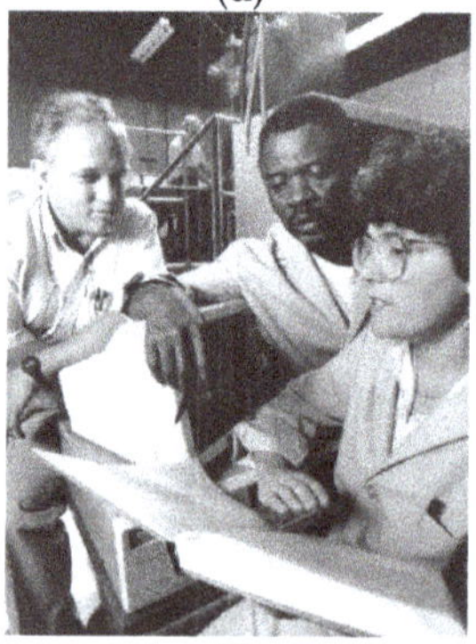

(**f**)

Dr Sally Rigden and Mr Gabriel Gwanmesia with the ultrasonic interferometer.

(**g**)

Figure 7. Stony Brook High Pressure Laboratory circa 1986. (**a**) 1000-ton Kennedy-Getting press installed in High Pressure Laboratory after inheritance from the laboratory of William LeNoble in the Department of Chemistry. (**b**) Gabriel Gwanmesia calibrating the girdle-anvil apparatus in the Harwood press as part of his M.S. thesis. (**c**) The Stony Brook High Pressure Laboratory in 1986 with the USSA-2000, the 500-ton Harwood press with girdle-anvil and the 1000-ton Kennedy-Getting press installed and operational. Reproduced with permission from Liebermann [1], published by Elsevier B.V., 2014. (**d**) Graduate students and postdocs of Bob in High Pressure Lab circa 1988. L. to R. Bob, Yves Bertran (France), Gabriel Gwanmesia (Cameroon), Ren Lu (China), François Guyot (France), Anne Remsberg (USA), Yanbin Wang and Xing Liu (China). Reproduced with permission from Liebermann [1], published by Elsevier B.V., 2014. (**e**) Co_2SiO_4. Residual core of olivine phase (magenta) surrounded by a rim of large golden brown single-crystals of spinel [20]. (**f**) Sally Rigden from the ANU with Gabriel Gwanmesia and the author in the High-Pressure Lab in 1991. Reproduced with permission from Liebermann [1], published by Elsevier B.V., 2014. (**g**) Gabriel Gwanmesia and Sally Rigden in Ian Jackson's laboratory at the ANU.

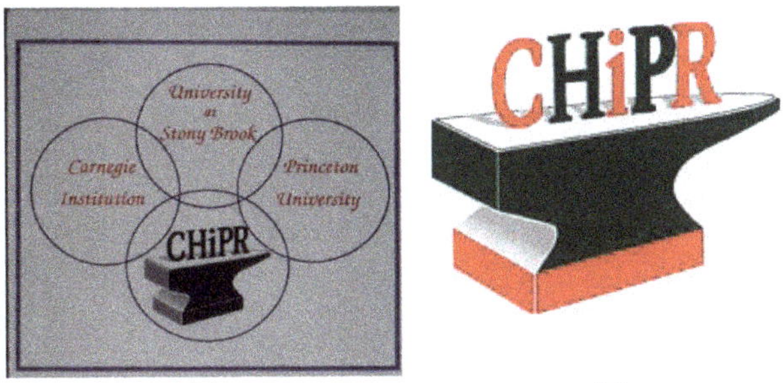

(**a**) (**b**)

Figure 8. Founding of CHiPR in 1991. Triumvirate of institutions for CHiPR: Stony Brook University; Carnegie Institution of Washington, and Princeton University and official CHiPR logo (using a blacksmith's anvil as a base). Reproduced with permission from Liebermann [1], published by Elsevier B.V., 2014.

One of the new initiatives in my research group in the CHiPR era began in 1994 when Baosheng Li, then a graduate student and now a Professor in the Department of Geosciences, developed techniques to incorporate ultrasonic interferometric measurements of sound velocities in situ in the Kennedy-Getting high-pressure apparatus in collaboration with Ian Jackson (my first graduate student and subsequently the Director of the Research School of Earth Sciences at the Australian National University; see Figure 9ab). This pioneering development provided the opportunity to conduct sound velocity experiments to pressures of 10 GPa [30] and was exploited over the next few years by many graduate students (Tony Cooke [31], Lucy Flesch [32], Yegor Sinelnikov [33], Jun Liu [34], Kenneth Darling [35]) and postdocs (Ganglin Chen [36], Frédéric Decremps [37,38], Jennifer Kung [39], Gabriel Gwanmesia [40]) for such studies.

At the nearby Brookhaven National Laboratory, the National Synchrotron Light Source (NSLS) was built in the early 1980s. Charlie Prewitt opened our eyes to the research opportunities of such synchrotron X-ray facilities (Figure 10a). In the early 1990s at the beginning of the CHiPR era, Don Weidner and his team moved SAM-85 from the campus to the NSLS and installed it on the superconducting beamline (X17B2). Don recruited Mike Vaughan (Stony Brook Ph.D. 1979) back from the University of Chicago to energize the MPI multi-anvil operation at the NSLS (Figure 10b).

Ultrasonic interferometry was first adapted for use in conjunction with synchrotron X-rays at the X17B2 hutch by my research team (led by Baosheng Li). In a typical experiment with a polycrystalline sample, three types of data are collected: (1) X-ray diffraction from the sample and the pressure standard, (2) ultrasonic waveforms using transfer function method, and (3) X-ray radiographic image of the sample. Data processing of these raw data results in the determination of pressure, unit cell volume, compressional and shear wave velocities at all pressure and temperature (T). Such acoustic experiments have been conducted by many of my students and postdocs over the past 15 years. These simultaneous ultrasonic interferometry and X-radiation measurement techniques which we developed at the NSLS [41–43] have now been implemented at many major synchrotron sources in the world, including SPring-8 in Japan, the Advanced Photon Source in the U.S., and DESY in Germany.

(a) (b)

Figure 9. (**a**) Baosheng Li and Ian Jackson using the Kennedy-Getting press to develop their new experiments extending ultrasonic interferometric measurements of sound velocities to 10 GPa in 1994. (**b**) Baosheng Li and Kennedy-Getting press with Walker-type module in High Pressure Lab in 1994. Reproduced with permission from Liebermann [1], published by Elsevier B.V., 2014.

(a) (b)

Figure 10. (**a**) Charlie Prewitt at X7 beamline of the new NSLS in early 1980s (from newsletter of State University of New York). (**b**) SAM-85 installed on the superconducting wiggler beamline (X17B2) at NSLS with Don Weidner and Mike Vaughan.

During the first quarter century of my research programs at Stony Brook, we were extremely lucky to have access to the personnel and equipment resources of the departmental machine shop and electronics support facility, which the founding fathers (no women on the faculty in the 1960s) of the Department had wisely incorporated into the staffing plan. The Machine Shop, which originally included Bob Muller in addition to Fred Gwinner and Al Catalano, continued to provide excellent service under Paul Hoversen with Ed Vorisek and Herb Schay and later Carey Koleda and Ricky Palencia. For this entire period, Bill Huebsch and Ben Vitale, who had been recruited from Brookhaven National Lab by the founding Chair Oliver Schaeffer, lent their considerable expertise in all things electrical and electronic to our research programs (from major projects such as the High Pressure Lab to the MPI's experimental program at the NSLS to minor needs such as installing antennas on the roof of the Health Sciences Center to receive the signals from our seismic stations on Shelter Island and Lloyd's Neck). These experimental programs simply could not have achieved their national and international stature without the dedicated services provided by these extraordinary staff members. Unfortunately, the Department and the University have recently decided to close these support facilities with the likely consequence that it will be difficult or impossible to establish and maintain such world-class experimental programs in the future.

All of the 25 NSF Science and Technology Centers established in the late 1980s and early 1990s were required to wind-down operations in Year #10 and close at the end of Year #12; for CHiPR, this end came in February 2002; at Stony Brook, we marked this occasion with a group photo which we sent as a thank you to Dan Weill, the Program Director for the Instrumentation and Facilities Program in the NSF Division of Earth Sciences (see Figure 11).

Figure 11. Group photo of faculty, staff and graduate students at Stony Brook in early 2002 as thank you to Dan Weill of the Instrumentation and Facilities Program in Earth Sciences at NSF.

In anticipation of "sunset" of CHiPR, we began to develop ideas for an expanded mineral physics initiative to expand access of high-pressure facilities to broader community, especially for graduate students, postdocs and junior faculty at U.S. academic institutions.

A proposal was submitted in August 2001 on behalf of 18 universities and national laboratories in the U.S. to establish a Consortium for Materials Properties Research in Earth Sciences (COMPRES); funding approved for 5 years commencing in 2002, with Stony Brook University as the headquarters of COMPRES.

COMPRES headquarters subsequently moved to the University of Illinois at Urbana-Chapaign in 2010 and then to the University of New Mexico in 2015. Additional details of the evolution of

COMPRES may be found in the Bass, Tempe and Long-Range Planning reports (Figure 12a–c); see also [44].

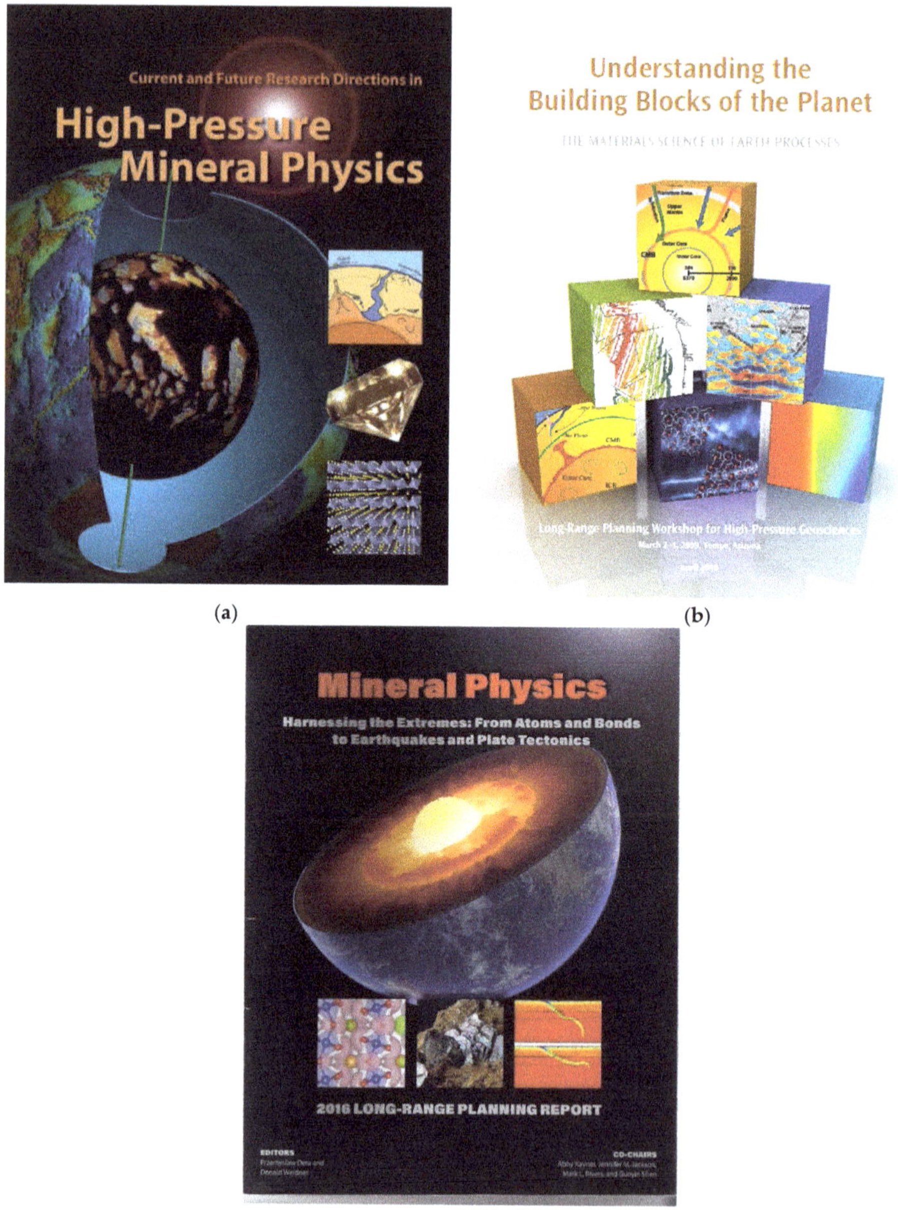

(**a**) (**b**)

(**c**)

Figure 12. (**a**) Frontispiece of Bass Report of COMPRES (edited by Jay Bass, 2004). (**b**) Frontispiece of Tempe Report of COMPRES (edited by Quentin Williams, 2010). Reproduced with permission from Liebermann [1], published by Elsevier B.V., 2014. (**c**) Frontispiece of Long Range Planning Report (edited by Przemyslaw Dera and Donald Weidner, 2016).

5. Bob as an Administrator at Stony Brook

My administrative career while at Stony Brook has been characterized by a series of serendipitous events; each of these has involved finding something good or useful to do even though I was not specifically looking for it.

The first of these events occurred in 1982–1983: I had been encouraged to apply for a position as Program Director for Geophysics in EAR at the NSF. I decided not to apply as I was under consideration for Chair of the Department of Earth and Space Sciences; when the Astronomy faculty vetoed my appointment as Chair, I was free to go on sabbatical leave to France in September 1983 to work with Jaoul and Poirier and their colleagues in Orsay and Paris (see above).

After a 3-month stay in Japan, which included the "shopping trip" describe above, I became the Director of the new High-Pressure Laboratory and later the Co-Director of CHiPR (1991). In 1993, I was a candidate to become the Director of the Research School of Earth Sciences at the Australian National University; however, I withdrew my candidacy after a receiving a generous retention offer from Stony Brook organized by Don Lindsley and Don Weidner.

In 1997, I became Chair of the newly-named Department of Geosciences and found that to be one of the most rewarding administrative experiences of my life; we hired new faculty (including Lianxing Wen and Troy Rasbury), recruited many outstanding graduate students and expanded our research programs.

When Paul Armstrong (Dean of the College of Arts and Sciences) departed for Brown University in 2000, Provost Robert McGrath asked me to serve as Interim Dean. I did so with every intention of only staying a year or so but became somewhat addicted to the challenges of the College leadership and so applied for the permanent job. When I was not chosen to be the new Dean, I decided to take another sabbatical in France, and was fortunate that Olivier Jaoul and his team had now moved to the Université Paul Sabatier in southern France (see Figure 13), where I spent another sabbatical leave in 2002–2003.

Figure 13. Author receiving degree of Docteur Honoris Causa at Université Paul Sabatier from Olivier Jaoul in 2004. Reproduced with permission from Liebermann [1], published by Elsevier B.V., 2014.

On my return to Stony Brook in 2003, I assumed the role of President of COMPRES and held that position until 2010. I then "retired" from administrative duties and returned to teaching at the graduate level (GEO 556 and 607) and helping to advise two groups of graduate students: (a) Three African-American women who were pursuing M.S. degrees in Geosciences Instrumentation under the supervision of Lars Ehm (Associate Professor in the Department of Geosciences) with the objective of qualifying for positions as Science Associates at the national laboratories of the U.S. Department of

Energy (such as Brookhaven National Lab) [45]; (b) Three Chinese women who were working with my colleague Baosheng Li (Professor in the Department of Geosciences) on projects to utilize ultrasonic interferometry in conjunction with synchrotron X-radiation to measure sound velocities of minerals at high pressures and temperatures [46–48]. All six of these graduate students took my Solid-State Geophysics course in the Fall 2013 semester (see Figure 14), which was my last semester of teaching at Stony Brook as I formally retired in 2014 and was then appointed as a Research Professor in the Department of Geosciences and the Mineral Physics Institute.

Figure 14. Graduate students in GEO 556: Solid-State Geophysics in Fall 2013 semester, with Bob teaching his last class at Stony Brook. Left to right: Xintong Qi, Ting Chen, Xuebing Wang, Bob, Adairé Heady, Ashley Thompson, Melissa Sims.

6. For Additional Details

Liebermann, R.C. Bob-san and High Pressure Science and Technology in Japan: A 40+-Year History. *Rev. High Press. Sci. Technol.* **2011**, *21*, 115–126.

Liebermann, R.C. Multi-anvil, high-pressure apparatus: A half century of development and progress. *High Press. Res.* **2011**, *31*, 493–532.

Liebermann, R.C.; Prewitt, C.T. From Airlie House in 1977 to Lake Morey in 2007: 30 Years of evolution of mineral physics. *Phys. Earth Planet. Inter.* **2014**, *228*, 36–45.

Liebermann, R.C. The Role of Serendipity in My Career in Mineral Physics: 1968 to 2013. *Phys. Earth Planet. Inter.* **2014**, *228*, 307–323.

Liebermann, R.C. History of Japan-US/High Pressure Mineral Physics Seminars: 1976 to 2012. *Phys. Earth Planet. Inter.* **2014**, *233*, 1–35.

Funding: This paper was written with support from a research grant to Baosheng Li and the author from the National Science Foundation (EAR-1524078).

Acknowledgments: I owe an enormous debt to many colleagues and friends for their support and collaboration over the past 43 years at Stony Brook, in addition to the graduate students and postdoctoral associates and visiting scientists cited above. In particular, I wish to express my appreciation and thanks to Don Weidner, Charlie Prewitt, John Parise Lars Ehm and Baosheng Li on the faculty of the Department of Geosciences and Michael Vaughan, Jiuhua Chen, and Jianzhong Zhang on the research faculty of the MPI.

Conflicts of Interest: The author declares no conflict of interest.

Appendix A

Appendix A.1 GEO 556: Solid-State Geophysics (Originally ESS 556)

From the Spring 1977 to Fall 2013 semesters, Bob has taught this course 14 times and had the privilege of introducing the following graduate students and postdocs to the wonders of mineral physics (some of whom audited the course and others who took it twice!).

Sp77: Doug Anderson, Elizabeth Hauser, and Robin Reichlin
—listed as GEO 607.
Audits: Alan Kafka, Hubert King, Louise Levien, Michael Vaughan.
Sp78: Jay Bass, Ben Burton, Paula Davidson, Barbara Leitner, Susan Narbut.
First time listed as ESS 556.
F79: Andy Au, Jay Bass, Ben Burton, Ann Singer.
F82: Hwai-kuo Chen, John Kandelin, Tom Ruubel, Michael Simpson, Ann Sirinides.
Sp85: Gabriel Gwanmesia, Virginia Haniford, Jaidong Ko, Xing Liu, Amir Yeganeh-haeri, Jianzhong Zhang.
F87: Ren Lu, George Moussouris, Rosemary Pacalo, Kim Pacanovsky, Anne Remsberg, Yanbin Wang.

Audit: Gabriel Gwanmesia.
F89: Yves Bertran, Yuan Liu, Yue Meng, Woosun Yang, Yusheng Zhao.
F91: Jishan Jin, Baosheng Li, Wenlu Zhu, Yong Zou.
F93: Reggie Minault, Bingming Shen-tu, Yegor Sinelnikov, Xiang Xia.
F95: Tony Cooke, Lucy Flesch, Rui Li, Jun Liu, Yujin Wu, Hui Zhao.
Audit: Frédéric Béjina.
F97: Corne Kreemer, Haibin Su
Audits: Ganglin Chen, Jiuhua Chen, Hiroki Kagi, Jianzhong Zhang.
SP00: Stan Adler, Kenneth Darling, Elliot Klein, Li Li, Laura Rossier, Veronika Vajdova
F11: Yu Chen, Yuan-yuan Liu, Guangrui.Qian, Gina Scherbenko, Terry-Ann Suer.
Audit: Matt Jacobsen.
F13: Ting Chen, Adairè Heady, Xintong Qi, Ashley Thompson, Melissa Sims, XuebingWang.
Audits: Xuefei Li, Yongtao Zou.

Appendix A.2 Theses Supervised by Bob Liebermann at ANU and SBU

Appendix A.2.1 Australian National University—Research School of Earth Sciences

Ian Jackson (1976)—Ph.D.
Subject: Phase equilibria and elastic properties in silicate analogue systems: studies of melting and polymorphic phase transformations.

Leonie Jones (1976)—Ph.D.
Subject: High-temperature elastic properties of fluoride and oxide analogues.

State University of New York at Stony Brook:
Department of Earth and Space Sciences (1976–1997)
Department of Geosciences (1997–2019)

Barbara J. Leitner (1979)—M.S.
Subject: Elasticity of Single Crystal Pyrope

Ellyn A. Schlesinger (1979)—M.S.

Subject: A Seismic Noise Survey of Long Island, New York

Susan M. Narbut (1979)—M.S.
Subject: Phase equilibria of the disproportionation reaction Ca_2SnO_4 (strontium plumbate) $CaSnO_3$ (perovskite) plus CaO (rocksalt)

Peter J. Lellis (1980)—B.S. Honors
Subject: Isostructural systematics of elasticity in silicate minerals

Ann E. Sirinides (1983)—M.S.
Subject: Elastic Properties of two-phase Aggregates

Richard J. Wilkinson (1983)—M.S.
Subject: The Long Island Sound Earthquake of October 21, 1981: Location, Focal Mechanism, Intensity Survey

Thomas Ruubel (1984)—M.S.
Subject: Laboratory measurements of velocities in rocks from the Ramapo fault system

Kirk Maasch (1986)—B.S. Honors
Subject: Acoustic and static compression experiments on the elastic behavior of hematite.

Gabriel Gwanmesia (1987)—M.S.
Subject: Pressure calibration in a girdle-anvil and a DIA-type pressure apparatus at room temperature (25 °C) and high temperature (1000 °C)

Appendix A.2.2 Theses Supervised (cont.)

Anne R. Remsberg (1990)—Ph.D.
Subject: A study of the polymorphic phase transformations in Co_2SiO_4

Ren Lu (1990)—M.S.
Subject: Study of kinetic rates of transformation between garnet and perovskite phases of $CaGeO_3$

Yanbin Wang (1991)—Ph.D
Subject: Electron microscopy and X-ray diffraction studies on structural phase transitions in $MgSiO_3$ perovskite.

Gabriel Gwanmesia (1991)—Ph.D
Subject: High pressure elasticity for the beta and spinel polymorphs of Mg_2SiO_4 and composition of the transition zone of the Earth's mantle.

Appendix A.2.3 Yves Bertran at Université Paris XI [Olivier Jaoul]

Baosheng Li (1993)—M.S.
Subject: Polycrystalline Stishovite: Hot-pressing and elastic properties

Baosheng Li (1996)—Ph.D.
Subject: Ultrasonic Measurements of the Elastic Wave Velocities of Olivine and Beta Polymorphs of Mg_2SiO_4 at Mantle Transition Zone P and T and Geophysical Implications

Lucy C. Flesch (1997)—M.S.
Subject: Sound velocities in polycrystalline $MgSiO_3$-orthopyroxene to 10 GPa at room temperature

Jun Liu (1997)—M.S.

Subject: Calorimetry study of the coesite-stishovite transformation and calculation of the phase boundary

Yegor Sinelnikov (1997)—M.S.
Subject: Elasticity of $CaTiO_3$-$CaSiO_3$ perovskites

Joseph A. Cooke (1997)—M.S.
Subject: Ultrasonic measurements of the elastic wave velocities of $Mg_3Al_2Si_3O_{12}$ pyrope garnet to 9 GPa at room temperature

Yegor Sinelnikov (1998)—Ph.D.—Check This Date?
Subject: Elasticity of $MgSiO_3$-perovskite at high pressure and temperature by ultrasonic interferometry

Appendix A.2.4 Theses Supervised (cont.)

Jun Liu (2001)—Ph.D.
Subject: Elasticity of Pyrope-Majorite Garnets at High Pressures and Temperatures and Implications for Geophysics

Kenneth Darling (2002)—M.S.
Subject: Ultrasonic Measurements of the Elastic Wave Velocities of Polycrystalline (Mg,Fe) San Carlos Olivine

References

1. Liebermann, R.C. The Role of Serendipity in My Career in Mineral Physics: 1968 to 2013. *Phys. Earth Planet. Inter.* **2014**, *228*, 307–323. [CrossRef]
2. Anderson, O.L. *Equations of State of Solids for Geophysics and Ceramic Science*; Oxford University Press: Oxford, UK, 1995.
3. Weidner, D.J.; Carleton, H.R. Elasticity of coesite. *J. Geophys. Res.* **1997**, *82*, 1334–1346. [CrossRef]
4. Bass, J.D.; Liebermann, R.C.; Weidner, D.J.; Finch, S.J. Elastic properties from acoustic and volume compression experiments. *Phys. Earth Planet. Inter.* **1981**, *25*, 140–158. [CrossRef]
5. Leitner, B.J.; Weidner, D.J.; Liebermann, R.C. Elasticity of single crystal pyrope and implications for garnet solid solution series. *Phys. Earth Planet. Inter.* **1980**, *22*, 111–121. [CrossRef]
6. Sasaki, S.; Prewitt, C.T.; Liebermann, R.C. The crystal structure of CaGeO3 perovskite and the crystal chemistry of the GdFeO3-type perovskites. *Am. Mineral.* **1983**, *68*, 89–98.
7. Poirier, J.P.; Liebermann, R.C. On the activation volume for creep and its variation with depth in the Earth's lower mantle. *Phys. Earth Planet. Inter.* **1984**, *35*, 283–293. [CrossRef]
8. Liebermann, R.C.; Maasch, A.K. Acoustic and static compression experiments on the elastic behavior of hematite. *J. Geophys. Res.* **1986**, *91*, 4651–4656. [CrossRef]
9. Liebermann, R.C. Impact of technology on high-pressure geophysics. In *The Impact of Technology on Geophysics*; a study conducted by the Impact of Technology Panel of the Geophysics Study Committee of the Geophysics Research Board; National Research Council, National Academy of Sciences: Washington, DC, USA, 1979.
10. Houlier, B.; Jaoul, O.; Abel, F.; Liebermann, R.C. Oxygen and silicon self-diffusion in natural olivine at T = 1300 °C. *Phys. Earth Planet. Inter.* **1988**, *50*, 240–250. [CrossRef]
11. Bertran-Alvarez, Y.; Jaoul, O.; Liebermann, R.C. Fe-Mg interdiffusion in single crystal olivines at very high pressure and controlled oxygen fugacity: Technological advances and initial data at 7 GPa. *Phys. Earth Planet. Inter.* **1992**, *70*, 102–118. [CrossRef]
12. Jaoul, O.; Bertran-Alvarez, Y.; Liebermann, R.C.; Price, G.D. Fe-Mg interdiffusion in olivine up to 9 GPa at T = 600–900 °C. Experimental data and comparison with defect calculations. *Phys. Earth Planet. Inter.* **1995**, *89*, 199–218. [CrossRef]
13. Raterron, P.; Bejina, F.; Doukhan, J.C.; Jaoul, O.; Liebermann, R.C. Olivine/Fe-metal equilibrium under high pressure: An ATEM investigation. *Phys. Chem. Miner.* **1998**, *25*, 285–493. [CrossRef]
14. Béjina, F.; Raterron, P.; Zhang, J.; Jaoul, O.; Liebermann, R.C. Activation volume of silicon diffusion in San Carlos olivine. *Geophys. Res. Lett.* **1997**, *24*, 2597–2600. [CrossRef]

15. Béjina, F.; Jaoul, O.; Liebermann, R.C. Activation volume for Si diffusion in San Carlos olivine: Implications for upper mantle rheology. *J. Geophys. Res.* **1999**, *104*, 25529–25542. [CrossRef]
16. Ingrin, J.; Liebermann, R.C. Deviatoric stress in a girdle-anvil type high pressure apparatus: Effect on the quartz-coesite phase transformation. *Phys. Earth Planet. Inter.* **1989**, *54*, 378–385. [CrossRef]
17. Wang, Y.; Liebermann, R.C.; Boland, J.N. Olivine as an in situ piezometer in high pressure apparatus. *Phys. Chem. Miner.* **1988**, *15*, 493–497. [CrossRef]
18. Guyot, F.; Gwanmesia, G.D.; Liebermann, R.C. An olivine to beta phase transformation mechanism in Mg2SiO4. *Geophys. Res. Lett.* **1991**, *18*, 89–92. [CrossRef]
19. Martinez, I.; Wang, Y.; Guyot, F.; Liebermann, R.C.; Doukhan, J.-C. Microstructures and iron partitioning in (Mg,Fe)SiO3 perovskite-(Mg,Fe)O Magnesiowustite assemblages: An analytical transmission electron microscopy study. *J. Geophys Res.* **1997**, *102*, 5265–5280. [CrossRef]
20. Remsberg, A.R.; Boland, J.N.; Gasparik, T.; Liebermann, R.C. Mechanism of the olivine-spinel transformation in Co_2SiO_4. *Phys. Chem. Miner.* **1988**, *15*, 498–506. [CrossRef]
21. Wang, Y.; Guyot, F.; Liebermann, R.C. Electron microscopy of (Mg,Fe)SiO_3 perovskite: Evidence of structural phase transitions and implications for the lower mantle. *J. Geophys. Res.* **1992**, *97*, 12327–12347. [CrossRef]
22. Gwanmesia, G.D.; Liebermann, R.C. Polycrystals of high-pressure phases of mantle minerals: Hot-pressing and characterization of physical properties. In *High Pressure Research: Applications to Earth and Planetary Sciences;* Syono, Y., Manghnani, M.H., Eds.; Terra Scientific Publishing Co., Tokyo/American Geophysical Union: Washington, DC, USA, 1992; pp. 117–135.
23. Gwanmesia, G.D.; Li, B.; Liebermann, R.C. Hot pressing of polycrystals of high pressure phases of mantle minerals in multi-anvil apparatus. *PAGEOPH* **1993**, *141*, 467–484. [CrossRef]
24. Gwanmesia, G.D.; Rigden, S.; Jackson, I.; Liebermann, R.C. Pressure dependence of elastic wave velocity for β-Mg_2SiO_4 and the composition of the Earth's mantle. *Science* **1990**, *250*, 794–797. [CrossRef]
25. Rigden, S.M.; Gwanmesia, G.D.; FitzGerald, J.D.; Jackson, I.; Liebermann, R.C. Spinel elasticity and seismic structure of the transition zone of the mantle. *Nature* **1991**, *354*, 143–145. [CrossRef]
26. Rigden, S.M.; Gwanmesia, G.D.; Liebermann, R.C. Elastic wave velocities of a pyrope-majorite garnet to 3 GPa. *Phys. Earth Planet. Inter.* **1994**, *84*, 35–44. [CrossRef]
27. Li, B.; Rigden, S.M.; Liebermann, R.C. Elasticity of stishovite at high pressure. *Phys. Earth Planet. Inter.* **1996**, *96*, 113–127. [CrossRef]
28. Antao, S.M.; Jackson, I.; Li, B.; Kung, J.; Chen, J.; Hassan, I.; Liebermann, R.C.; Parise, J.B. High-temperature elasticity of magnesioferrite spinel. *Phys. Chem. Miner.* **2007**, *34*, 345–350. [CrossRef]
29. Kung, J.; Jackson, I.; Liebermann, R.C. High-temperature elasticity of polycrystalline orthoenstatite ($MgSiO_3$). *Am. Mineral.* **2011**, *96*, 577–585. [CrossRef]
30. Li, B.; Jackson, I.; Gasparik, T.; Liebermann, R.C. Elastic wave velocity measurement in multi-anvil apparatus to 110 GPa using ultrasonic interferometry. *Phys. Earth Planet. Inter.* **1996**, *98*, 79–91. [CrossRef]
31. Chen, G.; Cooke, J.A., Jr.; Gwanmesia, G.D.; Liebermann, R.C. Elastic Wave Velocities of $Mg_3Al_2Si_3O_{12}$-Pyrope Garnet to 10 GPa. *Am. Mineral.* **1999**, *84*, 384–388. [CrossRef]
32. Flesch, L.C.; Li, B.; Liebermann, R.C. Sound velocities in polycrystalline $MgSiO_3$-orthopyroxene to 10 GPa at room temperature. *Am. Mineral.* **1998**, *83*, 444–450. [CrossRef]
33. Sinelnikov, Y.D.; Chen, G.; Neuville, D.R.; Liebermann, R.C. Ultrasonic shear velocities of $MgSiO_3$-perovskite at high pressure and temperature and lower mantle composition. *Science* **1998**, *281*, 677–679. [CrossRef]
34. Liu, J.; Chen, G.; Gwanmesia, G.D.; Liebermann, R.C. Elastic wave velocities of pyrope-majorite garnets ($Py_{62}Mj_{38}$ and $Py_{50}Mj_{50}$) to 9 GPa. *Phys. Earth Planet. Inter.* **2000**, *120*, 153–163. [CrossRef]
35. Darling, K.L.; Gwanmesia, G.D.; Kung, J.; Li, B.; Liebermann, R.C. Ultrasonic measurements of the sound velocities in polycrystalline San Carlos olivine in multi-anvil, high-pressure apparatus. *Phys. Earth Planet. Inter.* **2004**, *143–144*, 19–31. [CrossRef]
36. Chen, G.; Li, B.; Liebermann, R.C. Selected elastic moduli of single crystal olivines from ultrasonic experiments to mantle pressures. *Science* **1996**, *272*, 979–980. [CrossRef] [PubMed]
37. Decremps, F.; Li, B.; Liebermann, R.C. Single-crystal elasticity of ZnO at high pressure. In Proceedings of the International Conference on High Pressure Science and Technology (AIRAPT-17), Honolulu, HI, USA, 25–30 July 1999; Volume 2, pp. 810–813.
38. Decremps, F.; Zhang, J.; Li, B.; Liebermann, R.C. Pressure-induced softening of shear modes in ZnO. *Phys. Rev. B* **2001**, *63*, 224105. [CrossRef]

39. Kung, J.; Li, B.; Weidner, D.J.; Zhang, J.; Liebermann, R.C. Elasticity of $(Mg_{0.83}\ Fe_{0.17})O$ ferropericlase at high pressure: Ultrasonic measurements in conjunction with X-radiation techniques. *Earth Planet. Sci. Lett.* **2002**, *203*, 557–566. [CrossRef]
40. Gwanmesia, G.D.; Zhang, J.; Darling, K.; Kung, J.; Li, B.; Wang, L.; Neuville, D.; Liebermann, R.C. Elasticity of polycrystalline pyrope $(Mg_3Al_2Si_3O_{12})$ to 9 GPa and 1000 °C. *Phys Earth Planet. Inter.* **2006**, *155*, 179–190. [CrossRef]
41. Liebermann, R.C.; Chen, G.; Li, B.; Gwanmesia, G.D.; Chen, J.; Vaughan, M.T.; Weidner, D.J. Sound Velocity Measurements in Oxides and Silicates at Simulatneous High Pressures and Temperatures using Ultrasonic Techniques in Multi-Anvil Apparatus in Conjunction with Synchrotron X-radiation Determination of Equation of State. *Rev. High Press. Sci. Technol.* **1998**, *7*, 75–78. [CrossRef]
42. Li, B.; Liebermann, R.C. Indoor seismology by probing the Earth's interior by using sound velocity measurements at high pressures and temperatures. *Proc. Natl. Acad. Sci. USA* **2007**, *104*, 9145–9150. [CrossRef]
43. Li, B.; Liebermann, R.C. Study of the Earth's interior using measurements of sound velocities in minerals by ultrasonic interferometry. *Phys. Earth Planet. Inter.* **2014**, *233*, 135–153. [CrossRef]
44. Liebermann, R.C.; Prewitt, C.T. From Airlie House in 1977 to Lake Morey in 2007: 30 Years of evolution of mineral physics. *Phys. Earth Planet. Inter.* **2014**, *228*, 36–45. [CrossRef]
45. Liebermann, R.C.; Ehm, L.; Gwanmesia, G.D. Creating career paths for African-American students in Geosciences. *EOS Earth Space Sci. News* **2016**, *97*, 9–11. [CrossRef]
46. Chen, T.; Gwanmesia, G.D.; Wang, X.; Zou, Y.; Liebermann, R.C.; Michaut, C.; Li, B. Anomalous elastic properties of coesite at high pressure and implications for the upper mantle X-discontinuity. *Earth Planet. Sci. Lett.* **2015**, *412*, 42–51. [CrossRef]
47. Wang, X.; Chen, T.; Qi, X.; Zou, Y.; Kung, J.; Yu, T.; Wang, Y.; Liebermann, R.C.; Li, B. Acoustic travel time gauges for in-situ determination of pressure and temperature in multi-anvil apparatus. *J. Appl. Phys.* **2015**, *118*, 065901. [CrossRef]
48. Chen, T.; Liebermann, R.C.; Zou, Y.; Li, Y.; Qi, X.; Li, B. Tracking silica in Earth's upper mantle using new sound velocity data for coesite to 5.8 GPa and 1073K. *Geophys. Res. Lett.* **2017**, *44*, 7757–7765. [CrossRef]

minerals

MDPI

Review

Complex Geophysical Investigations under Extreme P,T–Conditions at Zentralinstitut für Physik der Erde (ZIPE) (1970–1990)

Hans J. Mueller [1,*] and Heiner Vollstädt [2]

1 Karlsruhe Institute of Technology, Institute of Applied Geosciences, Adenauerring 20b, building 50.40, D-76131 Karlsruhe, Germany
2 Vollstädt–Diamant GmbH, Kiefernweg 7, D-14554 Seddiner See, Germany; info@vollstaedt.com
* Correspondence: hans-joachim.mueller@rcftp.de; Tel.: +49-721-608-45221

Received: 13 February 2020; Accepted: 13 April 2020; Published: 6 May 2020

Abstract: The development of the geophysical high pressure research in the former German Democratic Republic (GDR) is described here. The GDR was a German state established in 1949 at the territory of the Soviet occupation zone. The different experimental investigations under extreme pressure and temperature conditions and their industrial applications, including the pilot manufacture of synthetic diamonds are explained. A review of the research topics pursued including experiments on lunar material and Earth core/mantle material is described.

Keywords: high pressure; physical properties; partial melting; lunar samples; diamond synthesis

1. Introduction

High pressures and high temperatures are two of the most important physical parameters in characterizing the physics of materials. In chemistry research and production lines under higher pressures and temperatures are also important; but in this field the volumes are bigger and pressures over 1 GPa are rarely applied.

In the mid-1970s when the author began working at the Zentralinstitut für Physik der Erde (ZIPE), scientists could only read international journals at the institute's library. Based on these data we could fill out order cards for special printed copies. Any personal contact with the authors was not allowed. Some of these special printed copies became personal guides to the scientific. For us in the discipline, perhaps the most valuable were undoubtedly the papers by Birch (1960/1961), Christensen (1965), Manghnani et al. (1974) and Kern (1978) [1–5]. Everything that was described in this literature could also be possible at Potsdam–pressure, geologically targeted samples, ultrasonic measurements, material interpretation of seismic data. We have to realize the world was divided into two different scientific regions at that time. That means any international published result could be possible by inaccessible materials and equipment. Consequently this "translation" to the "eastern" market was the first step of commencing scientific work. The publication of our own research results only happened at meetings and in the journals of countries of the Council for Mutual Economic Assistance (COMECON—an "eastern" analogue of the European Economic Community– EEC). During the 1970s, the conditions and methods of ZIPE experiments became comparable to those in the western countries and achieved-especially with the new "French press" (see later) – even higher conditions. The ZIPE department V had close relations with geoscientific institutions inside COMECON mainly at Kiev, Moscow, Novosibirsk, Apatity and Prague, including mutual visits. Later in the mid-1980s, the publication in "western" scientific journals was allowed and became customary. Consequently, the years from 1989—the year of the peaceful revolution—to 1992—the year of establishing the

GeoForschungsZentrum Potsdam—were, for the ZIPE scientific staff, the first time to be involved in the worldwide science community.

It was one of the peculiarities of scientific research in the German Democratic Republic (GDR) that the main part of high pressure research was conducted in the field of geosciences. From the mid-1960s to the early 1970s, the Institute for Geodynamik in Jena and the Geomagnetic Institute in Potsdam (integrated in the later Central Institute for Physics of the Earth) and the Karl Marx University in Leipzig, initiated the first experimental high pressure research. The background and basis for these research activities were polymorphic [6–9].

In Potsdam, at the Geomagnetic Institute, a rock magnetic department was established under the aegis of Frölich. The original theoretical and experimental investigations on the behaviour of the Earth's magnetic field and its geological development over time were very soon followed by the first rock-physical investigations under higher pressure and temperature conditions. The influence of Stiller and Vollstädt was particularly noticeable in this development; the latter was able to provide the mineralogical component to the mainly geophysically oriented group (see Table 1).

Table 1. The status of high pressure research at different institutions of the German Democratic Republic (GDR) in 1967.

Institution	VEB Geophysik Leipzig	department for geophysical survey and geology of Karl-Marx-University Leipzig	Central Institute for Physics of the Earth, Potsdam	Central Institute for Physics of the Earth, Potsdam Physics section, Dep. of Crystallography of Humboldt-University Berlin
equipment	200 t press	100 t press	200 t press, 50 t press, 2.5 t press	high pressure device, pressure chamber up to 40–60 kbar, uniaxial with small internal space
	ultrasonic equipment 0.3–1 MHz	ultrasonic equipment, high pressure hamber up to 8 kbar in test	ultrasonic equipment 1–4 MHz, high resistance measurement devices, strain gauges measurement devices, pressure chamber up to 2 kbar in test	polarisation microscopy, X-ray and spectroscopic devices
recent research	comparison of sound velocities of in-situ measurements with laboratory tests under uniaxial pressure	hysteresis and anisotropy of p-wave velocities under uniaxial pressure	deformation and electrical resistivity under uniaxial pressure	
research projects	pressure chamber up to 2 kbar		pressure chamber up to 10 kbar with heater, pressure chamber up to 30 kbar with heater, pressure chamber up to 100 kbar with heater	further development above chamber (heating)

In Jena, a strong theoretical group was established in the Institute for Geodynamics under the aegis of Uhlmann, which took particular interest in the condition of the deeper interior of the Earth and the material behaviour under extreme pressure and temperature conditions [10–13]. There too, under the influence of Stiller and against the background of theoretical investigations, a high-pressure group with a geophysical orientation similar to that in Potsdam was formed, i.e., mainly at conditions of the lower crust and upper mantle. However, reaching higher pressures was not given up. Later at Potsdam a new high pressure laboratory was established for performing experiments with the newly developed split sphere high pressure chamber (see later).

Organizational changes within the field of geosciences of the Academy of Sciences of the GDR led to the fusion of these regionally separated research activities and to the later concentration of the experimental activities at Potsdam (see Figures 1–4).

At the end of 1970 the first successful synthesis of industrial diamonds was made at ZIPE, for the time being without economic consequences. In 1981, the industrial pilot production was started in cooperation with district industry. By 1989, all hard material research and pilot production in Oranienburg were discontinued.

Figure 1. "Wood cutters" rest—von Faber.

Figure 2. Construction site of the first high pressure hall with a pit for the triaxial press (6 January 1971).

Figure 3. Assembling the triaxial press.

Figure 4. Prof. Vollstädt and his biggest press at that time.

2. Methods and Equipment of Measurement

2.1. Petrophysical Properties of Rocks and Minerals under Extreme P,T–Conditions

In autumn 1973, department V of ZIPE consisted of two buildings—a basement two-storey office part with lecture hall and the high-pressure laboratory, consisting of a hall, two offices and a large preparation room. In the hall there were two hydraulic presses—500 t and 600 t maximum load. For the latter, there was a removable carriage with a lateral punch ring allowing forces up to a maximum of 400 t. When combined this resulted in a triaxial system; i.e., a compression system in which a cube-shaped sample could be compressed in three directions. The world-famous laboratory of Prof. Kern in Kiel had the same principle. A piston heater was also developed in order to be able to experimentally simulate the temperature, which increased with depth, during the high-pressure experiments. After moving out the triaxial insert, a 600 t press with a yoke adjustable via two spindles was available [14–17] (see Figures 5 and 6).

Figure 5. 600 t press with triaxial device.

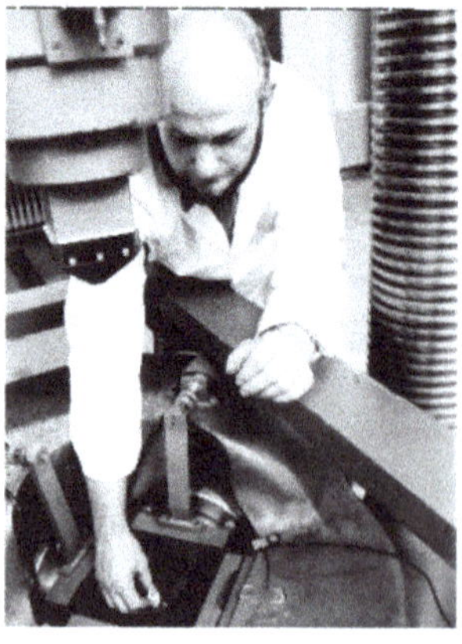

Figure 6. Sample insertion at the triaxial press.

The 500 t press had a manually adjustable yoke. Two classical oil pressure chambers up to 0.5 GPa maximum pressure, inner diameter 50 mm, as well as two piston-cylinder devices with different diameters (10 and 17 mm) of the ring system up to 1.75 GPa were available. Lead sheeting was used as a quasi-hydrostatic pressure transfer medium in which the rock samples were embedded. Ceramic ultrasonic transducers for compressional waves were installed in the upper and lower part of the column guide frames. For the oil pressure chambers there were initially no fixed measuring set-ups available. The task of the first author's (Mueller) final thesis was to develop an attenuation measurement for P-waves. The solution was a spring-loaded transmission method. Since the elastic coupling to the sample is critical in this case and the oil infiltration of samples should be avoided, all rock samples were covered by an air-drying lacquer. The ultrasonic transducers were separated from the sample only by a thin copper foil. The damping values were determined relative to a metallic specimen of known attenuation or by measuring with two samples of different lengths. In the course of the expansion of the high-pressure methods for measuring the elastic properties, a statistical measuring principle was established; i.e., first, the elastic wave velocities of all samples of a rock up to 0.5 GPa and room temperature were measured. On this basis, representative samples for the high-pressure, high-temperature experiments were then conducted, which were then finally incorporated into interpretation of deep seismic data. For this purpose, a special measuring set-up with two pairs of barium-titanate-zirconate ultrasonic transducers for elastic compression and shear waves was developed (see Figures 7 and 8). Measuring electronics, two measuring instruments for electric cables with preamplifiers, were utilized. In compact form, the instruments contained all the components (pulse generator, delay unit, oscilloscope) which are required for an ultrasonic measurement. Later a special ultrasonic measurement device from the company Krompholz was obtained [18–23].

Figure 7. Internal set-up of the 0.5 GPa oil pressure chamber.

These complex physical laboratory investigations also included measurements of the electrical conductivity and thermal properties of rocks under ambient and extreme pressure and temperature conditions. The results showed that with increasing pressure e.g., the electrical resistance increases, whereby the increase is more significant for sediments than for magmatic rocks.

Other parameters, such as magnetic quantities (magnetic anisotropy, direction of magnetization, Curie temperature), were also the subject of laboratory investigations and were used to clarify the magnetic rock behaviour down to the pressures of the deeper Earth crust (about 1 GPa and 750 °C).

The existing high-pressure measurements of elastic properties were essentially limited to crustal pressures; i.e., hydrostatic 0.5 GPa and quasi-hydrostatic 0.6 GPa. Thus, if the conversion factor of pressure in GPa to depth in km was 30, this corresponded to depths less than 20 km. It was then decided to purchase an oil pressure chamber up to a maximum pressure of 1.2 GPa. For this purpose, a two-layer shrunk jacketed vessel was built. Following the example of the existing 0.5 GPa chamber, the pressure chamber stood on the fixed lower punch with the electrical feedthroughs. The long upper punch was used for compression (Figures 7 and 8). The inner structure, the actual measuring cell, was connected to the lower punch with cables. Since there were problems with the compression path in practical test operation, the measuring assembly was connected to the lower punch with a plug. For this purpose, the measuring set-up was fed from above with a specially manufactured key and connected to the lower punch. However, there were problems with the pressure-transmitting medium. The hydraulic oil, which had been used up to 0.5 GPa, could no longer be used. From about 0.7 to 0.8 GPa its viscosity increased so much that cables were torn. Therefore, a mixture of ethanol and methanol was used as a new pressure transmitting medium. As expected, the viscosity behaviour was excellent, but the lubricating effects on the metal seals were not. Therefore, up to 10% hydraulic oil was added as lubricant. The measuring set-up was designed for measuring the elastic wave velocities of longitudinal and shear waves up to 750 °C [24–36].

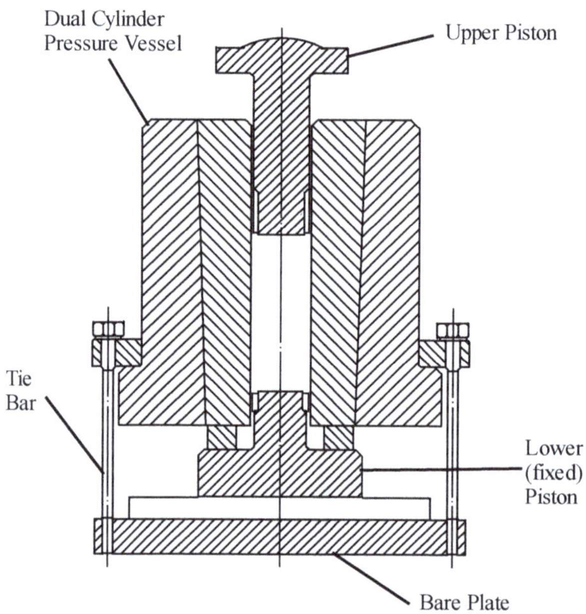

Figure 8. Sketch of the 1.2 GPa oil pressure chamber.

In 1977, there was an opportunity for scientific institutions to order major purchases on the world market, if this could remedy a massive shortage in the sense of scientific progress. Department V applied for a high-performance gas pressure chamber and was awarded the contract. Harwood, USA and Basset Bretagne Loire, France were shortlisted. Harwood offered a gas pressure chamber operated with argon up to 1.4 GPa. From Basset came a bid for a helium chamber up to 2.5 GPa (Figures 9 and 10). Therefore, the French offer was chosen. According to reports the selling price was 1.5 Mio Valuta Mark. In autumn 1977 a very large box arrived at Potsdam and was placed on the ramp of the high pressure laboratory, only about 10 m away from its future installation site. The following spring, two installers from the manufacturer arrived for the installation. The construction consisted of a 500 t press on top of which the pressure chamber, a four-shell vessel with 41 mm inner diameter, was placed. In addition, all the necessary pumps were combined on one platform. There was a low-pressure high performance pump for the hydraulic oil for the test preparation and post processing. This was used, for example, to push the lower piston with its seal set to its position immediately below the lateral gas inlet. Then, there were two plunger high-pressure pumps to move the lower piston into the pressure chamber after the automatic stopping the gas pressure pumps to reach a maximum pressure of 2.5 GPa. The third set of pumps was a two-stage high-pressure diaphragm compressor, which compressed the working gas from 10 MPa to 350 MPa and filled it into the chamber via the lateral gas supply. The third assembly was a large control cabinet containing all nova Swiss valves, electrical switches and display instruments for operating the press (Figure 10). For safety reasons, the entire high-pressure equipment, with the exception of this control cabinet, was enclosed in angle steel protective walls covered with 5 mm deep-drawing sheet steel. If 1.3 GPa was sufficient as the final pressure, high-purity argon could be used. However, the pressure chamber was designed for high-purity helium. "High purity" was necessary because hydrogen in particular would have diffused into the inner wall of the pressure chamber, which would have made it brittle. Extreme cleanliness, i.e., the removal of all residues of the surface coating of the metal sealing rings from the last experiment from the inner surface, was necessary. For this purpose, there was a platinum wire furnace, which was further developed in department V with an ultrasonic measuring set-up and in its last version allowed measurements of the propagation velocity of elastic compressional and shear waves up to 2.5 GPa and 1600 °C (Figure 11). To prevent infiltration of the rock samples by the compressed gas, the samples were welded into a stainless steel casing. However, there were also experiments on electrical conductivity and investigation of the deposition of diamond from the gas phase at extreme pressures. In the experiments for the material interpretation of deep seismic profiles, these gas pressure experiments up to 2.5 GPa and 1600 °C formed the upper end after preliminary tests up to 0.5 GPa and 1.2 GPa/500 °C oil pressure. Research carried out at a later date confirmed the unique commercial possibilities on a world scale for this pressure chamber. During the 13 years of operation of the high performance gas pressure chamber in the ZIPE, there was not a single accident [37–52].

Figure 9. 2.5 GPa gas pressure vessel.

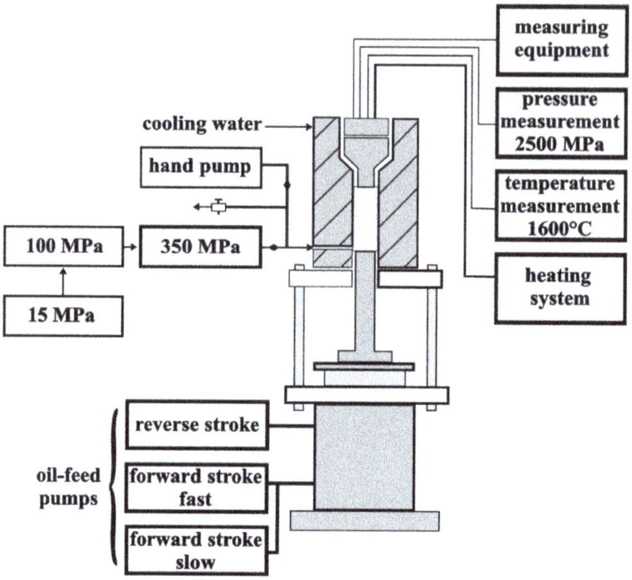

Figure 10. Scheme of the 2.5 GPa gas pressure vessel.

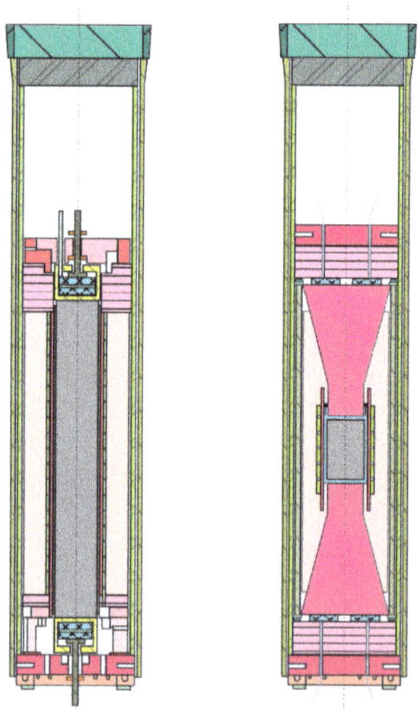

Figure 11. Two different measurement devices with sample encapsulation for the 2.5 GPa gas pressure vessel.

In the early 1980s, a new building for department V was also urgently required. This was a two-storey building with no basement and a high-pressure hall with hand washbasins at every room. This accommodated a Czech isostatic press; i.e., an extremely large oil pressure chamber. It could compress about 20 l of oil to 0.4 GPa. For safety reasons, the whole equipment was assembled below ground level. Therefore, the first part of the new building was the "basement", i.e., a room sized space with massive concrete walls. Only after assembling the press in this pit was the whole building established above. The purpose was to operate the split-sphere high pressure chamber (Figure 12) developed by the company itself following the publication of Kawaii and Endo (1970) [52]. Six steel anvils were combined to provide an outside sphere; the inner part is cube-shaped. It was filled by eight tungsten carbide anvils with a small octahedron sample in the center. After assembling the whole sphere was imbedded in a polysiloxane rubber jacket to prevent oil infiltration. The split-sphere chamber is the moulding of the later multi-anvil chambers. For a first test of the new chamber, the help of the Institute for Marine Research in Warnemünde was called upon. There was a 0.2 GPa plant working with water as pressure transmitting medium. The test was successful, but did not yet reach the target pressure of 1 Mbar. Additionally, in the mid-1980s a 1000 t press and an up-to-date 400 t press (see Figure 13) were installed in the new high-pressure hall. However, the new building with thc high pressure hall with the subsurface installation had another advantage. Now the high pressure staff had a party room (see Figures 14 and 15)

Figure 12. Split-sphere high pressure chamber.

Figure 13. New 400 t press.

Figure 14. The party at Vollstaedt's habilitation (left, central Mueller, right Waesch).

Figure 15. The colleagues celebrate this habilitation.

2.2. Electrical and Structural Relations in the Fe-Ni-S System at High Pressures and Temperatures and the State of the Earth Interior

The theoretical investigations of the state and structural behaviour of the deeper interior of the Earth required an extended electrical technique, which allowed a reliable extrapolation for an interpretation of the material behaviour to the outer core of the Earth (see Figure 16). The results were published among other things at the meeting of the European High Pressure Research Group at

Potsdam (see Figure 17). The material system investigated consisted mainly of the elements relevant to the Earth's core: iron, nickel and manganese in combination with Sulphur [53–71].

Figure 16. Bridgman device for electrical investigations.

Figure 17. Prof. Stiller gives a talk about the Earth core structure at the European High Pressure Research Group 1987 in Potsdam.

The verification of the pressure and temperature stability of the various phases of the elements mentioned or their sulphides and the determination of possible phase transitions under the extreme conditions required the development and manufacture of special high-pressure equipment.

The diamond anvil apparatus (also known as "Squeezer" in ZIPE) and the high-pressure apparatus "Belt" proved to be suitable. Department V especially concentrated on latter, which was later used for industrial diamond synthesis [72–77].

Figure 18 shows a diamond anvil apparatus developed and manufactured at the Central Institute for Physics of the Earth at Potsdam. Phase transitions in the range of more than 10 GPa could be produced in a very small space. Temperatures of more than 1000 °C could be reached by a special laser heating technology. The advantage of this high-pressure technology was that no additional press was required.

Figure 18. The squeezer (diamond anvil apparatus).

For these experiments the Fe-Ni-S-system was selected. X-ray diffraction measurements up to 21 GPa were carried out and phase transitions at 0.4 and 5.0 GPa were detected. In comparison with density calculations, derived from shock wave experiments, the Fe-Ni-S-system could be a candidate to discuss the composition of the inner Earth.

The "Belt" type high-pressure chamber (Figure 19) was a further development of a known variant with regard to a split support ring system and reached pressures of up to 8 GPa at temperatures of up to 1800 °C. For the "Belt" suitable presses according to their size were necessary. The advantage was the relatively large working volume and the fact that the design of this chamber enabled quasihydrostatic conditions, which led to optimum pressure and temperature gradients during the experiments.

2.3. Phase Transitions at High Pressure and Temperature–the System Graphite-Diamond

After the successful synthesis of industrial diamonds in the USA and Sweden in 1956, in the Soviet Union in 1961 and in the ČSSR (Czechoslovak Socialist Republic) in 1965 on the basis of various high-pressure technologies, the first tentative proposals to include this economically important technology in the research spectrum were made in the GDR as early as 1966. Such ideas were not supported by leading industrial companies (e.g., VEB Carl Zeiß Jena) or government officials (VEB – nationally owned company). Politicians referred to the agreements within the framework of the COMECON, according to which the Soviet Union, as a producer of industrial diamonds, was responsible for the entire COMECON sector.

Figure 19. Split belt ring system.

At the Central Institute for Physics of the Earth, Potsdam, high-pressure technology had been so developed at 1978 that nothing gets in the way of initial attempts at diamond synthesis. Contrary to clear instructions from the Presidium of the Academy of Sciences of the GDR, successful synthesis experiments to prove the pressure and temperature parameters were carried out at Potsdam in 1979. A few years later these results were also protected by patents. Figure 20 shows the investigation of the first successful synthesis experiment [78–81].

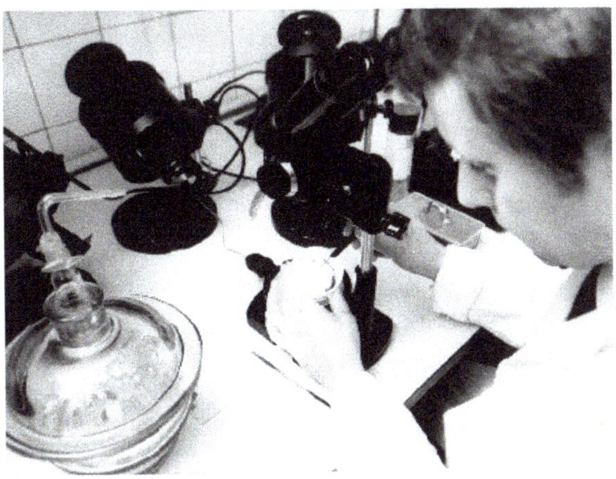

Figure 20. Inspection of the first synthetic diamonds by Vollstaedt.

In 1981, the GDR's Academy of Sciences (Zentralinstitut für Physik der Erde, Potsdam, ZIPE, in cooperation with the Zentralinstitut für Festkörperphysik und Werkstoffwissenschaften, Dresden) was surprisingly commissioned to begin research into the production of super-hard materials as part of a state plan topic. This basic research, starting with the synthesis of diamond, was also carried out for the synthesis of the superhard materials cubic bornitride (CBN) and polycrystalline diamond (PKD) and only ended with the transformation of the ZIPE.

Only shortly after the start of the basic research mentioned, companies were founded for the pilot production. Until the year 1989 more than 50,000 carats, diamond powder in different grain sizes were produced here and successfully used in the glass industry as well as in tool manufacturing. These achievements and successes are therefore especially important for as there have been a number of difficulties and mistakes in the accomplishment of the tasks. Materials important in the production process for superhard materials were not available on the GDR market. The usual standard material for the sealing pressure transmission–pyrophyllite-could only be obtained for foreign currencies, as well as the larger high-pressure high-temperature press technology.

After extensive field work to extract pyrophyllite in Vietnam (supported by the Institute for Raw Materials Research, Dresden), a lucky "substitute" was found from the company's own deposits. A deposit near Bockau/Erzgebirge described as pyrophyllite in older works was paragonite. A site survey by the second author revealed the presence of paragonite. After intensive investigations, this material, which was the unused vein material for emery mining at the beginning of the 20th century, proved to be a suitable sealing material for diamond synthesis [82]. Even special processing of the material made it possible to improve the properties of the diamond synthesis process (patent application).

The final decision for the production at the industrial partner Collective Combine Erdöl/Erdgas at Gommern came too late. The import of high-pressure technology and the establishment of diamond production in Gommern did not occur. In addition, the third high pressure hall of department V had a very short life. Here the two northern Korean wound frame presses and a US MTS stiff press were accommodated. During a research visit of the coauthor to the National Institute of Inorganic Materials, Tsukuba, Japan (1989), the cubic phase of aluminum nitride could be detected for the first time (Figure 21).

Figure 21. Vollstaedt in Tsukuba, Japan (central, left Kanda, right Fukunaga).

2.4. *Planetary Investigations and Special Experiments–Measurements on Lunar Material*

In 1970, 1972 and 1976 the Russian automatic spacecrafts Luna 16, 20 and 24 transported ca. 400 g of material from the lunar surface to the Earth for scientific investigations. The Central Institute for Physics of the Earth in Potsdam received a small part of this lunar regolith (ca. 3 g) for complex mineralogical experiments, together with other Institutes and Universities (see Figures 22 and 23).

Figure 22. Handing over the lunar material (standing Prof. Klare, President of Academy of Sciences, left Prof. Stiller).

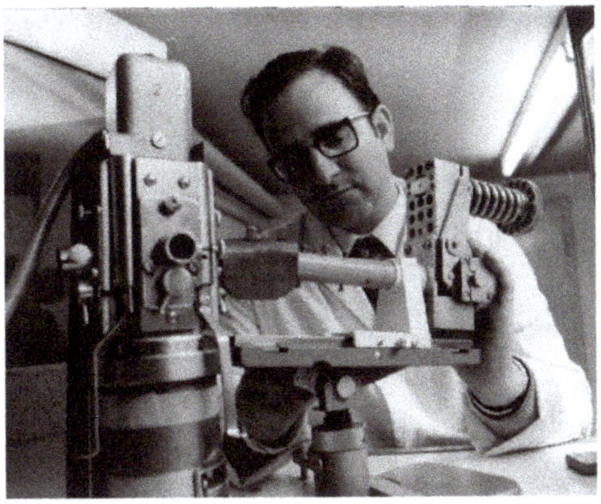

Figure 23. Waesch at the squeezer.

The main tasks for these investigations were:

- the determination of the composition and the properties of the lunar regolith and the lunar rocks, helping to explain the lunar history and the planetary evolution, especially the origin of the moon, and
- to develop experimental methods for using very small quantities of lunar materials (mg) to get very exact results.

To determine the composition of the minerals, their structural and chemical behaviour X-ray-spectroscopic investigations have been used. The silicic components of the different rock types were determined.

Petrological investigations on small rock fragments to describe fracture structures together with measurements of the electrical conductivity of the regolith were carried out [83–90].

For the investigation of the very small regolith particles (see Figures 24 and 25) a special high voltage microscope (1 MV) and a transmission electron microscope (up to 100 kV) was used. The most interesting results were:

- The potassium content of the lunar rocks is higher than in the terrestrial rocks.
- The composition of the pyroxene-group of lunar rocks corresponds to the rock composition on Earth.
- The main elements, O, Si, Fe, Al, Mg have similar contents in all lunar samples. Only lunar 16 has higher iron content.

- Lunar 20 has high cobalt content caused by meteorite bombardments.
- Complicated fracture surfaces (Figure 25) on regolith particles could be detected, fragment size (0.02–0.2 mm).

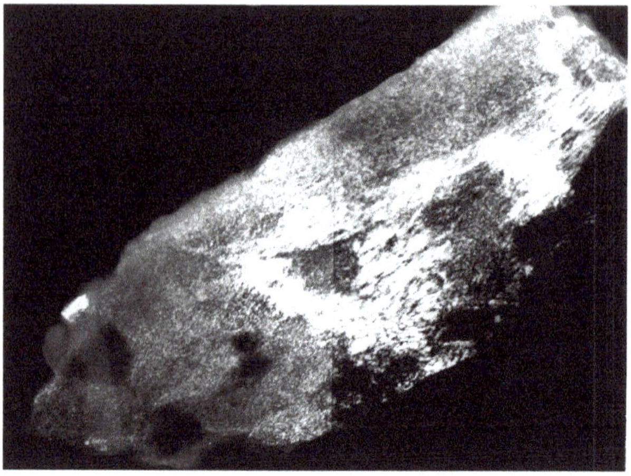

Figure 24. Fragment of anorthite (HEM 18.00).

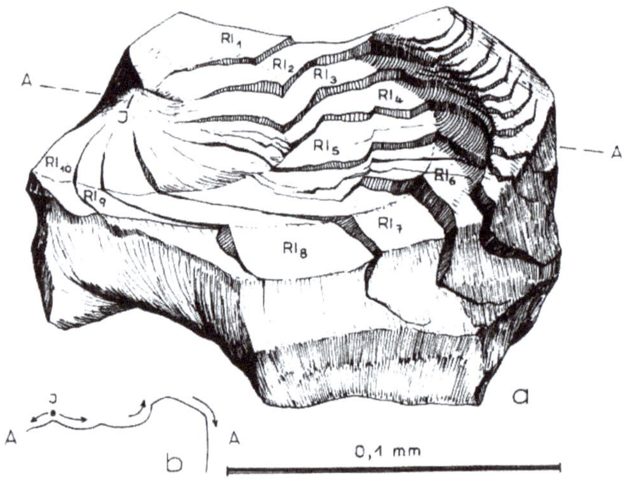

Figure 25. Fracture documentation of a small regolith fragment.

3. Summary

In a history of about 25 years, the department V of ZIPE with about 40 to 60 employees had made it possible to participate in practically all high-pressure directions relevant for geophysics. The new buildings of the GeoForschungsZentrum Potsdam (GFZ) were established at the place where the three high pressure halls with offices and machine shop had taken over the place. In contact with the demolition or slightly later all high pressure technology (with the only exception of the MTS-press) and all developments were scrapped after the reunification of Germany without necessity. All general high-pressure research was discontinued. At GFZ, three different sections are involved with special aspects of high pressure.

Author Contributions: The paper is the result of a joint effort. All authors have read and agreed to the published version of the manuscript.

Funding: This research received no external funding.

Acknowledgments: The authors highly acknowledge the patience and kindness of all editors just as well especially the support of one of the reviewers.

Conflicts of Interest: The authors declare no conflict of interest.

References

1. Birch, F. The velocity of compressional waves in rocks to 10 kilobars, part 1. *J. Geophys. Res.* **1960**, *65*, 1083–1102. [CrossRef]
2. Birch, F. The velocity of compressional waves in rocks to 10 kilobars, part 2. *J. Geophys. Res.* **1961**, *66*, 2199–2224. [CrossRef]
3. Christensen, N.I. Compressional wave velocities in metamorphic rocks at pressures to 10 kilobars. *J. Geophys. Res.* **1965**, *70*, 6147–6164. [CrossRef]
4. Manghnani, M.H.; Ramananantoandro, R.; Clark, S.P., Jr. Compressional and shear wave velocities in granulite facies rocks and eclogites to 10 kbar. *J. Geophys. Res.* **1974**, *79*, 5427–5446. [CrossRef]
5. Kern, H. Effect of high-low quarts transition on compressional and shear wave velocities in rocks under high pressure. *Phys. Chem. Miner.* **1979**, *4*, 161–171. [CrossRef]
6. Stiller, H.; Vollstädt, H.; Faber, J. *Hochdruckphysikalische Untersuchungen im Zentralinstitut für Physik der Erde*; Veröff. KAPG-Arbeitsgruppe 1.11; Veröff. ZIPE: Potsdam, Germany, 1974; pp. 22–28.
7. Stiller, H.; Vollstädt, H. Festkörper unter hohen Drücken und Temperaturen. *Veröff. ZIPE* **1972**, *17*, 7–32.
8. Stiller, H.; Vollstädt, H. *Feste Körper unter hohen Drücken und Temperaturen*; Sitzungsberichte AdW; Springer: Berlin, Germany, 1975; H.-1N (20.11.75).
9. Stiller, H.; Vollstädt, H. Festkörperphysik und Untersuchung des Erdinnern. *Z. Geol. Wiss.* **1981**, *11*, 1333–1341.
10. Vollstädt, H. Hochdruckverfahren in Wissenschaft und Technik. *Fert. Betr.* **1976**, *26*, 307.
11. Vollstädt, H. Geophysikalische Hochdruckforschungen und Diamantsynthese in der DDR. *Schr. Geowiss.* **2011**, *18*, 33–40.
12. Vollstädt, H.; Recht, H. Hochdruckforschung in der Werkstoffwissenschaft. *Wiss. Fortschr.* **1985**, *35*, 313–315.
13. Stiller, H.; Vollstädt, H. Hohe Drücke. In *Kleine Enzyklopädie, Struktur der Materie*; Bibliographisches Institut: Leipzig, Germany, 1982; pp. 632–846.
14. Müller, H.J. Die Messung der Dämpfung Elastischer Longitudinalwellen in Gesteinen unter Druck—Erarbeitung und Erprobung Eines Meßplatzes. Ph.D. Thesis, Humboldt-University, Berlin, Germany, 1974; 170p.
15. Müller, H.J.; Raab, S.; Seipold, U. Apparaturnye i metodićeskie voprosy izućenija skorosti i zatuchanija uprugich voln v sisteme FeS i gornych porodach.—(instrumental and methodical issues of investigations of the propagation velocity and the attenuation of elastic waves in the FeS-system and in rocks). In *Issledovanija Fizićeskich Svojstv Mineral'nogo Vešćestva Zemli pri Vysokich Termodinamićeskich Parametrach*; Naukova Dumka: Kiew, Ukraine, 1977; pp. 39–42.
16. Müller, H.J.; Raab, S.; Seipold, U. Uprugaja anizotropija gornych porod.—(elastic anisotropy of rocks). In *Issledovanija Fizićeskich Svojstv Mineral'nogo Vešćestva Zemli pri Vysokich Termodinamićeskich Parametrach*; Naukova Dumka: Kiew, Ukraine, 1977; pp. 73–76.
17. Müller, H.J.; Raab, S.; Seipold, U. Equipment for anisotropy investigations of rock samples under high pressures. In *Theoretical and Experimental Investigations of Physical Properties of Rocks and Minerals under Extreme p, T-Conditions*; Stiller, H., Volarovich, M.P., Eds.; Akademie-Verlag: Berlin, Germany, 1979; pp. 15–20.
18. Seipold, U.; Engler, R.; Gutzeit, W.; Müller, H.J. Petrophysikalische Untersuchungen an Granuliten unter hohen thermodynamischen Bedingungen. *Freib. Forsch.* **1979**, *350*, 33–39.
19. Seipold, U.; Müller, H.J. Ultraschallmessungen an Gesteinen mit dem Zählfrequenzmesser Tscha 3-34. *Neue Bergbautech.* **1979**, *6*, 347–349.
20. Engler, R.; Müller, H.J. Elastische Eigenschaften klüftiger Gesteine unter einaxialer Belastung. *Gerlands Beitr. Geophys.* **1980**, *3*, 217–222.

21. Seipold, U.; Müller, H.J.; Raab, S. Das Druck- und Temperaturverhalten von Gesteinen und dessen Bedeutung für die Interpretation von Anomalien. *Z. Geol. Wiss.* **1980**, *8*, 517–523.
22. Müller, H.J. Elastische Eigenschaften. In *Geophysikalische Methodik—Paläozoiden*; Res. Report (G4) ZIPE; Central Institute for Physics of the Earth: Potsdam, Germany, 1980.
23. Janssen, C.; Müller, H.J. Analyse paläotektonischer Spannungen auf der Grundlage von. Ultraschallmessungen an Horizontalstylolithen. *Z. Angew. Geol.* **1981**, *1*, 607–610.
24. Stiller, H.; Lebedev, T.S.; Mundt, W.; Hurtig, E.; Franck, S.; Müller, H.J. *Physikalische Eigenschaften Planetarer Materie bei Hohen Drücken und Temperaturen und Möglichkeiten der Komplexinterpretation von Beobachtungsdaten*; KAPG-Monogr.; Naukova Dumka: Kiew, Ukraine, 1984.
25. Müller, H.J. Elastische Eigenschaften von Krustengesteinen. In *Elastische und thermische Eigenschaften von Krustengesteinen*; Res. Report (G4) ZIPE; Central Institute for Physics of the Earth: Potsdam, Germany, 1984; pp. 53–108.
26. Müller, H.J. *Konzeption zur Nutzung der Isostatischen Presse für Petrophysikalische Messungen*; Res. Report ZIPE; Central Institute for Physics of the Earth: Potsdam, Germany, January 1985.
27. Kopf, M.; Müller, H.J.; Gottesmann, B. Correlation between pyroxene content and v_p and v_s under high pressure. In *Physical Properties of the Mineral System of the Earth's Interior*; Kapička, A., Kropaček, V., Pros, Z., Eds.; Institute Geophysics: Prague, Czech Republic, 1985; pp. 168–172.
28. Müller, H.J. *Petrophysikalische Messungen an Krustengesteinen*; Res. Report, "Methodik Petrologie" (G4 8/85), ZIPE; Central Institute for Physics of the Earth: Potsdam, Germany, 1985.
29. Müller, H.J. *Abschluß der Entwicklungs- und Fertigungsarbeiten zur Messung der Petroakustischen Parameter unter Höchsten Temperaturen mit Hilfe der Hydrostatischen 2,5 GPa-Anlage*; Res. Report ZIPE; Central Institute for Physics of the Earth: Potsdam, Germany, 1985; pp. 45–50.
30. Müller, H.J. *Die Experimentelle Erfassung von Zusammenhängen Zwischen den Elastischen und Stofflichen Eigenschaften unter den Druck-Temperatur-Bedingungen der Unterkruste im Südteil der DDR*; KAPG-Monogr.; Central Institute for Physics of the Earth: Potsdam, Germany, 1986.
31. Müller, H.J. *Investigation of Elastic Properties*; IASPEI—National Report; Central Institute for Physics of the Earth: Potsdam, Germany, 1986.
32. Müller, H.J. *Komplexe Hp- und HT- Petrophysikalische Untersuchungen an Gesteinen der Kruste und des Oberen Mantels*; Res. Report, (G3 5/1987, G4 11/1987), ZIPE; Central Institute for Physics of the Earth: Potsdam, Germany, 1987; pp. 35–45.
33. Müller, H.J. *Die Untersuchung der Elastischen Eigenschaften von Gesteinen des sächsischen Granulitgebirges unter den Bedingungen Hoher Drücke und Hoher Temperaturen*; Central Institute for Physics of the Earth: Potsdam, Germany, 1987; pp. 24–30.
34. Müller, H.J.; Schmädicke, E. Relation between the elastic properties and the structural behaviour of high-metamorphic rocks under the thermodynamical conditions of the upper mantle. In *Veröff. des Forschungsbereiches Geo—und Kosmoswissenschaften/High Pressure Geosciences and Material Synthesis*; Central Institute for Physics of the Earth: Potsdam, Germany, 1988; pp. 42–45.
35. Müller, H.J. High Pressure Investigation of the Elastic Properties under Partial Melting Conditions. In *Proceedings of the XXVIII. EHPRG, Bordeaux, France, 8–13 July 1990*; Special Edition High Press. Res.: Didcot, UK, 1990.
36. Müller, H.J.; Raab, S. Elastische Eigenschaften von Erzgebirgs-Graniten. *Z. Geol. Wiss.* **1993**, *21*, 207–216.
37. Müller, H.J.; Massonne, H.J. DFG-progress report. In *Hochdruck-Experimentelle Petrophysikalische Untersuchungen zum Krustenbau des Erzgebirges*; Deutsche Forschungsgemeinschaft: Frankfurt (Main), Germany, 1993.
38. Müller, H.J.; Raab, S. The velocities of elastic P- and S-waves in high-grade metamorphic rocks under high pressures and temperatures. *Geophys. J. Internat* **1995**, *17*, 78–85.
39. Behr, H.-J.; Dürbaum, H.-J.; Bankwitz, P.; Bankwitz, E.; Benek, R.; Berger, H.-J.; Brause, H.; Conrad, W.; Förste, K.; Frischbutter, A.; et al. Crustal structure of the Saxothuringian Zone: Results of the deep seismic profile MVE-90 (East). *Z. Geol. Wiss.* **1994**, *22*, 647–669.
40. Mueller, H.J. Modelling the lower crust by simulation of the in situ conditions: An example from the Saxonian Erzgebirge. *Phys. Earth Planet. Int.* **1995**, *92*, 3–15. [CrossRef]

41. Kern, H.; Müller, H.J. *Petrophysikalisch-Mineralogische Untersuchungen zur Teufenvariation der Granite des Erzgebirges als Beitrag zur Stofflichen Interpretation des Profils DEKORP 3-MVE*; DFG-final report; Deutsche Forschungsgemeinschaft: Frankfurt (Main), Germany, 1997.
42. Tietz, O.; Schwab, M.; Bankwitz, P.; Müller, H.J. *Komplexe Geologische Untersuchungen am Gemischten Bodegang und Seinen Xenolithen an den Gewitterklippen bei Thale (Harz)*; DFG final report; Deutsche Forschungsgemeinschaft: Frankfurt (Main), Germany, 1995.
43. Raab, S.; Hoth, P.; Huenges, E.; Mueller, H.J. Role of sulfur and carbon in the electrical conductivity of the middle crust. *J. Geophys. Res.* **1998**, *103*, 9681–9689. [CrossRef]
44. Seipold, U.; Mueller, H.J.; Tuisku, P. Principal differences in the high pressure behaviour between thermal and elastic properties of rocks. *Phys. Chem. Earth* **1998**, *23*, 357–360. [CrossRef]
45. Mueller, H.J.; Massonne, H.-J. Experimental High Pressure Investigation of Partial Melting in Natural Rocks and their Influence on v_p and v_s. *Phys. Chem. Earth Part A* **2001**, *26*, 325–332. [CrossRef]
46. Mueller, H.J.; Roetzler, K.; Schilling, F.R.; Lathe, C.; Wehber, M. Techniques for measuring the elastic wave velocities of melts and partial molten systems under high pressure conditions. *J. Phys. Chem. Sol.* **2010**, *71*, 1109–1117. [CrossRef]
47. Stiller, H.; Vollstädt, H. Influence of pressure and joints on the velocity of elastic waves in jointy rocks in the case of Earth quakes. In *Physical Properties of Rocks and Minerals under Extreme P-T-Conditions*; Akademie Verlag: Berlin, Germany, 1979; Volume 9, pp. 177–194.
48. Stiller, H.; Vollstädt, H.; Seipold, U. Investigations of thermal and elastic properties of rocks by means of a cubic press. *Phys. Earth Planet. Int.* **1978**, *17*, 31–34. [CrossRef]
49. Stiller, H.; Vollstädt, H.; Seipold, U. Complex physical measurements of several rocks under pressure. *Publ. Inst. Phys. Pol. Acad. Sci.* **1980**, *9*, 239–246.
50. Stiller, H.; Vollstädt, H.; Seipold, U.; Raab, S.; Volarovich, M.; Bajuk, E.; Efimova, G. Investigations of elastic properties and polymorphism of calcite by three-axial stress to 20 kbar. In *Physical Properties of Rocks at High Thermodynamical Paraneters*; The Institute of Physics of the Earth: Moscow, Russia, 1978; pp. 174–175.
51. Vollstädt, H. Contributions of petrophysical investigations to the interpretation of seismological models of the lithosphere. In *Proceedings of the UNESCO-Seminar, Mexico city, Mexico, 26 July–8 August 1982*; UNESCO: Berlin, Germany, 1982.
52. Vollstädt, H.; Budzinski, H. *Hochdruckphysik und die Struktur der Lithosphäre*; Veröff. ZIPE; Central Institute for Physics of the Earth: Potsdam, Germany, 1981.
53. Kawai, N.; Endo, S. The generation of ultrahigh hydrostatic pressures by a split sphere apparatus. *Rev. Sci. Instr.* **1970**, *41*, 1178–1181. [CrossRef]
54. Volarovitsch, M.-P.; Bayuk, E.I.; Yefimova, G.A.; Stiller, H.; Vollstädt, H.; Seipold, U.; Raab, S. A study on elastic properties and polymorphism of calcite under three-axial loading up to 20 kbar. *Phys. Earth* **1979**, *11*, 8–12.
55. Stiller, H.; Seipold, U.; Vollstädt, H. Electrical. thermoelectrical and elastic properties of transition metals chalcogenides under extreme p-T-conditions. In *Physical Properties of Rocks at High Pressures and Temperatures*; Inst. Geophys.; Ukrainian SSR, Naukova Dumka: Kiev, Ukraine, 1978.
56. Seipold, U.; Ullner, H.A.; Vollstädt, H. Thermal effects at the B1-B2-transition of Potassium Bromide. *Physica* **1986**, *139*, 269–271. [CrossRef]
57. Seipold, U.; Vollstädt, H.; Gutzeit, W. Determination of the thermal diffusivity of some rocks at high pressure. In Proceedings of the XIII. AIRAPT-Conference, Uppsala, Sweden, 17–22 August 1981; pp. 755–759, Part 2: Uppsala.
58. Frölich, F.; Vollstädt, H. Zur geophysikalischen Tiefenerkundung—Die Druck-Temperaturauswirkungen auf das magnetische und elektrische Gesteinsverhalten. *Ber. Ges. Geol. Wiss. B Miner. Lagerstättenf* **1970**, *15*, 307–312.
59. Stiller, H.; Vollstädt, H.; Seipold, U.; Wäsch, R. Electrical and thermal investigations of rocks and minerals under extreme p-T-conditions. *Pure Appl. Geophys.* **1976**, *114*, 262–272. [CrossRef]
60. Stiller, H.; Seipold, U.; Vollstädt, H. Electrical, thermoelectrical and elastic properties of transition metal chalcogenides under extreme PT conditions. In *Physikal. Eigenschaften d. Gesteine unter hohen Drücken und Temperaturen*; Sammelband, 4. Allunionskonferenz; Tbilissi, Hrsg.: Inst. f. Geophys; Ukrainian SSR, Naukova Dumka: Kiev, Ukraine, 1974; pp. 198–200.

61. Stiller, H.; Seipold, U.; Vollstädt, H. *Elektrische und Thermische Eigenschaften von einigen Mineralen und Gesteinen unter Extremen Bedingungen*; National Committee Geodesy Geophysics: Berlin, Germany, 1975; Volume III, p. 36ff.
62. Stiller, H.; Seipold, U.; Vollstädt, H. Electrical, thermoelectrical and elastic properties of transition metal chalcogenides under extreme p-T-conditions. In *Physikal. Eigenschaften d. Gesteine unter hohen Drücken und Temperaturen*; Naukova Dumka: Kiev, Ukraine, 1978; pp. 35–51.
63. Stiller, H.; Seipold, U.; Vollstädt, H. Thermal properties of rocks under high p-T-conditions. In Proceedings of the XVI'th General Assembly IWGG, Grenoble, France, 17–18 September 1975.
64. Vollstädt, H.; Seipold, U.; Gutzeit, W. Zur Messung der Temperaturleitfähigkeit unter extremen Bedingungen. *Geofys. Sb.* **1972**, *47*, 13–16.
65. Stiller, H.; Wagner, F.C.; Vollstädt, H. Zum Verhalten elastischer Wellen bei Bruchvorgängen in Gesteinen. *Abh. AdW DDR Abt. Math. Naturwiss. Tech.* **1978**, *1*, 111–121.
66. Stiller, H.; Wagner, F.C.; Vollstädt, H. Petrophysical causes and quantitative results of v_p changes before earthquakes. In *High Pressure and Temperature Studies of Physical Properties of Rocks and Minerals*; Naukowa Dumka: Kiew, Ukraine, 1977; pp. 86–192.
67. Stiller, H.; Wagner, F.C.; Vollstädt, H. *Wellengeschwindigkeitsänderungen in brüchigen Gesteinen unter Druck und Zusammenhänge mit Erscheinungen vor Erdbeben*; 7. Wiss. Jahrestagung des ZIPE, Tagungsband, (Heft: "Berichte über Arb. d. ZIPE und seiner NAN"); Central Institute for Physics of the Earth: Potsdam, Germany, 1975; p. 56.
68. Stiller, H.; Wagner, F.C.; Vollstädt, H. Die Druckabhängigkeit der Wellengeschwindigkeiten in brüchigen Gesteinen und ihre Beziehungen zu Erscheinungen vor Erdbeben. *Fiz. Zemli* **1980**, *1*, 52–61.
69. Stiller, H.; Wagner, F.C.; Vollstädt, H.; Hurtig, E. Physical processes in earthquake sources, interpretation possibilities on the base of laboratory experiments. In Proceedings of the IASPEI-Symposium, Bad Honnef, Germany, 26–29 September 1978.
70. Stiller, H.; Wagner, F.C.; Vollstädt, H. A two-phase model for the description of the influence of cracks on the v_p and v_s wave velocities in dry and saturated rock samples. *Tectonophysics* **1977**, *43*, 181–197. [CrossRef]
71. Stiller, H.; Vollstädt, H.; Franck, S. Present problems of high pressure geophysics. In *Proceedings of the VII. AIRAPT-Conference*; Plenum Press: New York, NY, USA, 1980.
72. Stiller, H.; Vollstädt, H.; Franck, S. State of material inside the Earth and some implications for geodynamics. Proceedings part I. 4. Int. In Proceedings of the Symposium Geodesy and Physics of the Earth, Karl-Marx-Stadt, Germany, 12–17 May 1980.
73. Vollstädt, H.; Stiller, H.; Kraft, A. Experimental high pressure investigations at the Fe-Ni-S-system and some geophysical implications. In Proceedings of the EGS-ESC-Symposium, Budapest, Hungary, 21–29 August 1980.
74. Vollstädt, H.; Kraft, A.; Seipold, U. *Zum Verhalten von FeS unter extremen p-T-Bedingungen*; Veröff. ZIPE; Central Institute for Physics of the Earth: Potsdam, Germany, 1974; Volume 21, pp. 55–71.
75. Kraft, A.; Stiller, H.; Vollstädt, H. The monosulfid solid solution in the Fe-Ni-S system: Relationship to the Earth's core on the basis of experimental high-pressure investigations. *Phys. Earth Plan. Int.* **1982**, *27*, 255–262. [CrossRef]
76. Fukunaga, O.; Nakano, S.; Maki, J.; Vollstädt, H.; Lorenz, H. Kinetics of phase transformation from hBN to cBN in the presence of Mg_5BN_3 catalyst. *High Press. Res.* **1990**, *6*, 947–949. [CrossRef]
77. Gerlach, U. Beiträge zum Mechanismus der Diamantsynthese. Ph.D. Thesis, Technol University Dresden, Dresden, Germany, 1987.
78. Gerlach, U.; Vollstädt, H. On diamond formation at static pressure: I. Reaction model of the nucleation process. *J. Hard. Mater.* **1990**, *22*, 45–58.
79. Lorenz, H. Zur Kinetik und den Mechanismen der katalytischen Phasenumwandlung des Bornitrids bei hohen Drücken und Temperaturen. Ph.D. Thesis, AdW ZIPE, Potsdam, Germany, 1991.
80. Gerlach, U.; Vollstädt, H.; Fukunaga, O.; Koinzer, M.; Kanda, H. Influence of the nucleation behaviour on the synthesis of diamond. *High Press. Res.* **1990**, *5*, 929–931. [CrossRef]
81. Gerlach, U.; Risse, G.; Vollstädt, H. On solid state diamond nucleation at static pressure. In Proceedings of the XI. AIRAPT-Conference, Kiew, Ukraine, 12–17 July 1987; Volume 1, pp. 385–390.
82. Raab, S.; Ullrich, B.; Vollstädt, H. Mechanical and thermal investigations of Paragonite. In Proceedings of the XXV. EHPRG-Conference, Potsdam, Germany, 25–27 August 1987; pp. 132–135.

83. Wäsch, R.; Vollstädt, H. Mineralogische Kennzeichnung und Klassifizierung des Luna 16-Materials. *Gerlands Beitr. Geophs.* **1973**, *82*, 445–453.
84. Stiller, H.; Vollstädt, H.; Wäsch, R.; Bankwitz, P.; Bankwitz, E.; Wagner, F.C.; Schön, J. Results of special mechanical analyses of Luna XVI Material, NASA sc. Techn. Inform. Office, Washington D.C. 1977. In Proceedings of the Tagungsband der sowjetisch-amerikanischen Konferenz über Kosmochemie des Mondes und der Planeten, Moscow, Russia, May 1975; Volume 1, pp. 23–40.
85. Stiller, H.; Vollstädt, H.; Wäsch, R.; Kraft, A. Microscopical and X-Ray Investigations of Lunar and Analogous Terrestrial Materials under extreme Conditions. In Proceedings of the Sammelband der V. Internat. Konferenz Hochdruckphysik und Technologie, Moskau, Russia, May 1975.
86. Stiller, H.; Vollstädt, H.; Wäsch, R. *Results of Special Mechanical Analyses of Luna XVI Material*; Kosmochemie des Mondes und der Planeten; Geology and Geophysics: Moscow, Russia, 1975; pp. 170–175.
87. Wäsch, R.; Vollstädt, H.; Kraft, A. Untersuchungen zur Identifizierung und zum Strukturverhalten lunarer Minerale sowie zur elektrischen Leitfähigkeit des Luna 16- und Luna 20-Regoliths in Abhängigkeit von Druck und Temperatur. *Gerlands Beitr. Geophys.* **1976**, *86*, 449–468.
88. Raidt, N.; Wäsch, R.; Vollstädt, H. Charakterisierung des Luna 16- und Luna 20-Regoliths mit Hilfe des Rasterelektronenmikroskops. *Gerlands Beitr. Geophs.* **1976**, *85*, 441–448.
89. Frölich, F.; Vollstädt, H. *Veränderungen des Materialverhaltens unter planetaren Bedingungen*; Veröff. ZIPE; Central Institute for Physics of the Earth: Potsdam, Germany, 1979; Volume 57, pp. 1–177.
90. Vetter, J. *Elektronenoptische Untersuchung der Luna-16-Mondbodenprobe*; Interner Bericht IFE Halle; Institute Electrophysics: Halle, Germany, 1974; pp. 1–65.

minerals

Editorial

Cornell's Role in Developing Synchrotron Radiation for Mineral Physics

William A. Bassett

Department of Earth and Atmospheric Sciences, Cornell University, Ithaca, NY 14853, USA; Wab7@cornell.edu

Received: 7 May 2020; Accepted: 28 May 2020; Published: 1 June 2020

Abstract: The newly invented diamond anvil cell (DAC) in 1960, and the newly constructed Cornell High Energy Synchrotron Source (CHESS) in 1979 were a perfect match, as CHESS could provide such an intense X-ray beam with such extraordinary properties that a whole new approach to mineral physics research became possible. The very high intensity of the X-ray beam from CHESS made it possible to make real-time observations of crystal structures during phase transitions for the first time. For instance, the olivine-spinel transition, important for understanding deep focus earthquakes can be shown to take place first by the displacive shift of oxygen layers supporting shear stress as most likely earthquake trigger followed by the diffusion of the cations to their positions in the spinel structure. X-ray emission spectra of high-pressure, high-temperature samples also made it possible to determine phase compositions, as well as the structures of complex ions in solution.

Keywords: high pressure; high temperature; diamond anvil cell; olivine-spinel transition; complex ions in solution

---

1. Introduction

The extreme intensity of X-rays that can be packed into a tiny beam is just the thing for high-pressure, high-temperature research using the diamond anvil cell (DAC). The appeal of the DAC had always been that so much could be learned, as long as analytical techniques exist that can be used for microscopic samples. I took a position in the Geology Department at Cornell University in 1978. It had been 17 years since my post-doc appointment at Brookhaven National Laboratory on Long Island. When I was there, I had watched the construction of the Alternating Gradient Synchrotron being built for electron-positron collision experiments. When the workmen had left for the day, I explored the partially constructed new facility. I was especially taken by the massive shielding made of salvaged old naval ship decking. When I inquired about its need, I was told it was to shield against synchrotron radiation, something that seemed to be simply a nuisance. I had never paid any more attention to that kind of radiation until the day I arrived at Cornell University for an interview and a tour of campus with Geology's chairman, Jack Oliver, and was shown the Cornell High Energy Synchrotron Source (CHESS) just being built on the campus. Figure 1 gives a simplified layout and outlines the basic principles involved in generating X-ray beams that can be used for research. When its capabilities were explained to me, and I learned that synchrotron radiation had acquired its new reputation, I realized that its intense, small X-ray beams and fast X-ray detectors might be the next ideal step for research with the DAC. Fortunately, I was offered the position and immediately started designing the equipment that would be needed to take advantage of its properties. Perhaps diffraction patterns could be made in less time than one week, as was needed on a conventional X-ray source, such as one that I had used earlier at the University of Rochester. Maybe data on the effects of high pressures and temperatures could be collected much more rapidly. The director of CHESS, Bob Batterman, and the beamline scientists were very generous and allowed some of us to make preliminary runs on CHESS before it was officially opened for users. The beamline scientists were extremely helpful and had my students and me up and running smoothly in a few days. These runs were so successful that I knew merging the DAC with synchrotron radiation would be a fascinating line of research for a long time to come. I learned many years later from Bob Batterman at his retirement that I was, in fact, the first person to formally apply for beamtime at CHESS when it was opened up to general users. The prospect of being able to collect a diffraction pattern at high pressures and temperatures in a few seconds rather than a week was irresistible. Later, when a visual camera with a strong lens system was added, visual changes could be correlated with the X-ray observations, and the X-ray beam could be seen for aiming when fluorescent diamond anvils were used. It was possible to watch phase transitions and reactions proceed in real time under either hydrostatic or deviatoric stress.

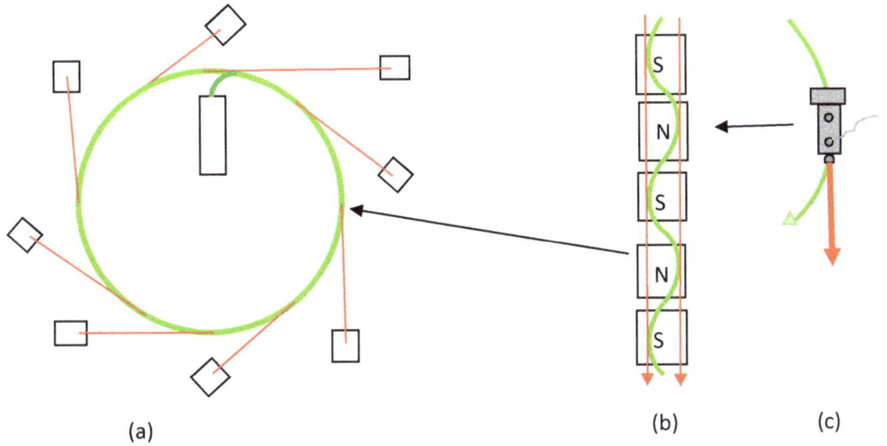

Figure 1. (**a**) Diagram of a synchrotron storage ring in which an electron beam (green) is traveling clockwise. (**b**) Every time the electron beam passes through a vertical magnetic field above the north end of a "C-shaped" magnet (N) and below the south end of the "C-shaped" magnet (S), the magnetic field causes the electron beam to be steered to the right. Every time the electron beam passes through the inverted magnetic field, it is steered to the left. Much more intense beams (shown as red arrows) can be produced if the electron beam passes through a series of alternating magnetic fields in a straight section of the ring, so that the beams produced at all the crests and troughs are added together. These are called wigglers or undulators. (**c**) Every time the beam is deflected, some of its energy is converted to an electromagnetic beam shining in a forward direction, like the headlight on a locomotive (c). Pipes leading off tangentially from the storage ring at the bending magnets, wigglers and undulators carry the X-ray beams to workstation hutches (rectangles) for experiments. Where there is no magnetic field, the electrons travel in a straight line and produce no electromagnetic beam.

In addition to X-ray diffraction spectroscopic techniques, such as emission spectroscopy, not only made it possible to determine the compositions of individual phases, but also made it possible to determine the valence states and structures of complex ions in aqueous solution. As beamlines were equipped with specialized instrumentation, researchers happily paid for travel to the nearest synchrotron source or the place equipped with the best instrumentation for their research. Even if numerous trips to a synchrotron source were needed for one research project, it always seemed like a desirable choice. Being on the same campus with CHESS, however, meant that my students and I had no need to travel, and would sometimes be given beamtime when other groups had to cancel. Although CHESS was not the first synchrotron source, it was early enough so that a number of specialized beamlines were developed there. A number of good sites can be found by digital searching for "synchrotron radiation sources world wide".

2. The Amazing Phenomenon of Synchrotron Radiation

Figure 1 is a simplified layout for a synchrotron storage ring for producing X-ray beams for research experiments.

If the electron-beam energy is high, the frequency of the light is high, e.g., X-rays. Although synchrotron sources can and do produce lower-energy radiation, such as visible light and infrared, these high-energy X-ray beams have been the most valuable for mineral physics. It was found that much more intense beams could be produced if the electron beam passed through a wiggler or undulator so that a series of X-ray beams produced at each bend in the electron beam could be added together. CHESS was instrumental in perfecting them, and today they are the preferred sources for much of synchrotron research. The crest and trough of each wiggle produces an electromagnetic beam; the series of crests and troughs combine to produce a much greater intensity shown by the red arrows. (c) Shows the comparison between the synchrotron phenomenon and a locomotive's headlight. The electric vector of the generated X-rays is highly oriented by the magnetic fields, so that the polarization is horizontal, a property especially valuable for our research using X-ray spectroscopy.

3. Results of Mineral Physics Research Using Synchrotron Radiation

One of the most interesting and valuable uses of synchrotron radiation for mineral physics was the ability to observe crystal structure change in real time. One of our most successful uses of this capability was the olivine-spinel phase transition which could have important implications for deep focus earthquakes. It had been pointed out by Green, et al. [1] and Burnley, et al. [2], that rocks are expected to be so ductile at the depths where deep-focus quakes originate, that a mechanism other than brittle failure must cause them. They suggest that, rather than brittle failure, the olivine-spinel phase transition made the most sense, and that shear stress might be an important mechanism. It occurred to me that we might be able to study the relationship between shear stress and the phase transition, if we were to watch the olivine-spinel transition by X-ray diffraction as the transition was taking place—something that synchrotron radiation would make possible. It might show that shear strain caused by shear stress is an even more intimate part of the transition by promoting the transition that, in turn, increasing the shear stress elsewhere on a fault plane driving additional transition and, in turn, additional phase transition. Perhaps shear strain rather than brittle failure could be shown to trigger failure by a shear mechanism at other points in a fault plane.

My student, Mike Furnish, and I [3] thought that if we could very slowly increase pressure at a temperature expected at the depth of deep-focus earthquakes, we might learn more about the olivine-spinel phase transition and what triggers it. We designed a gear system that could very slowly increase the pressure as we turned the hand crank on a remote selsyn driver that turned the selsyn motor in the experimental hutch that increased the pressure as the crank was turned. That ability to increase the pressure or halt for an interesting observation in the diffraction pattern was exactly what we needed. We chose fayalite as our sample, mounted it without a gasket and made a series of diffraction patterns over quite a long time, as we slowly increased stress in the form of both isotropic stress (pressure) and shear stress, as a result of the extrusion of sample from between the anvil faces. Finally, we saw a subtle change and slowed progress even further and ultimately halted progress. What we observed amazed us; it was completely unexpected, too good to be true. The transition proceeded by two steps. There were several intermediate diffraction patterns and so we were certain it was real. First, we could clearly see that the oxygen atoms were moving by displacement from their positions in the olivine structure to their positions in the spinel structure while the cations remained in their olivine positions until given more time or more pressure. Then, they slowly migrated by diffusion to the spinel structure positions.

We made a model of the olivine structure out of plastic balls that would let us slide the layers of oxygen atoms over each other. We already suspected that shear stress played a role because of the displacive nature of hexagonal close-packed structure to cubic close-packed structure; we just never expected it to be so visible in our data.

4. A Different Experiment with the Same Conclusions

Pam Burnley joined us as a post-doc. We turned once more to the olivine-spinel transition, as she already had familiarity with it, and found yet another way of showing the connection between shear stress and the olivine-spinel transition. We ran several different olivine compositions in a diamond anvil cell with no gasket, so they would be subjected to radial shear stress as they extruded [4,5]. Observations of the extruding sample gave a clear picture of the amount of shear strain as pressure and/or temperature increased. The central pressure was well above the transition pressure for each of the samples. This time we used only a microscope and heated gradually, until color change indicated the high-pressure phase. The high-pressure phase appeared as a ring in every run for every sample. We concluded that this was another sure indication that shear triggered the high-pressure phase. The outer portion of the sample was at too low a pressure and while the center was well above the transition pressure it experienced less shear (Figure 2).

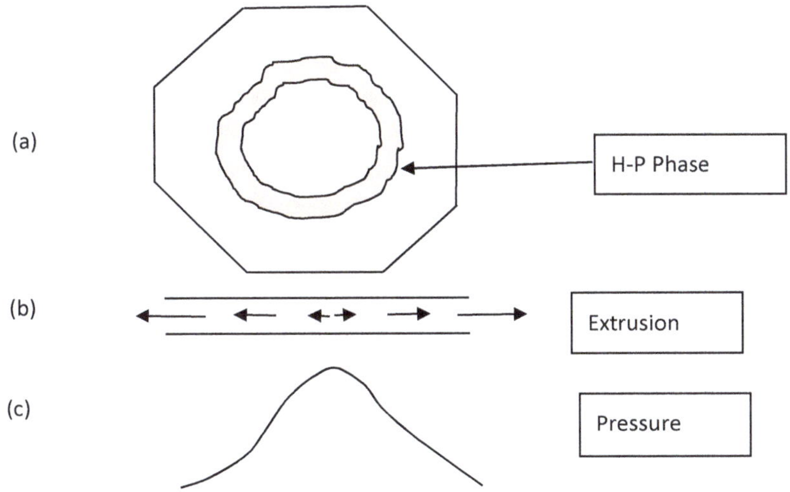

Figure 2. Olivine structure sample squeezed between diamond anvils with no gasket (**a**) View down through the upper diamond anvil. (**b**) The distribution of shear stress due to extrusion. (**c**) The distribution of pressure. In every run for every composition the spinel phase made its first appearance where the shear stress, not the pressure, was greatest. Of course, this did not give any direct information about crystal structures and the intermediate structure, but it did illustrate that all the samples with olivine structure showed the same behavior [4,5].

5. Advanced Photon Source (APS) and Fluid Studies at High Pressures and Temperatures (1995)

As synchrotron sources gained in popularity, the new ones took advantage of technology developed at the earlier ones, so they could add many new analytical techniques with new specialized instrumentation. And so it was with CHESS and the Advanced Photon Source (APS) planned for the Argonne National Laboratory in Illinois. A number of us from Cornell were glad to attend planning sessions and contribute to the design of experimental hutches at APS with an even greater variety of new instrumentation. My colleagues and I made a number of preliminary experimental runs at CHESS to develop methods for studying fluid aqueous solutions, but then realized that what we wanted to do could be done even more easily with the newly developed X-ray spectrometric instruments at APS. For our research on fluid samples at high pressures and temperatures we traveled to Argonne to use the facilities at APS. Figure 3 shows a diamond anvil cell designed specifically for passing an X-ray beam "between" the upper and lower diamond anvils perpendicular to the load axis of the cell [6].

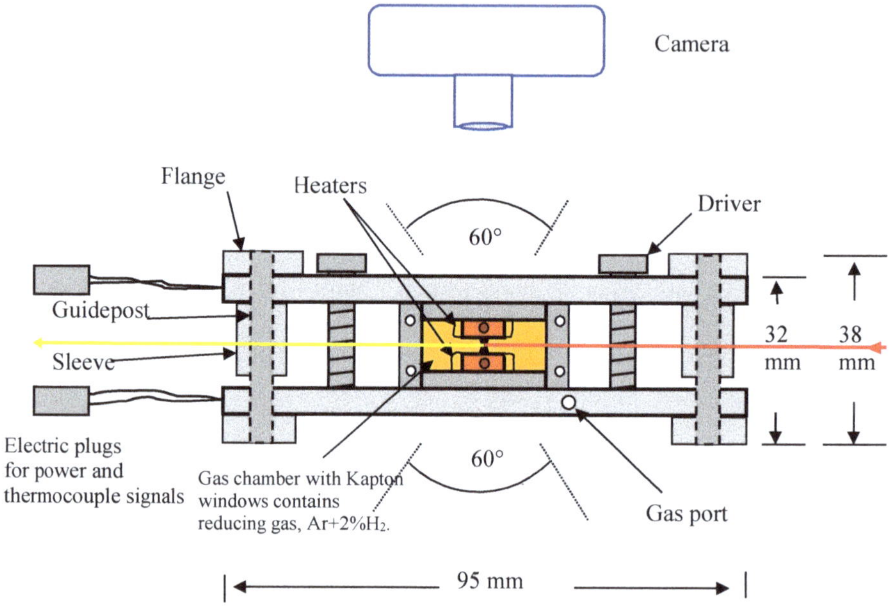

Figure 3. Diagram of diamond anvil cell model HDAC-5, suitable for spectrometric analysis because it provides side access for the X-ray beam to enter and exit the sample, while both the camera and the spectrometer have access to the sample at 90° to the X-ray beam. The main X-ray beam enters from the right at 45° to the page normal and the emitted X-rays to be analyzed exit to the left at 45° to the page normal. The camera is above looking through the upper anvil.

Figure 4 is an image of the setup for seeing the X-ray beam so we can aim it through our fluid or mixed-phase samples. Sometimes we use a metal gasket in addition to the recess for the sample; other times we just pressed the two anvil faces together so tightly that the sample cannot leak at the high sample pressures produced by heating (Figure 4). The choice depended on whether we needed to increase the pressure by squeezing the gasket to reduce sample volume, or whether the pressures resulting from heating were sufficient. We would usually avoid using a gasket when possible so that we could minimize contamination of the sample. The X-ray beam enters "between" the gasket and anvil face or "between" the two anvil faces when no gasket is used. The notches are there to minimize attenuation by diamond. Emitted X-rays from the sample pass out through the notch on the left, where they can be spectrometrically analyzed. The left notch, because it is at a 90° angle to the incident beam, yields the clearest signal for determining absorption, because emission is a direct function of the absorption, while Rayleigh and other forms of scattering not indicative of absorption are minimized, due to the highly polarized nature of the impinging X-ray beam, as described in Figure 2. We used this method to collect spectra that can be used to produce diagrams of complex ions, like the examples shown in Figure 5 [7]. This is valuable information for determining the mobility and reactivity of elements in solution as a function of concentration, pressure and temperature.

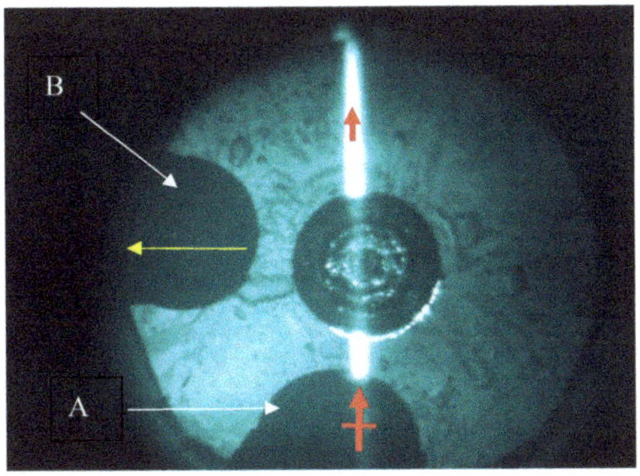

Figure 4. View of the X-ray beam entering and exiting the central, round fluid sample chamber in the lower diamond as seen through the unaltered upper diamond anvil. The entering X-ray beam (longer red arrow with a cross line to indicate polarization orientation, i.e., the electric vector). Diamond was removed from the lower anvil by laser ablation to form the two notches (A and B), to minimize attenuation of the X-rays as they pass through diamond, and is the reason that the beam itself does not cause fluorescence as it passes through notch A before it enters the fluorescent diamond anvil, where it appears white due to over exposure, and its less intense scattered blue fluorescence illuminates the anvil's features. A very faint fluorescent line shows its course through the fluid sample in the circular recess in the center of the lower anvil face. The shorter red arrow shows the direct beam exiting the sample chamber, and the yellow arrow shows the very weak emitted beam directed to the detector, where its spectrum can be analyzed [7].

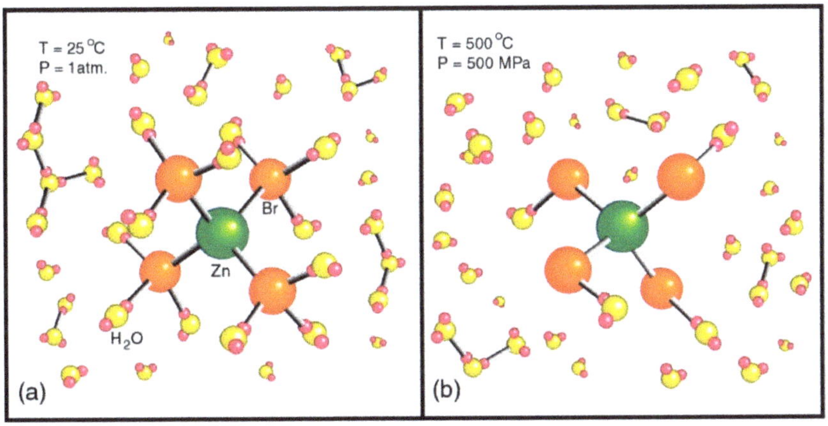

Figure 5. (**a**) This hydration shell surrounding a $ZnBr_4^{2-}$ ion is typical of the structures we found for the hydrated molecules of a range of compositions. (**b**) At 500 °C and 500 MPa, the sample density, although above the critical point, is only slightly less than the liquid phase. Some of the outermost H_2O molecules become detached from of $ZnBr_4^{2-}$ ion. Other compounds in solution such as rare-earth bromides have very different complex configurations and responses to temperature [7].

As interests in applications of synchrotron radiation to mineral physics research increased, it became clear that the understanding of many samples involves the compositions and properties of fluids and the structures of the ions in them. Virtually all high-pressure-high-temperature natural fluids and supercritical dense vapors are solutions in which minerals are carried, and are therefore very important to understanding hydrothermal, magmatic and nearly all metamorphic processes. Compositions of solutions at high pressure and temperature can be determined by emission spectra.

However, it seems odd to speak of the structures of ions in solution, a job that diffraction cannot do, and yet we find that the structures of ions that are complexes undergo significant changes with pressure and temperature using X-ray spectroscopy. Diffraction cannot provide that information, but the extraordinary details found close to the X-ray absorption edges of the heavier elements contain much information about the structures of complex ions at high pressures and temperatures. An explanation of the principles behind XAFS are beyond the scope of this paper but can be found by digital searching for "what is XAFS".

Ironically, those absorption details are more easily and accurately determined by the emission spectra than by absorption spectra. Diamond anvils with recesses in the anvil faces like that shown in Figure 4 can be used to collect absorption spectra by means of the emission spectra. There is an important reason for this; the X-ray beam from a synchrotron is highly polarized horizontally for the reason explained in Figure 2; all of the X-rays are produced by electrons following curved paths that are horizontal. The X-rays that are of value for interpreting ion structures are emitted, and therefore not polarized, while the polarized X-rays in the beam are not Rayleigh scattered by the sample, because the detector is aligned with the beam's electric vector. This geometry greatly reduces interference with the emitted X-rays that contain the information for determining the structures of ions in solution.

6. The Importance of Complex Ionic Structures

The distribution of charge over the surface of an ion can make the difference between solubility and insolubility especially if some of the complex ion contains other ions. When that occurs, the charge on the complex ion can even change sign + to − or − to +. Complexing can also affect the polarity of the complex ion, and finally, it can affect the size of the ion. For these reasons, the structures of complex ions can have profound effects on the dissolving, transport and precipitation of minerals. All of these can have profound effects on mineral behavior in solution. With the effects that pressure and temperature on the structures of complex ions it is expected that depth within Earth's interior has profound effects on hydrothermal, magmatic and metamorphic processes.

Acknowledgments: The author wishes to thank Bob Liebermann for suggesting this paper as part of the special issue *"Mineral Physics—in Honor of Orson Anderson"* in *Minerals*. He is also grateful for the help given him by Bob Mayanovic who was so important in guiding our research on complex ions in aqueous solutions. Everyone, from the directors to the beamline scientists, associated with CHESS at Cornell University and APS at Argonne National Laboratory made conducting research at their facilities so smooth-functioning and so rewarding that it was always a pleasure working there. The author also wishes to thank Elise Skalwold for her help locating publications.

Conflicts of Interest: The author declares no conflict of interest.

References

1. Green, H.W.; Young, T.E.; Walker, D.; Scholtz, C.H. Anticrack-associated faulting at very high pressure in natural olivine. *Nature* **1990**, *248*, 720–722. [CrossRef]
2. Burnley, P.; Green, H.W.; Prior, D.J. Faulting associated with the olivine to spinel transformation in Mg2GeO4 and its implications for deep-focus earthquakes. *J. Geophys. Res.* **1991**, *96*, 425–443. [CrossRef]
3. Furnish, M.D.; Bassett, W.A. Investigation of the mechanism of the olivine-spinel transition in fayalite by synchrotron radiation. *J. Geophys. Res.* **1983**, *88*, 10333–10341. [CrossRef]
4. Burnley, P.C.; Bassett, W.A.; Wu, T.-C. Diamond anvil cell study of the transformation mechanism from olivine to spinel phase in Co2SiO4, Ni2SiO4, and Mg2SiO4. *J. Geophys. Res.* **1995**, *100*, 17715–17723. [CrossRef]
5. Wu, T.-C.; Bassett, W.A.; Burnley, P.C.; Weathers, M.S. Shear-promoted phase transitions in Fe2SiO4 and Mg2SiO4 and the mechanism of Deep arthquakes. *J. Geophys. Res.* **1993**, *98*, 19767–19776. [CrossRef]

6. Bassett, W.A.; Shen, A.H.; Bucknum, M.; Chou, I.-M. A new diamond anvil cell for hydrothermal studies to 2.5 GPa and from −190 to 1200 °C. *Rev. Sci. Instrum.* **1993**, *64*, 2340–2345. [CrossRef]
7. Mayanovic, R.A.; Anderson, A.J.; Bassett, W.A.; Chou, I.-M. Hydrogen bond breaking in aqueous solutions near the critical point. *Chem. Phys. Lett.* **2001**, *336*, 212–218. [CrossRef]

minerals

MDPI

Article

Phase Relations in MAFSH System up to 21 GPa: Implications for Water Cycles in Martian Interior

Chaowen Xu [1,*] **and Toru Inoue** [1,2,3]

[1] Geodynamics Research Center, Ehime University, 2-5 Bunkyo-cho, Matsuyama 790-8577, Japan; toinoue@hiroshima-u.ac.jp

[2] Department of Earth and Planetary Systems Science, Hiroshima University, 1-3-1 Kagamiyama, Higashi-Hiroshima 739-8526, Japan

[3] Hiroshima Institute of Plate Convergence Region Research (HiPeR), Hiroshima University, Higashi-Hiroshima, Hiroshima 739-8526, Japan

* Correspondence: dkchaowen@126.com; Tel.: +81-050-3699-0952

Received: 3 August 2019; Accepted: 14 September 2019; Published: 16 September 2019

Abstract: To elucidate the water cycles in iron-rich Mars, we investigated the phase relation of a water-undersaturated (2 wt.%) analog of Martian mantle in simplified $MgO\text{-}Al_2O_3\text{-}FeO\text{-}SiO_2\text{-}H_2O$ (MAFSH) system between 15 and 21 GPa at 900–1500 °C using a multi-anvil apparatus. Results showed that phase E coexisting with wadsleyite or ringwoodite was at least stable at 15–16.5 GPa and below 1050 °C. Phase D coexisted with ringwoodite at pressures higher than 16.5 GPa and temperatures below 1100 °C. The transition pressure of the loop at the wadsleyite-ringwoodite boundary shifted towards lower pressure in an iron-rich system compared with a hydrous pyrolite model of the Earth. Some evidence indicates that water once existed on the Martian surface on ancient Mars. The water present in the hydrous crust might have been brought into the deep interior by the convecting mantle. Therefore, water might have been transported to the deep Martian interior by hydrous minerals, such as phase E and phase D, in cold subduction plates. Moreover, it might have been stored in wadsleyite or ringwoodite after those hydrous materials decomposed when the plates equilibrated thermally with the surrounding Martian mantle.

Keywords: high pressure; high temperature; Martian interior; water storage; water transport

---

1. Introduction

Water is an important volatile material that affects the physical and chemical properties of planetary interiors, such as those of Earth and Mars. Water transportation and storage are crucially important components of the water cycle, strongly affecting geodynamic processes. On Earth, several studies have indicated that some hydrous minerals can hold and transport water to the deep Earth by cold subducting slabs [1–3]. The so-called dense hydrous magnesium silicates (DHMSs) in $MgO\text{-}SiO_2\text{-}H_2O$ (MSH) system, such as phase A ($Mg_7Si_2O_{14}H_6$), phase E ($Mg_{2.3}Si_{1.25}O_6H_{2.4}$), superhydrous phase B ($Mg_{10}Si_3O_{18}H_4$), phase D ($MgSi_2O_6H_2$), and phase H ($MgSiO_4H_2$), are considered to be important carriers of subducted water from mantle transition zone down to the middle part of the lower mantle [4–9]. By contrast, the major minerals—wadsleyite and ringwoodite—in the Earth's mantle transition zone (MTZ), might act as a large water reservoir because they might hold several oceans' masses of H_2O [10].

The existence of water on Mars has long been controversial. Recent studies of topographic features, for example, the northern plains, sedimentary deposits, and valley networks [11–13] and the detection of subsurface ice, as well as various hydrous minerals in Lyot crater, suggest the existence of an ancient Martian ocean on the surface [14–16]. The convecting Martian mantle may be hydrated when reacting with overlying hydrous crust, bringing water into the deep interior [17]. Therefore, similarly to an Earth-like planet, some hydrous minerals might exist in cold region of iron-rich Mars, and wadsleyite and ringwoodite might also hold a huge amount of water in the Martian interior, as it was argued for the present-day Earth.

Several studies have identified phase relations in the MSH and $MgO-Al_2O_3-SiO_2-H_2O$ (MASH) systems, with the observation of various hydrous minerals at *P–T* conditions related to the cold subduction slabs [3,4,6,7,18,19]. Nevertheless, few data are available for hydrous iron-bearing systems: data for iron-rich systems, such as Mars, are rarely reported [5]. A better understanding of the phase relations in $MgO-Al_2O_3-FeO-SiO_2-H_2O$ (MAFSH) system might help to elucidate the geodynamic processes associated with the deepwater cycles of Mars. Therefore, we determined the phase relations in iron-rich MAFSH system between 15 GPa and 21 GPa to systematically ascertain the stability of DHMSs, and further estimate the possible water transportation into the Martian interior by subducting processes.

2. Materials and Methods

High-pressure and high-temperature experiments were conducted at Geodynamics Research Center (GRC), Ehime University, using a Kawai-type 1000 ton multi-anvil apparatus. Tungsten carbide cubes with truncation edge length (TEL) of 4 mm were used in combination with Co-doped MgO-octahedra of 10 mm edge length (10/4 assemblage). Preformed pyrophyllite gaskets were used between the cubes, and $LaCrO_3$ was used as the heater. A gold sample capsule was used in the cell assemblage. Pressures were calibrated at room temperature (25 °C) by diagnostic changes in the electrical resistances of ZnTe (9.6 and 12 GPa), ZnS (15.5 GPa), GaAs (18.3 GPa), and GaP (23 GPa) induced by the semiconductor-metal phase transitions at high pressures. The temperature was monitored using a $W_{97}Re_3$-$W_{75}Re_{25}$ thermocouple. The electromotive force (EMF) was not corrected for the effects of pressure. The sample was compressed to the desired pressure. Then, the oil pressure was held constant. Subsequently, alternating current (AC) power was supplied to the heater in the furnace assemblage. After heating for 40–240 min, the power was stopped by shutting off the electric power supply. Samples were recovered after releasing pressure slowly during 12 hrs. The recovered run products were mounted in epoxy resin and were polished for phase identification and chemical composition analysis.

The simplified Martian composition by Dreibus and Wänke [20], which is an analog of the primitive Martian mantle composition corresponding to mantle + crust, was adopted. The oxide mixture of MgO, Al_2O_3, SiO_2, and FeO was prepared in appropriate proportions. FeO was put in the reduced furnace at 1000 °C for 24 h before mixing to ensure that the ferrous ion was used. We added 2 wt.% H_2O in the form of $Mg(OH)_2$. The chemical compositions are presented in Table 1. To create a reduced environment, we inserted some Mo foil in a gold capsule before the starting material was encapsulated into the capsule. The phase assemblages were identified using a micro-focus X-ray diffractometer (MicroMax-007HF; Rigaku Corp., Tokyo, Japan) with Cu Kα radiation. The obtained data were processed using 2PD software, which can display and process two-dimensional data, including smoothing, background correction, and 2D to 1D conversion. The micro-textures and composition were obtained using a field emission scanning electron microscope (FESEM, JSM7000F, JEOL, Akishima-shi, Japan) combined with an energy-dispersive X-ray spectrometer (EDS, X-MaxN, Oxford Instruments, Plc., Abingdon, UK). The Fe_2SiO_4, Mg_2SiO_4, and Al_2O_3 were used as standards in the EDS analyses. Working parameters of 15 kV, 1 nA, and collection times of 30–50 s were used. The chemical composition was analyzed using EDS. We used software (Aztec ver. 2.4, Oxford Instruments Nanotechnology Tools Ltd., Abingdon, UK) to process EDS data. The Raman spectrum was obtained using laser Raman spectrometer (NRS-5100gr) to identify some of the recovered phases, with 532 nm laser excitation. Laser power applied to the sample was 10 mW. The Raman spectra were obtained from a linear baseline, and peak characteristics were carried out using the commercial software package. The EDS measurements suggested that we obtained the homogeneous composition of Mo where it appeared. On the other hand, we calculated the chemical formula of garnet obtained in this study. The result showed that only the Fe^{2+} in the formula could make the charge balance. Therefore, we created a reduced environment in the sample chamber by using Mo foil.

Table 1. Chemical composition (wt.%) of starting materials.

Composition	MgO	Al_2O_3	FeO	SiO_2	H_2O	Total
MAFSH	30.2	3.5	19.9	44.4	2	100

MAFSH, in simplified MgO-Al_2O_3-FeO-SiO_2-H_2O composition.

3. Results and Discussion

3.1. Phase Relations

Experimental conditions and the results are presented in Table 2 and Figure 1. Nominally anhydrous phases, such as clinopyroxene (Cpx), garnet (Gt), wadsleyite (Wd), and ringwoodite (Rw), were presented in quenched samples (Figure 2). Based on experimental conditions, different hydrous phases were observed with increased pressure. Phase E (PhE) was observed in the low-temperature region between 15–16.5 GPa to coexist with Gt and Wd or Rw, as presented in Figure 2a. It became unstable at 16.5 GPa with a temperature higher than 1100 °C. However, the thermal stability limit of PhE in this study was about 100 °C higher than that reported in iron-free hydrous peridotite system [21]. Phase D (PhD) remained as a major hydrous phase at the present pressure range from 18 GPa to 21 GPa below 1100 °C (Figure 2b). With increased temperature, PhD decomposed to Rw, stishovite (St), and Gt. It seems that PhD in MAFSH system has the same stability region, as reported for a MASH system [21]. Therefore, it was expected that the PhD had a positive pressure-temperature stability slope, as presented in Figure 1. The high iron content in MAFSH system might inhibit the superhydrous phase B (SuB) formation, leading to the disappearance of SuB in the whole pressure range.

An earlier report described that the loop in the Wd-Rw boundary shifted towards higher pressure by the effect of water [22] or towards lower pressure with increased iron content [23]. In this study, Wd was stable up to 16 GPa at 1200 °C; then it transformed to Rw, as identified by Raman spectrum. The Rw formation was also found shifting to low pressure, which was observed at 16 GPa and 1200 °C. Therefore, this loop was influenced more strongly by the higher iron content than by the effects of H_2O in the hydrous Martian mantle.

Table 2. Experimental conditions and results.

Pressure (GPa)	Temperature (°C)	Time (min)	Phase
15	900	240	Gt, Wd, PhE
15	1100	240	Gt, Wd, Cpx
15	1250	120	Gt, Wd, Cpx
15	1450	90	Gt, Wd, Cpx
16	1200	120	Gt, Wd, Rw, Cpx
16.5	1100	120	Gt, Rw, PhE, St
16.5	1300	90	Gt, Rw, Cpx
18	1000	240	Rw, PhD
18	1200	120	Gt, Rw, St
18	1550	40	Gt, Rw, St, Melt
19.5	1400	40	Gt, Rw, St
21	900	240	Rw, PhD, St
21	1050	240	Rw, PhD, h-Fe
21	1250	120	Gt, Rw, St
21	1500	40	Gt, Rw, St

Gt, garnet; Cpx, clinopyroxene; Wd, wadsleyite; Rw, ringwoodite; St, stishovite; PhE, phase E; PhD, phase D; h-Fe, iron-rich hydrous phase.

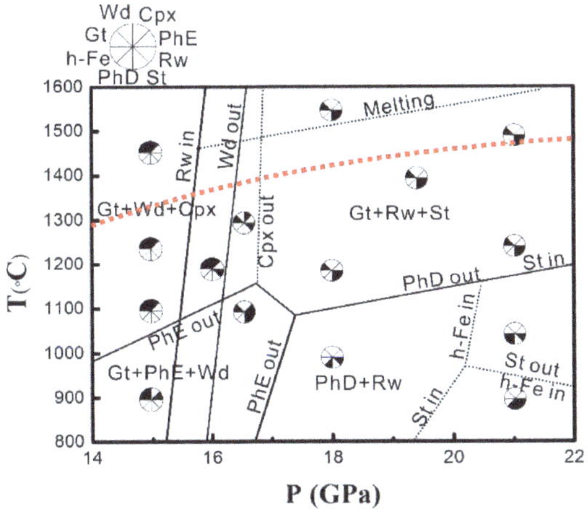

Figure 1. Phase relations in the system MgO-Al_2O_3-FeO-SiO_2 with 2% H_2O. Solidus lines are obtained according to the quenched samples combined with reported phase relation in CaO-MgO-Al_2O_3-SiO_2-pyrolite with 2% H_2O [24]. Dashed black lines are the proposed phase boundaries based on phase assemblages in recovered samples due to the limited data point at these regions. The dashed red line is proposed areotherm [17,25]. Gt, garnet; Cpx, clinopyroxene; Wd, wadsleyite; Rw, ringwoodite; St, stishovite; PhE, phase E; PhD, phase D; h-Fe, iron-rich hydrous phase.

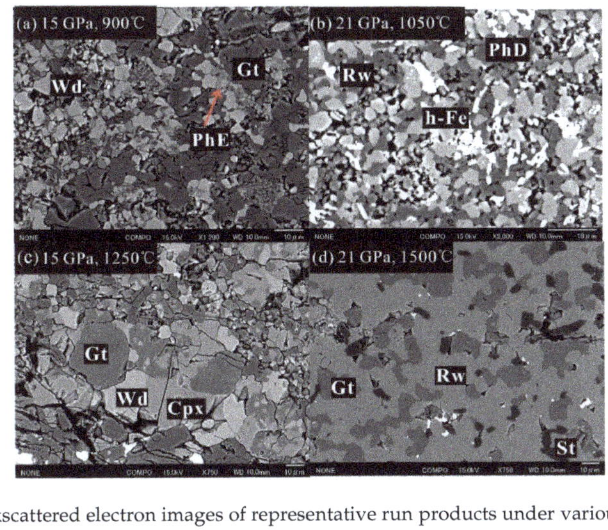

Figure 2. Backscattered electron images of representative run products under various pressure and temperature conditions: (**a**) 15 GPa and 900 °C; (**b**) 21 GPa and 1050 °C; (**c**) 15 GPa and 1250 °C; (**d**) 21 GPa and 1500 °C. Gt, garnet; Cpx, clinopyroxene; Wd, wadsleyite; Rw, ringwoodite; St, stishovite; PhE, phase E; PhD, phase D; h-Fe, iron-rich hydrous phase.

Cpx was observed at 16.5 GPa and 1300 °C, which further transformed to Gt at higher pressures. Stishovite (St) was found to be coexisting with Rw at temperatures higher than 1100 °C. Some amount of ferrous oxide was detected at 21 GPa and 1050 °C. At 21 GPa and 1000 °C, we found iron-rich hydrous phase, based on analyzing the deficit of weight total. Both, the total weight and diffraction pattern, were similar to εFeOOH; however, the diffraction peaks were slightly shifted compared with εFeOOH, probably because of the incorporation of Mg due to the similar ionic radius of $^{VI}Mg^{2+}$ (0.72 Å) and $^{VI}Fe^{2+}$ (0.78 Å). It was also a little strange that this phase did not contain Si and Al. Perhaps, the solubility of Al and Si decreases when Mg is included. More work is needed to clarify this issue.

3.2. Mineral Chemistry in DHMSs, Wadsleyite, and Ringwoodite

The measured chemical compositions of the phases in the experiments are presented in Table 3. Several phases exhibited broad compositional variations. PhE had a composition of 4.5–3.1 wt.% for Al_2O_3 and 12.8–8.5 wt.% for FeO, at elevated pressures and temperatures from 15 GPa and 900 °C to 16.5 GPa and 1100 °C. With increasing pressure from 18 GPa to 21 GPa, the Al_2O_3 content decreased from 7.9 to 4.7 wt.%. However, FeO remained almost stable at around 4.3 wt.%. The water contents in PhE and PhD were both estimated for 13 wt.% on average, based on the deficit from the EDS weight total (each phase was calculated by summing all the deficit of weight total in Table 3, and then divided by counted number). PhD in the MAFSH system generally had low amounts of FeO < 4.4 wt.%. The Al_2O_3 contents were 3.5–7.9 wt.%, whereas the FeO contents exhibited a small variation of 3.9–4.8 wt.% throughout the samples quenched under various pressure and temperature conditions (Table 3).

Table 3. Representative mineral compositions.

P (GPa)	*T* (°C)	Phase	MgO	Al_2O_3	SiO_2	FeO	Total
21	1500	Gt	26.48 (54)	13.27 (142)	48.63 (84)	12.33 (39)	100.71 (23)
		Rw	36.20 (29)	0	38.11 (32)	25.42 (45)	99.73 (66)
		St	0	1.45 (46)	100.68 (81)	0.77 (6)	102.90 (52)
	1250	Gt	25.01 (60)	15.21 (142)	46.27 (85)	12.34 (37)	98.84 (33)
		Rw	35.72 (33)	0	37.49 (23)	24.95 (43)	98.15 (58)
		St	38.20	0.55 (38)	99 (63)	0.78 (33)	100.33 (83)
	1050	Rw	41.60 (98)	0	39.27 (62)	15.99 (27)	96.86 (88)
		PhD	20.24 (42)	4.71 (71)	56.11 (85)	4.79 (34)	85.86 (145)
		h-Fe	7.25 (33)	0	0.87 (60)	81.74 (106)	89.87 (120)
	900	Rw	43.57 (86)	1.33 (7)	40.54 (73)	12.11 (87)	96.22 (143)
		PhD	29.36 (78)	3.50 (48)	52.95 (90)	3.93 (48)	89.74 (81)
		St	1.53 (71)	0	99.08 (78)	1.15 (27)	101.72 (58)
19.5	1400	Gt	26.02 (39)	12.76 (65)	47.78 (60)	12.21 (39)	98.78 (92)
		Rw	35.26 (13)	0	36.96 (19)	25.55 (34)	97.77 (44)
		St	0	0.96 (47)	98.42 (58)	0.72 (32)	100.10 (81)
18	1550	Gt	27.78 (29)	11.04 (37)	49.70 (30)	12.45 (80)	100.45 (37)
		Rw	34.88 (37)	2.05 (18)	37.77 (30)	26.70 (24)	99.35 (58)
		St	0	1.54 (58)	99.90 (97)	0.62 (36)	102.05 (63)
		Melt	20.65	1.62	16.45	16.57	55.31
18	1200	Gt	25.88 (90)	12.61 (99)	47.96 (67)	12.41 (64)	98.86 (76)
		Rw	35.55 (35)	0	37.56 (34)	25.61 (47)	98.71 (79)
		St	0	0.58 (23)	99.49 (71)	0.70 (40)	100.77 (82)
18	1000	Rw	32.89 (61)	0	36.98 (46)	27.86 (81)	97.73 (57)
		PhD	20.48 (75)	7.86 (59)	54.68 (73)	4.57 (46)	87.59 (53)
16.5	1300	Gt	25.42 (32)	11.23 (48)	47.18 (29)	12.76 (57)	96.59 (50)
		Rw	28.89 (32)	0	34.88 (27)	32.21 (25)	95.98 (64)
		Cpx	33.79 (24)	19.19 (60)	55.66 (53)	6.98 (47)	96.44 (95)
16.5	1100	Gt	25.76 (80)	12.81 (36)	47.64 (77)	14.74 (57)	100.94 (63)
		Rw	34.59 (57)	0	37.72 (17)	26.67 (63)	98.97 (37)
		PhE	38.42 (75)	3.12 (20)	37.82 (18)	8.47 (28)	87.82 (57)
		St	1.38 (67)	0.76 (64)	96.75 (75)	1.14 (20)	100.03 (27)

Table 3. *Cont.*

P (GPa)	*T* (°C)	Phase	MgO	Al_2O_3	SiO_2	FeO	Total
16	1200	Gt	24.28 (47)	13.42 (37)	46.67 (33)	14.26 (83)	98.63 (44)
		Rw	28.36 (42)	0	35.76 (39)	33.43 (59)	97.55 (69)
		Wd *	-	-	-	-	-
		Cpx	34.99 (51)	20.70 (54)	56.73 (58)	6.52 (37)	98.25 (37)
15	1450	Gt	26.61 (71)	11.35 (25)	49.48 (61)	13.10 (75)	100.54 (82)
		Wd	34.43 (35)	0	37.35 (33)	27.61 (46)	99.40 (81)
		Cpx	34.68 (74)	0	58.09 (41)	7.56 (70)	100.34 (82)
15	1250	Gt	27.39 (13)	11 (54)	49.79 (37)	13.06 (34)	101.24 (40)
		Wd	30.91 (34)	0	36.54 (18)	32.11 (47)	99.55 (46)
		Cpx	35.80 (39)	0	58.17 (57)	6.65 (35)	100.61 (67)
15	1100	Gt	25.19 (57)	12.32 (28)	47.52 (64)	13.08 (61)	98.11 (81)
		Wd	31.03 (96)	0	36.49 (80)	30.23 (94)	97.76 (46)
		Cpx	35.32 (70)	30.91 (34)	56.57 (72)	6.27 (81)	98.16 (13)
15	900	Gt	27.61 (31)	12.45 (79)	44.08 (83)	17.22 (96)	101.35 (74)
		Wd	27.91 (46)	0	36.15 (50)	35.08 (27)	99.14 (84)
		PhE	34.01 (84)	4.54 (26)	35.92 (47)	12.79 (54)	87.27 (79)

* Measuring the chemical composition by EDS was difficult because of the small crystal size. This phase was identified by Raman spectrum. Gt, garnet; Cpx, clinopyroxene; Wd, wadsleyite; Rw, ringwoodite; St, stishovite; PhE, phase E; PhD, phase D; h-Fe, iron-rich hydrous phase.

The chemical composition changes in Rw and Gt that occur with increasing temperature at some different pressures are shown in Figure 3. The MgO and SiO_2 contents decreased concomitantly with increasing temperature and then increased slightly in Rw at 16 and 21 GPa (Figure 3a). However, the opposite trend was observed for FeO content. At 18 GPa, opposite trends were observed in MgO, SiO_2, and FeO contents below 1250 °C compared to pressures at 16 and 21 GPa. Both Wd and Rw have near-stoichiometric bulk composition. However, the (Mg + Fe)/Si ratio of Wd and Rw was lower than 2, indicating incorporation of H^+. The H_2O contents in Rw were greater than those in Wd, based on deficit total weight estimation. Generally, the Al_3O_2 and SiO_2 contents in Gt exhibited an opposite tendency because of Tschermak substitution ($Mg^{2+} + Si^{4+} = 2Al^{3+}$). The Al_3O_2 content decreased, and the SiO_2 content increased concomitantly with increasing temperature. However, with increasing pressure, the Al_3O_2 content increased, and the SiO_2 content decreased (Figure 3b). The FeO content remained fundamentally unchanged in all quenched samples, except under the condition of 15 GPa and 900 °C (Table 3).

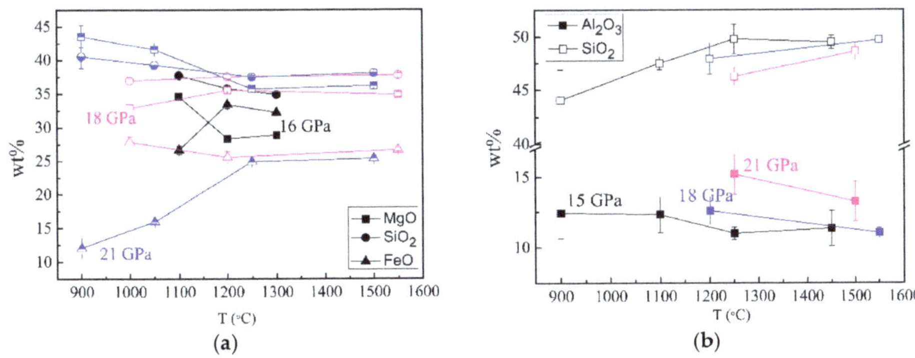

Figure 3. Chemical composition changes in ringwoodite (**a**) and garnet (**b**) with increasing temperature at different pressures.

3.3. Stability and Water Contents of Hydrous Phases in Iron-Rich Martian Mantle

Several reports of petrological studies have described that dense hydrous magnesium silicates (DHMSs) remain stable in the hydrous pyrolite mantle compositions along a cold subducting slab [1,3,21]. In the present study, we observed that the stability regions of DHMSs in the hydrous iron-rich Martian mantle (2 wt.% H_2O) were generally consistent with those obtained in CaO-MgO-Al_2O_3-SiO_2 (CMAS) pyrolite with 2 wt.% of H_2O and a water-saturated MSH system. However, some differences were apparent.

The stability region of PhE partially overlapped with Wd or Rw below 17 GPa in the low-temperature field. Although Wd could hold up to 3 wt.% of H_2O in its crystal structure [24], it only accommodated approx. 0.8 wt.% of H_2O in a water-undersaturated condition at 15 GPa and 900 °C. The water content increased drastically to approx. 2.2 wt.% after PhE decomposed at elevated temperature 1100°C and 15 GPa. It subsequently decreased to approx. 0.6 wt.% with the temperature increased to 1450 °C. The water content in Wd was consistent with those proposed in CMAS pyrolite with 2 wt.% of H_2O at 15 GPa and 1450 °C [24], indicating that Wd also has large water storage capacity within the stability field of the Martian mantle.

Reportedly, SuB appears in CMAS or MHS systems at pressures higher than 17 GPa [24], but we observed PhD instead of SuB at 18–21 GPa, which means that much water might be held in the low-temperature region because of the higher water solubility in PhD. It is expected that PhD might transport water to the deepest part of the Martian interior because PhD was reported to be stable up to 44 GPa, which corresponds to a depth of 1250 km [6].

Hydrous Rw appeared at pressures higher than 16.5 GPa, which exhibited a wider stability region than that in hydrous pyrolite. It seemed readily apparent that water contents in Rw decreased concomitantly with increasing temperature, as shown in Table 2. We observed trace amounts of melt at 18 GPa and 1550 °C, suggesting that it was very close to the wet solidus in MAFS system under a water-unsaturated condition. We also observed some amount of iron-rich hydrous phase at 21 GPa and 1050 °C, implicating it is a potential water carrier in low-temperature regions of the hydrous iron-rich Martian mantle. Because of its greater density than its surrounding materials, it might bring water to the Martian core.

4. Implications

The existence of water on Mars has long been controversial. Some evidence has shown that water disappeared from the Mars surface after its formation [17,26,27]. Although the whereabouts of the water have long been debated, some parts of the hydrated crust of Mars may have been brought into the deep interior by the convecting mantle. We inferred that DHMSs might act as an important water carrier in the deep region of Fe-rich Mars, although the temperature profile of the Mars interior remains unclear. Several models have been proposed to constrain the structure of the Martian interior based on geophysical observations and high-pressure petrological studies. Then, they suggested the pressure of the core-mantle boundary as 19–25 GPa [25,28,29]. Our results demonstrated that DHMSs could be expected to transport water even to the iron-rich Martian core in the cold region in this pressure range.

PhE and PhD were potentially relevant DHMSs in the Martian transition zone in a simplified MAFSH model. The DHMSs would dehydrate completely if the temperature of the surrounding Martian mantle was higher than their stability limits (Figure 2). The released water would be stored in Wd or Rw. These phases might subsequently act as a large water reservoir in the Martian mantle, as in the Earth's mantle. The results of the present phase relation in iron-rich Martian mantle were fundamentally consistent with data for hydrous pyrolite, except for the SuB absence in the present study. In addition, it is noteworthy that much more water could be held in the Martian mantle than in Earth's mantle because the phase transition pressure of the loop in both olivine (Ol)–Wd and Wd–Rw boundary shifted towards lower pressure [22,24].

Furthermore, it seems that the thermal structure of Mars is key to understand water cycles in a deeper region; however, the areotherm of Mar's is still under debate [17,25,28]. In this study, we adopted a model areotherm, as shown in Figure 1, which was determined according to the present day core-mantle boundary temperature of ~1400 °C suggested by Hauck and Phillips [17]. In this model, Mars was slowly cooling down from 4.5 Ga to the present whose areotherm was still at least higher than the dehydration temperature of PhE and PhD, as shown in Figure 1. During the evolution of early Mars, much of the water was possibly lost because of low melting temperatures of DHMSs, hydrous Wd, and Rw compared with areotherm [17]. In addition, water is an incompatible and volatile component in a solid-melt system, which easily causes partial melting of Martian mantle. Water also affects the thermal evolution of the planet. The generated magma may bring the released water to the shallow region, and help conduct and transfer heat more efficiently. This process may help early Martian mantle quickly cool down than a dry system. By contrast, this may become a barrier for water transportation at least to some critical depth during some part of Martian history. However, the accumulated water in the shallow region may hydrate some part of the crust. Unlike the Earth system in which water is recycled by subducting slabs, the mantle of Mars might be convected, but the crust is stagnant and not subducted [17]. The mantle might be hydrated by reaction with the overlying hydrous crust. Eventually, the water should be restored in Wd and Rw during mantle convection due to their high thermal stability region than that of DHMSs.

Our present result indicated that both DHMSs and nominally anhydrous minerals, Wd and Rw, have the potential to accommodate a certain amount of water, elucidating model geodynamic processes associated with the deepwater cycles of Mars.

Author Contributions: Conceptualization, C.X.; data curation, C.X.; writing (original draft preparation), C.X.; writing (review and editing), T.I.; supervision, T.I.; funding acquisition, C.X. and T.I.

Funding: This research was funded by JSPS KAKENHI Grant Numbers 18J12511 for C.X. and 26247073, 15H05828, and 18H03740 for T.I. In addition, C.X. was supported by Research Fellowships of the Japan Society for the Promotion of Science (JSPS) for Young Scientists (DC2).

Acknowledgments: The authors are grateful to Takeshi Arimoto for the preparation of starting materials.

Conflicts of Interest: The authors declare no conflict of interest.

References

1. Gasparik, T. The role of volatile in the transition zone. *J. Geophys. Res.* **1993**, *98*, 4287–4299. [CrossRef]
2. Ohtani, E.; Touma, M.; Litasov, K.; Kubo, T.; Suzuki, A. Stability of hydrous phases and water storage capacity in the transitional zone and lower mantle. *Phys. Earth Planet. Inter.* **2001**, *124*, 105–117. [CrossRef]
3. Komabayashi, T.; Omori, S. Internally consistent thermodynamic dataset for dense hydrous magnesium silicates up to 35 GPa, 1600 °C: Implications for water circulation in the Earth's deep mantle. *Phys. Earth Planet. Int.* **2006**, *156*, 89–107. [CrossRef]
4. Litasov, K.; Ohtani, E.; Sano, A.; Suzuki, A.; Funakoshi, K. Wet subduction versus cold subduction. *Geophys. Res. Lett.* **2005**, *32*, 1–5. [CrossRef]
5. Ghosh, S.; Schmidt, M. Melting of phase D in the lower mantle and implications for recycling and storage of H_2O in the deep mantle. *Geochim. Cosmochim. Acta* **2014**, *145*, 72–88. [CrossRef]
6. Nishi, M.; Irifune, T.; Tsuchiya, J.; Tange, Y.; Nishihara, Y.; Fujino, K.; Higo, Y. Stability of hydrous silicate at high pressures and water transport to the deep lower mantle. *Nat. Geosci.* **2014**, *7*, 224–227. [CrossRef]
7. Pamato, M.; Myhill, R.; Ballaran, T.; Frost, D.; Heidelbach, F.; Miyajima, N. Lower-mantle water reservoir implied by the extreme stability of a hydrous aluminosilicate. *Nat. Geosci.* **2015**, *8*, 75–79. [CrossRef]
8. Walter, M.J.; Thomson, A.R.; Wang, W.; Lord, O.T.; Ross, J.; McMahon, S.C.; Baron, M.A.; Melekhova, E.; Kleppe, A.K.; Kohn, S.C. The stability of hydrous silicates in Earth's lower mantle: Experimental constraints from the systems $MgO–SiO_2–H_2O$ and $MgO–Al_2O_3–SiO_2–H_2O$. *Chem. Geol.* **2015**, *418*, 16–29. [CrossRef]
9. Liu, X.; Matsukage, K.; Nishihara, Y.; Suzuki, T.; Takahashi, E. Stability of the hydrous phases of Al-rich phase D and Al-rich phase H in deep subducted oceanic crust. *Am. Mineral.* **2019**, *104*, 64–72. [CrossRef]
10. Inoue, T. Effect of water on melting phase relations and melt composition in the system Mg_2SiO_4-$MgSiO_3$-H_2O up to 15 GPa. *Phys. Earth Planet. Inter.* **1994**, *85*, 237–263. [CrossRef]

11. Cardenas, B.; Mohrig, D.; Goudge, T. Fluvial stratigraphy of valley fills at Aeolis Dorsa, Mars: Evidence for base-level fluctuations controlled by a downstream water body. *GSA Bull.* **2018**, *130*, 484–498. [CrossRef]
12. Chan, N.; Perron, J.; Mitrovica, J.; Gomez, N. New Evidence of an Ancient Martian Ocean from the Global Distribution of Valley Networks. *J. Geophys. Res.* **2018**, *123*, 2138–2150. [CrossRef]
13. Ivanov, M.; Erkeling, G.; Hiesinger, H.; Bernhardt, H.; Reiss, D. Topography of the deuteronilus contact on mars: Evidence for an ancient water/mud ocean and long-wavelength topographic readjustments. *Planet. Space Sci.* **2017**, *144*, 49–70. [CrossRef]
14. Byrne, S.; Dundas, C.; Kennedy, M. Distribution of mid-latitude ground ice on Mars from new impact craters. *Science* **2009**, *325*, 1674–1676. [CrossRef] [PubMed]
15. Conway, S.J.; Balme, M.R.; Kreslavsky, M.A.; Murray, J.B.; Towner, M.C. The comparison of topographic long profiles of gullies on Earth to gullies on Mars: A signal of water on Mars. *Icarus* **2016**, *253*, 189–204. [CrossRef]
16. Pan, L.; Ehlmann, B. Aqueous Processes from Diverse Hydrous Minerals in the Vicinity of Amazonian-Aged Lyot Crater. *J. Geophys. Res.* **2018**, *123*, 1618–1648. [CrossRef]
17. Hauck, S.A.; Phillips, R.J. Thermal and crustal evolution of Mars. *J. Geophys. Res. Planets* **2002**, *107*. [CrossRef]
18. Frost, D.J.; Fei, Y.W. Stability of phase D at high pressure and high temperature. *J. Geophys. Res.* **1998**, *103*, 7463–7474. [CrossRef]
19. Ohira, I.; Ohtani, E.; Sakai, T.; Miyahara, M.; Hirao, N.; Ohishi, Y.; Nishijima, M. Stability of a hydrous δ-phase, AlOOH-$MgSiO_2(OH)_2$, and a mechanism for water transport into the base of lower mantle. *Earth Planet. Sci. Lett.* **2014**, *401*, 12–17. [CrossRef]
20. Dreibus, G.; Wänke, H. Mars, a volatile-rich planet. *Icarus* **1987**, *71*, 225–240. [CrossRef]
21. Ohtani, E.; Litasov, K.; Hosoya, T.; Kubo, T.; Kondo, T. Water transport into the deep mantle and formation of a hydrous transition zone. *Phys. Earth Planet. Inter.* **2004**, *143–144*, 255–269. [CrossRef]
22. Litasov, K.D.; Ohtani, E.; Sano, A. Influence of water on major phase transitions in the Earth's mantle. In *Earth Deep Water Cycle*; Jacobsen, S.D., van der Lee, S., Eds.; Geophysical Monograph Series; Wiley Blackwell: Hoboken, NJ, USA, 2006; Volume 168, pp. 95–111.
23. Akaogi, M.; Ito, E.; Navrotsky, A. Olivine-modified spinel-spinel transitions in the system Mg_2SiO_4–Fe_2SiO_4: Calorimetric measurements, thermochemical calculation, and geophysical application. *J. Geophys. Res.* **1989**, *94*, 15671–15685. [CrossRef]
24. Litasov, K.; Ohtani, E. Stability of various hydrous phases in CMAS pyrolite-H_2O system up to 25 GPa. *Phys. Chem. Miner.* **2003**, *30*, 147–156. [CrossRef]
25. Bertka, C.M.; Fei, Y.W. Mineralogy of Martian interior up to core-mantle boundary pressures. *J. Geophys. Res.* **1997**, *102*, 5251–5264. [CrossRef]
26. Baker, V.R. Water and the Martian landscape. *Nature* **2001**, *412*, 228–236. [CrossRef] [PubMed]
27. Masson, P.; Carr, M.H.; Costard, F.; Greeley, R.; Hauber, E.; Jaumann, R. Geomorphologic evidence for liquid water. *Space Sci. Rev.* **2001**, *96*, 333–364. [CrossRef]
28. Zharkov, V.N.; Gudkova, T.V. Seismic model of Mars: Effects of hydration. *Planet. Space Sci.* **2014**, *104*, 270–278. [CrossRef]
29. Raevskiy, S.N.; Gudkova, T.V.; Zharkov, V.N. Diagnostic possibilities of body waves for studying the interior structure of Mars. *Izv-Phys. Solid Eart.* **2015**, *51*, 143–155. [CrossRef]

minerals

MDPI

Article

High-Pressure and High-Temperature Phase Transitions in Fe_2TiO_4 and Mg_2TiO_4 with Implications for Titanomagnetite Inclusions in Superdeep Diamonds

Masaki Akaogi *, Taisuke Tajima, Masaki Okano and Hiroshi Kojitani

Department of Chemistry, Gakushuin University, Mejiro, Toshima-ku, Tokyo 171-8588, Japan
* Correspondence: masaki.akaogi@gakushuin.ac.jp; Tel.: +81-3-5904-9377; Fax: +81-3-5992-1029

Received: 8 September 2019; Accepted: 3 October 2019; Published: 6 October 2019

Abstract: Phase transitions of Mg_2TiO_4 and Fe_2TiO_4 were examined up to 28 GPa and 1600 °C using a multianvil apparatus. The quenched samples were examined by powder X-ray diffraction. With increasing pressure at high temperature, spinel-type Mg_2TiO_4 decomposes into MgO and ilmenite-type $MgTiO_3$ which further transforms to perovskite-type $MgTiO_3$. At ~21 GPa, the assemblage of $MgTiO_3$ perovskite + MgO changes to 2MgO + TiO_2 with baddeleyite (or orthorhombic I)-type structure. Fe_2TiO_4 undergoes transitions similar to Mg_2TiO_4 with pressure: spinel-type Fe_2TiO_4 dissociates into FeO and ilmenite-type $FeTiO_3$ which transforms to perovskite-type $FeTiO_3$. Both of $MgTiO_3$ and $FeTiO_3$ perovskites change to $LiNbO_3$-type phases on release of pressure. In Fe_2TiO_4, however, perovskite-type $FeTiO_3$ and FeO combine into calcium titanate-type Fe_2TiO_4 at ~15 GPa. The formation of calcium titanate-type Fe_2TiO_4 at high pressure may be explained by effects of crystal field stabilization and high spin–low spin transition in Fe^{2+} in the octahedral sites of calcium titanate-type Fe_2TiO_4. It is inferred from the determined phase relations that some of Fe_2TiO_4-rich titanomagnetite inclusions in diamonds recently found in São Luiz, Juina, Brazil, may be originally calcium titanate-type Fe_2TiO_4 at pressure above ~15 GPa in the transition zone or lower mantle and transformed to spinel-type in the upper mantle conditions.

Keywords: phase transition; high pressure; Mg_2TiO_4; Fe_2TiO_4; spinel; ilmenite; perovskite; lithium niobate; calcium titanate; diamond inclusion

1. Introduction

Titanomagnetite ($Fe_{3-x}Ti_xO_4$), one of the important magnetic minerals in igneous and metamorphic rocks, is a solid solution between Fe_3O_4 magnetite ($x = 0$) and Fe_2TiO_4 ulvöspinel ($x = 1$), formed by replacement of Fe^{3+} by Ti^{4+}. Fe_2TiO_4 ulvöspinel has an inverse spinel structure where tetrahedral sites are occupied by Fe^{2+} and octahedral sites are occupied randomly by Fe^{2+} and Ti^{4+} [1,2]. Mg_2TiO_4 qandilite is one of endmembers of spinel solid solution in the system Fe_2TiO_4–Mg_2TiO_4–Fe_3O_4–$MgFe_2O_4$, and is important in industrial ceramics. Mg_2TiO_4 qandilite also has the inverse spinel structure [3]. High-pressure phase transitions of A_2BO_4 and AB_2O_4 spinels have received much attention in geophysics and mineral physics, because they provide valuable information on post-spinel transitions of spinel-structured Mg_2SiO_4 ringwoodite which occurs at 660 km depth in the mantle and on stability of aluminous phases with the AB_2O_4 stoichiometry in the deep mantle.

Phase transitions of Fe_2TiO_4 spinel at room temperature and high pressure have been studied in detail [2,4–6]. The studies indicated that at room temperature Fe_2TiO_4 cubic spinel (space group *Fd*-3*m*) first changes into tetragonal spinel (*I*4$_1$/*adm*) at 9 GPa due to a Jahn-Teller effect of Fe^{2+} in the tetragonal site. At 12 GPa, the tetragonal spinel transforms to $CaTi_2O_4$-type structure (*Cmcm*), in which octahedral sites are occupied by disordered Fe^{2+} and Ti^{4+} and eight-fold sites by Fe^{2+}. The $CaTi_2O_4$-type Fe_2TiO_4 further transforms at 53 GPa to another high-pressure polymorph (*Pmma*) with ordered Fe^{2+} and Ti^{4+} in the octahedral sites. Yamanaka et al. [4] reported that the high spin-low spin transition in Fe^{2+} in the octahedral sites of the $CaTi_2O_4$-type Fe_2TiO_4 started to occur above about 15 GPa at room temperature and that all of Fe^{2+} in the octahedral sites were in the low spin state in the *Pmma* phase, while Wu et al. [6] reported that the spin transition in Fe^{2+} initiated at around 40 GPa.

In contrast, studies on high-pressure and high-temperature transitions of Fe_2TiO_4 and Mg_2TiO_4 spinels have been limited. Phase transitions of the spinels were examined only up to 5 GPa at 800–1600 °C by Akimoto and Syono [7]. They reported that both Fe_2TiO_4 and Mg_2TiO_4 spinels decompose into ATiO$_3$ ilmenite and AO with rock-salt structure (A = Mg, Fe). The phase transitions at high temperature and high pressure above ~5 GPa, however, have not been well examined so far. Changes in elastic properties associated with the phase transitions in Fe_2TiO_4 and Mg_2TiO_4 shown above were investigated by Liebermann et al. [8]. The physical basis of elastic properties at high pressure and high temperature is described in detail by Anderson [9].

In this study, we have investigated phase transitions in Fe_2TiO_4 and Mg_2TiO_4 up to 28 GPa and 1600 °C, using a Kawai-type multianvil apparatus. We have found that the transition sequences in Fe_2TiO_4 at high pressure and high temperature are substantially different from those at high pressure and room temperature. Based on the results on Fe_2TiO_4 and Mg_2TiO_4, we discuss the difference in the transition behaviors between Fe_2TiO_4 and Mg_2TiO_4 from a crystal–chemical point of view. In addition, we discuss stability and possible origin of Fe_2TiO_4-rich titanomagnetites which were found as inclusions in superdeep diamonds from Juina area, Brazil [10,11].

2. Experimental Methods

Starting materials for high-pressure and high-temperature experiments of Mg_2TiO_4 and Fe_2TiO_4 were synthesized as follows. Spinel(Sp)-type Mg_2TiO_4 (qandilite) was synthesized from a 2:1 mixture (molar ratio) of MgO (>99.9% purity, Wako Co., Osaka, Japan) and TiO_2 (>99.9%, Wako Co.) by heating at 1300 °C for 20 h. Sp-structured Fe_2TiO_4 (ulvöspinel) was synthesized from a 1:1 mixture of Fe_2O_3 (>99.9%, Wako Co.) and TiO_2 by heating at 1270 °C for 27 h in controlled oxygen fugacity with mixed gasses of CO_2, H_2 and Ar (3:2:5 volume ratios). Powder X-ray diffraction measurements and composition analyses using a scanning electron microscope (SEM) with an energy-dispersive X-ray spectrometer (EDS) indicated that the synthesized samples were single-phase Sp-type Mg_2TiO_4 and Fe_2TiO_4 with the stoichiometric compositions. The lattice parameters determined by powder X-ray diffraction were a = 8.4398(1) Å for Mg_2TiO_4 Sp and a = 8.5375(3) Å for Fe_2TiO_4 Sp, which were in good agreement with those by Wechsler et al. [3] and Wechsler et al. [1], respectively. For high-pressure and high-temperature experiments of Fe_2TiO_4, 5 wt % metallic iron (>99.9 %, Johnson Matthey Co., London, UK) was added to the above synthesized Fe_2TiO_4 Sp to keep iron in the Fe_2TiO_4 samples in ferrous state.

High-pressure and high-temperature experiments were performed with the quench method using a Kawai-type 6-8 multianvil apparatus [12] at Gakushuin University. Tungsten carbide anvils with 2.5 mm truncated edge-length were used as the second-stage anvils. A 5 wt % Cr_2O_3-doped MgO octahedron of 7 mm edge length was used as the pressure medium. A cylindrical Re heater was placed in the central part of the octahedron. For the experiments of Mg_2TiO_4, the powdered starting material was directly put into the Re heater. For thermal insulation, a $LaCrO_3$ sleeve was placed between the heater and the MgO octahedron along with two $LaCrO_3$ plugs in both ends of the furnace. Two Pt discs were placed between the sample and the plugs to avoid any reaction between them. For the experiments of Fe_2TiO_4, the mixture of 95 wt % Fe_2TiO_4 and 5 wt % Fe was put into an Fe capsule which was placed in the central part of the Re heater. A BN sleeve was inserted between the Fe capsule and Re heater for electrical insulation. Temperature was measured at the central part of the outer surface of the furnace with a Pt/Pt-13%Rh thermocouple. Effect of pressure on emf of the thermocouple was ignored.

Pressure was calibrated at room temperature using pressure-fixed points: Bi I-II (2.55 GPa), Bi III-V (7.7 GPa), ZnS (15.6 GPa), and GaAs (18.3 GPa) in Ito [12] and GaP (23 GPa) by Dunn and Bundy [13]. The pressure was further corrected at 1200 and 1600 °C, using transition pressures of Mg_2SiO_4 olivine–wadsleyite–ringwoodite [14,15] and $MgSiO_3$ akimotoite–bridgmanite [16] and decomposition pressure of $Mg_3Al_2Si_3O_{12}$ pyrope to $MgSiO_3$-rich bridgmanite + Al_2O_3-rich corundum [16]. Uncertainties of pressure and temperature were estimated to be ±0.3 GPa and ±20 °C, respectively. In each run, pressure was raised to a targeted pressure of 11–28 GPa at a constant rate during about 2–4 h, and then temperature was raised to a targeted temperature of 1000–1600 °C at a rate of about 100 °C/min. The samples were kept for 60–120 min at the pressure-temperature conditions, and then quenched under pressure, slowly decompressed, and recovered to ambient conditions.

The recovered samples were pulverized and examined by a powder X-ray diffractometer (Rigaku, RINT2500, Tokyo, Japan) operated at 45 kV and 250 mA with a rotating Cr anode. Monochromatized CrKα radiation was used for identification of phases in the run products and determination of lattice parameters. High-purity Si powder was used to calibrate 2θ angle of the powder X-ray diffractometer. The lattice parameters were refined by the least squares method. By using some parts of run products which were mounted on slide-glass plates with epoxy resin and polished flat, phase identification and chemical analysis were made using a SEM (JEOL, JMS-6360, Tokyo, Japan) in combination with an EDS (SGX Sensortech Sirius SD-10133, High Wycombe, UK). The SEM was operated with acceleration voltage of 15 kV and probe current of 0.55 nA. Natural forsterite and synthetic fayalite and $MnTiO_3$ provided by JEOL were used as standard materials for Mg, Fe and Ti, respectively.

3. Results

3.1. Phase Transitions in Mg_2TiO_4

Table 1 summarizes results of phase transition experiments in Mg_2TiO_4 at 21–28 GPa and 1000–1600 °C. The previous study showed that at pressure below 1–2 GPa spinel (Sp)-type Mg_2TiO_4 first dissociates into ilmenite (Ilm)-type $MgTiO_3$ + MgO periclase (Pc) [7]. Figure 1a shows a powder X-ray diffraction pattern of the run product (No. O161021) quenched from 21 GPa and 1600 °C. The diffraction pattern exhibits MgO Pe and a phase with very similar diffraction peaks to $MgTiO_3$ Ilm (space group *R*-3). However, some diffraction peaks such as 003 and 101 of $MgTiO_3$ Ilm in 2θ range of 20–30° are absent. Almost all the peaks were indexed with hexagonal symmetry, and refined cell parameters were a = 5.0774(5) Å, c = 13.716(2) Å, giving c/a = 2.701. This c/a value is considerably smaller than 2.750 of $MgTiO_3$ Ilm [3]. The diffraction data in Figure 1a and the lattice parameters are consistent with those of $LiNbO_3$ (Ln)-type $MgTiO_3$ (*R3c*) [17]. Linton et al. [17] reported that Ilm-type $MgTiO_3$ transforms at about 15–20 GPa to the perovskite (Pv) phase, which is converted to the Ln-type phase on decompression. Therefore, our results in Table 1 and Figure 1a reveal that $MgTiO_3$ Pv + MgO Pe are stable at 21 GPa and 1600 °C and that $MgTiO_3$ Pv transforms to the Ln-type on release of pressure.

Table 1. Results of high-pressure and high-temperature experiments in Mg_2TiO_4.

Run No.	Pressure (GPa)	Temperature (°C)	(min)	Run Product
O160915	21	1000	120	Pe, Ln
O161025	23	1000	90	Pe, α
O161017	21	1300	90	Pe, Ln
O170116	22.5	1300	50	Ln, Pe, α
O161128	23.5	1300	60	Pe, α, Ln(tr)
O160909	24.5	1300	60	Pe, α

Table 1. *Cont.*

Run No.	Pressure (GPa)	Temperature (°C)	(min)	Run Product
O160924	28	1300	90	Pe, α
O161106	28	1400	90	Pe, α, Ln(tr)
O170219	22	1500	90	Ln, Pe, α
O170202	25	1500	60	Pe, α
O170126	28	1500	60	Pe, α
O161021	21	1600	90	Pe, Ln
O161124	22	1600	60	Pe, Ln
O161209	23.5	1600	60	Ln, Pe, α
O160903	24.5	1600	60	Pe, α, Ln(tr)

Abbreviations are as follows: Ln, $LiNbO_3$-type $MgTiO_3$; Pe, MgO periclase; α, αPbO_2 type TiO_2; tr, trace.

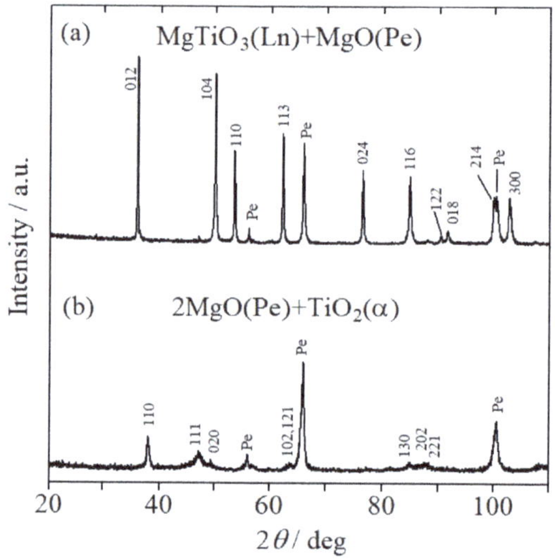

Figure 1. Powder X-ray diffraction patterns (Cr Kα) of run products of Mg_2TiO_4, (**a**) No. O161021 (21 GPa, 1600 °C) and (**b**) No. O160924 (28 GPa, 1300 °C). Miller indices in (**a**) are for Ln-type $MgTiO_3$, and those in (**b**) are for αPbO_2-type TiO_2. Ln: $LiNbO_3$-type, Pe: periclase, α: αPbO_2-type.

Figure 1b shows a powder X-ray diffraction pattern of the run product (No. O160924) quenched from 28 GPa and 1300 °C, in which diffraction peaks of MgO Pe and αPbO_2 type TiO_2 are observed. Broad diffraction peaks of αPbO_2 type TiO_2 in Figure 1b suggest that the phase was a retrograde transformation product from a high-pressure phase. Previous studies on TiO_2 indicated that at about 800–1000 °C TiO_2 rutile transforms to αPbO_2 type at about 6–8 GPa, which further transforms to baddeleyite (Bd)-type TiO_2 (akaogiite) [18,19] at about 15–20 GPa [20–22]. The Bd-type TiO_2 transforms to orthorhombic-I (OI) type TiO_2 at about 25–30 GPa at 1200–1500 °C [23,24]. These studies also showed that both of Bd- and OI-type TiO_2 phases back-transform to αPbO_2 type TiO_2 on release of pressure. Therefore, based on the experimental pressure range and the broad X-ray diffraction peaks, we conclude that the αPbO_2 type TiO_2 found in the run products of 28 GPa and 1300 °C was the retrograde transition product from Bd- or OI-type TiO_2 synthesized at the P, T conditions. This leads to the conclusion that $MgTiO_3$ Pv stable at 21 GPa decomposes to the constituent oxides at higher pressure.

Figure 2 shows the phase diagram of Mg_2TiO_4. Our experimental results at 21–28 GPa and 1000–1600 °C indicate that the transition boundary of $MgTiO_3$ Pv + MgO Pe to 2MgO Pe + Bd-type TiO_2 which is the decomposition boundary of $MgTiO_3$ Pv is located at 21–23 GPa and 1000–1600 °C with a small positive dP/dT slope. Figure 2 also includes the dissociation boundary of Mg_2TiO_4 Sp to $MgTiO_3$ Ilm + MgO Pe below ~1–2 GPa [7], as well as the Ilm-Pv transition boundary of $MgTiO_3$ at ~ 15–20 GPa [25]. The assemblage of MgO and TiO_2 is stable to at least 35 GPa at 1400 °C in our preliminary study on phase transitions in $MgTiO_3$ [25].

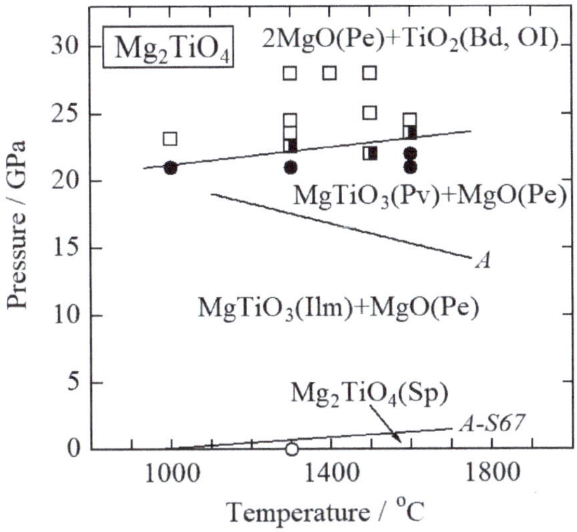

Figure 2. Phase diagram of Mg_2TiO_4. An open circle: Mg_2TiO_4(Sp), solid circles: $MgTiO_3$(Pv) + MgO(Pe), open squares: 2MgO(Pe) + TiO_2 (Bd, OI). A thick solid line: this study, a thin line with A-S67: Sp dissociation boundary to Ilm + Pe [7], a thin line with A: Ilm-Pv boundary [25]. Sp: spinel-type, Ilm: ilmenite-type, Pe: periclase, Pv: perovskite-type, Bd: baddeleyite-type, OI: orthorhombic I-type.

Molar volumes of high-pressure phases in the systems MgO-TiO_2 and FeO-TiO_2 are listed in Table 2. Using the data in Table 2, for the following transitions in Mg_2TiO_4:

$$Mg_2TiO_4(Sp) \rightarrow MgTiO_3(Ilm) + MgO(Pc) \rightarrow MgTiO_3(Pv) + MgO(Pc) \rightarrow 2MgO(Pc) + TiO_2(Bd) \rightarrow 2MgO(Pc) + TiO_2(OI),$$

we obtain molar volume changes at ambient conditions of −3.15, −1.55, −1.17 and −0.07 cm^3/mol, respectively, which are −7.0, −3.7, −2.9 and −0.2%, respectively.

3.2. *Phase Transitions in Fe_2TiO_4*

Results of high-pressure transition experiments in Fe_2TiO_4 are summarized in Table 3. Phase identification was made mostly by powder X-ray diffraction method. Metallic iron in the samples was identified by the SEM-EDS analysis and/or powder X-ray diffraction. Fe_2TiO_4 ulvöspinel first dissociates into Ilm-type $FeTiO_3$ and FeO wustite (Wu) at 4–5 GPa and 1000–1300 °C [7]. Figure 3a shows a powder X-ray diffraction pattern of the run product (No. M150115) synthesized at 14 GPa and 1100 °C. The diffraction pattern of Figure 3a is composed of FeO Wu, metallic iron, and the other phase whose diffraction peaks agree with those of Ln-type $FeTiO_3$ by Akaogi et al. [26]. The Ln-type $FeTiO_3$ phase is interpreted to be the retrograde transition product from Pv-type $FeTiO_3$ [27]. Natural occurrences of Ln-type $FeTiO_3$ were reported in shocked meteorites, and the new mineral was recently named wangdaodeite [28,29].

Figure 3b shows a powder X-ray diffraction pattern of the run product (No. M140710) synthesized at 18 GPa and 1300 °C. The diffraction pattern is similar to calcium titanate ($CaTi_2O_4$), and the peaks were indexed with orthorhombic symmetry (Figure 3b). The Miller indices were consistent with extinction rules of $CaTi_2O_4$(CT)-structured phase (*Cmcm*). The lattice parameters refined using twenty nine diffraction peaks of the Fe_2TiO_4 phase were a = 2.9473(5) Å, b = 9.6448(2) Å, c = 9.9085(2) Å, V = 281.663(7) Å^3. The cell parameters agree well with those extrapolated to ambient pressure of CT-type Fe_2TiO_4 at 38–50 GPa and room temperature by Yamanaka et al. [4]. These results indicate that the Fe_2TiO_4 phase in the run product synthesized at 18 GPa and 1300 °C is the CT-type phase. This is consistent with the results in the previous studies that Fe_2TiO_4 CT is quenchable at ambient conditions [5,30]. Compositions of coexisting Fe_2TiO_4 CT and $FeTiO_3$ Ln with very small grains of FeO Wu in the run product (No. M140626) at 15 GPa and 1300 °C were analyzed using the SEM-EDS: $Fe_{1.97(1)}Ti_{1.01(1)}O_4$ for CT-type phase and $Fe_{0.95(2)}Ti_{1.02(1)}O_3$ for Ln-type phase, indicating almost stoichiometric compositions within the analytical errors.

Table 2. Molar volumes of phases in the systems MgO-TiO_2 and FeO-TiO_2.

Comp.	Structure	V_0 (cm^3/mol)	Ref.	Comp.	Structure	V_0 (cm^3/mol)	Ref.
MgO	rock-salt	11.24	a	FeO	rock-salt	12.17	i
$MgTiO_3$	ilmenite	30.86	b	$FeTiO_3$	ilmenite	31.72	j
$MgTiO_3$	perovskite	29.31	c	$FeTiO_3$	perovskite	30.34	k
$MgTiO_3$	$LiNbO_3$	30.71	d	$FeTiO_3$	$LiNbO_3$	31.34	j
Mg_2TiO_4	spinel	45.25	e	Fe_2TiO_4	spinel	46.84	e
TiO_2	αPbO_2	18.41	f	Fe_2TiO_4	$CaTi_2O_4$	42.41	e
TiO_2	baddeleyite	16.90	g				
TiO_2	OI	16.83	h				

V_0: molar volume at ambient conditions. Ref. a: Hazen [31], b: Wechsler and Von Dreele [3], c: Linton et al. [17], d: Linton et al. [32], e: This study, f: Kojitani et al. [22], g: Al-Khatatbeh et al. [24], h: Nishio-Hamane et al. [33], i: McCammon [34], j: Akaogi et al. [26], k: Leinenweber et al. [27].

Table 3. Results of high-pressure and high-temperature experiments in Fe_2TiO_4.

Run no.	Pressure (GPa)	Temperature (°C)	Time (min)	Run Product *
M150109	12.5	1000	60	Ilm, Wu
M150107	12.5	1100	60	Ilm, Wu
M140806	13	1100	60	Ilm, Ln, Wu
M150115	14	1100	60	Ln, Wu
M141128	15.5	1100	60	Ln, Wu
M140707	16	1100	60	CT
M150218	17	1100	60	CT
M150603	20	1100	60	CT
M150226	23	1100	60	CT
M150516	24	1100	60	CT
M150127	12.5	1200	60	Ilm, Ln, Wu
M150416	18	1200	60	CT
M150707	11	1300	60	Ilm, Wu
M150725	12	1300	60	Ilm, Wu
M140728	13	1300	60	Ln, Wu
M150625	14	1300	60	Ln, Wu
M150202	15	1300	60	CT
M140626	15	1300	60	CT, Ln, Wu
M140701	18	1300	60	CT
M140710	18	1300	60	CT
M150227	20	1300	60	CT
M150606	22	1300	60	CT

* Metallic iron coexisted in the run products. Ilm: ilmenite-type $FeTiO_3$, Wu: FeO wustite, Ln: $LiNbO_3$-type $FeTiO_3$, CT: CT-type Fe_2TiO_4.

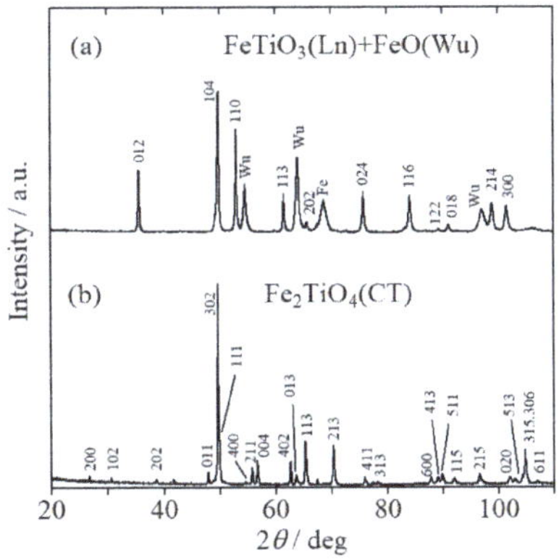

Figure 3. Powder X-ray diffraction patterns (Cr Kα) of run products of Fe_2TiO_4, (**a**) No. M150115 (14 GPa, 1100 °C) and (**b**) No. M140710 (18 GPa, 1300 °C). Miller indices in (**a**) are for Ln-type $FeTiO_3$, and those in (**b**) are for CT-type Fe_2TiO_4. Ln: $LiNbO_3$-type, Wu: wustite, Fe: metallic iron, CT: calcium titanate-type.

Figure 4 shows the phase diagram of Fe_2TiO_4. Our experimental results indicate that at 1000–1300 °C the assemblage of FeO Wu + $FeTiO_3$ Ln which was Pv at high P, T was observed in the run products synthesized at 12.5–15.5 GPa, while FeO Wu + $FeTiO_3$ Ilm was observed at 11–12.5 GPa. The Ilm-Pv transition boundary has a small negative slope. The Ilm-Pv transition pressure in this study is generally consistent with that obtained by extrapolation of the boundary determined at 500–900 °C by Ming et al. [35]. We found that the assemblage of $FeTiO_3$ Ln (Pv at high P, T) and FeO Wu changes to Fe_2TiO_4 CT in the run products at 15–16 GPa at 1100–1300 °C with a small negative slope boundary. Therefore, the stability field of $FeTiO_3$ Pv and FeO Wu is limited in a narrow pressure interval of about 2 GPa. This study indicates that Fe_2TiO_4 CT is stable up to at least 22–24 GPa at 1100–1300 °C. In our previous study in $FeTiO_3$ composition, we found that Fe_2TiO_4 CT coexists with OI-type TiO_2 up to about 33 GPa, above which it decomposes into FeO Wu + TiO_2 OI [26]. Therefore, the upper stability pressure of Fe_2TiO_4 CT is about 33 GPa.

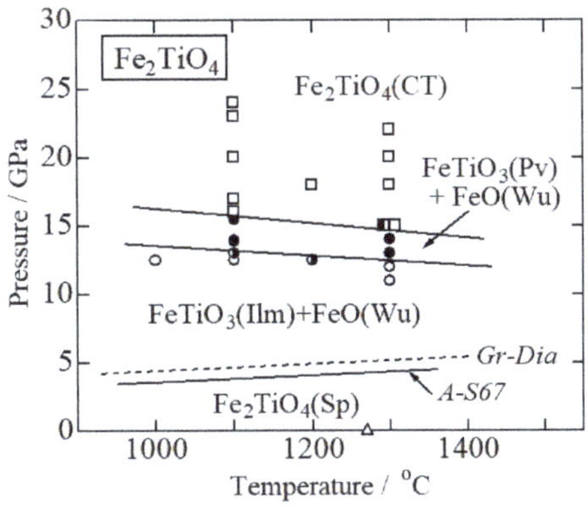

Figure 4. Phase diagram of Fe_2TiO_4. An open triangle: Fe_2TiO_4(Sp), open circles: $FeTiO_3$(Ilm) + FeO(Wu), solid circles: $FeTiO_3$(Pv) + FeO(Wu), open squares: Fe_2TiO_4(CT). Thick solid lines: this study, a thin line with A-S67: Sp dissociation boundary to Ilm + Wu [7], a thin dashed line with Gr-Dia: the graphite–diamond boundary [36]. Sp: spinel-type, Ilm: ilmenite-type, Wu: wustite, Pv: perovskite-type, CT: calcium titanate-type.

The results in Figure 4 combined with those by Akaogi et al. [26] indicate that the following transitions occur in Fe_2TiO_4 with increasing pressure:

$$Fe_2TiO_4(Sp) \rightarrow FeTiO_3(Ilm) + FeO(Wu) \rightarrow FeTiO_3(Pv) + FeO(Wu) \rightarrow Fe_2TiO_4(CT) \rightarrow 2FeO(Wu) + TiO_2(OI),$$

where molar volume changes at ambient conditions are −2.95, −1.38, −0.10 and −1.24 cm^3/mol, which are −6.3%, −3.1%, −0.2% and −2.9%, respectively, using the data in Table 2.

4. Discussion

Our results indicate that Fe_2TiO_4 Sp undergoes phase transitions to $FeTiO_3$ Ilm + FeO Wu and subsequently to $FeTiO_3$ Pv + FeO Wu at high pressure and high temperature (Figure 4). The decomposition of Fe_2TiO_4 Sp to $FeTiO_3$ Ilm + FeO Wu is consistent with the result by Akimoto and Syono [7]. However, at room temperature and high pressure, the transition behavior is different: Fe_2TiO_4 cubic Sp is distorted to tetragonal Sp at ~9 GPa due to the Jahn-Teller effect of Fe^{2+} at the tetrahedral site [2]. The difference may arise from kinetic hindrance to decomposition into $FeTiO_3$ Ilm + FeO Wu at room temperature. In the experiments at room temperature and high pressure, tetragonal Sp transforms to Fe_2TiO_4 CT at ~15 GPa [4,6]. The pressure is generally consistent with transition pressure to Fe_2TiO_4 CT at room temperature extrapolated from the results at 1000–1300 °C, considering the uncertainty of the boundary slope (Figure 4).

Here, we compare the transitions in Mg_2TiO_4 and Fe_2TiO_4. As shown in Figures 2 and 4, both of Sp-type Mg_2TiO_4 and Fe_2TiO_4 dissociate into ATiO$_3$ Ilm and AO (A = Mg, Fe) at pressures below ~5 GPa, and both of the ATiO$_3$ Ilm transform to ATiO$_3$ Pv at ~13–15 GPa. At higher pressure, however, the transition behaviors are different between Mg_2TiO_4 and Fe_2TiO_4. In Mg_2TiO_4, the assemblage of $MgTiO_3$ Pv + MgO Pe changes into 2MgO Pe + TiO_2 Bd due to decomposition of $MgTiO_3$ Pv. However, in Fe_2TiO_4 the assemblage of $FeTiO_3$ Pv + FeO Wu changes at ~15 GPa to CT-type Fe_2TiO_4 which is stable up to ~33 GPa.

Figure 5 illustrates the CT-type structure (*Cmcm*). The structure consists of double chains of edge-sharing octahedra running parallel to the a-axis, and tunnel spaces are formed by four corner-sharing double chains [37,38]. In the structure of CT-type Fe_2TiO_4, Fe^{2+} and Ti^{4+} are randomly distributed in the octahedral sites and only Fe^{2+} in the eight-fold sites in the tunnel spaces. Compared with Mg_2TiO_4, crystal-field effect of Fe^{2+} in the octahedral sites may stabilize the CT-type phase of Fe_2TiO_4 composition. Using effective ionic radii by Shannon [39], the difference of ionic radii between Fe^{2+} (0.78 Å in high-spin state) and Ti^{4+} (0.605 Å) in octahedral site is larger than that between Mg^{2+} (0.72 Å) and Ti^{4+}. However, the ionic radius of Fe^{2+} in low-spin state in the octahedral site is 0.61 Å, which is much closer to that of Ti^{4+} compared with Mg^{2+}. When we adopt that the high-spin to low-spin transition of Fe^{2+} in the octahedral site of Fe_2TiO_4 CT starts to occur at pressure of ~15–20 GPa at room temperature and proportion of low-spin Fe^{2+} increases with pressure [4], it is suggested that the very similar ionic radii of low-spin Fe^{2+} and Ti^{4+} is more favorable for the CT-type structure. Although the high-spin state and low-spin state would be mixed at high temperature such as 1000–1300 °C [40], it is likely that both of the crystal-field effect and the high-spin to low-spin transition of Fe^{2+} may facilitate the stability of Fe_2TiO_4 CT above ~15 GPa.

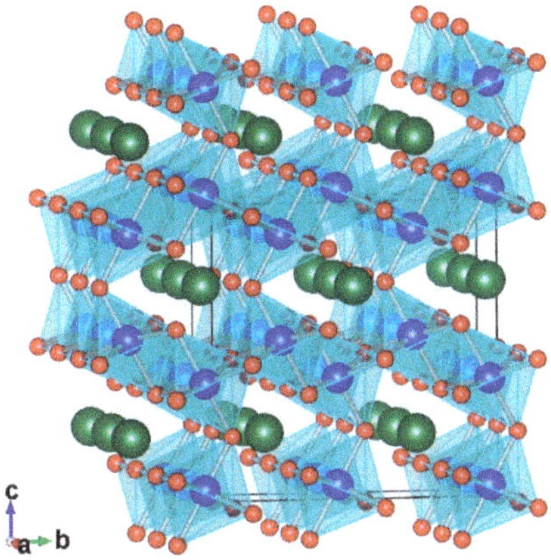

Figure 5. Crystal structure of $CaTi_2O_4$-type AB_2O_4. Small red spheres, middle blue ones and large green ones express oxygen, B and A cations, respectively. The structure was drawn by VESTA [41].

Sp-type titanomagnetites were found as inclusions in diamonds from some localities in Juina, Brazil, and the diamonds were interpreted to be derived from the deep mantle [10,11]. Two different occurrences of titanomagnetites as the diamond inclusions have been reported: one was a titanomagnetite-bearing mineral composite, and the other was separate single-phase inclusions of titanomagnetite. As the former-type inclusion, Walter et al. [10] discovered a ~30 μm-sized composite consisting mostly of orthopyroxene together with olivine and titanomagnetite (called "ulvöspinel" in their study) in a diamond from the Juina-5 kimberlite, Brazil. The analyzed composition of the titanomagnetite was approximately 36 mol % Fe_2TiO_4·36 mol % Fe_3O_4·28 mol % Mg_2TiO_4. Walter et al. [10] interpreted that the three-phase composite was originally a homogeneous Mg-rich bridgmanite in the lower mantle, and that the diamond was transported to the upper mantle at a depth range of ~150–200 km where immiscibility into the three phases occurred before eruption of the kimberlite. Based on the experimental data on $NaAlSiO_4$-$MgAl_2O_4$ [42,43], Walter et al. [10] estimated the depth range of ~150–200 km using nepheline–spinel composite inclusions which were interpreted to be originally calcium ferrite (CF)-type phase and hexagonal aluminous (NAL) phase in the lower mantle and the unmixing occurred at the depth range. Figures 1 and 3 and the results of Woodland et al. [44] indicate that the upper bound of stability field of Sp-type Mg_2TiO_4, Fe_2TiO_4 and Fe_3O_4 increases in the order of ~1, ~4 and ~10 GPa, respectively. Using the pressures, we roughly estimate the upper limit of stability field of the titanomagentite inclusion of the above composition to be ~5–6 GPa. The pressure is compatible with the estimated depth range of ~150–200 km from the CF and NAL inclusions.

The other kind of titanomagnetites was separate single-phase inclusions in superdeep diamonds. By a new combined method of synchrotron microtomography and single-crystal X-ray diffraction with fast, non-destructive methodology, Wenz et al. [11] discovered ~10–20 μm-sized inclusions of separate titanomagnetite crystals and powders in diamonds from São Luiz, Juina, Brazil. They observed that titanomagnetite inclusions were relatively abundant next to (Mg,Fe)O magnesio-wustite and $Fe_{2-x}Ti_xO_3$ titanohematite, and determined lattice parameters of the titanomagnetites embedded in the diamonds to be 8.511–8.405 Å by in situ X-ray diffraction method. Wenz et al. [11] assumed that compositions of the titanomagnetites were on the join Fe_2TiO_4-Fe_3O_4, because chemical analysis of the titanomagnetites was not made in the study. However, it is likely that the titanomagnetites contained minor amounts of Mg_2TiO_4 and $MgFe_2O_4$ components, as that found by Walter et al. [10]. At ambient conditions, cell parameters of Fe_2TiO_4 Sp and Fe_3O_4 Sp are 8.5375(3) Å (this study) and 8.3984(8) Å [4], respectively, while Mg_2TiO_4 Sp and $MgFe_2O_4$ Sp are 8.4398(1) Å (this study) and 8.391 Å [45], respectively. Therefore, there is large uncertainty in estimating compositions of the titanomagnetites in the diamonds only from the lattice parameters determined by the in situ measurements. It should also be considered that remnant pressure probably remained in the titanomagnetites embedded in the diamonds. When we use measured bulk modulus of Fe_2TiO_4 Sp of 121 GPa [8,46] or 147 GPa [47], the pressure dependence of cell parameter is calculated to be −0.023 or −0.019 Å/GPa, respectively.

We suggest, however, that the remnant pressure and possible presence of Mg_2TiO_4 and $MgFe_2O_4$ components, if any, decrease the lattice parameter of titanomagnetite from that in the Fe_2TiO_4–Fe_3O_4 system at 1 atm. When we assume that the titanomagnetites in the diamonds are Mg-free solid solutions in the system Fe_2TiO_4–Fe_3O_4 and the remnant pressure is negligibly small, we can estimate 81 mol % of Fe_2TiO_4 component for the titanomagnetite with the largest cell parameter (a = 8.511 Å). Because this is the lower bound for the Fe_2TiO_4 component, as discussed above, it is probable that the titanomagnetite composition was more Fe_2TiO_4-rich, possibly almost pure Fe_2TiO_4. Such titanomagnetites of very high Fe_2TiO_4 content probably could not be directly incorporated at a depth where the diamonds were formed, because the graphite–diamond transition boundary [36] is placed by ~1 GPa at higher pressure than the upper bound of the stability field of Fe_2TiO_4 Sp (Figure 4). Therefore, the above estimate would suggest that the Fe_2TiO_4-rich titanomagnetites were originally CT-type phases which are stable above ~15 GPa in the transition zone and the lower mantle conditions, and that they back-transformed to the Fe_2TiO_4-rich titanomagnetites in the diamonds after being transported to the upper mantle. The decomposition into $FeTiO_3$ Ilm (or Pv) + FeO Wu might have been kinetically hindered. Further experimental studies, particularly compositional analysis of the titanomagnetites, would be required to better clarify the origin of the titanomagnetites in the superdeep diamonds.

Author Contributions: T.T., M.O. and M.A. conducted the experiments; T.T., M.O., H.K. and M.A. analyzed the obtained data; M.A. and H.K. wrote the manuscript.

Funding: This work was supported in part by JSPS KAKENHI (grant nos. 25287145 and 17H02986 to M.A.) and by the MEXT-supported program of Gakushuin University for the Strategic Research Foundation at Private Universities.

Acknowledgments: I (M.A.) thank Robert C. Liebermann (Bob-san) for his suggestion to submit the paper to the Special Issue, Mineral Physics−In Memory of Orson Anderson, and for his encouragements on various occasions to me. We are grateful to T. Ishii, H. Yusa, D. Mori and Y. Inaguma for useful suggestions and discussion.

Conflicts of Interest: The authors declare no conflict of interest.

References

1. Wechsler, B.A.; Lindsley, D.H.; Prewitt, C.T. Crystal structure and cation distribution in titanomagnetites (Fe_{3-x},Ti_xO_4). *Am. Mineral.* **1984**, *69*, 754–770.
2. Yamanaka, T.; Mine, T.; Asogawa, S.; Nakamoto, Y. Jahn-Teller transition of Fe_2TiO_4 observed by maximum entropy method at high pressure and low temperature. *Phys. Rev.* **2009**, *B80*, 134120. [CrossRef]
3. Wechsler, B.A.; Von Dreele, R.B. Structure refinements of Mg_2TiO_4, $MgTiO_3$ and $MgTi_2O_5$ by time-of-flight neutron powder diffraction. *Acta Cryst.* **1989**, *B45*, 542–549. [CrossRef]
4. Yamanaka, T.; Kyono, A.; Nakamoto, Y.; Meng, Y.; Kharlamova, S.; Struzhkin, V.V.; Mao, H.K. High-pressure phase transitions of $Fe_{3-x}Ti_xO_4$ solid solution up to 60 GPa correlated with electronic spin transition. *Am. Mineral.* **2013**, *98*, 736–744. [CrossRef]
5. Wu, Y.; Wu, X.; Qin, S. Pressure-induced phase transition of Fe_2TiO_4: X-ray diffraction and Mössbauer spectroscopy. *J. Sold State Chem.* **2012**, *185*, 72–75. [CrossRef]
6. Xu, W.M.; Hearne, G.R.; Layek, S.; Levy, D.; Itie, J.-P.; Pasternak, M.P.; Rozenberg, G.K.; Greenberg, E. Site-specific spin crossover in Fe_2TiO_4 post-spinel under high pressure up to nearly a megabar. *Phys. Rev.* **2017**, *B96*, 045108. [CrossRef]
7. Akimoto, S.; Syono, Y. High-pressure decomposition of some titanate spinels. *J. Chem. Phys.* **1967**, *47*, 1813–1817. [CrossRef]
8. Liebermann, R.C.; Jackson, I.; Ringwood, A.E. Elasticity and phase equilibria of spinel disproportionation reactions. *Geophys. J. R. Astr. Soc.* **1977**, *50*, 553–586. [CrossRef]
9. Anderson, O.L. *Equations of State of Solids for Geophysics and Ceramic Science*; Oxford Univ. Press: New York, NY, USA, 1995; 405p.

10. Walter, M.J.; Kohn, S.C.; Araujo, D.; Bulanova, G.P.; Smith, C.B.; Gaillou, E.; Wang, J.; Steele, A.; Shirey, S.B. Deep mantle cycling of oceanic crust: Evidence from diamonds and their mineral inclusions. *Science* **2011**, *334*, 54–57. [CrossRef]
11. Wenz, M.D.; Jacobsen, S.D.; Zhang, D.; Regier, M.; Bausch, H.J.; Dera, P.K.; Rivers, M.; Eng, P.; Shirey, S.B.; Pearson, D.G. Fast identification of mineral inclusions in diamond at GSECARS using synchrotron X-ray microtomography, radiography and diffraction. *J. Synchrotron Rad.* **2019**, *26*, 1–6. [CrossRef]
12. Ito, E. Theory and Practice—Multianvil cells and high-pressure experimental methods. In *Mineral Physics*; Price, G.D., Ed.; Elsevier: Amsterdam, The Netherlands, 2007; Volume 2, pp. 197–230.
13. Dunn, K.J.; Bundy, F.P. Materials and techniques for pressure calibration by resistance-jump transitions up to 500 kilobars. *Rev. Sci. Instrum.* **1978**, *49*, 365–370. [CrossRef] [PubMed]
14. Morishima, H.; Kato, T.; Suto, M.; Ohtani, E.; Urakawa, S.; Utsumi, W.; Shimomura, O.; Kikegawa, T. The phase boundary between α- and β-Mg_2SiO_4 determined by in situ X-ray observation. *Science* **1994**, *265*, 1202–1203. [CrossRef] [PubMed]
15. Suzuki, A.; Ohtani, E.; Morishima, H.; Kubo, T.; Kanbe, Y.; Kondo, T. In situ determination of the phase boundary between wadsleyite and ringwoodite in Mg_2SiO_4. *Geophys. Res. Lett.* **2000**, *27*, 803–806. [CrossRef]
16. Fei, Y.; Van Orman, J.; Li, J.; Van Westrenen, W.; Sanloup, C.; Minarik, W.; Hirose, K.; Komabayashi, T. Experimentally determined post-spinel transformation boundary in Mg_2SiO_4 using MgO as an internal pressure standard and its geophysical implications. *J. Geophys. Res.* **2004**, *109*. [CrossRef]
17. Linton, J.A.; Fei, Y.; Navrotsky, A. The $MgTiO_3$-$FeTiO_3$ join at high pressure and temperature. *Am. Mineral.* **1999**, *84*, 1595–1603. [CrossRef]
18. El Goresy, A.; Chen, M.; Dubrovinsky, L.; Gillet, P.; Graup, G. An ultradense polymorph of rutile with seven-coordinated titanium from the Ries Crater. *Science* **2001**, *293*, 1467–1470. [CrossRef] [PubMed]
19. El Goresy, A.; Dubrovinsky, L.; Gillet, P.; Graup, G.; Chen, M. Akaogiite: An ultra-dense polymorph of TiO_2 with the baddeleyite-type structure, in shocked garnet gneiss from the Ries Crater, Germany. *Am. Mineral.* **2010**, *95*, 892–895. [CrossRef]
20. Akaogi, M.; Kusaba, K.; Susaki, J.; Yagi, T.; Matsui, M.; Kikegawa, T.; Yusa, H.; Ito, E. High-pressure high -temperature stability of αPbO_2-type TiO_2 and $MgSiO_3$ majorite: Calorimetric and in situ x-ray diffraction studies. In *High-Pressure Research: Application to Earth and Planetary Sciences*; Syono, Y., Manghnani, M.H., Eds.; Am. Geophys. Union: Washington, DC, USA, 1992; pp. 447–455.
21. Tang, J.; Endo, S. P-T boundary of α-PbO_2 type and baddeleyite type high-pressure phases of titanium dioxide. *J. Am. Ceram. Soc.* **1993**, *76*, 796–798. [CrossRef]
22. Kojitani, H.; Yamazaki, M.; Kojima, M.; Inaguma, Y.; Mori, D.; Akaogi, M. Thermodynamic investigation of the phase equilibrium boundary between TiO_2 rutile and its α-PbO_2-type high-pressure polymorph. *Phys. Chem. Min.* **2018**, *45*, 963–980. [CrossRef]
23. Dubrovinskaia, N.A.; Dubrovinsky, L.S.; Ahuja, R.; Prakapenka, V.B.; Dmitriev, V.; Weber, H.P.; Osorio-Guillen, J.M.; Johansson, B. Experimental and theoretical identification of a new high-pressure TiO_2 polymorph. *Phys. Rev. Lett.* **2001**, *87*, 275501. [CrossRef]
24. Al-Khatatbeh, Y.; Lee, K.K.M.; Kiefer, B. High-pressure behavior of TiO_2 as determined by experiment and theory. *Phys. Rev.* **2009**, *B79*, 134114. [CrossRef]
25. Akaogi, M.; Arai, S.; Abe, K.; Kojitani, H. Unpublished work. 2018.
26. Akaogi, M.; Abe, K.; Yusa, H.; Ishii, T.; Tajima, T.; Kojitani, H.; Mori, D.; Inaguma, Y. High-pressure high-temperature phase relations in $FeTiO_3$ up to 35 GPa and 1600 °C. *Phys. Chem. Min.* **2017**, *44*, 63–73. [CrossRef]
27. Leinenweber, K.; Utsumi, W.; Tsuchida, Y.; Yagi, T.; Kurita, K. Unquenchable high-pressure perovskite polymorphs of $MnSnO_3$ and $FeTiO_3$. *Phys. Chem. Min.* **1991**, *18*, 244–250. [CrossRef]
28. Dubrovinsky, L.; El Goresy, A.; Gillet, P.; Wu, X.; Simionivici, A. A novel natural shock-induced high- pressure polymorph of $FeTiO_3$ ilmenite with the Li-niobate structure from the Ries crater, Germany. *Meteorit. Planet. Sci.* **2009**, *44*, A64.
29. Xie, X.; Gu, X.; Yang, H.; Chen, M.; Li, K. Wangdaodeite. IMA 2016-007. CNMNC Newsletter, No. 31. *Mineral. Mag.* **2016**, *80*, 691–697.
30. Nishio-Hamane, D.; Zhang, M.; Yagi, T.; Ma, Y. High-pressure and high-temperature phase transitions in $FeTiO_3$ and a new dense $FeTi_3O_7$ structure. *Am. Mineral.* **2012**, *97*, 568–572. [CrossRef]

31. Hazen, R.M. Effects of temperature and pressure on the cell dimension and X-ray temperature factors of periclase. *Am Mineral.* **1976**, *66*, 266–271.
32. Linton, J.A.; Fei, Y.; Navrotsky, A. Complete Fe-Mg solid solution in lithium niobate and perovskite structures in titanates at high pressures and temperatures. *Am. Mineral.* **1997**, *82*, 639–642. [CrossRef]
33. Nishio-Hamane, D.; Shimizu, A.; Nakahira, R.; Niwa, K.; Sano-Furukawa, A.; Okada, T.; Yagi, T.; Kikegawa, T. The stability and equation of state for the cotunnite phase of TiO_2 up to 70 GPa. *Phys. Chem. Min.* **2010**, *37*, 129–136. [CrossRef]
34. McCammon, C.A. Effect of pressure on the composition of the lower mantle end member Fe_xO. *Science* **1993**, *259*, 66–68. [CrossRef]
35. Ming, L.C.; Kim, Y.H.; Uchida, T.; Wang, Y.; Rivers, M. In situ X-ray diffraction study of phase transitions of $FeTiO_3$ at high pressures and temperatures using a large-volume press and synchrotron radiation. *Am. Mineral.* **2006**, *91*, 120–126. [CrossRef]
36. Kennedy, C.S.; Kennedy, G.C. The equilibrium boundary between graphite and diamond. *J. Geophys. Res.* **1976**, *81*, 2467–2470. [CrossRef]
37. Yamanaka, T.; Uchida, A.; Nakamoto, Y. Structural transition of post-spinel phases $CaMn_2O_4$, $CaFe_2O_4$, and $CaTi_2O_4$ under high pressure up to 80 GPa. *Am. Mineral.* **2008**, *93*, 1874–1881. [CrossRef]
38. Ishii, T.; Kojitani, H.; Tsukamoto, S.; Fujino, K.; Mori, D.; Inaguma, Y.; Tsujino, N.; Yoshino, T.; Yamazaki, D.; Higo, Y. High-pressure phase transitions in $FeCr_2O_4$ and structure analysis of new post-spinel $FeCr_2O_4$ and $Fe_2Cr_2O_5$ phases with meteoritical and petrological implications. *Am. Mineral.* **2014**, *99*, 1788–1797. [CrossRef]
39. Shannon, R.D. Revised effective ionic radii and systematic studies of interatomic distances in halides and chalcogenides. *Acta Cryst.* **1976**, *A32*, 751–767. [CrossRef]
40. Oganov, A.R. Thermodynamics, phase transitions, equations of state, and elasticity of minerals at high pressures and temperatures. In *Mineral Physics*, 2nd ed.; Price, G.D., Ed.; Elsevier: Amsterdam, The Netherlands, 2015; Volume 2, pp. 179–201.
41. Momma, K.; Izumi, F. VESTA 3 for three-dimensional visualization of crystal, volumetric and morphology data. *J. Appl. Crystallogr.* **2011**, *44*, 1272–1276. [CrossRef]
42. Akaogi, M.; Tanaka, A.; Kobayashi, M.; Fukushima, N.; Suzuki, T. High-pressure transformations in $NaAlSiO_4$ and thermodynamic properties of jadeite, nepheline, and calcium ferrite-type phase. *Phys. Earth Planet. Inter.* **2002**, *130*, 49–58. [CrossRef]
43. Ono, A.; Akaogi, M.; Kojitani, H.; Yamashita, K.; Kobayashi, M. High-pressure phase relations and thermodynamic properties of hexagonal aluminous phase and calcium-ferrite phase in the systems $NaAlSiO_4$-$MgAl_2O_4$ and $CaAl_2O_4$-$MgAl_2O_4$. *Phys. Earth Planet. Inter.* **2009**, *174*, 39–49. [CrossRef]
44. Woodland, A.B.; Frost, D.J.; Trots, D.M.; Klimm, K.; Mezouar, M. In situ observation of the breakdown of magnetite (Fe_3O_4) to Fe_4O_5 and hematite at high pressures and temperatures. *Am. Mineral.* **2012**, *97*, 1808–1811. [CrossRef]
45. O'Neill, H.S.C.; Annersten, H.; Virgo, D. The temperature dependence of the cation distribution in magnesioferrite ($MgFe_2O_4$) from powder XRD structural refinements and Mössbauer spectroscopy. *Am. Mineral.* **1992**, *77*, 725–740.
46. Syono, Y.; Fukai, Y.; Ishikawa, Y. Anomalous elastic properties of Fe_2TiO_4. *J. Phys. Soc. Jpn.* **1971**, *31*, 471–476. [CrossRef]
47. Xiong, Z.; Liu, X.; Shieh, S.R.; Wang, F.; Wu, X.; Hong, X.; Shi, Y. Equation of state of a synthetic ulvöspinel, $(Fe_{1.94}Ti_{0.03})Ti_{1.00}O_{4.00}$, at ambient temperature. *Phys. Chem. Miner.* **2015**, *42*, 171–177. [CrossRef]

minerals

MDPI

Article

Validation of Digital Rock Physics Algorithms

Rongrong Lin [1,*] and Leon Thomsen [1,2]

[1] Department of Earth and Atmospheric Sciences, University of Houston, Houston, TX 77204, USA; lathomsen@uh.edu
[2] Delta Geo LLC, Houston, TX 77002, USA
* Correspondence: rlin5@uh.edu

Received: 23 September 2019; Accepted: 29 October 2019; Published: 31 October 2019

Abstract: With a detailed microscopic image of a rock sample, one can determine the corresponding 3-D grain geometry, forming a basis to calculate the elastic properties numerically. The issues which arise in such a calculation include those associated with image resolution, the registration of the digital numerical grid with the digital image, and grain anisotropy. Further, there is a need to validate the numerical calculation via experiment or theory. Because of the geometrical complexity of the rock, the best theoretical test employs the Hashin–Shtrikman result that, for an aggregate of two isotropic components with equal shear moduli, the bulk modulus is uniquely determined, independent of the micro-geometry. Similarly, for an aggregate of two isotropic components with a certain combination of elastic moduli defined herein, the Hashin–Shtrikman formulae give a unique result for the shear modulus, independent of the micro-geometry. For a porous, saturated rock, the solid incompressibility may be calculated via an "unjacketed" test, independent of the micro-geometry. Any numerical algorithm proposed for digital rock physics computation should be validated by successfully confirming these theoretical predictions. Using these tests, we validate a previously published staggered-grid finite difference damped time-stepping algorithm to calculate the static properties of digital rock models.

Keywords: digital rock physics; Hashin–Shtrikman; finite-difference; staggered grid

1. Introduction

It is now technically possible to construct a high-resolution 3-D image of a real rock sample, and to use this to compute the elastic properties of that rock sample, numerically (e.g., in [1]). The present work examines some of the issues which can arise in this program. Some general remarks on common issues are followed by defining procedures for validating any proposed numerical algorithm, and an application to one particular algorithm.

1.1. Common Issues

Any digital image of a rock sample is defined on a discrete imaging grid, and the subsequent numerical calculation is also conducted on a discrete numerical grid. Typically, the imaging grid is Cartesian, and very fine. The numerical grid is also typically Cartesian, and may be coarser, to save computation effort. However, the numerical resolution should be sufficiently fine that the contents of each numerical cell may be treated as uniform.

Furthermore, it is known that very tiny geometric features, like cracks and grain-contacts, can have a disproportionate effect on the overall elastic response, depending upon the micro-geometry. For example, the elastic effect of a crack depends, among other factors, on its ratio of thickness/length [2]. It is clearly important that the imaging grid, and the numerical grid, have sufficient spatial resolution to fully describe such features, in the sense described above [3].

Then, elastic properties are assigned to the grains. Of course, this means that, in addition to grain geometry, the mineralogy for each grain in the sample image must be determined, this may or may not be straightforward, depending on the sample.

Further, most minerals are anisotropic, and this creates another fundamental difficulty. For some minerals, the internal crystalline symmetry influences the external shape; for others not so much. For example, clay minerals often have the external shape of platelets. For such minerals, assigning the appropriate stiffness tensor and orientation to the grains is straight-forward.

For other minerals, the external shape of a grain may not indicate the orientation of its crystalline axes. In such a case, it may not be clear (without X-ray diffraction studies) how the anisotropic axes of any particular grain are oriented. In general, in some places within the rock, stiff axes will be juxtaposed against soft faces, but not in others. Therefore, the rock will be effectively heterogeneous, in the elastic sense, even if it were macro-homogeneous and mono-mineralic. This may be called "orientational heterogeneity", as opposed to "compositional heterogeneity"; most real rocks have both.

In an energetic depositional environment, the orientations of the grains will be partially or fully randomized, and so it is tempting to assume that the orientation of the anisotropic axes of such minerals is random. It may be that, if the micro-anisotropy is indeed randomly oriented, the anisotropic stress tensor in each grain can be replaced by its isotropic average, but this has not been demonstrated, either theoretically or numerically.

Any digital rock physics calculation should address these issues of resolution and anisotropy, which may affect its accuracy, even when the micro-geometry is realistic.

1.2. Measures of Isotropic Elasticity

In a physical test on a real rock, the stress is applied externally, and the deformation is measured also on the exterior surface. Hence the incompressibility can be measured from the surface average of interior compressional stress/strain for a cubic sample with edge L as:

$$K_{Surf} = -p\left\langle \frac{normal\ displacement}{L} \right\rangle_{surf}^{-1} \tag{1}$$

where p is the applied pressure, and $\langle\ \rangle_{surf}$ indicates a surface average. Similarly, the shear modulus can be measured from the surface average of interior shear stress/strain as:

$$\mu_{Surf} = \tau_{shear}\left\langle \frac{shear\ displacement}{L} \right\rangle_{surf}^{-1} \tag{2}$$

where τ_{shear} is the applied shear stress. These same averages can be calculated numerically from a digital rock image, recognizing the issues discussed above.

However, from that same image, the volume average of interior incompressibility may also be calculated as:

$$K_{Vol} = -\frac{\langle pressure \rangle_{vol}}{\langle dilatation \rangle_{vol}} \tag{3}$$

where $\langle\ \rangle_{vol}$ indicates an average over the volume. Similarly, the volume average of interior shear modulus may be calculated as:

$$\mu_{Vol} \equiv \frac{\langle shear\ stress \rangle_{vol}}{\langle shear\ strain \rangle_{vol}} \tag{4}$$

These volume averages are different in principle from the surface averages, and may be numerically different, depending on the accuracy of the approximations discussed above.

2. Methods

2.1. Validating A Digital Algorithm for Non-Porous Rocks

In view of the several approximations mentioned above, and others not mentioned here (since they are dependent on particular techniques), it is important to validate any numerical algorithm against a theoretical solution. For application to certain elastic problems with simple geometry (e.g., a borehole in a uniform stressed medium), theoretical solutions are available. However, for a realistic rock-physics problem, the complexity of geometry precludes any general theoretical solution.

It is possible to construct a rock physics model with an idealized micro-geometry for which a theoretical solution exists (e.g., for ellipsoidal cracks [3]). However, this is not useful for a realistic rock. The overall elasticity of a realistic rock is, in general, not theoretically calculable because the microgeometry is so complicated. Instead, theoretical upper and lower bounds on the bulk elastic moduli have been proposed. Those derived by Hill [4] are not absolute bounds, since they depend upon a micro-geometric assumption (non-correlation of stress and strain, at the micro level), which might not be true.

However, strict bounds on the overall elasticity were derived by Hashin and Shtrikman [5], independent of these issues, and furthermore are closer together than are the Hill "bounds". For solid composites of two isotropic constituents (with bulk moduli K_i and shear moduli μ_i and volume fractions f_i), the lower (−) and upper (+) H–S (Hashin - Shtrikman) bounds for the bulk modulus K and the shear modulus μ are

$$K_- \equiv K_1 + \frac{f_2}{\frac{1}{K_2-K_1} + \frac{f_1}{K_1+4G_1/3}} \leq K \leq K_2 + \frac{f_1}{\frac{1}{K_1-K_2} + \frac{f_2}{K_2+4G_2/3}} \equiv K_+ \tag{5}$$

$$\mu_- \equiv \mu_1 + \frac{f_2}{\frac{1}{\mu_2-\mu_1} + \frac{f_1}{H_1}} \leq \mu \leq \mu_2 + \frac{f_1}{\frac{1}{\mu_1-\mu_2} + \frac{f_2}{H_2}} \equiv \mu_+ \tag{6}$$

In Equation (6), the combination of moduli

$$H_i = \frac{5\mu_i(K_i + 4\mu_i/3)}{2(K_i + 2\mu_i)} \tag{7}$$

for each constituent appears in the bounds for the shear modulus. Here, it is assumed (following [5]) that $K_1 \leq K_2$ and also that $\mu_1 \leq \mu_2$; this correlation is common. If it should happen that $K_1 \geq K_2$ while $\mu_1 \leq \mu_2$, then the roles of K_+ and K_-, defined in Equation (5), are reversed. If it should happen that $K_1 \leq K_2$ while $\mu_1 \geq \mu_2$, then the roles of μ_+ and μ_-, defined in Equation (6), are reversed.

The bounds on the bulk modulus (Equation (5)) may be combined to show the difference:

$$K_+ - K_- = \frac{4f_1f_2(\Delta K)^2}{3\left(K_1 + 4\mu_1/3 + f_1\Delta K\right)\left(K_2 + 4\mu_2/3 - f_2\Delta K\right)}\Delta\mu \tag{8}$$

where $\Delta K \equiv K_2 - K_1$ and $\Delta\mu \equiv \mu_2 - \mu_1$. Hence, the bounds on $\boldsymbol{K}$ coincide exactly in the special case where the two minerals have identical shear modulus ($\Delta\mu = 0$), in which case the bulk modulus is given exactly by

$$K_{HS} \equiv K_-^* = K_+^* = \frac{(4f_1K_1\mu + 4f_2K_2\mu + 3K_2K_1)}{(3f_1K_2 + 3f_2K_1 + 4\mu)} \tag{9}$$

(where μ is the common shear modulus, of the two constituents and of the aggregate) for any random microgeometry of the two isotropic constituents. (The special case of Equation (9) with $\mu = 0$ yields the Reuss average [4] bulk modulus, the correct result for such a suspension.)

The generalization of Equation (5) for composites of more than two isotropic constituents was given by Hashin and Shtrikman [5]. For such composites, the upper and lower bounds on K coincide, yielding a unique result, if the shear moduli are equal for all constituents.

Of course, few if any physical aggregates have constituents with identical shear moduli. However, although Equation (9) is not a useful result for physical aggregates, it is a very useful result for numerical analyses, since it poses a necessary criterion for the validity of any digital rock physics numerical algorithm: any numerical algorithm which is proposed to calculate the elasticity of a realistic 3D geometry should first be validated by verifying that it satisfies Equation (9), when equal isotropic shear moduli are assigned to the grains; this may be called the "K-test".

Similarly, the bounds on the shear modulus in Equation (6) may be combined to show the difference:

$$\mu_+^* - \mu_-^* = \frac{f_1 f_2 (\Delta\mu)^2}{(H_1 + f_1\Delta\mu)(H_2 - f_2\Delta\mu)}[H_2 - H_1 - \Delta\mu] \tag{10}$$

Hence, the bounds on the shear moduli coincide exactly, in the special case where

$$H_2 - \mu_2 = H_1 - \mu_1 \tag{11}$$

For composites of more than two isotropic constituents, the upper and lower bounds on μ coincide, yielding a unique result, if the condition of Equation (11) holds for all constituents.

The condition of Equation (11) serves as a calculation of any one of the four constituent moduli, when the other three are specified. For example, it implies that

$$K_2 = \frac{4\mu_2(H_1 - \mu_1 - 2\mu_2/3)}{(3\mu_2 - 2H_1 + 2\mu_1))} \tag{12}$$

For composites of two constituents which confirm Equation (11), the shear modulus is given exactly by

$$\mu_{HS} = \mu_- = \mu_+ = \frac{f_1\mu_1 H_2 + f_2\mu_2 H_1}{f_1 H_1 + f_2 H_2} \tag{13}$$

independent of the microgeometry of the two isotropic constituents.

Of course, few if any physical aggregates have constituents which confirm Equation (11). However, although Equations (12) and (13) are not a useful result for physical aggregates, they are a very useful result for numerical analyses, since they pose another necessary criterion for the validity of any digital rock physics numerical algorithm: any numerical algorithm which is proposed to calculate the elasticity of a realistic 3D geometry should first be validated by verifying that it satisfies Equation (13) when the constituent moduli are constrained by Equation (11); this may be called the "μ-test".

2.2. *Validating A Digital Algorithm for Porous, Saturated Rocks*

Modeling a rock with fluid-filled porosity involves additional considerations. The theory of poro-elasticity (e.g., in [6,7]) requires consideration of the pore fluid properties, the frequency of excitation, and the hydraulic condition (e.g., open or closed) of the rock, all of which are outside of the present scope. However, it also requires consideration of K_S, the average solid modulus of the grains of the rock; this topic can be addressed here.

Love [8] (see especially Sections 121 and 123 (iii, iv)) proved that, for any shape of a homogeneous isotropic solid, its elastic response to a uniform increase of pore pressure (on all of its surfaces, internal and external) is independent of that shape, with each linear dimension decreasing proportionally, so that its shape is preserved. Hence the bulk modulus of the solid shape is the intrinsic bulk modulus of that solid. The proof is valid for any homogeneous isotropic shape, if the pore space is fully connected hydraulically, so that the fluid pressure is uniform.

This theorem may be applied to a representative volume element of a porous rock in a (hydraulically open) "unjacketed compression" experiment, wherein the increase in external pressure on a rock is balanced by an equal increase in internal pore pressure. If the porosity is fully connected on the time scale of the compression, the pore pressure will be uniform throughout, regardless of the complexity of the geometry. In such a test, the solid is compressed on all sides with the same additional pressure, and Love's theorem applies.

A straightforward extension of Love's proof to the case of a *heterogeneous* isotropic solid then concludes that, in an unjacketed test on such a rock with fully connected porosity, the solid modulus is approximately the intrinsic average modulus of the grains, as calculated above.

2.3. A Particular Algorithm

The tests proposed above are the most general conclusions of the present work. Successful passage of these tests should be demonstrated for any digital rock physics project proposed in the future. A failure in any test might be due to a shortcoming in the numerical algorithm applied, or to any of the other issues mentioned in Section 1.1, which may be different in each project. Below, we apply these tests to a particular algorithm, in a context where none of these other issues occur.

A staggered-grid finite-difference damped time-stepping algorithm has been proposed by Lin et al. [9,10] ("LFZ") for this digital rock physics problem. The LFZ algorithm produces a solution to the static elastic problem by solving the corresponding elasto-dynamic wave equation, with attenuation artificially added to damp out the kinetic energy (while keeping the strain energy) so that the dynamic solution converges to the static solution. The advantage of the algorithm is its computational efficiency: the time-marching scheme does not require the inversion of a large matrix, as in the static approach. The computational effort in such an inversion can be so significant that approximations permitting a 2D inversion of a 3D model have been proposed [11]. Therefore, the computational efficiency of the LFZ algorithm offers a significant advantage.

The method is based upon the first-order hyperbolic system of Virieux [12]: the equation of motion for continua:

$$\rho(\vec{x})\frac{\partial v_i(\vec{x},t)}{\partial t} = \frac{\partial \tau_{ij}(\vec{x},t)}{\partial x_j} \tag{14}$$

and the linear isotropic constitutive relation:

$$\frac{\partial \tau_{ij}}{\partial t} = \lambda(\vec{x})\delta_{ij}\frac{\partial v_k}{\partial x_k} + \mu\,(\vec{x})\left(\frac{\partial v_i}{\partial x_j} + \frac{\partial v_j}{\partial x_i}\right) \tag{15}$$

where v_i is the particle velocity vector (a function of space and time), ρ is the density, $\vec{x}$ is the position vector, and $\lambda = K - 2\mu/3$ is the Lame parameter. However, LFZ augmented the equation of motion (14) with a velocity-damping term:

$$\rho(\vec{x})\frac{\partial v_i(\vec{x},t)}{\partial t} = \frac{\partial \tau_{ij}(\vec{x},t)}{\partial x_j} - d\rho(\vec{x})v_i(\vec{x},t) \tag{16}$$

where the constant artificial damping coefficient d was chosen to efficiently damp out the kinetic energy, for the study of static elastic problems. The physical realism of the damping is immaterial; it is just an artifice to attenuate the kinetic energy. With the inertial terms damped out, the resulting deformation is independent of the material density.

These equations incorporate local heterogeneity of the elastic moduli and density explicitly (hence no need to consider local boundary conditions), although the model analyzed herein (described below) has piecewise uniform elasticity within each numerical cell. The dynamic algorithm (Equations (15) and (16)) is implemented on a staggered Cartesian numerical grid, similar to that proposed by Virieux [12], shown in Figure 1.

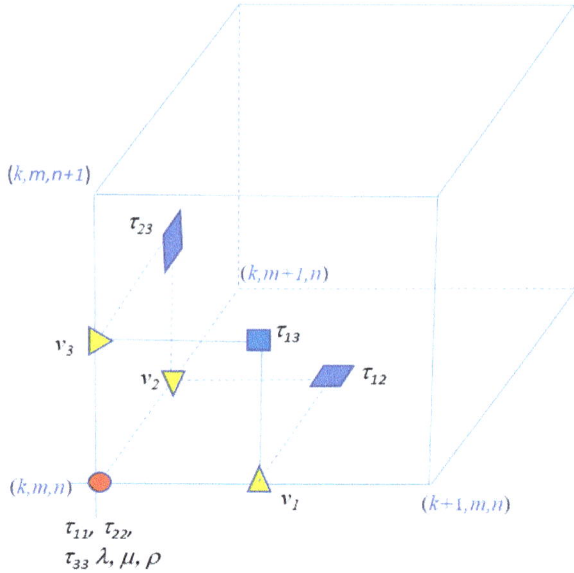

Figure 1. The staggered grid, showing various locations of evaluations.

The derivatives are discretized by centered finite differences, with the various quantities evaluated at different positions on the staggered grid, as shown in the Figure 1. The stress components are evaluated at t time units; the velocity components are evaluated at (t + 1/2) time units. As the computation approaches its asymptotic (static) limit, these time differences become irrelevant. This pattern of staggering arises naturally from the centering of the finite derivatives.

Where a grid node separates domains of different elasticity, the elastic moduli and density at that node are assigned appropriate average values, depending on the properties of its eight nearest neighbors in three dimensions. This procedure partially smooths the spatial variation of physical properties, with an accuracy which depends upon the resolution of the grid.

3. Results

3.1. Application of the LFZ Algorithm to Non-Porous Rocks

The Hashin–Shtrikman tests are here applied to the LFZ algorithm. In order to avoid problems caused by mismatch between the numerical grid and the model grid, the calculation was performed for the model shown in Figure 2, with cubic grains. Although this is not a realistic model, with this geometry, any inaccuracies in the calculation can be attributed to other features of the algorithm. To ensure uniform application of external stress, the model shown was surrounded by a uniform jacket (four cells thick) with the properties of water (for applied pressure) or of an average solid (for applied shear stress).

In the model, the cells are randomly assigned to one or another of two isotropic constituents. For the H–S "K-test" in Section 2.1, the elastic parameters are set as K_1 = 13.564 GPa, K_2 = 8.564 GPa, $\mu_1 = \mu_2$ = 4.586 GPa, $f_1 = f_2$ = 0.5. For this model, the theoretical H–S value from Equation (9) is K_{HS} = 10.700 GPa. The surface-average incompressibility from Equation (1) is K_{Surf} =10.689 GPa (an error of −0.1%); the volume-average incompressibility from Equation (3) is K_{Vol} = 10.685 GPa (an error of −0.1%).

The calculation was repeated, with the same model, but with a numerical grid much finer (200 × 200 × 200 cells), so that each cubical grain was represented by 64 cells, instead of by 1. The surface-average incompressibility from Equation (1) is K_{Surf} = 10.691 GPa (an error of −0.1%).

The volume-average incompressibility from Equation (2) is K_{Vol} = 10.690 GPa (an error of −0.1%). In this instance (with the numerical grid aligned with the model), the increased resolution made little difference in accuracy.

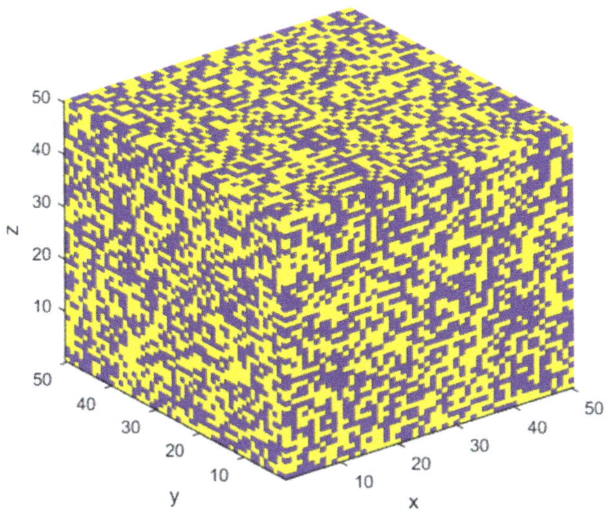

Figure 2. The model has cubic grains of two isotropic constituents (50/50 mix), randomly distributed. The boundaries of the grains are exactly aligned with the Cartesian numerical grid (50 × 50 × 50 cells), so that the contents of each cell are uniform, within that cell.

Then, with the same model geometry, the cells are assigned, for the H–S "μ-test", the elastic parameters are K_1 = 8.564 GPa, μ_1 = 3.236, μ_2 = 3.886 GPa, K_2 = 4.012 GPa (from Equation (12)), $f_1 = f_2$ = 0.5. For this model, the theoretical H–S value from Equation (13) is μ_{HS} = 3.546 GPa. The surface-average shear modulus from Equation (2) is μ_{Surf} = 3.544 GPa (precise to three significant figures). The volume-average shear modulus from Equation (4) is μ_{Vol} = 3.546 GPa (precise to four significant figures). The incompressibility is not uniquely determined with these parameters.

3.2. Application of the LFZ Algorithm to Porous, Saturated Rocks

To numerically test the procedures of Section 2.2, the algorithm of LFZ was applied to the model of Figure 3. The previous model (Figure 2) is penetrated by straight interconnected channels of water (K_{water} = 2.25 GPa) as depicted; the porosity is about 35%. The figure shows the inhomogeneous case, with solid matrix composed 50/50 of the two minerals specified in the "K-test" above. Again, this model is not realistic, however, with this geometry, any inaccuracies in the calculation can be attributed to other features of the algorithm. Since a shell of water surrounds this model completely, and pressure is applied to the exterior of this fluid shell, the solid grains are exposed on all sides to the same pressure.

For a model like that of Figure 3, but with homogeneous solid (with solid incompressibility K_2 = K_1 = 13.564 GPa), the surface-average incompressibility (simulating an unjacketed experiment) from Equation (1) is K_{Surf} = 13.543 GPa (an error of −0.2%). The calculated solid-volume-average incompressibility from Equation (2) is K_{Vol} = 13.564 GPa (precise to five significant figures).

For the inhomogeneous model of Figure 3, the surface-average incompressibility (simulating an unjacketed experiment from Equation (1)) of the solid portion only is K_{Surf} = 10.642 GPa, (an error of −0.5%). The volume-average incompressibility from Equation (2) of the solid portion only is K_{Vol} = 10.650 GPa, (an error of −0.5%). These differences are a measure of the accuracy of the extension of Love's theorem to the case of inhomogeneous solid, as well as the accuracy of the LFZ algorithm.

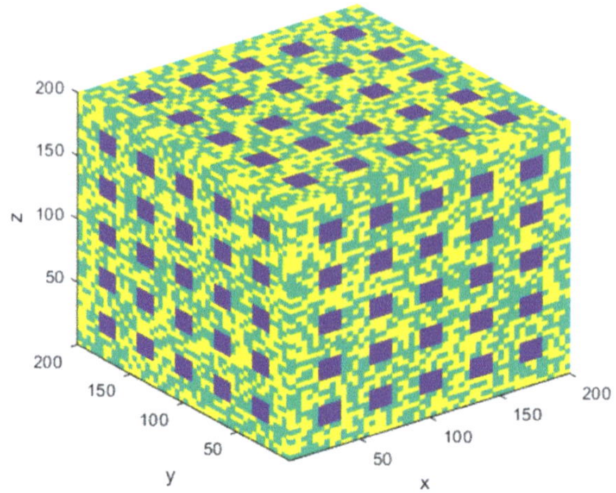

Figure 3. The previous heterogeneous model, penetrated with straight channels of water, all aligned with the numerical grid (200 × 200 × 200 cells).

There is no equivalent process for determining the average shear modulus μ_S of the grains of the porous rock. If the H–S "μ-test" were applied to a composite with one constituent fluid ($\mu_1 = 0$), the required K_2 from Equation (13) would be negative (hence un-physical), so the H–S "μ-test" is not applicable.

4. Discussion

We briefly discussed here some of the important issues related to digital rock physics computations. These include issues related to resolution (of the imaging grid and the numerical grid), to registration of the two grids, and to macro- and micro-anisotropy. A primary issue is the need to validate the numerical algorithm with a theoretical solution. For the complicated geometry of a digital rock, suitable quantitative validation procedures for any numerical computation of isotropic solid rock elasticity were presented here.

H–S theory [5] provides rigorous bounds for the bulk and shear moduli of an isotropic aggregate of isotropic components. It is well known that for two-constituent aggregates, wherein both constituents have the same shear moduli, the bounds on the bulk modulus of the aggregate coincide, so that its bulk modulus is uniquely determined, regardless of the internal geometry. Although this case is vanishingly rare in real rock, this "K-test" provides a useful validation procedure for any numerical calculation of digital solid rock physics.

A second consequence of the H–S bounds [5] is that if the moduli of the two constituents are constrained by the present Equation (11) above, the bounds on the shear modulus of the aggregate coincide, so that its shear modulus is uniquely determined, regardless of the internal geometry. Although this case is vanishingly rare in real rock, this "μ-test" provides a second useful validation procedure for any numerical calculation of digital solid rock physics.

For a porous rock with homogeneous solid, the bulk modulus may be measured (and calculated) under the condition of equal external pressure and pore pressure. A theorem by Love [8] proves that, in this ("unjacketed") condition, the bulk modulus of the aggregate is that of the solid itself. By extension, for a rock with inhomogeneous solid phase, the bulk modulus is approximately the average modulus of the solid itself. This provides a useful validation for any numerical calculation of digital rock physics for porous rocks.

These validation tests were applied to the LFZ algorithm [9,10]. The algorithm calculates the static modulus as the asymptotic limit of the corresponding elasto-dynamic equations, with artificial damping, thus avoiding the necessity to invert a large numerical matrix. The algorithm smooths the discontinuities in material properties at each grain boundary; the consequent errors are smaller with higher resolution of the numerical grid.

For the non-porous inhomogeneous solid, the algorithm reproduces the H–S results within a fraction of a percent. For the porous rock, with either homogeneous or inhomogeneous solid, the algorithm reproduces the Love result for the solid, within a fraction of a percent. We conclude that the LFZ algorithm is well-validated for this model, by all these tests. These inaccuracies would presumably be larger for a realistic model, with grains not conforming exactly to the numerical grid, but such errors would be different for each model and each algorithm, so that no generalizations are possible for such errors.

More generally, we recommend that these tests should be applied to any numerical algorithm which may be proposed to calculate the elasticity of any rock model. Of course, these tests, although necessary, are insufficient for a complete validation, as they do not consider all potential sources of inaccuracy. Nonetheless, only after an algorithm has passed these necessary validation tests should it be applied to a case with realistic grain parameters.

Author Contributions: Conceptualization, R.L. and L.T.; Methodology, R.L.; Software, R.L.; Validation, R.L.; Formal Analysis, L.T.; Writing—Original Draft Preparation, L.T.; Writing—Review and Editing, R.L.; Visualization, R.L.; Supervision, L.T.

Funding: This study was supported by the University of Houston, and Delta Geo, LLC.

Acknowledgments: Author R.L. appreciates the UH Xfrac group for providing computing hardware. Author L.T. received his first understanding of mineral physics from Orson Anderson, in 1967.

Conflicts of Interest: The authors declare no conflicts of interest.

References

1. Madonna, C.J.; Almqvist, B.S.G.; Saenger, E.H. Digital rock physics: Numerical prediction of pressure-dependent ultrasonic velocities using micro-CT imaging. *Geophys J. Int.* **2012**, *189*, 1475–1482. [CrossRef]
2. Budiansky, B.; O'Connell, R.J. Elastic moduli of a cracked solid. *Int. J. Solids Struct.* **1976**, *12*, 81–97. [CrossRef]
3. Arns, C.H.; Knackstedt, M.A.; Pinczewski, W.V.; Garboczi, E.J. Computation of linear elastic properties from microtomographic images: Methodology and agreement between theory and experiment. *J. Geoph.* **2002**, *67*, 1396–1405. [CrossRef]
4. Hill, R. The elastic behavior of a crystalline aggregate. *Proc. Phys. Soc.* **1952**, *65*, 439. [CrossRef]
5. Hashin, Z.; Shtrikman, S. Note on a variational approach to the theory of composite elastic materials. *J. Frankl. Inst.* **1961**, *271*, 336–341. [CrossRef]
6. Biot, M.A. General Theory of three-dimensional consolidation. *J. Appl. Phys.* **1941**, *12*, 155–164. [CrossRef]
7. Brown, R.J.S.; Korringa, J. On the dependence of the elastic properties of a porous rock on the compressibility of the pore fluid. *Geophysics* **1975**, *40*, 608–616. [CrossRef]
8. Love, A.E.H. *A Treatise on the Mathematical Theory of Elasticity*, 4th ed.; Cambridge University Press: Cambrige, UK, 2013.
9. Lin, R.; Fang, X.; Zheng, Y. A quasistatic finite-difference method for calculating the effective elastic moduli of digital rock models. In *SEG Technical Program Expanded Abstracts 2018*; Society of Exploration Geophysicists: Tulsa, OK, USA, 2018; pp. 3703–3707.
10. Lin, R.; Fang, X.; Gan, Y.; Zheng, Y. A damped dynamic finite difference approach for modeling static stress-strain fields. *Pure Appl. Geophys.* **2019**, *176*, 1–15. [CrossRef]
11. Saxena, N.; Mavko, G. Estimating elastic moduli of rocks from thin sections: Digital rock study of 3D properties from 2D images. *Comput. Geosci.* **2016**, *88*, 9–21. [CrossRef]
12. Virieux, J. *P-SV* wave propagation in heterogeneous media: Velocity-stress finite difference method. *Geophysics* **1986**, *51*, 889–901. [CrossRef]

minerals

MDPI

Article

Equations of State of Simple Solids (Including Pb, NaCl and LiF) Compressed in Helium or Neon in the Mbar Range

Agnès Dewaele

Commissariat à l'Energie Atomique, Direction des Applications Militaires, DIF, F-91297 Arpajon, France; agnes.dewaele@cea.fr

Received: 14 October 2019; Accepted: 1 November 2019; Published: 5 November 2019

Abstract: The equations of state measured under ambient temperature in the Mbar range are reviewed, focusing on experiments using diamond anvils cells with a quasi-hydrostatic pressure transmitting medium (helium or neon) and coupled with X-ray diffraction. Equations of state (EoS) parameters are listed with an unified pressure metrology for all data. This metrology is based on the efforts made in the 2000s to update the ruby luminescence pressure scale, after the collection of original data. To complete this database, unpublished *P-V* data for lead (Pb), sodium chloride (NaCl) and lithium fluoride (LiF) are also provided with the same metrology. Systematic effects of the pressure metrology on the EoS parameters are discussed.

Keywords: diamond anvil cell; lead; sodium chloride; lithium fluoride; high pressure

1. Introduction

Ambient temperature equations of state (EoS) $V(P, T = 300\text{ K})$ describe the evolution of bonding in solids under high compression. In static high-pressure devices such as diamond anvil cells (DACs), the unit cell volume V is measured with X-ray diffraction (XRD) and the pressure P is estimated using a pre-calibrated gauge. In the 2000s, the calibration of the most widely used gauge in DACs, ruby luminescence, has been updated on the basis of new measurements made in quasi-hydrostatic pressure transmitting media [1–5]. It has been suggested that the historical calibration by Mao et al. (hereafter noted as Mao 86) [6] underestimates the pressure by some 8% at 150 GPa, as represented in Figure 1. As a result, the EoS measured under quasi-hydrostatic compression published before, or during the period of ruby metrology updates have to be corrected.

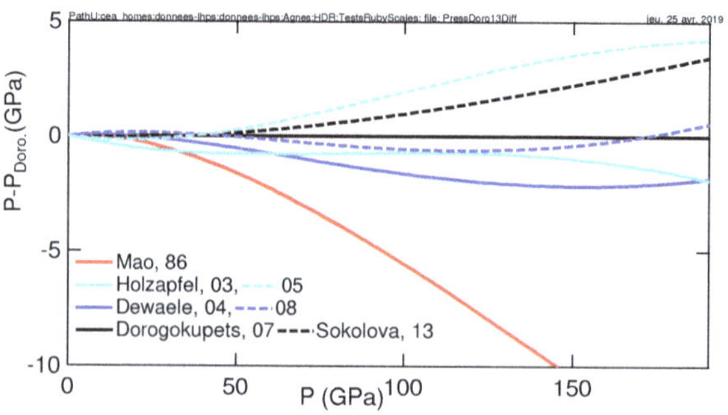

Figure 1. Comparison between calibrations of the ruby luminescence gauge [1–7]. The reference calibration is [4].

I present here EoS parameters for several simple solids (metals-alkali halides-diamond) measured with similar techniques in the Mbar range, between 2003 and 2012, which originally considered various calibrations of the ruby luminescence gauge. Here, a unified ruby pressure metrology (based on [4]) is used. Corrections to published EoS, subsequent to the update of the ruby scale, are listed in Table 1. In addition, *P*-*V* data are provided for some additional compounds: hcp-Pb, NaCl and LiF. NaCl and LiF are standards for static and dynamic compression measurements, and Pb behavior under high compression is typical of post-transition metals, and also used as a standard. These measurements are presented in the third section and compared with literature data. In the last section, the EoS parameters are listed in Table 1 and discussed.

Table 1. Rydberg–Vinet [8] equations of state (EoS) parameters at 300 K for elements and compounds obtained by fitting of the *P*-*V* data. P_{Mao} (P_{Dor}) indicates that the ruby calibration has been taken from [4,6]. W gauge has been calibrated against ruby in [2]. PTM: pressure transmitting medium. *V* in Å^3/at or /formula unit, *P* and K_0 in GPa. The numbers in bold have been fixed during the fit. The number in parenthesis indicate fit error bars (95% confidence level). They are of the same value in the third column as in the second column and are not reported. The reference for the published data is provided in the last column.

Element or Compound	P_{Mao} [6] V_0, K_0, K_0'	P_{Dor} [4] V_0, K_0, K_0'	*P* Domain	PTM	*P* Gauge	Ref.
Au	16.983(20), 166.4(2.0), 5.47(6)	16.986, 163.4, 6.04	0–131	He	ruby	[9]
Pt	15.099(25), 273.4(2.5), 4.83(8)	15.098, 270.8, 5.50	0–95	He	ruby	[2]
Cu	11.810(15), 135.3(1.5), 4.91(6)	11.81, 133.1, 5.38	0–155	He	ruby	[2]
Ta	18.020(18), 197.9(3.7), 3.17(10)	18.019, 196.1, 3.64	0–90	He	ruby	[2]
Al	16.573(19), 76.32(1.5), 4.16(6)	16.584, 74.2, 4.52	0–155	He	ruby	[2]
W	15.862(16), 298.3(4.1), 3.82(11)	15.858, 298.6, 4.37	0–155	He	ruby	[2]
Co	11.077(12), 197.0(3.2), 3.85(20)	11.077, 194.85, 4.36	0–66	He	ruby	[5]
Ag	17.070(16), 100.2(1.6), 5.70(9)	17.088, 96.6, 6.22	0–124	He	ruby	[5]
Mo	15.569(21), 270.3(3.9), 3.34(12)	15.567, 269.3, 3.87	0–124	He	ruby	[5]
Ni	10.954(18), 177.5(2.4), 4.83(9)	10.952, 176, 5.322	0–157	He	ruby	[5]
Zn	15.147(19), 64.3(1.2), 5.30(10)	15.155, 62.2, 5.705	0–157	He	ruby	[5]
Be	8.133(5), 115.2(1.1), 2.94(5)	8.134, 113.4, 3.29	0–95	He	ruby	[10]
Pb-hcp	28.063(55), 71.8(2.1), 4.40(8)	28.058, 70.0, 4.77	13–131	He	ruby	this work
Re	14.737(20), 350.5(8.0), 3.98(17)	14.734, 350.5, 4.62	0–144	He	W	[11]
ϵ-Fe	11.209(50), 164.5(7.9), 4.96(16)	11.177, 168.4, 5.33	17–204	He, Ne	W	[5]
C	5.673(8), 446.8(5.0), 3.01(60)	5.672, 448.9, 3.66	0–151	He, Ne	ruby	[12,13]
KCl-B2	55.98(76), 15.4(2.5), 5.75(15)	56.86, 13.1, 6.21	0–165	He	ruby	[14]
KBr-B2	64.37(81), 14.5(2.5), 5.58(20)	65.29, 12.44, 6.01	0–165	He	ruby	[14]
NaCl-B1	44.90(12), 24.0(8), 5.09(6)	44.93, 23.4, 5.29	0–35	He	ruby	this work
NaCl-B2	**42.3**, 24.0(1.2), 5.37(20)	**42.3**, 22.664, 5.735	37–155	He	ruby	this work
LiF	16.371(30), 64.6(1.4), 4.62(60)	16.391, 62.3, 5.01	0–109	He	ruby	this work

2. Methods

The experimental methods are similar for all measurements. Membrane diamond anvil cells with diamonds culets ranging from 400 μm to 100 × 300 (central flat × bevel) μm diameters are used. The sample, a grain smaller than 5 μm, is placed close (a few μm) to a pressure gauge and compressed in helium or neon pressure transmitting medium. It is analyzed with angular-dispersive monochromatic X-ray diffraction (XRD), with an X-ray spot size ranging from 6 × 8 μm to 2 × 3 μm, on high pressure beamlines of the European Synchrotron Radiation Facility (ID30, ID27, ID09). The diffraction geometry (sample to detector distance, detector position and angle) is calibrated using a reference sample. The volume is estimated from fitting of 3 to 8 diffraction lines of the sample in each XRD spectrum, yielding a relative precision of $\sim 5 \times 10^{-4}$. The pressure is estimated from the measurement of the luminescence of a ruby gauge, or the XRD signal of an X-ray gauge, here W (tungsten). This X-ray gauge has been calibrated against ruby using the data from [2]. Up to 3 samples have been placed in the same pressure chamber in order to ensure same pressure for calibration cross-check purpose. The pressure is increased with steps of 1 GPa to 4 GPa, with sufficient stabilization time to collect the data with negligible pressure drift (less than 0.5 GPa, estimated by measuring the pressure before and after X-ray exposure). This requires wait time of up to 30 min for low pressure (below 20 GPa) points, and of less than 10 min for highest pressure points. The data collection time is typically 5 min, 1 min for ruby pressure measurement and another 2 min for the X-ray diffraction exposure (including beam search), and 2 min to check the pressure after exposure. In order to diminish the measurement bias due to pressure gradients (which can reach 0.4 GPa at 50 GPa [15]), the gauge was located less than 4 microns away from the sample.

The *P*-*V* data have been fitted with a Rydberg–Vinet [8] form:

$$P = 3K_0(1-x)x^{-2}\exp\left(\frac{3}{2}(K_0'-1)(1-x)\right), \tag{1}$$

with

$$x = \left(\frac{V}{V_0}\right)^{\frac{1}{3}},$$

the compression V_0 volume under ambient conditions, K_0 the bulk modulus of the material and K_0' its pressure derivative under ambient conditions. The fitted V_0 is within experimental error bars of measured volume under ambient conditions, when available. For the same set of parameters (V_0, K_0, K_0'), Rydberg–Vinet pressure increases less than Birch–Murnaghan pressure or H02 pressure [16] under extreme compression, it is therefore considered as the most suited for soft solids such as rare-gas solids. It is used here for metals, diamond and alkali halides and its extrapolation beyond the pressure range where the data have been fitted is not fully justified. However, it has been observed that the extrapolation of gold EoS with a Rydberg–Vinet formulation beyond the measured compression range agrees well with measurements carried out later up to 600 GPa [17]. This possibility should be confirmed on other materials. Unless specified, (V_0, K_0, K_0') have been left as free parameters to fit the *P*-*V* data.

3. EoS of Pb, NaCl and LiF

3.1. Pb

Lead adopts a fcc phase under ambient conditions and transforms to a hcp phase by cold compression to ~14 GPa [18]; hcp-Pb partially transforms to a bcc phase above 107 GPa [19]. bcc-Pb is also observed below the melting line around 44 GPa and above, suggesting a negative hcp-bcc Clapeyron slope [20]. *P*-*V* data published in the 1990s have been collected with samples compressed directly between the diamond anvils [19,21,22], while Kuznetsov et al. [18] measurements have been performed with a sample compressed in a pressure transmitting medium (NaCl), which was also used as a pressure gauge with Brown scale [23]. The different experimental conditions and techniques yield different scatter of the *P*-*V* data (Figure 2). Due to their large scatter and a likely bias due to non-hydrostatic compression [24], I have not taken into account the data of [19,21,22] to obtain EoS parameters.

Here, Pb has been compressed in He up to 131 GPa without observing any formation of a bcc phase, indicating that non-hydrostatic compression helps inducing phase transformations, as already noticed for iron [25]. The current (14–131 GPa, see Table 2) and Kuznetsov et al. data have been merged to determine EoS parameters listed in Table 1.

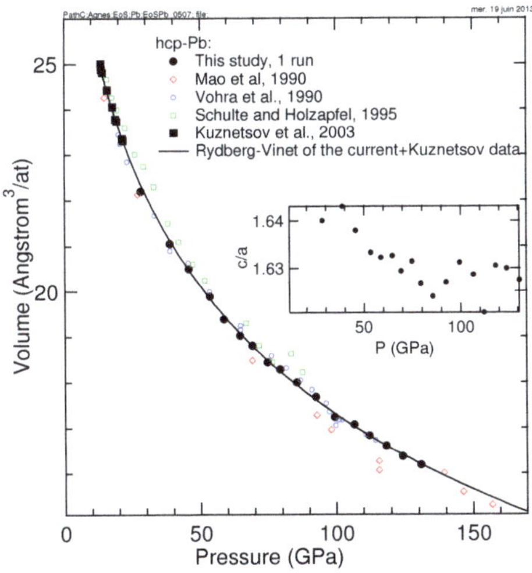

Figure 2. Volume of hcp-Pb volume measured in this study and from the literature [19,21,22]. Inset: c/a ratio measured in this study.

Table 2. Pressure P (from ruby luminescence, with calibration from [4]) and volume V (measured with X-ray diffraction) for hcp-Pb. The data are listed in the order they have been taken.

P (GPa)	V (Å^3/at)
28.16	22.2010
38.74	21.0521
45.62	20.5023
53.51	19.8897
58.77	19.3907
64.69	19.0305
69.24	18.8032
74.79	18.4456
79.34	18.2875
85.48	17.9984
92.47	17.6854
99.5	17.2410
106.6	17.0756
112.2	16.8330
118.3	16.6038
124.2	16.3813
130.8	16.1940

3.2. NaCl

The compression curve of NaCl has been measured in three runs, using helium as pressure medium, reaching 155 GPa (see Figure 3 and Table 3). The low pressure B1 phase transformed to the denser high pressure B2 phase at a pressure between 27 and 36 GPa.

In the literature, EoS measured under non-hydrostatic compression can be found; in particular, two studies in the multi-Mbar range [26,27]. Gold and platinum have been used as X-ray pressure calibrants, respectively, in Ono et al. and Sakai et al. studies. We have taken the calibrations of these gauges from [4] to plot the points in Figure 3. With this calibration, both Ono et al. and Sakai et al. measurements agree correctly with the current data up to 155 GPa. This suggests that non-hydrostatic stress remains weak in NaCl under pressure, as noted by Sakai et al. The P-V data collected using platinum as an X-ray pressure gauge up to 110 GPa in [28] also agree with the current one (the Birch–Murnaghan parameters from Table 1 in [28] have been considered, because there is a likely typographic error in Vinet parameters). The calibration of platinum gauge used by Fei et al. diverges by less than 0.2% with [4] in that range.

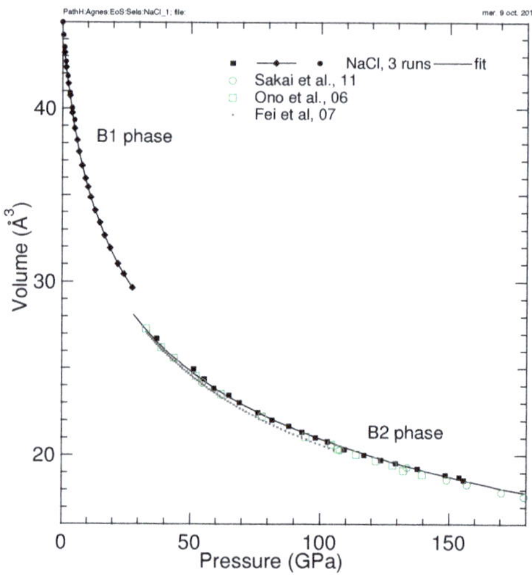

Figure 3. Volume of NaCl measured in this study and in the literature [26–28].

Table 3. Pressure *P* (from ruby luminescence, with calibration from [4]) and volume *V* (measured with X-ray diffraction) for NaCl (B1 and B2 phases). Each pair of columns corresponds to one experimental run. The data are listed in the order they have been taken.

P (GPa)	*V* ($Å^3$/fu) (B1)	*P* (GPa)	*V* ($Å^3$/fu) (B1)	*P* (GPa)	*V* ($Å^3$/fu) (B2)
0.868	43.2022	0.319	44.2471	37.06	26.7309
1.46	42.3505	0.768	43.5143	51.22	24.9611
2.29	41.4177	1.37	42.6836	55.45	24.3764
2.96	40.7222	2.11	41.8422	59.18	23.8456
3.85	39.7304	2.91	40.8914	65.03	23.4403
4.88	38.8363	0.0171	44.9814	69.06	23.0081
5.87	38.1616			76.24	22.4425
6.68	37.4870			81.68	22.0179
7.87	36.6937			88.07	21.6523
9.12	35.9584			93.12	21.3184
10.2	35.4483			98.24	21.0017
11.3	34.8669			102.8	20.7833
13.0	34.0953			109.5	20.3174
14.8	33.3907			117.1	20.0107
16.7	32.6566			123.7	19.7202
18.9	31.9352			129.3	19.5347
21.9	31.0243			137.7	19.2101
24.3	30.4328			148.5	18.8551
27.6	29.6501			153.9	18.7279
				155.6	18.5570

The current *P*-*V* data, plotted in black in Figure 3, have been used to determine the EoS parameters listed in Table 1. The value of V_0 for the B2 phase has been fixed to 42.3 $Å^3$/at, 5.7% denser than the B1 phase (the relative volume difference measured around 27 GPa), to prevent overfitting.

3.3. LiF

The compression curve of LiF has been measured in three runs (see Table 4). The samples were single crystals, compressed in helium. LiF remained in the B1 phase in the scanned pressure range (0.7–109 GPa). Its compression behavior (see fitting parameters for these three runs in Table 1) agrees well with one recent study up to 92 GPa [29], but not with another one to 37 GPa [30]. The disagreement is more obvious when this EoS is extrapolated in the Mbar range (see Figure 4). It can be noted that in Liu et al. [30], the fitted bulk modulus is 12% higher than its ultrasonic value, 64.3 GPa [31]; the current fitted bulk modulus agrees with the ultrasonic one.

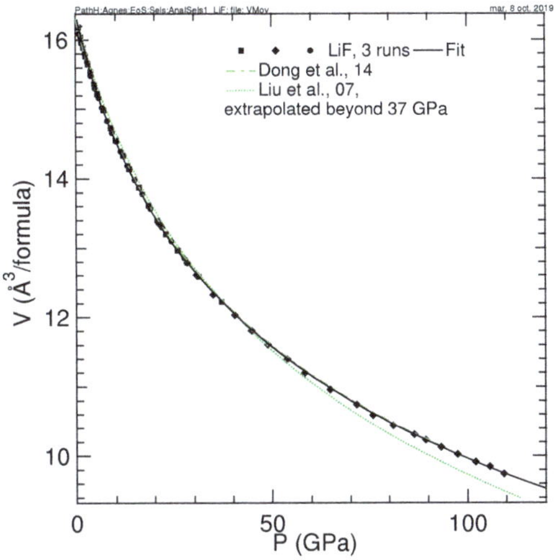

Figure 4. Volume of LiF measured in this study and in the literature [29,30].

Table 4. Pressure P (from ruby luminescence, with calibration from [4]) and volume V (measured with X-ray diffraction) for LiF. Each pair of columns corresponds to one experimental run. The data are listed in the order they have been taken.

P (GPa)	V (Å^3/fu)	P (GPa)	V (Å^3/fu)	P (GPa)	V (Å^3/fu)
0.722	16.181	21.3	13.3466	0.868	16.0879
1.32	16.040	28.3	12.7874	1.46	15.9574
1.95	15.8910	30.6	12.6060	2.29	15.8032
2.50	15.7684	34.8	12.3274	2.96	15.6871
3.14	15.6498	40.4	12.0328	3.85	15.4849
3.81	15.5367	44.7	11.8066	4.88	15.3078
4.75	15.3677	48.8	11.6103	5.87	15.1685
5.54	15.2234	53.8	11.3973	6.68	15.0231
7.03	14.9857	58.2	11.1980	7.87	14.8317
8.85	14.7288	64.8	10.9605	9.12	14.6591
10.1	14.5479	71.7	10.7381	10.2	14.5370
11.9	14.3310	75.8	10.5914	11.3	14.3930
13.5	14.1368	81.0	10.4369	13.0	14.1842
15.9	13.8710	86.4	10.3086	14.8	13.9848
18.4	13.6116	89.3	10.2328	16.7	13.7792
20.7	13.3988	93.2	10.1232	18.9	13.5672
22.9	13.2003	97.4	10.0220	21.9	13.2994
25.9	12.9683	102.1	9.91426	24.3	13.1015
28.4	12.7853	105.7	9.84496	27.6	12.8561
31.2	12.5880	109.3	9.74253		
37	12.2217				

4. Equation of State Parameters

The EoS parameters (V_0, K_0, K'_0) listed in Table 1 have been obtained for two calibrations of the ruby gauge: Mao 86 [6] and Dorogokupets 07 [4]. This last calibration has been used for Be [10], Re [11], U [32], KCl [14] and KBr [14]. Therefore, no correction of the original publications EoS parameters is needed for these compounds. They have been listed in Table 1 to provide an extended comparison EoS parameters for the two calibrations. To establish their calibration, Mao et al. have compressed a ruby up to 80 GPa together with metals (Cu, Ag) used as primary X-ray gauges in DACs, in argon pressure transmitting medium. Cu and Ag gauges were calibrated using reduced shock waves equations of state. The calibration of [4] has been established by the same method, but using several metals (Au, Pt, Ta, W, Cu, Al [2]) compressed together with ruby in a more hydrostatic helium pressure medium [24].

Table 1 allows extracting trends in the differences between EoS parameters for the two calibrations. Not surprisingly, V_0 are identical when the compound volume could be measured close to ambient conditions. The bulk moduli K_0 are close within a few percents, because the pressures for the two calibrations are close at moderate compression; the slightly higher value of K_0 obtained with Mao 86 calibration is due to the fact that the pressure is slightly higher up to 9 GPa, although not noticeably with the scale of Figure 1. The parameter which differs the most is K_0': it is higher with Dorogokupets calibration than Mao calibration, because the pressure is higher in the Mbar range, the domain that places the most constraints on the value of K_0'. The difference reaches 0.65 (out of 3.01, a 22% increase) for the most incompressible element, diamond. The relative variation of K_0' is smaller for softer solids such as aklali halides (K_0' of B2-NaCl higher by 7% for Dorogokupets scale than Mao 86 scale). It has already been noted that K_0' is more sensitive to the pressure scale for incompressible solids [1], which can easily been explained by a derivation of the Rydberg–Vinet EoS formulation. Reciprocally, a precise measurement of K_0' for a hard material would provide a tight constraint on the high pressure metrology.

Funding: The author acknowledges the European Synchrotron Radiation Facility for provision of X-ray diffraction beamtime.

Acknowledgments: I thank F. Occelli, M. Mezouar and M. Hanfland for experimental help.

Conflicts of Interest: The author declares no conflict of interest.

References

1. Holzapfel, W. Refinement of the ruby luminescence pressure scale. *J. Appl. Phys.* **2003**, *93*, 1813–1818. [CrossRef]
2. Dewaele, A.; Loubeyre, P.; Mezouar, M. Equations of state of six metals above 94 GPa. *Phys. Rev. B* **2004**, *70*, 094112–094119. [CrossRef]
3. Holzapfel, W.B. Progress in the realization of a practical pressure scale for the range 1–300 GPa. *High Press. Res.* **2005**, *25*, 187–196. [CrossRef]
4. Dorogokupets, P.I.; Oganov, A.R. Ruby, metals, and MgO as alternative pressure scales: A semiempirical description of shock-wave, ultrasonic, X-ray, and thermochemical data at high temperatures and pressures. *Phys. Rev. B* **2007**, *75*, 024115. [CrossRef]
5. Dewaele, A.; Torrent, M.; Loubeyre, P.; Mezouar, M. Compression curves of transition metals in the mbar range: Experiments and projector augmented-wave calculations. *Phys. Rev. B* **2008**, *78*, 104102–104114. [CrossRef]
6. Mao, H.-K.; Xu, J.; Bell, P. Calibration of the ruby pressure gauge to 800 kbar under quasi-hydrostatic conditions. *J. Geophys. Res.* **1986**, *91*, 4673–4676. [CrossRef]
7. Sokolova, T.S.; Dorogokupets, P.I.; Litasov, K.D. Self-consistent pressure scales based on the equations of state for ruby, diamond, MgO, B2-NaCl, as well as au, pt, and other metals to 4 mbar and 3000 k. *Russ. Geol. Geophys.* **2013**, *54*, 181–199. [CrossRef]
8. Vinet, P.; Ferrante, J.; Smith, J.R.; Rose, J.H. A universal equation of state for solids. *J. Phys. Condens. Matter* **1986**, *19*, L467. [CrossRef]
9. Takemura, K.; Dewaele, A. Isothermal equation of state for gold with a He-pressure medium. *Phys. Rev. B* **2008**, *78*, 104119–104131. [CrossRef]
10. Lazicki, A.; Dewaele, A.; Loubeyre, P.; Mezouar, M. High pressure-high temperature phase diagram and the equation of state of beryllium. *Phys. Rev. B* **2012**, *86*, 174118–174128. [CrossRef]
11. Anzellini, S.; Dewaele, A.; Occelli, F.; Loubeyre, P.; Mezouar, M. Equation of state of rhenium and application for ultra high pressure calibration. *J. Appl. Phys.* **2014**, *115*, 043511. [CrossRef]

12. Occelli, F.; Loubeyre, P.; Letoullec, R. Properties of diamond under hydrostatic pressures up to 140 GPa. *Nat. Mater.* **2003**, *2*, 151–154. [CrossRef] [PubMed]
13. Dewaele, A.; Datchi, F.; Loubeyre, P.; Mezouar, M. High pressure high temperature equations of state of neon and diamond. *Phys. Rev. B* **2008**, *77*, 094106–094114. [CrossRef]
14. Dewaele, A.; Belonoshko, A.B.; Garbarino, G.; Occelli, F.; Bouvier, P.; Hanfland, M.; Mezouar, M. High pressure-high temperature equation of state of KCl and KBr. *Phys. Rev. B* **2012**, *85*, 214105–214112. [CrossRef]
15. Klotz, S.; Chervin, J.-C.; Munsch, P.; Le Marchand, G. Hydrostatic limits of 11 pressure transmitting media. *J. Phys. D Appl. Phys.* **2009**, *42*, 075413. [CrossRef]
16. Holzapfel, W. Equations of State for Ideal and Real Solids Under Strong Compression. *Europhys. Lett.* **1991**, *16*, 67. [CrossRef]
17. Dewaele, A.; Loubeyre, P.; Occelli, F.; Marie, O.; Mezouar, M. Toroidal diamond anvil cell for detailed measurements under extreme static pressures. *Nat. Commun.* **2018**, *9*, 2913. [CrossRef] [PubMed]
18. Kuznetsov, A.; Dmitriev, V.; Dubrovinsky, L.; Prakapenka, V.; Weber, H.P. Fcc-hcp phase boundary in lead. *Solid State Commun.* **2002**, *122*, 125–127. [CrossRef]
19. Mao, H.K.; Wu, Y.; Shu, J.F.; Hu, J.Z.; Hemley, R.J.; Cox, D.E. High pressure phase transition and equation of state of lead to 238 GPa. *Solid State Commun.* **1990**, *74*, 1027–1029. [CrossRef]
20. Dewaele, A.; Mezouar, M.; Guignot, N.; Loubeyre, P. Melting of lead under high pressure using second-scale times resolved X-ray diffraction. *Phys. Rev. B* **2007**, *76*, 144106. [CrossRef]
21. Vohra, Y.K.; Ruoff, A.L. Static compression of metals mo, pb and pt to 272 GPa: Comparison with shock data. *Phys. Rev. B* **1990**, *42*, 8651–8653. [CrossRef] [PubMed]
22. Schulte, O.; Holzapfel, W.B. Equation-of-state behavior for different phases of lead under strong compression. *Phys. Rev. B* **1995**, *52*, 12636–12639. [CrossRef] [PubMed]
23. Brown, J.M. The NaCl pressure standard. *J. Appl. Phys.* **1999**, *86*, 5801–5809. [CrossRef]
24. Dewaele, A.; Loubeyre, P. Pressurizing conditions in helium-pressure-transmitting medium. *High Press. Res.* **2007**, *27*, 419–429. [CrossRef]
25. Boehler, R.; Bargen, N.V.; Chopelas, A. Melting, thermal expansion, and phase transitions of iron at high pressures. *J. Geophys. Res.* **1990**, *95*, 21731–21736. [CrossRef]
26. Ono, S.; Kikegawa, T.; Ohishi, Y. Structural property of CsCl-type sodium chloride under pressure. *Solid State Commun.* **2006**, *137*, 517–521. [CrossRef]
27. Sakai, T.; Ohtani, E.; Hirao, N.; Ohishi, Y. Equation of state of the NaCl-b2 phase up to 304 GPa. *J. Appl. Phys.* **2011**, *109*, 084912. [CrossRef]
28. Fei, Y.; Ricolleau, A.; Frank, M.; Mibe, K.; Shen, G.; Prakapenka, V. Toward an internally consistent pressure scale. *Proc. Natl. Acad. Sci. USA* **2007**, *104*, 9182–9186. [CrossRef]
29. Dong, H.; Dorfman, S.M.; Holl, C.M.; Meng, Y.; Prakapenka, V.B.; He, D.; Duffy, T.S. Compression of lithium fluoride to 92 GPa. *High Press. Res.* **2014**, *34*, 39–48. [CrossRef]
30. Liu, J.; Dubrovinsky, L.; Ballaran, T.B.; Crichton, W. Equation of state and thermal expansivity of LiF and NaF. *High Press. Res.* **2007**, *27*, 483–489. [CrossRef]
31. Simmons, G.; Wang, H. *Single Crystal Elastic Constants and Calculated Aggregate Properties: A Handbook*; The MIT Press: Cambridge, UK, 1971.
32. Dewaele, A.; Bouchet, J.; Occelli, F.; Hanfland, M.; Garbarino, G. Refinement of the equation of state of α-uranium. *Phys. Rev. B* **2013**, *88*, 134202. [CrossRef]

minerals

MDPI

Article

Hydrogen Effect on the Sound Velocities of Upper Mantle Omphacite

Wade Mans [1,2,*], Jin S. Zhang [1,2,*], Ming Hao [1,2], Joseph R. Smyth [3], Dongzhou Zhang [4,5], Gregory J. Finkelstein [5,6] and Przemyslaw Dera [5]

1 Department of Earth and Planetary Sciences, University of New Mexico, Albuquerque, NM 87131, USA; minghao@unm.edu
2 Institute of Meteoritics, University of New Mexico, Albuquerque, NM 87131, USA
3 Department of Geological Sciences, University of Colorado, Boulder, CO 80309, USA; smyth@colorado.edu
4 GeoSoiEnviroCARS, University of Chicago, Argonne National Laboratory, Argonne, IL 60439, USA; dzhang@hawaii.edu
5 Hawaii Institute of Geophysics and Planetology, University of Hawaii at Manoa, Honolulu, HI 96822 USA; gjfinkel@princeton.edu (G.J.F.); pdera@hawaii.edu (P.D.)
6 Department of Geosciences, Princeton University, Princeton, NJ 08544, USA
* Correspondence: wmans@unm.edu (W.M.); jinzhang@unm.edu (J.S.Z.)

Received: 7 October 2019; Accepted: 6 November 2019; Published: 8 November 2019

Abstract: Clinopyroxene (Cpx) is commonly believed to be the best structural water (hydrogen) carrier among all major upper mantle nominally anhydrous minerals (NAMs). In this study, we have measured the single-crystal elastic properties of a Cpx, a natural omphacite with ~710 ppm water at ambient pressure (*P*) and temperature (*T*) conditions. Utilizing the single-crystal X-ray diffraction (XRD) and electron microprobe data, the unit cell parameters and density were determined as a = 9.603(9) Å, b = 8.774(3) Å, c = 5.250(2) Å, β = 106.76(5)°, V = 255.1(4) Å^3, and ϱ = 3.340(6) g/cm^3. We performed Brillouin spectroscopy experiments on four single crystals along a total of 52 different crystallographic directions. The best-fit single-crystal elastic moduli (C_{ij}s), bulk and shear moduli were determined as: C_{11} = 245(1) GPa, C_{22} = 210(2) GPa, C_{33} = 249.6(9) GPa, C_{44} = 75.7(9) GPa, C_{55} = 71.2(5) GPa, C_{66} = 76(1) GPa, C_{12} = 85(2) GPa, C_{13} = 70(1) GPa, C_{23} = 66(2) GPa, C_{15} = 8.0(6) GPa, C_{25} = 6(1) GPa, C_{35} = 34.7(6) GPa, and C_{46} = 8.7(7) GPa, K_{S0} = 125(3) GPa, and G_0 = 75(2) GPa, respectively. Compared with the anticipated elastic properties of an anhydrous omphacite with the same chemical composition, our results indicate that the incorporation of ~710 ppm structural water has no resolvable effect on the aggregate elastic properties of omphacite, although small differences (up to ~9 GPa) were observed in C_{13}, C_{25}, C_{44}, and C_{66}.

Keywords: elasticity; omphacite; clinopyroxene; eclogite; structural water; nominally hydrous minerals; seismic velocities

1. Introduction

Quantifying the water content in the Earth's upper mantle through seismic observations requires the knowledge of how structural water content affects the elastic properties for various nominally anhydrous minerals (NAMs) [1,2]. Whether the upper mantle is universally or locally hydrated is still controversial [3,4], although the water carried down by the subducting slabs is considered a major source of the water in the Earth's interior [5–8]. Subduction is one of the main driving forces of the mantle convection and is responsible for many geological processes in the Earth's interior [9,10]. The basaltic slab crust transforms into eclogite at depths greater than ~100 km and it remains denser than the ambient mantle down to ~600 km depths [11]. As the solid solution between diopside (Di, $CaMgSi_2O_6$) and jadeite (Jd, $NaAlSi_2O_6$), omphacite is the major mineral phase that constitutes up to

80 vol% of eclogite. The subducting slabs carry the surface water into the Earth's interior primarily in the form of structural water; thus, it is important to know how much structural water can be stored in the major mineral phases in the subducting slabs. The maximum solubility of H_2O—depending on temperature (T) and pressure (P)—in pure Di ranges from 121–568 ppm, but in aluminous Di, this value increases up to 2500 ppm [12]. Natural Clinopyroxene (Cpx) samples can host as high as 2000 ppm water in their crystal structures [12,13], much higher than all other major upper mantle minerals such as olivine, orthopyroxene, or garnet. As a type of Cpx, which can retain the highest amount of water in its structure among all upper mantle NAMs, omphacite is likely both an important water carrier as well as a potential water reservoir in the Earth's interior [5,6,14–17].

Structural water (hydrogen) is incorporated into the Cpx crystal structure through cation vacancies primarily on the M2 site [13,18]. Previous experimental studies on olivine, another NAM commonly found in the ambient upper mantle, have suggested that the incorporation of hydrogen into the NAMs decreases both the P-wave and S-wave velocities (V_p and V_s) [19–23]. This softening effect of structural water can potentially affect our understanding of the upper mantle seismic structures and mineralogical composition, as well as the total water budget of the Earth's interior [6,21–25]. In addition, there is a growing interest in the mineral physics and petrology communities in using various minerals' thermoelastic parameters to establish new elastic geobarometers for interpreting various deep geological processes that happened in the past [26,27]. Unfortunately, most previous investigations are restricted to hydrous olivine and its high-P polymorphs [21–23,28–30], and no experimental sound velocity measurements have previously been made for hydrous Cpx. Therefore, in this study, we performed single-crystal Brillouin spectroscopy experiments on a hydrous omphacite sample with ~710 ppm water in order to study the possible structural water effect on the single-crystal elastic properties of omphacite.

2. Materials and Methods

The omphacite crystals were hand-selected from the natural omphacite sample SBB-46 from the South African Bobbejaan mine. Four different crystals were double-side polished to less than 30 μm thickness. Under optical examination, all crystals were inclusion- and scratch-free. The chemical composition $(Na_{0.396}K_{.005})(Mg_{0.537}Ca_{0.550}Fe_{0.072}Cr_{0.022}Ti_{0.007}Mn_{0.001})Al_{0.433}Si_{1.971}O_6$ (simplified to $Di_{59.1}Jd_{40.9}$) and the ~710 ppm hydrogen content of SBB-46 was determined in [13].

In order to determine the unit cell parameters and crystal orientation, we carried out the single-crystal XRD experiments for the 2 crystals a and b at ambient P–T condition at experimental station 13-BM-C, GeoSoilEnviroCARS (GSECARS), Advanced Photon Source, Argonne National Laboratory. The remaining 2 crystals c and d were measured at ambient P–T condition in the X-Ray Atlas Diffraction Lab at the University of Hawai'i at Manoa.

At GSECARS, the X-ray beam was monochromated to 28.6 keV with the beam size of ~12 μm × 18 μm determined at the full width at half maximum. The omphacite crystal was placed in an empty DAC, which had an opening angle of ±38°. Two separate detector positions were used by rotating a MAR165 Charge Coupled Device detector on a rotational arm [31]. The first detector angle was perpendicular to the incident X-Ray beam, while the second detector position was offset from the first position by 20° around the horizontal axis. Sample-to-detector distance and detector tilting were calibrated in the Dioptas program [32] using NIST standard LaB_6 powder. The diffraction images were collected in both wide-angle images that covered the whole ±38° opening angle range and 1° step-segments with 1 s/° exposure time. At the X-Ray Atlas Diffraction Lab in University of Hawai'i, a Bruker D8 Venture XRD diffractometer with Incoatec IμS 3.0 AgKα microfocus source, Helios focusing optics, Photon II detector was used for collecting the single-crystal diffraction images at a wide range of scattering angles. Diffraction data from both experiments was processed using Bruker APEX III software. For the needs of this study, full structure refinement was not performed, and only unit cell parameters were refined.

We performed the Brillouin spectroscopy experiments on all 4 double-polished SBB-46 omphacite single crystals at the High-*P* Laser Spectroscopy Laboratory at University of New Mexico. The light source was a 532-nm 300-mW single-mode diode-pumped solid-state laser. The experiment utilized a 50° symmetric forward scattering geometry, with the scattering angle precisely calibrated to be 50.42(5)° using the Corning 7980 standard silica glass [33,34]. V_p and V_s were determined for all 4 pre-oriented samples at ambient *P*–*T* condition for a total of 52 crystallographic directions. Each sample was measured at 13 different χ angles (0°, 30°, 60°, 120°, 150°, 180°, 195°, 225°, 255°, 285°, 315°, 345° and 360°) along the 360° azimuth to account for any possible geometric errors with an average collection time of 10 min per spectrum. The resulting Brillouin spectra all have very high signal-to-noise ratios (Figure 1).

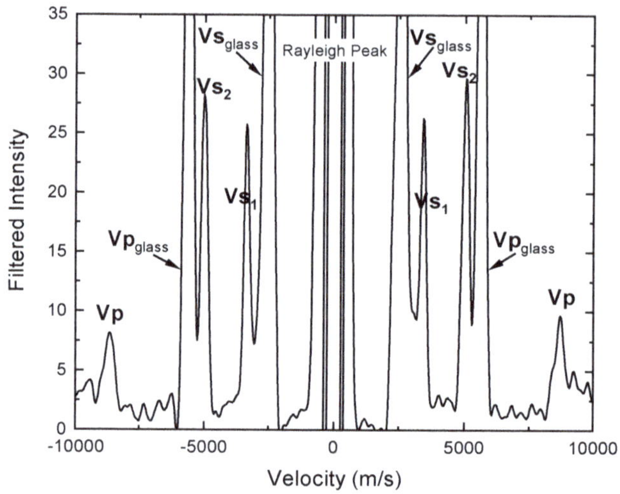

Figure 1. Typical Brillouin spectrum of omphacite crystal *d* at $\chi = 0°$.

3. Results and Discussion

The single-crystal XRD measurements for all four crystals yielded the averaged unit cell parameters of the SBB-46 omphacite of a = 9.603(9) Å, b = 8.774(3) Å, c = 5.250(2) Å, β = 106.76(5)°, V = 255.1(4) cm^3. The planes normal of the four crystals were determined to be: (0.9131 −0.4076 0.0018) for crystal *a*, (−0.2742 0.9457 −0.1748) for crystal *b*, (0.4646 0.1359 −0.8750) for crystal *c*, and (−0.6860 −0.5145 0.5145) for crystal *d* (Figure 2). The density at ambient condition was then calculated as $\varrho_0 = 3.340(6)$ g/cm^3. Omphacite crystals are known to assume one of two different symmetries, depending on the crystallization conditions: the disordered high-temperature phase crystallizes with space group C2/c, whereas the cation-ordered phase crystallizes with space group P2/n [35]. At ambient pressure, the order–disorder transition takes place at 725 °C [36]. The omphacite SBB-46 had a C2/c symmetry based on its high *T* thermal history. The difference in cation ordering was not found to have a resolvable effect on the elastic properties beyond experimental uncertainties [37].

The single-crystal C_{ij}s for the SBB-46 hydrous omphacite under ambient conditions were calculated through the least-squares inversion using the Christoffel equation. The best-fit C_{ij} model was: C_{11} = 245(1) GPa, C_{22} = 210(2) GPa, C_{33} = 249.6(9) GPa, C_{44} = 75.7(9) GPa, C_{55} = 71.2(5) GPa, C_{66} = 76(1) GPa, C_{12} = 85(2) GPa, C_{13} = 70(1) GPa, C_{23} = 66(2) GPa, C_{15} = 8.0(6) GPa, C_{25} = 6(1) GPa, C_{35} = 34.7(6) GPa, and C_{46} = 8.7(7) GPa. The root-mean-square residual was less than 53 m/s between the modeled and observed velocities (Figure 2). The K_{S0} and G_0 were calculated as 125(3) and 75(2) GPa from the single-crystal C_{ij}s under the Voigt-Reuss-Hill averaging scheme [38].

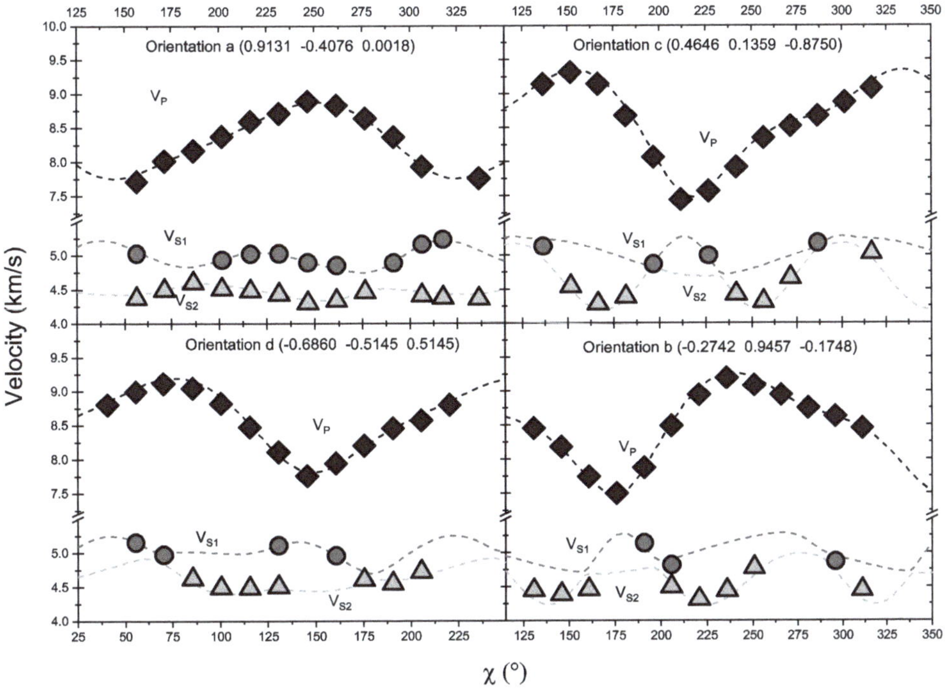

Figure 2. Acoustic velocities measured versus laboratory χ angles for the four orientations of SBB-46. Diamonds, circles, and triangles represent experimentally determined V_p, V_{s1}, and V_{s2}, respectively. The dashed lines were calculated from the best fit single-crystal C_{ij} model. Errors were smaller than the symbols.

The single-crystal elastic properties of anhydrous omphacite at ambient conditions had been measured [39–44] and systematically analyzed within a wide compositional range [44]. As shown in Figure 3, within the Di–Jd solid solution, most elastic moduli showed a close-to-ideal linear mixing trend except C_{13} and C_{23}. The C_{33}, C_{55}, and C_{35} presented in Ref. [44] lay outside the trends determined from other measurements, which can be explained by the high Tschermak content of the sample (12 mol%). The bulk and shear elastic moduli (K_s and G), as well as most of the single-crystal C_{ij}s of the hydrous omphacite SBB-46, were well within the 95% confidence interval determined by all of the previous measurements of anhydrous omphacite samples within the Di–Jd solid solution (Table 1). A few single-crystal elastic moduli such as C_{13}, C_{25}, C_{44}, and C_{66} showed a small decrease within the Di–Jd solid solution (up to ~9 GPa), deviating outside of the 95% confidence interval, which might have been caused by the incorporation of water into its crystal structure (Table 1). However, the decrease of these single-crystal elastic moduli caused by hydration is unlikely to produce a strong seismic anomaly. For example, the ~7% decrease of C_{44} and C_{66} caused by ~710 ppm water transforms to about 3.5% reduction in [010] polarized Vs propagating along the [001] and [100] directions, respectively. Considering the 50–70 vol% of omphacite in the eclogite rock, as well as the elastically isotropic nature of the garnet, it is difficult for the decrease of V_s in the eclogite rock in a realistic scenario to exceed 1%, unless near-perfect alignment of the omphacite crystals takes place in the form of SL-type fabrics [45–48]. In addition, the aggregate V_p and V_s did not show any structural water-induced softening either (Figure 4). Therefore, it is unlikely that the hydration of omphacite crystals would affect the seismic properties of the subducted slab crust.

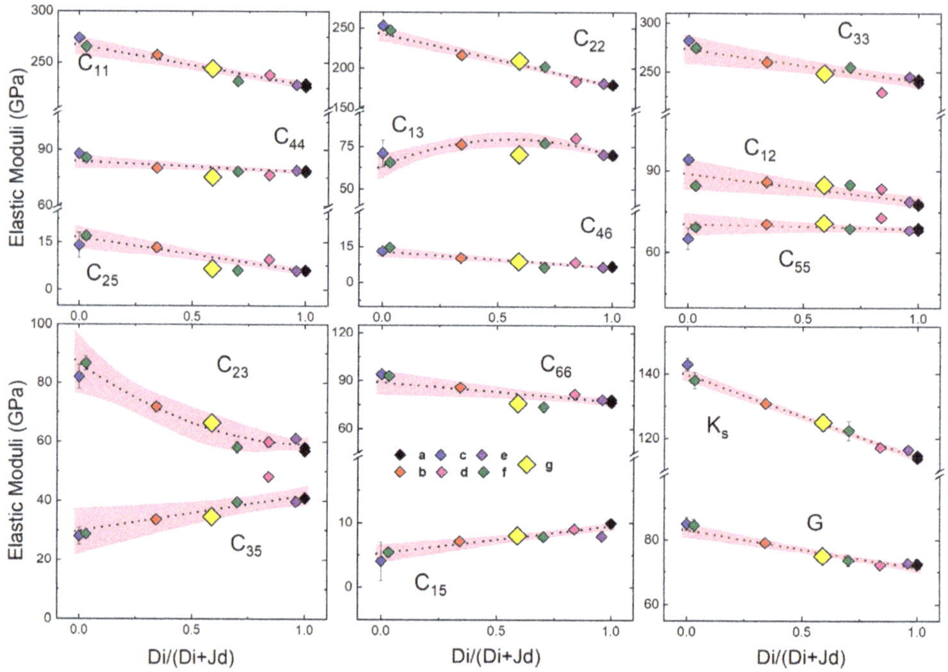

Figure 3. Single-crystal elastic properties of SBB-46 compared with previous studies on anhydrous omphacites within the Di–Jd solid solution as measured by (**a**) Sang et al., 2011 [39], (**b**) Bhagat et al., [40], (**c**) Kandelin and Weidener, 1988 [41], (**d**) Collins and Brown, 1998 [42], (**e**) Isaak and Ohno, 2003 [43], (**f**) Hao et al., 2019 [44]. The results are compared with the hydrous omphacite measured in (**g**) this study. Dashed lines show the compositional dependence of $C_{ij}s$, K_s, and G. Shaded regions represent the 95% confidence intervals determined in [44].

Table 1. Elastic moduli of the hydrous omphacite sample measured in this study and the anhydrous omphacite with the same chemical composition predicted from Hao et al., 2019 [44]. The uncertainties for the anhydrous omphacite are defined by the 95% confidence intervals shown in Figure 3.

Elastic Moduli	Anhydrous Di59.1Jd40.9	SBB-46 Di59.1Jd40.9 ~710 ppm Water
C_{11} (GPa)	244(4)	245(1)
C_{22} (GPa)	205(4)	210(2)
C_{33} (GPa)	254(6)	249.6(9)
C_{44} (GPa)	81(2)	75.7(9)
C_{55} (GPa)	69(2)	71.2(5)
C_{66} (GPa)	82(2)	76(1)
C_{12} (GPa)	83(2)	85(2)
C_{13} (GPa)	79(4)	70(1)
C_{23} (GPa)	64(6)	66(2)
C_{15} (GPa)	7.7(6)	8.0(6)
C_{25} (GPa)	10(2)	6(1)
C_{35} (GPa)	37(3)	34.7(6)
C_{46} (GPa)	9(1)	8.7(7)
Ks (GPa)	125(1)	125(3)
G (GPa)	76(1)	75(2)

We also calculated three elastic anisotropy indices for the hydrous omphacite SBB-46 and plotted the data along with other anhydrous omphacite samples in the Di–Jd solid solution (Figure 4): (1) The

Universal Anisotropy Index (A^U) which described the overall elastic anisotropy for any materials with arbitrary symmetry, (2) V_p azimuthal anisotropy Index (A^{Vp}) which was the maximum difference in V_p along all crystallographic directions divided by the aggregate V_p, and (3) the polarization anisotropy Index for V_s (D^{Vs}) which measured the maximum velocity difference between two orthogonally polarized V_s that propagated in the same direction. Among these indices, only the A^U for SBB-46 was slightly lower than the expected value for its anhydrous counterpart. This again suggested that the effect of ~710 ppm water on the aggregate elastic properties of omphacite was very small if not negligible. It was also worth noting that Ref. [44] again lay outside the trends determined from all of the other studies probably due to the high Tschermak content (12 mol%) of the sample. This observation is also important for the recent development of elastic geobarometry [26,27]. The elastic anisotropy seems to be a major factor that needs to be taken into account for calculating the *P*s for various geological processes, in particular when the rock has experienced high-grade metamorphism and eclogite facies are found [49–51].

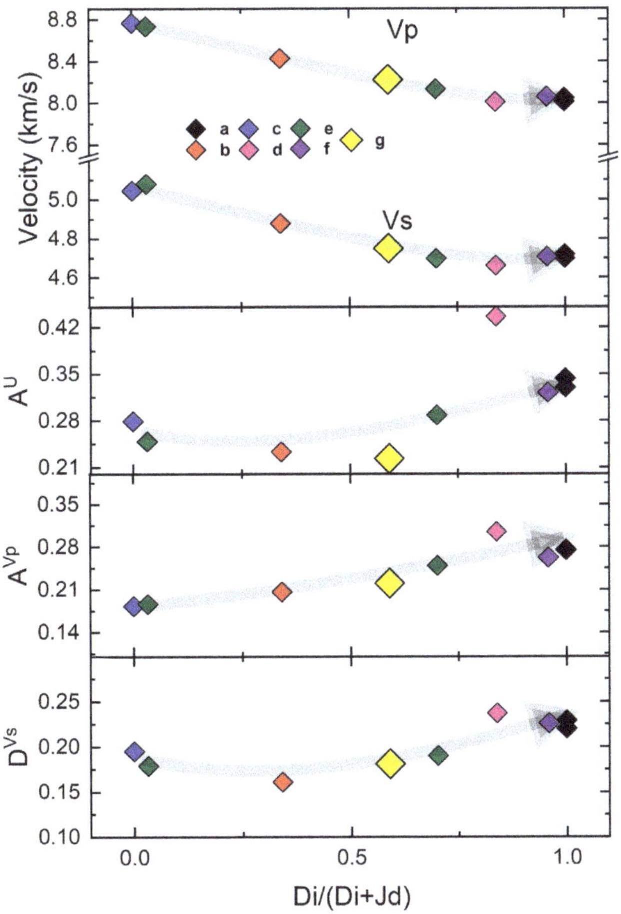

Figure 4. The V_p and V_s, A^U, A^{Vp}, and D^{Vs} of SBB-46 compared with previous studies on anhydrous omphacites (adapted from Hao et al., 2019 [44]) within the Di–Jd solid solution as measured by (**a**) Sang et al., 2011 [39], (**b**) Bhagat et al., [40], (**c**) Kandelin and Weidener, 1988 [41], (**d**) Collins and Brown, 1998 [42], (**e**) Isaak and Ohno, 2003 [43], (**f**) Hao et al., 2019 [44], and (**g**) this study.

4. Conclusions

For the first time, we have experimentally explored the effect of structural water on the single-crystal elastic properties of a hydrous omphacite sample SBB-46. Most of the single-crystal elastic moduli of the hydrous omphacite SBB-46 were well within the 95% confidence interval determined by all the previous measurements of anhydrous omphacite samples within the Di–Jd solid solution. The small decrease of the shear elastic moduli C_{13}, C_{25}, C_{44}, and C_{66} is unlikely to result in a strong seismic anomaly unless the eclogite has experienced extremely high strain and a near-perfect alignment of most omphacite crystals takes place. Based on this study, low to moderate amounts of structural water (~710 ppm) are unlikely to have significant impact on the seismic velocities of the subducted oceanic crust. Further studies of hydrous omphacite at a much higher water concentration (e.g., 1800 ppm [13]) and at high *P*–*T* conditions might provide additional insights.

Author Contributions: Conceptualization, J.S.Z. and W.M.; methodology, J.S.Z., J.R.S., G.J.F., P.D., D.Z.; software, J.S.Z.; validation, J.S.Z. and M.H.; formal analysis, W.M., M.H., J.R.S., G.J.F., P.D.; resources, J.S.Z., J.R.S., G.J.F., P.D., D.Z.; data curation, W.M., J.S.Z., G.J.F., P.D., D.Z.; writing—original draft preparation, W.M.; writing—review and editing, J.S.Z. and everybody else; visualization, W.M., J.S.Z., and M.H.; supervision, J.S.Z.; project administration, J.S.Z.; funding acquisition, J.S.Z.

Funding: This research was funded by National Science Foundation, grant number EAR 1646527 (J.S.Z.). The high-P laser lab was funded by the startup (J.S.Z) at University of New Mexico and National Science Foundation, grant number EAR 1646527 (J.S.Z.). P.D. was supported by the National Science Foundation grant number EAR 1722969. Portions of the X-ray diffraction work were conducted using the X-ray Atlas instrument at the University of Hawaii, funded by NSF grant EAR 1541516. Beamline 13BMC is supported by COMPRES, the Consortium for Materials Properties Research in Earth Sciences under NSF Cooperative Agreement EAR 1661511, and GSECARS funded by NSF (EAR-1634415), and Department of Energy (DOE)—GeoSciences (DE-FG02-94ER14466). This research used resources of the APS, a U.S. DOE Office of Science User Facility operated for the DOE Office of Science by Argonne National Laboratory under contract DE-AC02-06CH11357.

Acknowledgments: The authors thank Zhiyuan Ren and Wenyi Zhou for the help with the experiments.

Conflicts of Interest: The authors declare no conflict of interest. The funders had no role in the design of the study; in the collection, analyses, or interpretation of data; in the writing of the manuscript, or in the decision to publish the results.

References

1. Jacobsen, S.D. Effect of Water on the Equation of State of Nominally Anhydrous Minerals. In *Water in Nominally Anhydrous Minerals*; Walter de Gruyter GmbH: Berlin, Germany, 2018; pp. 321–342. [CrossRef]
2. Liu, L.; Du, J.; Zhao, J.; Liu, H.; Gao, H.; Chen, Y. Elastic Properties of Hydrous Forsterites under High Pressure: First-Principle Calculations. *Phys. Earth Planet. Inter.* **2009**, *176*, 89–97. [CrossRef]
3. Karato, S.I. Mapping Water Content in the Upper Mantle. In *Geophysical Monograph Series*; Blackwell Publishing Ltd.: Hoboken, NJ, USA, 2004; Volume 138, pp. 135–152. [CrossRef]
4. Thompson, A.B. Water in the Earth's Upper Mantle. *Nature* **1992**, *358*, 295–302. [CrossRef]
5. Katayama, I.; Nakashima, S.; Yurimoto, H. Water Content in Natural Eclogite and Implication for Water Transport into the Deep Upper Mantle. *Lithos* **2006**, *86*, 245–259. [CrossRef]
6. Ohtani, E.; Litasov, K.; Hosoya, T.; Kubo, T.; Kondo, T. Water Transport into the Deep Mantle and Formation of a Hydrous Transition Zone. *Phys. Earth Planet. Inter.* **2004**, *143*, 255–269. [CrossRef]
7. Skogby, H. Water in Natural Mantle Minerals I: Pyroxenes. *Rev. Mineral. Geochem.* **2006**, *62*, 155–167. [CrossRef]
8. Xia, Q.K.; Liu, J.; Kovács, I.; Hao, Y.T.; Li, P.; Yang, X.Z.; Chen, H.; Sheng, Y.M. Water in the Upper Mantle and Deep Crust of Eastern China: Concentration, Distribution and Implications. *Natl. Sci. Rev.* **2019**, *6*, 125–144. [CrossRef]
9. Rapp, R.P.; Irifune, T.; Shimizu, N.; Nishiyama, N.; Norman, M.D.; Inoue, T. Subduction Recycling of Continental Sediments and the Origin of Geochemically Enriched Reservoirs in the Deep Mantle. *Earth Planet. Sci. Lett.* **2008**, *271*, 14–23. [CrossRef]
10. Liu, L.; Zhang, J.S. Differential Contraction of Subducted Lithosphere Layers Generates Deep Earthquakes. *Earth Planet. Sci. Lett.* **2015**, *421*, 98–106. [CrossRef]

11. Irifune, T.; Sekine, T.; Ringwood, A.E.; Hibberson, W.O. The Eclogite–Garnetite Transformations at High Pressure and Some Geophysical Implications.Pdf. *Earth Planet. Sci. Lett.* **1986**, *77*, 245–256. [CrossRef]
12. Gavrilenko, P. Water Solubility in Diopside. Ph.D. Thesis, Bayerisches Geoinstitut, University Bayreuth, Bayreuth, Germany, 2008; p. 139.
13. Smyth, J.R.; Bell, D.R.; Rossman, G.R. Incorporation of hydroxyl in upper-mantle clinopyroxenes. *Nature* **1991**, *351*, 732–735. [CrossRef]
14. Wen, S.; Zhendong, Y.; Bolin, C.; Kai, Y. Role of Water in Deformed Omphacite in UHP Eclogite from the Dabie Mountains, Eastern China. *Acta Geol. Sin. Engl. Ed.* **2003**, *77*, 320–325. [CrossRef]
15. Weis, F.A.; Ros, L.; Reichart, P.; Skogby, H.; Kristiansson, P.; Dollinger, G. Hydrogen Concentration Analysis in Clinopyroxene Using Proton–Proton Scattering Analysis. *Phys. Chem. Miner.* **2018**, *45*, 669–678. [CrossRef]
16. Weis, F.A.; Skogby, H.; Troll, V.R.; Deegan, F.M.; Dahren, B. Magmatic Water Contents Determined through Clinopyroxene: Examples from the Western Canary Islands, Spain. *Geochem. Geophys. Geosys.* **2015**, *16*, 2127–2146. [CrossRef]
17. Warren, J.M.; Hauri, E.H. Pyroxenes as Tracers of Mantle Water Variations. *J. Geophys. Res. Solid Earth* **2014**, *119*, 1851–1881. [CrossRef]
18. Bromiley, G.D.; Keppler, H. An Experimental Investigation of Hydroxyl Solubility in Jadeite and Na-Rich Clinopyroxenes. *Contrib. Mineral. Petrol.* **2004**, *147*, 189–200. [CrossRef]
19. Ghosh, S.; Ohtani, E.; Litasov, K.D.; Suzuki, A.; Dobson, D.; Funakoshi, K. Effect of Water in Depleted Mantle on Post-Spinel Transition and Implication for 660 km Seismic Discontinuity. *Earth Planet. Sci. Lett.* **2013**, *371–372*, 103–111. [CrossRef]
20. Ghosh, S.; Ohtani, E.; Litasov, K.D.; Suzuki, A.; Dobson, D.; Funakoshi, K. Corrigendum to "Effect of Water in Depleted Mantle on Post-Spinel Transition and Implication for 660 Km Seismic Discontinuity" [Earth Planet. Sci. Lett. 371–372 (2013) 103–111]. *Earth Planet. Sci. Lett.* **2013**, *382*, 85–86. [CrossRef]
21. Jacobsen, S.D.; Jiang, F.; Mao, Z.; Duffy, T.S.; Smyth, J.R.; Holl, C.M.; Frost, D.J. Effects of Hydration on the Elastic Properties of Olivine. *Geophys. Res. Lett.* **2008**, *35*. [CrossRef]
22. Mao, Z.; Jacobsen, S.D.; Jiang, F.; Smyth, J.R.; Holl, C.M.; Duffy, T.S. Elasticity of Hydrous Wadsleyite to 12 GPa: Implications for Earth's Transition Zone. *Geophys. Res. Lett.* **2008**, *35*. [CrossRef]
23. Mao, Z.; Lin, J.F.; Jacobsen, S.D.; Duffy, T.S.; Chang, Y.Y.; Smyth, J.R.; Frost, D.J.; Hauri, E.H.; Prakapenka, V.B. Sound Velocities of Hydrous Ringwoodite to 16GPa and 673K. *Earth Planet. Sci. Lett.* **2012**, *331–332*, 112–119. [CrossRef]
24. Wang, W.; Walter, M.J.; Peng, Y.; Redfern, S.; Wu, Z. Constraining Olivine Abundance and Water Content of the Mantle at the 410-Km Discontinuity from the Elasticity of Olivine and Wadsleyite. *Earth Planet. Sci. Lett.* **2019**, *519*, 1–11. [CrossRef]
25. Xu, J.; Zhang, D.; Fan, D.; Zhang, J.S.; Hu, Y.; Guo, X.; Dera, P.; Zhou, W. Phase Transitions in Orthoenstatite and Subduction Zone Dynamics: Effects of Water and Transition Metal Ions. *J. Geophys. Res. Solid Earth* **2018**, *123*, 2723–2737. [CrossRef]
26. Angel, R.J.; Nimis, P.; Mazzucchelli, M.L.; Alvaro, M.; Nestola, F. How Large Are Departures from Lithostatic Pressure? Constraints from Host–Inclusion Elasticity. *J. Metamorph. Geol.* **2015**, *33*, 801–813. [CrossRef]
27. Angel, R.J.; Mazzucchelli, M.L.; Alvaro, M.; Nestola, F. EosFit-Pinc: A Simple GUI for Host-Inclusion Elastic Thermobarometry. *Am. Mineral.* **2019**, *102*, 1957–1960. [CrossRef]
28. Maxisch, T.; Ceder, G. Elastic Properties of Olivine Lix FePO4 from First Principles. *Phys. Rev. B-Condens. Matter Mater. Phys.* **2006**, *73*, 174112. [CrossRef]
29. Sinogeikin, S.V.; Katsura, T.; Bass, J.D. Sound Velocities and Elastic Properties of Fe-Bearing Wadsleyite and Ringwoodite. *J. Geophys. Res. Solid Earth* **1998**, *103*, 20819–20825. [CrossRef]
30. Inoue, T.; Weidner, D.J.; Northrup, P.A.; Parise, J.B. Elastic Properties of Hydrous Ringwoodite (γ-Phase) in Mg_2SiO_4. *Earth Planet. Sci. Lett.* **1998**, *160*, 107–113. [CrossRef]
31. Zhang, D.; Dera, P.K.; Eng, P.J.; Stubbs, J.E.; Zhang, J.S.; Prakapenka, V.B.; Rivers, M.L. High Pressure Single Crystal Diffraction at PX^2. *J. Vis. Exp.* **2017**, *119*, e54660. [CrossRef]
32. Prescher, C.; Prakapenka, V.B. DIOPTAS: A Program for Reduction of Two-Dimensional X-Ray Diffraction Data and Data Exploration. *High Press. Res.* **2015**, *35*, 223–230. [CrossRef]
33. Zhang, J.S.; Bass, J.D.; Taniguchi, T.; Goncharov, A.F.; Chang, Y.Y.; Jacobsen, S.D. Elasticity of Cubic Boron Nitride under Ambient Conditions. *J. Appl. Phys.* **2011**, *109*, 063521. [CrossRef]

34. Zhang, J.S.; Bass, J.D.; Zhu, G. Single-Crystal Brillouin Spectroscopy with CO_2 Laser Heating and Variable Q. *Rev. Sci. Instrum.* **2015**, *86*, 063905. [CrossRef] [PubMed]
35. Oberti, R.; Caporuscio, F.A. Crystal chemistry of clinopyroxenes from mantle eclogites: A study of the key role of the M2 site population by means of crystal-structure refinement. *Am. Mineral.* **1991**, *76*, 1141–1152.
36. Fleet, M.E.; Henznrnc, C.T.; Bancroft, G.M.; Aldridge, L.P. Omphacite Studies, I. The P2ln-C2 /c Transformation. *Am. Mineral.* **1978**, *63*, 1100–1106.
37. Hao, M.; Zhang, J.S.; Pierotti, C.E.; Ren, Z.; Zhang, D. High-Pressure Single-Crystal Elasticity and Thermal Equation of State of Omphacite and Their Implications for the Seismic Properties of Eclogite in the Earth's Interior. *J. Geophys. Res. Solid Earth* **2019**, *124*, 2368–2377. [CrossRef]
38. Hill, R. Elastic Properties of Reinforced Solids: Some Theoretical Principles. *J. Mech. Phys. Solids* **1963**, *5*, 357–372. [CrossRef]
39. Sang, L.; Vanpeteghem, C.B.; Sinogeikin, S.V.; Bass, J.D. The Elastic Properties of Diopside, $CaMgSi_2O_6$. *Am. Mineral.* **2011**, *96*, 224–227. [CrossRef]
40. Bhagat, S.S.; Bass, J.D.; Smyth, J.R. Single-Crystal Elastic Properties of Omphacite-C2/c by Brillouin Spectroscopy. *J. Geophys. Res.* **1992**, *97*, 6843–6848. [CrossRef]
41. Kandelin, J.; Weidener, D.J. The single-crystal elastic properties of jadeite. *Phys. Earth Planet. Inter.* **1988**, *3*, 251–260. [CrossRef]
42. Brown, J.M.; Collins, M.D. Elasticity of an Upper Mantle Clinopyroxene. *Phys. Chem. Miner.* **1998**, *26*, 7–13. [CrossRef]
43. Isaak, D.G.; Ohno, I. Elastic Constants of Chrome-Diopside: Application of Resonant Ultrasound Spectroscopy to Monoclinic Single-Crystals. *Phys. Chem. Miner.* **2003**, *30*, 430–439. [CrossRef]
44. Hao, M.; Pierotti, C.E.; Tkachev, S.; Prakapenka, V.; Zhang, J.S. The Single-Crystal Elastic Properties of the Jadeite-Diopside Solid Solution and Their Implications for the Composition-Dependent Seismic Properties of Eclogite. *Am. Mineral.* **2019**, *104*, 1016–1021. [CrossRef]
45. Bascou, J.; Barruol, G.; Vauchez, A.; Mainprice, D.; Egydio-Silva, M. EBSD-Measured Lattice-Preferred Orientations and Seismic Properties of Eclogites. *Tectonophysics* **2001**, *342*, 61–80. [CrossRef]
46. Bascou, J.; Tommasi, A.; Mainprice, D. Plastic Deformation and Development of Clinopyroxene Lattice Preferred Orientations in Eclogites. *J. Struct. Geol.* **2002**, *24*, 1357–1368. [CrossRef]
47. Ulrich, S.; Mainprice, D. Does Cation Ordering in Omphacite Influence Development of Lattice-Preferred Orientation? *J. Struct. Geol.* **2005**, *27*, 419–431. [CrossRef]
48. Zhang, J.; Green, H.W.; Bozhilov, K.N. Rheology of Omphacite at High Temperature and Pressure and Significance of Its Lattice Preferred Orientations. *Earth Planet. Sci. Lett.* **2006**, *246*, 432–443. [CrossRef]
49. Mazzucchelli, M.L.; Reali, A.; Morganti, S.; Angel, R.J.; Alvaro, M. Elastic Geobarometry for Anisotropic Inclusions in Cubic Hosts. *Lithos* **2019**, *350–351*, 105218. [CrossRef]
50. Alvaro, M.; Mazzucchelli, M.L.; Angel, R.J.; Murri, M.; Campomenosi, N.; Scambelluri, M.; Nestola, F.; Korsakov, A.; Tomilenko, A.A.; Marone, F.; et al. Fossil Subduction Recorded by Quartz from the Coesite Stability Field. *Geology* **2020**, *48*, G46617.1. [CrossRef]
51. Bonazzi, M.; Tumiati, S.; Thomas, J.B.; Angel, R.J.; Alvaro, M. Assessment of the Reliability of Elastic Geobarometry with Quartz Inclusions. *Lithos* **2019**, *350–351*, 105201. [CrossRef]

minerals

MDPI

Article

Cracked, Porous Rocks and Fluids: Moon and Earth Paradox

Jan V.M. Borgomano *, Jérôme Fortin and Yves Guéguen

Laboratoire de Géologie, Ecole Normale Supérieure/CNRS, UMR8538, PSL Research University, 75005 Paris, France; fortin@geologie.ens.fr (J.F.); gueguen@geologie.ens.fr (Y.G.)
* Correspondence: borgomano@geologie.ens.fr

Received: 14 October 2019; Accepted: 7 November 2019; Published: 9 November 2019

Abstract: Elastic wave velocities are key parameters in geosciences. In seismology at a large scale, or in seismic exploration at a more local and shallower scale, they were the main source of information for a long time. At the time of the Apollo mission, Anderson explained the unexpected result of very low velocities in Moon surface rocks by an intense cracking resulting from meteoritic impacts. Yet, it was also known that the Q factor was high. This could appear as a paradox. In the shallow layers of the Earth, rocks are porous. These shallow layers are of major importance in the Earth since they contain fluids. This is why velocities are higher and Q values lower in the Earth's shallow layers than in the Moon's shallow layers. Cracks have a determining effect on elastic properties because they are very compliant. Fluids also play a key role. Combining poroelasticity and effective elasticity, two independent theories much developed since the time of the Apollo mission, makes it possible to revisit the contrasting results observed in the Moon case and in the Earth case. Experimental results obtained on cracked synthetic glass show that dry cracks result in a strong decrease in velocity. On the other hand, saturated porous limestones exhibit a strong frequency-dependent attenuation when thermally cracked. The presence of fluid is the key factor.

Keywords: rock physics; pores; cracks; fluids; effective medium theory; poroelasticity

1. Introduction

In general, the heterogeneity of crustal rocks is mainly the result of variable mineral composition and of the presence of pores and cracks. At all scales, from local to regional, rocks are heterogeneous. The influence of heterogeneity in rock, as long as it remains moderate, can often be handled by considering that the medium is statistically homogeneous on the local scale. That means that a representative volume element (RVE) exists and that any part of the system with a volume much larger than the RVE has identical physical properties.

Within this assumption, use of effective elasticity in order to predict elastic properties and, hence, elastic wave velocities is of direct interest. In shallow conditions, rocks contain pores and cracks. These "defects" were identified for a long time as having a major influence on elastic properties.

When fluids are present, frequency dependence is expected, and the combined use of effective elasticity and poroelasticity allows accounting for it. This can be applied to the results described a long time ago by Anderson [1], who reported that the average sound velocity of Moon rocks was close to provolone cheese and very low in comparison to those found on Earth (Figure 1).

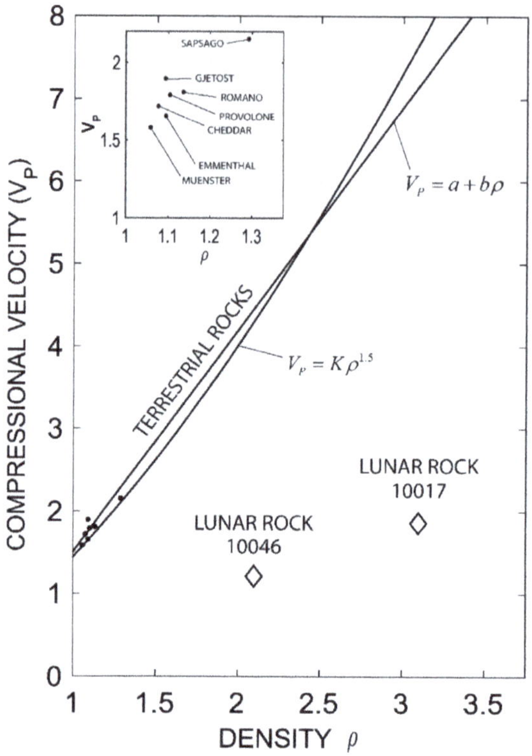

Figure 1. Comparison between sound velocities for lunar rocks and for various earth materials, from Schreiber and Anderson [1].

2. Effective Elasticity of Cracked Rocks

Effective elasticity allows one to derive the elastic moduli for a dry or saturated cracked rock in terms of the parameters defined below. Let us consider the simple, but interesting case of an isotropic distribution of three-dimensional (3D) identical circular cracks of radius r and aspect ratio ζ (thickness/diameter). The non-cracked matrix is assumed to be isotropic. Three groups of parameters are needed: (1) E_0, Young modulus (or equivalently K_0, the bulk modulus, or G_0, the shear modulus) and υ_0 Poisson ratio of the isotropic matrix; (2) K_f, the fluid bulk modulus (the fluid is taken below as liquid water); (3) crack density $\rho = nr^3/V$, if an REV of volume V contains n cracks of radius r (this non-dimensional parameter is expected to vary between 0 and 1, but the model under the assumption of non-interactive cracks is accurate only if ρ is low enough). Because cracks are very compliant, the effect of dry cracks on the modulus is very strong. The bulk modulus is decreased in the dry case (compared to the intact matrix modulus), and so is the shear modulus. Within the approximation of very thin cracks ($\zeta < 10^{-3}$) [2,3], the following relationships exist:

$$\frac{K_0}{K_{dry}} = 1 + \frac{16\left(1-\nu_0^2\right)}{9(1-2\nu_0)}\rho \text{ and } \frac{G_0}{G_{dry}} = 1 + \frac{32(1-\nu_0)(5-\nu_0)}{45(2-\nu_0)}\rho, \quad (1)$$

where K_{dry} stands for the dry cracked rock bulk modulus (same notation for G). Using a standard value $\upsilon_0 = 0.25$, the above equations become $K_0/K_{dry} = 1+ 3.3\rho$ and $G_0/G_{dry} = 1 + 1.45\rho$.

For a water-saturated rock, the effect is less important for the bulk modulus, and more important for the shear modulus. Note that, in (high-frequency) effective elasticity, fluid phases are considered to be immobile. This means that different cracks may experience different fluid pressures since fluid has no time to move. This justifies the name "unrelaxed" for the effective saturated moduli. Experimentally, the appropriate measurements are ultrasonic measurements, because they are very-high-frequency ones, so that the "unrelaxed" situation is obtained. Then, with the same approximations as above [2,3], the following relationships exist:

$$\frac{K_0}{K_{ur}} = 1 \text{ and } \frac{G_0}{G_{ur}} = 1 + \frac{32(1-\nu_0)}{15(2-\nu_0)}\rho, \quad (2)$$

where K_{ur} stands for the unrelaxed saturated cracked rock bulk modulus (same notation for G). Using a standard value $\upsilon_0 = 0.25$, the above equation becomes $G_0/G_{ur} = 1 + 0.91\rho$.

Saturation strongly modifies the dry results; the unrelaxed saturated bulk modulus has, under the assumption of thin cracks, the same value as that of the intact matrix, but the saturated shear modulus is decreased and different from the dry case. Using values reported by Anderson [1], (dry rock P-wave velocity of the order of 2 km/s for lunar rocks instead of 6 km/s for Earth rocks), it is straightforward to check that ρ should be higher than 1. The crack density is so high that the above model under the approximation of non-interactive cracks is no longer quantitative, but the implication is that the cracking is very intense. A very useful (but unavailable) complementary set of data would be the velocity variation of lunar rocks with pressure. Since cracks close under pressure, one expects a strong decrease in crack density at higher pressure, down to a crack density range where the above model would be quantitatively applicable [2,3]. Differential [4,5] or self-consistent schemes [6] may also be used beyond the limits of applicability of the non-interaction approximation.

As Anderson pointed out, this shows that lunar rocks are highly cracked and dry. If not dry, they would exhibit a much higher bulk modulus and, hence, a much higher P-wave velocity.

3. Effective Elasticity of Porous Rocks

The situation is very different for terrestrial rocks. They can contain cracks, but with a much lower crack density. Most importantly, in the crust, sedimentary rocks are porous and not dry. Pores are not very compliant because they are approximately round-shaped. Then, the bulk modulus and the shear modulus can be expressed approximately, in the dry case usually, in terms of porosity Φ [2,3] as follows:

$$\frac{K_0}{K_{dry}} = 1 + \frac{3(1-\nu_0)}{2(1-2\nu_0)}\Phi \text{ and } \frac{G_0}{G_{dry}} = 1 + \frac{15(1-\nu_0)}{7-5\nu_0}\Phi. \quad (3)$$

Using a standard value $\upsilon_0 = 0.25$, the above equations become $K_0/K_{dry} = 1 + 2.25\Phi$ and $G_0/G_{dry} = 1 + 1.9\Phi$.

For the saturated case, unrelaxed moduli are determined as follows:

$$\frac{K_0}{K_{ur}} = 1 \text{ and } \frac{G_0}{G_{ur}} = 1 + \frac{15(1-\nu_0)}{7-5\nu_0}\Phi. \quad (4)$$

In contrast with the previous case (cracks), saturation does not modify the shear modulus compared to the dry case. However, the most important result is that the effect of pores is less important than that of cracks.

4. Poroelasticity

As before, we consider a cracked isotropic rock, with an isotropic distribution of identical cracks. This is a particular case of a porous medium, with a low porosity. The cracks are assumed to be connected and water-saturated. Pores can be present as well. Then, porosity is higher, and cracks and pores are assumed to be connected. For simplification, we consider below a case where only cracks are present.

4.1. Relaxed (Isobaric) Moduli

The assumption is that the fluid is isobaric within an RVE (a reasonable assumption at seismic frequencies that are low enough). This is a different case from the one examined just above. Poroelasticity applies [3]. The situation we are interested in is such that there is no macroscopic fluid flow between RVEs, but the fluid pressure varies from one RVE to another. The bulk and shear moduli are called in that case the undrained bulk and shear moduli K_u and G_u. They can be measured under any condition where there is not enough time for the macroscopic flow to occur from one RVE to another. Then, the following relationships apply:

$$\frac{K_0}{K_u} = 1 \text{ and } \frac{G_0}{G_{dry}} = 1 + \frac{32(1-\nu_0)(5-\nu_0)}{45(2-\nu_0)}\rho \text{ and } \frac{G_{dry}}{G_u} = 1. \tag{5}$$

This simple result means that cracks are "invisible" in that case for the bulk modulus. The cracks affect only the shear modulus.

4.2. Unrelaxed Moduli

If wavespeed measurements are performed at ultrasonic frequencies, typical of laboratory experiments, the RVE is not isobaric. This case is out of the validity range of poroelasticity since fluid pressure is variable within an RVE. Of course, it is well within the validity range of effective elasticity as described above (Equation (2)). Above a certain critical frequency, any experimental measurement of wavespeeds corresponds to this non-isobaric situation. There exists a transition domain between regimes of "low" (relaxed) and "high" (unrelaxed) frequency.

A key question regards the critical frequency f_c value. In fluid-saturated cracked rocks, the effect is called "squirt flow". The critical frequency of "squirt flow" (or local fluid flow) f_c is obtained by calculating the time needed for local fluid motion between two neighbor cracks, as due to local pressure gradients [7].

$$f_c \sim \frac{\zeta^3 E_0}{20\eta}, \tag{6}$$

where, typically, the crack aspect ratio ζ is ~10^{-3}, Young's modulus E_0 is ~70 GPa, and water viscosity η is ~10^{-3} Pa·s, resulting in f_c ~3.5 kHz. This means that, typically, in cracked saturated rocks, one expects to get the "unrelaxed" moduli from ultrasonic data, and the "relaxed" ones from seismic data.

5. Q Factor

It results from the above that, in a saturated cracked medium, wavespeeds are expected to be frequency-dependent. If we use a viscoelastic framework [8], the frequency dependence effect is accompanied by dispersion and attenuation, following the Kramers–Kronig relationships [9].

In the simple case considered above, the dissipation is in shear only. Using the previous parameter values, one gets the following:

$$\frac{\Delta G}{G} = 0.54\rho, \tag{7}$$

where ΔG stands for the maximum variation of G (from the unrelaxed state to the relaxed one, in a water-saturated cracked rock).

In the approximation of the simple linear viscoelastic body, $Q^{-1}{}_{max} = 0.5\ \Delta G/G$, so that the squirt flow effect is expected to produce an attenuation peak on shear waves at f_c of

$$Q^{-1}_{max} = 0.27\rho, \tag{8}$$

and, similarly, will also produce some attenuation for the P-wave ($Q^{-1}{}_{max} = 0.09\rho$).

For a crack density of 0.2 (likely to be lower than that for Moon rocks, but in a range where the above models are approximately valid), one gets an attenuation peak of $Q^{-1}{}_{max}$ = 0.05 for the shear wave ($Q^{-1}{}_{max}$ = 0.02 for the P-wave). This is a high value, possible only if fluids are present.

The above model can be extended to porous rocks. In that case, it can be shown that crack–pore flow also induces a strong bulk effect [10].

In the situation of dry cracks, no dissipation related to fluid flow is expected and Equation (1) would apply both at low and high frequencies. In that case, $\Delta G/G = 0$ and $\Delta K/K = 0$, which results in $Q^{-1}{}_{max} = 0$, regardless of the crack density. This explains the apparent paradox between Moon (high crack density, low attenuation) and Earth rocks (low crack density, high attenuation).

6. Experimental Data

Convenient examples to illustrate the above predictions are given by data obtained on glass and Indiana limestone. A homogenous glass can be quenched, resulting in a connected network of cracks (Figure 2a) [11]. As with most sedimentary rocks, Indiana limestone is porous (Figure 2b). Its crack content can be modified by thermal cracking, but its porosity is a characteristic of the rock that cannot be modified easily. The fact that pores are present is important. It implies that squirt flow can take place not only from crack to crack like in the glass, but as also from crack to pore, resulting in bulk dispersion together with shear dispersion [10]. Here, we present a glass sample that was quenched from a temperature of 300 °C, and an Indiana limestone that was heated up to 500 °C for an hour prior to a natural cooling.

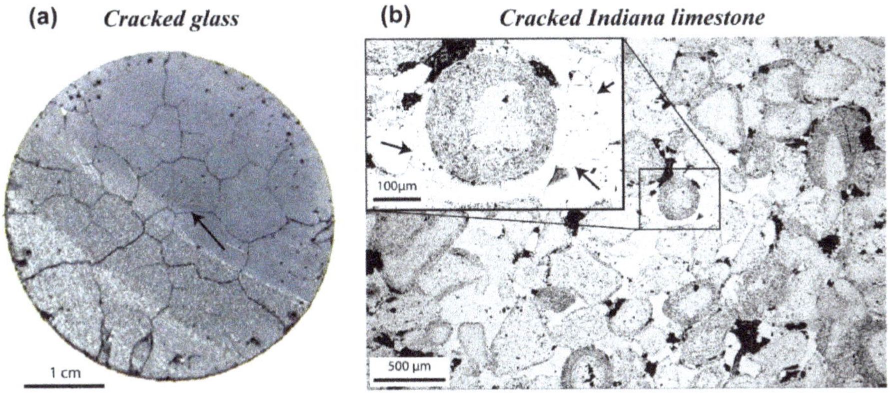

Figure 2. (**a**) Surface photography of cracked glass [12]. (**b**) SEM photomicrograph of Indiana limestone [13]. Arrows show examples of thermally induced cracks.

As seen on the microstructures (Figure 2), the glass was composed of a network of cracks that could be of a millimetric to centimetric scale (Figure 2a), whereas, in the Indiana limestone, the cracks were essentially induced in the homogeneous intergranular cement, with sizes around 100 µm (Figure 2b). The crack porosity of the thermally treated (TT) glass was measured around 0.24% using a mercury porosimeter [14]. Overall, the Indiana sample bears a total porosity of Φ = 11.4%, measured by the triple-weight method [13], essentially distributed between intragranular micropores and some intergranular mesopores (Figure 2b).

Permeability of the cracked glass was measured using the pulse decay method with argon and water [14], and it was found to decrease strongly from 8×10^{-17} to 4×10^{-21} m^2 with an increase in effective pressure from 2 to 20 MPa (Figure 3a). This is characteristic of crack closure with pressure, since the hydraulic conductivity is solely controlled by crack porosity. On the other hand, the permeabilities of the intact and the cracked Indiana limestones, measured with the steady-state flow rate method with water, exhibited a much smaller variation with effective pressure, decreasing from an average of 2.3 × 10^{-17} to 1.4 × 10^{-17} m^2 (Figure 3b). We can note that the results for the intact and cracked limestone differed negligibly, leading to the belief that the pores were the main contributor to the hydraulic conductivity, with a very small contribution of the microcracks (Figure 3b).

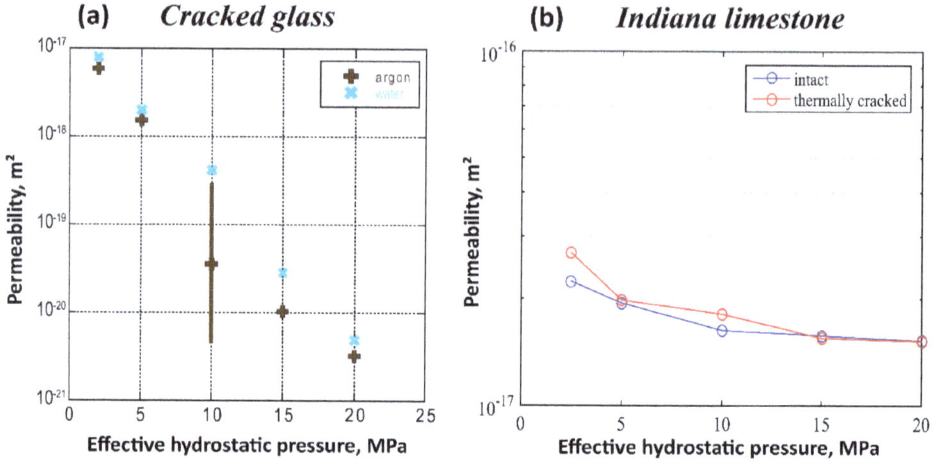

Figure 3. (**a**) Permeability as a function of effective pressure for the cracked glass [14] and (**b**) for the intact and cracked Indiana limestone [13].

Results of the elastic/acoustic properties are presented in Figure 4. For the glass sample, ultrasonic P- and S-wave velocities were measured over a range of effective pressures from 2.5 to 20 MPa [14] (Figure 4a). The intact sample exhibited no pressure dependence. For the cracked sample in dry conditions, the P- and S-wave velocities increased with pressure, consistently with crack closure. However, the water-saturated cracked sample exhibited no pressure dependence, and velocities were close to the intact sample (Figure 4a). This is consistent with the high-frequency unrelaxed regime when crack-to-crack squirt flow occurs (Equation (2)).

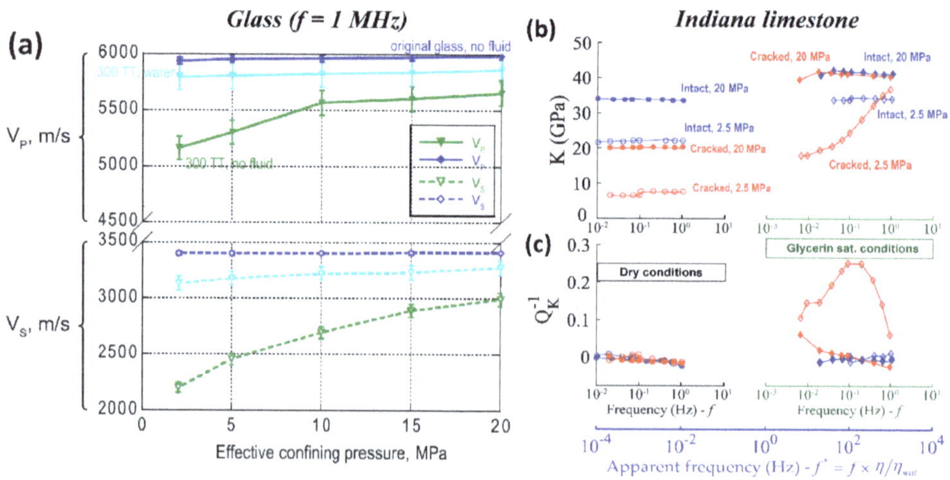

Figure 4. (**a**) Ultrasonic P- and S-wave velocities on the glass sample, intact and thermally cracked (TT) [14]. (**b**) Bulk dispersion and (**c**) attenuation of the intact and cracked Indiana limestone, at effective pressures of 2.5 and 20 MPa. The samples were tested in dry and fluid-saturated conditions. "Apparent frequency" is frequency normalized by the fluid's viscosity, using water as a reference [15].

The Indiana limestone was investigated using a specific apparatus at École normale supérieure (ENS) Paris that measures the frequency dispersion/attenuation of elastic moduli over a broad frequency range (0.004–100 Hz) using the forced-oscillation method [13,15]. Here, the results of the bulk modulus dispersion and attenuation (Q_K^{-1}) are presented (Figures 4b and 4c, respectively), as induced from pure hydrostatic conditions over an experimental frequency range of 0.004–1 Hz and effective pressures of 2.5 and 20 MPa. The frequency was normalized by the dynamic viscosity of the saturating fluid/gas (η), using water as a reference, in order to visualize the results in terms of "apparent frequency" ($f^* = f \times \eta/\eta_{wat}$). With the use of glycerin η_{gly} = 1000 mPa·s) instead of water (η_{wat} = 1 mPa·s), one can investigate the bulk dispersion/attenuation in the apparent seismic frequency range (4–1000 Hz) of a water-saturated rock (Figure 4b,c), as long as solely viscous-driven fluid-flow processes occur such as squirt flow (Equation (6)).

Both the intact and cracked Indiana samples were non-dispersive in the absence of fluid and showed no attenuation (Figure 4b,c); the glycerin-saturated intact sample also showed no attenuation. On the other hand, the glycerin-saturated cracked sample at low effective pressure (2.5 MPa) exhibited a strong bulk dispersion (Figure 4b) in the seismic range, associated with a strong attenuation peak (Figure 4c) around 100 Hz. This dispersion/attenuation was absent for a high effective pressure (20 MPa), related to crack closure. Therefore, the observed dispersion was likely to be related to crack-to-pore squirt flow, with a transition between the relaxed (<10 Hz) and unrelaxed regimes (>1000 Hz). Similarly to the glass sample, the "unrelaxed" elastic properties of the cracked limestone had low sensibility to the effective pressure, contrarily to the dry case.

One may ask if the cut-off frequency of the squirt flow may be predicted. In order to use the prediction given by Equation (6), one must obtain an estimate of the crack aspect ratio (ζ). The skeleton parameters of the glass were given by the intact sample (E_0 = 84 GPa, υ_0 = 0.27), and, for the Indiana limestone, one can take the properties of calcite in carbonates (E_0 = 83 GPa, υ_0 = 0.32, [16]). The pressure closure of a crack is given by $P_{close} = E_0\pi\zeta/\left(4(1-\nu_0{}^2)\right)$ [17], and the valu was found to be around 20 MPa for both the glass and the Indiana limestone, which would give an aspect ratio around $\zeta \approx 2.8 \times 10^{-4}$ for both samples. Therefore, the predicted cut-off frequency for squirt flow (Equation (6)) would be around 92 Hz, under water-saturated conditions ($\eta = 10^{-3}$ Pa·s), which seems in agreement with the results obtained on the cracked limestone (Figure 4b,c).

One interesting alternative method to determine ζ is to use the pressure dependence of the permeability. In a cracked medium, permeability may be calculated as follows [18]:

$$k = k_0 e^{-aP}, \tag{9}$$

where k_0 is the unconfined permeability. Moreover, the aperture (w) of the crack varies with pressure according to Equation (10) [19].

$$w \sim w_0\left(1 - \frac{P}{E_0\zeta}\right), \tag{10}$$

where w_0 is the unconfined aperture. If we assume that the variations of k and w^3 with pressure are proportional, from Equations (9) and (10), we obtain $a = 3/(E_0\zeta)$. This method applies only to a cracked medium and not to a mixture of cracks and pores. For the glass sample, the slope of the permeability versus pressure in Figure 3a (considering the log scale) gives us $a \approx 1.8 \times 10^{-7}$, from which we finally obtain $\zeta = 3/(E_0 a) \approx 2 \times 10^{-4}$, consistent with the previous method.

7. Conclusions

Simple models of isotropic cracked dry rocks show that the elastic wave velocities can decrease substantially if crack density is high. However, P-wave velocities are not really affected in the saturated case (because only the shear modulus is modified). Shear attenuation is predicted if cracked rocks are saturated. This fits well with Moon surface rocks, where a strong velocity decrease is expected for dry cracked rocks, but no strong attenuation.

In the Earth case, rocks are often saturated but crack density is low, implying a small velocity decrease. Attenuation and frequency-dependent behavior are expected because of the fluid and the possible presence of pores. Experimental results documented the squirt-flow effect.

Effective elasticity and poroelasticity provide a satisfactory theoretical background to explain these contrasting situations, and revisit the results obtained long ago by Anderson.

Author Contributions: Conceptualization, J.V.M.B., J.F. and Y.G.; validation, J.V.M.B.; formal analysis, J.V.M.B.; investigation, J.V.M.B.; data curation, J.V.M.B.; writing—original draft preparation, J.V.M.B. and Y.G.; writing—review and editing, J.F.; visualization, J.V.M.B.; supervision, J.F. and Y.G.; project administration, J.F.; funding acquisition, J.F.

Funding: This project was supported by TOTAL, under project number FR00007429.

Conflicts of Interest: The author declares no conflicts of interest.

References

1. Schreiber, E.; Anderson, O.L. Properties and composition of lunar materials: Earth analogies. *Science* **1970**, *168*, 1579–1580. [CrossRef] [PubMed]
2. Fortin, J.; Guéguen, Y.; Schubnel, A. Effect of pore collapse and grain crushing on ultrasonic velocities and Vp/Vs. *J. Geophys. Res.* **2007**, *112*, B08207. [CrossRef]
3. Guéguen, Y.; Kachanov, M. Effective elastic properties of cracked and porous rocks. In *Mechanics of Crustal Rocks, CISM Courses and Lectures*; Leroy, Y., Lehner, F., Eds.; Springer: Vienna, NY, USA, 2011; Volume 533, pp. 73–126.
4. Berryman, J.G.; Pride, S.R.; Wang, H.F. A differential scheme for elastic properties of rocks with dry or saturated cracks. *Geophys. J. Int.* **2002**, *151*, 597–611. [CrossRef]
5. Hashin, Z. The differential scheme and its application to cracked materials. *J. Mech. Phys. Solids* **1988**, *36*, 719–734. [CrossRef]
6. Budiansky, B.; O'Connell, R.J. Elastic moduli of a cracked solid. *Int. J. Solids Struct.* **1976**, *12*, 81–97. [CrossRef]
7. O'Connell, R.J.; Budiansky, B. Viscoelastic properties of fluid-saturated cracked solids. *J. Geophys. Res.* **1977**, *82*, 5719–5735. [CrossRef]
8. Pimienta, L.; Fortin, J.; Guéguen, Y. Effect of fluids and frequencies on Poisson's ratio of sandstone samples. *Geophysics* **2016**, *81*, D183–D195. [CrossRef]
9. O'Connell, R.J.; Budiansky, B. Measurements of dissipation in viscoelastic media. *Geophys. Res. Lett.* **1978**, *5*, 5–8. [CrossRef]
10. Adelinet, M.; Fortin, J.; Guéguen, Y. Dispersion of elastic moduli in a porous-cracked rock. *Tectonophysics* **2011**, *503*, 173–181. [CrossRef]
11. Mallet, C.; Fortin, J.; Guéguen, Y.; Bouyer, F. Effective elastic properties of cracked solids: An experimental investigation. *Int. J. Fract.* **2013**, *182*, 275–282. [CrossRef]
12. Mallet, C.; Fortin, J.; Guéguen, Y.; Bouyer, F. Brittle creep and subcritical crack propagation in glass submitted to triaxial conditions. *J. Geophys. Res. Solid Earth* **2015**, *120*, 879–893. [CrossRef]
13. Borgomano, J. Dispersion of the Elastic Moduli of Saturated Carbonate Rocks: Experimental Study and Modelisation. Ph.D. Thesis, PSL Research University, Paris, France, 2018. Available online: https://tel.archives-ouvertes.fr/tel-01892045v2 (accessed on 30 June 2019).
14. Ougier-Simonin, A.; Guéguen, Y.; Fortin, J.; Schubnel, A.; Bouyer, F. Permeability and elastic properties of cracked glass under pressure. *J. Geophys. Res. Solid Earth* **2011**, *116*, 1–12. [CrossRef]
15. Borgomano, J.V.M.; Pimienta, L.; Fortin, J.; Guéguen, Y. Dispersion and attenuation measurements of the elastic moduli of a dual-porosity limestone. *J. Geophys. Res. Solid Earth* **2017**, *122*, 2690–2711. [CrossRef]
16. Mavko, G.; Mukerji, T.; Dvorkin, J. *The Rock Physics Handbook: Tools for Seismic Analysis of Porous Media*; Cambridge University Press: Stanford, CA, USA, 2009.
17. Walsh, J.B. The effect of cracks on the compressibility of rock. *J. Geophys. Res.* **1965**, *70*, 381–389. [CrossRef]

18. Brace, W.F.; Walsh, J.B.; Frangos, W.T. Permeability of granite under high pressure. *J. Geophys. Res.* **1968**, *73*, 2225–2236. [CrossRef]
19. Jaeger, J.C.; Cook, N.G.; Zimmerman, R. *Fundamentals of Rock Mechanics*; Blackwell Pub.: Malden, MA, USA, 2007.

minerals

MDPI

Article

Effect of Carbon on the Volume of Solid Iron at High Pressure: Implications for Carbon Substitution in Iron Structures and Carbon Content in the Earth's Inner Core

Jing Yang [1,*], Yingwei Fei [1], Xiaojun Hu [2], Eran Greenberg [3] and Vitali B. Prakapenka [3]

1 Geophysical Laboratory, Carnegie Institution of Washington, 5251 Broad Branch Road, N.W., Washington, DC 20015, USA; yfei@carnegiescience.edu
2 School of Science, Wuhan University of Technology, Wuhan 430070, China; xjhu@whut.edu.cn
3 Center for Advanced Radiation Sources, University of Chicago, Chicago, IL 60637, USA; erangre@gmail.com (E.G.); prakapenka@cars.uchicago.edu (V.B.P.)
* Correspondence: jyang@carnegiescience.edu

Received: 15 October 2019; Accepted: 16 November 2019; Published: 20 November 2019

Abstract: Understanding the effect of carbon on the density of hcp (hexagonal-close-packed) Fe-C alloys is essential for modeling the carbon content in the Earth's inner core. Previous studies have focused on the equations of state of iron carbides that may not be applicable to the solid inner core that may incorporate carbon as dissolved carbon in metallic iron. Carbon substitution in hcp-Fe and its effect on the density have never been experimentally studied. We investigated the compression behavior of Fe-C alloys with 0.31 and 1.37 wt % carbon, along with pure iron as a reference, by in-situ X-ray diffraction measurements up to 135 GPa for pure Fe, and 87 GPa for Fe-0.31C and 109 GPa for Fe-1.37C. The results show that the incorporation of carbon in hcp-Fe leads to the expansion of the lattice, contrary to the known effect in body-centered cubic (bcc)-Fe, suggesting a change in the substitution mechanism or local environment. The data on axial compressibility suggest that increasing carbon content could enhance seismic anisotropy in the Earth's inner core. The new thermoelastic parameters allow us to develop a thermoelastic model to estimate the carbon content in the inner core when carbon is incorporated as dissolved carbon hcp-Fe. The required carbon contents to explain the density deficit of Earth's inner core are 1.30 and 0.43 wt % at inner core boundary temperatures of 5000 K and 7000 K, respectively.

Keywords: earth's core; light elements; interstitial carbon; X-ray diffraction

---

1. Introduction

Carbon (C) has been considered as one of the light elements to account for the density deficit of the Earth's core based on emerging evidence from cosmological, seismic, geochemical, and mineral physics data [1–4]. If the Earth's inner core were crystallized from a carbon-bearing core, either iron (Fe) carbide or Fe with interstitial carbon would be a candidate for the solid inner core. Whether the inner core crystallizes into an Fe carbide or Fe with interstitial carbon phase depends on the eutectic composition at inner core boundary (ICB) as well as the bulk carbon content in the Earth's core [2,5,6]. For an Fe carbide core, the Fe_7C_3 phase has been proposed as a leading candidate because its properties at high pressure can reconcile with several seismic observations, including matching the observed density, sound velocity and Poisson's ratio [4,7,8]. On the other hand, thermodynamic calculations of the Fe-C system at the ICB condition based on experimental data at high pressure suggest that if the bulk carbon content is less than 2.24 wt % in the core, a metallic Fe core with interstitial carbon, instead of Fe carbide, would be expected [2].

Interstitial carbon under solubility limit in its host lattice would not change the crystal structure of the host metal. Although it is still debated that whether Fe in the inner core would be in a body-centered cubic (bcc) [9,10] or hexagonal close-packed structure (hcp), the hcp-Fe is commonly accepted as the stable phase in the core [11–13]. The elastic properties of hcp-Fe can also provide a plausible explanation for the observed elastic anisotropy of the inner core [14,15]. The interstitial carbon in Fe would have an effect on the density and elastic properties of Fe at high pressure. Theoretical results show that interstitial carbon in hcp-Fe increases the compressional wave velocity, decreases the shear wave velocity, with enhanced seismic anisotropy. Furthermore, only 1–2.5 wt % of carbon is needed to match the density deficit of the inner core depending on the geotherm profile of the core [16]. However, there have been no experimental data on the effect of carbon on the density and elastic properties of hcp-Fe at high pressure. The current modeling of the Fe-C phase diagram at high pressure assumes that the behavior of carbon in the hcp-Fe structure is similar to that in the bcc- or fcc-Fe [2,5]. Such an assumption needs to be further examined by experiments in the stability field of hcp-Fe. Recent melting experiments on the Fe-C binary system showed that carbon content in the solid Fe remains 1 wt % up to 255 GPa [6]. The authors estimated the density of Fe with 1 wt % carbon by assuming a linear density change between pure Fe, Fe_3C and Fe_7C_3 because of the lack of data on the density of Fe with interstitial carbon. Such a linear interpolation could lead to a large uncertainty in the estimated carbon content.

In this study, we directly determined the density of Fe with dissolved carbon at high pressure to evaluate the relationship between the carbon content and the inner core density deficit. We have determined the equations of state of pure Fe and two Fe-C alloys with 0.31 and 1.37 wt % interstitial carbon up to 135 GPa. The experiments with multiple samples in the same sample chamber under the neon pressure medium provide high precision data that allow systematic evaluation of the carbon effect on the elastic properties and density of hcp-Fe. Using the new data together with previously published thermoelastic parameters for hcp-Fe, we placed constraint on the maximum carbon content in the inner core by comparing the density profiles between our mineral physics model and the seismic observation.

2. Experimental Method

2.1. Sample Synthesis

We used the commercial standard reference materials from National Institute of Standards and Technology (NIST), 19H basic electric steel (Fe with 0.2 wt % carbon, Fe-0.2C) and 5m Cast Iron (Fe with 2.59 wt % carbon, Fe-2.59C), and pure Fe powder (Alfa Aesar, Haverhill, MA, USA, 99.5%) as starting materials for the synthesis of Fe-C alloys with 0.31 wt % carbon (denoted as Fe-0.31C) and 1.37 wt % carbon (denoted as Fe-1.37C). It should be noted that the carbon content in the commercial NIST reference materials were measured in bulk, and they were not homogeneous in the micron scale, which needed further high pressure-temperature treatments. The Fe-0.31C sample was homogenized and sintered from the NIST Fe-0.2C, while Fe-1.37C sample was synthesized from mixed pure Fe and Fe-2.59C with a 1:1 ratio. The starting materials were loaded into an MgO capsule with crushable MgO spacers and sintered using piston cylinder apparatus at 1 GPa and 1150 °C at the Geophysical Laboratory. The temperature was measured with a type-C (W-5%Re vs. W-26%Re) thermocouple for all the experiments. Samples were heated at a rate of 100 °C/min to the set temperature and kept for at least 24 hours to ensure the sample homogeneity.

2.2. Carbon Analytical Procedure

The synthesized samples with a cylinder shape were cut into two pieces using a diamond saw. For each sample, one of the pieces was mounted into resin epoxy and polished with 0.05 μm Al_2O_3 gel on a lapidary wheel using water as lubricant. After a desirable polished surface was achieved, samples were cleaned using ethanol and then coated with Ir with a thickness of 100 nm. The samples were analyzed using electron probe microanalyzer (EMPA, JEOL 8530F, Akishima, Tokyo) at the Geophysical

Laboratory with 12 keV accelerating voltage and 30 nA current. The EMPA is equipped with an LDE6L crystal for high-precision carbon analysis. We also employed cold finger for all the carbon analyses, which reduced the carbon contamination in the vacuum chamber by half compared to that of without cold finger [17,18]. The standards were also coated with Ir with a thickness of 100 nm to keep X-ray absorptions uniform. The on-peak time was 10 second and off-peak time was 5 s for the carbon analysis. NIST pure Fe, synthetic Fe_3C (6.7 wt % C) and Fe_7C_3 (8.43 wt % C) were used as standards for carbon. The typical peak to background ratio for the C Kα was about 14.

The Fe_3C standard was synthesized in a piston cylinder apparatus using a half inch assembly with $BaCO_3$ pressure cell, crushable MgO spacers, and graphite furnace at the Geophysical Laboratory. A pure Fe wire ran in presence of excess carbon was loaded in a graphite capsule and kept at 2 GPa and 1200 °C for 51 h. The Fe_7C_3 standard was synthesized using a multi-anvil apparatus at the University of Michigan [8], with the mixture of fine Fe and excess graphite powers loaded into an MgO capsule and heated to 1200 °C at 14 GPa for 7 h.

2.3. Diamond Anvil Cell and Synchrotron X-ray Diffraction Experiements

For X-ray diffraction (XRD) experiments, high-pressure conditions up to 135 GPa were generated using a symmetric diamond anvil cell (DAC) with a pair of 350 μm flat or 150–250 μm beveled culet diamonds. The bulk samples were cut into pieces with dimensions of approximately 25 μm × 30 μm × 7 μm or 10 μm × 10 μm × 5 μm using a tungsten carbide knife. Rhenium gaskets were pre-indented to a thickness of 40 μm or 25 μm using the opposing diamonds. Holes of 180 μm or 80 μm were drilled in the center of the pre-indented area. Three samples (pure Fe, Fe-0.31C, and Fe-1.37C) were loaded in the sample chamber, separated with spaces filled with neon pressure medium. A small Au flake was also loaded as a pressure calibrant. Neon served as an additional pressure calibrant and a hydrostatic pressure medium [19].

Angle-dispersive synchrotron XRD measurements were performed at 13-IDD GSECARS beamline at the Advanced Photon Source (APS), Argonne National Laboratory (ANL). The incident X-ray beam had a monochromatic wavelength of 0.3344 Å and a full width at half maximum of 5 μm × 5 μm. At each pressure point, the X-ray diffraction images of pure Fe, Fe-0.31C, Fe-1.37C and Au were recorded for an exposure time of 20–50 s. Data were collected at an interval of 1–2 GPa up to 45 GPa for the first run (1st RUN) and 10–15 GPa interval up to 135 GPa for the second run (2nd RUN). The XRD images were integrated using the software Dioptas [20]. The diffraction patterns were analyzed using the Crystallography Data Analysis software (GSAS) [21] to obtain the lattice parameters at different pressures.

3. Results

Iron adopts different packing types upon heating or compression: bcc at ambient condition, fcc upon heating, and hcp upon compression. The speciation of carbon in Fe depends on carbon concentration, pressure, and temperature. The carbon solubility in the bcc-Fe was relatively low compared to that of the case in fcc-Fe due to the large distance between Fe atoms. The maximum carbon solubility in bcc-Fe was less than ~0.022 wt % at ambient pressure and temperature. In fcc-Fe, the maximum carbon solubility was as high as 2.14 wt % at ambient pressure [22], and slightly decreased with increasing pressure [23]. For example, at 1 GPa and 1150 °C, which was the synthesis condition in this study, the maximum carbon dissolved in fcc-Fe is ~2.0 wt %. By a rapid quenching of carbon dissolved fcc-Fe, the carbon stayed in the interstitial site of the lattice to form a bct (body-centered tetragonal) structure, which is a deformed bcc structure [24]. In order to obtain a homogeneous sample with a high carbon concentration, the quenched samples with a bct structure were synthesized and loaded into the diamond anvil cells for further high-pressure experiments. In this study, the results showed that the bct structure transformed to hcp structure at ~14–16 GPa at room temperature, which was slightly higher than the pressure of the bcc to hcp phase transition.

Back scattered electron (BSE) images of our polished samples (Figure S1) showed chemical homogeneity without any quench textures for the entire recovered sample, which was important to extract a tiny sample with the same composition for the DAC experiments. The inhomogeneous samples and samples with quenched texture as shown in BSE images were excluded from further analysis and DAC experiments (Figure S2). We randomly sampled at least 40 points on each sample for the carbon analysis. With the insertion of the cold finger, measurements of the pure Fe standard for carbon always yielded ~0.2 wt % carbon, which was taken as the background and corrected accordingly for the carbon analysis of the alloys. The background corrected counts for the carbon Kα peak showed a linear relationship with the concentrations of carbon in the standards. We used this calibration curve to determine the carbon contents in the samples, yielding 0.31(± 0.03) wt % and 1.37(± 0.06) wt % carbon for the synthesized samples used in the DAC experiments (Figure S1).

We loaded pure Fe, Fe-0.31C, and Fe-1.37C into one sample chamber to obtain high-precision data for comparison. XRD patterns were measured for each sample with a 1–2 GPa interval in the first run. The XRD patterns for the C-bearing samples were consistent with a bct-Fe structure based on peaks broadening comparing to those of pure Fe (Figure 1c). The bct structure was a slightly distorted bcc structure with a diverged *c* axis. It is usually difficult to measure the *c* axis from powder diffraction data because of peak overlap [24]. Figure 1a shows the (110) peak of bcc phase for pure Fe (Fe-0C), Fe-0.31C, and Fe-1.37C at 2.3 GPa, which had the highest relative peak intensity in the XRD pattern. In Figure 1a,c, the intensity of each spectrum was normalized to one to better compare the peak shifts. Furthermore, the full XRD patterns are shown in Figure 1c,d. Generally, the peaks shifted to the higher angle side with increasing pressure and reducing lattice parameters. We only calculated the lattice constant *a* for bct-Fe using (110) and (200) peaks since other peaks used to calculate *c* axis were difficult to be identified (Figure 1c). At 2.3 GPa, comparing the peak positions of Fe (Fe-0C), Fe-0.31C, and Fe-1.37C, the Fe-1.37C had the smallest *a*, and Fe-0C—the largest *a* lattice parameter (Figure 1a). This trend continued up to 14.4 GPa before the hcp phase transition (Figure S3).

We observed the structure transformation in Fe-0C, Fe-0.31C, and Fe-1.37C to hcp structure started at ~16 GPa and was completed at ~20 GPa. Figure 1b shows the (002) and (101) peaks of iron in its hcp structure for Fe-0C, Fe-0.31C and Fe-1.37C at 32.5 Gpa. The (111) peak of solid neon is also shown. We used the neon (111) peak as a primary pressure calibrant above 16 GPa in order to compare the subtle difference in pressure within the sample chamber. With solidification of neon at high pressure, the samples were under a quasi-hydrostatic condition that supports slight deviatoric stress. The measured pressure gradient from the center to the edge within the sample chamber was within 2 GPa at the highest experimental pressure, indicated by the small shift of the neon peaks. In Figure 1b, the slight shift in neon (111) peak among the diffraction patterns of Fe-0C, Fe-0.31C, and Fe-1.37C clearly showed a non-hydrostatic condition in the samples chamber. Fe-1.37C was under the highest pressure (33.1 GPa), and Fe-0C, under the lowest pressure (31.7 GPa). In the meantime, we found that the lattice parameters for hcp-Fe-1.37C were the largest, while the lattice parameters for hcp-Fe-0C were the smallest. The results indicated that the unit cell volume in the hcp structure increased with increasing carbon content under the same pressure (Figure 2a).

Tables S1 and S2 list the compression data for pure Fe, Fe-0.31C and Fe-1.37C in bct and the hcp structure. The pressure vs. volume data for hcp-Fe were used to fit to the third-order Birch–Murnaghan (3BM) equation of state (EoS) [25]:

$$P = \frac{3K_0}{2}\left[\left(\frac{V_0}{V}\right)^{\frac{7}{3}} - \left(\frac{V_0}{V}\right)^{\frac{5}{3}}\right]\left\{1 + \frac{3}{4}(K_0' - 4)\left[\left(\frac{V_0}{V}\right)^{\frac{2}{3}} - 1\right]\right\} \tag{1}$$

where P is the pressure (GPa); V_0 and V are the unit-cell volumes (Å^3) at ambient conditions and high pressures, respectively; K_0 and K_0' are the bulk modulus (GPa) and its pressure derivative at 300 K, respectively. The fitting results for hcp-Fe were shown in Table 1 and the calculated isothermal compression curves for pure Fe, Fe-0.31C and Fe-1.37C in hcp phase are shown in Figure 2.

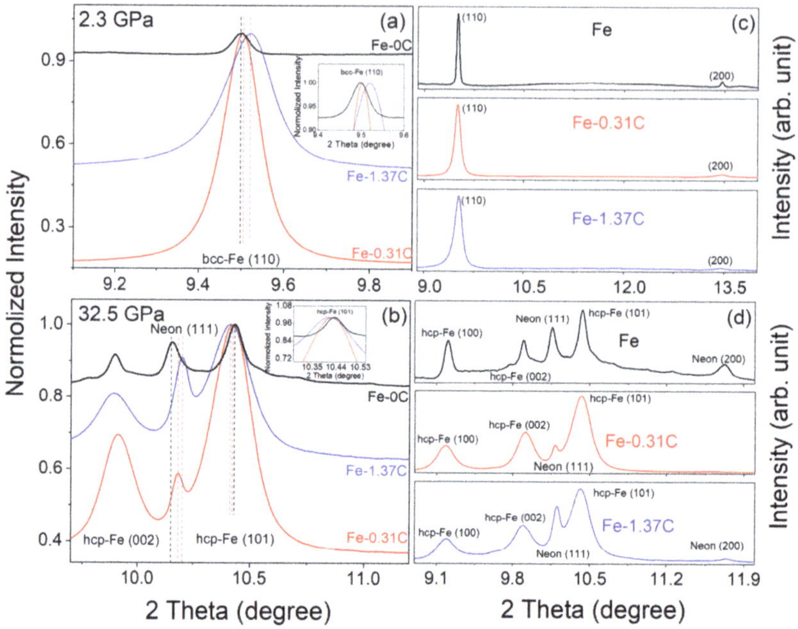

Figure 1. Representative X-ray diffraction (XRD) patterns for Fe-C samples in (**a**,**c**) selected peaks and full XRD patterns of the body-centered tetragonal (bct) structure at 2.3 GPa and (**b**,**d**) selected peaks and full XRD pattern of hcp structure at 32.5 GPa, using Au as first-order pressure calibrant. Insert in (**a**) shows the zoom-in of the (110) peak of body-centered cubic (bcc)-Fe and (**b**) zoom-in of (101) and (002) peaks of hcp-Fe. Fe-0C (pure Fe), Fe-0.31C, Fe-1.37C, and Au are loaded in the same sample chamber using neon as the pressure medium. Black, red and blue curves are for Fe-0C, Fe-0.31C and Fe-1.37C, respectively. Neon (111) peak shift in each pattern for Fe-0C, Fe-0.31C and Fe-1.37C under 32.5 GPa indicates non-hydrostatic pressure in the sample chamber, with Fe-0C under lowest pressure, and Fe-1.37C under highest pressure. Vertical dash lines are plotted to show the center of each peak and for comparison of peak shift.

Figure 2a shows the compression curves for hcp-Fe, Fe-0.31C and Fe-1.37C up to ~135 GPa. The trend of P–V relation of pure Fe was in a good agreement with selected previous results [26–28] up to 135 GPa. The slightly offset of the absolute values of the volume for pure Fe may have been caused by using a different pressure calibrant or pressure medium in each experimental study. Because of the covariance between V_0, K_0 and K_0', determination of V_0 of the high-pressure phase was difficult. In previous experimental studies, V_0 differed from 22.15 to 22.43 Å^3, K_0' differed from 4.5 to 5.8, and K_0 varied from 156 to 202 GPa [26–32]. Here we used fixed K_0' = 4.79 [27] for EoS fitting to systematically compare the EoS parameters for pure Fe, Fe-0.31C, and Fe-1.37C in this study. The volume at ambient V_0 calculated from 3BM EoS is 22.24(4), 22.26(3), and 22.44(3) Å^3 for pure Fe, Fe-0.31C and Fe-1.37C, respectively. Without constraining the K_0', we obtained V_0 = 22.26(6), K_0 = 169.7(5.2) and K_0' = 5.19(16) for pure Fe, V_0 = 22.37(4), K_0 = 168.9 (4.8) and K_0' = 5.19(14) for Fe-0.31C, and V_0 = 20.37(6), K_0 = 182.2(4.2) and K_0' = 4.68(17) for Fe-1.37C. The fitting results were summarized in Table 1. The fitting results showed that Fe with more interstitial carbon would have a larger unit cell volume at the same pressure up to ~135 GPa. The magnitude of increase remained the same over the pressure range of the study because the compression curves were almost parallel (Figure 2a). Figure S4 shows the unit cell volumes of the hcp phase at ambient conditions as a function of carbon content. Contrary to the volume–carbon content relation for the bcc phase in previous studies [2], the unit cell volume of hcp-Fe increased with increasing carbon content. This is the first experimental determination of the effect of carbon on the volume in the hcp structure that invalidates the common assumption that hcp and bcc iron have similar effect with respect to the interstitial carbon on the unit cell volume [2] as shown in Figure S4. The assumption may have significant effect on the calculated phase relations in the Fe-C system at high pressure, particularly in terms of the carbon solubility at very high pressure.

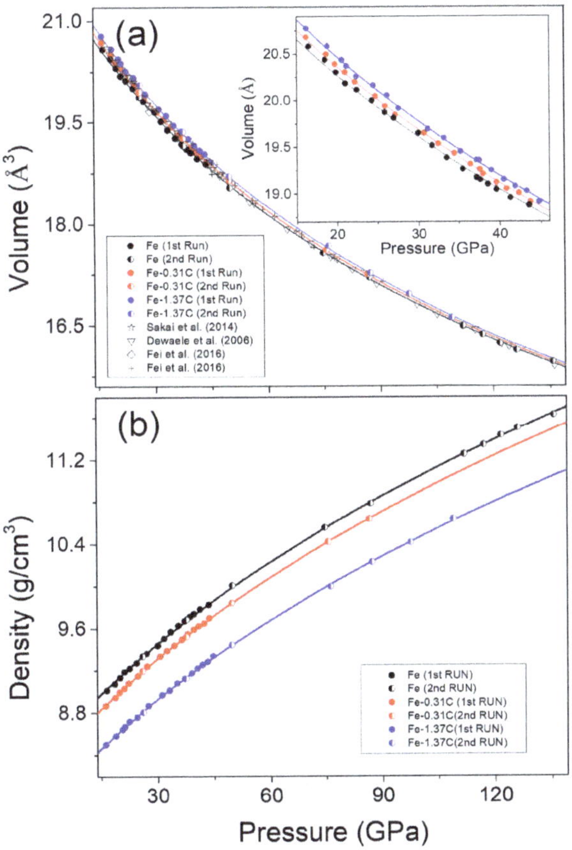

Figure 2. Compression curves for Fe-C samples. (**a**) Pressure and volume relation (P–V) of hcp-Fe with 0, 0.31 and 1.37 wt % of carbon; (**b**) pressure and density relation (P–ρ) of hcp-Fe with 0, 0.31 and 1.37 wt % of carbon. Au was used as the pressure calibrant below 15 GPa, while neon peaks were used as the pressure calibrant above 15 GPa. Solid symbols represent data from the 1st RUN and half-open symbols represent data from the 2nd RUN; the P–V and P–ρ relations were fitted using a third-order Birch–Murnaghan equation of state using combined data from 1st and 2nd RUN; black, red and blue curves are for Fe-0C, Fe-0.31C and Fe-1.37C, respectively.

The effect of interstitial carbon on the density is critical to estimate the carbon content in the Earth's inner core. The density as a function of pressure for hcp-Fe-C alloys are shown in Figure 2b. The results show that the density of Fe with 1.37 wt % of carbon was 5.6% smaller than that of the pure Fe with hcp structure. Due to the volume expansion, interstitial carbon in the hcp structure enhanced the density contrast between the pure iron and Fe-C alloy. First principle calculations using the projector augmented wave (PAW) method also showed a density decrease with additional interstitial carbon in the Fe lattice. Huang et al. (2005) reported that the density of Fe with 1.3 wt % of carbon was 8.7% smaller than that of the pure Fe [33], while others yielded a smaller density decrease [16,34].

For example, Caracas (2017) showed that 1 wt % added carbon would decrease the density of hcp-Fe by about 2.8%, while Li et al. (2018) showed that density of Fe with 1.4 wt % carbon dropped by 2.4% [16,34].

Table 1. Elastic parameters for hcp-structure of Fe and Fe-C alloys in this and previous studies.

Compositions	V_0 (Å^3)	ρ (g/cm^3)	K_0 (GPa)	K_0'
This study				
Fe	22.24(4)	8.392	187.3(1.3)	4.79(fixed)
Fe	22.26(6)	8.332	169.7(5.2)	5.19(16)
Fe-0.31C	22.26(3)	8.242	182.8(1.0)	4.79(fixed)
Fe-0.31C	22.37(4)	8.197	168.9(4.8)	5.19(14)
Fe-1.37C	22.44(3)	7.881	177.6(1.2)	4.79(fixed)
Fe-1.37C	20.37(6)	7.895	182.2(4.2)	4.68(17)
Calculation (Fe)				
Non-magnetic, Steinle-Neumann et al., (1999) [35]	20.45	9.069	292	4.4
Steinle-Neumann et al., (1999) [35]	21.10	8.790	209	5.2
Vočadlo et al., (2009) [36]	20.86(4)	8.891	237.8(4)	4.8
Non-magnetic, Huang et al., (2005) [33]	20.52	9.0385	289	4.46
Non-magnetic, Sha and Cohen (2010) [37]	20.18	9.191	296	4.4
Calculation (Fe-1.3C)				
Huang et al., 2005 [33]	21.46	8.254	291	4.34
Experiments (Fe)				
Fei et al. (2016) [27]	22.427(fixed)	8.270(fixed)	172.7(1.4)	4.79
Fei et al. (2016) [27]	22.185	8.360(24)	191.4(5.3)	4.52(8)
Yamazaki et al. (2012) [31]	22.15	8.373	202(7)	4.5(2)
Mao et al. (1990) [32]	22.35(2)	8.298	165(4)	5.33(90)
Dubrovinsky et al. (2000) [30]	22.40	8.280	156(4)	5.81
Boehler et al. (2008) [29]	22.46(4)	8.258	160(6)	5.6(2)
Dewaele et al. (2006) [28]	22.43	8.269	165(fixed)	4.97(4)

4. Discussion

4.1. Carbon Substitution Mechanism

As shown in the BSE images as well as the EMPA results (Figure S1), the carbon concentration in Fe samples was distributed homogenously, which indicated no other carbon-bearing phases existing in these samples. Thus, the carbon atoms added into Fe must be as an either interstitial or substitutional solid solution, which mostly depends on the atomic size factor. The interstitial solid solution is commonly found in Fe at ambient conditions because the atomic radius of Fe is ~1.7 times larger than the radius of carbon atoms. Carbon atoms fill the holes in the Fe lattice, occupying either octahedral sites (O-sites) or tetrahedral sites (T-sites) [22]. It has been suggested that interstitial carbon energetically prefers to occupy the O-sites in bcc-Fe instead of T-site at ambient conditions [38], where the O-site is nearly twice larger than the T-site. Upon compression, the carbon atom may change its atomic environment in order to become energetically favorable. Theoretical calculations show that when the pressure exceeds 300 GPa, the interstitial carbon atoms in bcc-Fe energetically occupy the T-site instead of the O-site [33], which, however, has not been validated by experiments yet. Changing of the carbon occupancy-site may result in a change of physical properties of the carbon-bearing host. For example, carbon atoms would exert tensile strains on the surrounding crystal lattice to contract the lattice if the carbon atoms occupy the holes that are larger than atomic radius of the carbon atoms; conversely, if the carbon atoms are larger than the holes of the host Fe, the carbon atoms would impose compressive strains in their vicinity, which would expand the volume of the lattice [22]. Increasing interstitial carbon content results in a decrease in unit-cell volume of bcc-Fe [2], which is consistent with the metallurgy study [22]. In this study, we showed that carbon expands both the *a*-axis and *c*-axis as well as the unit-cell volume of the hcp-Fe. The experimental results in this study, for the first time, revealed that carbon in hcp-Fe and bcc-Fe have an opposite effect on molar volume. The results may imply a change in the local environment for carbon atoms. One explanation could be that the carbon atoms enter the smaller T-site in hcp-Fe upon compression, resulting in the expansion of the unit cell because of the imposed compressive strains in the vicinity.

Another explanation for the carbon-induced volume change in hcp-Fe is that carbon atoms substitute the Fe atoms to form a substitutional solid solution instead of an interstitial solid solution. The substitutional solid solution commonly occurs in Fe-Si alloys. Both theoretical and experimental results show that Fe-Si alloys with hcp structure always have a larger unit-cell volume comparing to that of the pure hcp-Fe [39,40]. On the other hand, theoretical calculation shows that the hcp phase containing ~6 at. % of substitutional carbon becomes more stable than that with the same amount of interstitial carbon at pressures above 280 GPa [33]. However, there is no report for the volume difference between the interstitial and substitutional carbon-bearing hcp-Fe. Further experimental analysis to detect the local environment of carbon in hcp-Fe have to be done to better understand the substitution mechanism of carbon in the hcp-Fe at high pressure.

Interstitial atoms in solid Fe would lock the dislocation movements and enhance the strength of the solid Fe, which has been extensively studied in the steel industry [41]. The ability of a metal to plastically deform depends on the ability of dislocations to move. In bcc-Fe, increasing interstitial carbon content increases the hardness and strength of the Fe in the carbon content range of 0.35–1.86 wt %. Both tensile and compressive strain field interactions between dislocations and interstitial atoms result in the restriction of dislocation movements. The environmental change of carbon in Fe, due to the phase transition from bcc to hcp could affect the movement of the dislocation, which would have significant implications for the deformation of Earth's inner core. Thus, further studies on the flow of dislocation for Fe with interstitial or substitutional carbon at high pressure would be necessary for the interpretation of the seismic anisotropy of Earth's interior.

4.2. Anisotropic Axial Compressibility in the hcp-Fe Phase

As shown in Figure 3a, the axial compression data for a and c axes were fitted by the modified Birch–Murnaghan EoS equations:

$$P = \frac{3K_x}{2}\left[\left(\frac{x_0}{x}\right)^7 - \left(\frac{x_0}{x}\right)^5\right]\left\{1 + \frac{3}{2}(K_x' - 4)\left[\left(\frac{x_0}{x}\right)^2 - 1\right]\right\} \quad (2)$$

where x_0 and x represent the lattice constant a or c (Å) at ambient and high-pressure conditions, K_x and K_x' represent the axis compressibility (GPa) and its pressure derivative at 300 K, respectively. To avoid the tradeoff between K_x and K_x', we fixed K_x' at 4 in order to compare the K_x systematically with carbon effects. The fitting yielded *Ka* for Fe-0C, Fe-0.31C and Fe-1.37C were 232.5(3.6), 223.0(2.7) and 210.4(3.7) GPa, respectively. The result showed that the a axis becomes more compressible with the increasing carbon content. For the c axis, the fitting yielded *Kc* for Fe-0C, Fe-0.31C and Fe-1.37C was 224.6(3.3), 236.0(4.8) and 238.8(3.9) GPa, respectively, indicating that Fe-1.37C had the least compressible c axis and Fe-0C had the most compressible c axis (Figure 3b). The results showed that the alloying carbon can increase the anisotropy of the hcp-Fe crystal structure.

It has been observed that strong seismic anisotropy exists in the Earth's inner core, with longitudinal waves travelling ~3% faster along the polar axis than in the equatorial plane [42], which can be attributed to the preferred orientation of hcp-Fe with strong single-crystal elastic anisotropy. The question is widely debated on whether the a axis or c axis of hcp-Fe is parallel to the Earth's equatorial plane. Some studies reported that preferential alignment of the a axis of hcp-Fe in the equatorial plane is the most plausible explanation for the observed compressional wave travel time anomalies or seismic anisotropy in the inner core [43,44]. However, polycrystalline plasticity simulations for pure hcp-Fe found that initially random polycrystal become preferentially oriented with their c axes parallel to the equatorial plane, which can explain the observed seismic anisotropy [45]. Our results showed that $Ka > Kc$ for pure Fe with the a axis being 3.4% less compressible than the c axis, which is also consistent with previous theoretical calculations with $C_{11} < C_{33}$ for hcp-Fe since C_{11} and C_{33} associated purely with compressional stresses and strains along the *a* and *c* axis, respectively [35]. With increasing carbon content to 1.37 wt %, the axial compressibility reversed to $Kc > Ka$ with the *c* axis being 13.5% less compressible than the *a* axis, indicating that interstitial carbon in hcp-Fe could affect the direction of fast travelling waves in the Earth's inner core. It implied that if Fe with interstitial carbon presents in the Earth's core, the crystals preferentially oriented with the *a* axis parallel to the equatorial plane would reconcile the seismic observation, which behaves in an opposite way compared to that of the pure Fe. On the other hand, carbon can enhance seismic anisotropy as increasing carbon content (Figure 2b) with the fastest axis of compressional wave velocity propagation along the *c* axis. This contradicts the theoretical calculations, which showed that the fastest axis of compressional wave velocity propagation tilted away from the *c* axis of Fe [16]. Further experiments should investigate the deformation mechanism in the hcp-Fe with interstitial carbon and the effects of interstitial substituted carbon on the seismic anisotropy in the Earth's core to resolve the contradiction.

The *c*/*a* axial ratio of preferentially aligned crystals of hcp-Fe can be also related to the anisotropy of the Earth's inner core [46]. In theoretical calculations, the *c*/*a* axial ratio serves as an input to calculate the elastic moduli or to model the core anisotropy, which has previously been done in the case of pure hcp-Fe [35,36]. Previous studies have shown that the *c*/*a* ratio in hcp-Fe is closely related to pressure, temperature, and composition, which always increases with increasing temperature, but this effect is strongest at low pressure, and becomes weaker with increasing pressure. Some studies reported a trend of the *c*/*a* ratio of hcp-Fe decreasing with increasing pressure [26–28,31], while others observed a weak trend or even no trend [32,47–49]. The value of *c*/*a* ratio typically ranges from 1.60 to 1.61 over experimental pressures. Our results show that the *c*/*a* ratio for pure Fe was ~1.600–1.605 with a decreasing trend up to 135 GPa (Figure 3d), which is consistent with some of the previous studies. There have been studies on the effect of light elements on the *c*/*a* ratio of the hcp-Fe [46]. For example, adding 2.8 mol % sulfur to Fe-Ni alloy lowers the *c*/*a* ratio by ~0.005 [50]. In this study, it has been observed that carbon also has an effect on the *c*/*a* ratio of the hcp-Fe. The *c*/*a* ratio for Fe-0.31C and Fe-1.37C is ~1.603–1.607 and 1.607 to 1.610, respectively, and adding 1.37 wt % of carbon in the hcp-Fe would increase the *c*/*a* ratio by 0.005–0.007 at high pressures. The experimentally determined *c*/*a* ratio for pure Fe, Fe-0.31C and Fe-1.37C as a function of pressure has been linearly fitted to examine the pressure effect (Figure 3d). The results show that contrary to pure Fe, the *c*/*a* ratio of Fe-C had an increasing trend with increasing pressure, which implied an enhanced anisotropy of hcp-Fe with increasing carbon content.

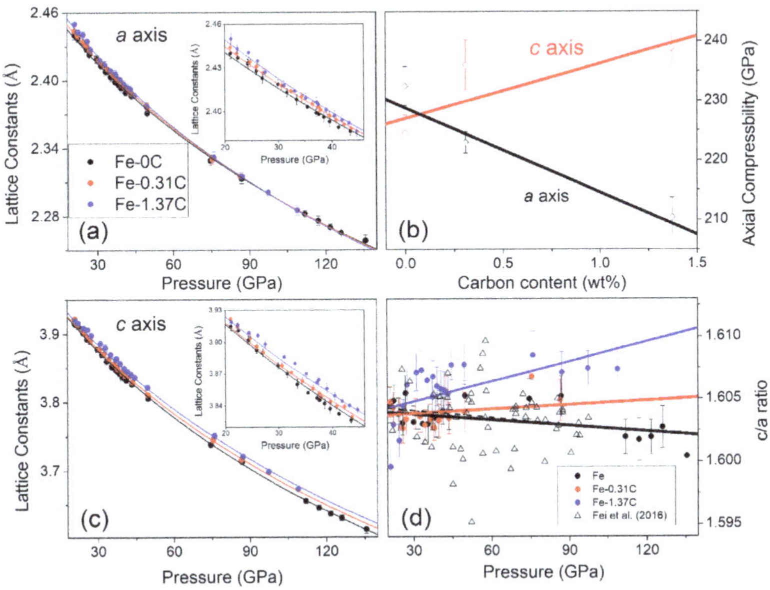

Figure 3. Compression of lattice constants *a* and *c* of the hcp-Fe with 0, 0.31 and 1.37 wt % of carbon. (**a**,**c**) The lattice constants of the *a* and *c* axes as a function of pressure; the experimental data were fitted by modified Birch–Murnaghan equation of state (EoS, see text for more detail); black, red and blue curves are for Fe-0C, Fe-0.31C and Fe-1.37C, respectively; inserts in (**a**,**c**) shows the zoomed plots from 20–45 GPa collected from 1st RUN; (**b**) the axial compressibility as function of carbon content; lines are linearly fitting of the data; (**d**) *c*/*a* ratio of hcp-Fe with 0, 0.31, and 1.37 wt % of carbon as a function of pressure; lines in (**b**,**d**) are linear fittings of the data.

4.3. Density Deficit of Earth's Inner Core

Incorporation of carbon in metallic iron can significantly reduce the density of hcp-Fe, providing a solution for the observed density deficit of the core if carbon is the dominant light element in the Earth's core. Due to the limited pressure and temperature range of the experimental data, thermoelastic modelling and extrapolation are needed to determine the carbon effect on the density under core conditions. Using the derived elastic moduli, the density of the hcp-Fe, Fe-0.31C, and Fe-1.37C at 300 K together with previously determined thermoelastic parameters, we can model the density of carbon-bearing Fe along an adiabatic geotherm by determining the thermal pressures based on the Debye model:

$$P_{th} = P - P_{300\mathrm{K}} = \frac{\gamma_{vib}}{V}\int_{300}^{T} C_{V,vib}dT + \frac{\gamma_e}{V}\int_{300}^{T} C_{Ve}dT \quad (3)$$

where $P_{300\mathrm{K}}$ is the reference pressure at 300 K from this study, V is the unit cell volume (Å^3), γ_{vib} is the vibrational Grüneisen parameter, γ_e is the electronic Grüneisen parameter, $C_{V,vib}$ is the vibrational specific heat (Jkg^{-1}K^{-1}) and C_{Ve} is the electronic specific heat (Jkg^{-1}K^{-1}). Because there are no measured thermoelastic parameters for carbon-bearing hcp-Fe, we adopted the parameters (Table S3) from the pure hcp-Fe derived from combined static *P*–*V*–*T* data up to 200 GPa and 1795 K [27] and dynamic Hugoniot data [51] for the calculations. To test the reliability of the modelling, we calculated the density profile for pure-Fe along the adiabatic geotherm with a fixed ICB temperature of 6000 K (T_{ICB} = 6000 K), which is comparable to that reported by Fei et al. (2016) within the experimental uncertainties [27] (Figure 4). The calculated density profile for Fe-1.37C plotted below that of the

Preliminary Reference Earth Model (PREM) for the inner core, indicating that less than 1.37 wt % carbon in the inner core was required to explain the density deficit.

With the established pressure–density relations for pure Fe, Fe-0.31C and Fe-1.37C along the geotherm with the fixed T_{ICB} of 5000, 6000 and 7000 K, we can estimate the carbon content that can reconciled the density deficit in the Earth's inner core depending on the thermal profile of the core. Using the Debye thermoelastic model (f) as described in Equation (3), we calculated the density of pure Fe (d_{Fe}), Fe-0.31C ($d_{Fe\text{-}0.31C}$), and Fe-1.37C ($d_{Fe\text{-}1.37C}$) as a function of pressure alone the geotherm:

$$d_{Fe} = f_{Fe}(P, T), \quad (4)$$

$$d_{Fe\text{-}0.31C} = (1 - 0.31\%)d_{Fe} + 0.31\%\Delta d_{0.31}, \quad (5)$$

$$d_{Fe\text{-}1.37C} = (1 - 1.37\%)d_{Fe} + 1.37\%\Delta d_{1.37} \quad (6)$$

where $\Delta d_{0.31}$ and $\Delta d_{1.37}$ represent density deviation from the density profile of pure Fe for Fe-0.31C and Fe-1.37C, respectively, which represent the carbon effect on the density of hcp-Fe. Using the PREM density profile (d_{PREM}) as the constraint, we performed least-squares regression to determine the carbon content through:

$$d_{PREM} = (1 - x)d_{Fe} + x\Delta d_x \quad (7)$$

where x is the predicted carbon content in the Earth inner core. The results yielded the carbon content x = 1.30(8)% with T_{ICB} = 5000 K, x = 0.95(5)% with T_{ICB} = 6000 K, and x = 0.43(13)% with T_{ICB} = 7000 K, which provided the best match to the density profile of PREM at inner core condition (Figure 4). Our results showed that on the basis of density constraint alone, incorporation of 0.43–1.30 wt % carbon in the inner core is sufficient to account for the inner core density deficit depending on the thermal profile. Previous results on the melting of Fe-C binary system showed that the carbon solubility in the solid Fe would remain a constant value of 1 wt % up to 255 GPa [6]. If this is the case and carbon is the only light element in the Earth's inner core, ~6000 K would be a good estimation of the inner core boundary temperature.

Considering the carbon partitioning between liquid and solid Fe, $D_C^{liquid/solid} \approx 2$, in the Fe-C binary system at inner core boundary conditions [2], the outer core would contain ~0.86–2.60 wt % carbon based solely on partitioning data. The determined carbon content in the outer core would be less than the recently determined eutectic composition of Fe-C system, which is 3 wt % carbon at the inner core boundary [6]. Thus, we would expect a hcp-Fe phase with dissolved carbon instead of Fe carbide crystallize at the inner core boundary. Previous studies have suggested that 1.7–3.8 wt % carbon is necessary to account for the density deficit [52,53] in the Earth's outer core, which is comparable with our estimations for the carbon content in the Earth's outer core. The amount of carbon would be reduced if other light elements also exist in the inner core. Sound velocity could provide additional constraint on the carbon content in the core. Theoretical calculations show that interstitial carbon increases compressional seismic wave velocity, while decreases shear wave velocity [16]. On the other hand, Li et al. (2018) inferred that the decreasing sound velocities and density of hcp-Fe with dissolved carbon cannot simultaneously match PREM, which has not been validated experimentally [34]. Thus, the sound velocity measurement of hcp-Fe with dissolved carbon, not only iron carbides, at high pressures would be essential for understanding the role of carbon in the solid inner core.

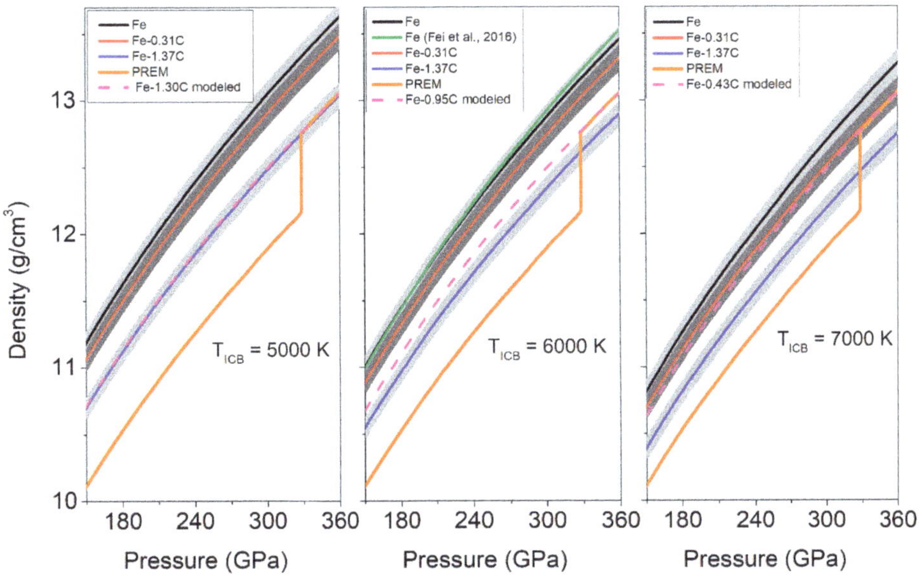

Figure 4. The calculated densities of hcp-Fe with 0, 0.31 and 1.37 wt % of carbon along the adiabatic geotherm with fixed temperatures of 5000, 6000 and 7000 K at the inner core boundary (T_{ICB}), compared with density profile of the core from PREM. Black, red and blue solid lines represent the density profile of pure Fe, Fe-0.31C and Fe-1.37C, respectively. The grey shaded area represents the propagated uncertainties for extrapolated density. The green line is the literature data for pure Fe along the adiabatic geotherm with a fixed temperature of 6000 K at the inner core boundary (Fei et al., 2016 [27]). The orange line represents the density profile of PREM seismic model. The dashed magenta line represent the modeled densities of Fe contains 1.30(7) 0.95 (5) and 0.43(10) wt % carbon at 5000, 6000, and 7000 K of T_{ICB} that can reconcile the density profile of the PREM.

Supplementary Materials: The following are available online at http://www.mdpi.com/2075-163X/9/12/720/s1, Figure S1: Carbon analysis for recovered Fe-C samples by electron microprobe; Figure S2: Back scattered electron images for unsuccessful synthesis runs; Figure S3: The length of lattice constant a as a function of pressure for Fe, Fe-0.31C, and Fe-1.37C; Figure S4: Unit cell volume of bcc- and hcp-Fe as a function of carbon content at ambient condition; Table S1: Compression data for pure-Fe, Fe-0.31C and Fe-1.37C in bct structure using Au as the pressure calibrant; Table S2: Compression data for pure-Fe, Fe-0.31C and Fe-1.37C in hcp phases using Ne as the pressure calibrant; Table S3: Thermoelastic parameters used for modeling the density of hcp-Fe, Fe-0.31C and Fe-1.37C along an adiabatic geotherm with T_{ICB} = 5000, 6000 and 7000 K.

Author Contributions: Conceptualization, J.Y. and Y.F.; methodology, J.Y., Y.F., E.G. and V.B.P.; formal analysis, J.Y. and X.H.; writing, J.Y. and Y.F.; all authors contribute review and editing.

Funding: This work was funded by an National Science Foundation (NSF) grant (EAR-1619868 to Y.F.). GSECARS was supported by the National Science Foundation (EAR-0622171) and U.S. Department of Energy (DE-FG02-94ER14466) under contract DE-AC02-06CH11357. APS was supported by Department of Energy Basic Energy Sciences Program (DOE-BES), under contract DE-AC02-0611357.

Acknowledgments: The authors thank Zhixue Du for their assistance with the X-ray diffraction experiments at 13ID-D, GSECARS. The authors also thank Megan Duncan and Emma Bullock for their guidance of carbon analysis.

Conflicts of Interest: The authors declare no conflict of interest.

References

1. Buchwald, V.F. *Handbook of Iron Meteorites: Their History, Distribution, Composition, and Structure*; University of California Press: Berkeley, CA, USA, 1975.

2. Fei, Y.; Brosh, E. Experimental study and thermodynamic calculations of phase relations in the Fe–C system at high pressure. *Earth Planet. Sci. Lett.* **2014**, *408*, 155–162. [CrossRef]
3. Wood, B.J. Carbon in the core. *Earth Planet. Sci. Lett.* **1993**, *117*, 593–607. [CrossRef]
4. Prescher, C.; Dubrovinsky, L.; Bykova, E.; Kupenko, I.; Glazyrin, K.; Kantor, A.; McCammon, C.; Mookherjee, M.; Nakajima, Y.; Miyajima, N. High Poisson's ratio of Earth's inner core explained by carbon alloying. *Nat. Geosci.* **2015**, *8*, 220. [CrossRef]
5. Lord, O.; Walter, M.; Dasgupta, R.; Walker, D.; Clark, S. Melting in the Fe–C system to 70 GPa. *Earth Planet. Sci. Lett.* **2009**, *284*, 157–167. [CrossRef]
6. Mashino, I.; Miozzi, F.; Hirose, K.; Morard, G.; Sinmyo, R. Melting experiments on the Fe–C binary system up to 255 GPa: Constraints on the carbon content in the Earth's core. *Earth Planet. Sci. Lett.* **2019**, *515*, 135–144. [CrossRef]
7. Chen, B.; Li, Z.; Zhang, D.; Liu, J.; Hu, M.Y.; Zhao, J.; Bi, W.; Alp, E.E.; Xiao, Y.; Chow, P. Hidden carbon in Earth's inner core revealed by shear softening in dense Fe_7C_3. *Proc. Natl. Acad. Sci. USA* **2014**, *111*, 17755–17758. [CrossRef]
8. Liu, J.; Li, J.; Ikuta, D. Elastic softening in Fe_7C_3 with implications for Earth's deep carbon reservoirs. *J. Geophys. Res. Solid Earth* **2016**, *121*, 1514–1524. [CrossRef]
9. Belonoshko, A.B.; Ahuja, R.; Johansson, B. Stability of the body-centred-cubic phase of iron in the Earth's inner core. *Nature* **2003**, *424*, 1032. [CrossRef]
10. Belonoshko, A.B.; Lukinov, T.; Fu, J.; Zhao, J.; Davis, S.; Simak, S.I. Stabilization of body-centred cubic iron under inner-core conditions. *Nat. Geosci.* **2017**, *10*, 312. [CrossRef]
11. Tateno, S.; Hirose, K.; Ohishi, Y.; Tatsumi, Y. The structure of iron in Earth's inner core. *Science* **2010**, *330*, 359–361. [CrossRef]
12. Mao, H.-k.; Shu, J.; Shen, G.; Hemley, R.J.; Li, B.; Singh, A.K. Elasticity and rheology of iron above 220 GPa and the nature of the Earth's inner core. *Nature* **1998**, *396*, 741. [CrossRef]
13. Stixrude, L.; Cohen, R. High-pressure elasticity of iron and anisotropy of Earth's inner core. *Science* **1995**, *267*, 1972–1975. [CrossRef] [PubMed]
14. Karato, S.-I. Inner core anisotropy due to the magnetic field—Induced preferred orientation of iron. *Science* **1993**, *262*, 1708–1711. [CrossRef] [PubMed]
15. Song, X. Anisotropy of the Earth's inner core. *Rev. Geophys.* **1997**, *35*, 297–313. [CrossRef]
16. Caracas, R. The influence of carbon on the seismic properties of solid iron. *Geophys. Res. Lett.* **2017**, *44*, 128–134. [CrossRef]
17. Dasgupta, R.; Walker, D. Carbon solubility in core melts in a shallow magma ocean environment and distribution of carbon between the Earth's core and the mantle. *Geochim. Cosmochim. Acta* **2008**, *72*, 4627–4641. [CrossRef]
18. Robaut, F.; Crisci, A.; Durand-Charre, M.; Jouanne, D. Practical aspects of carbon content determination in carburized steels by EPMA. *Microsc. Microanal.* **2006**, *12*, 331–334. [CrossRef]
19. Fei, Y.; Ricolleau, A.; Frank, M.; Mibe, K.; Shen, G.; Prakapenka, V. Toward an internally consistent pressure scale. *Proc. Natl. Acad. Sci. USA* **2007**, *104*, 9182–9186. [CrossRef]
20. Prescher, C.; Prakapenka, V.B. DIOPTAS: A program for reduction of two-dimensional X-ray diffraction data and data exploration. *High Press. Res.* **2015**, *35*, 223–230. [CrossRef]
21. Toby, B.H. EXPGUI, a graphical user interface for GSAS. *J. Appl. Crystallogr.* **2001**, *34*, 210–213. [CrossRef]
22. Callister, W.D.; Rethwisch, D.G. *Materials Science and Engineering: An Introduction*; John Wiley & Sons: New York, NY, USA, 2007; Volume 7.
23. Walker, D.; Dasgupta, R.; Li, J.; Buono, A. Nonstoichiometry and growth of some Fe carbides. *Contrib. Mineral. Petrol.* **2013**, *166*, 935–957. [CrossRef]
24. Lu, Y.; Yu, H.; Cai, X.; Rong, Y.; Sisson, R.D. Martensite lattice parameter measured by modern X-ray Diffraction in Fe-C alloy. In Proceedings of the 23rd International Federation of Heat Treatment and Surface Engineering Congress (IFHTSE 2016), Savannah, GA, USA, 18–21 April 2016.
25. Birch, F. Finite elastic strain of cubic crystals. *Phys. Rev.* **1947**, *71*, 809. [CrossRef]
26. Sakai, T.; Takahashi, S.; Nishitani, N.; Mashino, I.; Ohtani, E.; Hirao, N. Equation of state of pure iron and $Fe_{0.9}Ni_{0.1}$ alloy up to 3 Mbar. *Phys. Earth Planet. Inter.* **2014**, *228*, 114–126. [CrossRef]
27. Fei, Y.; Murphy, C.; Shibazaki, Y.; Shahar, A.; Huang, H. Thermal equation of state of hcp-iron: Constraint on the density deficit of Earth's solid inner core. *Geophys. Res. Lett.* **2016**, *43*, 6837–6843. [CrossRef]

28. Dewaele, A.; Loubeyre, P.; Occelli, F.; Mezouar, M.; Dorogokupets, P.I.; Torrent, M. Quasihydrostatic equation of state of iron above 2 Mbar. *Phys. Rev. Lett.* **2006**, *97*, 215504. [CrossRef] [PubMed]
29. Boehler, R.; Santamaría-Pérez, D.; Errandonea, D.; Mezouar, M. Melting, density, and anisotropy of iron at core conditions: New X-ray measurements to 150 GPa. *J. Phys. Conf. Ser.* **2008**, *121*, 022018. [CrossRef]
30. Dubrovinsky, L.; Saxena, S.; Tutti, F.; Rekhi, S.; LeBehan, T. In situ X-ray study of thermal expansion and phase transition of iron at multimegabar pressure. *Phys. Rev. Lett* **2000**, *84*, 1720. [CrossRef]
31. Yamazaki, D.; Ito, E.; Yoshino, T.; Yoneda, A.; Guo, X.; Zhang, B.; Sun, W.; Shimojuku, A.; Tsujino, N.; Kunimoto, T. P-V-T equation of state for ε-iron up to 80 GPa and 1900 K using the Kawai-type high pressure apparatus equipped with sintered diamond anvils. *Geophys. Res. Lett.* **2012**, *39*. [CrossRef]
32. Mao, H.; Wu, Y.; Chen, L.; Shu, J.; Jephcoat, A.P. Static compression of iron to 300 GPa and $Fe_{0.8}Ni_{0.2}$ alloy to 260 GPa: Implications for composition of the core. *J. Geophys. Res. Solid Earth* **1990**, *95*, 21737–21742. [CrossRef]
33. Huang, L.; Skorodumova, N.; Belonoshko, A.B.; Johansson, B.; Ahuja, R. Carbon in iron phases under high pressure. *Geophys. Res. Lett.* **2005**, *32*. [CrossRef]
34. Li, Y.; Vočadlo, L.; Brodholt, J.P. The elastic properties of hcp-Fe alloys under the conditions of the Earth's inner core. *Earth Planet. Sci. Lett.* **2018**, *493*, 118–127. [CrossRef]
35. Steinle-Neumann, G.; Stixrude, L.; Cohen, R.E. First-principles elastic constants for the hcp transition metals Fe, Co, and Re at high pressure. *Phys. Rev. B* **1999**, *60*, 791. [CrossRef]
36. Vočadlo, L.; Dobson, D.P.; Wood, I.G. Ab initio calculations of the elasticity of hcp-Fe as a function of temperature at inner-core pressure. *Earth Planet. Sci. Lett.* **2009**, *288*, 534–538. [CrossRef]
37. Sha, X.; Cohen, R. First-principles thermal equation of state and thermoelasticity of hcp Fe at high pressures. *Phys. Rev. B* **2010**, *81*, 094105. [CrossRef]
38. Williamson, G.; Smallman, R. X-ray evidence for the interstitial position of carbon in α-iron. *Acta Crystallogr.* **1953**, *6*, 361–362. [CrossRef]
39. Komabayashi, T.; Pesce, G.; Morard, G.; Antonangeli, D.; Sinmyo, R.; Mezouar, M. Phase transition boundary between fcc and hcp structures in Fe-Si alloy and its implications for terrestrial planetary cores. *Am. Mineral. J. Earth Planet. Mater.* **2019**, *104*, 94–99. [CrossRef]
40. Fischer, R.A.; Campbell, A.J.; Caracas, R.; Reaman, D.M.; Heinz, D.L.; Dera, P.; Prakapenka, V.B. Equations of state in the Fe-FeSi system at high pressures and temperatures. *J. Geophys. Res. Solid Earth* **2014**, *119*, 2810–2827. [CrossRef]
41. Goldschmid, H.J. *Interstitial Alloys*; Springer: Berlin, Germany, 2013.
42. Morelli, A.; Dziewonski, A.M.; Woodhouse, J.H. Anisotropy of the inner core inferred from PKIKP travel times. *Geophys. Res. Lett.* **1986**, *13*, 1545–1548. [CrossRef]
43. Lincot, A.; Merkel, S.; Cardin, P. Is inner core seismic anisotropy a marker for plastic flow of cubic iron? *Geophys. Res. Lett.* **2015**, *42*, 1326–1333. [CrossRef]
44. Antonangeli, D.; Merkel, S.; Farber, D.L. Elastic anisotropy in hcp metals at high pressure and the sound wave anisotropy of the Earth's inner core. *Geophys. Res. Lett.* **2006**, *33*. [CrossRef]
45. Buffett, B.; Wenk, H.-R. Texturing of the Earth's inner core by Maxwell stresses. *Nature* **2001**, *413*, 60. [CrossRef] [PubMed]
46. Fischer, R.A.; Campbell, A.J. The axial ratio of hcp Fe and Fe–Ni–Si alloys to the conditions of Earth's inner core. *Am. Mineral.* **2015**, *100*, 2718–2724. [CrossRef]
47. Jephcoat, A.P.; Mao, H.; Bell, P.M. Static compression of iron to 78 GPa with rare gas solids as pressure-transmitting media. *J. Geophys. Res. Solid Earth* **1986**, *91*, 4677–4684. [CrossRef]
48. Ma, Y.; Somayazulu, M.; Shen, G.; Mao, H.-K.; Shu, J.; Hemley, R.J. In situ X-ray diffraction studies of iron to Earth-core conditions. *Phys. Earth Planet. Inter.* **2004**, *143*, 455–467. [CrossRef]
49. Ono, S.; Kikegawa, T.; Hirao, N.; Mibe, K. High-pressure magnetic transition in hcp-Fe. *Am. Mineral.* **2010**, *95*, 880–883. [CrossRef]
50. Sakai, T.; Ohtani, E.; Kamada, S.; Terasaki, H.; Hirao, N. Compression of $Fe_{88.1}Ni_{9.1}S_{2.8}$ alloy up to the pressure of Earth's inner core. *J. Geophys. Res. Solid Earth* **2012**, *117*. [CrossRef]
51. Brown, J.; Fritz, J.; Hixson, R. Hugoniot data for iron. *J. Appl. Phys.* **2000**, *88*, 5496–5498. [CrossRef]

52. Nakajima, Y.; Imada, S.; Hirose, K.; Komabayashi, T.; Ozawa, H.; Tateno, S.; Tsutsui, S.; Kuwayama, Y.; Baron, A.Q. Carbon-depleted outer core revealed by sound velocity measurements of liquid iron–carbon alloy. *Nat. Commun.* **2015**, *6*, 8942. [CrossRef]
53. Morard, G.; Andrault, D.; Antonangeli, D.; Nakajima, Y.; Auzende, A.; Boulard, E.; Cervera, S.; Clark, A.; Lord, O.; Siebert, J. Fe–FeO and Fe–Fe_3C melting relations at Earth's core–mantle boundary conditions: Implications for a volatile-rich or oxygen-rich core. *Earth Planet. Sci. Lett.* **2017**, *473*, 94–103. [CrossRef]

minerals

MDPI

Article

Viscoelastic Behaviour from Complementary Forced-Oscillation and Microcreep Tests

Ian Jackson

Research School of Earth Sciences, Australian National University, Canberra, ACT 0200, Australia; Ian.Jackson@anu.edu.au

Received: 12 October 2019; Accepted: 19 November 2019; Published: 21 November 2019

Abstract: There is an important complementarity between experimental methods for the study of high-temperature viscoelasticity in the time and frequency domains that has not always been fully exploited. Here, we show that the parallel processing of forced-oscillation data and microcreep records, involving the consistent use of either Andrade or extended Burgers creep function models, yields a robust composite modulus-dissipation dataset spanning a broader range of periods than either technique alone. In fitting this dataset, the alternative Andrade and extended Burgers models differ in their partitioning of strain between the anelastic and viscous contributions. The extended Burgers model is preferred because it involves a finite range of anelastic relaxation times and, accordingly, a well-defined anelastic relaxation strength. The new strategy offers the prospect of better constraining the transition between transient and steady-state creep or, equivalently, between anelastic and viscous behaviour.

Keywords: viscoelasticity; anelasticity; creep function; forced-oscillation methods; microcreep methods

1. Introduction

Viscoelastic relaxation at small strain amplitudes is of geophysical significance as the cause of the attenuation and frequency-dependent wavespeeds (dispersion) of seismic waves within the Earth's deep interior [1–3]. For example, Minster and Anderson [4] demonstrated the potentially important link between dislocation creep and seismic-wave attenuation, and Karato [5] showed how the temperature sensitivity of wavespeeds is enhanced by viscoelastic relaxation.

In a very influential contribution, Goetze [6] reviewed the extensive literature concerning internal friction in metals, measured mainly with resonance (torsional pendulum) techniques, and emphasized the need for high-temperature laboratory measurements on geological materials at sub-Hz teleseismic frequencies rather than the MHz frequencies of conventional laboratory wave-propagation methods. This challenge has been addressed mainly with superior sub-resonant forced-oscillation techniques in multiple laboratories worldwide (e.g., [7–12]).

Concerning such methods for the study of high-temperature viscoelasticity, it is well known that there is, in principle, a valuable complementarity between time-domain and frequency- (or period-) domain methods (e.g., [13]). In practice, time-domain microcreep studies have been used to constrain the value of steady-state viscosity for use in modelling the results of sub-resonant forced oscillation studies [10,11]. In our laboratory, consistency between forced-oscillation and microcreep data has been used as evidence of linearity of mechanical behaviour [14], and microcreep studies have been used to provide a qualitative or quantitative indication of the extent to which the inelastic strain is recoverable on removal of the applied stress [15,16]. However, we have not yet fully exploited the valuable complementarity between the forced-oscillation and microcreep methods. The purpose of this investigation is to explore how results obtained in our laboratory from the two methods might best be combined to take advantage of the complementarity.

If a steady stress is suddenly applied at time $t = 0$, i.e., $\sigma(t) = H(t)$, where H is the Heaviside step function, the resulting strain is specified by the creep function $J(t)$. Provided only that the mechanical behaviour is linear, the strain $\varepsilon(t) = \varepsilon_0\exp(i\omega t - \delta)$ associated with a stress $\sigma(t) = \sigma_0\exp(i\omega t)$, that is sinusoidally time-varying with angular frequency ω, may be calculated by superposition of the strains resulting from a series of consecutive step-function changes in stress each of infinitesimal amplitude. In this way, it is established that

$$\varepsilon(t) = J^*(\omega)\sigma(t) \tag{1}$$

where the (complex) dynamic compliance $J^*(\omega)$ is related to the creep function $J(t)$ by (e.g., [13,17]).

$$J^*(\omega) = i\omega \int_0^{\infty} J(\xi)\exp(-i\omega\xi)d\xi \tag{2}$$

In the representation of linear viscoelastic behaviour, the alternative Andrade and extended Burgers creep functions $J(t)$, respectively,

$$J(t) = J_U + \beta t^n + t/\eta \tag{3}$$

and

$$J(t) = J_U\{1 + \Delta \int_{-\infty}^{\infty} D(\ln\tau)[1 - \exp(-t/\tau)]d\ln\tau\} + t/\eta \tag{4}$$

are widely used. In Equations (3) and (4), J_U is the unrelaxed compliance, β is the coefficient of the term representing the transient creep varying as the fractional n-th power of time t, and η is the steady-state viscosity. For the Burgers model, the parameter Δ is the anelastic relaxation strength associated with the distribution $D(\ln\tau)$ of anelastic relaxation times.

The distribution $D(\ln\tau)$ of anelastic relaxation times is commonly prescribed by separately normalised contributions $D_B(\ln\tau)$ and $D_P(\ln\tau)$, appropriate for monotonic dissipation background and a superimposed peak, respectively, along with the associated modulus dispersion:

$D_B(\ln\tau) = \alpha\tau^{\alpha}/(\tau_H^{\alpha} - \tau_L^{\alpha})$, for $\tau_L \le \tau \le \tau_H$, and zero elsewhere [4,18].

$$D_P(\ln\tau) = \sigma^{-1}(2\pi)^{-1/2}\exp\{-[\ln(\tau/\tau_P)/\sigma]^2\}. \tag{5}$$

In Equation (5), τ_L and τ_H are respectively the lower and upper limits of the distribution of anelastic relaxation times τ, and α the τ-exponent in the distribution $D_B(\ln\tau)$. The parameters τ_P and σ define the centre and width, respectively, of the log-normal distribution of relaxation times given by $D_P(\ln\tau)$. The duration of transient creep involving grain-boundary sliding has been identified with the Maxwell time $\tau_M = \eta J_U$ [19]. Accordingly, consistent with insights from more recent micromechanical modelling of grain-boundary sliding [20] and our recent practice [16], we here set $\tau_H = \tau_M$.

The real and negative imaginary parts of the associated dynamic compliance $J^*(\omega) = J_1(\omega) - iJ_2(\omega)$ for the Andrade model are

$$\begin{aligned} J_1(\omega) &= J_U + \beta\Gamma(1 + n)\omega^{-n}\cos(n\pi/2) \\ J_2(\omega) &= \beta\Gamma(1 + n)\omega^{-n}\sin(n\pi/2) + 1/\eta\omega. \end{aligned} \tag{6}$$

The corresponding quantities for the extended Burgers model are

$$\begin{aligned} J_1(\omega) &= J_U\{1 + \Delta \int_{-\infty}^{\infty} D(\ln\tau)d\ln\tau/(1 + \omega^2\tau^2)\} \\ J_2(\omega) &= \omega J_U\Delta \int_{-\infty}^{\infty} \tau D(\ln\tau)d\ln\tau/(1 + \omega^2\tau^2) + 1/\eta\omega. \end{aligned} \tag{7}$$

In equivalent parameterisations, the Maxwell time $\tau_M = \eta J_U$ takes the place of viscosity η. The (stiffness) modulus $M(\omega)$ and strain energy dissipation $Q^{-1}(\omega)$ then follow as

$$M(\omega) = [J_1^2(\omega) + J_2^2(\omega)]^{-1/2} \text{ and } Q^{-1}(\omega) = J_2(\omega)/J_1(\omega). \quad (8)$$

Accordingly, it is seen that, in principle, the creep function $J(t)$ and the dynamic compliance $J^*(\omega)$ contain identical information concerning the mechanical behaviour.

In practice, however, there are two important qualifications. Forced-oscillation measurements of shear modulus G (or Young's modulus E) and associated strain-energy dissipation Q^{-1} involve comparison of stress and strain signals of prescribed frequency or period (T_o) that are intensively sampled—even at short periods. In contrast, microcreep records, of much longer duration, are normally much more sparsely sampled. Thus, in our laboratory, the forced-oscillation and microcreep records are routinely sampled at frequencies of $128/T_o$ and 1 Hz, respectively. Accordingly, forced-oscillation data are expected to better resolve the behaviour at short timescales, whereas microcreep records will better resolve the behaviour at long timescales. Furthermore, our microcreep testing protocol involving successive switching of the torque between steady values 0, +L, 0, −L, and 0 has the capacity to distinguish between recoverable (and therefore anelastic) and permanent (viscous) strains, as demonstrated below. Forced-oscillation records comprising multiple complete cycles of oscillation have no such capacity to distinguish between anelastic and viscous behaviour.

2. Materials and Methods

2.1. The Processing of Experimental Forced-Oscillation Data

In our laboratory, shear modulus G and dissipation Q^{-1} are measured at each prescribed oscillation period T_o by comparing the complex torsional compliances (rad $(\text{Nm})^{-1}$) measured on complementary experimental assemblies containing, respectively, the cylindrical specimen of interest and a control specimen of known, nearly elastic, shear modulus [9]. Calculation of the compliance differential between the two assemblies serves to eliminate the contribution to the overall compliance from ceramic/steel torsion rods in series mechanically with the specimens. Accordingly, the compliance differential is the difference in compliance between the metal-jacketed specimen and a similarly jacketed control specimen of either polycrystalline alumina or a sapphire crystal. The compliances of the respective assemblies are routinely corrected for the small perturbation caused by interaction between the gas (argon) pressure medium and the central plate moving between the pair of relatively closely spaced outer fixed plates of the capacitance displacement transducers [21]. The force exerted by the gas medium on the central plate perturbs the observed compliance both by bending the lever arms on which the transducer central plates are mounted and by contributing to the torque. If the space between two parallel circular plates of radius R is occupied by gas (argon) of density ρ and viscosity η, the force $F(t)$ on the plate with normal displacement $d(t) = d_0\exp(i\omega t)$ is

$$F(t) = -\{\pi\rho\omega^2 R^4/8h[(b/2)^{-1}\tanh(b/2) - 1]\} \times d(t) \quad (9)$$

with

$$b = (1 + i)h/\delta, \text{ and boundary layer thickness } \delta = (2\eta/\rho\omega)^{1/2}. \quad (10)$$

The compliance differential is also corrected for any small differences between the dimensions and jacketing of the two specimens. Established procedures for the routine processing of such forced-oscillation data for each of a series of oscillation periods T_o, approximately logarithmically equally spaced at each temperature T, are outlined in previous publications [9,22]. All such G and Q^{-1} data obtained at relatively high temperatures and oscillation periods of 1–1000 s with $\log Q^{-1} > -2.2$ are then usually fitted by a non-linear least-squares procedure to an extended Burgers creep function model (e.g., [23]).

2.2. The Processing of Complementary Torsional Microcreep Records

Torsional microcreep tests in our laboratory, conducted on the same assembly as for the forced-oscillation measurements, involve the application of a torque of amplitude 0, +L, 0, −L, and 0 for successive time intervals typically of 2000 s duration each [4]. The first segment is used to estimate and correct for any linear drift, leaving a four-segment record of 8000 s duration within which the torque is switched at times t_i, $(i = 1, 2, 3, 4)$. The switching of the torque can be modelled as the superposition of Heaviside step functions of appropriate sign s_i (+1 at t_1 and t_4, −1 at t_2 and t_3) with rise times of order 1 s.

The motion of the central transducer plate relative to the pair of fixed plates following each such switching of the steady torque is well approximated by the expression

$$d(t) = d_0[1 - \exp(-\kappa t) + \gamma t]. \tag{11}$$

The force exerted by the gas medium on the moving plate is found from an analysis similar to that of [21] to be

$$F(t) = -3\pi R^4\eta\gamma d_0/2h^3 - \pi\kappa^2\rho R^4 d_0\exp(-\kappa t)/\{8h[(b/2)^{-1}\tan(b/2) - 1]\} \tag{12}$$

with

$$b = h/\delta, \text{ and boundary layer thickness now } \delta = (\eta/\kappa\rho)^{1/2}. \tag{13}$$

In Equation (12), the first term on the right-hand side is the force associated with the steady rate $d_0\gamma$ of plate separation, whereas the second term relates to the transiently enhanced rate $\kappa d_0\exp(-\kappa t)$ of plate separation. Representative second-segment microcreep data have been fitted to Equation (11), allowing calculation of and correction for the influence of the forces (Equation (12)) exerted on the moving transducer plates. The resulting perturbations are of order 0.001 μm at $t = 1$ s—negligible when compared with displacement amplitudes of 50–100 μm. By comparison, the correction for forced-oscillation displacement amplitudes of similar magnitude at the shortest period of 1.28 s is of order 0.1 μm—large enough to justify their routine correction.

Accordingly, the raw microcreep data are first processed to obtain a quantity termed the instantaneous torsional compliance $S_{spc}(t)$, being the time-dependent twist (radian) per unit torque (Nm) for the specimen assembly containing the polycrystalline specimen sandwiched between torsion rods within the enclosing metal jacket. Subtraction of the corresponding twist per unit torque for the reference assembly, $S_{ref}(t)$, in which a control specimen (either Lucalox^{TM} polycrystalline alumina or sapphire) of known properties is substituted for the specimen, eliminates the unwanted contribution from the steel and alumina torsion rods. The difference $S_{rel}(t)$ is thus the twist of the jacketed specimen relative to that of the jacketed control specimen, yet to be corrected for any (usually minor) differences in geometry. This difference signal $S_{rel}(t)$ is fitted to a function $S_{fit}(t)$ that is the superposition of the responses to each of the torque switching episodes, prescribed by the appropriate creep function $J(t')$:

$$S_{fit}(t) = \Sigma_{i=1,k}\, s_i J(t_i') \tag{14}$$

where $t_i' = t - t_i$ is the time elapsed since the i-th switching of the torque, for time t belonging to the k–th ($k = 1, 4$) segment of the four-segment record.

We have explored the option of weighting the fit by specifying an uncertainty in $S_{rel}(t)$ proportional to $\log t_j'$, t_j' being the time elapsed since the most recent (k–th) switching of the torque. The effect is to weight the data early in each segment relatively more heavily—in order to strengthen the connection with the forced-oscillation data. For the same reason, it may be desirable, as explored in the Results section, to fix certain parameters in the creep function model at values constrained by the forced oscillation data. For reasons of parametric economy, the Andrade creep function (Equation (3)) has thus far been preferred over the extended Burgers alternative (Equation (4)) for use in Equation (14) [16].

The Andrade creep function $J(t)$ associated with this time-domain fit is then Laplace transformed to obtain the corresponding complex dynamic compliance at selected periods T_o (those of the forced-oscillation experiments within the range 1–1000 s, and similarly chosen periods within the range 1000–10,000 s). The further processing of the microcreep data then proceeds exactly as for the forced-oscillation data. For the purpose of reconciling the results of the forced-oscillation and microcreep tests, the critical step in the processing of the microcreep record is therefore the fitting of the differential compliance $S_{rel}(t)$ to Equation (14).

A normally very small correction is then applied to the Laplace transform of $S_{rel}(t)$ for any differences in geometry between the two assemblies, to obtain the relative dynamic torsional compliance $S_{rl}(T_o)$. $S_{rl}(T_o)$ is added to the dynamic torsional compliance $S_{jc}(T_o)$ calculated *a priori* for the jacketted control specimen from forced-oscillation data for the relevant jacket metal, along with data concerning the viscoelasticity of polycrystalline alumina, or appropriate temperature-dependent elastic properties of single-crystal sapphire. The result is an estimate of the dynamic torsional compliance $S_{js}(T_o)$ of the jacketted specimen. The reciprocal of S_{js}, the torsional stiffness (Nm rad^{-1}), is then corrected for the stiffness of the jacket (inclusive of any foil wrapper used to control redox conditions), and inverted to obtain the (complex) dynamic torsional compliance $S_s(T_o)$ of the bare cylindrical specimen from which its shear modulus G and dissipation Q^{-1} are calculated. At each stage of this process, conducted within the period domain, the dynamic torsional compliance, fitted to an Andrade creep function, allows the construction of an associated virtual microcreep record by superposition of the responses to the successive episodes of torque switching (through Equation (14)). The results for $S_{rl}(t)$, $S_{jc}(t)$, $S_{js}(t)$ and $S_s(t)$ are intended to provide a clear indication of the relative contributions of the control specimen and jacket (inclusive of any liner) to the observed behaviour.

Concerning the recoverability of the strain, we first examine the use of the Andrade creep function in Equation (14) for the fit to the instantaneous torsional compliance within the second segment of the four-segment record as follows

$$S_{fit}(t) = \Sigma_{i=1,2}\, s_i – J(t_i') = \beta[(t - t_1)^n - (t - t_2)^n] + (t_2 - t_1)/\eta. \tag{15}$$

For $t >> t_i$ ($i = 1, 2$), this expression becomes approximately

$$S_{fit}(t) \approx n\beta(t_2 - t_1)t^{n-1} + (t_2 - t_1)/\eta. \tag{16}$$

Thus, as $t \to \infty$, the first term in Equation (16) with n typically ~1/3 becomes vanishingly small, indicating that the strain associated with the βt^n term in the Andrade creep function is ultimately recoverable.

Accordingly, the fraction f_R of the total inelastic strain $\beta t^n + t/\eta$ at time t that is ultimately recoverable following removal of the applied torque is [16].

$$f_R = \beta t^n/(\beta t^n + t/\eta) = 1/(1 + t^{1-n}/\beta\eta) \tag{17}$$

Such an estimate can be made for the Andrade model $S_{fit}(t)$ fitted to the compliance difference $S_{rel}(t)$ between the raw microcreep records right through to the Andrade model fitted to the final $S_s(T_o)$ data—with closely consistent results concerning the fraction of recoverable strain.

For the alternative extended Burgers model, the fraction f_R of the inelastic strain that is recoverable varies with elapsed time t as

$$f_R = \Delta\int_{-\infty}^{\infty} D(\ln\tau)[1 - \exp(-t/\tau)d\ln\tau/[\Delta\int_{-\infty}^{\infty} D(\ln\tau)[1 - \exp(-t/\tau)]d\ln\tau + t/\tau_M] \tag{18}$$

3. Results

3.1. Reconciliation of Forced-Oscillation and Microcreep Data with the Andrade Model

Our established procedure for the processing of forced-oscillation data was applied to representative results, obtained at 1200 °C and 200 MPa, for a specimen assembly containing an $Ni_{70}Fe_{30}$-sleeved Ti-doped Fo_{90} olivine specimen and for a reference assembly containing a similarly sleeved sapphire control specimen [23]. The fact that the forced-oscillation data best resolve the behaviour at relatively short periods was highlighted by the experience of fitting the Andrade model to forced-oscillation data spanning the period range 1–1000 s. It proved possible to refine the values of the parameters J_U, n, and β, but for the viscosity η, only a minimum value near 0.4×10^6 GPa.s compatible with the data could be established. The resulting period-dependent values of the shear modulus G and dissipation Q^{-1}, thus derived from forced-oscillation data, will later be compared with the same quantities inferred from the microcreep records.

The previously described procedure for the processing of microcreep data [16] was applied to the complementary microcreep record (Figure 1) for the same specimen and reference assemblies under the same conditions (1200 °C and 200 MPa). The viscoelastic behaviour, calculated a priori, for the jacketed sapphire control specimen ('jc' in Figure 1b) and the behaviour of the reference assembly as a whole ('ref' in Figure 1a) reveal a close approach to perfectly elastic behaviour. Small departures from the elastic ideal are attributable to the high-temperature viscoelasticity of the metallic materials of jacket and liner, and of the ceramic torsion rods.

The Andrade creep function model was fitted through Equation (14) to the compliance differential $S_{rel}(t)$ between the two assemblies, with ('w') and without ('u') the weighting described in the Materials and Methods section, and with ('f') and without (' ') the exponent n fixed at the value provided by the Andrade fit to the forced-oscillation data. Overall, the resulting 'u', 'uf', 'w', and 'wf' fits represent $S_{rel}(t)$ almost equally well. Accordingly, only the 'wf' fit is displayed in Figure 1.

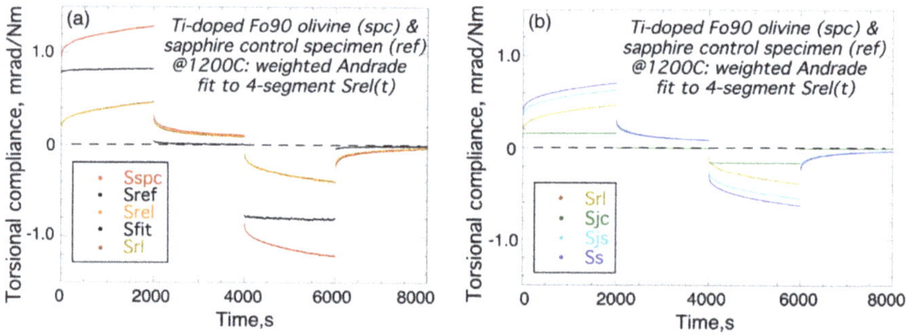

Figure 1. The processing of the four-segment microcreep record for a Ti-doped synthetic olivine polycrystal tested at 1200 °C [23]. (**a**) Instantaneous torsional compliances associated with raw microcreep data for the specimen (S_{spc}) and reference (S_{ref}) assemblies, along with their difference (S_{rel}) fitted with weighting described in the text to an Andrade model (S_{fit}) with exponent n from the Andrade fit to the period-dependent shear modulus and dissipation measured in the parallel forced-oscillation tests (i.e., the 'wf' model). The curve labelled S_{rl} represents the Andrade fit to the differential dynamic torsional compliance corrected in the period domain for minor geometrical differences between the two assemblies. Note that the S_{rel}, S_{fit}, and S_{rl} curves are very nearly coincident, but that the residuals $S_{rel}(t)$-$S_{fit}(t)$ for both the 'wf' and the other fits are plotted in Figure 2. (**b**) Virtual microcreep records representing further stages in the processing of the microcreep data as described in the text: S_{rl} as in panel (**a**), S_{jc} for the jacketed sapphire control specimen, S_{js} for the jacketed specimen, and S_s for the bare olivine specimen.

The optimal Andrade model (S_{fit}), thus fitted to $S_{rel}(t)$ was transformed to the period (T_o) domain for all steps in the further processing, as described previously. $S_{fit}(T_o)$ was first corrected for minor geometrical differences between the two assemblies (S_{rl}), and then added to the torsional compliance (S_{jc}) for the jacketed control specimen (Figure 1b). The result (S_{js}) is thus a robust estimate of the behaviour of the jacketed specimen which then requires a modest correction for the contribution of the enclosing NiFe liner and steel jacket—modelled as mild steel—to isolate the behaviour (S_s) of the bare specimen. Although the processing of $S_{fit}(t)$, to obtain ultimately the shear modulus and dissipation, was conducted within the period domain, at each step a virtual four-segment microcreep record (respectively labelled S_{rl}, S_{jc}, S_{js}, and S_s) was calculated by superimposing the appropriately time-delayed versions of the Andrade creep function fitted to the period-dependent complex compliance $S_i(T_o)$ with i = 'rl', 'jc', 'js', and 's' (Figure 1b).

For the alternative 'u', 'uf', 'w', and 'wf' Andrade models, the RMS misfits range narrowly between 0.67×10^{-5} rad/Nm and 0.72×10^{-5} rad/Nm—amounting to ~6% of the unrelaxed compliance J_U. The associated residuals displayed in Figure 2 are dominated by offsets of order 10^{-5} rad/Nm between successive segments of the record and by noise, typically of 5–10 s period. Such noise on the raw displacement time series is attributed to imperfect discrimination by the pairs of displacement transducers against flexural disturbance of the assembly. Parallel experiments, in which the displacement transducers are configured for measurement of the flexural rather than torsional response, indicate that the convective motions of the dense gas pressure medium within the internal furnace result in flexural disturbance of the specimen assembly.

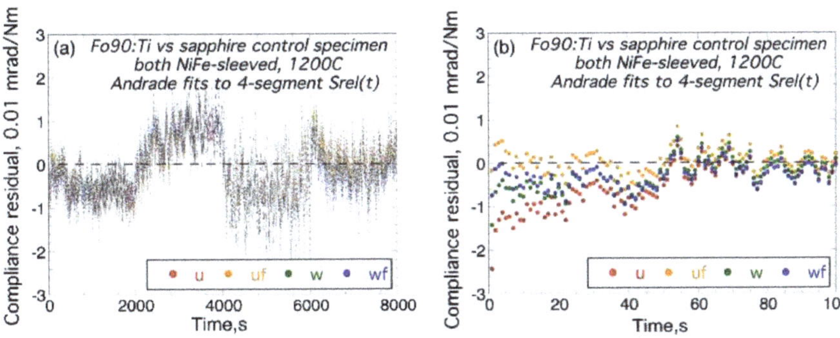

Figure 2. The residual $S_{rel}(t) - S_{fit}(t)$ versus time t for the alternative 'u', 'uf', 'w', and 'wf' Andrade models described in the text for (**a**) the entire four-segment record, and (**b**) the first 100 s of the first segment. Note that the 'wf' fit and the results of its use in further processing are plotted in Figure 1.

Divergence between the alternative Andrade models at short times (Figure 2b) reflects the strong covariance between the Andrade model parameters J_U, n, and β with correlation coefficients > 0.9 in magnitude. The somewhat divergent behaviour amongst the various Andrade creep functions for t < 50 s (Figure 2b) highlights the limited capacity of microcreep data to resolve mechanical behaviour at relatively short periods.

The complex dynamic compliance for the bare specimen, ultimately derived from the microcreep data, provides estimates through Equations (6) and (8) of the shear modulus G and dissipation Q^{-1}, which are compared in Figure 3 with the corresponding quantities from forced-oscillation data obtained for the same experimental assemblies at the same temperature. As expected, there is a broad consistency between the results obtained from the data acquired in the period (or frequency) and time domains—especially at the longer periods. The divergence at short periods between the results derived from forced-oscillation data and microcreep records is evidently reduced by according more weight to the $S_{rel}(t)$ observations soon after each torque switching event (Figure 3, 'w')—more so than by using the forced-oscillation data to constrain the value of the exponent n for the fits to the microcreep $S_{rel}(t)$ (Figure 3, 'uf' and 'wf').

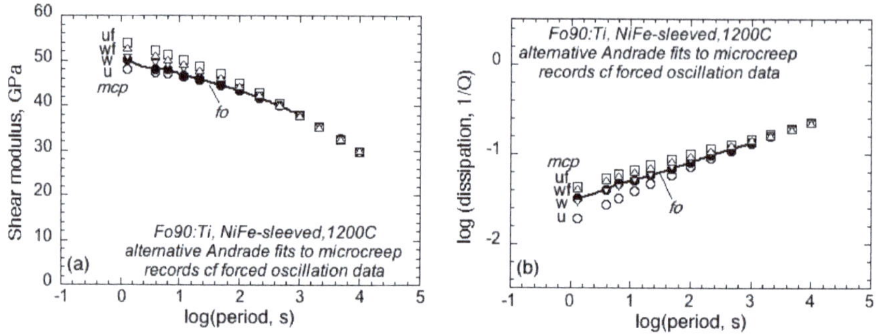

Figure 3. A comparison of (**a**) shear modulus G and (**b**) associated strain-energy dissipation Q^{-1} for the NiFe-sleeved synthetic Ti-doped Fo_{90} olivine specimen at 1200 °C—derived from forced-oscillation data (solid symbols and curve labelled 'fo') and from the four-segment microcreep record (open symbols, appropriately labelled, representing the results obtained with the alternative 'u', 'uf', 'w', and 'wf' Andrade fits to $S_{rel}(t)$.

Accordingly, the shear modulus and dissipation deriving from forced-oscillation data and weighted microcreep records have been combined as follows for fitting an Andrade model to the composite dataset. The 1 Hz sampling of the microcreep record becomes more intensive than that ($128/T_o$) for forced-oscillation for periods greater than 100 s. Accordingly, modulus and dissipation inferred from the 'w' fit to the microcreep $S_{rel}(t)$ records for the chosen periods > 200 s are used to supplement the forced-oscillation data for periods within the range 1–1000 s. The forced-oscillation and microcreep datasets thus overlap for periods of 200–1000 s, beyond which period, information derives solely from the microcreep records of 8000 s total duration. The optimal Andrade model, indistinguishable from the Burgers model discussed below and shown in Figure 4, adequately represents the composite dataset spanning four decades in period.

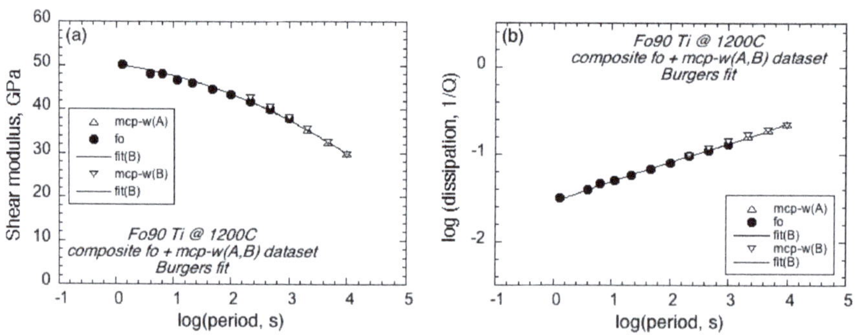

Figure 4. A comparison of shear modulus G and associated strain-energy dissipation Q^{-1} for the NiFe-sleeved synthetic Ti-doped Fo_{90} olivine derived from forced-oscillation data (solid symbols) and the microcreep records (open symbols). The labels 'mcp-w(A)' and 'mcp-w(B)', respectively, relate to microcreep results deriving from the alternative weighted Andrade and extended Burgers fits to the four-segment differential compliance $S_{rel}(t)$ record (Figures 1 and 5) without forced-oscillation constraint on the exponent n. The curves represent the optimal extended Burgers fit to the corresponding combined dataset as explained in the text. (**a**) G and (**b**) $\log Q^{-1}$ versus log (period, T_o, s) at 1200 °C.

3.2. Reconciliation of Forced-Oscillation and Microcreep Data with the Extended Burgers Model

The previously described processing with the Andrade creep function of complementary forced-oscillation data and microcreep records for the NiFe-sleeved Ti-doped Fo_{90} olivine yields a composite (G, Q^{-1}) dataset spanning four decades in period that is well represented by the final optimal Andrade model (Figure 4). However, because the Burgers model is regarded as physically more transparent than the Andrade model, if parametrically less economical, it is of interest to assess the feasibility of using the extended Burgers model throughout the parallel processing of forced-oscillation data and microcreep records. Because the focus here is on insight into the relatively long-period behavior at our highest routinely accessed temperature, we will use only the distribution $D_B(\ln\tau)$ of anelastic relaxation times (Equation (5)) corresponding to the high-temperature background.

That possibility is examined with reference to the same 1200 °C olivine dataset in Figure 5. As for the Andrade model, alternative extended Burgers fits to $S_{rel}(t)$ were explored—with ('w') and without ('u') weighting, and with ('f') and without (' ') forced-oscillation constraint of the parameter α. Again, the alternative models diverge significantly only for times < 50 s, and such divergence was minimised by weighting, more so than through imposition of a forced-oscillation constraint on α. The extended Burgers model 'w', with parameters J_U, Δ, α, and τ_M (but not τ_L) each refined within a modest uncertainty, evidently provides a satisfactory fit to the $S_{rel}(t)$ record (Figure 5). The RMS misfit of 0.72×10^{-5} rad/Nm is comparable with that for the Andrade mode.

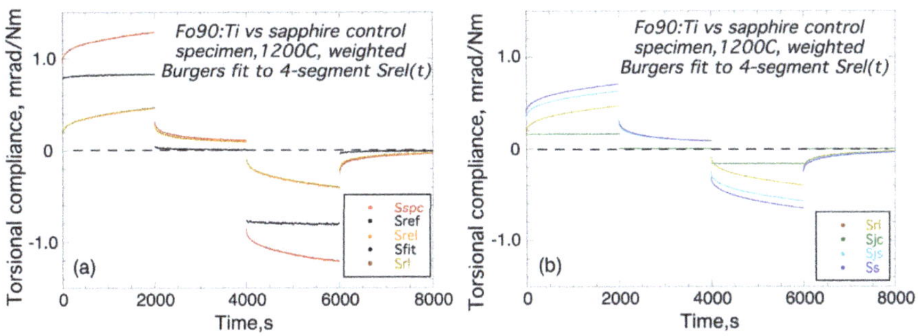

Figure 5. As for Figure 1, but now involving (**a**) the use of the extended Burgers rather than Andrade model for a weighted ('w') fit of $S_{rel}(t)$, and (**b**) in representation by virtual microcreep records of the derived quantities $S_{rl}(T_o)$, $S_{jc}(T_o)$, $S_{js}(T_o)$, and finally $S_s(T_o)$. Import revised figures.

The values of shear modulus and dissipation deriving from the alternative 'u', 'uf', 'w', and 'wf' fits of the extended Burgers model to the four-segment $S_{rel}(t)$ record are compared with the corresponding quantities from forced-oscillation data in Figure 6. As for the Andrade fit to $S_{rel}(t)$, the results are broadly consistent, especially at long periods. At short periods, the microcreep results diverge significantly from the forced oscillation data, but less so for the weighted fits.

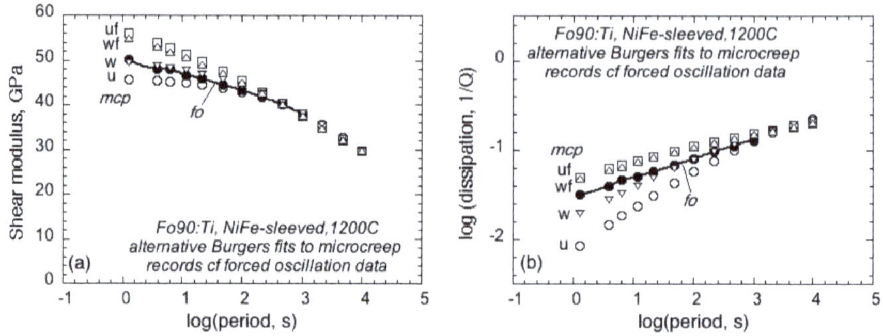

Figure 6. As for Figure 3, but now involving the use of the extended Burgers rather than Andrade model for the alternative fits of $S_{rel}(t)$. (**a**) Shear modulus G and (**b**) associated strain-energy dissipation Q^{-1} vs log (period, s).

The values of G and Q^{-1} for periods of 200–10,000 s derived from the microcreep records processed with the Burgers model ('w': weighted but with α unconstrained by forced–oscillation data) are combined with those from the forced-oscillation data to construct a composite ($G(T_o)$, $Q^{-1}(T_o)$) dataset. The optimal fit of the extended Burgers model to the composite dataset is compared with the constraining data in Figure 4. As for the Andrade model, the extended Burgers creep function provides a satisfactory representation of the composite dataset spanning four decades in period.

The values of the various parameters for the optimal Andrade and extended Burgers models fitted to the composite (G, Q^{-1}) datasets are assembled in Table 1. The values of the unrelaxed compliance J_U for the two models, and the parameters n and α, each approximating $\partial \log Q^{-1}/\partial \log T_o$, are consistent within the formal uncertainties. The Maxwell relaxation times $\tau_M = \eta J_U$ are 1.9×10^4 s and 1.2×10^4 s for the Andrade and Burgers models, respectively. The longer Maxwell time (higher viscosity) for the Andrade model reflects the fact that the distribution of anelastic relaxation times extends to ∞ for the Andrade model. Accordingly, more (less) of inelastic strain is treated as anelastic (viscous) for the Andrade creep function than for the Burgers model. As the 'transient' creep term βt^n in the Andrade creep function is unbounded as $t \to \infty$, there is no finite anelastic relaxation strength. Accordingly, we compare the anelastic relaxation strength $\Delta = 1.2 \pm 0.2$ for the extended Burgers model with the value (1.6) of $\beta t^n/J_U$ evaluated at $t = \tau_M$ for the Andrade model. This comparison confirms the impression from comparison of the Maxwell times (or viscosities) that the Andrade model partitions the inelastic strain more strongly into the anelastic contribution than does the extended Burgers model.

Table 1. Parameters of the optimal Andrade and extended Burgers creep function models fitted to the composite datasets comprising $N = 16$ (G, Q^{-1}) pairs for a Ti-doped Fo_{90} specimen tested at 1200 °C [15]. For the Burgers fit, log τ_L was fixed at −2.5. The assigned *a priori* uncertainties are $\sigma(G)/G = 0.03$ and $\sigma(\log Q^{-1}) = 0.05$. The quantities in parentheses are formal uncertainties in the last decimal place.

Andrade				
J_U 10^{-1} GPa^{-1}	n	β 10^{-2} GPa^{-1}s^{-n}	η, 10^6 GPa.s	$(\chi^2/2N)^{1/2}$
0.186(3)	0.25(2)	0.26(2)	1.0(6)	0.27
Burgers				
J_U, GPa	α	Δ	log(τ_M,s)	$(\chi^2/2N)^{1/2}$
0.191(2)	0.26(2)	1.2(2)	4.1(2)	0.35

Viscosities of order 10^{15} Pa s (Table 1) have thus been inferred in this study from torsional forced-oscillation tests at maximum shear strain amplitudes of (0.9–1.3) × 10^{-5} for periods of 1–1000 s, along with complementary microcreep experiments reaching somewhat larger strains (~2.5×10^{-5}). It is appropriate to compare such viscosities with those measured at the much larger stresses and strains of steady-state diffusional creep. At the same shear stress (0.44 MPa), temperature (1200 °C), and average grainsize (25 μm), the diffusional-creep flow law of Faul and Jackson[24] established in compressive testing to strains of order 10% for similar synthetic sol-gel-derived olivine predicts a shear viscosity $\eta = 5 \times 10^{16}$ Pa s. The lower viscosities of the tests at 10^{-5} strain amplitude may reflect transient rather than steady-state creep [24]. The effect of the constraint $\tau_H = \tau_M$ on the inferred value of η in the extended Burgers model will be examined in future work involving the systematic application of the method described here.

3.3. Recoverability of the Inelastic Strain

Of the inelastic strain at time t following the application of a steady torque, the fraction f_R that is ultimately recoverable following removal of the applied torque has been evaluated through (Equations (17) and (18)) for the Andrade and extended Burgers models, respectively. Consistent with the conclusion in the previous section that more (less) of the inelastic strain is attributed to the anelastic (viscous) terms in the Andrade creep function, $f_R(t)$ is consistently higher for the Andrade than for the extended Burgers model (Figure 7).

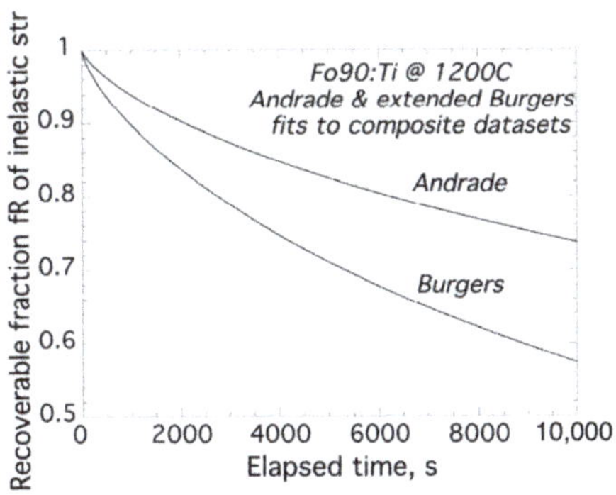

Figure 7. Variation with elapsed time of the fraction of inelastic strain that is ultimately recoverable on removal of the applied torque for the optimal Andrade and extended Burgers models (Table 1) fitted to the respective composite (G, Q^{-1}) datasets.

4. Discussion and Conclusions

It has been demonstrated that complementary forced-oscillation and microcreep data can be processed with either the Andrade or extended Burgers creep function models to yield consistent composite [$G(T_o)$, $Q^{-1}(T_o)$] datasets. Such a parallel processing of forced-oscillation data for periods of 1–1000 s and microcreep records of 10,000 s total duration serves to extend from three to four decades the effective range of periods over which the modulus and dissipation are constrained. Despite such consistency between the composite (G, Q^{-1}) datasets, the Andrade and extended Burgers models differ substantially in their attribution of strain between the transient and steady-state terms in the respective creep functions. Because the transient creep and associated anelastic strain extend to infinite time in the Andrade model, relatively more of the inelastic strain is attributed to the anelastic term, with a correspondingly higher viscosity and longer Maxwell time than for the Burgers model. The extended Burgers model is preferred for its greater clarity in prescribing a finite range of anelastic relaxation times and accordingly a finite anelastic relaxation strength.

The method developed and demonstrated here has the potential for routine application in the parallel processing of forced-oscillation data and microcreep records obtained in our laboratory with substantial benefit in better constraining microstrain viscoelastic behaviour at relatively long periods. Improved understanding of viscoelastic behaviour is critical not only for the interpretation of long-period (normal-mode) seismological models[25], but also for the interpretation of strain-energy dissipation at the even longer periods/timescales of tidal forcing[26,27] and glacial rebound [25–27].

Funding: This research received no external funding.

Acknowledgments: It is a pleasure and an honour to dedicate this article to the memory of Orson Anderson. Our research community is heavily indebted to Orson for his key role in founding and fostering the still relatively young discipline of Mineral Physics. As one of Orson's academic grandchildren, I am personally grateful to Orson for my introduction, through his student Bob Liebermann, to the laboratory measurement of elastic wave speeds with ultrasonic techniques. Orson served in 1976 as one of the external examiners of my ANU PhD thesis and twenty years later, we co-edited a special volume of *Physics of the Earth and Planetary Interiors* entitled *Thermoelastic Properties of Deep Mantle Phases*. Over the decades, we enjoyed many vigorous and fruitful discussions of diverse topics in mineral physics. Orson will long be remembered and sorely missed. Three anonymous reviewers are thanked for helpful comments on an earlier version of this manuscript.

Conflicts of Interest: The authors declare no conflict of interest.

References

1. Jackson, D.D.; Anderson, D.L. Physical mechanisms of seismic wave attenuation. *Rev. Geophys. Space Phys.* **1970**, *8*, 1–63. [CrossRef]
2. Randall, M.J. Attenuative dispersion and frequency shifts of the earth's free oscillations. *Phys. Earth Planet. Inter.* **1976**, *12*, P1–P4. [CrossRef]
3. Kanamori, H.; Anderson, D.L. Importance of physical dispersion in surface wave and free oscillation problems-Review. *Rev. Geophys. Space Phys.* **1977**, *15*, 105–112. [CrossRef]
4. Minster, J.B.; Anderson, D.L. A model of dislocation-controlled rheology for the mantle. *Philos. Trans. R. Soc. Lond.* **1981**, *299*, 319–356. [CrossRef]
5. Karato, S. Importance of anelasticity in the interpretation of seismic tomography. *Geophys. Res. Lett.* **1993**, *20*, 1623–1626. [CrossRef]
6. Goetze, C. A brief summary of our present-day understanding of the effect of volatiles and partial melt on the mechanical properties of the upper mantle. In *High-Pressure Research: Applications in Geophysics*; Manghnani, M.H., Akimoto, S., Eds.; Academic Press: New York, NY, USA, 1977; pp. 3–23.
7. Berckhemer, H.; Kampfmann, W.; Aulbach, E.; Schmeling, H. Shear modulus and Q of forsterite and dunite near partial melting from forced oscillation experiments. *Phys. Earth Planet. Inter.* **1982**, *29*, 30–41. [CrossRef]
8. Guéguen, Y.; Darot, M.; Mazot, P.; Woirgard, J. Q^{-1} of forsterite single crystals. *Phys. Earth Planet. Inter.* **1989**, *55*, 254–258. [CrossRef]
9. Jackson, I.; Paterson, M.S. A high-pressure, high temperature apparatus for studies of seismic wave dispersion and attenuation. *Pure Appl. Geophys.* **1993**, *141*, 445–466. [CrossRef]
10. Gribb, T.T.; Cooper, R.F. Low-frequency shear attenuation in polycrystalline olivine: Grain boundary diffusion and the physical significance of the Andrade model for viscoelastic rheology. *J. Geophys. Res.* **1998**, *103*, 27267–27279. [CrossRef]
11. McCarthy, C.; Takei, Y.; Hiraga, T. Experimental study of attenuation and dispersion over a broad frequency range: 2. The universal scaling of polycrystalline materials. *J. Geophys. Res.* **2011**, *116*, B09207. [CrossRef]
12. Li, L.; Weidner, D.J. Effect of dynamic melting on acoustic velocities in a partially molten peridotite. *Phys. Earth Planet. Inter.* **2013**, *222*, 1–7. [CrossRef]
13. Nowick, A.S.; Berry, B.S. *Anelastic Relaxation in Crystalline Solids*; Academic Press: New York, NY, USA, 1972.
14. Jackson, I. Dynamic compliance from torsional creep and forced oscillation tests: An experimental demonstration of linear viscoelasticity. *Geophys. Res. Lett.* **1993**, *20*, 2115–2118. [CrossRef]
15. Jackson, I.; Fitz Gerald, J.D.; Faul, U.H.; Tan, B.H. Grainsize sensitive seismic wave attenuation in polycrystalline olivine. *J. Geophys. Res.* **2002**, *107*, 2360. [CrossRef]
16. Barnhoorn, A.; Jackson, I.; Fitz Gerald, J.D.; Kishimoto, A.; Itatani, K. Grain-size sensitive viscoelastic relaxation and seismic properties of polycrystalline MgO. *J. Geophys. Res. Solid Earth* **2016**, *121*, 4955–4976. [CrossRef]

17. Jackson, I. Properties of rocks and minerals-physical origins of anelasticity & attenuation in rock. In *Treatise on Geophysics*, 2nd ed.; Schubert, G., Ed.; Elsevier: Amsterdam, the Netherlands, 2015; Volume 2, pp. 539–571. [CrossRef]
18. Kampfmann, W.; Berckhemer, H. High temperature experiments on the elastic and anelastic behaviour of magmatic rocks. *Phys. Earth Planet. Inter.* **1985**, *40*, 223–247. [CrossRef]
19. Raj, R. Transient behaviour of diffusion-induced creep and creep rupture. *Metall. Trans. A* **1975**, *6A*, 1499–1509. [CrossRef]
20. Morris, S.J.S.; Jackson, I. Diffusionally-assisted grain-boundary sliding and viscoelasticity of polycrystals. *J. Mech. Phys. Solids* **2009**, *57*, 744–761. [CrossRef]
21. Jackson, I.; Paterson, M.S. Shear modulus and internal friction of calcite rocks at seismic frequencies: Pressure, frequency and grainsize dependence. *Phys. Earth Planet. Inter.* **1987**, *45*, 349–367. [CrossRef]
22. Jackson, I.; Faul, U.H. Grainsize-sensitive viscoelastic relaxation in olivine: Towards a robust laboratory-based model for seismological application. *Phys. Earth Planet. Inter.* **2010**, *183*, 151–163. [CrossRef]
23. Cline, C.J., II; Faul, U.H.; David, E.C.; Berry, A.J.; Jackson, I. Redox-influenced seismic properties of upper-mantle olivine. *Nature* **2018**, *555*, 355–358. [CrossRef]
24. Faul, U.H.; Jackson, I. Diffusion creep of dry, melt-free olivine. *J. Geophys. Res.* **2007**, *112*, B04204. [CrossRef]
25. Lekić, V.; Matas, J.; Panning, M.; Romanowicz, B. Measurement and implications of frequency dependence of attenuation. *Earth Planet. Sci. Lett.* **2009**, *282*, 285–293. [CrossRef]
26. Nimmo, F.; Faul, U.H.; Garnero, E.J. Dissipation at seismic and tidal frequencies in a melt-free Moon. *J. Geophys. Res.* **2012**, *117*, E09005. [CrossRef]
27. Lau, H.C.P.; Faul, U.H. Anelasticity from Seismic to Tidal Timescales: Theory and Observations. *Earth Planet. Sci. Lett.* **2019**, *508*, 18–29. [CrossRef]

minerals

MDPI

Article

Thermal Equation of State of Fe_3C to 327 GPa and Carbon in the Core

Suguru Takahashi [1], Eiji Ohtani [1,*], Daijo Ikuta [1], Seiji Kamada [1], Tatsuya Sakamaki [1], Naohisa Hirao [2] and Yasuo Ohishi [2]

1 Department of Earth Science, Graduate School of Science, Tohoku University, 6-3 Aza-Aoba, Aramaki, Aoba-ku, Sendai 980-8571, Japan

2 Japan Synchrotron Radiation Research Institute, 1-1-1 Kouto, Sayo 679-5198, Japan

* Correspondence: eohtani@tohoku.ac.jp

Received: 16 October 2019; Accepted: 26 November 2019; Published: 30 November 2019

Abstract: The density and sound velocity structure of the Earth's interior is modeled on seismological observations and is known as the preliminary reference Earth model (PREM). The density of the core is lower than that of pure Fe, which suggests that the Earth's core contains light elements. Carbon is one plausible light element that may exist in the core. We determined the equation of state (EOS) of Fe_3C based on in situ high-pressure and high-temperature X-ray diffraction experiments using a diamond anvil cell. We obtained the P–V data of Fe_3C up to 327 GPa at 300 K and 70–180 GPa up to around 2300 K. The EOS of nonmagnetic (NM) Fe_3C was expressed by two models using two different pressure scales and the third-order Birch–Murnaghan EOS at 300 K with the Mie–Grüneisen–Debye EOS under high-temperature conditions. The EOS can be expressed with parameters of V_0 = 148.8(±1.0) Å^3, K_0 = 311.1(±17.1) GPa, K_0' = 3.40(±0.1), γ_0 = 1.06(±0.42), and q = 1.92(±1.73), with a fixed value of θ_0 = 314 K using the KBr pressure scale (Model 1), and V_0 = 147.3(±1.0) Å^3, K_0 = 323.0(±16.6) GPa, K_0' = 3.43(±0.09), γ_0 = 1.37(±0.33), and q = 0.98(±1.01), with a fixed value of θ_0 = 314 K using the MgO pressure scale (Model 2). The density of Fe_3C under inner core conditions (assuming P = 329 GPa and T = 5000 K) calculated from the EOS is compatible with the PREM inner core.

Keywords: iron carbide; Fe_3C; equation of state; high-pressure and high-temperature; inner core; in situ X-ray diffraction

1. Introduction

The density of the Earth's interior has been determined using the preliminary reference Earth model (PREM) [1]. The Earth's core is composed mainly of iron. The density of iron at specific pressure and temperature conditions has been determined in high-pressure and high-temperature experiments and from theoretical calculations (e.g., [2,3]). These studies implied that the density of pure hcp-Fe is higher than that obtained from seismological models such as PREM [1]. The density deficit is estimated to be ~10% for the outer core [4–7] and 2–5% for the inner core [8,9]. A recent study estimated the density of hcp-Fe to be 13.8–14.0 g/cm^3 [10,11] under inner core conditions, which corresponds to a density deficit of the inner core of 3.5–5.1%. Therefore, light elements may be contained within the Earth's inner core. Carbon is one candidate for these light elements, and Fe_3C is thought to be a constituent of the inner core (e.g., [12]). There are several works on the phase relations of the Fe–C system. Liu et al. [13] reported that Fe_3C is stable up to 150 GPa and that it melts incongruently into Fe_7C_3 and liquid at high temperatures in this pressure range and decomposes to Fe_7C_3 and Fe at higher pressures and temperatures. However, recent experiments report contradicting results. For example, Mashino et al. [14] reported Fe_3C is stable up to 250 GPa, and they did not find any evidence for decomposition of Fe_3C at around 150 GPa, as was reported by Liu et al. [13]. More recently, Takahashi et al. [15] studied the stability of Fe_3C up to and above 300 GPa and showed that it was stable up to this pressure, whereas it melted incongruently into Fe_7C_3 and liquid under the pressure and temperature conditions studied. The stability of Fe_3C is also supported by a recent shock experiment by Hu et al. [16]. They reported no evidence for decomposition of Fe_3C to Fe_7C_3 and Fe in the pressure range from 80 GPa to 248 GPa along the Hugoniot, which is consistent with the results of Mashino et al. [14] and Takahashi et al. [15]. Therefore, Fe_3C is an important potential constituent of the inner core.

The compression behavior of Fe and Fe alloys is related to their magnetic properties. The transition of magnetic Fe_3C to a nonmagnetic (NM) phase is estimated to occur at 70 GPa [17]. However, based on X-ray emission spectroscopy data [18] and the Mössbauer spectroscopy data [19], it has been revealed that this magnetic transition finishes at 20–25 GPa. Although there is a discrepancy in the absolute pressure of this transition to an NM phase, Fe_3C will be NM under inner core conditions. The compression behavior of NM Fe_3C up to core pressure conditions was reported by Sata et al. [20], but these experiments were only carried out at room temperature. Litasov et al. [21] determined the compression behavior of ferromagnetic (FM) Fe_3C up to about 30 GPa and estimated the density of NM Fe_3C under inner core conditions by assuming that the compression behavior of NM Fe_3C is similar to that of FM Fe_3C, which they determined. Very recently, Hu et al. [16] conducted shock experiments of Fe_3C and estimated the density of Fe_3C under inner core conditions. However, it is not well resolved whether their data correspond to NM Fe_3C. Presently, there are no high-temperature static compression experiments on NM Fe_3C.

To precisely estimate the density of NM Fe_3C in the inner core, it is necessary to understand the compression behavior of NM Fe_3C under high-pressure and high-temperature conditions. In this study, we measured the compression behavior of NM Fe_3C using in situ X-ray diffraction (XRD) experiments in a double-sided, laser-heated diamond anvil cell and determined the equation of state (EOS) to estimate the amount of carbon in the inner core.

2. Materials and Methods

2.1. Sample Preparation

The starting material was a powdered sample of Fe_3C, which was synthesized from a mixture of iron and graphite powders that was heated to 1273 K at 3 GPa for 12 h using a 3000 ton, Kawai-type multianvil apparatus installed at Tohoku University, Japan. The high pressure was generated using a symmetric-type diamond anvil cell and a membrane diamond anvil cell. The culet sizes of the diamond anvils were 100 and 150 μm, respectively. A tungsten gasket was indented to a thickness of 25–40 μm, and a hole with a diameter of 30–50 μm was drilled into the indented gasket to form a sample chamber. A foil from the synthesized Fe_3C sample was made using a cold-compression technique. KBr and MgO were used as the pressure marker [22,23] and the thermal insulator/pressure medium, respectively, for the high-pressure and high-temperature experiments up to 181.7 GPa. The pressure marker of KBr was sandwiched by the Fe_3C layers, and then the KBr and Fe_3C layers were sandwiched between MgO. We sandwiched the Fe_3C foil between the SiO_2 powder of the thermal insulator/pressure medium for the both the experiments above 298.9 GPa and those at room temperature.

2.2. In Situ XRD Experiments at the BL10XU Beamline at Spring-8

The in situ XRD experiments were conducted at the BL10XU beamline at the Spring-8 facility (Hyogo, Japan) [24]. The sample was heated using a double-sided laser heating technique [25] employing an SPI fiber laser (λ = 1.070 μm). The shape of the fiber laser beam was adjusted to a flattop beam using the beam-shaping system at the BL10XU beamline. This technique enabled us to decrease the temperature gradient across the sample and achieve a homogenous heated area with a diameter of approximately 30 μm. The temperature was determined by fitting the emission spectra from the surface of the heated sample to Planck's radiation law as a grey body formula using the typical wavelength range between 600 and 800 nm. The wavelength of the monochromatic X-ray beam and the distance between the sample and the X-ray detector were calibrated using XRD patterns from CeO_2 based on the double-cassette method. The typical wavelength of the X-rays was 0.4142(2)–0.4157(1) Å. The X-ray beam was collimated to a diameter of 15–20 μm. An imaging plate (IP) detector (RAXIS-IV; Rigaku, Tokyo, Japan) was used to collect the angle-dispersive XRD data. The exposure times for the IP were either 1 or 3 min. Each integrated XRD pattern, along with the 2θ angle (i.e., a one-dimensional XRD profile), was analyzed using an IP analyzer and PD Indexer software package [26].

The experimental pressures were determined using the EOS of KBr (B2) [22] and that of MgO [23] for the high-pressure and high-temperature experiments, whereas the compression curve of pyrite SiO_2 at 300 K was used for determination of pressure for the experiments exceeding 298.9 GPa [27]. The temperature distribution in the pressure medium in the double-heated diamond anvil cell (DAC) was evaluated from 3-dimensional numerical modeling [28]. This model calculation indicates that with a thickness of 1–2 μm, the temperature difference between the inner and anvil surfaces in the MgO layer is around 100–200 K at 2000 K, which is equivalent to the temperature uncertainty in the present experiment. Therefore, the temperature of the MgO layer in the DAC was considered to be the same as the sample temperature. The pressure and temperature conditions for the present experiments, based on the KBr pressure scale [22] and MgO pressure scale [23], are shown in Table S1 in Supplementary Materials.

3. Results

3.1. X-ray Diffraction Patterns of Fe_3C

In situ X-ray diffraction experiments were carried out from 71 GPa up to 182 GPa, and samples were heated up to around 2000 K at each pressure condition. The compression behavior of Fe_3C above 298.9 GPa at 300 K was obtained after quenching the sample from high temperatures above 1500 K, releasing stress during solid state compression. Its experimental pressures were determined by the equation of state of pyrite-type SiO_2 [27]. Previous studies show that the magnetic transition of Fe_3C to a nonmagnetic phase occurs at 20–25 GPa [13,19]. Therefore, Fe_3C is considered to be nonmagnetic under the conditions of the present experiments. Our results are summarized in Table S1. XRD patterns at pressures of 113–116 GPa and high temperatures are shown in Figure 1. XRD patterns of Fe_3C, KBr, and MgO were observed at all present experimental conditions. Fe_3C with the cementite structure was stable at all present pressure–temperature conditions.

3.2. Compression Behavior and the Thermal Equation of State of Fe_3C

The compressional data at 300 K was fitted to the third-order Birch–Murnaghan equation of state (3BM EOS) as follows:

$$P(V,\ 300) = \frac{3}{2}K_0\left[\left(\frac{V_0}{V}\right)^{\frac{7}{3}} - \left(\frac{V_0}{V}\right)^{\frac{5}{3}}\right]\left\{1 - \frac{3}{4}\left(4 - K_0'\right)\left[\left(\frac{V_0}{V}\right)^{\frac{2}{3}} - 1\right]\right\} \quad (1)$$

where, V_0, K_0, and $K_0{}'$ are the zero-pressure volume, the bulk modulus, and its pressure derivative, respectively. To investigate the thermoelastic parameters of Fe_3C, the Mie–Grüneisen–Debye (MGD) EOS can be fitted to P–V–T data obtained at 71–326 GPa and 300–2450 K. The MGD EOS is represented as follows (e.g., [29]):

$$P(V,\ T) = P(V,\ 300) + \Delta P_{th} \quad (2)$$

where the first and second terms denote the room temperature 3BM EOS (Equation (1)) pressure and the thermal pressure, respectively. The thermal pressure is expressed by the following expressions:

$$\Delta P_{th} = (\gamma/V)\ [E(T,\ \theta_D) - E(300,\ \theta_D)] \quad (3)$$

$$E = 9nRT\left(\frac{T}{\theta_D}\right)^3 \int_0^{\frac{\theta_D}{T}} \frac{t^3}{e^t - 1} dt \quad (4)$$

where E is internal thermal energy, γ is the Grüneisen parameter, θ_D is the Debye temperature, n is the number of atoms per formula unit, and R is the gas constant. The Debye temperature and the Grüneisen parameter are expressed as a function of volume:

$$\gamma = \gamma_0 \left(\frac{V}{V_0}\right)^q \quad (5)$$

$$\theta_D = \theta_0 \exp\left(\frac{\gamma_0 - \gamma}{q}\right) \quad (6)$$

where q is the volume dependence on the Grüneisen parameter. By fitting five parameters (V_0, K_0, K_0', γ_0 and q) with the Debye temperature θ_0 = 314 K [21], the following parameters were determined for NM-Fe_3C for the model using the pressure scale of KBr: V_0 = 148.8(1.0) Å^3, K_0 = 311.1(±17.1) GPa, K_0' = 3.40(±0.10), γ_0 = 1.06(±0.42), and q = 1.92(±1.73) [22] (Model 1). Our compression curve of Fe_3C at 300 K and high temperatures (Model 1) is shown in Figure 2a,b. Using the MgO pressure scale [23], we obtained the following parameters for NM Fe_3C: V_0 = 147.3(±1.0) Å^3, K_0 = 323(±16.6) GPa, K_0' = 3.43(±0.09), γ_0 = 1.37(±0.33), and q = 0.98(±1.01), with the Debye temperature of 314 K (Model 2). The compression curve of Fe_3C at 300 K and high temperatures (Model 2) is shown in Figure 2c,d.

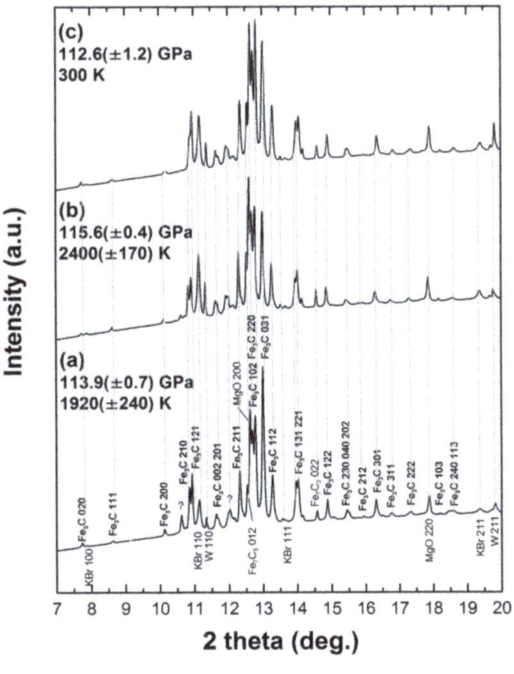

Figure 1. X-ray diffraction (XRD) patterns of Fe_3C at (**a**) 113.9(±0.7) GPa and 1920(±240) K, (**b**) 115.6(±0.4) GPa and 2400(±170) K, and (**c**) 112.8(±1.2) GPa and 300 K. XRD peaks were assigned to Fe_3C—KBr as a pressure marker, MgO as a thermal insulator, and W as a gasket. A few minor peaks derived from Fe_7C_3 were observed. 2 theta, a diffraction (Bragg) angle.

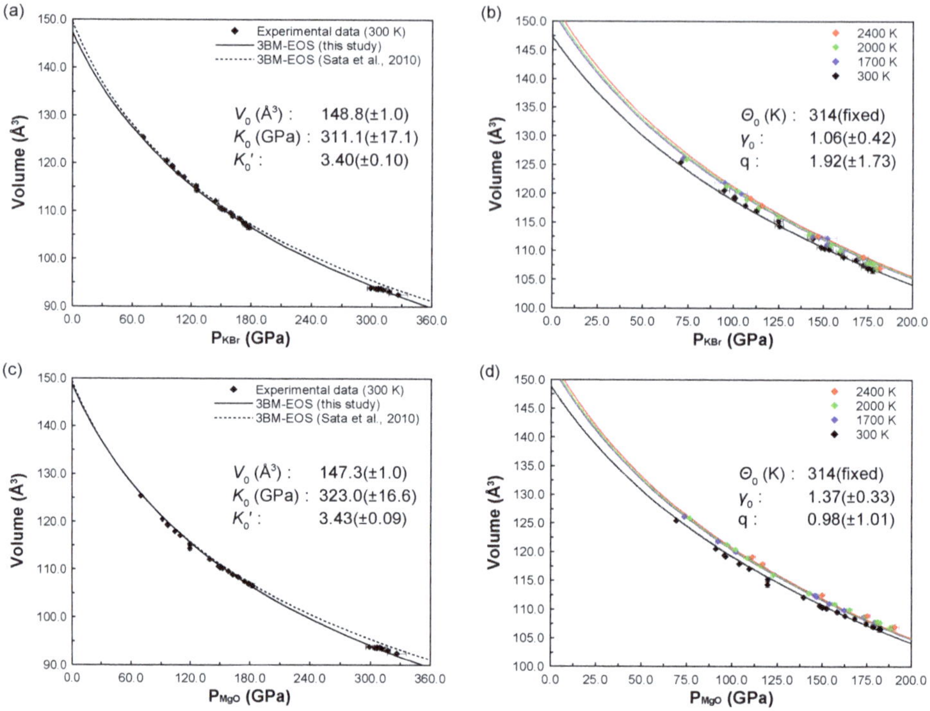

Figure 2. Volume compression of NM Fe_3C at 300 K and high temperatures. The compression at 300 K (**a**) and high temperatures based on the KBr pressure scale [22] (**b**) (Model 1). That at 300 K (**c**) and high temperatures based on the MgO pressure scale [23] (**d**) (Model 2). The pressure values above 290 GPa at 300 K in (**a**,**c**) are based on the equation of state of pyrite SiO_2 [27]. Black solid diamond symbols show the volume at high pressure and 300 K. Solid and dashed lines in (**a**,**c**) show the compression curves from this study and the study of Sata et al. [20], respectively. Black, blue, green, and red diamond symbols in (**b**,**d**) show the volume data set at around 1700(±100) K, 2000(±200) K, and 2400(±200) K and high pressure, respectively. Black, blue, green, and red curves in (**b**,**d**) show the compression curves at 300 K, 1700 K, 2000 K, and 2400 K, respectively. The detailed volume data at high pressure and temperature are given in Table S1. The fitting parameters for the thermal equation of state (Models 1 and 2) are shown in each figure.

The values of K_0 and K_0' obtained in this study are shown in Table 1, together with those of the previous works. These values are close to the NM Fe_3C reported by Sata et al. [20]. On the other hand, the K_0 of NM Fe_3C is higher than that of ferromagnetic and paramagnetic (PM) Fe_3C, and the K_0' of the NM Fe_3C is lower than that of FM and PM Fe_3C [17,20,21,30,31]. Ono and Mibe [32] reported a significant reduction in b-axis associated to the phase transition from FM to NM Fe_3C at 55 GPa, although this discontinuity was assigned to the transition from PM to NM Fe_3C by Litasov et al. [21]. However, since our experiments were performed at pressures above 69 GPa (at which NM-Fe_3C is stable) and no magnetic transition exists in the present pressure range, we did not observe the discontinuity in any axis at high pressure. In this work, the MGD EOS parameters of our models (Models 1 and 2), together with those determined by Litasov et al. [21], are shown in Table 2.

Table 1. Fitting parameters of thermal equations of state of Fe_3C.

V_0 (Å^3)	K_0 (GPa)	K_0'	EOS	Phase	Reference
148.8 ± 1.0	311.1 ± 17.1	3.40 ± 0.10	3BM	NM	Model 1 (based on KBr EOS)
147.3 ± 1.0	323 ± 16.6	3.43 ± 0.09	3BM	NM	Model 2 (based on MgO EOS)
149.46	290	3.76	3BM	NM	Sata et al. [20]
148.9	317	4.3	3BM	NM	Vočadlo et al. [17]
155.3	174	4.8	3BM	-	Li et al. [30]
155.2	175	5	Vinet	PM	Litasov et al. [21]
154.42	194	4.6	Vinet	PM	Litasov et al., MGD [21]
155.26	175.4	5.1	3BM	FM	Scott et al. [31]
155.4	167	6.7	Vinet	FM	Ono and Mibe [32]

Table 2. Fitting parameters of MGD EOS for Fe_3C.

θ_0 (K)	γ_0	q	Phase	Reference
314 *	1.06 ± 0.42	1.92 ± 1.73	NM	Model 1 (based on KBr EOS)
314 *	1.37 ± 0.33	0.98 ± 1.01	NM	Model 2 (based on MgO EOS)
314	2.15	−0.03	PM	Litasov et al. [21]

* Fixed following Litasov et al. [21].

4. Discussion

Although there are ambiguities concerning the thermal state of the core, the temperatures at the inner core boundary (ICB) and at the center of the Earth are in the range of 5000–6000 K (e.g., [33]). The density of Fe_3C at the pressure of the ICB (329 GPa) has been calculated to be 12.83(±0.67) g/cm^3 at 5000 K and 12.79(±0.73) g/cm^3 at 6000 K, respectively. The density of Fe_3C at the pressure of the center of the core (364 GPa) has been calculated to be 13.21(±0.70) g/cm^3 at 5000 K and 13.18(±0.76) g/cm^3 at 6000 K, respectively. PREM data give a density of 12.76 g/cm^3 and 13.09 g/cm^3 at the ICB (329 GPa) and the center of the core (364 GPa), respectively.

Figure 3a shows a comparison of the density of Fe [10], NM Fe_3C at 300 K and 5000 K, and the PREM density [1] as a function of pressure. The thermal expansion of NM Fe_3C with temperature is small compared to hcp-Fe. Our results indicate that the density of Fe_3C is compatible with seismological data on the inner core, assuming T = ~5000 K at the ICB pressure. The present density values for NM Fe_3C at 5000 K are different from those estimated by Litasov et al. [21] based on FM Fe_3C. The estimation of the high-temperature density based on FM Fe_3C could be different from that of the real NM Fe_3C, although the recent density estimation of Fe_3C based on shock compression [16] is close to the estimation by Litasov et al. [21]. In Figure 3b, we show the density of NM Fe_7C_3 at 5000 K, as described by Chen et al. [34] using thermal parameters for the PM phase [35]. The extrapolated densities of NM Fe_3C are slightly greater than those of Fe_7C_3. However, we cannot quantitatively evaluate the difference in the present accuracy of the experiments.

We concluded that the density of the Earth's inner core can be explained both by the Fe_3C and Fe_7C_3 density data at 5000~6000 K. Fe and Ni are thought to be major components in the Earth's core; however, the Earth's core may also contain carbon and other elements. We need to investigate the effects of Ni and the other elements on the density and sound velocity to form a conclusive argument regarding the composition of the inner core.

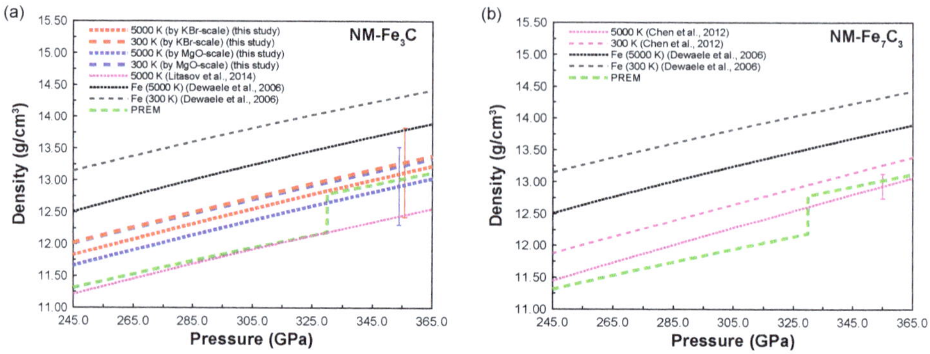

Figure 3. Pressure and density curves of NM Fe_3C (present), NM Fe_7C_3 [34], and hcp-Fe [10] at 300 K and 5000 K. The preliminary reference Earth model (PREM) density under the core conditions [1] is also shown in the figure. (**a**) The density curves for NM Fe_3C at 300 K and 5000 K using the KBr scale (Model 1) [22] and the MgO scale (Model 2) [23] are shown as red-dashed and red-dotted curves, and blue-dashed and blue-dotted curves, respectively. The uncertainties of the density of the two models extrapolated to the inner core pressure and 5000 K are shown in this figure. The PREM density in the core is shown as a green dashed curve. The density curve at 5000 K estimated by Litasov et al. [21] is shown as a pink dotted curve. (**b**) The density curves for NM Fe_7C_3 at 300 K and 5000 K are shown as a pink dashed curve and a pink dotted curve, respectively. The uncertainty of the density [34] at 5000 K is shown in this figure. The PREM density is shown as a green dashed curve.

Supplementary Materials: The following are available online at http://www.mdpi.com/2075-163X/9/12/744/s1, Table S1: Experimental conditions, volumes, and lattice parameters of Fe_3C, KBr, MgO and SiO_2.

Author Contributions: E.O. and S.T. devised the project. S.T. performed the sample preparation and managed the experimental plan. All authors participated in the experiments. S.T. and D.I. analyzed the data and performed the calculations. S.T. and E.O. wrote the manuscript with input from all authors.

Funding: This research was funded by Ministry of Education, Culture, Science, Sport, and Technology of Japanese Government, Kakenhi grant (JP22000002) and by Japan Society for the Promotion of Science, Kakenhi grant (JP15H05748) to Eiji Ohtani. Suguru Takahashi was supported by the Research Fellowship for Young Scientists of Japan Society for the Promotion of Science.

Acknowledgments: We thank Takeshi Sakai, Takanori Sakairi, Fumiya Maeda, Nanami Suzuki, and Akio Suzuki for their technical assistance and useful discussions. The in situ XRD experiments were performed with the approval of the Japan Synchrotron Radiation Research Institute (JASRI) (Proposal Nos 2013B0104, 2014A0104, and 2014B0104).

Conflicts of Interest: The authors declare no conflict of interest.

References

1. Dziewonski, A.M.; Anderson, D.L. Preliminary reference Earth model. *Phys. Earth Planet. Inter.* **1981**, *25*, 297–356. [CrossRef]
2. Dubrovinsky, L.S.; Dubrovinskaia, N.; Narygina, O.; Kantor, I.; Kuznetzov, A.; Prakapenka, A.B.; Vitos, L.; Johansson, B.; Mikhaylushkin, A.S.; Simak, S.I.; et al. Body-centered cubic iron-nickel alloy in Earth's core. *Science* **2007**, *316*, 1880–1883. [CrossRef] [PubMed]
3. Vočadlo, L.; Alfe, D.; Gillan, M.J.; Price, G.D. The properties of iron under core conditions from first principles calculations. *Phys. Earth Planet. Inter.* **2003**, *140*, 101–125. [CrossRef]
4. Birch, F. Elasticity and constitution of the Earth's interior. *J. Geophys. Res.* **1952**, *57*, 2199–2224. [CrossRef]
5. Birch, F. Density and composition of mantle and core. *J. Geophys. Res.* **1964**, *69*, 4377–4388. [CrossRef]
6. Brown, J.M.; McQueen, R.G. Phase transitions, Grüneisen parameter, and elasticity for shocked iron between 77 GPa and 400 GPa. *J. Geophys. Res.* **1986**, *91*, 7485–7494. [CrossRef]
7. Anderson, W.W.; Ahrens, T.J. An equation of state for liquid iron and implications for the Earth's core. *J. Geophys. Res.* **1994**, *99*, 4273–4284. [CrossRef]

8. Mao, H.K.; Wu, Y.; Chen, L.C.; Shu, J.F. Static compression of iron to 300 GPa and $Fe_{0.8}Ni_{0.2}$ alloy to 260 GPa: Implications for composition of the core. *J. Geophys. Res.* **1990**, *95*, 21737–21742. [CrossRef]
9. Dubrovinsky, L.S.; Saxena, S.K.; Tutti, F.; Rekhi, S.; LeBehan, T. In-situ X-ray study of thermal expansion and phase transition of iron at multimegabar pressure. *Phys. Rev. Lett.* **2000**, *84*, 1720–1723. [CrossRef]
10. Dewaele, A.; Loubeyre, P.; Occelli, F.; Mezouar, M.; Dorogokupets, P.I.; Torrent, M. Quasihydrostatic equation of state of iron above 2 Mbar. *Phys. Rev. Lett.* **2006**, *97*, 215504–215507. [CrossRef]
11. Sakai, T.; Takahashi, S.; Naoya Nishitani, N.; Mashino, I.; Ohtani, E.; Hirao, N. Equation of state of pure iron and $Fe_{0.9}Ni_{0.1}$ alloy up to 3 Mbar. *Phys. Earth Planet. Inter.* **2014**, *228*, 114–126. [CrossRef]
12. Wood, B.J. Carbon in the core. *Earth Planet. Sci. Lett.* **1993**, *117*, 593–607. [CrossRef]
13. Liu, J.; Lin, J.F.; Prakapenka, V.B.; Prescher, C.; Yoshino, T. Phase relations of Fe_3C and Fe_7C_3 up to 185 GPa and 5200 K: Implication for the stability of iron carbide in the Earth's core. *Geophys. Res. Lett.* **2016**, *43*, 12415–12422. [CrossRef]
14. Mashino, I.; Miozzi, F.; Hirose, K.; Morard, G.; Sinmyo, R. Melting experiments on the Fe-C binary system up to 255 GPa: Constraints on the carbon content in the Earth's core. *Earth Planet. Sci. Lett.* **2019**, *515*, 135–144. [CrossRef]
15. Takahashi, S.; Ohtani, E.; Sakai, T.; Kamada, S.; Ozawa, S.; Sakamaki, T.; Ito, Y.; Hirao, N.; Ohishi, Y. Phase and melting relations of Fe_3C to 300 GPa and carbon in the core. In *Carbon in Earth's Interior*; AGU Monograph Volume; Manning, C.E., Lin, J.-F., Mao, W., Eds.; American Geophysical Union: Washington, DC, USA, 2019; Chapter 3; pp. 25–36.
16. Hu, X.; Fei, Y.; Yang, J.; Cai, Y.; Ye, S.; Qi, M.; Liu, F.; Zhang, M. Phase stability and thermal equation of state of iron carbide Fe_3C to 245 GPa. *Geophys. Res. Lett.* **2019**. [CrossRef]
17. Vočadlo, L.; Brodholt, J.; Dobson, D.P.; Knight, K.; Marshall, W.; Price, G.D.; Wood, I.G. The effect of ferromagnetism on the equation of state of Fe_3C studied by first-principles calculations. *Earth Planet. Sci. Lett.* **2002**, *203*, 567–575. [CrossRef]
18. Lin, J.F.; Struzhkin, V.V.; Mao, H.K.; Hemley, R.J.; Chow, P.; Hu, M.Y.; Li, J. Magnetic transition in compressed Fe_3C from X-ray emission spectroscopy. *Phys. Rev. B* **2004**, *70*, 212405. [CrossRef]
19. Prescher, C.; Dubrovinsky, L.; McCammon, C.; Glazyrin, K.; Nakajima, Y.; Kantor, A.; Merlini, M.; Hanfland, M. Structurally hidden magnetic transitions in Fe_3C at high pressures. *Phys. Rev. B* **2012**, *85*, 10402. [CrossRef]
20. Sata, N.; Hirose, K.; Shen, G.; Nakajima, Y.; Ohishi, Y.; Hirao, N. Compression of FeSi, Fe_3C, $Fe_{0.95}O$, and FeS under the core pressures and implication for light element in the Earth's core. *J. Geophys. Res.* **2010**, *115*, B09204. [CrossRef]
21. Litasov, K.D.; Sharygin, I.S.; Dorogokupets, P.I.; Shatskiy, A.; Gavryushkin, P.N.; Sokolova, T.S.; Ohtani, E.; Li, J.; Funakoshi, K. Thermal equation of state and thermodynamic properties of iron carbide Fe_3C to 31 GPa and 1473 K. *J. Geophys. Res.* **2013**, *118*, 5274–5284. [CrossRef]
22. Dewaele, A.; Belonoshko, A.B.; Garbarino, G.; Occelli, F.; Bouvier, P.; Hanfland, M.; Mezouar, M. High-pressure–high-temperature equation of state of KCl and KBr. *Phys. Rev. B* **2012**, *85*, 214105. [CrossRef]
23. Kono, Y.; Irifune, T.; Higo, Y.; Inoue, T.; Barnhoon, A. P–V–T relation of MgO derived by simultaneous elastic wave velocity and in situ X-ray measurements: A new pressure scale for the mantle transition region. *Phys. Earth Planet. Inter.* **2010**, *183*, 196–211. [CrossRef]
24. Ohishi, Y.; Hirao, N.; Sata, N.; Hirose, K.; Takata, M. Highly intense monochromatic X-ray diffraction facility for high-pressure research at SPring-8. *High Press. Res.* **2008**, *28*, 163–173. [CrossRef]
25. Shen, G.; Mao, H.K.; Hemley, R.J. Laser-heated Diamond Anvil Cell Technique: Double-sided Heating with Multimode Nd:YAG Laser, in Advanced Materials '96. In Proceedings of the 3rd NIRIM International Symposium on Advanced Materials, Tsukuba, Japan, 4–8 March 1996; pp. 149–152.
26. Seto, Y.; Nishio-Hamane, D.; Nagai, T.; Sata, N. Development of a software suite on X-ray diffraction experiments. *Rev. High Press. Sci. Technol.* **2010**, *20*, 269–276. [CrossRef]
27. Kuwayama, Y.; Hirose, K.; Sata, N.; Ohishi, Y. Pressure-induced structural evolution of pyrite-type SiO_2. *Phys. Chem. Miner.* **2011**, *38*, 591–597. [CrossRef]
28. Rainey, E.S.G.; Hernlund, J.W.; Kavner, A. Temperature distributions in the laser-heated diamond anvil cell from 3-D numerical modeling. *J. Appl. Phys.* **2013**, *114*, 204905. [CrossRef]
29. Jackson, I.; Rigden, S.M. Analysis of *P–V–T* data: Constraints on the thermoelastic properties of high-pressure minerals. *Phys. Earth Planet. Inter.* **1996**, *96*, 85–112. [CrossRef]

30. Li, J.; Mao, H.K.; Fei, Y.; Gregoryanz, E.; Eremets, M.; Zha, C.S. Compression of Fe_3C to 30 GPa at room temperature. *Phys. Chem. Miner.* **2002**, *29*, 166–169. [CrossRef]
31. Scott, H.P.; Williams, Q.; Knittle, E. Stability and equation of state of Fe_3C to 73 GPa: Implications for carbon in the Earth's core. *Geophys. Res. Lett.* **2001**, *28*, 1875–1878. [CrossRef]
32. Ono, S.; Mibe, K. Magnetic transition of iron carbide at high pressures. *Phys. Earth Planet. Inter.* **2010**, *180*, 1–6. [CrossRef]
33. Terasaki, H.; Kamada, S.; Sakai, T.; Ohtani, E.; Hirao, N.; Ohishi, Y. Liquidus and solidus temperature of a Fe-O-S alloy up to the pressures of the outer core: Implication for the thermal structure of the Earth's core. *Earth Planet. Sci. Lett.* **2011**, *232*, 379–392. [CrossRef]
34. Chen, B.; Gao, L.; Lavina, B.; Dera, P.; Alp, E.E.; Zhao, J.; Li, J. Magneto-elastic coupling in compressed Fe_7C_3 supports carbon in Earth's inner core. *Geophys. Res. Lett.* **2012**, *39*, L18301. [CrossRef]
35. Nakajima, Y.; Takahashi, E.; Sata, N.; Nishihara, Y.; Hirose, K.; Funakoshi, K.; Ohishi, Y. Thermoelastic property and high-pressure stability of Fe_7C_3: Implication for iron-carbide in the Earth's core. *Am. Mineral.* **2011**, *96*, 1158–1165. [CrossRef]

minerals

MDPI

Article

Evaluating the Role of Iron-Rich (Mg,Fe)O in Ultralow Velocity Zones

Vasilije V. Dobrosavljevic *, Wolfgang Sturhahn and Jennifer M. Jackson

Seismological Laboratory, California Institute of Technology, Pasadena, CA 91125, USA; wolfgang@gps.caltech.edu (W.S.); jackson@gps.caltech.edu (J.M.J.)

* Correspondence: vasilije@caltech.edu

Received: 12 November 2019; Accepted: 6 December 2019; Published: 8 December 2019

Abstract: The composition of ultralow velocity zones (ULVZs) remains an open question, despite advances in both seismology and experimental work. We investigate the hypothesis of iron-rich (Mg,Fe)O (magnesiowüstite) as a cause of ULVZ seismic signatures. We report new quasi-hydrostatic X-ray diffraction measurements to constrain the equation of state of $(Mg_{0.06}Fe_{0.94})O$ with fit parameters $V_0 = 9.860 \pm 0.007$ Å^3, $K_{0T} = 155.3 \pm 2.2$ GPa, $K'_{0T} = 3.79 \pm 0.11$, as well as synchrotron Mössbauer spectroscopy measurements to characterize the high-pressure magnetic and spin state of magnesiowüstite. We combine these results with information from previous studies to calculate the elastic behavior at core–mantle boundary conditions of magnesiowüstite, as well as coexisting bridgmanite and calcium silicate perovskite. Forward models of aggregate elastic properties are computed, and from these, we construct an inverse model to determine the proportions of magnesiowüstite that best reproduce ULVZ observations within estimated mutual uncertainties. We find that the presence of magnesiowüstite can explain ULVZ observations exhibiting 1:2 V_P:V_S reduction ratios relative to the Preliminary Reference Earth Model (PREM), as well as certain 1:3 V_P:V_S reductions within estimated uncertainty bounds. Our work quantifies the viability of compositionally distinct ULVZs containing magnesiowüstite and contributes to developing a framework for a methodical approach to evaluating ULVZ hypotheses.

Keywords: (Mg,Fe)O; ultralow velocity zones; core–mantle boundary; equations of state; finite strain analysis; inverse model; X-ray diffraction; synchrotron Mössbauer spectroscopy

---

1. Introduction

The boundary layer separating the iron-dominant liquid outer core from the silicate-rich mantle is a region of great complexity, where extreme contrasts in material properties promote the persistence of multiscale structural heterogeneities (e.g., [1–3]). The thermo-elastic variations at the lowermost mantle play a dominant role in the evolutionary history of the Earth through regulation of heat flow and consequent influence over the dynamics of both the mantle and the core. Despite the centrality of this region in the development of the solid Earth system, many open questions remain regarding the characteristics, origins, and dynamic interactions of observed heterogeneities, such as large thermochemical piles, ultralow velocity zones, subducted former oceanic material, and small seismic scatterers. As individual observational studies of such features and experimental investigations into candidate compositions continue to develop, synthesis of results from seismology and mineral physics provides quantitative and systematic avenues for revealing new insights into this complex region.

Advances in seismological observation over the past several decades have provided increasing evidence for the existence of ultralow velocity zones (ULVZs)—Small (<50 km in height) localized seismic anomalies at the base of the mantle that are characterized by significant reductions in velocities (up to 25% and 50% for P and S waves, respectively) and likely increases in density (up to 20%)

relative to the surrounding mantle ([4–6], see also [7] for a recent review). To date, less than 20% of the core–mantle boundary (CMB) has been explored by seismic studies investigating ULVZs, yet the existing observations of ULVZs reveal a large variety of elastic properties, locations, and structural geometries (e.g., [7]). Nevertheless, the growing set of seismic studies observing diverse ULVZ properties can inform us of the causes of such significant velocity reductions. Whether all ULVZs share common characteristics and/or origins and what dynamical links may exist between them and other multi-scale features like slabs or slab debris, large thermochemical piles, and upwelling mantle plumes is the subject of current multidisciplinary research.

The very existence of ULVZs poses the question of their origin and dynamics and has attracted a variety of proposed explanations, including an iron-bearing layer of FeO and FeSi [8,9], iron-enriched postperovskite (Mg,Fe)SiO_3 [10], subducted banded iron formations [11], silicate sediments from the core [12], melt within subducted oceanic crust [13–15], slab-derived metallic melt [16], and partial melting of mantle material [17,18]. Reports of preferential iron partitioning into ferropericlase coexisting with bridgmanite or postperovskite [19] have motivated recent high-pressure experimental studies on iron-rich compositions of (Mg,Fe)O. These materials show remarkably low sound velocities [20,21], and dynamic modeling work on assemblages containing this material could reproduce the velocities and topography of some observed ULVZs [22]. In addition, recent seismic and dynamic studies have suggested that compositionally distinct origin may be necessary to explain many ULVZs [23,24]. However, quantitative comparisons of such proposed features with seismic observations remain largely unexplored, creating difficulty in evaluating the likelihood of proposed explanations.

With this study, we aim to develop a quantitative approach to assess the role of iron-rich (Mg,Fe)O in ULVZs. First, we present new measurements of the compressional behavior of $(Mg_{0.06}Fe_{0.94})O$ in a helium pressure medium at ambient temperatures using synchrotron X-ray diffraction and produce a well-constrained equation of state for this material, which had not been done previously [25]. We combine these results with sound velocities and thermo-elastic information from previous studies [21,25,26] to calculate the densities and velocities of iron-rich (Mg,Fe)O compositions at CMB pressure–temperature conditions. Using recent thermoelasticity measurements of bridgmanite and calcium silicate perovskite, we calculate seismic properties of coexisting mineral assemblages containing iron-enriched (Mg,Fe)O and build a linear mixing model that combines uncertainty estimates from both mineral physics and seismic observations to invert for the best-fit concentrations of iron-rich (Mg,Fe)O for select ULVZ observations. While an inversion approach has been used for other applications, such as constraining the composition of the bulk lower mantle [27], this study applies a similar approach to constraining the compositions of ULVZs. Our study quantifies the viability of iron-rich (Mg,Fe)O to account for certain classes of ULVZ seismic observations and, in doing so, contributes to developing a framework for a systematic evaluation of proposed origins of ULVZs.

2. Materials and Methods

A sample of polycrystalline $(Mg_{0.058(1)}Fe_{0.942(1)})O$, hereafter referred to as Mw94, mixed with NaCl powder, was synthesized at ambient pressure conditions using 95% enriched ^{57}Fe and MgO powders, with ferric iron content capped at 5% by conventional Mössbauer spectroscopy [25]. The powdered sample was loaded inside a rhenium gasket between two beveled diamond anvils of 250 μm diameter within a symmetric diamond-anvil cell. The sample chamber also contained two ruby spheres for use as pressure markers by measurement of their pressure-dependent fluorescence spectra. The chamber was then loaded with compressed helium gas at 170 MPa, explicitly chosen as a pressure-transmitting medium to minimize the nonhydrostatic behavior of the pressure environment (e.g., [26]), using the gas-loading system at Caltech.

2.1. X-Ray Diffraction

High-pressure X-ray diffraction experiments were performed at Beamline 12.2.2 of the Advanced Light Source (ALS) of Lawrence Berkeley National Laboratory (Berkeley, CA, USA). The LaB_6 standard was used to calibrate the sample to detector distance. Diffraction patterns at each pressure point were collected using a high-resolution image plate (MAR345) with an incident X-ray energy of 25 keV and a size of about 10 μm × 10 μm (full width at half maximum). Diffraction patterns were collected at a series of 27 compression points across the pressure range of 1.5 to 88.7 GPa. Intervals between pressure steps were limited to ~2 GPa for the range below 30 GPa, in order to finely sample the compressional behavior of the material in the low-pressure regime, a critical step for tightly determining the material's equation of state.

Measurements of the ruby fluorescence spectra were performed immediately before and after each diffraction measurement, using the pressure scale of [28]. The cell was allowed to rest for 15 min on average after pressure increases, in order to reduce pressure creep during the collection of diffraction patterns. The uncertainty in pressure at each step was estimated as the difference of the pressures measured for the two rubies before and after each diffraction measurement. Diffraction patterns were radially integrated from the 2D image plate using the Dioptas software [29]. Selected diffraction patterns are shown in Figure S1. Peaks for Mw94 and NaCl, as well as for ruby and rhenium, were identified and fit using the GSAS-II software [30] to derive the unit cell volume for Mw94 at each pressure point. The unit cell volumes of NaCl were used as a secondary pressure gauge [31] when present in the diffraction patterns and confirmed pressures determined by ruby fluorescence.

2.2. Synchrotron Mössbauer Spectroscopy

We complement our diffraction data with synchrotron Mössbauer spectroscopy (SMS) performed at Sector 3-ID-B of the Advanced Photon Source (APS) at Argonne National Laboratory (Lemont, IL, USA), using top-up mode of the storage ring with 24 bunches separated by 153 ns. X-rays were focused to an area of about 12 μm × 18 μm and tuned to 14.4125 keV, a nuclear transition energy of ^{57}Fe. A high-resolution monochromator provided an energy bandwidth of 1 meV [32]. A time window of 25 to 125 ns after excitation was used to observe nuclear resonant forward scattering and to fit the data. We performed SMS measurements at three compression points (95.4, 97.0, and 116.4 GPa) on the identical sample used for the diffraction measurements, in order to determine the magnetic ordering and spin state. Pressures were measured before and after SMS data collection using ruby fluorescence [28] and Raman spectroscopy, using the high-frequency edge of the diamond Raman band [33].

3. Experimental Results

The unit cell volume per atom of Mw94 at each compression point was calculated using the following hkl reflections: 111, 200, 220, 311, and 222 for B1; 003, 101, 102, 104, 110, and 105 for the rhombohedral phase (Figure 1 and Figure S2, and Table S1). The splitting of 111 and 220 B1 peaks was observed at 34.5 GPa, suggesting a rhombohedral distortion of the cubic lattice. We bracketed this structural transition from the B1-cubic phase to a rhombohedral phase between 30.1 and 34.5 GPa. The material remains in the rhombohedral phase up to at least 88.7 GPa. We observed a change in the trend of the unit-cell volume as a function of pressure at the two compression points prior to the structural transition (26.8 GPa, 30.1 GPa) and excluded these data in fitting the equation of state.

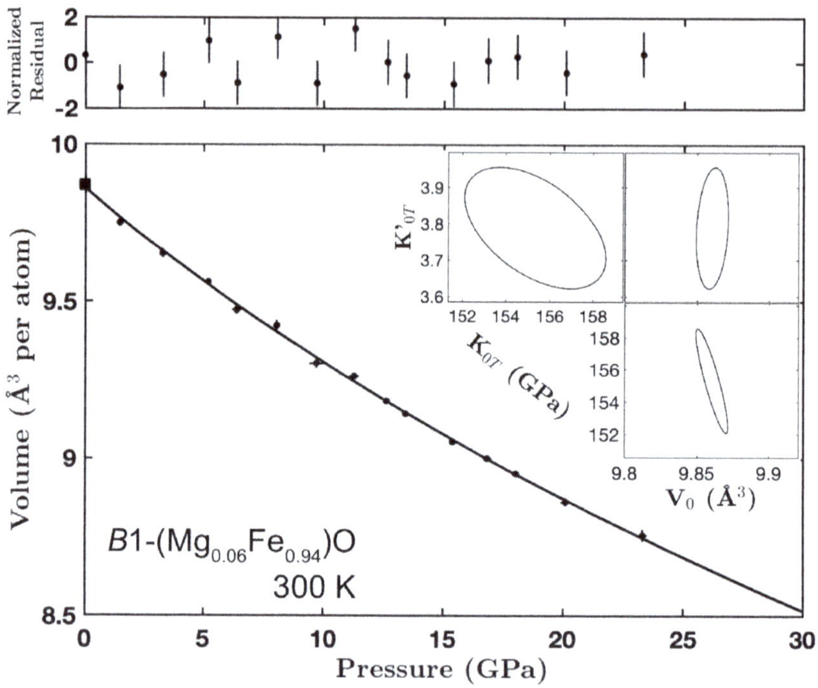

Figure 1. Unit-cell volumes of B1-($Mg_{0.058(1)}Fe_{0.942(1)}$)O (Mw94) (solid black circles) and equation of state (solid black line) fit using MINUTI [34]. The uncertainty in pressure at each step was estimated as the difference of the pressures measured by the two rubies before and after each diffraction measurements. We include a zero-pressure measurement of Mw94 (solid black square) from a previous study [25] for comparison. Normalized residuals are shown on the top panel. Error ellipses demonstrate the correlations between fit parameters.

3.1. Equation of State (B1 Phase)

The pressure–volume data for the B1-cubic phase and for the rhombohedral phase were fit separately with two distinct equations of state using the MINUTI software package [34]. We used a third-order Burch–Murnaghan equation of state and varied the fit parameters V_0, K_{0T}, and K'_{0T} representing the zero-pressure unit cell volume per atom, the isothermal zero-pressure bulk modulus, and its pressure derivative, respectively. Due to large tradeoffs between K_{0T} and K'_{0T} and motivated by previous reports of K'_{0T} values for similar compositions (see [26]), we performed the fit using an initial value of 3.8 for K'_{0T} with a prior window of 0.3; given the reduced χ^2 value of 0.70 ± 0.23 using this prior and an equivalent value of χ^2 = 0.74 ± 0.25 with no prior, we find the prior information to be a good fit to the data. For the B1-cubic phase, the best fit parameters are V_0 = 9.860 ± 0.007 Å^3 /atom, K_{0T} = 155.3 ± 2.2 GPa, K'_{0T} = 3.79 ± 0.11 (Table 1), which are significantly more compatible with the trends of K_{0T} as a function of iron concentration presented in [26] for the (Mg,Fe)O solid solution, compared with the results from an X-ray diffraction study of Mw94 from the same bulk sample [25]. These results underscore the need to perform measurements using a helium pressure medium in the low pressure regime, which were absent in the experimental range of Wicks et al. [25]. While the choice of neon or helium as the pressure medium can have a significant effect on elastic properties and transition pressure ranges for the (Mg,Fe)O system, the effect may be different for other phases with different crystal structures, such as (Mg,Fe)SiO_3 and $CaSiO_3$.

Table 1. Isothermal equation of state parameters for the B1 and rhombohedral phases of Mw94, with thermal equation of state models for three iron-rich compositions of Mw.

Parameters	B1–Mw94	r–Mw94	Mw94 Model [a]	Mw84 Model [b]	Mw78 Model [c]
V_0 (Å^3/atom)	9.860 ± 0.007	9.59 ± 0.10	9.860 ± 0.007	9.79 ± 0.04	9.86 ± 0.02
K_{0T} (GPa)	155.3 ± 2.2	217 ± 19	155.3 ± 2.2	155.3 ± 2.2	148 ± 3
K'_{0T}	3.79 ± 0.11	2.06 ± 0.22	3.79 ± 0.11	3.79 ± 0.11	4.09 ± 0.12
θ_0 (K)	—	—	426	426	426
γ_0	—	—	1.17	1.17	1.17
q	—	—	0.5	0.5	0.5
Reduced χ^2	0.70 ± 0.23	1.37 ± 0.37	—	—	—

[a] γ_0 and q taken from previously reported values for Mw94 [25]. [b] Identical to Mw94 model except for V_0 taken from Wicks et al. [21]. [c] Non-thermal parameters from Finkelstein et al. [26].

3.2. Equation of State (Rhombohedral Phase)

We fit a third-order Burch–Murnaghan isothermal equation of state to the pressure–volume data for the rhombohedral phase and found best fit parameter values V_0 = 9.59 ± 0.10 Å^3/atom, K_{0T} = 217 ± 19 GPa, K'_{0T} = 2.06 ± 0.22, with a reduced χ^2 value of 1.37 ± 0.37 (data and fit with residuals is shown in Figure S2). While Wicks et al. [25] bracketed the transition pressure for the cubic to rhombohedral transition between 13 and 24 GPa for a sample of Mw94 in a neon pressure medium, we observed a transition pressure in the range of 30.1 to 34.5 GPa for Mw94 in a helium medium. It is likely that a steep increase of non-hydrostaticity at ~20 GPa caused by neon compared to helium [26] induced the rhombohedral distortion at a lower pressure.

3.3. Synchrotron Mössbauer Spectroscopy

In order to explore the possibility of a high-pressure ferrous-iron spin transition in Mw94, we performed synchrotron Mössbauer spectroscopy (SMS) on Mw94 at three compression points (95.4 ± 0.2, 97.0 ± 0.5, and 116.4 ± 0.5 GPa). Pressures were determined by Raman spectroscopy using the high-frequency edge of the diamond Raman band [33], with uncertainties estimated as the difference of pressures before and after SMS measurements and of pressures at two different locations on the anvil. At the two lower pressures, we observed fast oscillations in the time spectra that are caused by magnetic ordering in the sample. At 116.4 GPa, a significant fraction of the fast oscillations was no longer present (Figure S3), suggesting an absence of magnetic ordering that could indicate a high-spin to low-spin transition occurring in the Fe^{2+} sites of Mw94. In order to evaluate the possibility of a spin transition, we used the CONUSS software version 2.2.0 [35] to fit the 116.4 GPa spectrum. We found that three distinct Fe^{2+} environments are required to fit the spectrum. One site is characteristic of low-spin Fe^{2+} and the other two sites are characterized by high-spin Fe^{2+}-like sites: one with no magnetic ordering and one with a magnetic hyperfine field. The best-fit model (Table S2) shows that 37 ± 2% of the iron atoms have transitioned into the low-spin state at 116.4 GPa, with 3% of the iron atoms retaining the magnetic ordering in a high-spin state, and the remaining fraction indicative of a broad high-spin Fe^{2+}-like site with no magnetic ordering. The pressure of the Fe^{2+} spin transition suggested by this result is consistent with the trend of increasing spin transition pressure with increasing iron content in (Mg,Fe)O (e.g., [20,36]). It should be noted that Mw94 is in the rhombohedral phase at the P–T conditions of these SMS measurements, so the reduced symmetry of the crystal structure compared to the B1 phase may affect the pressure range of the spin transition. The temperature dependence of the spin state in Fe-rich (Mg,Fe)O at CMB conditions has not been thoroughly studied, although the higher temperatures of Earth's mantle would likely drive the spin transition to higher pressures than those expected within the mantle (e.g., [37,38]). Taken together, these results suggest that, unlike for ferropericlase [39], the fraction of low-spin magnesiowüstite would either be relatively low or negligible at CMB conditions.

4. Modeling Iron-Rich (Mg,Fe)O in the Lowermost Mantle

4.1. Calculating Iron-Rich (Mg,Fe)O Elasticity at CMB Conditions

Mw94 has been reported to remain in the B1-cubic phase along the mantle geotherm up to CMB conditions [25]. We thus combined our results with those from previous studies in order to calculate the densities and sound velocities of iron-rich (Mg,Fe)O compositions at the pressure and temperature of the CMB. We first calculated the bulk sound velocity for Mw94 at 135.8 GPa (CMB pressure given in PREM) and 300 K using our isothermal equation of state. We then used this value and the Debye velocity V_D = 4.27 ± 0.09 km/s reported by Wicks et al. [21] for an identical composition at the same pressure–temperature condition in order to calculate a V_P and V_S for Mw94 at 135.8 GPa and 300 K, following the procedure taken by Wicks et al. [21]. We applied the temperature corrections reported by Wentzcovitch et al. [40] for MgO at CMB conditions to our results and thereby calculated the sound velocities for Mw94 at 135.8 GPa and 3800 K. In addition, we incorporated the thermal parameters reported by Wicks et al. [25] for Mw94 in order to calculate the density of Mw94 at 135.8 GPa and 3800 K. We then repeated this procedure for Mw84 by using the same equation of state except with a zero-pressure volume V_0 = 9.79 ± 0.04 Å^3 /atom as reported by Wicks et al. [21], and for Mw78 using the isothermal equation of state reported by Finkelstein et al. [26]. In doing so, we computed V_P, V_S, and density with uncertainties estimated from experimental reports (Table S3) for the behavior of three iron-rich compositions of (Mg,Fe)O at the pressure–temperature conditions of the CMB. We note that while the equation of state for these iron-rich compositions is taken from measurements of the B1 phase that exists at CMB conditions, the measurements reported for the Debye velocity at lower mantle pressures [21] were performed on the rhombohedral phase. Nevertheless, these values reflect the best available understanding of this material's shear elastic behavior. It is also important to note that Wicks et al. found no discernible compositional dependence of the Debye velocity for Mw84, Mw94, and FeO [21], such that the effect of composition appears primarily in the equations of state used for extrapolation (Table 1).

4.2. Forward Modeling

Having calculated sound velocities and densities for three iron-rich compositions of (Mg,Fe)O at the pressure–temperature conditions of the CMB, we next investigated the likelihood of the presence of this material in ULVZs given the range of seismic observations. In order to do so, we first constructed a forward linear mixing model for calculating the sound velocities and densities of mineral aggregates containing iron-rich (Mg,Fe)O mixed with bridgmanite (Br) and calcium silicate perovskite (CaPv). The iron content of Br was determined by the partitioning of iron between Mw and Br using a K_D value of 0.03 [21]. For CaPv, we used recent ultrasonic interferometry experiments reported by Gréaux et al. [41] and extrapolated to CMB conditions using finite strain analysis [42]. For modeling Br properties, we used a combination of theoretical and experimental constraints (e.g., [40,43–45]) (see Table S3). We calculated properties for aggregates containing concentrations of Mw ranging from 0% to 50% combined with a mixture of Br and CaPv where the concentration of CaPv was controlled by fixing it at 10% of Br concentration, with an uncertainty of 5%. By computing bulk and shear moduli for the minerals we were mixing, we could compute Voigt and Reuss bounds for the aggregate elastic properties in order to determine the range of velocity reductions that can be accounted for by the presence of iron-rich (Mg,Fe)O. The Hill averages for P and S wave velocity reductions relative to PREM for aggregates containing Mw94, Mw84, and Mw78 are shown in Figure 2a, as well as markers indicating density increases relative to PREM. Voigt and Reuss bounds for all three compositions of Mw are shown in Figure S4, with markers indicating Mw concentrations.

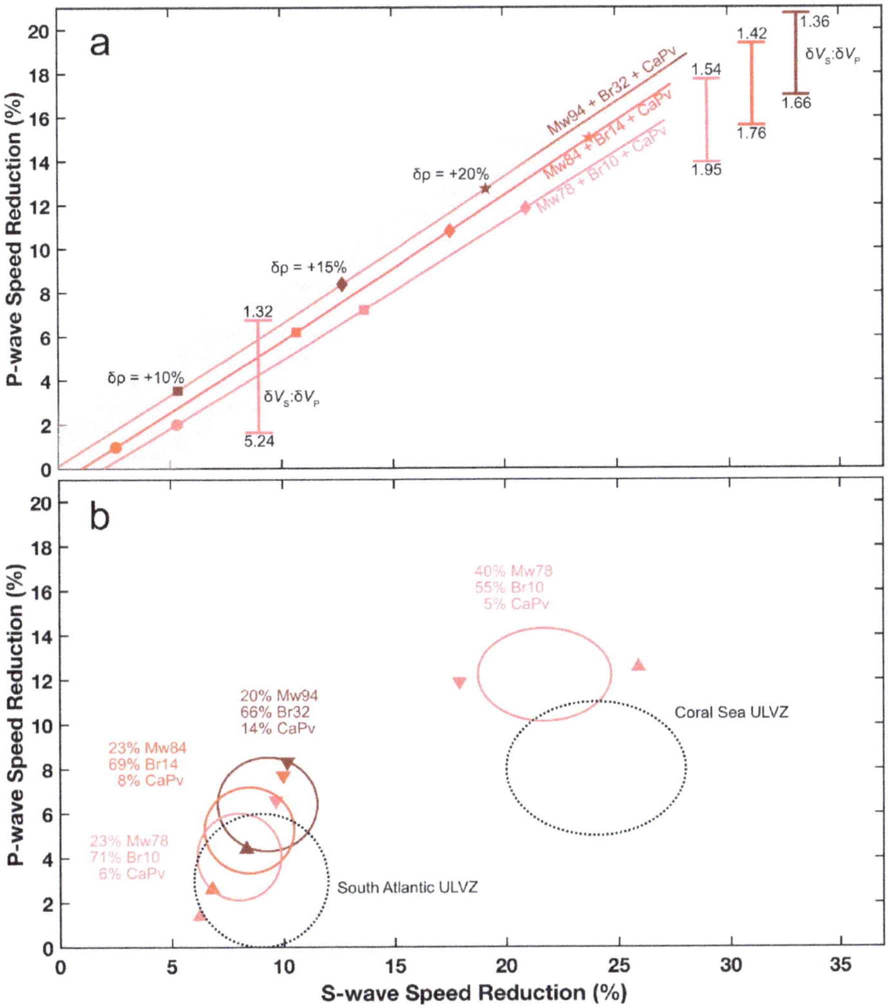

Figure 2. Seismic signatures of iron-rich (Mg,Fe)O in the lowermost mantle relative to PREM [46]. (**a**) Forward models (Hill average) of three Mw compositions with Mw concentrations ranging from 8.3% (Mw94), 10.2% (Mw84), and 10.6% (Mw78) up to 50%, mixed with bridgmanite and calcium perovskite at a constant Br:CaPv ratio of 10:1. Symbols indicate density increases relative to PREM (circle 5%, square 10%, diamond 15%, star 20%). Vertical bars indicate the range of δV_P:δV_S for a given Mw composition at a given δV_S. (**b**) Best-fit results of the inverse model for two ULVZ observations: South Atlantic [47] and Coral Sea [48]. Ellipses represent uncertainties for observations (black dotted lines) and uncertainties at the 68% level for Hill average of best-fit Mw-containing assemblages (solid lines), with corresponding concentrations and compositions noted. Downward and upward triangles indicate best-fit assemblage properties for Voigt and Reuss mixing bounds, respectively. The complete inversion results, including the Voigt and Reuss bounds, are reported in Table 2.

Table 2. Inversion results for two ULVZ observations showing best-fit assemblages of Mw mixed with Br and CaPv with properties as a percentage relative to PREM at the CMB. See text and Supporting Information for details.

Result	δV_P (%)[a]	δV_S (%)	$\delta\rho$ (%)	Mw#	X_{Mw} (%)	X_{Br} (%)	X_{CaPv} (%)	Reduced χ^2
ULVZ: S. Atlantic [b]	−3 ± 3	−9 ± 3	+10 ± 10					
Fit: Voigt	−8.3 ± 3.4	−10.2 ± 3.9	+17.2 ± 2.6	94	29.0 ± 11	59.9 ± 13	11.1	0.58 ± 0.54
Reuss	−4.4 ± 2.3	−8.3 ± 2.0	+8.2 ± 0.6	94	11.6 ± 1.8	72.1 ± 5.6	16.3	0.09 ± 0.21
Hill	−6.4 ± 2.1	−9.3 ± 1.7	+12.7 ± 2.2	94	20.3 ± 5.6	66.0 ± 7.1	13.7	
Fit: Voigt	−7.7 ± 2.9	−10.0 ± 3.1	+15.3 ± 1.7	84	34.1 ± 8.6	64.5 ± 9.9	1.4	0.32 ± 0.40
Reuss	−1.6 ± 2.6	−6.1 ± 2.6	+2.4 ± 0.6	84	11.4 ± 2.7	73.0 ± 9.2	15.6	0.22 ± 0.33
Hill	−5.3 ± 1.9	−8.5 ± 2.0	+9.7 ± 0.9	84	22.7 ± 4.5	68.7 ± 6.7	8.6	
Fit: Voigt [c]	−6.6 ± 2.6%	−9.7 ± 2.7	+12.8 ± 1.3	78	34.4 ± 7.7	63 ± 11	2.6	0.15 ± 0.27
Reuss[c]	−1.4 ± 2.9	−6.3 ± 2.6	+2.8 ± 0.7	78	11.8 ± 4.1	80 ± 19	8.2	0.28 ± 0.37
Hill	−4.1 ± 1.9	−8.0 ± 1.9	+7.8 ± 0.8	78	23.1 ± 4.4	71 ± 11	5.9	
ULVZ: Coral Sea [d]	−8 ± 3	−24 ± 4	+8 ± 6					
Fit: Voigt [e]	−11.9 ± 3.6	−17.9 ± 5.2	+18.3 ± 2.4	78	46.7 ± 14	48.4 ± 17	4.9	3.5 ± 1.1
Reuss [e]	−12.6 ± 2.1	−25.9 ± 2.4	+12.7 ± 1.5	78	34.1 ± 4.0	61.1 ± 6.3	7.9	0.41 ± 0.37
Hill	−12.2 ± 2.1	−21.7 ± 3.0	+15.5 ± 1.3	78	40.4 ± 7.3	54.7 ± 9.1	4.9	

Note: Uncertainties are reported at the 68% level. Unless otherwise noted, all inversions use a prior of X_{Br} = 0.7 with a window of 0.2. [a] Velocity and density changes are reported as a percentage relative to PREM at the CMB. [b] [47]. [c] Prior: X_{Br} = 0.7 with a window of 0.4. [d] [48]. [e] Priors: X_{Br} = 0.6 and X_{Mw} = 0.4, with windows of 0.1 for each.

4.3. Inverse Modeling

While the forward model demonstrates that a wide array of observed velocity reductions could be caused by the presence of iron-rich (Mg,Fe)O, it does not offer a direct quantitative assessment of the compatibility of ULVZ observations with hypothetical mineral aggregates bearing iron-rich (Mg,Fe)O. To that end, we additionally constructed an inverse linear mixing model that can more comprehensively compare seismic observations with experimental results from mineral physics. The inputs to the inverse model were the bulk and shear moduli and density (computed from reported seismic velocities and density) of an observed ULVZ with estimated uncertainties, as well as these properties for the minerals that were being mixed (e.g., Mw94 + Br32 + CaPv, where Br32 represents ($Mg_{0.68}Fe_{0.32}$)SiO_3), with errors propagated from the relevant experimental results. The model minimizes the difference between the target assemblage elastic properties and the hypothetical assemblage properties, with properties weighted according to the inverse of their observational and experimental uncertainties during the fitting procedure. In this way, the model computes individual mineral concentrations for the assemblage that best fits the observational target, following either the Voigt or Reuss formulation of aggregate mixing. Thus, for a given ULVZ observation, we can compute the concentration (X, in percent) of an iron-rich composition of (Mg,Fe)O for the aggregate that would provide the closest fit of seismic wave velocities and density to the observation. The uncertainty in the resulting (Mg,Fe)O concentration includes uncertainty estimates from both the seismic observation of the ULVZ and from the elastic properties of the constituent minerals. By using the χ^2 value to evaluate the quality of the fit, we can compare best-fit assemblages of various mineral compositions to determine which set of minerals with which set of concentrations can best explain a ULVZ observation. More details of the inverse modeling approach can be found in Appendix A.

In order to demonstrate the applicability of this approach, we evaluated two ULVZ seismic observations chosen specifically because the studies report constraints on both V_P and V_S, as well as density and estimates of uncertainties (Table 2). For the inversions, we report the best-fit assemblages containing magnesiowüstite compositions that result in the lowest χ^2 values (Table 2). As part of the inversion process, we can use priors on mineral concentrations to test whether these observed ULVZ properties are compatible with an iron-enriched pyrolitic composition, or whether a non-pyrolitic composition is required to explain the observation. We additionally make use of priors to account for the tradeoffs in concentration between Br and CaPv, which exhibit velocities much closer to PREM than Mw. To that end, for the ULVZ underneath the South Atlantic [47], we impose a prior on the Br concentration of 0.7 with a prior window of 0.2, both due to the trade-off with CaPv and in order to maintain pyrolitic proportions, while for the Mw78 inversion a wider prior window of 0.4 can be used to produce reasonable results. The results demonstrate that an iron-enriched pyrolitic composition is compatible with the South Atlantic ULVZ observation. In contrast, the inversion for the selected seismic observation of a ULVZ beneath the Coral Sea [48] requires more magnesiowüstite than a pyrolitic model would suggest, necessitating a different set of priors. These results suggest that two different formation scenarios may be required for these two ULVZ observations.

The modeling results shown in Figure 2 demonstrate that the presence of iron-rich (Mg,Fe)O results in velocity reductions relative to PREM close to the 1:2 V_P:V_S ratio. It can additionally be seen that the inversion results suggest that ULVZ observations exhibiting a 1:3 V_P:V_S reduction ratio can also be explained by the presence of iron-rich (Mg,Fe)O, within the estimated uncertainty bounds. To help clarify this result, the vertical bars on the forward models (Figure 2a and Figure S4) demonstrate the wide range of reduction ratios, particularly at low S-wave speed reductions, that can be produced by the presence of Mw within uncertainty bounds. It is additionally worth noting that the aggregate elastic properties of the inverted ULVZs could in fact be closer to the Reuss or Voigt bounds than the Hill average, depending on the microstructure of the mineral phases. While the observed 1:3 V_P:V_S reduction ratio is commonly attributed to the presence of partial melt (e.g., [47,49]), the compatibility of solid iron-enriched magnesiowüstite with this reduction ratio, as demonstrated by our results, underscores the necessity of applying a quantitative approach to evaluating ULVZ hypotheses.

By using an inverse linear mixing model, we have shown that the presence of iron-rich (Mg,Fe)O is a viable hypothesis for explaining the seismic anomalies observed within certain ULVZs. The strength of our approach lies in the fact that it incorporates estimated uncertainties from both seismic observations and results from mineral physics, thus permitting a synthesis of information from both fields of study in a quantitative way. We have therefore contributed to the development of a framework for evaluating the likelihood of proposed hypotheses for ULVZs that can be expanded and applied in a systematic way to the growing set of ULVZ observations, in order to develop a comprehensive understanding of heterogeneities in the lowermost mantle.

Supplementary Materials: The following are available online at http://www.mdpi.com/2075-163X/9/12/762/s1, Figure S1: Examples of Integrated XRD Patterns for Mw94, Figure S2: Unit-Cell Volumes and Equation of State for Rhombohedral Mw94, Figure S3: SMS Spectra of Mw94, Figure S4: Forward Modeling the Seismic Signatures of Mw-bearing Aggregates, Table S1: Pressure-Volume Measurements for Mw94, Table S2: Hyperfine Parameters of Mw94 from SMS Measurements, Table S3: Seismic Velocities and Densities of Relevant Phases at CMB Conditions.

Author Contributions: Conceptualization, J.M.J. and V.V.D.; methodology, J.M.J., V.V.D. and W.S.; software, W.S.; validation, J.M.J., V.V.D. and W.S.; formal analysis, V.V.D.; investigation, J.M.J. and V.V.D.; resources, J.M.J. and W.S.; data curation, J.M.J.; writing—original draft preparation, V.V.D.; writing—review and editing, J.M.J., V.V.D. and W.S.; funding acquisition, J.M.J.

Funding: We thank NSF-CSEDI-EAR-1161046 and NSF-EAR-CAREER-0956166 for support of this research. Operations at Sector 3 (APS) and beamline 12.2.2 (ALS) are partially supported by COMPRES. This research used resources of the Advanced Photon Source and of the Advanced Light Source, which are DOE Office of Science User Facilities under contracts DE-AC02-06CH11357 and DE-AC02-05CH11231, respectively.

Acknowledgments: We thank Christine Beavers, as well as Gregory Finkelstein and Natalia Solomatova, for their help with the diffraction experiments. We thank Don Helmberger, Christine Thomas, and Zhongwen Zhan for valuable discussions. We thank June K. Wicks for synthesis of the sample.

Conflicts of Interest: The authors declare no conflict of interest.

Appendix A

We develop an inverse mixing model with the purpose to determine the mixture of individual minerals with aggregate properties that best match a set of target properties (e.g., velocity reductions and density increase relative to PREM for a given ULVZ observation). The mixture contains $(N+1)$ individual minerals, e.g., (Mg,Fe)O magnesiowüstite (Mw), (Mg,Fe)SiO_3 bridgmanite (Br), $CaSiO_3$

calcium silicate perovskite (CaPv). The concentrations $p_j \geq 0$ of the minerals in the mixture are normalized by

$$\sum_{j=1}^{N+1} p_j = 1 \tag{A1}$$

The minerals have properties ϕ_{jk}, where j and k are the number of mineral types and properties, respectively. In addition to the density, properties of interest for our inversions are the bulk modulus and shear modulus or their inverse, dependent on Voigt or Reuss mixing, respectively. We calculate these moduli from the compressional and shear velocities. Average properties of the aggregate are then

$$\langle\phi_k\rangle = \sum_{j=1}^{N+1} p_j\phi_{jk} = \phi_{N+1,k} + \sum_{j=1}^{N} p_j\left(\phi_{j,k} - \phi_{N+1,k}\right) \tag{A2}$$

where the concentration p_{N+1} is eliminated by use of $\sum_{j=1}^{N+1} p_j = 1$. The measure for matching these average properties and the target properties Φ_k is defined as

$$M\left(\{p_j\}\right) = \sum_{k=1}^{n} w_k(\langle\phi_k\rangle - \Phi_k)^2 + \sum_{j=1}^{N} \frac{\left(p_j - P_j\right)^2}{\delta^2 P_j} \tag{A3}$$

where n is the number of properties. In order to produce a physically meaningful solution to the inversion, we can introduce priors P_j on the mineral concentrations with uncertainties $\delta^2 P_j$. The best matching is achieved for the smallest measure. The weights w_k are taken as

$$w_k = \left(\delta\Phi_k^2 + \sum_{j=1}^{N+1} p_j^2 \delta\phi_{jk}^2\right)^{-1} \tag{A4}$$

where $\delta\Phi_k^2$ and $\delta\phi_{jk}^2$ are the variances of target properties Φ_k and mineral properties ϕ_{jk}, respectively. Because n (number of properties: elastic moduli and density) exceeds the dimensionality of the parameter space N (total minerals minus one), the measure never takes its smallest possible value, zero, but can only be minimized by finding a local minimum of it. The optimal concentrations satisfy N non-linear equations given by

$$\frac{\partial}{\partial p_j} M\left(\{p_j\}\right) = 0 \tag{A5}$$

The non-linearity is caused by the concentration dependence of the weights w_k. If the solution does not satisfy $p_j \geq 0$ for all concentrations, the local minimum of the measure is located outside of the truncated N dimensional parameter space.

References

1. Lay, T. Deep Earth Structure: Lower Mantle and D. In *Treatise on Geophysics*, 2nd ed.; Elsevier: Amsterdam, The Netherlands, 2015; Volume 1, pp. 683–723. [CrossRef]
2. Lay, T.; Helmberger, D.V. A Lower Mantle S-wave Triplication and the Shear Velocity Structure of D''. *Geophys. J. R. Astron. Soc.* **1983**, *75*, 799–837. [CrossRef]
3. Cobden, L.; Thomas, C.; Trampert, J. Seismic Detection of Post-Perovskite inside the Earth. In *The Earth's Heterogeneous Mantle: A Geophysical, Geodynamical, and Geochemical Perspective*; Springer: New York, NY, USA, 2015; pp. 391–440. [CrossRef]
4. Garnero, E.J.; Helmberger, D.V. Further Structural Constraints and Uncertainties of a Thin Laterally Varying Ultralow-Velocity Layer at the Base of the Mantle. *J. Geophys. Res. Solid Earth* **1998**, *103*, 12495–12509. [CrossRef]

5. Thorne, M.S.; Garnero, E.J. Inferences on Ultralow-Velocity Zone Structure from a Global Analysis of SPdKS Waves. *J. Geophys. Res. Solid Earth* **2004**, *1098*, 1–22. [CrossRef]
6. McNamara, A.K.; Garnero, E.J.; Rost, S. Tracking Deep Mantle Reservoirs with Ultra-Low Velocity Zones. *Earth Planet. Sci. Lett.* **2010**, *299*, 1–9. [CrossRef]
7. Yu, S.; Garnero, E.J. Ultralow Velocity Zone Locations: A Global Assessment. *Geochem. Geophys. Geosystems* **2018**, *19*, 396–414. [CrossRef]
8. Manga, M.; Jeanloz, R. Implications of a Metal-Bearing Chemical Boundary Layer in D″ for Mantle Dynamics. *Geophys. Res. Lett.* **1996**, *23*, 3091–3094. [CrossRef]
9. Knittle, E.; Jeanloz, R. Earth's Core-Mantle Boundary: Results of Experiments at High Pressures and Temperatures. *Science* **1991**, *251*, 1438–1443. [CrossRef]
10. Mao, W.L.; Mao, H.K.; Sturhahn, W.; Zhao, J.; Prakapenka, V.B.; Meng, Y.; Shu, J.; Fei, Y.; Hemley, R.J. Iron-Rich Post-Perovskite and the Origin of Ultralow-Velocity Zones. *Science* **2006**, *3123*, 564–565. [CrossRef]
11. Dobson, D.P.; Brodholt, J.P. Subducted Banded Iron Formations as a Source of Ultralow-Velocity Zones at the Core-Mantle Boundary. *Nature* **2005**, *434*, 371–374. [CrossRef]
12. Buffett, B.A.; Garnero, E.J.; Jeanloz, R. Sediments at the Top of Earth's Core. *Science* **2000**, *290*, 1938–1942. [CrossRef]
13. Andrault, D.; Pesce, G.; Bouhifd, M.A.; Bolfan-Casanova, N.; Heńot, J.M.; Mezouar, M. Melting of Subducted Basalt at the Core-Mantle Boundary. *Science* **2014**, *244*, 895. [CrossRef]
14. Ohtani, E.; Maeda, M. Density of Basaltic Melt at High Pressure and Stability of the Melt at the Base of the Lower Mantle. *Earth Planet. Sci. Lett.* **2001**, *344*, 892–895. [CrossRef]
15. Pradhan, G.K.; Fiquet, G.; Siebert, J.; Auzende, A.L.; Morard, G.; Antonangeli, D.; Garbarino, G. Melting of MORB at Core-Mantle Boundary. *Earth Planet. Sci. Lett.* **2015**, *431*, 247–255. [CrossRef]
16. Liu, J.; Li, J.; Hrubiak, R.; Smith, J.S. Origins of Ultralow Velocity Zones through Slab-Derived Metallic Melt. *Proc. Natl. Acad. Sci. USA* **2016**, *113*, 5547–5551. [CrossRef]
17. Williams, Q.; Garnero, E.J. Seismic Evidence for Partial Melt at the Base of Earth's Mantle. *Science* **1996**, *273*, 1528–1530. [CrossRef]
18. Berryman, J.G. Seismic Velocity Decrement Ratios for Regions of Partial Melt in the Lower Mantle. *Geophys. Res. Lett.* **2000**, *27*, 421–424. [CrossRef]
19. Sinmyo, R.; Hirose, K.; Nishio-Hamane, D.; Seto, Y.; Fujino, K.; Sata, N.; Ohishi, Y. Partitioning of Iron between Perovskite/Postperovskite and Ferropericlase in the Lower Mantle. *J. Geophys. Res. Solid Earth* **2008**, *113*. [CrossRef]
20. Wicks, J.K.; Jackson, J.M.; Sturhahn, W. Very Low Sound Velocities in Iron-Rich (Mg,Fe)O: Implications for the Core-Mantle Boundary Region. *Geophys. Res. Lett.* **2010**, *37*, 1–5. [CrossRef]
21. Wicks, J.K.; Jackson, J.M.; Sturhahn, W.; Zhang, D. Sound Velocity and Density of Magnesiowüstites: Implications for Ultralow-Velocity Zone Topography. *Geophys. Res. Lett.* **2017**, *44*, 2148–2158. [CrossRef]
22. Bower, D.J.; Wicks, J.K.; Gurnis, M.; Jackson, J.M. A Geodynamic and Mineral Physics Model of a Solid-State Ultralow-Velocity Zone. *Earth Planet. Sci. Lett.* **2011**, *303*, 193–202. [CrossRef]
23. Li, M.; McNamara, A.K.; Garnero, E.J.; Yu, S. Compositionally-Distinct Ultra-Low Velocity Zones on Earth's Core-Mantle Boundary. *Nat. Commun.* **2017**, *8*, 177. [CrossRef]
24. Brown, S.P.; Thorne, M.S.; Miyagi, L.; Rost, S. A Compositional Origin to Ultralow-Velocity Zones. *Geophys. Res. Lett.* **2015**, *42*, 1039–1045. [CrossRef]
25. Wicks, J.K.; Jackson, J.M.; Sturhahn, W.; Zhuravlev, K.K.; Tkachev, S.N.; Prakapenka, V.B. Thermal Equation of State and Stability of (Mg0.06Fe0.94) O. *Phys. Earth Planet. Inter.* **2015**, *249*, 28–42. [CrossRef]
26. Finkelstein, G.J.; Jackson, J.M.; Sturhahn, W.; Zhang, D.; Ercan Alp, E.; Toellner, T.S. Single-Crystal Equations of State of Magnesiowüstite at High Pressures. *Am. Mineral.* **2017**, *102*, 1709–1717. [CrossRef]
27. Matas, J.; Bass, J.; Ricard, Y.; Mattern, E.; Bukowinski, M.S.T. On the Bulk Composition of the Lower Mantle: Predictions and Limitations from Generalized Inversion of Radial Seismic Profiles. *Geophys. J. Int.* **2007**, *170*, 764–780. [CrossRef]
28. Dewaele, A.; Datchi, F.; Loubeyre, P.; Mezouar, M. High Pressure-High Temperature Equations of State of Neon and Diamond. *Phys. Rev. B Condens. Matter Mater. Phys.* **2008**, *77*, 094106. [CrossRef]
29. Prescher, C.; Prakapenka, V.B. DIOPTAS: A Program for Reduction of Two-Dimensional X-Ray Diffraction Data and Data Exploration. *High Press. Res.* **2015**, *35*, 223–230. [CrossRef]

30. Toby, B.H.; Von Dreele, R.B. GSAS-II: The Genesis of a Modern Open-Source All Purpose Crystallography Software Package. *J. Appl. Crystallogr.* **2013**, *46*, 544–549. [CrossRef]
31. Fei, Y.; Ricolleau, A.; Frank, M.; Mibe, K.; Shen, G.; Prakapenka, V. Toward an Internally Consistent Pressure Scale. *Proc. Natl. Acad. Sci. USA* **2007**, *104*, 9182–9186. [CrossRef]
32. Toellner, T.S. Monochromatization of Synchrotron Radiation for Nuclear Resonant Scattering Experiments. *Hyperfine Interact.* **2000**, *125*, 3–28. [CrossRef]
33. Akahama, Y.; Kawamura, H. Pressure Calibration of Diamond Anvil Raman Gauge to 410 GPa. *J. Phys. Conf. Ser.* **2010**, *215*, 1. [CrossRef]
34. Sturhahn, W. MINUTI Open Source Software, Version 2.0.0. Available online: http://www.nrixs.com (accessed on 10 January 2018).
35. Sturhahn, W. CONUSS and PHOENIX: Evaluation of Nuclear Resonant Scattering Data. *Hyperfine Interact* **2000**, *125*, 149–172. [CrossRef]
36. Solomatova, N.V.; Jackson, J.M.; Sturhahn, W.; Wicks, J.K.; Zhao, J.; Toellner, T.S.; Kalkan, B.; Steinhardt, W.M. Equation of State and Spin Crossover of (Mg,Fe)O at High Pressure, with Implications for Explaining Topographic Relief at the Core-Mantle Boundary. *Am. Mineral.* **2016**, *101*, 1084–1093. [CrossRef]
37. Sturhahn, W.; Jackson, J.M.; Lin, J.F. The Spin State of Iron in Minerals of Earth's Lower Mantle. *Geophys. Res. Lett.* **2005**, *32*, 1–5. [CrossRef]
38. Tsuchiya, T.; Wentzcovitch, R.M.; da Silva, C.R.S.; de Gironcoli, S. Spin Transition in Magnesiowüstite in Earth's Lower Mantle. *Phys. Rev. Lett.* **2006**, *96*, 198501. [CrossRef]
39. Wu, Z.; Wentzcovitch, R.M. Spin Crossover in Ferropericlase and Velocity Heterogeneities in the Lower Mantle. *Proc. Natl. Acad. Sci. USA* **2014**, *111*, 10468–10472. [CrossRef]
40. Wentzcovitch, R.M.; Wu, Z.; Carrier, P. First Principles Quasiharmonic Thermoelasticity of Mantle Minerals. *Rev. Mineral. Geochem.* **2010**, *71*, 99–128. [CrossRef]
41. Gréaux, S.; Irifune, T.; Higo, Y.; Tange, Y.; Arimoto, T.; Liu, Z.; Yamada, A. Sound Velocity of $CaSiO_3$ Perovskite Suggests the Presence of Basaltic Crust in the Earth's Lower Mantle. *Nature* **2019**, *565*, 218–221. [CrossRef]
42. Duffy, T.S.; Anderson, D.L. Seismic Velocities in Mantle Minerals and the Mineralogy of the Upper Mantle. *J. Geophys. Res.* **1989**, *94*, 1895–1912. [CrossRef]
43. Li, B.; Zhang, J. Pressure and Temperature Dependence of Elastic Wave Velocity of $MgSiO_3$ Perovskite and the Composition of the Lower Mantle. *Phys. Earth Planet. Inter.* **2005**, *151*, 143–154. [CrossRef]
44. Dorfman, S.M.; Meng, Y.; Prakapenka, V.B.; Duffy, T.S. Effects of Fe-Enrichment on the Equation of State and Stability of $(Mg,Fe)SiO_3$ Perovskite. *Earth Planet. Sci. Lett.* **2013**, *361*, 249–257. [CrossRef]
45. Wolf, A.S.; Jackson, J.M.; Dera, P.; Prakapenka, V.B.; Al, W.E.T. Solid Earth The Thermal Equation of State of (Mg, Fe) SiO_3 Bridgmanite (Perovskite) and Implications for Lower Mantle Structures. *J. Geophys. Res.* **2015**, *120*, 7460–7489. [CrossRef]
46. Dziewonski, A.M.; Anderson, D.L. Preliminary Reference Earth Model. *Phys. Earth Planet. Inter.* **1981**, *25*, 297–356. [CrossRef]
47. Simmons, N.A.; Grand, S.P. Partial Melting in the Deepest Mantle. *Geophys. Res. Lett.* **2002**, *29*, 1552. [CrossRef]
48. Rost, S.; Garnero, E.J.; Williams, Q. Fine-Scale Ultralow-Velocity Zone Structure from High-Frequency Seismic Array Data. *J. Geophys. Res.* **2006**, *111*, B09310. [CrossRef]
49. Yuan, K.; Romanowicz, B. Seismic Evidence for Partial Melting at the Root of Major Hot Spot Plumes. *Science* **2017**, *357*, 1–5. [CrossRef]

minerals

MDPI

Article

Ultimate Mechanical Properties of Forsterite

Karine Gouriet [1,*], Philippe Carrez [1] and Patrick Cordier [1,2]

[1] Univ. Lille, CNRS, INRA, ENSCL, UMR 8207-UMET-Unité Matériaux et Transformations, F-59000 Lille, France; Philippe.Carrez@univ-lille.fr (P.C.); patrick.cordier@univ-lille.fr (P.C.)

[2] Institut Universitaire de France, 1 rue Descartes, F-75005 Paris, France

* Correspondence: karine.gouriet@univ-lille.fr

Received: 21 November 2019; Accepted: 12 December 2019; Published: 14 December 2019

Abstract: The ultimate mechanical properties, as characterized here by the ideal strengths of Mg_2SiO_4 forsterite, have been calculated using first-principles calculations and generalized gradient approximation under tensile and shear loading. The ideal tensile strengths (ITS) and ideal shear strengths (ISS) are computed by applying homogeneous strain increments along high-symmetry directions ([100], [010], and [001]) and low index shear plane ((100), (010), and (001)) of the orthorhombic lattice. We show that the ultimate mechanical properties of forsterite are highly anisotropic, with ITS ranging from 12.1 GPa along [010] to 29.3 GPa along [100], and ISS ranging from 5.6 GPa for simple shear deformation along (100) to 11.5 GPa for shear along (010).

Keywords: forsterite; ultimate mechanical properties; ideal tensile strength; ideal shear strength

1. Introduction

Olivine $(Mg,Fe)_2SiO_4$ is a mineral of prominent importance since it is a major component of the diffuse interstellar medium and of protoplanetary disks around young stars [1]. Olivine dust in the interstellar medium appears to be almost entirely amorphous, whereas the spectra of protoplanetary disks also show evidence of crystallinity. In the solar system, olivine is found in comets [2], chondritic and nonchondritic meteorites [3,4], and in the mantle of terrestrial planets. On Earth, olivine is the main constituent of the upper mantle [5] and its transformation under pressure to wadsleyite and ringwoodite is the main cause of the observed seismic discontinuities at 410 and 520 km depths. Olivine glass is very difficult to obtain from the melt and requires extreme cooling rates [6]. The first report of olivine glass in 1977 by [7] was related to shock experiments of a single crystal of San Carlos olivine. A few years later, [8] reported evidence of fayalite olivine glass formed after heating in a diamond anvil cell. In 1990, [9,10] presented evidence of amorphization of fayalite pressurized above 39 GPa and 35 GPa respectively. Occurrence of pressure induced amorphization of Mg-rich olivines was further documented by [11–13]. Although the role of pressure was generally put forward as the cause for amorphization, the influence of non-hydrostatic stresses was highlighted by [12,14]. Pressure-induced amorphization is usually described as a kinetically preferred transformation resulting from frustration in reaching the high-pressure equilibrium crystalline state. This transformation questions the mechanical stability of crystalline solids.

Here, we propose a first investigation of the mechanical stability of olivine based on elasticity. The elastic properties of solids do not just address their response to small strains. In pure hydrostatic compression, a solid is not prone to failure, and elastic properties deviate significantly from a linear response. The equation of state which describes the evolution of the volume in response to a hydrostatic compression is a fundamental parameter for the minerals of the interior of the Earth. For any other solicitation, the elastic response of solids is difficult to assess to large stresses since the presence of defects leads to failure and plastic deformation. The recent advances in theoretical methods and computation make it possible to calculate the elastic behavior of solids submitted to homogeneous deformation at large strains, until the point of instability (the ideal strength) is reached [15–18]. Ideal strength is thus the critical stress above which a defect-free crystal reaches its mechanical instability and undergoes a spontaneous structural transformation (to another phase possibly amorphous). The method consists therefore in the computation of the crystal energy and stresses under the application of successive deformation increments in order to access to the corresponding stress–strain curves. The instability corresponds to the maximum of stress which also corresponds to an inflection point in the free energy curve as the Cauchy stress is related to the derivative of the free energy as a function of strain [19].

In this study, we focus on the elastic properties and mechanical stability of pure, iron-free Mg_2SiO_4 forsterite. The ideal tensile strengths (ITS) and ideal shear strengths (ISS) are computed from a first-principles method along high-symmetry directions [100], [010], and [001] and for homogenous shear of (100), (010), and (001) planes (here given with respect to the *Pbnm* space group of forsterite).

2. Materials and Methods

In this work, the derivation of the anisotropic ideal strength is performed according to the recent ADAIS (version 1.0.0, Beihang University, China) free open source software written by Zhang and co-workers [20], which allows for an automatic implementation of homogenous deformation to standard first-principles VASP calculations [21]. Thus, all simulations correspond to calculations based on the density functional theory (DFT) using a plane wave basis set and the projector augmented wave method (PAW) [22]. Exchange–correlation energy is accounted by employing the Perdew-Wang (PW91), gradient-corrected functional (GGA) [23]. For all the calculations, plane-wave basis set expansion was limited using a kinetic energy cutoff of 520 eV, known to ensure an adequately atomic force convergence in forsterite [24]. Regarding the k-point sampling, throughout this study, we use a single grid of 6 × 4 × 6 according to a Monkhorst and Pack scheme [25] corresponding to 18 k-points per forsterite unit cell.

As mentioned in the introduction, once a Mg_2SiO_4 unit cell has been fully relaxed, tensile or shear tests are performed by applying incremental homogeneous strain, i.e., atomic layers of the crystal are uniformly displaced along the tensile or shear direction (Figure 1). At each deformation state, a relaxation of both the cell shape and the atomic positions is performed until all the components of the stress tensor are brought to zero, except for one corresponding to the applied stress condition. In practice, we verify that structural relaxation allows for residual stresses of the order of a few MPa at the most.

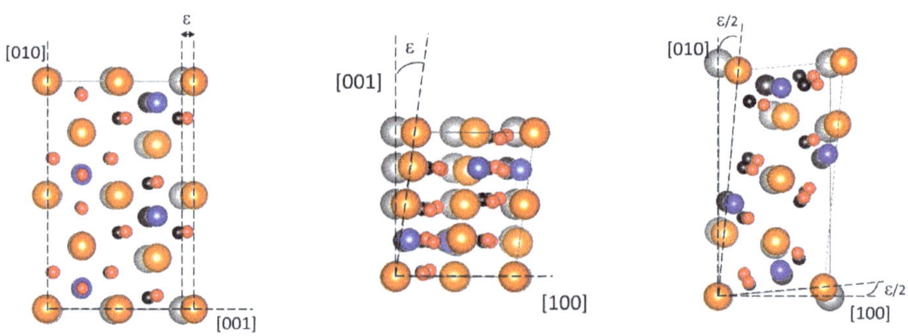

(**a**) (**b**) (**c**)

Figure 1. Illustration of the various loading conditions applied in this study (**a**) tensile deformation along [001] where $\varepsilon_{zz} = \varepsilon$, (**b**) simple shear [001](100) where $\varepsilon_{xz} = \varepsilon/2$ and $\varepsilon_{zx} = 0$, or (**c**) pure shear deformation in (001), where $\varepsilon_{xy} = \varepsilon_{yx} = \varepsilon/2$. The colored image represents a deformed state compared with the undeformed, reference structure in grey. Mg is in orange, Si in blue, O in red.

3. Results

3.1. Ground State Properties

Before calculating the ultimate mechanical properties of Mg_2SiO_4, we optimized the equilibrium structure. A unit cell has been built and relaxed for the *Pbnm* configuration, giving rise to the equilibrium lattice parameters *a*, *b*, and *c*. The results are displayed in Table 1, where they are compared with available data (both theoretical and experimental). It is shown that the calculations predict the correct Mg_2SiO_4 ground state structure.

Table 1. Crystallographic data for forsterite at 0 K and 0 GPa compared with calculated data.

	a (Å)	*b* (Å)	*c* (Å)	*V* (Å³)
This study	4.79	10.27	6.03	296.63
Calculated GGA [24]	4.79	10.28	6.02	296.43
Calculated GGA [26]	4.79	10.28	6.04	297.68
Calculated GGA [27]	4.71	10.15	5.96	284.92
Calculated LDA [28]	4.64	9.99	6.07	281.67
Experimental [29]	4.75	10.19	5.98	289.58

3.2. Ideal Strength in Tension and in Shear

From the optimized unit cell containing 28 atoms, the ideal strengths are determined from an incremental application of strain of 0.005. Three tensile directions have been tested and for the shear deformation, we investigated both simple and pure shear modes.

3.2.1. Tensile Tests

The evolutions of the total energy as a function of strain are reported in Figure 2a for tensile experiments performed along [100], [010] and [001]. As expected, one observes first, for the three tensile tests, a parabolic evolution corresponding to elastic energy storing which corresponds to the initial linear portion of the stress–strain curves (Figure 2b). The elastic anisotropy of the structure is readily visible from the slope of the stress–strain curves. Pulling along the [100] axis leads to the largest Young's modulus (Table 2). The same conclusion arises from the computation of the Poisson ratios. Indeed, since during tensile tests, transverse directions are fully relaxed, we are able to determine the corresponding Poisson ratios (summarized in Table 3) according to the variation of the transverse lattice parameters. Except for the [100] tensile experiment, the computed Poisson ratio are close to 0.25, i.e., in agreement with what can be deduced from Voigt–Reuss averaging. Thus, the [100] direction exhibits a peculiar behavior in tension which may be related to the distorted hexagonal closed packing of the oxygen sublattice as pointed out by [30].

For the three tests, we observe an inflection point on the energy curve which corresponds, by convention, to the maximum (or ideal) tensile strength (ITS). With the corresponding stress–strain curves shown in Figure 2b, one may notice the remarkable agreement between the Cauchy stress derivation (solid line) obtained by derivation of the energy and the stress state (according to the Hellmann–Feynman theorem) of the strained volume (symbols). Whatever the investigated pulling directions, the instability in tension is reached for rather comparable strains (11–16%). However, the magnitudes of the ideal stresses are strongly dependent on the crystallographic direction showing, again, the anisotropy of the forsterite crystal structure. Pulled along the [100] axis, forsterite can indeed sustain the highest stress, i.e., 29.3 GPa and 13% of strain, whereas the instability is reached for a stress equal to 12.1 GPa, at a critical strain of 11% along [010] and for 15.9 GPa at 16% along the [001] axis. The behavior just after the stability limit is also contrasted. For [100] tensile loading, stress drops rapidly above the critical strain. For loading along [001], stress decreases progressively after the maximum. The case of [010] loading attracts attention since after the critical strain, one observes a first softening stage followed by a strain stiffening which allows the structure to sustain the stress and reach further strains.

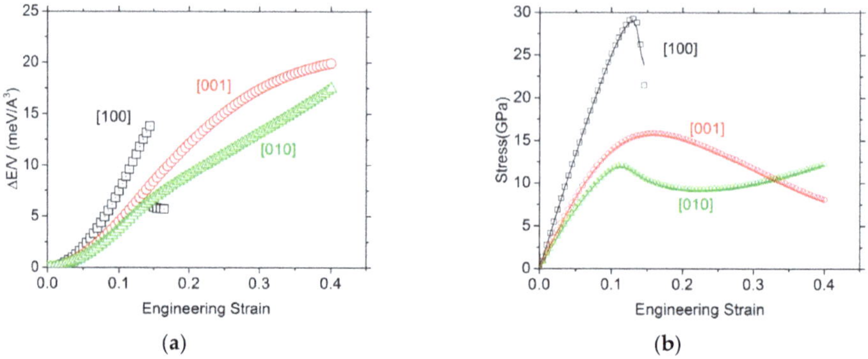

Figure 2. (**a**) Evolution of the strain energy as a function of the engineering strain; (**b**) Stress as a function of the engineering strain. The tensile directions are [100] (black empty square), [001] (red empty circle), and [010] (green empty triangle). The instability occurs at the inflection point of the strain energy–strain curves, or the maximum of the stress–strain curves. In (b), the lines correspond to the Cauchy stress evaluations from the derivative of the energy curves shown in (a).

Table 2. Ideal stresses (and associated engineering strains) determined in this study under tensile, pure and simple shear loading. For tensile and simple shear tests, we report also the Young's modulus and Poisson ratio. The normalized stresses are the ideal stresses divided by the elastic modulus (Young's modulus in tension and shear modulus in shear).

Tensile Tests	[100]		[010]		[001]	
ITS (GPa)	29.3		12.1		15.9	
Corresponding strain (%)	13.0		11.5		16	
Young's modulus (GPa)	274.4		153.2		170.9	
Normalized stress	0.10		0.08		0.09	
Pure shear tests	**(100)**		**(010)**		**(001)**	
ISS (GPa)	5.6		11.8		8.7	
Corresponding strain (%)	18.5		26.5		18.5	
Simple shear tests	**[010](001)**	**[001](010)**	**[100](001)**	**[001](100)**	**[010](100)**	**[100](010)**
ISS (GPa)	6.2	5.3	13.4	11.2	9.0	8.5
Corresponding strain (%)	20	18	29.5	26	20	18
Shear modulus (GPa)	58.7 (i.e., C_{44})	58.7 (i.e., C_{44})	73.7 (i.e., C_{55})	73.7 (i.e., C_{55})	73.0 (i.e., C_{66})	73.0 (i.e., C_{66})
Normalized stress	0.10	0.09	0.18	0.15	0.12	0.11

Table 3. Poisson ratio determined in this study under tensile tests for strains below 5%.

Tensile Tests	**[100]**	**[010]**	**[001]**
$\upsilon_{[100]}$	-	0.13	0.14
$\upsilon_{[010]}$	0.23	-	0.29
$\upsilon_{[001]}$	0.20	0.24	-

3.2.2. Shear Tests

Nine ideal shear deformation tests have been performed within this study. Six experiments correspond to simple shear and three to pure shear. The evolutions of the energy as a function of the engineering strain are shown Figure 3, and the corresponding stress–strain curves are shown Figure 4. The initial slopes of the stress–strain curves give the shear moduli C_{ii} (with i = 4, 5, or 6 in Voigt notations). In Figure 4, the results are presented in three groups, each corresponding to a pure shear deformation test associated with the two related simple shear experiments. The three groups naturally emerge from Figure 3. A first set, involving [100](001), [001](100) simple shear deformation and pure shear in (010), corresponds to the highest energy curves. On the opposite side, applying [010](001) or [001](010) simple shear deformation or pure shear in (100) corresponds to the smallest energy increase. In between, the last set of experiments corresponds to simple shear [100](010) and [010](100), and pure shear in (001).

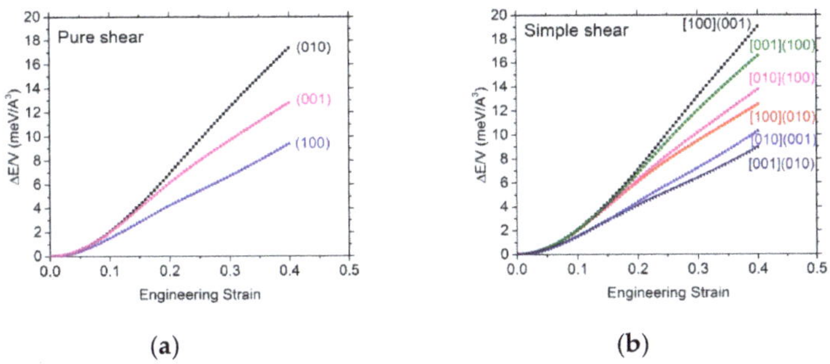

Figure 3. Variation of the total energy as a function of the engineering strain calculated using a unit cell of forsterite (**a**) under pure shear deformation (**b**) under simple shear deformation, for the [100](001) (square), [100](010) (circle), [010](001) (up triangle), [010](100) (down triangle), [001](100) (diamond) and [001](010) (left triangle) shear deformations.

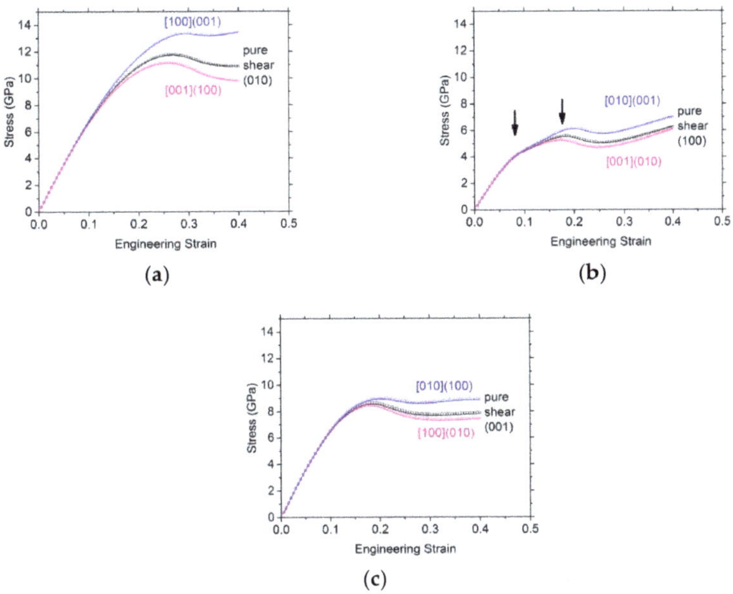

Figure 4. Stress as a function of the engineering strain calculated using a unit cell of forsterite under shear deformation for (**a**) simple shear along [100](001) and [001](100) and pure shear in (010), (**b**) simple shear along [010](001) and [001](010) and pure shear in (100), and (**c**) simple shear along [100](010) and [010](100) and pure shear in (001). Arrows in (b) mark the occurrence of Mg–O bonds breaking as described in the discussion section.

The correspondence between the different configurations tested within each group appears clearly on the stress–strain curves (Figure 4). Within each group, all curves superimpose within the elastic regime. According to the slopes of the stress–strain curves, we find C_{44} = 58.7 GPa, C_{55} = 73.7 GPa and C_{66} = 73 GPa in agreement with the elastic properties of olivine found experimentally [31,32] or numerically [26]. Moreover, in each group, the pure shear stress–strain curve lies in between the simple shear stress–strain curves until the instability is reached. At the ISS, the stress differences between simple or pure shear tests are within a few GPa. Critical strains and ISS are summarized in Table 1.

The largest ISS (Figure 4a) corresponds to the first group of experiments reported above, with [100](001) and [001](100) simple shears and pure shear in (010) with ISS between 11.2 and 13.4 GPa. For the other shear deformation experiments, the instability of the structure occurs at lower stress and lower strain. The softest group (shown Figure 4b), corresponds to [010](001) and [001](010) simple shear deformations, and pure shear in (100) with ISS between 5.3 and 6.2 GPa. It is worth noting that this shear mode exhibits a more complex behavior with the instability preceded by an inflection in the stress–strain curves, followed by significant strain stiffening.

4. Discussion and Conclusions

The main objective of this work is to investigate the mechanical response of forsterite to applied strains until it becomes mechanically unstable. Before reaching this point, the crystal is strained in the linear elastic regime and all energy–strain curves first exhibit a parabolic regime. This allows us to determine elastic moduli which, with the calculated lattice parameters, validate our calculations.

The ITS, defined as the first maximum of the stress–strain curve, are 29.3, 12.1, and 15.9 GPa along the [100], [010], and [001] directions respectively. With a 2.4 ratio between the extreme values, the ITS illustrate the anisotropy of orthorhombic forsterite which follows quite well the elastic behavior since, normalized by the Young's modulus, all ITS values are of the order of 0.1 (Table 2). It is striking however, that loading along different directions leads to contrasted behaviors at the instability. To understand the origin of the differences of the stress–strain curves, we follow the evolution of the bond lengths as a function of strain. The tensile direction which corresponds to the higher stiffness is [100]. The stress drops abruptly after the maximum. Figure 5a shows how the Mg–O bond lengths evolve for the two magnesium sites in forsterite, called Mg^1 and Mg^2. It appears that the stress maximum corresponds to a divergence of the Mg^2–O bond. The authors of [33] show that the strength of metallic bond in oxides correlate with their lengths. Above 2.5 Å, the Pauli strength decreases drastically, and the bond loses its strength. This is what occurs at the inflection point of the energy curve when forsterite is strained along [100]. Loading forsterite in tension along [010] and [100] leads to different behaviors. The case of tension along [010] is interesting since after the maximum, the stress first decreases before stabilizing and progressively stiffening up to an engineering strain of 0.4. Again, the origin of this behavior is found in the bond distances as shown in Figure 5b. After a first increase from 2.2 to ca. 2.35 Å, the Mg^2–O bond length decreases to recover its original value in the strain interval 0.2–0.3 before increasing again. On both Figure 5a,b, one can see that the SiO_4 tetrahedra are not affected by the loading since the Si–O distances remain almost constant. This is a general observation for all solicitations investigated here which illustrates the stiffness of the ionocovalent Si–O bond.

It must be noted that care is needed when discussing the behavior beyond the instability since the system may evolve with structural modifications which may be constrained by the choice of the system size [34], especially when bonds are broken, leading to structural reconstruction [35] like in the [100] tensile test.

In shear we observe, as in tension, that the ISS follow quite well the elastic anisotropy with normalized values (Table 2) in between 0.1 and 0.18. The authors of [18] have compiled the ISS of several simple metals and ceramics. Most metals exhibit normalized ISS above 0.005 and below 0.15 when oxides (MgO, CaO) are slightly above 0.15. Normalized ISS of covalent materials (C, SiC, Si_3N_4) are around 0.2. Our results on forsterite are consistent with this general pattern since the resistance of the structure depends on the Mg–O bonds mostly.

For all tests, we observe that the pure shear is bracketed by the two related simple shears. The shear tests presented in Figure 4b present several interesting features. They correspond to the weakest shear directions with ISS of the order of 6 GPa. Also the stress–strain curves exhibit a change of slope (at ca. 10%, marked by the left arrow on Figure 4b) before the instability (marked by the right arrow) followed by a significant stiffening after the instability. To analyze these features, we compute the Mg–O bond lengths. Results corresponding to the [010](001) shear test are presented on Figure 6. At the first change of slope (just before 10% strain), one observes (Figure 6b) that one Mg^2 shows a bond divergence (with O^9). A second similar feature (bond breakage between Mg^2 and O^1) is responsible for the instability. However, in parallel, several Mg–O bonds show their distances decrease significantly (below 2Å) which result in the significant stiffening observed. Such features, with new bond formation or reorganization under large strain, has also been reported in various crystalline solids (for instance in cementite Fe_3C) leading to a strong strain-stiffening effect [36] as observed here in forsterite.

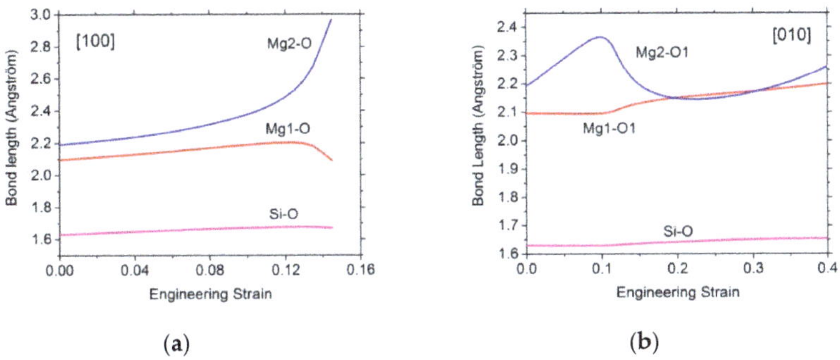

Figure 5. Typical bond lengths evolution as a function of strain for tensile deformation along (**a**) [100] and (**b**) [010]. Mainly because of the ionocovalency of the bond, Si–O bonds are rather unaffected by strain, whereas one may notice the differential behavior between Mg^1 and Mg^2 sites.

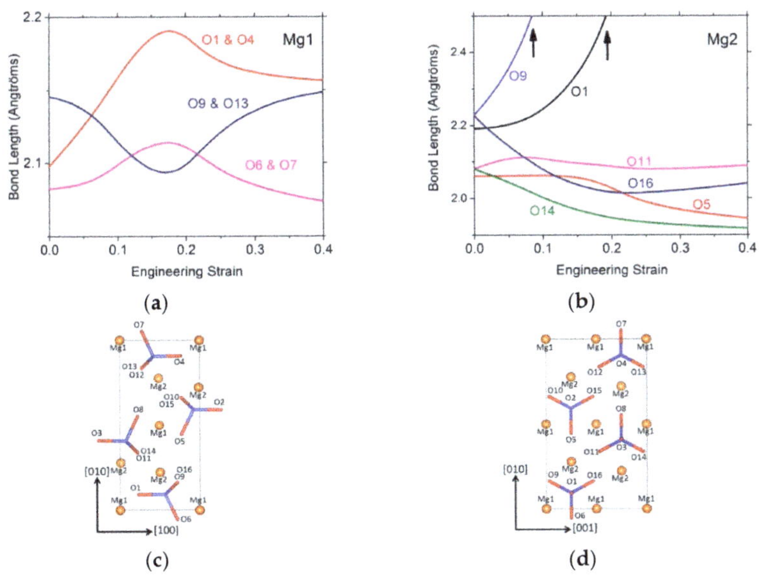

Figure 6. Mg–O bond lengths for both (**a**) Mg^1 and (**b**) Mg^2 sites, as a function of the applied strain for [010](001) simple shear deformation. In (b), the two arrows highlight the breaking of Mg–O bonds at, respectively, first inflexion of the stress–strain curve and the instability as quoted in Figure 4b. The oxygen numbering refers to the labeled atoms shown in (**c**) and (**d**) within the unit cell of forsterite.

Olivine is not stable at high pressure. At mantle temperatures, $(Mg_{0.9}, Fe_{0.1})_2SiO_4$ olivine transforms into wadsleyite at ca. 13 GPa and then to ringwoodite at ca. 18 GPa. At room temperature, these reconstructive phase transformations are kinetically hindered and the olivine structure can be further compressed until it collapses to an amorphous phase above ca. 40 GPa [9,14]. The onset of pressure induced amorphization varies depending on the composition, the type of loading (static, dynamic), but also, although this is less constrained, on non-hydrostaticity as pointed out by [12] and [37]. Here, we characterize the limit of mechanical stability of forsterite without confining pressure and under tensile and shear loading. We show that the onset of instability can be as low as 5–6 GPa for some shear conditions (Table 2). Such deviatoric stress conditions can be reached in nanoindentation, which can significantly facilitate high pressure phase transitions and lower the pressure threshold. Evidence for amorphization has been reported under contact loading in silicon [38] and in boron carbide [39]. Nanoindentation has been performed recently on olivine by [40,41], however, no microstructural investigation was conducted to show a possible amorphization. Such characterizations should provide a test for our theoretical predictions.

Author Contributions: Conceptualization, P.C. (Patrick Cordier); methodology, P.C. (Philippe Carrez); validation, P.C. (Philippe Carrez); investigation, K.G.; writing—original draft preparation, K.G.; writing—review and editing, P.C. (Philippe Carrez) and P.C. (Patrick Cordier); funding acquisition, P.C. (Patrick Cordier).

Funding: This research was funded by European Research Council (ERC) under the European Union's Horizon 2020 Research and Innovation Programme under grant agreement No 787198—TimeMan.

Conflicts of Interest: The authors declare no conflict of interest.

References

1. Fogerty, S.; Forrest, W.; Watson, D.M.; Sargent, B.A.; Koch, I. Silicate Composition of the Interstellar Medium. *Astrophys. J.* **2016**, *830*, 71. [CrossRef]
2. Zolensky, M.E.; Zega, T.J.; Yano, H.; Wirick, S.; Westphal, A.J.; Weisberg, M.K.; Weber, I.; Warren, J.L.; Velbel, M.A.; Tsuchiyama, A. Mineralogy and petrology of comet 81P/Wild 2 nucleus samples. *Science* **2006**, *314*, 1735–1739. [CrossRef] [PubMed]
3. Brearley, A.J.; Jones, R.H.; Papike, J.J. (Eds.) Chondritic meteorites. Planetary Materials. In *Review in Mineralogy*; Mineralogical Society of America: Chantilly, VA, USA, 1998; Volume 36, p. C1.
4. Mittlefehldt, D.W.; McCoy Timothy, J.; Goodrich, C.A.; Kracher, A. Non-chondritic meteorites from asteroidal bodies. *Rev. Miner. Geochem.* **1998**, *36*, 4.1–4.195.
5. Ringwood, A.E. Phase transformations and their bearing on the constitution and dynamics of the mantle. *Geochim. Cosmochim. Acta* **1991**, *55*, 2083–2110. [CrossRef]
6. Richet, P.; Leclerc, F.; Benoist, L. Melting of forsterite and spinel, with implications for the glass transition of Mg_2SiO_4 liquid. *Geophys. Res. Lett.* **1993**, *20*, 1675–1678. [CrossRef]
7. Jeanloz, R.; Ahrens, T.; Lally, J.S.; Nord, G.L.; Christie, J.M.; Heuer, A.H. Shock-Produced Olivine Glass: First Observation. *Science* **1977**, *197*, 457–459. [CrossRef]
8. Lacam, A.; Madon, M.; Poirier, J.P. Olivine glass and spinel formed in a diamond anvil high-pressure cell. *Nature* **1980**, *288*, 155–157. [CrossRef]
9. Williams, Q.; Knittle, E.; Reichlin, R.; Martin, S.; Jeanloz, R. Structural and electronic properties of Fe_2SiO_4-fayalite at ultrahigh pressures: Amorphization and gap closure. *J. Geophys. Res.* **1990**, *95*, 21549–21563. [CrossRef]
10. Richard, G.; Richet, P. Room-temperature amorphization of fayalite and high-pressure properties of Fe_2SiO_4 liquid. *Geophys. Res. Lett.* **1990**, *17*, 2093–2096. [CrossRef]
11. Guyot, F.; Reynard, B. Pressure-induced structural modifications and amorphization in olivine compounds. *Chem. Geol.* **1992**, *96*, 411–420. [CrossRef]
12. Andrault, D.; Bouhifd, M.A.; Itie, J.P.; Richet, P. Compression and amorphization of $(Mg,Fe)_2SiO_4$ olivines: An X-ray diffraction study up to 70 GPa. *Phys. Chem. Min.* **1995**, *22*, 99–107. [CrossRef]
13. Finkelstein, G.J.; Dera, P.K.; Jahn, S.; Oganov, A.R.; Holl, C.M.; Meng, Y.; Duffy, T.S. Phase transitions and equation of state of forsterite to 90 GPa from single-crystal X-ray diffraction and molecular modeling. *Am. Miner.* **2014**, *99*, 35–43. [CrossRef]

14. Santamaria-Perez, D.; Thomson, A.; Segura, A.; Pellicer-Torres, J.; Manjon, F.J.; Corà, F.; McColl, K.; Wilson, M.; Dobson, D.; McMillan, P.F. Metastable structural transformations and pressure-induced amorphization in natural $(Mg,Fe)_2SiO_4$ olivine under static compression: A Raman spectroscopic study. *Am. Miner.* **2016**, *101*, 1642–1650. [CrossRef]
15. Li, W.; Wang, T. Ab initio investigation of the elasticity and stability of aluminium. *J. Phys. Condens. Matter.* **1998**, *10*, 9889–9904. [CrossRef]
16. Roundy, D.; Krenn, C.R.; Cohen Marvin, L.; Morris, J.W., Jr. Ideal Shear Strengths of fcc Aluminum and Copper. *Phys. Rev. Lett.* **1999**, *82*, 2713–2716. [CrossRef]
17. Ogata, S.; Kitagawa, H. Ab initio tensile testing simulation of aluminum and aluminum nitride ceramics based on density functional theory. *Comput. Mater. Sci.* **1999**, *15*, 435–440. [CrossRef]
18. Ogata, S.; Li, J.; Hirosaki, N.; Shibutani, Y.; Yip, S. Ideal shear strain of metals and ceramics. *Phys. Rev. B* **2004**, *B70*, 104404. [CrossRef]
19. Yip, S.; Li, J.; Tang, M.; Wang, J. Mechanistic aspects and atomic-level consequences of elastic instabilities in homogeneous crystals. *Mater. Sci. Eng.* **2001**, *A317*, 236–240. [CrossRef]
20. Zhang, S.H.; Fu, Z.H.; Zhang, R.F. ADAIS: Automatic Derivation of Anisotropic Ideal Strength via high-throughput first-principles computations. *Comput. Phys. Commun.* **2019**, *238*, 244–253. [CrossRef]
21. Kresse, G.; Furthmüller, J. Efficient iterative schemes for ab initio total-energy calculations using a plane-wave basis set. *Phys. Rev.* **1996**, *B54*, 11169. [CrossRef]
22. Perdew, J.P.; Wang, Y. Accurate and simple analytic representation of the electron-gas correlation energy. *Phys. Rev.* **1992**, *B45*, 13244–13249. [CrossRef] [PubMed]
23. Wang, Y.; Perdew, J.P. Correlation hole of the spin-polarised electron-gas, with exact small-wave-vector and high-density scaling. *Phys. Rev.* **1991**, *B44*, 13298–13307. [CrossRef] [PubMed]
24. Hernandez, E.R.; Brodholt, J.; Alfe, D. Structural, vibrational and thermodynamic properties of Mg_2 SiO_4 and $MgSiO_3$ minerals from first-principles simulations. *Phys. Earth Planet. Inter.* **2015**, *240*, 1–24. [CrossRef]
25. Monkhorst, H.J.; Pack, J.D. Special points for Brillouin-zone integrations. *Phys. Rev.* **1976**, *B13*, 5188–5192. [CrossRef]
26. Durinck, J.; Legris, A.; Cordier, P. Influence of crystal chemistry on ideal plastic shear anisotropy in forsterite: First principle calculations. *Am. Miner.* **2005**, *90*, 1072–1077. [CrossRef]
27. Brodholt, J. Ab initio calculations on point defects in forsterite (Mg_2SiO_4) and implications for diffusion and creep. *Am. Miner.* **1997**, *82*, 1049–1053. [CrossRef]
28. Brodholt, J.; Patel, A.; Refson, K. An ab initio study of the compressional behavior of forsterite. *Am. Miner.* **1996**, *81*, 257–260. [CrossRef]
29. Fujino, K.; Sasaki, S.; Takeuchi, Y.; Sadanaga, R. X-ray determination of electron distributions in forsterite, fayalite, and tephroite. *Acta Cryst.* **1981**, *B37*, 513–518. [CrossRef]
30. Poirier, J.-P. On the slip systems of olivine. *J. Geophys. Res.* **1975**, *80*, 4059–4061. [CrossRef]
31. Webb, S.L. The elasticity of the upper mantle orthosilicates olivine and garnet to 3 GPa. *Phys. Chem. Miner.* **1989**, *16*, 684–692. [CrossRef]
32. Zha, C.-S.; Duffy, T.S.; Downs, R.T.; Mao, H.-K.; Hemley, R.J.; Weidner, D.J. Single-Crystal Elasticity of the α and β of Mg_2SiO_4 Polymorphs at High Pressure. In *Properties of Earth and Planetary Materials at High Pressure and Temperature*; Manghnani, M.H., Yagi, T., Eds.; American Geophysical Union: Washington, DC, USA, 1998; Geophysical Monograph; Volume 101, pp. 9–16.
33. Gibbs, G.V.; Hill, F.C.; Boisen, M.B.; Downs, R.T. Power law relationships between bond length, bond strength and electron density distributions. *Phys. Chem. Min.* **1998**, *25*, 585–590. [CrossRef]
34. Jahnátek, M.; Hafner, J.; Krajčí, M. Shear deformation, ideal strength, and stacking fault formation of fcc metals: A density-functional study of Al and Cu. *Phys. Rev. B* **2009**, *B79*, 224103. [CrossRef]
35. Garvik, N.; Carrez, P.; Cordier, P. First-principles study of the ideal strength of Fe_3C cementite. *Mater. Sci. Eng.* **2013**, *A572*, 25–29. [CrossRef]
36. Jiang, C.; Srinivasan, S.G. Unexpected strain-stiffening in crystalline solids. *Nature* **2013**, *496*, 339–342. [CrossRef] [PubMed]
37. Machon, D.; Meersman, F.; Wilding, M.C.; Wilson, M.; McMillan, P.F. Pressure-induced amorphization and polyamorphism: Inorganic and biochemical systems. *Prog. Mater. Sci.* **2014**, *61*, 216–282. [CrossRef]
38. Saka, H.; Shimatani, A.; Suganuma, M.S. Transmission electron microscopy of amorphization and phase transformation beneath indents in Si. *Philos. Mag. A* **2002**, *82*, 1971–1981. [CrossRef]

39. Ge, D.; Domnich, V.; Juliano, T.; Stach, E.A.; Gogotsi, Y. Structural damage in boron carbide under contact loading. *Acta Mater.* **2004**, *52*, 3921–3927. [CrossRef]
40. Kranjc, K.; Rouse, Z.; Flores, K.M.; Skemer, P. Low temperature plastic rheology of olivine determined by nanoindentation. *Geophys. Res. Lett.* **2015**, *43*, 176–184. [CrossRef]
41. Kumamoto, K.M.; Thom, C.A.; Wallis, D.; Hansen, L.N.; Armstrong, D.E.J.; Warren, J.M.; Goldsby, D.L.; Wilkinson, A.J. Size effects resolve discrepancies in 40 years of work on low-temperature plasticity in olivine. *Sci. Adv.* **2017**, *3*, e1701338. [CrossRef]

minerals

MDPI

Article

Ab Initio Thermoelasticity of Liquid Iron-Nickel-Light Element Alloys

Hiroki Ichikawa † **and Taku Tsuchiya** *,†

Geodynamics Research Center, Ehime University, Matsuyama 790-8577, Japan; takutsuchiya@gmail.com
* Correspondence: tsuchiya.taku.mg@ehime-u.ac.jp
† These authors contributed equally to this work.

Received: 8 October 2019; Accepted: 8 January 2020; Published: 9 January 2020

Abstract: The earth's core is thought to be composed of Fe-Ni alloy including substantially large amounts of light elements. Although oxygen, silicon, carbon, nitrogen, sulfur, and hydrogen have been proposed as candidates for the light elements, little is known about the amount and the species so far, primarily because of the difficulties in measurements of liquid properties under the outer core pressure and temperature condition. Here, we carry out massive ab initio computations of liquid Fe-Ni-light element alloys with various compositions under the whole outer core P, T condition in order to quantitatively evaluate their thermoelasticity. Calculated results indicate that Si and S have larger effects on the density of liquid iron than O and H, but the seismological reference values of the outer core can be reproduced simultaneously by any light elements except for C. In order to place further constraints on the outer core chemistry, other information, in particular melting phase relations of iron light elements alloys at the inner core-outer core boundary, are necessary. The optimized best-fit compositions demonstrate that the major element composition of the bulk earth is expected to be CI chondritic for the Si-rich core with the pyrolytic mantle or for the Si-poor core and the $(Mg,Fe)SiO_3$-dominant mantle. But the H-rich core likely causes a distinct Fe depletion for the bulk Earth composition.

Keywords: ab initio molecular dynamics; high-pressure thermoelasticity; outer core chemistry

1. Introduction

The earth's core is thought to be composed of Fe-Ni alloy including substantially large amounts of light elements. These light elements account for observed density deficits of ~10% for the liquid outer core and ~5% for the solid inner core [1–7]. Determination of the light element (LE) composition of the outer core (OC) has long been one of the central research topics in the deep earth sciences. The density (ρ) and adiabatic bulk (K_T) and shear (K_S) moduli of iron and iron-LE alloys are key to interpreting seismological observations and then constructing a compositional model of the core [5,8]. However, those of the liquid states at the OC pressure (P) and temperature (T) (from ~136 to ~329 GPa and from ~4000 to ~6000 K) are still limitedly clarified experimentally. So far, static experiments have been performed up to less than 100 GPa [9–11]. Higher-P behavior of liquid iron was on the other hand investigated by shock wave experiments in multi-Mbar condition [2,12–16]. The temperature, however, changes along the principal Hugoniot and dramatically increases with increasing pressure to more than 8000 K at the pressure of the inner core (IC)-OC boundary (P_{ICB}) of ~329 GPa, which is far higher than the expected actual ICB temperature (T_{ICB}) of ~5000–6000 K [17–22]. Experimental determination of thermoelasticity of liquid iron alloys in the whole P, T condition of the earth's OC thus remains technically impractical.

In contrast, ab initio molecular dynamics (AIMD) simulations have been widely applied to clarify ρ and P-wave velocity (V_P) of liquid iron and iron-LE alloys at the OC conditions in order to constrain the OC composition by interpreting seismological observations. These parameters for the Fe-O, Fe-Si, Fe-S, Fe-C, Fe-Ni, and Fe-Si-O system were calculated [23–25]. However, the data points in these studies were limited; two particular compositions of $Fe_{0.82}Si_{0.10}O_{0.08}$ and $Fe_{0.79}Si_{0.08}O_{0.13}$ only were considered [24] and two particular pressures of the core-mantle boundary (CMB) and ICB only were considered [23]. In particular, in the latter, empirical pressure corrections of 10 GPa and 8 GPa were adopted at the CMB and ICB respectively, though the optimized OC compositions are essentially sensitive to these corrections. Meanwhile, some studies have been performed throughout the whole OC P, T conditions for pure Fe [7,26], Fe-S [27], and Fe-H [28]. However, different formulations were employed to model their thermal equations of state, making a quantitative comparison of the reported thermoelasticity not easy.

In this study, ab initio MD simulations are performed for binary and ternary Fe-Ni-LE alloys with several different LE and Ni fractions from ~100 to ~450 GPa and from 4000 to 8000 K. Equations of state (EoS) and thermoelasticity are then analyzed for each alloy through the same internally consistent way [7]. Using modeled thermoelasticity, we optimize light element compositions for each alloy as a function of the T_{ICB} and discuss the possible OC composition.

2. Results and Discussion

2.1. Effects of LE on the Thermoelasticity of Liquid Iron

Calculations with several different LE concentrations clarify systematic trends on the effects of LEs on the thermoelasticity of liquid Fe. Incorporations of LEs always decrease ρ but increase V_P (Table 1), but trends are different depending on the type of LE. It is found that incorporations of larger Si and S atoms have only marginal effects on the volume (volume per atom), then the EoS is nearly unchanged (Figure S1). In contrast, incorporations of smaller O, C, and in particular H atoms reduce the volume considerably in the whole OC P range. These are related to the fact that the Fe-Si and S alloys are so-called substitutional-type, while the Fe-O, C, and H alloys are interstitial-type as recognized generally in lower P condition.

Table 1. Effects of light element (LE) incorporation on V_P and ρ of liquid Fe calculated at the P_{CMB} and 4000 K and at the P_{ICB} and 5300 K. X_{LE} represents the fraction of LE in atom%.

P, *T* Condition		O	Si	S	C	*H*
$P = P_{CMB}$ $T = 4000$ K	$\frac{\partial \ln V_P}{\partial X_{LE}}$	0.05(1)	0.13(1)	0.06(1)	0.16(1)	0.02(1)
	$\frac{\partial \ln \rho}{\partial X_{LE}}$	−0.34(1)	−0.51(1)	−0.41(1)	−0.30(1)	−0.24(1)
	$\frac{\partial \ln V_P}{\partial \ln \rho}$	−0.14(1)	−0.26(1)	−0.16(1)	−0.54(1)	−0.10(1)
$P = P_{ICB}$ $T = 5300$ K	$\frac{\partial \ln V_P}{\partial X_{LE}}$	0.09(1)	0.21(1)	0.16(1)	0.20(1)	0.07(1)
	$\frac{\partial \ln \rho}{\partial X_{LE}}$	−0.31(1)	−0.48(1)	−0.38(1)	−0.31(1)	−0.21(1)
	$\frac{\partial \ln V_P}{\partial \ln \rho}$	−0.29(1)	−0.44(1)	−0.42(1)	−0.63(1)	−0.35(1)

Because of these volume reductions, ρ variations associated with the O and H incorporations are smaller than those expected from the small masses. As a result, the effects of Si and S incorporations on ρ are larger than those of O and H incorporations (Table 1). These behaviors are consistent with a recent study reporting structural and dynamical properties of Fe-LE alloy liquids [29]. A similar tendency is seen in V_P, but the systematics is less pronounced since the effects of LEs on ρ and K_S are partially cancelled. Perturbation ratios ($\partial \ln V_P/\partial \ln \rho$) are sometimes referred to discuss the chemical heterogeneity in Earth's deep interior [30,31]. In the present cases, absolute values of this ratio are always smaller than 1, indicating that the effects of LE incorporations are always much larger in ρ than in V_P.

2.2. Optimized Compositions

Misfits in ρ and V_P between the Fe-Ni-X liquid alloys and the preliminary reference earth model (PREM) [32] are then evaluated as $\sum\left[\left(\frac{\rho-\rho_{PREM}}{\rho_{PREM}}\right)^2+\left(\frac{V_P-V_{P_{PREM}}}{V_{P_{PREM}}}\right)^2\right]$ (Figure 1), where the summation is taken over the whole OC pressure range. It is clearly demonstrated that the misfits are sensitive to the LE concentration and temperature but not so sensitive to the Ni concentration. The best-fit compositions along two adiabats (T_{ICB} = 5000 K and 6500 K) with three different Ni/(Fe + Ni) ratios, which can be defined by the minima of misfits, are listed in Table 2 with misfits and the ρ and V_P of best-fit compositions along two adiabats are shown in Figure 2.

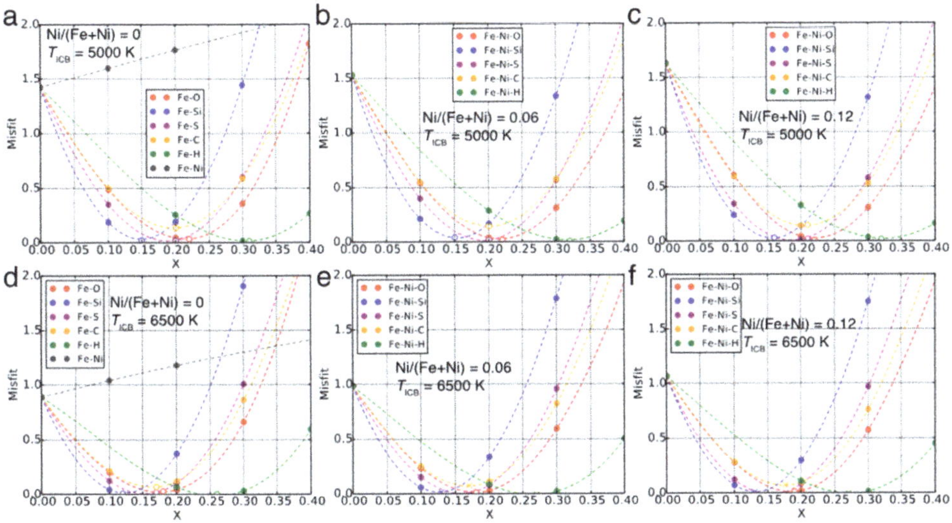

Figure 1. The misfit from the value of the preliminary reference earth model (PREM) as a function of atomic fraction of LEs (X_{LE}). Three different Ni fractions, 0 (**a**,**d**), 0.06 (**b**,**e**), and 0.12 (**c**,**f**) and two different T_{ICB} of 5000 K (**a**–**c**) and 6500 K (**d**–**f**) are examined. Filled circles are the results of molecular dynamics (MD) and dashed lines are the cubic spline interpolations. The atomic fractions at the minima correspond to the best-fit LE concentrations. Open circles indicate the misfits obtained from MD with the best-fit compositions, which are in good agreement with the minima of spline interpolations. Errors in the misfits originated in the fitting procedures are comparable to the size of symbols.

Figures 1 and 2 and Table 2 indicate that among the best-fit compositions, the misfit of the Fe(-Ni)-C system is distinctly large. This suggests that carbon could be eliminated from the major LE in the OC, though there is possibility that the situation could change in ternary or higher-order multicomponent systems. In contrast, the misfits of the best-fit compositions of the other LEs have marginal differences, which are almost indistinguishable from each other within the computational uncertainty (shaded regions in Figure 2). There have been some previous studies which constrained the OC composition within the similar manner, taking two LEs into account, suggesting an oxygen-depleted OC [12] or oxygen-rich OC (3.7 wt. % O, 1.9 wt. % Si) [23]. However, according to the results of the present study, the difference between the misfits of best-fit composition models (Figure 1) is very small except for C, indicating that the information of ρ and V_P are insufficient to determine the OC composition uniquely. Therefore, some other information, e.g., melting phase relations, partitioning behavior between solids and liquids, the bulk earth (BE) compositional property and so on, are quite helpful to place further constraints on the LE composition, but all of these are not well understood at the moment.

Table 2. Best-fit compositions of binary and ternary alloys at T_{ICB} = 5000 K and 6500 K. Misfit, Mg/Si, and Mg/Fe represent a misfit in V_P and ρ from the PREM, Mg/Si, and Mg/Fe ratios expected to the bulk earth with the pyrolytic mantle, respectively. Errors from the fitting procedures are represented in parentheses.

T_{ICB}	Best-Fit Composition	Misfit(×10^{-2})	Mg/Si	Mg/Fe
5000 K	$Fe_{0.78}O_{0.22}$	1.8(1)	1.25(1)	1.03(1)
	$Fe_{0.85}Si_{0.15}$	2.7(1)	1.06(1)	1.04(1)
	$Fe_{0.81}S_{0.19}$	1.6(1)	1.25(1)	1.08(1)
	$Fe_{0.80}C_{0.20}$	11.2(1)	1.25(1)	1.01(1)
	$Fe_{0.70}H_{0.30}$	1.9(1)	1.25(1)	0.97(1)
	$Fe_{0.73}Ni_{0.05}O_{0.22}$	0.8(1)	1.25(1)	1.10(1)
	$Fe_{0.80}Ni_{0.05}Si_{0.15}$	1.3(1)	1.06(1)	1.10(1)
	$Fe_{0.76}Ni_{0.05}S_{0.19}$	0.6(1)	1.25(1)	1.14(1)
	$Fe_{0.75}Ni_{0.05}C_{0.20}$	11.8(1)	1.25(1)	1.07(1)
	$Fe_{0.64}Ni_{0.04}H_{0.32}$	1.1(1)	1.25(1)	1.03(1)
	$Fe_{0.69}Ni_{0.09}O_{0.22}$	0.7(1)	1.25(1)	1.15(1)
	$Fe_{0.74}Ni_{0.10}Si_{0.16}$	1.4(1)	1.05(1)	1.17(1)
	$Fe_{0.71}Ni_{0.10}S_{0.19}$	1.2(1)	1.25(1)	1.21(1)
	$Fe_{0.7}Ni_{0.09}C_{0.21}$	10.4(1)	1.25(1)	1.13(1)
	$Fe_{0.6}Ni_{0.08}H_{0.32}$	0.9(1)	1.25(1)	1.09(1)
6500 K	$Fe_{0.82}O_{0.18}$	4.8(1)	1.25(1)	1.02(1)
	$Fe_{0.88}Si_{0.12}$	1.7(1)	1.09(1)	1.02(1)
	$Fe_{0.85}S_{0.15}$	0.8(1)	1.25(1)	1.05(1)
	$Fe_{0.84}C_{0.16}$	7.7(1)	1.25(1)	1.00(1)
	$Fe_{0.74}H_{0.26}$	0.1(1)	1.25(1)	0.97(1)
	$Fe_{0.77}Ni_{0.05}O_{0.18}$	0.9(1)	1.25(1)	1.08(1)
	$Fe_{0.82}Ni_{0.05}Si_{0.13}$	1.8(1)	1.08(1)	1.08(1)
	$Fe_{0.79}Ni_{0.05}S_{0.16}$	0.2(1)	1.25(1)	1.12(1)
	$Fe_{0.78}Ni_{0.05}C_{0.17}$	7.5(1)	1.25(1)	1.06(1)
	$Fe_{0.69}Ni_{0.04}H_{0.27}$	2.3(1)	1.25(1)	1.02(1)
	$Fe_{0.71}Ni_{0.1}O_{0.19}$	2.2(1)	1.25(1)	1.15(1)
	$Fe_{0.77}Ni_{0.1}Si_{0.13}$	1.3(1)	1.08(1)	1.15(1)
	$Fe_{0.74}Ni_{0.1}S_{0.16}$	0.6(1)	1.25(1)	1.18(1)
	$Fe_{0.72}Ni_{0.1}C_{0.18}$	7.7(1)	1.25(1)	1.13(1)
	$Fe_{0.64}Ni_{0.09}H_{0.27}$	1.0(1)	1.25(1)	1.09(1)

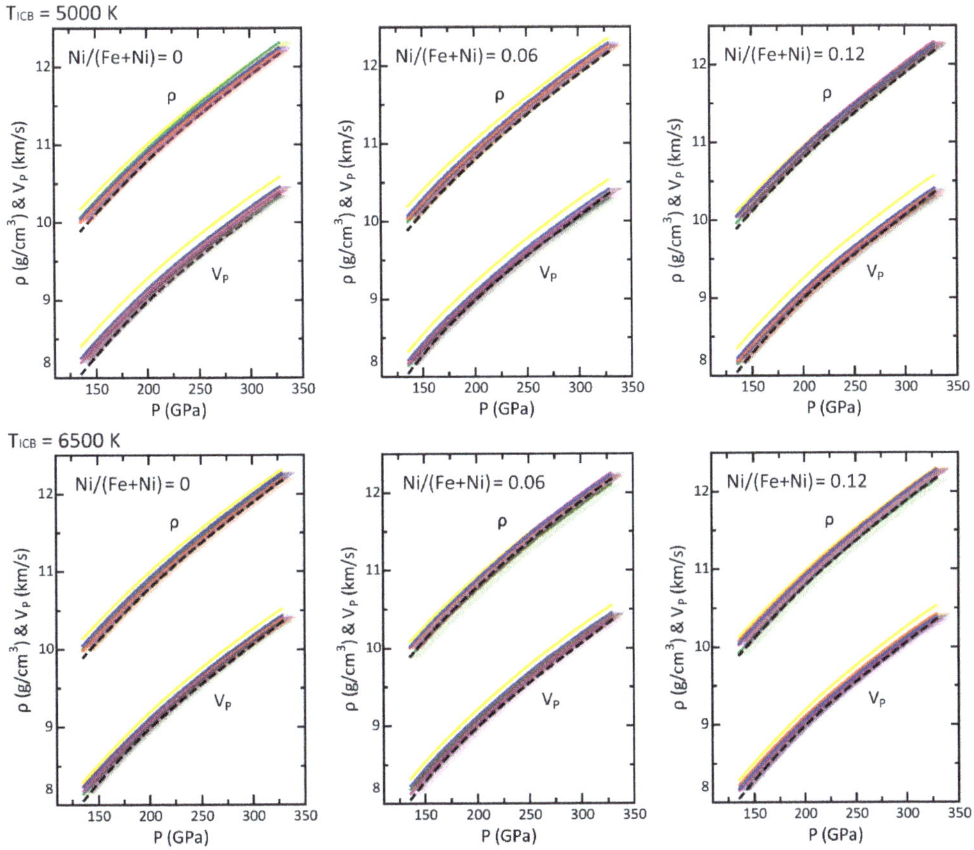

Figure 2. ρ and V_P for the best-fit compositions along two adiabats with T_{ICB} = 5000 K and 6500 K (red, Fe-Ni-O; blue, Fe-Ni-Si; purple, Fe-Ni-S; yellow, Fe-Ni-C; green, Fe-Ni-H). Solid lines correspond to the results determined from raw thermoelasticity data and shaded regions correspond to the uncertainties in pressure with +10 GPa [33]. Dashed lines indicate the PREM values [32]. The ρ and V_P of almost all alloys overlap, indicating that these data only are insufficient to determine the OC composition uniquely.

The best-fit compositions vary depending on the setting of T_{ICB} since the amount of LEs required to reproduce the PREM decreases for higher T (Table 2). However, the misfit values are insensitive to temperature without any systematic variations, meaning that it is difficult to constrain T_{ICB} through this optimization. Based on the calculated results at two different T_{ICB}, the best-fit compositions are represented as a function of T_{ICB} within the first-order as follows:

$$X_O\ (\text{atom\%}) = -2.60 \times 10^{-3}\ T_{ICB}(\text{K}) + 35.0 \text{ for O,} \quad (1)$$

$$X_{Si}\ (\text{atom\%}) = -1.80 \times 10^{-3}\ T_{ICB}(\text{K}) + 24.4 \text{ for Si,} \quad (2)$$

$$X_S\ (\text{atom\%}) = -2.27 \times 10^{-3}\ T_{ICB}(\text{K}) + 30.6 \text{ for S,} \quad (3)$$

and

$$X_H\ (\text{atom\%}) = -3.33 \times 10^{-3}\ T_{ICB}(\text{K}) + 48.3 \text{ for H.} \quad (4)$$

These are found to change only marginally when Ni is incorporated. Here, the Fe(-Ni)-C systems are eliminated since they show larger misfits than others.

Since the T_{ICB} should correspond to the freezing temperature of the OC liquid, melting phase relations of the Fe-LEs systems at the P_{ICB} are quite important to place further constraints on the OC composition. There are however almost no available data with enough quality at the moment. Some experiments, though all were conducted at substantially lower pressures than P_{ICB}, suggest that the eutectic temperature of Fe-FeS system is more than 1000 K lower than the melting temperature of pure Fe [19,34], while solidus or eutectic temperatures are not quite different (within a few 100 K) in the Fe-FeSi [35] and Fe-O systems [36]. A large drop in the melting temperature might also be expected in the Fe-H system [37]. The T_{CMB} is usually thought to be ~4000 K [38,39] and its adiabatic extrapolation leads to ~5200 K for the T_{ICB} [7]. This might be ~1000 K lower than the T_M of pure Fe at the P_{ICB}, suggesting S and H as the potential LEs in the OC. But even larger temperature drops could exist in the ternary or quaternary systems, so it is hard to exclude Si and O from the LE candidates based on this discussion.

Another point is the density jump across the ICB ($\Delta\rho_{ICB}$), which is reported seismologically to reach ~4.7% [32]. This observed $\Delta\rho_{ICB}$ is however too large to be reconciled simply by the solid–liquid transition of pure Fe[41]. Partitioning of LEs between solid and liquid phases is therefore thought to be required, namely LEs dissolving in the OC should strongly prefer liquid to solid. Again, nothing can be conclusive before the melting phase relations of the Fe-LEs systems are clarified at P_{ICB}, but extrapolations of experimental knowledge obtained at lower pressures suggest that the strong partitioning occurs in the Fe-O system [36] but not in the Fe-S [19,34], Fe-Si [35,40], and Fe-H systems [37].

In this study, we select the PREM model as a reference P-wave velocity and density of the OC. It is however well-known that the velocity structure of the earth's interior depends on the reference model. For example, the AK135 model [41,42] has the P wave velocity different from the PREM model in particular at the uppermost and lowermost outer core. Although contrasts between the AK135 and PREM reach ~0.11 km/s and ~0.065 km/s at the uppermost and lowermost part respectively, these make no significant changes in the insights obtained from our analyses.

2.3. Bulk Earth Composition

The BE composition can be affected by the LE compositions of the OC. We next examine what major element composition of the BE are lead from each best-fit composition for the OC (Table 2). In this modeling, the mantle composition is assumed to be the pyrolytic [43], and the IC and crust are ignored because of its negligibly small masses. The chemistry of the mantle, in particular of the lower mantle, is still under debate, but some recent studies similarly suggested the pyrolytic one might be reasonable [44,45].

Table 2 shows the Mg/Si and Mg/Fe ratios of the BE expected from the best-fit compositions for the core combined with the pyrolytic composition for the mantle, which are calculated using the total mass of the earth, weight and atomic % of major elements (Mg, Fe, Si, and O) in the pyrolytic model for the mantle and the optimized composition models for the core. CI chondrite, one of the major candidates of earth's building block, is known to have the Mg/Si and Mg/Fe ratios of ~1.05 and ~1.23, respectively [43]. Table 2 shows that this Mg/Si ratio is achieved only when Si is the major LE in the OC, but no case can explain the Mg/Fe ratio. Enstatite (EH) chondrite is another candidate of earth's building block, which is known to have the Mg/Si and Mg/Fe ratios of ~0.79 and ~0.91. Very Si-rich core and mantle are required to explain this small Mg/Si and Table 2 shows that such composition is incompatible with the observations of the OC. An Mg/Si value similar to CI chondrite, ~1.03, is proposed in the OCCAM model [46] with an Mg/Fe ratio of ~1.12. These ratios are close to the values expected for our Si-bearing best-fit compositions. In summary, Si is a geochemically plausible candidate for the major LEs in the core. But if the lower mantle is assumed to be $MgSiO_3$-dominant, the Mg/Si ratios of the BE expected with the best-fit composition for the core decreases by ~0.2. Then, the Mg/Si and Mg/Fe ratios of all the best-fit compositions except for Si and H-bearing cases match the ratios of CI chondrite and OCCAM model. In this case, Si-rich and H-rich OC with $MgSiO_3$-dominant lower mantle lead to a too Si-rich and Fe-poor BE composition, respectively.

3. Conclusions

Ab initio thermoelasticity of Fe-Ni-LEs alloy liquids in the whole OC P, T condition indicates that all the LEs have the effects to decrease ρ and increase V_P of pure Fe, but the effects are counterintuitively larger for the Si and S incorporations than for the O, C, and H incorporations. Any best-fit alloy composition except the C-rich case can reproduce the ρ and V_P of the actual OC in the comparable level, so that the information of ρ and V_P only are insufficient to determine the OC composition uniquely. Melting phase relations and LE partitioning in the Fe-Ni-LE systems at the P_{ICB} are therefore essential to place a tighter constraint on the OC chemistry. The Si-rich best-fit composition for the core with an assumption of the pyrolytic mantle predicts the CI chondritic BE composition, but the O and S-rich best-fit compositions for the core with the $MgSiO_3$-dominant mantle also lead to the similar chemistry for the BE. The H-rich best-fit composition however causes a distinct deficit of Fe for the BE. In future studies, it might be important to investigate correlations between LEs in higher-order multicomponent systems, which are ignored in this study.

4. Methods

4.1. *Ab Initio Molecular Dynamics Simulations*

To determine the P-V-T equation of state (EoS) of liquid iron-light element alloys, total internal energy (E) and total pressure (P) are calculated by means of the AIMD technique within the canonical (NVT) ensemble in the same manner as our previous study [7] using a PWSCF code [47] for electronic structure with an original implementation of the constant temperature molecular dynamics (MD) module [48]. The simulations are performed on binaries and ternaries, $(Fe\text{-}Ni)_{1-X}O_X$; $(Fe\text{-}Ni)_{1-X}Si_X$; $(Fe\text{-}Ni)_{1-X}S_X$; $(Fe\text{-}Ni)_{1-X}C_X$; $(Fe\text{-}Ni)_{1-X}H_X$, at different atomic fractions ($X_O \leq 0.5, X_{Si} \leq 0.3, X_S \leq 0.3, X_C \leq 0.3, X_H \leq 0.4$). Three Ni/(Fe + Ni) ratios of 0, 0.06 (consistent with the geochemically modeled value) [43], and 0.12 are examined.

The Newton's equation of motion is numerically integrated by using the velocity Verlet algorithm with time steps of 1 fs (10^{-15} s) for the Fe-LE systems, which is the same for previous studies [7,17,33,49], and 0.5 fs for the Fe-H system. Some results (pure Fe and $Fe_{1-X}O_X$ systems) are, in part, already reported in the previous studies [7,49]. MD cells basically contain 50 atoms as in our previous study [7] but 100 atoms for the optimized compositions, and T is controlled by the kinetic energy scaling method. The validity of the cell size with 50–100 atoms for liquid iron can be seen in previous calculations [17,33], where a minor variation of the melting temperature of iron (~100 K) was found with changing the cell size from 67 to 980 atoms. Thermodynamic properties of liquid iron were also found to be sufficiently converged for this cell size.

For electronic structure calculations, we apply the generalized gradient approximation (GGA) [50] to the exchange correlation functional instead of the local density approximation (LDA). This is essential since many previous studies reported that GGA shows significant improvements over LDA when it comes to correctly describing ground-state properties and compression behaviors for iron [51,52]. The ultrasoft pseudopotential and plane-wave basis set are used to describe electronic structures. Here, an electronic configuration of $3s^23p^63d^{6.5}4s^14p^0$ is pseudized with a sufficiently small core radius of 2.0 a.u. for Fe; $2s^22p^4$, with a core radius of 1.5 a.u. for O; $3s^23p^4$, with a core radius of 1.7 a.u. for S; $3s^23p^23d^0$, with a core radius of 1.4 a.u. for Si; $2s^22p^2$, with a core radius of 1.1 a.u. for C; $1s^1$, with a core radius of 0.8 a.u. for H; and $3s^23p^63d^84s^24p^0$ with a core radius of 2.0 a.u. for Ni by the Vanderbilt scheme [53] with non-linear core corrections [54]. We apply a kinetic energy cutoff of 50 Ry and spin polarization is not taken into account. These conditions are already well tested in our previous calculations [7,49,55] and are fairly similar to those in calculations by other groups [24]. Liquids in principle have no periodic structure, thus the Γ point only is sampled in our simulations. All the MD simulations are conducted in P,T condition from ~80 to ~500 GPa and from 4000 to 8000 K (Figure S1), which covers the whole P,T range of the core. Standard deviations in calculated T and P are found to be typically ~50 K and ~3 GPa at 5000 K and ~130 GPa and ~120 K and ~6 GPa at 8000 K and ~400 GPa, respectively.

4.2. EoS Analysis of Liquid Iron Alloys

The calculated E-P-V-T relations of liquid iron alloys are analyzed using a single EoS model basically identical to the one in our previous study [7]. For the isothermal part at a reference temperature T_0, we used the Vinet (Morse-Rydberg) Equation,

$$P_{T_0}(V) = 3K_{T0}\left(\frac{V}{V_0}\right)^{-\frac{2}{3}}\left[1-\left(\frac{V}{V_0}\right)^{\frac{1}{3}}\right]\exp\left\{\frac{3}{2}\left(K'_{T_0}-1\right)\left[1-\left(\frac{V}{V_0}\right)^{\frac{1}{3}}\right]\right\} \tag{5}$$

Here, K_{T_0} and K'_{T_0} are the isothermal bulk modulus and its pressure derivative at zero pressure at T_0. The internal thermal energy is represented by a second-order polynomial of temperature with a volume dependent second-order coefficient,

$$E_{th}(V,T) = 3nk_B\left[T + e_0\left(\frac{V}{V_0}\right)^g T^2\right] \tag{6}$$

where k_B is the Boltzmann constant and n is the number of atoms per formula unit. The first term corresponds to the phonon energy (atomic contribution), while the second term represents the electronic contribution.

The thermal pressure is linked with the internal thermal energy by Grüneisen parameter γ as in the following Equation,

$$P_{th}(V,T) = \frac{1}{V}\int \gamma(V,T)dE_{th}(V,T). \tag{7}$$

We employ the highest temperature of 8000 K in the present calculations as a reference temperature T_0 in order to constrain the reference isotherm as tightly as possible within the broad pressure range. Although in the previous study [7], the following functional form for $\gamma(V)$

$$\gamma(V) = \gamma_0\left\{1 + a\left[(V/V_0)^b - 1\right]\right\} \tag{8}$$

was employed, we realized that three additional parameters γ_0, a, and b make the least-square analyses less stable and less systematic. Instead, in this study, we assume the γ to be constant for each composition. The previous study [7] reported that the variation of γ of pure Fe is 0.2 only from 100 to 400 GPa and from 4000 and 7000 K and we confirmed that this small variation of γ does not affect the results of analyzed thermoelasticity. Consequently, the present EoS model requires six parameters in total (V_0, K_{T_0}, K'_{T_0}, γ, e_0, and g) to calculate P at a given V, T. These parameters are determined by least squares analyses on the datasets obtained from the AIMD calculations. The derived EoS parameters for best-fit compositions are summarized in Table S1.

Derivative quantities of EoS such as K_T and α are obtained based on the thermodynamic definitions as $(\partial P/\partial V)_T = -K_T/V$ and $(\partial P/\partial T)_V = \alpha K_T$, respectively. K_T is then converted to K_S as

$$\frac{K_S}{K_T} = 1 + \alpha\gamma T, \tag{9}$$

and adiabatic temperature gradient is computed using the relationship,

$$\left(\frac{\partial T}{\partial P}\right)_S = \frac{\alpha VT}{C_P} = \frac{\gamma T}{K_S}. \tag{10}$$

V_P is then calculated for each composition (Figure S1) as $V_P = \sqrt{\frac{K_S}{\rho}}$ along two different adiabats explained below.

4.3. Adiabats

The adiabatic temperature profile is calculated numerically by integrating Equation (7) from the ICB pressure. ρ and V_P are calculated along the adiabats anchored by two possible ICB temperatures: T_{ICB} = 5000 K and T_{ICB} = 6500 K. The former T_{ICB} is found to give ~3700 K at 136 GPa, which is close to a proposed core-mantle boundary temperature [38,39]. The latter corresponds to the melting temperature (T_M) of pure iron at 329 GPa [17,18], which would be close to the upper bound of ICB temperature since T_M of iron-LE alloys are expected in general to be lower than the T_M of pure Fe.

Supplementary Materials: The following are available online at http://www.mdpi.com/2075-163X/10/1/59/s1, Figure S1: The calculated P-V-T data of iron alloys with fitted EoS. Filled red circles represent the data used for EoS analysis, Table S1: EoS parameters for the best-fit composition models.

Author Contributions: H.I. and T.T. conducted the ab initio calculations. Both authors analyzed the results and wrote the manuscript. All authors have read and agreed to the published version of the manuscript.

Funding: This research was supported by X-ray Free Electron Laser Priority Strategy Program (MEXT), and KAKENHI JP15H05834, JP26800237, JP17K05638, JP17H06457, and JP26287105.

Acknowledgments: We thank M. Ohsumi for helping with the analyses, C. Shiraishi for helping with the data plotting, and S. Kaneshima and in particular T. Ohtaki for the helpful discussion.

Conflicts of Interest: The authors declare no conflicts of interest.

References

1. Birch, F. Density and composition of mantle and core. *J. Geophys. Res.* **1964**, *69*, 4377–4388. [CrossRef]
2. Brown, J.M.; McQueen, R.G. Phase transitions, Grüneisen parameter, and elasticity for shocked iron between 77 GPa and 400 GPa. *J. Geophys. Res. Solid Earth* **1986**, *91*, 7485–7494. [CrossRef]
3. Jephcoat, A.; Olson, P. Is the inner core of the Earth pure iron? *Nature* **1987**, *325*, 332–335. [CrossRef]
4. Uchida, T.; Wang, Y.; Rivers, M.L.; Sutton, S.R. Stability field and thermal equation of state of e-iron determined by synchrotron X-ray diffraction in a multianvil apparatus. *J. Geophys. Res. Solid Earth* **2001**, *106*, 21799–21810. [CrossRef]
5. Anderson, O.L.; Isaak, D.G. Another look at the core density deficit of Earth's outer core. *Phys. Earth Planet. Inter.* **2002**, *131*, 19–27. [CrossRef]
6. Dewaele, A.; Loubeyre, P.; Occelli, F.; Mezouar, M.; Dorogokupets, P.I.; Torrent, M. Quasihydrostatic Equation of State of Iron above 2 Mbar. *Phys. Rev. Lett.* **2006**, *97*, 215504. [CrossRef] [PubMed]
7. Ichikawa, H.; Tsuchiya, T.; Tange, Y. The P-V-T equation of state and thermodynamic properties of liquid iron. *J. Geophys. Res. Solid Earth* **2014**, *119*, 240–252. [CrossRef]
8. Poirier, J.-P. Light elements in the Earth's outer core: A critical review. *Phys. Earth Planet. Inter.* **1994**, *85*, 319–337. [CrossRef]
9. Morard, G.; Siebert, J.; Andrault, D.; Guignot, N.; Garbarino, G.; Guyot, F.; Antonangeli, D. The Earth's core composition from high pressure density measurements of liquid iron alloys. *Earth Planet. Sci. Lett.* **2013**, *373*, 169–178. [CrossRef]
10. Sanloup, C.; Guyot, F.; Gillet, P.; Fiquet, G.; Mezouar, M.; Martinez, I. Density measurements of liquid Fe-S alloys at high-pressure. *Geophys. Res. Lett.* **2000**, *27*, 811–814. [CrossRef]
11. Sanloup, C.; Van Westrenen, W.; Dasgupta, R.; Maynard-Casely, H.; Perrillat, J.-P. Compressibility change in iron-rich melt and implications for core formation models. *Earth Planet. Sci. Lett.* **2011**, *306*, 118–122. [CrossRef]
12. Huang, H.; Fei, Y.; Cai, L.; Jing, F.; Hu, X.; Xie, H.; Zhang, L.; Gong, Z. Evidence for an oxygen-depleted liquid outer core of the Earth. *Nature* **2011**, *479*, 513–516. [CrossRef]
13. Huang, H.; Hu, X.; Jing, F.; Cai, L.; Shen, Q.; Gong, Z.; Liu, H. Melting behavior of Fe-O-S at high pressure: A discussion on the melting depression induced by, O and S. *J. Geophys. Res.* **2010**, *115*, B05207. [CrossRef]
14. Huang, H.; Wu, S.; Hu, X.; Wang, Q.; Wang, X.; Fei, Y. Shock compression of Fe-FeS mixture up to 204 GPa. *Geophys. Res. Lett.* **2013**, *40*, 687–691. [CrossRef]

15. Zhang, Y.; Sekine, T.; He, H.; Yu, Y.; Liu, F.; Zhang, M. Shock compression of Fe-Ni-Si system to 280 GPa: Implications for the composition of the Earth's outer core. *Geophys. Res. Lett.* **2014**, *41*, 4554–4559. [CrossRef]
16. Zhang, Y.; Sekine, T.; He, H.; Yu, Y.; Liu, F.; Zhang, M. Experimental constraints on light elements in the Earth's outer core. *Sci. Rep.* **2016**, *6*, 22473. [CrossRef]
17. Alfè, D. Temperature of the inner-core boundary of the Earth: Melting of iron at high pressure from first-principles coexistence simulations. *Phys. Rev. B* **2009**, *79*. [CrossRef]
18. Anzellini, S.; Dewaele, A.; Mezouar, M.; Loubeyre, P.; Morard, G. Melting of iron at Earth's inner core boundary based on fast X-ray diffraction. *Science* **2013**, *340*, 464–466. [CrossRef]
19. Kamada, S.; Ohtani, E.; Terasaki, H.; Sakai, T.; Miyahara, M.; Ohishi, Y.; Hirao, N. Melting relationships in the Fe–Fe_3S system up to the outer core conditions. *Earth Planet. Sci. Lett.* **2012**, *359*, 26–33. [CrossRef]
20. Ma, Y.; Somayazulu, M.; Shen, G.; Mao, H.-K.; Shu, J.; Hemley, R.J. In situ X-ray diffraction studies of iron to Earth-core conditions. *Phys. Earth Planet. Inter.* **2004**, *143*, 455–467. [CrossRef]
21. Nguyen, J.H.; Holmes, N.C. Melting of iron at the physical conditions of the Earth's core. *Nature* **2004**, *427*, 339–342. [CrossRef] [PubMed]
22. Terasaki, H.; Kamada, S.; Sakai, T.; Ohtani, E.; Hirao, N.; Ohishi, Y. Liquidus and solidus temperatures of a Fe–O–S alloy up to the pressures of the outer core: Implication for the thermal structure of the Earth's core. *Earth Planet. Sci. Lett.* **2011**, *304*, 559–564. [CrossRef]
23. Badro, J.; Côté, A.S.; Brodholt, J.P. A seismologically consistent compositional model of Earth's core. *Proc. Natl. Acad. Sci. USA* **2014**, 1–4. [CrossRef] [PubMed]
24. Pozzo, M.; Davies, C.; Gubbins, D.; Alfè, D. Transport properties for liquid silicon-oxygen-iron mixtures at Earth's core conditions. *Phys. Rev. B* **2013**, *87*, 14110. [CrossRef]
25. Brodholt, J.; Badro, J. Composition of the low seismic velocity E′ layer at the top of Earth's core. *Geophys. Res. Lett.* **2017**, *44*, 8303–8310. [CrossRef]
26. Vočadlo, L.; Alfè, D.; Gillan, M.J.; Price, G.D. The properties of iron under core conditions from first principles calculations. *Phys. Earth Planet. Inter.* **2003**, *140*, 101–125. [CrossRef]
27. Umemoto, K.; Hirose, K.; Imada, S.; Nakajima, Y.; Komabayashi, T.; Tsutsui, S.; Baron, A.Q.R. Liquid iron-sulfur alloys at outer core conditions by first-principles calculations. *Geophys. Res. Lett.* **2014**, *41*, 6712–6717. [CrossRef]
28. Posner, E.S.; Steinle-Neumann, G. Mass transport and structural properties of binary liquid iron alloys at high pressure. *Geochem. Geophys. Geosyst.* **2019**, *20*, 3556–3568. [CrossRef]
29. Umemoto, K.; Hirose, K. Liquid iron-hydrogen alloys at outer core conditions by first-principles calculations. *Geophys. Res. Lett.* **2015**, *42*, 7513–7520. [CrossRef]
30. Karato, S.; Karki, B.B. Origin of lateral variation of seismic wave velocities and density in the deep mantle. *J. Geophys. Res.* **2001**, *106*, 21771–21783. [CrossRef]
31. Wentzcovitch, R.M.; Tsuchiya, T.; Tsuchiya, J. $MgSiO_3$ post-perovskite at D" conditions. *Proc. Natl. Acad. Sci. USA* **2006**, *103*, 543–546. [CrossRef] [PubMed]
32. Dziewonski, A.M.; Anderson, D.L. Preliminary reference Earth model. *Phys. Earth Planet. Inter.* **1981**, *25*, 297–356. [CrossRef]
33. Alfè, D.; Price, G.; Gillan, M. Iron under Earth's core conditions: Liquid-state thermodynamics and high-pressure melting curve from ab initio calculations. *Phys. Rev. B* **2002**, *65*. [CrossRef]
34. Mori, Y.; Ozawa, H.; Hirose, K.; Sinmyo, R.; Tateno, S.; Morard, G.; Ohishi, Y. Melting experiments on Fe–Fe_3S system to 254 GPa. *Earth Planet. Sci. Lett.* **2017**, *464*, 135–141. [CrossRef]
35. Fischer, R.A.; Campbell, A.J.; Reaman, D.M.; Miller, N.A.; Heinz, D.L.; Dera, P.; Prakapenka, V.B. Phase relations in the Fe–FeSi system at high pressures and temperatures. *Earth Planet. Sci. Lett.* **2013**, *373*, 54–64. [CrossRef]
36. Seagle, C.T.; Heinz, D.L.; Campbell, A.J.; Prakapenka, V.B.; Wanless, S.T. Melting and thermal expansion in the Fe-FeO syustem at high pressure. *Phys. Earth Planet. Inter.* **2008**, *265*, 655–665. [CrossRef]
37. Shibazaki, Y.; Terasaki, H.; Ohtani, E.; Tateyama, R.; Nishida, K.; Funakoshi, K.; Higo, Y. High-pressure and high-temperature phase diagram for $Fe_{0.9}Ni_{0.1}$-H alloy. *Phys. Earth Planet. Inter.* **2014**, *228*, 192–201. [CrossRef]
38. Kawai, K.; Tsuchiya, T. Temperature profile in the lowermost mantle from seismological and mineral physics joint modeling. *Proc. Natl. Acad. Sci. USA* **2009**, *106*, 22119–22123. [CrossRef]

39. Nomura, R.; Hirose, K.; Uesugi, K.; Ohishi, Y.; Tsuchiyama, A.; Miyake, A.; Ueno, Y. Low core-mantle boundary temperature inferred from the solidus of pyrolite. *Science* **2014**, *343*, 522–525. [CrossRef]
40. Ozawa, H.; Hirose, K.; Yonemitsu, K.; Ohishi, Y. High-pressure melting experiments on Fe–Si alloys and implications for silicon as a light element in the core. *Earth Planet. Sci. Lett.* **2016**, *456*, 47–54. [CrossRef]
41. Kennett, B.L.N.; Engdahl, E.R.; Buland, R. Constraints on seismic velocities in the Earth from traveltimes. *Geophys. J. Int.* **1995**, *122*, 108–124. [CrossRef]
42. Montagner, J.-P.; Kennett, B.L.N. How to reconcile body-wave and normal-mode reference earth models. *Geophys. J. Int.* **1996**, *125*, 229–248. [CrossRef]
43. McDonough, W.F.; Sun, S.-S. The composition of the Earth. *Chem. Geol.* **1995**, *120*, 223–253. [CrossRef]
44. Cottaar, S.; Heister, T.; Rose, I.; Unterborn, C. BurnMan: A lower mantle mineral physics toolkit. *Geochem. Geophys. Geosyst.* **2014**, *15*, 1164–1179. [CrossRef]
45. Wang, X.; Tsuchiya, T.; Hase, A. Computational support for a pyrolitic lower mantle containing ferric iron. *Nat. Geosci.* **2015**, *8*, 556–559. [CrossRef]
46. Fitoussi, C.; Bourdon, B.; Wang, X. The building blocks of Earth and Mars: A close genetic link. *Earth Planet. Sci. Lett.* **2016**, *434*, 151–160. [CrossRef]
47. Giannozzi, P.; Baroni, S.; Bonini, N.; Calandra, M.; Car, R.; Cavazzoni, C.; Ceresoli, D.; Chiarotti, G.L.; Cococcioni, M.; Dabo, I.; et al. QUANTUM ESPRESSO: A modular and open-source software project for quantum simulations of materials. *J. Phys. Condens. Matter* **2009**, *21*, 395502–395521. [CrossRef] [PubMed]
48. Usui, Y.; Tsuchiya, T. Ab initio two-phase molecular dynamics on the melting curve of SiO_2. *J. Earth Sci.* **2010**, *21*, 801–810. [CrossRef]
49. Ichikawa, H.; Tsuchiya, T. Atomic transport property of Fe-O liquid alloys in the Earth's outer core P, T condition. *Phys. Earth Planet. Inter.* **2015**, *247*, 27–35. [CrossRef]
50. Perdew, J.P.; Burke, K.; Ernzerhof, M. Generalized gradient approximation made simple. *Phys. Rev. Lett.* **1996**, *77*, 3865–3868. [CrossRef]
51. Asada, T.; Terakura, K. Cohesive properties of iron obtained by use of the generalized gradient approximation. *Phys. Rev. B* **1992**, *46*, 13599–13602. [CrossRef] [PubMed]
52. Stixrude, L.; Cohen, R.E.; Singh, D.J. Iron at high pressure: Linearized-augmented-plane-wave computations in the generalized-gradient approximation. *Phys. Rev. B* **1994**, *50*, 6442–6445. [CrossRef] [PubMed]
53. Vanderbilt, D. Soft self-consistent pseudopotentials in a generalized eigenvalue formalism. *Phys. Rev. B* **1990**, *41*, 7892–7895. [CrossRef] [PubMed]
54. Louie, S.G.; Froyen, S.; Cohen, M.L. Nonlinear ionic pseudopotentials in spin-density-functional calculations. *Phys. Rev. B* **1982**, *26*, 1738–1742. [CrossRef]
55. Tsuchiya, T.; Fujibuchi, M. Effects of Si on the elastic property of Fe at Earth's inner core pressures: First principles study. *Phys. Earth Planet. Inter.* **2009**, *174*, 212–219. [CrossRef]

minerals
MDPI
Article

The Bridgmanite–Akimotoite–Majorite Triple Point Determined in Large Volume Press and Laser-Heated Diamond Anvil Cell

Britany L. Kulka [1], Jonathan D. Dolinschi [1], Kurt D. Leinenweber [2], Vitali B. Prakapenka [3] and Sang-Heon Shim [1,*]

[1] School of Earth and Space Exploration, Arizona State University, Tempe, AZ 85287, USA; bkulka@asu.edu (B.L.K.); natinschi@gmail.com (J.D.D.)
[2] Eyring Materials Center, Arizona State University, Tempe, AZ 85287, USA; kurtl@asu.edu
[3] GeoSoilEnviroCars, University of Chicago, Chicago, IL 60439, USA; prakapenka@cars.uchicago.edu
* Correspondence: shdshim@asu.edu

Received: 11 December 2019; Accepted: 9 January 2020; Published: 15 January 2020

Abstract: The bridgmanite–akimotoite–majorite (Bm–Ak–Mj or BAM) triple point in $MgSiO_3$ has been measured in large-volume press (LVP; COMPRES 8/3 assembly) and laser-heated diamond anvil cell (LHDAC). For the LVP data, we calculated pressures from the calibration provided for the assembly. For the LHDAC data, we conducted in situ determination of pressure at high temperature using the Pt scale at synchrotron. The measured temperatures of the triple point are in good agreement between LVP and LHDAC at 1990–2000 K. However, the pressure for the triple point determined from the LVP is 3.9 ± 0.6 GPa lower than that from the LHDAC dataset. The BAM triple point determined through these experiments will provide an important reference point in the pressure–temperature space for future high-pressure experiments and will allow mineral physicists to compare the pressure–temperature conditions measured in these two different experimental methods.

Keywords: triple point; bridgmanite; akimotoite; majorite; large-volume press; laser-heated diamond anvil cell

1. Introduction

Accurate determination of pressure (P) and temperature (T) is essential for laboratory experiments to contribute to the geophysical understandings of the deep interiors of Earth, other planets in the solar system, and exoplanets. As Orson Anderson demonstrated through his important work, thermal equations of state (EOS) of the standard materials can be used for estimating pressures from measurements of volume and temperature in high-pressure experiments [1].

The development of third-generation synchrotron facilities in the late 1990s and early 2000s enabled in situ measurements of phase boundaries and physical properties, allowing mineral physicists to take advantage of the thermal EOS established by Anderson and others. However, early experiments found that results from different pressure scales and different experimental techniques differed by 2–3 GPa for the important phase boundaries near the 660-km discontinuity [2–6]. Intense efforts have been made since then [7–9], but the discrepancy for some important phase boundaries still remain unresolved [10,11]. Other potential sources of the discrepancy have been investigated. For example, pressure effects on the thermocouple electromotive force (emf) calibration are important for large-volume press (LVP) experiments [12–14]. Spectroradiometry has been the standard method for temperature measurements in laser heating, but optical effects from diamond anvils and thermal gradients could potentially introduce artifacts in the measured temperature [15–18]. While the accuracies of spectroradiometry through LHDAC [19] and thermocouple emf [13,14] have been separately investigated, temperatures from spectroradiometry and thermocouples have never been cross examined at high pressure to our knowledge.

Comparing phase boundaries provides an opportunity to examine the pressure and temperature scales in high-pressure apparatuses. However, in such an effort, it is difficult to separate pressure effects and temperature effects [10]. It is desirable to have a reference point in the $P-T$ space for such comparison. At pressures between 20 and 24 GPa, the triple point exists between bridgmanite (Bm), akimotoite (Ak), and majorite (Mj) in $MgSiO_3$. Not only does understanding of the triple point provide a new opportunity to compare pressures and temperatures measured in different techniques but also the triple point itself is important in geophysics for understanding the origin of the seismic discontinuity structures near 660-km depths. Although the 660-km discontinuity has been related mainly to the post-spinel transition in Mg_2SiO_4 for many decades [20], it has been well known that the phase boundaries in $MgSiO_3$ can exist at depths very close to the 660-km discontinuity and therefore affect the complex discontinuity structures at the bottommost mantle transition zone [21,22]. In this paper, we report the triple point between Bm, Ak, and Mj (BAM) in pure $MgSiO_3$ measured in both LVP and LHDAC. We compare the pressure and temperature of the BAM triple point from those two separate measurements and discuss possible sources of discrepancy between LVP and LHDAC.

2. Methods

2.1. Large-Volume Press (LVP)

High-pressure LVP experiments were conducted in the 1100-ton multi-anvil press at the Eyring Materials Center at Arizona State University (ASU). The LVP consists of eight second-stage anvils surrounding the cell assembly, six first-stage anvils surrounding the second-stage anvils, and a support ring surrounding the first-stage anvils. The second-stage anvils are tungsten carbide, have a 3.0-mm truncation on each corner, and were used with injection-molded ceramic octahedra with 8.0-mm edges [23]. All LVP experiments reported here were conducted with the COMPRES 8/3 assembly (more information on the assembly can be found in Walker et al. [24] and Leinenweber et al. [23]).

Each assembly used a cylindrical Re-capsule made from a 3×6 mm^2 sheet of rhenium foil that was filled with pure $MgSiO_3$ glass synthesized using the containerless laser levitation method [25,26]. This capsule was place inside of an MgO sleeve within the Re-furnace and $LaCrO_3$ sleeve. The Re-furnace and $LaCrO_3$ sleeve were placed inside the octahedron. The completed assembly was placed on top of four second-stage anvils with four other second-stage anvils placed on top once inside of the LVP.

Pressures for each run were raised at a rate of 400 psi/h, and temperatures were raised to target temperatures at a constant rate of 100 K/min. Heating of the assemblies lasted 30–60 min to reach chemical equilibrium. Temperature was monitored just above the capsule in the assembly by a W5Re-W%26Re (type C) thermocouple. The run was held at target temperature for the heating duration until quenched. Pressure began being pumped down immediately after quenching at a rate of 400 psi/h. Once at ambient conditions, the assembly was recovered for analysis (Table 1).

Table 1. Large-volume press runs performed in this study: Estimated uncertainty for pressure is 0.5 GPa. Estimated uncertainty without the pressure effects on thermocouple is 5 K. We assigned an uncertainty of ±50 K for the runs where we estimated temperature through applied voltage calibrated for thermocouple. *T*: temperature, *P*: pressure, O.P.: oil pressure, *F*: force in metric tons, *t*: heating duration.

Run ID	*T* (K)	O.P. (psi)	*P* (GPa)	*F*(ton)	*t* (min)	Product
BB1434JD	2173 ± 5	5000	20.5	515	30	Bm
BB1438JD	2023 ± 5	4500	20.0	463	30	Bm
BB1469BK	1873 ± 5	4937	22.0	611	60	Bm
BB1470BK	1873 ± 5	4365	20.0	449	60	Ak
BB1472BK	2073 ± 50	4103	19.0	421	30	Mj
BB1475BK	1873 ± 5	4892	21.0	504	60	Bm
BB1476BK	1875 ± 50	4609	20.5	475	60	Bm
BB1477BK	1973 ± 5	4025	19.0	414	30	Mj
BB1478BK	1923 ± 50	4192	19.5	432	30	Ak + Mj
BB1479BK	1948 ± 5	4344	20.0	447	30	Ak

For the pressure calculation, we used the calibration for COMPRESS 8/3 assembly reported by Leinenweber et al. [23], including the reported thermal effects. In that study, in situ X-ray diffraction was acquired in LVP at the Advanced Photon Source (APS), Argonne National Lab. The oil pressure was calibrated for the capsule pressure using tungsten thermal EOS [27] (Table 1). The LVP at beamline 13-IDD of the GeoSoilEnviroCARS sector uses a split module (T-cup) while the press at ASU uses a Walker module. These and other factors could result in different friction effects in the two different presses. The equivalence of the forces on the two presses was confirmed by using a steel cube fitted with a strain gauge and by measuring the strain on the cube at the two different installations.

In order to understand the thermal gradient in the sample chamber, we conducted a calculation on the cell assembly we used through the approach presented in Hernlund et al. [28] using thermal conductivity data valid to 1300 K. The temperature calculation showed a 65-K temperature difference between the coldest (thermocouple position) and the warmest spots in the sample. If the calculation is extrapolated to 2000 K, the difference could be about 110 K, although it is uncertain. For the runs with single phase observation (all but one), the temperature values represent the lower bound for the stability of the observed phases, and therefore, the runs can confirm the stability of the observed phases up to approximately 100 K higher temperatures than the values presented in Table 1 based on our modeling. However, because the thermal gradient modeling can be affected by uncertainty sources different from our experiment, we will focus on the experimentally measured temperatures and observed phases.

Recovered LVP samples were sliced axially and were carefully polished using alumina sandpaper. For phase identification, Raman spectroscopy was used for all LVP products in 1D point and 2D scanning modes at ASU. Laser wavelength was 532.22 nm, and the beam size was 2 μm. Measurements were taken for 1000 accumulations of 1-s exposures each (16 min total) on each spot at 30–45 mW of laser power.

2.2. Laser-Heated Diamond Anvil Cell (LHDAC)

The same $MgSiO_3$ glass used in LVP was loaded in diamond-anvil cell (DAC) #1 in LHDAC experiments (Table 2). The sample powder was mixed with 10 wt% Pt powder, which served as the pressure calibrant and laser coupler for heating. Pressure was calculated by combining the measured unit-cell volume and temperature with the thermal EOS by Dorogokupets and Dewaele [29]. The powder mixture was pressed into a foil which was 100–200 μm in size and 15–20 μm thick. The foil was loaded into a laser-drilled hole in a pre-indented rhenium gasket using a micromanipulator (Axis Pro SS, Micro Support Co., Ltd., Shizuoka, Japan). The foil was supported by 5 spacers that consisted of <15 μm pure sample grains on each side of the diamond culet. Argon gas (Ar) was loaded as a pressure medium and thermal insulator in a gas loading system (GLS1500, Sanchez Technologies, Frépillon, France) at 1450 bar pressure at ASU.

Table 2. Experimental runs in a Laser-Heated Diamond Anvil Cell (LHDAC) in this study. DAC: diamond-anvil cell, PM: pressure medium, SM: starting material, XE: X-ray energy, Gl: $MgSiO_3$ glass. Other abbreviations are the same as in Table 1.

DAC ID	Setting	Spot ID	*P* (GPa)	*T* (K)	Product
#1	PM: Ar	1	20.2	1506	Ak
	SM: Gl	2	21.3	1897	Mj
	XE: 37 keV	3	24.1	1903	Ak
		4	21.0	1610	Ak
		5	21.4	2056	Mj
		6	22.0	1531	Ak
		7	23.2	1870	Ak
		8	30.3	2110	Bm
		9	26.4	2097	Bm
		10	24.5	1717	Bm
		11	27.2	1884	Bm
		12	23.5	1717	Ak
		13	29.2	1706	Bm
		14	24.3	2123	Bm
		15	27.1	2096	Bm
		16	26.3	1992	Bm
		17	27.0	2021	Bm
		18	25.5	1934	Bm
		19	21.4	1541	Ak
		20	22.5	1897	Mj
		21	23.6	2050	Mj
		22	22.0	1637	Ak
#2	PM: NaCl	1	16.8	1616	Mj
	SM: Ak	2	17.5	1761	Mj
	XE: 30 keV	3	22.4	2019	Mj
		4	20.1	1964	Mj
		5	24.0	2349	Mj
		6	20.6	1931	Mj
		7	24.0	2136	Mj
		8	20.2	1922	Mj
		9	21.2	1905	Mj
		10	26.1	1809	Bm
		11	21.7	1798	Mj
		12	22.8	1804	Ak
		13	22.8	2004	Mj
		14	27.5	1835	Bm
		15	29.3	2003	Bm
#3	PM: NaCl	1	26.1	1625	Bm
	SM: Ak	2	27.0	1600	Bm
	XE: 30 keV	3	24.6	1900	Bm
		4	25.3	1831	Bm
		5	25.1	1908	Bm
		6	24.6	1600	Bm
		7	26.9	2126	Bm

For measurements in DACs #2 and #3, we used Ak as a starting material. The Ak sample was synthesized in the LVP at ASU. For sufficient amount of akimotoite sample, we conducted the synthesis in a 10/5 COMPRES assembly with Re foil capsule filled with $MgSiO_3$ glass. The synthesis conditions were 22 GPa (corresponding to a ram force of 720 tonnes) and 1473 K for 1 h. After quenching, the capsule was cut in half with a diamond wire saw and analyzed with Raman to ensure pure Ak product. The Ak sample was removed from the capsule and ground in an agate mortar under acetone for 20 min. The sample powder was mixed with 10 wt% Pt powder, which once again served as the pressure calibrant [29] and laser coupler. The mixture was again pressed into a foil of the same size as the $MgSiO_3$ sample. For these samples, we used NaCl as a pressure medium, which was dried at 100 °C for 24 h. Thin foils of NaCl were loaded into the sample chamber above and below the sample foil. All samples were compressed in symmetric-type DAC utilizing type Ia standard design 400 μm diamond anvils.

LHDAC data was acquired using monochromatic XRD in DAC at sector 13-IDD of the GeoSoilEnviroCARS (GSECARS) at the Advanced Photon Source (APS) utilizing the double-sided laser heating system. The laser beams were coaxially aligned with the X-ray beam to measure diffraction patterns on the heating spot. The typical beam diameters for the X-ray beam and laser heating spot are 5 μm and 20 μm, respectively.

The 2D diffraction images were acquired using a Dectris Pilatus detector at GSECARS 13-IDD and then integrated into 1D diffraction patterns using calibration parameters obtained from the LaB_6 standard in DIOPTAS software [30]. Diffraction images were exposed for 5–10 s and were collected before, during, and after heating. The diffraction images were analyzed, and phases were identified using PeakPo software [31]. Pt peaks were fitted with pseudo-Voigt profile function to obtain the peak positions in PeakPo. Pressure was calculated with pytheos [32]. The unit-cell volume of Pt was calculated with the 111 and 200 peaks and some with the 220 peak as well.

3. Results

3.1. *Large-Volume Press*

Raman spectroscopy was used to identify the recovered samples. Bm was identified by the strong modes at 383 and 501 cm^{-1} (Figure 1), and a weaker mode at 542 cm^{-1} also appeared clearly in spectra. Ak was identified by the strong modes at 480 and 799 cm^{-1} along with a few other weaker modes. Mj was identified by the strong modes at 602 and 931 cm^{-1}. The peak positions for all three phases are in good agreement with those reported in the literature [33–36].

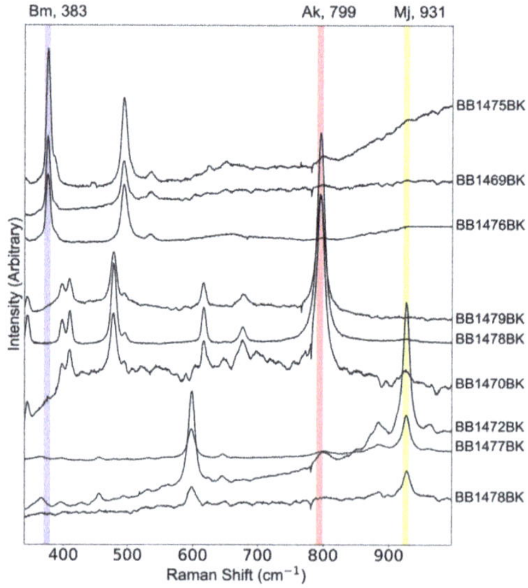

Figure 1. Raman spectra of the samples recovered from the Large-Volume Press (LVP) runs: The vertical lines highlight the key peaks for the identification of Bm (blue) at 383 cm^{-1}, Ak (red) at 799 cm^{-1}, and Mj (yellow) at 931 cm^{-1}. An LVP run ID is provided to each spectrum (Table 1).

For sample BB1478BK, we identified both Ak and Mj phases located at different areas within the sample: Ak was found within the cooler region of the sample (outer area close to capsule), and Mj was found within the warmer region of the sample (center and equatorial region of capsule). This sample with two phases allows us to tightly constrain the Ak–Mj phase boundary (and therefore the BAM triple point).

We inspected the samples under a stereomicroscope. The Bm samples are white and light-gray, and semi-translucent with 5–10 μm crystals. The Ak samples are dark-gray with light-gray speckled throughout and with 2–5 μm crystals. The Mj samples are gray to light-gray in color with the middle of the sample being lighter in color and with <5 μm crystals.

We chose to conduct LVP experiments at a narrow *P*–*T* field to make high-resolution determination of the BAM triple point. The observations of both Mj and Ak in one sample (BB1478BK) provide a tight constraint on the location of the Ak–Mj boundary (Figure 2a). The observations of Ak at 19.5 GPa and 1923 K and of Mj at 19 GPa and 1973 K indicate that the Ak–Mj boundary temperature should not exceed ∼2000 K at 19–20 GPa. The observations of Ak at 20 GPa and 1873–1948 K and Bm at 20.5–22 GPa and 1873–2023 K suggest that the Ak–Bm boundary should be located between these pressures in the LVP experiments. In this determination, we adopted the Clapeyron slopes of the three boundaries from Ishii et al. [37]. Combining these constraints with other data points, the $P-T$ condition for the triple point is determined to be 19.9 ± 0.4 GPa and 2000 ± 50 K in our LVP experiments on $MgSiO_3$. We attempted to fit the data with different Clapeyron slopes [4]. Because of the narrow $P-T$ region of our study and the better agreement in the Clapeyron slopes of these boundaries in the literature, our determined $P-T$ conditions for the triple point are not sensitive to the choice of the Clapeyron slopes of the boundaries.

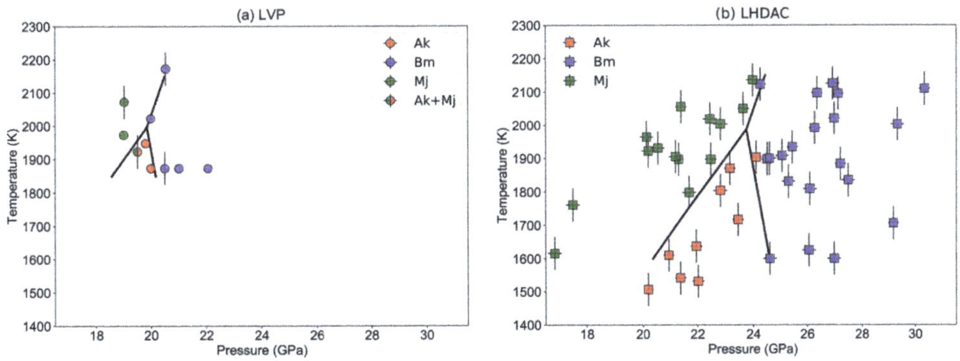

Figure 2. Data points from our (**a**) LVP and (**b**) LHDAC experiments: The red, green, and blue symbols represent Ak, Mj, and Bm, respectively. The estimated phase boundaries and triple point are shown. We fixed the Clapeyron slopes of the involved phase boundaries to those reported in Ishii et al. [37] and then adjusted the location of the phase boundaries to fit our data points. Our LVP triple point is located at 2000 ± 50 K and 19.9 ± 0.4 GPa. Our LHDAC triple point is located at 1990 ± 100 K and 23.8 ± 0.6 GPa.

3.2. Laser-Heated Diamond Anvil Cell

Three DAC were used to acquire data for the BAM triple point. Cell #1 loaded with the $MgSiO_3$ glass starting material was used to determine the Mj–Bm phase boundary through rapid heating into the Mj field and then by slowly increasing temperature to gain pressure through thermal pressure effects until the first Bm peaks were observed (Figure 3a). In order to avoid forming Ak, rapid heating was conducted because Ak could remain as a metastable phase and could overlap with the Bm and Mj diffraction peaks. Rapid heating was accomplished by aiming the heating laser beam at a previously heated spot. Once the desired temperature was reached, the sample was rapidly translated by moving the LHDAC such that the heating laser was aimed at an adjacent unheated spot and thus reached target temperature quickly without a slow laser power ramp up. Once the Mj peaks are observed, the temperature was increased slowly until the first Bm peaks form. The location in $P-T$ space where this happens was used to determine the Mj–Bm phase boundary (Figure 2b).

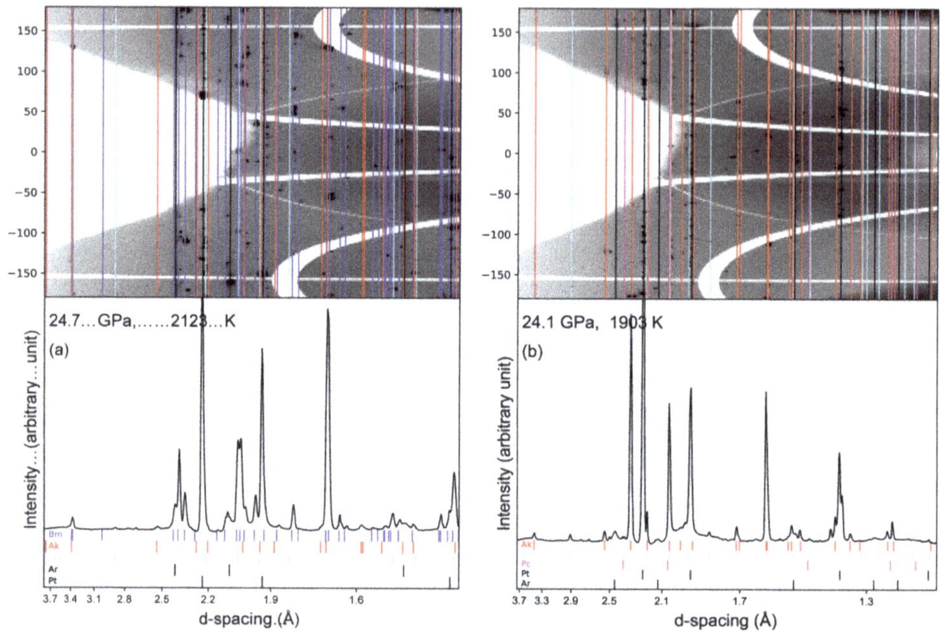

Figure 3. Two-dimensional diffraction images of LHDAC data (top) with integrated 1D diffraction patterns (bottom): (**a**) the formation of Bm, shown in blue, across the Mj/Bm phase boundary and (**b**) the stability of Ak and absence of Bm. We provide $P-T$ conditions of the diffraction measurements. The colored vertical bars in the 2D image and the colored ticks in the 1D pattern show the diffraction peak positions of different phases. We found some weak diffraction features possibly from Ak and Stv (stishovite) in Figure 3a and Stv and Pc (periclase) in Figure 3b. They are likely formed as transient phases due to kinetics and differential elemental diffusion during phase transition.

The experiments with cells #2 and #3 loaded with the Ak starting material were designed to constrain the boundaries between Ak and Mj and between Ak and Bm. Temperature was raised slowly until the first observation of the Bm or Mj diffraction peaks.

Mj was mainly identified through the 040, 323, and 431 peaks. The 444 and 046 peaks were observable in some patterns, and these helped to further confirm the presence of Mj. Ak was identified with the $10\bar{2}$, 104, 110, $20\bar{4}$, and $11\bar{6}$ peaks. The 113 peak was sometimes used, but it overlaps enough with the Mj 125 peak that it was less useful. Finally, Bm was identified with the 111, 200, 120, 210, 022, 122, 121, 023, and 221 peaks. Higher 2θ peaks were used to confirm the presence of Bm, if available. Similar to our LVP analysis, the BAM triple point was constrained by moving the three phase boundaries in the $P-T$ space while we fixed the Clapeyron slopes of the boundary to those reported in Ishii et al. [37]. The $P-T$ condition for the BAM triple point from our LHDAC dataset is 23.8 $\pm$ 0.6 GPa and 1990 $\pm$ 100 K.

4. Discussion

The $P-T$ conditions of the Bm–Ak–Mj (BAM) triple point that we obtained are 19.9 $\pm$ 0.4 GPa and 2000 $\pm$ 50 K for LVP and 23.8 $\pm$ 0.6 GPa and 1990 $\pm$ 100 K for LHDAC (Figure 4). Despite the fact that they can be biased by different systematic error sources (such as pressure effects on thermocouple emf calibration in LVP and optical effects on spectroradiometry through diamond anvils in LHDAC), a remarkable agreement was found in temperature for the triple point from both techniques. To our knowledge, this is the first direct cross examination of the two temperature measurement techniques. Such an agreement may not necessarily be applicable for pressures much lower or much higher than the range in which we conducted our measurements, i.e., 20–24 GPa, because some of the perceived systematic error sources could be pressure dependent, such as pressure effects on thermocouple emf calibration. However, for the pressure range of the mantle transition zone, our result provides important experimental confirmation for comparing temperature measurements from LVP and LHDAC experiments.

Our results suggest that pressure calibration is the most important issue to resolve in comparing LVP and LHDAC datasets of a 3.9 GPa difference. So far, LHDAC studies have reported systematically higher pressures for the phase boundaries in the mantle transition zone by 2–3 GPa compared with LVP studies [2–4,6,10]. Our new results reported here also confirm the trend but with a greater magnitude. The difference is particularly important to resolve because the COMPRES 8/3 assembly and its pressure calibration [23] have been widely used in high-pressure studies.

The calibration for the 8/3 assembly was conducted through in situ measurements using the tungsten EOS [27] in Leinenweber et al. [23]. For our LHDAC experiments, we chose to use the thermal EOS of Pt by Dorogokupets and Dewaele [29]. This pressure scale is known to yield better agreements with the Au and MgO scales according to Ye et al. [11] for a wide pressure range. In order to ensure the consistency between LHDAC and LVP results, it would be useful to examine the agreements between the W scale and the Pt scale at in situ high $P-T$. However, the high shear strength of tungsten can be a potential issue for the accurate determination on its EOS [38,39]. Therefore, the COMPRES 8/3 assembly can be calibrated using other pressure standards. Pt, Au, and MgO could be good candidates as they have been used widely in LHDAC. However, despite the agreements over a larger pressure range, the MgO, Au, and Pt scales by Dorogokupets and Dewaele [29] are different by $\sim$2 GPa at 20–40 GPa and high temperatures according to Ye et al. [11]. Therefore, an important challenge still remains for the thermal EOS of important materials at the pressure range for the mantle transition zone. It is notable that some of the standard materials (particularly Au) have low melting temperatures compared with the mantle geotherm at pressures of $\leq$ 30 GPa, potentially causing significant anharmonic effects in their EOS or even abnormal premelting behaviors.

The seismic properties of the 660-km discontinuity are in general agreement with those of the post-spinel transition [40]. Ishii et al. [37] showed that the Ak–Bm boundary should be close to the post-spinel boundary within 1 GPa [37]. Figure 4 shows that the Ak–Bm and Mj–Bm boundaries in $MgSiO_3$ measured by LHDAC are closer to the $P-T$ conditions expected for the 660-km discontinuity. Does this mean that LHDAC yields more reliable results for the location of the phase boundary? This approach is not desirable as the experimental methods should be able to address the question of whether the mantle phase boundaries are indeed the source of the 660-km discontinuity rather than the other way around.

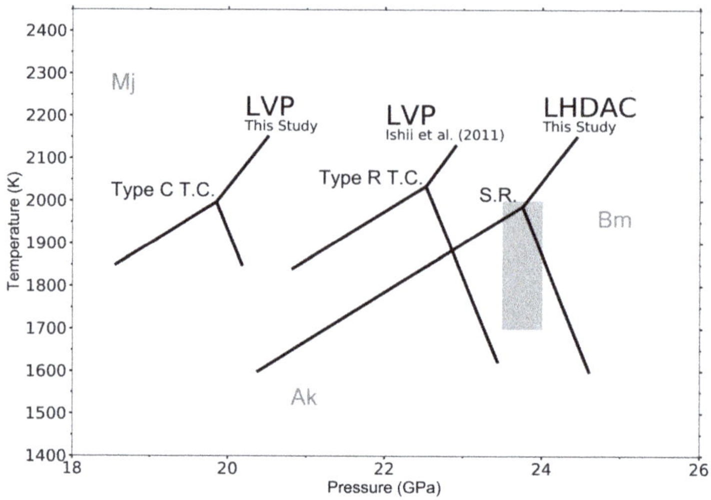

Figure 4. The phase boundaries between bridgmanite, akimotoite, and majorite and the triple point between them: The gray rectangular area represents pressure–temperature conditions expected for the 660-km discontinuity [41–44].

Some former LVP studies have measured the Ak–Mj, Mj–Bm, and Ak–Bm boundaries in pure $MgSiO_3$ and inferred the triple point [4,37,45]. However, the reported $P-T$ conditions for the boundaries and the triple point do not agree with each other: the discrepancy can be as large as 2 GPa in pressure and 300 K in temperature among the LVP measurements. In terms of temperature, our results are in best agreement with the most recent report by Ishii et al. [37]. They reported stability of Ak up to 1973 K at 22.3 GPa, and the triple point in their phase diagram can be inferred to be 2035 $\pm$ 60 K, which is in agreement with our LVP and LHDAC results on the temperature of the Bm–Ak–Mj (BAM) triple point within 100 K. This agreement is encouraging in that at least the recent studies converge on the temperature of the BAM triple point within 100 K even between different high-pressure techniques (LVP and LHDAC) and between different temperature measurement techniques (thermocouple W5%Re-W26%Re in our LVP study; Pt/Pt-13%Rh in the LVP study by Ishii et al. [37]; and spectroradiometry in our LHDAC study). As efforts are being made for enhancing the accuracy of thermocouple emf calibrations at high pressures [13,14], it remains to be seen if future calibration work on the thermocouples used in this study and in Ishii et al. [37] can find further improvement in the agreement.

The pressure inferred for the BAM triple point in Ishii et al. [37] is between our LVP and LHDAC results, located approximately in the middle (Figure 4). They calibrated pressures at high temperatures based on previously reported boundaries in Mg_2SiO_4, $MgSiO_3$, and $MgAl_2O_4$ (see Ishii et al. [37] for references) which are all different in pressure calculation methods. If we were to use the same Ak–Bm transition pressure point (at 1873 K and 22.3 GPa) that is used in Ishii et al. [37] as an internal calibration point, our multi-anvil BAM triple point would lie at almost the same P and T as that in Ishii et al. [37]. Although it would depend on experimental setup, including the sample geometry and anvil materials, possible pressure change during heating is an important factor to consider for improving pressure estimation in LVP experiments [46]. In order to further gain insight into the differences between the DAC and LVP pressures and to possibly close the gap, it would be worthwhile to make detailed measurements on the BAM triple point in situ using the same sample and Pt pressure standard that was used in the LHDAC.

5. Conclusions

We have established the location of the phase boundaries involving bridgmanite, akimotoite, and majorite with both large-volume press (LVP) and laser-heated diamond anvil cell (LHDAC). With the use of these two techniques, we were able to determine the location of the bridgmanite–akimotoite–majorite (BAM) triple point where these three phases are at equilibrium. The results indicate that the temperature of the BAM triple point established with the two techniques are in agreement: W5%Re-W26%Re thermocouple used in LVP and spectroradiometry in LHDAC constrain the temperature of the BAM triple point between 1990 K and 2000 K. Our result is also in good agreement with the temperature measured by Ishii et al. [37] despite the fact that they used a different thermocouple (Pt-Pt13%Rh). This result now enables direct comparison of the experimental results at the $P-T$ conditions of the mantle transition zone for temperature. Furthermore, the BAM triple point $P-T$ condition reported here can be used as a reference point for calibrating pressure and temperature in high-pressure experiments. Unlike conventional "fixed-pressure points" that are often used in LVP calibration, the BAM triple point does not need an assumption of zero thermal effects on such points. Our study also reveals that the pressure measurement is the main source for the discrepancy between LVP studies and between LVP and LHDAC results. This result now calls for efforts in improving pressure measurement techniques in LVP and LHDAC.

Author Contributions: Conceptualization, S.-H.S. and K.D.L.; methodology, V.B.P.; software, S.-H.S.; validation, B.L.K. and J.D.D.; formal analysis, B.L.K. and J.D.D.; investigation, B.L.K. and J.D.D.; resources, S.-H.S.; data curation, B.L.K., J.D.D., and S.-H.S.; writing—original draft preparation, B.L.K., J.D.D., and S.-H.S.; writing—review and editing, K.D.L. and S.-H.S.; visualization, B.L.K. and J.D.D.; supervision, S.-H.S.; funding acquisition, S.-H.S. All authors have read and agreed to the published version of the manuscript.

Funding: The work has been supported by the NSF (EAR1725094) and NASA (80NSSC18K0353). D.S. was supported partially by the Keck Foundation (PI: P. Buseck). The results reported herein benefit from collaborations and information exchange within NASA's Nexus for Exoplanet System Science (NExSS) research coordination network sponsored by NASA's Science Mission Directorate. Synchrotron measurements were conducted at the Advanced Photon Source, a Department of Energy (DOE) Office of Science User Facility operated for the DOE Office of Science by Argonne National Laboratory under Contract DE-AC02-06CH11357. The synchrotron X-ray diffraction experiments were conducted at GSECARS (University of Chicago, Sector 13), Advanced Photon Source (APS). GSECARS is supported by the NSF-Earth Science (EAR-1128799) and DOE-GeoScience (DE-FG02-94ER14466).

Acknowledgments: We thank four anonymous reviewers and the editors. We acknowledge the use of facilities within the Eyring Materials Center at Arizona State University.

Conflicts of Interest: The authors declare no conflict of interest.

Abbreviations

The following abbreviations are used in this manuscript:

P	Pressure
T	Temperature
LVP	Large-volume press
LHDAC	Laser-heated diamond anvil cell
Ak	Akimotoite
Mj	Majorite
Bm	Bridgmanite
BAM	Bridgmanite–Akimotoite–Majorite

References

1. Anderson, O.L. The Grüneisen ratio for the last 30 years. *Geophys. J. Int.* **2000**, *143*, 279–294. [CrossRef]
2. Irifune, T.; Nishiyama, N.; Kuroda, K.; Inoue, T.; Isshiki, M.; Utsumi, W.; Funakoshi, K.i.; Urakawa, S.; Uchida, T.; Katsura, T.; et al. The postspinel phase boundary in Mg_2SiO_4 determined by in situ X-ray diffraction. *Science* **1998**, *279*, 1698–1700, [CrossRef] [PubMed]

3. Shim, S.H.; Duffy, T.S.; Shen, G. The post-spinel transformation in Mg_2SiO_4 and its relation to the 660-km seismic discontinuity. *Nature* **2001**, *411*, 571. [CrossRef] [PubMed]
4. Hirose, K.; Komabayashi, T.; Murakami, M.; Funakoshi, K.I. In situ measurements of the majorite-akimotoite-perovskite phase transition boundaries in $MgSiO_3$. *Geophys. Res. Lett.* **2001**, *28*, 4351–4354. [CrossRef]
5. Katsura, T.; Yamada, H.; Shinmei, T.; Kubo, A.; Ono, S.; Kanzaki, M.; Yoneda, A.; Walter, M.J.; Ito, E.; Urakawa, S.; et al. Post-spinel transition in Mg_2SiO_4 determined by high $P-T$ in situ X-ray diffractometry. *Phys. Earth Planet. Inter.* **2003**, *136*, 11–24. [CrossRef]
6. Fei, Y.; Van Orman, J.; Li, J.; Van Westrenen, W.; Sanloup, C.; Minarik, W.; Hirose, K.; Komabayashi, T.; Walter, M.; Funakoshi, K.I. Experimentally determined postspinel transformation boundary in Mg_2SiO_4 using MgO as an internal pressure standard and its geophysical implications. *J. Geophys. Res. Solid Earth* **2004**, *109*, B02305. [CrossRef]
7. Fei, Y.; Ricolleau, A.; Frank, M.; Mibe, K.; Shen, G.; Prakapenka, V. Toward an internally consistent pressure scale. *Proc. Nat. Acad. Sci. USA* **2007**, *104*, 9182–9186. [CrossRef]
8. Sinogeikin, S.; Bass, J.; Prakapenka, V.; Lakshtanov, D.; Shen, G.; Sanchez-Valle, C.; Rivers, M. Brillouin spectrometer interfaced with synchrotron radiation for simultaneous X-ray density and acoustic velocity measurements. *Rev. Sci. Instrum.* **2006**, *77*, 103905. [CrossRef]
9. Li, B.; Kung, J.; Uchida, T.; Wang, Y. Simultaneous equation of state, pressure calibration and sound velocity measurements to lower mantle pressures using multi-anvil apparatus. In *Advances in High-Pressure Technology for Geophysical Applications*; Elsevier: Amsterdam, The Netherlands , 2005; pp. 49–66.
10. Ye, Y.; Gu, C.; Shim, S.H.; Meng, Y.; Prakapenka, V. The postspinel boundary in pyrolitic compositions determined in the laser-heated diamond anvil cell. *Geophys. Res. Lett.* **2014**, *41*, 3833–3841. [CrossRef]
11. Ye, Y.; Prakapenka, V.; Meng, Y.; Shim, S.H. Intercomparison of the gold, platinum, and MgO pressure scales up to 140 GPa and 2500 K. *J. Geophys. Res. Solid Earth* **2017**, *122*, 3450–3464. [CrossRef]
12. Getting, I.; Kennedy, G. Effect of pressure on the emf of chromel-alumel and platinum-platinum 10% rhodium thermocouples. *J. Appl. Phys.* **1970**, *41*, 4552–4562. [CrossRef]
13. Nishihara, Y.; Fuke, K.; Tange, Y.; Higo, Y. Determination of pressure effect on thermocouple electromotive force using multi-anvil apparatus. *High Press. Res.* **2016**, *36*, 121–139. [CrossRef]
14. Nishihara, Y.; Doi, S.; Kakizawa, S.; Higo, Y.; Tange, Y. Effect of pressure on temperature measurements using WRe thermocouple and its geophysical impact. *Phys. Earth Planet. Inter.* **2019**, *298*, 106348. [CrossRef]
15. Boehler, R. High-pressure experiments and the phase diagram of lower mantle and core materials. *Rev. Geophys.* **2000**, *38*, 221–245. [CrossRef]
16. Shen, G.; Rivers, M.L.; Wang, Y.; Sutton, S.R. Laser heated diamond cell system at the Advanced Photon Source for in situ X-ray measurements at high pressure and temperature. *Rev. Sci. Instrum.* **2001**, *72*, 1273–1282. [CrossRef]
17. Kavner, A.; Panero, W.R. Temperature gradients and evaluation of thermoelastic properties in the synchrotron-based laser-heated diamond cell. *Phys. Earth Planet. Inter.* **2004**, *143*, 527–539. [CrossRef]
18. Deng, J.; Du, Z.; Benedetti, L.R.; Lee, K.K. The influence of wavelength-dependent absorption and temperature gradients on temperature determination in laser-heated diamond-anvil cells. *J. Appl. Phys.* **2017**, *121*, 025901. [CrossRef]
19. Lin, J.F.; Sturhahn, W.; Zhao, J.; Shen, G.; Mao, H.K.; Hemley, R.J. Absolute temperature measurement in a laser-heated diamond anvil cell. *Geophys. Res. Lett.* **2004**, *31*, L14611. [CrossRef]
20. Bullen, K. The problem of the Earth's density variation. *Bull. Seismol. Soc. Am.* **1940**, *30*, 235–250.
21. Ai, Y.; Zheng, T.; Xu, W.; He, Y.; Dong, D. A complex 660 km discontinuity beneath northeast China. *Earth Planet. Sci. Lett.* **2003**, *212*, 63–71. [CrossRef]
22. Day, E.A.; Deuss, A. Reconciling *PP* and $P'P'$ precursor observations of a complex 660 km seismic discontinuity. *Geophys. J. Int.* **2013**, *194*, 834–838. [CrossRef]
23. Leinenweber, K.D.; Tyburczy, J.A.; Sharp, T.G.; Soignard, E.; Diedrich, T.; Petuskey, W.B.; Wang, Y.; Mosenfelder, J.L. Cell assemblies for reproducible multi-anvil experiments (the COMPRES assemblies). *Am. Mineral.* **2012**, *97*, 353–368. [CrossRef]
24. Walker, D.; Carpenter, M.; Hitch, C. Some simplifications to multianvil devices for high pressure experiments. *Am. Mineral.* **1990**, *75*, 1020–1028.
25. Tangeman, J.A.; Phillips, B.L.; Navrotsky, A.; Weber, J.R.; Hixson, A.D.; Key, T.S. Vitreous forsterite (Mg_2SiO_4): Synthesis, structure, and thermochemistry. *Geophys. Res. Lett.* **2001**, *28*, 2517–2520. [CrossRef]

26. Shim, S.H.; Catalli, K. Compositional dependence of structural transition pressures in amorphous phases with mantle-related compositions. *Earth Planet. Sci. Lett.* **2009**, *283*, 174–180. [CrossRef]
27. Hixson, R.; Fritz, J. Shock compression of tungsten and molybdenum. *J. Appl. Phys.* **1992**, *71*, 1721–1728. [CrossRef]
28. Hernlund, J.; Leinenweber, K.; Locke, D.; Tyburczy, J.A. A numerical model for steady-state temperature distributions in solid-medium high-pressure cell assemblies. *Am. Mineral.* **2006**, *91*, 295–305. [CrossRef]
29. Dorogokupets, P.; Dewaele, A. Equations of state of MgO, Au, Pt, NaCl-B1, and NaCl-B2: Internally consistent high-temperature pressure scales. *High Press. Res.* **2007**, *27*, 431–446. [CrossRef]
30. Prescher, C.; Prakapenka, V.B. DIOPTAS: A program for reduction of two-dimensional X-ray diffraction data and data exploration. *High Press. Res.* **2015**, *35*, 223–230. [CrossRef]
31. Shim, S.H. PeakPo—A python software for X-ray diffraction analysis at high pressure and high temperature. *Zenodo* **2017**. Available online: https://zenodo.org/record/3376238#.Xhn22S2ZM1I (accessed on 14 January 2020).
32. Dan Shim, S.-H. Pytheos—A Python Tool Set For Equations Of State. 2017. Available online: https://zenodo.org/record/802392#.Xhn2iC2ZM1I (accessed on 14 January 2020).
33. Williams, Q.; Jeanloz, R.; McMillan, P. Vibrational spectrum of $MgSiO_3$ perovskite: Zero-pressure Raman and mid-infrared spectra to 27 GPa. *J. Geophys. Res. Solid Earth* **1987**, *92*, 8116–8128. [CrossRef]
34. Gillet, P.; Guyot, F.; Wang, Y. Microscopic anharmonicity and equation of state of $MgSiO_3$-perovskite. *Geophys. Res. Lett.* **1996**, *23*, 3043–3046. [CrossRef]
35. Rauch, M.; Keppler, H.; Häfner, W.; Poe, B.; Wokaun, A. A pressure-induced phase transition in $MgSiO_3$-rich garnet revealed by Raman spectroscopy. *Am. Mineral. J. Earth Planet. Mater.* **1996**, *81*, 1289–1292.
36. Okada, T.; Narita, T.; Nagai, T.; Yamanaka, T. Comparative Raman spectroscopic study on ilmenite-type $MgSiO_3$ (akimotoite), $MgGeO_3$, and $MgTiO_3$ (geikielite) at high temperatures and high pressures. *Am. Mineral.* **2008**, *93*, 39–47. [CrossRef]
37. Ishii, T.; Kojitani, H.; Akaogi, M. Post-spinel transitions in pyrolite and Mg_2SiO_4 and akimotoite–perovskite transition in $MgSiO_3$: Precise comparison by high-pressure high-temperature experiments with multi-sample cell technique. *Earth Planet. Sci. Lett.* **2011**, *309*, 185–197. [CrossRef]
38. Millett, J.; Gray III, G.; Bourne, N. Measurement of the shear strength of pure tungsten during one-dimensional shock loading. *J. Appl. Phys.* **2007**, *101*, 033520. [CrossRef]
39. He, D.; Duffy, T.S. X-ray diffraction study of the static strength of tungsten to 69 GPa. *Phys. Rev. B* **2006**, *73*, 134106. [CrossRef]
40. Lebedev, S.; Chevrot, S.; van der Hilst, R.D. Seismic evidence for olivine phase changes at the 410-and 660-kilometer discontinuities. *Science* **2002**, *296*, 1300–1302. [CrossRef]
41. Dziewonski, A.M.; Anderson, D.L. Preliminary reference Earth model. *Phys Earth Planet. Inter.* **1981**, *25*, 297–356. [CrossRef]
42. Brown, J.; Shankland, T. Thermodynamic parameters in the Earth as determined from seismic profiles. *Geophys. J. Int.* **1981**, *66*, 579–596. [CrossRef]
43. Kennett, B.L.; Engdahl, E.; Buland, R. Constraints on seismic velocities in the Earth from traveltimes. *Geophys. J. Int.* **1995**, *122*, 108–124. [CrossRef]
44. Katsura, T.; Yoneda, A.; Yamazaki, D.; Yoshino, T.; Ito, E. Adiabatic temperature profile in the mantle. *Phys. Earth Planet. Inter.* **2010**, *183*, 212–218. [CrossRef]
45. Sawamoto, H. Phase diagram of $MgSiO_3$ at pressures up to 24 GPa and temperatures up to 2200 °C: Phase stability and properties of tetragonal garnet. *High Press. Res. Miner. Phys.* **1987**, *209*, 219.
46. Ishii, T.; Huang, R.; Fei, H.; Koemets, I.; Liu, Z.; Maeda, F.; Yuan, L.; Wang, L.; Druzhbin, D.; Yamamoto, T.; et al. Complete agreement of the post-spinel transition with the 660-km seismic discontinuity. *Sci. Rep.* **2018**, *8*, 6358. [CrossRef] [PubMed]

minerals

MDPI

Article

Compressional Behavior of Hydrous Orthoenstatite: Insight into the Nature of LVZ under Continental Plate

Florian Tian-Siang Hua [1], Przemyslaw Dera [2] and Jennifer Kung [1,*]

[1] Department of Earth Sciences, National Cheng Kung University, Tainan 701, Taiwan; gn012421567333@gmail.com

[2] Hawaii Institute of Geophysics and Planetology, School of Ocean and Earth Science and Technology, University of Hawaii at Manoa, Honolulu, HI 96822, USA; pdera@hawaii.edu

* Correspondence: jkung@mail.ncku.edu.tw; Tel.: +886-6-275-7575

Received: 7 December 2019; Accepted: 12 January 2020; Published: 16 January 2020

Abstract: A suite of hydrous orthoenstatite crystals were synthesized at 5–7 GPa and 1100–1300 °C, corresponding to the mid upper mantle conditions in continental regions. The synthetic crystals presented a clear, inclusion-free, and euhedral form with a size range from 100 to a few hundred microns. The Al- and water content of crystals were less than 2 wt. % and ranging from ~500 ppm to 1000 ppm, respectively, characterized by Raman and IR spectroscopy, electron microscopy, and SIMS. The analysis shows that the capability of water incorporation for this suite of hydrous orthoenstatite is correlated to the Al-content in the crystal structure. To understand how the detailed crystal structure reflects the influence of water and Al, single crystal X ray diffraction analysis was performed for this suite of hydrous orthoenstatite. By comparing the results obtained at ambient conditions, we find that for low-level of Al-content, <2 wt. %, the Al^{3+} cation tends to occupy one of tetrahedral sites (TB) only. Analysis of the X ray diffraction results under high pressure indicates that the elasticity of orthoenstatites is insensitive to the presence of low-level water and Al. We use this finding to evaluate the velocity profile at the mid upper mantle of continental regions to compare with seismic observation. The comparison indicates that the cause of the low velocity zone in continental regions originates from the geotherm profile rather than the effect of water on the elasticity of mantle phases.

Keywords: hydrous pyroxene; high pressure; X ray diffraction; compressional behavior; low velocity zone

1. Introduction

The Mg-rich orthoenstatite (OEn) phase is one of the major mineral phases in lherzolite and harzburgite xenoliths, which put constraints on the composition of the lithosphere in continental regions [1]. To shed light on the physical state of the upper mantle, the elasticity of the anhydrous OEn phase has been investigated under high pressure and high temperature [2–10]. In terms of crystal chemistry, the pyroxene phase is a "nominally anhydrous mineral (NAM)". Field evidence shows that the OEn phase often contains some amount of "water" within the crystal structure, up to 600–800 ppm, usually incorporated as hydroxyl, OH^- [11–13]. It is well-known that the crystal chemistry (e.g., H^+ concentration and minor elements) of mineral phases could have a different influence on the physical properties (e.g., [14–18]). Recent experiments indicate that the phase boundary of monoclinic enstatite (low-pressure phase to high-pressure phase) is noticeably shifted by the water content [19,20]. On the other hand, the phase transition in the hydrous orthoenstatite with low water-content at high pressure is only shifted by less than 1 GPa (from Pbca to $P2_1/c$) [21].

The experimental results also indicate that the presence of high Al-content in the synthetic orthoenstatite can enhance the water concentration up to 7500 ppm at low-pressure, but the Al content presented in orthopyroxene structure is depressed by the pressure effect [22–24]. The results of X-ray single-crystal diffraction experiments suggest that the site occupancy of H^+ varies with the concentration of Al^{3+}, as the H^+ is located at the Si^{4+} site when the Al-content is fairly low. In contrast, H^+ is located at both M and Si^{4+} sites when the Al-content was high in orthoenstatite [23,25]. To shed light on the effect of H^+ incorporation in different crystallographic sites on elasticity, the compressional behavior of the hydrous OEn phase has to be investigated.

In the upper mantle, a low-velocity zone (LVZ) is characterized by low seismic velocity, high seismic attenuation and high electrical conductivity. Some potential causes of LVZ have been proposed, including: (1) change of geothermal gradient, (2) the presence of small amounts of the melt, and (3) the hydrogen present in the structure of NAMs [22,26–30]. In the wet mantle, the "water" can either be incorporated into the mantle phases to affect the elasticity, or exist as a hydrous fluid, which reduces the solidus of rock and may produce melt if the geotherm intersects with solidus [27,31,32]. The above factors have an effect on each other in the wet upper mantle. Among the mantle phases, the pyroxene phase is capable of carrying the highest water content at the mantle conditions [13,33]. As mentioned above, previous measurements on orthoenstatite were focused on anhydrous orthoenstatite. In order to evaluate the possible causes of LVZ in the continental regions, it is important to have access to the physical properties of hydrous pyroxene around the depth of LVZ. In this study, we investigated the compressional behavior of the orthoenstatite phase with crystal chemistry (Al^{3+} and water contents) under the mid upper mantle conditions.

2. Materials and Methods

Single-crystal synthesis was carried out in a multi-anvil large volume high pressure apparatus installed at National Cheng Kung University (Tainan, Taiwan). The stoichiometry of the starting mixture of oxides, SiO_2, MgO, $Mg(OH)_2$, and $Al(OH)_3$, was chosen to correspond to the composition of pure end-member or aluminous orthopyroxene. The $MgSiO_3$ glass was used to synthesize anhydrous orthoenstatite, hereafter "OEn", (R638). The synthesis was carried out at the pressure of 5–7 GPa and temperature of 1100–1300 °C. Experimental details are compiled in Table 1. The synthesized single crystals presented clear, inclusion-free and euhedral form (Figure 1). To identify the synthesized phase and verify the presence of $[OH^-]$ bonds, Raman spectroscopy measurements were carried out with the wavenumber range extended up to 4000 cm^{-1}. The detailed characterization of $[OH^-]$ bands were performed by FTIR carried out at the Magmas and Volcanoes Laboratory of Clermont Auvergne University (Clermont-Ferrand, France). Figure 2 shows the Raman and IR spectra in the $[OH^-]$ vibrational region, collected from two crystals from runs R623 and R624 in this study. Water content was measured using the SIMS technique, with Cameca IMS-1280 ion microprobe installed at School of Ocean and Earth Science and Technology, University of Hawaii at Manoa (Honolulu, HI, USA), with the calibration curves constructed using standards employed in previous studies [34,35]. Chemical composition analyses were carried out by the electron microprobe in energy dispersive mode with FEI Quanta 250 electron microscopy installed at the Department of Earth Sciences, National Cheng Kung University.

Table 1. Experimental conditions of crystal synthesis and products.

Run		Pressure (GPa)	Temperature (°C)	Crystal Size Long-Axis Lengths (μm)	Run Products
Anhydrous (intended)					
	R638	5.5	1300	50–150	OEn
	R626	6	1200	100–400	OEn
Hydrous					
	R598	5	1200	100–300	OEn
	R545	5	1400	100–300	OEn
	R623	5.5	1200	500–1000	OEn, Coe [1]
Al-bearing hydrous					
	R624	5.5	1200	400–700	OEn
	R608	7	1200	200–400	OEn, Ol [2]

[1] Coe, coesite; [2] Ol, olivine.

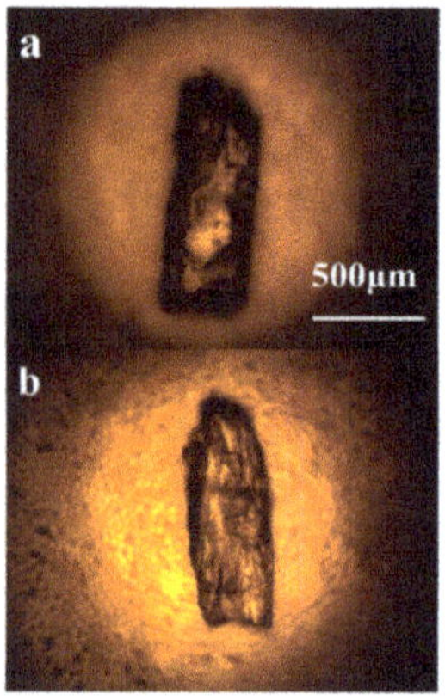

Figure 1. Samples of (**a**) hydrous Mg#100 (R623) and (**b**) Al-bearing OEn (R624).

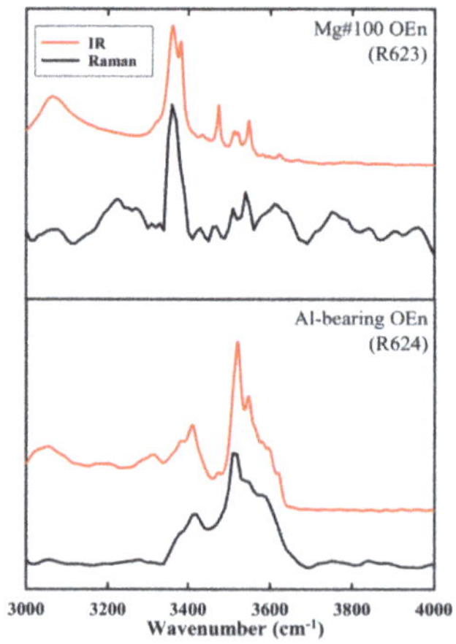

Figure 2. Polarized Raman and IR spectra (E parallel c axis) of hydrous OEn crystals in this study. The black and red lines are the data collected from Raman and FTIR spectroscopy, respectively.

X-ray single-crystal diffraction was performed at ambient conditions and at high pressure on a Bruker D8 Venture diffractometer equipped with IμS 3.0 Microfocus Source with Ag Kα radiation and a fixed-chi type goniometer installed at X-ray Atlas lab of the University of Hawaii at Manoa (Honolulu, HI, USA). Diffraction images were collected using PHOTON II detector. The tilting and rotation of the detector and the sample-to-detector distance were calibrated using NIST standard LaB_6 powder. BX90 diamond anvil cells equipped with two 400-μm diameter culet diamonds were used for high pressure experiments. Rhenium gaskets were preindented to ~40 μm thicknesses and then ~180 μm diameter holes were laser-drilled at the center of gasket as sample chamber. Methanol-ethanol (4:1) mixture was loaded into the chamber as a pressure medium. Two ~10 μm ruby spheres were also loaded into the sample chamber to measure the pressure.

Diffraction images were analyzed using the APEX3 software package (v2018.1, Bruker AXS, Madison, WI, USA), including XPREP and SHELXL. Diffraction data were integrated using SAINT+ (Bruker-AXS) after exclusion of the non-sample diffraction spots and then corrected for absorption using the ABSORB software. Crystal symmetry and space group were determined by XPREP. Crystal structures were refined from the intensity data using the SHELXL software (2014/7, Bruker AXS, Madison, WI, USA). The cation occupancy parameters were constrained according to chemical analysis and charge balance. The balance formulas are shown in the following:

$$X_{obs} = 4\,X_{Si} + 3\,X_{Al} + 2\,X_{Mg} - 2\,X_{O} \text{ (charge balance)} \quad (1)$$

$$X_{obs} = X_{SiA} + X_{SiB} \text{ (chemical analysis)} \quad (2)$$

$$X_{obs} = X_{M1} + X_{M2} \text{ (chemical analysis)} \quad (3)$$

where X is molar number in chemical formula and X_{obs} is the value based on the chemical analysis. The instrumental errors were corrected according to NIST standard LaB_6 powder following previous described procedure [21,36]. For high pressure experiments, the positions and intensities of ~400–600 unique peaks were extracted from the raw diffraction patterns at each pressure point. Dynamic masks were employed to correct for the regions on the detector shaded by the diamond-anvil cell. The rest of the data processing was carried out in a similar fashion to the data processed at ambient conditions.

3. Results and Discussion

In this study, we intended to synthesize anhydrous OEn by using mixed oxides (MgO and SiO_2) and $MgSiO_3$ glass for high pressure study. However, the Raman measurements showed that these "anhydrous phase" specimens (R626 and R638, starting with oxides and $MgSiO_3$ glass) contained $[OH^-]$ signals. By comparing the integrated area of $[OH^-]$ bands of specimen R624, the water content of which was measured by SIMS, we estimated the crystals from R626 and R638 runs to contain ~300 ppm of water. Therefore, the specimens synthesized in this study are all hydrous phases. Chemical analysis of synthetic OEn is summarized in Table 2 and indicates that the synthetic pure Mg end-member specimens incorporated a small amount of aluminum (less than 0.34 Al_2O_3 wt. %), which were contaminated during the preparation of starting powder. This suite of low-Al content specimens, including R545, R598, R623, R626, and R638, will be referred to as Mg#100. The other two specimens, R608 and R624, contain Al_2O_3 at around 1.5–1.9 wt. % and will be referred as the Al-bearing OEn. The SIMS results indicate that the Al-bearing group contains about 800–1000 ppm of H_2O, in contrast to Mg#100, which H_2O content was around 400–500 ppm or less (Table 2). Figure 3a presents the relation of Al- and H_2O- content of synthetic OEns in this study and [23], illustrating the Al-content having a strong influence of water incorporation in OEn structure, with higher Al concentration resulting in higher water content. The following discussion of water-dependence of the crystal structure will be highlighted by the influence of Al-replacement in the hydrous phase. Furthermore, based on the chemical analysis of Al-bearing OEn and the crystal-structure refinement, we observed that with less than 2 wt. % of Al_2O_3, the Al^{3+} cation only substitutes for Si^{4+} in tetrahedral site and prefers to occupy the SiB (i.e., TB) sites rather than the SiA (i.e., TA) sites, which is indicated as "AlB" occupancy in Supplementary Table S2 (R624 and R608).

Table 2. Chemical compositions of synthesized orthoenstatite.

Phase	Run No.	MgO (wt. %)	SiO_2 (wt. %)	Al_2O_3 (wt. %)	Water (ppm)
Hydrous Mg#100	R545	40.39 ± 0.30	61.13 ± 0.49	0.13 ± 0.03	319 ± 40
	R598	38.60 ± 0.05	58.47 ± 0.09	0.25 ± 0.07	515 ± 75
OEn	R623	38.30 ± 0.56	59.51 ± 0.74	0.09 ± 0.07	417 ± 18
	R626 *	38.66 ± 0.18	60.59 ± 0.33	0.34 ± 0.05	300–400 *
	R638 *	38.36 ± 0.18	60.78 ± 0.35	0.22 ± 0.04	300–400 *
Hydrous Al-bearing	R608	38.08 ± 0.13	58.56 ± 0.27	1.56 ± 0.09	978 ± 117
OEn	R624	37.69 ± 0.79	58.73 ± 1.15	1.92 ± 0.66	849 ± 69

* Both R626 and R638 were intended to be the anhydrous OEn, but obvious OH^- bands were observed by Raman and IR measurements. The concentration of water was estimated by comparing the area of OH^- bands with those of R608. The details were described in text.

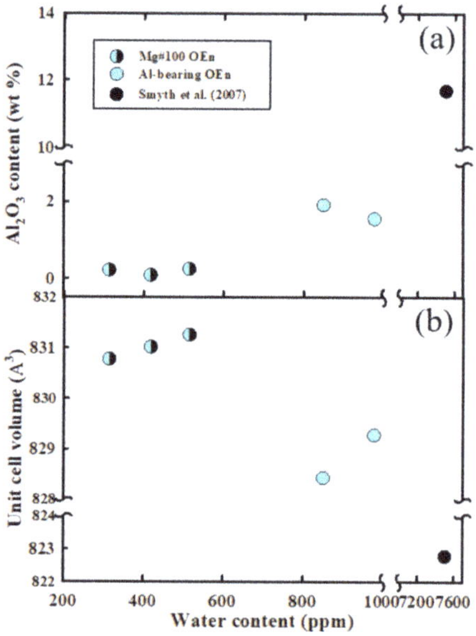

Figure 3. The relations between Al-content (**a**), cell volume (**b**) and water content among the investigated OEn crystals from this study and [23]. The half-filled and filled blue symbols are hydrous pure and Al-bearing OEn from this study, respectively; the black symbol is higher-aluminous OEn from the study of [23]. The error bars are smaller than the size of symbols. The details were described in text.

The X-ray diffraction result on the synthetic enstatite crystals clearly indicated orthorhombic symmetry with space group Pbca. X-ray diffraction measurements at ambient conditions were performed on six synthetic OEn specimens. Detailed refinement information is listed in Supplementary Tables S1 and S2. The cell volumes of the synthetic OEn investigated in this study (shown in Figure 3b, blue and half-filled blue circles) are distributed into two groups, corresponding to a detectable volume reduction caused by the Al-content incorporating in the structure. Compared with the anhydrous OEn from Ref. [37] (V_0 = 831.4 Å^3), the cell volumes of low-level water content OEn (less than 500 ppm, half-filled blue symbols in Figure 3b) are comparable within uncertainty, between 830.5 and 831.5 Å^3. At higher water-content (~800–1000 ppm, blue symbols in Figure 3b), the measured cell volumes show systematically lower value (<830 Å^3), compared with those of the hydrous Mg#100 (>830 Å^3) specimens. The trend of the cell volume of hydrous OEn reduced by the water content observed in this study fits the earlier observation of high water-content Al-bearing specimen (11.7 wt. % of Al_2O_3 with 7500 ppm of water) presented by [23] (shown as black circle in Figure 3b). The relation of lengths of crystallographic axes and Al-content is illustrated in Figure 4. Within 2 wt. % Al_2O_3, the *b*-axial length presents a systematic decrease by more than 0.3%. In contrast, the *a*- and *c*-axial lengths remain roughly unchanged. When the OEn contains high Al_2O_3 content (~11.7 wt. %) as shown in [23], the *b*-axial length is dramatically shortened by 1%, while *a* and *c*-axes change only by ±0.1% (black symbols in Figure 4). Examination of the changes in unit cell parameters as a function of Al-content clearly shows that the reduction of cell volume is controlled by the shortening of *b*-axial length.

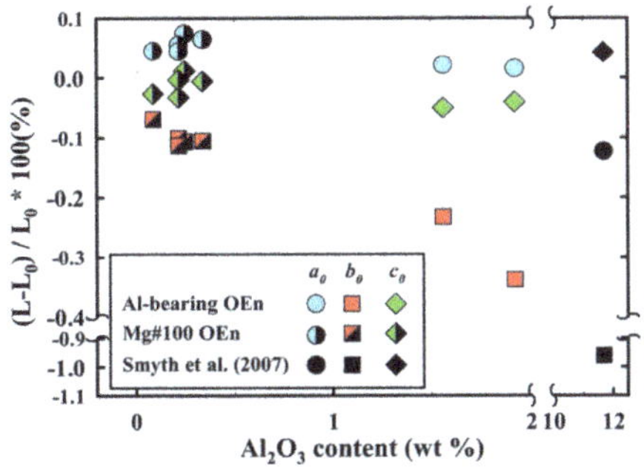

Figure 4. The relative difference of cell parameters of investigated hydrous OEn as a function of Al-content. The circle, diamond and square represent *a*, *b* and *c* axis, respectively. The half-filled colored symbols represent the composition of Mg#100 and full colored to be Al-bearing crystals. The black symbols are the data from [23]. The cell parameter of *a*, *b*, and *c* in this figure were compared to those of anhydrous OEn presented in [37]. The error bars are smaller than the size of symbols.

Pyroxene crystal structure contains two different polyhedra, SiO_4^{4-} tetrahedra, marked as T(A, B), and M(1,2) octahedra. Through the detailed structural analysis of Mg#100 and Al-bearing OEn (Supplementary Tables S2 and S3), we note that the trend of tetrahedral volume of B-chain (V_{TB}, diamond symbol) as function of Al-content presents a positive change (Figure 5a), whereas that of A-chain (V_{TA}, circle symbols) remains unchanged, including the results of high Al-content OEn from the study of [23] (black symbols). In the case of the octahedral volumes of M1 and M2 sites ($V_{M1,2}$), a decreasing trend presents as the Al-containing OEn is increasing in general, shown in Figure 5b. Furthermore, in the high Al-content OEn [23], the reduction of V_{M1} (the black circle in Figure 5b) is almost two times as large as that of V_{M2} (black diamonds in Figure 5b). The observed volume change as a function of Al-content can be explained by the analysis of crystal chemistry and the refined structural parameters obtained in this study and [23] (Supplementary Tables S2 and S3). When the Al-content is as low as <2 wt. % Al_2O_3, the larger Al^{3+} cation (ionic radius 0.53Å) tends to occupy the TA site rather than the TB site by replacing small Si^{4+} cation (ionic radius 0.40Å) (see item "AlB" in Supplementary Table S2). In contrast, when the Al-content is as high as ~11.7 wt. % of Al_2O_3 [23], the Al^{3+} cation occupies not only the tetrahedral site (TA) but also octahedral M1 site, replacing larger cation, like Mg^{2+} (ionic radius 0.72Å), resulting in M1 volume reduction by ~3.5%. On the other hand, the observation of minor volume reduction (~1%) in M1 and M2 sites for the low Al-content specimens in this study (<2.0 wt. %, blue filled symbols) is related to the introduction of vacancies in both sites, noted in Supplementary Table S2. The M2 volume reduction observed in [23] (black diamond in Figure 5b) is also caused by the vacancy incorporation.

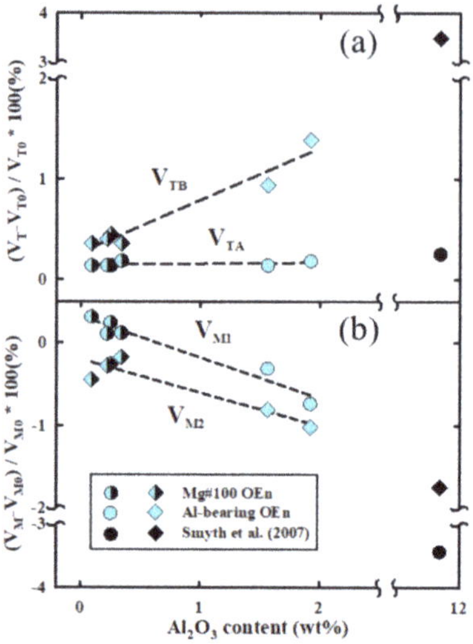

Figure 5. The relative difference of tetrahedral (TA and TB sites, (**a**)) and octahedral (M1 and M2 sites, (**b**)) volumes among investigated hydrous OEn as a function of Al-content. The half-filled colored symbols represent Mg#100 and full colored to be Al-bearing composition. The black symbols are the data from [23]. The polyhedral volumes in this figure were compared to those of anhydrous OEn presented in [37]. The error bars are smaller than the size of symbols.

Two high-pressure X ray diffraction experiments were performed on hydrous Mg#100 OEn (R598, 515 ppm of water) and hydrous Al-bearing OEn (R608, 1.56 wt. % of Al_2O_3 with 978 ppm of water) up to 15 GPa. The diffraction data were successfully indexed using the orthorhombic *Pbca* unit cell up to 13.2 and 10.0 GPa for Mg#100 OEn and Al-bearing OEn, respectively (Supplementary Table S4). The refinement details for Al-bearing OEn at high pressure are listed in Supplementary Table S5. Beyond the pressures mentioned above, the data became difficult to resolve for both runs due to a known phase transition at high pressure, from the Pbca phase to the $P2_1/c$ phase, which was confirmed in previous studies [36,38]. Our high-pressure diffraction data suggest that the transition pressure for the Mg#100 phase and the Al-bearing phase were above 13.0 and 10.0 GPa, respectively.

Figure 6 shows the volume compression behavior of hydrous Mg#100 and Al-bearing OEn from this study, compared with previous studies on anhydrous Mg#100 [37] and hydrous Mg#100 with 600 ppm H_2O [21]. All four datasets show similar compression behavior in orthoenstatite structure within *Pbca* stability field. This indicates that, below 1000 ppm, water/low Al-content in OEn has a small effect on the compression behavior of volume. As for the axial compressibility (Figure 7), all four OEn crystals with different crystal chemistry show the *b*-axis to be the most compressible, decreasing by about 3.5% at 12 GPa, while those *a*- and *c*-axes shorten between 2.0% and 2.5%, respectively. Taking a close look of the axial compression behavior (Figure 7), one can notice that the trends of b/b_0 and c/c_0 of the hydrous phases from this study (Mg#100 and Al-bearing) and [21] (Mg#100) follow those of anhydrous Mg#100 of [37] (empty symbols). The trends for the a-axis compressibility for the hydrous Al-bearing phase (with ~1000 ppm water) in this study and the hydrous Mg#100 (with ~620 ppm water) from [21] (red dashed line in Figure 7a) present a detectable difference from our hydrous Mg#100 (with ~520 ppm water) and the anhydrous Mg#100 from [37] (black dashed line in Figure 7a). The minor difference of compressibility in *a*-axis measured from our Al-bearing phase and Mg#100 in [21] does not affect the volume compressibility shown in Figure 6. At ambient conditions, the polyhedral volumes of the investigated Al-bearing OEn show a minor difference from those of anhydrous OEn from [37] (V_{TB}, M1 and M2; Supplementary Table S3b). Under high pressure, the compressibility of tetrahedral voids (TA and TB) and octahedral voids (M1 and M2) of Al-bearing phase are comparable to those of anhydrous phase within uncertainty (Figure 8a,b).

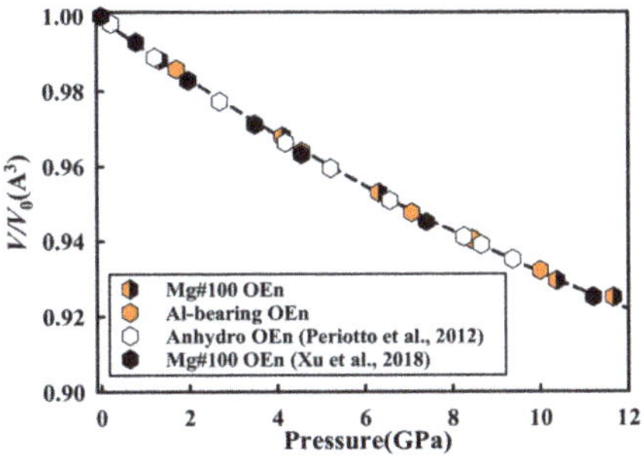

Figure 6. Pressure evolution of the normalized unit-cell volume of anhydrous and hydrous enstatite. The half-filled and filled orange symbols represent the hydrous #Mg100 and Al-bearing OEn from this work, respectively; the empty hexagonal symbols for data of anhydrous OEn from [37]; the filled hexagonal symbols for data of hydrous OEn from [21]. The dashed line is the third-order Birch-Murnaghan equation of state (EoS) fit from anhydrous OEn from [37]. The error bars of the data points are smaller than the size of symbols.

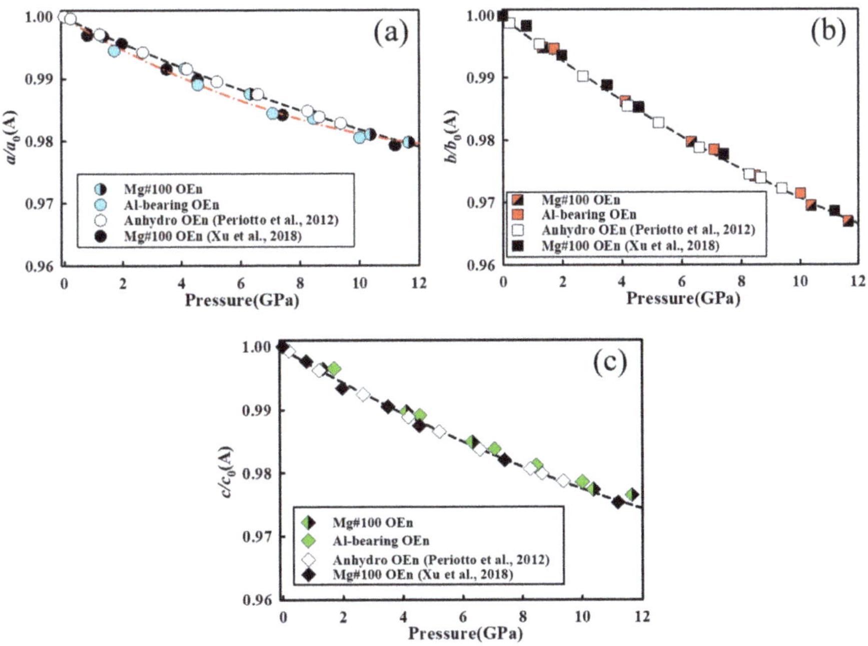

Figure 7. Pressure evolution of normalized cell parameters of anhydrous and hydrous enstatite for (**a**) a/a_0, (**b**) b/b_0 and (**c**) c/c_0. The half-filled and filled colored symbols represent the hydrous Mg#100 and Al-hydrous OEn from this work; the empty symbols for data of anhydrous OEn from [37]; the black symbols for data of hydrous OEn from [21]. The dashed black line is the fit of third-order Birch-Murnaghan equation of state (EoS) from [37] and the red dashed line in (**a**) represents the fit of Al-bearing OEn from this work and [21]. The error bars of the data points are smaller than the size of symbols.

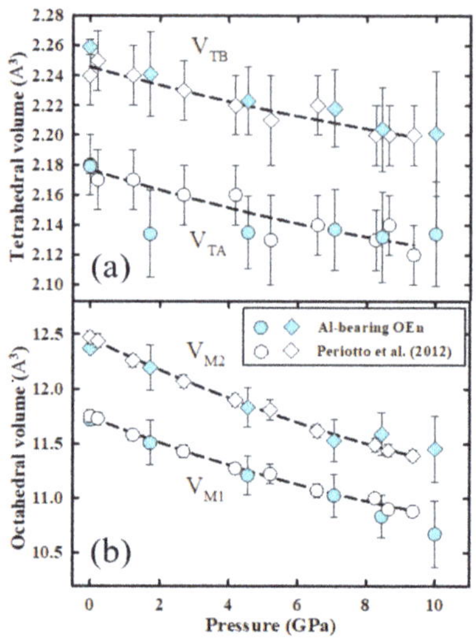

Figure 8. Evolution of tetrahedral (**a**) and octahedral (**b**) volumes as a function of pressure. The blue symbols represent the data of hydrous Al-bearing OEn from this work, and the empty symbols are the data from [37].

We investigated the synthetic OEn single crystals with less than 2 wt. % of Al_2O_3 incorporating water content from ~500 ppm to near 1000 ppm, corresponding to the upper limit of hydration at mid-upper mantle conditions (depth of 150–250 km). The results of single-crystal X ray diffraction indicate that the detailed crystal structure of hydrous Al-doped OEn shows only minor difference from the anhydrous Mg#100 OEn at ambient conditions, and that compressional behavior for this series studied OEn is insensitive to current crystal chemistry (Figures 6–8). The P-V results of this study suggest that the derived bulk modulus of hydrous low Al-bearing OEn is similar to that of anhydrous Mg#100 phase (Figure 6). The analysis of [39] indicated that the major elastic moduli, Cij (i = j, i = 1–6), of pyroxene are strongly correlated to the distortion of M1 and M2 octahedra. Our results on the compression of tetrahedra and octahedra do not show much difference from those of anhydrous Mg#100 phase. Therefore, we draw a conclusion that the elasticity of the studied OEn is insensitive to this level of Al- and water-content.

4. Geophysical Implications

In this study, the water content in our synthetic OEn phases was within the range of 500–1000 ppm, close to that observed in natural samples [33] and experimental values observed at the pressure of 5–7 GPa, temperature of 1100–1400 °C [22,24], corresponding to the depth of 200 ± 50 km. The OEn phase is a major mineral phase in addition to olivine, in continental petrological models. Here, we employ the current findings to explore the velocity profile of continental regions, where the LVZ is observed around 200–250 km depth [40]. For simplicity, we model by using only olivine and OEn phases in the weight ratios from 60:40 to 80:20, respectively, for the petrological compositions ranging from harzburgite to lherzolite. In the following illustration, the water content of the olivine phase is set to be at a maximum level of 1200 ppm, equilibrating with ~900 ppm H_2O in Al-bearing OEn phase using the partitioning model of [24] (D_{water} ~0.78) at conditions of 7.5 GPa, 1175 °C (Supplementary Table S6).

Among the upper mantle phases, olivine has been investigated most extensively, regarding how the elasticity is affected by a wide range of water content within the structure [41]. The elasticity of the hydrate olivine, with ~9000 ppm of water, are 0.6% and 0.4% for P and S waves, respectively, lower than those of anhydrous olivine. The water contents in our current investigated hydrous OEn phases is about the level of 1000 ppm, about one-ninth of samples in Ref. [41]. From the comparison of compressional behavior between anhydrous and hydrous OEn, we expect that the current water-content would not affect the elasticity too much. The compressibility study of hydrous diopside [42], with ~600 ppm of water, provide some evidence for that. Without available elasticity measured by optic-acoustic measurements (i.e., Brillouin scattering or ultrasonic measurement) for hydrous OEn phase, we use the reduction factor established from the hydrous olivine phase summarized by [41] to scale down the elastic velocities of hydrous OEn investigated in this study. The tested P-T conditions are set at shallow upper mantle (less than 250 km depth), and thus, a linear relation of V_S as function pressure and temperature is employed. Here, we only use the shear wave data to compare the S wave seismic models from [40],

$$V_S = V_{S(0)} + (dV_S)/dP \cdot \Delta P + (dV_S)/dT \cdot \Delta T \quad (4)$$

where $V_{S(0)}$ are S-wave velocity at ambient condition or high pressure-high temperature conditions, respectively, and dV_S/dP and dV_S/dT are the pressure and temperature derivatives of the velocity of olivine or enstatite, respectively. The thermoelastic parameters used in this study are presented in Table 3 [43–45]. Due to the measurements of the thermoelastic parameters for both hydrous olivine and OEn phases is incomplete, those parameters tabulated in Table 3 are from anhydrous phases. We tested the velocities of olivine and OEn under upper mantle conditions using the thermoelastic parameters in Table 3 along 100 Ma geotherm to compare with those calculated in [28]. The calculated S wave velocity profiles of olivine and OEn shared the same feature with those shown in Figure 4 of [28], displaying a velocity minimum around 150 km depth with Vs less 3% higher (the illustration is not shown in this study). This demonstrates that the simplified Equation (4) generates the same feature as reported in the previous study. We computed the pure olivine and OEn phases along the continental platform geotherm (solid black line in Figure 9, T_p = 1350 °C, the mantle potential temperature). As clearly shown in Figure 10, the shear velocities of hydrous-olivine and -OEn (the blue dashed lines next to the green and orange lines, respectively) are insensitive to the water content at the level of ~1000 ppm. The blue shed area is the resultant shear velocity profile of the continental petrological models (from harzburgite to lherzolite) calculated using a simple linear equation with different ratio in weight % of olivine and OEn. The calculated shear velocity profile for the continental area presents a broad low-velocity channel around the depth of 200 km, compared with the seismic models of Australia (yellow dashed line) and Canadian (red dashed line) areas where the LVZ presented below 200 km depth [40]. The difference of velocity minimum between our calculation and those seismic models is less than 1.7%, which can be reduced by introducing the low-velocity Ca-rich pyroxene into mineralogical assemblage models. The above modeling thermoelastic parameters are determined for end-member compositions, forsterite (Mg_2SiO_4) and enstatite ($MgSiO_3$). To consider the iron effect on elasticity of olivine and OEn phases, the S wave velocity may reduce by ~3% for the modeled mineral assemblies when the Mg number for both phases remain roughly the same at upper mantle condition, Mg# = 92 ± 2. Ref. [14] pointed out that the elastic velocities of Al-bearing (5 wt. %) OEn is approximately 3% faster than those of Mg#100. At the upper mantle, the Al-content in OEN is the function of depth. To consider the effect of Al-content on the S wave velocity in Figure 10, we expect the increase of the magnitude of velocity drop from the depth of 100 km to 200 km. However, we do not know the elasticity behavior of hydrous Al-bearing OEn, when high Al-content may contain higher water in the crystal structure at 'wet' mantle condition. Considering the ratio of OEn phase decreasing upon the pressure increasing at upper mantle condition, the blue shaded zone in Figure 10 also gives the velocity trend, shifting towards left-side of area. The influence of crystal chemistry of NAM phases most likely is to show on the magnitude of velocity drop, not the depth of the velocity minimum channel and the width of the channel. The current geotherm model of the continental platform in Figure 9 would not satisfy the velocity minimum channel observed below 200 km depth, shown in the seismic models [40]. We also computed the shear velocity with harzburgite composition model (60 wt. % olivine: 40 wt. % OEn) along an averaged cratonic geotherm derived from heat flow of

40 mW/m², suggested by [46], shown as black dashed line in Figure 9. The low-temperature geotherm results in a fast velocity drop (thick gray dashed line) and a deeper velocity minimum located at the depth of mantle adiabat, 250 km. By comparing different gradients of geotherms, we illustrate the depth of low-velocity channel strongly associated with the depth where the geotherm gradient turns from conductive mode (shallower than the empty red cross along the continental platform geotherm in Figure 9) to the condition of the mantle adiabat (at a depth of 250 km). Under the constraint of the seismic feature, the thermal gradient of the cratonic areas could be set between the continental platform and averaged cratonic regions in Figure 9, while considering the effect of thermal profile only.

Table 3. Elastic properties of olivine and orthoenstatite at high pressure and high temperature.

	Olivine		Orthoenstatite
	Anhydrous	Hydrous (0.9 wt. %)	Anhydrous
$V_{S(0)}$ (km/s)	5.03 [a]	5.01 [b]	4.88 [c]
$\delta V_S/\delta P$ (km/s/GPa)	0.032 [d]	0.029 [b]	0.035 [e]
$\delta V_S/\delta T$ (km/s/K)	−0.00037 [f]		−0.00041 [c]

a: [43], b: [41], c: [10], d: [44], e: [9], f: [45].

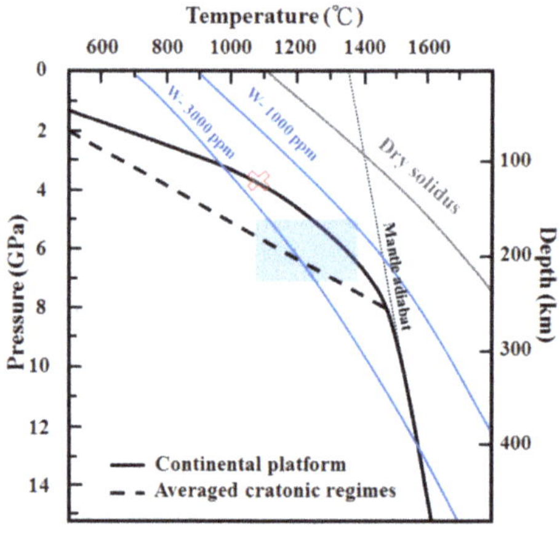

Figure 9. Solidus and thermal gradient as a function of depth discussed in this study. The continental geothermal gradient is from [32] and the averaged cratonic geotherm is from [46]. The dry and wet solidus are from [47,48], respectively. The geotherm above empty red cross represents the "conductive" geotherm and the mantle adiabat starts at a depth of 250 km at the upper mantle. The blue shaded area is the P-T conditions of synthetic orthoenstatite crystals in this study.

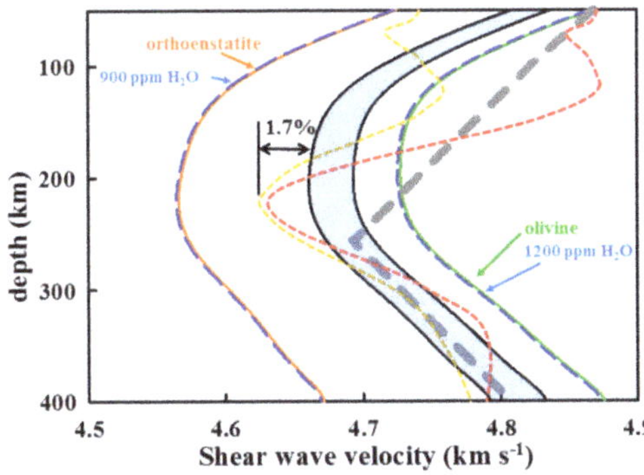

Figure 10. Shear wave velocity of olivine and OEn, and modeled continental petrological compositions computed along different geotherms as a function of depth in the continental area. The green and orange curves represent the velocity of anhydrous olivine and OEn phases. The long-dashed blue lines are the velocities of hydrous OEn and olivine. The blue shaded zone represents the shear velocity computed from different mineralogical assembly using different weight ratios of olivine and OEn. The thick grey dashed line represents the shear velocity computed from the low-temperature geotherm [46]. The short-dashed lines are reprinted from the shear velocity model of [40], including Australia craton (yellow) and Canadian craton (red). The details are described in the text.

The water content of the measured OEn in this study (~1000 ppm) would be the upper bound under the upper mantle conditions according to the observation of natural mantle samples [13,33]. We use the amount of ~1000 ppm water in the OEn phase as a constraint to model the water content in the olivine phase utilizing the partitioning coefficient at a pressure of 7.5 GPa, temperature of ~1200 °C determined by [24], resulting in the estimated bulk water content to be 1000~1200 ppm for the petrology models employed in this study. As shown in Figure 9, the continental platform geotherm (solid black line) and the mantle solidus with 1000 ppm H_2O (blue shaded area) could not intersect at the depth of 200 km or deeper. In this case, the origin of LVZ has little to do with partial melting at this depth, when the potential temperature was set around 1350 °C.

We demonstrated that the major influence of the depth and shape of LVZ in the continental area would be related to the character of the geothermal gradient, especially the depth segment from "conductive" mantle transforming to "adiabatic" mantle if the thermal effect is the major consideration. On the other hand, under such a cool continental geotherm, shown in Figure 9, if the cause of LVZ involves the melt, then the water content has to be higher than 1500 ppm at least in the mantle rock carried by the hydrous NAM minerals or free fluid. To investigate whether partial melting is involved in the origin of LVZ, the results of electrical conductivity measurements would need to provide additional constraints [49,50].

Supplementary Materials: The following are available online at http://www.mdpi.com/2075-163X/10/1/71/s1, Tables S1 to S6.

Author Contributions: Conceptualization, J.K.; methodology, J.K., P.D., F.T.-S.H.; validation, J.K., F.T.-S.H.; formal analysis, F.T.-S.H., P.D.; resources, J.K., P.D.; data curation, J.K., P.D., F.T.-S.H.; writing-original draft preparation, F.T.-S.H., J.K.; writing-review and editing, J.K., P.D., F.T.-S.H.; visualization, F.T.-S.H., J.K.; supervision, J.K.; project administration, J.K.; funding acquisition, J.K. All authors have read and agreed to the published version of the manuscript.

Funding: The research grants were funded by Ministry of Science and Technology, grant numbers MOST 106(107,108)-M-2116-010(080,015) (to J.K.). The visiting grants for F.H. was supported by the grants of MOST (to J.K.), Higher Education Sprout Project, Ministry of Education to the Headquarters of University Advancement at National Cheng Kung University(NCKU) and National Science Foundation grant EAR 1722969 (to P.D.). High-pressure facility at NCKU was partially supported by the grant of "Geochemical Precious Instrument Platform Project: Southern Branch Ministry of Science and Technology". X-ray diffraction work was conducted using the X-ray Atlas instrument at the University of Hawaii, funded by NSF grant EAR 1541516 (to P.D.).

Acknowledgments: The authors are grateful to Gary Huss and Kazuhide Nagashima for the assistance of SIMS measurement and Gregory Finkelstein for the starting of diffraction work. ChingChien Li is thanked for IR measurement and sample R545 synthesis.

Conflicts of Interest: The authors declare that none has any conflict of interest.

References

1. McDonough, W.F. Constraints on the composition of the continental lithospheric mantle. *Earth Planet. Sci. Lett.* **1990**, *101*, 1–18. [CrossRef]
2. Frisillo, A.L.; Barsch, G.R. Measurement of single-crystal elastic-constants of bronzite as a function of pressure and temperature. *J. Geophys. Res.* **1972**, *77*, 6360–6384. [CrossRef]
3. Webb, S.L.; Jackson, I. The pressure-dependence of the elastic-moduli of single-crystal ortho-pyroxene ($Mg_{0.8}Fe_{0.2}$) SiO_3. *Eur. J. Mineral.* **1993**, *5*, 1111–1119. [CrossRef]
4. Flesch, L.M.; Li, B.S.; Liebermann, R.C. Sound velocities of polycrystalline $MgSiO_3$-orthopyroxene to 10 GPa at room temperature. *Am. Mineral.* **1998**, *83*, 444–450. [CrossRef]
5. Angel, R.J.; Jackson, J.M. Elasticity and equation of state of orthoenstatite, $MgSiO_3$. *Am. Mineral.* **2002**, *87*, 558–561. [CrossRef]
6. Jackson, J.M.; Palko, J.W.; Andrault, D.; Sinogeikin, S.V.; Lakshtanov, D.L.; Wang, J.Y.; Bass, J.D.; Zha, C.S. Thermal expansion of natural orthoenstatite to 1473 K. *Eur. J. Mineral.* **2003**, *15*, 469–473. [CrossRef]
7. Jackson, J.M.; Sinogeikin, S.V.; Carpenter, M.A.; Bass, J.D. Novel phase transition in orthoenstatite. *Am. Mineral.* **2004**, *89*, 239–244. [CrossRef]
8. Jackson, J.M.; Sinogeikin, S.V.; Bass, J.D. Sound velocities and single-crystal elasticity of orthoenstatite to 1073 K at ambient pressure. *Phys. Earth Planet. Inter.* **2007**, *161*, 1–12. [CrossRef]
9. Kung, J.; Li, B.S.; Uchida, T.; Wang, Y.B.; Neuville, D.; Liebermann, R.C. In situ measurements of sound velocities and densities across the orthopyroxene → high-pressure clinopyroxene transition in $MgSiO_3$ at high pressure. *Phys. Earth Planet. Inter.* **2004**, *147*, 27–44. [CrossRef]
10. Kung, J.; Jackson, I.; Liebermann, R.C. High-temperature elasticity of polycrystalline orthoenstatite ($MgSiO_3$). *Am. Mineral.* **2011**, *96*, 577–585. [CrossRef]
11. Skogby, H. Water in natural mantle minerals I: Pyroxenes. In *Water in Nominally Anhydrous Minerals*; Book Series: Reviews in Mineralogy & Geochemistry; Mineralogical Soc America: Chantilly, VA, USA, 2006; Volume 62, pp. 155–167.
12. Beran, A.; Libowitzky, E. Water in natural mantle minerals II: Olivine, garnet and accessory minerals. In *Water in Nominally Anhydrous Minerals*; Book Series: Reviews in Mineralogy & Geochemistry; Mineralogical Society America: Chantilly, VA, USA, 2006; Volume 62, pp. 169–191.
13. Demouchy, S.; Bolfan-Casanova, N. Distribution and transport of hydrogen in the lithospheric mantle: A review. *Lithos* **2016**, *240*, 402–425. [CrossRef]
14. Chai, M.; Brown, J.M.; Slutsky, L.J. The elastic constants of an aluminous orthopyroxene to 12.5 GPa. *J. Geophys. Res. Solid Earth* **1997**, *102*, 14779–14785. [CrossRef]
15. Frost, D.J.; Dolejs, D. Experimental determination of the effect of H_2O on the 410-km seismic discontinuity. *Earth Planet. Sci. Lett.* **2007**, *256*, 182–195. [CrossRef]
16. Poe, B.T.; Romano, C.; Nestola, F.; Smyth, J.R. Electrical conductivity anisotropy of dry and hydrous olivine at 8 GPa. *Phys. Earth Planet. Inter.* **2010**, *181*, 103–111. [CrossRef]
17. Mao, Z.; Li, X.Y. Effect of hydration on the elasticity of mantle minerals and its geophysical implications. *Sci. China-Earth Sci.* **2016**, *59*, 873–888. [CrossRef]
18. Chang, Y.Y.; Hsieh, W.P.; Tan, E.; Chen, J.H. Hydration-reduced lattice thermal conductivity of olivine in Earth's upper mantle. *Proc. Natl. Acad. Sci. USA* **2017**, *114*, 4078–4081. [CrossRef]
19. Withers, A.C.; Hirschmann, M.M. H_2O storage capacity of $MgSiO_3$ clinoenstatite at 8–13 GPa, 1100–1400 degrees C. *Contrib. Mineral. Petrol.* **2007**, *154*, 663–674. [CrossRef]
20. Jacobsen, S.D.; Liu, Z.X.; Ballaran, T.B.; Littlefield, E.F.; Ehm, L.; Hemley, R.J. Effect of H_2O on upper mantle phase transitions in $MgSiO_3$: Is the depth of the seismic X-discontinuity an indicator of mantle water content? *Phys. Earth Planet. Inter.* **2010**, *183*, 234–244. [CrossRef]

21. Xu, J.G.; Zhang, D.Z.; Fan, D.W.; Zhang, J.S.; Hu, Y.; Guo, X.Z.; Dera, P.; Zhou, W.G. Phase transitions in orthoenstatite and subduction zone dynamics: Effects of water and transition metal ions. *J. Geophys. Res. Solid Earth* **2018**, *123*, 2723–2737. [CrossRef]
22. Mierdel, K.; Keppler, H.; Smyth, J.R.; Langenhorst, F. Water solubility in aluminous orthopyroxene and the origin of Earth's asthenosphere. *Science* **2007**, *315*, 364–368. [CrossRef]
23. Smyth, J.R.; Mierdel, K.; Keppler, H.; Langenhorst, F.; Dubrovinsky, L.; Nestola, F. Crystal chemistry of hydration in aluminous orthopyroxene. *Am. Mineral.* **2007**, *92*, 973–976. [CrossRef]
24. Ferot, A.; Bolfan-Casanova, N. Water storage capacity in olivine and pyroxene to 14 GPa: Implications for the water content of the earth's upper mantle and nature of seismic discontinuities. *Earth Planet. Sci. Lett.* **2012**, *349*, 218–230. [CrossRef]
25. Takeda, H. Tetrahedral sizes of orthopyroxenes and silicon-aluminum ordering. *Am. Mineral.* **1973**, *58*, 1096–1097.
26. Bell, D.R.; Rossman, G.R. Water in Earth's Mantle—The role of nominally anhydrous minerals. *Science* **1992**, *255*, 1391–1397. [CrossRef] [PubMed]
27. Hirschmann, M.M. Water, melting, and the deep Earth H_2O cycle. *Ann. Rev. Earth Planet. Sci.* **2006**, *34*, 629–653. [CrossRef]
28. Stixrude, L.; Lithgow-Bertelloni, C. Mineralogy and elasticity of the oceanic upper mantle: Origin of the low-velocity zone. *J. Geophys. Res. Solid Earth* **2005**, *110*. [CrossRef]
29. Kawakatsu, H.; Kumar, P.; Takei, Y.; Shinohara, M.; Kanazawa, T.; Araki, E.; Suyehiro, K. Seismic evidence for sharp lithosphere-asthenosphere boundaries of oceanic plates. *Science* **2009**, *324*, 499–502. [CrossRef]
30. Karato, S.I. On the origin of the asthenosphere. *Earth Planet. Sci. Lett.* **2012**, *321*, 95–103. [CrossRef]
31. Green, D.H.; Hibberson, W.O.; Kovacs, I.; Rosenthal, A. Water and its influence on the lithosphere-asthenosphere boundary. *Nature* **2010**, *467*, 448–497. [CrossRef]
32. Condie, K.C. The Mantle. In *Earth as an Evolving Planetary System*, 3rd ed.; Condie, K.C., Ed.; Elsevier Ltd.: Amsterdam, The Netherlands, 2016; Volume 3, pp. 89–133.
33. Ingrin, J.; Skogby, H. Hydrogen in nominally anhydrous upper-mantle minerals: Concentration levels and implications. *Eur. J. Mineral.* **2000**, *12*, 543–570. [CrossRef]
34. Peslier, A.H.; Luhr, J.F.; Post, J. Low water contents in pyroxenes from spinel-peridotites of the oxidized, sub-arc mantle wedge. *Earth Planet. Sci. Lett.* **2002**, *201*, 69–86. [CrossRef]
35. Shimizu, K.; Ushikubo, T.; Hamada, M.; Itoh, S.; Higashi, Y.; Takahashi, E.; Ito, M. H_2O, CO_2, F, S, Cl, and P_2O_5 analyses of silicate glasses using SIMS: Report of volatile standard glasses. *Geochem. J.* **2017**, *51*, 299–313. [CrossRef]
36. Finkelstein, G.J.; Dera, P.K.; Duffy, T.S. Phase transitions in orthopyroxene (En_{90}) to 49 GPa from single-crystal X-ray diffraction. *Phys. Earth Planet. Inter.* **2015**, *244*, 78–86. [CrossRef]
37. Periotto, B.; Balic-Zunic, T.; Nestola, F.; Katerinopoulou, A.; Angel, R.J. Re-investigation of the crystal structure of enstatite under high-pressure conditions. *Am. Mineral.* **2012**, *97*, 1741–1748. [CrossRef]
38. Dera, P.; Finkelstein, G.J.; Duffy, T.S.; Downs, R.T.; Meng, Y.; Prakapenka, V.; Tkachev, S. Metastable high-pressure transformations of orthoferrosilite Fs_{82}. *J. Geophys. Res. Solid Earth* **2013**, *221*, 15–21. [CrossRef]
39. Duffy, T.S.; Vaughan, M.T. Elasticity of enstatite and its relationship to crystal-structure. *J. Geophys. Res. Solid Earth Planets* **1988**, *93*, 383–391. [CrossRef]
40. Romanowicz, B. The thickness of tectonic plates. *Science* **2009**, *324*, 474–476. [CrossRef]
41. Mao, Z.; Jacobsen, S.D.; Jiang, F.; Smyth, J.R.; Holl, C.M.; Frost, D.J.; Duffy, T.S. Velocity crossover between hydrous and anhydrous forsterite at high pressures. *Earth Planet. Sci. Lett.* **2010**, *293*, 250–258. [CrossRef]
42. Gavrilenko, P.; Ballaran, T.B.; Keppler, H. The effect of Al and water on the compressibility of diopside. *Am. Mineral.* **2010**, *95*, 608–616. [CrossRef]
43. Suzuki, I.; Anderson, O.L.; Sumino, Y. Elastic properties of a single-crystal forsterite Mg_2SiO_4, up to 1,200 K. *Phys. Chem. Miner.* **1983**, *10*, 38–46. [CrossRef]
44. Darling, K.L.; Gwanmesia, G.D.; Kung, J.; Li, B.S.; Liebermann, R.C. Ultrasonic measurements of the sound velocities in polycrystalline San Carlos olivine in multi-anvil, high-pressure apparatus. *Phys. Earth Planet. Inter.* **2004**, *143*, 19–31. [CrossRef]
45. Isaak, D.G. high-temperature elasticity of iron-bearing olivines. *J. Geophys. Res. Solid Earth* **1992**, *97*, 1871–1885. [CrossRef]

46. Cammarano, F.; Romanowicz, B. Insights into the nature of the transition zone from physically constrained inversion of long-period seismic data. *Proc. Natl. Acad. Sci. USA* **2007**, *104*, 9139–9144. [CrossRef] [PubMed]
47. Katz, R.F.; Spiegelman, M.; Langmuir, C.H. A new parameterization of hydrous mantle melting. *Geochem. Geophys. Geosys.* **2003**, *4*, 1073. [CrossRef]
48. Green, D.H. Experimental petrology of peridotites, including effects of water and carbon on melting in the Earth's upper mantle. *Phys. Chem. Miner.* **2015**, *42*, 95–122. [CrossRef]
49. Dai, L.D.; Karato, S.I. Electrical conductivity of orthopyroxene: Implications for the water content of the asthenosphere. *Proc. Jpn. Acad. Series B-Phys. Biol. Sci.* **2009**, *85*, 466–475. [CrossRef]
50. Jones, A.G.; Fullea, J.; Evans, R.L.; Muller, M.R. Water in cratonic lithosphere: Calibrating laboratory-determined models of electrical conductivity of mantle minerals using geophysical and petrological observations. *Geochem. Geophys. Geosys.* **2012**, *13*. [CrossRef]

minerals

MDPI

Article

Riesite, a New High Pressure Polymorph of TiO_2 from the Ries Impact Structure

Oliver Tschauner [1,*], Chi Ma [2], Antonio Lanzirotti [3] and Matthew G. Newville [3]

1 Department of Geoscience, University of Nevada Las Vegas, Las Vegas, NV 89154, USA
2 Division of Geology and Planetary Sciences, California Institute of Technology, CA 91125, USA; chi@gps.caltech.edu
3 GeoSoilEnviroCARS, University of Chicago at the Advanced Photon Source, Argonne National Laboratory, Argonne, IL 60367, USA; lanzirotti@uchicago.edu (A.L.); Newville@uchicago.edu (M.G.N.)
* Correspondence: oliver.tschauner@unlv.edu

Received: 14 December 2019; Accepted: 16 January 2020; Published: 18 January 2020

Abstract: This paper describes riesite, a new high-pressure polymorph of TiO_2 from the Ries impact structure, Germany. Riesite occurs in shock-induced melt veins within xenoliths of bedrock in suevite. It is structurally closely related to srilankite from which it differs by having two distinct cation sites rather than one and through its monoclinic symmetry. It is indicative that riesite forms only upon release from the shock state upon back transformation from akaogiite.

Keywords: high-pressure mineral; high pressure; mineral physics; impacts

1. Introduction

Rutile is a common accessory mineral in many rocks of the crust. At low pressure, its stability field is constrained by transformation into anatase at high temperature and to brookite at intermediate pressures and temperatures. At high pressure, rutile undergoes a sequence of transitions to phases isotypic with scrutinyite (α-PbO_2) [1–3], baddeleyite [4,5], and cottunite ($PbCl_2O$) [6,7]. Baddeleyite-type TiO_2 is an approved mineral with the name akaogiite [8]. The boundaries between these polymorphs depend on temperature but also on composition; the solubility of Zr in rutile increases with increasing temperature [9], but the boundary between rutile and scrutinyite-type TiO_2 is shifted to lower pressures with increasing Zr-content [10]. Nb and Ta exert a similar effect, but the structural relations between rutile- and post-rutile phases are more complex than for Zr because the incorporation of Nb and Ta in rutile requires coupled substitution, which results in cation ordering under most conditions. Ixiolite is scrutinyite-type—Fe-Ti-Nb-Ta-oxide [11]. Naturally occurring scrutinyite-type TiO_2 is also an approved mineral with the name srilankite [12]. Most srilankites in terrestrial rocks contain about 40–60 mol % ZrO_2 [12–14]. Endmember or near to endmember srilankite has been reported from impact sites [15,16] and from ultrahigh pressure metamorphic rocks [17]. Unfortunately, srilankite from those occurrences has been labeled as "TiO_2-II" as in the earlier experimental studies on high-pressure polymorphs of TiO_2 [1,4,6], while it should properly be labeled as srilankite, a mineral known since 1983 [12]. The fact that srilankite was originally reported with chemical formula $ZrTi_2O_6$ but with scrutinyite-structure corresponding to a structure formula $(Ti,Zr)O_2$ [12] has certainly contributed to the confusion. Natural occurrences, as well as the majority of experimental studies, indicate a sequence of pressure-induced transformations: rutile → srilankite → akaogiite with transition pressures 10–15 and 20–25 GPa at 300 K, respectively (2,3,7). This range of phase transitions and the abundance of TiO_2 as accessory phases in the continental crust makes high-pressure polymorphs of TiO_2 good pressure markers for impact events [14,15,18]. This includes studies of the evolution of particular impact sites [14,15] as well as the assessment of the frequency of impacts in the ancient geologic past because rutile is a common detrital mineral and its high-pressure polymorphs may be conserved in detrital grains as well, similar to reidite that has been found in detritral zircon grains [19].

Here we describe structure, composition, and occurrence of a new high-pressure polymorph of TiO_2, riesite. Riesite has been approved as a new mineral species with number IMA-2015-110 [20].

2. Materials and Methods

Riesite was found in a thin section (ZLN114c) from a xenolith of garnet-sillimanite restite with shock-melt veins that was trapped in suevite. This xenolith and the mineralogy of the shock-melt vein have been described in an earlier study [21], to impact-related melt veins which carry high-pressure minerals whose formation requires several GPa pressure or more. This is in distinction to impact-related pseudotachylites which do not contain such high-pressure minerals. The high-pressure polymorphs of common accessories like rutile and zircon permit distinction between both types of shock-induced melt veins. In the current case, the melt vein contains majorite-rich garnet [21], a jadeite-rich clinopyroxene, and accessory akaogiite, and reidite (the high-pressure polymorph of zircon). Akaogiite and reidite had previously been reported from the Ries [8,22,23]. Entrapped garnet clasts contain stishovite, which probably formed through direct solid state transformation of quartz inclusions [21]. Since an extensive discussion of the petrography and mineralogy of the xenolith and on the composition of the shock melt vein has been presented by Stähle et al. [21] we will not recapitulate these findings here and instead will focus on the observation and characterization of riesite. Within the shock melt vein, we observed clasts of TiO_2 and $FeTiO_3$. These clasts are opaque in the optical microscope but yield a high back scatter electron ('BSE') intensity region in the scanning electron microscope ('SEM') (Figure 1a,b). At higher magnification, we find that the titanium dioxide clasts are composed of grains of sub-micrometer in diameter (Figure 1c).

Electron backscatter diffraction ('EBSD') reveals that these grains are not from a phase with rutile structure, although the composition is close to pure TiO_2 (Table 1). In one of the cases examined, a fine grained dense aggregate of this phase encloses a highly deformed rutile kernel (see below). Quantitative elemental microanalyses of these clasts were carried out at Caltech using a JEOL 8200 electron microprobe operated at 15 kV and 20 nA in focused beam mode. Analyses were processed with the CITZAF correction procedure [24]. Ten point analyses of type riesite were averaged and the results are given in Table 1.

Table 1. Analytical data for riesite. Average of ten point analyses. The empirical formula (based on 2 O atoms pfu) is $(Ti_{0.997}Fe_{0.005})O_2$.

Constituent	Wt %	Range	SD	Probe Standard
TiO_2	99.25	98.98–99.62	0.19	TiO_2
FeO	0.42	0.33–0.53	0.07	Fayalite
CaO	0.03	0.02–0.04	0.01	Anorthite
Total	99.70	-	-	-

We collected X-ray microdiffraction data at the undulator beamline 13-IDE (GSECARS, APS, Argonne National Laboratory) using a primary beam of wavelength at 0.6199 Å, monochromatized by a two-crystal Si monochromator. The X-ray beam was focused to 2×3 μm^2 by vertical and horizontal Kirkpatrick-Baez mirrors of 200 mm focal length. A MAR165 CCD area detector was used for collecting diffraction data in forward scattering geometry. The thin section that contains riesite was initially examined by X-ray fluorescence mapping with same spatial resolution of 2×3 μm^2 in order to identify Ti-rich clasts in the shock-melt vein of section ZLN114c. Clasts with rutile composition were then examined by X-ray diffraction mapping in 2 μm steps horizontally and vertically through the focused X-ray beam over the selected region of the clasts and diffraction patterns recorded at each step. The diffraction pattern images were corrected for diffuse scattering from the glass slide of the thin section using background image subtraction in Fit2D [25]. Then the patterns were corrected for geometric distortion from detector tilt using the GSE-ADA analysis software [26] and integrated using Fit2D [25]. Riesite was observed together with ilmenite, rutile, and akaogiite in several transformed or partially transformed rutile clasts. Initially we interpreted riesite as Ti-srilankite, but soon we found that deviations of observed peak intensities and their 2ϑ angles from the modelled pattern of TiO_2-srilankite reflect an actual reduction in crystal structure symmetry (Figure 2a,b). The possible sub-group related structures were examined. The more promising model structures were used for LeBail extraction of structure factor moduli and examined by simulated annealing [27,28] without symmetry bias (in space group *P1*). After convergence, automatic symmetry search resulted consistently in a monoclinic distortion of the srilankite structure into subgroup *P2/c* in setting *P1 2/c 1*.

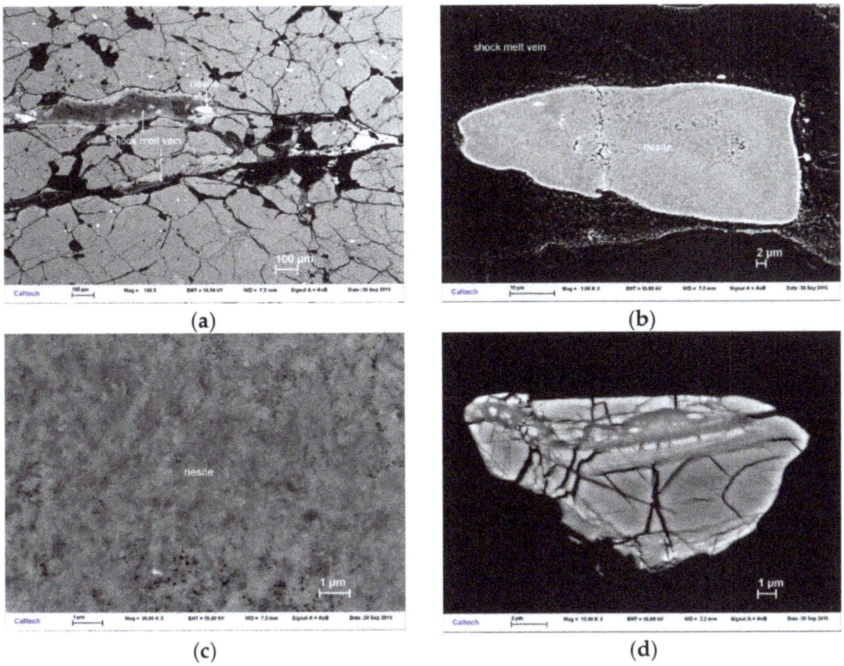

Figure 1. Field Emission secondary electron microscope (SEM) back scatter electron (BSE) images of riesite type material at different magnification: (**a**) Overview image showing the location of a riesite clast within the shock melt vein which cuts through the ZNL114c thin section. (**b**) The same clast of type riesite at higher magnification. Riesite is pseudomorph after rutile, replacing a former single crystal through a fine-grained aggregate. The enhanced brightness at the border between the riesite aggregate and the shock melt vein is result of electrostatic charging. (**c**) Area of the same clast of riesite, shown at high magnification. A bright crystallite of submicrometer dimensions is Zr-rich. Type riesite contains no significant Zr. (**d**) Reidite lamellae in a zircon clast trapped in the shock melt vein of ZLN114c. Reidite was identified by electron backscatter diffraction (EBSD).

For powder data in many cases (and in this particular case), indexation cannot be well conducted independent from structure modeling. Thus, we did not rely on forward modeling of monoclinic cells but used the full set of integrated intensities of the new TiO_2-phase for reversed Monte Carlo (rMC) modeling (local optimization) without symmetry bias (that is: in space group *P1*). rMC converged to an F-based refinement factor R_F of less than 7% and subsequent automatic symmetry search gave consistently the riesite cell and space group (*P2/c*, SG number 13) for different sets of intensities obtained through LeBail extractions from diffraction patterns from different locations and paragenesis: riesite and rutile (Figure 2a,b), as well as patterns with riesite, ilmenite, and akaogiite (Figure 2c). Thus, the choice of space group and cell dimensions is robust and is based on a large statistical set of data obtained from several patterns of riesite-bearing sample material. We point out that the srilankite-model gives an R_F worse by factor of two than the monoclinic structure. EBSD patterns match better to the monoclinic structure with a mean angular deviation of 0.38° than to the orthorhombic structure with a mean angular deviation of 0.47° (Figure 2d). The main difference between both structures is in the site symmetry and fractional atomic coordinates of Ti. Since EBSD does not directly probe these parameters, it is not very sensitive to their differences.

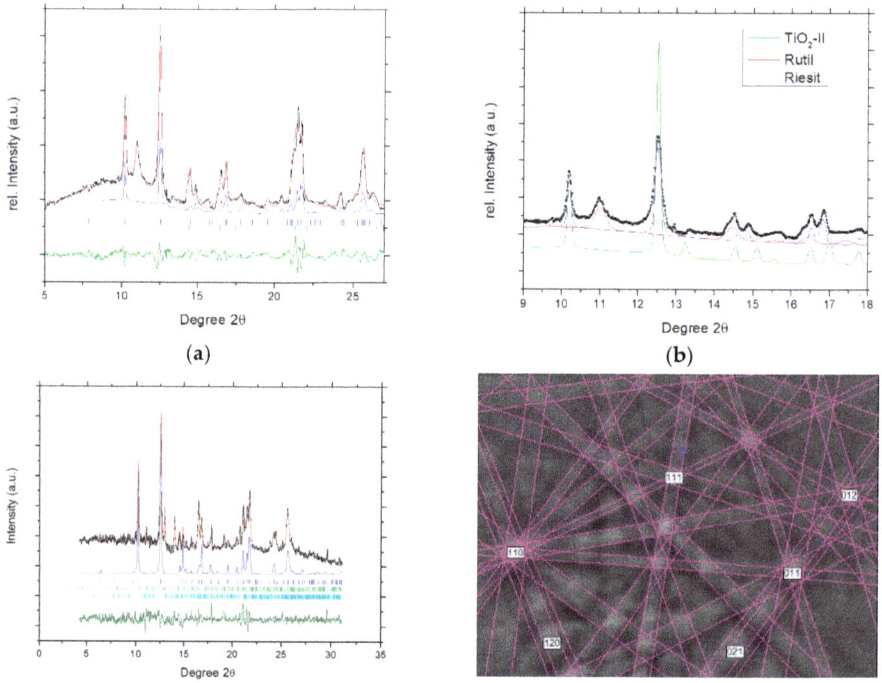

Figure 2. X-ray diffraction and EBSD data of riesite: (**a**) Representative X-ray diffraction pattern of type riesite. Black crosses = observed data, red line: Rietveld refined pattern of riesite plus rutile. The pattern represents ~70 mass% riesite and 30 mass% rutile. Blue line: Contribution of riesite to the Rietveld-refined pattern. (**b**) The same pattern between 9 and 19° 2ϑ with the calculated pattern of Ti-endmember srilankite added (green line). The same profile parameters and scale factor as for riesite were used. Ti-Srilankite clearly mismatches the observed pattern. (**c**) Rietveld refinement of diffraction pattern from a different clast of riesite, coexisting with ilmenite and akaogiite. (**d**) Observed and fitted EBSD pattern of type riesite. The mean angular deviation of the fit was 0.38°.

3. Results

We used this monoclinic distorted structure for structure refinement with the Rietveld method using Powdercell [29] and GSAS [30]. Pseudovoigt peak profiles were used with Gaussian terms $U = 728$, $V = 8.1$, $W = 7.3$ and Lorentzian terms $Lx = 6.3$ and $Ly = 8.1$ (in GSAS). Rietveld Refinement converged to a weighted refinement factor R_{wp} of 6.8% (Figure 2a), whereas the orthorhombic Ti-srilankite structure converges to not better than 11.1% (Figure 2b). As a further refinement measure we obtained a profile-based refinement factor R_p of 5.1% and $\chi^2 = 11.2$ for 1534 observations. R_{F^2} was 9.6% and R_F was found to be 7.8%. LeBail refinement of riesite converged to an R_p of 5.0%. The pattern used for final structure refinement of riesite (Figure 2a) contained 71 mass% riesite and 29 mass% rutile.

An additional refinement of riesite from an aggregate with ilmenite and remnant rutile (Figure 2c) gave an R_p of 5.7% and a R_{wp} of 7.8%.

4. Discussion

We place the new mineral riesite into its context with other high-pressure polymorphs of rutile. Riesite assumes space group *P2/c*, which is a direct subgroup of the space group of srilankite, *Pnma*. However, in riesite the cations reside on different Wykhoff sites than in srilankite. A direct group-subgroup relation maps the cations from srilankite onto Wyckoff-sites 2e and 2f in the monoclinic structure. However, our structure analysis converged persistently to a structure with the cations on site 4g rather than 2e and 2f. In other words, the present phase is not a product of a simple distortive lattice relaxation of Ti-rich srilankite into a monoclinic structure but reflects a significant cation sublattice shift relative to the direct subgroup structure of srilankite of between 5% and 22% (Ti1 coordinate y/b):

$$\text{Ti2: } \Delta(x|y|z) = 0.022|0.018|0.041$$

$$\text{Ti1: } \Delta(x|y|z) = 0.038|0.040|0.015$$

Thus, despite the small deviation of the beta angle from 90 degrees, the change in axes length generates overall splitting of powder diffraction lines and the difference in cation lattice causes changes in the observed intensities that make the powder patterns of riesite distinct from srilankite. Figure 2b shows clearly that the observed pattern cannot be matched by the orthorhombic srilankite cell and structure.

This fundamental difference between riesite and srilankite brings us to the formation mechanism of riesite. Riesite assumes a space group that is a direct subgroup of the space group of srilankite. At first glance this suggests riesite to form upon a distortive transition that is guided by this group-subgroup relation as one of the Landau-criteria. As we already showed above, this is not the case: The continuous transition from direct group-subgroup transformation *Pnma* → *P2/c* with index 2 gives the following mapping of Wyckoff sites:

$$\text{Ti: } 4c \rightarrow 2e + 2f$$

$$\text{O: } 8d \rightarrow 4g + 4g$$

However, in riesite the cations clearly reside on sites 4g with half occupancy (Table 2). Internal cation shift between riesite and monoclinic distorted srilankite is marked with changes of atomic position between 5% (mostly) and 22% (Ti1 coordinate y/b). In other words, there is no direct group-subgroup relation that maps the cation sublattice of srilankite onto that of riesite.

Table 2. Atom coordinates of riesite. The structure is monoclinic and assumes space group number. 13, *P12/c1* with unit cell dimensions a = 4.519(3) Å, b = 5.503(8) Å, c = 4.888(2) Å, β = 90.59(8)°, V = 121.5(1) Å^3. Site fractional occupancies (SFO) and isothermal displacement factors B_{iso} are given. For comparison, the unit cell of endmember srilankite (TiO_2-II) is 4.5318(7), 5.5019(7), 4.9063(6) Å, respectively, with all angles 90°, and the fractional atomic coordinates are 0, 0.1704(3), $\frac{1}{4}$ (Ti) and 0.2716(6), 0.3814(7), 0.4142(7) (O) [31].

Atom	Wyckoff	*x/a*	*y/b*	*z/c*	SFO	B_{iso} (Å^2)
Ti1	4g	0.041(2)	0.142(4)	0.268(2)	0.47(3)	0.6(1)
Ti2	4g	0.51(1)	0.311(7)	0.78(1)	0.53(1)	0.7(1)
O1	4g	0.28(1)	0.36(1)	0.436(7)	1 [1]	0.9(2)
O2	4g	0.224(1)	0.086(6)	0.937(7)	1 [1]	0.9(2)

[1] Fixed based on chemical analysis.

However, space group and Wyckoff site symmetries of the riesite structure can be obtained, through relation to akaogiite.

$$P1\ 2_1/c\ 1(\text{akaogiite}) \rightarrow P1\ c\ 1(\text{transient}) \rightarrow P1\ 2/c\ 1(\text{riesite})$$

which maps all cations and anions onto sites 4g with partial occupancy. In riesite, there is no indication for partially occupied O-sites (as obtained through this mapping), but the cation sites are clearly partially occupied and of site symmetry 4g. We emphasize that riesite and akaogiite are distinct structures. Riesite is not a variety of akaogiite. The group-subgroup-chain $P1\ 2_1/c\ 1 \rightarrow P1\ c\ 1 \rightarrow P1\ 2/c\ 1$ involves a direct group-subgroup mapping from $P2_1/c$ onto Pc. However, the mapping of Pc into $P2/c$ is klassengleich not translationsgleich, whence it involves a sublattice-shift. Keeping the atomic fractional coordinates of akaogiite, this group-subgroup chain results in a fluorite-like arrangement, whereof in riesite half of the Ti-sublattice is shifted along [110] (in compliance with the klassengleiche mapping from Pc into $P2/c$).

Akaogiite is a polymorph of TiO_2 that occurs at higher pressure than srilankite. According to our group-theoretical analysis above, riesite assumes a structural state in between srilankite and akaogiite. One may expect riesite to form in a pressure-temperature regime intermediate between srilankite and akaogiite but it has not been reported from experiments. The fact that the riesite structure involves low-symmetric partially occupied cation sites is indicative for formation through a different process than static compression. The sublattice disorder and the structural relation between akaogiite and riesite suggest that riesite is the product of retrograde transformation of akaogiite upon rapid decompression and cooling. In this case, riesite does not have a thermodynamic stability field but forms at elevated temperature under similar pressures as srilankite in dependence of the path (rapid decompression of akaogiite at elevated temperature). In fact, the riesite type material has been found in exactly this environment: a shock melt vein in an impact-metamorphosed bedrock fragment. The density of riesite is 4.37 ± 0.11 g/cm^3 (Table 2), which is in between rutile (4.25 g/cm^3) and srilankite (4.38 g/cm^3 [12]), but below that of akaogiite at reference conditions (4.72 g/cm^3 [8]).

In general, the occurrence of riesite along with akaogiite is expected to be common in impact-metamorphism of rutile-bearing rock. In these environments, riesite indicates that peak pressures have been in the stability field of akaogiite above 20 to 25 GPa. Occurrence of Ti-endmember srilankite instead of riesite indicates that the peak pressures were below 20 GPa.

In the Ries xenolith ZLN114, riesite occurs as complete or partial replacement of rutile clasts trapped within a shock melt vein composed of majoritic garnet and jadeite. In partially transformed clasts, we find akaogiite, riesite, and very fine-grained rutile at the outer rim of the clasts, akaogiite further inside, and residual, highly strained rutile crystallites in the kernel of larger clasts. Reidite (Figure 1d) and stishovite are other high-pressure minerals identified in clasts adjacent to the type material of the new mineral.

In an earlier study of similar shock-melt veins in amphibolite xenoliths from Ries suevite, Stähle et al. [23] estimated peak shock pressure up to 17–20 GPa based on garnet barometry. However, the present paper is the first assessment of shock pressures for a xenolith of garnet-cordierite-sillimanite restite. The observation of riesite and akaogiite is consistent with peak pressure 20–25 GPa. The presence of reidite (Figure 1d) is also consistent with pressures above 12 GPa, if temperatures were above 1200 K [32]. Lower temperature shifts the transition from zircon to reidite to higher pressure [32]. We note that the shock-driven direct conversion along the Hugoniot of zircon and on experimental time scales of less than one ms occurs above 60 GPa [33]. Observation of liebermannite at the rim of the shock melt vein and majoritic garnet in the vein indicates that peak shock pressure was not above 20–22 GPa [34]. We note that this pressure range is markedly below the pressure in the isobaric core of the Ries impact event. Xenoliths like ZLN114 that were trapped in suevite record different peak-shock pressures and release paths depending on their original position relative to the isobaric core and their entrapment in suevite [35].

Author Contributions: O.T. collected X-ray diffraction and –fluorescence data, analyzed crystal structures, and contributed to the interpretation of data and writing of the manuscript, C.M. conducted chemical analysis, EBSD analysis, collected FE-SEM images, and contributed to the interpretation of data and writing of the manuscript, A.L. and M.G.N. set up experiment for X-ray diffraction and –fluorescence analysis, and contributed to the interpretation of data and writing of the manuscript. All authors have read and agree to the published version of the manuscript.

Funding: This work was supported by DOE Award DESC0005278, NSF EAR-1128799, DE-FG02-94ER14466, NSF Grants EAR-0318518, and DMR-0080065. Portions of this work were performed at GeoSoilEnviroCARS (The University of Chicago, Sector 13), Advanced Photon Source (APS), Argonne National Laboratory. GeoSoilEnviroCARS is supported by the National Science Foundation – Earth Sciences (EAR – 1634415) and Department of Energy- GeoSciences (DE-FG02-94ER14466). This research used resources of the Advanced Photon Source, a U.S. Department of Energy (DOE) Office of Science User Facility operated for the DOE Office of Science by Argonne National Laboratory under Contract No. DE-AC02-06CH11357.

Acknowledgments: We would like to thank Robert C. Liebermann for his suggestion to submit this paper to the Special Issue of *Minerals* in memory of Orson Anderson. We also thank V. Stähle and D. Stöffler for providing the thinsection ZLN114c which contains type riesite and we thank the Institut für Geowissenschaften, University of Heidelberg and in particular M. Trieloff for permitting us to examine this section, which is part of their collection. The publication fees for this article were supported by the UNLV University Libraries Open Article Fund.

Conflicts of Interest: The authors declare no conflict of interest. The funders had no role in the design of the study; in the collection, analyses, or interpretation of data; in the writing of the manuscript, or in the decision to publish the results.

References

1. Jamieson, J.C.; Olinger, B. High-pressure polymorphism of titanium dioxide. *Science* **1968**, *151*, 893–895. [CrossRef] [PubMed]
2. Withers, A.C.; Essene, E.J.; Jhang, Y. Rutile/TiO_2II phase equilibria. *Contrib. Mineral. Petrol.* **2003**, *145*, 199–204. [CrossRef]
3. Kojitani, H.; Yamazaki, M.; Kojima, M.; Inaguma, Y.; Mori, D.; Akaogi, M. Thermodynamic investigation of the phase equilibrium boundary between TiO_2 rutile and its alpha-PbO_2-type high-pressure polymorph. *Phys. Chem. Min.* **2018**, *45*, 963–980. [CrossRef]
4. Mammone, J.F.; Nicol, M.; Sharma, S.K. Raman-spectra of TiO_2-II, TiO_2-III, SnO_2, and GeO_2 at high-pressure. *J. Phys. Chem. Sol.* **1981**, *42*, 379–384. [CrossRef]
5. Sato, H.; Endo, S.; Sugiyama, M.; Kikegawa, T.; Shimomura, O.; Kusaba, K. Baddeleyite-type high-pressure phase of TiO_2. *Science* **1991**, *251*, 786–788. [CrossRef] [PubMed]
6. Dubrovinskaia, N.A.; Dubrovinsky, L.S.; Ahuja, R.; Prokopenko, V.B.; Dmitriev, V.; Weber, H.P.; Johansson, B. Experimental and theoretical identification of a new high-pressure TiO_2 polymorph. *Phys. Rev. Lett.* **2001**, *87*, 275501. [CrossRef]
7. Nishio-Hamane, D.; Shimizu, A.; Nakahira, R.; Niwa, K.; Sano-Furukawa, A.; Okada, T.; Kikegawa, T. The stability and equation of state for the cotunnite phase of TiO_2 up to 70 GPa. *Phys. Chem. Min.* **2010**, *37*, 129–136. [CrossRef]
8. El Goresy, A.; Dubrovinsky, L.; Gillet, P.; Graup, G.; Chen, M. Akaogiite: An ultra-dense polymorph of TiO_2 with the baddeleyite-type structure, in shocked garnet gneiss from the Ries Crater, Germany. *Am. Min.* **2010**, *95*, 892–895. [CrossRef]
9. Ferry, J.M.; Watson, E.B. New thermodynamic models and revised calibrations for the Ti-in-zircon and Zr-in-rutile thermometers. *Contr. Min. Petr.* **2007**, *154*, 429–437. [CrossRef]
10. Akaogi, M.; Horiuchi, N.; Ishii, T.; Kojitani, H. High-pressure phase relations in the system TiO_2-ZrO_2 to 12 GPa: Stability of αPbO_2-type srilankite solid solutions of $(Ti_{1-x}, Zr_x)O_2$ ($0 \le x \le 0.6$). *Phys. Chem. Min.* **2012**, *39*, 797–802. [CrossRef]
11. Nickel, E.H.; Rowland, J.F.; McAdam, R.C. Ixiolite—A columbite substructure. *Am. Min.* **1963**, *48*, 961–979.
12. Willgallis, A.; Siegmann, E.; Hettiaratchi, T. Srilankite, a new Zr-Ti-oxide mineral. *Neues Jahrbuch für Mineralogie Monatshefte* **1983**, *4*, 151–157.
13. Wang, L.P.; Essene, E.J.; Zhang, Y.X. Mineral inclusions in pyrope crystals from Garnet Ridge, Arizona, USA: Implications for processes in the upper mantle. *Contr. Min. Petr.* **1999**, *135*, 164–178. [CrossRef]
14. Varlamov, D.A.; Garanin, V.K.; Kostrovitskii, S.I. The exotic high-titanium minerals as inclusions within granets from low crust and mantle xenoliths. *Dokl. Akad. Nauk.* **1995**, *345*, 364–366.

15. El Goresy, A.; Chen, M.; Gillet, P.; Dubrovinsky, L.; Graup, G.; Ahuja, R. A natural shock-induced dense polymorph of rutile with alpha-PbO_2 structure in the suevite from the Ries crater in Germany. *Earth Planet. Sci. Lett.* **2001**, *192*, 485–495. [CrossRef]
16. Chen, M.; Gu, X.P.; Xie, X.D.; Yin, F. High-pressure polymorph of TiO_2-II from the Xiuyan crater of China. *Chin. Sci. Bul.* **2013**, *58*, 4655–4662. [CrossRef]
17. Zhang, R.Y.; Liou, J.G.; Ernst, W.G. The Dabie-Sulu continental collision zone: A comprehensive review. *Gondw. Res.* **2009**, *16*, 1–26. [CrossRef]
18. Tschauner, O. High-pressure minerals. *Am. Mineral.* **2019**, *104*, 1701–1731. [CrossRef]
19. Cavosie, A.J.; Erickson, T.M.; Timms, N.E. Nanoscale records of ancient shock deformation: Reidite ($ZrSiO_4$) in sandstone at the Ordovician Rock Elm impact crater. *Geology* **2015**, *43*, 315–318. [CrossRef]
20. Tschauner, O.; Ma, C. Riesite, IMA 2015-110a. CNMNC Newsletter No. 35, February 2017: p 213. *Min. Mag.* **2017**, *81*, 209–213.
21. Staehle, V.; Altherr, R.; Nasdala, L.; Ludwig, T. Ca-rich majorite derived from high-temperature melt and thermally stressed hornblende in shock veins of crustal rocks from the Ries impact crater (Germany). *Contr. Min. Petr.* **2011**, *161*, 275–291. [CrossRef]
22. Erickson, T.M.; Pearce, M.A.; Reddy, S.M.; Timms, N.E.; Cavosie, A.J.; Bourdet, J.; Nemchin, A.A. Microstructural constraints on the mechanisms of the transformation to reidite in naturally shocked zircon. *Contr. Min. Petr.* **2017**, *172*, 6. [CrossRef]
23. Stähle, V.; Altherr, R.; Nasdala, L.; Trieloff, M.; Varychev, A. Majoritic garnet grains within shock-induced melt veins in amphibolites from the Ries impact crater suggest ultrahigh crystallization pressures between 18 and 9 GPa. *Contr. Min. Petr.* **2017**, *172*, 86. [CrossRef]
24. Armstrong, J.T. CITAF—A package of correction programs for the quantitative electron microbeam X-ray analysis of thick polished materials, thin-films, and particles. *Microbeam Anal.* **1995**, *4*, 177–200.
25. Hammersley, A.P.; Svensson, S.O.; Hanfland, M.; Fitch, A.N.; Hausermann, D. Two-dimensional detector software: From real detector to idealised image or two-theta scan. *High. Pres. Res.* **1996**, *14*, 235–248. [CrossRef]
26. Dera, P.; Zhuravlev, K.; Prakapenka, V.; Rivers, M.L.; Finkelstein, G.J.; Grubor-Urosevic, O.; Downs, R.T. High pressure single-crystal micro X-ray diffraction analysis with GSE_ADA/RSV software. *High. Pres. Res.* **2013**, *33*, 466–484. [CrossRef]
27. Putz, H.; Schon, J.C.; Jansen, M. Combined method for ab initio structure solution from powder diffraction data. *J. Appl. Cryst.* **1999**, *32*, 864–870. [CrossRef]
28. Altomare, A.; Camalli, M.; Cuocci, C.; Giacovazzo, C.; Moliterni, A.; Rizzi, R. EXPO2009: Structure solution by powder data in direct and reciprocal space. *J. Appl. Cryst.* **2009**, *42*, 1197–1202. [CrossRef]
29. Kraus, W.; Nolze, G. POWDER CELL—A program for the representation and manipulation of crystal structures and calculation of the resulting X-ray powder patterns. *J. Appl. Cryst.* **1996**, *29*, 301–303. [CrossRef]
30. von Dreele, R.B.; Larson, A.C. *General Structure Analysis System (GSAS)*; Los Alamos National Laboratory Report LAUR; Los Alamos National Laboratory: Los Alamos, NM, USA, 2004.
31. Grey, I.E.; Li, C.; Madsen, I.C.; Braunshausen, G. TiO_2-II. Ambient pressure and structure refinement. *Mat. Res. Bull.* **1988**, *23*, 743–753. [CrossRef]
32. Timms, N.E.; Erickson, T.M.; Pearce, M.A.; Cavosie, A.J.; Schmieder, M.; Tohver, E.; Wittmann, A. A pressure-temperature phase diagram for zircon at extreme conditions. *Earth-Science Rev.* **2017**, *165*, 185–202. [CrossRef]
33. Leroux, H.; Reimold, W.U.; Koeberl, C.; Hornemann, U.; Doukhan, J.C. Experimental shock deformation in zircon: A transmission electron microscopic study. *Earth. Planet. Sci. Lett.* **1999**, *169*, 291–301. [CrossRef]
34. Ma, C.; Tschauner, O.; Beckett, J.R.; Rossman, G.R.; Prescher, C.; Prakapenka, V.B.; MacDowell, A. Liebermannite, KAlSi3O8, a new shock-metamorphic, high-pressure mineral from the Zagami Martian meteorite. *Meteorit. Planet. Sci.* **2018**, *53*, 50–61. [CrossRef]
35. Artemieva, N.A.; Wuennemann, K.; Krien, F.; Reimold, W.U.; Stöffler, D. Ries crater and suevite revisited-Observations and modeling Part II: Modeling. *Meteorit. Planet. Sci.* **2013**, *48*, 590–627. [CrossRef]

minerals

MDPI

Article

X-ray and Neutron Study on the Structure of Hydrous SiO_2 Glass up to 10 GPa

Satoru Urakawa [1,*], Toru Inoue [2,3], Takanori Hattori [4], Asami Sano-Furukawa [4], Shinji Kohara [5,6], Daisuke Wakabayashi [7], Tomoko Sato [2], Nobumasa Funamori [7] and Ken-ichi Funakoshi [8]

1 Department of Earth Sciences, Okayama University, Okayama 700-8530, Japan
2 Department of Earth and Planetary Systems Science, Hiroshima University, Higashi-Hiroshima 739-8526, Japan; toinoue@hiroshima-u.ac.jp (T.I.); tomokos@hiroshima-u.ac.jp (T.S.)
3 Geodynamics Research Center, Ehime University, Matsuyama 790-8577, Japan
4 J-PARC Center, Japan Atomic Energy Agency, Tokai 319-1195, Japan; hattori.takanori@jaea.go.jp (T.H.); sanoasa@post.j-parc.jp (A.S.-F.)
5 Research Center for Advanced Measurement and Characterization, National Institute for Materials Science (NIMS), Sayo, Hyogo 679-5148, Japan; KOHARA.Shinji@nims.go.jp
6 Diffraction and Scattering Division, Japan Synchrotron Radiation Research Institute, Sayo, Hyogo 679-5198, Japan
7 Institute of Materials Structure Science, High Energy Accelerator Research Organization (KEK), Tsukuba 305-0801, Japan; daisuke.wakabayashi@kek.jp (D.W.); nobumasa.funamori@kek.jp (N.F.)
8 Neutron Science and Technology Center, Comprehensive Research Organization for Science and Society, Tokai 319-1106, Japan; k_funakoshi@cross.or.jp
* Correspondence: urakawa@okayama-u.ac.jp

Received: 15 December 2019; Accepted: 16 January 2020; Published: 20 January 2020

Abstract: The structure of hydrous amorphous SiO_2 is fundamental in order to investigate the effects of water on the physicochemical properties of oxide glasses and magma. The hydrous SiO_2 glass with 13 wt.% D_2O was synthesized under high-pressure and high-temperature conditions and its structure was investigated by small angle X-ray scattering, X-ray diffraction, and neutron diffraction experiments at pressures of up to 10 GPa and room temperature. This hydrous glass is separated into two phases: a major phase rich in SiO_2 and a minor phase rich in D_2O molecules distributed as small domains with dimensions of less than 100 Å. Medium-range order of the hydrous glass shrinks compared to the anhydrous SiO_2 glass by disruption of SiO_4 linkage due to the formation of Si–OD deuterioxyl, while the response of its structure to pressure is almost the same as that of the anhydrous SiO_2 glass. Most of D_2O molecules are in the small domains and hardly penetrate into the void space in the ring consisting of SiO_4 tetrahedra.

Keywords: hydrous silica glass; medium-range order; first sharp diffraction peak; phase separation; small angle X-ray scattering; X-ray diffraction; neutron diffraction; high pressure

1. Introduction

Silica glass is the most fundamental, fully polymerized network glass whose structure and properties under pressure have long been of interest as an important analog material of silicate magma [1–9]. For the structural aspect, SiO_2 glass is densified by compression with the change in medium-range order in the pressure region up to about 10 GPa [2,4,8]. Namely, reduction of the Si–O–Si bond angle between SiO_4 tetrahedra shrinks the ring and squeezes the interstitial void, leading to permanent densification of SiO_2 glass. SiO_2 glass becomes fully densified glass by room temperature compression to the pressure of 9 to 13 GPa, of which the density is about 20% higher than ordinary SiO_2 glass at ambient conditions [8].

Water in silicate glass has a large effect on the structure as well as on the properties such as viscosity and glass transition temperature [10,11]. Water can react with SiO_2 glass to form hydroxyl species (Si–OH), disrupting the linkage of SiO_4 tetrahedra. As a result, the medium-range order of the SiO_2 glass is also affected by the addition of water in which the size of the ring of the SiO_2 glass containing water becomes smaller than that of ordinary SiO_2 glass [12]. The molecular water, as well as the hydroxyl, is also present in the silicate glasses. It is known that the molecular water becomes the dominant water species in the silicate glasses with increasing total water content [10,11]. However, it is not yet well understood how the molecular water is incorporated into the silicate glasses.

In this study, the hydrous SiO_2 glass containing 13 wt.% of D_2O was synthesized under high-pressure and high-temperature conditions and small angle X-ray scattering (SAXS), X-ray diffraction (XRD), and neutron diffraction (ND) have been applied in order to investigate the short-range order, medium-range order, and phase separation of this hydrous glass. In addition to this, in-situ high-pressure XRD and ND measurements of this hydrous glass have been conducted up to about 10 GPa to clarify the pressure response of its structure. Effects of the pressure and the water dissolution on the medium-range order of the SiO_2 glass and the state of molecular water in the SiO_2 glass are also discussed.

2. Hydrous SiO_2 Glass Sample

The glass sample with a composition of SiO_2–13wt.% D_2O was synthesized using a Kawai-type high-pressure apparatus driven by the 3000 ton press installed at GRC, Ehime University. Powdered quartz enclosed in a Pt capsule together with 13 wt.% D_2O was melted at 3 GPa and 1873 K for 30 min, then it was quenched to room temperature by turning off the electric power supply. The quenching rate was approximately 1000 K/s. Subsequently, the applied pressure was released slowly in several hours, and then the sample was recovered in ambient conditions (Figures S1 and S2). Since we have not measured the content of heavy water in SiO_2 glass, the 13 wt.% D_2O content is a nominal value. The recovered glass is optically clear and homogeneous. Density of this hydrous SiO_2 glass was determined to be 2.239 g/cm^3 by Archimedes' method. This value is slightly higher than the density of ordinary SiO_2 glass (2.20 g/cm^3), although it contains a large amount of heavy water. This means that this high-pressure hydrous SiO_2 glass is an intermediately densified glass.

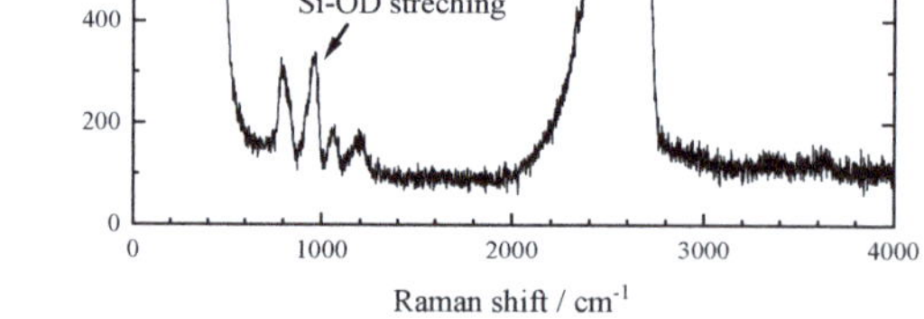

Figure 1. Raman spectrum of the hydrous SiO_2 glass at ambient conditions. Spectrum is unpolarized.

Raman spectroscopy analysis has been applied to clarify the species of water dissolved in the SiO_2 glass. Raman spectrum shows that the hydrous SiO_2 glass has some additional peaks except those assigned to silica glass (Figure 1). Those are a sharp peak around 960 cm^{-1} and a broad peak from 2100 to 2750 cm^{-1}. The former is assigned to the Si–OD stretching vibrations and the latter to OD stretching vibrations in D_2O molecules and Si–OD groups [13,14]. Heavy water in the hydrous SiO_2 glass is thought to take two states, OD deuterioxyl and D_2O molecule. In general, the amount of OH hydroxyls dissolved in the silicate glass increases with the total H_2O content, but reaches a limit of about 3 wt.% [10,11]. Thus, this hydrous SiO_2 glass may contain about 10 wt.% of the heavy water as molecular species.

3. Experiments

3.1. Small Angle X-ray Scattering at Ambient Conditions

SAXS experiments were performed at the BL18C of the Photon Factory (PF) of the High Energy Accelerator Research Organization (KEK), Japan. Incident X-rays were monochromatic X-rays of 15.28 keV with a beam diameter of 35 micrometers, and scattered X-rays were detected in Q-range from 0.02 to 4.0 $Å^{-1}$ with an image plate. The detail of experimental setup is described elsewhere [15].

3.2. Angle-Dispersive X-ray Diffraction at Ambient Conditions

Angle-dispersive XRD experiment at ambient conditions was carried out using two-axis diffractometer at the BL04B2 beamline of SPring-8, Japan. We acquired diffraction data up to $Q = 25$ $Å^{-1}$ by angle dispersive method using the monochromatic X-rays of 61.37 keV. The detail of the experimental setup is described elsewhere [16,17].

3.3. High-Pressure X-ray Diffraction

XRD experiments on the hydrous glass during compression were performed using the DIA-type cubic press MAX80 installed at the AR-NE5C beamline of the PF. Tungsten carbide anvils and a boron-epoxy pressure transmitting medium were used. A powdered sample was pressed into a pellet with a diameter of 2 mm and enclosed in a boron nitride (BN) capsule. Sample pressure was determined from the volume of the NaCl pressure marker [18]. XRD profiles were acquired by an energy-dispersive method in a transmitting geometry with an intrinsic Ge detector at room temperature and pressures of up to 9.6 GPa. White X-rays up to 120 keV were used and the data was collected at 11 fixed angles ranging from 3° to 30° to cover a wide Q range. Background intensities were also measured at each angle using an empty cell.

3.4. High-Pressure Neutron Diffraction

ND experiments of the hydrous SiO_2 glass under pressure were conducted by the time-of-flight (TOF) method combined with a multi-anvil press ATSUHIME [19] at the PLANET beamline [20] of the spallation neutron source of the Materials and Life Science Experimental Facility (MLF) at the J-PARC, Japan. A multi-anvil 6-6 type high-pressure apparatus was used [21]. The second-stage Ni-bound WC anvils were used. Clumps of SiO_2–13wt.%D_2O glasses were directly packed in the ZrO_2 pressure medium. The size of the hydrous SiO_2 glass sample was 4.7 mm in diameter and 6.7 mm in height. Sample pressures were calculated on the basis of the pressure-load calibration curves which were determined in separated runs beforehand. The diffracted neutrons were detected by a pair of 90° detector banks consisting of ^3He position sensitive detectors equipped with receiving radial collimators. The data were acquired for 15–33 h at the proton beam power of 300 kW. ND profiles were acquired at room temperature and with the same pressure conditions as those of XRD experiments. Diffraction profiles of a vanadium pellet in a high-pressure cell and an empty cell were also acquired for the correction of scattering intensity.

4. Results and Discussion

4.1. Phase Separation of Hydrous SiO_2 Glass

It is known that the silicate glass with high water content undergoes the glass-in-glass phase separation at a low temperature [11]. Due to the limitation of mutual solubility, it separates into a silica-rich and a water-rich phase. We performed the SAXS measurements on several parts of the hydrous SiO_2 glass to clarify the possible phase separation. The hydrous SiO_2 glass clearly shows significant scattering intensity compared with anhydrous glass (Figure 2). The scattering intensity, however, shows a difference depending on location; some regions have strong scattering intensity, while others only show very weak scattering intensity. This indicates that the distribution of scattering entities in the hydrous glass is heterogeneous in the dimension on the order of from μm to mm. The SAXS patterns of the hydrous glass has a broad peak at Q = 0.05–0.1 $Å^{-1}$, which indicates the existence of an average distance frequently realized between neighboring scattering entities. The average distance is estimated to be about 100 Å from the length scale ($2\pi/Q$) corresponding to the position of the peaks. It is, therefore, considered that the size of the scattering entities is less than 100 Å. The similar SAXS pattern was observed in hydrated Na-silicate glass, and it was interpreted by the glass-in-glass phase separation [11]. In our hydrous SiO_2 glass, similar phase separation may occur. When combined with the results of the XRD and ND described below, the most likely candidate for this scattering entities in hydrous SiO_2 glass is a phase rich in molecular D_2O. Although this hydrous glass is optically homogeneous, it is considered to be the mixture of SiO_2-rich glass part and D_2O-rich domain with the dimensions of less than 100 Å.

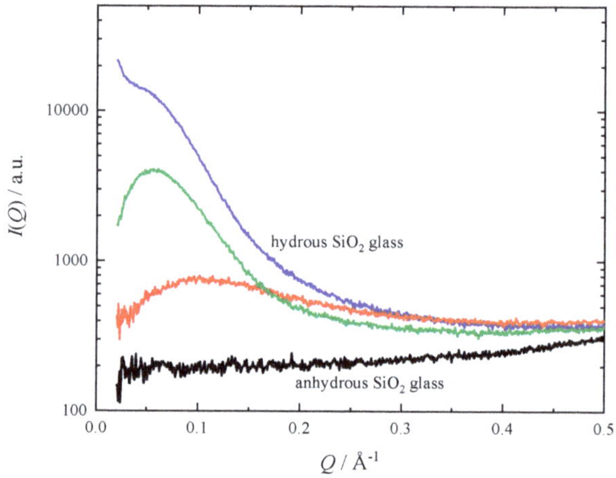

Figure 2. Small angle X-ray scattering intensity from the hydrous SiO_2 glass at ambient conditions. Scattering intensity from anhydrous SiO_2 glass (black curve) is also shown for comparison. Three curves were observed at locations from a few hundred micrometers to a millimeter apart.

4.2. Comparison with Dry SiO_2 Glass at Ambient Conditions

We compare the structure of the hydrous glass and the anhydrous glass at atmospheric pressure using the results of XRD and ND. The structure factors $S(Q)$ and the total correlation functions $T(r)$ for the hydrous and the anhydrous glass are shown in Figure 3. The $S(Q)$ and $T(r)$ for the anhydrous SiO_2 glass were reported by Kohara et al. [22] for X-ray and Hannon [23] for neutron. Here, we consider what the $S(Q)$ and the $T(r)$ of the hydrous SiO_2 glass represent. XRD and ND measurements were performed on the bulk hydrous SiO_2 glass. As shown in the results of the SAXS measurements, the hydrous glass is separated into two phases. Therefore, the $S(Q)$ and $T(r)$ obtained from XRD and ND are the sum of contributions from the SiO_2-rich glass part and the D_2O-rich domains. Hereafter, we discuss the structure of the hydrous SiO_2 glass based on this view.

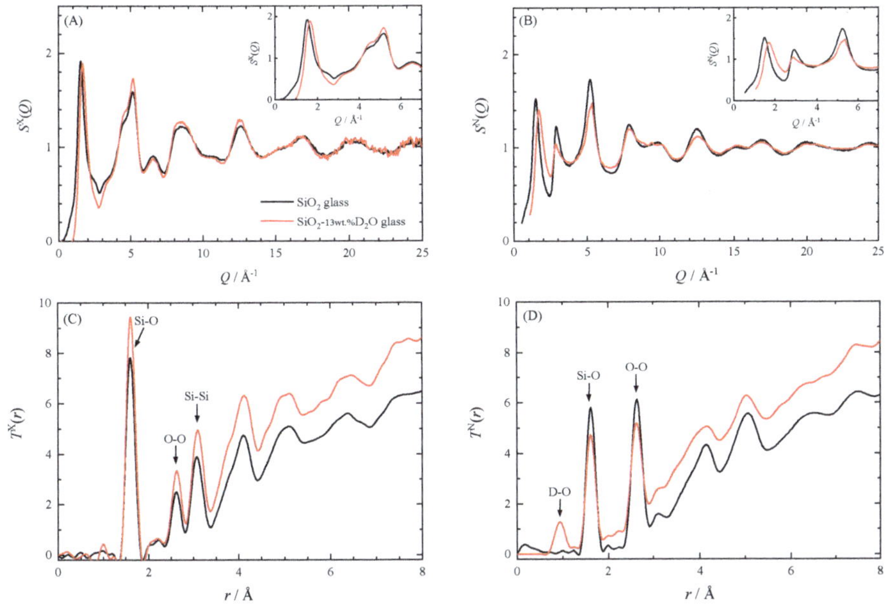

Figure 3. X-ray and neutron diffraction data of the hydrous SiO_2 glass and the anhydrous SiO_2 glass at ambient conditions. (**A**) The X-ray structure factor $S^X(Q)$, (**B**) the neutron structure factor $S^N(Q)$, (**C**) the X-ray total correlation function $T^X(r)$, and (**D**) the neutron total correlation function $T^N(r)$.

As discussed below, both $S(Q)$ and $T(r)$ show that the short-range order composed of SiO_4 tetrahedra of the hydrous SiO_2 glass is almost the same as that of anhydrous glass, while the medium-range order related to the ring of the SiO_4 tetrahedra is diminished by addition of water. The first sharp diffraction peak (FSDP) observed at $Q \sim 1.5$–1.7 Å^{-1} in both the $S^X(Q)$and $S^N(Q)$ of hydrous glass clearly shifts to the higher-Q side than those of anhydrous glass. The second peak is observed at $Q = 2.9$ Å^{-1} only in neutron data. In the other peaks, the intensity and the position of $S^X(Q)$ are almost the same between the hydrous and the anhydrous glass, whereas the $S^N(Q)$ shows a difference. This is due to the very small X-ray atomic scattering factor of D and the relatively large neutron scattering length of D. The similarity of $S^X(Q)$ means the Si–O correlation between the hydrous and anhydrous glass is similar. The X-ray total correlation function $T^X(r)$ of hydrous glass shows almost the same peaks positions as those of anhydrous glass, but the height of $T^X(r)$ of hydrous glass is higher than that of anhydrous glass because of its higher number density. The Si coordination number of the hydrous glass calculated from the area of the first peak at 1.6 Å is about 3.9, which is almost the same as that for the anhydrous glass. On the other hand, the neutron total correlation function $T^N(r)$ of hydrous glass has a different shape from that of anhydrous glass because of the presence of peaks due to the OD deuterioxyl and the D_2O molecules. The peak at about 0.93 Å for hydrous glass corresponds to the D–O distance. In the $T^N(r)$, it is difficult to identify peaks originating from deuterium other than this peak, but the effect of overlapping peaks is recognized. Since the Si-O correlation typically located around 1.6 Å in the $T^N(r)$ is superimposed on the intramolecular D–D correlation (1.56 Å) and the intermolecular O–D correlation (1.92 Å) of D_2O [24], the coordination number of Si cannot be determined.

FSDP of $S(Q)$ is thought to be related to the formation of medium-range order, although its origin has been still under debate [25–29]. The position of FSDP, Q_1, corresponds to the length scale of periodicity in real space l_1 ($=2\pi/Q_1$). Thus, the shift of the FSDP toward high-Q side by dissolution of water means the shrinkage of the medium-range order. This is attributed to the decrease of the size of the SiO_4 ring by breaking SiO_4 linkages with the OH (OD) group. In ordinary SiO_2 glass, the six-membered ring of SiO_4 tetrahedra is most frequent followed by the five-membered ring [30], but the population of a smaller ring, such as five- and four-membered rings, may increase by water dissolution. The shift of the FSDP toward high-Q side due to water dissolution has been reported for rhyolitic glasses containing more than 70 wt.% of SiO_2 [31]. The displacement of the FSDP of $S^X(Q)$ in rhyolitic glass is about 1.5 Å^{-1} by the 7.5 wt.% dissolution of H_2O, which is comparable to that of our hydrous silica glass containing 13 wt.% of D_2O.

On the other hand, the FSDP of densified SiO_2 glasses is known to be on the high-Q side compared to ordinary SiO_2 glass [32]. This is considered to correspond to the shrinkage of the medium-range order, which is related to a reduction of Si–O–Si bond angle associated with the reduction of interstitial voids by compression [4]. Our hydrous SiO_2 glass was prepared by quenching from the melt at 3 GPa and is thought to partially retain the structure of intermediately densified glass. Therefore, the position of FSDP in our hydrous glass can be attributed to the effects of both water addition and densification.

As revealed by SAXS, XRD, and ND, the hydrous glass synthesized in this study is a mixture of SiO_2-rich glass parts and a D_2O-rich domain. By assuming that the structure of each phase is the same as that for pure SiO_2 glass and pure liquid D_2O, the $S(Q)$ for hydrous glass can be calculated from those for pure phases [22,33–35] (the detailed method is described in Appendix B). Comparison of the $S^X(Q)$ and the $S^N(Q)$ of the hydrous SiO_2 with those calculated from the previously reported $S(Q)$ are shown in Figure 4. They showed good agreement except the position of FSDP of the $S^X(Q)$ and the $S^N(Q)$, and the height of the second peak of the $S^N(Q)$. The difference may be attributed to the shrinkage of the medium-range order of the SiO_2 glass by dissolution of water.

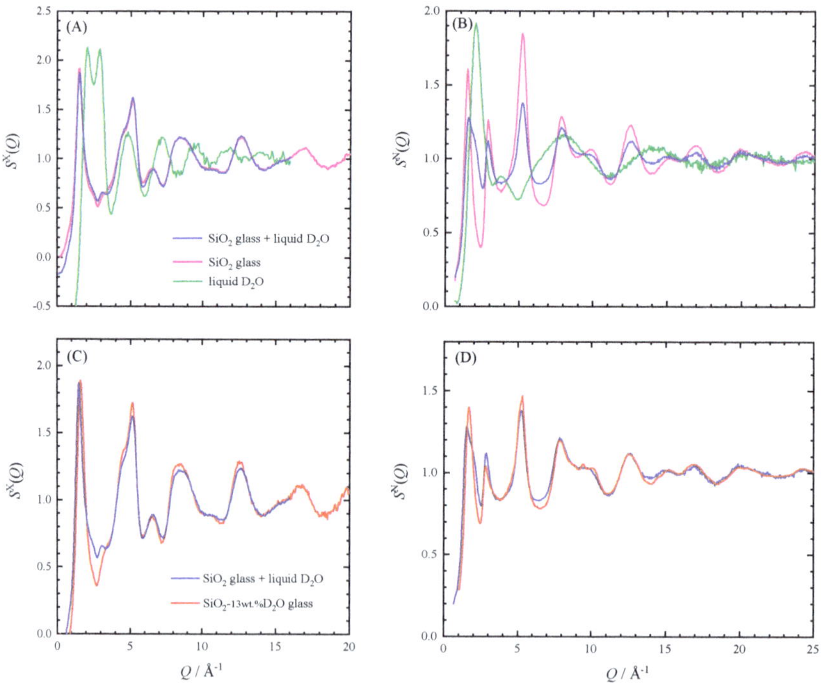

Figure 4. (**A**) $S^X(Q)$'s of SiO_2 glass and liquid D_2O at ambient conditions [22,34], and that simulated for the hydrous glass using them. (**B**) Corresponding $S(Q)$ for neutron, $S^N(Q)$ [33,35]. (**C**) Comparison of the observed and simulated $S^X(Q)$s for the hydrous SiO_2 glass. (**D**) Comparison of the observed and simulated $S^N(Q)$s.

4.3. Hydrous SiO_2 Glass under Pressure

We consider the response of the structure of the hydrous SiO_2 glass to pressure. The $S(Q)$ and $T(r)$ for the hydrous glass during compression to about 10 GPa are shown in Figure 5. In XRD, the $S^X(Q)$ of hydrous SiO_2 glass changes with pressure in the same way as anhydrous SiO_2 glass reported by Inamura et al. [4], except for the position of the FSDP. This means that the response to the pressure of the short-range order of the hydrous glass is the same as that of the anhydrous glass. This is consistent with the structure deduced from SAXS showing that the hydrous SiO_2 glass is mainly composed of relatively dry parts rich in SiO_2.

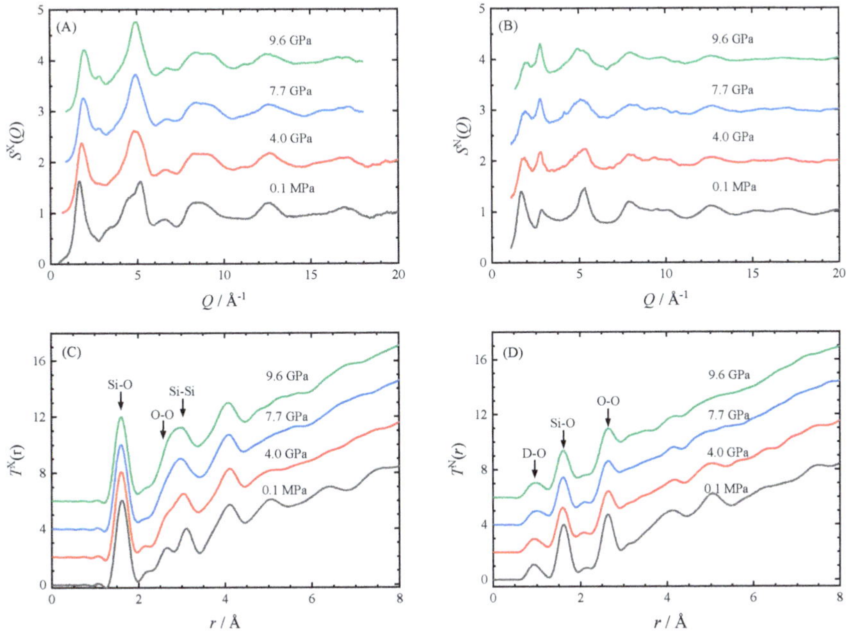

Figure 5. X-ray and neutron diffraction data of the hydrous SiO_2 glass with 13 wt.% D_2O during room temperature compression to 10 GPa. (**A**) The X-ray structure factor $S^X(Q)$, (**B**) the neutron structure factor $S^N(Q)$, (**C**) the X-ray total correlation function $T^X(r)$, and (**D**) the neutron total correlation function $T^N(r)$.

Figure 6 compares the pressure dependence of the position of FSDP of hydrous SiO_2 glass obtained from both XRD and ND with that of anhydrous glass [4,9]. The FSDP of $S^X(Q)$ and $S^N(Q)$ shifts in parallel to the high-Q side in proportion to the pressure up to 10 GPa. Those shifts are also parallel to the FSDP of ordinary anhydrous glass. In addition, the intensity of the FSDP of the hydrous glass decreases with pressure in the same way as anhydrous glass for both the $S^X(Q)$ and $S^N(Q)$ [4,9]. Shrinkage of the medium-range order is also found in $T(r)$. The Si–O distance is almost constant with up to 10 GPa, but the Si–Si distance decreases with increasing pressure (Figure 5). Thus, the Si–O–Si bond angle decreases, and the network linkage of the SiO_4 tetrahedra is distorted with increasing pressure. These show that even though the medium-range order of the hydrous SiO_2 glass is partly disrupted by OD deuterioxyl, it shrinks with pressure just like the anhydrous glass.

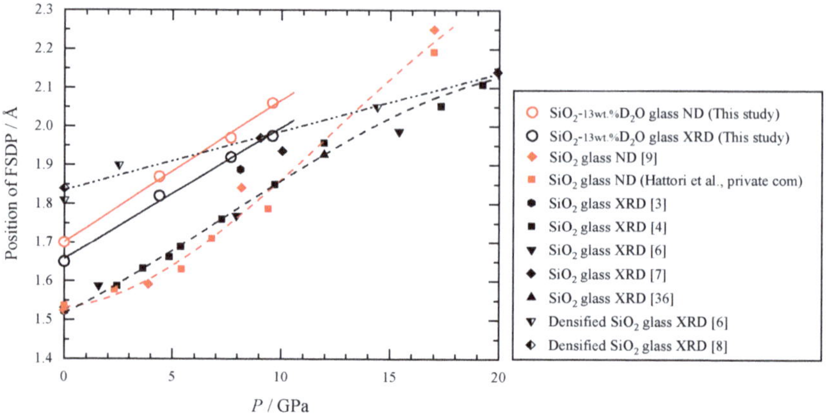

Figure 6. Comparison of the position of the first sharp diffraction peak (FSDP) of $S^X(Q)$ and $S^N(Q)$ between the hydrous and anhydrous SiO_2 glass [3,4,6–8,36].

We estimate the density of hydrous SiO_2 glass under pressure in order to derive the total correlation function $T(r)$. Since the FSDP of the hydrous glass changes with pressure at almost the same rate as that of the anhydrous glass, and the shape of the structure factors resemble each other except the FSDP, we assume that the densities of hydrous glass and anhydrous glass would change in the same rate with pressure. We evaluate the $T(r)$ using the densities of hydrous glass at high pressures, which is estimated from the relationships between density and pressure of ordinary SiO_2 glass shown in Figure 1 of Wakabayashi et al. [8]. The coordination number of Si calculated from $T^X(r)$ is about four up to 10 GPa, which is consistent with that of anhydrous SiO_2 glass. This shows that our estimation of high-pressure density of hydrous SiO_2 glass is fairly good, because the similarity of the short-range order between hydrous glass and anhydrous glass is expected.

What happens to the D_2O-rich phase in the hydrous SiO_2 glass during compression? Unfortunately, neutron diffraction data does not give us any information except nearly constant D–O distance in $T^N(r)$ up to 10 GPa. When evaluated using the partial molar volume of H_2O at 3 GPa [37], at which the hydrous SiO_2 glass was synthesized, 10 wt.% D_2O is equivalent to about 14% in volume. It is interesting that such an amount of molecular D_2O does not greatly affect the compression behavior of SiO_2 glass.

4.4. *Molecular Water in the Hydrous SiO2 Glass*

The SAXS strongly suggests that most of the D_2O molecules may form the separate phase in small domains distributed in the hydrous SiO_2 glass. Can all D_2O molecules in the hydrous glass be incorporated into the small domains? Here, we consider the possibility that D_2O molecules are contained in the voids characteristic of the ring structure made by connecting SiO_4 tetrahedra.

Sato et al. [38] showed that when He penetrates into SiO_2 glass, SiO_2 glass becomes rigid. Since the interstitial voids in the SiO_2 glass are occupied by He, the voids are not squeezed even when pressure is applied and the bulk modulus of the SiO_2 glass with He becomes larger than that of the ordinary SiO_2 glass. It is shown that the FSDP position of the SiO_2 glass saturated with He does not change up to 10 GPa. If D_2O molecules penetrate the interstitial voids formed by the SiO_4 linkage, the bulk modulus of the hydrous glass should be larger than that of ordinary glass, as shown in the case of SiO_2 glass with He. The position of the FSDP of our hydrous glass, however, changes with pressure in the same way as the anhydrous glass (Figure 6), and its bulk modulus is estimated to be almost the same as that of the anhydrous glass. Thus, it is considered that the interstitial void space in the hydrous SiO_2 glass is almost empty, and D_2O does not penetrate into the SiO_4 network. It is supported from void size: The interstitial void in the SiO_4 network of SiO_2 glass is too small to accommodate

D_2O molecule in it. In the size distribution model of interstitial voids in silica glass estimated from the solubility site density of He and Ne dissolved in cristobalite, about 10% of voids can contain He and only 6% of voids can contain Ne [39]. According to Zhang and Xu [40], the molecular diameter of H_2O is 2.74 Å (that of D_2O is also expected to be about the same), which is larger than He (2.16 Å) and Ne (2.42 Å), so that it seems difficult to incorporate the H_2O (or D_2O) molecules in the interstitial voids of SiO_2 glass. Further, the void size in the hydrous glass is smaller than that of anhydrous glass due to the reduction of size of the SiO_4 ring, and it is more difficult for the H_2O (or D_2O) molecule to be incorporated into the interstitial void.

5. Conclusions

SAXS, XRD and ND show that the SiO_2–13wt.% D_2O composition glass, which was synthesized by quenching from melt at 3 GPa is separated into two phases: SiO_2-rich glass phase and D_2O-rich minor phase were distributed as small domains with dimensions of less than 100 Å. Both the $S^X(Q)$ and the $S^N(Q)$ of the hydrous SiO_2 glass can be reproduced from those of the SiO_2 glass and liquid D_2O. Medium-range order and short-range order derived from XRD and ND mainly reflect the structure of the SiO_2-rich glass part. The FSDP of hydrous glass shifts to the higher-Q side than those of anhydrous glass at ambient conditions. This corresponds to the shrinkage of the medium-range order associated with the decrease of the size of SiO_4 rings by breaking the SiO_4 linkage with the OD group, as well as by distorting the network linkage of SiO_4 with a pressure of 3 GPa during glass synthesis. The response of the structure of the SiO_2-rich glass part to pressure is almost the same as that of the anhydrous SiO_2 glass. The FSDP of the hydrous glass shifts to the high-Q side with increasing pressure in parallel with that of the anhydrous glass. The Si–O–Si bond angle decreases with pressure, although the coordination number of Si is about four at the pressures of from 0.1 MPa to 10 GPa. Thus, the medium-range order shrinks with pressure associated with distortion of network linkage of SiO_4 tetrahedra. Most of D_2O molecules are in the small domains of the D_2O-rich phase and hardly penetrate into the void space in the ring consisting of SiO_4 tetrahedra of SiO_2-rich glass part.

Supplementary Materials: The following are available online at http://www.mdpi.com/2075-163X/10/1/84/s1, Figure S1: Cell assemble for the synthesis of hydrous SiO_2 glass at 3 GPa and 1873 K, Figure S2: Recovered Pt capsule and hydrous SiO_2 glass quenched from the conditions of 3 GPa and 1873 K.

Author Contributions: Conceptualization, S.U. and T.I.; formal analysis, S.U., T.H., S.K.; investigation, S.U., T.I., T.H., A.S.-F., S.K., D.W., T.S., N.F., K.-i.F.; writing—original draft preparation, S.U.; All author contribute review and editing. All authors have read and agreed to the published version of the manuscript.

Funding: This research was in part supported by the JSPS KAKENHI Grant (19GS0205, 18K03805), the Earthquake Research Institute Joint Usage/Research Program (2010-G-01, 2011-G-05, 2012-G-02), and the Joint Usage/Research Center PRIUS, Ehime University.

Acknowledgments: We thank Y. Katayama and T. Kimura for help with the ND experiments at J-PARC, and K. Mibe and M. Kanzaki for help with the synthesis of hydrous SiO_2 glass. We are grateful to A. Zeidler and P. S. Salmon for providing us with high-pressure neutron diffraction data for SiO_2 glass. We are also grateful to two anonymous reviewers for their thoughtful formal reviews. The XRD experiments were conducted under approval of PF Proposal Advisory Committee (2011G652, 2017G135) and the JASRI (2014B1480), and the ND experiments were performed with the approval of the Neutron Science Proposal Review Committee of J-PARC, MLF (2013I0011).

Conflicts of Interest: The authors declare no conflict of interest.

Appendix A

Appendix A.1 Data Reduction of X-ray and Neutron Diffraction

Appendix A.1.1 Energy Dispersive X-ray Diffraction at High Pressures

We used a Faber-Ziman definition for X-ray structure factor $S^X(Q)$ which is defined as follows.

$$S^X(Q) = (i^X_{coh}(Q) - \{\langle f(Q)^2\rangle - \langle f(Q)\rangle^2\})/\langle f(Q)\rangle^2 \quad \text{(A1)}$$

where $i_{coh}{}^{X}(Q)$ is a coherent scattering intensity per atom, $\langle f(Q)^2\rangle = \sum_i c_i f_i^2(Q), \langle f(Q)\rangle^2 = (\sum_i c_i f_i(Q))^2$, c_i and $f_i(Q)$ are an atomic fraction and an atomic scattering factor of the ith atom, respectively.

The $S^{X}(Q)$ was evaluated by the procedure developed by Funakoshi [41]. Energy profiles of XRD pattern obtained at various Bragg angles can be expressed by the following equation.

$$I_{ob}^{X}(E,\theta) = NA(E,\theta)P(E,\theta)I_0^{X}(E)\left[i_{coh}^{X}(E,\theta) + i_{inc}^{X}(E,\theta) + i_m^{X}(E,\theta)\right] + I_{BG}^{X}(E,\theta) \quad \text{(A2)}$$

where N is the number of atom, $A(E,\theta)$ is the absorption factor, $P(E,\theta)$ is the polarization factor, $I_0{}^{X}(E)$ is the intensity of incident X-rays. $i_{inc}(E,\theta)^{X}$ and $i_m(E,\theta)^{X}$ are incoherent and multiple scattering intensity per atom, respectively. $I_{BG}{}^{X}(E,\theta)$ is background intensity. At first, $I_{BG}{}^{X}(E,\theta)$'s were subtracted from $I_{ob}{}^{X}(E,\theta)$'s. As the energy spectrum of incident X-rays from the synchrotron source $I_0{}^{X}(E)$ is unknown, we evaluated simultaneously both $I_0{}^{X}(E)$ and $i_{coh}{}^{X}(E,\theta)$'s from $I_{ob}{}^{X}(E,\theta)$'s by simulation using the Monte Carlo method [41]. Here, we ignored the differences of absorption and polarization effects and multiple scattering terms among data taken with different Bragg angles. The incoherent scattering term was corrected using the formulation of Hajdu [42] and Pálinkás [43]. Then coherent scattering intensities at each Bragg angle were unified to obtain $I_{coh}{}^{X}(Q)$. The Faber–Ziman total structure factor $S^{X}(Q)$ was evaluated from Equation (A1) in which the atomic scattering factor was calculated using the approximation formula given by Doyle and Turner [44].

Appendix A.1.2 Time-of-Flight Neutron Diffraction at High Pressures

Structure factor for neutron $S^{N}(Q)$ was evaluated from a ND profile obtained by the TOF method combined with a multi-anvil press. Observed ND profile is described as follows.

$$I_{ob}^{N}(\lambda,\theta) = NA(\lambda,\theta)I_0^{N}(\lambda)\left[i_{coh}^{N}(\lambda,\theta) + i_{inc}^{N}(\lambda,\theta) + i_m^{N}(\lambda,\theta)\right] + I_{BG}^{N}(\lambda,\theta) \quad \text{(A3)}$$

where N is the number of atoms, $A(\lambda,\theta)$ is the absorption factor, $I_0{}^{N}(\lambda)$ is the intensity of incident neutron, $i_{coh}{}^{N}(\lambda,\theta)$, $i_{inc}{}^{N}(\lambda,\theta)$, and $i_m{}^{N}(\lambda,\theta)$ are coherent, inelastic, incoherent, and multiple scattering cross section per one atom, respectively. $I_{BG}{}^{N}(\lambda,\theta)$ is background intensity. Observed scattering intensities $I_{ob}{}^{N}(\lambda,\theta)$ are normalized by $I_0{}^{N}(\lambda)$ obtained from the scattering intensity of a vanadium pellet after background subtraction and absorption correction. Absorption correction was performed according to the method of Paalman and Pings [45]. Then, incoherent scattering and multiple scattering terms were corrected. The incoherent term was evaluated by a composition average of incoherent scattering length b_{inc} of the elements, which are listed in the table of Sears [46]. The incoherent scattering for D atom was significantly deviated from the value listed in the table due to the inelastic effect. Therefore, we corrected it by empirical method employed by Kameda et al. [47] (as described below). The multiple scattering term was corrected by the method of Blech and Averbach [48]. Thus, we calculated the structure factor $S^{N}(Q)$ from the corrected intensity $I_{cor}{}^{N}(\lambda,\theta)$, which is the sum of $i_{coh}{}^{N}(\lambda,\theta)$ and $i_{ine}{}^{N}(\lambda,\theta)$.

Inelastic effect for incoherent scattering of deuterium has a small but non-negligible effect on the total scattering intensity of the hydrous SiO_2 glass. The effect was corrected based on the empirical method by Kameda et al. [48] in which incoherent scattering was estimated from the self term of the liquid null-H_2O using a known O–O correlation. The applicability of this method at PLANET has been confirmed from the good coincidence of the $S(Q)$ for D_2O water with that reported so far in the literature (Hattori et al., private com).

Definition of Faber–Ziman total structure factor for the $S^{N}(Q)$ is as follows.

$$S^{N}(Q) = (i_{coh}^{N}(Q) - \{\langle b\rangle^2 - \langle b\rangle^2\})/\langle b\rangle^2 \quad \text{(A4)}$$

where $\langle b^2\rangle = \sum_i c_i b_i^2$, $\langle b\rangle^2 = (\sum_i c_i b_i)^2$, c_i, and b_i are an atomic fraction and a coherent scattering length of the ith atom [46], respectively.

Appendix A.1.3 Fourier Analysis

The $S(Q)$ data in reciprocal space can be transformed into total correlation function $T(r)$ in real space by means of a Fourier transform.

$$T(r) = 4\pi r\rho_0 + \frac{2}{\pi}\int_{Q_{min}}^{Q_{max}} M(Q)Q\{S(Q)-1\}\sin rQdQ \tag{A5}$$

where ρ_0 is a number density of atoms and $M(Q)$ is a modification function. We used a Lorch function as $M(Q)$ [49].

Appendix B

Faber–Ziman Structure Factor of a Two-Phase Mixture can be Written as Follows for X-ray

$$\begin{aligned} S^X_{A+B}(Q) = X_A(\langle f(Q)\rangle^2_A/\langle f(Q)\rangle^2_{A+B})S^X_A(Q) + X_B(\langle f(Q)\rangle^2_B/\langle f(Q)\rangle^2_{A+B})S^X_B(Q) \\ +X_A(\langle f(Q)^2\rangle_A - \langle f(Q)\rangle^2_A)/\langle f(Q)\rangle^2_{A+B} + X_B(\langle f(Q)^2\rangle_B - \langle f(Q)\rangle^2_B)/\langle f(Q)\rangle^2_{A+B} \\ -(\langle f(Q)^2\rangle_{A+B} - \langle f(Q)\rangle^2_{A+B})/\langle f(Q)\rangle^2_{A+B} \end{aligned} \tag{A6}$$

where X_A and X_B are the mole fraction of phase A and B, respectively. Here, each value satisfies the condition of $X_A + X_B = 1$. Equation (A6) can be applied to the neutron structure factor $S^N(Q)$ by replacing an atomic scattering factor with a coherent scattering length of neutron.

References

1. Bridgman, P.W. Effects of very high pressures on glass. *J. Appl. Phys.* **1953**, *24*, 405–413. [CrossRef]
2. Hemley, R.J.; Mao, H.K.; Bell, P.M.; Mysen, B.O. Raman spectroscopy of SiO_2 glass at high pressure. *Phys. Rev. Lett.* **1986**, *57*, 747–750. [CrossRef] [PubMed]
3. Meade, C.; Hemley, R.J.; Mao, H.K. High-pressure X-ray diffraction of SiO_2 glass. *Pys. Rev. Lett.* **1992**, *69*, 1387–1390. [CrossRef] [PubMed]
4. Inamura, Y.; Katayama, Y.; Ustumi, W.; Funakoshi, K. Transformations in the intermediate-range structure of SiO_2 glass under high pressure and temperature. *Phys. Rev. Lett.* **2004**, *93*, 015501. [CrossRef]
5. Sato, T.; Funamori, N. Sixfold-coordinated amorphous polymorph of SiO_2 under high pressure. *Phy. Rev. Lett.* **2008**, *101*, 255502. [CrossRef]
6. Benmore, C.J.; Soignard, E.; Amin, S.A.; Guthrie, M.; Shastri, S.D.; Lee, P.L.; Yarger, J.L. Structural and topological changes in silica glass at pressure. *Phys. Rev. B* **2010**, *81*, 054105. [CrossRef]
7. Sato, T.; Funamori, N. High-pressure structural transformation of SiO_2 glass up to 100 GPa. *Phys. Rev. B* **2010**, *82*, 184102. [CrossRef]
8. Wakabayashi, D.; Funamori, N.; Sato, T.; Taniguchi, T. Compression behavior of densified SiO_2 glass. *Phys. Rev. B* **2011**, *84*, 144103. [CrossRef]
9. Zeidler, A.; Wezka, K.; Rowlands, R.F.; Whittaker, D.A.J.; Salmon, P.S.; Polidori, A.; Drewitt, J.W.E.; Klotz, S.; Fischer, H.E.; Wilding, M.C.; et al. High-pressure transformation of SiO_2 glass from a tetrahedral to an octahedral network: A joint approach using neutron diffraction and molecular dynamics. *Phys. Rev. Lett.* **2014**, *113*, 13501. [CrossRef]
10. Stolper, E. Water in silicate glasses: An infrared spectroscopic study. *Contrib. Mineral. Petrol.* **1982**, *81*, 1–17. [CrossRef]
11. Tomozawa, M. Water in glass. *J. Non-Cryst. Solids* **1985**, *73*, 197–204. [CrossRef]
12. Zotov, N.; Keppler, H.; Hannon, A.C.; Soper, A.K. The effect of water on the structure of silicate glasses—A neutron diffraction study. *J. Non-Cryst. Solids* **1996**, *202*, 153–163. [CrossRef]
13. Van der Steen, G.H.A.M.; van den Boom, H. Raman spectroscopic study of hydrogen-containing vitreous silica. *J. Non-Cryst. Solids* **1977**, *23*, 279–286. [CrossRef]
14. McMillan, P.F.; Remmele, R.L. Hydroxyl sites in SiO_2 glass: A note on infrared and Raman spectra. *Am. Mineral.* **1986**, *71*, 772–778.

15. Sato, T.; Funamori, N.; Wakabayashi, D.; Nishida, K.; Kikegawa, T. Coexistence of two states in optically homogeneous silica glass during the transformation in short-range order. *Phys. Rev. B* **2018**, *98*, 144111. [CrossRef]
16. Kohara, S.; Suzuya, K.; Kashihara, K.; Matsumoto, N.; Umesaki, N.; Sakai, I. A horizontal two-axis diffractometer for high-energy X-ray diffraction using synchrotron radiation on bending magnet beamline BL04B2 at SPring-8. *Nucl. Instrum. Methods Phys. Res.* **2001**, *A467–A468*, 1030–1033. [CrossRef]
17. Kohara, S.; Suzuya, K. High-energy X-ray diffraction studies of disordered materials. *Nucl. Instrum. Methods Phys. Res.* **2003**, *B199*, 23–28. [CrossRef]
18. Decker, D.L. High-pressure equation of state for NaCl, KCl, and CsCl. *J. Appl. Phys.* **1971**, *42*, 3239–3244. [CrossRef]
19. Sano-Furukawa, A.; Hattori, T.; Arima, H.; Yamada, A.; Tabata, S.; Kondo, M.; Nakamura, A.; Kagi, H.; Yagi, T. Six-axis multi-anvil press for high-pressure, high-temperature neutron diffraction experiments. *Rev. Sci. Instrum.* **2014**, *85*, 113905. [CrossRef]
20. Hattori, T.; Sano-Furukawa, A.; Arima, H.; Komatsu, K.; Yamada, A.; Inamura, Y.; Nakatani, T.; Seto, Y.; Nagai, T.; Utsumi, W.; et al. Design and performance of high-pressure PLANET beamline at pulsed neutron source at J-PARC. *Nucl. Instrum. Methods Phys. Res.* **2015**, *A780*, 55–67. [CrossRef]
21. Nishiyama, N.; Wang, Y.; Sanehira, T.; Irifune, T.; Rivers, M.L. Development of the multi-anvil assembly 6-6 for DIA and D-DIA type high-pressure apparatuses. *High Press. Res.* **2008**, *28*, 307–314. [CrossRef]
22. Kohara, S.; Ito, M.; Suzuya, K.; Inamura, Y.; Sakurai, Y.; Ohishi, Y.; Takata, M. Structural studies of disordered materials using high-energy X-ray diffraction from ambient to extreme conditions. *J. Phys. Condens. Matter* **2007**, *19*, 506101. [CrossRef]
23. Hannon, A.C. Unpublished GEM Datra Examples: Silica Glass, Diffraction Data for Vitreous SiO_2, Oxide Glass Data. 1990. Available online: https://www.isis.stfc.ac.uk/Pages/Oxide-Glass-Data.aspx (accessed on 3 December 2019).
24. Kameda, Y.; Amo, Y.; Usui, T.; Umebayashi, Y.; Ikeda, K.; Otomo, T. Neutron diffraction study on partial pair correlation functions of water at ambient temperature. *Bull. Chem. Soc. Jpn.* **2018**, *91*, 1586–1595. [CrossRef]
25. Elliot, S.R. Medium-range structural order in covalent amorphous solids. *Nature* **1991**, *354*, 445–452. [CrossRef]
26. Gaskell, P.H. Medium-range structure in glasses and low-Q structure in neutron and X-ray scattering data. *J. Non-Cryst. Solids* **2005**, *351*, 1003–1013. [CrossRef]
27. Crupi, C.; Carini, G.; González, M.; D'Angelo, G. Origin of the first sharp diffraction peak in glasses. *Phys. Rev. B* **2015**, *92*, 134206. [CrossRef]
28. Zeidler, A.; Salmon, P.S. Pressure-driven transformation of the ordering in amorphous network-forming materials. *Phys. Rev. B* **2016**, *93*, 214204. [CrossRef]
29. Onodera, Y.; Kohara, S.; Tahara, S.; Masuno, A.; Inoue, H.; Shiga, M.; Hirata, A.; Tsuchiya, K.; Hiraoka, Y.; Obayashi, I.; et al. Understanding diffraction patterns of glassy, liquid and amorphous materials via persistent homology analyses. *J. Ceram. Soc. Jpn.* **2019**, *127*, 853–863. [CrossRef]
30. Pasquarello, A.; Car, R. Identification of Raman defect lines as signatures of ring structures in vitreous silica. *Phys. Rev. Lett.* **1998**, *80*, 5145–5147. [CrossRef]
31. Zotov, N.; Yanev, Y.; Epelbaum, M.; Konstantinov, L. Effect of water on the structure of rhyolite glasses—X-ray diffraction and Raman spectroscopy studies. *J. Non-Cryst. Solids* **1992**, *142*, 234–246. [CrossRef]
32. Susman, S.; Volin, K.J.; Price, D.L.; Grimsditch, M.; Rino, J.P.; Kalia, R.K.; Vashishta, P.; Gwanmessia, G.; Wang, Y.; Libermann, R.C. Intermediate-range order in permanently densified vitreous SiO_2: A neutron-diffraction and molecular-dynamics study. *Phys. Rev. B* **1991**, *43*, 1994–1997. [CrossRef] [PubMed]
33. Onodera, Y.; Takimoto, Y.; Hijiya, H.; Taniguchi, T.; Urata, S.; Inaba, S.; Fujita, S.; Obayashi, I.; Hiraoka, Y.; Kohara, S. Origin of the mixed alkali effect in silicate glass. *NPG Asia Mater.* **2019**, *11*, 75. [CrossRef]
34. Hart, R.T.; Benmore, C.J.; Neuefeind, J.; Kohara, S.; Tomberli, B.; Egelstaff, P.A. Temperature dependence of isotropic quantum effects in water. *Phys. Rev. Lett.* **2005**, *94*, 047801. [CrossRef] [PubMed]
35. Kameda, Y.; Uemura, O. The intramolecular structure of oxonium ion in concentrated aqueous deuterochloric acid solutions. *Bull. Chem. Soc. Jpn.* **1992**, *65*, 2021–2028. [CrossRef]
36. Funamori, N.; Sato, T. A cubic boron nitride gasket for diamond-anvil experiments. *Rev. Sci. Instrum.* **2008**, *79*, 053903. [CrossRef]
37. Sakamaki, T. Density of hydrous magma. *Chem. Geol.* **2017**, *475*, 135–139. [CrossRef]

38. Sato, T.; Funamori, N.; Yagi, T. Helium penetrates into silica glass and reduces its compressibility. *Nat. Commun.* **2011**, *2*, 345. [CrossRef]
39. Shacckelford, J.F.; Masaryk, J.S. The interstitial structure of vitreous silica. *J. Non-Cryst. Solids* **1978**, *30*, 127–139. [CrossRef]
40. Zhang, Y.; Xu, Z. Atomic radii of noble gas elements in condensed phases. *Am. Mineral.* **1995**, *80*, 670–675. [CrossRef]
41. Funakoshi, K. Energy-dispersive X-ray Diffraction Study for Alkali Silicate Melts Using Synchrotron Radiation under High Pressure and Temperature. Ph.D. Thesis, Tokyo Institute of Technology, Tokyo, Japan, March 1997.
42. Hajdu, F. Analytic approximation for incoherent scattering X-ray intensities. *Acta Crystallogr.* **1971**, *A27*, 73–74. [CrossRef]
43. Pálinkás, G. Analytic approximation for the incoherent X-ray intensities of the atoms from Ca to Am. *Acta Crystallogr.* **1973**, *A29*, 10–12. [CrossRef]
44. Doyle, P.A.; Turner, P.S. Relativistic Hartree-Fock X-ray and electron scattering factors. *Acta Crystallogr.* **1968**, *A24*, 390–397. [CrossRef]
45. Paalman, H.H.; Pings, C.J. Numerical evaluation of X-ray absorption Factors for cylindrical samples and annular sample cells. *J. App. Phys.* **1962**, *33*, 2635–2639. [CrossRef]
46. Sears, V.F. Neutron scattering lengths and cross sections. *Neutron News* **1992**, *3*, 26–37. [CrossRef]
47. Kameda, Y.; Sasaki, M.; Usuki, T.; Otomo, T.; Itoh, K.; Suzuya, K.; Fukunaga, T. Inelasticity effect on neutron scattering intensities of the null-H_2O. *J. Neutron Res.* **2003**, *11*, 153–163. [CrossRef]
48. Blech, I.A.; Averbach, B.L. Multiple scattering of neutrons in vanadium and copper. *Phys. Rev.* **1965**, *137*, A113–A116. [CrossRef]
49. Lorch, E. Neutron diffraction by germania, silica and radiation-damaged silica glasses. *J. Phys. C Solid State Phys.* **1969**, *2*, 229–237. [CrossRef]

minerals

MDPI

Article

Axial Compressibility and Thermal Equation of State of Hcp Fe–5wt% Ni–5wt% Si

Eric Edmund [1,2], Francesca Miozzi [1], Guillaume Morard [1,3], Eglantine Boulard [1], Alisha Clark [1,4], Frédéric Decremps [1], Gaston Garbarino [5], Volodymyr Svitlyk [5], Mohamed Mezouar [5] and Daniele Antonangeli [1,*]

1 Sorbonne Université, Muséum National d'Histoire Naturelle, UMR CNRS 7590, Institut de Minéralogie, de Physique des Matériaux et de Cosmochimie, IMPMC, 75005 Paris, France; eric.edmund@hpstar.ac.cn (E.E.); francesca.miozzi@upmc.fr (F.M.); guillaume.morard@univ-grenoble-alpes.fr (G.M.); eglantine.boulard@upmc.fr (E.B.); alisha.clark@colorado.edu (A.C.); frederic.decremps@upmc.fr (F.D.)

2 Centre for High Pressure Science & Technology Advanced Research (HPSTAR), Shanghai 201203, China

3 Univ. Grenoble Alpes, Univ. Savoie Mont Blanc, CNRS, IRD, IFSTTAR, ISTerre, 38000 Grenoble, France

4 Geological Sciences, University of Colorado, Boulder, CO 80309-0399, USA

5 ESRF, F-38043 Grenoble, France; gaston.garbarino@esrf.fr (G.G.); svitlyk@esrf.fr (V.S.); mezouar@esrf.fr (M.M.)

* Correspondence: daniele.antonangeli@upmc.fr

Received: 4 December 2019; Accepted: 17 January 2020; Published: 23 January 2020

Abstract: Knowledge of the elastic properties and equations of state of iron and iron alloys are of fundamental interest in Earth and planetary sciences as they are the main constituents of telluric planetary cores. Here, we present results of X-ray diffraction measurements on a ternary Fe–Ni–Si alloy with 5 wt% Ni and 5 wt% Si, quasi-hydrostatically compressed at ambient temperature up to 56 GPa, and under simultaneous high pressure and high temperature conditions, up to 74 GPa and 1750 K. The established pressure dependence of the c/a axial ratio at ambient temperature and the pressure–volume–temperature (P–V–T) equation of state are compared with previous work and literature studies. Our results show that Ni addition does not affect the compressibility and axial compressibility of Fe–Si alloys at ambient temperature, but we suggest that ternary Fe–Ni–Si alloys might have a reduced thermal expansion in respect to pure Fe and binary Fe–Si alloys. In particular, once the thermal equations of state are considered together with velocity measurements, we conclude that elements other than Si and Ni have to be present in the Earth's inner core to account for both density and seismic velocities.

Keywords: iron alloys; nickel; silicon; high pressure; high temperature; equation of state; Earth's core; telluric planetary cores

1. Introduction

Different lines of evidence, from the analysis of meteorite compositions and Earth's differentiation models to the comparison of shock compression measurements with seismic observations, have put forward the notion that iron (Fe) is the main constituent of the Earth's core [1,2]. Cosmochemical arguments have also suggested that some weight percent (wt%) of nickel (Ni) is alloyed to iron [3], and, since Birch's pioneering work [1], it has been clear that elements lighter than an Fe–Ni alloy are present in the liquid outer core [2]. The density mismatch in the case of the solid inner core is less than that of the outer core, and though the fact that pure Fe is too dense in respect to seismic models has been well-established, the accurate quantification of this density deficit is still an object of active research (e.g., [4,5]).

Concerning inner core density, the effect of Ni is generally considered minor. Indeed, based on results obtained at high pressure and ambient temperature, Ni alloying seems to marginally affect the compressibility curves [6–8] and compressional sound velocities [9–11] of pure Fe. Possible effects of Ni on the shear sound velocity have been suggested [12], but calculations have indicated that these too should become negligible at inner core conditions [13]. However, studies of the elasticity of Fe–Ni alloys at simultaneous high-pressure, high-temperature conditions are very limited [9,13]. Face-centered cubic (fcc) Fe–Ni alloys are known to exhibit an anomalously low (almost zero) thermal expansion over a wide temperature range (the Invar effect) [14,15]. Anomalous compressibility has also been observed for certain ranges of pressure as a consequence of a pressure-induced Invar effect [16]. The degree to which such effects extend to the more moderate Ni concentration expected for the Earth's core (about 5 wt% [17]), and in the hexagonal close-packed (hcp) structure, which is expected to be stable at inner core pressure and temperature conditions [18,19], remain to be addressed. Additionally, the comparative effect of nickel and light elements on the elasticity and equation of state of a ternary alloy needs further investigation, with only few experimental studies in literature that have been limited to ambient temperature [7,8,10,12]. In particular, amongst the light elements that have been proposed to enter into the inner core composition, silicon has recently received much attention [7,8,10,19–23], as both core differentiation models [24,25] and isotopic arguments [26] have supported its presence at a small level of wt%.

The c/a axial ratio is another material property of interest for the characterization of the Earth's inner core, as this relates to elastic anisotropy in materials with an hcp structure. Recently, starting from a survey of previous work, the effect of pressure, temperature, and composition on the axial ratio of iron and alloys in the Fe–Ni–Si system have been parameterized [27]. However, the determination of axial ratios in static compression experiments is highly sensitive to the pressurizing conditions of the sample (e.g., [4,6]). Due to the challenges inherent in the generation of high static pressures under quasi-hydrostatic conditions, data on iron alloys became available only recently [8,21,23]. It follows that the direct determination of the axial ratios of iron alloys is critical to validate and improve such models to more adequately reflect actual material properties at the conditions of planetary interiors.

We thus carried out X-ray diffraction (XRD) measurements on a ternary Fe–Ni–Si alloy at high pressure and under simultaneous high pressure and high temperature conditions. Data have been used to establish the pressure dependence of the c/a axial ratio and the pressure–volume–temperature (P–V–T) equation of state. A comparison of the obtained results with previous work indicates that Ni does not affect the compressibility and axial compressibility of Fe–Si alloys at ambient temperature, but we suggest that Ni-bearing Fe–Si alloys might have a reduced thermal expansion. Hence, the effects of nickel should be considered when modeling the Earth's inner core or, more generally, iron-rich planetary cores.

2. Materials and Methods

An Fe–Ni–Si alloy with 5 wt% Ni and 5 wt% Si, hereafter referred to as Fe–5Ni–5Si, was synthesized by an ultra-rapid quench method at the Institut de Chimie et des Matériaux de Paris-Est, Paris, France, starting from Fe (99.9%, GoodFellow), Ni (99.9%, GoodFellow), and Si (99.9%, GoodFellow), following a protocol detailed in [28]. Sample composition and homogeneity were studied with an electron microprobe (Camparis center, Sorbonne Université, Paris, France) and with scanning electron microscopy (Institut de Minéralogie, de Physique des Matériaux et de Cosmochimie, IMPMC, Paris, France) analysis. The homogeneity of the sample was confirmed below the 1 μm scale, and the measured composition was observed within 0.3 (Ni and Si) and 0.7 (Fe) wt% of the nominal values.

High pressure was generated by using Le Toullec-type membrane-driven diamond anvil cells (DAC) equipped with Re gaskets and beveled diamonds with a culet size of 150/300 μm. Samples were scraped from a large, thin ribbon, and specimen dimensions were chosen such that there was no bridging between the sample and the diamonds, nor was there contact between the gasket and the sample. Selected samples were 5–7 μm thick for ambient temperature measurements and 1–3 μm thick for measurements at high temperature.

For room temperature measurements, the Fe–Ni–Si alloy was loaded alongside Pt as a pressure calibrant [29], with neon as a pressure transmitting medium so to ensure quasi-hydrostatic compression over the pressure range of interest.

For measurements at high temperature, the Fe–Ni–Si alloy was sandwiched between dry KCl disks that served as a pressure transmitting medium and to thermally and chemically insulate the sample from the diamonds. KCl was also used as the pressure calibrant by employing the thermal equation of state (EoS) reported in [30] and the temperature correction for KCl outlined in [31]. After loading the DAC, the assembly was left open to dry in a vacuum oven at 130 °C for several hours, after which the DAC was closed. This practice ensured that the moisture content was minimal within the sample chamber.

Angle-dispersive XRD measurements were performed at beamline ID27 at the European Synchrotron Radiation Facility (ESRF). The monochromatic radiation ($\lambda = 0.3738$ Å) was focused to approximately $3 \times 3\ \mu m^2$, horizontal x vertical full-with-half-maximum (FWHM). Diffraction patterns were collected on a MarCCD camera, with collection times of 30–60 s. Samples were heated on both sides by two continuous Nd:YAG fiber lasers (TEM00), each one supplying up to 110 W. Hot spots were approximately 20 μm in diameter, much larger than the FWHM of the focused X-ray beam. Temperatures were measured on both sides before and after XRD data collection, as well as on one side during data collection, by the spectroradiometric method while using a Planck fit of the observed blackbody radiation from the center of the heating spot. While absolute errors in temperature can be estimated on the order of 150 K, the measured temperature was seen to vary by less than 30 K during pattern collection (averaged over 3–5 measurements per diffraction pattern).

Diffraction images were calibrated against a CeO_2 standard and then radially integrated by using the Dioptas image processing software [32]. The integrated diffraction pattern was then analyzed by use of Le Bail fits in the software Jana 2006 [33].

3. Results

3.1. X-Ray Diffraction at High Pressure and Ambient Temperature

X-ray diffraction patterns were collected at ambient temperature in the stability field of the hcp structure between 21 and 56 GPa. A typical two-dimensional diffraction image and the corresponding integrated diffraction pattern are shown in Figure 1. All observed peaks are indexed as either belonging to sample or to the Ne pressure transmitting medium. Pressure was assessed from the measured volume of the Pt calibrant (collected independently from that of the sample by translating the cell a few microns from the sample position, before and after sample measurement). Aside from an intrinsic 1–2% error in absolute pressure due to the systematic effects of pressure calibration, the primary source of quantifiable error in pressure measurement was due to the pressure drift between the calibrant and sample measurements. Errors due to volume determination are very difficult to quantify, as the calibrant EoS is often the outcome of combined studies, and heavily parameterized with the use of results from diffraction as well as other techniques (e.g., ambient pressure ultrasonics). The Le Bail method of fitting power diffraction patterns while using the standard crystallography software typically leads to an underestimation of uncertainties when the sample background is large, as in the case of studies employing diamond anvil cells [34]. Differences between determined pressures before and after sample measurements were less than 1 GPa at all pressures, and errors in pressure were generally less than 2%. The observation of the near-hydrostatic peak ratios for the hcp Fe–5Ni–5Si sample and the minimal misfit of the Le Bail fit to both the sample and calibrant diffraction patterns indicated that the effects of deviatoric stress were negligible for the present dataset. Due to the weak intensity of the Ne reflections at the sample and calibrant measurement positions, as well as their overlap with strong sample and Pt reflections, the volume of Ne could not reliably be used to determine the magnitude of pressure gradients within the cell, which were expected to be small [35].

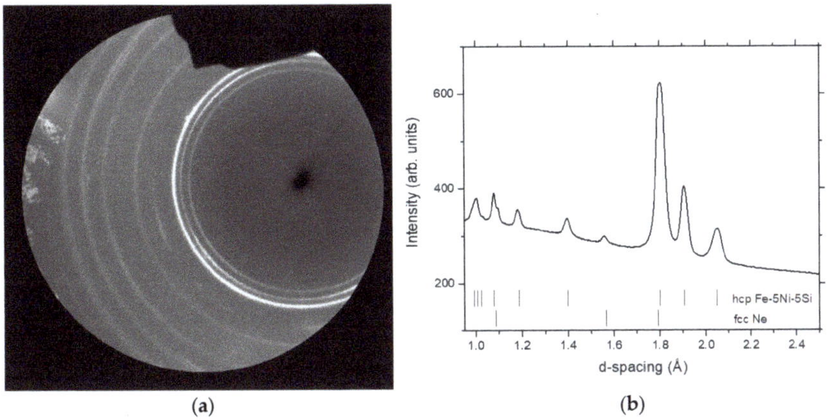

Figure 1. Example of the collected data at high pressure and ambient temperature. (**a**) 2D diffraction image and (**b**) integrated diffraction pattern of Fe–5Ni–5Si at 55.7 GPa.

The obtained compression curve is reported in Figure 2 along with the literature results for pure Fe [4] and alloys in the Fe–Ni–Si system [8,21,23] that were compressed by using either He or Ne as a pressure transmitting medium. All data are generally consistent within mutual uncertainties, thus highlighting that the compressibility of iron at 300 K is not significantly modified over the here-considered pressure range by addition of either Ni or Si at a small wt% level. Accordingly, the presence of Ni and Si affect the density of an alloy essentially by slightly increasing it, when Ni is present, or decreasing it, when Si is present; according to their difference in atomic mass in respect to Fe.

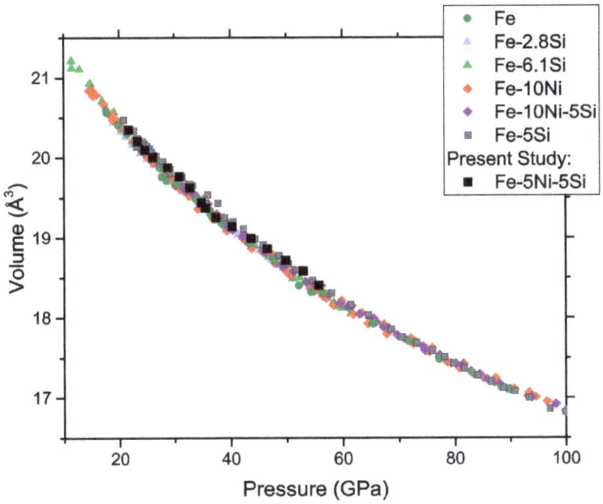

Figure 2. Isothermal compression curve of hexagonal close-packed (hcp) Fe–5Ni–5Si at ambient temperature compared to those of Fe [4] and Fe–Si alloys with 2.8 [21], 5 [23] and 6.1 wt% Si [21], an Fe–Ni alloy with 10 wt% Ni [8] and an Fe–Ni–Si alloy with 10 wt% Ni and 5 wt% Si [8], all in the hcp structure.

Conversely, as illustrated in Figure 3, both Si and Ni have a significant effect on the axial c/a ratio of the alloys, increasing it in respect to that of pure Fe while only moderately affecting its pressure dependence. As such, both Si and Ni are expected to qualitatively affect, in the same way, the elastic anisotropy of the alloy at ambient temperature. The pressure dependence of the c/a ratio of pure Fe and the effects on the c/a ratio when alloyed with 10 wt% Ni and when alloyed with 5 wt% Si were estimated by following the parameterization proposed in [27] (Figure 3). It stands clear that this model overestimates the c/a ratio of hcp Fe at low pressure, and it also overestimates its reduction with increasing pressure. On the other hand, the effect of Si and Ni inclusion are underestimated.

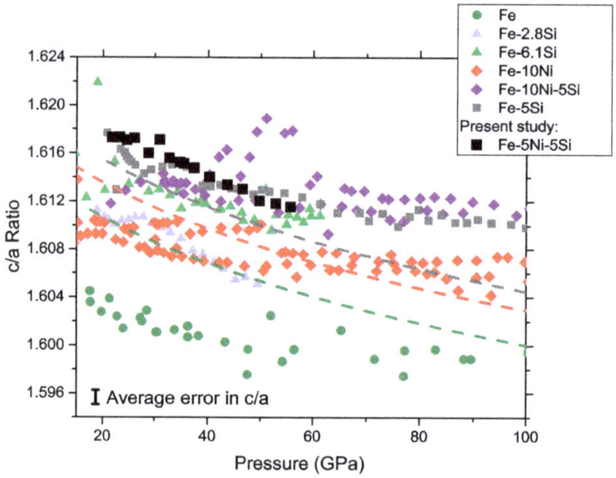

Figure 3. Axial c/a ratio of hcp alloys in the Fe–Ni–Si system at ambient temperature as a function of pressure. Legend is same as in that in Figure 2. Dashed lines are the expected c/a ratio according to the parameterization proposed in [27] for Fe (dark green), Fe–10Ni (red), and Fe–5Si (gray).

3.2. X-Ray Diffraction at High Pressure and High Temperature

X-ray diffraction patterns were collected under simultaneous high pressure and high temperature conditions along two isotherms, at 1450 K, upon compression up to about 74 GPa, and at 1750 K upon decompression. Temperatures varied by less than 50 K along each isotherm (1σ variation of 40 K for the 1450 K isotherm and 20 K for the 1750 K isotherm). A typical two-dimensional diffraction image and the corresponding integrated diffraction pattern are shown in Figure 4. All observed peaks are indexed either belong to the sample or to KCl. Pressure was increased (or decreased) step-wise by using a gas-driven membrane while the sample was maintained at high temperature via laser heating. The relative alignment of the focused X-ray spot, laser-heating system, and cell assembly was checked every 10–20 GPa. The application of high temperatures resulted in the relaxation of deviatoric stress in both the sample and the pressure-transmitting medium, becoming effectively negligible for high-temperature runs in the present study (quasi-hydrostatic conditions). This also ensured a comparable stress-state in both the compression and decompression runs. As previously mentioned in Section 3.1, standard crystallography softwares tend to underestimate error bars. Reported error bars for high temperature data were the determined by refitting the diffraction patterns with PDIndexer.

A single fcc phase was observed along the 1450 K isotherm up to 31–35 GPa, pressure above which this phase transformed into the hcp structure. Remnant traces of fcc diffraction lines were observed to 56 GPa. Along the 1750 K isotherm, an fcc phase was observed to crystallize upon decompression to pressures below 40 GPa. The entire pressure–volume–temperature (P–V–T) data set measured for the hcp structure in this study, including data at ambient temperature, is reported in Figure 5.

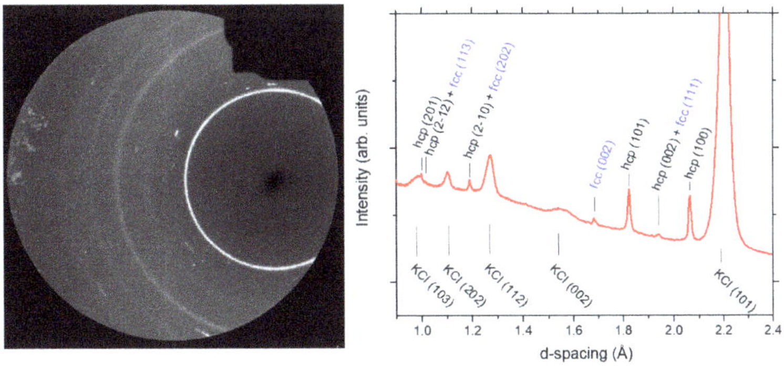

(a) (b)

Figure 4. Example of the collected data at high pressure and high temperature. (**a**) 2D diffraction image and (**b**) integrated diffraction pattern of Fe–5Ni–5Si at 52 GPa and 1470 K.

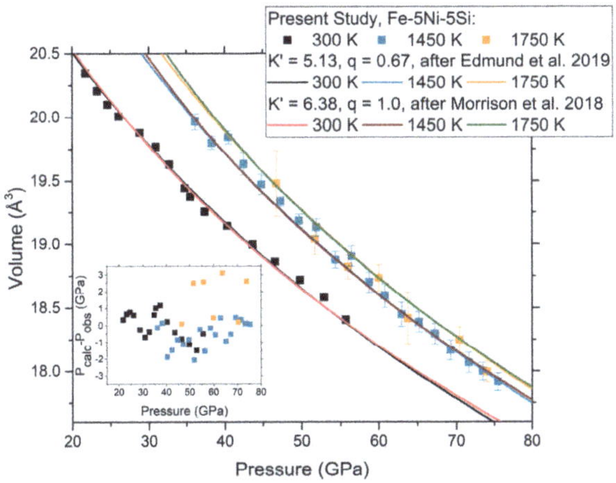

Figure 5. Isothermal compression curves of hcp Fe–5Ni–5Si at 300, 1450 and at 1750 K. Color-coded solid lines are the thermal equation of state fit to the experimental data, with the values of K′ and q fixed according to either Edmund et al., 2019 [23] or Morrison et al., 2018 [8]—cases 1 and 3, specifically. Inset: residuals in pressure for the present dataset (colors correspond to the respective isotherms of the main figure). See main text for more details.

Acquired data were used to establish a thermal equation of state (P–V–T EoS) following the formalism presented in [23]. Briefly, the ambient temperature component of the P–V–T thermal model was based on a Rydberg–Vinet equation of state

$$P_{300\mathrm{K}}(\mathrm{V}) = 3K_0\left(\frac{1-\eta}{\eta^2}\right)\exp\left[\frac{3}{2}(K'-1)(1-\eta)\right] \tag{1}$$

where $\eta = \left(\frac{V}{V_0}\right)^{\frac{1}{3}}$, V is the unit cell volume, V_0 is the ambient pressure unit cell volume, K_0 is the bulk modulus, and K′ is its pressure derivative at ambient pressure and temperature. Minor differences in fitted elastic parameters can arise when using a different EoS formalism (e.g., Birch–Murnaghan vs. Rydberg–Vinet). The Rydberg–Vinet formalism was chosen here because it is generally accepted to better reflect the compressional behavior of metals at high pressures compared to other finite-strain equations of state [36,37].

The thermal pressure includes vibrational (P_{vib}) and electronic and anharmonic (P_{el+anh}) terms:

$$P(V,T) = P_{300K}(V) + P_{vib}(V,T)|_{300}^{T} + P_{el+anh}(V,T)|_{300}^{T} \quad (2)$$

The vibrational term of the thermal pressure can be written as:

$$P_{vib}(V,T) = \frac{9R\gamma_{vib}}{V}\left[\frac{\theta_D}{8} + T\left(\frac{T}{\theta_D}\right)^3 \int_{300}^{\theta_D/T} \frac{x^3}{\exp(x)-1}dx\right] \quad (3)$$

where γ_{vib} is the vibrational Grüneisen parameter, θ_D is the Debye temperature, and R is the ideal gas constant.

The volume dependence of the vibrational Grüneisen parameter and Debye temperature are, respectively, given by:

$$\frac{\gamma_{vib}}{\gamma_{vib,0}} = \left(\frac{V}{V_0}\right)^q \quad (4)$$

$$\theta_D = \theta_{D,0} \exp\left[\left(\gamma_{vib,0} - \gamma_{vib}\right)/q\right] \quad (5)$$

where q is an exponent that characterizes the volume dependence.

As both P_{el} and P_{anh} scale as T^2, a single term was used in the fitting procedure:

$$P_{el+anh}(V,T) = \frac{\gamma_e}{V}\beta_0\left(\frac{V}{V_0}\right)^k T^2 \quad (6)$$

where γ_e is the electronic Grüneisen parameter, β_0 is the electronic heat capacity, and k is an exponent that characterizes the volume dependence.

Though the variation from study to study across the literature in terms of data points at ambient temperature is small (see Figure 2), the reported values for V_0, K_0 and K′ vary up to about 3%, 40% and 35%, respectively [4,8,21,23]. As all these studies were performed employing Ne or He as a pressure transmitting medium and with pressure calibrations that are generally in close agreement (e.g., [38,39]), we argue that such discrepancies likely arise primarily from technical challenges that are associated with very high pressures generation and the strong covariance between V_0, K_0 and K′ [5,8]. In regards to the former, at Mbar pressures, it is expected that all pressure-transmitting media develop non-negligible stress gradients across the sample chamber [38,40] that may represent a systematic bias among these studies and strongly contribute to the differences between the available quasi-hydrostatic EoS for the Fe–Ni–Si system (e.g., [4,8,23]). In regard to the latter, the lack of direct measurements on the V_0 of hcp iron and iron alloys, these being high-pressure phases, directly reflects on the ambient temperature EoS parameters, which are particularly sensitive to the investigated pressure range, the employed pressure metrology, and the stress-state of both the sample and the calibrant. The fitting of only the 300 K data of the present study with a fixed K′ value systematically resulted in values of V_0 and K_0 that were in agreement with those reported in studies over a similar pressure range (e.g., [21]) but were higher in respect to studies that used the same value of K′ but covered a larger pressure range. The simultaneous refinement of both ambient and high temperature datasets provides a closer agreement with the V_0 and K_0 reported by studies that extended to higher pressures [8,23], likely due to the expanded pressure range of the high T data and the cancellation of experimental errors across the measured isotherms. Given the uncertainty in K′ for the Fe–Si alloys, q was kept fixed because it is related to the variation of K′ with temperature [41] and strongly dependent on the fixed value of K′ when left as a free parameter. In view of the overall limited pressure and temperature range covered in this study and in order to better highlight the effects due to Ni, the values for K′, q and θ_D were either fixed to those established for Fe–5Si [23] (cases 1 and 2) or to those proposed for Fe–10Ni–5Si [8] (cases 3 and 4). Furthermore, given the considerable uncertainties over the role of anharmonic effects in the thermal EoS of iron and iron alloys, Table 1 includes fits to the present dataset with (cases 1 and 3) or without (cases 2 and 4) the incorporation of an anharmonic term. When an anharmonic term was incorporated, following previous studies, γ_e, β_0 and k were fixed to 2, 3.2 and 1.34, respectively [23]. In particular, value of β_0 was chosen in between previously reported values of 3.9 [5] and ~1.7 [4]—it must be noted, however, that the latter employed a different parameterization which led to small differences in pressure dependence, as it separately treated lattice and electronic anharmonicity. As highlighted in Figure 5, these fitting strategies similarly accounted for experimental data (as a matter of fact, the fits are virtually indistinguishable over the investigated pressure and temperature range).

Table 1. Pressure–volume–temperature equation of state (P–V–T EoS) fitting parameters for hcp Fe–5Ni–5Si (cases 1–4) and reported thermoelastic parameters for Fe–5Si [23] and Fe–10Ni–5Si [8]. For cases 1–4, when errors are not given, the parameter was kept fixed (see main text).

Case#	V_0 (Å^3)	K_0 (GPa)	K′	θ_D (K)	$\gamma_{vib,0}$	q	γ_e	β_0($10^{-6}\cdot cm^3\cdot mol^{-1}\cdot J\cdot K^{-1}$)	k
1	22.61 ± 0.08	160 ± 5	5.13	422 [1]	1.66 ± 0.09	0.67	2	3.2	1.34
2	22.61 ± 0.07	160 ± 5	5.13	422 [1]	2.13 ± 0.08	0.67	-	-	-
3	22.97 ± 0.10	125 ± 5	6.38	417 [1]	1.82 ± 0.09	1	2	3.2	1.34
4	22.97 ± 0.09	125 ± 4	6.38	417 [1]	2.30 ± 0.09	1	-	-	-
[23]	22.59	163	5.13	422	1.73	0.67	2	3.2	1.34
[8]	22.84	125	6.38	417	2	1	†	†	†

[1] Literature values for θ_D in hcp Fe and Fe–Si alloys ranged between 417 and 422 K [5,8,23,42,43]. Actual values within this range do not affect fit quality or other fitting parameters.† Reference [8] employed anharmonic thermal pressure corrections after reference [4].

When the P–V–T EoS determined here for hcp Fe–5Ni–5Si is compared to the P–V–T EoS of hcp Fe–5Si [23] (Figure 6), the effects of the inclusion of 5 wt% Ni are found to be very minor on compressibility (P–V curves of Fe–5Ni–5Si and Fe–5Si are almost parallel along the three isotherms), while the Ni-bearing Fe–Si alloys appear to have a somewhat reduced thermal expansion. Indeed, the volumes of Fe–5Ni–5Si along the high temperature isotherms are systematically below those of Fe–5Si (although the difference is small). All data presented in this study may be found in the Supplementary Information.

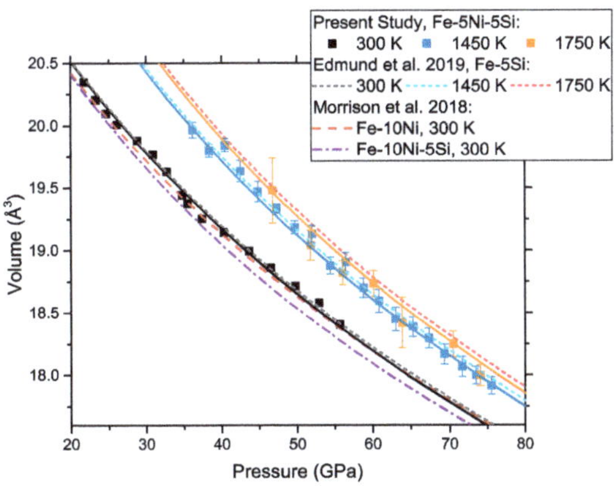

Figure 6. Comparison of the isothermal compression curves of hcp Fe–5Ni–5Si at 300, 1450 and 1750 K with literature results on Fe–Si, Fe–Ni and Fe–Ni–Si alloys. Color-coded solid lines are the thermal equation of state fit to the experimental data (see Figure 5 and discussion in the main text). P–V–T relations for hcp Fe–5Si [23] are also reported as dotted curves (same temperature–color code). Dashed and dash-dotted lines are the P–V relation established at 300 K for hcp Fe–10Ni and hcp Fe–10Ni–5Si, respectively [8]. While compression curves at ambient temperature do not show differences, the high-temperature compression curves of Fe–5Ni–5Si are parallel but systematically below those of Fe–5Si, suggesting a reduced thermal expansion.

4. Discussion

Seismic studies have argued for an elastically anisotropic inner core [44–46], and this feature has been commonly ascribed to the preferential alignment of comprising crystals (e.g., [47,48]). The elastic anisotropy of the hcp-structured materials in the meridian a–c plane is largely governed by the ratio of the elastic moduli C_{33}/C_{11} which, in turn, can be related to the axial compressibility and, hence, to the c/a axial ratio. In agreement with previous measurements [8], both nickel and silicon are individually observed to increase the c/a axial ratio in binary Fe–Si and Fe–Ni alloys in respect of pure iron, while the effect on its pressure dependence is less evident (Figure 3). Interestingly, we note that Ni addition does not significantly modify the c/a ratio of the Fe–Si alloys. As illustrated in Figure 3, Fe–5Ni–5Si and Fe–10Ni–5Si have the same c/a ratio as Fe–5Si. As already pointed out, the recently proposed parameterization of the effects of pressure, temperature and composition on the c/a axial ratio of the hcp Fe–Ni–Si alloys [27] fails to reproduce the values measured in quasi-hydrostatically compressed Fe, binary Fe–Si and Fe–Ni alloys, and ternary Fe–Ni–Si alloys.

To quantitatively correlate the c/a ratio with C_{33}/C_{11} and elastic anisotropy is not a straightforward exercise (see, for instance, discussion in [8]). Calculations pointed out the importance of the c/a ratio to derive the correct elastic moduli from stress–strain methods [49,50]. As the c/a ratio in Fe–Ni, Fe–Si and Fe–Ni–Si alloys is closer to the ideal value for hcp structure of 1.633 (Figure 3), we might expect the elastic anisotropy to be reduced in respect to pure Fe, but direct measurements remain necessary to substantiate this speculation.

Concerning inner core density, the current results do not allow for a reliable extrapolation to inner core conditions. Further measurements over an extended pressure and temperature range are planned. Nonetheless, a few considerations can be made. The negligible effect of nickel and silicon at a small wt% level on the compressibility of pure iron argues for substitutional alloys where Ni and/or Si randomly replace Fe in the crystal lattice without significantly modifying it. Thus, the main effect on density (and the only effect at ambient temperature) is the scaling according to relative atomic masses. On the other hand, thermal expansion seems to be smaller in hcp Fe–Ni–Si alloys than in pure hcp-Fe [5] and hcp-Fe–Si alloys [23]. On a qualitative ground, this is not surprising because fcc Fe–Ni alloys are known for their anomalously low thermal expansion. In hcp alloys and for moderate Ni concentration in the range expected for the Earth's core, such effect is much reduced but appears to still be present. Accordingly, nickel acts to increase the density of the alloy at inner core conditions by both the increased atomic mass of Ni in respect to Fe and because Ni-bearing alloys have comparatively smaller volume at core temperatures due to the reduced thermal expansion.

Based on these arguments, the effects of nickel vs. silicon alloying are qualitatively different in many aspects. Though both Ni and Si do not change the compressibility of Fe (Figures 2 and 6), Ni increases density without significantly affecting compressional sound velocity [9–11,13], while Si decreases density and increases compressional sound velocity [10,20,23]. No composition within the binary Fe–Ni or Fe–Si system is expected to match both seismically observed velocities and densities at inner core conditions [13,20,23]. In particular, an Fe–Si alloy with 5 wt% Si can match the density of the Earth's inner core for all reasonable core temperatures, but its velocities remain too high in respect to seismological observations [23]. The addition of Ni would call for more Si to account for the inner core density, pushing the velocity of this ternary Fe–Ni–Si alloy even further away. Fe–Si alloys whose velocities are expected to get close to seismological observations are too dense at relevant pressure and temperatures [20,23]. The addition of Ni would make the ternary Fe–Ni–Si alloy even denser. We can then conclude that no Fe–Ni–Si alloy can account for the seismically observed physical properties of the inner core and that elements other than Si and Ni have to be present in the Earth's inner core.

Supplementary Materials: The following are available online at http://www.mdpi.com/2075-163X/10/2/98/s1.

Author Contributions: Conceptualization, G.M. and D.A.; formal analysis, E.E. and F.M.; funding acquisition, F.D. and D.A.; investigation, E.E., F.M., G.M., E.B., A.C., G.G., V.S., M.M. and D.A.; project administration, D.A.; supervision, G.M., F.D. and D.A.; writing—original draft, E.E. and D.A.; writing—review and editing, F.M., G.M., E.B. and V.S. All authors have read and agreed to the published version of the manuscript.

Funding: This work was supported by the Investissements d'Avenir programme (reference ANR-11-IDEX-0004-02) and more specifically within the framework of the Cluster of Excellence MATriaux Interfaces Surfaces Environnement (MATISSE) led by Sorbonne Université. This project has received funding from the European Research Council (ERC) under the European Union's Horizon 2020 research and innovation programme (Grant agreement No. 724690). F.M. has received funding from the European Research Council (ERC) under the European Union's Horizon 2020 research and innovation Programme (Grant agreement 670787).

Acknowledgments: The authors wish to thank Jeroen Jacobs for technical assistance at the ESRF. Yoann Guarnelli and Paraskevas Parisiadis are acknowledged for their support at IMPMC. Femtosecond laser micro-machining at the Institut de Minéralogie, de Physique des Matériaux et de Cosmochimie (IMPMC), Paris, has been developed and realized by the "Cellule Projet" with the financial support of ANR 2010-JCJC-604-01.

Conflicts of Interest: The authors declare no conflict of interest. The funders had no role in the design of the study; in the collection, analyses, or interpretation of data; in the writing of the manuscript, or in the decision to publish the results.

References

1. Birch, F. Elasticity and constitution of the Earth's interior. *J. Geophys. Res.* **1952**, *57*, 227–286. [CrossRef]
2. Poirier, J.P. Light elements in the Earth's outer core: A critical review. *Phys. Earth Planet. Inter.* **1994**, *85*, 319–337. [CrossRef]
3. Li, J.; Fei, Y. Experimental constraints on core composition. In *Treatise on Geochemistry*; Elsevier: New York, NY, USA, 2007; pp. 1–31.
4. Dewaele, A.; Loubeyre, P.; Occelli, F.; Mezouar, M.; Dorogokupets, P.I.; Torrent, M. Quasihydrostatic equation of state of iron above 2 Mbar. *Phys. Rev. Lett.* **2006**, *97*, 215504. [CrossRef] [PubMed]
5. Fei, Y.; Murphy, C.; Shibazaki, Y.; Shahar, A.; Huang, H. Thermal equation of state of hcp-iron: Constraint on the density deficit of the Earth's solid inner core. *Geophys. Res. Lett.* **2016**, *43*, 6837–6843. [CrossRef]
6. Mao, H.K.; Wu, Y.; Chen, L.C.; Shu, J.F. Static compression of iron to 300 GPa and $Fe_{0.8}Ni_{0.2}$ alloy to 260 GPa: Implications for composition of the core. *J. Geophys. Res.* **1990**, *95*, 21737–21742. [CrossRef]
7. Asanuma, H.; Ohtani, E.; Sakai, T.; Terasaki, H.; Kamada, S.; Hirao, N.; Ohishi, Y. Static compression of $Fe_{0.83}Ni_{0.09}Si_{0.08}$ alloy to 374 GPa and $Fe_{0.93}Si_{0.07}$ alloy to 252 GPa: Implications for the Earth's inner core. *Earth Planet. Sci. Lett.* **2011**, *310*, 113–118. [CrossRef]
8. Morrison, R.A.; Jackson, J.M.; Sturhahn, W.; Zhang, D.; Greenberg, E. Equation of state and anisotropy of Fe-Ni-Si alloys. *J. Geophys. Res. Solid Earth* **2018**, *123*, 4647–4675. [CrossRef]
9. Kantor, A.P.; Kantor, I.Y.; Kurnosov, A.V.; Kuznetsov, A.Y.; Dubrovinskaia, N.A.; Krisch, M.; Bossak, A.A.; Dimitriev, V.P.; Urusov, V.S.; Dubrovinsky, L.S. Sound wave velocities of fcc Fe-Ni alloy at high pressure and temperature by mean of inelastic X-ray scattering. *Phys. Earth Planet. Inter.* **2007**, *164*, 83–89. [CrossRef]
10. Antonangeli, D.; Siebert, J.; Badro, J.; Farber, D.L.; Fiquet, G.; Morard, G.; Ryerson, F.J. Composition of the Earth's inner core from high-pressure velocity measurements in Fe-Ni-Si alloys. *Earth Planet. Sci. Lett.* **2010**, *295*, 292–296. [CrossRef]
11. Wakamatsu, T.; Ohta, K.; Yagi, T.; Hirose, K.; Ohishi, Y. Measurements of sound velocity in iron-nickel alloys by femtosecond laser pulses in a diamond anvil cell. *Phys. Chem. Miner.* **2018**, *45*, 589–595. [CrossRef]
12. Morrison, R.A.; Jackson, J.M.; Sturhahn, W.; Zhao, J.; Toellner, T.S. High pressure thermoelasticity and sound velocities of Fe-Ni-Si alloys. *Phys. Earth Planet. Inter.* **2019**, *294*, 106268. [CrossRef]
13. Martorell, B.; Brodholt, J.; Wood, I.G.; Vočadlo, L. The effect of nickel on the properties of iron at the conditions of Earth's inner core: Ab initio calculations of seismic wave velocities of Fe-Ni alloys. *Earth Planet. Sci. Lett.* **2013**, *365*, 143–151. [CrossRef]
14. Guillaume, C.E. Recherches sur les aciers au nickel. Dilatations aux temperatures elevees; resistance electrique. *CR Acad. Sci.* **1897**, *125*, 18.
15. Van Schilfgaarde, M.; Abrikosov, I.A.; Johansson, B. Origin of the Invar effect in iron-nickel alloys. *Nature* **1999**, *400*, 46–49. [CrossRef]
16. Dubrovinsky, L.; Dubrovinskaia, N.; Abrikosov, I.A.; Vennström, M.; Westman, F.; Carlson, S.; van Schilfgaarde, M.; Johansson, B. Pressure-induced invar effect in Fe-Ni alloys. *Phys. Rev. Lett.* **2001**, *86*, 4851–4854. [CrossRef]
17. McDonough, W.F.; Sun, S.S. Composition of the Earth. *Chem. Geol.* **1995**, *120*, 223–253. [CrossRef]
18. Tateno, S.; Hirose, K.; Komabayashi, T.; Ozawa, H.; Ohishi, Y. The structure of Fe-Ni alloy in Earth's inner core. *Geophys. Res. Lett.* **2012**, *39*, L12305. [CrossRef]

19. Tateno, S.; Kuwayama, Y.; Hirose, K.; Ohishi, Y. The structure of Fe-Si alloy in Earth's inner core. *Earth Planet. Sci. Lett.* **2015**, *418*, 11–19. [CrossRef]
20. Antonangeli, D.; Morard, G.; Paolasini, L.; Garbarino, G.; Murphy, C.A.; Edmund, E.; Decremps, F.; Fiquet, G.; Bosak, A.; Mezouar, M.; et al. Sound velocities and density measurements of solid hcp-Fe and hcp-Fe-Si(9wt.%) alloy at high pressure: Constraints on the Si abundance in the Earth's inner core. *Earth Planet. Sci. Lett.* **2018**, *482*, 446–453. [CrossRef]
21. Kamada, S.; Suzuki, N.; Maeda, F.; Hirao, N.; Hamada, N.; Hamada, M.; Ohtani, E.; Masuda, R.; Mitsui, T.; Ohishi, Y.; et al. Electronic properties and compressional behavior of Fe-Si alloys at high pressure. *Am. Mineral.* **2018**, *103*, 1959–1965. [CrossRef]
22. Komabayashi, T.; Pesce, G.; Morard, G.; Antonangeli, D.; Sinmyo, R.; Mezouar, M. Phase transition boundary between fcc and hcp structures in Fe-Si alloys and its implications for terrestrial planetary cores. *Am. Mineral.* **2019**, *104*, 94–99. [CrossRef]
23. Edmund, E.; Antonangeli, D.; Decremps, F.; Miozzi, F.; Morard, G.; Boulard, E.; Clark, A.N.; Ayrinhac, S.; Gauthier, M.; Morand, M.; et al. Velocity-density systematics of Fe-5wt%Si: Constraints on Si content in the Earth's inner core. *J. Geophys. Res. Solid Earth* **2019**, *124*, 3436–3447. [CrossRef]
24. Siebert, J.; Badro, J.; Antonangeli, D.; Ryerson, F.J. Terrestrial accretion under oxidizing conditions. *Science* **2013**, *339*, 1194–1197. [CrossRef] [PubMed]
25. Fischer, R.A.; Nakajima, Y.; Campbell, A.J.; Frost, D.J.; Harries, D.; Langenhorst, F. High pressure metal-silicate partitioning of Ni, Co, V, Cr, Si and O. *Geochim. Cosmochim. Acta* **2015**, *167*, 177–194. [CrossRef]
26. Fitoussi, C.; Bourdon, B.; Kleine, T.; Oberli, F.; Reynolds, B.C. Si isotope systematics of meteorites and terrestrial peridotites: Implications for Mg/Si fractionation in the solar nebula and for Si in the Earth's core. *Earth Planet. Sci. Lett.* **2009**, *287*, 77–85. [CrossRef]
27. Fischer, R.A.; Campbell, A.J. The axial ration of hcp Fe and Fe-Ni-Si alloys to the conditions of Earth's nner core. *Am. Mineral.* **2015**, *100*, 2718–2724. [CrossRef]
28. Morard, G.; Andrault, D.; Guignot, N.; Siebert, J.; Garbarino, G.; Antonangeli, D. Melting of Fe-Ni-Si and Fe-Ni-S alloys at magabar pressure: Implications for the core-mantle boundary temperature. *Phys. Chem. Miner.* **2011**, *38*, 767–776. [CrossRef]
29. Dorogokupets, P.I.; Dewaele, A. Equation of state of MgO, Au, Pt, NaCl-B1 and NaCl-B2: Internally consistent high-temperature pressure scale. *High Press. Res.* **2012**, *27*, 431–446. [CrossRef]
30. Dewaele, A.; Belonoshko, A.B.; Garbarino, G.; Occelli, F.; Bouvier, P.; Hanfland, M.; Mezouar, M. High-pressure high-temperature equation of state of KCl and KBr. *Phys. Rev. B* **2012**, *85*, 214105. [CrossRef]
31. Campbell, A.J.; Danielson, L.; Righter, K.; Seagle, C.T.; Wang, Y.; Prakapenka, V.B. High pressure effects on the iron-iron oxide and nickel-nickel oxide oxygen fugacity buffers. *Earth Planet. Sci. Lett.* **2009**, *286*, 556–564. [CrossRef]
32. Prescher, C.; Prakapenka, V.B. DIOPTAS: A program for reduction of two-dimensional X-ray diffraction data and data exploration. *High Press. Res.* **2015**, *35*, 223–230. [CrossRef]
33. Petříček, V.; Dusek, M.; Palatinus, L. Crystallographic computing system JANA2006: General features. *Z. Krist. Cryst. Mater.* **2014**, *229*, 345–352. [CrossRef]
34. Angel, R.J.; Equation of state. High-temperature and high-pressure crystal chemistry. *Rev. Mineral. Gepchemistry* **2000**, *41*, 445–520.
35. Klotz, S.; Chervin, J.C.; Munsch, P.; Le Marchand, G. Hydrostatic limit of 11 pressure transmitting media. *J. Phys. D Appl. Phys.* **2009**, *42*, 075413. [CrossRef]
36. Dewaele, A.; Loubeyre, P.; Mezouar, M. Equation of state of six metals above 94 GPa. *Phys. Rev. B* **2004**, *70*, 094112. [CrossRef]
37. Dewaele, A.; Torrent, M.; Loubeyre, P.; Mezouar, M. Compression curves of transition metals in the Mbar range: Experiments and projector augmented-wave calculations. *Phys. Rev. B* **2008**, *78*, 104102. [CrossRef]
38. Dorfman, S.M.; Prakapenka, V.B.; Meng, Y.; Duffy, T.S. Intercomparison of pressure standards (Au, Pt, Mo, MgO, NaCl and Ne) to 2.5 Mbar. *J. Geophys. Res.* **2012**, *117*, B08210. [CrossRef]
39. Ye, Y.; Prakapenka, V.B.; Meng, Y.; Shim, S.-H. Intercomparison of the gold, platinum and MgO pressure scales up to 140 GPa and 2500 K. *J. Geophys. Res. Solid Earth* **2017**, *122*, 3450–3464. [CrossRef]
40. Dewaele, A.; Loubeyre, P. Pressurizing conditions in helium-pressure-transmitting-medium. *High Press. Res.* **2007**, *27*, 419–429. [CrossRef]

41. Bina, C.R.; Hellfrich, G.R. Calculation of Elastic Properties from Thermodynamic Equation of State Principles. *Annu. Rev. Earth Sci.* **1992**, *20*, 527–552. [CrossRef]
42. Fischer, R.A.; Campbell, A.J.; Caracas, R.; Reaman, D.M.; Dera, P.; Prakapenka, V.B. Equation of state and phase diagram of Fe-16Si alloy as a candidate component of Earth's core. *Earth Planet. Sci. Lett.* **2012**, *357–358*, 268–276. [CrossRef]
43. Fischer, R.A.; Campbell, A.J.; Caracas, R.; Reaman, D.M.; Heinz, D.L.; Dera, P.; Prakapenka, V.B. Equation of state in the Fe-FeSi system at high pressures and temperatures. *J. Geophys. Res. Solid Earth* **2014**, *119*, 2810–2827. [CrossRef]
44. Attanayake, J.; Cormier, V.F.; De Silva, S.M. Uppermost inner core seismic structure—New insights from body waveform inversion. *Earth Planet. Sci. Lett.* **2014**, *385*, 49–58. [CrossRef]
45. Deuss, A. Heterogeneity and anisotropy of Earth's inner core. *Annu. Rev. Earth Planet. Sci.* **2014**, *42*, 103–126. [CrossRef]
46. Lythgoe, K.H.; Deuss, A.; Rudge, J.F.; Neufeld, J.A. Earth's inner core: Innermost inner core or hemispherical variations? *Earth Planet. Sci. Lett.* **2014**, *385*, 181–189. [CrossRef]
47. Antonangeli, D.; Merkel, S.; Farber, D.L. Elastic anisotropy of in hcp metals at high pressure and the sound wave anisotropy of the Earth's inner core. *Geophys. Res. Lett.* **2006**, *33*, L24303. [CrossRef]
48. Lincot, A.; Cardin, P.; Deguen, R.; Merkel, S. Multiscale model of global inner-core anisotropy induced by hcp alloy plasticity. *Geophys. Res. Lett.* **2016**, *43*, 1084–1091. [CrossRef]
49. Gannarelli, C.M.S.; Alfé, D.; Gillan, M.J. The axial ration of hcp iron at the conditions of the Earth's inner core. *Phys. Earth Planet. Inter.* **2005**, *152*, 67–77. [CrossRef]
50. Vočadlo, L.; Dobson, D.P.; Wood, I.G. Ab initio calculations of the elasticity of hcp-Fe as a function of temperature at inner-core pressure. *Earth Planet. Sci. Lett.* **2009**, *288*, 534–538. [CrossRef]

minerals

MDPI

Article

Study on the High-Pressure Behavior of Goethite up to 32 GPa Using X-Ray Diffraction, Raman, and Electrical Impedance Spectroscopy

Ruilian Tang [1,2], Jiuhua Chen [3,*], Qiaoshi Zeng [1], Yan Li [4], Xue Liang [3], Bin Yang [1] and Yu Wang [5]

1 Center for High Pressure Science and Technology Advanced Research, Changchun 130012, China; ruilian.tang@hpstar.ac.cn (R.T.); zengqs@hpstar.ac.cn (Q.Z.); yangbin@hpstar.ac.cn (B.Y.)
2 School of Materials Science and Engineering, Changchun University of Science and Technology, Changchun 130022, China
3 Center for the Study of Matter at Extreme Conditions, Department of Mechanical and Materials Engineering, Florida International University, Miami, FL 33199, USA; xlian002@fiu.edu
4 State Key Laboratory of Superhard Materials, College of Physics, Jilin University, Changchun 130012, China; liyan2012@jlu.edu.cn
5 State Key Laboratory of Inorganic Synthesis and Preparative Chemistry, College of Chemistry, Jilin University, Changchun 130012, China; wangyu@jlu.edu.cn
* Correspondence: chenj@fiu.edu

Received: 2 January 2020; Accepted: 20 January 2020; Published: 23 January 2020

Abstract: Goethite is a major iron-bearing sedimentary mineral on Earth. In this study, we conducted in situ high-pressure x-ray diffraction, Raman, and electrical impedance spectroscopy measurements of goethite using a diamond anvil cell (DAC) at room temperature and high pressures up to 32 GPa. We observed feature changes in both the Raman spectra and electrical resistance at about 5 and 11 GPa. However, the x-ray diffraction patterns show no structural phase transition in the entire pressure range of the study. The derived pressure-volume (P-V) data show a smooth compression curve with no clear evidence of any second-order phase transition. Fitting the volumetric data to the second-order Birch–Murnaghan equation of state yields V_0 = 138.9 ± 0.5 Å^3 and K_0 = 126 ± 5 GPa.

Keywords: goethite; high-pressure; Raman spectroscopy; x-ray diffraction; electrical impedance spectroscopy; electrical resistance; phase transition

1. Introduction

Although both goethite (α-FeOOH) and hematite (Fe_2O_3) are major iron-bearing sedimentary minerals, goethite has received less attention in the research of Earth's deep interior due to its instability at high temperatures. Upon heating, goethite decomposes into hematite and water [1,2]. The recent discovery of a pyrite-structure phase in FeOOH at lower mantle pressure and temperature conditions has led to more research on goethite [3–6]. There have been a few studies on its behavior at high pressures. In addition to its three other naturally occurring forms (akaganeite (β-FeOOH), lepidocrocite (γ-FeOOH), and feroxyhyte (δ-FeOOH)), two high-pressure phases (ε-FeOOH and pyrite-FeOOH) have been reported experimentally through heating at high pressures above 5 GPa [1,2,7] and 90 GPa [4–6,8], respectively. Before transforming to pyrite-FeOOH, ε-FeOOH experiences a second-order transition at 43 GPa with symmetry change from *P21nm* to *Pnnm*, accompanied by a sluggish high-spin to low-spin transition [9]. The spin transition also occurs in goethite at 45 GPa as an isostructural first-order phase transition [10,11]. Infrared and Raman spectroscopy indicate a few changes in the pressure range up to 45 GPa [10,12,13]. In a recent study, Liu et al. [12] reported two pressure-induced phase transitions at 7 and 20 GPa through Raman spectroscopy and electrical conductivity studies and attributed these two transitions to the structural transformation to ε-FeOOH (at 7 GPa) and an isostructural second-order transition (at 20 GPa). P-V equation of state (EOS) studies of goethite at room temperature are also very diverse; the measured bulk modulus (K_0) of ambient pressure and temperature ranges from 86 to 148 GPa [1,10,14–18]. Here, we report our high-pressure experimental studies of goethite using x-ray diffraction, Raman spectroscopy, and electrical impedance spectroscopy (EIS) up to 32 GPa in order to comprehend the discrepancies among observations of the high-pressure behavior of goethite from different studies in this pressure range. While we found similar phenomena in Raman spectroscopy and electrical impedance experiments to those reported by Liu et al. [12], the x-ray diffraction shows no structural phase transition in the pressure range studied. The analysis of the volumetric data as a function of pressure indicates that there is unlikely to be any second-order transition with a discontinuity of compressibility. The origin of the change in Raman spectra and electric conductivity remains unclear, although a sluggish magnetic state transition from antiferromagnetic (AFM) to ferromagnetic (FM) may be an explanation.

2. Materials and Methods

The starting material in our experiments was goethite (α-FeOOH) from Alfa Aesar with ≥99% purity. All measurements of x-ray diffraction (XRD), Raman, and electrical impedance spectroscopy (EIS) were conducted using a diamond anvil cell (DAC) (manufacture, city, country) with a 400 μm anvil culet and rhenium metal gasket for high-pressure generation. Inert neon gas and a mixture of 16:3:1 methanol-ethanol-water was used for pressure transmission in the XRD and Raman experiments, respectively, and no pressure transmission medium was used in EIS experiments. The goethite sample was loaded into a chamber 200 μm in diameter, together with small ruby spheres. The sample pressure was determined through in situ fluorescence measurement and the ruby pressure scale [19]. In situ x-ray diffraction was conducted at HPCAT of Advanced Photon Source (APS) (Lemont, IL, USA) with a wavelength of 0.4066 Å. The collected x-ray diffraction patterns were processed using the General Structure Analysis System (GSAS) software (manufacture, city, country) [20]. The P-V equations of state (EOS) were derived using EosFit7-GUI [21]. Raman spectroscopy was performed using a Micro-Raman Spectroscopy (inVia Reflex, Renishaw, Wotton-under-Edge, United Kingdom) System. A 532 nm diode-pumped solid-state laser was used to excite the sample, and Raman spectra were collected using a charge-coupled device (CCD)-spectrograph in the 150–5000 cm^{-1} range.

In the EIS measurement, a time (*t*)-dependent sinusoidal voltage was applied to the sample inside the DAC (Figure 1):

$$E(t) = E_0 \sin(\omega t), \tag{1}$$

where ω is the angular frequency ($\omega = 2\pi f$, where f is the frequency). For a linear (or pseudolinear) sample, the current response to a sinusoidal voltage is a sinusoid at the same frequency, but shifted in phase by $\phi(\omega)$, which is a function of frequency:

$$I(t) = I_0(\omega) \sin(\omega t + \phi(\omega)). \tag{2}$$

The impedance of the sample is defined by the ratio of E(t)/I(t):

$$Z(\omega) = E(t)/I(t) = (E_0/I_0(\omega)) \sin(\omega t)/\sin(\omega t + \phi(\omega)). \tag{3}$$

when the applied voltage and its current response are expressed as a complex number,

$$E(t) = E_0 \exp(j\omega t) \text{ and } I(t) = I_0(\omega) \exp[j(\omega t + \phi(\omega))],$$

the impedance is expressed by

$$\begin{aligned} Z(\omega) &= [E_0 \exp(j\omega t)]/[I_0(\omega) \exp(j\omega t + j\phi(\omega))] \\ &= (E_0/I_0(\omega)) \exp(-j\phi(\omega)) = Z_0(\omega)\,[\cos(\phi(\omega)) - (j)\sin\phi(\omega)] \\ &= Z_R(\omega) - Z_I(\omega)\, j \end{aligned} \tag{4}$$

where $Z_0(\omega) = E_0/I_0(\omega)$ is the magnitude of impedance and $Z(\omega)$ is composed of a real part ($Z_R(\omega) = Z_0(\omega)\cos(\phi(\omega))$) and an imaginary part ($-Z_I(\omega) = -Z_0(\omega)\sin(\phi(\omega))$).

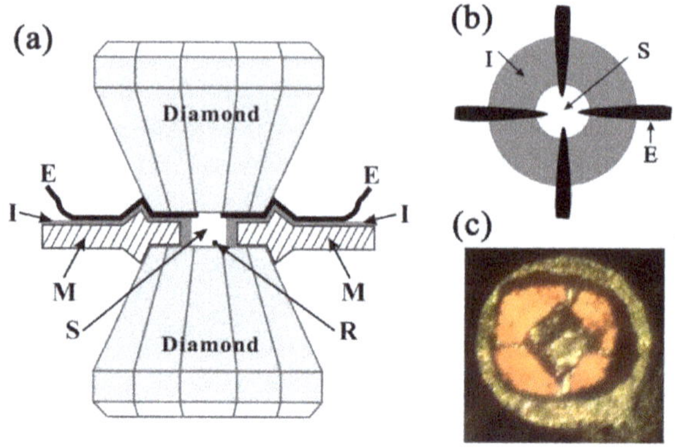

Figure 1. Sketch of the sample inside a diamond anvil cell (DAC) for electrical impedance spectroscopy. E: electrode, I: electrical insulation layer, M: metal gasket, S: sample chamber, and R: ruby sphere for pressure calibration. (**a**) Side view of the cell sketch; (**b**) top view of the gasket ring and electrodes; (**c**) sample in the cell under a microscope.

A simple sample system can be considered as an R-C circuit, i.e., a bulk resistance (*R*) and bulk capacitance (*C*) in parallel (Figure 2a). The impedance (*Z*) of this circuit follows

$$\frac{1}{Z(\omega)} = \frac{1}{R} + \frac{1}{1/(j\omega C)} \tag{5}$$

therefore,

$$Z(\omega) = \frac{R}{1 + j\omega CR} = \frac{R}{1 + (\omega C)^2} - j\frac{\omega CR^2}{1 + (\omega C)^2}. \tag{6}$$

let

$$Z_R(\omega) = \frac{R}{1 + (\omega C)^2} \text{ and } Z_I(\omega) = \frac{\omega CR^2}{1 + (\omega C)^2}, \tag{7}$$

and eliminating ω from Z_R and Z_I, we get the relation between Z_R and Z_I:

$$\left(Z_R(\omega) - \frac{R}{2}\right)^2 + Z_I(\omega)^2 = \left(\frac{R}{2}\right)^2. \tag{8}$$

Hence, the bulk resistance R can be obtained from the radius of the $Z_I(\omega)$ vs. $Z_R(\omega)$ semicircle (i.e., Nyquist plot [22]) of the EIS experiment with $Z(\omega) = Z_R(\omega) - Z_I(\omega)j$ (Figure 2b). In the Nyquist plot, the frequency (ω) does not appear in the equation, but the Z_R axis is the direction of decreasing ω. When $\omega = 0$ (i.e., direct current), the impedance only has the real part with Z = R. As ω increases, $Z_R(\omega)$ decreases from R and $Z_I(\omega)$ increases from 0, following the semicircle. At the maximum point of the semicircle, $Z_R(\omega_m) = Z_I(\omega_m) = R/2$. Incorporating this into Equation (7), we get $\omega_m RC = 1$. Therefore, $\omega_m = 1/(RC) = 1/\tau$ ($\tau = RC$ is the time constant or relaxation time of the system).

In a polycrystalline sample, an equivalent circuit can be composed through a simplified series two-phase model [23], taking the grain interior (intra-grain) as phase 1 and the grain boundary (inter-grain) as phase 2 (Figure 2c). In reality, the bulk capacitance (C) is replaced by the constant phase element (CPE) with $Z_{CPE} = Z_C/(j\omega)^n$, where $0 \le n \le 1$. When n = 0, Z_C = R (pure resistance); when n = 1, Z_C = 1/C (pure capacitance). The Nyquist plot for the polycrystalline sample consists of two semicircles (Figure 2d). For porous polycrystals, the resistance of the grain boundary is larger and, therefore, its $Z_I(\omega)$ vs. $Z_R(\omega)$ semicircle has a larger radius than that of the grain interior, according to Equation (8).

The EIS measurements were conducted using a Solartron 1260 Phase-Gain-Analyzer equipped with a Solartron 1296 dielectric interface. A sinusoidal voltage signal with a 1.0 V magnitude was applied to the sample in a DAC (Figure 1). The signal frequency ranged from 0.1 Hz to 10 MHz.

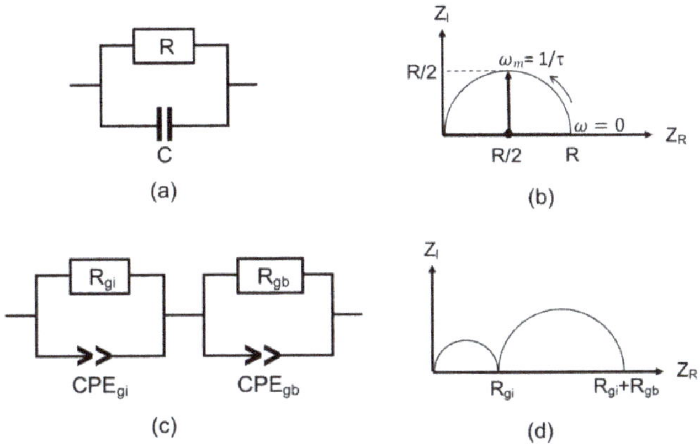

Figure 2. Equivalent circuits for (**a**) single crystal and (**c**) polycrystal samples, and their corresponding Nyquist plots (**b**,**d**), respectively. Impedance $Z(\omega) = Z_R(\omega) - Z_I(\omega)$ j, where $Z_I(\omega)$: negative of R: bulk resistance, CPE: constant phase element, subscript gi: grain interior contribution, and subscript gb: grain boundary contribution.

3. Results

3.1. Raman Spectroscopy

Figure 3a shows the in situ Raman spectra of goethite in the wavenumber range 150–1300 cm^{-1} during compression up to 32 GPa at room temperature. The spectra above 1300 cm^{-1} up to 5000 cm^{-1} are not shown as there is no significant feature detected in this wavenumber range. The spectrum near ambient pressure is consistent with data previously reported. The assignments of the observed Raman peaks are listed in Table 1. There are some unassigned weak peaks, i.e., 209, 399, 420, 887, 1020, and 1117 cm^{-1} (at ambient conditions). These peaks were also present in previously reported Raman spectra [10,12,24–26]. Peaks above 800 cm^{-1} are associated with OH bonding, and, therefore, are difficult to observe in some Raman experiments. The total number of Raman active modes calculated using the SMODES module of ISOTROPY Software Suite (Version 1.2.4, Brigham Young University, Provo, USA) [27] is 24, which is consistent with that of diaspore (α-AlOOH), an isostructural mineral of goethite with the space group *Pbnm* [28]. Upon compression, the Raman peak at 302 cm^{-1} shows an obvious splitting at 5.2 GPa, and the main peak at about 391 cm^{-1} loses one of its two shoulder peaks on the high wavenumber side at about the same pressure. These shoulder peaks are theoretically predicted by the density functional theory (DFT) calculation for goethite [29]. A new peak at about 260 cm^{-1} appears when the pressure is increased from 10.6 to 11.7 GPa, and the peak at 302 cm^{-1} further yields a small shoulder peak. Both of these new peaks are very weak. When the positions of all the observed peaks are plotted as a function of pressure (Figure 3b,c), more features become visible. As the pressure increases, all peaks show smooth blueshifts as a function of pressure due to shortening of the bond distances. At about 5 GPa, where the peak at 302 cm^{-1} splits, discontinuities in the peak shift as a function of pressure occurring at most peaks, and the slopes of the shifts also change across this pressure. Distinctly, the new peak split at 302 cm^{-1} shows a redshift in contrast to all other peaks. At 11.7 GPa, where two weak peaks appear (at 260 and 302 cm^{-1}), only subtle changes in the slopes of the shifts are observed. Such a subtle change appears to occur at 20 GPa for the ν_4 and ν_5 peaks. Upon decompression to ambient pressure, the original Raman spectrum of the starting sample is recovered, indicating that all the changes during compression are reversible.

Table 1. Assignments of the observed Raman peak of goethite at ambient conditions.

Label	ν_1	ν_2	ν_3	ν_4	ν_5	ν_6
Wavenumber	250	302	391	486	556	687
Assignment [30]	Fe–O stretching symmetric	Fe–O Hbending symmetric	Fe–O–Fe/–OH stretching symmetric	Fe–OH stretching asymmetric	Fe–OH stretching asymmetric	Fe–O stretching symmetric

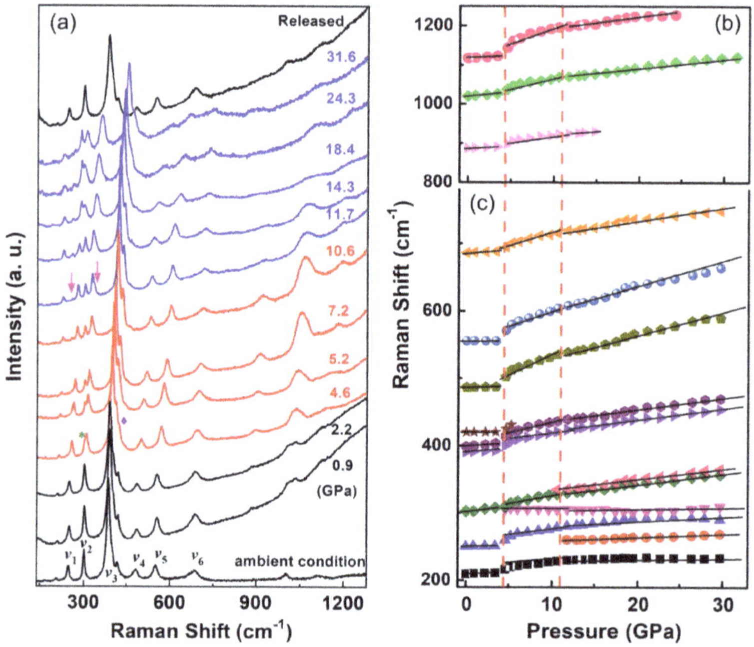

Figure 3. (**a**) Selected Raman spectra of goethite during compression (pressures as labeled) and decompressed back to 0.2 GPa. Assignments for peaks labeled with ν_i are listed in Table 1. The star and arrows indicate where a new peak emerges, and the diamond indicates peak disappearance. (**b**,**c**) are positions of the Raman peak as a function of pressure. Vertical lines indicate where some major feature changes occur.

3.2. Electrical Impedance Spectroscopy

In situ EIS measurements were conducted up to 31 GPa at room temperature. Figure 4 shows the Nyquist plots of the EIS data collected at different pressures. Resistances of grain-interior and grain-boundary contributions were derived from the first and second semicircles of the Nyquist plots and are shown in Figure 5, together with some of the previously reported results [10,12]. Since the starting sample was pre-pressed into a pallet to minimize the void space in the grain boundaries, the resistance contribution from the grain boundary started at a value comparable to that from the grain interior. Upon compression, the grain boundaries were further reduced, and, therefore, the grain-boundary resistance became much smaller than the grain-interior resistance (about one order of magnitude smaller at the peak pressure, 31 GPa). The pressure influences on the grain-interior resistance and grain-boundary resistance are well-correlated. Two significant changes can be recognized in the total resistance as a function of pressure. Upon compression, the total resistance initially decreases with pressure. At about 5 GPa, the resistance begins to increase with pressure and then decreases with pressure again at 11 GPa. These changes are also well-correlated with those observed in Raman spectroscopy.

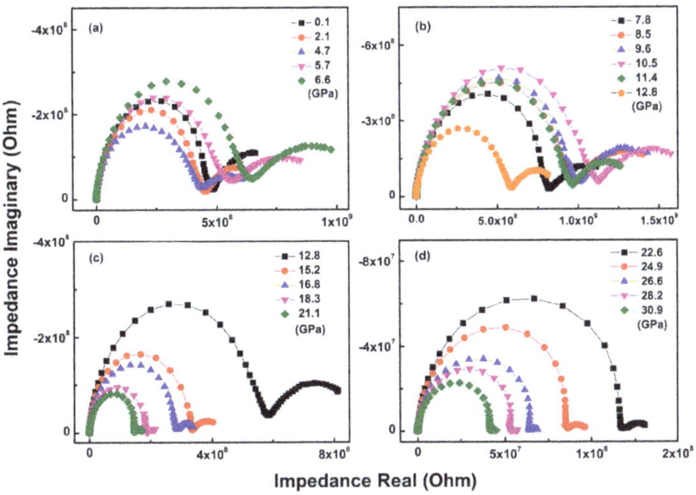

Figure 4. The Z_I vs. Z_R Nyquist plots of electrical impedance spectroscopy for the goethite sample inside a diamond anvil cell at different pressures, as labeled. The first semicircle of each data is the contribution from the grains (R_g), and the second semicircle is that from the grain boundaries (R_{gb}). The total resistance of the sample $R = R_g + R_{gb}$.

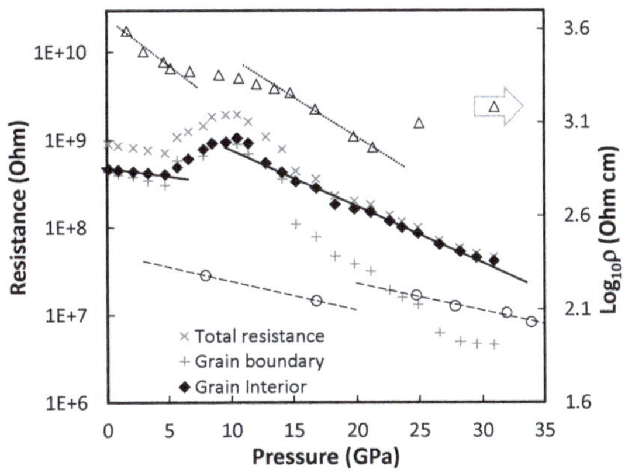

Figure 5. Measured grain-interior resistance (R_{gi}: solid diamonds), grain-boundary resistance (R_{gb}: pluses), and total resistance ($R = R_g + R_{gb}$: crosses) of the goethite sample at different pressures using electrical impedance spectroscopy. Triangles and open circles are the resistivity and resistance from Liu et al. [12] and Xu et al. [10], respectively.

3.3. In Situ X-Ray Diffraction

Figure 6 shows the sequence of in situ synchrotron x-ray diffraction patterns at high pressures up to 32 GPa at room temperature. No diffraction feature change was observed in the entire pressure range, except for the peak shift due to compression. A peak showing a dramatic shift in contrast to the rest of the diffraction pattern was obtained from the solidified neon pressure transmission medium.

The volumetric data (Figure 7) derived from the diffraction patterns were fitted to the third-order Birch–Murnaghan equation of state (BM EOS):

$$P = 3K_0 f(1 + 2f)^{5/2}(1 + (3/2)(K' - 4)f), \quad (9)$$

where K_0 and K' are the bulk modulus at ambient pressure and its pressure derivative, respectively, and $f = ((V_0/V)^{2/3} - 1)/2$ is the Eulerian strain. The EOS fitting yields $V_0 = 138.8 \pm 0.8$ Å^3, $K_0 = 129 \pm 19$ GPa, and $K' = 3.8 \pm 1.2$.

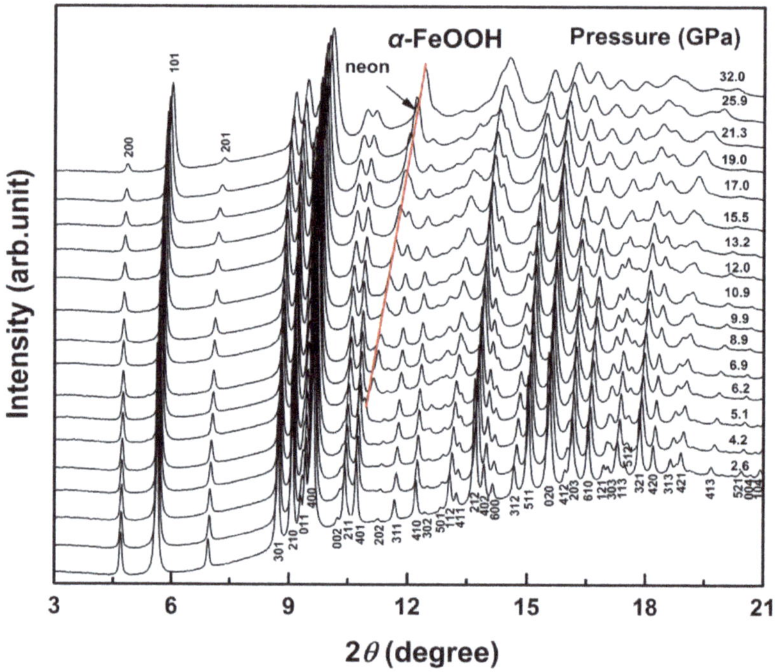

Figure 6. In situ synchrotron x-ray diffraction patterns of goethite at high pressures, as labeled, up to 32 GPa, during compression. The x-ray wavelength is 0.4066 Å. The Miller index of goethite structure is given at each peak. The red line indicates the track of the diffraction peak from the neon pressure transmission medium.

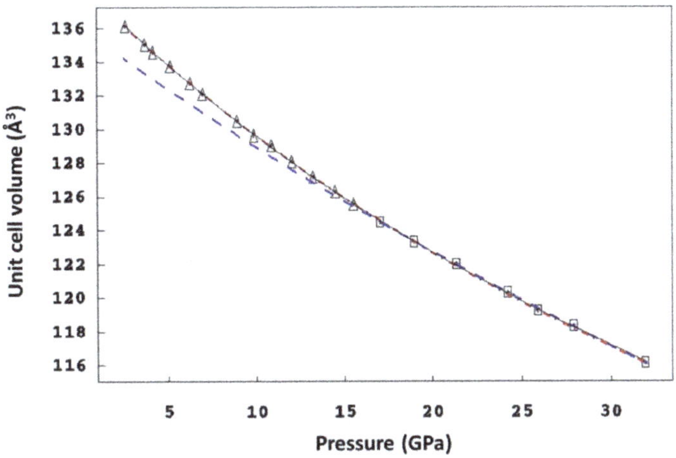

Figure 7. Volumetric data (triangles and squares) of goethite as a function of pressure derived from the x-ray diffraction. Errors are smaller than the size of the symbols. The red dot-dashed line, blue dashed line, and thin solid line represent the equation of state (EOS) fitting of the data from 0 to 16 GPa (triangles), 16 to 32 GPa (squares), and 0 to 32 GPa.

4. Discussion

Although solidification of the pressure medium (mixture of 16:3:1 methanol-ethanol-water) at about 11 GPa in the Raman experiments may give rise to the change in the slope of peak-shift as a function of pressure, the change in the EIS experiment at the same pressure is independent of the solidification because no pressure medium was used in the EIS measurement. Similar phenomena have been observed in other in situ high-pressure Raman spectroscopy studies [10,12]. Xu et al. [10] resolved the peak at 300 cm^{-1} as three peaks at ambient pressure, whereas we recognized only one peak initially, a splitting at about 5 GPa, and a second splitting at about 11 GPa. On the other hand, they did not observe the shoulder peak on the high wavenumber side of the main peak at 400 cm^{-1} at ambient pressure. Instead, they observed a splitting of the peak at 5 GPa and changes in the slope of peak-shift with pressure at 16 GPa in some of the peaks (i.e., at 250 and 300 cm^{-1}). No discontinuity of peak-shift with pressure was reported throughout the entire pressure range (up to 40 GPa) they studied. In their Raman study, Liu et al. [12] did not focus their attention on weak peaks, such as the shoulder peak from the second splitting of the peak at 300 cm^{-1} and the shoulder peaks on the high wavenumber side of the main peak at 400 cm^{-1}. Their observation of the peak-shift discontinuities with pressure at 7 GPa is approximately consistent with our current study (5 GPa). The second slope-change they reported is at a higher pressure (20 GPa) compared to what we observed in this study (11 GPa), except for the ν_4 and ν_5 peak, which change their slopes at 20 GPa. Discrepancies among these in situ high-pressure Raman spectroscopy results seem to originate from the instrumental resolution, stress environment, and experimentalist's attention.

A comparison of our EIS data with the resistance data from Xu et al. [10] and the resistivity (reciprocal of conductivity) data from Liu et al. [12] in the same pressure range is shown in Figure 5. Since the resistance value highly depends on the geometries of the sample and electrodes, the data from Xu et al. [10] are about one order of magnitude lower than the result of this study. However, the trend of reducing resistance with pressure is consistent with the current study, and a notable discontinuity of resistance as a function of pressure can be recognized in their data (dashed lines), although the pressure step in their experiment is too large to unveil any details due to their primary interest in the higher pressure range. The resistivity data as a function of pressure from Liu et al. [12] show two changes at pressures similar to those observed in this study. Nevertheless, we did not observe the resistance increase at a higher pressure (20 GPa) reported by Liu et al. [12], and neither did Xu et al. [10].

Although both Raman spectroscopy (Figure 3) and EIS (Figure 5) indicate feature changes at about 5 and 11 GPa in this study, the observation of in situ x-ray diffraction (Figure 6) rules out the possibility of any first-order structural transition at these pressures. Xu et al. [10] reported a change in the bulk modulus at 16 GPa, based on their x-ray diffraction data, from 120 to 197 GPa. When we break our volumetric data into two groups at 16 GPa and fit the two datasets separately to the third-order BM EOS, we get V_0 = 138.8 Å^3, K_0 = 129 GPa, and K' = 3.7 for data below 16 GPa, and V_0 = 136.1 Å^3, K_0 = 175 GPa, and K' = 2.1 for data above 16 GPa. An increase in bulk modulus appears to be present across the breaking pressure. However, the confidence ellipses (at a 1σ level) in K_0 and K' (Figure 8) of the two datasets indicate that the fitting results of K_0 and K' in each dataset are highly correlated and both datasets can be fitted almost equally well with K_0 and K' located in the intersection area of the two ellipses [31], e.g., K' = 4. In addition, fitting the entire data with single EOS yields a much smaller confidence ellipse lying within the intersection area of the two ellipses, indicating an improvement in the quality of fitting. Therefore, we fitted the two datasets using the second-order BM EOS, and got V_0 = 138.8 Å^3 and K_0 = 127 GPa, and V_0 = 139.0 Å^3 and K_0 = 125 GPa for the data below and above 16 GPa, respectively. The difference between the results of the two datasets is negligible. Therefore, the current volumetric data from x-ray diffraction indicates no second-order phase transition. The final fitting of the entire data into the second-order BM EOS yields V_0 = 138.9 ± 0.5 Å^3 and K_0 = 126 ± 5 GPa. To validate the final choice of the entire data for the EOS fitting, we made a plot of normalized stress $F = P/[3f(1+2f)^{5/2}]$ vs. Eulerian strain f, i.e., f-F plot (Figure 9). Substituting P in Equation. (9) with F, the third-order BM EOS is expressed as

$$F = K_0 + (3/2)K_0(K' - 4)f. \tag{10}$$

A linear fit of the f-F plot (Figure 9) yields K_0 = 125.6 GPa and K'= 4.07, which indicates that the second-order BM EOS is an appropriate estimate. A comparison of the EOS parameters for goethite from published reports is given in Table 2.

Table 2. Comparison of EOS parameters for goethite.

Reference	V_0	K_0	K' *	Method **	Pressure Medium ***	Pressure Calibrant	Pressure Range (GPa)
This study	138.9(5)	126(5)	4	DAC, AD	Ne	Ruby	0–32
Suzuki [16]	138.96(7)	85.9(15)	12.6(8)	LVP, ED	NaCl	NaCl	0–7.55
Xu et al. [10]	138.4(3)	120(3)	4	DAC, AD	He, Ne	Au	0–16
Gleason et al. [1]	138.75(2)	140.3(37)	4.6(4)	DAC, AD	ME(4:1)	Au	0–29.4
Kim et al. [14]	138.8	131.1(58)	4	LVP, ED	ME(4:1)	NaCl	0–9.54
Nagai et al. [15]	140.45(1)	111(2)	4	DAC, AD	ME(4:1)	Ruby	0–24.5
Kim et al. [18]	139.3	147.9	4	DAC, ED	MEW(16:3:1)	MgO	0–27.6
Tunega [32]	138.70	93.5	4	spDFT	-	-	-
	140.86	118.7	4	spDFT+U	-	-	-
Guo et al. [33]	141.49	114.1	-	PW-GGA+U	-	-	-
Otte et al. [34]	144.4	108.8	5.9	GGA+U	-	-	-

* K' = 4 indicates a fitting with the second-order Birch–Murnaghan equation of state (BM EOS). ** DAC: diamond anvil cell experiment; LVP: large volume press experiment; AD/ED: angle/energy dispersive x-ray diffraction; spDFT: spin-polarized density functional theory; U: the on-site Coulomb repulsion parameter in density functional theory (DFT); PW: plane-wave basis; GGA: generalized gradient approximation in DFT. *** ME: mixture of methanol and ethanol; MEW: mixture of methanol, ethanol, and water.

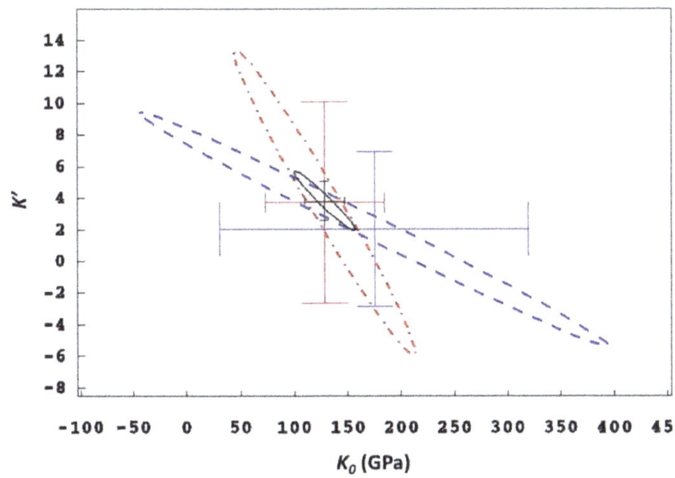

Figure 8. Confidence ellipses in K_0 and K' for the fitting of volume-pressure data below 16 GPa (dot-dashed red line), above 16 GPa (dashed blue line), and the entire pressure range up to 32 GPa (solid black line). Crosses indicate the center of each ellipse.

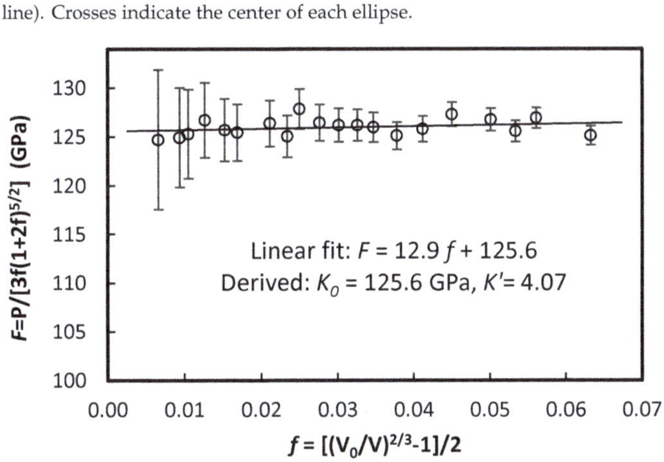

Figure 9. *f-F* plot based on the volume-pressure data of goethite. The solid line is a linear fit of the data.

Ruling out the possibilities of a first-order structural transition and a second-order compressibility transition, the origin for the changes observed in Raman spectroscopy and EIS remains unjustified [11]. A theoretical calculation [34] once predicted a pressure-induced structural phase transformation to the high-pressure ε-FeOOH at 6–7 GPa, followed by a high-spin to low-spin transition at 7.7 GPa, using first-principle generalized gradient approximation (GGA). However, experimental studies have indicated that the ε-FeOOH phase only forms at an elevated temperature [1,2], and the predicted spin transition is shifted to a much higher pressure (56 GPa) if the on-site Coulomb repulsion term (U) is included in the GGA calculation [34]. Xu et al. [10] experimentally observed high-spin to low-spin transition in goethite at 45 GPa. Therefore, the transformation to ε-FeOOH phase and the spin transition cannot be responsible for the changes observed in Raman spectroscopy and EIS at 5 and 11 GPa. The theoretical calculation using spDFT by Tunega [32] indicates that a high-spin ferromagnetic (FM) state of goethite may become energetically favorable at about 15 GPa, whereas the high-spin antiferromagnetic (AFM) state is stable at ambient pressure. The GGA+U calculation [34] shows that the FM state at high pressure has a wider bandgap over the AFM state, and, therefore, is expected to have a higher resistance. If we follow these theoretical results, we may speculate that the FM state (higher resistance phase) becomes stable at 5 GPa, and the transition from an AFM to FM state is a sluggish process. Following this speculation, the electrical resistance behavior of goethite from EIS (Figure 5) can be illustrated as a pure AFM state (lower resistance phase) before 5 GPa, a pure FM state (higher resistance phase) above 11 GPa, and a transition stage from AFM to FM between 5 and 11 GPa, resulting in a slow increase of resistance. The pressure-induced AFM to FM transition has been experimentally observed in many other materials, e.g., Au_2Mn [35], EuTe [36], USb_2 [37], $SrFeO_3$, and $CaFeO_3$ [38]. The electrical resistance behavior observed in USb_2 is very similar to what we observed here in goethite. Upon compression, the resistance decreases within the AFM phase, increases during the AFM-FM transition, and finally decreases again in the FM phase [37]. The AFM-FM transition pressure assumed here (5 GPa) in goethite is lower than that of the spDFT prediction (15 GPa). Since the major changes in Raman spectra at 11 GPa (delayed to 20 GPa for ν_4 and ν_5 peaks) are exhibited on the slope of the peak shifts with pressure, they are consistent with the AFM-FM transition assumption as well. Further theoretical simulation of the influence of AFM-FM transition on phonon vibrations will be interesting. No apparent crystallographic signature is observed to accompany this transition in our x-ray diffraction data. Suzuki [16] reported an anomaly in the compression behavior along the unit-cell axes at about 5 GPa, but this was not observed in either our compression data (Figure 10) or other published data (except for the effect of pressure-medium solidification) [1,3,10,14,15].

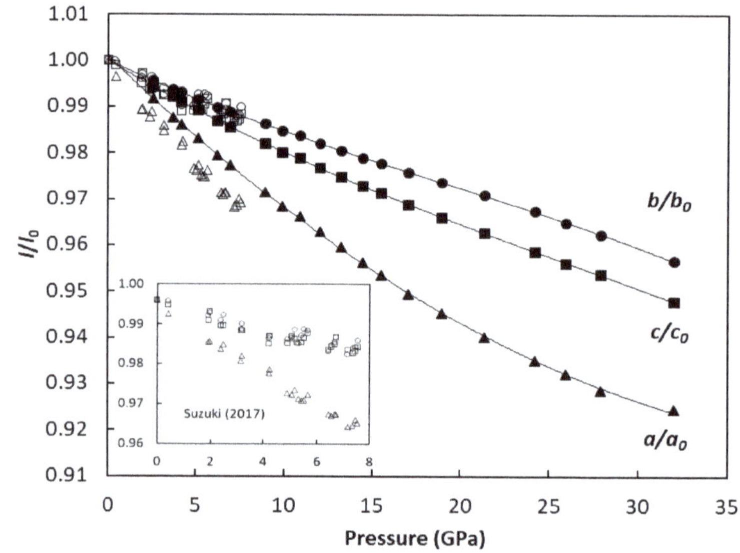

Figure 10. Normalized unit-cell parameters of goethite as a function of pressure with a_0 = 4.602 Å, b_0 = 9.961 Å, and c_0 = 3.026 Å. Open symbols are from Suzuki [16].

FeOOH was recently found to be stable in a pyrite-type structure in deep mantle conditions [5,6]. If the mineral presents at shallower depths where goethite (a-FeOOH) is stable, its influence on the magnetic properties of Earth's mantle needs to be considered when understanding the magnetism of Earth's interior, which is a matter still the subject of much debate [39,40]. Further studies on this may be worthwhile.

Author Contributions: R.T., conceptualization, methodology, formal analysis, investigation, data curation, writing-original draft preparation, visualization; J.C., conceptualization, methodology, formal analysis, investigation, writing-original draft preparation, review and editing, visualization, supervision; Q.Z., formal analysis, investigation, resources, data curation, writing-review and editing; Y.L., investigation, resources, data curation, writing- review and editing; X.L., investigation, resources, data curation, writing-review and editing; B.Y., investigation, resources, data curation, writing-review and editing; Y.W., investigation, resources, writing-review and editing. All authors have read and agreed to the published version of the manuscript.

Funding: A portion of this work was performed at HPCAT (Sector 16), Advanced Photon Source (APS), Argonne National Laboratory. HPCAT operations were supported by DOE-NNSA's Office of Experimental Sciences. The Advanced Photon Source is a U.S. Department of Energy (DOE) Office of Science User Facility operated for the DOE Office of Science by Argonne National Laboratory under Contract No. DE-AC02-06CH11357. This research was supported in part by the NSAF Grant No. U1530402; the Youth science foundation of Changchun University of Science and Technology (Grant No. XQNJJ-2018-15); and the Jilin Province Science and Technology Development Program, China (20170520119JH). J.C acknowledges the support from NSF (EAR-1723185).

Acknowledgments: The authors thank Yue Meng for her technical support of the synchrotron beamline, Vadym Drozd and Jin Zhang for their insightful discussions on the project.

Conflicts of Interest: The authors declare no conflict of interest.

References

1. Gleason, A.E.; Jeanloz, R.; Kunz, M. Pressure-Temperature Stability Studies of Feooh Using X-Ray Diffraction. *Am. Mineral.* **2008**, *93*, 1882–1885. [CrossRef]
2. Voigt, R.; Will, G. The System Fe_2o_3-H_2o under High Pressures. *Neues Jahrb. Fur Mineral.* **1981**, *2*, 89–96.
3. Hu, Q.; Kim, D.Y.; Yang, W.; Yang, L.; Meng, Y.; Zhang, L.; Mao, H.K. Feo_2 and Feooh under Deep Lower-Mantle Conditions and Earth's Oxygen-Hydrogen Cycles. *Nature* **2016**, *534*, 241–244. [CrossRef]
4. Hu, Q.; Kim, D.Y.; Liu, J.; Meng, Y.; Yang, L.; Zhang, D.; Mao, W.L.; Mao, H.K. Dehydrogenation of Goethite in Earth's Deep Lower Mantle. *Proc. Natl. Acad. Sci. USA* **2017**, *114*, 1498–1501. [CrossRef]
5. Liu, J.; Hu, Q.; Kim, D.Y.; Wu, Z.; Wang, W.; Xiao, Y.; Chow, P.; Meng, Y.; Prakapenka, V.B.; Mao, H.-K.; et al. Hydrogen-Bearing Iron Peroxide and the Origin of Ultralow-Velocity Zones. *Nature* **2017**, *551*, 494–497. [CrossRef]
6. Nishi, M.; Kuwayama, Y.; Tsuchiya, J.; Tsuchiya, T. The Pyrite-Type High-Pressure Form of Feooh. *Nature* **2017**, *547*, 205–208. [CrossRef]
7. Suzuki, A. High-Pressure X-Ray Diffraction Study of E-Feooh. *Phys. Chem. Miner.* **2010**, *37*, 153–157. [CrossRef]
8. Mao, H.-K.; Hu, Q.; Yang, L.; Liu, J.; Kim, D.Y.; Meng, Y.; Zhang, L.; Prakapenka, V.B.; Yang, W.; Mao, W.L. When Water Meets Iron at Earth's Core–Mantle Boundary. *Natl. Sci. Rev.* **2017**, *4*, 870–878. [CrossRef]
9. Gleason, A.E.; Quiroga, C.E.; Suzuki, A.; Pentcheva, R.; Mao, W.L. Symmetrization Driven Spin Transition in E-Feooh at High Pressure. *Earth Planet. Sci. Lett.* **2013**, *379*, 49–55. [CrossRef]
10. Xu, W.; Greenberg, E.; Rozenberg, G.K.; Pasternak, M.P.; Bykova, E.; Boffa-Ballaran, T.; Dubrovinsky, L.; Prakapenka, V.; Hanfland, M.; Vekilova, O.Y.; et al. Pressure-Induced Hydrogen Bond Symmetrization in Iron Oxyhydroxide (Supplementary Materials). *Phys. Rev. Lett.* **2013**, *111*, 175501. [CrossRef] [PubMed]
11. Reagan, M.M.; Gleason, A.E.; Daemen, L.; Xiao, Y.; Mao, W.L. High-Pressure Behavior of the Polymorphs of Feooh. *Am. Mineral.* **2016**, *101*, 1483–1488. [CrossRef]
12. Liu, K.; Dai, L.; Li, H.; Hu, H.; Zhuang, Y.; Yang, L.; Pu, C.; Hong, M. Pressure-Induced Phase Transitions for Goethite Investigated by Raman Spectroscopy and Electrical Conductivity. *High Press. Res.* **2019**, *39*, 106–116. [CrossRef]
13. Williams, Q.; Guenther, L. Pressure-Induced Changes in the Bonding and Orientation of Hydrogen in Feooh-Goethite. *Solid State Commun.* **1996**, *100*, 105–109. [CrossRef]
14. Kim, Y.; Hwang, G.; Do, J. Compression Study of Goethite at Room Temperature. *J. Mineral. Soc. Korea* **2007**, *20*, 261–266.
15. Nagai, T.; Kagi, H.; Yamanaka, T. Variation of Hydrogen Bonded O ... O Distances in Goethite at High Pressure. *Am. Mineral.* **2003**, *88*, 1423–1427. [CrossRef]
16. Suzuki, A. Thermal Equation of State of Goethite (A-Feooh). *High Press. Res.* **2017**, *37*, 193–199. [CrossRef]
17. Suzuki, A. Pressure–Volume–Temperature Equation of State of E–Feooh to 11 Gpa and 700 K. *J. Mineral. Petrol. Sci.* **2016**, *111*, 420–424. [CrossRef]

18. Kim, Y.; Yi, Z. High Pressure X-Ray Diffraction Study on a Goethite Using Synchrotron Radiation. *J. Min. Soc. Korea* **1997**, *10*, 75–81.
19. Mao, H.K.; Xu, J.; Bell, P.M. Calibration of the Ruby Pressure Gauge to 800 Kbar under Quasi-Hydrostatic Conditions. *J. Geophys. Res.* **1986**, *91*, 4673–4676. [CrossRef]
20. Toby, B.H.; Von Dreele, R.B. Gsas-Ii: The Genesis of a Modern Open-Source All Purpose Crystallography Software Package. *J. Appl. Crystallogr.* **2013**, *46*, 544–549. [CrossRef]
21. Gonzalez-Platas, J.; Alvaro, M.; Nestola, F.; Angel, R. Eosfit7-Gui: A New Graphical User Interface for Equation of State Calculations, Analyses and Teaching. *J. Appl. Crystallogr.* **2016**, *49*, 1377–1382. [CrossRef]
22. Barsoukov, E.; Macdonald, J.R. *Impedance Spectroscopy: Theory, Experiment, and Applications*, 3rd ed.; John Wiley & Sons, Inc.: New Jersey, NJ, USA, 2018.
23. Bonanos, N.; Steele, B.C.H.; Butler, E.P. Applications of Impedance Spectroscopy: Characterization of Materials. In *Impedance Spectroscopy: Theory, Experiment, and Applications*, 3rd ed.; Barsoukov, E., Macdonald, J.R., Eds.; John Wiley & Sons, Inc.: New Jersey, NJ, USA, 2018; pp. 175–224.
24. Lafuente, B.; Downs, R.T.; Yang, H.; Stone, N. The Rruff Project. Available online: http://rruff.info/goethite/display=default/R120086 (accessed on 30 December 2019).
25. Dünnwald, J.; Otto, A. An Investigation of Phase Transitions in Rust Layers Using Raman Spectroscopy. *Corros. Sci.* **1989**, *29*, 1167–1176. [CrossRef]
26. De Faria, D.L.A.; Venâncio Silva, S.; de Oliveira, M.T. Raman Microspectroscopy of Some Iron Oxides and Oxyhydroxides. *J. Raman Spectrosc.* **1997**, *28*, 873–878. [CrossRef]
27. Stokes, H.T.; Hatch, D.M.; Campbell, B.J. Smodes, Isotropy Software Suite. Available online: http://stokes.byu.edu/iso/isotropy.php (accessed on 30 December 2019).
28. Delattre, S.; Balan, E.; Lazzeri, M.; Blanchard, M.; Guillaumet, M.; Beyssac, O.; Haussühl, E.; Winkler, B.; Salje, E.K.H.; Calas, G. Experimental and Theoretical Study of the Vibrational Properties of Diaspore (A-Alooh). *Phys. Chem. Miner.* **2012**, *39*, 93–102. [CrossRef]
29. Kubicki, J.D.; Paul, K.W.; Sparks, D.L. Periodic Density Functional Theory Calculations of Bulk and the (010) Surface of Goethite. *Geochem. Trans.* **2008**, *9*, 4. [CrossRef]
30. Legodi, M.A.; de Waal, D. The Preparation of Magnetite, Goethite, Hematite and Maghemite of Pigment Quality from Mill Scale Iron Waste. *Dyes Pigments* **2007**, *74*, 161–168. [CrossRef]
31. Angel, R.J. Equations of State. In *High-Temperature and High-Pressure Crystal Chemistry*; Hazen, R.M., Downs, R.T., Eds.; Mineralogical Society of America: Chantilly, VA, USA, 2000; Volume 41, pp. 35–60.
32. Tunega, D. Theoretical Study of Properties of Goethite (A-Feooh) at Ambient and High-Pressure Conditions. *J. Phys. Chem. C* **2012**, *116*, 6703–6713. [CrossRef]
33. Guo, H.; Barnard, A.S. Modeling the Iron Oxides and Oxyhydroxides for the Prediction of Environmentally Sensitive Phase Transformations. *Phys. Rev. B* **2011**, *83*, 094112. [CrossRef]
34. Otte, K.; Pentcheva, R.; Schmahl, W.W.; Rustad, J.R. Pressure-Induced Structural and Electronic Transitions in Feooh from First Principles. *Phys. Rev. B* **2009**, *80*, 205116. [CrossRef]
35. Wayne, R.C.; Smith, F.A. The Pressure Induced Metamagnetic Transition in Au2mn and the Pressure Depen-Dence of the Ferromagnetic Curie Temperature. *J. Phys. Chem. Solids* **1969**, *30*, 183–185. [CrossRef]
36. Ishizuka, M.; Kai, Y.; Akimoto, R.; Kobayashi, M.; Amaya, K.; Endo, S. Pressure-Induced Ferromagnetism in Eute. *J. Magn. Magn. Mater.* **1997**, *166*, 211–215. [CrossRef]
37. Jeffries, J.R.; Stillwell, R.L.; Weir, S.T.; Vohra, Y.K.; Butch, N.P. Emergent Ferromagnetism and T-Linear Scattering in Usb_2 at High Pressure. *Phys. Rev. B* **2016**, *93*, 184406. [CrossRef]
38. Kawakami, T.; Nasu, S. High-Pressure Mössbauer Spectroscopy of Perovskite High Valence Iron Oxides under External Magnetic Field. *J. Phys. Condens. Matter* **2005**, *17*, S789–S793. [CrossRef]
39. Wasilewski, P.J.; Thomas, H.H.; Mayhew, M.A. The Moho as a Magnetic Boundary. *Geophys. Res. Lett.* **1979**, *6*, 541–544. [CrossRef]
40. Kupenko, I.; Aprilis, G.; Vasiukov, D.M.; McCammon, C.; Chariton, S.; Cerantola, V.; Kantor, I.; Chumakov, A.I.; Rffer, R.; Dubrovinsky, L.; et al. Magnetism in Cold Subducting Slabs at Mantle Transition Zone Depths. *Nature* **2019**, *570*, 102–106. [CrossRef] [PubMed]

minerals

MDPI

Article

A New Reference for the Thermal Equation of State of Iron

Francesca Miozzi [1,*], Jan Matas [2], Nicolas Guignot [3], James Badro [4,5], Julien Siebert [4] and Guillaume Fiquet [1]

1 Sorbonne Université, UMR CNRS 7590, Muséum National d'Histoire Naturelle, Institut de Minéralogie, de Physique des Matériaux et de Cosmochimie, IMPMC, 75005 Paris, France; guillaume.fiquet@sorbonne-universite.fr
2 Laboratoire de Géologie de Lyon, Université de Lyon, UMR CNRS 5276, 69007 Lyon, France; jan.matas@cnrs.fr
3 Synchrotron SOLEIL, L'Orme des Merisiers, 91192 Gif-sur-Yvette, France; nicolas.guignot@synchrotron-soleil.fr
4 Université de Paris, Institut de physique du globe de Paris, CNRS, 1 rue Jussieu, 75005 Paris, France; badro@ipgp.fr (J.B.); siebert@ipgp.fr (J.S.)
5 École Polytechnique Fédérale de Lausanne, CH-1015 Lausanne, Switzerland
* Correspondence: francesca.miozzi@upmc.fr

Received: 23 December 2019; Accepted: 18 January 2020; Published: 24 January 2020

Abstract: The high-pressure, high-temperature behavior of iron was investigated to 140 GPa and 3500 K with in situ synchrotron X-ray diffraction. Iron samples were compressed in diamond-anvil cells and heated up with the double-sided laser-heating system installed at the high-pressure ID27 of the European Synchrotron Radiation Facility (ESRF). Three different structures, namely α-bcc, γ-fcc or ε-hcp Fe were identified as a function of pressure and temperature in the domain we explored. At pressures above 90 GPa, it is clearly shown that ε-iron is the single stable solid phase up to 160 GPa at high temperatures. The analysis of the P-V-T relationship allows us to propose a reliable experimental thermal equation of state (EoS) for iron. We also show that the addition of low pressure points to our EoS refinement yields more robust constrain on the determination of the reference volume V_0 of the ε-hcp structure, which has important implications on the final parametrization of the equation of state. The extrapolation of the proposed EoS to core pressure conditions indicates that a pure iron core would have an excess of density of 3% compared to the PREM density profile.

Keywords: equations of state; Mie-Grüneisen-Debye; iron; Earth's core

1. Introduction

Iron is considered to be the main constituent of the Earth's core. Cosmochemical abundances and iron meteorites also support the idea that iron is present in numbers of planetary cores. Since the first pioneering work [1], a wealth of data has been produced on iron structure at extreme pressure and temperature, elastic properties or transport properties [2–6], so as to yield a precise model of the structure, composition and dynamics of the Earth's core. A lot of studies focus on the determination of a phase diagram for iron at relevant pressure and temperature conditions [4,6–9]. Light elements have also been shown to be incorporated in the core (see [10] for a review), in order to match density and sound waves velocities obtained by seismological modelling. These light elements produce significant variations of density and elastic properties of the alloys. The nature and amount of light elements has been recently re-examined and models built on a comparison of elastic properties of alloys with those of the Earth's core [11,12]. These studies indicate that silicon (with ca 2–3 wt %) is the main alloying element in the Earth's inner core, and oxygen likely to be the main light element of the Earth's liquid outer core, in agreement with cosmochemical abundances [13,14]. This exercise, however, can only be done when reliable thermal equations of state are available. In this respect, because high pressure and temperature experiments on relevant metallic alloys are not available, most constraints arise from measurements carried out on pure iron. The latest measurements carried out on the structure of iron at pressure and temperature conditions of the Earth's center have shown the ε-hcp structure (defined hereafter as hcp) is favored [6], provided light elements or nickel content is not too large [15]. As for other properties, a wealth of ambient temperature and thermal equations of state has been reported for hcp Fe. These equations of state have been established either using first principles calculations [16–18]. Experiments were carried out at moderate pressures (20–30 GPa) and temperatures between 300 and 1500 K [19–21] in multi-anvils large volume devices. The investigated pressure range was extended up to 300 GPa with the use of diamond anvil cells [22–26], however, when reliable measurements are available to 200 GPa at room temperature, the thermal component of the equation of state is either calculated so as to provide a thermal equation of state [24] or constrained by a limited number of data points collected simultaneously at high pressure and temperature [25]. In the existing literature several sets of parameters of the ambient temperature equation of state (V_0, K_0, and K_0', see also Table 3 in [26]) and the thermal model (θ, γ, q) were proposed. In particular, V_0 has values that spans from 22.15 Å^3 [21] to 22.7 Å^3 [26], K_0 is comprised between 202 GPa [21] and 135 GPa [320] and K_0' between 4.5 [21] and 6 [20].

In this study, we conducted a series of experiments in order to establish a new experimental thermal equation of state of iron, built on a large number of experimental points collected simultaneously at high pressure and high temperature. Measurements of molar volume of hcp iron are reported to 140 GPa and temperatures up to 3500 K. These experiments have been performed at the high-pressure beamline ID27 of the ESRF. Combined with measurements obtained at room-temperature, it allows us to establish a reliable thermal equation of state that can be extrapolated to core conditions Our results are discussed in the light of measurements obtained for pure iron at core conditions [6] and compared to recent melting line data and reference thermal equations of state from the literature [20,24,25].

2. Methods

2.1. Samples

We prepared a mixture of fine grained iron (source 99.999% purity, 1–2 μm grain size) which was dispersed in a matrix of MgO (Prolabo, spectroscopic grade). MgO powder was first fired at temperatures above 1000 °C during several hours before being mixed with about 10% iron in weight and subsequently hot-pressed for 24 h at 1.5 GPa and 800 K in a piston cylinder under reducing conditions so as to obtain a compact composite polycrystalline sample, free from any iron oxides. This sample was then subsequently thinned down to a 15 μm thick wafer in which discs of 30 μm in diameter were shaped using femtosecond laser micromachining, and finally loaded in a diamond-anvil cell under a dry neon atmosphere in a 2000 bar gas vessel. In our experiments, neon acted as a pressure transmitting medium and thermal insulation. A combination of flat (culets 300 microns in size) and beveled diamonds (150–300 and 100–300 μm in size) were used for these experiments. Gaskets were made from a rhenium foil initially 200 microns in thickness, pre-indented to thicknesses of 30 to 20 μm before pressure chambers were drilled. A set of low-pressure measurements was also carried out using helium as a pressure transmitting medium so as to add solid constraints on the determination of the reference volume V_0 for the hcp structure. In this set of measurements, pressure measurements were provided by the fluorescence of ruby in perfect hydrostatic and quasi-hydrostatic conditions pressures up to 40 GPa and back down to 10 GPa.

2.2. High Pressure Laser Heating

Diamond-anvil cell samples kept at high-pressure within neon pressure medium were then heated up using two infrared lasers sources, composed of two vertically polarized high power–high stability

Nd-YAG diode pumped lasers from Spectra Physics. These lasers operate in the TEM00 mode and are able to generate a continuous total laser power of 80 W (40 W for each laser) at a wavelength of 1.064 μm. This very high power is largely sufficient to heat any metals or opaque materials well above 3000 K in the megabar regime. The IR beams were focused onto the sample through the diamonds at an angle of 16° from the normal of the diamond-anvil surface with aplanatic objectives (f = 120 mm). These lasers also provide an excellent power and beam pointing stability; the peak-to-peak power variation is indeed better than 0.2% in the 0.1–1000 Hz frequency domain; and the pointing variations are better than 100 μrad. These two parameters are crucial because the sample temperature and temperature distribution are directly correlated to their variation.

2.3. Temperature Measurement

As for temperature profile and peak temperatures measurements, we used optimized Schwarzschild-type reflecting microscope objectives totally free of chromatic aberrations. The two reflective objectives are vertically and horizontally mounted on high-precision translations. Temperatures are classically measured using the multi-wavelength spectral radiometry, where the emitted light is analyzed using a Jobin-Yvon CCD camera (model ATE-1024 × 256) mounted on an Acton spectrograph (model SP556i,). This spectrometer is equipped with a grating optimized for the temperature measurements. The grating is 150 gr/mm blazed at 500 nm and covers the wavelength domain 480–820 nm, with a central wavelength at 650 nm. The sample temperature is measured on an image spatially filtered by a pinhole positioned in front of the entrance slit to avoid any temperature errors induced by the spatial nonlinearity of the CCD camera. The spatial resolution is fixed by the pinhole entrance, and corresponds to a spatial resolution of about 4 μm on the sample.

In practice, the temperatures are measured using the multi-wavelength radiometry method. The collected thermal radiation is analysed over a wavelength domain of 500–800 nm and is fitted to a Planck law (Equation (1)), which expresses the spectral intensity $I(\lambda,T)$ as a function of the emissivity $\varepsilon(\lambda,T)$ and the temperature T:

$$I(\lambda,T) = \varepsilon(\lambda,T)c_1\lambda - 5[e(c_2/\lambda T) - 1] - 1 \tag{1}$$

where λ is the wavelength and c_1 and c_2 are two constants.

It is necessary to determine the system response $R(\lambda,T_{ref})$ at a fixed temperature T_{ref} in order to extract the correct temperature and emissivity. The system response has the following expression:

$$R(\lambda,T_0) = S(\lambda)\varepsilon(\lambda,T_{ref})c_1\lambda - 5[e(c_2/\lambda T_{ref}) - 1] - 1 \tag{2}$$

where $S(\lambda)$ is the normalized system response.

In our experiments, the normalized system response $S(\lambda)$ has been obtained from the measured intensities of a tungsten lamp that is calibrated to temperature (T_{ref} was set at 2650 K in this experiment). Details about our calibration lamp and procedures are provided in [27].

2.4. X-Ray Optics and Diffraction

The X-ray optical elements are composed of a nitrogen-cooled channel-cut Si(111) monochromator located in the optics hutch at 30 m from the source and a pair of multilayer mirrors in the KB geometry are installed in the experimental hutch to focus the monochromatic beam on the pressurized sample. As already mentioned, the quality of focusing optics at high X-ray energies is of primary importance in high-pressure experiments because of the very small sample volume. KB mirrors made of iridium–alumina multilayers deposited on silicon wafers have a maximum of 80% reflectivity at 30 keV (see [28] for more details). Large focal distances of 800 mm and 1200 mm are used for the horizontal and the vertical mirrors, respectively, in order to avoid serious loss of spatial resolution on the detector. A wavelength of 0.26472 Å (samarium k-edge @ 46.8 keV) has been chosen for this experiment. At such energy, the focal spot at sample location is 2.7 × 1.8 μm^2 (full width at half maximum). With such a spatial resolution, we are confident we can eliminate most radial temperature gradients. As iron particles dispersed in MgO platelets have grain size of 1–2 μm, axial temperature gradients are expected to be as reduced as possible. As a matter of a fact, X-ray diffraction pattern presented in Figure 1 does not show any peak broadening compared to ambient temperature data.

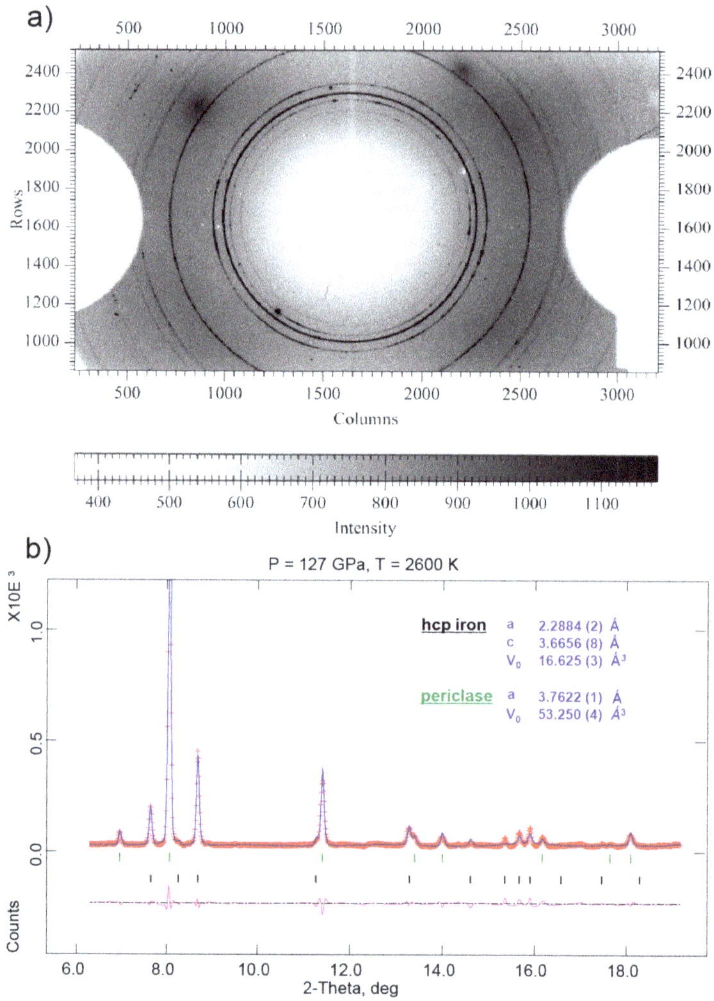

Figure 1. Example of the collected data at high pressure and high temperature. (**a**) Two dimensional (2D) diffraction image recorded on the MAR 345 image plate; (**b**) integrated diffraction pattern of hcp iron along with the Le Bail refinement with the General Structure Analysis System (GSAS) package, presented here after background subtraction.

The present data set has been collected in a former configuration of the ID27 beamline at the ESRF and is presently re-analyzed in this paper. X-ray images were recorded on a MAR345 image-plate scanner, with acquisition time varying between 15 and 120 s. Diffraction images were integrated with Fit2D software [29]. The integrated one-dimensional diffraction patterns were analyzed with the General Structure Analysis System (GSAS) software package [30] using the Le Bail method to refine lattice parameters (see Figure 1a,b).

2.5. Pressure Measurements

MgO was used as pressure calibrant in all runs, exception made for the room-temperature low-pressure experiments for which helium was used as a pressure medium. The volumes, obtained with a Le Bail fit made on the diffraction patterns, were subsequently used in a thermal equation of state with a Mie Grüneisen Debye formalism to obtain accurate pressure measurements at high-temperature. In particular, after careful evaluation of the results obtained employing two different MgO EoS [31,32] (see details and Figure S1 in Supplementary Materials) and following the discussion in [33] we opted for the one proposed by Speziale et al. [31]. The pressure measurements with the MgO thermal EoS is of course central to our exercise. We thus analyzed some recovered samples with TEM microscopy (JEOL 2100F, see details and Figure S2 in Supplementary Materials). to check the integrity of MgO and the absence of chemical reaction/diffusion between MgO and iron. These measurements, available in Supplementary Materials, show that iron diffusion into MgO never exceeds 1 at% when detected. For the room-temperature low-pressure measurements carried out using helium as pressure transmitting medium, we used the combination of two gauges: ruby with the hydrostatic scale (exponent 7.665 for the power law) [34] and $SrB_4O_7{:}Sm^{2+}$ [35].

2.6. Equation of State and Thermal Model

Several formalisms are classically proposed to express the variation of a volume with pressure. Here we used a Birch Murnaghan EoS at the third order, defined as:

$$P = \frac{3}{2}K_0\left[\left(\frac{V_0}{V}\right)^{\frac{7}{3}} - \left(\frac{V_0}{V}\right)^{\frac{5}{3}}\right]\left\{1+\frac{3}{4}[K_0' - 4]\left[\left(\frac{V_0}{V}\right)^{\frac{2}{3}} - 1\right]\right\} \quad (3)$$

With V the unit cell volume, V_0 the unit cell volume at ambient pressure, K_0 the bulk modulus, and K_0' its pressure derivative.

The high-temperature behaviour is described by the thermal pressure (ΔP_{th}).

$$P(V,T) = P(V,300K) + \Delta P_{th}(V,T) \quad (4)$$

The thermal pressure takes in account the vibrational energy of the lattice and is written as:

$$\Delta P_{th}(V,T) = \frac{\gamma(V)}{\gamma}[E_{th}(V,T), E_{th}(V,300K)] \quad (5)$$

where γ stands for the Grüneisen parameter, θ_D the Debye temperature and R is the ideal gas constant. The Grüneisen parameter (γ) is expressed as

$$\gamma(V) = \gamma_0\left(\frac{V}{V_0}\right)^q \quad (6)$$

with q corresponding to the logarithmic volume dependence of $\gamma(V)$.

The internal energy is defined as:

$$E_{th}(V,T) = 9R\left[\frac{\theta_D}{8} + T\left(\frac{T}{\theta_D}\right)\int_0^{\theta_{0,D}/T}\frac{x^3}{\exp(x)-1}dx\right] \quad (7)$$

The Debye temperature is a function of the Grüneisen parameter (γ)

$$\theta_D = \theta_{D,0}\exp\left[\frac{(\gamma_0 - \gamma(V))}{q}\right] \quad (8)$$

All refinements were carried out using EosFit-7c [36].

3. Results and Discussion

3.1. Iron Structure at High-Pressure and High-Temperature

The diffraction patterns recorded between 30 GPa and 140 GPa at temperatures varying from ambient temperature to 3400 K, display how the face centered cubic (fcc) structure is stable at low pressure and high temperature and further transforms into the hexagonal closed packed (hcp) structure above 60 GPa. The recorded transition is consistent with most recent experimental results [9]. The cell parameters for the hexagonal closed packed structure have a linear and monotonous evolution with temperature, with no kink or inversion in the axial compressibility of c and a axis as a function of pressure.

3.2. Room Temperature Equation of State

The ambient temperature EoS was refined with data points from both the P–V and P–V–T data sets. The collected volumes are in good agreement with the diamond anvil cell data recorded from previous studies [24,25] as well as with the multi anvil data from [20] (see Figure 2).

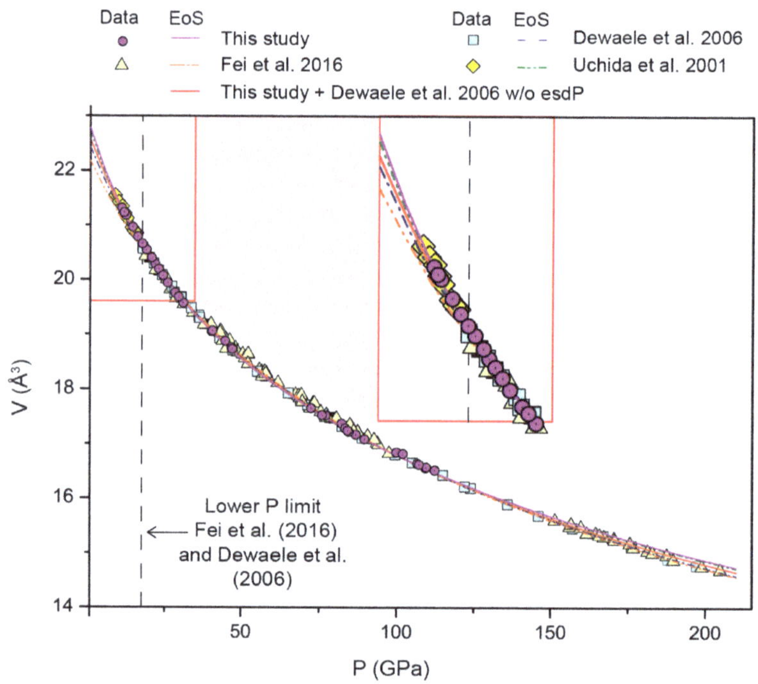

Figure 2. Compression data at room temperature with the fitted equations of state. Symbols represent the experimental points from different studies; lines are the fitted equations of state. The error bars on the data from the present study are within the dimension of the symbol. Relative to the present study, purple lines represent the equation of state obtained accounting for the errors on pressure, volume and temperature. The red line instead stands for the EoS obtained when the errors in pressure are not taken in account.

A Birch Murnaghan and a Vinet equations of state at the third order were fitted to the experimental data. Refined values are reported in Table 1. The results obtained with the two EoS formalisms are in agreement within the error bars.

Table 1. Pressure volume equations of state fit results. Values in parenthesis are the published error bars.

	This Study		This Study + Data from Dewaele et al. 2006 [24]		Fei et al. 2016 [25]	Dewaele et al. 2006 [24]		Uchida et al. 2001 [20]
			with esdP	w/o esdP				
Formalism	BM	Vinet	BM	BM		BM	Vinet	BM
V_0 (Å^3)	22.80 (2)	22.81	22.75 (1)	22.57 (1)	22.18	22.47	22.43	22.7 (3)
K_0 (GPa)	129 (6)	125 (5)	134 (5)	153 (3)	191 (5)	165 *	163 (8)	135 (19)
K_0'	6.2 (2)	6.5 (2)	6.1 (2)	5.3 (1)	4.5 (1)	4.97 (4)	5.4 (2)	6 (0.4)

* Fixed values; BM = Birch Murnaghan formalism; esdP = estimated standard deviation for pressures.

When compared with the existing equations of state determined with diamond anvil cell data only [20,21], our results indicate a lower K_0 and higher K_0' and V_0. We are instead in agreement within the error bars with the parameters proposed combining data from multi anvil press [20] and former DAC studies [24,25] (Figure 2). In all these studies the results from fitting procedures have to be considered as fitting parameters and should be used carefully when used to extrapolate materials properties.

We evaluated if the different pressure ranges of the available sets of data had an effect on the value of the EoS parameters. Excluding the points at lower pressures from our data set (in particular the points collected below 30 GPa in helium), the refinement of the EoS produces a set of parameters with a high V_0 and K_0' (respectively 23.31 Å^3 and 7.03) and low K_0 (104 GPa).

The outcome is different if we consider our full dataset (with the low pressure data) and include those from Dewaele and co-authors [24] in the refinement procedure, to extend the covered pressure range. In this case we obtain an equation of state consistent within the error bars with our results (i.e., 129 ± 6 GPa and 134 ± 5 GPa). However, the extrapolated EoS (solid purple line in Figure 2) shows a small offset above 150 GPa compared to the experimental data [24,25]. A better agreement between the volumes extrapolated and the high-pressure data is obtained when the uncertainties in pressures are excluded during the EoS refinement. The refinement then converges toward a higher K_0 and lower V_0 and K_0' (see Table 1—column w/o esdP). When plotted (red solid line in Figure 2), it reproduces the high-pressure evolution of the experimental volumes of reference [20]. This observation points out some limitations of the experimental approach and leaves some room for discussion about combining datasets using different pressure standards (MgO and ruby versus ruby and tungsten [20]), and about the way how uncertainties are taken into account in the refinement procedure.

It is noteworthy how a significant shift in the refined parameters is induced by a difference in the volume at zero pressure, resulting from the different pressure range examined. Indeed, the presence of our low pressure data collected with helium as pressure transmitting medium yields a much better constraint on the refinement of the V_0 of the hcp structure that has in turn a significant effect on the refined EoS and on the possibility of establishing material properties.

This is clearly shown in Figure 2, where different data sets and refined equations of state are reported. The lower pressure part is upscaled to enhance the different EoS and V_0 obtained, when data collection starts above 20 GPa only [24,25] or when the refinement includes data down to 10 GPa in hydrostatic conditions. Like other studies, the present results confirm the existence of a significant trade-off between the EoS parameters representing volume at zero pressure and the bulk modulus. It also points out that EoS fitting parameters obtained by different studies should be used with caution when extrapolating material properties outside the pressure and temperature range studied experimentally. The choice of V_0 has a strong effect on the values of the EoS, stronger than the addition of high-pressure points. It has therefore a significant effect on the extrapolation of the properties of iron. As a matter of a fact, if we choose to fix the reference volume V_0 at 22.43 Å^3 as in [24], the fit made to our own set of data then converges to a bulk modulus in perfect agreement with this latter set of measurements and refined ambient temperature EoS parameters.

Finally we tested the EoS obtained with the present dataset, and the one determined adding the values from [24] and determined without the uncertainties in pressure, against measurements carried out at inner core pressures [6,22,26]. This test is presented in Figure 3.

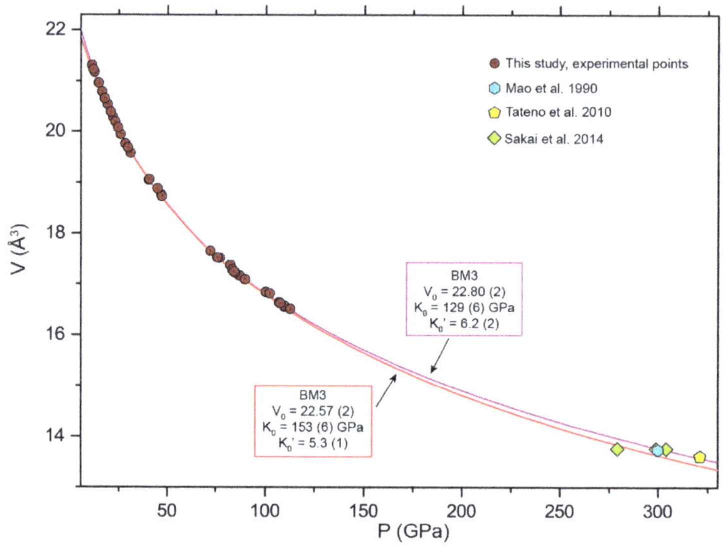

Figure 3. Extrapolation of the two room-temperature equations of state discussed at pressures compatible with the experimental data from [6]. The experimental data from the present study are added as well.

When compared to experimental data collected at the highest pressures [6,22,26], the extrapolation with the alternate EoS (K_0 = 153 GPa, K'_0 = 5.3, V_0 = 22.57 Å^3) produces smaller volumes, in agreement with the point obtained by Sakai et al. [26] using the P1 calibration (see details in [26]) but fails to reproduce other measurements. The extrapolation with the EoS obtained using only the dataset from the present study (K_0 = 129 GPa, K'_0 = 6.2, V_0 = 22.80 Å^3) perfectly matches the measurements published so far at conditions close to the actual pressure and temperature of the inner core / outer core boundary [6,22,26]. Accordingly, we retain the EoS obtained with our experimental data only as the best solution.

3.3. Thermal Model

P–V–T data were collected at temperatures comprised between 1300 and 3300 K and pressures between 45 and 135 GPa. We used the parameters refined with the Birch Murnaghan at 3rd order discussed above (K_0 = 129 GPa, K'_0 = 6.2, V_0 = 22.80 Å^3) as a reference for the room temperature EoS and the Mie Grüneisen Debye formalism for the thermal model.

A stepwise regression refinement procedure was used: we first refined the parameters of the thermal model only (θ_D, γ, q) with the others fixed (V_0, K_0, K_0'). We then determined also bulk modulus and first pressure derivative through an iterative procedure. We kept the V_0 fixed to the value obtained at room temperature and identified several solutions, either keeping θ_D fixed to a commonly accepted value (i.e., 420 K) or including it in the refined parameters. However, in the latter case, it was difficult to find a univocal solution and convergence was not always achieved. Conversely, with a fixed θ_D, the refinement cycles ran smoothly and produced accurate results as shown by the low chi^2 and pressure residuals.

The best fit to our experimental data (and preferred solution) converges to V_0 = 6.87 (2) cm^3/mol (V_0 = 22.81 $Å^3$), K_0 = 129 (1) GPa and K_0' = 6.24 (4) with θ_D = 420 K, γ_0 = 1.11 (1) and q = 0.3 (3), the resulting chi^2 is 1.79. Alternatively, we also provide the fit to our experimental data coupled with those of Dewaele and co-authors [24] (w/o esdP) with V_0 = 6.79 (1) cm^3/mol (V_0 = 22.57 $Å^3$), K_0 = 153 (1) GPa and K_0' = 5.39 (4), with θ_D = 420 K, γ_0 = 1.09 (1) and q = 0.26 (4).

The results are shown in Figure 4a, where the experimental data are reported along with the calculated isotherms as well as with the pressure residuals for the fit. The good agreement between the calculated isotherms and the experimental data and the symmetric distribution around 0 of the pressure residuals (Figure 4b) and their low value ($-3 < \Delta P < 3$ GPa) attest the quality of the fit.

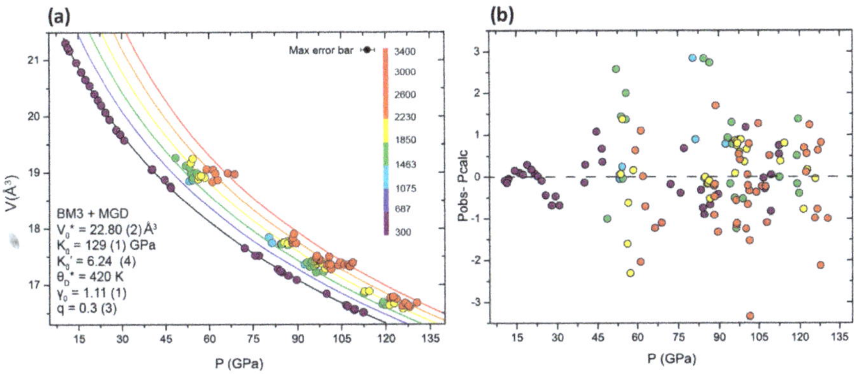

Figure 4. Results obtained fitting our data with the Mie Grüneisen Debye thermal model. (**a**) Experimental points with the isotherms calculated between 300 and 3500 K; (**b**) pressure residuals for the fit in GPa, colors correspond to the inset in (**a**).

As demonstrated by the refinements, the current dataset can be reliably fitted with a standard vibrational model. The consistency of the volume and axial compressibility data over the whole pressure and temperature range, and the absence of specific features such as volume jumps or inversions in the axial compressibility permit to exclude the presence of strong magnetic effects.

Previously determined thermal equations of state for iron also considered some additional terms to account for other contributions such as electronic or anharmonic contributions, in addition to the thermal pressure described with the standard vibrational model presented in Equations (5)–(7). To test the influence of these parameters, we used the parameterization proposed in [24] to compute the values of the two contributions for the present data set. The maximum contribution is obtained for the higher temperature data (i.e., 3399 K at 68 GPa) and amounts 1.3 GPa and 5.2 GPa for the electronic and anharmonic term respectively. If these values are subtracted from the total thermal pressure term (in order to keep the pure vibrational part only), and the data set fitted again using the Mie Grüneisen Debye model, we do not have a noticeable difference in the refined parameters.

To further test our results, the current thermal EoS is also plotted against high pressure and temperature experimental data available from literature. In Figure 5a the calculated isotherms are compared with the HP-HT data points from [25]. The present model satisfactorily describes this dataset, the main discrepancies arising from the scatter of the data themselves. Furthermore, the isotherms extrapolated at lower pressures also show a good agreement with the multi-anvil data [20] (Figure 5b).

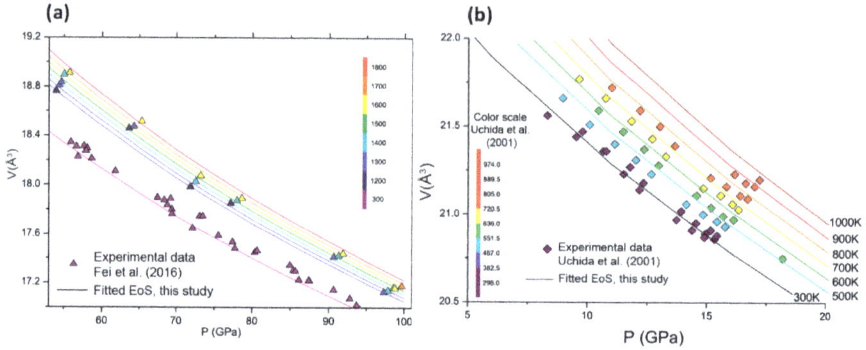

Figure 5. Volume over pressure plots of the isotherms from the present study compared with experimental data points from literature. (**a**) Data points from diamond anvil cell experiments [21]; (**b**) data points from multi anvil experiments [19].

Ultimately, the available parameterizations for the electronic and anharmonic terms used in [24,25] result in a significant difference in density once thermodynamic quantities are calculated from the volumes extrapolated at P-T conditions relevant for the inner core. Constraining these parameters with only one dataset thus appears to be very difficult.

4. Discussion

The thermal equation of state of iron has a critical role in the evaluation of the chemical and physical state of the Earth's solid inner core. As a matter of fact, such a thermal EoS is required to extrapolate the densities of iron at relevant P-T conditions and provides a reference for the comparison with models from seismology. The density deficit shown in the preliminary reference Earth model (i.e., the PREM [37]) with respect to the extrapolated densities of pure Fe, implies the incorporation of light elements in the solid inner core. It is therefore essential to assess a reference density profile for pure iron to identify the nature and the amount of such light elements. As such, the accuracy of the Fe thermal EoS, and deduced parameters representing material properties, has great influence on the interpretation of geophysical data.

We used the best matching thermal EoS to extrapolate the densities of iron at inner core pressures (330–360 GPa) and a temperature of 6000 K, 6500 K, as well as at room temperature. The results are shown along the PREM model and the values from previous studies [24,25] in Figure 6. At room temperature the densities calculated in the present study are lower than the values from Dewaele and co-authors [24] and in agreement with the experimental data from Mao et al. [22], Tateno et al. [6] and Sakai et al. [26]. The difference in the refined parameters lead to 1.8% density difference 300 GPa and 300 K between the two extrapolated EoS. As previously discussed, this difference has its origin in the incorporation of low-pressure data in the refinement, which allows us to have better constraints on the reference volume V_0, with consequences on the couple K_0 and K_0'.

At high temperature as well, the densities calculated in the present study are lower compared to those reported in literature. The higher difference is obtained when the present data are compared with those from [24], showing a 1.5% density difference at the ICB (inner core boundary).

The reduction is 0.5% when our results are compared to the values obtained by [25]. It is noteworthy that the shift in density between the present study and the one from Fei et al. [25] is similar to the one existing between their data and those of Dewaele and coauthors [24]. Hence, the difference between our results obtained using a simple standard vibrational model and another model including the electronic term, is of the same order as the difference induced by using two different parameterizations to account for the same electronic term.

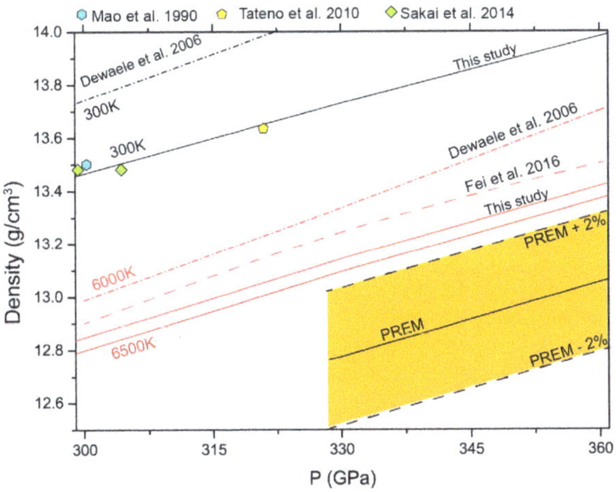

Figure 6. Density over pressure plot in the pressure range representative of the Earth's inner core. Densities calculated from the present thermal equation of state are compared those from literature. Densities from the seismological reference of the PREM are shown with a 2% error bar. For [29] only points at or above 300 GPa are shown.

According to our preferred EoS parameters, a pure iron core would be 3% denser than the density profile in PREM if the core temperature is assumed to be 6000 K, thus possibly revising the amount of light elements needed to match the seismological inferred densities. This density difference progressively diminishes if the temperature of the core is increased, though at 6500 K there is still a 2.5% difference and extreme temperatures would be needed to have a match with the seismological reference. As well, a different density of solid iron changes the estimate of the density jump between the liquid outer core and the solid inner core at the ICB. We estimate a 1.1% density contrast between solid and liquid iron [38] at the ICB. The present value is lower than the 2% difference predicted in [38] and increases the divergence from the seismological reference [37], proposed to be 4.5%. Accordingly, the estimates on the amount of light elements in the core need to be reassessed. In particular, the amount of oxygen, considered as the most elegant solution to account for the ICB density contrast due to its incompatibility with solid iron [12] might be higher than expected.

5. Conclusions

We presented an in situ X-ray diffraction study of iron at high pressure and temperature. The present data extends the temperature range previously investigated with static compression techniques. The results highlight that the study of the hexagonal close packed structure at the lowest possible pressure is a critical step to constrain the refinement of the ambient temperature equation of state, due to the trade-off between the V_0, K_0 and K_0'. The value of V_0 has indeed a strong incidence on the room-temperature EoS, with significant consequences when the EoS is extrapolated to core pressures. We propose a set of parameters refined with a 3rd order Birch Murnaghan equation and the thermal contribution is described with a standard Mie Grüneisen Debye vibrational model, without the need of additional electronic or anharmonic terms. Our thermal equation of state extrapolated at core pressures yields a +3% density difference for a pure iron core compared to the density profile of PREM, assuming a core temperature of 6000 K. Hence, this suggest to re-evaluate the amount of light elements in the Earth's core and/or the effects of light elements on the physical properties of iron. For instance the 1.1% density contrast between solid and liquid iron at the ICB, estimated from the present equation of state, suggests the presence of a higher amount of oxygen in the liquid outer core.

Supplementary Materials: The following are available online at http://www.mdpi.com/2075-163X/10/2/100/s1. Figure S1: Eos for iron and extrapolations made using different MgO pressure scales [31,32]; Figure S2: ATEM analyses on a FIB prepared recovered sample after diamond anvil cell experiment. The two data sets used for the refinement are attached as separate files.

Author Contributions: Investigation, G.F., J.B., J.S., N.G. and J.M.; formal analysis, F.M., G.F.; writing—original draft preparation, F.M. and G.F.; writing—review and editing, F.M., G.F., J.B., J.M., and N.G. All authors have read and agreed to the published version of the manuscript.

Funding: F.M. and G.F. have received funding from the European Research Council (ERC) under the European Union's Horizon 2020 research and innovation Programme, grant agreement No 670787 to Guillaume Fiquet

Acknowledgments: We thank Eugene Gregoryanz for his assistance during the X-ray data acquisition and Anne-Line Auzende for her help with the TEM examination of the recovered sample. Francesca Miozzi thanks G. Morard and R.J. Angel for fruitful discussions.

Conflicts of Interest: The authors declare no conflict of interest. The funders had no role in the design of the study; in the experiments reported; in the writing of the manuscript, or in the decision to publish the results.

References

1. Birch, F. Elasticity and constitution of the Earth's interior. *J. Geophys. Res.* **1952**, *57*, 227–286. [CrossRef]
2. Anderson, O.L. Mineral physics of iron and of the core. *Rev. Geophys.* **1995**, *33*, 429. [CrossRef]
3. Boehler, R. Temperatures in the Earth's core from melting-point measurements of iron at high static pressures. *Nature* **1993**, *363*, 534–536. [CrossRef]
4. Ma, Y.; Somayazulu, M.; Shen, G.; Mao, H.; Shu, J.; Hemley, R.J. In situ X-ray diffraction studies of iron to Earth-core conditions. *Phys. Earth Planet. Inter.* **2004**, *143*, 455–467. [CrossRef]
5. Fiquet, G. Sound Velocities in Iron to 110 Gigapascals. *Science* **2001**, *291*, 468–471. [CrossRef] [PubMed]
6. Tateno, S.; Hirose, K.; Ohishi, Y.; Tatsumi, Y. The structure of iron in earth's inner core. *Science* **2010**, *330*, 359–361. [CrossRef] [PubMed]
7. Yoo, C.S.; Akella, J.; Campbell, A.J.; Mao, H.K.; Hemley, R.J. Phase Diagram of Iron by in Situ X-ray Diffraction: Implications for Earth's Core. *Science* **1995**, *270*, 1473–1475. [CrossRef]
8. Dubrovinsky, L.S.; Saxena, S.K.; Tutti, F.; Rekhi, S.; LeBehan, T. In Situ X-Ray Study of Thermal Expansion and Phase Transition of Iron at Multimegabar Pressure. *Phys. Rev. Lett.* **2000**, *84*, 1720–1723. [CrossRef]
9. Morard, G.; Boccato, S.; Rosa, A.D.; Anzellini, S.; Miozzi, F.; Henry, L.; Garbarino, G.; Mezouar, M.; Harmand, M.; Guyot, F.; et al. Solving Controversies on the Iron Phase Diagram Under High Pressure. *Geophys. Res. Lett.* **2018**, *45*, 11074–11082. [CrossRef]
10. Poirier, J. Light elements in the Earth's outer core: A critical review. *Phys. Earth Planet. Inter.* **1994**, *85*, 319–337. [CrossRef]
11. Antonangeli, D.; Siebert, J.; Badro, J.; Farber, D.L.; Fiquet, G.; Morard, G.; Ryerson, F.J. Composition of the Earth's inner core from high-pressure sound velocity measurements in Fe–Ni–Si alloys. *Earth Planet. Sci. Lett.* **2010**, *295*, 292–296. [CrossRef]
12. Badro, J.; Cote, A.S.; Brodholt, J.P. A seismologically consistent compositional model of Earth's core. *Proc. Natl. Acad. Sci.* **2014**, *111*, 7542–7545. [CrossRef]
13. McDonough, W.F.; Sun, S.-S. The composition of the Earth. *Chem. Geol.* **1995**, *120*, 223–253. [CrossRef]
14. Allègre, C.J.; Poirier, J.-P.; Humler, E.; Hofmann, A.W. The chemical composition of the Earth. *Earth Planet. Sci. Lett.* **1995**, *134*, 515–526. [CrossRef]
15. Kuwayama, Y.; Hirose, K.; Sata, N.; Ohishi, Y. Phase relations of iron and iron–nickel alloys up to 300 GPa: Implications for composition and structure of the Earth's inner core. *Earth Planet. Sci. Lett.* **2008**, *273*, 379–385. [CrossRef]
16. Belonoshko, A.B. Equation of state for ϵ-iron at high pressures and temperatures. *Condens. Matter Phys.* **2010**, *13*, 23605–23615. [CrossRef]
17. Sha, X.; Cohen, R.E. First-principles thermal equation of state and thermoelasticity of hcp Fe at high pressures. *Phys. Rev. B* **2010**, *81*, 094105. [CrossRef]
18. Dorogokupets, P.I.; Dymshits, A.M.; Litasov, K.D.; Sokolova, T.S. Thermodynamics and Equations of State of Iron to 350 GPa and 6000 K. *Sci. Rep.* **2017**, *7*, 41863. [CrossRef]
19. Funamori, N.; Yagi, T.; Uchida, T. High-pressure and high-temperature in situ x-ray Diffraction study of iron to above 30 Gpa using MA8-type apparatus. *Geophys. Res. Lett.* **1996**, *23*, 953–956. [CrossRef]

20. Uchida, T.; Wang, Y.; Rivers, M.L.; Sutton, S.R. Stability field and thermal equation of state of ε-iron determined by synchrotron X-ray diffraction in a multianvil apparatus. *J. Geophys. Res. Solid Earth* **2001**, *106*, 21799–21810. [CrossRef]
21. Yamazaki, D.; Ito, E.; Yoshino, T.; Yoneda, A.; Guo, X.; Zhang, B.; Sun, W.; Shimojuku, A.; Tsujino, N.; Kunimoto, T.; et al. P-V-T equation of state for ε -iron up to 80 GPa and 1900 K using the Kawai-type high pressure apparatus equipped with sintered diamond anvils. *Geophys. Res. Lett.* **2012**, *39*, 2012GL053540. [CrossRef]
22. Mao, H.K.; Wu, Y.; Chen, L.C.; Shu, J.F.; Jephcoat, A.P. Static compression of iron to 300 GPa and Fe 0.8 Ni 0.2 alloy to 260 GPa: Implications for composition of the core. *J. Geophys. Res.* **1990**, *95*, 21737. [CrossRef]
23. Jephcoat, A.P.; Mao, H.K.; Bell, P.M. Static compression of iron T78 GPa with rare gas solids as pressure-transmitting media. In *Elastic Properties and Equations of State*; American Geophysical Union: Washington, DC, USA, 1988; Volume 91, pp. 524–531.
24. Dewaele, A.; Loubeyre, P.; Occelli, F.; Mezouar, M.; Dorogokupets, P.I.; Torrent, M. Quasihydrostatic Equation of State of Iron above 2 Mbar. *Phys. Rev. Lett.* **2006**, *97*, 215504. [CrossRef]
25. Fei, Y.; Murphy, C.; Shibazaki, Y.; Shahar, A.; Huang, H. Thermal equation of state of hcp-iron: Constraint on the density deficit of Earth's solid inner core. *Geophys. Res. Lett.* **2016**, *43*, 6837–6843. [CrossRef]
26. Sakai, T.; Takahashi, S.; Nishitani, N.; Mashino, I.; Ohtani, E.; Hirao, N. Equation of state of pure iron and Fe0.9Ni0.1 alloy up to 3Mbar. *Phys. Earth Planet. Inter.* **2014**, *228*, 114–126. [CrossRef]
27. Schultz, E.; Mezouar, M.; Crichton, W.; Bauchau, S.; Blattmann, G.; Andrault, D.; Fiquet, G.; Boehler, R.; Rambert, N.; Sitaud, B.; et al. High Pressure–High Temperature Monochromatic X-Ray Diffraction At the Esrf. *High Press. Res.* **2005**, *25*, 71–83. [CrossRef]
28. Mezouar, M.; Crichton, W.A.; Bauchau, S.; Thurel, F.; Witsch, H.; Torrecillas, F.; Blattmann, G.; Marion, P.; Dabin, Y.; Chavanne, J.; et al. Development of a new state-of-the-art beamline optimized for monochromatic single-crystal and powder X-ray diffraction under extreme conditions at the ESRF. *J. Synchrotron Radiat.* **2005**, *12*, 659–664. [CrossRef]
29. Hammersley, A.P. FIT2D: A multi-purpose data reduction, analysis and visualization program. *J. Appl. Crystallogr.* **2016**, *49*, 646–652. [CrossRef]
30. Larson, A.C.; Von Dreele, R.B.; Alamos, L. Gsas. *Rep. lAUR* **1994**, *1994*, 86–748.
31. Speziale, S.; Zha, C.-S.; Duffy, T.S.; Hemley, R.J.; Mao, H. Quasi-hydrostatic compression of magnesium oxide to 52 GPa: Implications for the pressure-volume-temperature equation of state. *J. Geophys. Res. Solid Earth* **2001**, *106*, 515–528. [CrossRef]
32. Tange, Y.; Nishihara, Y.; Tsuchiya, T. Unified analyses for P-V-T equation of state of MgO: A solution for pressure-scale problems in high P-T experiments. *J. Geophys. Res.* **2009**, *114*, B03208. [CrossRef]
33. Dorogokupets, P.I. P–V–T equations of state of MgO and thermodynamics. *Phys. Chem. Miner.* **2010**, *37*, 677–684. [CrossRef]
34. Mao, H.K.; Xu, J.; Bell, P.M. Calibration of the ruby pressure gauge to 800 kbar under quasi-hydrostatic conditions. *J. Geophys. Res.* **1986**, *91*, 4673. [CrossRef]
35. Datchi, F.; LeToullec, R.; Loubeyre, P. Improved calibration of the SrB4O7:Sm2+optical pressure gauge: Advantages at very high pressures and high temperatures. *J. Appl. Phys.* **1997**, *81*, 3333–3339. [CrossRef]
36. Gonzalez-Platas, J.; Alvaro, M.; Nestola, F.; Angel, R. EosFit7-GUI: A new graphical user interface for equation of state calculations, analyses and teaching. *J. Appl. Crystallogr.* **2016**, *49*, 1377–1382. [CrossRef]
37. Dziewonski, A.M.; Anderson, D.L. Preliminary reference Earth model. *Phys. Earth Planet. Inter.* **1981**, *25*, 297–356. [CrossRef]
38. Alfè, D.; Gillan, M.; Price, G. Composition and temperature of the Earth's core constrained by combining ab initio calculations and seismic data. *Earth Planet. Sci. Lett.* **2002**, *195*, 91–98. [CrossRef]

minerals

MDPI

Article

A Synchrotron Mössbauer Spectroscopy Study of a Hydrated Iron-Sulfate at High Pressures

Tyler Perez *,†, Gregory J. Finkelstein ‡, Olivia Pardo, Natalia V. Solomatova § and Jennifer M. Jackson *

Seismological Laboratory, California Institute of Technology, Pasadena, CA 91125, USA; gjfinkel@princeton.edu (G.J.F.); opardo@caltech.edu (O.P.); natalia.solomatova@ens-lyon.fr (N.V.S.)

* Correspondence: tperez11@jhu.edu (T.P.); jackson@gps.caltech.edu (J.M.J.)

† Now at Department of Earth and Planetary Sciences, Johns Hopkins University, Baltimore, MD 21218, USA.

‡ Now at Department of Geosciences, Princeton University, Princeton, NJ 08544, USA.

§ Now at École Normale Supérieure de Lyon, Laboratoire de Géologie, CNRS, UMR 5276, 69342 Lyon, France.

Received: 21 December 2019; Accepted: 4 February 2020; Published: 8 February 2020

Abstract: Szomolnokite is a monohydrated ferrous iron sulfate mineral, $FeSO_4 \cdot H_2O$, where the ferrous iron atoms are in octahedral coordination with four corners shared with SO_4 and two with H_2O groups. While somewhat rare on Earth, szomolnokite has been detected on the surface of Mars along with several other hydrated sulfates and is suggested to occur near the surface of Venus. Previous measurements have characterized the local environment of the iron atoms in szomolnokite using Mössbauer spectroscopy at a range of temperatures and 1 bar. Our study represents a step towards understanding the electronic environment of iron in szomolnokite under compression at 300 K. Using a hydrostatic helium pressure-transmitting medium, we explored the pressure dependence of iron's site-specific behavior in a synthetic szomolnokite powdered sample up to 95 GPa with time-domain synchrotron Mössbauer spectroscopy. At 1 bar, the Mössbauer spectrum is well described by two Fe^{2+}-like sites and no ferric iron, consistent with select conventional Mössbauer spectra evaluations. At pressures below 19 GPa, steep gradients in the hyperfine parameters are most likely due to a structural phase transition. At 19 GPa, a fourth site is required to explain the time spectrum with increasing fractions of a low quadrupole splitting site, which could indicate the onset of another transition. Above 19 GPa we present three different models, including those with a high- to low-spin transition, that provide reasonable scenarios of electronic environment changes of the iron in szomolnokite with pressure. We summarize the complex range of Fe^{2+} spin transition characteristics at high-pressures by comparing szomolnokite with previous studies on ferrous-iron bearing phases.

Keywords: iron-sulfates; hydrous phases; szomolnokite; high-pressure; synchrotron Mössbauer spectroscopy; spin-transition

1. Introduction

Historically, sulfate minerals have been studied mainly in the context of surface processes such as evaporitic deposits and hydrothermal systems [1] or in the context of mine tailings and wastes [2]. Sulfate salts play important roles in the cycling of metals and sulfates in terrestrial systems [3], and are thought to play important roles on the surface weathering processes on other planetary bodies. In particular, relatively large deposits of monohydrated sulfates have been detected on the surface of Mars using absorption spectroscopy [4,5], with Lichtenberg et al. [5] specifically preferring szomolnokite to explain certain signatures present in Aram Chaos. Talla and Wildner [6] perform a detailed spectroscopic study on the kieserite-szomolnokite solid-solution series under ambient and Martian temperature conditions, due to the high probability that intermediate compositions along this join exist on Mars. Additionally, Lane et al. [7] suggests that hydrous iron sulfates closely match

Mini-TES (Miniature Thermal Emission Spectrometer) and Mössbauer data from Martian soil analyzed by MER (Mars Exploration Rovers). Chou et al. [3] performed experiments on the stability of a variety of hydrated and anhydrous sulfates in Martian reaction environments. They suggest that hydrated sulfates could play important roles in the hydrologic cycle of Mars. There is also evidence for hydrated sulfates on the surface of Europa [8], and they have been investigated up to pressures of ~2.5 GPa as possible constituents of icy moon mantles [9]. Barsukov et al. [10] suggested that barium and strontium sulfates could possibly be stable in the crust in Venus, although it is not clear if they would be formed sub-surface as a result of dehydration processes.

Previous experimental studies on hydrated sulfates at elevated pressures include Fortes et al. [11] who used neutron powder diffraction of deuterated $MgSO_4{\cdot}11D_2O$ to explore phase transitions within the range $0.1 < P < 1000$ MPa and 150 to 280 K. $MgSO_4{\cdot}11D_2O$ is the deuterated analogue of meridianiite, which is triclinic with space group $P\bar{1}$ (Z = 2). They detected evidence of peritectic melting at 0.545 GPa and 275 K as well as a phase transition at 0.9 GPa and 240 K, decomposing into ice VI + $MgSO_4{\cdot}9D_2O$ which is monoclinic with space group $P2_1/c$ (Z = 4). Previous Mössbauer work on szomolnokite ($FeSO_4{\cdot}H_2O$) has focused largely on the effect of temperature on the hyperfine parameters at ambient pressure. In a recent comprehensive review paper, Dyar et al. [12] presented and summarized conventional Mössbauer spectra evaluations for a suite of iron-bearing sulfates. Alboom et al. [13] studied szomolnokite with energy domain Mössbauer spectroscopy from 4.2 K to 450 K and found a magnetic order-disorder transition at 29.6 ± 0.5 K. Giester et al. [14] used Mössbauer spectroscopy as well as X-ray diffraction to find that for $(Fe,Cu)SO_4{\cdot}H_2O$, there is a reduction in symmetry from monoclinic to triclinic beyond 20 mol % Cu and report a magnetic order transition between 15 K and 4.2 K at ambient pressure.

As discussed in Meusburger et al. [15], szomolnokite ($FeSO_4{\cdot}H_2O$) is isostructural to kieserite ($MgSO_4{\cdot}H_2O$), which is monoclinic and in space-group *C2/c*. The kieserite structure consists of corner sharing $[MO_4(H_2O)]^{6-}$ units, which run parallel to the crystallographic *c*-axis. Lattice parameters for szomolnokite obtained in this study and previous studies [14–17] are discussed below and given in Table S1. Meusberger et al. [15] studied the structural evolution of szomolnokite up to pressures of 9.2 GPa by means of X-ray diffraction, Fourier-transform infrared spectroscopy, and Raman spectroscopy. They found a transition from monoclinic phase in space group *C2/c* to triclinic phase in space group $P\bar{1}$. They find that the transition is ferroelastic and second order in thermodynamic character.

A recent study has characterized the magnetic and structural changes of jarosite, a hydroxylated iron sulfate, $KFe_3(OH)_6(SO_4)_2$, as a function of pressure (up to 40 GPa) using a variety of techniques, including synchrotron Mössbauer spectroscopy [18]. They described the iron environment with a single Fe^{3+} site which steadily increases in quadrupole splitting and steadily decreases in isomer shift as pressure increases. They found a continuous pressure induced phase transition that did not affect the spin state of the Fe^{3+}. They also reported a dramatic increase in magnetic ordering temperature as a function of pressure; up to 240 K at 40 GPa.

In this study, we use time-domain synchrotron Mössbauer spectroscopy to examine the high-pressure behavior of szomolnokite at the crystal chemical level. Mössbauer spectroscopy directly probes the local electronic environment of iron atoms within solids, which can provide valuable physical insight to transitions occurring in the examined phase under high pressures. Understanding the high-pressure behavior of szomolnokite will provide a step towards characterizing the behavior of complex hydrated minerals under compression and has implications for planetary interiors.

2. Materials and Methods

The szomolnokite powder used in this study was synthesized through a collaboration with Isoflex ($FeSO_4{\cdot}H_2O$, using 96% ^{57}Fe). For ambient pressure X-ray diffraction characterization, a powdered sample of szomolnokite contained in a Kapton tube was used. X-ray diffraction measurements were taken at beamline 12.2.2 at the Advanced Light Source (ALS) operating with a wavelength of 0.4972 Å and beam size of 20 μm^2 (full width at half maximum). The integrated ambient pressure X-ray diffraction pattern is shown in Figure 1 and was fit using Rietveld refinement in GSAS-II [19]. The resulting lattice parameters are in agreement with previously reported values [14–17] (Table S1).

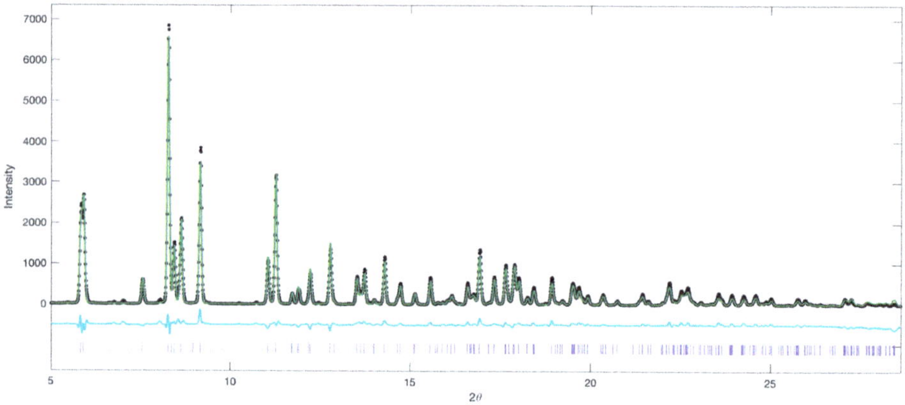

Figure 1. Integrated ambient pressure-temperature diffraction pattern with predicted reflections for szomolnokite. The background-corrected integrated pattern (black dots) is fit using Rietveld refinement in GSAS-II. The resulting fit is shown in green with residuals below (cyan). Predicted reflections for the monoclinic *C2/c* phase are shown in blue. The reflections highlighted in the grey region are chosen to calculate the ambient pressure lattice parameters: a = 7.09 Å, b = 7.55 Å, c = 7.78 Å, β = 118.65(2)°, volume = 365.2(1) $Å^3$. Comparison to previous studies are shown in Table S1.

For the synchrotron Mössbauer spectroscopy measurements, a powdered $^{57}FeSO_4{\cdot}H_2O$ sample (<5 μm thick) was loaded into a Princeton-design symmetric diamond anvil cell, assembled with two one-quarter-carat diamonds with 250 μm diameter culets, plus 50 μm bevels, mounted on tungsten-carbide seats. A rhenium gasket was indented to ~50 μm and drilled with a 130-μm hole, which was then loaded with the powdered szomolnokite sample. Two rubies were placed in the sample chamber to be used as pressure gauges. The cell was placed into a vacuum of ~10^{-5} Pa to remove any residual liquids before it was loaded with helium under ~0.17 GPa (25,000 pounds per square inch).

The hyperfine interactions, namely the quadrupole splitting (QS) and isomer shift (IS), were determined using time-domain synchrotron Mössbauer spectroscopy (SMS), a technique that probes the local electronic environment of ^{57}Fe atoms in the sample. The isomer shift is proportional to the s-electron density at the nucleus while the quadrupole splitting describes the asymmetry in the electric field gradient at the Mössbauer nucleus. Knowledge of both quantities provides constraints on the valence and spin state of the iron atoms in szomolnokite. The distribution of quadrupole splittings for a particular Mössbauer site (expressed as the full width at half maximum, FWHM, in units of mm/s) as well as site weight fraction was also determined in the fitting. The SMS measurements were conducted at Sector 3-ID-B of the Advanced Photon Source of Argonne National Laboratory. The storage ring was operated in 24 bunch mode, top-up, with 153 ns bunch separation. The focus size of the X-ray beam was about 15 × 15 μm^2. The time window used to evaluate the spectra was 23 ns to 129 ns after excitation. The pressure in the sample chamber was determined before and after the SMS measurement using the ruby fluorescence method [20] and the standard deviation between two ruby spheres proximal to the sample.

3. Results

The SMS spectra were fitted with version 2.2.0 of the CONUSS software [21], which uses a least-square algorithm to fit iron's hyperfine parameters in szomolnokite, as well as material properties such as effective thickness and the Lamb–Mössbauer factor. We estimated the sample thickness before compression to be <5 μm thick, using a calibrated binocular microscope. Although it is known that the Lamb–Mössbauer factor of solids generally increases with increasing pressure, there are no quantitative constraints for szomolnokite. In order to keep the number of fitting parameters at a minimum, the Lamb–Mössbauer factor was fixed at 0.6, and physical thickness was fixed at 3 μm at all compression points. The quadrupole splitting value for each site was fitted. In some cases, the distribution of quadrupole splittings for a particular Mössbauer site (expressed as the full width at half maximum, FWHM, in units of mm/s) could be determined in the fitting, whereas in some models incorporating additional sites, the FWHM was fixed. At 1 bar, the isomer shifts of sites 1 and 2 were determined in a CONUSS fitting procedure called, "dual fit", where the spectrum of the sample and the spectrum of the sample with a 10-μm thick stainless steel reference foil containing the natural ^{57}Fe enrichment level are fit simultaneously (Tables S2–S4). The best fit results at 1 bar agree reasonably well with the model from Dyar et al. [12], given that the highest weight fraction site falls in the same area in IS vs. QS space as the other szomolnokite samples (Figure 2; Table S5). The primary site also agrees with other ambient pressure Mössbauer studies on szomolnokite. However, the isomer shift of the second site (representing 9% of total iron content) required to fit the spectrum, is relatively low (~1 mm/s) compared with that of previously characterized szomolnokite samples. It is clear, though, that this second site is not associated with ferric iron, as its QS value is relatively high (2.07 ± 0.03 mm/s), and therefore, we have ruled out the possibility of any Fe^{3+} in our sample. We have also ruled out amorphous components as well as significant texture effects.

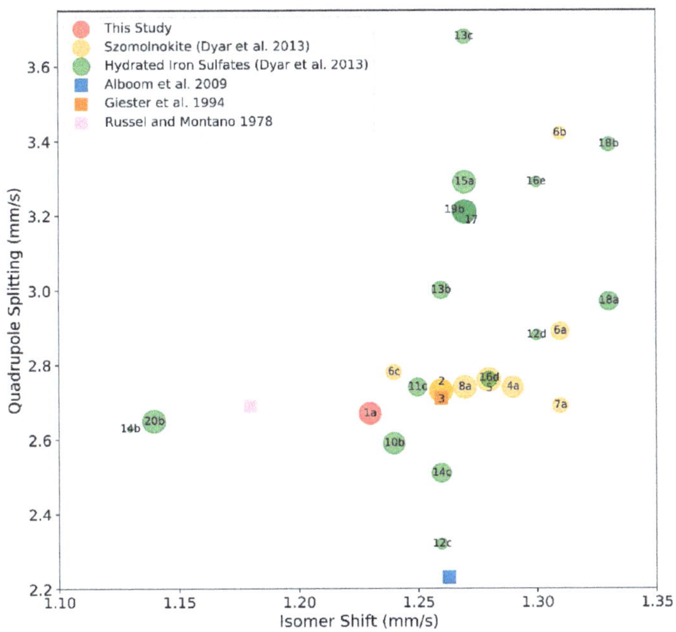

Figure 2. Isomer shift relative to bcc-iron versus quadrupole splitting at 0 GPa of ^{57}Fe-Mössbauer sites in a variety of hydrated iron sulfates. Circles with the same number represent different sites within a single sample. Labels come from [12–14,22] (see Table S5). Size of circle represents weight fraction of that site. The colored squares represent values from earlier Mössbauer studies of szomolnokite. The IS of site 1b from this study is out of the range of this figure at 0.89 ± 0.02 mm/s.

Although there is only one crystallographic site for iron in szomolnokite at ambient conditions [13–15], two Mössbauer sites are required to fit the data. As discussed in previous works on iron-bearing phases characterized by a single crystallographic site for ferrous iron, such as bridgmanite and (Mg,Fe)O ferropericlase [23–26], additional Fe^{2+}-like sites in the data evaluation of Mössbauer spectra may arise from differences in the next nearest-neighbor environments. Atomic-specific probes like Mössbauer spectroscopy are therefore capable of resolving such local-environment characteristics. At higher pressures, three or more sites are required to fit the spectra. This is not unreasonable if we look to a recent combined single-crystal X-ray diffraction study of szomolnokite [15], where they found two distinct crystallographic sites for iron in the triclinic structure at 7.3 GPa. A recent combined X-ray diffraction and SMS study on $Fe_2Si_2O_6$ ferrosilite also showed that additional iron sites are required to fit the SMS spectra. Specifically, at the structural phase transition of ferrosilite from *C*2/*c* to HP-*P*2$_1$/*c*, Solomatova et al. [27] found that four high-spin Mössbauer sites are required to fit the SMS spectra (M1a, M1b, M2a, M2b), although only two distinct crystallographic sites for iron could be resolved with single-crystal X-ray diffraction (M1 and M2).

The isomer shifts could not be determined as a function of pressure relative to stainless steel foil due to the fact that the sample was smaller than the X-ray focus size, leading to complications in interpreting the scattering effects around its edges with respect to the much larger stainless steel reference foil. We therefore did not attempt to interpret the isomer shift values from the reference foil with compression. Rather, we report the relative difference in isomer shift of the sites with respect to site 1.

The results indicate significant changes in the hyperfine parameters with increasing pressure. We explore three different best-fitting models over the compression range of our study. Although these models produce the same hyperfine parameters for the iron sites at low pressures, the models diverge at pressures above 19 GPa, as discussed below (Figure 3, Figure 4, and Figure S1, and Tables S2–S4).

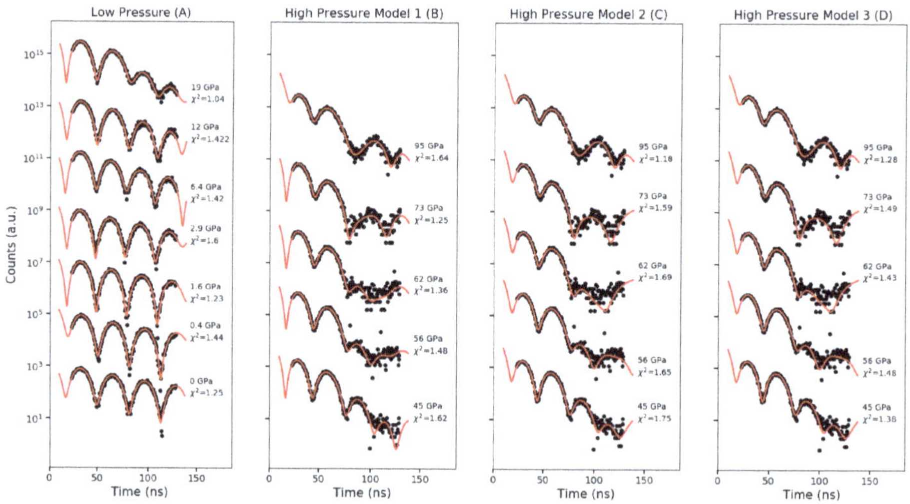

Figure 3. Representative synchrotron Mössbauer spectroscopy (SMS) spectra of szomolnokite: black circles are binned data and red curves are best fit results from CONUSS. (**A**) All models are identical from 0 to 19 GPa. (**B**) Model 1 features only 3 sites and assumes all the iron in the material remains in the high-spin state up to 95 GPa. (**C**) Model 2 applied to 19 to 95 GPa. This features an additional site, which could indicate a transition (increased distortion, structural, or high- to low-spin transition) that occurs between 19 and 45 GPa. (**D**) Model 3 applied to the higher pressures. This model features a gradual increase in presence of a low spin site from 19 to 95 GPa.

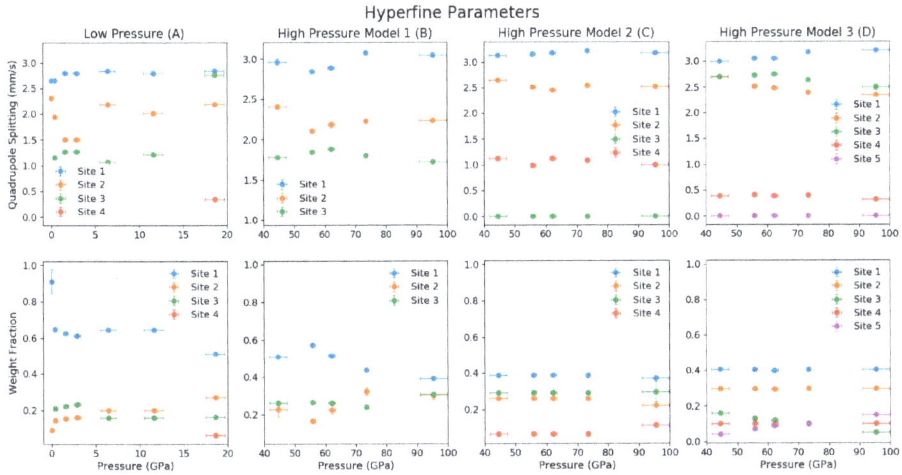

Figure 4. Fitted hyperfine parameters of szomolnokite as a function of pressure: quadrupole splitting and weight fractions of the iron sites. (**A**) Low pressure values, as all models are identical from 0 to 19 GPa. (**B**) Model 1 features only 3 sites from 45 to 95 GPa and assumes all the iron in the material remains in the high-spin state. (**C**) Values from Model 2, featuring a high- to low-spin transition in site 3 (quadrupole-splitting of 0 mm/s) between 19 and 45 GPa. (**D**) Values from Model 3 consider a gradual high-spin (site 3) to low-spin (site 5) transition from 45 to 95 GPa: site 3 (green) and site 5 (purple) trade off in weight fraction until site 3 has zero weight fraction and site 5 reaches a weight fraction around 0.2. Values are reported in Tables S2–S4.

Three Mössbauer sites were needed to fit the data in the 0.4 to 19 GPa compression range. In this pressure range, the quadrupole splitting fluctuates in each of the three sites (Figure 4). The weight fraction of the primary site rapidly decreases in this pressure range and the other two sites gain that weight fraction. There is a significant change at 19 GPa where the QS for site 3 increases to 2.7 mm/s, and the primary site weight fraction decreases. The isomer shift values relative to site 1 show large fluctuations in this compression range. The relative IS of site 2 increases sharply before dropping off slightly at 19 GPa (Figure S1). The relative IS of site 3 decreases at low compressions, then increases with pressure up to 19 GPa. These changes are likely related to a structural transition, as discussed in the next section.

Above 19 GPa, a fourth site with a low weight fraction and low quadrupole splitting was introduced to increase the quality of the fit. This site has a small but non-zero quadrupole splitting suggesting that this additional feature could be indicative of another transition. Therefore, above 19 GPa we consider three models that incorporate different transition scenarios, including a high- to low-spin-transition. Model 1 does not incorporate a low spin site (zero quadrupole-splitting) over the compression range of this study and removes the fourth site at 19 GPa. Model 1 shows a small decrease in the QS of sites 1 and 2 above 45 GPa, followed by a steady rise over the remaining pressure range, whereas the QS of site 3 is essentially invariant with compression. With fewer total parameters in Model 1 than the other models, the weight fractions of the sites could freely vary in the fitting procedure without running into correlation problems. We observe a small increase in weight percent of site 1 above 45 GPa followed by a steady decrease with pressure. The weight fraction of site 2 exhibits the opposite trend as site 1, while the weight percent of site 3 shows minimal fluctuations in this pressure range with a small increase at 95 GPa.

Model 2 contains four sites between 19 and 45 GPa, where site 4 has a low QS at 19 GPa (6% of the total iron) and steadily increases with increasing pressure and site 3 has zero QS at 45 GPa (low-spin ferrous iron, 30% of total iron) (Figure 4). This model would reflect a structural transition around

19 GPa and/or a high- to low-spin transition at 45 GPa. Above 45 GPa, the QS and weight fraction remain largely pressure-invariant in this model, except for slight increase in the weight fraction of site 4 at 95 GPa.

Model 3 introduces a fifth Mössbauer site as a low-spin site (zero quadrupole-splitting), which gradually increases in weight fraction (starting at ~4% of the total iron) at the expense of site 3 (high-spin in this model) between 45 GPa to 95 GPa. This model represents a gradual high- to low-spin transition beginning around 45 GPa. In this model, the QS of site 1 has a weak positive correlation with pressure, whereas the QS of sites 2 and 3 have a weak negative correlation with pressure; the QS of site 4 is independent of pressure within uncertainties.

The relative IS values and broadening (expressed as the FWHM of the QS in mm/s) for the iron sites in these models are plotted as a function of pressure in Figure S1. The relative IS of Model 1 shows sites 2 and 3 decreasing with pressure. In Model 2, the relative isomer shifts are fairly stable with respect to pressure, whereas in Model 3, the IS of site 4 decreases with increasing pressure up to 73.4 GPa. All models feature significant broadening of sites 1, 2, and 3 above 45 GPa (FWHM values range from around 0.3 to 0.6 mm/s) (Figure S1).

4. Discussion and Implications

The steep gradient in hyperfine parameters at low pressures (Figure 4; Tables S2–S4) suggests a structural change, as demonstrated by recent single-crystal X-ray diffraction results by Meusberger et al. [15]. At higher pressures, the three different scenarios fit the data well, as illustrated in Figure 3, resulting in an average reduced χ^2 of 1.39 for Model 1 (which does not incorporate a spin transition), 1.44 for Model 2 (sharp spin transition in one site at 45 GPa), and 1.37 for Model 3 (gradual spin transition from 45 to 95 GPa). We find that, while some sites show decreasing QS values, the QS values for several sites across the models increase with increasing pressure. This trend is distinct from the trend highlighted at room-pressure comparing a range of iron-bearing minerals [12], which found that the quadrupole splitting tends to have a positive correlation with bond length. An increase in the QS values with pressure can be correlated with the combination of bond-shortening and other complex changes influencing the local electronic environments crystal structure, including lattice distortion [23]. This suggests an increasingly distorted (non-cubic) local iron environment in the crystal structure. Above 19 GPa, the quadrupole splittings observed in the present study are relatively constant, suggesting an increasing resistance of the lattice against further distortion.

To gain a better understanding of the connections between terrestrial planetary interiors and their surfaces, the behavior of less common minerals should be evaluated. Their interaction with the bulk materials of a planet can create meaningful deviations from an assumed average behavior. Structural and spin transitions of individual minerals or phases would likely lead to changes in the physical and chemical properties of the bulk mantle. For example, the occurrence of spin transitions in iron-bearing materials of Earth's lower mantle has received increased attention in the last 15 years owing to their potential geophysical, geochemical, and geodynamical implications [28]. At room-temperature, (Mg,Fe)O ferropericlase and $FeSiO_3$ ferrosilite exhibit a similar broad spin crossover behavior with pressure [24,29], whereas the iron-bearing carbonate $FeCO_3$ siderite experiences a sharp spin transition [30]. In Figure 5, we schematically compare these transitions with szomolnokite, along with iron-bearing silicate glasses.

This study is among the first to experimentally observe the effect of pressure on the electronic environment of iron in szomolnokite, a hydrated iron sulfate. Although szomolnokite exists on the surface of Earth and likely on Mars, it is unknown if it exists at depth within either planet. Although it is also known that hydrated sulfates play important roles in the metal and hydrologic exchanges on the surface of terrestrial planets [3], we know less about the specific role that szomolnokite plays on the surfaces and interiors of planetary bodies, including icy satellites.

Our results suggest that there could be major structural transitions in szomolnokite at relatively low-pressure (Figures 3 and 4), corroborated by recent X-ray diffraction results [15]. The basalt-to-eclogite transition (>1.2 GPa, 400–1000 °C) [33] plays an important role in the context of plate tectonics and subducting slab dynamics on Earth, owing to the large density increase across this transition [33,34]. Water content, as well as the behavior of all volatiles, can have an important effect on processes closely related to plate tectonics [35–38]. Hydrated minerals (e.g., hydrated silicates and sulfates) could react with basalt, affecting the basalt-to-eclogite phase transition, thus, influencing processes related to plate dynamics on planetary bodies. Future measurements at a range of pressure and temperature conditions are needed in order to understand the extent to which sulfate phases exist at depth in various planetary bodies and their role in volatile cycling.

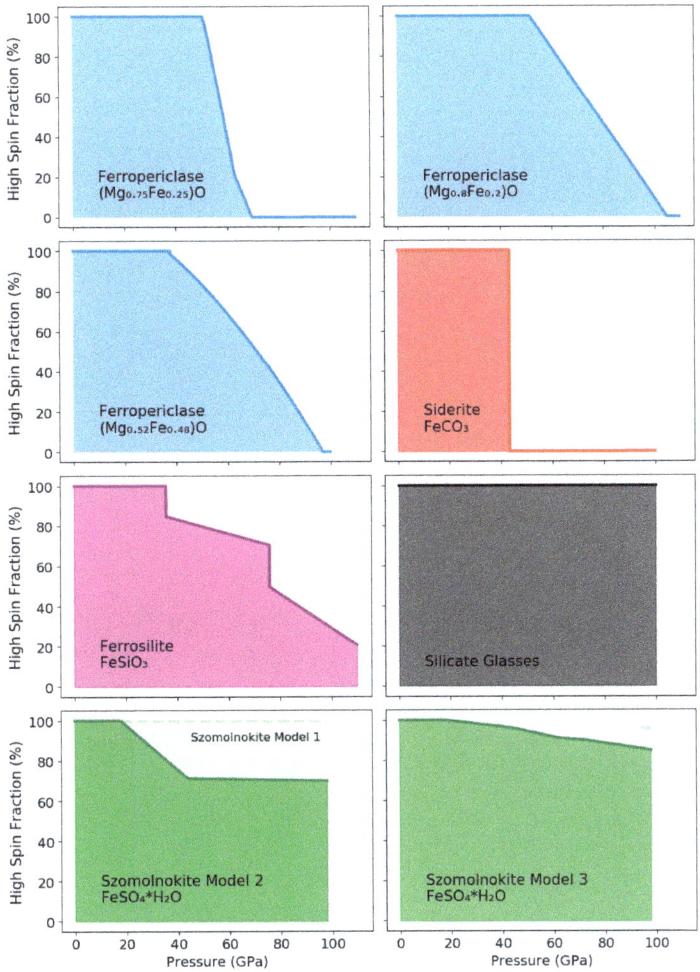

Figure 5. Schematic comparison of the fraction of high-spin ferrous iron (relative to low-spin ferrous iron) as a function of pressure at 300 K in various Fe^{2+}-bearing phases: $(Mg_{0.75}Fe_{0.25})O$ ferropericlase, [31], $(Mg_{0.8}Fe_{0.2})O$ ferropericlase [32], $(Mg_{0.52}Fe_{0.48})O$ ferropericlase [29], $FeCO_3$ siderite [30], $FeSiO_3$ ferrosilite [27], basaltic and rhyolitic glasses [24], and $FeSO_4{\cdot}H_2O$ szomolnokite (this study). For szomolnokite, we include Model 2 (where site 3 undergoes a high- to low-spin transition) and Model 3 (where a fifth low-spin site gradually increases its weight fraction over a broad range of pressures). Model 1 is shown in the background of Model 2, exhibiting no spin transition over the pressure range investigated.

Supplementary Materials: The following are available online at http://www.mdpi.com/2075-163X/10/2/146/s1, Table S1: Lattice parameters of szomolnokite at ambient pressure and temperature conditions; Table S2: Best fit hyperfine parameters for szomolnokite according to Model 1; Table S3: Best fit hyperfine parameters for szomolnokite according to Model 2; Table S4: Best fit hyperfine parameters for szomolnokite according to Model 3; Table S5: Hydrated iron-bearing sulfates and their experimentally determined quadrupole splitting and isomer shift values at ambient pressure and temperature; Figure S1: The relative IS and FWHM of szomolnokite for Models 1, 2, and 3.

Author Contributions: Conceptualization, J.M.J.; methodology, G.J.F., T.P., and J.M.J.; validation, T.P., G.J.F., O.P., N.V.S., and J.M.J.; formal analysis, T.P.; investigation, T.P. and J.M.J.; writing—original draft preparation, T.P., G.J.F., N.V.S., and J.M.J.; writing—review and editing, T.P. and J.M.J.; resources, data curation, and funding acquisition, J.M.J. All authors have read and agree to the published version of the manuscript.

Funding: We thank the W.M. Keck Institute for Space Studies and NSF-CSEDI-EAR-1161046 for support of this research. Operations at Sector 3 (APS) and beamline 12.2.2 (ALS) are partially supported by COMPRES. This research used resources of the Advanced Photon Source and of the Advanced Light Source, which are DOE Office of Science User Facilities under contracts DE-AC02-06CH11357 and DE-AC02-05CH11231, respectively.

Acknowledgments: We are thankful to Wolfgang Sturhahn, Rachel Morrison, and Thomas S. Toellner for helpful discussions. We are thankful to Bob Liebermann for handling our manuscript and to three anonymous reviewers for their helpful feedback and suggestions. We thank the Mr. and Mrs. Larson for their contribution to the Summer Undergraduate Research Fellowship program at Caltech, which supported part of this work. N.V.S. was partly funded by the European Research Council (ERC) under the European Union's Horizon 2020 research and innovation program (Grant Agreement Number 681818–IMPACT).

Conflicts of Interest: The authors declare no conflict of interest.

References

1. Klein, C.; Hurlbut, C.S., Jr. *Manual of Mineralogy*; Wiley: New York, NY, USA, 1985.
2. Hammarstrom, J.; Seal, R.; Meier, A.; Kornfeld, J. Secondary sulfate minerals associated with acid drainage in the eastern US: recycling of metals and acidity in surficial environments. *Chem. Geol.* **2005**, *215*, 407–431. [CrossRef]
3. Chou, I.-M.; Seal, R.R.; Wang, A. The stability of sulfate and hydrated sulfate minerals near ambient conditions and their significance in environmental and planetary sciences. *J. Asian Earth Sci.* **2013**, *62*, 734–758. [CrossRef]
4. Wendt, L.; Gross, C.; Kneissl, T.; Sowe, M.; Combe, J.-P.; LeDeit, L.; McGuire, P.C.; Neukum, G. Sulfates and iron oxides in Ophir Chasma, Mars, based on OMEGA and CRISM observations. *Icarus* **2011**, *213*, 86–103. [CrossRef]
5. Lichtenberg, K.A.; Arvidson, R.E.; Morris, R.V.; Murchie, S.L.; Bishop, J.L.; Remolar, D.F.; Glotch, T.D.; Dobrea, E.N.; Mustard, J.F.; Andrews-Hanna, J.; et al. Stratigraphy of hydrated sulfates in the sedimentary deposits of Aram Chaos, Mars. *J. Geophys. Res. Space Phys.* **2010**, *115*. [CrossRef]
6. Talla, D.; Wildner, M. Investigation of the kieserite–szomolnokite solid-solution series, (Mg,Fe)$SO_4 \cdot H_2O$, with relevance to Mars: Crystal chemistry, FTIR, and Raman spectroscopy under ambient and martian temperature conditions. *Am. Miner.* **2019**, *104*, 1732–1749. [CrossRef]
7. Lane, M.D.; Dyar, M.D.; Bishop, J.L. Spectroscopic evidence for hydrous iron sulfate in the Martian soil. *Geophys. Res. Lett.* **2004**, *31*. [CrossRef]
8. Hand, K.P.; Carlson, R.W. Europa's surface color suggests an ocean rich with sodium chloride. *Geophys. Res. Lett.* **2015**, *42*, 3174–3178. [CrossRef]
9. Nakamura, R.; Ohtani, E. The high-pressure phase relation of the $MgSO_4$–H_2O system and its implication for the internal structure of Ganymede. *Icarus* **2011**, *211*, 648–654. [CrossRef]
10. Barsukov, V.L.; Volkov, V.P.; Khodakovsky, I.L. The crust of Venus: Theoretical models of chemical and mineral composition. *J. Geophys. Res. Space Phys.* **1982**, *87*, A3–A9. [CrossRef]
11. Fortes, A.D.; Fernandez-Alonso, F.; Tucker, M.; Wood, I.G. Isothermal equation of state and high-pressure phase transitions of synthetic meridianiite ($MgSO_4 \cdot 11D_2O$) determined by neutron powder diffraction and quasielastic neutron spectroscopy. *Acta Crystallogr. Sect. B Struct. Sci. Cryst. Eng. Mater.* **2017**, *73*, 33–46. [CrossRef]
12. Dyar, M.D.; Breves, E.; Jawin, E.; Marchand, G.; Nelms, M.; O'Connor, V.; Peel, S.; Rothstein, Y.; Sklute, E.C.; Lane, M.D. What Lurks in the Martian Rocks and Soil? Investigations of Sulfates, Phosphates, and Perchlorates. Mössbauer parameters of iron in sulfate minerals. *Am. Mineral.* **2013**, *98*, 1943–1965. [CrossRef]

13. Van Alboom, A.; De Resende, V.G.; De Grave, E.; Gómez, J.A.M. Hyperfine interactions in szomolnokite ($FeSO_4{\cdot}H_2O$). *J. Mol. Struct.* **2009**, *924*, 448–456. [CrossRef]
14. Giester, G.; Lengauer, C.L.; Redhammer, G.J. Characterization of the $FeSO_4{\cdot}H_2O$-$CuSO_4{\cdot}H_2O$ solid-solution series, and the nature of poitevinite,(Cu, Fe)$SO_4{\cdot}H_2O$. *Can. Mineral.* **1994**, *32*, 873–884.
15. Meusburger, J.; Ende, M.; Talla, D.; Wildner, M.; Miletich, R. Transformation mechanism of the pressure-induced C2/c-to- $P\bar{1}$ transition in ferrous sulfate monohydrate single crystals. *J. Solid State Chem.* **2019**, *277*, 240–252. [CrossRef]
16. Wildner, M.; Giester, G. The crystal structures of kieserite-type compounds. I, Crystal structures of Me (II) SO_4. H_2O (Me = Mn, Fe, Co, Ni, Zn). *Neues Jahrbuch für Mineralogie Monatshefte* **1991**, *7*, 296–306.
17. Pistorius, W. Lattice Constants of $FeSO_4{\cdot}H_2O$ (artificial szomolnokite) and $NiSO_4{\cdot}H_2O$. *Bulletin des Sociétés Chimiques Belges* **1960**, *69*, 570–574. [CrossRef]
18. Klein, R.A.; Walsh, J.P.; Clarke, S.M.; Guo, Y.; Bi, W.; Fabbris, G.; Meng, Y.; Haskel, D.; Alp, E.E.; Van Duyne, R.P. Impact of pressure on magnetic order in jarosite. *J. Am. Chem. Soc.* **2018**, *140*, 12001–12009. [CrossRef]
19. Toby, B.H.; Von Dreele, R.B. GSAS-II: The genesis of a modern open-source all purpose crystallography software package. *J. Appl. Crystallogr.* **2013**, *46*, 544–549. [CrossRef]
20. Dewaele, A.; Torrent, M.; Loubeyre, P.; Mezouar, M. Compression curves of transition metals in the Mbar range: Experiments and projector augmented-wave calculations. *Phys. Rev. B* **2008**, *78*, 104102. [CrossRef]
21. Sturhahn, W. Open Source Software: CONUSS (COherent NUclear Scattering by Single crystals). Available online: www.nrixs.com (accessed on 21 December 2019).
22. Russell, P.E.; Montano, P.A. Magnetic hyperfine parameters of iron containing minerals in coals. *J. Appl. Phys.* **1978**, *49*, 1573–1575. [CrossRef]
23. Jackson, J.M.; Sturhahn, W.; Shen, G.; Zhao, J.; Hu, M.Y.; Errandonea, D.; Bass, J.D.; Fei, Y. A synchrotron Mossbauer spectroscopy study of (Mg,Fe)SiO_3 perovskite up to 120 GPa. *Am. Mineral.* **2005**, *90*, 199–205. [CrossRef]
24. Solomatova, N.V.; Jackson, J.M.; Sturhahn, W.; Rossman, G.R.; Roskosz, M. Electronic environments of ferrous iron in rhyolitic and basaltic glasses at high pressure. *J. Geophys. Res. Solid Earth* **2017**, *122*, 6306–6322. [CrossRef]
25. Fei, Y.; Virgo, D.; Mysen, B.; Wang, Y.; Mao, H. Temperature-dependent electron delocalization in (Mg, Fe) SiO_3 perovskite. *Am. Mineral.* **1994**, *79*, 826–837.
26. McCammon, C. Perovskite as a possible sink for ferric iron in the lower mantle. *Nature* **1997**, *387*, 694. [CrossRef]
27. Solomatova, N.V.; Alieva, A.; Finkelstein, G.J.; Sturhahn, W.; Baker, M.B.; Beavers, C.M.; Zhao, J.; Toellner, T.S.; Jackson, J.M. High-pressure single-crystal X-ray diffraction and synchrotron Mössbauer study of monoclinic ferrosilite. *Comptes Rendus Geosci.* **2019**, *351*, 129–140. [CrossRef]
28. Badro, J. Spin transitions in mantle minerals. *Annu. Rev. Earth Planet. Sci.* **2014**, *42*, 231–248. [CrossRef]
29. Solomatova, N.V.; Jackson, J.M.; Sturhahn, W.; Wicks, J.K.; Zhao, J.; Toellner, T.S.; Kalkan, B.; Steinhardt, W.M. Equation of state and spin crossover of (Mg,Fe)O at high pressure, with implications for explaining topographic relief at the core-mantle boundary. *Am. Mineral.* **2016**, *101*, 1084–1093. [CrossRef]
30. Lavina, B.; Dera, P.; Downs, R.T.; Prakapenka, V.; Rivers, M.; Sutton, S.; Nicol, M. Siderite at lower mantle conditions and the effects of the pressure-induced spin-pairing transition. *Geophys. Res. Lett.* **2009**, *36*. [CrossRef]
31. Lin, J.-F.; Gavriliuk, A.G.; Struzhkin, V.V.; Jacobsen, S.D.; Sturhahn, W.; Hu, M.Y.; Chow, P.; Yoo, C.-S. Pressure-induced electronic spin transition of iron in magnesiowustite-(Mg,Fe)O. *Phys. Rev. B* **2006**, *73*, 113107. [CrossRef]
32. Kantor, I.Y.; Dubrovinsky, L.; McCammon, C. Spin crossover in (Mg,Fe)O: A Mössbauer effect study with an alternative interpretation of X-ray emission spectroscopy data. *Phys. Rev. B* **2006**, *73*, 100101. [CrossRef]
33. Hacker, B.R. Eclogite formation and the rheology, buoyancy, seismicity, and H_2O content of oceanic crust. *Geophys. Monogr. -Am. Geophy. Union* **1996**, *96*, 337–346.
34. Bjørnerud, M.; Austrheim, H. Inhibited eclogite formation: The key to the rapid growth of strong and buoyant Archean continental crust. *Geology* **2004**, *32*, 765–768. [CrossRef]
35. Unterborn, C.T.; Hull, S.D.; Stixrude, L.P.; Teske, J.K.; Johnson, J.A.; Panero, W.R. Stellar Chemical Clues As To The Rarity of Exoplanetary Tectonics. *arXiv* **2017**, arXiv:1706.10282 [astro-ph].

36. Han, R.; Shimamoto, T.; Hirose, T.; Ree, J.H.; Ando, J.I. Ultralow Friction of Carbonate Faults Caused by Thermal Decomposition. *Science* **2007**, *316*, 878–881. [CrossRef]
37. Kessel, R.; Ulmer, P.; Pettke, T.; Schmidt, M.W.; Thompson, A.B. The water–basalt system at 4 to 6 GPa: Phase relations and second critical endpoint in a K-free eclogite at 700 to 1400 °C. *Earth Planet. Sci. Lett.* **2005**, *237*, 873–892. [CrossRef]
38. Korenaga, J. On the Likelihood of Plate Tectonics on Super-Earths: Does Size Matter? *Astrophys. J.* **2010**, *725*, L43–L46. [CrossRef]

minerals

MDPI

Article

Development of a Stress Sensor for In-Situ High-Pressure Deformation Experiments Using Radial X-Ray Diffraction

Jennifer Girard [1,*], Reynold E. Silber [1], Anwar Mohiuddin [1,2], Haiyan Chen [3] and Shun-ichiro Karato [1]

1 Department of Geology and Geophysics, Yale University, New Haven 06511 CT, USA; reynold.silber@yale.edu (R.E.S.); anwar.mohiuddin@yale.edu (A.M.); shun-ichiro.karato@yale.edu (S.-i.K.)
2 Intel Ronler Acres Campus, 2501 NE Century Boulevard, Hillsboro 97124 OR, USA
3 Mineral Physics Institute, Stony Brook University, Stony Brook 11794 NY, USA; haiyan.chen@stonybrook.edu
* Correspondence: jennifer.girard@yale.edu

Received: 30 December 2019; Accepted: 10 February 2020; Published: 13 February 2020

Abstract: We developed a stress sensor for in-situ deformation experiments using synchrotron radial X-ray diffraction. This stress sensor provided nearly diffraction-plane-independent stress that, when used in series with a sample, reduced the uncertainty of the average stress estimation acting on a sample. Here, we present the results of a study where pyrope was used as a stress sensor. Using a Deformation-DIA (D-DIA) high-pressure deformation apparatus, pyrope, olivine and alumina were deformed in the same run/cell assembly placed in series along the compression direction. Deformation experiments were conducted at pressures between 4 and 5 GPa and temperatures between 730 and 1273 K with strain-rates between 10^{-5} and 10^{-6} s^{-1}. Stresses estimated from various (hkl) planes in pyrope were nearly the same; i.e., pyrope is plastically isotropic with ≤10 % variation with (hkl). However, stresses from various (hkl) planes in olivine and alumina varied by approximately a factor of 3. Comparisons between average stresses inferred from pyrope and those from different diffraction planes in olivine and alumina showed that the average stress in these materials evolved from low-end stress, estimated from various (hkl) planes at small strain, to high-end stress at a large strain. This suggests that the rate-controlling slip system in these materials changes from the soft to the hard slip system with strain.

Keywords: high pressure; deformation; in-situ; D-DIA; stress sensor

---

1. Introduction

In a classic deformation experiment, one measures macroscopic strain and stress using a displacement transducer and a load cell [1,2]. From these measurements, one obtains a relationship between macroscopic stress and strain. The results provide key data sets to help us understand deformation of a material.

A load cell cannot be used under high-pressure conditions because a sample is surrounded by various materials and, thus, the load (or the torque), determined outside of a sample assembly, does not reflect directly the stress acting on a sample. Radial X-ray diffraction is a powerful tool for estimating the stress in a sample during plastic deformation, including deformation under high pressures (see e.g., [3–5]). This is possible using high-intensity X-rays generated by a synchrotron radiation facility. In this approach, diffracted X-rays from a sample are collected, at different orientations with respect to the macroscopic stress orientation, to estimate the stress. Strain (and strain rate) is estimated by collecting X-ray radiography images of the sample during deformation. A relationship between stress and strain is obtained and is essential to understanding the plastic properties of a material.

However, a fundamental challenge in this method of characterization of plastic properties is the fact that the relationship between the macroscopic stress and the stress estimated from X-ray diffraction is not simple. In this method, strain is measured by the displacement of the sample (i.e., macroscopic strain), but macroscopic stress (the average stress acting on a sample) is estimated from microscopic strain (the strain in individual grains) determined by X-ray diffraction. In this technique (called "radial X-ray diffraction"), one measures the variation in *d*-spacing of various lattice planes, $\frac{\Delta d_{hkl}}{d_{hkl}}$, and converts it to stress using the known elastic constants of a sample (see e.g., [6–8]). However, it is often noted that the estimated stress from different diffraction planes of a material differs substantially (sometimes by more than a factor of 2–3 [9,10]), and it is unclear how one can estimate the average (macroscopic) stress acting on a sample from a variety of stress values estimated from radial X-ray diffraction. Karato [11] developed a theory to explain this by the large variation in stress from different diffraction planes. Karato's theory shows that a large variation in stress is largely due to plastic anisotropy that affects the distribution of local stress in an aggregate. Plastic anisotropy is usually much larger than elastic anisotropy and can lead to a large variation in the estimated stress from lattice strain ($\frac{\Delta d_{hkl}}{d_{hkl}}$).

The purpose of this work is to develop an isotropic stress sensor from which the average stress acting on a sample is estimated with a smaller uncertainty than that estimated with previous stress sensors, and to test it under high-pressure and temperature conditions using olivine and alumina as samples.

In some previous studies, a stress sensor was used to estimate the stress acting on a sample when the sample itself cannot be used to estimate the stress during deformation (e.g., single-crystal deformation) (see e.g., [12–16]). In these cases, polycrystalline alumina was used as a stress sensor. However, these stress sensors have large plastic anisotropy, which led to large uncertainty in the stress estimation.

2. Materials and Methods

2.1. Sample Preparation and Analysis

The theoretical guideline for choosing a material for a stress sensor is simple: the (hkl) dependence of estimated stress from the radial X-ray diffraction comes from elastic and plastic anisotropy of a material [11]. Therefore, a good material as a stress sensor is a material with small elastic and plastic anisotropy. Most garnets are elastically almost isotropic (see e.g., [17]). Additionally, due to the high symmetry of slip systems, garnets are also plastically nearly isotropic (see e.g., [18,19]). Consequently, we chose an end-member of the garnet group with formula $Mg_3Al_2Si_3O_{12}$ (pyrope) as a sensor material.

Pyrope ($Mg_3Al_2Si_3O_{12}$) garnet was synthesized from an oxide mixture ($MgO–Al_2O_3–SiO_2$). The oxides were thoroughly mixed in an agate mortar and a ball roller mixer for 12 h. The mixture was then melted at the temperature (T) of 1973 K in a vertical tube furnace (Deltech Inc. manufactured, Denver, CO, USA), following the melting phase diagram proposed by Irifune and Ohtani [20]. The molten mixture was quenched by dropping it in distilled water, resulting in pyrope glass, which was subsequently ground and analyzed by X-ray diffraction (Rigaku miniflex 600, manufactured, Tokyo, Japan) to ensure the absence of unwanted crystalline phases due to inhomogeneous melting. The glass powder was hot pressed in a 25/17 cell assembly at the pressure (P) of ~3 GPa and 1573 K for 1 h to form fine grain polycrystalline pyrope following Irifune et al. [21] using a Kawaii-type large volume high-pressure apparatus (Try Engineering, Japan), available at Yale University.

San Carlos olivine crystals (San Carlos NM, CA, USA) ($(Mg_{0.91}, Fe_{0.09})_2SiO_4$), also referred to as Fo91, without any visible inclusions, were handpicked and ground to a powder using a hand-operated hydraulic piston press and then a ball mill. The powder was sorted to obtain fine-grained olivine with grain size less than ~1 μm using a sedimentation technique. The fine-grained powder was hot pressed in a 25/17 cell assembly at 3 GPa and 1073 K for 1 h. Recovered samples were polished and analyzed using a SEM (Scanning Electron Microscope XL30 ESEM-FEG, FEI, Hillsboro OR, USA) to confirm that the grain size was ≤ 1 μm.

Each hot-pressed sample was then sliced to the appropriate thickness (0.2–0.4 mm for pyrope and 1.3 mm for olivine; alumina is part of the cell assembly in D-DIA experiments), and core-drilled into 1 mm diameter cylinders for D-DIA uniaxial deformation experiments.

2.2. D-DIA High-Pressure Deformation Experiments

We performed two high-pressure deformation experiments using a D-DIA apparatus at the 6-BM-B white X-ray beamline at Advanced Photon Source (APS) in Argonne National Laboratory. These are referred to by the run number throughout the text, and also shown this way in Table 1 in the results section (i.e., San 430 and San 452).

Pressure was increased to a desired value by applying a load on all of the six anvils. Heating was usually achieved by using a graphite heater as shown in Figure 1A (used in run San 430). To minimize the temperature gradient, we used a stepped heater as shown in Figure 1B (used in run San 452). In addition, two lateral thermocouples were placed; one on top and one at the center of the cell, to monitor the possible temperature gradient in the cell assembly.

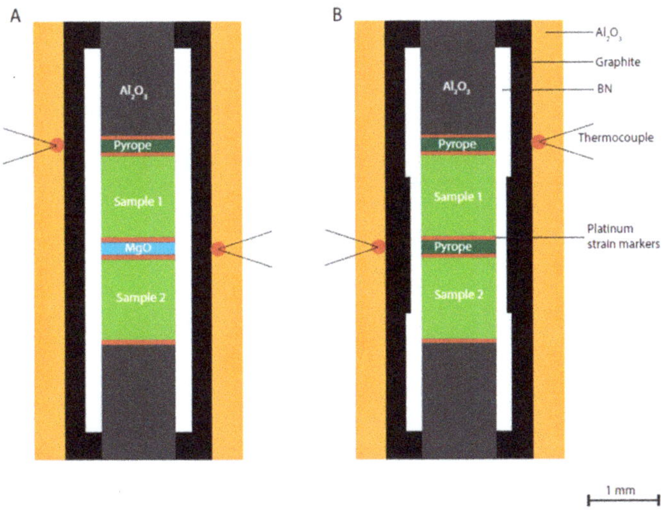

Figure 1. (**A**) Schematic of the traditional D-DIA cell assembly used in San 430 to test the pyrope stress sensor and synthesized fine grain olivine samples. (**B**) A D-DIA cell with a graphite stepped furnace developed to minimize the temperature gradient in the sample in San 452.

During the experiments, temperatures were estimated from thermocouple readings (for run San 430), and from a temperature calibration run (for the run San 452). A temperature calibration run was performed with thermocouples positioned at the center of the cell assembly, using NaCl as a sample at $P = 5$ GPa (the pressure was estimated from NaCl's equation of state). The pressure and temperature were also estimated using the equation of state of MgO and platinum (for run San 430), and using the equation of state of pyrope and platinum (for run San 452).

To deform samples, the top anvil was advanced relative to the bottom anvil under a nearly constant pressure. These samples were surrounded by materials that were X-ray transparent and thus stress and strain could be monitored during deformation. A schematic of our cell design is shown in Figure 1. The cell consisted of two machinable alumina pistons used to compress the samples (olivine stress sensor and pyrope stress sensor). The samples were sandwiched between platinum foils used as strain markers.

Table 1. Summary of deformation conditions for runs San 430 and San 452.

Run #	*P* (GPa)	*T* (K)	Stress (GPa) †			Strain Rate (s^{-1}) ‡		Total Strain (%) ‡
			Pyrope	San Carlos Olivine	Alumina	Pyrope	San Carlos Olivine	
San 430	4.4 ± 1 *	1273 ± 130 *	0.16 ± 0.01	0.12 ± 0.04	0.14 ± 0.04	$0.7 \pm 0.1 \times 10^{-6}$	$3.5 \pm 0.4 \times 10^{-6}$	14 ± 1.5
			0.27 ± 0.02	0.17 ± 0.06	0.230 ± 0.06	$2 \pm 0.2 \times 10^{-6}$	$7.9 \pm 0.8 \times 10^{-6}$	
San 452	4.8 ± 0.6 **	730 ± 90 **	3.6 ± 0.1	3.04 ± 0.7	3.05 ± 0.4	-	$3.7 \pm 0.4 \times 10^{-6}$	5 ± 0.6

* Obtained from the EOS (Equation of State) of Pt and MgO and confirmed with a thermocouple during the experiment (a 10 K difference between the thermocouple and the EOS estimate.). The uncertainty in *P-T* was calculated using the uncertainty on d-spacing resulting from peak fitting from platinum and from MgO. ** Temperature and pressure reported in the table are the average values with one standard deviation calculated from a power versus temperature calibration run and from the EOS of pyrope and platinum. † Stress values reported in the table are weighted average stress values at steady state condition (constant stress, constant strain rate). The error reported corresponds to one standard deviation of stress estimated for each (hkl) peak. ‡ Strain and strain rate uncertainties were calculated using the sample length measurement uncertainty (10%). Strain of pyrope in San 452 was too small to be measured accurately; therefore, strain rate could not be estimated.

Strain in each sample was measured by X-ray absorption images of strain markers collected periodically during a deformation experiment. The distance between strain markers was measured from the images at various times to estimate the strain rate [22].

Stress was estimated from energy dispersive X-ray diffraction, with a constant 2θ collected through a conical slit [23] using a set of 10 detectors along 10 different azimuthal angles (0°, 22.5°, 45°, 67.5°, 90°, 112.5°, 135°, 157.5°, 180°, and 270°).

The diffracted X-ray signals collected at each azimuthal angle were used to estimate the lattice strain which, was then used to calculate the differential stress on crystals during deformation, following Singh [8], which connects lattice strain to stress (Equation (1)).

$$d_{hkl}(\psi) \propto d^0_{hkl}\left\{1 + \frac{\sigma_u}{6M}\left(1 - 3cos^2\psi\right)\right\} \quad (1)$$

where d_{hkl} is the d-spacing for the lattice plane (hkl), d^0_{hkl} is the d-spacing in hydrostatic conditions for the lattice plane (hkl), M is the elastic modulus of the crystallographic orientation, ψ is the azimuthal angle, and σ_u is the uniaxial stress applied on the material. Note that Singh's theory [8] ignores the influence of plastic anisotropy, and, consequently, stress estimated from different diffraction planes (hkl) shows a large variation when plastic anisotropy is strong.

X-rays diffracted from the olivine sample, as well as the pyrope stress sensor and alumina piston, were collected throughout the deformation experiment (Figure 2), and stress was calculated using Equation (1) with the appropriate elastic constants [24–26]. An example of measured lattice strain (i.e., d-spacing) for different (hkl) peaks for pyrope, olivine and alumina is shown in Figure 3.

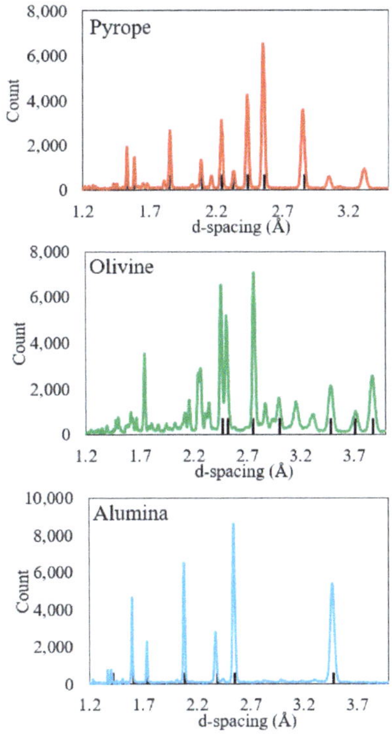

Figure 2. Example of the Energy dispersive X-ray diffraction spectra collected on each sample (pyrope, olivine and alumina) from run San 430 collected before the experiment, at room pressure and room temperature. The black lines indicate the diffraction peaks analyzed for stress calculation.

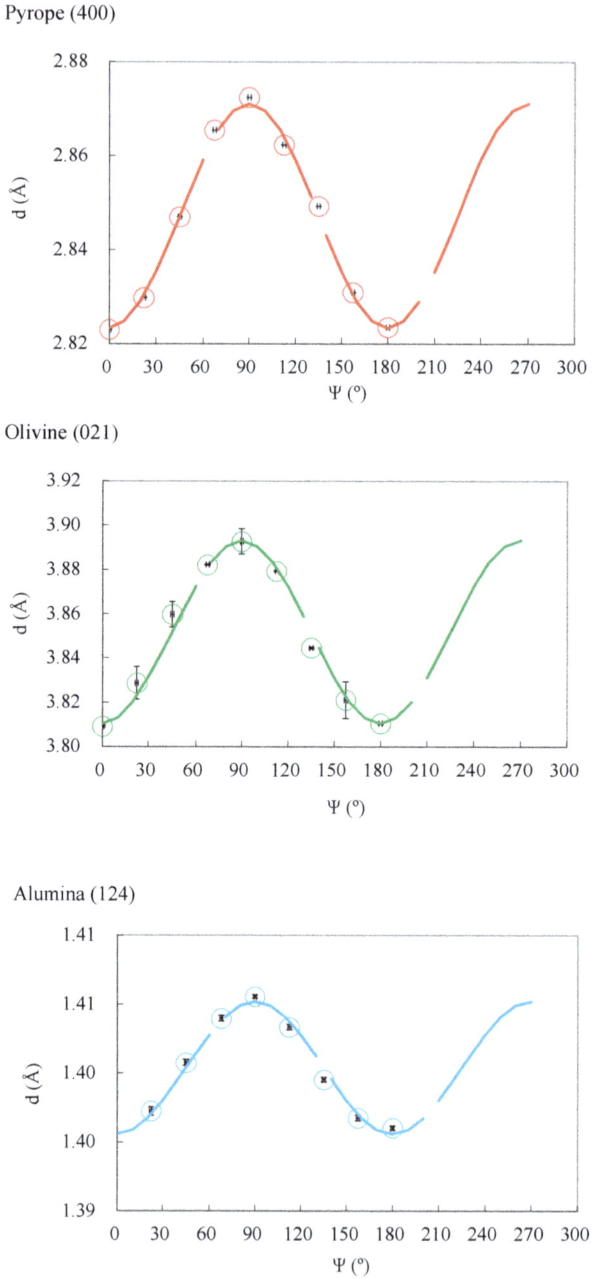

Figure 3. Example *d*-spacing versus azimuthal angle (ψ) for different (hkl) diffraction peaks used to estimate stresses for pyrope (400), Olivine (021) and alumina (124).

In our cell assembly (Figure 1), two samples as well as a stress sensor were stacked (+ alumina pistons). When friction along the edge of each sample was small, stress in each sample and in the stress sensor was the same. To evaluate the influence of friction, we used two stress sensors; one stress sensor at the top of the sample chamber and a second stress sensor in the center for run San 452. Stress σ_{hkl} for each (hkl) diffraction plane (for pyrope, alumina and olivine) is shown in Figure 4. Uncertainties of each individual σ_{hkl} were calculated from the uncertainty in peak fitting. These uncertainties were used as weights in weighted average stresses calculations and are reported in Table 1; stress uncertainty in Table 1 corresponds to one standard deviation.

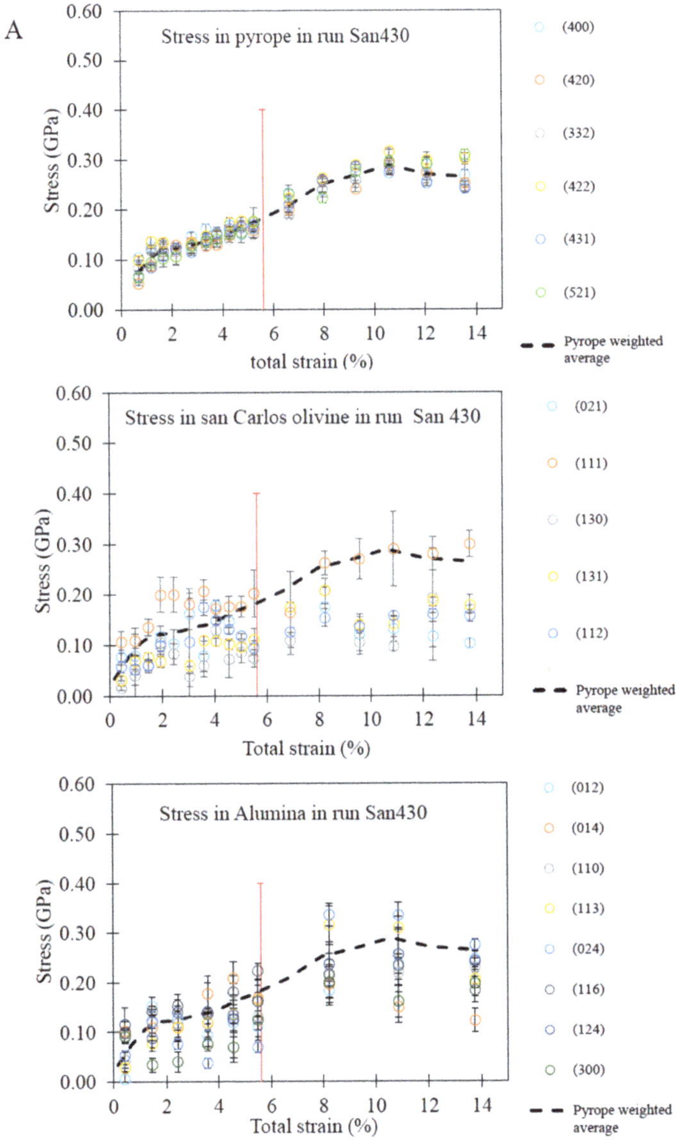

Figure 4. *Cont.*

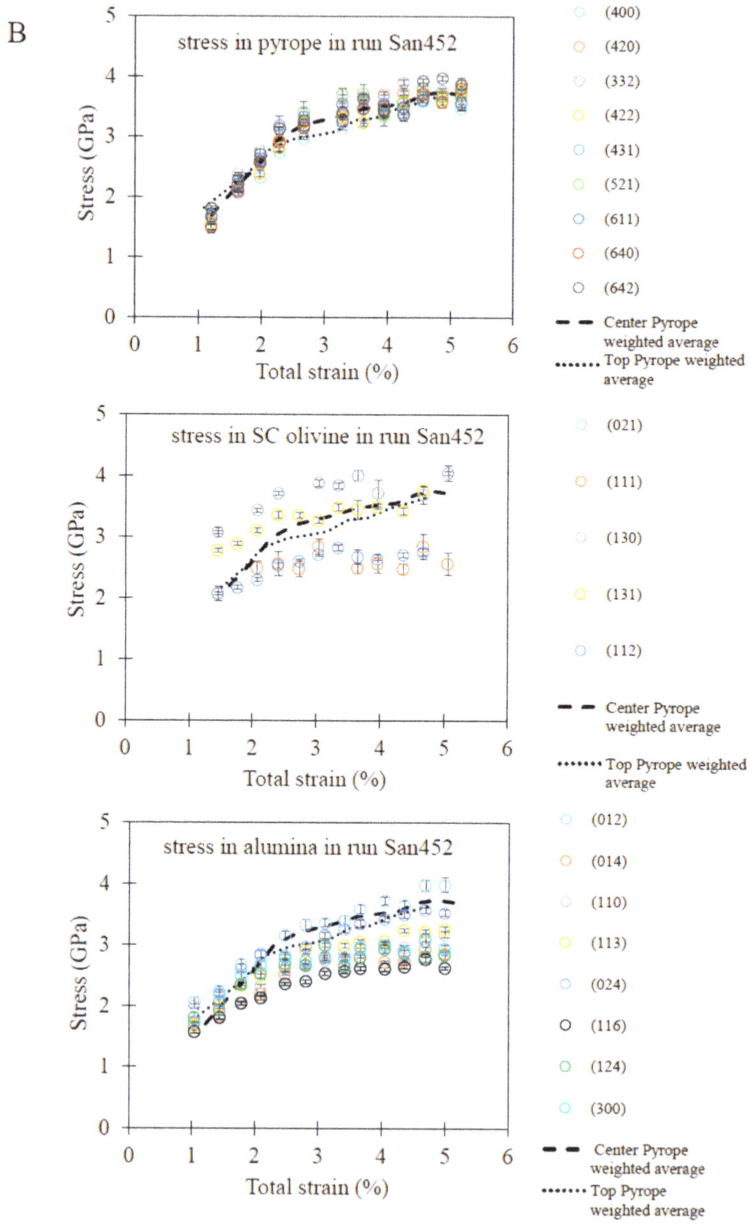

Figure 4. (**A**) Stress versus total strain for run San 430. The dashed line represents the weighted average stress estimate for pyrope. The red vertical line marks the time when the strain rate was increased. (**B**) Stress versus total strain for run San 452 estimated for various (hkl) diffraction peaks for pyrope, olivine and alumina.

3. Results and Discussion

Figure 4 shows the stress estimated from pyrope, olivine and alumina, collected in two different high-pressure uniaxial deformation experiments. One run (San 430; Figure 4a) is at P = 4.4 GPa under low stress, and another run (San 452; Figure 4b) is at P = 4.8 GPa. Due mainly to the differences in temperature (500 K difference), the stress levels are markedly different between these two runs.

We report stresses σ_{hkl} estimated from various (hkl) peaks from each sample with their respective uncertainties $\Delta\sigma_{hkl}$ obtained using peak fitting uncertainties. The dashed line represents the weighted average stress estimated from the pyrope stress sensor $\sigma_{(hkl)}$, where individual uncertainties $\Delta\sigma_{hkl}$ were used as weights. The red vertical line indicates the time at which the strain rate was changed. In San 430, a pyrope stress sensor was placed only at the top of the column, but in San 452, we placed two pyrope stress sensors (one on top of the deformation column and one in the center of the cell assembly). Figure 4b shows stress estimated from the pyrope at the center of the cell; the dotted line is the weighted average stress estimated from the pyrope's position at the top. Both stress estimates are similar and thus friction can be considered negligible.

We note that, because of its isotropic properties, variation in stress among different diffraction planes for pyrope was small with a maximum of 10% whereas, for olivine and alumina, stress values varied substantially with the diffraction planes (up to a factor of 3 or more).

Therefore, pyrope works as a good sensor to estimate the average stress acting on a sample. Consequently, a comparison of average stress estimated from the stress sensor and the stresses estimated from various diffraction planes in olivine provides a new insight into the role of various slip systems in deformation of a polycrystalline aggregate (see e.g., [27–29]). Our results show that, in these two runs, the average stress estimated from various diffraction planes in pyrope (represented by the dotted and dash lines) was in the upper part of the stress values estimated from different diffraction planes in olivine and alumina. This means that the strength of alumina and olivine aggregates are largely determined by their respective strong slip system(s) (since the largest stress is required to deform the aggregate), and therefore implies that deformation of each aggregate in these cases occurs at nearly homogeneous strain.

Finally, we compared the stress estimated in alumina in our study with Raterron et al. [30]. We confirm that the ordering of the σ_{hkl} is the same in both studies (e.g., stress estimated from the (012) diffraction plane is always the largest and the (116) diffraction plane always shows the smallest stress). However, in Raterron et al. [30], large anisotropy in stress estimated from alumina prevented an accurate stress estimation. Thus, they used EPSC (Elastic Plastic Self Consistent) modeling to calculate "true" stress. Their results showed that, at the end of deformation, EPSC modeling agreed with the lower end of stresses estimated in alumina. This is not consistent with our observations using a pyrope stress sensor. Since these two methods differ, it is difficult to explain where the variances come from. In our study, stress in pyrope was directly measured from X-ray diffraction but the "true" stress corresponding to the EPSC modeling was calculated following some assumptions. Therefore, some of the EPSC assumptions may need to be refined.

We also note that the average stress of a deforming material evolves from low end to high end with strain, suggesting that the slip system that controls the strength of an aggregate changes from the soft slip system at a small strain to the hard slip system at a large strain (this is clearly seen for run San 452 where the relative error in stress estimate is small compared to run San 430). This is consistent with an idea presented by Karato [31]. However, it is not clear if this is the case for more general cases. At higher strain, the stress–strain distribution may change. A stress sensor can be used to investigate the evolution of stress–strain distribution during plastic deformation. A good stress sensor material can be any material with low (or no) plastic anisotropy. This technique can also be applied to deformation of two-phase mixtures where the evolution of stress–strain with strain is an important issue in characterizing the processes of shear localization.

Author Contributions: J.G. and S.-i.K. conceived the idea of the study. J.G., A.M. and R.E.S. performed sample preparation and conducted experiments at Argonne National Laboratory with the help of H.C. J.G. analyzed the data and wrote the manuscript, which was edited by the co-authors. All authors have read and agreed to the published version of the manuscript.

Funding: This research was funded by National Science Foundation, NSF grant number EAR 1764271.

Acknowledgments: It is our pleasure to contribute to a special volume dedicated to Orson Anderson. Orson made a huge contribution to our community by introducing a rigorous physics approach to geological sciences. We hope that our work represents one of the contributions to follow the approach pioneered by Orson Anderson. The authors would like to thank COMPRES for partly supporting the facility 6-BM-B, where the data for this study were collected. We are also grateful to W. Samella and C. Fiederlein for helping with preparation of parts for 25/15 as well as the D-DIA cell assemblies used in this study.

Conflicts of Interest: The authors declare no conflict of interest.

References

1. Griggs, D. Deformation of rocks under high confining pressure: I experiment at room temperature. *J. Geol.* **1936**, *44*, 541–577. [CrossRef]
2. Paterson, M.S. A high-pressure, high-temperature apparatus for rock deformation. *Int. J. Rock Mech. Min. Sci.* **1970**, *7*, 517–526. [CrossRef]
3. Weidner, D.J. Rheological studies at high pressure. In *Ultrahigh-Pressure Mineralogy*; Hemley, R.J., Ed.; The Mineralogical Society of America: Washington, DC, USA, 1998; pp. 492–524.
4. Weidner, D.J.; Li, L. Measurements of stress using synchrotron x-rays. *J. Phys. Condens. Matter* **2006**, *18*, S1061–S1067. [CrossRef] [PubMed]
5. Weidner, D.J.; Vaughan, M.T.; Wang, L.; Long, H.; Li, L.; Dixon, N.A.; Durham, W.B. Precise stress measurements with white synchrotron x-rays. *Rev. Sci. Instrum.* **2010**, *81*, 013903. [CrossRef] [PubMed]
6. Singh, A.K. The lattice strain in a specimen (cubic system) compressed nonhydrostatically in an opposed anvil device. *J. Appl. Phys.* **1993**, *73*, 4278–4286. [CrossRef]
7. Uchida, T.; Funamori, N.; Yagi, T. Lattice strains in crystals under uniaxial stress field. *J. Appl. Phys.* **1996**, *80*, 739–746. [CrossRef]
8. Singh, A.K.; Balasingh, C.; Mao, H.-K.; Hemley, R.; Shu, J. Analysis of lattice strains measured under nonhydrostatic pressure. *J. Appl. Phys.* **1998**, *83*, 7567–7578. [CrossRef]
9. Li, L.; Weidner, D.J.; Chen, J.; Vaughan, M.T.; Davis, M.; Durham, W.B. X-ray strain analysis at high pressure: Effect of plastic deformation in mgo. *J. Appl. Phys.* **2004**, *95*, 8357–8365. [CrossRef]
10. Chen, J.; Li, L.; Yu, T.; Long, H.; Weidner, D.J.; Wang, L.; Vaughan, M.T. Does reuss and voigt bounds really bound in high-pressure rheology experiments? *J. Phys. Condens. Matter* **2006**, *18*, S1049–S1059. [CrossRef]
11. Karato, S. Theory of lattice strain in a material undergoing plastic deformation: Basic formulation and applications to a cubic crystal. *Phys. Rev.* **2009**, *79*, 214106. [CrossRef]
12. Girard, J.; Chen, J.; Raterron, P.; Holyoke, C.W. Deformation of periclase homogenous at high pressure and high temperature: Quantification of the effect of pressure on slip-system activities. *J. Appl. Phys.* **2012**, *111*, 112607. [CrossRef]
13. Girard, J.; Chen, J.; Raterron, P.; Holyoke, C.W. Hydrolytic weakening of olivine at mantle pressure: Evidence of [100](010) slip system softening from single-crystal deformation experiements. *Phys. Earth Planet. Inter.* **2013**, *216*, 12–20. [CrossRef]
14. Raterron, P.; Amiguet, E.; Chen, J.; Li, L.; Cordier, P. Experimental deformation of olivine single crystals at mantle pressures and temperatures. *Phys. Earth Planet. Inter.* **2009**, *172*, 74–83. [CrossRef]
15. Raterron, P.; Chen, J.; Li, L.; Weidner, D.J.; Cordier, P. Pressure-induced slip-system transition in forsterite: Single crystal rheological properties at mantle pressure and temperature. *Am. Mineral.* **2007**, *92*, 1436–1445. [CrossRef]
16. Amiguet, E.; Raterron, P.; Cordier, P.; Couvy, H.; Chen, J. Deformation of diopside single crystal at mantle pressure 1: Mechanical data. *Phys. Earth Planet. Inter.* **2009**, *177*, 122–129. [CrossRef]
17. Sinogeikin, S.V.; Bass, J.D. Elasticity of pyrope and majorite-pyrope solid solutions to high temperatures. *Earth Planet. Sci. Lett.* **2002**, *203*, 549–555. [CrossRef]
18. Rabier, J.; Garem, H.; Veyssière, P. Transmission electron microscopy determination of dislocation burgers vectors in plastically deformed yttrium iron garnet single crystals. *J. Appl. Phys.* **1976**, *47*, 4755–4758. [CrossRef]
19. Karato, S.; Wang, Z.; Liu, B.; Fujino, K. Plastic deformation of garnets: Systematics and implications for the rheology of the mantle transition zone. *Earth Planet. Sci. Lett.* **1995**, *130*, 13–30. [CrossRef]
20. Irifune, T.; Ohtani, E. Melting of pyrope $Mg_3Al_2Si_3O_{12}$ up to 10 Gpa possible of pressure induced structural change in pyrope melt. *J. Geophys. Res.* **1986**, *91*, 9357–9366. [CrossRef]

21. Irifune, T.; Kawakami, K.; Arimoto, T.; Ohfuji, H.; Kunimoto, T.; Shinmei, T. Pressure-induced nano-crystallization of silicate garnets from glass. *Nat. Commun.* **2016**, *7*, 13753. [CrossRef]
22. Vaughan, M.; Chen, J.; Li, L.; Weidner, D.; Li, B. Use of x-ray imaging techniques at high-pressure and temperature for strain measurements. In Proceedings of the AIRAPT-17, Honolulu, HI, USA, 25–30 July 1999; Manghnani, M.H., Nellis, W.J., Nicol, M.F., Eds.; Universities Press: Hyderabad, India, 2000; pp. 1097–1098.
23. Durham, W.B.; Weidner, D.J.; Karato, S.-i.; Wang, Y. New developments in deformation experiments at high pressure. In *Plastic Deformation of Minerals and Rocks*; Karato, S.-i., Wenk, H.-R., Eds.; MSA: Washington, DC, USA, 2002; Volume 51, pp. 21–49.
24. Hu, Y.; Wu, z.; Dera, P.; Bina, C.R. Thermodynamic and elastic properties of pyrope at high pressure and high tempratue by first principles calculation. *J. Geophys. Res. Solid Earth* **2016**, *121*, 9. [CrossRef]
25. Liu, W.; Li, B. Thermal equation of state of $(Mg_{0.9}Fe_{0.1})_2SiO_4$ olivine. *Phys. Earth Planet. Inter.* **2006**, *157*, 188–195. [CrossRef]
26. Wang, W.; Wu, Z. Elasticity of corundum at high pressures and temperatures: Implications for pyrope decomposition and al-content effect on elastic properties of bridgmanite. *J. Geophys. Res. Solid Earth* **2018**, *123*, 2. [CrossRef]
27. Kocks, U.F. The relation between polycrystal deformation and single-crystal deformation. *Metall. Mater. Trans. B* **1970**, *1*, 1121–1143. [CrossRef]
28. Kocks, U.F.; Canova, G.R. *How Many Slip Systems, and Which?* Hansen, N., Horsewell, A., Leffers, T., Lilholt, H., Eds.; Risø DTU National Laboratory for Sustainable Energy: Roskilde, Denmark, 1981.
29. Hutchinson, J.W.; Hill, R. Bounds and self-consistent estimates for creep of polycrystalline materials. *Proc. R. Soc. Lond. A* **1976**, *348*, 101–127.
30. Raterron, P.; Merkel, S.; Holyoke, C.W. Axial temperature gradient and stress measurements in the deformation-dia cell, using alumina pistons. *Rev. Sci. Instrum.* **2013**, *84*, 043906. [CrossRef]
31. Karato, S.-I. *Micro-Physics of Post Glacial Rebound, Retrospective Collection*; Trans Tech Publication: Stafa-Zuerich, Switzerland, 1998; pp. 351–364.

minerals

MDPI

Article

X-ray Laue Microdiffraction and Raman Spectroscopic Investigation of Natural Silicon and Moissanite

Camelia Veronica Stan [1,*], Earl Francis O'Bannon III [2,*], Pavel Mukhin [3], Nobumichi Tamura [4] and Larissa Dobrzhinetskaya [5]

[1] NIF & Photon Science Directorate, Lawrence Livermore National Laboratory, PO Box 808, Livermore, CA 94551, USA
[2] Physical and Life Sciences Directorate, Lawrence Livermore National Laboratory, PO Box 808, Livermore, CA 94551, USA
[3] Independent Consultant in Ore Deposits, Tel Aviv 2067041, Israel; pavelmukhin4@gmail.com
[4] Advanced Light Source, Lawrence Berkeley National Laboratory, 1 Cyclotron Road, Berkeley, CA 94720, USA; ntamura@lbl.gov
[5] Earth and Planetary Sciences, University of California Riverside, 900 University Avenue, Riverside, CA 92521, USA; larissa@ucr.edu
* Correspondence: stan2@llnl.gov (C.V.S.); obannon2@llnl.gov (E.F.O.III)

Received: 30 January 2020; Accepted: 21 February 2020; Published: 25 February 2020

Abstract: Moissanite, SiC, is an uncommon accessory mineral that forms under low oxygen fugacity. Here, we analyze natural SiC from a Miocene tuff-sandstone using synchrotron Laue microdiffraction and Raman spectroscopy, in order to better understand the SiC phases and formation physics. The studied crystals of SiC consist of 4H- and 6H-SiC domains, formed from either, continuous growth or, in one case, intergrown, together with native Si. The native Si is polycrystalline, with a large crystal size relative to the analytical beam dimensions (>1–2 μm). We find that the intergrown region shows low distortion or dislocation density in SiC, but these features are comparatively high in Si. The distortion/deformation observed in Si may have been caused by a mismatch in the coefficients of thermal expansion of the two materials. Raman spectroscopic measurements are discussed in combination with our Laue microdiffraction results. Our results suggest that these SiC grains likely grew from an igneous melt.

Keywords: SiC; Laue diffraction; microdiffraction; natural silicon; moissanite; Si; synchrotron; polymorph; mineral deformation

---

1. Introduction

SiC may be the archetypal polytypic material (e.g., polymorphs only differ in the stacking sequence of identical sheets or structural units) with greater than 250 known synthetic polytypes, ~11 of which are reported as naturally occurring [1]. The polytypes are typically described with a number that refers to the number of layers that repeat along the stacking direction, and a letter that refers to the crystal system [2]. They generally lie on a continuum between the wurtzite (2H) and zinc-blende (3C) structures, depending on the relative interplanar layering of Si and C. Figure 1 shows the stacking sequence of three of the simplest naturally-occurring polytypes of SiC. Moissanite, the naturally occurring SiC mineral, is typically considered to be either, the 6H- or, to a lesser extent, 4H- and 15R-SiC structure types. Predicting which polytype forms under natural conditions is complicated, and currently there are several well-summarized theories to explain formation [3], ranging from thermodynamic, kinetic, and growth considerations. Moreover, transformation from one polytype to another is thought to occur by periodic slip around dislocations [4], diffusional rearrangement with the nucleation and expansion of stacking faults [5], and/or changes in Si/C ratios during crystallization [6].

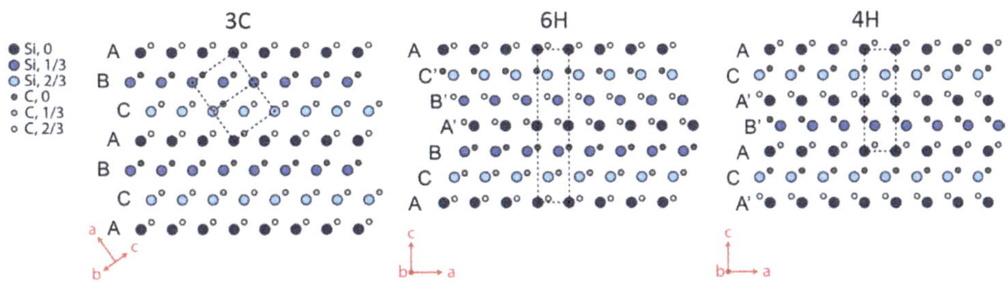

Figure 1. 3C-, 6H- and 4H-SiC structures can be interpreted through the orientation of alternating Si and C layers [7,8]. The unit cell in each case is represented by a dashed line. The layers are labelled based on relative stacking position and depth coming out of the page.

Naturally occurring moissanite (SiC) was first discovered within the impact rocks of the Canyon Diablo meteor crater of Arizona [9]. Moissanite was once considered a geological aberration [10]. Many synthetic routes have been reported (see review by Abderrazak and Hmida [11]), with variations in the resulting crystal properties, but none truly mimic geological conditions, which range from relatively low pressure (upper crust) to high pressure (mid to lower mantle), high temperature, and very low oxygen fugacity. For more than a century, natural occurrences of moissanite have been reported in the literature and these reports show that SiC origin may be broken up into three categories: (i) high-, ultra-high pressure environments, such as upper- and lower mantle and mantle transition zone, and even core–mantle boundary; (ii) ambient/low-pressure environments, such as metamorphic, magmatic rocks and hydrothermal processes recorded in both continental and oceanic crust; or (iii) formation of SiC and native metals during lightning strikes in the ophiolitic rocks exposed on the Earth's surface [12–20]. All require temperature ranged from 700–800 °C to 2500 °C, and extremely low oxygen fugacity, e.g. ~8 to <12 log-bar units [21], or 6 to 8 log-bar units [22] below the iron-wüstite (IW) buffer. In the last two decades many more well-documented, including in-situ, finds of natural moissanite have been reported [12–18,23–43]. These reports suggest that moissanite may be a much more common accessory mineral than previously thought. More analytical data and measurements from well-documented *in situ* natural SiC are necessary in formulating a less incontrovertible mechanism of moissanite formation under geological conditions.

Here, we analyze naturally-occurring SiC in a tuff-sandstone, focusing on the understanding of the mechanism of SiC formation, and on the intrinsic relationship between SiC and Si, recorded in one of the studied moissanite grains. We employ X-ray Laue microdiffraction (μXRD) and Raman spectroscopy to investigate the microstructure, distortion, and phase relations in both SiC and Si.

2. Materials and Methods

2.1. Sample Description

The sample is a lithified tuff-sandstone (also referred to as tuffite) related to the Miocene age Lower Basalt Formation in the Yizre'el Valley of Kishon River, Israel. The rock includes fragments of local sediments mixed with the pyroclastic material deposited during volcanic eruption (e.g., Baer et al., [44]). The research sample consists of the hypidiomorphic crystals of picroilmenite, magnetite and ulvospinel intermixed with eroded round crystals of quartz and anorthite, all of which are cemented with secondary Ca, Fe, Mg-hydrous carbonates, chlorite and serpentine (see detailed description in Dobrzhinetskaya et al., [41]). The thin section used for this research is a piece of the same sample, described in Dobrzhinetskaya et al., [41] as tuffite. However, it can be now identified more precisely as a tuff-sandstone, due to presence of a microfossil (Figure 2a). The rock was cut by a diamond mini-saw and mounted into an epoxy disc followed by polishing with 50 μm, 15 μm, and 3 μm size corundum pastes and ultrasonic rinsing in distilled water after each step of polishing. The SiC crystals were identified with the aid of optical microscopy and Raman spectroscopy. Three grains of moissanite, labelled SiC1, SiC2—containing native Si inclusions, and SiC3, were chosen for analysis (Figure 2a–d).

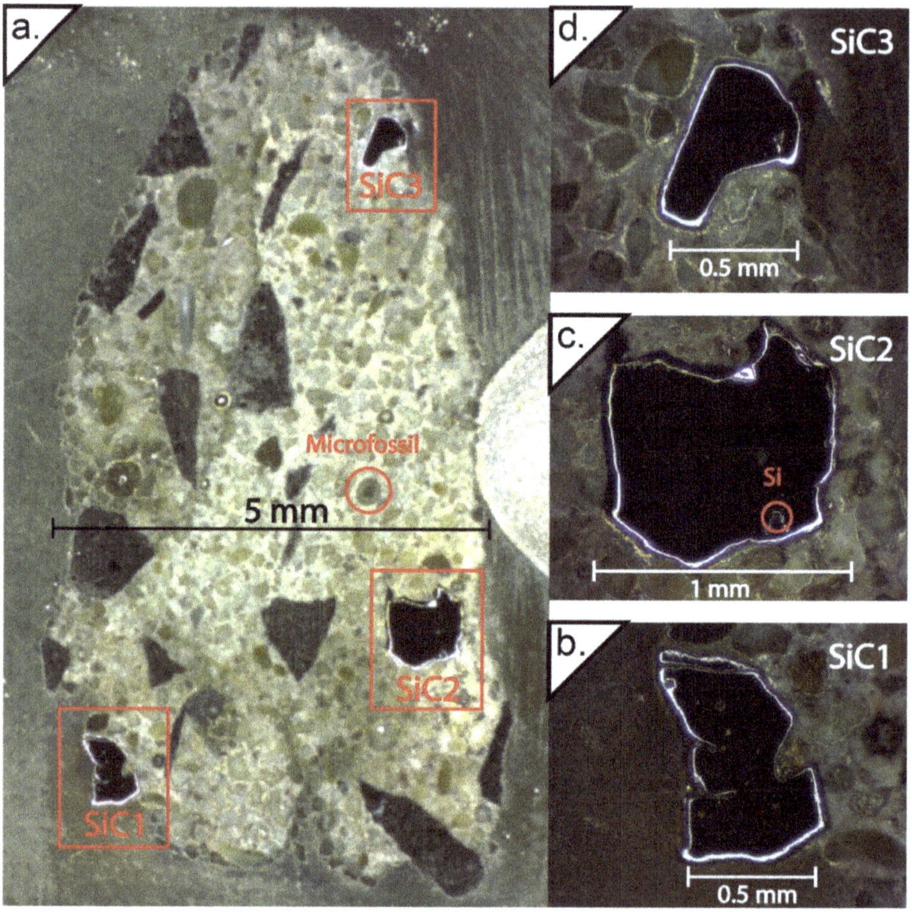

Figure 2. (**a**) The microscope image of the sample clearly shows the location of the SiC grains (**b**–**d**), which have different reflectivities from the surrounding rock matrix. The red circle indicates the location of a microfossil (Figure S1). Locations of SiC grains in a. and the Si inclusion in SiC2 (**c**)are indicated.

2.2. Data Collection

X-ray fluorescence (XRF) and μXRD were collected at beamline 12.3.2 of the Advanced Light Source at Lawrence Berkeley National Laboratory. A white light beam (5–24 keV) was focused to an area of ~1 μm × 1 μm using Kirkpatrick-Baez mirrors. The sample was placed onto a high-precision translational stage at 45° relative to the incident beam and raster scanned. XRF data was collected using a Vortex-EM XRF detector at a 20 μm × 20 μm resolution. Scanning was performed at a rate of 0.5 s/pixel. The emission lines of Cr, Fe, Ni, Ca, and Ti were used to determine crystal grain locations. Since the XRF detector is only sensitive in the 2000–20,000 eV range, Si (Kα = 1740 eV) and C (Kα = 277 eV) cannot directly be detected, so the SiC location was determined through the absence of any signals.

Microdiffraction mapping was performed on SiC1, SiC2, and SiC3 grains (Table 1), allowing some margin around each crystal to account for the penetration depth of the beam (~100 μm in SiC) and the 45° geometry of the sample relative to the incident beam. Data was collected in 90° geometry by a Pilatus 1M detector using a 0.5 s exposure time. The sample-to-detector distance was calibrated using an unstrained synthetic Si sample. Further information about the experimental setup can be found in Kunz et al. [45] and Stan et al. [46]. μXRD data analysis was performed using XMAS [45,47], the super-computing facilities at the National Energy Research Scientific Computing Center (NERSC).

Table 1. Measurement conditions for microdiffraction.

Grain	Map Dimensions (μm^2)	Pixel Dimension (μm)
SiC1	798 × 972	6
SiC2	1064 × 1080	8
SiC2 [a]	382 × 202	2
SiC3	665 × 755	5

[a] The second map was performed over the area where Si was present.

Raman maps of the Si inclusion in SiC2 were collected on a Horiba LabRam Evolution Raman spectrometer with a focal length of 800 mm. Spectra were collected from 450–900 cm^{-1} and the colors were assigned based on the first order Raman mode of Si and the manifold of TO modes for 4H- and 6H-SiC centered near 775 cm^{-1}. Spectra were collected in a backscattered geometry with an excitation wavelength of 532 nm and an 1800 lines/mm grating, which results in a spectral resolution of ~1 cm^{-1}. An Olympus BXFM-ILHS microscope with a 50× long working distance objective was used to focus the laser beam to a ~2 μm spot size onto the sample. Maps were collected using a 2 μm step size. A synthetic, unstrained Si chip was used to calibrate the Raman spectrometer. The peaks were fit using non-linear least squares to background-subtracted pseudo-Voight or Gaussian line profiles using the program Igor Pro (Version 8.0.4.2, WaveMetrics, Inc., Lake Oswego, OR, USA).

3. Results

Three grains of SiC (size ~0.5–1 mm) were chosen for detailed studies. XRF measurements corresponded well with the known shape and dimensions of each grain, as expected from optical microscope images of the sample (Figure 2).

3.1. SiC1

Initial μXRD measurements indicates that the sample consists of a strongly diffracting single crystal. Initial indexing attempts focused on the 4H-SiC polytype, which we anticipated based on previous TEM and Raman spectroscopic work [41]. Most of the scanned area could be indexed as either 4H- or 6H-SiC, with more than 30 peaks indexed for both polytypes (Figure 3). It is expected in general that 6H-SiC would have more diffraction peaks than 4H-SiC over the same angular range, due to the difference in the c-axis length between the two phases. This can easily be seen in the relative number of indexed peaks (Figure 3a,b).

The grain contains an area of overlap where both unit cells can be indexed (Figures 3 and 4). On the left-hand side of the grain, all observed peaks can be indexed as 4H-SiC only (Figure 4a), whereas on the right-hand side, all peaks can be indexed as 6H-SiC only (Figure 4c). In the center, the diffraction pattern contains peaks from both, and both can be indexed, with some peaks overlapping for both unit cells (Figure 4b).

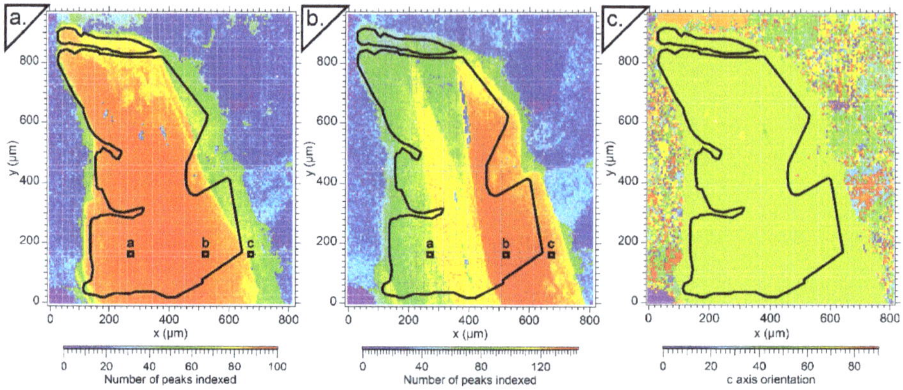

Figure 3. Grain can be indexed as both (**a**) 4H-SiC; and (**b**) 6H-SiC; (**c**) c-axis orientation with respect to the sample normal is also shown, to demonstrate the intergrowth of the two crystal types along the same stacking direction. The black outline indicates the surface expression of the sample, based on microscopy images. The squares labelled a, b, and c correspond to the diffraction pattern locations from Figure 4.

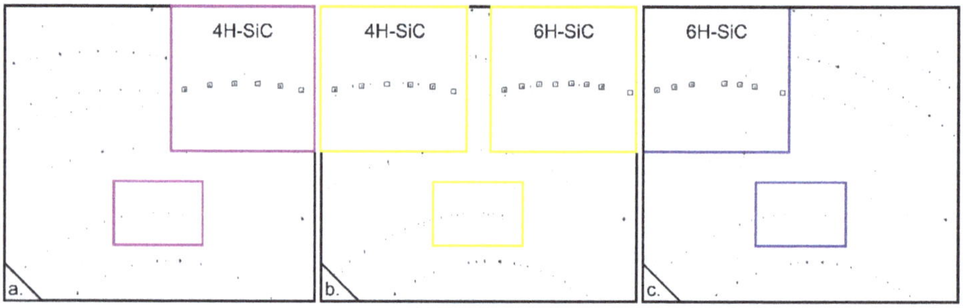

Figure 4. Diffraction patterns from SiC1 can be indexed as either the 4H- or 6H-SiC. (**a**). This part of the sample indexes solely as 4H-SiC. The inset shows squares around all successfully identified peaks. No additional peaks remain. (**b**). This area can be indexed as both 4H- and 6H-SiC. The two insets show that all peaks are identified as being part of one or the other stacking type. (**c**). This pattern can be fully indexed as 6H-SiC, as seen in the inset.

3.2. *SiC2*

The bulk of this grain consists of a strongly diffracting single crystal that indexes as 4H-SiC (Figure 5). The lower right-hand portion of the SiC2 sample contains both, the region where an Si inclusion was found, as well as a more poorly indexed region that is better fit as a 6H-SiC (Figure 5b). Although, a second Si inclusion was identified by optical microscopy and scanning electron microscopy (SEM) [41], the diffraction data were of insufficient quality to index it. We chose the mixed 4H/6H-SiC area for more detailed mapping and analyzed it with 2 μm × 2 μm resolution. The bulk 4H-SiC crystal pattern is visible in this entire area. However, single crystal diffraction, taking the form of continuous lines, can also be observed (Figure 6). We can fit the pattern using a 6H-SiC unit cell, which provides a more accurate fit than the 4H-SiC pattern of the rest of the crystal body. This can be seen when comparing Figure 6e,f, where all maxima can be captured with a 6H-SiC fit except for a select few, which correspond to 4H-SiC. However, most of the grain is polycrystalline in this area, with a few discrete orientations dominating (Figure 5c).

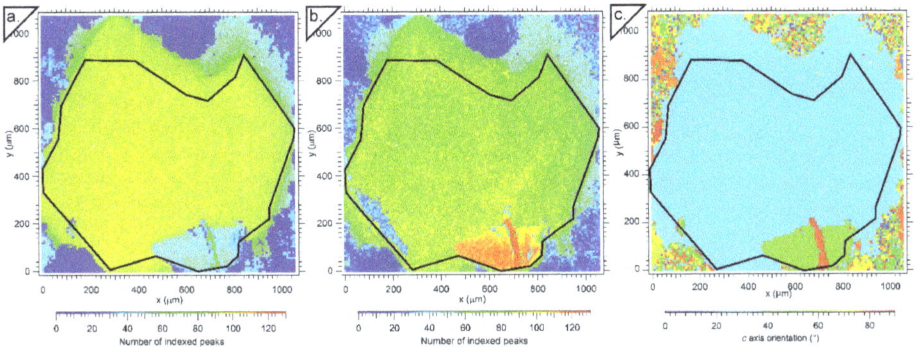

Figure 5. The SiC2 grain indexed as (**a**). 4H-SiC and (**b**). 6H-SiC. (**c**). The c-axis orientation with respect to the sample normal shows several different orientations are present in this sample. The black outline indicates the surface expression of the sample, based on microscopy images.

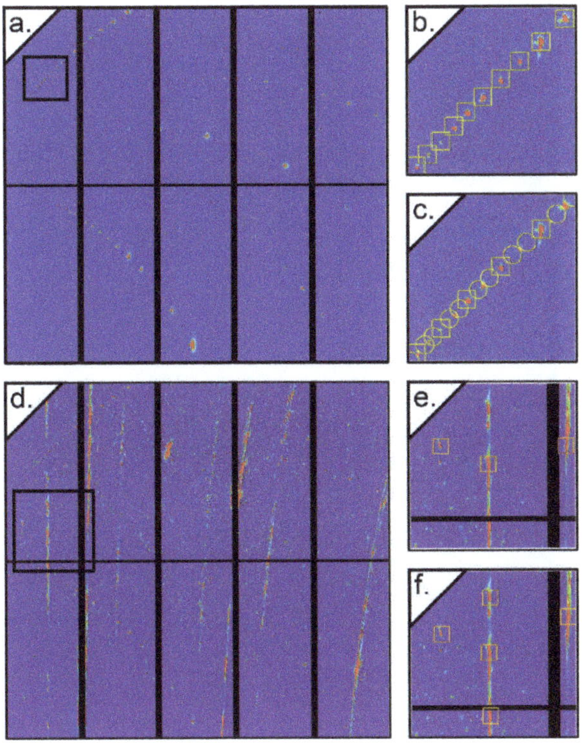

Figure 6. (**a**). This pattern is representative of the SiC2 sample in the 4H-SiC-indexed region. (**b**). The yellow squares indicate peaks that were identified and indexed as 4H-SiC. All peaks can be fit with this unit cell. (**c**). The same pattern from a. is indexed as 6H-SiC. The yellow circles show peaks that are predicted to occur, but are not found (with signal-to-noise ratio > 5). (**d**). This pattern is representative of the polycrystalline region of the SiC2 sample (**e**) and (**f**). are representative fits to the 4H- and 6H-SiC unit cells, respectively. Boxes in (**a**). and (**d**). outline enlarged areas from (**b**)., (**c**)., (**e**) and (**f**).

A native Si inclusion was initially identified by visual inspection due to differences in its reflectivity relative to the surrounding SiC. This inclusion occurs within the mixed 4H-/6H-SiC region of the sample. By scanning through the diffraction patterns, we find one area with broad, deformed peaks (Figure 7) that can be indexed as diamond structured ($Fd\bar{3}m$) Si. At least 3 distinct crystallographic orientations can be indexed, indicating that the native Si inclusion is polycrystalline, but with large grain sizes (Figure 7b–d). These grains overlap for some part of the sample. A {113} reflection is the most intense in all three crystallites. Examination of this peak, however, indicates that it is likely that many more subgrains exist, as evidenced by the existence of several overlapping maxima (Figure 7e–g). This is also indicative of a plastic deformation that is significant enough to cause a subgrain boundary formation. Although, an area of only ~36 μm × ~30 μm is exposed to the surface, we can track the persistence of the {113} reflection in these three main Si grains for at least 340 μm horizontally and 590 μm vertically. This area of persistence is roughly coincident with where 6H-SiC is present, but with inhomogeneous diffraction intensity suggesting varying emplacement depths and/or grain thickness in Si.

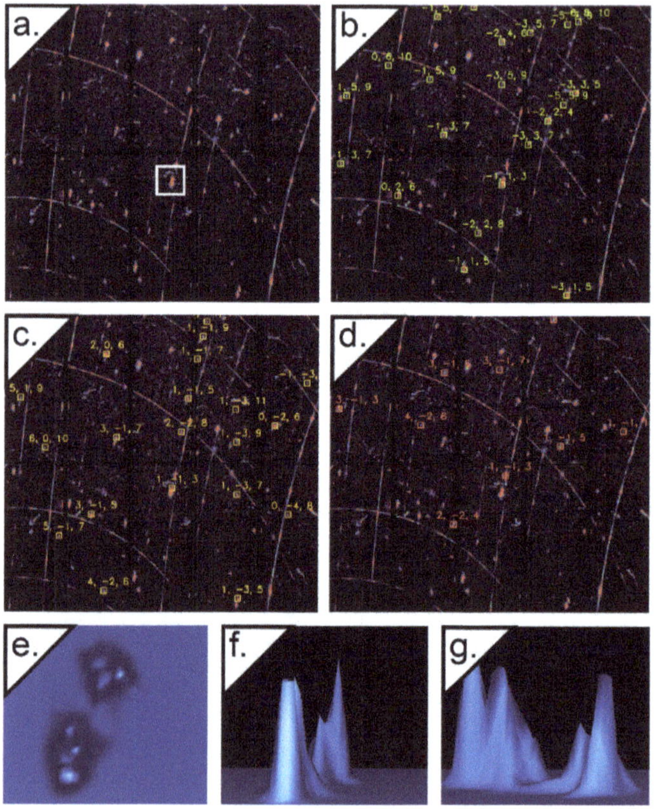

Figure 7. (**a**). This is a representative pattern taken in the region of exposed Si inclusion. (**b**)., (**c**)., and (**d**). show indexation fits for three different Si orientations (~33.5, 15.5, and 24.8° *c*-axis orientation with respect to sample normal, respectively). (**e**). This is a magnified view of the {113} reflection outlined by a white box in (**a,f**). and (**g**). show three-dimensional (3D) projections of the peaks from e. in vertical, and horizontal transects, respectively.

A Raman map of the Si inclusion is shown in Figure 8. The 4H- and 6H-SiC can also be identified from the Raman spectra, but were not the focus of this measurement. The single Raman mode of silicon is due to the first-order Raman scattering of the longitudinal optical (LO) and the transverse optical (TO) phonon modes, which are degenerated at the Γ-point [48]. Its room pressure and temperature position has been reported to range from 519.5 ± 0.8 cm^{-1} to 523.0 ± 1.0 cm^{-1} [49–52]. The large range of reported frequencies is likely due to many factors, such as instrument calibration, tensile, or compressive stress state of the Si sample. More recently it has been shown that the Raman mode of un-stressed Si is observed between 520.5 and 520.7 cm^{-1} [48,53]. In this sample, the Si Raman mode ranges from 522.3 to 525.2 cm^{-1}.

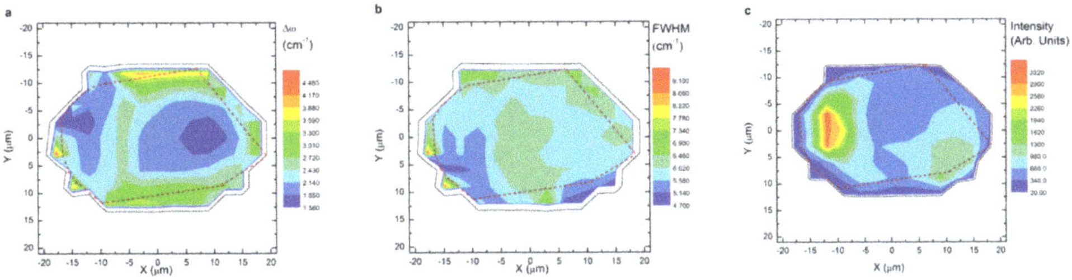

Figure 8. Raman maps of the Si inclusion. (**a**). $\Delta\omega = \omega_{observed} - \omega_{unstrained\ Si}$. a value of 520.7 cm^{-1} was used for the unstrained Si. (**b**). Full width at half maximum (FWHM) of the first order Si Raman mode. (**c**). Intensity of the first order Si Raman mode. The red dashed line outlines the surface expression of the Si inclusion.

3.3. SiC3

Like SiC1, the SiC3 grain is made of two overlapping domains, which easily index to 4H-SiC or 6H-SiC (Figure 9). The overlap region is like that of SiC1 (Figure 4), where peaks from both orientations can be observed, suggesting that the two structures are intercalated. A second SiC orientation is observed on the left-hand side of the grain, which visually appears to coincide with the presence of both a slight fracture in the grain as well as the presence of a metal silicide inclusion [41]. The data were insufficient to determine the mineral type of the silicide.

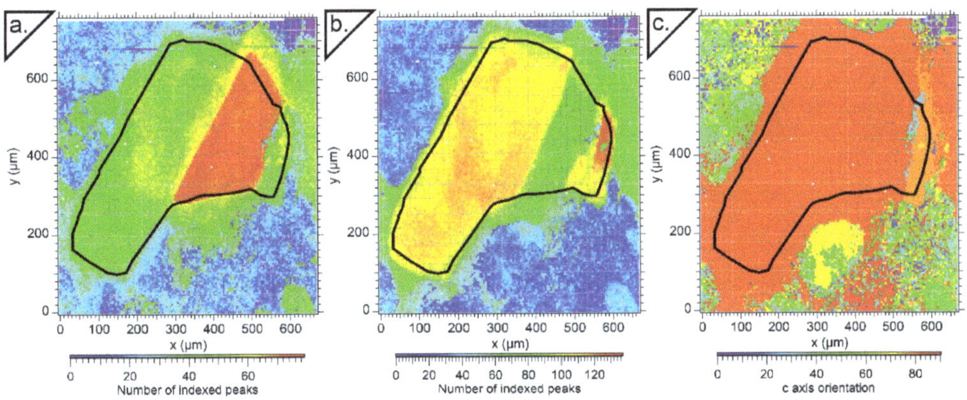

Figure 9. The SiC3 grain indexed as (**a**). 6H-SiC and (**b**). 4H-SiC. (**c**). The c-axis orientation with respect to the sample normal shows several different orientations are present in this sample. The black outline indicates the surface expression of the sample, based on microscopy images.

4. Discussion

4.1. SiC Formation—Geological Context

The three crystals of SiC in the studied sample are characterized as xenocrysts transported together with other pyroclastic materials by the volcanic vent related to Miocene intraplate alkali basalt activities [41]. Together with other pyroclastic materials, they mixed with ashes and local sediments containing microfauna (Figure S1) to become a part of the tuffite/tuffo-sandstone formation. Dobrzhinetskaya et al. [41] hypothesized that SiC, presented in the 4H-SiC and 6H-SiC polytypes, was formed at depth ~60–100 km, through the reaction of SiO_2 (melt) with highly reducing fluids (H_2O–CH_4–H_2–C_2H_6, as proposed by Zhang and Duan, [54]). This is supported by well-known reactions which could assist the formation of both SiC and Si:

$$SiO_2\ (\text{melt}) + CH_4 \rightarrow SiC + 2H_2O \tag{1}$$

$$2CH_4 + H_2O \rightarrow 4H_2 + 2CO \rightarrow C + CO_2 + 4H_2 \tag{2}$$

$$SiO_2 + C \rightarrow Si + CO_2 \tag{3}$$

According to Dobrzhinetskaya et al. [41], the ultra-reduced fluid, originating from a mantle hot spot, could pass through an alkaline basalt magma reservoir and interact with the SiO_2 available from the walls of the crustal rocks surrounding this magmatic reservoir. This would lead to the formation of SiC, and this process could be accompanied by the reduction of metal-oxides to native metals, alloys, and silicides that could have originated from droplets of an immiscible melt fraction trapped by SiC during its crystallization.

4.2. Stress/Strain Relationships and Crystal Grain Formation

The three SiC grains fall into two distinct categories, based on the spatial distribution of the 4H- and 6H- unit cells. In the case of SiC1 and SiC3, there is an orientational relationship between the 4H- and 6H-SiC regions, where both unit cells can be described by the same orientation matrix. By traversing a sample, we can observe a 4H-SiC region, a mixed region where some lattice planes are shared, and finally a 6H-SiC region. The intercalation and lack of reorientation suggests that the sample is a single grain. Studies suggest that the degree of hexagonality of SiC varies with the Si/C ratio (Haase et al. [6] and references contained therein). Therefore, grains with variable structure, such as SiC1 and SiC3, could have grown from a slowly cooling Si/C melt slightly enriched in Si, where SiC growth would have led to C depletion in the melt and an increased Si/C ratio. In such a scenario, 4H-SiC would be formed at first, and continuous grain growth would lead to a less hexagonal (containing some cubic ABC stacking planes) structure over time as the surrounding melt changed, explaining the transition region between the two crystal types and the final 6H-SiC structure.

In the case of SiC2, there is no such relationship in the orientation of 4H-SiC and 6H-SiC, nor between SiC and Si. It is more likely that the 4H-SiC grain, which is visible in the diffraction patterns where 6H-SiC is also present, formed independently during different melting events from the 6H-SiC, and that these single crystals became intergrown or cemented together by remaining liquid Si at a later time.

The differential stress and strain in both SiC and Si were calculated from the indexed Laue diffraction map. In the 4H-SiC region, the stress/strain across the crystal was homogeneous for a given tensor, so we focus the discussion mainly on the part of the map that covers the exposed Si region and the entire 6H-SiC region. The strain in this area for all 3 crystal types is presented in Figure 10a–c. The overall strain in the 4H-SiC region of the sample is low, except for some tensile differential strain in the z direction of the sample; this observation may be due to the release of pressure as a result of cutting and polishing the sample. The Si is under compressive differential strain in the XY plane and tensile differential strain in the Z direction. The strain state seems not to depend on the crystallographic orientation of any of the 3 phases.

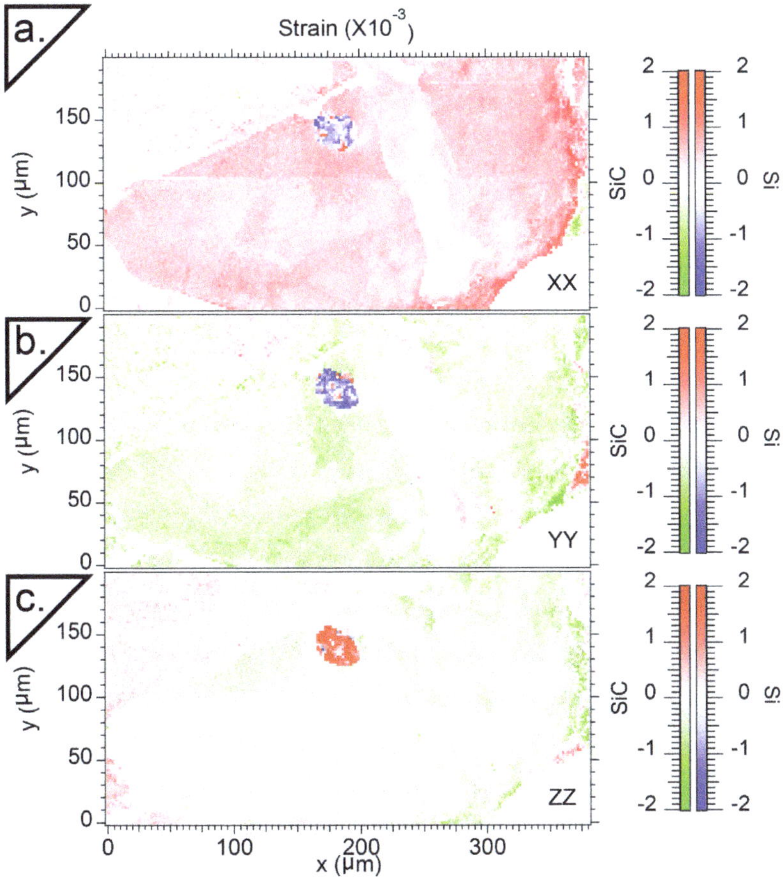

Figure 10. Strain maps of the mixed 4H-/6H-SiC region and Si inclusion in the (**a**) x (horizontal) (**b**) y (vertical) and (**c**) z (out of plane) directions.

A possible explanation for this may be that greater plastic deformation and differential strain in Si is due to a differential in the coefficient of thermal expansion (CTE) for the two materials, where SiC = 4.36×10^{-6} K^{-1} [55] and Si = 2.63×10^{-6} K^{-1} [56]. In prior studies of dislocation generation between materials with contrasting CTE, it was found that the dislocation density increased relative to the pure materials when subjected to similar temperature conditions [57,58]. We assume that the Si inclusion formed while the sample was at high temperature, which is reasonable as this SiC sample likely formed at a depth of 60–100 km (~2 GPa) and a temperature range of 1000–1600 °C [41]. Then the factor of ~2 differences in their CTE is likely the cause of the higher dislocation densities associated with the Si and SiC in the 6H-SiC region. In essence, SiC would contract at a faster rate than Si, leading to greater strain in the Si. We map the dislocation densities in both, assuming a {001}<110> slip system in SiC (Figure 11). This assumption is borne out by the presence of diffraction "lines" in the single-crystal Laue diffraction of SiC, which are indicative of stacking faults and are direct evidence of the activation of this slip system. Overall, Si displays a much greater density of dislocations than SiC. This is supported by observations in the diffraction, which suggest that the Si is under more plastic deformation at every point where both are present.

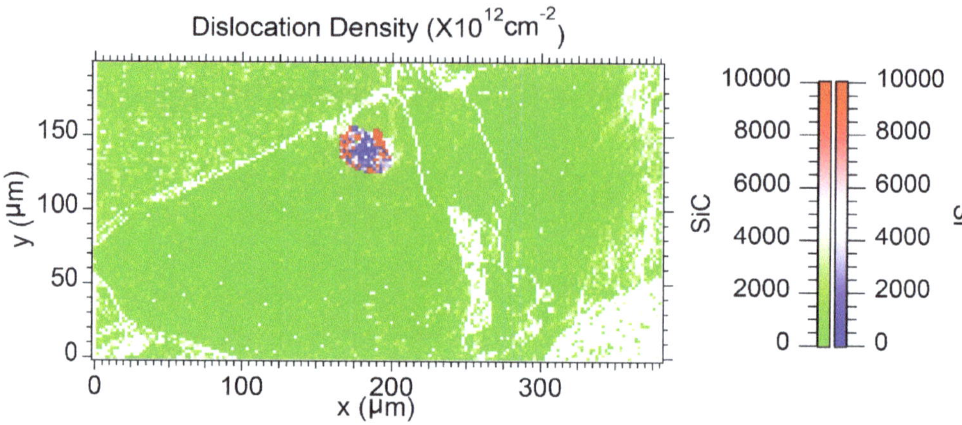

Figure 11. Geometrically necessary dislocation density of the mixed 4H-/6H-SiC region and Si inclusion. Notably the dislocation density of the Si inclusion is high at the edges and low in the middle and the SiC grain shows a very low dislocation density.

4.3. SiC/Si Relationship

One of the goals of our study was to understand the formation of a native silicon inclusion, such as that found in SiC2, which requires extremely reducing conditions of formation [21]. Due to the wide use of silicon in the semiconductor industry the effects of tensile and compressive stresses, hydrostatic pressure, temperature, and polishing on the first order Raman mode position and width are well-characterized [48,50,52,53,59–66]. It is well-known that tensile and compressive stress affect the Raman line by a redshift, and blueshift, respectively [61,62]. Anastassakis et al. [59] reported a splitting of the first order Raman mode of Si under uniaxial stress along the [001] or [111] direction. Thus, the observed shift in the first order Raman mode of our Si inclusion (Figure 8a) may be evidence of residual compressive stress. Most of the grain shows a shift of ~1.5–2.7 cm^{-1} (Figure 8a) from the ambient pressure value of 520.7 cm^{-1}. Taking the pressure shift of the first order Si Raman mode reported by Weinstein and Piermarini [52] this could suggest that the Si grain records a residual pressure of ~0.3–0.5 GPa.

The FWHM of this peak ranges from ~4–9 cm^{-1} (Figure 8b). The room pressure and temperature FWHM of unstrained Si has previously been reported to range from ~3.2–7.0 cm^{-1} depending on the laser power [50,60,63,64]. Fitting very low intensity modes can be difficult and can result in larger errors in peak position and FWHM. However, even disregarding the very large and small observed FWHMs, the majority of the Raman spectra of the Si inclusion have FWHM of ~5.5–7.7 cm^{-1}. Weinstein and Piermarini [52] report that there is no appreciable change in the peak shape with applied pressure to ~10 GPa. Hence, our observed peak widths and shapes did not unambiguously indicate that our inclusion recorded any residual pressure.

Not surprisingly, the largest mode shifts are observed near the edge of the Si inclusion, and the smallest shifts near the center of the inclusion. These shifts also correspond to regions of largest FWHM and lowest intensity (Figure 8c). We also observe some asymmetry to the higher wave-number side of the observed Si Raman modes primarily in the spectra collected near the edge of the Si inclusion, suggesting a uniaxial component to the residual stress. This is consistent with the diffraction measurements discussed above, where plastic deformation was clearly observed as asymmetrical streaking in three-dimensional (3D) peak shape (Figure 7e–g). Furthermore, a lower Raman intensity is observed at the edges of the inclusion, which can be explained by considering the penetration depth of the Raman measurements. The penetration depth of a laser is dependent on the material absorption coefficient, α, which is wavelength- and temperature-dependent, but for a constant α, intensity decays exponentially with depth, according to the Beer-Lambert Law. For a 532 nm laser the penetration depth in Si is ~0.7 μm [67]. The lowest intensities are from just outside where the Si inclusion is exposed to the surface, and these spectra also show peaks that are from SiC (Figure 8c).

Nazzareni et al. [43] report natural SiC from a peralkaline syenite from the Azores. They identified an Si inclusion with Raman spectroscopy and report that the first order Raman mode of Si was observed to be at 519.65 cm^{-1}, and the position of the synthetic Si first order Raman mode is 520.72 cm^{-1}. As discussed above and by Nazzareni et al. [43], the residual pressure of the Si inclusion was calculated using the pressure shift of the first order Raman mode of Si, reported by Weinstein and Piermarini [52]. In contrast to our results, Nazzareni et al. [43] report that the first order Raman mode of their Si inclusion is observed to be lower than the ambient pressure value, suggesting that it records a negative residual pressure. It should be noted that, in the absence of a detailed compositional analysis of the Si inclusion, reported by Nazzareni et al. [43], the negative shift could be due to impurities such as Fe, which, for example, has been reported in native silicon from the Luobusa Ophiolite [68]. Much of our Si inclusion is still enclosed within SiC, and only the upper part of the inclusion was exposed to the surface during cutting. Therefore, we can hypothesize that the positive shift of the Raman mode of our Si inclusion probably reflects some residual pressure similar but opposite to the conclusions reached by Nazzareni et al. [43].

A considerable amount of work has been reported on subsurface damage in Si, due to polishing and grinding [66,69]. It is possible that our observed shift in the first order Si Raman mode is a result of damage caused during sample polishing. Shifts of ~1 cm^{-1} are observed in the core of the exposed Si inclusion while larger shifts are observed at the edge of the exposed inclusion. Zhang et al. [66] report that the position of the first order Si Raman mode for several Si wafers that were polished with different grain sizes ranging from ~70–~4 μm. The ground wafer surfaces mainly present compressive states that are dependent on the grain sized used for polishing and shifts up to 4 cm^{-1} from the ambient value are observed. The main reason for the generation of residual stresses is due to phase transitions. Zarudi and Zhang [69] discuss subsurface damage of single-crystal silicon due to grinding and polishing. They show that the depth of damage is related to the particle size of the abrasives used, and that the damage can be removed with polishing (about 10 h). Phase transitions to α-Si, Si-III, and Si-XII are also observed in polished Si samples and the phases that are observed is dependent on the grain size of the polish that was used [66]. We do not observe any peaks from α-Si, Si-III, or Si-XII in our spectra. The stress regime in the near-surface environment of the polished and exposed Si inclusion isn't simple, and likely grades from mostly compressive near the edges to more tensile in the middle. It is difficult to deconvolve the signals from, (1) actual residual pressure that may be recorded by the grain, due to thermal expansion differences, (2) artefacts from polishing the sample, and (3) the exposure of the Si inclusion to the surface on one side. To verify these statements, it will be necessary to conduct additional Raman experiments of synthetic Si crystals before, and after, polishing.

5. Conclusions

We performed μXRD and Raman spectroscopy studies of three grains of natural SiC and found that each of them consists of two polytypes of SiC, 4H- and 6H-SiC. These SiC polytypes were found to be either, intergrown during a single growth event (SiC1 and SiC3), or cemented together after the initial formation of SiC crystals (SiC2) with polycrystalline, more deformed Si. There is no straightforward explanation for which SiC polytype will form first, and current theories, include impurity arguments, kinetic, and growth considerations, as well as thermodynamics. In a natural system the liquid will have a particular Si/C ratio, as well as some quantity of impurities. This liquid will cool at some rate and a certain SiC polytype will crystallize. Since we observe both 4H- and 6H-SiC, and if we assume a closed system crystallization of SiC, crystal growth will change the Si/C ratio. It has been proposed that the Si/C ratio will determine which type of polytype will form [6]. This hypothesis aims to explain the natural zoned SiC that was found in metamorphic rocks from Bulgaria [42]. Thus, we propose that for SiC1 and SiC3 one nucleation event led to grain formation through a change in the Si/C ratio, likely caused by C depletion during grain growth, and in the case of SiC2 two different nucleation events likely occurred.

Despite its high hardness, SiC deforms plastically and polygonizes at a high temperature [70]. Dislocations in (0001), dislocations with $\left[1\bar{1}20\right]$ vectors, pileups formed by slip, and dislocation walls formed by climb, were all reported by Amelinckx et al. [70]. Dislocations in microdiamonds have also been reported [71,72]. Kvasnytsya and Wirth [72] report a high density of curved dislocations in one of their microdiamond samples from a meteorite impact area and suggest a thermally activated process where the diamond experienced a period of thermal annealing after deformation. Dobrzhinetskaya et al. [71] found dislocations in microdiamonds from ultra-high-pressure metamorphic terranes and assigned them to dislocations of growth, e.g., during the nucleation process, the volume may have been under differential stress. The observed low dislocation density in the examined SiC crystals suggests that the samples may not have remained at high temperatures for an extended period, nor did they undergo any high temperature deformation at any time after their initial crystallization. While these new results still cannot place any constraint on the depth of formation of these grains, they do support their high-temperature origin which was reported earlier by Dobrzhinetskaya et al. [41]. Given the wide stability field of SiC, from the shallow mantle to the deep Earth, one should be cautious when interpreting formation conditions of natural SiC.

Supplementary Materials: The following are available online at http://www.mdpi.com/2075-163X/10/3/204/s1, Figure S1: (**a**). Optical microscope image of the tuffite sample containing 3 SiC crystals. The red outline indicates the location of the microfossil/microfauna. (**b**). Microfossil/microfauna as seen at larger magnification.

Author Contributions: C.V.S., E.F.O.III, data collection and interpretation, writing, and editing; N.T., data collection and interpretation; P.M., resources and sample preparation; L.D., conception, data interpretation, writing, and editing. All authors have read and agreed to the published version of the manuscript.

Funding: EFO was partially supported by NSF through EAR-1620423 and acknowledges helpful discussions with Q. Williams UCSC. A portion of this work was performed under the auspices of the US Department of Energy by Lawrence Livermore National Laboratory under Contract No. DE-AC52-07NA27344. This research used resources of the National Energy Research Scientific Computing Center, a DOE Office of Science User Facility supported by the Office of Science of the U.S. Department of Energy under Contract No. DE-AC02-05CH11231. The Advanced Light Source is supported by the Director, Office of Science, Office of Basic Energy Sciences, Materials Sciences Division, of the U.S. Department of Energy under Contract No. DE-AC02-05CH11231 at Lawrence Berkeley National Laboratory and University of California, Berkeley, California.

Conflicts of Interest: The authors declare no conflict of interest.

References

1. Cheung, R. *Silicon Carbide Microelectromechanical Systems for Harsh Environments*; Imperial College Press: London, UK, 2006; ISBN 978-1-86094-624-0.
2. Ramsdell, L.S. Studies on silicon carbide. *Am. Mineral.* **1947**, *32*, 64–82.
3. Jepps, N.W.; Page, T.F. Polytypic transformations in silicon carbide. *Prog. Cryst. Growth Charact.* **1983**, *7*, 259–307. [CrossRef]
4. Powell, J.A.; Will, H.A. Low-temperature solid-state phase transformations in 2H silicon carbide. *J. Appl. Phys.* **1972**, *43*, 1400–1408. [CrossRef]
5. Jagodzinski, H. Polytypism in SiC crystals. *Acta Cryst.* **1954**, *7*, 300. [CrossRef]
6. Haase, V.; Kirschstein, G.; List, H.; Ruprecht, S.; Sangster, R.; Schröder, F.; Töpper, W.; Vanecek, H.; Heit, W.; Schlichting, J.; et al. *Si Silicon: System Si-C. SiC: Natural Occurrence. Preparation and Manufacturing Chemistry. Special Forms. Manufacture. Electrochemical Properties. Chemical Reactions. Applications. Ternary and Higher Systems with Si and C*; Gmelin Handbook of Inorganic Chemistry; Springer-Verlag Berlin Heidelberg: Berlin, Germany, 1985; Volume B3, ISBN 978-3-662-06994-3.
7. Burdick, C.L.; Owen, E.A. The structure of carborundum determined by X-rays. *J. Am. Chem. Soc.* **1918**, *40*, 1749–1759. [CrossRef]
8. Thibault, N.W. Morphological and structural crystallography and optical properties of silicon carbide (SiC). *Am. Mineral.* **1944**, *29*, 327–362.
9. Moissan, H. Nouvelles recherches sur la météorité de Cañon Diablo. *Comptes rendus* **1904**, *139*, 773–786.

10. Milton, C.; Vitaliano, D.B. Moissanite SiC, a geological aberration. In Proceedings of the 98th Annual Meeting of the Geological Society of America, Orlando, FL, USA, 14 October 1985; p. 665.
11. Abderrazak, H.; Hmida, E.S.B.H. Silicon Carbide: Synthesis and Properties. In *Properties and Applications of Silicon Carbide*; InTech: Rijeka, Croatia, 2011; p. 361.
12. Lyakhovich, V.V. Origin of accessory moissanite. *Int. Geol. Rev.* **1980**, *22*, 961–970. [CrossRef]
13. Marshintsev, V.K. Nature of silicon carbide in kimberlite rocks of Yakutiya. *Mineral. Zhurnal* **1990**, *12*, 17–26.
14. Leung, I.; Guo, W.; Friedman, I.; Gleason, J. Natural occurrence of silicon carbide in a diamondiferous kimberlite from Fuxian. *Nature* **1990**, *346*, 352. [CrossRef]
15. Qi, X.; Yang, J.; Xu, Z.; Bai, W.; Zhang, Z.; Fang, Q. Discovery of moissanite in retrogressive eclogite from the Pre-pilot Hole of the Chinese Continental Scientific Drilling Project (CCSD-PP2) and its geological implication. *Acta Petrol. Sin.* **2007**, *23*, 3207–3214.
16. Lee, J.-S.; Yu, S.-C.; Tung, S.-F.; Bai, W.-J.; Yang, J.-S.; Fang, Q.-S.; Zhang, Z. The crystal structure of natural 33R moissanite from Tibet. *Z. für Krist.–Cryst. Mater.* **2009**, *221*, 213–217. [CrossRef]
17. Kaminsky, F. Mineralogy of the lower mantle: A review of 'super-deep' mineral inclusions in diamond. *Earth-Sci. Rev.* **2012**, *110*, 127–147. [CrossRef]
18. Shiryaev, A.A.; Griffin, W.L.; Stoyanov, E. Moissanite (SiC) from kimberlites: Polytypes, trace elements, inclusions and speculations on origin. *Lithos* **2011**, *122*, 152–164. [CrossRef]
19. Ballhaus, C.; Wirth, R.; Fonseca, R.O.C.; Blanchard, H.; Pröll, W.; Bragagni, A.; Nagel, T.; Schreiber, A.; Dittrich, S.; Thome, V.; et al. Ultra-high pressure and ultra-reduced minerals in ophiolites may form by lightning strikes. *Geochem. Perspect. Lett.* **2017**, *5*, 42–46. [CrossRef]
20. Ballhaus, C.; Blanchard, H.; Fonseca, R.O.C.; Bragagni, A. Reply 2 to Comment on "Ultra-high pressure and ultra-reduced minerals in ophiolites may form by lightning strikes". *Geochem. Perspect. Lett.* **2018**, *8*, 8–10. [CrossRef]
21. Mathez, E.A.; Fogel, R.A.; Hutcheon, I.D.; Marshintsev, V.K. Carbon isotopic composition and origin of SiC from kimberlites of Yakutia, Russia. *Geochim. Cosmochim. Ac.* **1995**, *59*, 781–791. [CrossRef]
22. Golubkova, A.; Schmidt, M.W.; Connolly, J.A.D. Ultra-reducing conditions in average mantle peridotites and in podiform chromitites: a thermodynamic model for moissanite (SiC) formation. *Contrib. Mineral. Petrol.* **2016**, *171*, 41. [CrossRef]
23. Ohrenschall, R.D.; Milton, C. The occurrence of moissanite (silicon carbide) in sediments. *J. Sediment. Res.* **1931**, *1*, 96–99.
24. Kaminskiy, F.V.; Bukin, V.J.; Potapov, S.V.; Arkus, N.G.; Ivanova, V.G. Discoveries of silicon carbide under natural conditions and their genetic importance. *Int. Geol. Rev.* **1969**, *11*, 561–569. [CrossRef]
25. Ross, J. *Kimberlites and Related Rocks*; John Wiley & Sons: Hoboken, NJ, USA, 1989; ISBN 978-0-86793-384-0.
26. Leung, I.S. Silicon carbide cluster entrapped in a diamond from Fuxian, China. *Am. Mineral.* **1990**, *75*, 1110–1119.
27. Wilding, M.C.; Harte, B.; Harris, J.W. Evidence for a deep origin of São Luis diamonds. In *CPRM Special Publication 2/91, Proceedings of the Extended Abstracts 5th International Kimberlite Conference*; CPRM: Brasília, Brazil, 1991; pp. 456–458.
28. Svisero, D.P. Distribution and origin of diamonds in Brazil: An overview. *J. Geodyn.* **1995**, *20*, 493–514. [CrossRef]
29. Di Pierro, S.; Gnos, E.; Grobety, B.H.; Armbruster, T.; Bernasconi, S.M.; Ulmer, P. Letters. Rock-forming moissanite (natural α-silicon carbide). *Am. Mineral.* **2003**, *88*, 1817–1821. [CrossRef]
30. Zhang, Z. Native gold and native copper grains enclosed by olivine phenocrysts in a picrite lava of the Emeishan large igneous province, SW China. *Am. Mineral.* **2006**, *91*, 1178–1183. [CrossRef]
31. Klein-BenDavid, O.; Wirth, R.; Navon, O. Micrometer-scale cavities in fibrous and cloudy diamonds—A glance into diamond dissolution events. *Earth Planet Sc. Lett.* **2007**, *264*, 89–103. [CrossRef]
32. Xu, S.; Wu, W.; Xiao, W.; Yang, J.; Chen, J.; Ji, S.; Liu, Y. Moissanite in serpentinite from the Dabie Mountains in China. *Mineral. Mag.* **2008**, *72*, 899–908. [CrossRef]
33. Trumbull, R.B.; Yang, J.-S.; Robinson, P.T.; Di Pierro, S.; Vennemann, T.; Wiedenbeck, M. The carbon isotope composition of natural SiC (moissanite) from the Earth's mantle: New discoveries from ophiolites. *Lithos* **2009**, *113*, 612–620. [CrossRef]
34. Yusupov, R.G.; Stanley, C.J.; Welch, M.D.; Spratt, J.; Cressey, G.; Rumsey, M.S.; Seltmann, R.; Igamberdiev, E. Mavlyanovite, Mn5Si3: A new mineral species from a lamproite diatreme, Chatkal Ridge, Uzbekistan. *Mineral. Mag.* **2009**, *73*, 43–50. [CrossRef]

35. Fritsch, E.; Toledo, V.; Matlins, A. Record-size natural moissanite crystals discovered in Israel. *Gems Gemol.* **2014**, *50*, 160–161.
36. Liang, F.; Xu, Z.; Zhao, J. In-situ moissanite in dunite: deep mantle origin of mantle peridotite in Luobusa ophiolite, Tibet. *Acta Geol. Sin. Engl. Ed.* **2014**, *88*, 517–529. [CrossRef]
37. Liu, Y.; He, D.; Gao, C.; Foley, S.; Gao, S.; Hu, Z.; Zong, K.; Chen, H. First direct evidence of sedimentary carbonate recycling in subduction-related xenoliths. *Sci. Rep.* **2015**, *5*, 11547. [CrossRef] [PubMed]
38. Xu, X.; Yang, J.; Robinson, P.T.; Xiong, F.; Ba, D.; Guo, G. Origin of ultrahigh pressure and highly reduced minerals in podiform chromitites and associated mantle peridotites of the Luobusa ophiolite, Tibet. *Gondwana Res.* **2015**, *27*, 686–700. [CrossRef]
39. Yang, J.; Meng, F.; Xu, X.; Robinson, P.T.; Dilek, Y.; Makeyev, A.B.; Wirth, R.; Wiedenbeck, M.; Cliff, J. Diamonds, native elements and metal alloys from chromitites of the Ray-Iz ophiolite of the Polar Urals. *Gondwana Res.* **2015**, *27*, 459–485. [CrossRef]
40. Di Pierro, S.; Gnos, E. Ca-Al-silicate inclusions in natural moissanite (SiC). *Am. Mineral.* **2016**, *101*, 71–81. [CrossRef]
41. Dobrzhinetskaya, L.; Mukhin, P.; Wang, Q.; Wirth, R.; O'Bannon, E.; Zhao, W.; Eppelbaum, L.; Sokhonchuk, T. Moissanite (SiC) with metal-silicide and silicon inclusions from tuff of Israel: Raman spectroscopy and electron microscope studies. *Lithos* **2018**, *310–311*, 355–368. [CrossRef]
42. Machev, P.; O'Bannon, E.F.; Bozhilov, K.N.; Wang, Q.; Dobrzhinetskaya, L. Not all moissanites are created equal: New constraints on moissanite from metamorphic rocks of Bulgaria. *Earth Planet. Sci. Lett.* **2018**, *498*, 387–396. [CrossRef]
43. Nazzareni, S.; Nestola, F.; Zanon, V.; Bindi, L.; Scricciolo, E.; Petrelli, M.; Zanatta, M.; Mariotto, G.; Giuli, G. Discovery of moissanite in a peralkaline syenite from the Azores Islands. *Lithos* **2019**, *324–325*, 68–73. [CrossRef]
44. Baer, G.; Aharon, L.; Heimann, A.; Shaliv, G.; Agnon, A. The Nahal Tavor vent: Interplay of Miocene tectonics, dikes, and volcanism in the Lower Galilee, Israel. *Isr. J. Earth Sci.* **2006**, *55*, 1–16. [CrossRef]
45. Kunz, M.; Tamura, N.; Chen, K.; MacDowell, A.A.; Celestre, R.S.; Church, M.M.; Fakra, S.; Domning, E.E.; Glossinger, J.M.; Kirschman, J.L.; et al. A dedicated superbend X-ray microdiffraction beamline for materials, geo-, and environmental sciences at the advanced light source. *Rev. Sci. Instrum.* **2009**, *80*, 035108. [CrossRef]
46. Stan, C.V.; Tamura, N. Synchrotron X-ray Microdiffraction and Fluorescence Imaging of Mineral and Rock Samples. *JoVE* **2018**, in press. [CrossRef]
47. Tamura, N.; MacDowell, A.A.; Spolenak, R.; Valek, B.C.; Bravman, J.C.; Brown, W.L.; Celestre, R.S.; Padmore, H.A.; Batterman, B.W.; Patel, J.R. Scanning X-ray microdiffraction with submicrometer white beam for strain/stress and orientation mapping in thin films. *J. Synchrotron Radiat.* **2003**, *10*, 137–143. [CrossRef]
48. Merlen, A.; Sangar, A.; Torchio, P.; Kallepalli, L.N.D.; Grojo, D.; Utéza, O.; Delaporte, P. Multi-wavelength enhancement of silicon Raman scattering by nanoscale laser surface ablation. *Appl. Surf. Sci.* **2013**, *284*, 545–548. [CrossRef]
49. Russell, J.P. Raman scattering in silicon. *Appl. Phys. Lett.* **1965**, *6*, 223–224. [CrossRef]
50. Parker, J.H.; Feldman, D.W.; Ashkin, M. Raman Scattering by Silicon and Germanium. *Phys. Rev.* **1967**, *155*, 712–714. [CrossRef]
51. Uchinokura, K.; Sekine, T.; Matsuura, E. Raman scattering by silicon. *Solid State Commun.* **1972**, *11*, 47–49. [CrossRef]
52. Weinstein, B.A.; Piermarini, G.J. Raman scattering and phonon dispersion in Si and GaP at very high pressure. *Phys. Rev. B* **1975**, *12*, 1172–1186. [CrossRef]
53. Poborchii, V.; Tada, T.; Kanayama, T. Study of stress in a shallow-trench-isolated Si structure using polarized confocal near-UV Raman microscopy of its cross section. *Appl. Phys. Lett.* **2007**, *91*, 241902. [CrossRef]
54. Zhang, C.; Duan, Z. A model for C–O–H fluid in the Earth's mantle. *Geochim. et Cosmochim. Acta* **2009**, *73*, 2089–2102. [CrossRef]
55. Li, Z.; Bradt, R.C. Thermal Expansion of the Hexagonal (6H) Polytype of Silicon Carbide. *J. Am. Ceram. Soc.* **1986**, *69*, 863–866. [CrossRef]
56. Watanabe, H.; Yamada, N.; Okaji, M. Linear Thermal Expansion Coefficient of Silicon from 293 to 1000 K. *Int. J. Thermophys.* **2004**, *25*, 221–236. [CrossRef]
57. Arsenault, R.J.; Shi, N. Dislocation generation due to differences between the coefficients of thermal expansion. *Mater. Sci. Eng.* **1986**, *81*, 175–187. [CrossRef]

58. Vogelsang, M.; Arsenault, R.J.; Fisher, R.M. An *in situ* HVEM study of dislocation generation at Al/SiC interfaces in metal matrix composites. *MTA* **1986**, *17*, 379–389. [CrossRef]
59. Anastassakis, E.; Pinczuk, A.; Burstein, E.; Pollak, F.H.; Cardona, M. Effect of static uniaxial stress on the Raman spectrum of silicon. *Solid State Commun.* **1970**, *8*, 133–138. [CrossRef]
60. Hart, T.R.; Aggarwal, R.L.; Lax, B. Temperature Dependence of Raman Scattering in Silicon. *Phys. Rev. B* **1970**, *1*, 638–642. [CrossRef]
61. Cerdeira, F.; Buchenauer, C.J.; Pollak, F.H.; Cardona, M. Stress-Induced Shifts of First-Order Raman Frequencies of Diamond- and Zinc-Blende-Type Semiconductors. *Phys. Rev. B* **1972**, *5*, 580–593. [CrossRef]
62. Campbell, I.H.; Fauchet, P.M. The effects of microcrystal size and shape on the one phonon Raman spectra of crystalline semiconductors. *Solid State Commun.* **1986**, *58*, 739–741. [CrossRef]
63. Kouteva-Arguirova, S.; Arguirov, T.; Wolfframm, D.; Reif, J. Influence of local heating on micro-Raman spectroscopy of silicon. *J. Appl. Phys.* **2003**, *94*, 4946–4949. [CrossRef]
64. Georgi, C.; Hecker, M.; Zschech, E. Effects of laser-induced heating on Raman stress measurements of silicon and silicon-germanium structures. *J. Appl. Phys.* **2007**, *101*, 123104. [CrossRef]
65. Yang, Y.; Munck, K.D.; Teixeira, R.C.; Swinnen, B.; Verlinden, B.; Wolf, I.D. Process induced sub-surface damage in mechanically ground silicon wafers. *Semicond. Sci. Technol.* **2008**, *23*, 075038. [CrossRef]
66. Zhang, Y.; Wang, D.; Gao, W.; Kang, R. Residual stress analysis on silicon wafer surface layers induced by ultra-precision grinding. *Rare Met.* **2011**, *30*, 278–281. [CrossRef]
67. Xu, Z.; He, Z.; Song, Y.; Fu, X.; Rommel, M.; Luo, X.; Hartmaier, A.; Zhang, J.; Fang, F. Topic Review: Application of Raman Spectroscopy Characterization in Micro/Nano-Machining. *Micromachines* **2018**, *9*, 361. [CrossRef] [PubMed]
68. Robinson, P.T.; Bai, W.-J.; Malpas, J.; Yang, J.-S.; Zhou, M.-F.; Fang, Q.-S.; Hu, X.-F.; Cameron, S.; Staudigel, H. Ultra-high pressure minerals in the Luobusa Ophiolite, Tibet, and their tectonic implications. *Geol. Soc. Lond. Spec. Publ.* **2004**, *226*, 247–271. [CrossRef]
69. Zarudi, I.; Zhang, L. Subsurface damage in single-crystal silicon due to grinding and polishing. *J. Mater Sci. Lett.* **1996**, *15*, 586–587. [CrossRef]
70. Amelinckx, S.; Strumane, G.; Webb, W.W. Dislocations in Silicon Carbide. *J. Appl. Phys.* **1960**, *31*, 1359–1370. [CrossRef]
71. Dobrzhinetskaya, L.F.; Green, H.W.; Bozhilov, K.N.; Mitchell, T.E.; Dickerson, R.M. Crystallization environment of Kazakhstan microdiamond: evidence from nanometric inclusions and mineral associations. *J. Metamorph. Geol.* **2003**, *21*, 425–437. [CrossRef]
72. Kvasnytsya, V.M.; Wirth, R. Nanoinclusions in microdiamonds from Neogenic sands of the Ukraine (Samotkan' placer): A TEM study. *Lithos* **2009**, *113*, 454–464. [CrossRef]

minerals

MDPI

Article

The Elastic Properties of β-Mg_2SiO_4 Containing 0.73 wt.% of H_2O to 10 GPa and 600 K by Ultrasonic Interferometry with Synchrotron X-Radiation

Gabriel D. Gwanmesia [1,2,*], Matthew L. Whitaker [2,3], Lidong Dai [4], Alwin James [5], Haiyan Chen [2,3], Richard S. Triplett [2,3] and Nao Cai [2]

1 Division of Physics, Engineering, Mathematics & Computer Science, Delaware State University, Dover, DE 19901, USA
2 Mineral Physics Institute, Stony Brook University, Stony Brook, NY 11794, USA; Matthew.whitaker@stonybrook.edu (M.L.W.); Haiyan@bnl.gov (H.C.); Richard.triplett@stonybrook.edu (R.S.T.); Cainao2013@gmail.com (N.C.)
3 Department of Geosciences, Stony Brook University, Stony Brook, NY 11794, USA
4 Key Laboratory for High Temperature and High-Pressure Study of the Earth's Interior, Institute of Geochemistry, Chinese Academy of Sciences, Guiyang 550002, China; dailidong@vip.gyig.ac.cn
5 Department of Chemistry, Stony Brook University, Stony Brook, NY 11794, USA; alwin.james@stonybrook.edu
* Correspondence: ggwanmesia@desu.edu

Received: 21 January 2020; Accepted: 21 February 2020; Published: 26 February 2020

Abstract: We measured the elastic velocities of a synthetic polycrystalline β-Mg_2SiO_4 containing 0.73 wt.% H_2O to 10 GPa and 600 K using ultrasonic interferometry combined with synchrotron X-radiation. Third-order Eulerian finite strain analysis of the high P and T data set yielded K_{so} = 161.5(2) GPa, G_o = 101.6(1) GPa, and $(\partial K_s/\partial P)_T$ = 4.84(4), $(\partial G/\partial P)_T$ = 1.68(2) indistinguishable from K_{so} = 161.1(3) GPa, G_o = 101.4(1) GPa, and $(\partial K_s/\partial P)_T$ = 4.93(4), $(\partial G/\partial P)_T$ = 1.73(2) from the linear fit. The hydration of the wadsleyite by 0.73 wt.% decreases K_s and G moduli by 5.3% and 8.6%, respectively, but no measurable effect was noted for $(\partial K_S/\partial P)_T$ and $(\partial G/\partial P)_T$. The temperature derivatives of the K_s and G moduli from the finite strain analysis $(\partial K_S/\partial T)_P$ = −0.013(2) GPaK^{-1}, $(\partial G/\partial T)_P$ = −0.015(0.4) GPaK^{-1}, and the linear fit $(\partial K_S/\partial T)_P$ = −0.015(1) GPaK^{-1}, $(\partial G/\partial T)_P$ = −0.016(1) GPaK^{-1} are in agreement, and both data sets indicating the $|(\partial G/\partial T)_P|$ to be greater than $|(\partial K_S/\partial T)_P|$. Calculations yield $\Delta V_{p(\alpha\text{-}\beta)}$ = 9.88% and $\Delta V_{S(\alpha\text{-}\beta)}$ = 8.70% for the hydrous β-Mg_2SiO_4 and hydrous α-Mg_2SiO_4, implying 46–52% olivine volume content in the Earth's mantle to satisfy the seismic velocity contrast $\Delta V_s = \Delta V_P$ = 4.6% at the 410 km depth.

Keywords: elasticity; hydrous wadsleyite; equation of state; ultrasonic interferometry; synchrotron X-radiation; high temperature; high pressure; mantle composition

1. Introduction

Wadsleyite [β-(Mg, Fe)$_2$SiO$_4$] is a high-pressure polymorph of olivine [α-(Mg, Fe)$_2$SiO$_4$] stable from 410 to 525 km in the Earth's transition zone. Wadsleyite can incorporate varying amounts of water up to 3.3% as hydroxyl (OH) groups in the structure depending on the pressure and temperature and phase conditions [1–4]. The hydration of wadsleyite takes place by two H atoms substituting for one octahedral Mg in the structure [1,4–11] that affects the elasticity when compared to the anhydrous phase, and most notably at high pressure (P) and temperature (T) due to the variability of the H atomic radius.

Previous studies [9,12–14] have suggested that the Earth's transition zone could be a potential water reservoir, given the abundance of wadsleyite and ringwoodite in the region. The seismic velocity jumps at the 410-km depth have been associated with changes in the elastic wave speeds resulting from the transformation of olivine to wadsleyite. Still, the depth of the phase transition is reduced by OH incorporation into the olivine and wadsleyite structures.

The elastic properties of hydrous wadsleyite, and particularly the pressure (P) and temperature (T) derivatives of the elastic moduli, are essential for providing tighter constraints on the olivine content of the Earth's mantle by comparing laboratory elasticity data with the seismic velocity jumps at the 410-km depth. The data are also essential for constraining the water content in the wadsleyite phase that constitutes about 60% of the mineral assemblage in the transition zone region.

Data are already available for anhydrous Mg, and Mg-Fe wadsleyite at high pressure and room temperature [15–17], high pressure and high temperature [18–20], and, high temperature and room pressure [21–24], but the elasticity data for the hydrous wadsleyite are sparse compared to the anhydrous wadsleyite. A few static compression studies [6,25] have yielded information on the isothermal bulk modulus (K_T) and its derivative (K_T') for OH-bearing wadsleyite. However, the static compression studies do not provide information on the shear modulus. Secondly, fitting such P-V-T data to an equation of state requires a significant trade-off between the bulk modulus and its derivative [26].

Brillouin scattering measurements [27] on single-crystal wadsleyite containing 0.37, 0.84, and 1.66 wt.% water at room pressure (P) and room temperature (T) show that both the bulk (K_{so}) and shear (G_o) moduli decrease linearly with water content in the wadsleyite. Measurements of the elasticity of wadsleyite containing 0.84 wt.% H_2O to 12 GPa at room T [28] yielded the P-wave and S-wave velocities that are 2.7% and 3.6%, respectively, lower than the corresponding data for the anhydrous wadsleyite [17]. The study also concluded that the effect of OH substitution in wadsleyite on the pressure derivatives of the bulk and shear moduli is immeasurable. There is no consistent effect on the pressure derivatives of K and G due to the hydration of wadsleyite from comparing the Brillouin scattering data on single-crystal hydrous Fe-bearing wadsleyite containing 0.24 wt.% H_2O (Fe = 0.112) measured at high P and room T [29], with similar measurements carried out for wadsleyite containing 1.93 wt.% H_2O (Fe = 0.112) [30], and combining the Fe-bearing wadsleyite data with the static compression studies and the Brillouin scattering data on end-member wadsleyite with 0.84 wt.% H_2O [28] still does not reveal a clear trend regarding the variation of K and G with OH content in wadsleyite.

Temperature is an important variable, in addition to the pressure that affects geophysical properties, including the elastic wave velocities in the Earth's mantle. However, compared to the limited data currently available for the anhydrous wadsleyite [18,20–24], there is as yet no study to investigate the effect of temperature on the elasticity of hydrous wadsleyite. Ultrasonic and Brillouin scattering techniques are well-established for investigating the elasticity of materials. The Brillouin scattering measures the elasticity of the single-crystal as a function of the crystallographic direction within the sample. However, the utilization of the technique to study the hydrous material at high T is challenging compared to the pressure study. The ultrasonic technique employs either dense synthetic polycrystalline or relatively large single-crystal specimens. However, the main difficulty is to make suitable high-acoustic polycrystalline specimens or relatively large single-crystals of the hydrous high-pressure phase containing structurally-bound water. In addition to the sample difficulty, the OH retention in the hydrous sample during the high P and T ultrasonic studies is an essential factor to consider in the studies.

We report the synthesis of a dense isotropic polycrystalline specimen of hydrous wadsleyite (β-Mg_2SiO_4) containing 0.73 wt.% H_2O, and the ultrasonic elasticity data for the sample measured to 10 GPa and temperatures to 600 K using ultrasonic interferometry techniques combined with synchrotron X-ray diffraction. We compare our data with those of previous studies of Mg and Mg-Fe bearing hydrous wadsleyite, evaluate the effect on the elasticity due to OH incorporation in the wadsleyite, and discuss the impact of the temperature derivatives of the elasticity of wadsleyite on the olivine content of the Earth's transition zone.

2. Materials and Methods

2.1. Sample Synthesis and Characterization

The synthetic polycrystalline specimen of hydrous wadsleyite (β-Mg_2SiO_4) used in this study was hot-pressed in the 2000-ton uniaxial split-cylinder apparatus (USSA-2000) of the Kawai-type [31,32] at the Stony Brook High-Pressure laboratory, under conditions of 15.5 GPa and 1000 °C for 3 h within a 14/8 mm cell assembly.

The starting mixture consisted of pure forsterite (α-Mg_2SiO_4) and brucite [$Mg(OH_2)$] calculated to yield 0.778 wt.% H_2O, with additional silica (SiO_2) to maintain chemical charge balance due to the breakdown of brucite. Before weighing, we dried the α-Mg_2SiO_4 and the SiO_2 powders at 1000 °C for 24 h and the $Mg(OH)_2$ at 350 °C for 24 h. The mixture was ground several times to a fine-grained homogenous powder in an agate mortar under alcohol. The powder mixture was dried in an oven at 150 °C for 24 h and then densely packed in an $Au_{75}Pd_{25}$ capsule previously annealed at 900 °C. The capsule was crimped, cold-sealed in the air, and placed inside a NaCl sleeve inside a graphite resistance furnace. The NaCl insulates the capsule from the graphite furnace and also provides a pseudo-hydrostatic environment for the sample during heating at high pressure. A previous study [33] has demonstrated that NaCl loses most of its shear strength at about 300 °C, thus minimizing non-hydrostatic stress around the sample.

The hot-pressing procedure used in the study is similar to those described in detail by previous investigators [34–36]. We first increased the pressure to 1 GPa, and then the sample was preheated to 200 °C to relax the NaCl around the sample. Subsequently, we increased the pressure slowly to 15.5 GPa in 13 h, and then the temperature was increased to 1000 °C in 15 min. After maintaining the sample at the P and T conditions for 3 h, the temperature was rapidly decreased to 200 °C in 10 min and then maintained at the T condition throughout the depressurization that took 15 h.

The recovered sample was cylindrical, about 2.7 mm in diameter and 2.2 mm long. Before further characterization, we ground both ends of the sample roughly flat and collected X-ray diffraction spectra of the surfaces on a Rigaku Ultima IV diffractometer (Tube voltage—40 kV, Tube current—30 mA, Cu Kα λ = 1.54059 Å, 10–70° 2θ, 0.02° step size, 0.4 min/step) in Bragg-Brentano geometry with a D-tex Ultra solid-state detector. The spectra confirmed the specimen to be single-phase wadsleyite (β-Mg_2SiO_4).

The bulk density of the sample was determined by Archimedes' method using distilled water in which we add a few drops of an organic solvent to reduce surface tension on the sample and immersed parts. We corrected the measured bulk density for the effect of air buoyancy on the sample weight in air and also the change in the density of the fluid with temperature. We obtained ρ = 3.435 (5) g/cc, which is in excellent agreement with the theoretical X-ray density (ρ = 3.436 g/cc) based on the single-crystal X-ray systematics developed [6] that report the variation of the density of wadsleyite with H_2O content.

We carried out a scanning electron microscopy (SEM) examination of both a fractured and a polished surface of the wadsleyite specimen using an LEO-1550 FEG SEM with electron dispersive X-ray spectroscopy (EDAX) operating at the high tension of 20 kV. The samples were sputter-coated with gold for the analysis. We observed, as evident in the photomicrograph of the fractured surface of the sample in Figure 1a, that the synthetic polycrystalline wadsleyite sample is homogeneous and fine-grained. The average grain size of the sample is about 3–5 μm. Examination of the polished surface of the sample (Figure 1b) reveals the grains to be well-equilibrated and well-developed with sharp straight edges meeting at high angles.

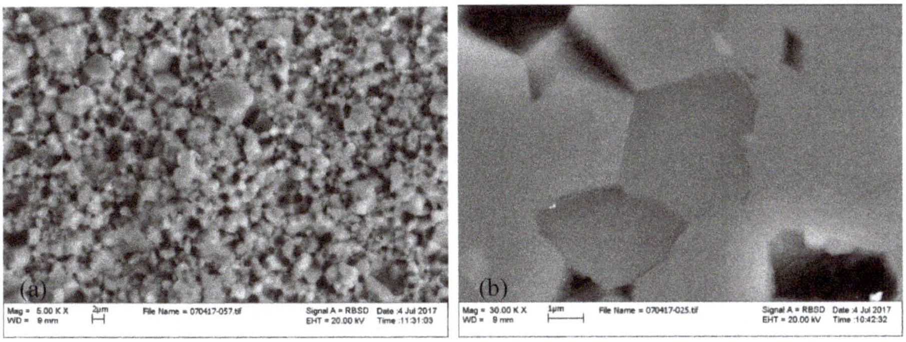

Figure 1. Scanning electron microscopy (SEM) micrographs of the cracked (**a**) and polished (**b**) surface of the hydrous β-Mg_2SiO_4 specimens. The average grain size of the specimen is about 2–5 μm.

The water content of the polycrystalline hydrous wadsleyite (β-Mg_2SiO_4) sample was measured using the Fourier transform infrared spectroscopy (FTIR) technique at the Key Laboratory of High-Temperature and High-Pressure Study of the Earth's Interior, Institute of Geochemistry, Chinese Academy of Sciences, Guiyang, China. Previous studies [37,38] have described in detail the set-up and experimental procedure of the FTIR technique.

We carried out the spectra measurements using a Fourier transform vacuum infrared spectroscopy (FTIR) spectrometer (Vertex-70V and Hyperion-1000 infrared microscope). We collected infrared spectra of the original polycrystalline sample from 350 to 8000 cm^{-1} wavenumbers, as well as the recovered specimen from the high pressure and high-temperature ultrasonic studies.

The specimens were doubly polished to a thickness of about 60 μm for the IR analysis. The IR absorption of the sample was measured using unpolarized radiation with a Mid-IR light source, a CaF_2 beam splitter, and an MCT detector with a 100 μm × 100 μm aperture. We collected 512 scans for each spectrum in the analysis. In Figure 2, we have overlapped the infrared spectra acquired for a sliced piece of the original sample and the remaining sample recovered from the ultrasonic velocity measurements to demonstrate that there was no change in the water content during the high P and T ultrasonic studies; the data agree within the uncertainties (10%) of the FTIR measurements. The Paterson calibration [39] was adopted to precisely determine the water content from FT-IR absorption data using,

$$C_{OH} = \frac{B_i}{150\xi} \int \frac{K(v)}{(3780 - v)} dv \tag{1}$$

where C_{OH} is the molar concentration of hydroxyl (ppm wt. of H_2O or H/106 Si), *Bi* is the density factor (4.08 cm × 104 cm H/106 Si), ξ is the orientation factor (1/3), and $K(v)$ is the absorption coefficient at wavenumber v in cm^{-1}. The integration was from 3000 to 3750 cm^{-1}. The analysis yielded identical water contents of 0.73 (7) wt.% for both specimens. The listed uncertainty of 10% is mainly due to errors in the sample thickness measurement, in addition to the current unavailability of a standard water content calibration for hydrous wadsleyite.

For ultrasonic studies, both ends of a sliced cylindrical piece of the sample were ground and polished flat and parallel to within 1/4λ of visible light, using 9, 6, 3, 1, and $\frac{1}{4}$ μm diamond compounds in succession. The polished sample, having a length of 1.017 mm, was cored to a 2 mm diameter for the ultrasonic studies. For internal consistency and data comparison, we also prepared a sample of anhydrous β-Mg_2SiO_4 that we have measured using the same the high P and T technique described subsequently for the hydrous sample.

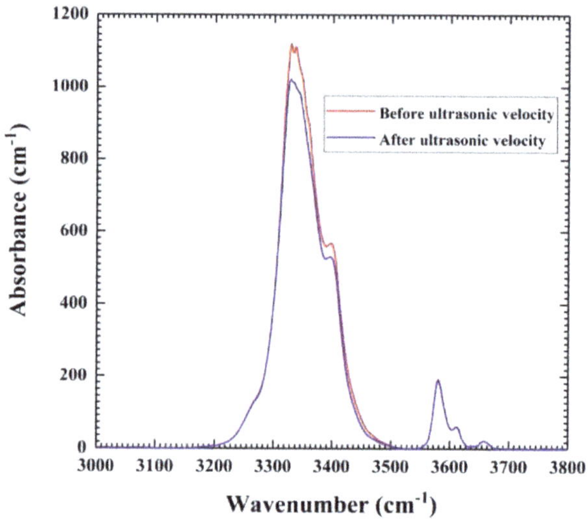

Figure 2. Fourier Transform Infrared (FTIR) absorption spectra for the hydrous β-Mg_2SiO_4 polycrystalline sample taken before and after the high P and T ultrasonic studies.

The synchrotron X-ray diffraction pattern for the hydrous wadsleyite (β-Mg_2SiO_4) containing 0.73 (7) wt.% of H_2O used in this study is shown in Figure 3 and compared with one taken with the press open (ambient condition) at the end of the high P and T ultrasonic experiments.

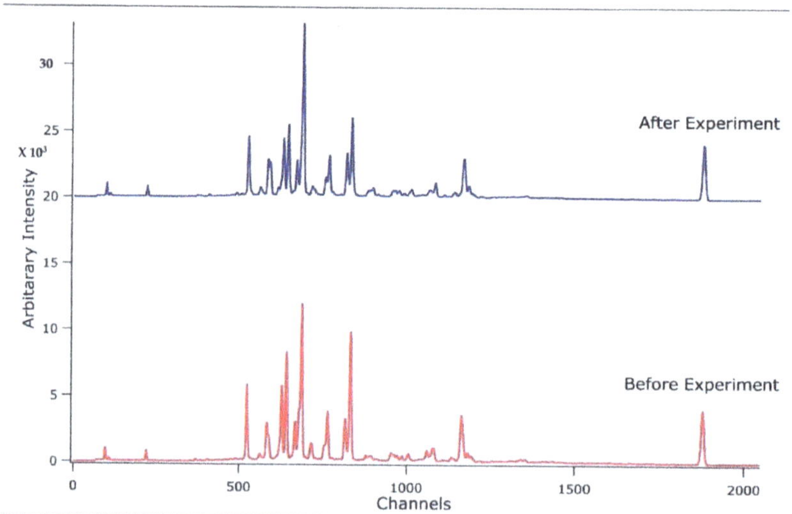

Figure 3. Comparison of X-ray diffraction patterns for the hydrous wadsleyite taken at ambient conditions (open press) before and after the high P and T ultrasonic experiments.

The spectra are sharp and show no uncharacteristic peaks, indicating that there is no significant residual stress in both the original and the recovered specimens, and also that the wadsleyite phase did not change during the high P and T experiments. We conducted the ultrasonic measurements below 600 K to avoid the dehydration of the hydrous β-Mg_2SiO_4 and also to avoid a possible back-transformation of the sample to the low P and T olivine phase.

2.2. Elasticity Measurements at High Pressure and High Temperature

The high P and T ultrasonic velocity measurements of the hydrous wadsleyite sample were carried out by ultrasonic interferometry techniques. The pressure generating device was a 250-ton hydraulic DIA-type multi-anvil apparatus equipped with a DDIA module installed at the 6-BM-B beamline at the advanced photon light source (APS) of the Argonne National Laboratory and inter-phased with in-situ X-ray diffraction and X-radiographic techniques. The experimental setup and the directly integrated acoustic system combined with pressure experiments (DIASCOPE) for data acquisition are described in detail in a previous study [40]. The DIASCOPE is an adapted ultrasonic interferometry technique for fast acoustic travel time measurements. A previous study [41] provides a detailed description of the main features and pressure-generating mechanism of the DIA-type high-pressure apparatus. The cell assembly and the acoustic piezoelectric transducer/cubic carbide anvil arrangement used in this study are the same as illustrated in Figure 1 of a previous study [42], and as described in other previous studies [43]. We utilized a dual-mode 10° Y-cut $LiNbO_3$ transducer capable of generating and receiving P and S waves simultaneously. An alumina buffer rod of length 2.0 mm and 2.0 mm in diameter ground and polished flat on both ends is inserted between the specimen and the tungsten carbide (WC) anvil, allowing high-frequency acoustic signals (20–70 MHz) to propagate to and from the sample. A BN sleeve houses the sample supported at the far end by a NaCl disc; the NaCl also serves as an in-situ pressure marker. A 2 μm gold Au foil was inserted at the top and bottom of the specimen to improve the mechanical coupling at the interfaces between the specimen and cell components as the sample was pressurized and to enhance the transmission and reception of the acoustic signal. The two Au foils delineate the specimen in the X-radiographic images acquired for determining the specimen length at each P-T condition [44,45]. A W/Re 3%–W/Re 25% thermocouple located at the interface between the sample and the NaCl in-situ pressure standard monitors the sample temperature.

We have designed the pressure-temperature (P-T) path of the ultrasonic experiments used in this study and shown in Figure 4 to minimize deviatoric stress in the sample. We initially compressed the sample to about 2 GPa. We then heated it to 600 K, where we collected travel times of acoustic compressional (P) and shear (S) waves, X-ray diffraction of the sample, the NaCl in-situ pressure standard, and a sample image. Data were subsequently collected at 100 K intervals as we decreased the temperature down to ambient temperature along the isobar. We then increased the pressure moderately and repeated the procedure until six cycles of data collection were completed up to the peak pressure of 9.8 GPa, followed by four data collecting sequences on decompression. At each P and T point, we waited for about 3 min to equilibrate the pressure and temperature within the sample before acquiring the data.

The P and T path in the current study differs from that of previous standard studies [16,19,20,46,47] in which the sample is first compressed to the maximum pressure, followed by heating and subsequent data collection during cooling along an isobaric path. Temperatures in the traditional studies were relatively high to relax stress accumulated in the sample during compression to high pressure. However, given the limited temperature of 600 K in the current study, we minimized the stress in the sample throughout the experiment by heating the sample to the peak temperature at each pressure step of the operation before the data acquisition.

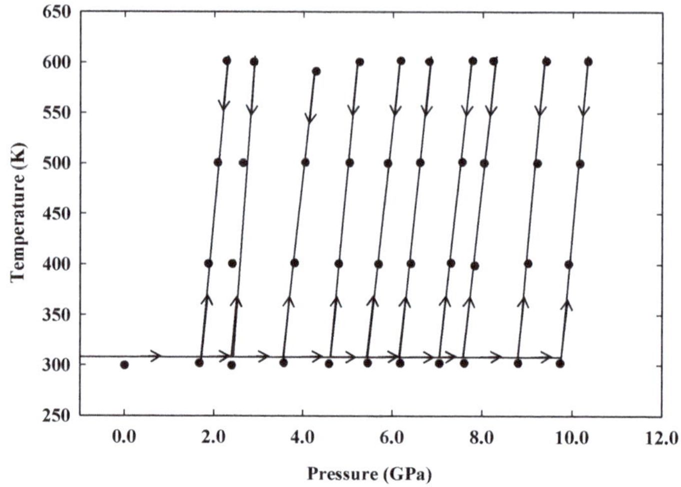

Figure 4. Pressure-temperature (P-T) path for the compression and decompression cycles, and the heating and data acquisition points in the ultrasonic experiments.

3. Results and Discussions

3.1. Data Acquisition and Analysis

We determine the P and S wave travel times by the pulse-echo-overlap (PEO) method [40] with a standard deviation of 0.2 ns (0.2%) and 0.5 ns (0.1%) for the P and S waves, respectively. A correction 1.437 ns and 0.272 ns was applied to all the P and S wave travel times, respectively, for the effect of the 2 μm Au bond between the sample and the buffer rod, using a previously established procedure [48].

The two-way corrected P and S wave travel times are tabulated in Table 1 and plotted as a function of pressure in Figure 5a,b, showing the travel times of both the P and S waves to decrease linearly and steadily with pressure along the isotherms.

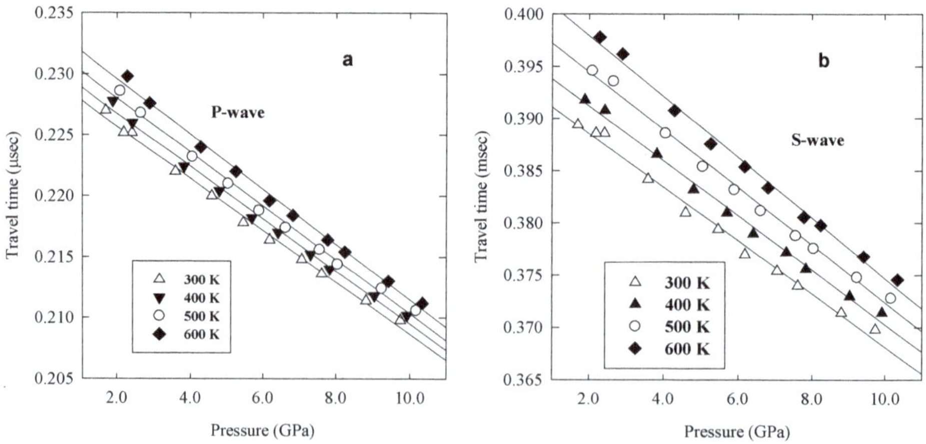

Figure 5. Travel times of acoustic compressional (P) (**a**) and shear (S) (**b**) wave velocities as a function of pressure. Lines are linear regression to the data along the experimental isotherms.

Table 1. Experimental ultrasonic and P-V-T data for hydrous wadsleyite from this study. Two- way travel times have 1σ of 0.2 ns (0.2%) for the P and 0.5 ns (0.1%) S waves. The uncertainties are length (0.1%), velocities about 1%, less than 1.5% in the derived moduli. V_o = 538.6 (2) and l_o = 1072 mm.

P (GPa)	T (K)	V (Å^3)	L (mm)	ρ (g/cc)	T_P (μsec)	T_s (μsec)	V_P (km/s)	V_s (km/s	L (GPa)	K (GPa)	G (GPa)
0.01	299	538.6	1.072	3.436							
2.41	299	532.1	1.068	3.478	0.2252	0.3886	9.48	5.49	312.7	172.7	105.0
2.28	601	537.0	1.071	3.446	0.2298	0.3978	9.32	5.38	299.4	166.2	99.9
2.09	500	535.7	1.070	3.455	0.2286	0.3946	9.36	5.42	302.8	167.3	101.6
1.88	400	534.5	1.069	3.462	0.2278	0.3918	9.39	5.46	305.1	167.6	103.2
1.69	301	533.4	1.069	3.470	0.2270	0.3894	9.41	5.49	307.5	168.2	104.5
5.26	600	528.0	1.065	3.505	0.2220	0.3876	9.59	5.49	322.6	181.5	105.8
5.04	500	526.9	1.064	3.512	0.2210	0.3854	9.63	5.52	325.8	182.9	107.1
4.81	400	525.9	1.064	3.519	0.2204	0.3832	9.65	5.55	327.7	183.2	108.4
4.59	301	525.0	1.063	3.525	0.2200	0.3810	9.66	5.58	329.1	182.8	109.7
6.82	600	523.6	1.062	3.535	0.2184	0.3834	9.72	5.54	334.3	189.6	108.5
6.61	501	522.5	1.061	3.542	0.2174	0.3812	9.76	5.57	337.6	191.2	109.8
6.42	400	521.5	1.061	3.548	0.2170	0.3790	9.77	5.60	339.0	190.8	111.1
6.19	302	520.6	1.060	3.555	0.2164	0.3770	9.80	5.62	341.1	191.3	112.4
8.24	601	519.2	1.059	3.564	0.2154	0.3798	9.83	5.58	344.6	196.8	110.8
8.04	500	518.1	1.058	3.572	0.2144	0.3776	9.87	5.61	348.1	198.4	112.2
7.84	398	517.3	1.058	3.577	0.2140	0.3756	9.88	5.63	349.6	198.3	113.5
7.61	302	516.4	1.057	3.584	0.2136	0.3740	9.90	5.65	351.1	198.4	114.5
9.41	601	516.1	1.057	3.586	0.2130	0.3768	9.92	5.61	353.1	202.7	112.8
9.23	500	515.1	1.056	3.593	0.2124	0.3748	9.95	5.64	355.3	203.2	114.1
9.03	401	514.2	1.056	3.599	0.2118	0.3730	9.97	5.66	357.6	203.8	115.3
8.80	302	513.5	1.055	3.604	0.2114	0.3714	9.98	5.68	359.1	204.0	116.3
10.35	601	513.4	1.055	3.605	0.2112	0.3746	9.99	5.63	359.8	207.3	114.4
10.17	500	512.7	1.054	3.610	0.2106	0.3728	10.01	5.66	362.0	208.0	115.5
9.92	400	511.9	1.054	3.615	0.2102	0.3714	10.03	5.68	363.6	208.3	116.5
9.74	302	511.2	1.053	3.620	0.2098	0.3698	10.04	5.70	365.1	208.4	117.5
7.78	601	520.3	1.060	3.557	0.2164	0.3806	9.79	5.57	341.2	194.1	110.3
7.55	501	519.7	1.059	3.561	0.2156	0.3788	9.83	5.59	343.9	195.3	111.4
7.31	401	519.0	1.059	3.566	0.2152	0.3772	9.84	5.61	345.3	195.4	112.4
7.06	301	518.3	1.058	3.570	0.2148	0.3754	9.85	5.64	346.7	195.4	113.5
6.18	601	525.4	1.063	3.522	0.2196	0.3854	9.68	5.52	330.2	187.3	107.2
5.89	500	524.5	1.063	3.529	0.2188	0.3832	9.71	5.55	332.9	188.2	108.5
5.70	400	523.8	1.062	3.533	0.2182	0.3810	9.74	5.58	334.8	188.4	109.8
5.46	302	522.9	1.061	3.539	0.2178	0.3794	9.75	5.60	336.3	188.5	110.8
4.29	591	530.6	1.067	3.488	0.2240	0.3908	9.52	5.46	316.3	177.8	103.9
4.05	500	529.8	1.066	3.493	0.2232	0.3886	9.55	5.49	318.8	178.6	105.2
3.82	401	528.9	1.065	3.499	0.2224	0.3866	9.58	5.51	321.3	179.5	106.3
3.58	302	527.9	1.065	3.506	0.2220	0.3842	9.59	5.54	322.6	179.0	107.7
2.89	600	534.9	1.070	3.460	0.2276	0.3962	9.40	5.40	305.6	171.1	100.9
2.65	500	534.1	1.069	3.465	0.2268	0.3936	9.43	5.43	307.9	171.6	102.2
2.42	400	533.0	1.068	3.472	0.2260	0.3908	9.45	5.47	310.3	171.9	103.8
2.18	300	531.8	1.067	3.480	0.2252	0.3886	9.48	5.49	312.7	172.7	105.0
−0.14	300	538.9	1.072	3.434							

The sample chamber for ultrasonic measurements in the cubic anvil DIA-type apparatus is composed of an alumina buffer rod and other cylindrical parts. These components impose axial deviatoric stress in the cell assembly that could cause vertical stress, σ_1 to be different from the lateral stress $\sigma_2 = \sigma_3$, as observed in previous studies [33,41]. Previous ultrasonic experiments in the multi-anvil press with synchrotron radiation [18–20,46,47] only measured vertical stress σ_1. However, recent replacement of the lateral WC anvil with a sintered diamond has afforded measurement of the lateral stress $\sigma_2 = \sigma_3$, in addition to the vertical stress. The pressure listed in Table 1 is the weighted average of the three components given by $P = (2\sigma_1 + \sigma_3)/3$, yielding a more precise pressure than previously possible. The axial stress was generally higher than the lateral stress by up to 0.4 GPa by heating the sample to 600 K. However, the difference diminished linearly with decreasing temperature, and ultimately reversed signs around ambient temperature where the vertical stress became moderately higher than the lateral stress. The analysis shows that while the temperature derivatives of the elastic moduli are sensitive to the stress differences between σ_1 and $\sigma_2 = \sigma_3$, the pressure derivatives of the elastic moduli are only moderately affected by the difference.

The sample volume in Table 1 is also a weighted average of data from the two detectors given by $V = (2V_1 + V_3)/3$, where V_1 and $V_2 = V_3$ is the sample volumes measured by the axial and lateral detectors, respectively.

The sample length was determined from the unit cell volume data using the relation $l = l_o\ (V/V_o)^{1/3}$, where l and V are the length and weighted unit cell volume, respectively; subscript zero represents ambient values. We observed considerable scatter in the lengths determined from the X-radiographic method, which we ascribed to the relatively small temperature interval (100 K) between measurements. We found changes in successive specimen lengths to be close to the precision (0.2–0.4%) of the ranges determined from the pixels data. In Table 1, we tabulated the specimen pressures calculated using the equation of state of NaCl [49] with a precision of 0.8%. We compute the listed sample densities in Table 1 from the unit-cell parameters from the X-ray data using the relationship $\rho = \rho_o(V_o/V)$, where ρ and V are density and weighted unit cell volume, respectively; subscript zero represents ambient values.

The travel time data were combined with the lengths to calculate the P and S velocities at high P and T. Uncertainty in the travel time, when combined with uncertainty in specimen length (0.1%), yields uncertainty in the velocity of about 1%. The P and S wave velocities are given in Table 1 and used to calculate the longitudinal $L = \rho V_P = (K_s + 4G/3)$, shear ($G = \rho V_S{}^2$), and bulk ($K_s = L - 4G/3$) moduli, also listed in Table 1.

We have plotted the P and S wave velocities from this study as a function of pressure (Figure 6a,b), and the elastic moduli K_s and G are also plotted as a function of pressure in Figure 7a,b, respectively. The acoustic velocities, as well as the moduli data, are observed to increase systematically and linearly over the entire pressure range for each isotherm.

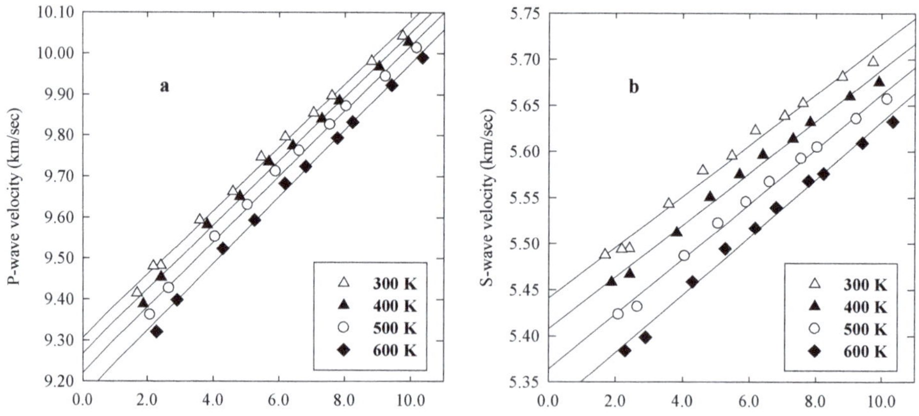

Figure 6. Elastic compressional (P) **(a)** and shear (S) **(b)** wave velocities as a function of pressure. The line is a linear regression to the data along the experimental isotherm.

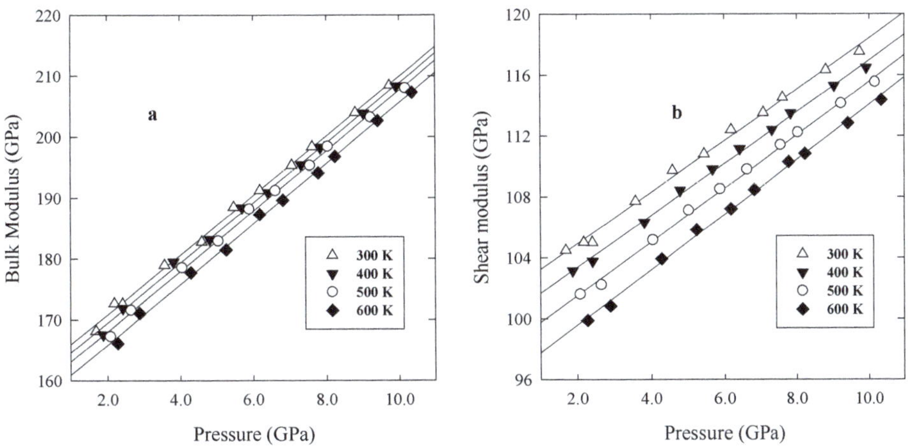

Figure 7. Elastic bulk (K) (**a**), and shear (G) (**b**) moduli as a function of pressure. The line is a linear regression to the data along the experimental isotherm.

For consistency in extrapolating to the high pressures of the transition zone, we fit the shear (G) and compressional ($Mp = K + 4G/3$) moduli, where K is the bulk modulus to functions of Eulerian strain {ε ($\varepsilon = [1 - (\rho/\rho_o)^{2/3}]/2$} of the third order:

$$G = (1 - 2\varepsilon)^{5/2}(M_1 + M_2\varepsilon) \tag{2}$$

$$M_p = (1 - 2\varepsilon)^{5/2}(L_1 + L_2\varepsilon) \tag{3}$$

The coefficients of the polynomials are related to the bulk (K_s) and shear (G_o) moduli and their pressure derivatives at zero pressure as follows:

$$M_1 = G_o \tag{4}$$

$$M_2 = 5G_o - 3K_sG_o' \tag{5}$$

$$L_1 = K_s + (4/3)G_o' \tag{6}$$

$$L_2 = 5(K_s + 4G_o'/3) - 3K_s(K_s' + 4G_o'/3) \tag{7}$$

where K_s, and G_o, are the elastic moduli and K_s', and G_o' and their pressure derivatives, respectively. Previous ultrasonic studies have shown that the data collected at the end of the cooling cycles at 300 K (cooled data) are generally under pseudo-hydrostatic conditions due to cell relaxation resulting from heating. By fitting the 300 K data from the cooling cycles to Equations (2) and (3), we obtain K_s = 161.1(4) GPa, G_o = 101.6(2) GPa, K_s' = 4.89(7), G_o' = 1.69(4) and from linear regression of the cooled data, we obtain K_s = 161.6(9) GPa, G_o = 101.7(2) GPa, K_s' = 5.03(1), G_o' = 1.79(4), where numbers in parenthesis represent the fitting uncertainty.

Following the previous procedure [46], we have treated each modulus as a linear function of pressure and temperature by fitting the P-T-K_S and P-T-G data sets separately to the linear Equation of the form:

$$M\,(P,T) = M_o + (\partial M/\partial P)_T(P - P_o) + (\partial M/\partial T)_P(T - T_o) \tag{8}$$

Where M_o is the ambient pressure and temperature K_s or G_o, and $(\partial M/\partial T)_P$ and $(\partial M/\partial P)_T$ are the temperature and pressure derivatives, respectively, of the elastic modulus M. In fitting, we exclude all data acquired before the sample was heated after each pressure increase because the data are likely to have residual stress caused by the cold compression. Fitting the entire P-T-K_s and P-T-G data sets separately to Equation (3) yielded: K_s = 161.1(3) GPa, G_o = 101.4(1) GPa, K_s' = 4.93(4), G_o' = 1.73(2), $(\partial K_s/\partial T)_P = -1.5(1)$ GPaK^{-1}, and $(\partial G/\partial T)_P = -1.6(1)$ GPaK^{-1}. The numbers in parenthesis represent the fitting uncertainty.

3.2. Ambient Elastic Bulk (K_s) and Shear (G_o) Moduli

In Table 2, we present the adiabatic bulk (K_s) and shear (G_o) moduli and their pressure derivatives $(\partial K_s/\partial P)_T$, and $(\partial G/\partial P)_T$, respectively, for the hydrous wadsleyite (β-Mg_2SiO_4) containing 0.73(7) wt.% H_2O and compare the data with those from previous studies of a hydrous β-Mg_2SiO_4, as well as Mg- and Fe-bearing anhydrous wadsleyite. Data for the anhydrous β-Mg_2SiO_4 polycrystalline sample measured in this study by the identical procedure are also listed in the table to validate the interferometry method in the DIA-type apparatus used for the measurements and to serve as an internally consistent data set for direct comparison with the hydrous wadsleyite data.

Our K_s and G_o data from the four analytical techniques are consistent and in excellent agreement with the mutual uncertainty of the data. We have selected our finite strain results for comparison with previous studies. Our finite strain K_s = 161(5) GPa and G_o = 101.6(2) GPa for the hydrated sample are 5.2% and 8.7%, respectively, lower than the corresponding values for anhydrous β-Mg_2SiO_4 (Table 2) from this study. We observe a comparable reduction in the K_s and G_o by comparing the values for the hydrous β-Mg_2SiO_4 from this study with results from other previous studies for anhydrous wadsleyite [16,18,21,50]. An earlier investigation of the elasticity of a single crystal of wadsleyite (β-Mg_2SiO_4) containing 0.37–1.66 wt.% H_2O by Brillouin spectroscopy [27] observed both K_s and G_o to decrease linearly with increasing water content in wadsleyite. The linear relationships given by Equations (4) and (5) of the study [27], correlating the bulk and shear moduli respectively, with water content in wadsleyite predict K_s and G_o of 161.4 GPa and 106.0 GPa, respectively, for hydrous β-Mg_2SiO_4 containing 0.73 wt.% H_2O as measured in this study. Whereas our K_s = 161.4 GPa is identical with the predicted value, the G_o = 101.4 GPa from this study is 4% smaller than predicted by Equation (5) of the Brillouin scattering study [27]. Our data show a stronger elasticity effect on G compared to Ks caused by the hydration in contrast to Brillouin scattering measurements on single-crystal wadsleyite [27,28] that report a comparable decrease in Ks and G of 7.6% and 7.0%, respectively, by dissolving one weight percent water in wadsleyite.

First principle calculations [51] report that one weight percent of water dissolved in wadsleyite reduces the Ks and G by ~5.6% and ~6%, respectively, which though also comparable, are out of the range of the Brillouin scattering data. The variability in the current data on the effect of wadsleyite hydration on the elasticity is mainly due to uncertainty in the water content measurements, generally around 10% [27,28]. Currently, standard water-content calibrations are available only for the olivine and ringwoodite phases, but not the wadsleyite phase. Resolving the magnitude and the relative contributions of the water effect on K_s and G requires the establishment of a conventional water content calibration for wadsleyite, as wells as systematic measurements of OH-bearing wadsleyite samples.

3.3. Pressure Derivatives of Elastic Moduli

As shown in Table 2, the pressure derivative of the adiabatic bulk modulus, K_s' = 4.84(2) for hydrous β-Mg_2SiO_4 with 0.73 wt.% H_20 is identical to K_s' = 4.81(2) for our anhydrous β-Mg_2SiO_4 and also in agreement with previously reported K_s' = 4.56(23) [19], within the mutual data uncertainties. The wadsleyite specimens used in this study have been hot-pressed by similar techniques and measured at high pressure using similar P and T paths and ultrasonic procedures.

The K_s' value for the hydrous wadsleyite is moderately higher than previous ultrasonic data for β-Mg_2SiO_4 (K_s' = 4.5(1) [16]). The Brillouin Scattering data (K_s' = 4.3(2)) for single-crystal β-Mg_2SiO_4 [17] measured to 15 GPa at room T. Our K_s' is also higher than the aggregate modulus K_s' = 4.5 from the first principle calculations [51] for β-$(Mg_{0.875}Fe_{0.125})_2SiO_4$) wadsleyite. The K_s' = 4.1(1) for single-crystal wadsleyite containing 0.84 wt.% H_2O measured to 12 GPa at room T by Brillouin spectroscopy [28], and K_S' = 4.13(8) reported for Fe-bearing single-crystal wadsleyite from Brillouin scattering studies [29] are in excellent agreement, but both are lower than the other published data.

Figure 8 is a plot of the elastic *K* and *G* moduli collected at the end of each cooling cycle (300 K) for the hydrous and anhydrous β-Mg_2SiO_4 as a function pressure. The line is the linear regression of the compression data. The overlap among the compression and decompression data yields a robust pressure dependence of the elastic *K* and *G* moduli for each composition. The similarity of the K_S' and G_o' for the hydrous and the anhydrous β-Mg_2SiO_4 phases shown in Figure 8 suggests that the pressure derivatives of the elastic moduli are insensitive to the OH content of the wadsleyite, at least up to 0.73 wt.% of the study. For comparison, we also plotted the Brillouin scattering single-crystal measurements carried out to 12 GPa and room T for containing 0.84 wt.% H_2O [28]. We note that whereas the Brillouin scattering shear modulus data further validates the insensitivity to the OH content in wadsleyite, the slope of the bulk modulus with pressure (K_S') is slightly lower than from the current study, as shown in Table 2.

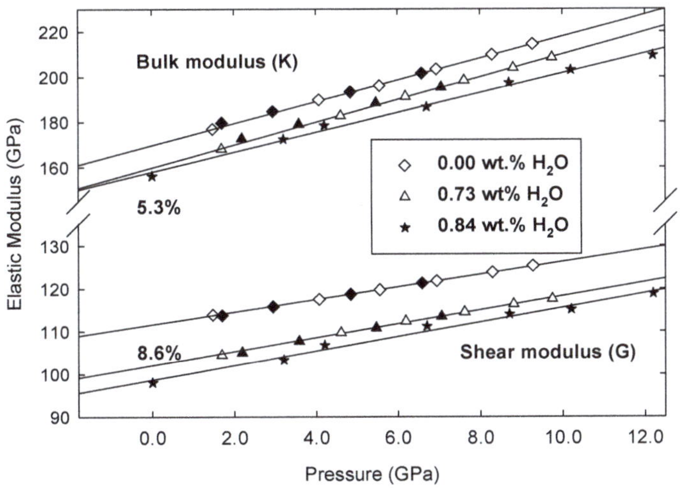

Figure 8. Elastic bulk (*K*) and shear (*G*) moduli at 300 K for hydrous (triangle) and anhydrous (diamond) β-Mg_2SiO_4 as a function of pressure. Open and filled symbols are data acquired on compression and decompression, respectively. Lines are linear fit to the compression data. Stars are Brillouin scattering single-crystal data [27]. Indicated are 5.3% and 8.6% decreases in *Ks* and *Go*, respectively, due to hydration of wadsleyite by 0.73%.

The G_o'= 1.55(1) from the current study of anhydrous β-Mg_2SiO_4 is in excellent agreement within the mutual data uncertainties with previous anhydrous studies G_o' = 1.4(2) [17], G_o' = 1.5(1) by [18] and in reasonable agreement with G_o'= 1.4(1) for hydrous wadsleyite containing 0.84 wt.% H_2O [28]. Our G_o' is also in close agreement with G_o' = 1.6 for β-$(Mg_{0.875}Fe_{0.125})_2SiO_4)$ from first-principle calculations [51]. However, as shown in Table 2, the G_o'= 1.68(4) for the hydrous β-Mg_2SiO_4 containing 0.73 wt.% H_2O from this study is marginally higher than the ultrasonic data for anhydrous wadsleyite [16,18,19] and the Brillouin scattering data for anhydrous wadsleyite [17] but in excellent agreement only with G_o' = 1.64(4) from the single-crystal Brillouin scattering measurement of Fe/(Mg + Fe) = 0.112(2) wadsleyite containing 0.2492 wt.% H_2O [29].

Table 2. Thermo-elastic properties of hydrous wadsleyite (0.73 wt.% H_2O). [a]: Linear fit of entire data set to $M\ (P,T) = M_o + (\partial M/\partial P)_T(P - P_o) + (\partial M/\partial T)_P(T - T_o)$: M = Elastic modulus ($K$ or G); [b]: third-order finite strain fit to entire data set; [c]: third-order finite strain fit to data set at the end of cooling cycles; [d]: linear regression of data set at the end of the cooling cycles. (1) This study. UI—ultrasonic interferometry; BS—Brillouin scattering; RUS—resonance ultrasound spectroscopy; RS—resonance sphere technique.

OH Content wt.%	K_{S0} (GPa)	G (GPa)	$(\partial K_s/\partial P)_T$	$(\partial G/\partial P)_T$	$(\partial K_S/\partial T)_P \times 10^{-2}$ GPaK^{-1}	$(\partial G/\partial T)_P \times 10^{-2}$ GPaK^{-1}	Method	Ref.
				β-$(Mg)_2SiO_4$				
0.73	161.1(3) [a] 161.5(2) [b] 161.1(4) 161.6(9)	101.4(1) [a] 101.6 (2) [b] 101.6(2) 101.7(2)	4.93(4) [a] 4.84(2) [b] 4.89(7) [c] 5.03(1) [d]	1.73 (2) [a] 1.68(4) [b] 1.69(4) [c] 1.79(4) [d]	−1.5(1) [a] −1.3(1) [b]	−1.6(1) [a] −1.5(2) [b]	UI	[1]
0.84	160.3(7)	105.3(6)	4.1(1)	1.4(1)			BS	[28]
0.00	170.6(2) [a] 170.3(2) [b]	111.5(1) [a] 111.3(1) [b]	4.76(3) [a] 4.81(3) [b]	1.53(1) [a] 1.55(1) [b]	−1.7(1) [a] −1.8(1) [b]	−1.60(3) [a] −1.7(1) [b]	UI	[1]
0.00	170(2)	115(2)	4.3(2)	1.4(2)			BS	[17]
0.00	170(2)	108(1)	4.5(1)	1.6(1)			UI	[16]
0.00	173(1)	113(1)	4.2(1)	1.5(1)	−1.2(1)	−1.7(1)	UI	[18]
0.00	170.7(11)	111.6(5)	4.56(23)	1.75(9)	−1.29(7)	−1.58(10)	UI	[19]
0.00	170.2(19)	113.9(7)			−1.71(5)	−1.57(3)	RUS	[21]
				β-$(Mg/Fe)_2SiO_4$				
$Mg_{(0.87)}Fe_{(0.13)}$	175.4(7)	108.0(4)	4.11(11)	1.56(5)	−1.35(10)	−1.44(8)	UI	[20]
$Mg_{(0.91)}Fe_{(0.09)}$	165.70	105.66			−1.6(3)	−1.2(1)	RS	[23]
$Mg_{(0.91)}Fe_{(0.09)}$	165.72(6)	105.43(1)			−1.75(3)	−1.59(1)	RUS	[24]
$Mg_{(0.92)}Fe_{(0.08)}$	170.8(1.2)	108.9(4)			−1.75(7)	−1.55(6)	RUS	[22]

The investigators of the single-crystal wadsleyite elasticity containing 0.84 wt.% H_2O at 12 GPa and room T [28] concluded that K_s' and G_o' for the hydrous wadsleyite phase is indistinguishable from those of the anhydrous phase, citing the overlap of the Ks' and G_o' from their study with the anhydrous data [17]. The K_s' values for the hydrous (0.73 wt.% H_2O) and the anhydrous wadsleyite samples (Table 2; also see Figure 8) are identical, and there is also a close agreement between the G_o' for the two compositions. We, therefore, conclude that the pressure derivatives of K and G are independent of the H_2O content of wadsleyite (at least up to the 0.73 wt.% of this study). The agreement of the K_s' and G_o' from this study with Mg and Mg-Fe wadsleyite data from previous studies further strengthens our assumption.

3.4. Temperature Derivatives of Elastic Moduli

We have tabulated values for $(\partial K_s/\partial T)_P$ and $(\partial G/\partial T)_P$ for the hydrous wadsleyite with 0.73(7) wt.% water and the anhydrous β-Mg_2SiO4 from this study in Table 2 and compared the results with data from previous studies on the anhydrous Mg and Mg-Fe wadsleyite. As observed in the table, the $(\partial K_s/\partial T)_P$ and $(\partial G/\partial T)_P$ results for the hydrous β-Mg_2SiO_4 obtained from fitting the entire data set linearly (see Equation (3)) and the third-order finite strain method to all the data is consistent and in good agreement within the mutual uncertainties of the analysis techniques. For consistency, we select the finite strain results for extrapolation to mantle conditions and comparison with previous data.

Our $(\partial K_s/\partial T)_P = -1.8(1) \times 10^{-2}$ GPaK^{-1} for the anhydrous β-Mg_2SiO_4 from this study is in excellent agreement with $(\partial K_s/\partial T)_P = -1.71(5) \times 10^{-2}$ GPaK^{-1} obtained on measurements of anhydrous polycrystalline β-Mg_2SiO_4 using resonance ultrasound spectroscopy techniques [21]. The data is also in excellent agreement with anhydrous Fe-bearing wadsleyite data [22–24]. The current $(\partial K_s/\partial T)_P$ values for anhydrous Mg- and Mg-Fe wadsleyites, including the data for the anhydrous β-Mg_2SiO_4 from this study, are higher in magnitude than the $(\partial K_s/\partial T)_P = -1.3(1) \times 10^{-2}$ GPaK^{-1} for our hydrated wadsleyite. The magnitude of the temperature derivative of the adiabatic bulk modulus $|(\partial K_s/\partial T)_P|$

for the anhydrous β-Mg_2SiO_4 from the current study is 28% greater than for the hydrous β-Mg_2SiO_4 containing 0.73 wt.% H_2O. However, our $(\partial K_s/\partial T)_P = -1.3(1) \times 10^{-2}$ GPaK^{-1} for the hydrated wadsleyite is in excellent agreement with the ultrasonic $(\partial K_s/\partial T)_P = -1.2 \times 10^{-2}$ GPaK^{-1} and $(\partial K_s/\partial T)_P = -1.29(7) \times 10^{-2}$ GPaK^{-1} for anhydrous β-Mg_2SiO_4 [18,19], the $(\partial K_s/\partial T)_P = -1.35(10) \times 10^{-2}$ GPaK^{-1} from ultrasonic measurement of anhydrous $Mg_{(0.83)}Fe_{(0.13)}$ wadsleyite [20].

The $(\partial G/\partial T)_P = -1.5(2) \times 10^{-2}$ GPaK^{-1} for the hydrous wadsleyite (0.73 wt.% H_2O) from the third-order finite strain analysis overlaps within the mutual data uncertainty with the $(\partial G/\partial T)_P = -1.7(1) \times 10^{-2}$ GPaK^{-1} for the anhydrous β-Mg_2SiO_4 in this study. Moreover, the current data set is in agreement with the published ultrasonic and resonance ultrasound spectroscopy $(\partial G/\partial T)_P$ measurements for anhydrous β-Mg_2SiO_4 [18,19,21], and the Fe-bearing wadsleyite [20,22,24] measurements, within the mutual data uncertainties. The resonance spectroscopy results of $(\partial G/\partial T)_P = -1.2(1) \times 10^{-2}$ GPaK^{-1} for anhydrous $Mg_{(0.91)}Fe_{(0.09)}$ wadsleyite [23] is the only data that falls outside the tight band of all current $(\partial G/\partial T)_P$ measurements. Thus, whereas the effect on $\partial K_s/\partial T)_P$ due to Fe or OH substitution in wadsleyite is still variable, the effect on $(\partial G/\partial T)_P$ is only minimal.

As observed in Table 2, our $(\partial K_s/\partial T)_P$ and $(\partial G/\partial T)_P$ data sets from the linear fitting and third-order finite strain analysis are consistent, and each shows $|(\partial G/\partial T)_P|$ to be moderately higher than $|(\partial K_S/\partial T)_P|$. We are currently analyzing elasticity data for wadsleyite containing 0.26 wt.%, 0.53 wt.%, and 1.00 wt.% H_2O, respectively, and the preliminary results seem to corroborate the current observation of the $|(\partial G/\partial T)_P|$ being higher than $|(\partial K_s/\partial T)_P|$ for the 0.26 wt.% and 0.53 wt.% wadsleyite samples. Only the $(\partial K_s/\partial T)_P$ and $(\partial G/\partial T)_P$ for the 1.00 wt.% H_2O wadsleyite shows $|(\partial K_s/\partial T)_P| > |(\partial G/\partial T)_P|$, as also observed for the anhydrous β-Mg_2SiO_4 in this study, as well as many previous studies of Mg- and Mg-Fe wadsleyite [21–24]. The $(\partial K_s/\partial T)_P$ and $(\partial G/\partial T)_P$ of −0.164(5) GPa/K and −0.130(3) GPa/K, respectively, from the ultrasonic measurements of San Carlos olivine to 8 GPa and 1073 K [19] exhibit similar patterns to those of most mantle phases where $|(\partial K_s/\partial T)_P|$ is greater than $|(\partial G/\partial T)_P|$. Apart from our current data for the hydrous wadsleyite, only two previous ultrasonic high P and T studies on anhydrous β-Mg_2SiO_4 [18,19] and $Mg_{(0.87)}Fe_{(0.13)}$ wadsleyite [20] also observe $|(\partial G/\partial T)_P|$ to be greater $|(\partial K_S/\partial T)_P|$. It is noteworthy that the anhydrous β-Mg_2SiO_4 data [18,19] are not two separate studies. Whereas the earlier study [18] used the linear function to fit the data, the latter [19] reanalyzed the data using the finite strain method. A previous study [27] has remarked on the difficulty of synthesizing a nominally anhydrous wadsleyite without a hydroxyl; despite the care taken, the sample still contained at least 50 ppm wt.% H_20. The temperature derivative results for the β-Mg_2SiO_4 [18,19] and the $Mg_{(0.87)}Fe_{(0.13)}$ wadsleyite [20] could reflect minor H_2O in the samples if the starting materials were completely dry before the hot-pressing. A detailed examination of the β-Mg_2SiO_4 and [18,19] and $Mg_{(0.87)}Fe_{(0.13)}$ wadsleyite [20] samples would have been useful in explaining the $(\partial K_s/\partial T)_P$ and $(\partial G/\partial T)_P$ observations of the studies. Nonetheless, based on the current $(\partial K_s/\partial T)_P$ and $(\partial G/\partial T)_P$ data for the hydrous wadsleyite (0.73 wt.% H_2O) and the preliminary data for wadsleyite containing 0.26 wt.% and 0.53 wt.%, we infer that the observation showing the bulk modulus ($(\partial K_s/\partial T)_P$) being affected more than shear modulus ($(\partial G/\partial T)_P$) by temperature could be an intrinsic structural property of the wadsleyite hydration for relatively small OH content up to 0.73 wt.% H_2O and if so, could have important implications for evaluating the orthosilicate content of the Earth's mantle.

We assumed orthorhombic symmetry in the X-ray analysis of all the wadsleyite specimens of this study. However, some previous studies [52,53] observed a monoclinic distortion of hydrous wadsleyite. The monoclinic distortion is associated with cation vacancy stacking disorder of distinct modules [52] or poly-synthetic twinning [3]. Another study [7] has found only samples with water content greater than 0.5 wt.% to display the monoclinic symmetry. Further studies of the hydrous wadsleyite symmetry and cause would enhance understanding of the temperature derivatives elastic moduli phenomena of wadsleyite observed in this study.

Examination of all current T-derivatives data shows that the effect of Fe substitution or OH incorporation in wadsleyite on $(\partial G/\partial T)_P$ is immeasurable. Previous resonance ultrasound spectroscopy measurements of anhydrous β-Mg_2SiO_4 [21] concluded that there is no measurable difference in

$(\partial K_S/\partial T)_P$ and $(\partial G/\partial T)_P$ due to Fe substitution in the anhydrous wadsleyite up to X_{Fe} = 0.09 from observation of agreement of reported $(\partial K_S/\partial T)_P$ and $(\partial G/\partial T)_P$ for Mg and Mg-Fe wadsleyite [18–21]. By including our hydrous wadsleyite data, we find the conclusion for $(\partial G/\partial T)_P$ to be strengthened, but less robust for the current $(\partial K_S/\partial T)_P$ measurements.

The worldwide 410-km seismic velocity discontinuities in the Earth's upper mantle are generally attributed to the phase transition of the olivine (α) to the wadsleyite (β) phase. Previous studies have estimated the olivine volume of the mantle by comparing the seismic P and S wave velocity contrasts (ΔV_p; ΔV_s) across the 410-km discontinuity caused by the phase change, with the corresponding laboratory velocity contrast between the olivine (α) and wadsleyite (β) phases [15,18–20,53].

We have applied our new elasticity data for the hydrous wadsleyite (0.73 wt.% H_2O) and adapted the analytical approach of previous studies [18–20], to determine the P and S wave velocity contrasts between the α and β phases at the 410-km P and T conditions. We use the third-order finite strain approach of previous studies [18–20] to calculate the P and S wave velocities of the hydrous wadsleyite from this study, the anhydrous α-$Mg_{(0.9Fe0.1)_2}SiO_4$ [19] and other mantle phases at the 410-km discontinuity. We applied the iron-partitioning data between the α- and the β-phases [54] and performed the calculations to 14 GPa along 1673 K-foot adiabat. We calculated the elasticity of the pyrolite lithology using the Voight-Reuss-Hill (VRH) averaging scheme. In addition to the anhydrous α-$Mg(_{0.9Fe0.1})_2SiO_4$, we also calculate the P and S wave velocities at 14 GPa using the Brillouin scattering elasticity data for hydrous olivine containing 0.8–0.9 wt.% H_2O measured at room P and T [55] and room T and 14 GPa [56]. For the calculations, we follow the same approach to correct for the effect of Fe on the elastic bulk (*K*) and shear (*G*) moduli and densities of the α and β phases, as reported in the previous ultrasonic analysis [18–20] used to evaluate the olivine content of the mantle.

The velocity contrasts from finite strain analysis of the hydrous β-Mg_2SiO_4 data from this study, and the hydrous α-Mg_2SiO_4 [55,56] data are $\Delta V_{p(\alpha\text{-}\beta)}$ = 9.88% and $\Delta V_{S(\alpha\text{-}\beta)}$ = 8.70%, implying 46–52% olivine volume content for the isochemical Earth's mantle when compared with the average regional body wave seismic velocity contrasts $\Delta V_P = \Delta V_S$ = 4.6% [57–60]. In contrast, comparing the P and S wave velocity contrasts between our new hydrous wadsleyite and the anhydrous α-$Mg(_{0.9Fe0.1})_2SiO_4$ [19] yields $\Delta V_{p(\alpha\text{-}\beta)}$ = 13.0% and $\Delta V_{S(\alpha\text{-}\beta)}$ = 14.70%, corresponding to a significantly lower olivine volume content (39.4–36%) for the mantle. The velocity contrasts for the α-β phase transition investigated using the thermoelasticity data β-$Mg(_{0.885Fe0.13})_2SiO_4$ and α-$Mg(_{0.9Fe0.1})_2SiO_4$ [19] yielded $\Delta V_{p(\alpha\text{-}\beta)}$ = 10.9% and $\Delta V_{S(\alpha\text{-}\beta)}$ = 12.2%, indicating that 38–56% olivine to satisfy the seismic discontinuity of 4.9% for the P wave and 4.6% for the S wave. A previous study [55] observes that water incorporation in the olivine decreases the velocity contrast between the olivine and wadsleyite phases at the 410-km depth, causing an increase in the predicted olivine volume content for the mantle.

The range of the velocity contracts across the 410-km discontinuity from body wave studies is 3.8% to 4.9% for P waves and 4% to 5% for S waves [61]. We obtain a higher olivine volume content (50%) for the P wave that is close to the S-wave value (52%) by adopting the seismic velocity contrasts (ΔV_P = 4.9%; ΔV_S = 4.6%) used by Liu and co-workers [20]. A comparison of the derived olivine volume content from the anhydrous α-$Mg(_{0.9Fe0.1})_2SiO_4$ [19] and the hydrous olivine [55,56] underscores the significance of our new data on the T-derivatives of the elastic moduli for the hydrous wadsleyite, as well as the importance of extending the measurements to the hydrous olivine and hydrous ringwoodite phases of $(Mg, Fe)_2SiO_4$.

Figure 9 is an adaptation of Figure 3 from the previous ultrasonic measurements of the elastic moduli of β-$(Mg)_2SiO_4$ carried in a liquid pressure piston-cylinder apparatus to 3 GPa at room temperature [15]. The figure displays the trade-off of $|(\partial G/\partial T)_P|$ and $|(\partial K_S/\partial T)_P|$ for the wadsleyite phase compatible with the orthosilicate content of an iso-chemical mantle model, calculated along a 1575 K-foot adiabat; the pressure dependence of the elastic moduli K'_s = 4.8(2) and G_o' = 1.7(1) used in the calculations are identical, within the mutual uncertainties of the data with values for the hydrous β-$(Mg)_2SiO_4$, with 0.73 wt.% H_2O and the anhydrous β-$(Mg)_2SiO_4$ measured in this study. The study assumes that the seismic velocity contrasts $\Delta V_p = \Delta V_s$ = 4.6% at 410-km in the Earth's mantle, as reported for regional seismic Earth model studies [57–60]. The calculations predicted that the values of $(\partial K_S/\partial T)_P$ and $(\partial G/\partial T)_P$ of the orthosilicate content for an iso-chemical model mantle would be those with $|(\partial G/\partial T)_P|$ being greater than $|(\partial K_S/\partial T)_P|$. We note that the predicted orthosilicate content is much less sensitive to temperature change than to uncertainties in the seismic velocity contrasts, as shown in Figure 9.

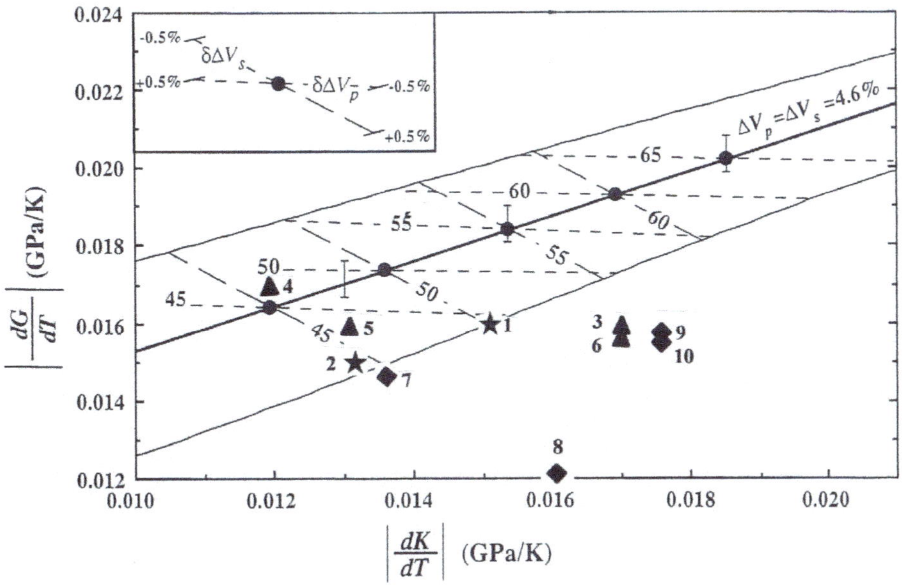

Figure 9. Absolute temperature derivatives for the elastic bulk (K) and shear (G) moduli for Mg_2SiO_4 and (Mg, Fe)$_2$SiO$_4$ wadsleyite and inferred composition (orthosilicate content) of the mantle adapted from Figure 3 [15]. The heavy solid line represents calculations at a potential temperature of 1300 °C with a seismic velocity contrast for both P and S waves of 4.6% at the 410 km discontinuity. Vertical bars show the effect of changing the temperature by ±100 °C. The outlying solid lines represent the effect of a ±0.5% change in ΔV_P or ΔV_S, as indicated in the insert. The lines labeled orthosilicate content (percent) correspond to $(\delta\Delta V_P) = \pm 0.5\%$ at constant ΔV_S (lines with short dashes with slope ~ 0) and to $(\delta\Delta V_S) = \pm 0.5\%$ at constant ΔV_P (lines with long dashes with slope ~−3/4). Also shown are measured values of these parameters; star, from this study, for hydrous β-(Mg)$_2$SiO$_4$, (0.73 wt.% H_2O); 1. the linear fit results to data (Equation (3)); 2. The finite strain result; triangle, anhydrous β-Mg_2SiO_4; 3. from this study; 4. [18]; 5. [19]; 6. [21]; diamond, anhydrous (Mg,Fe)$_2$SiO$_4$, 7. [20]; 8. [23]; 9. [24]; 10. [22].

In Figure 9, we plot the data for $|(\partial G/\partial T)_P|$ and $|(\partial K_S/\partial T)_P|$ obtained from the finite strain analysis (1) and the linear fit to the entire data set (2) from this study (star). For comparison, we also plot our new data for the anhydrous β-(Mg)$_2$SiO$_4$, (3) as well as high T data for anhydrous Mg- and Mg-Fe-bearing wadsleyite from previous studies [18–24]. The triangle and the diamonds are the Mg end-member and the Fe-bearing wadsleyite data, respectively. Our two new data sets predict about 45–50% orthosilicate content for the mantle, in excellent agreement with 46–52% from comparing the seismic velocity contrasts ($\Delta V_p = \Delta V_s = 4.6\%$) with the corresponding contrasts between the α and the β phases ($\Delta V_{p(\alpha\text{-}\beta)} = 9.88\%$; $\Delta V_{S(\alpha\text{-}\beta)} = 8.70\%$) for hydrated α-(Mg)$_2SiO_4$ and hydrated β-(Mg)$_2$SiO$_4$ at the 410-km depth obtained from the finite strain analysis. All current high-temperature data on anhydrous Mg and Mg-Fe wadsleyite, including our new β-(Mg)$_2$SiO$_4$, data (3) lead to velocity contrasts with ratio $\Delta V_p/\Delta V_s$ significantly different from unity for which there is no compelling seismological evidence [15].

The two ultrasonic data for anhydrous β-$(Mg)_2SiO_4$ [18,19], and anhydrous $Mg_{(0.87)}Fe_{(0.13)}$ wadsleyite [20] also fall within the predicted range of preferred orthosilicate content of the Earth's mantle. However, as noted above, given that there is no information on the sample characterization used in the study, and given that our anhydrous β-$(Mg)_2SiO_4$ data from this study also lie out of the range, we presume that the wadsleyite sample used in the study might have contained some structural water.

Figure 9 shows that the inferred $|(\partial G/\partial T)_P|$ and $|(\partial K_S/\partial T)_P|$ values for the wadsleyite phase consistent with the isochemical model mantle are those for which $|(\partial G/\partial T)_P|$ is higher than $|(\partial K_S/\partial T)_P|$. It also indicates that as the temperature derivatives increase along the curve, the inferred olivine volume content of the model mantle also increases from ~40% to 70%. The inferred olivine content is only minimally affected by changing the temperature at the zero-pressure of the appropriate adiabat but strongly influenced by perturbating the seismic relative velocity contrast ($\Delta V_P/V_P \neq \Delta V_S/V_S$). For a given K and G of the olivine phase, the decrease in K and G of the wadsleyite caused by hydration, as observed in this study, leads to increases in the inferred olivine content of the mantle, while temperature derivatives decrease the content. As noted above, preliminary data for wadsleyite containing 0.26 wt.% and 0.53 wt.% H_2O, show $|(\partial G/\partial T)_P|$ to be higher than $|(\partial K_s/\partial T)_P|$. Additionally, we also observe that the $(\partial G/\partial T)_P$ and $(\partial K_S/\partial T)_P$ values move towards higher olivine content with decreasing water OH content in the wadsleyite for constant pressure derivatives of the elastic moduli. Completion of the data analysis on the current OH-wadsleyite samples and any necessary additional elasticity measurements of hydrous wadsleyite could reveal the trend of the mantle olivine content variation with the elastic properties of wadsleyite.

4. Conclusions

We measured the compressional (P) and shear (S) wave velocities for a synthetic polycrystalline wadsleyite containing 0.73 wt.% H_2O to 10 GPa and 600 K by ultrasonic interferometry techniques combined with synchrotron X-radiation. The adiabatic bulk (Ks) and shear (G) moduli and their pressure and temperature derivatives are determined using the third-order finite strain Equation of state and also by fitting (K_s) and (G) to linear combinations of pressure and temperature. The results for the elastic bulk (K_s) and shear (G) moduli and their pressure derivatives (K_{so} = 161.5(2) GPa, G_o = 101.6(1) GPa, and $(\partial K_s/\partial P)_T$ = 4.84(4), $(\partial G/\partial P)_T$ = 1.68(2)), from the finite strain analysis are indistinguishable from (K_{so} = 161.1(3) GPa, G_o = 101.4(1) GPa, and $(\partial K_s/\partial P)_T$ = 4.93(4), $(\partial G/\partial P)_T$ = 1.73(2)) from the linear fit. Hydration of wadsleyites by 0.73 wt.% H_2O leads to a decrease in the (K_s) and (G) moduli of 5.3% and 8.6%, respectively, but does not significantly affect $(\partial K_s/\partial P)_T$ and $(\partial G/\partial P)_T$. Previous Brillouin scattering measurements of wadsleyite containing 0.84 wt.% H_2O also concluded that $(\partial K_s/\partial P)_T$ and $(\partial G/\partial P)_T$ are independent of the wadsleyite hydration. The temperature derivatives of the (K_s) and (G) moduli from the finite strain and the linear analysis yielded $(\partial K_S/\partial T)_P$ = −0.013(2) $GPaK^{-1}$, $(\partial G/\partial T)_P$ = −0.015(0.4) $GPaK^{-1}$, and $(\partial K_S/\partial T)_P$ = −0.015(1) $GPaK^{-1}$, $(\partial G/\partial T)_P$ = −0.016(1) $GPaK^{-1}$, respectively. The two data sets are consistent, and both also reveal the magnitude of the temperature derivative of the shear modulus to be greater than of the bulk modulus {$|(\partial G/\partial T)_P| > |(\partial K/\partial T)_P|$}, in contrast with data for anhydrous β-Mg_2SiO_4 from this study, as well as most previous high T studies of anhydrous Mg- and Mg-Fe-bearing wadsleyite. The new observation on $(\partial K_S/\partial T)_P$ and $(\partial G/\partial T)_P$ has implications for assessing the olivine volume content of the mantle.

Comparison of the body wave seismic velocity contrasts ($\Delta V_s = \Delta V_P = 4.6\%$) with the corresponding contrasts between the hydrous β-Mg_2SiO_4 phase from this study and the hydrous α-Mg_2SiO_4 ($\Delta V_{p(\alpha\text{-}\beta)}$ = 9.88% and $\Delta V_{S(\alpha\text{-}\beta)}$ = 8.70%) calculated for the 410-km depth by third-order finite strain technique yields 46–52% volume content in the Earth's mantle. However, in comparison, the contrast for the hydrous β-Mg_2SiO_4 and anhydrous α-$Mg_{(0.9}Fe_{0.1)}{}_2SiO_4$ [19] yields $\Delta V_{p(\alpha\text{-}\beta)}$ = 13.0% and $\Delta V_{S(\alpha\text{-}\beta)}$ = 14.70%, corresponding to a significantly lower olivine volume content (39.4–36%) for the mantle. The combined effect of OH and Fe on the high P and T elasticity of the wadsleyite and olivine phases need to be understood to improve the assessment of the olivine volume content of the Earth's mantle.

Author Contributions: Conceptualization, G.D.G.; Methodology, G.D.G. and M.L.W.; Software, M.L.W.; Validation, G.D.G., M.L.W., and A.J.; Formal Analysis, G.D.G., M.L.W., A.J, L.D., and N.C.; Investigation, G.D.G., M.L.W., R.S.T., L.D., and A.J.; Resources, H.C., R.S.T., and N.C.; Writing—original draft preparation, G.D.G.; Writing—review and editing, G.D.G., N.C., A.J., M.L.W., L.D., H.C., and R.S.T.; Project administration, G.D.G., M.L.W., and H.C. All authors have read and agreed to the published version of the manuscript.

Funding: The National Science Foundation supported this research under grant EAR-1417024 to GDG. Use of the Advanced Photon Source, Argonne National Laboratory, was supported by the U.S. Department of Energy, Office of Science, Office of Basic Research, under contract No. DE-AC02-06CH11357. The use of the 6-BM-B beamline was supported by the Consortium for Materials Properties Research in Earth Sciences (COMPRES) under NSF cooperative agreement No. EAR-01-35554. COMPRES supported the cell assemblies under cooperative agreement No. EAR-1661511.

Acknowledgments: Gabriel Gwanmesia dedicates this manuscript to O.L Anderson, who served as the external member of his Ph.D. committee at Stony Brook University in 1991 when he defended his thesis on the high-pressure elasticity of wadsleyite and ringwoodite, for avocating his work and for the eminent and estimable academic "*grandfather*" that he exemplified. We thank Jim Quinn, Director of Laboratories, in the Materials Science and Chemical Engineering laboratory at Stony Brook University for the SEM examination of the samples. We thank Baosheng Li for the use of the Stony Brook High-Pressure facility for the sample synthesis. This is Mineral Physics Institute Publication No. 511.

Conflicts of Interest: The authors declare no conflicts of interest.

References

1. Inoue, T.; Yurimoto, H.; Kudoh, Y. Hydrous modified spinel, $Mg_{1.75}SiH_{0.5}O_4$: A new water reservoir in the mantle transition region. *Geophys. Res. Lett.* **1995**, *22*, 117–120. [CrossRef]
2. Kohlstedt, D.L.; Keppler, H.; Rubie, D.C. The solubility of water in α, β, and γ phases of $(Mg, Fe)_2SiO_4$. *Contrib. Mineral. Petrol.* **1996**, *123*, 345–357. [CrossRef]
3. Kudoh, Y.; Inoue, T. Mg-vacant structural modules and dilution of the symmetry of hydrous wadsleyite, β-$Mg_{2-X}SiH_{2X}$ with $0.00 \leq X \leq 0.25$. *Phys. Chem. Miner.* **1999**, *28*, 232–241.
4. Smyth, J.R. β-Mg_2SiO_4: A potential host for water in the mantle? *Am. Mineral.* **1987**, *72*, 1051–1055.
5. Deon, F.; Koch-Muller, M.; Rhede, D.; Gottschalk, M.; Wirth, R.; Thomas, S.M. Location and quantification of hydroxyl in wadsleyite: New insights. *Am. Mineral.* **2010**, *95*, 312–322. [CrossRef]
6. Holl, C.M.; Smyth, J.R.; Jacobsen, S.D.; Frost, D.J. Effects of hydration on the structure and compressibility of wadsleyite, β-(Mg_2SiO_4). *Am. Mineral.* **2008**, *93*, 598–607. [CrossRef]
7. Jacobsen, S.D.; Demouchy, S.; Frost, D.J.; Boffa-Ballaran, T.; Kung, J. A systematic study of OH in hydrous wadsleyite from polarized FTIR spectroscopy and single-crystal X-ray diffraction: Oxygen sites for hydrogen storage in the Earth's interior. *Am. Mineral.* **2005**, *90*, 61–70. [CrossRef]
8. Kudoh, Y.; Inoue, T.; Arashi, H. Structure and crystal chemistry of hydrous wadsleyite, $Mg_{1.75}SiHO_{0.5}O_4$: Possible hydrous magnesium silicate in the mantle transition zone. *Phys. Chem. Miner.* **1996**, *23*, 461–469. [CrossRef]
9. Smyth, J.R. A crystallographic model of hydrous wadsleyite (β-Mg_2SiO_4): An ocean in the Earth's interior? *Am. Mineral.* **1994**, *79*, 1021–1024.
10. Smyth, J.R.; Kawamoto, T. Wadsleyite II: A new high-pressure hydrous phase in the peridotite-H_2O system. *Earth Planet. Sci. Lett.* **1997**, *146*, E9–E16. [CrossRef]
11. Ye, Y.; Brown, D.A.; Smyth, J.R.; Panero, W.R.; Jacobson, S.D.; Chang, Y.Y. Compressibility and thermal expansion of hydrous ringwoodite with 2,5(3) wt.% H_2O. *Am. Mineral.* **2012**, *97*, 573–582. [CrossRef]
12. Bercovici, D.; Karato, S. Whole-mantle convection, and the transition zone water filter. *Nature* **2003**, *425*, 39–44. [CrossRef] [PubMed]
13. Pearson, D.G.; Brenker, F.E.; Nestola, F.; Mcneill, J.; Nasdala, L.; Hutchison, M.T.; Matveev, S.; Mather, K.; Silversmit, G.; Schmitz, S.; et al. Hydrous mantle transition zone indicated by ringwoodite included within diamond. *Nature* **2014**, *507*, 221–224. [CrossRef] [PubMed]
14. Khan, A.; Shankland, T.J. A geophysical perspective on mantle water content and melting: Inverting electromagnetic sounding data using laboratory-based electrical conductivity profiles. *Earth Planet. Sci. Lett.* **2012**, *317*, 27–43. [CrossRef]
15. Gwanmesia, G.D.; Rigden, S.; Jackson, I.; Liebermann, R.C. Pressure dependence of elastic wave velocity for β-Mg_2SiO_4 and the composition of the Earth's mantle. *Science* **1990**, *250*, 794–797. [CrossRef] [PubMed]

16. Li, B.; Gwanmesia, G.D.; Liebermann, R.C. Sound velocities of olivine and beta polymorphs of Mg_2SiO_4 at Earth's transition zone pressures. *Geophys. Res. Lett.* **1996**, *23*, 2259–2262. [CrossRef]
17. Zha, C.S.; Duffy, T.S.; Mao, H.K.; Downs, R.T.; Hemley, R.J.; Weidner, D.J. Single-crystal elasticity of beta-Mg_2SiO_4 to the pressure of the 410 km seismic discontinuity in the Earth's mantle. *Earth Planet. Sci. Lett.* **1997**, *147*, E9–E15. [CrossRef]
18. Li, B.; Liebermann, R.C.; Weidner, D.J. P-V-V_P-V_S-T measurements on wadsleyite to 7 GPa and 873 K: Implications for the 410-km seismic discontinuity. *J. Geophys. Res.* **2001**, *106*, 30579–30591. [CrossRef]
19. Liu, W.; Kung, J.; Li, B. Elasticity of San Carlos olivine to 8 GPa and 1073 K. *Geophys. Res. Lett.* **2005**, *32*, L16301. [CrossRef]
20. Liu, W.; Kung, J.; Li, B.; Nishiyama, N.; Wang, Y. Elasticity of $(Mg_{0.87}Fe_{0.13})_2SiO_4$ wadsleyite to 12 GPa and 1073 K. *Phys. Earth Planet. Int.* **2009**, *174*, 98–104. [CrossRef]
21. Isaak, D.G.; Gwanmesia, G.D.; Falde, D.; Davis, M.G.; Triplett, R.S.; Wang, L. The elastic properties of β-Mg_2SiO_4 from 295 to 660 K and implications for the composition of the Earth's upper mantle. *Phys. Earth Planet. Int.* **2007**, *162*, 22–31. [CrossRef]
22. Isaak, D.G.; Gwanmesia, G.D.; Davis, M.G.; Stafford, C.S.; Stafford, A.M.; Triplett, R.S. The temperature dependence of Fe-bearing wadsleyite. *Phys. Earth Planet. Int.* **2010**, *182*, 107–112. [CrossRef]
23. Katsura, T.; Mayama, N.; Shouno, K.; Sakai, M.; Yoneda, A.; Suzuki, I. Temperature derivatives of the elastic moduli of $(Mg_{0.91}Fe_{0.09})_2SiO_4$ modified spinel. *Phys. Earth Planet. Inter.* **2001**, *124*, 163–166. [CrossRef]
24. Mayama, N.; Suzuki, I.; Saito, T. Temperature dependence of elastic moduli of β-$(Mg\text{-}Fe)_2SiO_4$. *Geophys. Res. Letts.* **2004**, *31*, L019247. [CrossRef]
25. Yusa, H.; Inoue, T. Compressibility of hydrous wadsleyite (β-phase) in Mg_2SiO_4 by high-pressure x-ray diffraction. *Geophys. Res. Lett.* **1997**, *24*, 1831–1834. [CrossRef]
26. Yagi, T.; Uchiyama, Y.; Akaogi, M.; Ito, E. Isothermal compression curve of $MgSiO_3$ tetragonal garnet. *Phys Earth Planet Int.* **1992**, *74*, 1–7. [CrossRef]
27. Mao, Z.; Jacobsen, S.D.; Jiang, F.; Smyth, J.R.; Holl, C.; Frost, D.J.; Duffy, T.S. Single-crystal elasticity of wadsleyites, β-Mg_2SiO_4, containing 0.37–1.66 wt.% H_2O. *Earth Planet. Sci. Lett.* **2008**, *268*, 540–549. [CrossRef]
28. Mao, Z.; Jacobsen, S.D.; Jiang, F.; Smyth, J.R.; Holl, C.M.; Duffy, T.S. Elasticity of hydrous wadsleyite to 12 GPa: Implications for Earth's transition zone. *Geophys. Res. Lett.* **2008**, *35*, L21305. [CrossRef]
29. Buchen, J.; Marquardt, H.; Speziale, S.; Kwazoe, T.; Ballaran, T.B.; Kurnosov, A. High-pressure single-crystal elasticity of wadsleyite and seismic signature of water in the shallow transition zone. *Earth Planet. Sci. Lett.* **2018**, *498*, 77–87. [CrossRef]
30. Mao, Z.; Jacobsen, S.D.; Frost, D.J.; McCammon, C.A.; Hauri, E.H.; Duffy, T.S. Effect of Hydration on the single-crystal elasticity of Fe-bearing wadsleyite to 12 GPa. *Am. Mineral.* **2011**, *96*, 1606–1612. [CrossRef]
31. Kawai, N.; Endo, S. The generation of ultrahigh hydrostatic pressures by a split-sphere apparatus. *Rev. Sci. Instrum.* **1970**, *41*, 1178–1181. [CrossRef]
32. Kawai, N.; Togaya, M.; Onodera, A. A new device for pressure vessels. *Proc. Japan Acad.* **1973**, *49*, 623–626. [CrossRef]
33. Weidner, D.J.; Wang, Y.; Vaughan, M.T.; Ko, J.; Liu, X.; Yeganeh-Haeri, A.; Pacalo, R.E.; Zhao, Y. Characterization of stress, pressure, and temperature in SAM-85, a DIA-type high-pressure apparatus. In *High-Pressure Research: Application to the Earth and Planetary Sciences*; Syono, Y., Manghnani, M.H., Eds.; TERRA PUB, Tokyo/American Geophysical Union: Washington, DC, USA, 1992; pp. 13–17.
34. Gwanmesia, G.D.; Liebermann, R.C.; Guyot, F. Hot-pressing and characterization of polycrystals of β-Mg_2SiO_4 for acoustic velocity measurements. *Geophys. Res. Lett.* **1990**, *17*, 1331–1334. [CrossRef]
35. Gwanmesia, G.D.; Li, B.; Liebermann, R.C. Polycrystals of high-pressure phases of mantle minerals: Hot pressing and characterization of physical properties. In *High-pressure Research: Application to Earth and Planetary Sciences*; Syono, Y., Manghnani, M., Eds.; American Geophysical Union: Washington, DC, USA, 1992; Volume 117, p. 135.
36. Gwanmesia, G.D.; Li, B.; Liebermann, R.C. Hot-pressing of polycrystals of high-pressure phases of mantle minerals in multi anvil apparatus. *PAGEOPH* **1993**, *141*, 467–484. [CrossRef]
37. Dai, L.D.; Hu, H.Y.; Li, H.P.; Wu, L.; Hui, K.S.; Jiang, J.J.; Sun, W.Q. Influence of temperature, pressure, and oxygen fugacity on the electrical conductivity of dry eclogite, and geophysical implications. *Geochem. Geophys. Geosyst.* **2016**, *17*, 2394–2407. [CrossRef]

38. Hu, H.Y.; Dai, L.D.; Li, H.P.; Hui, K.S.; Sun, W.Q. Influence of dehydration on the electrical conductivity of epidote and implications for high conductivity anomalies in subduction zones. *J. Geophys. Res.* **2017**, *122*, 2751–2762. [CrossRef]
39. Paterson, M.S. The determination of hydroxyl by infrared absorption in quartz, silicate glass, and similar materials. *Bull Mineral.* **1982**, *105*, 20–29.
40. Whitaker, M.L.; Baldwin, K.J.; Huebsch, W.R. DIASCoPE: Directly integrated acoustic system combined with pressure experiments- A new method for fast acoustic velocity measurements at high pressure. *Rev. Sci. Instr.* **2017**, *88*, 034901. [CrossRef]
41. Weidner, D.J.; Wang, Y.; Vaughan, M.T. Yield strength at high pressure and temperature. *Geophys. Res. Lett.* **1994**, *21*, 753–756. [CrossRef]
42. Liebermann, R.C.; Chen, G.; Li, B.; Gwanmesia, G.D.; Chen, J.; Vaughan, M.T.; Weidner, D.J. Sound velocity measurements in oxides and silicates at simultaneous high pressures and temperatures using ultrasonic techniques in multi-anvil apparatus in conjunction with synchrotron X-radiation determination of the equation of state. *Rev. High-Press. Sci. Tech.* **1998**, *7*, 75–78. [CrossRef]
43. Li, B.; Chen, K.; Kung, J.; Liebermann, R.C.; Weidner, D.J. Ultrasonic measurement using the transfer function method. *J. Phys. Condens. Matter.* **2002**, *14*, 11337–11342.
44. Kung, J.; Li, B.; Weidner, D.J.; Zhang, J.; Liebermann, R.C. Elasticity of ($Mg_{0.83}$, $Fe_{0.17}$)O ferropericlase at high pressure: Ultrasonic measurements in conjunction with X-radiation techniques. *Earth Planet. Sci. Lett.* **2002**, *203*, 227–566. [CrossRef]
45. Li, B.; Kung, J.; Liebermann, R.C. Modern techniques in measuring elasticity of earth materials at high pressure and high temperature using ultrasonic interferometry in conjunction with synchrotron X-radiation in multi-anvil apparatus. *Phys. Earth Planet Inter.* **2004**, *143–144*, 559–574. [CrossRef]
46. Gwanmesia, G.D.; Zhang, J.; Darling, K.; Kung, J.; Li, B.; Wang, L.; Neuville, D.; Liebermann, R.C. Elasticity of polycrystalline pyrope ($Mg_3Al_2Si_3O_{12}$) to 9 GPa and 1000 °C. *Phys. Earth Planet. Int.* **2006**, *155*, 179–190. [CrossRef]
47. Gwanmesia, G.D.; Wang, W.; Heady, A.; Liebermann, R.C. Elasticity and sound velocities of polycrystalline garnet ($Ca_3Al_2Si_3O_{12}$) at simultaneous high pressures and high temperatures. *Phys. Earth Planet. Int.* **2014**, *228*, 80–87. [CrossRef]
48. Niesler, H.; Jackson, I. Pressure derivatives of elastic wave velocities from ultrasonic interferometric measurements on jacketed polycrystals. *J. Acoust. Soc. Am.* **1989**, *86*, 1573–1585. [CrossRef]
49. Decker, D.L. High-pressure equation of state for NaCl, KCl, and CsCl. *J. Appl. Phys.* **1971**, *42*, 3244–3329. [CrossRef]
50. Sawamoto, H.; Weidner, D.J.; Sasaki, S.; Kumazawa, M. Single-crystal elastic properties of the modified-spinel (beta) phase of magnesium orthosilicate. *Science* **1984**, *224*, 749–751. [CrossRef]
51. Nunez-Valdez, M.; Wu, Z.; Yu, Y.G.; Wentzcovitch, R.M. Thermal elasticity of $(Fe_xMg_{1-x})_2SiO_4$ olivine and wadsleyite. *Geophys. Res. Lett.* **2013**, *40*, 290–294.
52. Smyth, J.R.; Kawamoto, T.; Jacobsen, S.D.; Swope, R.J.; Hervig, R.J.; Hollway, J.R. Crystal structure of monoclinic hydrous wadsleyite [β-$(Mg\text{-}Fe)_2SiO_4$]. *Am. Mineral.* **1997**, *82*, 270–275. [CrossRef]
53. Duffy, T.S.; Anderson, D.L. Seismic velocities in mantle minerals and the mineralogy of the upper mantle. *J. Geophys. Res.* **1989**, *94*, 1895–1912. [CrossRef]
54. Irifune, T.; Isshiki, M. Iron-partitioning in a pyrolite mantle and nature of the 410-km seismic discontinuity. *Nature* **1998**, *92*, 702–705. [CrossRef]
55. Jacobsen, S.D.; Jiang, F.; Mao, Z.; Duffy, S.D.; Smyth, J.R.; Holl, C.M.; Frost, D.J. Effects of hydration on the elastic properties of olivine. *Geophys. Res. Lett.* **2008**, *35*, L14303. [CrossRef]
56. Mao, Z.; Jacobsen, S.D.; Jiang, F.; Smyth, J.R.; Holl, C.M.; Frost, D.J. Velocity crossover between hydrous and anhydrous forsterite at high pressures. *Earth Planet. Sci. Lett.* **2010**, *293*, 250–258. [CrossRef]
57. Walck, M.C. The P-wave upper mantle structure beneath an active spreading center: The Gulf of California. *Geophys. J. R. Astr. Soc.* **1984**, *76*, 697–723. [CrossRef]
58. Grand, S.; Helmberger, D. Upper mantle shear structure of North America. *Geophys. J. R. Astr. Soc.* **1984**, *76*, 339–438. [CrossRef]
59. Le Fevre, L.V.; Helmberger, D.V. Upper mantle P velocity structure of the Canadian Shield. *J Geophys. Res.* **1989**, *94*, 17729–17765.

60. Nolet, G.; Grand, S.P.; Kennett, B.L.N. Seismic heterogeneity in the upper mantle. *J. Geophys. Res.* **1994**, *99*, 23753–23766. [CrossRef]
61. Nolet, G.; Wortel, M.J.R. Encyclopedia of Geophysics. James, D.A., Ed.; Van Nostrand Reinhold: New York, NY, USA, 1989; p. 775.

minerals

MDPI

Article

Thermal Analysis, Compressibility, and Decomposition of Synthetic Bastnäsite-(La) to Lanthanum Oxyfluoride

Richard L. Rowland II [1,*], Barbara Lavina [2,†], Kathleen E. Vander Kaaden [3], Lisa R. Danielson [4] and Pamela C. Burnley [5,*]

1 Static High Pressure Team, Explosive Science and Shock Physics (M-9), Los Alamos National Laboratory, Los Alamos, NM 87544, USA
2 High Pressure Science and Engineering Center, University of Nevada Las Vegas, Las Vegas, NV 89154, USA; Blavina@anl.gov
3 Astromaterials Research and Exploration Science Division (ARES), Jacobs, NASA Johnson Space Center, Mail Code XI3, Houston, TX 77058, USA; Kathleen.E.VanderKaaden@NASA.Gov
4 Space and Remote Sensing (ISR-2), Los Alamos National Laboratory, Los Alamos, NM 87544, USA; LDanielson@LANL.Gov
5 High Pressure Science and Engineering Center and Geoscience Department, University of Nevada Las Vegas, Las Vegas, NV 89154, USA
* Correspondence: rlrowland@lanl.gov (R.L.R.II); Pamela.Burnley@unlv.edu (P.C.B.)
† Current address: X-ray Science Division, Argonne National Laboratory, Argonne, IL 60439, USA.

Received: 30 December 2019; Accepted: 21 February 2020; Published: 27 February 2020

Abstract: Understanding basic material properties of rare earth element (REE) bearing minerals such as their phase stability and equations of state can assist in understanding how economically viable deposits might form. Bastnäsite is the most commonly mined REE bearing mineral. We synthesized the lanthanum-fluoride end member, bastnäsite-(La) ($LaCO_3F$), and investigated its thermal behavior and decomposition products from 298 K to 1173 K under ambient pressure conditions through thermogravimetric analysis, differential scanning calorimetry, evolved gas analysis, and high temperature powder X-ray diffraction. We also investigated the compressibility of bastnäsite-(La) via single crystal X-ray diffraction in diamond anvil cells at an ambient temperature up to 11.3 GPa and from 4.9 GPa to 7.7 GPa up to 673 K. At ambient pressure, bastnäsite-(La) was stable up to 598 K in air, where it decomposed into CO_2 and tetragonal γ-LaOF. Above 948 K, cubic α-LaOF is stable. High temperature X-ray diffraction data were used to fit the Fei thermal equation of state and the thermal expansion coefficient α_{298} for all three materials. Bastnäsite-(La) was fit from 298 K to 723 K with $V_0 = 439.82$ Å^3, $\alpha_{298} = 4.32 \times 10^{-5}$ K^{-1}, $a_0 = -1.68 \times 10^{-5}$ K^{-1}, $a_1 = 8.34 \times 10^{-8}$ K^{-1}, and $a_2 = 3.126$ K^{-1}. Tetragonal γ-LaOF was fit from 723 K to 948 K with $V_0 = 96.51$ Å^3, $\alpha_{298} = 2.95 \times 10^{-4}$ K^{-1}, $a_0 = -2.41 \times 10^{-5}$ K^{-1}, $a_1 = 2.42 \times 10^{-7}$ K^{-1}, and $a_2 = 41.147$ K^{-1}. Cubic α-LaOF was fit from 973 K to 1123 K with $V_0 = 190.71$ Å^3, $\alpha_{298} = -1.12 \times 10^{-5}$ K^{-1}, $a_0 = 2.36 \times 10^{-4}$ K^{-1}, $a_1 = -1.73 \times 10^{-7}$ K^{-1}, and $a_2 = -17.362$ K^{-1}. An ambient temperature third order Birch–Murnaghan equation of state was fit with $V_0 = 439.82$ Å^3, $K_0 = 105$ GPa, and K′ = 5.58.

Keywords: Bastnäsite; equation of state; rare earth element

1. Introduction

Rare earth elements (REE), the elements with atomic number 57 to 71, are the first on the periodic table to begin populating F-block orbitals. Due to this fact, they exhibit unique electronic properties and thus are sought after for various technological applications for which other metals are not suitable. They are considered rare, not because of their relative abundance in the Earth's crust, but because they are generally only found as trace impurities instead of primary cations in minerals [1,2]. Relative crustal abundances of REEs are higher than noble metals such as gold or platinum, and some REEs are more common than base metals like lead [2,3]. The limited number of minerals that form with high REE content are rarely found in high enough concentrations to make economically viable deposits [2]. Currently, China controls > 90% of the world's supply of REEs [1], presenting the potential for a supply problem. Understanding basic material properties of REE bearing minerals such as how they respond to changes in pressure and temperature can assist in understanding how economically viable deposits form.

Having the general formula of (Ce,La,Nd,Y)CO_3(F,OH), bastnäsite is the most common mineral mined for REEs [4]. It occurs mainly in carbonatites and is usually found in association with calcite, dolomite, and barite [5–7], but also occurs in other geologic settings important for REE-bearing minerals [8]. Bastnäsite-(La) is the lanthanum fluoride end member ($LaCO_3F$) [2,4,9–14]. This work explores the ambient pressure thermal behavior of bastnäsite-(La) and the oxyfluorides it decomposes to using thermogravimetric analysis, differential scanning calorimetry, evolved gas analysis, and high temperature X-ray powder diffraction. We also explore the behavior of bastnäsite-(La) at high pressure via single crystal X-ray diffraction in a diamond anvil cell.

Structure and Decomposition of Bastnäsite-(La) to Lanthanum Oxyfluoride

Bastnäsite-(La) ($LaCO_3F$) exhibits hexagonal symmetry in the P-62c space group. Each unit cell contains six formula units of alternating layers of carbonate anions and layers of lanthanum cations and fluorine anions (Figure 1). The lanthanum-fluoride layers are parallel to the *a*- axis while the carbonate anion groups between them are locally parallel to the *c*- axis [4,9,10]. With the application of sufficient energy in the form of heat, bastnäsite-(La) decomposes via the decarbonation reaction $LaCO_3F$ + (heat) = LaOF + CO_2 [7,15]. At ambient conditions, γ-LaOF exhibits tetragonal symmetry in the P4/nmm space group with two formula units per unit cell [16–20]. At high temperature, α-LaOF is stable with cubic symmetry in the Fm-3m space group with four formula units per unit cell [21–25]. β-LaOF is reported as stable at ambient pressure and temperature exhibiting rhombohedral symmetry in the R-3m space group with six formula units per unit cell [16,20,24]. This structure was not encountered during this investigation.

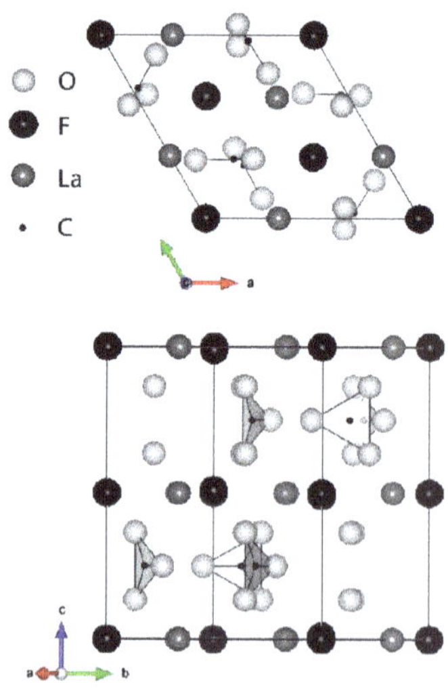

Figure 1. Structure of bastnäsite-(La) generated in the software VESTA [26], (**a**) as viewed down the *c*-axis, (**b**) as viewed down the *a**- axis [4]. In both images, the trigonal planar polyhedral for the CO_3^{2-} ions are highlighted.

2. Materials and Methods

2.1. Synthesis

Bastnäsite-(La) was synthesized via precipitation from an aqueous solution at room pressure and temperature using the method described by Janka and Schleid [15]. Aqueous solutions of each reagent were made by mixing powdered lanthanum nitrate ($La(NO_3)_3{\cdot}6H_2O$, Sigma Aldrich 203548-100G, >99% purity), sodium bicarbonate ($NaHCO_3$, Sigma Aldrich S6014-25G, >99% purity), and sodium fluoride (NaF, Sigma Aldrich 201154-5G, >99% purity) with deionized water in separate beakers. The solutions were combined in a larger beaker. Bastnäsite-(La) precipitated from the mixture of solutions immediately, with sodium nitrate ($NaNO_3$) remaining in the solution. The fluid was decanted, and the bastnäsite-(La) precipitate was washed in deionized water and centrifuged to remove residual sodium nitrate. The powder was then dried in a Fischer Scientific model 289A Isotemp Vacuum oven under vacuum at 488 K for at least 12 h.

Phase identification was completed by powder X-ray diffraction in a PANalytical X'Pert PRO and photoaccoustic infrared spectroscopy on a Digilab FTS-7000 Fourier transfer infrared (FTIR) spectrometer. Rietveld structure refinement using X'pert Highscore Plus software was used to determine that the bastnäsite-(La) starting material contained < 10% LaF_3 impurity. No OH peaks were detected by FTIR between 3600 cm^{-1} and 3400 cm^{-1}.

The powder X-ray diffraction peaks from the synthesized bastnäsite-(La) are wide due to the small particle size, so the material was next annealed in a Griggs modified piston cylinder apparatus [27] using NaCl as the pressure transmitting medium. Samples were placed in a platinum jacket, surrounded by graphite, and sealed in a copper capsule, then subjected to pressures between 0.25 GPa and 1.0 GPa and temperatures from 973 K to 1123 K [28]. See Figure 2 for a comparison of diffraction patterns before and after annealing.

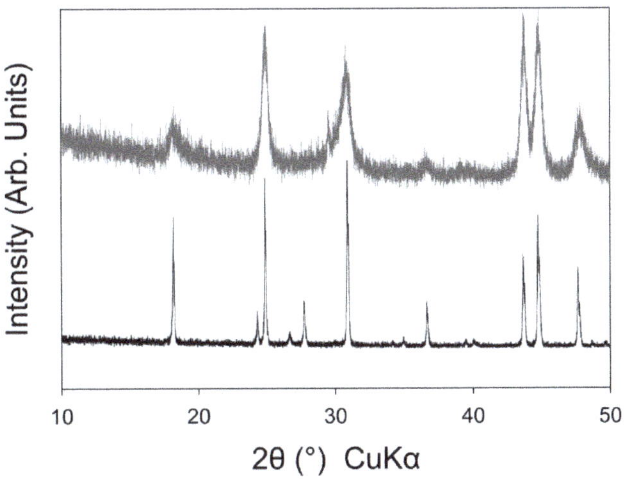

Figure 2. Powder XRD patterns of synthesized bastnäsite-(La) starting material. Top pattern is after washing and drying, bottom pattern is after annealing in the Griggs modified piston cylinder apparatus.

2.2. Themogravimetric Analysis and Differential Scanning Calorimetry

Thermogravimetric analysis (TGA), differential scanning calorimetry (DSC), and evolved gas analysis (EGA, Pfeiffer Vacuum Thermostar. Aßlar, Germany.) were used to determine the temperature at which bastnäsite-(La) begins to decompose and to determine the temperatures of phase transformations in the decomposition products. Powdered samples of at least 5 mg were heated in an alumina ceramic crucible in a Netzsch STA449 F1 Jupiter DSC/TGA apparatus under dry N_2 at 1 atm pressure at a programmed rate of 5°/min from 313 K to 1223 K. Nitrogen was used as a carrier gas because it was unreactive with the sample during heating. A second empty alumina crucible was used as the reference. It is noted that the Netzsch Jupiter DSC/TGA apparatus contains a heat-flux style calorimeter, thus exothermic reactions produce a measured increase in heat flow. A Pfeiffer Vacuum Thermostar Quadrupole mass spectrometer was used to record select masses between 1–100 AMU throughout the temperature ramp.

2.3. High Temperature Powder X-ray Diffraction

High temperature powder X-ray diffraction was used to determine unit cell volume as a function of temperature and to determine which phases were present at different temperatures. Measurements were made in air on a PANalytical X'Pert Pro MPD Diffractometer fitted with an Anton Paar XRK 900 thermal reactor stage and an X'celerator (2.02° 2θ) detector. A cobalt X-ray source (Co $K\alpha_1$ λ = 1.78901 Å) was used in angle dispersive mode from 4.0980 to 79.9814° 2θ with a step size of 0.0170° at 50.1650 s per step. Diffraction patterns were collected at 298 K, and in 50 degree increments from 323 K to 473 K, and in 25 degree increments from 473 K to 1173 K. A final pattern was collected at 298 K after the sample was allowed to cool. Data analysis was completed using QualX 2.0 [29], UnitCellWin64 [30] and EosFit7GUI [31].

2.4. Single Crystal X-ray Diffraction

In situ single crystal synchrotron X-ray diffraction was performed at Sector 16 HPCAT of the Advanced Photon Source at Argonne National Laboratory. The pressure cells used were four-post diamond anvil cells (DAC) with 700 μm diameter culets and laser-cut rhenium gaskets with a 380 μm diameter laser-cut gasket hole sample chamber [32,33]. Single crystals of the annealed bastnäsite-(La) ~20 μm in diameter were placed in the sample chamber. Ruby spheres along with gold or copper powder were included in the sample chamber as pressure standards. The diamond cells were gas loaded [34] with helium, neon, or argon as the pressure transmitting media. Angle dispersive X-ray diffraction images were captured on a 2048 × 2048 pixel MAR CCD detector with a monochromatic beam of 30 keV energy X-rays (λ = 0.3738 nm) while rotating the DAC ±30 degrees around the vertical axis; allowing for between 50 and 300 bastnäsite-(La) reflections to be measured. The ruby luminescence pressure scale [35] was used to estimate pressure during data collection and the equations of state for gold [36] and copper [37] were used for final pressure determination. Heated DAC experiments were conducted in a vacuum chamber with kapton and mylar windows. The DAC was equipped with resistive heating wires wound around the gasket between the anvils and around the exterior of the DAC housing to generate and maintain sample chamber temperatures up to 673 K. Thermocouples were placed in contact with the diamonds to measure the sample temperature. For the heated experiments, argon was used as the pressure transmitting media, and copper (Cu) as the pressure standard. Data analysis was completed using the software packages Fit2D [38], MDI Jade (Materials Data Inc), GSE_ADA/RSV [39], and EosFit7GUI [31].

3. Results

3.1. Thermogravimetric Analysis and Differential Scanning Calorimetry

Thermogravimetric analysis and mass spectrometry CO_2 detection curves for annealed and unannealed bastnäsite-(La) and LaOF are displayed in Figure 3. Samples containing bastnäsite-(La) lost 1.0% to 1.3% between 313 K and 598 K due to adsorbed water. From 598 K to 778 K, 13.0% to 17.4% was lost due to the decomposition of bastnäsite-(La) to γ-LaOF and CO_2. Since the LaOF sample was generated by the decomposition of bastnäsite-(La), the minor deflection of its mass loss curve at the bastnäsite-(La) decomposition point is likely caused by minor amounts of remaining bastnäsite-(La). Otherwise, the curve for LaOF does not show a significant change in slope, indicating that the LaOF itself did decompose.

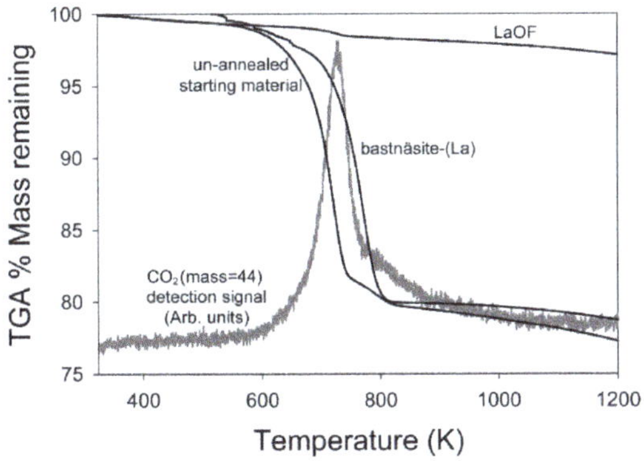

Figure 3. Thermogravimetric analysis (TGA) curves for annealed bastnäsite-(La), unannealed starting material, and LaOF from 325 K to 1223 K. CO_2 detection from the Pfeiffer Vacuum Thermostar Quadrupole mass spectrometer in light gray. Mass loss curves for the starting material and annealed bastnäsite-(La) begin to steepen at 600 K, at the same time CO_2 detection begins. The mass loss curve for LaOF does not show a significant change in slope, indicating that the decomposition reaction did not occur in that sample.

The differential scanning calorimetry curves for annealed and unannealed bastnäsite-(La) (Figure 4) exhibit endotherms between 598 K and 778 K, and the mass spectrometer's CO_2 detection peaked at 730 K. All samples exhibited exotherms at 948 K and endotherms at 1123 K, indicative of phase transitions.

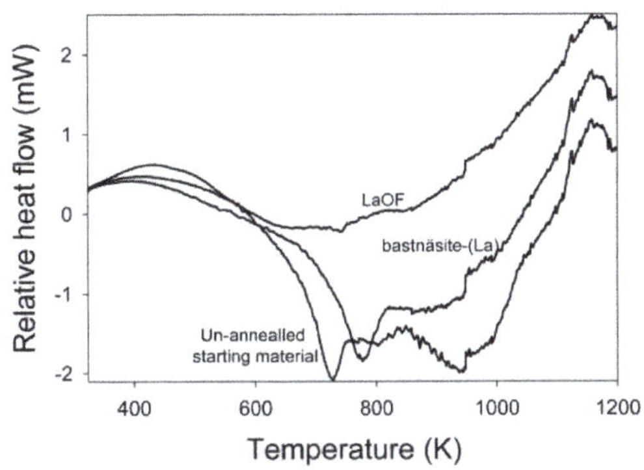

Figure 4. Differential scanning calorimetry (DSC) curves for annealed bastnäsite-(La), unannealed starting material, and LaOF from 325 K to 1223 K. Endotherms can be seen on both the annealed bastnäsite-(La), and unannealed starting material curves from the range near 598 K to 773 K, indicating the decomposition reaction. The LaOF curve does not exhibit this, since it did not decompose. Exotherms at 948 K and endotherms at 1123 K are evident on all three curves, indicating phase transformations.

3.2. High Temperature Powder X-ray Diffraction

Figure 5 displays a selection of the high temperature powder XRD patterns encompassing the bastnäsite-(La) decomposition reaction. Bastnäsite-(La) is the dominant species in the X-ray diffraction patterns from 298 K to 723 K. Diffraction peaks for γ-LaOF begin to show up in the patterns above 598 K. By 773 K, bastnäsite-(La) is no longer present in the patterns, and γ-LaOF is the only phase present. Above 948 K, only α-LaOF is present. γ-LaOF was the only species present in the pattern collected at 298 K after the sample was allowed to cool following the collection of the final high temperature pattern. β-LaOF was not evident in any of the patterns. See Tables 1–3 for the measured lattice parameters and volumes of all three materials. Figures 6–8 display measured unit cell volumes as a function of temperature for all three materials.

Table 1. Measured lattice parameters and volume for bastnäsite-(La) from 298 to 723 K.

Temperature (K)	*a* (Å)	Uncertainty	*c* (Å)	Uncertainty	V (Å^3)	Uncertainty	*a*/*c* Ratio
298	7.1867	0.00085	9.8328	0.00083	439.82	0.185	0.7309
323	7.1897	0.00085	9.8347	0.00083	440.26	0.185	0.7311
373	7.1954	0.00085	9.8394	0.00083	441.18	0.185	0.7313
423	7.1992	0.00085	9.8438	0.00083	441.84	0.185	0.7313
473	7.2044	0.00085	9.8486	0.00083	442.69	0.186	0.7315
498	7.2074	0.00085	9.8504	0.00083	443.14	0.186	0.7317
523	7.2096	0.00085	9.8539	0.00083	443.57	0.186	0.7316
548	7.2125	0.00085	9.8564	0.00083	444.04	0.186	0.7318
573	7.2159	0.00085	9.8572	0.00083	444.50	0.186	0.7320
598	7.2172	0.00085	9.8598	0.00083	444.76	0.186	0.7320
623	7.2205	0.00086	9.8626	0.00083	445.30	0.186	0.7321
648	7.2238	0.00086	9.8669	0.00083	445.91	0.187	0.7321
673	7.2266	0.00086	9.8669	0.00083	446.25	0.187	0.7324
698	7.2296	0.00086	9.8730	0.00083	446.90	0.187	0.7323
723	7.2339	0.00086	9.8731	0.00083	447.43	0.187	0.7327

Table 2. Measured lattice parameters and volumes for γ-LaOF at 298 K, and from 723 K to 948 K.

Temperature (K)	*a* (Å)	Uncertainty	*c* (Å)	Uncertainty	V (Å^3)	Uncertainty
298	4.0798	0.00041	5.798	0.0020	96.51	0.118
723	4.1242	0.00045	5.855	0.0030	99.59	0.130
748	4.1192	0.00041	5.873	0.0020	99.65	0.119
773	4.1192	0.00041	5.873	0.0020	99.65	0.119
798	4.1213	0.00042	5.862	0.0020	99.57	0.119
823	4.1230	0.00042	5.862	0.0020	99.64	0.119
848	4.1246	0.00042	5.867	0.0020	99.80	0.119
873	4.1276	0.00042	5.864	0.0020	99.91	0.119
898	4.1293	0.00042	5.860	0.0020	99.92	0.119
923	4.1286	0.00050	5.866	0.0022	99.99	0.122
948	4.1311	0.00042	5.859	0.0020	100.00	0.119

Table 3. Measured lattice parameters and volumes for α-LaOF from 973 K to 1173 K.

Temperature (K)	*a* (Å)	Uncertainty	V (Å^3)	Uncertainty
298	5.756 *	0.003 *	190.71 *	
973	5.8424	0.00049	199.42	0.138
998	5.8443	0.00049	199.62	0.138
1023	5.8476	0.00050	199.95	0.138
1048	5.8489	0.00050	200.08	0.138
1073	5.8500	0.00050	200.20	0.138
1098	5.8526	0.00050	200.47	0.138
1123	5.8538	0.00050	200.60	0.138
1148	5.8531	0.00050	200.52	0.138
1173	5.8511	0.00050	200.32	0.138

* values for ambient temperature (298 K) from [22], because this structure was not recoverable to room temperature by quenching in the XRK-900 reactor stage.

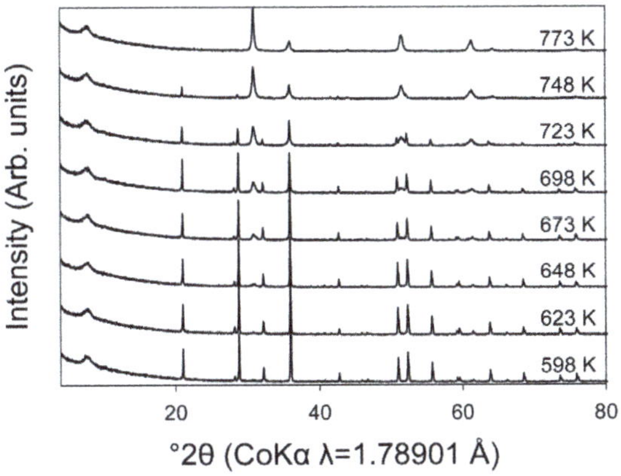

Figure 5. High temperature powder X-ray diffraction patterns from 598 K to 723 K. Bottom pattern is 598 K, and the patterns go up in 25° temperature increments. Bottom-most pattern exhibits only bastnäsite-(La). Top-most pattern exhibits only γ-LaOF. The middle six patterns exhibit both species as the decomposition reaction progresses from bastnäsite-(La) to γ-LaOF.

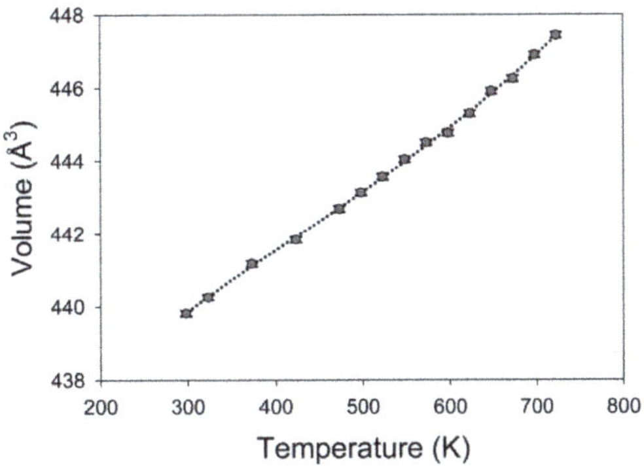

Figure 6. Measured unit cell volumes for bastnäsite-(La), as a function of temperature from 298 K to 723 K. The dotted black line is the fit for the Fei thermal equation of state [40].

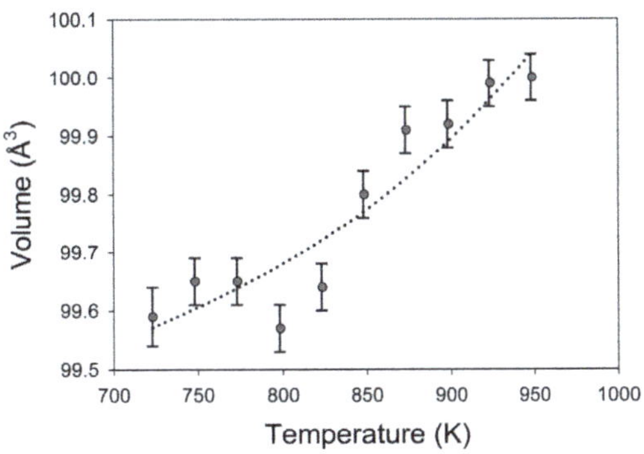

Figure 7. Measured unit cell volumes for tetragonal γ-LaOF as a function of temperature from 723 K to 948 K. The dotted black line is the fit for the Fei thermal equation of state [40].

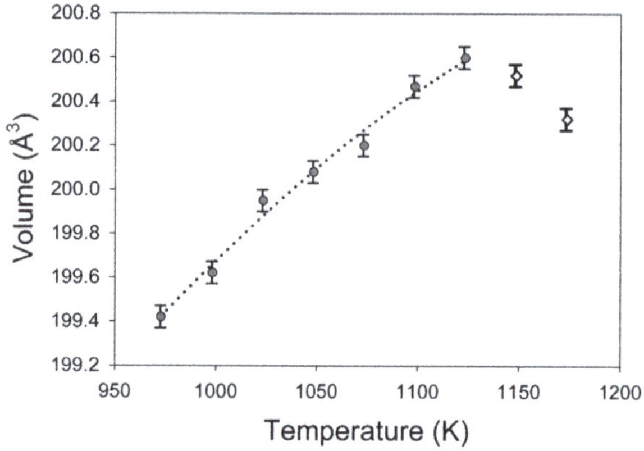

Figure 8. Measured unit cell volumes for cubic α-LaOF as a function of temperature from 973 K to 1123 K. Gray-filled circles represent data used to fit the thermal equation of state in this temperature range. White diamonds represent temperature points 1148 K and 1173 K, which are above the temperature that coincides with a contraction in unit cell volume and an exotherm on the DSC curve. The dotted black line is the fit for the Fei thermal equation of state [40].

3.3. *Single Crystal X-ray Diffraction*

At ambient temperature, 16 X-ray diffraction patterns of bastnäsite-(La) were collected from ambient pressure to 11.3 GPa. Over this pressure range, the *a*- crystallographic axis contracts from 7.187 Å to 6.974 Å, and the *c*- crystallographic axis contracts from 9.830 Å to 9.593 Å, with a corresponding unit cell volume contraction from 439.82 Å^3 to 404.06 Å^3. Table 4 lists the measured lattice parameters, calculated unit cell volumes, and crystallographic axis ratios for each pressure point. The unit cell volume data were fit to a third order Birch–Murnaghan equation of state [31,41] with V_o = 439.82 Å^3, K_o = 105 GPa, and K′ = 5.58. Figure 9 illustrates the unit cell volume measurements superimposed over the calculated equation of state.

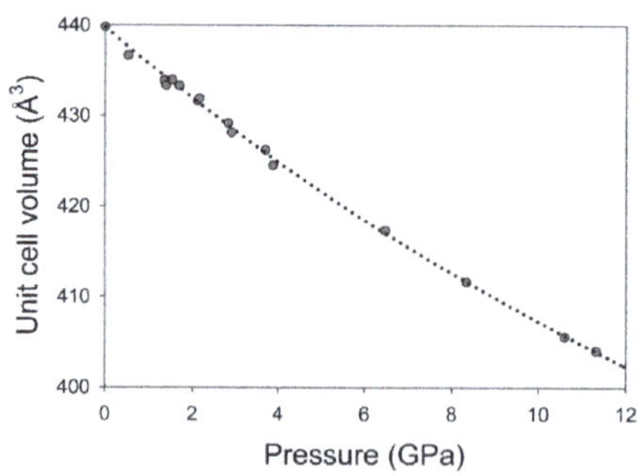

Figure 9. Pressure-volume data measured for bastnäsite-(La) fit to the 3rd order Birch–Murnaghan equation of state [41]. V_0 = 439.82 Å^3, K_0 = 105 GPa, K′ = 5.58. Gray circles are measured volumes from Table 4, and dotted line represents equation of state fit to data.

Over the temperature range of 347 K to 673 K, nine X-ray diffraction patterns were collected between 4.9 GPa and 7.7 GPa. Over this range of conditions, the *a*- crystallographic axis varies between 7.057 Å and 7.086 Å, and the *c*- crystallographic axis varies between 9.730 Å and 9.685 Å, with corresponding variation in unit cell volumes between 423.11 Å^3 and 418.32 Å^3. Table 5 lists the measured lattice parameters, calculated unit cell volumes, and crystallographic axis ratios for each pressure and temperature point.

Table 4. Bastnäsite-(La) measured lattice parameters, unit cell volumes, *a*/*c* crystallographic axis ratios, pressure transmitting media, and pressure standards from 0 GPa to 11.3 GPa at ambient temperature.

Pressure (GPa)	*a* (Å)	Uncertainty	*c* (Å)	Uncertainty	V (Å^3)	Uncertainty	*a*/*c* Ratio	Pressure Media	Pressure Standard
0.00	7.187	0.002	9.833	0.040	439.82	0.044	0.7309	n/a	n/a
0.53	7.169	0.003	9.811	0.003	436.68	0.009	0.7307	Ne	Au
1.35	7.148	0.003	9.806	0.003	433.91	0.008	0.7289	Ne	Au
1.39	7.159	0.001	9.763	0.003	433.32	0.006	0.7333	He	Cu
1.53	7.158	0.001	9.780	0.003	433.96	0.006	0.7319	He	Cu
1.69	7.150	0.002	9.787	0.004	433.30	0.008	0.7306	He	Cu
2.11	7.144	0.004	9.766	0.006	431.64	0.013	0.7315	Ne	Au
2.15	7.145	0.001	9.769	0.003	431.88	0.005	0.7314	He	Cu
2.82	7.129	0.001	9.751	0.003	429.18	0.005	0.7311	He	Cu
2.90	7.117	0.002	9.761	0.002	428.15	0.006	0.7291	Ne	Au
3.69	7.111	0.001	9.734	0.002	426.23	0.004	0.7305	He	Cu
3.87	7.095	0.003	9.738	0.003	424.53	0.009	0.7286	Ne	Au
6.47	7.051	0.003	9.694	0.003	417.38	0.009	0.7274	Ne	Au
8.34	7.014	0.003	9.663	0.003	411.65	0.009	0.7259	Ne	Au
10.60	6.973	0.003	9.632	0.003	405.59	0.009	0.7239	Ne	Au
11.31	6.974	0.004	9.593	0.003	404.06	0.011	0.7270	Ne	Au

Table 5. Bastnäsite-(La) measured lattice parameters, unit cell volumes, and *a*/*c* crystallographic axis ratios from 4.9 GPa to 7.7 GPa pressure and from 347 K to 673 K temperature.

Pressure (GPa)	Temp (K)	*a* (Å)	Uncertainty	*c* (Å)	Uncertainty	V (Å^3)	Uncertainty	*a*/*c* Ratio
4.9	347	7.086	0.001	9.730	0.002	423.11	0.005	0.7283
5.1	324	7.084	0.002	9.721	0.002	422.47	0.006	0.7287
5.8	373	7.076	0.002	9.696	0.002	420.44	0.006	0.7298
6.5	423	7.070	0.002	9.699	0.003	419.85	0.007	0.7289
6.7	473	7.060	0.002	9.691	0.002	418.32	0.006	0.7285
6.7	523	7.063	0.002	9.697	0.003	418.93	0.007	0.7284
7.0	573	7.057	0.002	9.713	0.003	418.91	0.007	0.7266
7.3	623	7.065	0.001	9.694	0.002	419.01	0.005	0.7288
7.7	673	7.064	0.002	9.685	0.002	418.53	0.006	0.7294

4. Discussion

From TGA/DSC and high temperature powder XRD data, it is apparent that at ambient pressure, the decarbonation temperature of bastnäsite-(La) is about 598 K. Above this temperature, the decomposition reaction of bastnäsite-(La) into γ-LaOF and CO_2 begins. Since our experiments were conducted without controlled pCO_2, this decarbonation point is not the equilibrium decomposition temperature; which at one bar pCO_2 is likely higher. The application of pressure significantly increases the stability of bastnäsite-(La), as was observed in our recrystallization experiments at 1073 K and 0.22 GPa (after five hours) and at 1123 K and 0.71 GPa after 21 h [28]. However, as these experiments were not reversed, a phase diagram for bastnäsite-(La) cannot yet be established. At 948 K, tetragonal γ-LaOF undergoes a first order structural phase transition reaction to cubic α-LaOF. The decomposition temperature for bastnäsite-(La) found in this study agrees with previous values reported by Janka and Schlied [15] who used similar methods, but did not report behavior for temperatures above 823 K. While the sample material cooled back down to 298 K, it transformed to γ-LaOF, the stable structure of lanthanum oxyfluoride at ambient conditions.

Table 6 lists the fitted thermal expansion coefficients for bastnäsite-(La), γ-LaOF, and α-LaOF over the temperature ranges measured in the high temperature powder X-ray diffraction experiments. Over the temperature ranges measured, all three species exhibit a trend of positive thermal expansion

(Figures 6–8), except between 1123 K and 1173 K, where the α-LaOF unit cell contracts. There is an endotherm in the DSC curve at 1123 K that could indicate a phase transition, but the X-ray diffraction patterns at these temperatures do not indicate a structural change, and the mass spectrometer did not detect any compounds outgassing at these temperatures; further studies are necessary to determine what this endotherm and coincident thermal contraction represent.

Table 6. Calculated thermal expansion coefficients for bastnäsite-(La), γ-LaOF, and α-LaOF.

Material	Dimension	Temp. Range (K)	$\alpha_{298} \times 10^{-5}$	$a_0 \times 10^{-5}$	$a_1 \times 10^{-7}$	a_2
bastnäsite-(La)	$a(a_0 = 7.1867$ Å)	298–723	1.73	−1.08	0.373	1.502
	$c(c_0 = 9.8328$ Å)	298–723	0.951	0.247	0.116	0.319
	$V(V_0 = 439.82$ Å^3)	298–723	4.32	−1.68	0.834	3.126
γ-LaOF	$a(a_0 = 4.0798$ Å)	723–948	13.5	−17.9	1.90	22.879
	$c(c_0 = 5.798$ Å)	723–948	0.179	0.1785	−0.229	8.154
	$V(V_0 = 96.51$ Å^3)	723–948	29.5	−24.1	2.42	41.147
α-LaOF	$a(a_0 = 5.756$ Å)	973–1123	−0.844	8.46	−0.625	−6.606
	$V(V_0 = 190.71$ Å^3)	973–1123	−1.12	23.6	−1.73	−17.362

Note: Reference temperature for all fits is 298 K.

Thermal analysis has not been performed on any other fluorocarbonate minerals, so a direct comparison of the properties of bastnäsite-(La) measured in this investigation to other materials with similar structure is not possible. Others have investigated the thermal expansion of more common carbonates including aragonite ($CaCO_3$), strontianite ($SrCO_3$), cerrusite ($PbCO_3$), witherite ($BaCO_3$) [42], calcite ($CaCO_3$), and magnesite ($MgCO_3$) [43]. A comparison of the thermal equations of state for bastnäsite-(La) and these materials is given in Figure 10. Compared with these more common carbonates, bastnäsite-(La) experiences more thermal expansion for a given increase in temperature than the trigonal R-3c carbonates (calcite and magnesite), and less thermal expansion for a given increase in temperature than the orthorhombic Pmcn carbonates (aragonite, cerrusite, strontianite, and witherite). Additionally, as shown in Table 6, bastnäsite-(La)'s *a*- and *c*- crystallographic axes expand at different rates; all of the other carbonates listed above also exhibit some degree of anisotropic thermal expansion. Overall, this indicates that bastnäsite-(La)'s thermal expansion behavior is similar to that of other carbonate minerals.

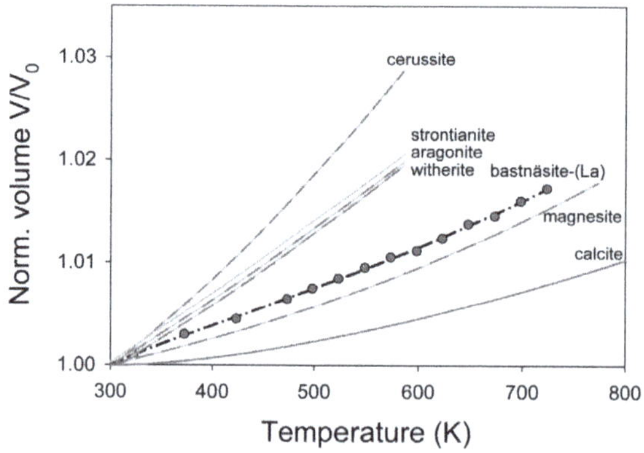

Figure 10. Comparisons of ambient pressure thermal equations of state for different carbonate minerals. Gray circles represent bastnäsite-(La) data from this study, and the dotted black line represent the fitted Fei equation of state [40] for bastnäsite-(La). Thermal equations of state are shown for calcite ($CaCO_3$), magnesite ($MgCO_3$) [43], witherite ($BaCO_3$), cerrusite ($PbCO_3$), strontianite ($SrCO_3$), and aragonite ($CaCO_3$) [42].

There are no previous studies of tetragonal REE-bearing oxyfluorides, but Achary et al. [24] synthesized the rhombohedral oxyfluorides of five REEs (La, Nd, Sm, Eu, and Gd), and used high temperature powder XRD to investigate their thermal expansion and phase transformation to cubic structures at high temperature. Their calculation and fitting of thermal expansion coefficients used the data for both rhombohedral and cubic phases, so the resulting thermal equations of state are not suitable to compare directly to that for α-LaOF derived in this study. However, using their data for just the cubic phases, we were able to fit thermal expansion coefficients using EOSFit7GUI [31]. Table 7 lists the fitted coefficients, and Figure 11 illustrates the thermal equations of state compared to that of α-LaOF measured in this study. The thermal expansion behavior we measured for α-LaOF is essentially the same as what we derived from Achary et al.'s data [24].

Table 7. Calculated thermal expansion coefficients for cubic REE-oxyfluorides.

Material	Temp. Range (K)	V_0 (Å^3)	α_{298} $\times 10^{-5}$	a_0 $\times 10^{-5}$	a_1 $\times 10^{-7}$	a_2
α-LaOF †	973–1123	190.71	−1.12	23.6	−1.73	−17.362
α-LaOF *	823–1077	196.1	0.855	0.0017	0.286	0.0001
NdOF *	823–1075	183.8	0.268	0.0007	0.0895	0.0001
SmOF *	811–1075	176.2	1.4	1.4	0.00028	0
EuOF *	815–1077	168.2	4.49	4.49	0.00052	0.0001
GdOF *	913–1067	172.6	0.461	0.0044	0.1530	0

Note: Reference temperature for all fits is 298 K; † this study; * data from Achary, et. al [24].

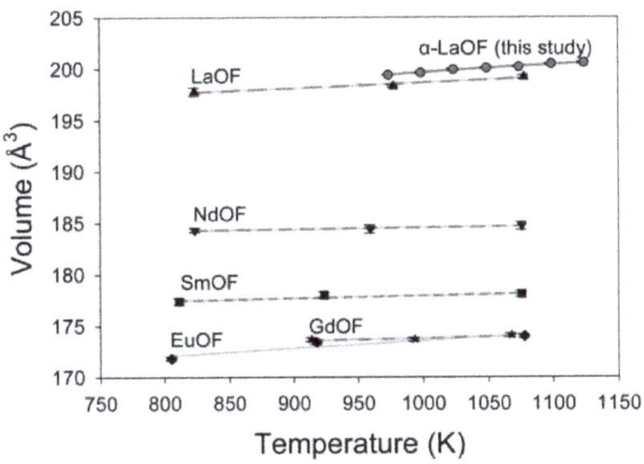

Figure 11. Comparisons of ambient pressure thermal equations of state for cubic REE-OF compounds from this study and using data from Achary et al. [24] (solid symbols). Gray circles are data for cubic α-LaOF from this study.

The anisotropy of bastnäsite-(La) is enhanced under compression; the *a*- crystallographic axis contracts more than the *c*- crystallographic axis. The changing ratio of *a*/*c* is given in Table 4. Similarly, the *a*- crystallographic axis is also more expansive during heating (Table 1). As illustrated in Figure 1, the planar carbonate ions are arranged in the structure such that they lie on planes containing the *c*-crystallographic axis. Others [44–46] have observed rigid body behavior of CO_3^{2-} anions in other carbonates at high pressure. Thus, we suggest that most of the expansion and contraction of the structure is accommodated by the La-O and La-F bonds.

There are no compressibility data fit to equations of state or any bulk moduli measured from other methods (e.g., ultrasonic methods) in the literature for bastnäsite or any other fluorocarbonate minerals, so a direct comparison of the properties of bastnäsite-(La) measured in this investigation to materials with similar structure is not possible. However, our results are consistent with recent work on natural materials that shows that the bastnäsite structure is stable up to 25 GPa at room temperature [47].

Anderson and Nafe [48] found that when comparing many compounds' bulk moduli versus their specific ionic volumes, multiple trends emerged. Figure 12 displays their data, with additional data for carbonate minerals from Knittle [49], Merlini [50], and Xu [51]. Anderson and Nafe [48] identified a sulfide-selenide-telluride trend, an oxide trend, a fluorite trend, and an alkali-halide trend. Rhodochrosite, dolomite, ankerite, calcite, strontianite, and witherite form a separate carbonate trend (yellow line in Figure 12), but the hydroxycarbonates azurite and malachite do not lie on this trend. The values measured for bastnäsite-(La) in this study plot in a location consistent with the carbonate trend, the fluorite trend, as well as the oxide trend. Work on other fluorocarbonates needs to be completed in order to determine which of these trends bastnäsites follow, or if fluorocarbonates have a separate trend. If the speculation above that bastnäsite-(La)'s compressional anisotropy is related to rigid body behavior and orientation of the carbonate ions is correct, it is then likely that bastnäsites would follow the carbonate trend.

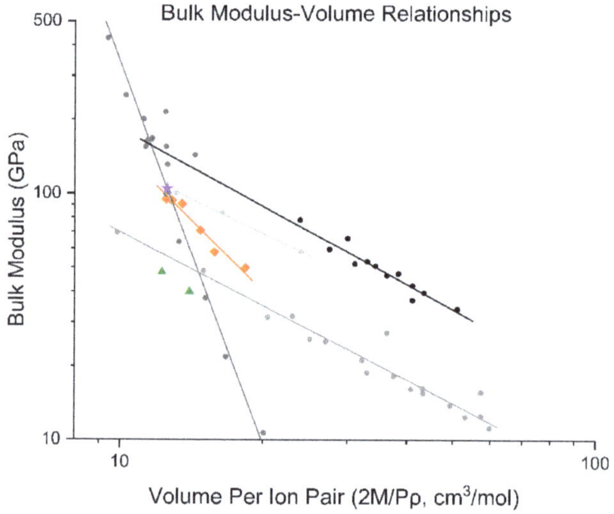

Figure 12. Bulk modulus versus specific volume for simple oxides, alkali halides, fluorites, sulfides, selenides, tellurides, and carbonates. Small circles include data from [48] and [49]. Large purple star is the value for bastnäsite-(La) measured in this study ($2M/P\rho = 12.60$ cm^3/mol, $K_o = 105$ GPa), large orange diamonds are carbonates from [49], and green triangles are malachite and azurite from [50,51]. Dark gray line is the oxide trend, black line is the sulfide, selenide, and telluride trend, light gray line is the fluorite trend, medium gray line is the alkali-halide trend, and orange line is proposed carbonate trend.

While some of our measurements were taken at elevated temperature and pressure (Table 5), there is not enough data to properly fit a high temperature and pressure equation of state. However, these data can be used to evaluate whether our isothermal and thermal equations of state can be combined to predict bastnäsite-(La)'s unit cell volume at high temperature and pressure. Thus, for each pressure temperature point measured in our heated DAC experiments, we calculated the value for V_o (at $P = 0$) using our values for the thermal equation of state [40], and then used our Birch–Murnaghan equation of state [41] to calculate the pressure required to produce the observed unit cell volume. The pressures calculated in this way agree well with those measured from the copper pressure standard. Table 8 lists the measured temperatures, unit cell volumes, and pressures, and the calculated pressures. Since the temperatures and volumes from the elevated temperature DAC experiments were not used in the fitting of Fei's thermal equation of state or in the fitting of the Birch–Murnaghan equation of state, these data show that the combination of both equations is a reasonable approximation for modeling changes in volume due to both pressure and temperature simultaneously.

Table 8. List of heated DAC experiments comparing measured pressure using the Cu standard versus calculated pressures using the Fei thermal EOS [40] parameters. Difference is calculated by subtracting the measured pressure from the calculated pressure. Error is calculated by dividing the difference by the measured pressure.

Temp (K)	V (Å^3)	Measured P (GPa)	Calculated P (GPa)	Difference	Error
347	423.11	4.90	5.03	0.13	0.03
324	422.47	5.09	5.11	0.02	0.00
373	420.44	5.78	6.02	0.24	0.04
423	419.85	6.46	6.46	0.00	0.00
473	418.32	6.74	7.23	0.49	0.07
523	418.93	6.71	7.28	0.57	0.08
573	418.91	6.97	7.57	0.60	0.09
623	419.01	7.33	7.84	0.51	0.07
673	418.53	7.69	8.33	0.64	0.08

5. Conclusions

Compressibility, thermal expansion and phase stability are fundamental thermodynamic properties of materials, and as such are significant in understanding how they interact with the geologic settings in which the minerals naturally occur. Bastnäsite-La and the lanthanum oxyfluoride polymorphs addressed above are important REE-bearing compounds, and thus studying them is useful for understanding how REE bearing compounds form and participate in chemical reactions with other compounds. Additionally, since bastnäsite is found in carbonatites, it is an important mineral to investigate in order to understand carbonated magmas and the formation of carbonatite-related ore deposits. Gaining insight into the thermodynamic behavior of bastnäsite and other fluorocarbonates could potentially lead to better methods of locating and processing REE ore in the future.

Author Contributions: Conceptualization, R.L.R.II, P.C.B. and B.L.; methodology, R.L.R.II, B.L., K.E.V.K.; formal analysis, R.L.R.II and B.L.; investigation, R.L.R.II; resources, P.C.B. and L.R.D.; data curation, P.C.B.; writing—original draft preparation, R.L.R.II; writing—review and editing, P.C.B., L.R.D., and K.E.V.K.; supervision, P.C.B.; project administration, P.C.B.; funding acquisition, P.C.B. All authors have read and agreed to the published version of the manuscript.

Funding: This research was sponsored in part by the National Nuclear Security Administration under the Stewardship Science Academic Alliances program through the DOE Cooperative Agreement #DE-NA0001982 as well as by National Science Foundation grant EAR-1220548.

Acknowledgments: The high-pressure work was conducted at HPCAT (Sector 16), Advanced Photon Source (APS), Argonne National Laboratory. HPCAT operations are supported by DOE-NNSA's Office of Experimental Sciences.The Advanced Photon Source, a U.S. Department of Energy (DOE) Office of Science User Facility operated for the DOE Office of Science by Argonne National Laboratory under Contract No. DE-AC02-06CH11357. Use of the COMPRES-GSECARS gas loading system was supported by COMPRES under NSF Cooperative Agreement EAR 11-57758 and by GSECARS through NSF grand EAR-1128799 and DOE grand DE-FG02-94ER1466. We thank Sergey Tkachev for his assistance with the gas loadings. This research also used resources and equipment at the National Aeronautics and Space Administration Lyndon B. Johnson Space Center's Astromaterials Research and Exploration Science division. We thank Brad Sutter of Jacobs JETS at NASA Johnson Space Center for assistance with the high temperature powder XRD data collection and Joanna V. Hogancamp of Geocontrols Systems and Jacobs JETS at NASA Johnson Space Center for assistance with TGA, DSC, and EGA data collection. The authors also wish to thank several anonymous reviewers for helpful suggestions.

Conflicts of Interest: The authors declare no conflict of interest. The funders had no role in the design of the study; in the collection, analyses, or interpretation of data; in the writing of the manuscript, or in the decision to publish the results.

References

1. Chen, Z. Global rare earth resources and scenarios of future rare earth industry. *J. Rare Earths* **2011**, *29*, 1–6. [CrossRef]
2. Long, K.R.; Van Gosen, B.S.; Foley, N.K.; Cordier, D. The Principal Rare Earth Elements Deposits of the United States: A Summary of Domestic Deposits and a Global Perspective. *US Geol. Surv. Sci. Investig.* **2010**, *2010*, 96.
3. Workman, R.K.; Hart, S.R. Major and trace element composition of the depleted MORB mantle (DMM). *Earth Planet. Sci. Lett.* **2005**, *231*, 53–72. [CrossRef]
4. Ni, Y.; Hughes, J.M.; Mariano, A.N. The atomic arrangement of bastnäsite-(Ce), $Ce(CO_3)F$, and structural elements of synchysite-(Ce), röntgenite-(Ce), and parisite-(Ce). *Am. Mineral.* **1993**, *78*, 415–418.
5. Olson, J.C.; Shawe, D.R.; Pray, L.C.; Sharp, W.N.; Hewett, D.F. Rare-Earth Mineral Deposits of the Mountain Pass District San Bernardino County California. *Geol. Surv. Prof. Pap.* **1954**, *261*, 4–15.
6. Williams-Jones, A.E.; Wood, S.A. A preliminary petrogenetic grid for REE fluorocarbonates and associated minerals. *Geochim. Cosmochim. Acta* **1992**, *56*, 725–738. [CrossRef]
7. Hsu, L.C. Synthesis and Stability of Bastnaesites in a Part of the System (Ce,La)-F-H-C-O. *Mineral. Petrol.* **1992**, *47*, 87–101. [CrossRef]
8. Schmandt, D.S.; Cook, N.J.; Ciobanu, C.L.; Ehrig, K.; Wade, B.P.; Gilbert, S.; Kamenetsky, V.S. Rare earth element fluorocarbonate minerals from the olympic dam Cu-U-Au-Ag deposit, South Australia. *Minerals* **2017**, *7*, 202. [CrossRef]
9. Oftedal, I. Zur Kristallstruktur von Bastnäsit, $(Ce,La)FCO_3$. *Zeitschrift fur Krist. Krist. Krist. Krist.* **1931**, *78*, 462.
10. Donnay, G.; Donnay, J.D.H. The Crystallography of Bastnaesite, Parisite, Roentgenite, and Synchisite. *Am. Mineral.* **1953**, *38*, 932–963.
11. Castor, S.B. Rare Earth Deposits of North America. *Resour. Geol.* **2008**, *58*, 337–347. [CrossRef]
12. Castor, S.B. The Mountain Pass Rare-Earth Carbonatite and Associated Ultrapotassic Rocks, California. *Can. Mineral.* **2008**, *46*, 779–806. [CrossRef]
13. Shivaramaiah, R.; Anderko, A.; Riman, R.E.; Navrotsky, A. Thermodynamics of bastnaesite: A major rare earth ore mineral. *Am. Mineral.* **2016**, *101*, 1129–1134. [CrossRef]
14. Gysi, A.P.; Williams-Jones, A.E. The thermodynamic properties of bastnäsite-(Ce) and parisite-(Ce). *Chem. Geol.* **2015**, *392*, 87–101. [CrossRef]
15. Janka, O.; Schleid, T. Facile Synthesis of Bastnaesite-Type $LaF[CO_3]$ and Its Thermal Decomposition to LaOF for Bulk and Eu^{3+}-Doped Samples. *Eur. J. Inorg. Chem.* **2009**, *2009*, 357–362. [CrossRef]
16. Woo, D.C.; Lee, M.-H.; Jung, W.-S. Synthesis and characterization of rhombohedral- and tetragonal-lanthanum oxyfluoride powders. *Ceram. Int.* **2013**, *39*, 1533–1538. [CrossRef]
17. Jacob, K.T.; Saji, V.S.; Waseda, Y. Lanthanum Oxyfluoride: Structure, Stability, and Ionic Conductivity. *Int. J. Appl. Ceram. Technol.* **2006**, *3*, 312–321. [CrossRef]
18. Fergus, J.W.; Chen, H.-P. Structure and Conductivity of Tetragonal and Rhombohedral Lanthanum Oxyfluoride Compounds. *J. Electrochem. Soc.* **2000**, *147*, 4696–4704. [CrossRef]
19. Shinn, D.B.; Eick, H.A. Phase Analyses of Lanthanide Oxide Fluorides. *Inorg. Chem.* **1969**, *8*, 232–235. [CrossRef]
20. Zachariasen, W.H. Crystal Chemical Studies of the 5f-series of Elements. XIV. Oxyfluorides, XOF. *Acta Crystallogr.* **1951**, *4*, 231–236. [CrossRef]
21. Pistorius, C.W.F.T. Effect of Pressure on the Rhombohedral/Cubic Transitions of Some Lanthanide Oxide Fluorides. *J. Less-Common Met.* **1973**, *31*, 119–124. [CrossRef]
22. Klemm, V.W.; Klein, H.-A. Lanthanoxyfluorid [Lanthanum Oxyfluoride]. *Zeitschrift fur Anorg. und Allg. Chemie* **1941**, *248*, 167–171, in Germany.
23. Mathews, M.D.; Tyagi, A.K.; Moorthy, P.N. Study of phase transition in REOF system by dilatometry (RE = La, Nd, Sm, Gd, Eu and Y). *Thermochim. Acta* **1997**, *298*, 165–167. [CrossRef]
24. Achary, S.N.; Ambekar, B.R.; Mathews, M.D.; Tyagi, A.K.; Moorthy, P.N. Study of phase transition and volume thermal expansion in a rare-earth (RE) oxyfluoride system by high-temperature XRD (RE = La, Nd, Sm, Eu and Gd). *Thermochim. Acta* **1998**, *320*, 239–243. [CrossRef]

25. Holtstam, D.; Grins, J.; Nysten, P. Håleniusite- (La) from the Bastnäs Deposit, Västmandland, Sweden: A new REE Oxyfluoride Mineral Species. *Can. Mineral.* **2004**, *42*, 1097–1103. [CrossRef]
26. Momma, K.; Izumi, F. VESTA 3 for three-dimensional visualization of crystal, volumetric and morphology data. *J. Appl. Crystallogr.* **2011**, *44*, 1272–1276. [CrossRef]
27. Tullis, T.E.; Tullis, J. Experimental Rock Deformation Techniques. *Geophys. Monogr.* **1986**, *36*, 297–324.
28. Rowland, R.L., II. *Phase Equilibria, Compressibility, and Thermal Analysis of Bastnaesite-(La)*; University of Nevada: Las Vegas, NV, USA, 2017.
29. Altomare, A.; Corriero, N.; Cuocci, C.; Falcicchio, A.; Moliterni, A.; Rizzi, R. QUALX2.0: A qualitative phase analysis software using the freely available database POW_COD. *J. Appl. Crystallogr.* **2015**, *48*, 598–603. [CrossRef]
30. Holland, T.J.B.; Redfern, S.A.T. Unit cell refinement from powder diffraction data: the use of regression diagnostics. *Mineral. Mag.* **1997**, *61*, 65–77. [CrossRef]
31. Angel, R.J.; Gonzalez-Platas, J.; Alvaro, M. EosFit7c and a Fortran module (library) for equation of state calculations. *Zeitschrift fur Krist.* **2014**, *229*, 405–419. [CrossRef]
32. Kantor, I.; Prakapenka, V.; Kantor, A.; Dera, P.; Kurnosov, A.; Sinogeikin, S.; Dubrovinskaia, N.; Dubrovinsky, L. BX90: A new diamond anvil cell design for X-ray diffraction and optical measurements. *Rev. Sci. Instrum.* **2012**, *83*. [CrossRef] [PubMed]
33. Hrubiak, R.; Sinogeikin, S.; Rod, E.; Shen, G. The laser micro-machining system for diamond anvil cell experiments and general precision machining applications at the High Pressure Collaborative Access Team. *Rev. Sci. Instrum.* **2015**, *86*. [CrossRef] [PubMed]
34. Rivers, M.; Prakapenka, V.B.; Kubo, A.; Pullins, C.; Holl, C.M.; Jacobsen, S.D. The COMPRES/GSECARS gas-loading system for diamond anvil cells at the Advanced Photon Source. *High Press. Res.* **2008**, *28*, 273–292. [CrossRef]
35. Mao, H.K.; Xu, J.; Bell, P.M. Calibration of the Ruby Pressure Gauge to 800 kbar Under Quasi-Hydrostatic Conditions. *J. Geophys. Res.* **1986**, *91*, 4673. [CrossRef]
36. Hirose, K.; Sata, N.; Komabayashi, T.; Ohishi, Y. Simultaneous volume measurements of Au and MgO to 140 GPa and thermal equation of state of Au based on the MgO pressure scale. *Phys. Earth Planet. Inter.* **2008**, *167*, 149–154. [CrossRef]
37. Wang, Y.; Zhang, J.; Xu, H.; Lin, Z.; Daemen, L.L.; Zhao, Y.; Wang, L. Thermal equation of state of copper studied by high P-T synchrotron X-ray diffraction. *Appl. Phys. Lett.* **2009**, *94*, 4. [CrossRef]
38. Hammersley, A.P. FIT2D: A multi-purpose data reduction, analysis and visualization program. *J. Appl. Crystallogr.* **2016**, *49*, 646–652. [CrossRef]
39. Dera, P.; Zhuravlev, K.; Prakapenka, V.; Rivers, M.L.; Finkelstein, G.J.; Grubor-Urosevic, O.; Tschauner, O.; Clark, S.M.; Downs, R.T. High pressure single-crystal micro X-ray diffraction analysis with GSE_ADA/RSV software. *High Press. Res.* **2013**, *33*, 466–484. [CrossRef]
40. Fei, Y. Thermal Expansion. In *Mineral Physics and Crystallography: A Handbook of Physical Constants*; Ahrens, T.J., Ed.; American Geophysical Union: Washington, DC, USA, 1995; pp. 29–44. ISBN 0-87590-852-7.
41. Birch, F. Finite Elastic Strain of Cubic Crystals. *Phys. Rev.* **1947**, *71*, 809–824. [CrossRef]
42. Ye, Y.; Smyth, J.R.; Boni, P. Crystal structure and thermal expansion of aragonite-group carbonates by single-crystal X-ray diffraction. *Am. Mineral.* **2012**, *97*, 707–712. [CrossRef]
43. Markgraf, S.A.; Reeder, R.J. High-temperature structure refinements of calcite and magnesite. *Am. Mineral.* **1985**, *70*, 590–600.
44. Ross, N.L.; Reeder, R.J. High-pressure structural study of dolomite and ankerite. *Am. Mineral.* **1992**, *77*, 412–421.
45. Ross, N.L. The equation of state and high-pressure behavior of magnesite. *Am. Mineral.* **1997**, *82*, 682–688. [CrossRef]
46. Zhang, J.; Reeder, R.J. Comparative compressibilities of calcite-structure carbonates: Deviations from empirical relations. *Am. Mineral.* **1999**, *84*, 861–870. [CrossRef]
47. Vennari, C.E.; Williams, Q. High-pressure Raman and Nd^{3+} luminescence spectroscopy of bastnäsite-(REE)CO_3F. *Am. Mineral.* **2019**, *104*, 1389–1401. [CrossRef]
48. Anderson, O.L.; Nafe, J.E. The Bulk Modulus-Volume Relationship for Oxide Compounds and Related Geophysical Problems. *J. Geophys. Res.* **1965**, *70*, 3951–3963. [CrossRef]

49. Knittle, E. Static Compression Measurements of Equation of State. In *Mineral physics and Crystallography: A Handbook of Physical Constants*; Ahrens, T.A., Ed.; American Geophysical Union: Washington, DC, USA, 1995; pp. 98–142.
50. Merlini, M.; Perchiazzi, N.; Hanfland, M.; Bossak, A. Phase transition at high pressure in $Cu_2CO_3(OH)_2$ related to the reduction of the Jahn–Teller effect. *Acta Crystallogr. Sect. B Struct. Sci.* **2012**, *B68*, 266–274. [CrossRef]
51. Xu, J.; Kuang, Y.; Zhang, B.; Liu, Y.; Fan, D.; Zhou, W.; Xie, H. High-pressure study of azurite $Cu_3(CO_3)_2(OH)_2$ by synchrotron radiation X-ray diffraction and Raman spectroscopy. *Phys. Chem. Miner.* **2015**, *42*, 805–816. [CrossRef]

minerals

MDPI

Article

Picosecond Acoustics Technique to Measure the Sound Velocities of Fe-Si Alloys and Si Single-Crystals at High Pressure

Eric Edmund †, Michel Gauthier, Daniele Antonangeli, Simon Ayrinhac, Silvia Boccato, Thibault Deletang, Marc Morand, Yiuri Garino, Paraskevas Parisiades and Frédéric Decremps *

Institut de Minéralogie de Physique des Matériaux et de Cosmochimie, Sorbonne Université— CNRS—Muséum National d'Histoire Naturelle, 4 Place Jussieu, F-75005 Paris, France; eric.edmund@hpstar.ac.cn (E.E.); michel.gauthier@sorbonne-universite.fr (M.G.); daniele.antonangeli@sorbonne-universite.fr (D.A.); simon.ayrinhac@sorbonne-universite.fr (S.A.); silvia.boccato@sorbonne-universite.fr (S.B.); thibault.deletang@etu.upmc.fr (T.D.); marc.morand@sorbonne-universite.fr (M.M.); yiuri.garino@sorbonne-universite.fr (Y.G.); Paraskevas.Parisiadis@upmc.fr (P.P.)

* Correspondence: frederic.decremps@sorbonne-universite.fr

† Current address: Center for High Pressure Science and Technology Advanced Research (HPSTAR), Shanghai 201203, China.

Received: 23 January 2020; Accepted: 19 February 2020 ; Published: 27 February 2020

Abstract: We describe here a time resolved pump-probe laser technique—picosecond interferometry—which has been combined with diamond anvil cells (DAC). This method enables the measurement of the longitudinal sound velocity up to Mbar pressure for any kind of material (solids, liquids, metals, insulators). We also provide a description of picosecond acoustics data analysis in order to determine the complete set of elastic constants for single crystals. To illustrate such capabilities, results are given on the pressure dependence of the acoustic properties for prototypical cases: polycrystal (hcp-Fe-5 wt% Si up to 115 GPa) and single-crystal (Si up to 10 GPa).

Keywords: picosecond acoustics; high pressure; sound velocity; elastic constant; iron alloys; silicon

1. Introduction

The study of the effects of pressure and temperature on the acoustic properties of condensed matter is well known to be particularly useful information in the framework of geophysics and physics. Among the most fascinating questions that still need to be addressed, the composition and structure of the planetary core [1], the role of electron-phonon interactions in the emergence of super-conducting phases [2], or the melting curve of alkali liquids [3], could be recalled.

Beyond such fundamental interest, the precise knowledge of the anisotropic distribution of sound velocities in a crystal and their evolution with pressure and temperature can have direct technological applications. The case of silicon is an obvious example for semiconducting technologies, where the knowledge of its complete set of elastic constants under constraints is employed to improve its mechanical stability [4].

Experimentally, access to the acoustic properties of materials submitted to extreme pressure and temperature conditions remains limited, mainly because of the severe constraints imposed on the sample volume by the pressure-generating device. With the notable exception of Brillouin scattering in diamond anvil cells (DACs), which allows measuring the energy of thermal phonons of a transparent sample [5], most studies on the elasticity or visco-elasticity of dense solids and liquids were done by conventional ultrasonics [6], limited to a few tens of GPa [7,8]. The technique of Inelastic X-ray scattering, compatible with the use of DACs, has paved the way for more ambitious studies and has helped to confirm the great relevance of acoustic phonon measurements in physics [9] and geophysics [10]. We emphasize here that this technique is nevertheless very time demanding and particularly suited to the study of the acoustic phonons far from the zone center. It may thus seem expensive (and somewhat less direct) to use such a powerful method to "simply" derive the speed of sound through the help of a dispersion model.

Thereafter, this article will focus on a review of a recent development, the application of picosecond acoustics in diamond anvils cell. Some of its capabilities will here be illustrated by a choice of two examples of experimental results: (i) the measurements of sound velocities of polycrystalline Fe-5 wt% Si and Fe-10 wt% Si at Mbar pressure and (ii) the determination of the complete set of elastic constants on an opaque single-crystalline sample (silicon) up to 10 GPa.

2. Measuring Techniques

Picosecond Acoustics in Diamond Anvil Cells: A Complementary Acoustic Technique for Sound Velocity Measurements at High Pressure and Temperature

The three most well-established types of experiment that allow the determination of the speed of sound in environments subject to extreme conditions are: (i) the ultrasound techniques; (ii) the inelastic neutrons scattering in the large volume press; or using a diamond anvils cell (DAC), (iii) the inelastic scattering of the light by phonons applied either in the visible (Brillouin) or X-ray (IXS) wavelength range. The capabilities, as well as the limitations of each technique are summarized below.

The measurement of sound speed by the ultrasonic technique, where ultrasonic waves are generated through the use of a piezoelectric transducer, may only be carried out for sample dimensions on the order of a few millimeters [6,11]. Therefore, the use of devices allowing large sample volumes (incompatible with the achievement of very high pressures) is required [7,8]. Some attempts have been made to adapt the ultrasonic technique at a very high frequency (on the order of the GHz) to apply the pulse-echo overlap technique in DACs [12], but complications due to interface contact between the sample and the diamond anvils limit the maximum pressure that can be reached, typically less than 15 GPa.

Brillouin scattering could only be applied to transparent materials [5]. In its simpler geometry, mostly used within a DAC, the knowledge of the sample index of refraction is required (including its pressure variation).

As a consequence, if one wished to directly measure the sound velocity of an opaque sample under pressures higher than 20 GPa, IXS has been the only adapted technique [10]. This is an extremely powerful technique as it allows the study of phonon dispersions in condensed matter at momentum transfers and energy transfers characteristic of collective atom motions in a wide range of materials. Disordered systems, opaque crystals, or samples only available in very small quantities could be studied. However, this technique necessitates the use of third-generation synchrotron light sources and long collection times—severely constraining the breadth and scope of such investigations.

In 1986, the first picosecond acoustic experiments were performed, where the authors reported the observation of acoustic oscillations during the measurement of the transitional transmissivity of semiconductor thin films [13]. This idea is schematically illustrated in Figure 1. With this technique, relative changes of the coefficient of electromagnetic reflectivity of the sample are measured as a function of the delay between probe (detection) and pump (generation) laser pulses.

A first ultrashort laser pulse with a duration of approximately 100 fs generates ps acoustic pulses within the sample. The generation process is a result of the photo-elastic coupling between the incident radiation and the material. The energy carried by the incident photons of the pump beam is absorbed by the sample, and electron-phonon coupling causes a local sharp heat pulse (dilatation). This rapid change in volume generates a longitudinal sound wave, the propagation of which is perpendicular to the surface of the sample. A second light pulse, shifted in time with respect to the pump through the use of a delay line, detects the arrival of the acoustic strain field at the surface of the sample. In the framework of a simple photo-elastic model [14,15], the time variation of the sample reflectivity $\delta r(t)$ of the probe is a function of the photo-elastic coefficient $\partial \tilde{n} / \partial \eta$ (where $\tilde{n}$ is the sample refractive index and η the stress generated by thermal expansion) and the surface displacement $u_0(t)$ along the propagation axis z. It can be expressed as:

$$\frac{\delta r(t)}{r_0} = 2ik_0 n_d \left\{ \int_0^{+\infty} \eta(z,t)dz + \left(\frac{\partial \tilde{n}}{\partial \eta}\right) \frac{2\tilde{n}}{n_d^2 - \tilde{n}^2} \int_0^{+\infty} \eta(z,t)e^{2ik_0\tilde{n}z}dz \right\} \quad (1)$$

where k_0 is the laser wave vector and n_d the refractive index of the material surrounding the sample (pressure transmitting medium or diamond anvil, depending on experimental design).

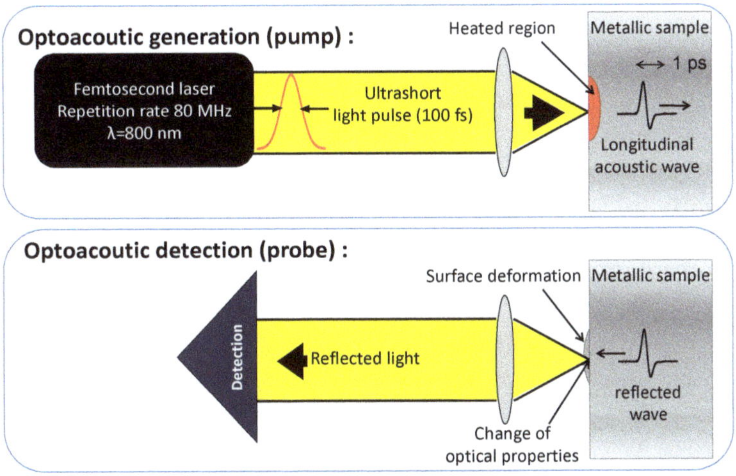

Figure 1. Femtosecond laser light pulse used here as a kind of ultimate "hammer" to generate an ultrashort (ps) acoustic strain. (**Top**) The pump is focused on one surface of the sample (focused spot of about 3 μm) through a diamond window and absorbed on a length scale of ξ ($\approx$ 10 nm). As soon as the pump laser pulse reaches the surface, it creates a sudden and small temperature increase ΔT (less than 10 K). (**Bottom**) The change of temperature relaxes by launching an acoustic strain field (with frequency typically on the order of GHz in a metallic liquid) propagating at the sound velocity v_l (the laser probe is here focused on the opposing face of the sample with respect to the pump beam).

In other words, after propagation across the sample, both thermal and acoustic effects alter the optical reflectivity sample in two ways: the photo-elastic effect and surface displacement (as the acoustic echo reaches the surface). The first modification contributes to the change in intensity of the reflectivity and its phase shift, whereas the second one only modifies the phase of the reflectivity signal.

Using this novel approach combined with DACs, we showed that such a laboratory setup circumvents all previous limitations [16–18]. Technical details related to the setup of such an experiment can be found in [19]. A schematic of the DAC/picosecond acoustics combination setup is given in Figure 2. Sound velocity extraction from the raw experimental data (the time dependence of the surface sample reflectivity) was described in detail in the following references [19–21].

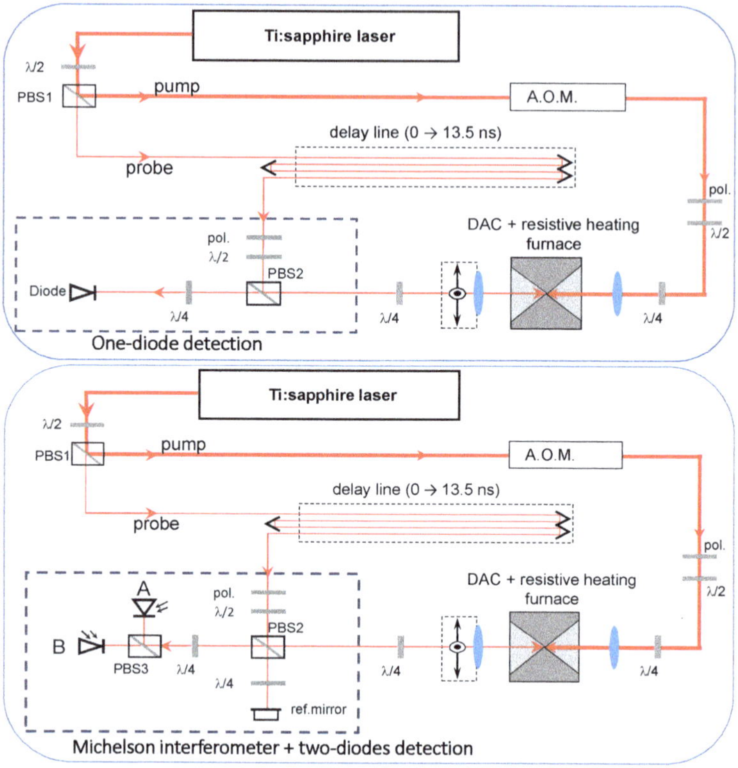

Figure 2. (**Top**) Schematic [19] of the diamond anvil cell (DAC)/picosecond acoustics setup (transmission geometry) within the reflectometry configuration (one diode detection). (**Bottom**) Interferometric setup (both the real and imagery part of the reflectivity are measured using a Michelson interferometer). PBS: polarizing beam-splitter, $\lambda/4$: quarter wave plate, $\lambda/2$: half-wave plate, pol.: linear polarizer, A.O.M.: acousto-optic modulator. A and B: diodes.

3. Measurement of the Sound Velocities of Polycrystalline Fe-Si Alloys at Mbar Pressures

Iron and iron alloys at extreme conditions have garnered significant interest due to their relevance to the Earth's deep interior. While there are strong constraints on the elastic properties of the Earth's interior via seismological investigations (e.g., [22]) providing constraints on density, longitudinal wave velocities and shear wave velocities, there is no direct chemical information regarding Earth's iron-rich core. Consequently, the chemical composition of the Earth's core is heavily debated, with the presence or absence of different light elements indicating the presence or absence of various geochemical cycles [23,24] or differing formation scenarios of the Earth itself (e.g., [25,26]).

One possible way to constrain the Si content of the inner core is the comparison between seismological data and experimental measurements, or calculations, of ρ, V_p, and V_s of candidate materials at pertinent P-T conditions. Thanks to the adaptation of picosecond acoustics to DAC previously described, it has been possible to make direct measurements of the compressional sound velocity of Fe-Si alloy with 5 wt% Si (Fe5Si) and 10 wt% Si (Fe10Si) up to more than 120 GPa.

Shown in Figure 3 are selected travel times collected during high pressure runs of Fe10Si before, during, and after the bcc-hcp transition. It is observed that well within the stability field of bcc or hcp Fe-Si alloys, the first acoustic echo is sharp and unambiguous. However, in the transition region, there is significant distortion of the shape and intensity of the initial rise of the time domain signal.

This likely arises because in the transition region, the sample exists as a mechanical mixture of two elastically distinct phases.

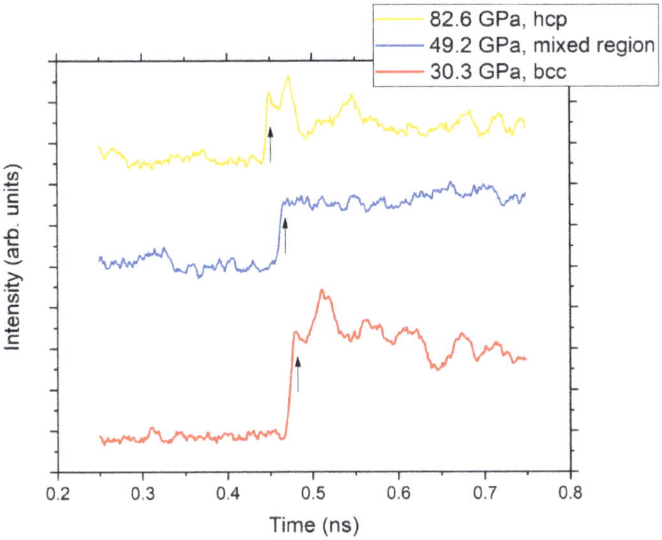

Figure 3. bcc-hcp transition observed by PA time domain measurements for Fe10Si. The signal shows a sharp, well-defined initial acoustic echo above and below the transition region observed by XRD [27].

It is seen that in each case, the bcc-hcp transformation results in some anomaly in the pressure evolution of the normalized travel time (Figure 4), which is somewhat different from alloy to alloy, due to complex changes in the texture, stress, thickness, and elasticity of the sample. Error bars in Figure 4 were estimated as the pressure difference between the last time domain measurement exhibiting a well-defined acoustic echo observed before the transition region in the case of the onset of the bcc-hcp transition, or the first time-domain measurement after the transition region, at the end of the bcc-hcp transition.

The pressure dependence of the sample thickness was taken into account using P-V EoS determined by X-ray diffraction, either after [28] for the bcc and hcp phases of Fe10Si or after [27,29] for Fe5Si. We relate the pressure evolution of the longitudinal sound velocity to the pressure evolution of thickness and acoustic travel time using the following equation:

$$e(t_0) = v_L \Delta t = v_L \left[t_0 + pT_{laser} - \tau\right] \tag{2}$$

where $e(t_0)$ is the sample thickness, p an integer that takes into account the delay due to the successive generation of echoes due to the the laser repetition rate (T_{laser} = 12.554 ns), τ the time at which the pump-probe coincidence occurs, and t_0 the emergence time of the acoustic wave at the surface of the sample.

In Equation (2), the thickness of the sample is assumed to vary purely with the hydrostatic variation of the unit cell volume of the sample. In general, stress exerted by the diamond anvils onto the sample chamber of a DAC leads to sample compression where the stress along the compression axis is equal to or greater than the stress perpendicular to the compression axis. The variation of these stress components and their influence upon elasticity measurements depends on many aspects of the experimental design, sample elasticity, and the choice of pressure-transmitting medium (e.g., [30]). While a hydrostatic stress (where the magnitude of the stress on the sample is equal in all directions) on a given sample can be achieved through the use of a liquid pressure-transmitting medium surrounding the sample, all conventional pressure-transmitting media crystallize into solids by about 12 GPa at ambient temperature [31–33], which leads to the development of differential stress across the sample and the possibility for sample deformation under compression at high pressures. The degree to which such effects take place is bounded by the elastic properties of the sample itself, and the elastic properties of the pressure transmitting medium. On the basis of the determined sound velocities and literature constraints on the shear strength of Ne [34], the pressure gradients that may develop in Ne or Ar are not sufficiently large to induce significant elastic or plastic deformation of the sample in the case of Fe or Fe-alloys relative to experimental uncertainties. Estimations of the Young's moduli of Fe, Fe5Si, and Fe10Si at the highest pressure data points used from these experiments are E = 890 GPa at 152 GPa, E = 810 GPa at 114 GPa, and E = 340 GPa at 36 GPa for these alloys, respectively [27,28,35]. Given that the shear strength of Ar is estimated to be about 3–6 GPa at 150 GPa [36,37] and the shear strength of Ne is less than 1–2 GPa over the pressure ranges investigated in [27,28], deviations from the assumption of purely volumetric variation in thickness are not significant in these systems relative to experimental uncertainties and scatter. However, future studies are needed to test this assumption for materials that undergo reconstructive phase transitions at very high pressures, as the rapid loss of shear or compressive strength in a sample under conditions where compression becomes strongly deviatoric may lead to plastic effects that would be important to constrain for more precise velocity determination at multi-Mbar pressures.

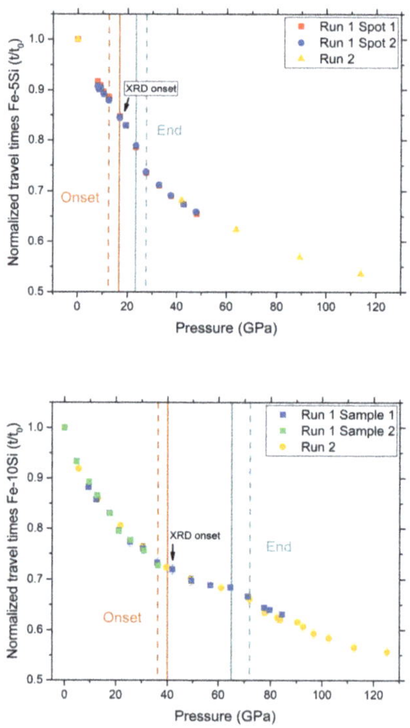

Figure 4. Normalized travel time t/t_0 vs P for different runs on Fe5Si (**up**) and Fe10Si (**down**). The bcc-hcp transition results in a significant reduction in the scatter of the measured travel times for Fe5Si. We also note that sample tilt within the sample chamber has a negligible effect on measured travel times due to the instrumental configuration used. Run numbers indicate the number of each experimental run for a given alloy, and Samples 1 and 2 for Fe10Si indicate two separate samples co-compressed in one experimental run.

Irrespective of such considerations, for Fe-5Si, we observed that the acoustic travel time of a sample varied by nearly 50% from ambient pressure to 1 Mbar, while changes in sample volume led to variations in sample thickness of about 10% over the same pressure range [28]. Consequently, the primary contribution to the variation of the sample acoustic travel time over this pressure range was due to the change in sound velocity, rather than the change in thickness. As a result, under strongly deviatoric conditions where the presented assumptions in thickness are violated, the main systematic errors would more likely arise due to variations in sound velocity from the development of sample texture. For the Fe-Si system, textures remained nearly constant over the studied pressure range as observed by XRD [27,28] and have been proposed as the primary explanation for the strong disagreement between ab initio molecular dynamics calculations (e.g., [38]), IXS [29,39,40], and NRIXS [41] investigations of the elastic properties of the Fe-Si system at high pressures. Indeed, due to the limitations on sample volume imposed by NRIXS and IXS, maintaining a good sample texture is challenging at Mbar pressures.

The sound velocities of Fe, Fe5Si, and Fe10Si are reported in Figure 5. We observed that following the transition region, a sharp rise in sound velocity was observed, with nonlinear variations up to about 30 GPa (Fe5Si) or 80 GPa (Fe10Si), across the bcc-hcp transition region.

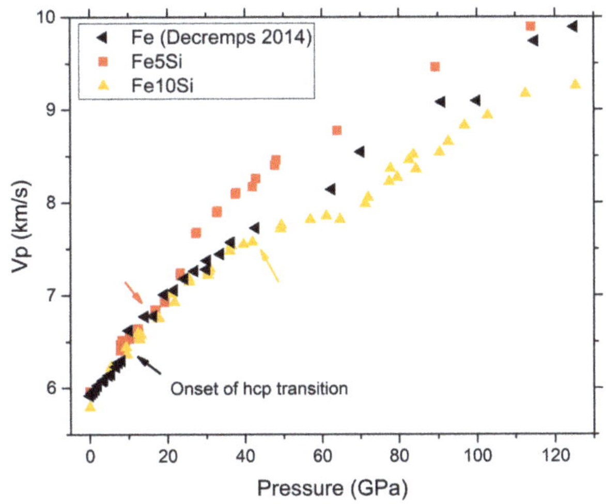

Figure 5. Vp vs. P for Fe (left triangles, labeled Fe [35]), Fe5Si (squares), and Fe10Si (up-triangles) compressed quasi-hydrostatically. While at the transition for Fe, there is a sharp increase in Vp, this does not occur in Fe5Si and Fe10Si, possibly due to a change in the bcc-hcp transition mechanism with Si addition [27]. Errors in Vp were about 2% up to 60 GPa, and up to 3% at higher pressures.

A critical discussion of these results can be found in [27,28].

Through the combination of a P-V-T equation of state and longitudinal sound velocities, it is possible to derive other thermodynamic quantities to place further constraints on core composition and properties. Adiabatic and isothermal bulk moduli can be related by the equation:

$$K_S = K_T(1 + \alpha\gamma_{th}T) \tag{3}$$

where K_S is the adiabatic bulk modulus, K_T is the isothermal bulk modulus, α is the coefficient of thermal expansion of the material, and γ_{th} is the thermodynamic Grüneisen parameter. It follows that knowing $V_p(\rho)$ and $K_S(\rho)$ allows for the determination of shear velocities (V_s) through the equation:

$$\frac{K_S}{\rho} = V_P^2 - \frac{4}{3}V_S^2 \tag{4}$$

For Fe-Si alloys at ambient temperature, adiabatic bulk moduli are only marginally different from isothermal bulk moduli under pressure (about 2% higher at 20 GPa and about 1% higher at 1.2 Mbar [28]). It follows that using either isothermal or adiabatic bulk moduli to derive V_s at ambient temperature results in similar extrapolations to core densities, albeit with isothermal moduli, resulting in a slight overestimation of V_s.

4. Propagation of the Acoustic Energy in a Silicon Single Crystal at High Pressure

4.1. "Phonon Focusing" at High Pressure: Experimental Principles and Data Analysis

In the preceding studies on Fe-Si alloys, the travel time determination was typically similar to what was usually done at MHz frequencies using the pulse-echo ultrasonic technique.

Because of the acoustic far field diffraction limit, the pump beam mainly generated a longitudinal acoustic strain field such that no shear wave could be detected by the probe. In order to circumvent this "shearless" limitation, we demonstrated in a previous study [42] that lateral compressive stresses (producing internal diffraction) could be produced through a minimization of the source area with respect to the characteristic acoustic wavelength. Similarly to the well-known phonon focusing technique [43], we developed a wavefront "imagery method" that could be used to extract the complete set of elastic constants as a function of pressure. The prototypical case of silicon single crystals was studied and described in the following, demonstrating the high accuracy of such method.

We here used a thin platelet single-crystal of silicon oriented along [100] (dimensions were about $70 \times 70\ \mu m^2$). As transducers, thin films (50 nm) of aluminum were sputtered on both sides of the single-crystal platelet. From data at ambient conditions [6] and the picosecond measurement of the travel time, we determined the thickness of the sample to be 42.2(1) µm. This sample was loaded within a DAC, using neon as the pressure transmitting medium and a ruby chip in order to determine the pressure.

For each pressure point, we then used the picosecond setup to image the acoustical surface energy of the sample, as described in [16,42]. The main issue here was the data analysis, as the determination of the elastic constants of a single-crystal through the measurement of its surface energy was a typical inverse problem. Apart from very few cases that could be analytically solved [44], a numerical procedure needed to be developed in order to extract with a high accuracy the complete set of elastic constants from the raw data. Such a procedure was particularly crucial within the context of high pressure data analysis.

In this work, we considered a large number N of experimental velocities $v_{\rm exp}^m$ in well-chosen directions. Indeed, by working on pre-aligned samples or with known orientations, experimental velocities could be measured in the plane (100) giving access to C_{11} and C_{44}, and (110) ($\theta = 45°$) for C_{12} (for a cubic symmetry). For each direction, the corresponding experimental velocity was the solution of the Christoffel equation.

Theoretical velocities $v_e^{\rm th}$ were calculated for several pump-probe delays (identical to the experimental delays) and in several directions for a given elastic tensor C_{ijkl}. Taking into account the laser characteristics (to determine the pump-probe delay), the density ρ and the thickness of the sample e and its initial input values C_{ijkl}^0 were approximated from the expected elastic constants. To estimate the agreement between the experimental and theoretical velocities, the χ^2 test was chosen, assuming that the experimental velocities followed a normal distribution, centered around theoretical velocities:

$$\chi^2 = \sum_{m=1}^{N} \frac{[v_{\rm th}^e(n_i, C_{ijkl}, \rho) - v_{\rm exp}^m]^2}{\sigma_m^2}. \tag{5}$$

where the parameter σ_m corresponds to the estimated uncertainty associated with the mth experimental velocity measurement. The procedure to optimize the elastic constants (schematized by a diagram giving in Figure 6) then consisted of minimizing the difference between the experimental and theoretical velocities by a least squares method.

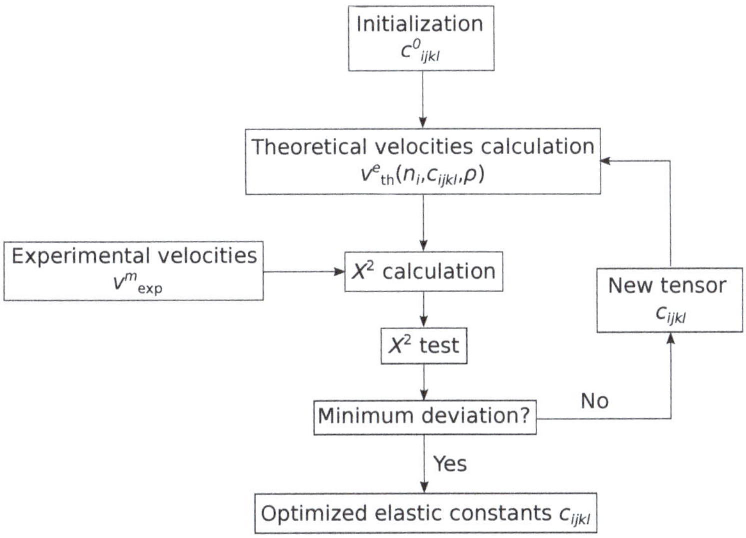

Figure 6. Calculation procedure for the determination of the complete set of elastic constants C_{ijkl} using experimental measurements of 3D phonon imaging patterns of single crystals.

The optimization procedure converged towards optimized elastic constants according to the initialization values C^0_{ijkl} given to the algorithm, but the latter succeeded in most cases (except in values much too higher/inferior compared to those expected) to find the expected values. The accuracy on the calculation of elastic constants by the inverse problem depended on the number of experimental velocities considered: the more there were, the more precise the calculation was.

4.2. *Ambient Conditions Elastic Constants*

Past work [45,46] at ambient conditions showed that pulling part of the diffraction effects, it was possible through a reduction of the pump laser spot to generate and detect transverse waves by picosecond acoustics. Following such a technique, the Christoffel equation was here solved for a set of wave vectors **k** lying within 45 degrees (cubic symmetry) around the [100] crystallographic direction of our silicon single crystal (ρ = 2.331 g·cm^{-3}). For the three acoustical polarizations, we calculated from ray theory [47] the set of elastic constants that produced the observed slowness curves and the wave front curves within a surface section in the plane (100) (see Figure 7).

The simulated process of the wave surfaces, where the elastic constants were the fitting parameters, well reproduced the experimental pattern and gave rise to an elastic tensor in very good agreement with previously published data [6,48]. We here emphasize that, for a given thermodynamical condition (ambient or high pressures), successive patterns at different pump-probe delays could be used to renew the fitting process in order to determine the complete set of elastic constants with an exceptional accuracy (see Figure 8).

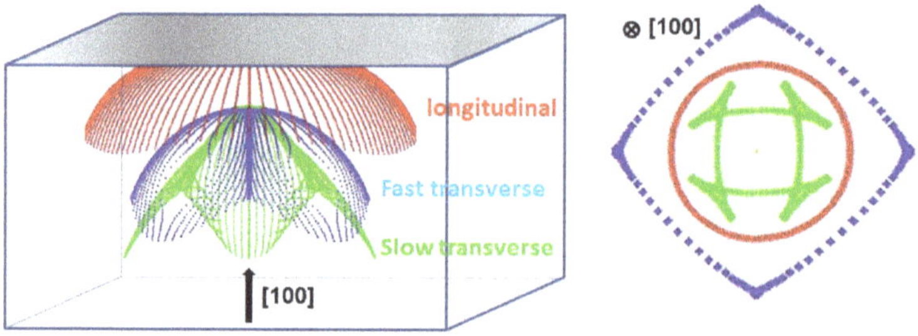

Figure 7. Group velocity surfaces at ambient conditions along the [100] symmetry axis of ZB-cubic silicon. The pump probe delay time here is 3.5 ns. Red, blue, and green dashed lines correspond respectively to the longitudinal, fast, and slow transversal group velocities. Right: surface wave sections along the (100) plane of a cubic zincblende structure)

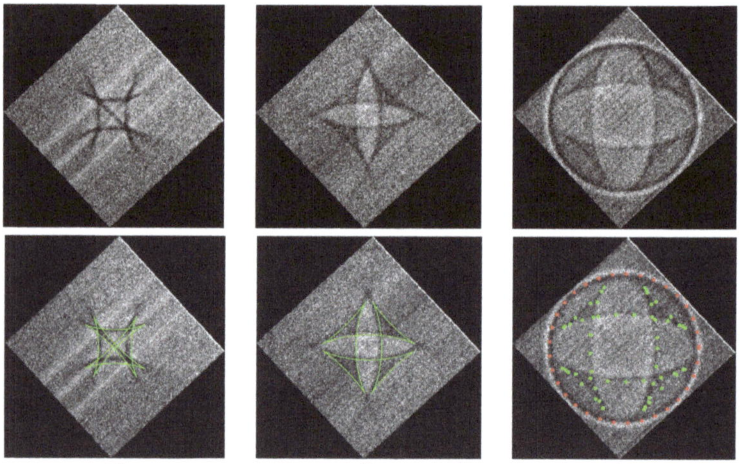

Figure 8. Top panel: Experimental phonon imaging pattern (100 × 100 μm^2) in the (100) plane of silicon at three different delay times: (**a**) 8.3 ns, (**b**) 9.6 ns, and (**c**) 10.7 ns. Lower panel: Comparison between simulation and experimental phonon imaging patterns using C_{11}= 165.7 GPa, C_{12}= 63.9 GPa, and C_{44} = 79.5 GPa.

4.3. High Pressure Dependence of the Complete Set of Elastic Constants

Using the procedure described previously and illustrated in Figure 6, we simulated for each high pressure point the 3D phonon imaging patterns for the three acoustic polarizations. Sound velocities and elastic constants of a Si cubic single crystal were then deduced from high pressure picosecond measurements in DAC. The high accuracy of such technique allowed fitting the C_{11}), C_{44}, and C_{12} high pressure data point with a second-order polynomial regression (data in GPa):

$C_{11}(P) = 165.7 + 4.73P - 0.09P^2$;
$C_{12}(P) = 63.60 + 5.78P - 0.12P^2$;
C_{44} (P) = 80 GPa (constant)

Finally, using the experimental values of C_{IJ}(p), the highly accurate and free-hypothesis equation of state B(p) was determined.

A second-order polynomial curve well reproduced the experimental pressure dependence of the bulk modulus with: B(P) = 97.60 + 5.74P − 0.16P^2, in excellent agreement with a fourth-order regression of the elastic energy with respect to the strain giving: B′ = dB/dP = 5.08 − 0.24P.

5. Conclusions

In conclusion, we showed that the adaptation of picosecond acoustics for measurements in diamond anvil cell opens a wide landscape of new experimental research opportunities. Complementary to the ongoing efforts focused on the use of large facilities, this new high pressure technique allows for determination of the sound velocity as a function of density over a pressure and temperature range one order of magnitude higher than what was previously attainable in both poly- and single-crystalline materials, including metallic samples.

In a first example, we showed that the compressional and shear properties of Fe-Si alloys were incompatible with seismic observations of Earth's inner core [28]. By combining measurements of the P-V-T EoS and compressional sound velocities of Fe5Si, it was shown that Fe-Si alloys whose velocities were expected to get close to seismological observations were too dense at relevant temperatures, and thus, significant quantities of other elements were needed to describe the elastic properties of Earth's inner core.

A second example showed that this technique is an adequate tool to extract at high pressure the complete set of elastic constants of single-crystalline silicon with an outstanding accuracy. This study demonstrated that picosecond acoustics in DAC enabled nonlinear acoustics, anisotropy, the equation of state, as well as structural stability measurement on single crystals, whatever the sample.

Finally, and from a more general point of view, we here emphasize that combined with laser heating, this technique would also open a way of determining the sound velocity under the thermodynamical conditions of the Earth's core (around three million atmospheres and 6000 K). Due to the shorter time scale of the experiments, the temperature stability of the laser was less technically challenging, and there was a lower chance for chemical reaction between the Fe-bearing sample and diamonds. Such measurements will be crucial to new and exciting advances in our comprehension of the properties of the Earth's deep interior.

Author Contributions: Software M.M. and Y.G.; investigation F.D., E.E., M.G., D.A., S.A., S.B., P.P.; data curation, F.D., E.E., M.G., S.A., T.D.; writing–original draft preparation, F.D.; writing–review and editing F.D. and E.E.; supervision F.D.; project administration, D.A.; funding acquisition, D.A. All authors have read and agreed to the published version of the manuscript.

Funding: This project received funding from the European Research Council (ERC) under the European Union's Horizon 2020 research and innovation program (Grant Agreement No. 724690).

Conflicts of Interest: The authors declare no conflict of interest.

References

1. Anderson, O.; Isaak, D. Another look at the core density deficit of Earth's outer core. *Phys. Earth Planet. Int.* **2002**, *131*, 19–27. doi:10.1016/S0031-9201(02)00017-1. [CrossRef]
2. Kong, Y.; Dolgov, O.; Jepsen, O.; Andersen, O. Electron-phonon interaction in the normal and superconducting states of MgB_2. *Phys. Rev. B* **2001**, *64*. doi:10.1103/PhysRevB.64.020501. [CrossRef]
3. Gregoryanz, E.; Degtyareva, O.; Somayazulu, M.; Hemley, R.J.; Mao, H.K. Melting of Dense Sodium. *Phys. Rev. Lett.* **2005**, *94*, 185502. [CrossRef] [PubMed]
4. Tang, Z.; Zhao, H.; Li, G.; Aluru, N.R. Finite-temperature quasicontinuum method for multiscale analysis of silicon nanostructures. *Phys. Rev. B* **2006**, *74*, 064110. doi:10.1103/PhysRevB.74.064110. [CrossRef]
5. Decremps, F.; Datchi, F.; Polian, A. Hypersonic velocity measurement using Brillouin scattering technique. Application to water under high pressure and temperature. *Ultrasonics* **2006**, *44*, E1495–E1498. [CrossRef]
6. McSkimin, H.; Andreacht, P. Elastic moduli of silicon vs hydrostatic pressure at 25 degrees C and −195.8 degrees C. *J. Appl. Phys.* **1964**, *35*, 2161. doi:10.1063/1.1702809. [CrossRef]

7. Li, B.; Gwanmesia, G.D.; Liebermann, R.C. Sound velocities of olivine and beta polymorphs of Mg_2SiO_4 at Earth's transition zone pressures. *Geophys. Res. Lett.* **1996**, *23*, 2259–2262. doi:10.1029/96GL02084. [CrossRef]
8. Gauthier, M.; Lheureux, D.; Decremps, F.; Fischer, M.; Itié, J.P.; Syfosse, G.; Polian, A. High-pressure ultrasonic setup using the Paris–Edinburgh press: Elastic properties of single crystalline germanium up to 6 GPa. *Rev. Sci. Instrum.* **2003**, *74*, 3712–3716. doi:10.1063/1.1593791. [CrossRef]
9. Krisch, M.; Loubeyre, P.; Ruocco, G.; Sette, F.; Cunsolo, A.; D'Astuto, M.; LeToullec, R.; Lorenzen, M.; Mermet, A.; Monaco, G.; et al. Pressure evolution of the high-frequency sound velocity in liquid water. *Phys. Rev. Lett.* **2002**, *89*. doi:10.1103/PhysRevLett.89.125502. [CrossRef]
10. Antonangeli, D.; Occelli, F.; Requardt, H.; Badro, J.; Fiquet, G.; Krisch, M. Elastic anisotropy in textured hcp-iron to 112 GPa from sound wave propagation measurements. *Earth Planet. Sci. Lett.* **2004**, *225*, 243–251. doi:10.1016/j.epsl.2004.06.004. [CrossRef]
11. Decremps, F.; Fischer, M.; Polian, A.; Sieskind, M. Elasticity of BaFCl single crystal under hydrostatic pressure. *Eur. Phys. J. B* **1998**, *5*, 7–13. doi:10.1007/s100510050412. [CrossRef]
12. Jacobsen, S.D.; Reichmann, H.J.; Spetzler, H. *Advances in High-Pressure Technology for Geophysical Applications*; Chen, J., Wang, Y., Duffy, T.S., Shen, G., Dobrzhinetskaya, L., Eds.; Cambridge University Press: Cambridge, UK, 2005; p. 25.
13. Maris, H. Picosecond ultrasonics. *Sci. Am.* **1998**, *278*, 86–89. [CrossRef]
14. Matsuda, O.; Larciprete, M.C.; Li Voti, R.; Wright, O.B. Fundamentals of picosecond laser ultrasonics. *Ultrasonics* **2015**, *56*, 3–20. doi:10.1016/j.ultras.2014.06.005. [CrossRef] [PubMed]
15. Wright, O.B.; Perrin, B.; Matsuda, O.; Gusev, V.E. Optical excitation and detection of picosecond acoustic pulses in liquid mercury. *Phys. Rev. B* **2008**, *78*, 024303. doi:10.1103/PhysRevB.78.024303. [CrossRef]
16. Decremps, F.; Belliard, L.; Gauthier, M.; Perrin, B. Equation of state, stability, anisotropy and nonlinear elasticity of diamond-cubic (ZB) silicon by phonon imaging at high pressure. *Phys. Rev. B* **2010**, *82*. doi:10.1103/PhysRevB.82.104119. [CrossRef]
17. Decremps, F.; Belliard, L.; Couzinet, B.; Vincent, S.; Munsch, P.; Le Marchand, G.; Perrin, B. Liquid mercury sound velocity measurements under high pressure and high temperature by picosecond acoustics in a diamond anvils cell. *Rev. Sci. Instrum.* **2009**, *80*. doi:10.1063/1.3160104. [CrossRef] [PubMed]
18. Decremps, F.; Belliard, L.; Perrin, B.; Gauthier, M. Sound velocity and absorption measurements under high pressure using picosecond ultrasonics in a diamond anvil cell: Application to the stability study of AlPdMn. *Phys. Rev. Lett.* **2008**, *100*. doi:10.1103/PhysRevLett.100.035502. [CrossRef]
19. Decremps, F.; Gauthier, M.; Ayrinhac, S.; Bove, L.; Belliard, L.; Perrin, B.; Morand, M.; Le Marchand, G.; Bergame, F.; Philippe, J. Picosecond acoustics method for measuring the thermodynamical properties of solids and liquids at high pressure and high temperature. *Ultrasonics* **2015**, *56*, 129–140. doi:10.1016/j.ultras.2014.04.011. [CrossRef]
20. Ayrinhac, S.; Gauthier, M.; Le Marchand, G.; Morand, M.; Bergame, F.; Decremps, F. Thermodynamic properties of liquid gallium from picosecond acoustic velocity measurements. *J. Phys. Condens. Matter* **2015**, *27*, 275103. [CrossRef]
21. Ayrinhac, S.; Gauthier, M.; Bove, L.E.; Morand, M.; Le Marchand, G.; Bergame, F.; Philippe, J.; Decremps, F. Equation of state of liquid mercury to 520 K and 7 GPa from acoustic velocity measurements. *J. Chem. Phys.* **2014**, *140*, 244201. [CrossRef]
22. Dziewonski, A.M.; Anderson, D.L. Preliminary reference Earth model. *Phys. Earth Planet. Int.* **1981**, *25*, 297–356. doi:10.1016/0031-9201(81)90046-7. [CrossRef]
23. Mao, H.K.; Hu, Q.; Yang, L.; Liu, J.; Kim, D.Y.; Meng, Y.; Zhang, L.; Prakapenka, V.B.; Yang, W.; Mao, W.L. When water meets iron at Earth's core–mantle boundary. *Natl. Sci. Rev.* **2017**, *4*, 870–878. doi:10.1093/nsr/nwx109. [CrossRef]
24. Dasgupta, R.; Hirschmann, M.M. The deep carbon cycle and melting in Earth's interior. *Earth Planet. Sci. Lett.* **2010**, *298*, 1–13. doi:10.1016/j.epsl.2010.06.039. [CrossRef]
25. Siebert, J.; Badro, J.; Antonangeli, D.; Ryerson, F.J. Terrestrial Accretion Under Oxidizing Conditions. *Science* **2013**, *339*, 1194–1197. doi:10.1126/science.1227923. [CrossRef] [PubMed]
26. Fischer, R.A.; Nakajima, Y.; Campbell, A.J.; Frost, D.J.; Harries, D.; Langenhorst, F.; Miyajima, N.; Pollok, K.; Rubie, D.C. High pressure metal-silicate partitioning of Ni, Co, V, Cr, Si, and O. *Geochim. Cosmochim. Acta* **2015**, *167*, 177–194. doi:10.1016/j.gca.2015.06.026. [CrossRef]

27. Edmund, E.; Antonangeli, D.; Decremps, F.; Morard, G.; Ayrinhac, S.; Gauthier, M.; Boulard, E.; Mezouar, M.; Hanfland, M.; Guignot, N. Structure and elasticity of cubic Fe-Si alloys at high pressures. *Phys. Rev. B* **2019**, *100*, 134105. doi:10.1103/PhysRevB.100.134105. [CrossRef]
28. Edmund, E.; Antonangeli, D.; Decremps, F.; Miozzi, F.; Morard, G.; Boulard, E.; Clark, A.N.; Ayrinhac, S.; Gauthier, M.; Morand, M.; et al. Velocity-Density Systematics of Fe-5wt%Si: Constraints on Si Content in the Earth's Inner Core. *J. Geophys. Res. Solid Earth* **2019**, *124*, 3436–3447. doi:10.1029/2018JB016904. [CrossRef]
29. Antonangeli, D.; Morard, G.; Paolasini, L.; Garbarino, G.; Murphy, C.A.; Edmund, E.; Decremps, F.; Fiquet, G.; Bosak, A.; Mezouar, M.; et al. Sound velocities and density measurements of solid hcp-Fe and hcp-Fe-Si (9 wt.%) alloy at high pressure: Constraints on the Si abundance in the Earth's inner core. *Earth Planet. Sci. Lett.* **2018**, *482*, 446–453. doi:10.1016/j.epsl.2017.11.043. [CrossRef]
30. Takemura, K. Evaluation of the hydrostaticity of a helium-pressure medium with powder X-ray diffraction techniques. *J. Appl. Phys.* **2001**, *89*, 662. doi:10.1063/1.1328410. [CrossRef]
31. Finger, L.W.; Hazen, R.M.; Zou, G.; Mao, H.K.; Bell, P.M. Structure and compression of crystalline argon and neon at high pressure and room temperature. *Appl. Phys. Lett.* **1981**, *39*, 892. doi:10.1063/1.92597. [CrossRef]
32. Loubeyre, P.; Besson, J.M.; Pinceaux, J.P.; Hansen, J.P. High-Pressure Melting Curve of ^{4}He. *Phys. Rev. Lett.* **1982**, *49*, 1172. doi:10.1103/PhysRevLett.49.1172. [CrossRef]
33. Klotz, S.; Chervin, J.C.; Munsch, P.; Marchand, G.L. Hydrostatic limits of 11 pressure transmitting media. *J. Phys. D Appl. Phys.* **2009**, *42*, 075413. [CrossRef]
34. Dorfman, S.M.; Prakapenka, V.B.; Meng, Y.; Duffy, T.S. Intercomparison of pressure standards (Au, Pt, Mo, MgO, NaCl and Ne) to 2.5 Mbar. *J. Geophys. Res. Solid Earth* **2012**, *117*, B08210. doi:10.1029/2012JB009292. [CrossRef]
35. Decremps, F.; Antonangeli, D.; Gauthier, M.; Ayrinhac, S.; Morand, M.; Le Marchand, G.; Bergame, F.; Philippe, J. Sound velocity of iron up to 152 GPa by picosecond acoustics in diamond anvil cell. *Geophys. Res. Lett.* **2014**, *41*, 1459–1464. doi:10.1002/2013GL058859. [CrossRef]
36. Mao, H.K.; Badro, J.; Shu, J.; Hemley, R.J.; Singh, A.K. Strength, anisotropy, and preferred orientation of solid argon at high pressures. *J. Phys. Condens. Matter* **2006**, *18*, 963. [CrossRef]
37. Marquardt, H.; Speziale, S.; Gleason, A.; Sinogeikin, S.; Kantor, I.; Prakapenka, V.B. Brillouin scattering and X-ray diffraction of solid argon to 65 GPa and 700 K: Shear strength of Argon at HP/HT. *J. Appl. Phys.* **2013**, *114*, 093517. doi:10.1063/1.4820578. [CrossRef]
38. Martorell, B.; Wood, I.G.; Brodholt, J.; Vocadlo, L. The elastic properties of hcp-Fe1-xSix at Earth's inner-core conditions. *Earth Planet. Sci. Lett.* **2016**, *451*, 89–96. doi:10.1016/j.epsl.2016.07.018. [CrossRef]
39. Mao, Z.; Lin, J.F.; Liu, J.; Alatas, A.; Gao, L.; Zhao, J.; Mao, H.K. Sound velocities of Fe and Fe-Si alloy in the Earth's core. *Proc. Natl. Acad. Sci. USA* **2012**, *109*, 10239–10244. doi:10.1073/pnas.1207086109. [CrossRef]
40. Sakairi, T.; Sakamaki, T.; Ohtani, E.; Fukui, H.; Kamada, S.; Tsutsui, S.; Uchiyama, H.; Baron, A.Q. Sound velocity measurements of hcp Fe-Si alloy at high pressure and high temperature by inelastic X-ray scattering. *Am. Mineral.* **2018**, *103*, 85. doi:10.2138/am-2018-6072. [CrossRef]
41. Lin, J.F.; Struzhkin, V.V.; Sturhahn, W.; Huang, E.; Zhao, J.; Hu, M.Y.; Alp, E.E.; Mao, H.k.; Boctor, N.; Hemley, R.J. Sound velocities of iron-nickel and iron-silicon alloys at high pressures. *Geophys. Res. Lett.* **2003**, *30*, 2112. doi:10.1029/2003GL018405. [CrossRef]
42. Decremps, F.; Belliard, L.; Perrin, B. Structure and elasticity of single-crystals by phonon imaging at high pressure. *Acta Crystallogr. A Found. Adv.* **2011**, *67*, C111–C112. doi:10.1107/S0108767311097273. [CrossRef]
43. Hurley, D.; Wolfe, J. Phonon Focusing in Cubic-Crystals. *Phys. Rev. B* **1985**, *32*, 2568–2587. doi:10.1103/PhysRevB.32.2568. [CrossRef] [PubMed]
44. Buchwald, V.; Davis, A. Surface waves in anisotropic elastic media. *Nature* **1961**, *191*, 899. doi:10.1038/191899a0. [CrossRef]
45. Audoin, B.; Bescond, C.; Deschamps, M. Measurement of stiffness coefficients of anisotropic materials from pointlike generation and detection of acoustic waves. *J. Appl. Phys.* **1996**, *80*, 3760–3771. doi:10.1063/1.363328. [CrossRef]
46. Rossignol, C.; Rampnoux, J.M.; Perton, M.; Audoin, B.; Dilhaire, S. Generation and Detection of Shear Acoustic Waves in Metal Submicrometric Films with Ultrashort Laser Pulses. *Phys. Rev. Lett.* **2005**, *94*, 166106. doi:10.1103/PhysRevLett.94.166106. [CrossRef] [PubMed]

47. Auld, B.A. *Acoustic Fields and Waves in Solids*; Malabar, F.L., Ed.; R.E. Krieger Publishing Company: Malabar, FL, USA, 1990; Volume I.
48. McSkimin, H.; Andreacht, P. Measurement of 3-order moduli of silicon and germmanium. *J. Appl. Phys.* **1964**, *35*, 3312. doi:10.1063/1.1713214. [CrossRef]

minerals

MDPI

Article

Melting Curve of Potassium Chloride from in situ Ionic Conduction Measurements

Dongyuan Zhou [1,2], Junjie Dong [1,3], Yanhan Si [1], Feng Zhu [1,4] and Jie Li [1,*]

1 Department of Earth and Environmental Sciences, University of Michigan, Ann Arbor, MI 48109, USA; zhoudy@mail.ustc.edu.cn (D.Z.); junjiedong@g.harvard.edu (J.D.); yanhans@umich.edu (Y.S.); fzhuum@umich.edu (F.Z.)

2 School of Earth and Space Sciences, University of Science and Technology of China, Hefei 230026, China

3 Department of Earth and Planetary Sciences, Harvard University, Cambridge, MA 02138, USA

4 Hawaii Institute of Geophysics and Planetology, University of Hawaii at Mānoa, Honolulu, HI 96822, USA

* Correspondence: jackieli@umich.edu

Received: 31 January 2020; Accepted: 5 March 2020; Published: 9 March 2020

Abstract: We report experimental constraints on the melting curve of potassium chloride (KCl) between 3.2 and 9 GPa from in situ ionic conduction measurements using a multi-anvil apparatus. On the basis of concurrent measurements of KCl and sodium chloride (NaCl) at 1 bar using the differential thermal analysis (DTA) method and Pt sphere marker, we show that the peak rate of increase in ionic current with temperature upon heating coincides with latent heat ledge and fall of Pt sphere, thus establishing the criterion for melting detection from ionic conduction measurements. Applying this criterion to high pressures, we found that the melting point of KCl rose steeply with increasing pressure to exceed 2443 ± 100 K at 9 GPa. Fitting the results of this study together with existing data at pressures below 4 GPa and above 20 GPa, we obtained the Simon's melting equation for KCl in the simple cubic B2 structure between 1.8 and 50 GPa: $T_m = 1323\left(\frac{P-1.87}{2.2(1)} + 1\right)^{\frac{1}{2.7(1)}}$, where T is in K and P is in GPa. Starting at 1 bar, the melting point of KCl increases at an average rate of ~150 K/GPa to cross that of Pt near 9 GPa. The highly refractory nature of KCl makes it a sensitive pressure calibrant for the large-volume pressure at moderate pressures and a potential sample container for experiments at moderate pressures and very high temperatures.

Keywords: melting criterion; high pressure; pressure calibration; pressure medium

1. Introduction

Potassium chloride (KCl) occurs in nature as sylvite and is widely used as a pressure transmitting medium and thermal insulating material in high-pressure experiments with the diamond-anvil cells [1]. The systematic melting behavior of alkali halides has been used to test theories on interatomic forces in solids and liquids, including long-range Coulomb interaction and short-range potential [2]. Existing data on the melting point of compressed KCl include piston-cylinder results at pressures up to 4 GPa [3] and several diamond-anvil cell data points at pressures above 20 GPa [2]. Because the maximum pressure of standard piston-cylinder apparatus is limited to ~5 GPa and the melting point of KCl rises steeply with pressure to approach or exceed the graphitization temperature of diamonds, experimental constraints on the melting curve of KCl at moderate pressures are challenging to obtain and are currently unavailable.

The high-pressure ionic conduction method was developed recently and applied to investigate the melting behavior of ionic compounds NaCl [4], sodium carbonate (Na_2CO_3), calcium carbonate ($CaCO_3$) [5], and barium carbonate ($BaCO_3$) [6]. In these studies, comparisons with separate experiments using platinum (Pt) sphere markers and literature data using differential thermal analysis (DTA) suggested that at a given pressure, melting could be detected by a steep rise in ionic conduction with increasing temperature. Given uncertainties associated with cross-experiment and cross-laboratory comparisons, it is desirable to rigorously test the criterion for melting detection through simultaneous observations and direct comparisons between conduction, latent heat effect, and sphere motion.

Previous ionic conduction experiments used a symmetric configuration where the tips of electrodes and junction of thermocouple were placed at the opposite ends of the sample chamber, presumably along an isotherm. In reality, symmetric positioning was not easily realized because of uncertainties in initial placement and deformation during compression and heating [5]. Moreover, the external thermocouple does not register the latent heat effect to accurately measure the sample temperature during melting, and therefore the data quality is compromised.

Here we performed concurrent DTA and Pt sphere marker experiments on NaCl and KCl at 1 bar to test the criterion for melting detection in ionic conduction observations. Applying modified configurations with single-side four-wire placements to reduce measurement errors and improve detection sensitivity, we determined the melting curve of KCl at high pressures. These results allowed us to cross-examine various criteria for melting detection and evaluate the applications of KCl as a pressure calibrant for large-volume press and as a material for thermal insulation, pressure transmission, and sample containment in high-pressure research.

2. Method

The starting material was high-purity crystalline KCl (Alfa Aesar 10839, 99.997%) or NaCl (Alfa Aesar 10862, 99.999%). The sample was kept in a vacuum oven at 400 K prior to the experiment, and directly loaded into Pt container or ground into fine powder and then packed into Pt container.

2.1. *Experiments at 1 bar*

Concurrent ionic conduction, differential thermal analysis (DTA), and Pt sphere marker measurements at the ambient pressure of 1 bar were carried out using a Petit box furnace. The DTA experiments used two type S (Pt-Rh) thermocouples with welded junctions, one near the crucible and the other embedded in the sample (Figure 1). Two Pt electrode wires were inserted into a four-bore alumina sleeve, along with the wires of the embedded thermocouple that measured the temperature of the sample near the tips of the electrodes. The conduction measurements circuit used 0.1 VAC to 1 VAC and 60 Hz excitation voltage and recorded current up to 900 μA using a FLUKE 289 multimeter [6]. A Mastech variac was used to adjust the excitation voltage. For the marker experiments, several Pt spheres with 100–300 μm diameters were placed on top of the sample. The furnace was heated at a rate of 0.2 to 1 K/s in most experiments, using an internal type K thermocouple, and cooled by shutting off the power. The temperatures of the thermocouples were monitored using a multi-channel DataQ device.

2.2. *Multi-Anvil Experiments*

Ionic conduction experiments were conducted at pressures up to 9 GPa using the 1000-ton Walker-type multi-anvil apparatus at the University of Michigan. Tungsten carbide cubic anvils with truncation edge length (TEL) of 8 mm (Fansteel) and 5 mm (Toshiba-Tungaloy F-grade) were used to generate high pressures. The 8-mm assembly used 646 Ceramacast octahedra with fins [7] and a single-side radial placement of a type C (W-Re) thermocouple and type C negative electrode wires in a four-bore alumina (Al_2O_3) tubing. The thermocouple junction and electrode tips were in contact with the sample and at the equator of a cylindrical-shaped rhenium (Re) heater where the temperature is the highest along the axis (Figure 2a). The 5-mm assembly was based on the COMPRES design and used a single-side axial configuration, where a type C (W-Re) thermocouple and a pair of Pt electrode wires were placed in a four-bore alumina tubing, also at the heater equator (Figure 2b). The axial assemblies were dried in the vacuum oven at 400 K for at least 24 h before loading into the multi-anvil apparatus.

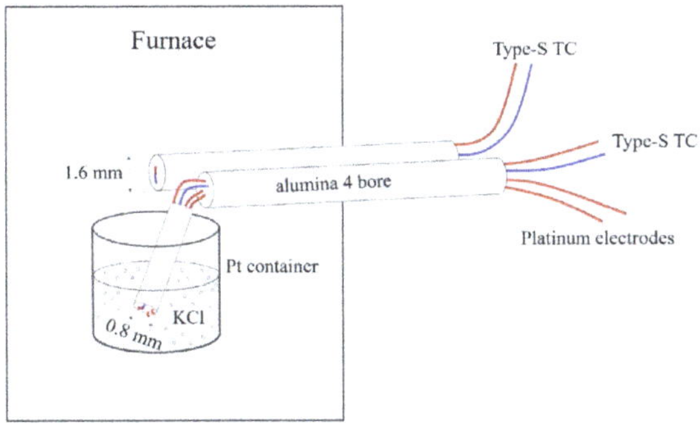

Figure 1. Configuration of concurrent differential thermal analysis (DTA) and ionic conduction experiments at 1 bar. The container was either a 10 mL Pt crucible or a cylindrical Pt container with 1.6 mm diameter and 3–5 mm height. The diameters of the type S thermocouples and Pt electrodes were 8 mil (0.20 mm) or 5 mil (0.13 mm).

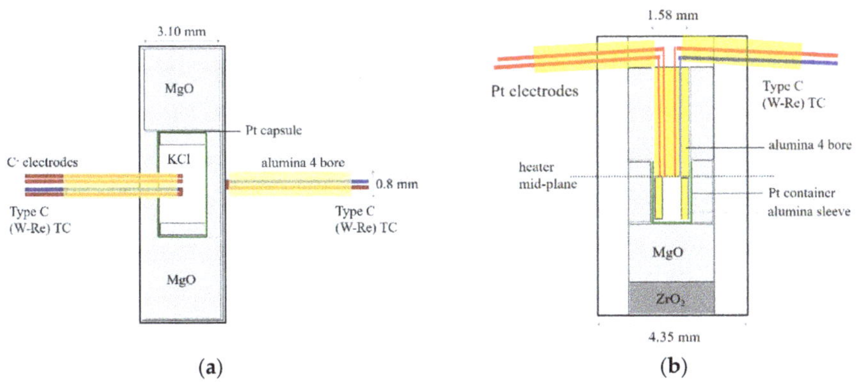

Figure 2. Configurations of multi-anvil conduction experiments with (**a**) radial placement and (**b**) axial placement of thermocouple and electrode wires.

The relationship between the sample pressure and oil load of the hydraulic press was calibrated through fixed point phase transitions and NaCl melting curve. With radial placement experiments (Figure 2a), we converted press loads to sample pressures using the NaCl-melting-based pressure calibration for 8-mm 646 octahedra [7]. The sample pressures in the axial placement experiments were calculated using the calibration for 5-mm COMPRES assembly [8], which was reproduced in this laboratory. The uncertainty in pressure measurement is estimated to be ±7%. This includes the precision of pressure calibration of ±5% estimated based on duplicate experiments, and systematic errors of ±5% arising from the effect of temperature on pressure calibration and pressure drift during heating and cooling [4]. The uncertainty in temperature measurement is estimated to be ±50 K, by considering the precision in the thermocouple calibration and the effects of pressure on the electromotive force (emf) of the thermocouples [9].

In a typical multi-anvil experiment, the sample was compressed to the first target press load, heated at the rate of ~1 K/s to cross the melting point by 50 to 200 K, then cooled to a few hundred degrees below the melting point to complete one heating and cooling cycle. The temperature was controlled using the embedded thermocouple. Multiple cycles were performed to check reproducibility and to improve the precision of melting detection at a reduced heating rate of 6 K/m, before further compression to the next target press load. After the final cycle, the sample was quenched to room temperature by shutting off the heating power, and then decompressed to room pressure. Temperature and multiple heating and cooling cycles may affect the pressure calibration, as discussed later in this paper.

The conduction measurement circuit was identical to that of the 1 bar experiments. External electromagnetic interference, including heating current and the pressure control motor, was less than a few μA and negligible compared with the ionic current through molten KCl.

Experimental products were recovered by grinding and polishing in oil or directly on SiC papers. An optical microscope was used to check the sample geometry and see if the Pt spheres remained separate or met at the bottom of the capsule.

3. Results

3.1. KCl at 1 bar

At 1 bar, our differential thermal analysis (DTA) measurements of KCl yielded a melting point of 1042 ± 2 K, where a latent heat ledge was observed upon heating and cooling (Table 1, Figure 3a). The width of the latent heat ledge varied with the heating or cooling rate and sample size. The Pt sphere experiments on KCl bracketed the melting point at 1 bar to 1042 ± 2 K. These results agree well with the reported value of 1043 K [3].

Table 1. Summary of 1 bar experiments

KCl Melting Temperature (K)			
#	**DTA**	**Current**	**Δ*T***
P072519			
1	1042	1042	0
2	1042	1042	0
3	1038	1043	−5
4	1039	1043	−4
5	1043	1043	0
6	1042	1042	0
7	1043	1043	0
NaCl KCl Melting Temperature (K)			
P071919_8 [1]			
1	1070	1070	0
2	1070	1071	−1
3	1071	1072	−1
4	1072	1072	0
5	1071	1072	−1
P071919_5A			
1	1081	1071	10
P071919_5B			
1	1075	1074	1
2	1075	1075	0
3	1074	1074	0
4	1073	1075	−2
P071919_5C			
2	1074	1074	0

refers to the sequence number of heating path; Measurements with unclear DTA signals were discarded; [1] 8 and 5 refer to wire diameters in 0.001 inch.

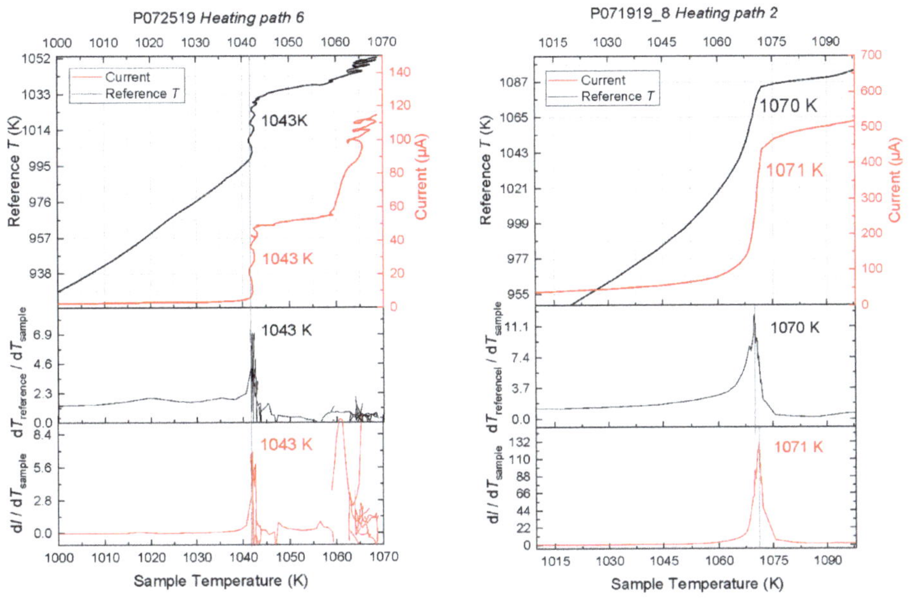

(a) **(b)**

Figure 3. Representative results of experiments on potassium chloride (KCl) (**a**) and sodium chloride (NaCl) (**b**) at 1 bar. In each figure, the three panels show the reference temperature (black) and the ionic current (red) and their rates of change as a function of sample temperature. The KCl data were recorded along the sixth heating path of experiment P072519 on KCl and yielded a melting point of 1042 ± 2 K. The NaCl data were recorded along the second heating path of experiment P071919 and yielded a melting point of 1070 ± 2 K. Note that the latent heat ledges appear as cliffs in these figures.

The latent heat ledge coincided with a similar ledge in the current-temperature curve (Figure 3a). Along a representative heating path, the ionic current remained below 3 μA at temperatures up to ~1000 K, climbed up slightly upon further heating to ~1038 K, and then rose steeply to more than 50 μA at constant sample temperature of ~1038 K. The rates at which the reference temperature or current increased with the sample temperature, calculated as the first derivatives of the recorded curves $dT_{reference}/dT_{sample}$ and dI/dT_{sample}, are nearly flat at temperatures up to ~1038 K. They increased steeply within 5 K and spiked at 1042 ± 2 K upon melting. Further heating brought these derivatives back to the baseline values. Another steep rise of the current occurred at ~1060 K. The origin of this rise is unclear and may be a result of melt spreading out along the wires. This rise can be distinguished from the melting signal because it was not accompanied by a steep rise in the temperature, as shown in the derivative plots. The maximum current varied with the composition of the electrode, whether it was copper or platinum, their size and geometry including immersion length and separation between the two electrodes, and it aged with time in repeated cycles.

3.2. NaCl at 1 bar

The DTA experiments on NaCl at 1 bar produced slightly tilted latent heat ledge (Figure 3b). Accordingly, the peak in $dT_{reference}/dT_{sample}$ spread over 10 K, likely because the heating rate was too low to produce the expected ledge. The melting temperature was assigned to 1070 ± 2 K, where the relative rate of increase in the reference temperature with respect to that of the sample temperature peaked. This peak was accompanied by a peak in the relative rate of increase in ionic current with respect to that of the sample temperature at 1071 ± 2 K. The current was below 50 μA at temperatures below 1033 K and rose steeply to >400 μA at 1073 K, as a result of melting. With further heating the ionic current increased at a much lower rate. These results agree well with the reported value of 1074 K within mutual uncertainties.

In both the KCl and NaCl experiments, the melting points measured by the DTA method were confirmed by the sinking of Pt spheres. Large clear crystals were recovered at ambient temperature in the containers. Furthermore, samples appeared on the crucible wall and along the thermocouple or electrode wires. We infer that melt climbed up the wall and wires under capillary force and that large crystals grew from melt upon cooling. We observed variations of the peak current with time and between experiments, and attribute some of the variations to the spreading of melt. The recovered samples were confirmed to be single phases according to inspections by optical microscopes.

3.3. *KCl at High Pressures*

From the high-pressure experiments with radial wire placement, we obtained the melting points of KCl at 3.2 to 5.5 GPa (Table 2). The melting temperature was also located at the steepest point of the current-temperature curve (Figure 4a). During heating, the press load drifted up because of thermal pressure. The reported sample pressures were calculated from the press loads where the steepest current rises took place.

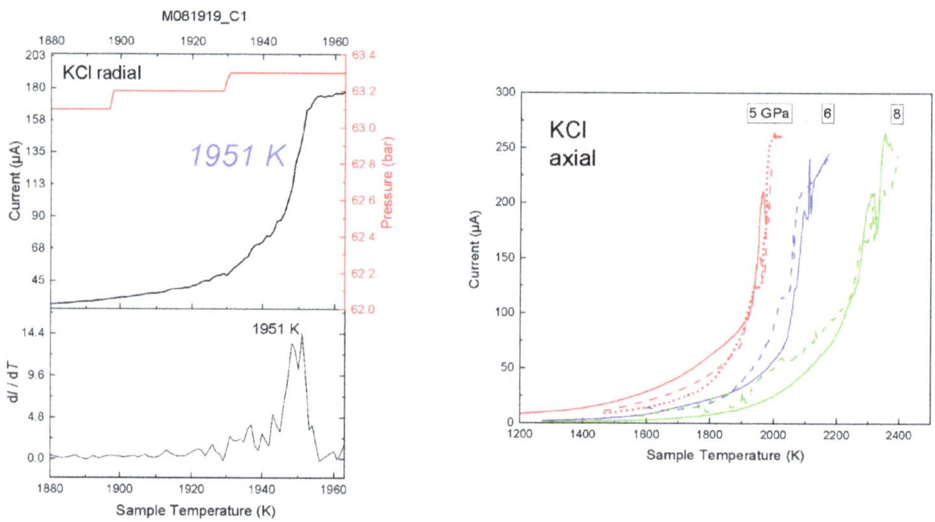

(a) (b)

Figure 4. Representative ionic conduction results of multi-anvil experiments on KCl. (**a**) Radial placement. The curve was recorded in experiment M081919_8. The press load (red bars) increased during heating. The steps in the press load reflect the resolution of the pressure transducer. Melting was located at the peak rate of current rise at 1951 ± 5 K. (**b**) Axial placement. All the current curves were recorded in experiment M080817_5. The solid, dashed, and dotted curves represent the first, second, and third heating cycles, respectively.

The high-pressure experiments yielded melting points of KCl at 3 to 8 GPa and a lower bound for the melting point at 9 GPa (Table 2). In repeated heating cycles, the steepest current rise occurred within 5–20 K, despite some sporadic excursions in the current-temperature curves (Figure 4b).

Table 2. Summary of multi-anvil experiments on KCl.

	Pressure, GPa	Melting T, K	Error, K
M081919_8 [1]	3.6	1669	±7
		1678	±8
		1668	±8
	4.7	1820	±5
		1814	±13
		1815	±12
	5.5	1951	±13
		1951	±9
M072619_8 [1]	3.2	1577	±5
M080817_5 [1]	5.0	1954	±15
		1975	±13
		1976	±12
	6.0	2073	±18
		2110	±29
	7.0	2173	±50
	8.0	2273	±43
		2314	±48
	9.0	≥2443 [2]	
M080417_5 [1]	4.4	1752	±22

Notes: [1] 8 and 5 refer to the truncation edge length in mm. The errors are estimated from the width of the peak in dI/dT and do not include additional errors as discussed in the text. [2] A lower bound for the melting point of KCl at 9 GPa is obtained from the melting point of Pt.

4. Discussion

4.1. Melting Curve of KCl

The melting temperatures of KCl measured in this study are in general agreement with the literature data (Figure 5). Where the pressure ranges overlap, the results of our radial-TC experiments agree with the existing data from piston-cylinder experiments [3,10], within uncertainties. Between 5 and 8 GPa, the results of our ionic conduction experiments are higher by 20–90 K, but marginally agree with the Simon–Glatzel melting curve [11] fitted through all the literature data on KCl in the simple cubic B2 structure between 1.8 and 50 GPa, which can be described by $T_m = 1323\left(\frac{P-1.87}{2.2(1)}+1\right)^{\frac{1}{2.7(1)}}$, where T is in K, P is in GPa, and the numbers in parentheses are uncertainties in the last digits.

Excluding the literature data at pressures above 20 GPa from diamond anvil cell (DAC) experiments [2], the fit becomes $T_m = 1323\left(\frac{P-1.87}{3.5(5)}+1\right)^{\frac{1}{1.8(2)}}$. This fit to the data from large volume apparatus alone diverge from the all-data fit at pressures above 10 GPa and differ by as much as 500 K at 25 GPa. The divergence confirms the notion that the Simon's equation is empirical and cannot be used for extrapolating melting curves beyond the experimental pressure range, although it is widely used for interpolation because the formulation is simpler than other melting laws and requires no knowledge of the equation-of-state of the solid phase. Indeed, a previous study on NaCl found that the Simon's law equation fit to data between 1 bar and 6.5 GPa overpredicts the melting temperature at 20 GPa by nearly 100 K [4]. On the other hand, more recent DAC data on the melting curve of water questioned the accuracy of data from piston-cylinder experiments [12]. Furthermore, the melting temperatures of iron determined in earlier DAC experiments were found to be overestimated by hundreds of degrees [13]. Therefore, independent measurements at pressures below 10 GPa and new data at pressures above 10 GPa are needed to validate the melting curve of KCl fitted to all the existing data.

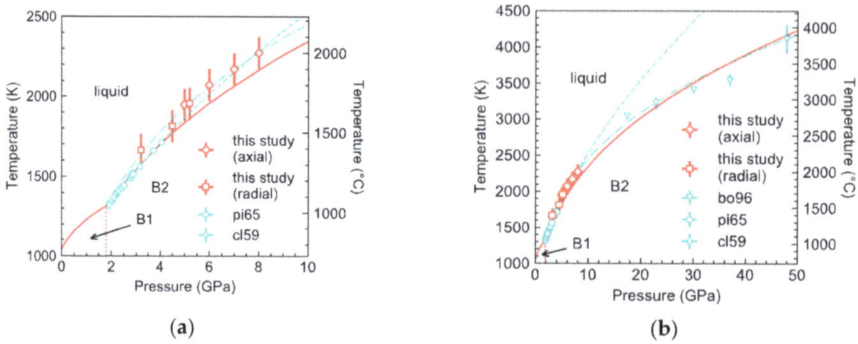

Figure 5. Melting curve of KCl up to 10 GPa (**a**) and up to 50 GPa (**b**). The red solid curve is a Simons–Glatzel fit to all the data, including the axial placement data (red diamonds) and radial placement data (red squares) from the ionic conduction measurements in this study, and the literature data using piston-cylinder apparatus ([10], green thin-diamonds, [3], green circles) and diamond anvil cells ([2], green triangles). The green dashed curve is a fit to all the data from large volume apparatus, including piston cylinder and multi-anvil press, and the green dot-dashed curve is a fit to all the literature data. The fits are weighted by uncertainties in pressure and temperature, estimated at ±100 K and ±0.5 GPa for our data. The red solid curve and green dashed curve are described by $T_m = 1323\left(\frac{P-1.87}{2.2(1)}+1\right)^{\frac{1}{2.7(1)}}$, and $T_m = 1323\left(\frac{P-1.87}{3.5(5)}+1\right)^{\frac{1}{1.8(2)}}$, respectively, where T is in K, P is in GPa, and the numbers in parentheses are uncertainties in the last digits.

4.2. Melting Detection from Ionic Conduction

The concurrent experiments at 1 bar demonstrate that ionic conduction measurements can be used to determine the melting temperature of ionic compounds. Where high-quality data were acquired at 1 bar, we observed simultaneous occurrences of latent heat ledge and peak rate of current rise with temperature, thus firmly establishing the criterion of melting detection from ionic conduction observations. The validity of this criterion is further supported by confirmation from the sinking of Pt sphere, movement of melt, and growth of large crystals from melt upon cooling.

The recorded current-temperature curves are not always of high quality, even at the ambient pressure. Random errors originated from several sources, including melt motion, cell deformation, and pressure drift with temperature. At 1 bar, the migration of melt along the wires or container wall altered the contact location and area of the electrodes, thus producing noise in the current reading. In the high-pressure experiments, components of the high-pressure assembly had diverse ranges of thermoelastic properties and porosities, and therefore responded to compression and heating differently. The resulting stress and deformation produced noise in the current and temperature signals. Furthermore, the current-temperature curve could have been broadened by pressure drift at high temperature. In some of our axial placement experiments, the current-temperature curves were not smooth and the rate of current rise with temperature fluctuated. As a result, the peak in the derivative plot was not as sharp and prominent as expected. Nevertheless, these data gave reasonable constraints on the melting temperature, which increased as a function of pressure, as expected.

Compared with the symmetric configuration used in previous conduction experiments (e.g., Dong et al., 2019 [6]), the single-side placement of thermocouple and electrodes reduce uncertainties in temperature measurements associated with temperature gradient. Furthermore, the placement of the thermocouple junction and electrode tips at the heater waist allows for detection of the first melt in the sample chamber. By limiting the amount of melt, this configuration minimizes heater deformation associated with melt movement, which may be significant before pore space is not fully closed. Heater deformation leads to perturbation of the temperature distribution and may cause shorting. With the thermocouple junction placed directly in the sample, the single-side placement has the additional advantage of accounting for the latent heat effect on the sample temperature. However, chemical contamination of the thermocouple could produce false signals. In particular, the W-Re based type C wires are known to be vulnerable to oxidation and they react with carbonate melts [5]. The type S wires are more chemical inert, but they are less refractory and do not work at temperatures beyond ~2300 K.

The axial and radial placement of thermocouple and electrodes each has its advantages and disadvantages. Installing the thermocouple requires similar levels of skills in the axial and radial placements, but the axial placement involves cutting two deep grooves in the octahedron and filling extra space with cement after installing the thermocouple, whereas the radial placement only involves drilling one bore through a fin of the octahedron. The radial placement ensures that the thermocouple junction and electrode tips are located at the hottest part of the heater to detect the first melt. Perforation in the heater causes larger temperature gradient and hot spots because each wire from the hot zone to the exterior acts as a heat sink that perturbs the thermal field. Moreover, melt may leak out from the perforation at the heater equator. However, the four-bore holder in the radial placement has a diameter of 0.80 mm so that the thermocouple junction and the electrode tips are less than 0.2 mm apart. This configuration allows reliable melting detection even in the presence of relatively large temperature gradients. Such a small holder can only fit 5 mil wires. The thinner wires lose less heat but are more fragile and break easily at a few GPa. In this study, we placed the second thermocouple just outside the heater in an attempt to observe the latent heat associated with melting in the radial experiments. These two wires can be left out in the radial placement experiment.

A successful conduction experiment requires the survival of the thermocouple and electrode wires and separation of the electrodes. In the KCl experiments, we used the W-Re-based type C wires, which are much stronger than the type S wires and therefore less likely to break or short, but they are more chemically reactive and cannot be used for carbonates such as $CaCO_3$ [5]. The Pt-Rh-based type S thermocouples are soft and prone to failure due to breaking or shorting. In the axial configuration, space is available to accommodate wires with 8 mil or even 10 mil diameters. If correctly placed, these wires are sufficiently thick to survive compression to 20 GPa, but the thicker electrodes short more easily. Shorting may occur upon compression, when the Pt container and malleable wires deform under stress. Chance of shorting increases at high temperature, when the wires soften and flow more readily. The movement of melt may also lead to shorting.

The ionic conduction method has a number of advantages over the differential thermal analysis (DTA) method. At 1 bar, the latent heat ledge may be missing if the sample is heated up too fast or too slow, but the steep rise in ionic current is not affected by the heating rate. The latent heat effect was not observed in the high-pressure experiments, possibly because the sample mass was too small, and the signal was overwhelmed by the thermal response of the surrounding materials in the high-pressure assembly. Thermopile amplification used for differential scanning calorimetry studies at high pressures [14,15] may allow us to observe the latent heat ledge associated with melting. Alternatively, we can apply the ionic conduction method to detect melting.

Compared with the ionic conduction method, applying the Pt sphere maker method at high pressures is labor-intensive. Theoretically the melting point at a given pressure can be bracketed by two experiments, but in reality, many more experiments are needed to locate the melting temperature, even if all are successful. Duplicate experiments may produce contradictory results because of uncertainties in sample placement in a temperature gradient. With limited width and length of the sample, Pt spheres could easily get stuck to the container wall or trapped in the thin layer of melt, and therefore produce no results.

Admittedly, ion conduction measurement at high pressure is challenging because it requires well controlled sample geometry. All the parts need to be aligned and closely fitted to ensure success. Our current success rate is about one third for users with average skills. However, this type of experiment is highly efficient when successful. Our study showed that melting produces a steep rise in ionic current with temperature, but we cannot exclude the possibility of an abrupt current rise resulting from other changes such as decomposition or chemical reaction. For these reasons, it is important to examine the experiment product for chemical changes, and to validate the melting points from conduction experiments by sphere marker experiments at selected pressures.

4.3. Application of KCl in High Pressure Research

The proposed melting curve of KCl informs its potential use as a pressure calibrant for large-volume press or material for pressure transmission, thermal insulation, and sample containment in high pressure research.

The melting curve of NaCl has been used to calibrate the relationship between press load and sample pressure in piston-cylinder and large-volume press. Such calibration revealed notable deviation from pressure calibration established at 1473 K in experiments using the 646-pressure medium [7]. At 7 GPa and ~2000 K, the thermal pressure reached as much as 2 GPa, which can be explained by the low thermal conduction of zirconia-based 646 ceramics. While NaCl works well as a pressure calibrant at pressures up to ~7 GPa, it loses resolution at higher pressures, where its melting curve becomes relatively flat (Figure 6). Between 1 bar and ~9 GPa, the melting point of KCl increases with pressure at an average rate of ~150 K/GPa. With such a large melting slope, KCl may be used as a high-resolution pressure calibrant at moderate pressures and high temperatures. Note that the reference melting curve of KCl carries an uncertainty of ±100 K and the measured melting temperature bears a comparable uncertainty of ±100 K with the current design of ionic conduction experiment including the uncertainties in temperature measurements, the peak width of dI/dT, and reproducibility among multiple heating cycles (Table 2). As a result, the uncertainty in pressure calibration is at least ±0.7 GPa, which can be improved when the KCl melting curve is more tightly constrained and a better design of ionic conduction experiment enables more accurate and precise melting detection.

Previous studies showed that pressure calibration of multi-anvil experiments is affected by temperature and heating cycle. Leinenweber et al. (2012) [8] found that the COMPRES 10/5 assembly experienced a pressure maximum during heating. At 15 to 20 GPa, the sample pressure varied by as much as 1.5 GPa between room temperature and 1900 K. More significantly, at 20–30 GPa, the COMPRES 8/3 assembly lost 5 to 10 GPa during heating between room temperature and 2300–2600 K [8]. On the other hand, the sample pressure increased by 2 GPa between 1500 and 2300 K at a press load of 1400 ton, after a temperature cycle at 680 ton [16]. Such large pressure variations with temperature and heating cycles suggest that some of the discrepancies in the measured melting temperatures between multiple heating cycles may originate from pressure change at a fixed press load, in addition to other sources such as melt movement and heater deformation. The reproducibility of the measured melting temperatures between repeated cycles and the agreement between our results and literature data on the melting curve of KCl suggests that in this study the deviations of sample pressures from the pressure calibrations are within the estimated uncertainties.

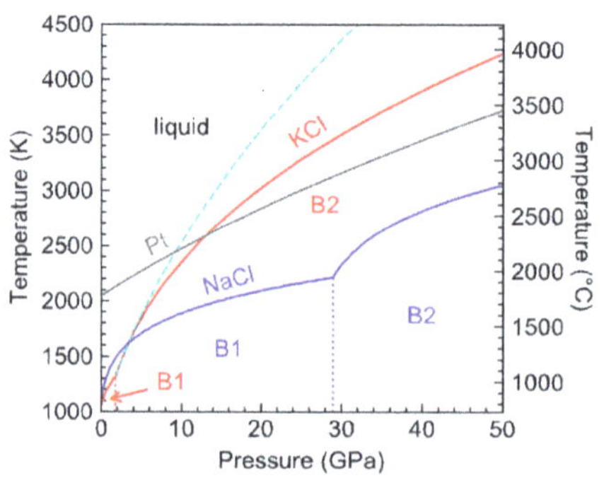

Figure 6. Phase diagrams of KCl and NaCl up to 50 GPa. The B1–B2 phase boundaries in KCl (red, [3,17]) and NaCl (blue, [18]) are plotted as dotted lines. The melting curve of Pt [19] intersects the Simons–Glatzel fit of KCl to the large volume data (green dashed curve) near 9 GPa.

We propose that the ionic conduction method may be used to quantify the effects of temperature and heating cycle on pressure calibration. Although in its current state this method still bears relatively large uncertainties and may have limited resolutions at high pressures, it offers a potential solution when synchrotron radiation is not available or opening an X-ray window in the cell is not desirable.

The divergence between the melting curves of NaCl and KCl is interesting (Figure 6). At 1 bar, they are both in the B1 face-centered cubic structure. As expected from the systematic relationship between the melting temperature and ionic radii [20], KCl melts at a lower temperature than NaCl and transforms into the simple cubic B2 structure at a lower pressure than NaCl. The B1–B2 transition is associated with a volume reduction, which may have led to steepening of the melting slope, as observed in the melting curve of SiO_2 upon the transition from high-quartz to coesite and from coesite to stishovite [21].

While NaCl remains in the B1 structure with small increase of melting temperature with pressure, the melting curve of KCl crosses that of NaCl near 3 GPa. Between 3 and 9 GPa, the melting point of KCl continues to rise steeply to cross that of Pt near 9 GPa (Figure 6). The highly refractory nature of KCl at moderate pressures may provide value for high-pressure research. In diamond-anvil cell experiments NaCl and KCl are widely used as pressure medium because they are transparent and have low shear moduli. They also serve as thermal insulating layer for laser-heating experiments. If the sample of interest is a melt, then a refractory medium or insulator may be preferred for stability.

Finding a suitable container is also important for high temperature experiments on materials synthesis phase equilibria and chemical reactions. For sample containment, KCl possesses a unique combination of characteristics: It is water-soluble, relatively soft, and electrically insulating in the solid state, and highly refractory at moderate pressure, therefore it is a promising option along with commonly used container materials such as graphite, MgO, BN, and various metals.

5. Conclusions

Through concurrent differential thermal analysis, platinum sphere marker, and ionic conduction measurements at 1 bar, we showed that melting of KCl or NaCl produces a prominent peak in the increase of ionic current with temperature; therefore, the first derivative of the current-temperature relation can be used to locate the melting point.

Applying this criterion for melting detection to ionic conduction measurements at high pressures, we determined the melting points of KCl at selected pressures between 3.5 and 8 GPa, which partially fill the gap between 4 GPa and 20 GPa in the literature data. Our results suggest that the melting temperature at 9 GPa exceeds that of Pt, and is at least 2400 ± 100 K. The Simon–Glatzel fit to the large volume data diverges from that through all the data including those from the DAC experiments. Considering the empirical nature of the Simon–Glatzel melting equation and large uncertainties in the early DAC results, further constraints are needed at pressures above 10 GPa to provide tighter constraints on the KCl melting curve.

Given the large dependence of melting temperature on pressure between 1 bar and 10 GPa, KCl may be used to calibrate large volume press at high temperatures in this pressure range. The highly refractory nature of KCl at 5 to 50 GPa makes it valuable candidate material for sample containment, pressure transmission, and thermal insulation at high temperatures and moderate pressures.

Author Contributions: Conceptualization, J.L.; methodology, D.Z., J.D., J.L; data analysis, D.Z., J.D., J.L.; investigation, D.Z., J.D., Y.S., F.Z., J.L.; writing—original draft preparation, J.L.; writing—review and editing, D.Z., J.D., Y.S., F.Z., J.L.; visualization, D.Z., J.D., J.L.; supervision, J.L.; All authors have read and agreed to the published version of the manuscript.

Funding: D.Z. acknowledges financial support from USTC's URP summer project. J.D. acknowledges support through a James Mills Peirce Fellowship from the Graduate School of Arts and Sciences at Harvard University. J.L. acknowledges support of USA National Science Foundation grant EAR 1763189.

Acknowledgments: We thank Robert Liebermann for encouraging and supporting our contribution to this special issue that honors Orson Anderson for his pioneering role in establishing mineral physics. We thank Dave Walker and two anonymous reviewers for providing valuable comments and suggestions, Skye Kaplan, Chengwei Zhang, and Ethan Gutierrez for their assistance with experiments and data archiving.

Conflicts of Interest: The authors declare no conflict of interest.

References

1. Dewaele, A.; Belonoshko, A.B.; Garbarino, G.; Occelli, F.; Bouvier, P.; Hanfland, M.; Mezouar, M. High-pressure–high-temperature equation of state of KCl and KBr. *Phys. Rev. B* **2012**, *85*, 214105. [CrossRef]
2. Boehler, R.; Ross, M.; Boercker, D.B. High-pressure melting curves of alkali halides. *Phys. Rev. B* **1996**, *53*, 556. [CrossRef] [PubMed]
3. Pistorius, C.W.F.T. Melting curves of the potassium halides at high pressures. *J. Phys. Chem. Solids* **1965**, *26*, 1543–1548. [CrossRef]
4. Li, Z.; Li, J. Melting curve of NaCl to 20 GPa from electrical measurements of capacitive current. *Am. Mineral.* **2015**, *100*, 1892–1898. [CrossRef]
5. Li, Z.; Li, J.; Lange, R.; Liu, J.; Militzer, B. Determination of calcium carbonate and sodium carbonate melting curves up to Earth's transition zone pressures with implications for the deep carbon cycle. *Earth Planet. Sci. Lett.* **2017**, *457*, 395–402. [CrossRef]
6. Dong, J.; Li, J.; Zhu, F.; FaRawi, R. Melting curve minimum of barium carbonate $BaCO_3$ near 5 GPa. *Am. Mineral.* **2019**, *104*, 671–678. [CrossRef]
7. Walker, D.; Li, J. Castable solid pressure media for multi-anvil devices. *Matter Radiat. Extrem.* **2020**, *5*, 018402. [CrossRef]
8. Leinenweber, K.D.; Tyburczy, J.A.; Sharp, T.G.; Soignard, E.; Diedrich, T.; Petuskey, W.B.; Mosenfelder, J.L. Cell assemblies for reproducible multi-anvil experiments (the COMPRES assemblies). *Am. Mineral.* **2012**, *97*, 353–368. [CrossRef]
9. Li, J.; Hadidiacos, C.; Mao, H.K.; Fei, Y.; Hemley, R.J. Behavior of thermocouples under high pressure in a multi-anvil apparatus. *High Press. Res.* **2003**, *23*, 389–401. [CrossRef]
10. Clark, S.P., Jr. Effect of pressure on the melting points of eight alkali halides. *J. Chem. Phys.* **1959**, *31*, 1526–1531. [CrossRef]
11. Simon, F.; Glatzel, G. Bemerkungen zur schmelzdruckkurve. *Z. Für Anorg. Und Allg. Chem.* **1929**, *178*, 309–316. [CrossRef]
12. Datchi, F.; Loubeyre, P.; LeToullec, R. Extended and accurate determination of the melting curves of argon, helium, ice (H_2O), and hydrogen (H_2). *Phys. Rev. B* **2000**, *61*, 6535. [CrossRef]
13. Anzellini, S.; Dewaele, A.; Mezouar, M.; Loubeyre, P.; Morard, G. Melting of iron at Earth's inner core boundary based on fast X-ray diffraction. *Science* **2013**, *340*, 464–466. [CrossRef] [PubMed]
14. Rapp, R.P.; Navrotsky, A. Differential scanning calorimetry in a piston-cylinder apparatus: Design and calibration. *Pure Appl. Geophys.* **1993**, *141*, 615–629. [CrossRef]
15. Secco, R.A. *High Pressure Science and Technology*; Schmidt, S.C., Shaner, J.W., Samara, G.A., Ross, M., Eds.; AIP: New York, NY, USA, 1994; Volume 2, p. 1593.
16. Fei, Y.; Van Orman, J.; Li, J.; van Westrenen, W.; Sanloup, C.; Minarik, W.; Hirose, K.; Komabayashi, T.; Walter, M.J.; Funakoshi, K.I. Experimentally determined postspinel transformation boundary in Mg_2SiO_4 using MgO as an internal pressure standard and its geophysical implications. *J. Geophys. Res. Solid Earth* **2004**, 109. [CrossRef]
17. Boehler, R.; Ross, M.; Boercker, D.B. Melting of LiF and NaCl to 1 Mbar: Systematics of ionic solids at extreme conditions. *Phys. Rev. Lett.* **1997**, *78*, 4589. [CrossRef]
18. An, Q.; Zheng, L.; Fu, R.; Ni, S.; Luo, S.N. Solid–liquid transitions of sodium chloride at high pressures. *J. Chem. Phys.* **2006**, *125*, 154510. [CrossRef] [PubMed]
19. Errandonea, D. High-pressure melting curves of the transition metals Cu, Ni, Pd, and Pt. *Phys. Rev. B* **2013**, *87*, 054108. [CrossRef]

20. Prewitt, C.T.; Downs, R.T. High-pressure crystal chemistry. *Rev. Mineral.* **1998**, *37*, 284–318.
21. Ghiorso, M. An equation of state for silicate melts. I. Formulation of a general model. *Am. J. Sci.* **2004**, *304*, 637–678. [CrossRef]

minerals

MDPI

Article

Thermoelastic Properties of $K_{0.7}Na_{0.3}AlSi_3O_8$ Hollandite and $NaAlSi_2O_6$ Jadeite: Implication for the Fate of the Subducted Continental Crust in the Deep Mantle

Steeve Gréaux [1,2,*], Youmo Zhou [1], Yoshio Kono [1], Akihiro Yamada [1,†], Yuji Higo [3] and Tetsuo Irifune [1,2]

1 Geodynamics Research Center, Ehime University, Matsuyama, Ehime 790-8577, Japan; youmo@sci.ehime-u.ac.jp (Y.Z.); kono@sci.ehime-u.ac.jp (Y.K.); yamada.ak@mat.usp.ac.jp (A.Y.); irifune@dpc.ehime-u.ac.jp (T.I.)

2 Earth-Life Science Institute, Tokyo Institute of Technology, Tokyo 152-8550, Japan

3 Japan Synchrotron Radiation Institute, SPring-8, Hyogo 679-5198, Japan; higo@spring8.or.jp

* Correspondence: greaux@sci.ehime-u.ac.jp

† Current address: Center for the Glass Science and Technology, The University of Shiga Prefecture, Hikone 522-8533, Japan.

Received: 31 January 2020; Accepted: 11 March 2020; Published: 13 March 2020

Abstract: The thermoelastic properties of $K_{0.7}Na_{0.3}AlSi_3O_8$ hollandite and $NaAlSi_2O_6$ jadeite, synthesized from a (K, Na)-felspar (microcline), were investigated by a combination of in situ energy dispersive synchrotron X-ray radiation and multi-anvil techniques at high pressure (P) and temperature (T) up to 21 GPa and 1700 K. The second-order phase transformation was found to occur in hollandite at ~16 GPa from tetragonal *I/4m* (hollandite-I) to monoclinic *I2/m* (hollandite-II), which confirms the previous report that the incorporation of Na in the hollandite structure decreases the transformation pressure. Fitting the pressure–volume–temperature data to the Birch–Murnaghan equation of state yielded estimates of the thermoelastic parameters for jadeite as well as the $K_{0.7}Na_{0.3}AlSi_3O_8$ hollandite-I and -II phases, which indicate that the incorporation of Na is likely to decrease the bulk moduli of both hollandite phases. The obtained thermoelastic parameters were combined with those of other mantle minerals reported previously to estimate the density of continental materials along an average mantle geotherm. Based on our results, continental crust and sediment become, respectively, 11% and 15% denser than the pyrolitic mantle at pressure >10 GPa, suggesting that once pulled down to the critical depth of ~300 km, the continental portions of the slab can subduct further into the deep mantle, down to the lowermost part of the mantle transition region.

Keywords: high pressure; thermal properties; in situ X-ray diffraction; liebermannite; jadeite; continental crust

1. Introduction

Geological observations suggest that the subduction of continental crust may have occurred through time and transported extensive amounts of continental crust and sedimentary materials into the Earth's mantle [1], which can be traced by the diamond inclusions of minerals such as K-rich hollandite [2,3]. At the mantle transition region (MTR) depths, petrological works have shown that continental crust materials transform to a mixture of garnet, stishovite, pyroxene, hollandite, and Ca-rich aluminosilicate (CAS) phase [4,5]. Numerical simulations on the basis of the density contrasts between continental crust and mantle components have predicted that when such transformations occur, continental crust is likely to sink even deeper into the Earth's mantle [6,7] where it could accumulate through time and form a thick and buoyant second continent at the bottom of the MTR [8]. Such a layer would have important implications for the location of alkali metals in the Earth, especially K^{40}, which is believed to play a key role in the Earth's internal heating and long-term evolution, by providing substantial amount of heat through its long radiogenic decay process [9,10].

The density estimates made by previous studies were, however, mostly based on assuming the thermoelastic properties of the endmembers of continental crust minerals [11,12]. To this day, there have been only a few works that addressed the effect of cationic substitutions on the elastic properties of those minerals under the relevant pressure and temperature conditions, which hampers precise estimates of the density of the continental crust in the deep mantle. (K,Na)$AlSi_3O_8$ hollandite is the solid solution in the $KAlSi_3O_8$ liebermannite [13] and $NaAlSi_3O_8$ lingunite [14] joint system, and jadeite is a Na-rich clinopyroxene. These two phases are very specific to continental crust compositions where they represent up to ~40 vol.% of the phase assemblages at high pressure, while they do not appear in pyrolite or mid-ocean ridge basalt (MORB) compositions at the MTR pressures [15–17]. Therefore, knowledge of their elasticity at high pressure (P) and temperature (T) is of importance for estimating the density of continental crust when it is subducted in the slab to the depths of the MTR. The thermoelastic properties of liebermannite have been reported previously by in situ measurements using multi-anvil and diamond anvil techniques [18–20], while those of (K,Na)$AlSi_3O_8$ hollandite have been poorly investigated. To this day, only one experimental study reported the compression curve for $K_{0.8}Na_{0.2}AlSi_3O_8$ hollandite at room T, based on which it was concluded that the incorporation of Na has no significant effect on the compressibility of the hollandite structure [21]. The elasticity of jadeite has been reported by several studies, although those works were limited to high T measurements at relatively low pressure [22,23] or high P measurements at room T [24,25], requiring large extrapolations when calculating the density of this phase under the high P and high T conditions of the mantle.

Here we investigated the thermoelastic properties of $K_{0.7}Na_{0.3}AlSi_3O_8$ hollandite and $NaAlSi_2O_6$ jadeite synthesized from a microcline feldspar by a combination of in situ synchrotron X-ray measurements and multi-anvil techniques at high pressures and high temperatures up to 21 GPa and 1700 K, respectively, at the beamline BL04B1 (SPring-8, Japan). This hollandite composition is comparable with those obtained in the phase relation studies on continental crust materials at high P up to ~22 GPa [4,5]. Our results of the thermoelastic properties of hollandite are compared with some previous studies, and the effect of Na on the compressibility of the hollandite structure is discussed. The density profiles of continental crust and sediment are estimated on the basis of a thermoelastic dataset that includes our new data on $K_{0.7}Na_{0.3}AlSi_3O_8$ hollandite and jadeite and compared to that of other mantle compositions.

2. Materials and Methods

2.1. Starting Material

We prepared a feldspar glass from a natural microcline (Nellore, Andhra Pradesh, India), which was melted at ~1800 K and quenched in cold water. Analysis of the resulting glass by a field emission scanning electron microscope (FE-SEM, JSM7000F, JEOL Ltd., Tokyo, Japan) equipped with an energy dispersive spectrometer (EDS, X-Max^N, Oxford Instruments plc, Oxon, UK) yielded the composition reported in Table 2. The microcline glass was powdered and cold-sealed in a gold capsule, then hot-pressed at 15 GPa and 1500 K for 30 min (OS2999, Table 1) using the 2000-ton Kawai-type multi-anvil press apparatus ORANGE2000 at the Geodynamics Research Center (Ehime University, Japan). The recovered samples appeared well sintered in a cylindrical shape (~2.0 mm diameter) and free of visible cracks. The texture analyses (Figure 1a,b) and chemical composition (Table 2) obtained from SEM–EDS analysis showed that the sample consisted of a multiphase aggregate of $K_{0.7}Na_{0.3}AlSi_3O_8$ liebermannite and $NaAlSi_2O_6$ jadeite intergrown with small grains (<1µm) of stishovite (Table 1).

Table 1. Experimental conditions and run products.

Run No.	P (GPa)	T (K)	Duration (h)	Phase Assemblages
OS2999	15.0	1500	0.5	Na-lieb, Jd, St
S2697	7.0–20.4	300–1700	*	Na-lieb, Jd, St
OD1667	21.4	1700	12	Na-lieb, Jd, St

* in situ measurements (see details in methods); Na-lieb, Na-bearing liebermannite; Jd, jadeite; St, Al-bearing stishovite.

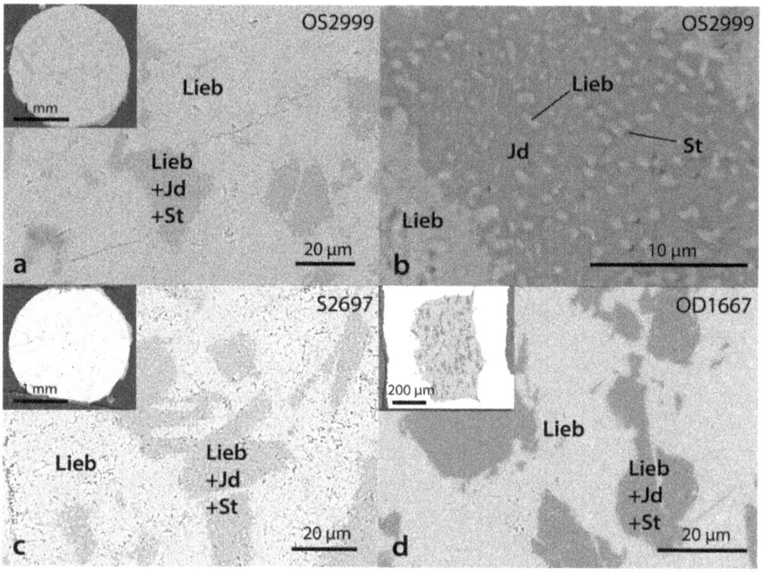

Figure 1. Electron image of (**a**) (**b**) the microcline sample before (OS2999) and (**c**) after (S2697) in situ measurements. (**d**) OD1667 show the result of an equilibrium experiment carried out for 12 h at ~21 GPa and 1700 K, corresponding to the highest pressure (P) and temperature (T) data of the in situ measurements (see Table 1). Abbreviations: Lieb, Na-rich liebermannite; Jd, jadeite; St, stishovite.

Table 2. Composition of starting material and run products.

	Glass Starting Material *	Hollandite			Jadeite + Stishovite		
Element (wt.%)		OS2999	S2697	OD1667	OS2999	S2697	OD1667
Na	1.48	2.4(2)	2.2(1)	1.5(1)	9.1(3)	9.2(1)	8.6(2)
K	9.96	10.1(3)	10.4(2)	10.9(2)	0.0(1)	0.0(0)	0.0(0)
Al	10.43	9.7(1)	9.6(1)	9.5(1)	10.6(4)	10.4(1)	10.1(1)
Si	30.15	30.6(2)	30.8(2)	31.4(1)	32.1(9)	32.0(2)	32.5(3)
O	46.40	46.4(4)	46.5(3)	47.0(1)	49.2(2)	48.9(2)	49.1(2)
Total	98.4	99.1(9)	99.4(7)	100.5(2)	100.1(7)	100.5(3)	100.4(4)
Element (p.f.u.)							
Na		0.28(2)	0.26(1)	0.18(1)	0.77(5)	0.78(1)	0.73(3)
K		0.71(2)	0.73(2)	0.76(2)	-	-	-
Al		0.99(1)	0.98(1)	0.96(1)	0.77(5)	0.75(1)	0.74(1)
Si		3.01(1)	3.02(1)	3.05(1)	2.23(6)	2.24(1)	2.27(1)
O		8	8	8	6	6	6
sum cations		4.99(1)	4.99(1)	4.95(1)	3.77(5)	3.78(1)	3.73(1)

- not detected by EDS analysis; * contains minor amount (i.e., value < 0.1 wt.%) of Ca, Fe, and Mg.

Micro-focused X-ray diffraction patterns collected on the recovered sample was well explained by an assemblage of Na-liebermannite (space group *I4/m*, Z = 2), jadeite (space group: *C2/c*, Z = 4), and stishovite (space group $P4_2/mnm$, Z = 2). Fitting of liebermannite peaks yielded the unit-cell parameters a = 9.3207(2) Å and c= 2.7193(1) Å; e.g., V = 236.2(2) Å^3, which is ~0.3%–0.6% smaller than that of the $KAl_2Si_3O_8$ endmember [18,19,26], due to the incorporation of Na, which decreases the unit-cell volume of the hollandite structure [21,26]. The jadeite unit-cell parameters a = 9.431(5) Å, b = 8.569(1) Å, c = 5.219(3) Å^3, and β = 107.43(3)° (e.g., V = 402.4(4) Å^3) are in good agreement with those of the $NaAlSi_3O_8$ endmember [22,24]. The stishovite unit-cell parameters a = 4.184(2) Å and c = 2.669(1) Å (e.g., V = 46.74(1) Å^3) is slightly higher than pure stishovite, which suggest the incorporation of ~1 wt.% Al_2O_3 [27], although the exact composition could not be measured by SEM–EDS due to the small size (<1 µm) of the stishovite grains. The unit-cell parameters and volumes of all the observed phases are shown in Table 3.

Table 3. Unit-cell volumes of Na-liebermannite and jadeite in the recovered samples.

	Na-Liebermannite			Jadeite		
	OS2999	S2697	OD1667	OS2999	S2697	OD1667
Unit-cell parameters						
a (Å)	9.3207(2)	9.3159(2)	9.3110(4)	9.431(5)	9.433(2)	9.405(1)
b (Å)	9.3207(2)	9.3159(2)	9.3110(4)	8.569(1)	8.563(1)	8.563(5)
c (Å)	2.7193(1)	2.7243(1)	2.7313(3)	5.219(3)	5.226(1)	5.227(8)
α (°)	90	90	90	90	90	90
β (°)	90	90	90	107.43(3)	107.52(1)	107.31(7)
γ (°)	90	90	90	90	90	90
V (Å^3)	236.2(2)	236.4(2)	236.8(4)	402.4(4)	402.6(1)	401.9(1)

2.2. *High Pressure Synchrotron Experiments*

X-ray diffraction measurements were carried out in a 1500-ton DIA-type multi-anvil press apparatus at the beamline BL04B1 at SPring-8 (Hyogo, Japan). High pressure was generated within so called 11/5 assemblies, equipped with a 20 µm thick cylindrical Re heater, which was transparent to high-energy X-rays. MgO was used to make a quasi-hydrostatic environment around the sample. A mixture of NaCl + BN + gold (10:1:0.1 wt.%) was placed at one end of the sample to enhance hydrostatic conditions and also served as a pressure standard using the EoS of NaCl and gold [28,29]. Temperature was monitored using a $W_{97}Re_3$–$W_{75}Re_{25}$ thermocouple (Type-D, Nilaco Corporation, Tokyo, Japan), which was maintained at high pressure, at the center of the cell and in contact with the pressure standard. Unit-cell volumes of the sample and pressure marker constituent phases were determined by energy dispersive X-ray diffraction (ED-XRD) at a fixed diffraction angle 2θ = 5.9757°, which was calibrated by the diffraction peaks of the gold standard at room P and T conditions. A multichannel Ge solid state detector was used to acquire photons in a range of 30–150 keV, which was calibrated with the characteristic fluorescence lines of ^{55}Fe, ^{57}Co, and ^{133}Ba, yielding the precision of the energy measurements within ±30 eV per channel. The polychromatic X-ray beam was collimated to 0.05 mm horizontally and 0.2 mm vertically, while its position relative to the sample and marker was determined before each measurement by X-ray radiography imaging techniques using a high-resolution CCD camera (C11440, Hamamatsu Photonics, Iwata, Japan).

2.3. *Data Reduction and EoS Procedure*

The *GSAS/EXPGUI* software package (v1.251, Argonne National Laboratory, Lemont, IL, USA) [30, 31] was used to analyze the diffraction data assuming the $BaMn_8O_{15}$ hollandite-type structure (space group *I4/m*, Z = 2) for the Na-liebermannite phase and the monoclinic structure (space group: *C2/c*, Z = 2) for the jadeite phase, respectively. The peak positions of Na-liebermannite and jadeite were refined simultaneously by a least-square fitting of the whole diffraction profile using the Le Bail technique [32] for energy dispersive spectra. The extracted unit-cell volumes for each experimental P and T conditions are summarized in Supplementary Materials Table S1.

The pressure-volume relations were determined by a least-square fitting of the room T unit-cell volumes $V_{P,300}$ of Na-liebermannite and jadeite as a function of the pressure (gold P-scale) to the third-order Birch–Murnaghan (BM) equation of state (EoS) with the general form given in Equation (1).

$$P(V) = \frac{3}{2}K_{0,T}\left[\left(\frac{V_{0,T}}{V_{P,T}}\right)^{7/3} - \left(\frac{V_{0,T}}{V_{P,T}}\right)^{5/3}\right] \times \left\{1 - \frac{3}{4}(K'_{0,T} - 4) \times \left[\left(\frac{V_{0,T}}{V_{P,T}}\right)^{2/3} - 1\right]\right\} \quad (1)$$

where the parameters $K_{0,T}$, $K'_{0,T}$, and $V_{0,T}$ are the isothermal bulk modulus, its pressure derivative, and the zero-pressure unit-cell volume, respectively. Because of the limited data in this study, fitting of all elastic parameters simultaneously generally yielded large errors due to the inaccuracy in determining $K'_{0,T}$ and $V_{0,T}$ from Equation (1). Therefore, we chose to fix $V_{0,T}$ to its experimental value at room P and T conditions, and constrain $K'_{0,T}$ to values varying from 3 to 5.

The thermal properties were subsequently determined by fitting all data at high P and T to Equation (1) while assuming the bulk modulus is a linear function of the zero-pressure bulk modulus $K_{0,300}$ and the first temperature derivative $(\partial K_{0,T}/\partial_T)_P$, e.g., Equation (2); the first pressure derivative is independent of the temperature, e.g., Equation (3); and the zero-pressure volume is an exponential function of the thermal expansion $\alpha_{0,T}$, e.g., Equation (4) and Equation (5).

$$K_{0,T} = K_{0,300} + \left(\frac{\partial K_{0,T}}{\partial T}\right)_P \times (T - 300) \quad (2)$$

$$K'_{0,T} = K'_{0,300} \quad (3)$$

$$V_{0,T} = V_{0,300} \times exp\left[\int_{300}^{T} \alpha_{0,x}dx\right] \quad (4)$$

$$\alpha_{0,T} = a_0 + b_0T \quad (5)$$

When fitting the thermal parameters, the room temperature parameters $K_{0,300}$ and $V_{0,300}$ were fixed to the values determined by the fitting of 300 K data by Equation (1).

3. Results

Figure 2 shows selected diffraction patterns collected in situ at high P and room T as well as high P and high T up to 20.4 GPa and 1700 K. At all P and T, our sample consisted of $K_{0.7}Na_{0.3}AlSi_3O_8$ hollandite (e.g., Na-liebermannite) and jadeite (Figure 2). Stishovite peaks, although visible at room conditions, had too low intensities, which did not permit us to calculate its unit-cell volume at high P and T. The compression curve of Na-liebermannite along the 300 K isotherm (Figure 3) shows that our unit-cell values were generally consistent with those of $K_{0.8}Na_{0.2}AlSi_3O_8$ hollandite reported by Boffa-Ballaran et al. [21], with differences that can be explained by the slightly higher Na-content in our hollandite phase. Similarly to Boffa-Ballaran et al. [21], we also observed a discontinuous volume decrease at P > 16 GPa, which they attributed to the hollandite to hollandite-II phase transition, although in our experiments, we could not observe clearly the broadening and splitting of diffraction peaks that generally characterize the transition [18,19,21] because of the low resolution of the present energy dispersive XRD method.

In order to assess that the volume decrease is related to the second order transition in the hollandite phase and not to an eventual reaction between liebermannite and jadeite at high P and T, we carried out an additional experiment (OD1667, Table 1) at 21.4 GPa and 1700 K, which corresponds to the highest P and T conditions where the density was measured in situ (Figure 2, Supplementary Materials Table S1). In OD1667, the microcline glass was compressed to the target pressure and annealed at 1700 K for 12 h, which is similar to the procedure for equilibrium experiments in the $KAlSi_3O_8$-$NaAlSi_3O_8$

system [26]. SEM–EDS analyses of the recovered sample showed Na-liebermannite and jadeite + stishovite, with rather similar texture and composition to the multiphase aggregate, before and after the in situ experiment (Figure 1 and Table 1). The slightly lower Na-content observed in OD1667's hollandite phase indicates that Na-content of liebermannite was unlikely to increase at P > 16 GPa, as a result of reaction with the coexisting jadeite, which is compatible with previous studies showing that the partitioning coefficient of Na in hollandite decreases with increasing P and T [26]. Analyses of the sample recovered after the in situ experiment (S2697, Table 1) were, however, quasi-identical to those of the starting aggregate (OS2999, Table 1) suggesting the kinetics of this reaction would be slower than the heating time required for our measurements. The volume decrease we observed at P > 16 GPa was likely to correspond to the hollandite to hollandite-II transition in $K_{0.7}Na_{0.3}AlSi_3O_8$ hollandite, which is consistent with Boffa-Ballaran et al.'s [21] observations of the transition in between 17 GPa and 20 GPa for $K_{0.8}Na_{0.2}AlSi_3O_8$ hollandite. Our results also indicated that substitution of Na into the $KAlSi_3O_8$ hollandite structure decreased the transition pressure with respect to that observed for the endmember at pressures higher than 20 GPa [19,33,34], which is generally consistent with conclusions of previous experimental and theoretical studies [21,35].

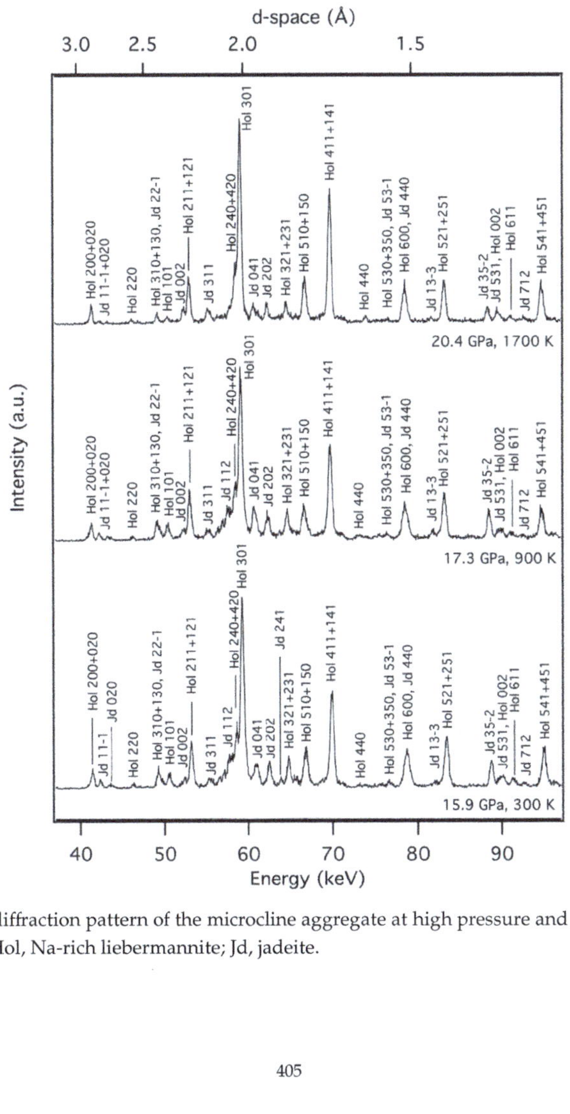

Figure 2. X-ray diffraction pattern of the microcline aggregate at high pressure and high temperature. Abbreviations: Hol, Na-rich liebermannite; Jd, jadeite.

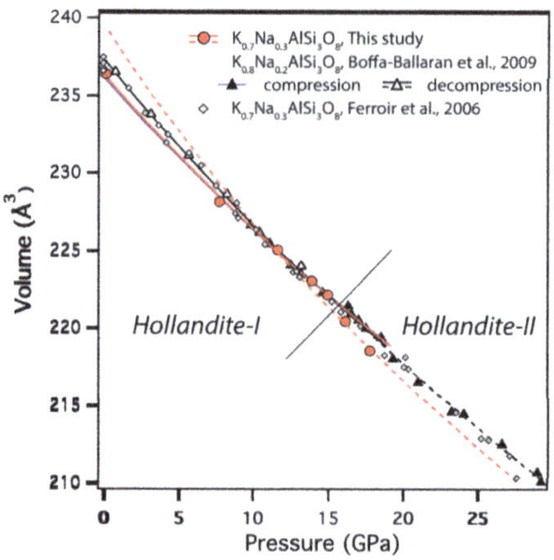

Figure 3. Unit-cell volume of $K_{0.7}Na_{0.3}AlSi_3O_8$ liebermannite as a function of pressure across the hollandite-I to hollandite-II transition. Red lines represent a fitting of hollandite-I (solid line) and hollandite-II (broken line) data by a Birch–Murnaghan EoS (Equation (1)). Black lines represent a fitting of Boffa-Ballaran et al.'s [21] data with (solid line) and without (broken line) decompression data.

3.1. P–V–T EoS of $K_{0.7}Na_{0.3}AlSi_3O_8$ Liebermannite

A least-square fitting of the room T Na-liebermannite unit-cell volume below 16 GPa to Equation (1) yielded the isothermal bulk modulus K_0 = 214(1) GPa and V_0 = 236.24(5) Å^3 (K' = 4), which was 6% to 18% higher than K_0 = 180–201 GPa reported for the $KAlSi_3O_8$ endmember [18,19,36]. Fitting of Equation (1) with K' fixed to 3 and 5 (Table 4) showed that such differences could not be explained by the tradeoff between K_0 and K′, and therefore suggested that incorporation of Na had a substantial effect on the compressibility of liebermannite. In contrast, we found our bulk modulus was ~8% higher than K_0 = 198 GPa for $K_{0.8}Na_{0.2}AlSi_3O_8$ hollandite [21], despite the two phases holding a similar Na-content. Fitting our data with V_0 = 237.3 Å^3 as proposed by Boffa-Ballaran et al. [21], however, gave a lower bulk modulus of K_0 = 195.4(4) GPa, which is more in agreement with their K_0 values and suggests that the discrepancy with Boffa-Ballaran et al. [21] lies in the determination of the room T zero-pressure volume.

Except for our study and Ferroir et al. [19], previous studies did not measure directly the unit-cell volume of their hollandite phase. Instead they derived their V_0 values from the fitting of the EoS, assuming the pressure derivative of the bulk modulus K' = 4 [18,21,36]. Comparison of previous V_0 values with a recent study by Zhou et al. [26], who carried out a precise investigation of the unit-cell volume of Na-bearing liebermannite as a function of Na-content, suggests that Boffa-Ballaran et al. [21] likely overestimated V_0 (Figure S1), which may be partly due to the use of decompression data in their EoS fitting. A second analysis of their high-pressure data, omitting decompression data, yielded K_0 = 217(4) GPa and V_0 = 236.1(2) Å^3 (K' = 4), which agrees well with our K_0 and V_0 (Table 4) and the Na-content dependence of V_0 proposed by Zhou et al. [26]. On the other hand, it is difficult to explain the differences between K_0 values of previous studies on the $KAlSi_3O_8$ endmember [18,19,36], which may be due not only to unconstrained V_0 but also limited data at high pressure [18,36]. Nevertheless, more recently, some theoretical works have favored higher K_0 values for the $KAlSi_3O_8$ endmember [34], such as K_0 = 201.4 GPa reported by Ferroir et al. [19]. Their value is ~9% smaller than our K_0 for $K_{0.7}Na_{0.3}AlSi_3O_8$, suggesting a substantial effect of Na-incorporation on the compressibility of liebermannite, which contrast with a previous conclusion by Boffa-Ballaran et al. [21], who proposed that Na does not affect the bulk compressibility of the hollandite structure.

Data at P > 16 GPa were used to investigate the compressibility of the $K_{0.7}Na_{0.3}AlSi_3O_8$ hollandite-II phase. Because the transition is reversible on release of pressure, the hollandite-II phase is unquenchable to ambient conditions and therefore there was no direct measurements of its zero-pressure volume. When we chose to fix V_0 = 239.4 Å^3 as proposed by Boffa-Ballaran et al. [21] for the $K_{0.8}Na_{0.2}AlSi_3O_8$ hollandite-II above 17 GPa, a least-square fitting of our hollandite-II room T unit-cell volume data yielded K_0 = 163.2(3) GPa (K' = 4). The difference between K_0 in the two studies may be partly related to the use of the tetragonal $BaMn_8O_{15}$ hollandite-type structure to fit our high-pressure data rather than the monoclinic cell *I2/m* used by Boffa-Ballaran et al. [21]. The reason we chose this structure is because the data were analyzed close to the transition pressure and therefore splitting of the peaks may not have been large enough to identify them as separate peaks. It is, however, worth noting that the fitting of hollandite-II data in Boffa-Ballaran et al. [21] also included data close to the transition, which may have led to overestimating the zero-pressure volume in their study. When we adopted a lower V_0 = 237.01 as proposed by a fitting of the combined $KAlSi_3O_8$ hollandite-II volume dataset of Hirao et al. [19] and Ferroir et al. [20], we obtained a larger K_0 = 188.2(4) GPa (K' = 4) Å^3, which is more compatible with bulk moduli proposed by experimental and theoretical works on the $KAlSi_3O_8$ endmember [20,34]. Our K_0 estimates for the $K_{0.8}Na_{0.2}AlSi_3O_8$ hollandite-II were lower than previous studies on the $KAlSi_3O_8$ endmember, suggesting that the incorporation of Na in the hollandite-II structure is likely to decrease the bulk modulus of the hollandite-II phase, although more high-pressure data on Na-bearing hollandites are necessary to constrain tightly the effect of Na on the compressibility of the hollandite-II structure.

Table 4. Thermoelastic parameters of hollandite-I and hollandite-II phases in the system $KAlSi_3O_8$–$NaAlSi_3O_8$.

	K_{T0} (GPa)	K_T'	V_0 (Å^3)	$\partial K_T/\partial T$ (GPa·K^{-1})	a_0 (10^{-5} K^{-1})	b_0 (10^{-8} K^{-2})	χ^2	Reference
Hollandite-I								
$K_{0.7}Na_{0.3}AlSi_3O_8$	220(1)	3*	236.20(5)				0.031	a
	214(1)	4*	236.24(5)				0.023	a
	207(1)	5*	236.27(5)				0.018	a
	211(2)	4*	236.24*	−0.043(20)	2.9(7)		0.122	a
$K_{0.8}Na_{0.2}AlSi_3O_8$	198(3)	4	237.3(2)					b
	217(4)	4*	236.1(2)					c
$KAlSi_3O_8$	183(3)	4	237.6(2)	−0.033(2)	3.32(5)	1.09(1)		d
$KAlSi_3O_8$	201.4(7)	4	237.01(33)					e
$KAlSi_3O_8$	180(3)	4	236.26(36)					f
Hollandite-II								
$K_{0.7}Na_{0.3}AlSi_3O_8$	163.2(3)	4	239.4*				0.008	a
	188.2(4)	4	237.01*				0.003	a
	160(2)	4	239.4*	−0.023(16)	3.2(11)		0.124	a
	184(3)	4	237.01*	−0.031(19)	3.3(12)		0.184	a
$K_{0.8}Na_{0.2}AlSi_3O_8$	174(7)	4	239.4(9)					b
$KAlSi_3O_8$	204(2)	4	237.01*					g

* fixed; a, this study; b [21]; c fitting of compression data from [21]; d [18]; e [19]; f [36] and g [20].

Figure 4 shows the unit-cell volume of Na-liebermannite up to ~21 GPa and 1700 K. A least-square fitting of all data below 16 GPa by Equations (1)–(5) yielded the isothermal bulk modulus K_{T0} = 211(2) GPa, its temperature derivative $(\partial K_T/\partial T)_P$ = −0.043(20) GPa·K^{-1}, and the thermal expansion α_T = 2.9(7) × 10^{-5} K^{-1} (K' = 4, V_0 = 236.24 Å^3) for the hollandite-I phase. Our values of $(\partial K_T/\partial T)_P$ and α_T were slightly different than that reported for the $KAlSi_3O_8$ endmember [18], but the difference falls within the uncertainties of the present measurement, suggesting that the substitution of Na may not have a significant effect on the temperature derivative of the bulk modulus of liebermannite. As for the hollandite-II, a similar fitting yielded K_{T0} = 160(2) GPa, $(\partial K_T/\partial T)_P$ = −0.023(16) GPa·K^{-1}, and α_T = 3.2(11) × 10^{-5} K^{-1} (K' = 4) for V_0 = 239.4 Å^3 fixed after Boffa-Ballaran et al. [21], while we obtained K_{T0} = 184(3) GPa, $(\partial K_T/\partial T)_P$ = −0.031(19) GPa·K^{-1}, and α_T = 3.3(12) × 10^{-5} K^{-1} (K' = 4) when we fixed V_0 = 237.01 Å^3 after Hirao et al. [20]. These are the first estimates of the thermal properties of Na-bearing hollandite-II phase, which allow for calculating the density of this phase at the P and T conditions of the deep mantle.

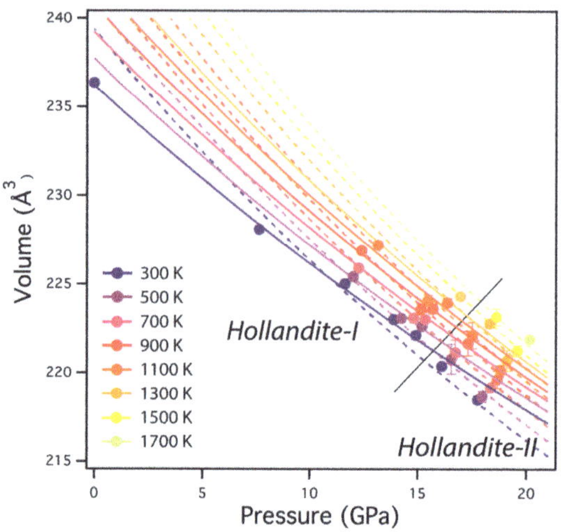

Figure 4. Unit-cell volume of $K_{0.7}Na_{0.3}AlSi_3O_8$ liebermannite as a function of pressure and temperature. The color-scale represents the data at temperatures varying from 300 K to 1700 K, respectively. Pressure and volume error bars are within the size of the symbols. Solid and broken lines represent the isothermal compression curves obtained from fitting all the data to a Birch–Murnaghan EoS (Equations (1)–(5)) for the hollandite-I (plain lines) and hollandite-II (broken lines) phases, respectively.

3.2. *P–V–T* EoS of Jadeite

The unit-cell volumes of jadeite underwent nonlinear decreases with no discontinuity in the compression curves up to the maximum pressure, as shown in Figure 5. Because of our limited data, V_0 was fixed to the value determined by ambient XRD on the starting sample (OS2999, Table 1). Fitting of the pressure–volume data at 300 K by Equation (1) yielded K_0 = 129.2(2) GPa (K' = 4, V_0 = 402.4 Å^3). The present K_0 value was in good agreement with the former study of Zhao et al. [22], while it was smaller than more recent studies [24,25]. The difference is, however, well explained by the tradeoff between K_0 and its first pressure derivative K' as shown by a subsequent fitting of our data with K' = 3.3, which yielded K_0 = 133.6(2) GPa, which is well consistent with the former studies of Posner et al. [24] and McCarthy et al. [25]. On the other hand, it is difficult to explain the high value of K_0 = 134.0(7) GPa reported by Nestola et al. [23] for K' = 4.4 (Table 5), although their unit-cell volume data as well as those of other studies [22,24,25] were well explained by our compression curve (Figure 5). The difference may lie in their determination of K' when fitting data in a relatively narrow pressure range, as pointed out by Zhao et al. [22] during fitting of their own data.

Figure 6 shows the unit-cell volume of Jadeite up to ~21 GPa and 1700 K. A least-square fitting of all data by Equations (1)–(5) yielded the isothermal bulk modulus K_{T0} = 127(1) GPa, its temperature derivative $(\partial K_T/\partial T)_P$ = –0.012(11) GPa·K^{-1}, and the thermal expansion α_T = 2.6(9) × 10^{-5} K^{-1} (K' = 4, V_0 = 402.4 Å^3), which are generally in good agreement with the former study of Zhao et al. [22]. We also noted that if we adopted K' = 3.3 as suggested by more recent studies [24,25], we found K_{T0} = 132(1) GPa, $(\partial K_T/\partial T)_P$ = –0.007(11) GPa·K^{-1}, and α_T = 2.3(8) × 10^{-5} K^{-1} (V_0 = 402.4 Å^3).

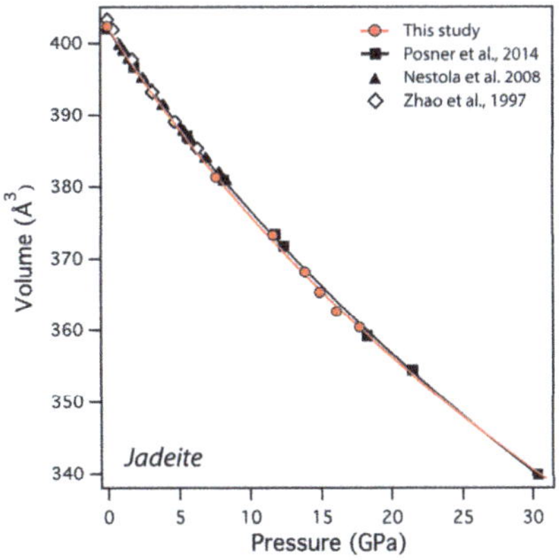

Figure 5. Unit-cell volume of Jadeite as a function of pressure. Data are fit with a Birch–Murnaghan equation (Equation (1)).

Table 5. Thermoelastic parameters of jadeite.

Reference	K_{T0} (GPa)	K_T'	V_0 (Å^3)	$\partial K_T/\partial T$ (GPa·K^{-1})	a_0 (10^{-5} K^{-1})	b_0 (10^{-8} K^{-2})	χ^2
This study	133.6(2)	3.3*	402.4*				0.014
	129.2(2)	4*	402.4*				0.015
	126.7(2)	4.4*	402.4*				0.017
	123.3(2)	5*	402.4*				0.023
	132(1)	3.3*	402.4*	−0.007(11)	2.3(8)		0.284
	127(1)	4*	402.4*	−0.012(11)	2.6(9)		0.259
Posner et al. [23]	136(3)	3.3(2)	402.5(4)				
McCarthy et al. [22]	136(1)	3.4(4)	402.03(2)				
Nestola et al. [21]	134.0(7)	4.4(1)	402.26(2)				
Zhao et al. [20]	127(5)	4	403	−0.014(5)	2.5(2)	0.2(2)	
	125(4)	5	403	−0.016(5)	2.6(2)	0.3(2)	

* fixed.

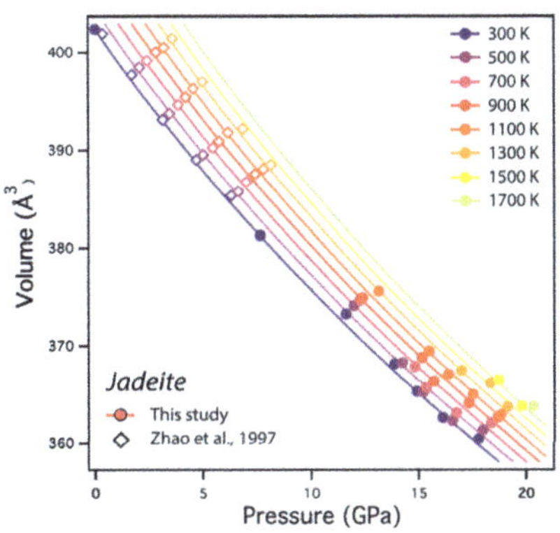

Figure 6. Unit-cell volume of jadeite as a function of pressure and temperature. Plain circles represent our volume data while open diamonds are data taken from Zhao et al. [22]. The color-scale represents the data at temperatures varying from 300 K to 1700 K, respectively. Pressure and volume error bars are within the size of the symbols. Solid lines represent the isothermal compression curves obtained from fitting all data to a Birch–Murnaghan EoS (Equations (1)–(5)).

4. Discussion

Implications for the Density of Subducted Continental Crust

We combined our thermoelastic data of hollandite and jadeite with those of other mantle minerals reported by previous experimental studies (Table S2) to estimate density of continental crust and sediment along an average mantle geotherm (Figure 4). Density of pyrolite, oceanic crust, and harzburgite compositions were also calculated and presented for comparison. We used the formalism of Equations (1)–(5) to calculate densities of each mineral as a function of pressure and temperature along an adiabatic mantle geotherm [37]. The density of the rock aggregates were calculated by a weighted average of the density of the constituent minerals, while the proportions of each phases at depth (Figure 7a) were taken from the phase equilibrium study of Irifune et al. [4]. In the shallower upper mantle, continental crust consists of clinopyroxene (40%), orthoclase (30%), coesite (25%), and kyanite (5%). Here we assume that the 40% clinopyroxene is jadeite, which is supported by the experimental result that the clinopyroxene in continental crust becomes more Na-rich with increasing pressure and exists as jadeite, essentially under the MTR conditions. With increasing pressure, clinopyroxenes also gradually enter into the garnet phase, resulting in an increase of the garnet proportions up to 30% at depths of ~400 km. At these depths, garnet coexists with pyroxene (15%), stishovite (25%), hollandite (25%), and kyanite (5%). The gradual exsolution of up to 10% CaPv from garnet occurs at ~550 km [38]. Kyanite is present down to a depth of ~450 km where it is replaced by 10% CAS phase. The phase transformations of coesite to stishovite and orthoclase to hollandite were fixed to 10 GPa (~300 km depth) while the hollandite phase transition from hollandite-I to hollandite-II was fixed to 17 GPa (~500 km depth).

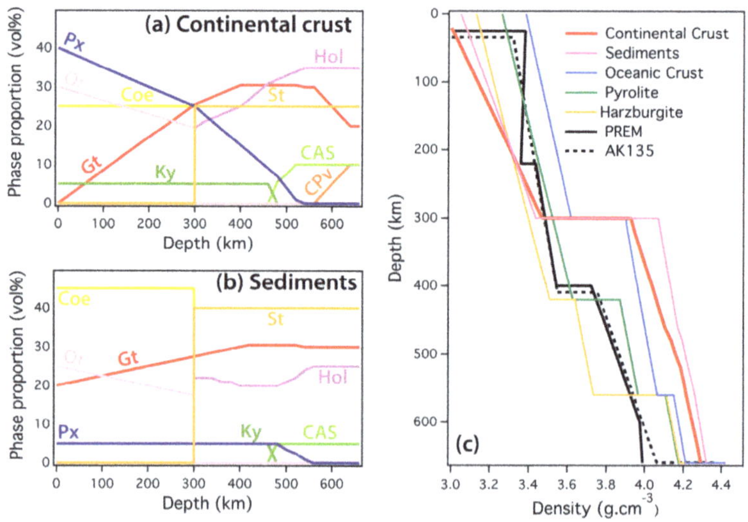

Figure 7. Phase proportions in (**a**) continental crust and (**b**) sediment as a function of depth, derived from Irifune et al. [4]. (**c**) Calculated density of continental crust and sediment as function of depth compared to that of other mineralogical models of pyrolite, oceanic crust (mid-ocean ridge basalt), and harzburgite, and global seismic models PREM [39] and AK135 [40]. Abbreviations: Px, pyroxene; Or, orthoclase; Coe, coesite; Gt, garnet; Ky, kyanite; St, stishovite; Hol, hollandite; CAS, Ca-rich aluminosilicate phase; CPv, $CaSiO_3$ perovskite.

Figure 7b shows that the densities of continental crust and sediment at depth < 200 km, are lower than any other lithologies, which would provide negative buoyancy to the slab and impede its subduction at the shallower upper mantle depths. We see, however, that both continental crust and sediment densities rapidly increase, due to the increase of the garnet proportions with pressure (Figure 7a), suggesting that their negative buoyancy is likely to be canceled by the density of the oceanic crust at depths > 200 km. The high-density of the oceanic crust component is generally considered to pull the slab downward, and thus could provide a way to transport broken fragments of continental blocks or sediments down to a depth of ~270 km [7,41]. When the continental crust and sediment reach a critical depth of ~300 km, their density would become, respectively, 11% and 15% greater than that of the pyrolitic mantle (Figure 7b) after the transformation of orthoclase to hollandite and coesite to stishovite at ~10 GPa (Figure 7a). Consequently, if the continental crust or the sediments would decouple from the main body of the slab at depths > 300 km, their high densities compared to the mantle are likely to favor the supply of continental materials to the deeper parts of MTR. In contrast, a slab breakoff before the critical depth would impede further subduction of both continental crust and sediment.

At depths > 660 km, previous studies proposed that the continental crust may be subducted into the lower mantle owing to its higher density compared to the surrounding mantle [4,5] when the temperature near the slab surface is lower than ~1700 K [5]. The final destination of the continental crust in the lower mantle is, however, still uncertain. Some studies suggested that continental crust may be partially molten when the slab is thermally equilibrated in the lower mantle [4], while others proposed that it may transform to a mineral assemblage stable at higher pressure if the subducted slab can remain cool enough in the lower mantle [5,42]. While recent laboratory studies favor the presence of oceanic crust in the uppermost lower mantle to explain globally the low-velocity profiles beneath subduction zones [43], the recycling of continental crust and sediment in the uppermost lower mantle may provide an alternative to explain locally some seismic observations that do not conform to global seismological models. Further investigation of the elasticity of continental crust constituent minerals at the pressure and temperature relevant to the uppermost lower mantle conditions are, however, necessary, and would provide an improved understanding on the recycling of continental crust materials in the deep Earth, with implications for the Earth's mantle internal heating and long-term evolution [9,10].

Supplementary Materials: The following are available online at http://www.mdpi.com/2075-163X/10/3/261/s1, Figure S1: Unit-cell volume of hollandite as a function of Na-content. The shaded area represents the uncertainties on the determination of the unit-cell volume of the $KAlSi_3O_8$ endmember on the basis of the data of Nishiyama et al. [18] and Ferroir et al. [19], which is translated to the other compositions., Table S1: Raw data for: Pressure (NaCl scale), Pressure (Au scale), Temperature (Kelvin), Unit-cell volume of Na-liebermannite (Angstrom cube), Unit-cell volume of Jadeite (Angstrom cube)., Table S2: Thermoelastic parameters of major mantle minerals.

Author Contributions: S.G. and T.I. conceived and designed the study. A.Y. prepared the starting glass materials. S.G. and Y.H. prepared the XRD measurement system combined with a multi-anvil apparatus and S.G. conducted experiments at SPring-8. Y.Z. conducted experiments at GRC. S.G., Y.Z. and Y.K. analyzed data. S.G. and Y.Z. wrote the manuscript. All authors discussed the content of the manuscript. All authors have read and agreed to the published version of the manuscript.

Funding: This research was carried with the support of a Grant-in-Aid for Young Scientists from the Japan Society for the Promotion of Science to S. Gréaux (Proposal No. 23740339) and the Joint Usage/Research Center PRIUS, Ehime University, Japan.

Acknowledgments: The authors thank Y. Tange, T. Kunimoto, Y. Zou, C. Zhou, and T. Arimoto for their assistance in the experiments at the beamline BL04B1 of SPring-8 (Project No. 2009A1300).

Conflicts of Interest: The authors declare no conflict of interest.

References

1. Yamamoto, S.; Senshu, H.; Rino, S.; Gondwana, S.O. 2009 Granite subduction: Arc subduction, tectonic erosion and sediment subduction. *Gondwana Res.* **2009**, *15*, 443–453. [CrossRef]
2. Bulanova, G.P.; Walter, M.J.; Smith, C.B.; Kohn, S.C.; Armstrong, L.S.; Blundy, J.; Gobbo, L. Mineral inclusions in sublithospheric diamonds from Collier 4 kimberlite pipe, Juina, Brazil: Subducted protoliths, carbonated melts and primary kimberlite magmatism. *Contrib. Mineral. Petro.* **2010**, *160*, 489–510. [CrossRef]

3. Cid, J.P.; Nardi, L.V.S.; Cid, C.P.; Gisbert, P.E.; Balzaretti, N.M. Acid compositions in a veined-lower mantle, as indicated by inclusions of (K,Na)-Hollandite + SiO_2 in diamonds. *LITHOS* **2014**, *196–197*, 42–53.
4. Irifune, T.; Ringwood, A.E.; Hibberson, W.O. Subduction of continental crust and terrigenous and pelagic sediments: An experimental study. *Earth Planet. Sci. Lett.* **1994**, *126*, 351–368. [CrossRef]
5. Ishii, T.; Kojitani, H.; Akaogi, M. High-pressure phase transitions and subduction behavior of continental crust at pressure–temperature conditions up to the upper part of the lower mantle. *Earth Planet. Sci. Lett.* **2012**, *357*, 31–41. [CrossRef]
6. Plank, T.; van Keken, P.E. Geodynamics: The ups and downs of sediments. *Nat. Geosci.* **2008**, *1*, 17–18. [CrossRef]
7. Ichikawa, H.; Kawai, K.; Yamamoto, S.; Kameyama, M. Supply rate of continental materials to the deep mantle through subduction channels. *Tectonophysics* **2013**, *592*, 46–52. [CrossRef]
8. Kawai, K.; Tsuchiya, T.; Tsuchiya, J.; Maruyama, S. Lost primordial continents. *Gondwana Res.* **2009**, *16*, 581–586. [CrossRef]
9. Helffrich, G.R.; Wood, B.J. The Earth's mantle. *Nature* **2001**, *412*, 501–507. [CrossRef]
10. Turcotte, D.L.; Paul, D.; White, W.M. Thorium-uranium systematics require layered mantle convection. *J. Geophys. Res.* **2001**, *106*, 4265–4276. [CrossRef]
11. Li, B.; Rigden, S.M.; Liebermann, R.C. Elasticity of stishovite at high pressure. *Phys. Earth Planet. Inter.* **1996**, *96*, 113–127. [CrossRef]
12. Li, B.; Neuville, D.R. Elasticity of diopside to 8 GPa and 1073K and implications for the upper mantle. *Phys. Earth Planet. Inter.* **2010**, *183*, 398–403. [CrossRef]
13. Ma, C.; Tschauner, O.; Beckett, J.R.; Rossman, G.R.; Prescher, C.; Prakapenka, V.B.; Bechtel, H.A.; MacDowell, A. Liebermannite, $KAlSi_3O_8$, a new shock-metamorphic, high-pressure mineral from the Zagami Martian meteorite. *Meteorit. Planet. Sci.* **2017**, *53*, 50–61. [CrossRef]
14. Liu, L.G.; El Gorsey, A. High-pressure phase transitions of the feldspars, and further characterization of lingunite. *Int. Geol. Rev.* **2007**, *49*, 854–860. [CrossRef]
15. Ringwood, A.E. Phase transformations and their bearing on the constitution and dynamics of the mantle. *Geochim. Cosmochim. Acta* **1991**, *55*, 2083–2110. [CrossRef]
16. Irifune, T.; Ringwood, A.E. Phase transformations in subducted oceanic crust and buoyancy relationships at depths of 600–800 km in the mantle. *Earth Planet. Sci. Lett.* **1993**, *117*, 101–110. [CrossRef]
17. Ishii, T.; Kojitani, H.; Akaogi, M. Phase relations and mineral chemistry in pyrolitic mantle at 1600–2200 °C under pressures up to the uppermost lower mantle: Phase transitions around the 660-km discontinuity and dynamics of upwelling hot plumes. *Phys. Earth Planet. Inter.* **2018**, *274*, 127–137. [CrossRef]
18. Nishiyama, N.; Rapp, R.P.; Irifune, T.; Sanehira, T.; Yamazaki, D.; Funakoshi, K.-I. Stability and P–V–T equation of state of $KAlSi_3O_8$-hollandite determined by in situ X-ray observations and implications for dynamics of subducted continental crust material. *Phys. Chem. Miner.* **2005**, *32*, 627–637. [CrossRef]
19. Ferroir, T.; Onozawa, T.; Yagi, T.; Merkel, S.; Miyajima, N.; Nishiyama, N.; Irifune, T.; Kikegawa, T. Equation of state and phase transition in $KAlSi_3O_8$ hollandite at high pressure. *Am. Mineral.* **2006**, *91*, 327–332. [CrossRef]
20. Hirao, N.; Ohtani, E.; Kondo, T.; Sakai, T.; Kikegawa, T. Hollandite II phase in $KAlSi_3O_8$ as a potential host mineral of potassium in the Earth's lower mantle. *Phys. Earth Planet. Inter.* **2008**, *166*, 97–104. [CrossRef]
21. Boffa Ballaran, T.; Liu, J.; Dubrovinsky, L.S.; Caracas, R.; Crichton, W. High-pressure ferroelastic phase transition in aluminosilicate hollandite. *Phys. Rev. B* **2009**, *80*, 214104. [CrossRef]
22. Zhao, Y.; Von Dreele, R.B.; Shankland, T.J.; Weidner, D.J.; Zhang, J.; Wang, Y.; Gasparik, T. Thermoelastic equation of state of jadeite $NaAlSi_2O_6$: An energy-dispersive Reitveld Refinement Study of low symmetry and multiple phases diffraction. *Geophys. Res. Lett.* **1997**, *24*, 5–8. [CrossRef]
23. Nestola, F.; Boffa Ballaran, T.; Liebske, C.; Bruno, M.; Tribaudino, M. High-pressure behaviour along the jadeite $NaAlSi_2O_6$–aegirine $NaFeSi_2O_6$ solid solution up to 10 GPa. *Phys. Chem. Miner.* **2006**, *33*, 417–425. [CrossRef]
24. Posner, E.S.; Dera, P.; Downs, R.T.; Lazarz, J.D.; Irmen, P. High-pressure single-crystal X-ray diffraction study of jadeite and kosmochlor. *Phys. Chem. Miner.* **2014**, *41*, 695–707. [CrossRef]
25. McCarthy, A.; Downs, R.; Thompson, R. Compressibility trends of the clinopyroxenes, and in-situ high-pressure single-crystal X-ray diffraction study of jadeite. *Am. Mineral.* **2008**, *93*, 198–209. [CrossRef]

26. Zhou, Y.; Irifune, T.; Ohfuji, H.; Shinmei, T.; Du, W. Stability region of $K_{0.2}Na_{0.8}AlSi_3O_8$ hollandite at 22 GPa and 2273 K. *Phys. Chem. Miner.* **2016**, *44*, 33–42. [CrossRef]
27. Lakshtanov, D.L.; Litasov, K.D.; Sinogeikin, S.V.; Hellwig, H.; Li, J.; Ohtani, E.; Bass, J.D. Effect of Al^{3+} and H^+ on the elastic properties of stishovite. *Am. Mineral.* **2007**, *92*, 1026–1030. [CrossRef]
28. Tsuchiya, T. First-principles prediction of the P-V-T equation of state of gold and the 660-km discontinuity in Earth's mantle. *J. Geophys. Res. Solid Earth* **2003**, *108*. [CrossRef]
29. Matsui, M.; Higo, Y.; Okamoto, Y.; Irifune, T.; Funakoshi, K.I. Simultaneous sound velocity and density measurements of NaCl at high temperatures and pressures: Application as a primary pressure standard. *Am. Mineral.* **2012**, *97*, 1670–1675. [CrossRef]
30. Larson, A.C.; Von Dreele, R.B. *GSAS General Structure Analysis System. Operation Manual*; LAUR 86-748; Los Alamos National Laboratory: Los Alamos, NM, USA, 2000; pp. 1–179.
31. Toby, B.H. EXPGUI, a graphical user interface for GSAS. *J. Appl. Crystallogr.* **2001**, *34*, 210–213. [CrossRef]
32. Le Bail, A.; Duroy, H.; Fourquet, J.L. Ab-initio structure determination of $LiSbWO_6$ by X-ray powder diffraction. *Mater. Res. Bull.* **1988**, *23*, 447–452. [CrossRef]
33. Sueda, Y.; Irifune, T.; Nishiyama, N.; Rapp, R.P.; Ferroir, T.; Onozawa, T.; Yagi, T.; Merkel, S.; Miyajima, N.; Funakoshi, K.-I. A new high-pressure form of $KAlSi_3O_8$ under lower mantle conditions. *Geophys. Res. Lett.* **2004**, *31*. [CrossRef]
34. Mookherjee, M.; Steinle-Neumann, G. Detecting deeply subducted crust from the elasticity of hollandite. *Earth Planet. Sci. Lett.* **2009**, *288*, 349–358. [CrossRef]
35. Kawai, K.; Tsuchiya, T. First-principles study on the high-pressure phase transition and elasticity of $KAlSi_3O_8$ hollandite. *Am. Mineral.* **2012**, *98*, 207–218. [CrossRef]
36. Zhang, J.; Ko, J.; Hazen, R.H.; Prewitt, C.T. High-pressure crystal chemistry of $KAlSi_3O_8$ hollandite. *Am. Mineral.* **1993**, *78*, 493–499.
37. Brown, J.M.; Shankland, T.J. Thermodynamic parameters in the Earth as determined from seismic profiles. *Geophys. J. Int.* **1981**, *66*, 579–596. [CrossRef]
38. Saikia, A.; Frost, D.J.; Rubie, D.C. Splitting of the 520-kilometer seismic discontinuity and chemical heterogeneity in the mantle. *Science* **2008**, *319*, 1515–1518.
39. Dziewonski, A.M.; Anderson, D.L. Preliminary reference Earth model. *Phys. Earth Planet. Inter.* **1981**, *25*, 297–356. [CrossRef]
40. Kennett, B.L.N.; Engdahl, E.R.; Buland, R. Constraints on seismic velocities in the Earth from traveltimes. *Geophys. J. Int.* **1995**, *122*, 108–124. [CrossRef]
41. Ichikawa, H.; Gréaux, S.; Azuma, S. Subduction of the primordial crust into the deep mantle. *Geosci. Front.* **2017**, *8*, 347–354. [CrossRef]
42. Ono, S.; Hirose, K.; Kikegawa, T.; Saito, Y. The compressibility of a natural composition calcium ferrite-type aluminous phase to 70 GPa. *Phys. Earth Planet.* **2002**, *131*, 311–318. [CrossRef]
43. Gréaux, S.; Irifune, T.; Higo, Y.; Tange, Y.; Arimoto, T.; Liu, Z.; Yamada, A. Sound velocity of $CaSiO_3$ perovskite suggests the presence of basaltic crust in the Earth's lower mantle. *Nature* **2019**, *565*, 218–221. [CrossRef] [PubMed]

minerals

MDPI

Article

Incorporation of Incompatible Strontium and Barium Ions into Calcite ($CaCO_3$) through Amorphous Calcium Carbonate

Ayaka Saito [1], Hiroyuki Kagi [1,*], Shiho Marugata [1], Kazuki Komatsu [1], Daisuke Enomoto [1], Koji Maruyama [1] and Jun Kawano [2]

1 Geochemical Research Center, Graduate School of Science, The University of Tokyo, 7-3-1 Hongo, Tokyo 113-0033, Japan; ayawaka921105@gmail.com (A.S.); marugata@eqchem.s.u-tokyo.ac.jp (S.M.); kom@eqchem.s.u-tokyo.ac.jp (K.K.); enomoto.daisuke@iri-tokyo.jp (D.E.); k6.maru@gmail.com (K.M.)

2 Department of Earth and Planetary Sciences, Faculty of Science, Hokkaido University, N10 W8, Kita-ku, Sapporo 060-0810, Japan; j-kawano@sci.hokudai.ac.jp

* Correspondence: kagi@eqchem.s.u-tokyo.ac.jp

Received: 31 January 2020; Accepted: 13 March 2020; Published: 17 March 2020

Abstract: Calcite is a ubiquitous mineral in nature. Heavy alkaline-earth elements with large ionic radii such as Sr^{2+} and Ba^{2+} are highly incompatible to calcite. Our previous study clarified that incompatible Sr^{2+} ions can be structurally incorporated into calcite through crystallization from amorphous calcium carbonate (ACC). In this study, we synthesized Sr-doped calcite with Sr/(Sr + Ca) up to 30.7 ± 0.6 mol% and Ba-doped calcite with Ba/(Ba + Ca) up to 68.6 ± 1.8 mol%. The obtained Ba-doped calcite samples with Ba concentration higher than Ca can be interpreted as Ca-containing barium carbonates with the calcite structure which have not existed so far because barium carbonate takes the aragonite structure. X-ray diffraction (XRD) patterns of the Sr-doped and Ba-doped calcite samples obtained at room temperature showed that reflection 113 gradually weakened with increasing Sr/(Sr + Ca) or Ba/(Ba + Ca) ratios. The reflection 113 disappeared at Ba/(Ba + Ca) higher than 26.8 ± 1.6 mol%. Extinction of reflection 113 was reported for pure calcite at temperatures higher than 1240 K, which was attributed to the rotational (dynamic) disorder of $CO_3{}^{2-}$ in calcite. Our Molecular Dynamics (MD) simulation on Ba-doped calcite clarified that the $CO_3{}^{2-}$ ions in Ba-doped calcites are in the static disorder at room temperature. The $CO_3{}^{2-}$ ions are notable tilted and displaced from the equilibrium position of pure calcite.

Keywords: calcite; amorphous calcium carbonate; disorder; MD simulation

1. Introduction

Calcium carbonate ($CaCO_3$) is a ubiquitous mineral in nature that has three anhydrous polymorphs: calcite, aragonite, and vaterite. Calcite has a trigonal structure and it is thermodynamically stable under ambient condition. Aragonite, an orthorhombic structure, is a high-pressure phase. Vaterite is a metastable phase and it is a rare mineral compared to calcite and aragonite. In addition to these three anhydrous polymorphs, amorphous calcium carbonate (hereafter amorphous calcium carbonate (ACC); $CaCO_3{\cdot}nH_2O$, $n < 1.5$) is known to exist [1]. In living organisms, ACC is mainly used as a precursor for the formation of crystalline calcium carbonate and either has an aragonite-like short-range order or a calcite-like short-range order. Apart from biominerals, synthetically prepared ACCs have calcite-like short range structures [2,3]. Since ACC is thermodynamically metastable, it is easily crystallized at high temperature or high pressure [4,5].

Calcium carbonate exists mostly as calcite and aragonite in the natural environment. It is known that the structures of carbonates depend on the ionic radii of divalent metal ions [6]. Divalent ions with an ionic radius smaller than that of Ca^{2+} (1.00 Å) form a carbonate with the calcite structure. On the other hand, divalent ions with an ionic radius larger than that of Ca^{2+} form a carbonate with the aragonite structure. The incorporation of impurity ions into structures of carbonates also depends on the radii of impurity ions. Carbonates with the calcite structure tend to capture impurity elements of which the ionic radius is smaller than that of calcium ion. In contrast, carbonates with the aragonite structure tend to capture impurity elements of which the ionic radius is larger than that of Ca^{2+}. From the above, divalent metal ions such as Sr^{2+} and Ba^{2+} are incompatible to calcite. Pingitore and Eastman [7] studied the behavior of trace elements, Sr^{2+} and Ba^{2+}, into calcite. Partition coefficients of Ba^{2+} and Sr^{2+} were $k^{Ba^{2+}}_{calcite} = 0.6 \pm 0.01$ and $k^{Sr^{2+}}_{calcite} = 0.05 \sim 0.14$, respectively. Therefore, $k^{Ba^{2+}}_{calcite} < k^{Sr^{2+}}_{calcite}$ is consistent with the size of ionic radii showing that Ba is more incompatible to calcite than Sr.

Matsunuma et al. reported that high concentration of incompatible Sr was captured into the structure of calcite by pressurizing ACC [8]. The unit-cell volume of calcite samples precipitated from a supersaturated solution showed no noteworthy increase. During crystal growth from the supersaturated solution, Sr is excluded from calcite because of its incompatibility. In contrast, the unit-cell volume of calcite samples obtained from crystallization from ACC notably increased with increasing Sr concentration in starting solutions [8]. Because ACC has significant flexibility in the structure, impurity ions with high incompatibility to calcite can be captured in ACC. Rapid crystallization from ACC may result in the introduction of highly incompatible ions into calcite. Not only being limited to calcite, impurity doping via amorphous state to a crystallite phase can be applied to other substances.

In this study, we verified the Sr/(Sr + Ca) in calcite prepared from Sr-doped ACC samples. Also, incorporation of more incompatible Ba^{2+} (1.35 Å) into calcite was investigated. Changes in properties of Sr-doped and Ba-doped calcite were investigated as well.

2. Experimental Procedures

2.1. Synthesis of Sr-Doped Calcite and Ba-Doped Calcite

Strontium-doped ACC samples were synthesized based on Matsunuma et al. [8]. Ice-cooled 0.1 M Na_2CO_3 and blended 0.1 M solutions of $CaCl_2$ and $SrCl_2$ with varying Sr/(Sr + Ca) from 0 to 50 mol% were mixed at 0 °C with the weight ratio of 1:1. The initial pH value of the mixed solution is approximately 11.5. White precipitates were obtained after mixing the solutions in the reaction vessel. The precipitates were immediately filtered using a membrane filter (0.45 μm) with vacuum filtration. Obtained precipitates were then washed with acetone and dried at 25 °C for 1 day in a vacuum desiccator evacuated with a diaphragm pump. The reasons why the sample solutions were kept at 0 °C and the obtained ACC samples were rinsed with acetone are to avoid crystallization of calcite. Synthesized Sr-doped ACC samples were pressurized using a hydraulic press in a tungsten carbide (WC) piston-cylinder of 4 mm inner diameter at 0.8 GPa for 10 min at room temperature for pressure-induced crystallization of calcite. After decompression, the samples were recovered and kept in a vacuum desiccator to remove water that emitted after pressure-induced crystallization. In addition, calcite samples were also prepared by heating Sr-doped ACC samples at 400 °C for 2 h.

Barium-doped ACC samples were also prepared by the same method as Sr-doped ACC samples, mixing ice-cooled 0.1 M Na_2CO_3 and blended solutions of $CaCl_2$ and $BaCl_2$ with varying Ba/(Ba + Ca) from 0 to 80 mol% with the weight ratio of 1:1. After the samples were dried for 1 day in a vacuum desiccator, Ba-doped ACC samples were heated for 2 h at 400 °C for heat-induced crystallization of calcite.

2.2. Sample Analysis

Powder X-ray diffraction (XRD) patterns of the obtained samples were measured using a silicon zero background plate and an X-ray diffractometer (Miniflex II, Rigaku Corp., Tokyo, Japan). The measurement conditions for XRD were 0.02° step, scanned region from 10° to 70° in 2θ with a scan rate of 1° per minute, Cu$K\alpha$ radiation operated at 15 mA and 30 kV. For refinement of lattice parameters of calcite samples, potassium chloride powder was mixed with the samples as an internal standard for lattice constant and Rietveld analysis was conducted using general structure analysis system (GSAS) software and EXPGUI [9,10]. Field emission Scanning electron microscope/Energy dispersive X-ray spectroscopy (SEM-EDS) analysis (JSM-7000, JEOL, Tokyo, Japan) was conducted to observe the morphology of impurity-doped calcite samples and to determine the Sr and Ba concentration in the synthesized samples. The pelletized calcite samples were fixed on a glass slide with carbon tapes and carbon coating was applied on the samples.

3. Molecular Dynamics (MD) Simulations of Ba-Doped Calcite

MD simulations of Ba-doped calcite were conducted to investigate the behavior of carbonate ions in Ba-doped calcite at an atomistic level. Because the ionic radius of Ba is larger than that of Sr, Ba-doped calcite should be more suitable to analyze the effect of captured divalent cation on the crystal structure with MD simulation. The interatomic potential function between two atoms (the i th and j th atoms) used in the present MD calculations is described as follows:

$$\phi_{ij}(r_{ij}) = \frac{z_i z_j e^2}{r_{ij}} + f_0(B_i + B_j)exp\frac{A_i + A_j - r_{ij}}{B_i + B_j} - \frac{C_i C_j}{r_{ij}^6} + D_{1ij}exp(-\beta_{1ij}r_{ij}) + D_{2ij}exp(-\beta_{2ij}r_{ij}).$$

The equation consists of the Coulombic interaction between point charges, short range repulsion, van der Waals attraction, and Morse potential terms. Here r_{ij} is the interatomic distance between the i th and j th atoms, $f_0 = 6.9511 \times 10^{-11}$ N is a constant, e is the electronic charge. z, A, B, and C are the parameters for each atomic species, and D_1, D_2, β_1, and β_2 are the parameters for the C–O pair.

MD simulations in this study were carried out with the MD program MXDTRICL [11]. The equations of motions were integrated by Verlet's algorithm with a time step of 0.5 fs. Periodic boundary conditions were applied on the MD basic cell. The temperatures and pressures were controlled by scaling particle velocities and simulating the cell parameters, respectively.

The potential parameters set for Ba-doped calcite was derived based on Kawano et al. (see Table 1): the same parameters for O, C, and Ca were used as these determined in Kawano et al. [12] for the phase transition in calcite at high temperatures, and the parameters for Ba was newly derived empirically to reproduce the crystal structure, thermal expansivity, and compressibility of witherite ($BaCO_3$ with aragonite structure) (see Supplementary Materials).

Table 1. Parameter sets for interatomic potential function for Molecular Dynamics (MD) simulations. Parameters for Ca, C, O, and O–C are from Kawano et al. [12]. Parameters for Ba was determined in this study.

Atoms	Z (e)	A (Å)	B (Å)	C (kcal$^{1/2}$ Å3 mol$^{-1/2}$)
O	−0.915	1.8836	0.1658	23.351
C	1.045	0.4638	0.0784	0
Ca	1.7	1.4466	0.1042	10.086
Ba	1.7	1.65	0.102	13
Atomic Pair	D1 (kJ mol^{-1})	β1 (Å^{-1})	D2 (kJ mol^{-1})	β2 (Å^{-1})
O–C	45735	5.14	−4936.066	2.57

The unit cell adopted in the MD simulation of Ba-doped calcite was composed of 72 crystallographic unit cells of calcite (a_{MD} = 6*a*, c_{MD} = 2*c*, in the hexagonal setting), containing 2160 atoms. Barium ions were randomly substituted to Ca ions in the calcite structure relaxed at 300 K and 1 atm to form initial structure of Ba-doped calcite with various Ba concentration. Atomic behavior was analyzed based on the results of calculations for 20–30 ps (= 40,000–60,000 steps), after preliminary annealing of the initial structure for at least 20 ps (= 40,000 steps) which is ensured to be long enough to equilibrate the system. Simulations at high temperature were started from the relaxed structure of Ba-doped calcite with the targeted Ba concentration, which were annealed at 300 K and 1 atom. The diffracted intensity was calculated from the structure factor that was directly obtained from the MD simulated atomic positions by using a program developed by Miyake et al. [13].

4. Results and Discussion

4.1. Increase of Lattice Parameters of Calcite Induced by Incorporation of Sr and Ba

Figure 1a displays XRD patterns of calcite samples obtained from Sr-doped ACC samples after pressure treatment at 0.8 GPa. Calcite was observed as the crystalline phase after pressure treatment on the ACC samples precipitated from solutions with Sr/(Sr + Ca) < 35 mol%. However, strontianite ($SrCO_3$ with the aragonite structure) was observed as a crystalline phase in addition to calcite after pressure treatment on ACC obtained from starting solutions with Sr/(Sr + Ca) > 40 mol%. In contrast, after heating treatment at 400 °C, no strontianite was observed from calcite samples crystallized from ACC obtained from starting solutions with Sr/(Sr + Ca) = 40 mol%. Figure 1b shows XRD patterns of calcite samples obtained by heating treatment on Sr-doped ACC samples. ACC samples precipitated from solutions with Sr/(Sr + Ca) < 40 mol% crystallized into calcite. In contrast, ACC samples from solutions with Sr/(Sr + Ca) = 40 mol% crystallized into strontianite and calcite in a similar way to the case of pressure-induced crystallization.

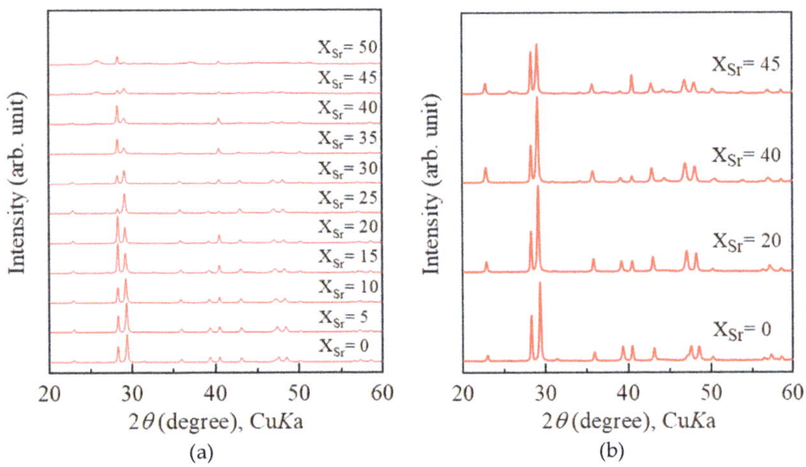

Figure 1. Powder X-ray diffraction patterns of calcium carbonate samples precipitated from solutions with various Sr concentrations. (**a**) Crystallization was induced by pressure treatment. (**b**) Powder X-ray diffraction patterns of calcium carbonate samples precipitated from solutions with various Sr concentrations. Crystallization was induced by heating treatment at 400 °C for 2 h. x_{Sr} stands for Sr/(Sr + Ca) of starting solutions.

The unit-cell parameters of calcite samples obtained from heating treatment and pressure treatment were obtained from Rietveld refinement on the XRD patterns. The obtained unit-cell volume data are presented in Figure 2.

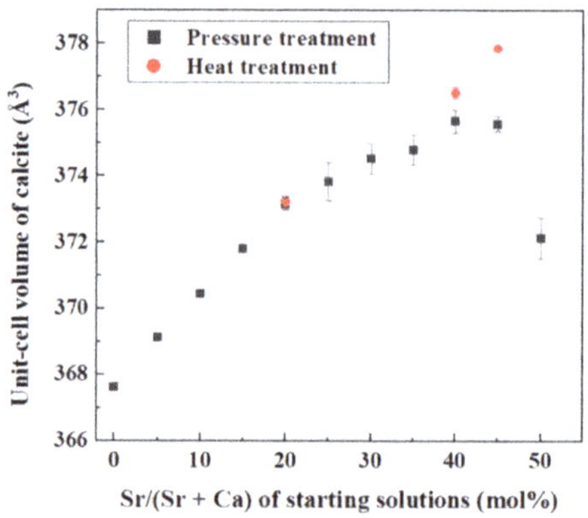

Figure 2. Unit-cell volume of calcite samples obtained from heating treatment and pressure treatment vs. Sr/(Sr + Ca) of starting solutions. The sample with the largest volume higher than 377 Å^3 contained strotntianite. The error bar was derived from standard deviation of unit cell volume estimated using general structure analysis system (GSAS).

Matsunuma et al. reported the unit-cell volume of calcite obtained by pressurization of ACC increased monotonically up to 372.5 Å^3 [8]. As shown in Figure 2, the unit-cell volume of calcite obtained by heating treatment on Sr-doped ACC increased up to 376.51 ± 0.18 Å^3, which extended to higher value than the previous study [8]. The unit-cell volume of calcite obtained from pressurization of ACC increased with increasing Sr concentration of starting solutions, but the unit-cell volume drops after the appearance of strontianite (see Figure 2). This indicates that Sr doped in the initial ACC was partitioned to strontianite and Sr concentration in calcite notably decreased.

Figure 3 displays powder XRD patterns of calcium carbonate samples obtained from heating treatment on Ba-doped ACC samples. ACC samples in the range of Ba/(Ba + Ca) = 0 to Ba/(Ba + Ca) = 70 mol% crystallized into exclusively calcite. Samples crystallized from solutions with Ba/(Ba + Ca) > 75 mol% contained witherite ($BaCO_3$ with the aragonite structure) in addition to calcite. Unit-cell parameters of calcite samples obtained from heating treatment on Ba-doped ACC were refined from the XRD patterns. The c/a axial ratio increased as the Ba concentration of starting solutions increased (data not shown). This trend also support that Ba is contained in the crystal structure of calcite [6,14]. Figure 4 shows that unit-cell volume of calcite crystallized from Ba-doped ACC monotonically increased with increasing Ba concentration in the starting solutions and the maximum unit-cell volume of Ba-doped calcite was 426.26 ± 0.08 Å^3. This result is much higher than that of Sr-doped calcite (376.51 ± 0.18 Å^3).

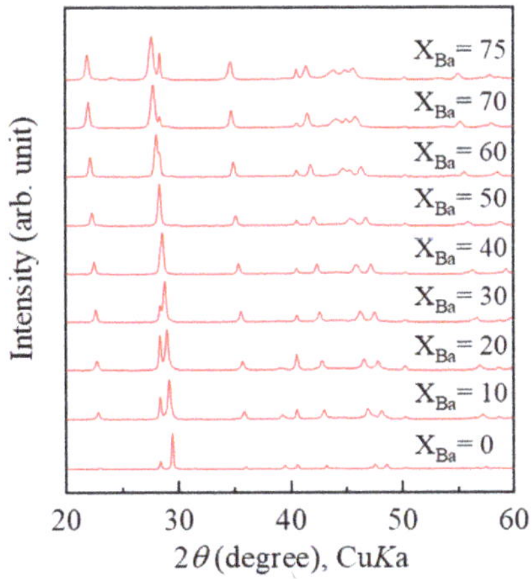

Figure 3. Powder X-ray diffraction patterns of calcium carbonate samples with various Ba concentrations obtained from heating treatment. Crystallization was induced by heating treatment at 400 °C for 2 h. x_{Ba} stands for Ba/(Ba + Ca) of starting solutions.

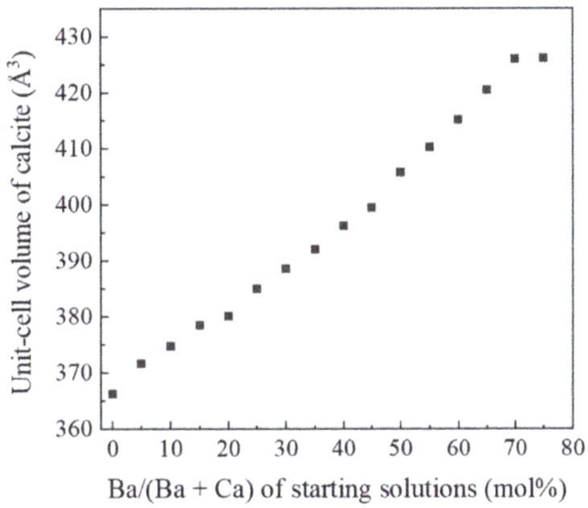

Figure 4. Unit-cell volume of calcite samples obtained from heating treatment vs Ba/(Ba + Ca) ratio of starting solutions.

4.2. Growth Texture of Sr-Doped Calcite and Ba-Doped Calcite

Figure 5a,b show SEM images of Sr-doped calcite obtained from pressure treatment and heating treatment, and Figure 5c is an SEM image of Ba-doped calcite obtained from the heating treatment. Calcite crystals obtained from pressure treatment have euhedral shapes and the grain size is in the order of 10 μm. In contrast, the grain size of calcite samples obtained from heating treatment was in the order of several tens of nanometers which is much smaller than those obtained from pressure treatment. In pressure-induced crystallization, coexisting fluid plays an important role in crystallization of calcite carbonate; dissolution-recrystallization from coexisting fluid can result in the crystallization of calcite with larger grain size [8,15]. On the other hand, the fine calcite crystals were obtained by heating treatment. During the heating process, the nucleation density and the nucleation rate were high. The contrastive growth textures imply the notable difference in the nucleation density and growth kinetics of calcite between the heating treatment and pressure-treatment.

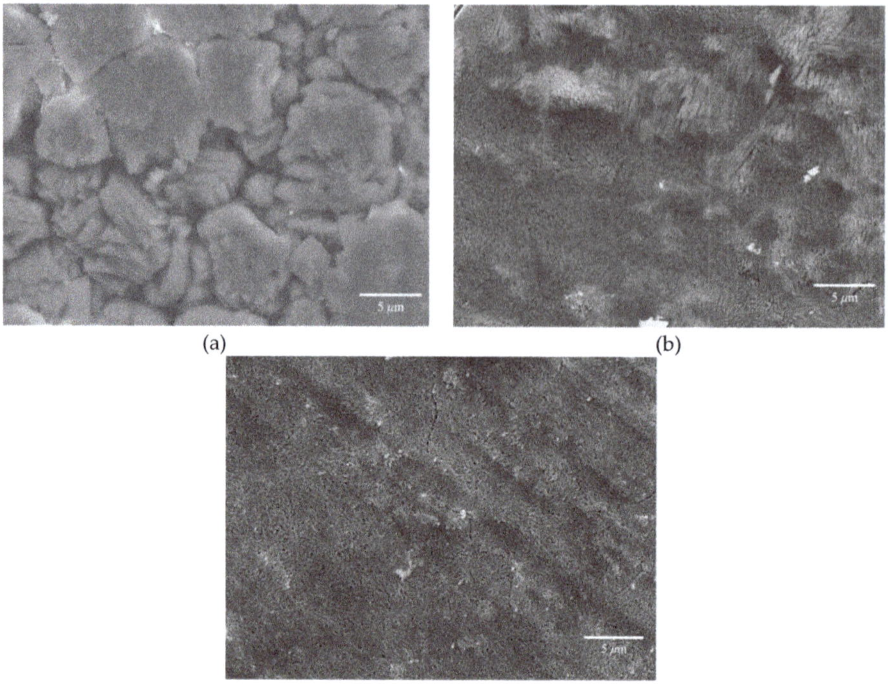

(a) (b) (c)

Figure 5. (**a**) SEM image of pressurized Sr-doped ACC sample (Sr/(Sr + Ca) = 0.05). (**b**) SEM image of heated Sr-doped ACC sample (Sr/(Sr + Ca) = 0.2). (**c**) SEM image of heated Ba-doped ACC sample (Ba/(Ba + Ca) = 0.45).

4.3. *Incorporation of Sr and Ba into Calcite Lattice*

Strontium concentration of calcite samples was determined from Energy-dispersive X-ray spectroscopy (EDS) measurements. The maximum Sr concentration of calcite obtained by heating Sr-doped ACC was 30.7 ± 0.6 mol% (see Table 2), which was larger than the pressurized sample (24.8 ± 0.4 mol%). These crystallized samples contained no strontianite as an impurity. Strontium concentration of natural samples such as corals and foraminiferas are less than 1 mol% [16,17] and the synthesized samples that were directly precipitated contain about 3 mol% of Sr at most [8]. It is obvious that the Sr concentration obtained in this study is remarkably higher.

Table 2. Sr/(Sr + Ca) of calcite samples determined from Energy-dispersive X-ray spectroscopy (EDS) measurements. The calcite samples listed in this table contained no strontianite ($SrCO_3$).

Sr/(Sr + Ca) in Starting Solutions (mol%)	Sr/(Sr + Ca) in Calcite Determined from EDS Measurements (mol%)	
	Pressure Treatment	Heat Treatment
5	5.1 ± 0.1	
10	9.5 ± 0.2	
15	14.2 ± 0.2	
20	19.4 ± 0.2	19.3 ± 0.7
40		30.7 ± 0.6

In the case of Ba incorporation, the maximum Ba concentration of heat-induced Ba-doped calcite was 68.6 ± 1.8 mol% (see Table 3). This result is much higher than the maximum Sr concentration in calcite. Barium carbonate ($BaCO_3$; witherite) takes an aragonite structure because the ionic radius of Ba^{2+} (1.35 Å) is much larger than Ca^{2+} (1.00 Å) and Sr^{2+} (1.18 Å). Ba-doped calcite obtained in this study has a calcite structure though they occupy more than half Ba^{2+} than Ca^{2+} in the crystal structure, which would be a Ca-doped $BaCO_3$ with a calcite structure. This study clarified that crystallization through ACC made it possible to synthesize such new material.

Table 3. Ba/(Ba + Ca) of calcite samples determined from EDS measurements.

Ba/(Ba + Ca) in Starting Solutions (mol%)	Ba/(Ba + Ca) in Calcite Determined from EDS Measurements (mol%)
5	8.2 ± 0.8
10	13.1 ± 1.1
15	17.6 ± 0.6
20	22.2 ± 1.0
25	26.8 ± 1.6
30	31.7 ± 1.4
35	36.5 ± 1.2
40	39.6 ± 1.7
45	43.0 ± 1.6
50	50.2 ± 1.6
55	53.2 ± 1.2
60	57.6 ± 1.3
65	63.3 ± 1.7
70	68.6 ± 1.8

4.4. Impurity-Induced Order–Disorder Phase Transition

Figure 6a,b display XRD patterns of Sr-doped calcite samples and Ba-doped calcite samples respectively. With increasing Sr concentration, 113 reflection gradually broadened and finally disappeared at Sr/(Sr + Ca) = 35 mol% as shown in Figure 6a. The same phenomenon was also observed in XRD patterns of Ba-doped calcite samples (see Figure 6b); 113 reflection gradually broadened and disappeared at Ba/(Ba + Ca) = 25 mol%.

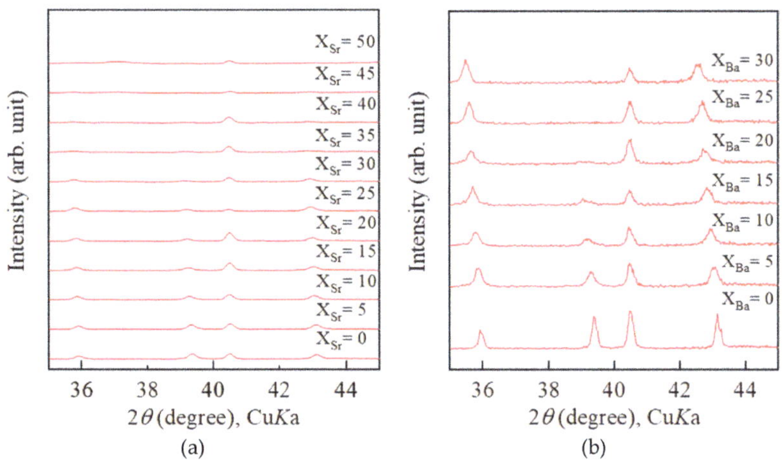

Figure 6. Disappearance of 113 reflection observed from powder X-ray patterns of (**a**) Sr-doped calcite samples and (**b**) Ba-doped calcite samples.

At room temperature, CO_3^{2-} in calcite vibrate at the original positions and the alternate layers of CO_3^{2-} are pointing in opposite directions. This is the ordered structure of calcite. In contrast, at high temperature, the CO_3^{2-} in calcite become disordered and leads to the order–disorder phase transition of calcite. This phase transition has been investigated in many studies [18–20]. Dove and Powell made neutron diffraction measurements on high temperature calcite and reported that the intensity of the 113 superlattice reflection falls rapidly on increasing the temperature and then disappears, showing that pure calcite has a phase transition from $R\bar{3}c$ to $R\bar{3}m$ at 1260 K [18]. Molecular dynamics (MD) simulation of order–disorder phase transition was conducted by Ferrario et al. and Liu et al. [19,20]. They concluded that the equilibrium positions of every CO_3^{2-} occupy the two orientations with equal probability. In addition to these MD simulation studies, Dove et al. studied neutron powder diffraction of calcite and suggested that the order–disorder phase transition is precipitated by the librational amplitude of the CO_3^{2-} [21]. However, the mechanism of this phase transition was still ambiguous. Ishizawa et al. carried out single-crystal X-ray diffraction experiments on high temperature calcite and determined the structure of high temperature calcite called phase V [22]. They reported that 113 reflection completely disappeared at 1275 K, which provides the $R\bar{3}m$ symmetry for crystal. In phase V, the oxygen atoms can freely migrate along undulated orbital, which means that carbonate ions are in rotational disorder.

In this study, the extinction of 113 reflection was observed at room temperature in contrast to the previous studies at high temperature. The phenomenon was induced by incorporation of the large incompatible elements which may suggest the rotational disorder of carbonate ions occurring at room temperature. This study is the first to find impurity-induced order–disorder phase transition at room temperature and the detailed mechanism needs to be investigated.

4.5. Molecular Dynamics (MD) Simulations of Ba-Doped Calcite

To elucidate the behavior of CO_3^{2-} ions, temperature-dependence of MD-simulated crystals was investigated. Kawano et al. indicated that the relative intensity of the 113–104 reflections decreases gradually upon increasing temperature and abruptly disappears between 1200 and 1250 K. The result is consistent with the experimental results on pure calcite and they suggested that the transition from $R\bar{3}c$ to $R\bar{3}m$ occurs at this temperature change. Above 1250 K, CO_3^{2-} ions sometimes 'flip' ±60○ from the original position and changes direction, which corresponds to the rotational (dynamical) disorder of CO_3^{2-} ions. In the same way, behaviors of $Ca_{0.75}Ba_{0.25}CO_3$ and $Ca_{0.5}Ba_{0.5}CO_3$ at high temperature were simulated. The results showed that rotational disorder of CO_3^{2-} ions occurred at 1050 K and 850 K for $Ca_{0.75}Ba_{0.25}CO_3$ and $Ca_{0.5}Ba_{0.5}CO_3$, respectively. The obtained results indicate that the transition temperature notably decreased with increasing Ba concentration in calcite. However, the transition temperatures are significantly higher than room temperature (300 K). This suggests that the rotational disorder associating with free flips of CO_3^{2-} ions does not occur at room temperature even for Ba-containing calcite samples.

Figure 7 shows probability distributions of rotation angle (θ) between the *a*-axis and the C–O bonds for $CaCO_3$, $Ca_{0.75}Ba_{0.25}CO_3$, and $Ca_{0.5}Ba_{0.5}CO_3$ at 300 K. These probability distributions were averaged for the all CO_3^{2-} ions in the simulated cells. All of the O atoms vibrate at the equilibrium positions of $R\bar{3}c$ symmetry at 60°, 180°, and 300° from the *a*-axis. With increasing Ba concentration in calcite, the width of the probability distribution increased. However, the probability between the peaks drops to zero. This means that no ±120° flipping of the CO_3 groups occur at 300 K.

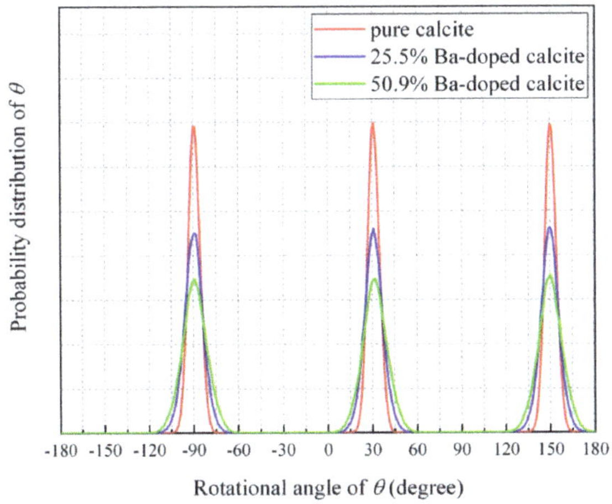

Figure 7. Average probability distributions of rotation angle (θ) between the *a*-axis and the C–O bonds for $CaCO_3$, $Ca_{0.5}Ba_{0.5}CO_3$, and $Ca_{0.5}Ba_{0.5}CO_3$ at 300 K.

Figure 8 shows probability distributions of rotation angle (θ) between the C–O bonds and the *a*-axis for $CaCO_3$, $Ca_{0.75}Ba_{0.25}CO_3$, and $Ca_{0.5}Ba_{0.5}CO_3$ at 300 K. This figure is focusing on one identical CO_3^{2-} ion for $CaCO_3$, $Ca_{0.75}Ba_{0.25}CO_3$, and $Ca_{0.5}Ba_{0.5}CO_3$. For pure calcite, three C–O bonds of the CO_3^{2-} ion distribute at −90°, 30°, and 150°. With increasing Ba concentration, and the three C–O bonds angularly displaces from the ideal angles (−90°, 30°, and 150°) and the widths of the distribution increased. Figure 9 shows probability distributions of the libration angle between the *a*–*b* plane and the CO_3^{2-} ion shown in Figure 8. For pure calcite, the libration angle distributes around 0°. With increasing Ba concentration, the libration angle of the carbonate ion increases and displaces from the equilibrium angle of pure calcite. The simulated results indicate that the involvement of Ba^{2+} ions to calcite induced the displacements of individual CO_3^{2-} ions in the rotational axis (θ) and librational axis (θ). Local structures surrounding a carbonate ion in 25.5% Ba-doped calcite and 50.9% Ba-doped calcite are shown in Figure 9c. The CO_3^{2-} ions in Ba-doped calcite samples vanishing 113 reflection take a static disorder.

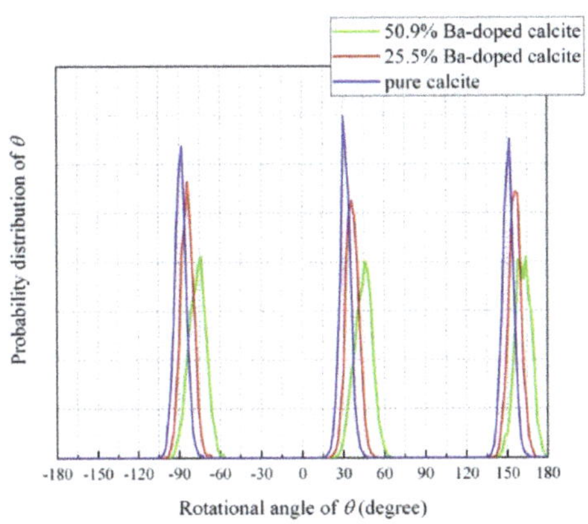

Figure 8. Probability distributions of rotation angle (θ) of one CO_3^{2-} ion between the C–O bonds and the *a*-axis for $CaCO_3$, $Ca_{0.5}Ba_{0.5}CO_3$, and $Ca_{0.5}Ba_{0.5}CO_3$ at 300 K.

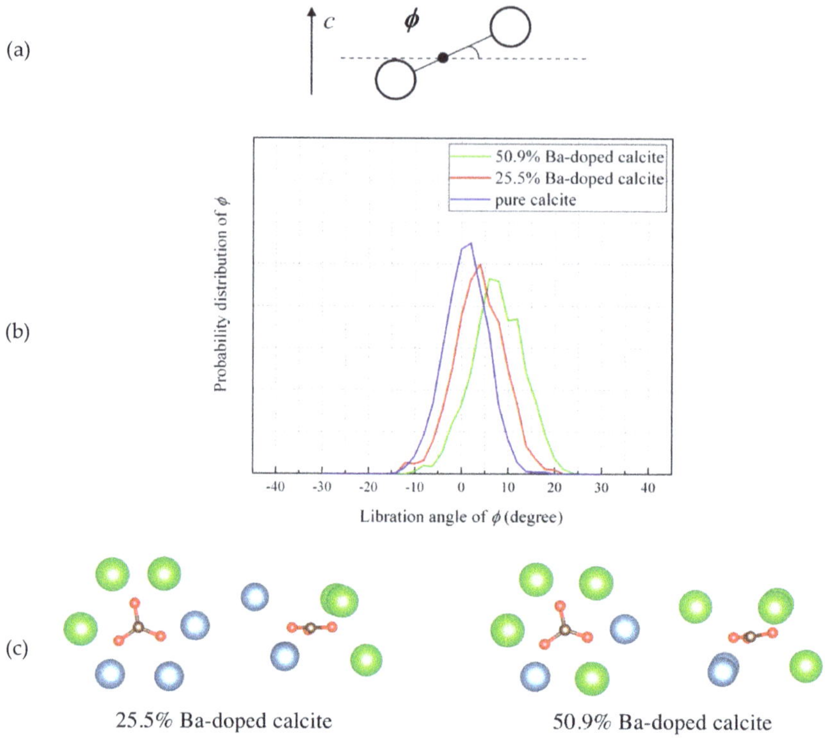

Figure 9. (**a**) Schematic drawing of CO_3^{2-} libration by an angle ϕ out of the *a*–*b* plane. (**b**) Probability distributions of the libration angle (ϕ between the *a*–*b* plane on the same CO_3^{2-} ion shown in Figure 8. (**c**) Local structures surrounding a carbonate ion in 25.5% Ba-doped calcite and 50.9% Ba-doped calcite.

Finally, X-ray diffraction patterns were obtained based on the simulated structure of Ba-doped calcite. As shown in Figure 10, the intensity of 113 reflection decreased with increasing Ba content in calcite and approached zero at Ba/(Ba + Ca) = 0.3. The simulated results are consistent with experimental results obtained in the present work although some difference can be seen between the simulation and the experimental results. The disappearance of 113 reflection in the Ba-doped and Sr-doped calcite samples are derived from the static disorder of CO_3^{2-} ions. This study first clarified the static disorder of CO_3^{2-} ions in calcium carbonate induced by large incompatible cations.

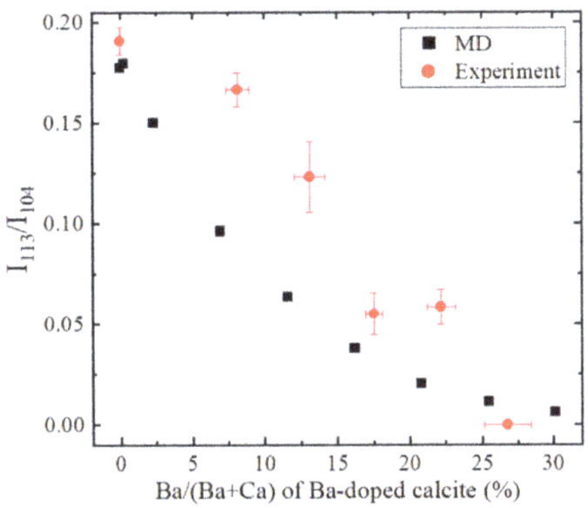

Figure 10. Ba concentration in calcite versus reflection intensities of 113 normalized by 104 (I_{113}/I_{104}) obtained from the MD simulation and experiments.

5. Conclusions

This study clarified that crystallization through an amorphous state can induce the structural incorporation of large and incompatible ions. The introduction of large incompatible ions to calcite induced a newly found phenomenon: the static disorder of carbonate ions. Moreover, the present study may open a window to develop a new material by doping incompatible elements. Specific conclusions obtained from this study are listed below.

(1) Crystallization from Sr-doped amorphous calcium carbonate and Ba-doped amorphous calcium carbonate resulted in the formation of calcite containing notably high concentrations of Sr and Ba. The maximum Ba concentration corresponded to the chemical formula of $Ba_{0.7}Ca_{0.3}CO_3$.

(2) With increasing Sr and Ba concentrations, the intensity of the 113 reflection of calcite decreased. The 113 reflection vanished at room temperature when Ba concentration was higher than 25 mol% ($Ca_{0.75}Ba_{0.25}CO_3$).

(3) The MD simulation indicated that the CO_3^{2-} ions in Ba-doped calcites are not in the rotational (dynamical) disorder but in the static disorder at room temperature. The CO_3^{2-} ions are tilted and angularly displaced from the equilibrium position of pure calcite.

Supplementary Materials: The following are available online at http://www.mdpi.com/2075-163X/10/3/270/s1, Figure S1: Ba/(Ba + Ca) molar ratios of starting solutions versus those of ACC samples precipitated from the solutions. Figure S2: (a) Temperature and (b) pressure dependence of the molar volume of MD-simulated and experimentally obtained $BaCO_3$. Black and red arrows in Figure S2(a) shows that phase transitions occurred at that temperature. Figure S3: Cell parameters of MD-simulated and the present experimental results of Ba-doped calcite.

Author Contributions: Conceptualization, A.S., H.K., K.K., and J.K.; Experiments, A.S., S.M., D.E., and K.M.; Methodology, A.S., K.K., and J.K.; Supervision, H.K.; Writing—original draft, A.S., H.K., and J.K. All authors have read and agreed to the published version of the manuscript.

Funding: The research was supported by JSPS KAKENHI Grant Numbers 18H05224 and JP18K18780.

Acknowledgments: The authors would like to thank Hiroshi Sakuma for valuable suggestions on the MD simulation. We are grateful to Takafumi Hirata and Yoshiki Makino for ICP-MS analysis and Hideto Yoshida for technical assistants for SEM-EDS analysis. Comments from three anonymous reviewers greatly helped us to improve this manuscript. We thank Robert C. Liebermann for providing us with an opportunity to contribute a paper to Special Issue of *Minerals* in memory of Orson Anderson.

Conflicts of Interest: The authors declare no conflict of interest.

References

1. Addadi, B.L.; Raz, S.; Weiner, S. Taking Advantage of Disorder: Amorphous Calcium Carbonate and Its Roles in Biomineralization. *Adv. Mater.* **2003**, *15*, 959–970. [CrossRef]

2. Cantaert, B.; Kuo, D.; Matsumura, S.; Nishimura, T.; Sakamoto, T.; Kato, T. Use of amorphous calcium carbonate for the design of new materials. *ChemPlusChem* **2017**, *82*, 107–120. [CrossRef] [PubMed]
3. Cartwright, J.H.E.; Checa, A.G.; Gale, J.D.; Gebauer, D.; Sainz-Díaz, C.I. Calcium carbonate polyamorphism and its role in biomineralization: How many amorphous calcium carbonates are there? *Angew. Chem. Int. Ed.* **2012**, *51*, 11960–11970. [CrossRef] [PubMed]
4. Koga, N.; Nakagoe, Y.; Tanaka, H. Crystallization of amorphous calcium carbonate. *Thermochim. Acta* **1998**, *318*, 239–244. [CrossRef]
5. Yoshino, T.; Maruyama, K.; Kagi, H.; Nara, M.; Kim, J.C. Pressure-Induced Crystallization from Amorphous Calcium Carbonate. *Cryst. Growth Des.* **2012**, *12*, 3357–3361. [CrossRef]
6. Veizer, J. Trace elements and isotopes in sedimentary carbonates. In *Carbonates: Mineralogy and Chemistry*; Reviews in Mineralogy; Reeder, R.J., Ed.; Mineralogical Society of America: Washington, DC, USA, 1983; Volume 11, pp. 265–299.
7. Pingitore, N.E.; Eastman, M.P. The experimental partitioning of Ba^{2+} into calcite. *Chem. Geol.* **1984**, *45*, 113–120. [CrossRef]
8. Matsunuma, S.; Kagi, H.; Komatsu, K.; Maruyama, K.; Yoshino, T. Doping incompatible elements into calcite through amorphous calcium carbonate. *Cryst. Growth Des.* **2014**, *14*, 5344–5348. [CrossRef]
9. Larson, A.C.; Von Dreele, R.B. *General Structure Analysis System (GSAS)*; Los Alamos National Laboratory: Los Alamos, NM, USA, 2000; pp. 86–748.
10. Toby, B.H. EXPGUI, a graphical user interface for GSAS. *J. Appl. Crystallogr.* **2001**, *34*, 210–213. [CrossRef]
11. Kawamura, K. MXDTRICL. Japan Chemical Program Exchange, #77, 1997.
12. Kawano, J.; Miyake, A.; Shimobayashi, N.; Kitamura, M.J. Molecular dynamics simulation of the rotational order–disorder phase transition in calcite. *Phys. Condens. Matter* **2009**, *21*, 95406. [CrossRef] [PubMed]
13. Miyake, A.; Hasegawa, H.; Kawamura, K.; Kitamura, M. Symmetry and its change in a reciprocal space of a quartz crystal simulated by molecular dynamics. *Acta Cryst.* **1998**, *A54*, 330–337. [CrossRef]
14. Mackenzie, F.T.; Bischoff, W.D.; Bishop, F.C.; Loijens, M.; Schoonmaker, J.; Wollast, R. Magnesian calcites: Low temperature occurrence, solubility and solid-solution behavior. In *Carbonates: Mineralogy and Chemistry*; Reviews in Mineralogy; Reeder, R.J., Ed.; Mineralogical Society of America: Washington, DC, USA, 1983; Volume 11, pp. 97–144.
15. Maruyama, K.; Kagi, H.; Inoue, T.; Ohfuji, H.; Yoshino, T. In situ observation of pressure-induced crystallization from amorphous calcium carbonate by time-resolved X-ray diffraction. *Chem. Lett.* **2015**, *44*, 434–436. [CrossRef]
16. Pingitore, N.E.; Lytle, F.W.; Davies, B.M.; Eastman, M.P.; Eller, P.G.; Larson, E.M. Mode of incorporation of Sr^{2+} in calcite: Determination by X-ray absorption spectroscopy. *Geochim. Cosmochim. Acta* **1992**, *56*, 1531–1538. [CrossRef]
17. Allison, N.; Austin, W.E.N. The potential of ion microprobe analysis in detecting geochemical variations across individual foraminifera tests. *Geochem. Geophys. Geosystems* **2003**, *4*, 1–9. [CrossRef]
18. Dove, M.T.; Powell, B.M. Neutron diffraction study of the tricritical orientational order/disorder phase transition in calcite at 1260 K. *Phys. Chem. Miner.* **1989**, *16*, 503–507. [CrossRef]
19. Ferrario, M.; Lynden-Bell, R.M.; McDonald, I.R. Structural fluctuations and the order–disorder phase transition in calcite. *J. Phys. Condens. Matter* **1994**, *6*, 1345–1358. [CrossRef]
20. Liu, J.; Duan, C.G.; Ossowski, M.M.; Mei, W.N.; Smith, R.W.; Hardy, J.R. Simulation of structural phase transition in $NaNO_3$ and $CaCO_3$. *Phys. Chem. Miner.* **2001**, *28*, 586–590. [CrossRef]
21. Dove, M.T.; Swainson, I.P.; Powell, B.M.; Tennant, D.C. Neutron powder diffraction study of the orientational order–disorder phase transition in calcite, $CaCO_3$. *Phys. Chem. Miner.* **2005**, *32*, 493–503. [CrossRef]
22. Ishizawa, N.; Setoguchi, H.; Yanagisawa, K. Structural evolution of calcite at high temperatures: Phase V unveiled. *Sci. Rep.* **2013**, *3*, 2832. [CrossRef] [PubMed]

minerals

MDPI

Article

Structure and Behavior of the Ni End-Member Schreibersite Ni_3P under Compression to 50 GPa

Sasithorn Chornkrathok [1,2,*], Dongzhou Zhang [2] and Przemyslaw Dera [2]

[1] Department of Earth Sciences, University of Hawai'i at Mānoa, Honolulu, HI 96822, USA
[2] Hawai'i Institute of Geophysics and Planetology, University of Hawai'i at Mānoa, Honolulu, HI 96822, USA; dzhang@hawaii.edu (D.Z.); pdera@hawaii.edu (P.D.)
* Correspondence: cs65@hawaii.edu; Tel.: +1-808-476-9793

Received: 15 February 2020; Accepted: 27 March 2020; Published: 30 March 2020

Abstract: To better understand the potential presence of light element alloys of Fe and Ni in the Earth's interior, the crystal structure and compressional behavior of the Ni-P binary compound, schreibersite (Ni_3P), have been investigated using synchrotron X-ray diffraction experiments. Both powder and two single-crystal samples of synthetic Ni_3P (in different orientations with respect to the loading axis of the diamond anvil cell) were compressed up to approximately 50 GPa at ambient temperature. The compressional data obtained for Ni_3P were fitted with a 3rd order Birch–Murnaghan equation of state. All data indicated that the c/a ratio of unit cell parameters remained approximately constant up to about 30 GPa but then increased progressively with pressure, exhibiting a second slight discontinuity at approximately 40 GPa. The changes in unit cell parameters at ~30 GPa and ~40 GPa suggested discontinuous changes in magnetic ordering. Moreover, the threshold of these subtle discontinuities is sensitive to the stress state and orientation of the crystal in the diamond anvil cell. This study is the first report on the compressional behavior of both powder and single-crystal schreibersite at high-pressure (up to 50 GPa). It offers insights into the effects of Ni_3P components on the compressional behavior of the Earth's core.

Keywords: schreibersite; high pressure; diamond anvil cell; synchrotron x-ray diffraction; single-crystal x-ray diffraction

1. Introduction

Knowledge of the chemical composition and physical properties of the Earth's interior comes mostly from seismic observations, geophysical modeling, direct observation of surface rocks, and the study of meteorites [1–3]. Unfortunately, at present, Earth's deep interior cannot be directly sampled due to engineering limitations, as the deepest hole that has been drilled thus far reached only approximately 12 km [4]. Consequently, experimental investigations at high pressures and high temperatures have played an essential role in studying the Earth's interior. The Earth's core is believed to be comprised primarily of an iron-nickel (Fe-Ni) alloy. Still, a light element (sulfur, phosphorus, oxygen, etc.) component appears to be required as well, based upon current models and observations [1–3,5]. From seismic evidence, we know that the density of Earth's core is approximately 10% lower than that of pure Fe at the estimated core pressure and temperature [2,5,6]. The current interpretation of this density deficit involves the presence of a significant amount of nickel, along with small amounts (up to 10%) of one or more light elements such as sulfur, phosphorus, oxygen, silicon, hydrogen, carbon, and potassium [1,3,7]. In addition to reducing the density, the presence of light elements in Fe-Ni alloys also affects the elastic properties and acoustic wave velocities of the core. For example, alloying of Fe with silicon increases both the compressional and shear wave velocities, whereas adding Ni decreases the compressional and shear wave velocities [8]. To better understand the evolution and formation of the Earth's core in terms of discrete material properties, experimentally measured elasticity and other physical parameters of potential alloys present in the Earth's core are required.

Among light elements suspected to be present in the core, phosphorus (P) is particularly interesting, because of its importance for life. P is present in the structure of living cells, helps to carry genetic information, is part of DNA, the fundamental building blocks of life, and contributes to adenosine triphosphate (ATP), the energy currency of life [9,10]. Although estimated concentration of P in the core is rather low (~0.2%) [3,11], phosphide minerals, especially Fe-Ni-P alloys that are found in iron-meteorites, such as iron monophosphide (FeP) [12,13], nickel monophosphide (NiP) [14], allabogdanite ($(Fe, Ni)_2P$ (orthorhombic)) [15], barringerite ($(Fe, Ni)_2P$) (hexagonal)) [16–18], schreibersite ($(Fe, Ni)_3P$ (tetragonal)) [19–21], nickel-phosphide ($(Ni, Fe)_3P$) (tetragonal)) [22,23], or melliniite ($(Ni, Fe)_4P$) [24], have long been studied to understand the formation of planetary bodies.

Natural schreibersite ($(Fe, Ni)_3P$), occurs as a minor phase in iron and stony-iron meteorites [19], where it usually shows a higher Fe than Ni content. Compositions with Ni content greater than 50% are called nickel phosphide [23]. Schreibersite has a metallic luster, a density of 7.0–7.3 g/cm^3, and hardness of 6.5–7 on the Mohs scale. It forms tetragonal crystals with perfect 001 cleavage and non-centrosymmetric space group I-4 [25]. In the schreibersite unit cell, there are four atomic positions in crystallographically non-equivalent 8g Wyckoff sites. Three of these are occupied by a metallic atom (Fe, Ni), while the fourth holds the P atom [25,26]. Although occurrences of schreibersite on the Earth's surface are rare, the implications of the possible presence of schreibersite in the deep interior might be important for the origin of life on Earth [10]. It is believed that meteoritic phosphide minerals, including schreibersite, could have been one of the primary sources of phosphorus on the early Earth [27,28]. Moreover, the concentration of phosphorus has a strong effect on the melting temperature in the Fe-P-S system [13].

The synthesis of minerals using high-pressure and temperature experimental techniques has given us insight into how these materials respond to the external forces that act on rocks at depth. Numerous previous studies have looked at alloys of Fe, Ni, and various light elements. Still, this research has focused primarily on Fe-rich compounds because Fe is the major component in the core. Pure Fe is stable in a hexagonal close-packed (hcp) crystal structure above ~10 GPa at 300 K [29,30], while pure Ni is stable in a face-centered cubic (fcc) structure at high pressure [31–33]. Alloying between iron and nickel leads to the stabilization of the fcc phase at high pressure and low temperature [30,32,33]. Another observation is that the value of the axial ratio (c/a) of the hcp-Fe phase is lower than the theoretical estimate. This has an effect on the longitudinal anisotropy of the hcp phase and might influence the seismic anisotropy of the Earth's core [32,34].

The possibility of forming alloys and solid compounds between Fe and light elements has stimulated a range of experimental studies that consider the stability, physical properties, and behavior of such phases, including schreibersite Fe end-member, Fe_3P, [19,35], as well as the isostructural Fe_3S [36–39]. Intriguing discontinuous behavior has been reported in Fe_3P, including, for example, a change in its c/a ratio on compression, attributed to changes in magnetic ordering [35]. A non-quenchable high-temperature phase of Fe_3P was discovered above 1600 K at 64 GPa [35]. The isostructural Fe_3S, is stable up to 220 GPa and 3300 K [39]. However, it also undergoes a magnetic collapse at 21 GPa, with the disordered magnetic moment affecting the elasticity and thermodynamics of this phase [36].

The Earth's core mainly consists of Fe, but the effects of Ni, the second major element, may also be significant, although they have not been investigated systematically. Constraining the properties of the Ni end-member of schreibersite is essential for understanding schreibersite's reactivity. It is crucial for proper modeling of solid solution phases with intermediate Fe/Ni ratios consistent with observations from meteorites. Thus, this investigation focuses on the study of the crystal structure and behavior of Ni end-member schreibersite (Ni_3P), specifically its crystal structure, at high pressures using synchrotron powder x-ray diffraction and synchrotron single-crystal X-ray diffraction.

2. Materials and Methods

2.1. Single Crystal X-Ray Diffraction at Ambient Conditions

The Ni_3P compound used in all our studies was synthesized by Dr. Nabil Boctor, at the Carnegie Institution for Science, from a stoichiometric mixture of elemental Ni and P in a molar ratio of 3:1, fused in a silica tube at 1000 °C and 1 atm in a high-temperature furnace. Ambient pressure measurements were conducted at the University of Hawai'i at Manoa X-ray Atlas Diffraction Laboratory, using a Bruker D8 Venture single crystal diffractometer equipped with a PHOTON-II CPAD detector and an Ag kα IμS microfocus source (0.56089 Å) Hawai'i. A synthetic single-crystal approximately 0.10 × 0.08 × 0.02 mm in size was mounted on a Kapton loop and then aligned with the center of the instrument using a goniometer head. The data collection was conducted using APEX3 software (Version 2018.1-0, Bruker AXS Inc., Madison, WI, USA) [40]. The atomic positions, bond lengths, and angles were determined with the SHELXL program (Version 2018, Göttingen, Germany) [41].

2.2. High-Pressure X-Ray Diffraction

High-pressure single-crystal X-ray diffraction experiments were conducted at beamline 13BM-C of Advanced Photon Source (APS), Argonne National Laboratory. Two separate experiments were performed on the synthetic Ni_3P. Experiment 1 used synchrotron powder X-ray diffraction. Finely ground powder of Ni_3P was loaded into the diamond anvil cell with 0.300 mm culet-size standard diamonds mounted on asymmetric backing plates (cubic boron nitride towards the X-ray source and tungsten carbide towards the detector). A piece of Re metal foil with an initial thickness of 0.255 mm was pre-indented to 0.039 mm. The sample chamber was made by drilling a 0.180 mm diameter, laser-cut hole in the pre-indented gasket. A piece of gold foil and a ruby sphere were placed in the sample chamber, next to the sample for pressure calibration [42–44]. The diamond anvil cell (DAC) was loaded to ~0.5 GPa with neon as a pressure medium using the GSECARS/COMPRES gas loading system [45]. A monochromatic incident beam with a wavelength of 0.434 Å (28.6 keV) was used. The incident beam was focused to a spot of approximately 0.015 by 0.015 mm with a pair of Kirkpatrick–Baez mirrors. A MAR165 charge-coupled device (CCD, Rayonix L.L.C, Evanston, IL, USA) detector for recording the diffraction images was placed approximately 180 mm from the sample. Lanthanum hexaboride (LaB_6) powder was used to calibrate the distance and the tilting of the detector. The total angular opening of the DAC was ±12°. The sample was kept motionless during the exposure, with a typical exposure time of 5 s. The data were integrated and converted to 1-dimensional diffraction patterns using the Dioptas program [46].

Experiment 2 employed the synchrotron single-crystal x-ray diffraction technique. Two synthetic single-crystals of Ni_3P, C1, and C2, with approximate sizes of 0.040 × 0.025 × 0.005 mm and 0.020 × 0.020 × 0.005 mm, respectively, were loaded into the DAC with standard diamonds (0.300 mm cutlets). The diamond anvils were mounted on asymmetric backing plates (cubic boron nitride towards the X-ray source and tungsten carbide towards the detector). Crystal C1 was oriented with the (001) direction perpendicular to the diamond's cutlet, and C2 was oriented with the (001) parallel to the cutlet. The Re metal foil with an initial thickness of 0.255 mm was pre-indented to 0.045 mm, and the sample chamber was prepared by laser-cutting a 0.180 mm diameter hole in the pre-indented gasket using laser cutting. Gold foil and two ruby spheres were placed in the sample chamber, next to the samples for pressure calibration [42–44]. The DAC was loaded to ~0.5 GPa with neon as a pressure medium using the GSECARS/COMPRES gas loading system [45]. The characteristics of the beam and the instrument in experiment 2 were the same as in experiment 1. The total angular range (φ) covered during the scans ranged from 57 to 123° (with a total angular opening of ±33°). A series of step and wide-step φ-exposures were collected. Step scans involved 1° angular increments, and wide-step scans used 9.8° angular increments. The sample was rotated at a constant speed during the exposure, with a typical exposure time of 1s/deg. The single-crystal X-ray diffraction data were analyzed with

GSE_ADA and Reciprocal Space Viewer (RSV), as described in Reference [47]. The refinement of the crystal structure was done using the program SHELXL [41].

3. Results

3.1. Single Crystal X-Ray Diffraction at Ambient Conditions

The unit cell parameters of the Ni end-member schreibersite (Ni_3P) measured in our single-crystal X-ray diffraction experiment at ambient pressure. The measurements were as follows: a = 8.9515(2) Å, c = 4.3859(1) Å, and V_0 = 351.44(8) Å^3, consistent with the tetragonal structure with I-4 space group [48–50]. The fractional atomic coordinates and interatomic distances (bonding distances) in Ni_3P at ambient pressure are summarized in Tables 1 and 2, respectively. The structure of Ni_3P determined in this experiment is shown in Figure 1. The final figure of merit of the structure refinement, R1 obtained in this experiment, was 0.0239.

Table 1. Fractional atomic coordinates of Ni_3P at ambient pressure and temperature (numbers in parentheses represent errors in the last digit).

Element	x	y	z
Ni	0.9228(1)	0.8887(1)	0.7608(1)
Ni	0.6353(1)	0.9675(1)	0.0219(1)
Ni	0.8313(1)	0.7802(1)	0.2458(1)
P	0.7139(1)	0.9517(1)	0.5192(2)

Table 2. Bonding interatomic distances in Ni_3P at ambient pressure and temperature.

Atom pair	Interatomic Distance (Å)
Ni_I–Ni_I	2.425(1), 2.710(1) × 2, 2.859(1)
Ni_I–Ni_{II}	2.649(1)
Ni_I–Ni_{III}	2.478(1), 2.592(1), 2.708(1), 2.732(1), 2.813(1)
Ni_{II}–Ni_{II}	2.491(1), 2.669(1), 2.669(1)
Ni_{II}–Ni_{III}	2.511(1), 2.544(1), 2.618(1)
Ni_{III}–Ni_{III}	2.687(1), 2.687(1)
Ni_I–P	2.222(1), 2.284(1)
Ni_{II}–P	2.296(1), 2.317(1), 2.318(1), 2.345(1)
Ni_{III}–P	2.214(1), 2.278(1), 2.337 (1)

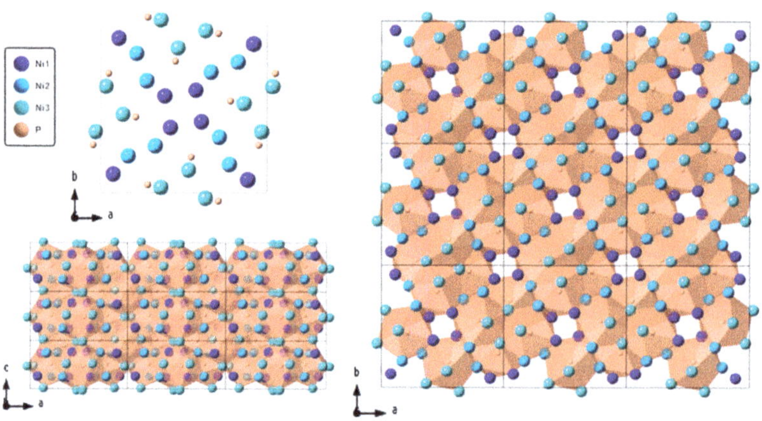

Figure 1. The crystal structure of Ni_3P at ambient pressure and temperature.

3.2. High-Pressure X-Ray Diffraction

The powder experiment was conducted in the pressure range of 0.5–46.0 GPa at ambient temperature. The unit cell parameters obtained from the powder data at the first pressure point are a = 8.945(1) Å, and c = 4.381(1) Å, consistent with the tetragonal structure. The crystal structure of Ni_3P remained tetragonal on compression to 46.0 GPa (Table 3). The ratio of the unit cell parameters (c/a) as a function of pressure remained constant with minor scatter around 0.4897(1) below 35.0 GPa. Above 35.0 GPa, the c/a ratio increased noticeably and approached 0.492(1) at 46.0 GPa (Figure 2). The normalized lattice parameters (a/a_0 and c/c_0) showed a decrease at a continuous rate for both parameters up to 46 GPa. Meanwhile, the slope of c changed discontinuously at 35 GPa (Figure 2), resulting in a change in the c/a ratio.

Table 3. Pressure dependence of the unit cell parameters of Ni_3P determined in the powder diffraction experiment.

Pressure (GPa)	*a* [Å]	*c* [Å]	Volume [A^3]
0.50(1)	8.945(1)	4.381(1)	350.5(1)
3.6(2)	8.901(1)	4.360(1)	345.5(1)
5.8(4)	8.874(1)	4.345(1)	342.2(1)
6.9(3)	8.853(4)	4.336(2)	339.9(4)
8.0(3)	8.842(1)	4.330(1)	338.5(1)
9.3(3)	8.824(1)	4.321(1)	336.5(1)
10.9 (4)	8.801(2)	4.311(2)	334.0(2)
12.9(5)	8.781(4)	4.301(4)	331.6(5)
15.0(5)	8.755(2)	4.288(2)	328.6(3)
16.5(5)	8.739(2)	4.280(1)	326.9(2)
19.0(5)	8.709(4)	4.267(3)	323.6(4)
21.5(5)	8.692(1)	4.257(1)	321.6(2)
21.8(1)	8.683(2)	4.253(1)	320.6(2)
24.0(4)	8.663(6)	4.242(3)	318.4(6)
25.8(4)	8.646(2)	4.233(2)	316.4(2)
27.7(4)	8.629(3)	4.225(2)	314.6(3)
29.5(5)	8.616(2)	4.218(1)	313.1(2)
31.9(5)	8.592(4)	4.207(3)	310.5(4)
33.4(4)	8.581(3)	4.200(1)	309.4(3)
35.2(3)	8.569(4)	4.194(2)	308.0(4)
36.6(4)	8.556(2)	4.189(2)	306.7(2)
38.5(4)	8.536(3)	4.180(1)	304.6(3)
41.0(2)	8.518(4)	4.174(3)	302.9(4)
42.2(2)	8.499(5)	4.170(3)	301.2(5)
44.1(1)	8.480(7)	4.165(5)	299.5(7)
46.0(2)	8.467(7)	4.163(7)	298.4(8)

The single crystal experiment was conducted over the pressure range 1.1–49.5 GPa at ambient temperature. Given the different orientation of the two sample crystals, the corresponding diffraction patterns were different, as shown in Figure 1. The unit cell parameters of the two samples are shown in Table 4. The unit cell parameters at 1.1 GPa confirmed that both single crystals were tetragonal, with a = 8.939(1) Å, and c = 4.378(2) Å for C1 and a = 8.936(1) Å, c = 4.375(1) Å for C2. Similar to the powder experiment, the c/a ratio was approximately constant at around 0.4899(3) below 30.0 GPa but started increasing significantly above that pressure, approaching 0.4928(4) at 49.5 GPa. For crystal C2, although the ratio of the cell parameter c/a below 25 GPa was constant at around 0.4897(2), it started to increase substantially above that pressure and reaches 0.4931(6) at 49.5 GPa, as shown in Figure 2. Moreover, between 40 and 45 GPa, both single crystals exhibited a slight change in the c/a ratio, which remained approximately constant at 0.4924(4). The normalized lattice parameters (a/a_0 and c/c_0) for the single crystal experiment revealed that there was a slight decrease in the slope of the c-axis at 30 GPa for C1 and 25 GPa for C2. This contributed to the change in the c/a ratio in both crystals, as shown in Figure 2.

Table 4. Pressure dependence of the unit cell parameters of Ni_3P determined in the single crystal experiment.

Pressure (GPa)	Crystal	*a* [Å]	*c* [Å]	Volume [A³]
1.1(1)	C1	8.939(1)	4.378(2)	349.8(2)
	C2	8.936(1)	4.375(1)	349.4(1)
3.1(2)	C1	8.911(1)	4.363(1)	346.4(1)
	C2	8.906(1)	4.362(1)	346.0(1)
5.1(3)	C1	8.883(1)	4.350(2)	343.3(2)
	C2	8.876(1)	4.348(1)	342.6(1)
7.4(4)	C1	8.853(1)	4.334(2)	339.7(1)
	C2	8.851(1)	4.334(1)	339.5(1)
10.2(5)	C1	8.823(1)	4.321(1)	336.4(1)
	C2	8.819(1)	4.318(1)	335.8(1)
11.9(3)	C1	8.799(1)	4.309(2)	333.6(2)
	C2	8.797(1)	4.307(2)	333.3(2)
14.3(1)	C1	8.768(1)	4.294(2)	330.1(2)
	C2	8.767(1)	4.293(2)	330.0(1)
16.2(2)	C1	8.749(1)	4.286(3)	328.1(2)
	C2	8.745(1)	4.282(2)	327.5(2)
18.5(8)	C1	8.727(1)	4.276(2)	325.7(2)
	C2	8.721(1)	4.271(2)	324.9(2)
20.2(6)	C1	8.706(1)	4.266(2)	323.4(2)
	C2	8.702(1)	4.262(3)	322.8(2)
22.5(5)	C1	8.685(1)	4.256(3)	321.0(2)
	C2	8.677(1)	4.250(3)	319.9(2)
25.0(6)	C1	8.664(1)	4.244(3)	318.6(3)
	C2	8.655(1)	4.243(3)	317.8(3)
27.0(7)	C1	8.643(1)	4.235(3)	316.3(2)
	C2	8.634(2)	4.234(4)	315.7(3)
29.1(5)	C1	8.623(1)	4.226(4)	314.3(3)
	C2	8.608(2)	4.223(4)	312.9(3)
30.0(3)	C1	8.613(1)	4.224(3)	313.4(3)
	C2	8.599(2)	4.223(1)	311.0(1)
32.3(8)	C1	8.594(1)	4.220(4)	311.7(3)
	C2	8.581(2)	4.216(4)	310.4(3)
33.3(3)	C1	8.579(1)	4.215(4)	310.2(3)
	C2	8.568(2)	4.211(4)	309.1(3)
35.2(8)	C1	8.565(1)	4.212(4)	309.0(3)
	C2	8.551(2)	4.204(5)	307.5(3)
38.0(7)	C1	8.539(2)	4.200(5)	306.2(3)
	C2	8.521(2)	4.193(5)	304.5(4)
39.4(4)	C1	8.529(1)	4.197(4)	305.3(3)
	C2	8.514(2)	4.191(5)	303.8(4)
40.6(3)	C1	8.519(1)	4.193(3)	304.3(3)
	C2	8.501(2)	4.184(5)	302.4(4)
43.3(5)	C1	8.497(1)	4.184(3)	302.1(2)
	C2	8.473(2)	4.173(5)	299.6(4)
45.2(1)	C1	8.479(1)	4.174(3)	300.1(2)
	C2	8.460(2)	4.166(5)	298.2(4)
47.4(3)	C1	8.457(1)	4.164(3)	297.8(3)
	C2	8.440(2)	4.159(6)	296.3(4)
49.5(3)	C1	8.441(1)	4.160(4)	296.4(3)
	C2	8.427(3)	4.156(6)	295.1(4)

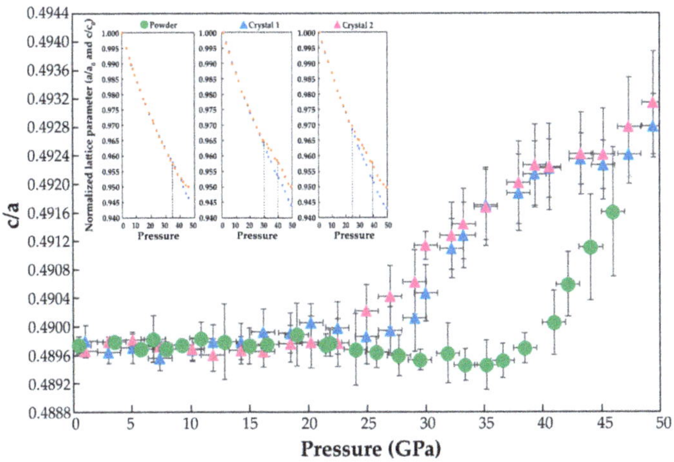

Figure 2. The c/a axial ratio of Ni_3P at different pressures at room temperature. The graphs at the top-left corner are the normalized lattice parameters (a/a_0 and c/c_0) of Ni_3P from this study.

The volume and pressure data from all experiments in this project were fit with a 3rd order Birch–Murnaghan equation of state (BM-EoS) using the EOS-FIT7 program [51]. Based on the independently-constrained value for the ambient unit cell volume (V_0 = 351.44(8) Å^3), for all data below the first transition of both powder and two single-crystal, the bulk modulus for schreibersite (K_{T0}) was 195(2) GPa and the first derivative of the bulk modulus (K_{T0}) was 4.5(2). The BM-EoS fit of powder data below 35 GPa yielded a bulk modulus of K_{T0} = 190(4) GPa, and its pressure derivative K'= 4.9(4). A discontinuity in incompressibility was observed at 35 GPa. For the single crystal experiment, the BM-EoS fit of C1 data below 30 GPa yielded the values K_{T0} = 199(5) GPa, K' = 4.7(4). Similarly, for C2, the volume data below 25 GPa were fit by 3rd BM-EoS, yielding K_{T0} = 194(5) GPa, K' = 4.7(8). The results of the BM-EoS fits in each sample are plotted in Figure 3. However, the data above the first discontinuity fit with the 2nd BM-EoS yielded V_0 = 352.22(2) Å^3, K_{T0} = 194(10) GPa.

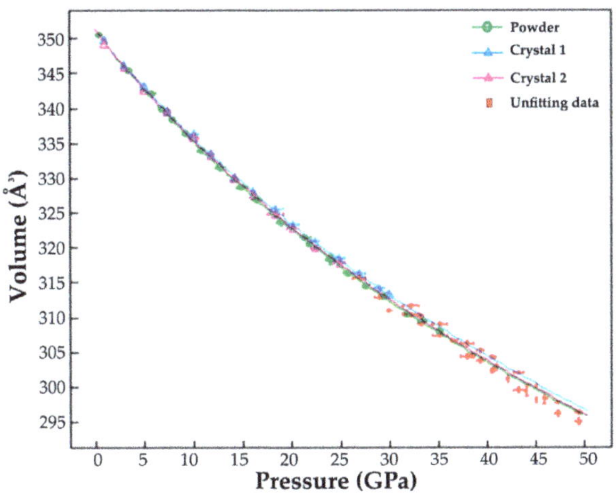

Figure 3. The volume-pressure curves represent the compression data of Ni_3P from high-pressure syn-XRD experiments.

The changes in bulk modulus as a function of pressure can be represented in the form of normalized pressure (F = P(3f(1 + 2f)$^{5/2}$) − 1) versus a Eulerian strain (f = 0.5((V_0/V)$^{2/3}$ − 1)) plot [52]. The F-f plot from all data indicates $K' > 4$ (Figure 4). The linear character of all three F-f plots suggested that the second pressure derivative of the bulk modulus (K'') was negligible. EoS fit confidence ellipses are used for a visual assessment of the quality of the correlation between K_{T0} and K' (Figure 5). The confidence ellipses from all our data on Ni_3P are shown in Figure 5, drawn at a 95.4% confidence level (2σ), and it indicates that all calculated K_{T0} and K' values from this study are internally consistent.

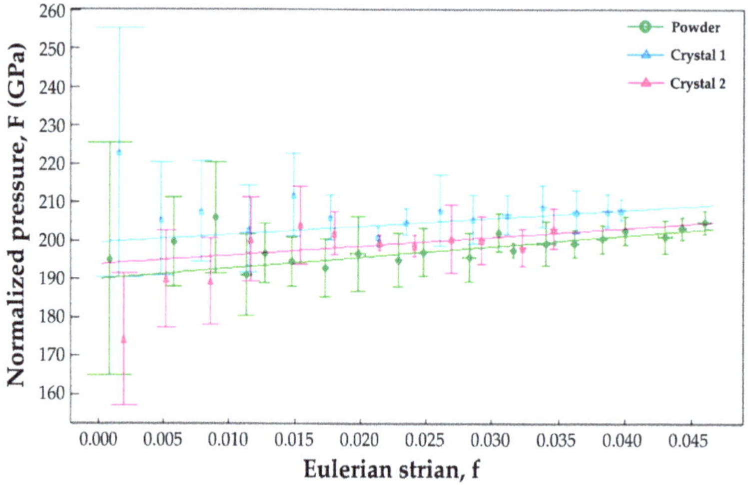

Figure 4. F-f plots for the Ni_3P compression below the first transition indicate that the second pressure derivative of the bulk modulus (K″) is negligible.

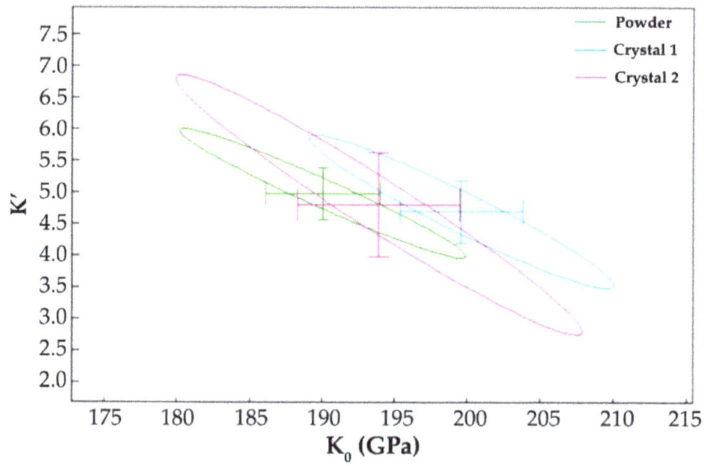

Figure 5. The confidence ellipses from all our data on Ni_3P are drawn at a 95.4% level, indicating that all calculated K_0 and K' values from this study are consistent.

4. Discussion

4.1. Crystal Structure at Ambient Condition.

The ambient unit cell parameters and interatomic distances of Ni_3P determined in this study are in good agreement with previous reports [48–50]. The atomic coordinates of Ni_3P determined in our single-crystal experiment represent the opposite absolute structure configuration of the mineral, which is a well-known phenomenon for molecules or crystals without a center of symmetry or mirror plane [53]. The space group of Ni_3P (I-4) does not include an inversion center (it is acentric). The fractional atomic coordinates from our experiments have been transformed to follow the convention used by previous studies [48–50]. However, despite differences in the absolute configuration, the structure of Ni_3P in this experiment is in excellent agreement with previously reported results.

4.2. Crystal Structure Evolution on Compression to 50 GPa

At ambient conditions, the unit cell parameters and volume of Ni_3P are slightly smaller than the ones of Fe_3P [19,35,54]. This shows that the unit cell parameters are controlled mainly by the cation size, as the ionic radius of nickel (~1.24 Å) is smaller than that of iron (~1.26 Å) [55].

In the powder sample, the c/a ratio starts to increase at 40 GPa; the same behavior is observed in single-crystal at 30 GPa in single crystal 1 and single-crystal 2 at 25 GPa (Figure 2). The experimental results indicate that the axial compressibility of Ni_3P is almost identical up to the transition point, above which the c-axis becomes less compressible. Differences in the transition pressure observed in the three experiments (powder, C1, and C2) indicate that the stress state and the orientation of the uniaxial stress component have a controlling effect over the transition point.

The sensitivity of some phase changes to uniaxial stress has been reported in the scientific literature. For example, first-principles calculations were used to predict the formation of a previously unknown high-pressure phase of SiO_2 with five-coordinated silicon that forms in response to stress applied to α-quartz, along a specific crystal orientation [56]. In SiO_2 α–cristobalite, two-phase changes to cristobalite II and cristobalite X-I are either observed [57] or suppressed [58] depending on the degree of hydrostaticity and stress rate. $CuGeO_3$ is one of the best-known examples of a material that is sensitive to uniaxial stress, with the ambient pressure phase going through two different series of phase changes depending on the degree of non-hydrostaticity [59].

In the three comparative experiments conducted in this study, the pressure medium and stress fields were very similar (Ne). Neon is a good soft quasi-hydrostatic medium, but at pressures above 20 GPa, it starts displaying signs of uniaxial stress [60]. The three samples in our experiments were oriented in different ways in the stress field of the diamond anvil cell, which featured an enhanced uniaxial component aligned with the loading axis. The two single crystals had either the [100] or [001] directions exposed explicitly to the uniaxial stress. In contrast, in the powder samples, individual grains had all possible orientations to the uniaxial stress. We do not know the exact dependence of the phase change transition point on the direction of the uniaxial stress. It is hypothetically possible, for example, that the strongest effect would be associated with the [110] or [011] directions, rather than the principal directions [100] or [001]. The observed transition pressures for the two single crystals are both lower than the transition point observed for the powder sample. It is worth noting that the nature of the discontinuity in the single crystal and powder experiments was different. Single crystals seemed to transform through an immediate single-crystal to the single-crystal mechanism, with no evidence of two-phase coexistence. In the powder sample, there was most likely a two-phase coexistence with populations of differently oriented powder grains undergoing the phase change at different pressures, and relative phase fractions changing gradually. However, this effect was difficult to quantify, given the continuous nature of changes in the unit cell parameters.

Very similar effects observed in Fe_3P and Fe_3S have been attributed to changes in magnetic ordering [35,37,38]; therefore, we assume that the nature of the discontinuity in Ni_3P is of magnetic origin. However, we did not directly measure magnetic properties to provide direct proof of this interpretation. We were also not able to conclusively verify the existence of the second transition due to the small number of pressure points collected above 40 GPa. Gu et al. [35] reported a magnetic collapse in Fe_3P around 40 GPa corresponding to a change of symmetry from I-4 to P4/mnc, based on first-principle calculations. The discontinuities in lattice parameters and volume as a function of pressure observed in our experiments at approximately 30 GPa and 40 GPa might be related to displacive phase changes. The change in the trend of the c/a ratio of Ni_3P in our experiments was also very similar to the magnetic transition and analogous c/a change observed in Cr_2O_3 [61,62]. The decreasing compressibility of the c-axis, contributing to the discontinuity in the c/a ratio trend, suggests that there are differences in the magnetic structure of schreibersite below and above the transition point [54].

The bulk modulus (K_{T0}) of Ni_3P from this experiment was in good agreement with the results of the first-principles calculations [63]. Moreover, the values of K_{T0} and K' of Ni_3P from our experiments were higher than values reported for isostructural compounds such as Fe_3P [19,35], Fe_3S [36–39], and Ni_3S [64]. This outcome suggested that the compressibility was controlled mainly by the cation size. Interestingly, K_{T0} and K' of Fe_3P and Fe_3S have quite similar values within the uncertainty, whereas K_{T0} and K' of Ni_3P are higher than Ni_3S. This indicates that changing the anion has a small effect on the compressibility behavior between Fe_3P and Fe_3S. However, the effect on the compressibility between Ni_3P and Ni_3S is stronger. The cause of the difference is that the synthetic Ni_3P in this paper was a stoichiometric compound, while Ni_3S in the earlier studies was non-stoichiometric. Moreover, the studies on Ni_3S were conducted only up to about 10 GPa, whereas we compressed our sample to approximately 50 GPa. If we focus only on the first 10 GPa of compressibility data, we also find that the trend of the c/a ratio of C1 is decreasing slightly over this range. Thus, to study the effect of the anion size between Ni_3P and Ni_3S, one might have to compress both samples to pressures higher than 10 GPa using samples as close to stoichiometric as possible.

4.3. Implications for Planetary Cores

To assess the effect the presence of Ni-rich schreibersite would have on the Earth's core, we compared the densities of pure iron, Fe_3P, and Ni_3P at pressures corresponding to the inner and outer core. It should be noted that these calculations were performed by extrapolating our results far beyond the range of conditions covered in our experiments based on a geothermal equation. The comparison showed that the density of Ni_3P from this experiment and Fe_3P from Gu et al. [35] was approximately 5% lower than the pure Fe at core pressure and ambient temperature. This result suggests that the presence of a significant amount of nickel and light elements such as phosphorus has the effect of lowering the density, consistent with the observed deficit in the density of the core as determined from seismic evidence [2,5,6]. Although our experiments did not include a study of the effects of high temperature on the unit cell parameter of Ni_3P, to a very simplified first approximation, we can compare the cold compressibility curves.

Gu et al., [35] argued that the discontinuities observed around 17 and 40 GPa were not only consistent with the calculated magnetic transitions but also probably produced discontinuous changes in the coefficient of thermal expansion. The thermal expansion coefficient $\alpha(p)$ decreases with increasing pressure. By analogy, we can assume that the compressional discontinuities observed in our experiments for Ni_3P at approximately 30 GPa and 40 GPa might also be related to discontinuities of thermal expansion and magnetic transitions. Thermal expansion decreases at high pressure, and with increasing bond energy, contributes to the high melting point. Therefore, at high pressure, most materials stay in the solid-state to much higher temperatures than at low pressure. At ambient pressure, Ni_3P has a melting point (~960 °C) lower than Fe_3P (~1100 °C) and pure iron (~1500 °C), which implies that addition of Ni_3P to natural schreibersite lowers the melting point [65–67].

The presence of Ni leads to loss of ferromagnetic order in Fe_3P because Ni causes a decrease of the magnetic moment [68,69]. Ni_3P is less compressible than pure hcp-Fe or iron phosphide phase, including Fe_3P [19,35]. Moreover, first-principles calculations on Ni_3P, Fe_3P, and pure hcp-Fe suggest that the shear modulus of Ni_3P is lower than Fe_3P and pure hcp-Fe by approximately 30% and 20%, respectively [63,70,71]. The higher incompressibility and lower shear modulus of Ni_3P are consistent with the decrease of shear wave velocity of Fe under pressure when adding Ni to Fe-compounds [8].

5. Conclusions

The results of our diamond anvil cell experiments suggest that the volume and unit cell parameters of Ni_3P experience two discontinuities, at approximately 30 GPa and 40 GPa, which may be associated with corresponding magnetic transitions. The results of this study also illustrate that the transition c-axis of Ni_3P becomes less compressible than the a-axis. The discontinuity is sensitive to the stress state and orientation of the crystal in the diamond anvil cell. Moreover, the experiments suggest that substituting Ni into Fe-bearing compounds results in higher incompressibility of the mineral.

Further investigations of Fe-Ni alloys, including more direct measurements of their magnetic properties, are needed to explain the observed transitions fully more fully. Moreover, to reliably model core conditions, additional measurements will need to be conducted at a higher pressure and high temperature.

Author Contributions: Conceptualization, S.C. and P.D.; methodology, S.C., D.Z., and P.D.; software, D.Z. and P.D.; validation, S.C., D.Z., and P.D.; formal analysis, S.C.; investigation, S.C., D.Z., and P.D.; resources, P.D.; data curation, S.C.; writing—original draft preparation, S.C.; writing—review and editing, S.C., D.Z., and P.D.; visualization, S.C.; supervision, P.D.; project administration, S.C.; funding acquisition, P.D. All authors have read and agreed to the published version of the manuscript.

Funding: This research was partly supported by the National Science Foundation, the U.S. Department of Energy, the W.M Keck Foundation, the U.S. Department of Agriculture, and the State of Illinois. Use of the APS was supported by the US Department of Energy, Office of Science, Office of Basic Energy Sciences, under Contract No. DE-AC02-06CH11357. The development of the X-ray Atlas instrument was funded by NSF EAR Infrastructure and Facilities grant 1541516.

Acknowledgments: We thank Nabil Boctor from the Carnegie Institution of Science for providing the synthesic sample. The X-ray diffraction work was conducted using the X-ray Atlas instrument at the University of Hawai'i. We are also grateful to the staff of GeoSoilEnviroCARS (Sector 13-BM-C), Advanced Photon Source, Argonne National Laboratory, for their assistance with data collection. We also thank Robert Rapp for help with editing the manuscript.

Conflicts of Interest: The authors declare no conflict of interest.

References

1. McDonough, W.F.; Sun, S.S. The composition of the Earth. *Chem. Geol.* **1995**, *120*, 223–253. [CrossRef]
2. Allègre, C.J.; Poirier, J.P.; Humler, E.; Hofmann, A.W. The chemical composition of the Earth. *Earth Planet. Sci. Lett.* **1995**, *134*, 515–526. [CrossRef]
3. Litasov, K.D.; Shatskiy, A.F. Composition of the Earth's core. *Russ. Geol. Geophys.* **2016**, *57*, 22–46. [CrossRef]
4. Kozlovsky, Y.A. The world's deepest well. *Sci. Am.* **1984**, *251*, 98–105. [CrossRef]
5. Birch, F. Elasticity and constitution of the Earth's interior. *J. Geophys. Res.* **1952**, *57*, 227–286. [CrossRef]
6. Li, J.; Fei, Y. Experimental constraints on core composition. *Treatise Geochem.* **2003**, *2*, 568.
7. Poirier, J.P. Light elements in the Earth's outer core: A critical review. *Phys. Earth Planet. Inter.* **1994**, *85*, 319–337. [CrossRef]
8. Lin, J.F.; Struzhkin, V.V.; Sturhahn, W.; Huang, E.; Zhao, J.; Hu, M.Y.; Alp, E.E.; Mao, H.K.; Boctor, N.; Hemley, R.J. Sound velocities of iron-nickel and iron-silicon alloys at high pressures. *Geophys. Res. Lett.* **2003**, *30*. [CrossRef]
9. Maciá, E.; Hernández, M.V.; Oró, J. Primary sources of phosphorus and phosphates in chemical evolution. *Orig. Life Evol. Biosph.* **1997**, *27*, 459–480. [CrossRef]
10. Bryant, D.E.; Greenfield, D.; Walshaw, R.D.; Johnson, B.R.; Herschy, B.; Smith, C.; Pasek, M.A.; Telford, R.; Scowen, I.; Munshi, T.; et al. Hydrothermal modification of the Sikhote-Alin iron meteorite under low pH geothermal environments. A plausibly prebiotic route to activated phosphorus on the early Earth. *Geochim. Cosmochim. Acta* **2013**, *109*, 90–112. [CrossRef]

11. McDonough, W.F. 3.16–Compositional model for the Earth's core. *Treatise Geochem (2nd edition).* **2014**, *3*, 559–577.
12. Gu, T.; Wu, X.; Qin, S.; Dubrovinsky, L. In situ high-pressure study of FeP: Implications for planetary cores. *Phys. Earth Planet. Inter.* **2011**, *184*, 154–159. [CrossRef]
13. Stewart, A.J.; Schmidt, M.W. Sulfur and phosphorus in the Earth's core: The Fe-P-S system at 23 GPa. *Geophys. Res. Lett.* **2007**, *34*. [CrossRef]
14. Dera, P.; Lazarz, J.D.; Lavina, B. Pressure-induced development of bonding in NiAs type compounds and polymorphism of NiP. *J. Solid State Chem.* **2011**, *184*, 1997–2003. [CrossRef]
15. Britvin, S.N.; Rudashevsky, N.S.; Krivovichev, S.V.; Burns, P.C.; Polekhovsky, Y.S. Allabogdanite, (Fe, Ni)$_2$P, a new mineral from the Onello meteorite: The occurrence and crystal structure. *Am. Mineral.* **2002**, *87*, 1245–1249. [CrossRef]
16. Brandstätter, F.; Koeberl, C.; Kurat, G. The discovery of iron barringerite in lunar meteorite Y-793274. *Geochim. Cosmochim. Acta* **1991**, *55*, 1173–1174. [CrossRef]
17. Buseck, P.R. Phosphide from meteorites: Barringerite, a new iron-nickel mineral. *Science* **1969**, *165*, 169–171. [CrossRef]
18. Dera, P.; Lavina, B.; Borkowski, L.A.; Prakapenka, V.B.; Sutton, S.R.; Rivers, M.L.; Downs, R.T.; Prewitt, C.T. Structure and behavior of the barringerite Ni end-member, Ni$_2$P, at deep Earth conditions and implications for natural Fe-Ni phosphides in planetary cores. *J. Geophys. Res. Solid Earth* **2009**, *114*. [CrossRef]
19. Geist, V.; Wagner, G.; Nolze, G.; Moretzki, O. Investigations of the meteoritic mineral (Fe, Ni)$_3$P. *Cryst. Res. Technol. J. Exp. Ind. Crystallogr.* **2005**, *40*, 52–64. [CrossRef]
20. Scott, H.P.; Huggins, S.; Frank, M.R.; Maglio, S.J.; Martin, C.D.; Meng, Y.; Santillán, J.; Williams, Q. Equation of state and high-pressure stability of Fe$_3$P -schreibersite: Implications for phosphorus storage in planetary cores. *Geophys. Res. Lett.* **2007**, *34*. [CrossRef]
21. He, X.J.; Guo, J.Z.; Wu, X.; Huang, S.X.; Qin, F.; Gu, X.P.; Qin, S. Compressibility of natural schreibersite up to 50 GPa. *Phys. Chem. Miner.* **2019**, *46*, 91–99. [CrossRef]
22. Britvin, S.N.; Kolomenskii, V.D.; Boldyreva, M.M.; Bogdanova, A.N.; Kretser, Y.L.; Boldyreva, O.N.; Rudashevskii, N.S. Nickelphosphide (Ni, Fe)$_3$P, the nickel analog of schreibersite. *ЗАПИСКИРОССИЙСКОГО МИНЕРАЛОГИЧЕСКОГО ОБЩЕСТВА* **1999**, *128*, 64–72.
23. Skála, R.; Drábek, M. Nickelphosphide from the Vicenice octahedrite: Rietveld crystal structure refinement of synthetic analogue. *Mineral. Mag.* **2003**, *67*, 783–792. [CrossRef]
24. Pratesi, G.; Bindi, L.; Moggi-Cecchi, V. Icosahedral coordination of phosphorus in the crystal structure of melliniite, a new phosphide mineral from the Northwest Africa 1054 acapulcoite. *Am. Mineral.* **2006**, *91*, 451–454. [CrossRef]
25. Skála, R.; Císařová, I. Crystal structure of meteoritic schreibersite: Determination of absolute structure. *Phys. Chem. Miner.* **2005**, *31*, 721–732. [CrossRef]
26. Pritekel, C. The crystal structure of meteoritic schreibersite: Refinement of the absolute crystal structure. Bachelor's Thesis, University of Colorado Boulder, Boulder, CO, USA, January 2015.
27. Pasek, M.A.; Lauretta, D.S. Aqueous corrosion of phosphide minerals from iron meteorites: A highly reactive source of prebiotic phosphorus on the surface of the early Earth. *Astrobiology* **2005**, *5*, 515–535. [CrossRef]
28. Gull, M.; Mojica, M.A.; Fernández, F.M.; Gaul, D.A.; Orlando, T.M.; Liotta, C.L.; Pasek, M.A. Nucleoside phosphorylation by the mineral schreibersite. *Sci. Rep.* **2015**, *5*, 17198. [CrossRef]
29. Hemley, R.J.; Mao, H.K. In situ studies of iron under pressure: New windows on the Earth's core. *Int. Geol. Rev.* **2001**, *43*, 1–30.
30. Kuwayama, Y.; Hirose, K.; Sata, N.; Ohishi, Y. Phase relations of iron and iron–nickel alloys up to 300 GPa: Implications for composition and structure of the Earth's inner core. *Earth Planet. Sci. Lett.* **2008**, *273*, 379–385. [CrossRef]
31. Huang, E.; Bassett, W.A.; Weathers, M.S. Phase relationships in Fe-Ni alloys at high pressures and temperatures. *J. Geophys. Res. Solid Earth* **1988**, *93*, 7741–7746. [CrossRef]
32. Lin, J.F.; Heinz, D.L.; Campbell, A.J.; Devine, J.M.; Shen, G. Iron-silicon alloy in Earth's core? *Science* **2002**, *295*, 313–315. [CrossRef] [PubMed]
33. Mao, W.L.; Campbell, A.J.; Heinz, D.L.; Shen, G. Phase relations of Fe–Ni alloys at high pressure and temperature. *Phys. Earth Planet. Inter.* **2006**, *155*, 146–151. [CrossRef]

34. Steinle-Neumann, G.; Stixrude, L.; Cohen, R.E.; Gülseren, O. Elasticity of iron at the temperature of the Earth's inner core. *Nature* **2001**, *413*, 57. [CrossRef] [PubMed]
35. Gu, T.; Fei, Y.; Wu, X.; Qin, S. High-pressure behavior of Fe_3P and the role of phosphorus in planetary cores. *Earth Planet. Sci. Lett.* **2014**, *390*, 296–303. [CrossRef]
36. Fei, Y.; Li, J.; Bertka, C.M.; Prewitt, C.T. Structure type and bulk modulus of Fe_3S, a new iron-sulfur compound. *Am. Mineral.* **2000**, *85*, 1830–1833. [CrossRef]
37. Lin, J.F.; Fei, Y.; Sturhahn, W.; Zhao, J.; Mao, H.K.; Hemley, R.J. Magnetic transition and sound velocities of Fe_3S at high pressure: Implications for Earth and planetary cores. *Earth Planet. Sci. Lett.* **2004**, *226*, 33–40. [CrossRef]
38. Seagle, C.T.; Campbell, A.J.; Heinz, D.L.; Shen, G.; Prakapenka, V.B. Thermal equation of state of Fe_3S and implications for sulfur in Earth's core. *J. Geophys. Res. Solid Earth* **2006**, *111*. [CrossRef]
39. Kamada, S.; Terasaki, H.; Ohtani, E.; Sakai, T.; Kikegawa, T.; Ohishi, Y.; Hirao, N.; Sata, N.; Kondo, T. Phase relationships of the Fe-FeS system in conditions up to the Earth's outer core. *Earth Planet. Sci. Lett.* **2010**, *294*, 94–100. [CrossRef]
40. Bruker. *APEX3 Crystallography Software Suite*; Bruker AXS Inc.: Madison, WI, USA, 2016.
41. Sheldrick, G.M. A short history of SHELX. *Acta Crystallogr. Sect. A Found. Crystallogr.* **2008**, *64*, 112–122. [CrossRef]
42. Heinz, D.L.; Jeanloz, R. The equation of state of the gold calibration standard. *J. Appl. Phys.* **1984**, *55*, 885–893. [CrossRef]
43. Mao, H.K.; Bell, P.M.; Shaner, J.T.; Steinberg, D.J. Specific volume measurements of Cu, Mo, Pd, and Ag and calibration of the ruby R1 fluorescence pressure gauge from 0.06 to 1 Mbar. *J. Appl. Phys.* **1978**, *49*, 3276–3283. [CrossRef]
44. Mao, H.K.; Xu, J.A.; Bell, P.M. Calibration of the ruby pressure gauge to 800 kbar under quasi-hydrostatic conditions. *J. Geophys. Res. Solid Earth* **1986**, *91*, 4673–4676. [CrossRef]
45. Rivers, M.; Prakapenka, V.B.; Kubo, A.; Pullins, C.; Holl, C.M.; Jacobsen, S.D. The COMPRES/GSECARS gas-loading system for diamond anvil cells at the Advanced Photon Source. *High Press. Res.* **2008**, *28*, 273–292. [CrossRef]
46. Prescher, C.; Prakapenka, V.B. DIOPTAS: A program for reduction of two-dimensional X-ray diffraction data and data exploration. *High Press. Res.* **2015**, *35*, 223–230. [CrossRef]
47. Dera, P.; Zhuravlev, K.; Prakapenka, V.; Rivers, M.L.; Finkelstein, G.J.; Grubor-Urosevic, O.; Tschauner, O.; Clark, S.M.; Downs, R.T. High pressure single-crystal micro X-ray diffraction analysis with GSE_ADA/RSV software. *High Press. Res.* **2013**, *33*, 466–484. [CrossRef]
48. Aronsson, B. The crystal structure of Ni_3P.(Fe_3P-Type). *Acta Chem. Scand.* **1955**, *9*, 137–140. [CrossRef]
49. Rundqvist, S.; Hassler, E.; Lundvik, L. Refinement of Ni_3P Structure. *Acta Chem. Scand.* **1962**, *16*, 242.
50. Jun, R.E.N.; Wang, J.G.; Li, J.F.; Li, Y.W. Density functional theory study on crystal nickel phosphides. *J. Fuel Chem. Technol.* **2007**, *35*, 458–464.
51. Gonzalez-Platas, J.; Alvaro, M.; Nestola, F.; Angel, R. EosFit7-GUI: A new graphical user interface for equation of state calculations, analyses and teaching. *J. Appl. Crystallogr.* **2016**, *49*, 1377–1382. [CrossRef]
52. Birch, F. Finite strain isotherm and velocities for single-crystal and polycrystalline NaCl at high pressures and 300 K. *J. Geophys. Res. Solid Earth* **1978**, *83*, 1257–1268. [CrossRef]
53. Flack, H.D. Chiral and achiral crystal structures. *Helv. Chim. Acta* **2003**, *86*, 905–921. [CrossRef]
54. Howard, J.W. Finite strain studies of single crystal Fe_3P under high pressures. Master's Thesis, University of Nevada Las Vegas, Las Vegas, NV, USA, 2010.
55. Greenwood, N.N.; Earnshaw, A. *Chemistry of the Elements*; Elsevier: Oxford, UK, 2012.
56. Badro, J.; Teter, D.M.; Downs, R.T.; Gillet, P.; Hemley, R.J.; Barrat, J.L. Theoretical study of a five-coordinated silica polymorph. *Phys. Rev. B* **1997**, *56*, 5797. [CrossRef]
57. Dove, M.T.; Craig, M.S.; Keen, D.A.; Marshall, W.G.; Redfern, S.A.T.; Trachenko, K.O.; Tucker, M.G. Crystal structure of the high-pressure monoclinic phase-II of cristobalite, SiO_2. *Min. Mag.* **2000**, *64*, 569–576. [CrossRef]
58. Dera, P.; Lazarz, J.D.; Prakapenka, V.B.; Barkley, M.; Downs, R.T. New insights into the high-pressure polymorphism of SiO_2 cristobalite. *Phys. Chem. Miner.* **2011**, *38*, 517–529. [CrossRef]
59. Dera, P.; Jayaraman, A.; Prewitt, C.T.; Gramsch, S.A. Structural basis for high-pressure polymorphism in $CuGeO_3$. *Phys. Rev. B* **2002**, *65*, 134105. [CrossRef]

60. Klotz, S.; Chervin, J.C.; Munsch, P.; Le Marchand, G. Hydrostatic limits of 11 pressure transmitting media. *J. Phys. D Appl. Phys.* **2009**, *42*, 075413. [CrossRef]
61. Dera, P.; Lavina, B.; Meng, Y.; Prakapenka, V.B. Structural and electronic evolution of Cr_2O_3 on compression to 55 GPa. *J. Solid State Chem.* **2011**, *184*, 3040–3049. [CrossRef]
62. Golosova, N.O.; Kozlenko, D.P.; Kichanov, S.E.; Lukin, E.V.; Liermann, H.P.; Glazyrin, K.V.; Savenko, B.N. Structural and magnetic properties of Cr_2O_3 at high pressure. *J. Alloy. Compd.* **2017**, *722*, 593–598. [CrossRef]
63. Zhao, D.; Zhou, L.; Du, Y.; Wang, A.; Peng, Y.; Kong, Y.; Sha, C.; Ouyang, Y.; Zhang, W. Structure, elastic and thermodynamic properties of the Ni–P system from first-principles calculations. *Calphad* **2011**, *35*, 284–291. [CrossRef]
64. Urakawa, S.; Matsubara, R.; Katsura, T.; Watanabe, T.; Kikegawa, T. Stability and bulk modulus of Ni_3S, a new nickel sulfur compound, and the melting relations of the system Ni-NiS up to 10 GPa. *Am. Mineral.* **2011**, *96*, 558–565. [CrossRef]
65. Lyman, T. *Metallography, Structures and Phase Diagrams, Metal Handbook*; American Society for Metals: Metals Park, OH, USA, 1973.
66. Okamoto, H. Fe-P (Iron-Phosphorus). *J. Phase Equilibria Diffus.* **2007**, *28*, 588. [CrossRef]
67. Kim, T.Y.; Son, H.J.; Lim, S.K.; Song, Y.I.; Park, H.S.; Suh, S.J. Electroless Nickel Alloy Deposition on SiO_2 for Application as a Diffusion Barrier and Seed Layer in 3D Copper Interconnect Technology. *J. Nanosci. Nanotechnol.* **2014**, *14*, 9515–9524. [CrossRef] [PubMed]
68. Gambino, R.J.; McGuire, T.R.; Nakamura, Y. Magnetic Properties of the Iron-Group Metal Phosphides. *J. Appl. Phys.* **1967**, *38*, 1253–1255. [CrossRef]
69. Goto, M.; Tange, H.; Tokunaga, T.; Fujii, H.; Okamoto, T. Magnetic properties of the $(Fe_{1-x}M_x)_3P$ compounds. *Jpn. J. Appl. Phys.* **1977**, *16*, 2175. [CrossRef]
70. Li, L.H.; Wang, W.L.; Hu, L.; Wei, B.B. First-principle calculations of structural, elastic and thermodynamic properties of Fe–B compounds. *Intermetallics* **2014**, *46*, 211–221. [CrossRef]
71. Wu, J.; Chong, X.; Zhou, R.; Jiang, Y.; Feng, J. Structure, stability, mechanical and electronic properties of Fe–P binary compounds by first-principles calculations. *RSC Adv.* **2015**, *5*, 81943–81956. [CrossRef]

minerals

MDPI

Article

High-Pressure Raman and Infrared Spectroscopic Study of Prehnite

Nancy L. Ross [1,*], Theresa A. Detrie [1,2] and Zhenxian Liu [3]

1 Department of Geosciences, Virginia Polytechnic Institute and State University, Blacksburg, VA 24061, USA; Theresa.Detrie@gmail.com
2 Trinity Industries, Inc., Dallas, TX 752017, USA
3 Department of Physics, University of Illinois at Chicago, Chicago, IL 60607, USA; zxliu@bnl.gov
* Correspondence: nross@vt.edu

Received: 20 January 2020; Accepted: 27 March 2020; Published: 31 March 2020

Abstract: High-pressure Raman and infrared spectra of a natural sample of prehnite, with a chemical composition of $Ca_2(Al_{0.74},Fe_{0.26})_2Si_3O_{10}(OH)_2$, are presented. Analyses of the spectra indicate that prehnite undergoes a reversible structural change between 6 and 8 GPa that is most likely associated with a subtle alteration in the orientation and/or deformation of the polyhedra comprising the framework of the structure. At pressures in excess of ~11 GPa, the high-pressure spectra indicate that prehnite undergoes a reversible phase transition involving the collapse of the framework structure.

Keywords: prehnite; infrared spectroscopy; Raman spectroscopy; high pressure; phase transition

1. Introduction

Prehnite is a hydrous calcium aluminosilicate mineral and a member of the prehnite-pumpellyite metamorphic facies, and is commonly associated with oceanic plate subduction zones, where hydrogen incorporation into mineral structures occurs [1]. The presence of prehnite is indicative of environmental conditions of 0.2–0.6 GPa and 473–623 K [2–4]. Prehnite and its associated minerals therefore provide valuable information pertaining to important geological processes, such as rock formation and water transportation, and for this reason, the evaluation of their physical and structural properties as a function of pressure and temperature is of importance.

The crystal structure of prehnite has been extensively investigated, including the location of the hydrogen positions with neutron diffraction [5–9]. The structure is comprised of corner-sharing $(Si,Al)O_4$ tetrahedra and AlO_6 octahedra, with any Fe^{3+} present substituting for Al^{3+}. When viewed along the [010] direction, the prehnite framework appears to consist of layers of octahedra, connected by three layers of tetrahedra (Figure 1). These $(Si,Al)O_4$ tetrahedra form a corkscrew arrangement that extends parallel to [010]. The tetrahedra that are directly bound to the octahedra are all SiO_4 units, however the intermediate layer of double tetrahedra is composed of both AlO_4 and SiO_4. The exact arrangement of these double tetrahedra is dependent on the space group to which the structure is assigned. If the structure is described in orthorhombic space group *Pncm*, then the average structure that is obtained is disordered, and the central atoms of the tetrahedra comprising the intermediate layer must be modelled as Si and Al in a 1:1 ratio (Figure 1). However, if the structure is refined in the acentric monoclinc space group, P2*cm*, then an ordered arrangement is obtained, in which half of these tetrahedra are AlO_4 and the remaining half are SiO_4 units. The apices of the octahedra are protonated, leading to possible hydrogen bonding between octahedral units that lie adjacent to each other in the [010] direction, as shown in Figure 1 where the bifurcated environment of the H and its next nearest oxygen atoms is indicated by dashed lines. The anionic charge on the framework is compensated for by Ca^{2+} ions, located within the channels of the framework that run parallel to [010].

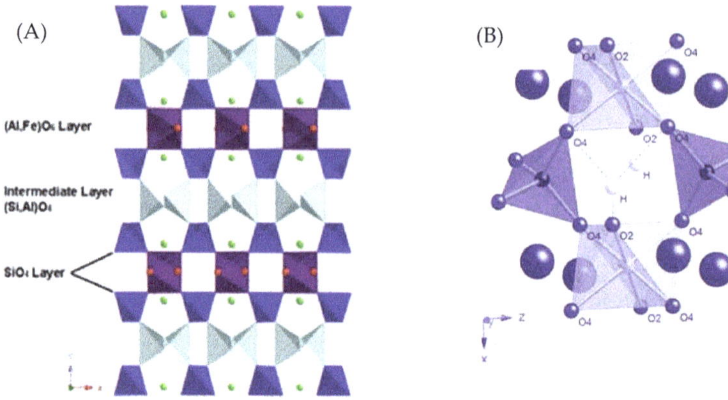

Figure 1. (**A**) The average structure of prehnite (*Pncm*), as viewed along [010] with (Al,Fe)O_6 (purple); SiO_4 (dark blue); ($Si_{0.5}$,$Al_{0.5}$)O_4 (light blue); Ca^{2+} (green spheres). (**B**) Hydrogen positions viewed down [010]. Both figures were generated from crystallographic data reported in [9].

In a high-pressure single-crystal X-ray diffraction study (HP-XRD) of prehnite, Detrie et al. [10] completed structure refinements up 9.8 GPa, that shed light on the compression mechanisms of prehnite. Above 8.7 GPa, they observed a change in the structure involving softening of the volume and *b* unit cell parameter, although the average structure was maintained across the transition. The driving force for the structural change in prehnite above 9 GPa was attributed to the reduction of the shear of the T2 tetrahedron, while the Si tetrahedron, the Al octahedron and the remainder of the Ca environment all remained essentially the same. The aim of this study was to elucidate the nature of this transition, by completing a high-pressure infrared and Raman study of prehnite. As described below, these data provide new insights into the phase transition and high-pressure behavior of prehnite.

2. Materials and Methods

The data were collected on a sample of prehnite from Mali, Africa, with a formula of $Ca_2(Al_{0.74},Fe_{0.26})_2Si_3O_{10}(OH)_2$. All infrared (IR) spectra were recorded at the National Synchrotron Light Source (NSLS), Brookhaven National Laboratory, on beamline U2A. Mid-IR spectra were collected with a Bruker IFS 66v/S vacuum Fourier transform interferometer (Bruker Optics, Karlsruhe, Germany). The instrument was equipped with a KBr beamsplitter (Bruker Optics, Karlsruhe, Germany), globular source, and a N_2 cooled mercury cadmium telluride (MCT) detector (Infrared Laboratories Inc., Tucson, AZ, USA). Far-IR measurements were performed with a modified Bruker IRscope II microscope (Bruker Optics, Karlsruhe, Germany), fitted with a mylar beamsplitter and SiB/Si bolometer (Infrared Laboratories Inc., Tucson, AZ, USA). Infrared spectra were first collected in air under ambient conditions and then the sample was loaded in a standard symmetric type diamond anvil cell (DAC), and compressed into a thin film of powder. A KBr or petroleum jelly was used as the pressure-transmitting medium during the high-pressure mid- or far-infrared experiment separately. These pressure media were used because it is transparent in the mid- or far-infrared region accordingly. The exact hydrostatic limit for this particular pressure medium is unknown, but it is assumed to be low based on studies of similar media, such as silicon oil (0.9 GPa) and glycerol (1.4 GPa) [11]. A fragment of ruby was included in the DAC as a pressure calibrant [12]. High pressure Raman spectra were also collected at the NSLS at beamline U2A. Ambient condition spectra were collected from a single crystal in air. For the high-pressure data collections, the crystal was loaded into a standard symmetric type DAC with the dominant (001) crystal face parallel to the diamond culets. Two ruby spheres were included in the DAC for pressure calibration [12]. The DAC was filled with a 4:1 methanol:ethanol solvent mixture, that acted as the pressure-transmitting medium; this mixture was chosen as it is known to remain hydrostatic up to 10 GPa [11].

3. Results and Discussion

3.1. IR and Raman Spectra of Prehnite Measured under Ambient Conditions

Table 1 summarizes the peaks derived from the ambient infrared spectra obtained by smoothing the spectra with a Savitsky–Golay function and then deconvoluting and fitting peaks with multiple Gaussian functions (using PeakFit [13]). There is acontinuous band of modes extending from 128 to 574 cm^{-1}, followed by a gap, then a band of peaks between 755 to 1150 cm^{-1}, followed by another gap before peaks between 3450 to 3500 cm^{-1} (O-H streching modes). The peak at 475(1) cm^{-1} is largely associated with the internal vibrational modes of the MO_6 octahedra, and the bands between 3453–3490 cm^{-1} are associated with O–H stretching vibrations. The peaks within the 976–1152 cm^{-1} range of the spectra reflect the internal vibrational modes of the SiO_4 tetrahedra. These assignments are based on the analyses of the infrared and Raman spectra of relevant silicate minerals [14–18]. However, the remainder of the vibrational bands in the spectra are complex and is comprised of multiple peaks that cannot be assigned unambiguously without a full lattice dynamical study, which is beyond the scope of the current contribution.

Table 1. Central energy values and frequency half width maximum (FHWM) values for principle Gaussian functions, comprising the fitted infrared and Raman data.

Infrared Peaks (cm^{-1})	Peak FHWM (cm^{-1})	Raman Peaks (cm^{-1})	Peak FHWM (cm^{-1})
128 (3)	27	112 (1)	21
146 (2)	21	138 (1)	27
213 (1)	26	161 (1)	16
243 (1)	16	216 (1)	17
296 (1)	27	317 (1)	18
343 (1)	33	350 (1)	27
377 (1)	21	381 (1)	25
423 (2)	55	464 (1)	21
475 (1)	35	492 (1)	22
505 (1)	36	518 (1)	20
542 (1)	38	541 (1)	14
574 (5)	44	606 (1)	26
755 (2)	47		
812 (1)	36		
876 (1)	33		
931 (9)	44		
976 (9)	40	940 (1)	30
998 (25)	62	985 (1)	16
1033 (19)	33		
1080 (13)	52	1074 (1)	28
1118 (16)	48		
1152 (24)	47		
3453 (1)	41	3456 (1)	26
		3475 (6)	18
3490 (1)	35	3419 (1)	13
		3494 (1)	13

Table 1 also includes the peaks obtained from fitting the Raman spectra. There is a continuous band of modes extending from 112 to 605 cm^{-1}, then a gap before modes between 940 to 1100 cm^{-1} (Si–O stretching modes), and then another gap before the modes between 3450 to 3500 cm^{-1} (O–H stretching modes).

3.2. High Pressure Infrared Spectra of Prehnite

The far- and mid-IR spectra of prehnite recorded over the 1 bar to 20 GPa range are shown in Figure 2. These plots show the evolution with pressure of the Ca and lattice vibrational modes, as well as modes associated with the translational motions of the octahedral and tetrahedra [14–16]. Examination of the spectra reveals a red-shift of the peak at ca. 540 cm^{-1}, accompanied by suppression of the peak intensity. The changes in this peak are gradual, but appear to commence at ca. 6.0 GPa and be completed by 8.2 GPa. A comparison of the high pressure far-IR spectra of prehnite with those of layered silicates, such as chrysotile, antigorite, talc, clinochlore, and montmorillonite [14–16] suggest that the peaks within the 400–700 cm^{-1} range are associated with the translational motions of the polyhedra, the bending/stretching of the P–O–P (P = polyhedron) links, and the M–OH vibrations.

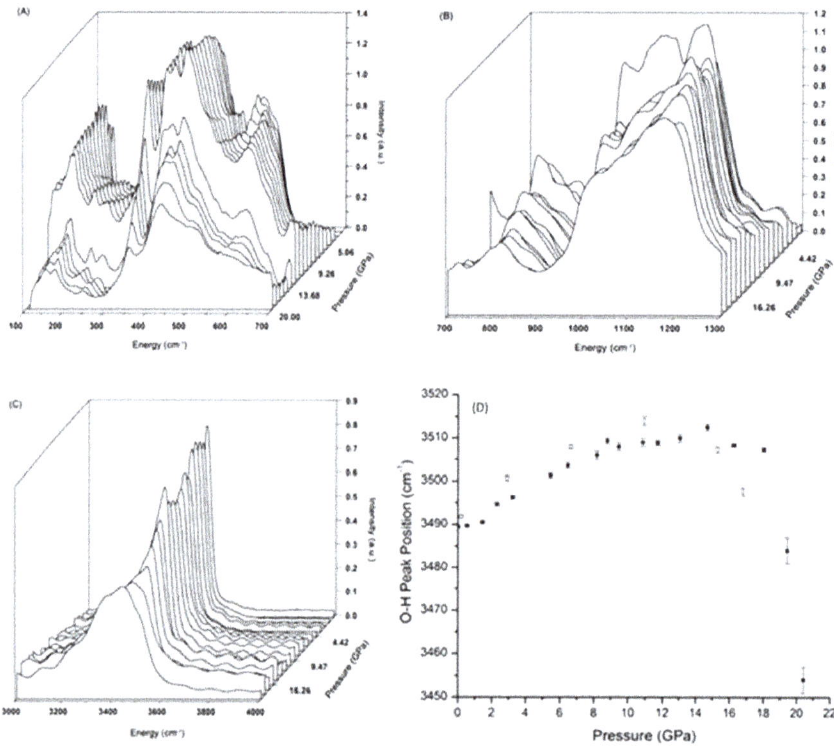

Figure 2. (**A**) Evolution with pressure of the far-IR spectra of prehnite over the 0–700 cm^{-1} range; (**B**) Over the 700–1300 cm^{-1} range; (**C**) Over the 3000–4000 cm^{-1} range. (**D**) Central position of the primary O–H vibrational peak, plotted as a function of pressure. Closed circles: data points measured during pressure increase; open circles: data points recorded during decompression.

Consequently, the changes observed between 6.0 and 8.2 GPa in the far-IR spectra, as evidenced by the red-shift of the 540 cm^{-1} peak, are likely to involve a subtle change in the relative orientations of the polyhedra that is insufficient to break the average crystal symmetry (P*ncm*) of the structure [9]. However, the mid-IR region of the high-pressure spectra associated with the O–H stretching modes (Figure 2C) do not exhibit any significant changes in the 6–8 GPa pressure range. This is further emphasized in Figure 2D, where the positions of the primary O–H vibration peaks are plotted as a function of pressure. No deviation of the curve is observable within this pressure range. Above 16 GPa, however, there is sudden red-shift, which is discussed below. In contrast to the IR data, examination of the corresponding peaks in the high-pressure Raman spectra between 6–8 GPa (Section 3.3) indicate that there is a change in the chemical environment of the O–H group associated with this phase transition.

The far-IR results are consistent with the HP-XRD results, that indicate the prehnite structure begins to soften above 8.7 GPa, in which subtle but significant structural changes occur [10]. The broad pressure range, and shift to lower pressure, reflected in the IR data relative to the HP-XRD data, may be due to the non-hydrostatic conditions under which these spectroscopic data were collected (e.g., [11]). Furthermore, the HP-XRD experiments indicated that the transition involves movement/distortion of the tetrahedra, as evidenced by alterations in the O–T–O and T–O–T (T = Si or Al) angles. The far-IR data concur with this finding.

The infrared spectra of prehnite over the 700–1300 cm^{-1} energy range are displayed in Figure 2B. These spectra contain information pertaining to the internal vibrational modes of both the octahedra and the tetrahedra, as well as complex interactive polyhedral modes. It is immediately apparent that the evolution of these spectra with pressure is more intricate than observed for the spectra recorded below 700 cm^{-1} (Figure 2A). This may be indicative of the internal vibrational modes of the polyhedra, and the complex coupled motions of these units, being more sensitive to pressure than the lattice modes and Ca–O vibrations. No abrupt changes are observed in the peaks from 900–1300 cm^{-1} over the 6–8 GPa pressure range. This may further suggest that the structural transition that occurs within this pressure range does not necessitate a change in the lengths, and hence strength, of the bonds within the polyhedra, a finding that is also consistent with the HP-XRD study [10].

A more dramatic change in the far-IR spectra is observed above 11 GPa (Figure 2A). The peaks in the high-pressure far-IR spectra display a dramatic change in shape (predominately broadening) and reduction in the intensity between 11.5 and 12.4 GPa. As these bands are principally associated with the lattice modes and polyhedral motions, these significant changes in the spectra are believed to be associated with the onset of the collapse of the prehnite framework. This framework collapse is fully reversible. Similar effects are seen in the IR spectra, shown in Figure 2B. At ca. 11 GPa, the bands begin to lose structure, broaden and weaken in intensity, coinciding with the onset of the framework collapse. In contrast, the O–H vibrational modes seem insensitive to the initial stages of this major structural change, as the peaks associated with these modes do not exhibit any notable changes until ca. 16 GPa, at which point there is a noticeable red-shift of the principle peak associated with the O–H vibrations (Figure 2D).

3.3. High Pressure Raman Spectra of Prehnite

The high-pressure Raman spectra for prehnite are presented in Figure 3. Many of the features that are apparent in the spectra measured under ambient conditions (Table 1) are obscured by the high background in the high-pressure spectra. Nonetheless, valuable information can still be gleaned, in particular from the evolution of the peak at ca. 520 cm^{-1} with pressure. This peak either arises from the internal vibrations of the MO_6 polyhedra, or from the T–O–T bending modes. As the pressure is increased from 6.73 GPa to 12.68 GPa, the splitting of this peak is amplified, yet as the pressure is decreased, the degree of splitting reduces, and below ca. 8.6 GPa, the peak becomes a single entity. This is strongly suggestive of an alteration of the prehnite structure with pressure. Furthermore, the merging of the split peak into a single peak below ca. 8.6 GPa coincides with the phase transition that is known to occur in prehnite at this pressure, as evidenced by the high-pressure IR spectra described above and the HP-XRD study [6]. As discussed above, the HP-XRD study of prehnite indicated that the structural phase transition at ca. 8.7 GPa is related to a distortion of the tetrahedra. Consequently, it is probable that the peak at ca. 520 cm^{-1} is due to T–O–T bending motions, and that in the prehnite phase that exists above 8.6 GPa, there is an increase in the number of tetrahedra in symmetrically distinct environments, that results in the splitting of this peak in the high pressure spectra. The peaks in the 850–950 cm^{-1} range of the spectra (Figure 3A) involve the internal vibrations, predominately T–O–T stretching, of the SiO_4 and AlO_4 tetrahedra. Unfortunately, the low resolution of the data and the interference of the methanol/ethanol solvent peak at ca. 1050 cm^{-1} prevent a conclusive evaluation of the peaks in this energy range.

Figure 3B,C show the evolution of the O–H vibration band with pressure and the position of the principle peak of this band, respectively. As the pressure increases, the position of the O–H vibration peak is, as expected, blue-shifted. However, at ca. 8.6 GPa, there is a noticeable red-shift, as evidenced by the sudden downturn in the data plotted in Figure 3C (closed circles); this provides further verification of a phase change occurring in the prehnite structure at ca. 8.6 GPa. Moreover, this phase transition is clearly accompanied by an alteration in the environment and weakening of the O–H bond. Such a change was not observed in the high-pressure mid-IR spectra of the sample (Figure 2), and this is presumably due to the different selection rules associated with the two techniques. The shift to lower energy of the O–H bond vibration is indicative of the weakening of this bond, which may suggest that the hydrogen bonding in which this bond participates is strengthened by the structural alterations that occur during the phase transition. This is not altogether surprising, as this electrostatic interaction is between the –OH group situated at the apex of the MO6 octahedra and the T–O–M bridging oxygen located on the opposite side of the cavity [2,3]. Therefore, it is feasible that the deformation of the tetratahedra that is believed to drive the phase transition results in the distance between the bridging oxygen atom and the hydrogen atom being shortened, and consequently the O–H··O hydrogen bond is strengthened. The phase transition at observed 8.6 GPa is reversible. This is indicated by the return of the O–H peak to its low-pressure position, as the pressure on the structure is alleviated (Figure 3C, open circles), and by the recombination below 8.6 GPa of the split peak at ca. 520 cm^{-1} (Figure 3B).

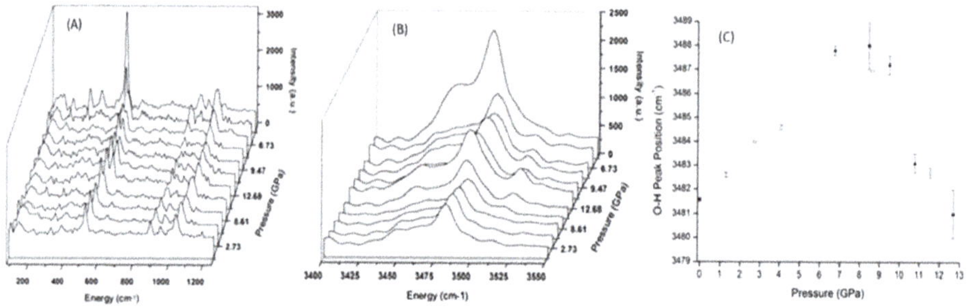

Figure 3. (**A**) Evolution with pressure of the Raman spectra of prehnite between 150–1250 cm^{-1}; (**B**) between 3400–3550 cm^{-1}; (**C**) central position of the primary O–H vibrational peak as a function of pressure. Closed circles: data points measured during pressure increase; open circles: data points recorded during decompression. All spectra have been corrected for background and smoothed by application of the Savitzty–Golay algorithm, to enable the spectral features to be observed.

4. Conclusions

The infrared and Raman spectra of prehnite, collected under ambient and high-pressure conditions, have allowed for an in-depth assessment of the evolution of the prehnite structure as a function of pressure. The vibrational data, combined with the previous HP-XRD study [10], indicate that the polyhedra comprising the prehnite framework undergo deformation/reorientation in response to pressure. Prehnite exhibits a reversible phase transition between 6–8 GPa, and the prehnite framework undergoes reversible collapse at pressures in excess of 11 GPa. It is expected that subtle reversible structural changes occur in other complex silicates, that are important as they will affect the elastic and thermodynamic properties of the material. The study also shows how high-pressure vibrational spectroscopic experiments complement HP-XRD studies, as they show details about changes in the local atomic environments and in the bonding of the structure with pressure.

Author Contributions: Conceptualization, N.L.R.; methodology, Z.L.; analysis, T.A.D.; experiments; T.A.D. and Z.L.; resources, N.L.R.; writing N.L.R.; writing draft, T.A.D.; review and editing, N.L.R.; supervision, N.L.R.; project administration, N.L.R.; funding acquisition, N.L.R. All authors have read and agreed to the published version of the manuscript.

Funding: This research was funded by the National Science Foundation (NSF), EAR-073892 and EAR-1118691. This research was also partially supported by COMPRES, the Consortium for Materials Properties Research in Earth Sciences, under NSF Cooperative Agreement EAR 01-35554.

Conflicts of Interest: The authors declare no conflict of interest.

References

1. Artioli, G.; Quartieri, S.; Deriu, A. Spectroscopic data on coexisting prehnite-pumpellyite and epidote-pumpellyite. *Can. Mineral.* **1995**, *33*, 67–75.
2. Liou, J.G. Synthesis and stability relations of prehnite, $Ca_2Al_2Si_3O_{10}(OH)_2$. *Amer. Mineral.* **1971**, *56*, 507–531.
3. Perkin, D., III; Westrum, E.F., Jr.; Essene, E.J. thermodynamic properties and phase relations of some minerals in the system $CaO–Al_2O_3–SiO_2–H_2O$. *Geochim. Cosmochim. Acta* **1980**, *44*, 61–84. [CrossRef]
4. Gottschalk, M. Internally consistent thermodynamic data for rock-forming minerals in the system $SiO_2–TiO_2–Al_2O_3–Fe_2O_3–CaO–MgO–FeO–K_2O–Na_2O–H_2O–CO_2$. *Eur. J. Miner.* **1997**, *9*, 175–223. [CrossRef]
5. Peng, S.-T.; Chou, K.-D.; Tang, Y.-C. The structure of prehnite. *Acta Chem. Sin.* **1959**, *25*, 56–63.
6. Papike, J.J.; Zoltai, T. Ordering of tetrahedral aluminium in prehnite. *Am. Miner.* **1967**, *52*, 974–984.
7. Akizuki, M. Al, Si order and the internal texture of prehnite. *Can. Miner.* **1987**, *25*, 707–716.
8. Balić-Žunić, T.; Šćavničar, S.; Molin, G. Crystal structure of prehnite from Komiža. *Eur. J. Miner.* **1990**, *2*, 731–734. [CrossRef]
9. Detrie, T.A.; Ross, N.L.; Angel, R.J.; Welch, M.D. Crystal chemistry and location of hydrogen atoms in prehnite. *Miner. Mag.* **2008**, *72*, 1163–1179. [CrossRef]
10. Detrie, T.A.; Ross, N.L.; Angel, R.J.; Diego Gatta, G. Equation of state and structure of prehnite to 9.8 GPa. *Eur. J. Miner.* **2009**, *21*, 561–570. [CrossRef]
11. Angel, R.J.; Bujak, M.; Zhao, J.; Gatta, D.; Jacobsen, S.D. Effective hydrostatic limits of pressure media for high-pressure crystallographic studies. *J. Appl. Cryst.* **2007**, *40*, 26–32. [CrossRef]
12. Mao, H.; Xu, J.; Bell, P.M. Calibration of the ruby pressure gauge to 800 kbar under quasi-hydrostatic conditions. *J. Geophys. Res.* **1986**, *91*, 4673–4676. [CrossRef]
13. SeaSolve Software Inc. *PeakFit v4.12 (1999–2003)*; SeaSolve Software Inc.: San Jose, CA, USA, 2000.
14. Šontevska, V.; Jovanovski, G.; Makreski, P. Minerals from Macedonia. Part XIX. Vibrational spectroscopy as identificational tool for some silicate minerals. *J. Mol. Struct.* **2007**, *834–836*, 318–327. [CrossRef]
15. Boev, B.; Jovanovski, G.; Makreski, P. Minerals from Macedonia. Part XX. Geological Setting, Lithologies, and Identification of the Minerals from Ržanovo Fe Ni deposit. *Turkish J. Earth Sci.* **2009**, *18*, 631–652.
16. McKeown, D.A.; Bell, M.I.; Etz, E.S. Raman spectra and vibrational analysis of the trioctahedral mica phlogopite. *Am. Miner.* **1999**, *84*, 970–976. [CrossRef]
17. Saniger, J.M. Al–O infrared vibrational frequencies of γ alumina. *Mater. Lett.* **1995**, *22*, 109–113. [CrossRef]
18. White, A.J.R.; Laukamp, C.; Stokes, M.A.; Legras, M.; Pejcic, B. Vibrational spectroscopy of epidote, pumpellyite and prehnite applied to low-grade regional metabasites. *Geochem. Expl. Env. Analy. (GEEA)* **2017**, *17*, 315–333. [CrossRef]

minerals

MDPI

Article

Stability and Solid Solutions of Hydrous Alumino-Silicates in the Earth's Mantle

Wendy R. Panero [1,*] and Razvan Caracas [2,3]

1 School of Earth Sciences, Ohio State University, Columbus, OH 43210, USA
2 CNRS, Laboratoire de Geologie de Lyon, Ecole Normale Supérieure de Lyon, 69364 Lyon, France; razvan.caracas@ens-lyon.fr
3 The Centre for Earth Evolution and Dynamics (CEED), University of Oslo, Postbox 1028 Blindern, N-0315 Oslo, Norway
* Correspondence: panero.1@osu.edu

Received: 14 February 2020; Accepted: 28 March 2020; Published: 8 April 2020

Abstract: The degree to which the Earth's mantle stores and cycles water in excess of the storage capacity of nominally anhydrous minerals is dependent upon the stability of hydrous phases under mantle-relevant pressures, temperatures, and compositions. Two hydrous phases, phase D and phase H, are stable to the pressures and temperatures of the Earth's lower mantle, suggesting that the Earth's lower mantle may participate in the cycling of water. We build on our prior work of density functional theory calculations on phase H with the stability, structure, and bonding of hydrous phases D, and we predict the aluminum partitioning with H in the Al_2O_3-SiO_2-MgO-H_2O system. We address the solid solutions through a statistical sampling of site occupancy and calculation of the partition function from the grand canonical ensemble. We show that each phase has a wide solid solution series between $MgSi_2O_6H_2$-$Al_2SiO_6H_2$ and $MgSiO_4H_2$-2δAlOOH + SiO_2, in which phase H is more aluminum rich than phase D at a given bulk composition. We predict that the addition of Al to both phases D and H stabilizes each phase to higher temperatures through additional configurational entropy. While we have shown that phase H does not exhibit symmetric hydrogen bonding at high pressure, we report here that phase D undergoes a gradual increase in the number of symmetric H-bonds beginning at ~30 GPa, and it is only ~50% complete at 60 GPa.

Keywords: hydrous mantle minerals; density functional theory; mineral disordering

1. Introduction

The storage and cycling of water in the Earth's deep interior can be greatly enhanced if hydrous silicates are stable under the high pressures and temperatures of the Earth's mantle. In particular, phases D and H are likely candidates for the storage and transport of water in the transition zone and deep mantle (e.g., [1,2]). Each phase is a hydrous magnesium silicate, $MgSi_2O_6H_2$ and $MgSiO_4H_2$, respectively. Due to the similarity in ion size, a common coupled substitution of Mg^{2+} + Si^{4+} = $2Al^{3+}$ predicts the stability of both aluminous solid solutions and equivalent aluminous phases, like aluminous phase D, $Al_2SiO_6H_2$, and aluminous phase H, otherwise referred to as δ-AlOOH.

Synthesis experiments of phase D in the absence of aluminum demonstrates that it is stable at pressures greater than 16 GPa, with increasing thermal stability with increasing pressure, such that under pressures of the base of the Earth's transition zone, phase D is stable to nearly 1700 K, suggesting it could even be stable in a mantle assemblage at moderately cool temperatures [3,4]. Phase D then breaks down to phase H with excess silica above 40–48 GPa [2,5], while phase D + brucite transforms to phase H at pressures as low as 33 GPa. Similarly, Walter et al. [6] demonstrate that increasing silica in a MgO-SiO_2-H_2O system increases the stability of phase H relative to phase D in equilibrium with bridgmanite [6].

Adding aluminum to the system increases the thermal stability of both phases D [5] and H [7]. Solid solutions in phase H are well documented, in which [8] demonstrated a complete solid solution series between $MgSiO_4H_2$ and δ-AlOOH, with more limited mixing with SiO_2. The entropy of mixing on this solid solution series is sufficient to explain an increase in stability by as much as 800 K. Phase D, in contrast, has a potentially incomplete solid solution series between the $MgSi_2O_6H_2$ and $Al_2SiO_6H_2$ end-member compositions, in which case crystal refinements of solid solutions show Al substituting for both Mg and Si but with partial occupancy of sites, a complexity not likely in the phase H system [9].

Multicomponent systems increase the pressure stability range of phase H [7] with a wide mutual stability region of bridgmanite, phase D, and phase H assemblage between 33 and 55 GPa [6], with some work suggesting that phase H can be stable as low as 25 GPa [10,11]. The partitioning of aluminum between phases D and H when in equilibrium, however, is unclear. Nishi et al. [5] demonstrate that Al preferentially partitions into phase H over phase D at 50 GPa and 1273 K by a factor of 6. In contrast, Bindi et al. [12] show under similar conditions that the aluminum instead favors phase D over phase H by nearly the same amount.

In this work, we focus on the stability, structure, and hydrogen bonding of hydrous phase D in the MgO-Al_2O_3-SiO_2-H_2O system, following the approach developed by [8]. Together with the results in [8], we examine the phase boundary between phases D and H, with implications on the partitioning of Al between the two phases.

2. Mineral Structures

For the $MgSi_2O_6H_2$ phase D composition, single crystal refinement [1] places both the Mg and Si in octahedral sites of the $P\bar{3}1m$ space group; the SiO_6 units form an edge-sharing sheet of octahedra in which one in three octahedra are vacant. The MgO_6 octahedra are linked by their corners to the silica layer above and below each vacant octahedral site in the silica layer. Hydrogen atoms bond to the underbonded corners of the silica octahedra in the c-direction, extending into the gaps between MgO_6 units (Figure 1a). As with most hydrous silicates, hydrogen atoms bond in one of two potential wells between neighboring oxygen atoms. This leads to a short OH bond and a longer H..O bond. With increasing pressure, the OH bond length increases at the expense of the H..O hydrogen bond as the two potential wells begin to merge, eventually leading to a symmetric hydrogen bond at about 40 GPa [13]. Such symmeterization is associated with a shift in the material properties.

The aluminum end-member of phase D, $Al_2SiO_6H_2$ can be thought of as a systematic substitution of two Al octahedra for pairs of Mg and Si octahedra, predicting an ordered arrangement of alternating Al and Si octahedra and an interlayer made up of AlO_6 units (Figure 1c). However, and with analogy to phase Egg [14] and δ-AlOOH-phase H solid solutions [8], the cations are likely disordered on octahedral sites. This is supported by a structural analysis of iron-bearing aluminous phase D [9]. Pure $Al_2SiO_6H_2$, however, may have additional disordering amongst all nominally vacant octahedral sites, which may explain the further thermal stability of this phase at intermediate pressures [15].

In contrast to phase D, phase H is a superstructure of the $CaCl_2$-structure, in which all octahedra are corner sharing, with an ordering of Mg and Si on the octahedra sites [8]. The fully aluminous structure, δ-AlOOH, is similar to phase H in which each octahedral site is filled with Al. The solid solutions lead to a complete disordering between the octahedral sites with a near-zero excess enthalpy of solution for pressures between 20 and 60 GPa. A consequence of the cation-site disorder is an ensemble average bonding environment in which OH..O bonds form an asymmetric, single-well potential at midmantle pressures. This is in contrast with the end-member compositions, in which the ordered arrangements of the charge distribution leads to an effective merging of two OH..O energy wells [2,8,13].

Figure 1. Relaxed crystal structures of phase D with composition (**a**) $MgSi_2O_6H_2$, (**b**) $Mg_5Al_6Si_{13}O_{48}H_{16}$, and (**c**) $Al_2SiO_6H_2$ at 20 GPa viewed normal to the (011) plane, in which MgO_6 units are yellow, SiO_6 units are dark blue, and AlO_6 units are light blue. H is pink, with OH bonds marked where bond lengths are less than 1.1Å.

3. Methods

We calculate the structure and energetics of Mg- and Al- phase D end-member compositions and $Mg_{1-x}Al_{2x}Si_{2-x}O_6H_2$ solid solutions. To do this, we calculate the energetics of at least 10 configurations in 10 GPa steps of $Mg_{1-x}Al_{2x}Si_{2-x}O_6H_2$ from 0 to 60 GPa for 9 compositions ranging from $x = 0$ to $x = 1$, for a total of 630 structural calculations. In each configuration, Al, Si, and Mg are randomly assigned to octahedral sites, and H atoms are placed in the stable arrangement as found in the ordered, $MgSi_2O_6H_2$ composition structure. Additionally, end-member compositions also include the fully ordered structure as an additional configuration at each pressure.

3.1. Calculations

We use the Vienna Ab-initio Simulation Package (VASP) package [16] to perform density functional theory calculations with the projector augmented wave method (PAW) [17] and generalized-gradient approximation (GGA) in the Perdew–Burke–Ernzerhof [18] formulation for the exchange-correlation part, with a kinetic energy cutoff of 800 eV. The electronic density is sampled in the reciprocal space on grids of regular and high-symmetry $2 \times 2 \times 2$ k-points to achieve better than 0.004 eV convergence in enthalpy. Each phase D composition is modeled in a $2 \times 2 \times 2$ supercell (88 atoms), generated in the same approach as in [8].

3.2. Thermodynamics

For each pressure and composition, we calculate the partition function, Z, as:

$$Z = \Sigma_i e^{\beta E_i} \tag{1}$$

where E_i is the calculated energy of the configuration i and β is $(k_B T)^{-1}$. The probability of each state, P_i, is therefore a function of both the energy difference relative to other states and the temperature, according to

$$P_i = (1/Z)\, e^{\beta E_i} \tag{2}$$

The ensemble average energy is then calculated directly from the weighted probabilities of each configuration

$$\langle E \rangle = \Sigma_i E_i P_i \tag{3}$$

and represents the expectation value of the energy of that composition.

We treat the configurational entropy, S_{config}, as a counting of microstates in which each octahedral site is independent; we neglect the effects of compositional variations in the vibrational component of the entropy, and work only with the "configurational" Gibbs free energy of the system, defined as

$$G(x,P,T) = (\langle E \rangle - TS_{config}) \tag{4}$$

where x is a compositional index representing the Al-component of phase D.

$$H_{mix}(x,P,T)) = (\langle E \rangle + PV) - (xH_A + (1-x)H_B) \tag{5}$$

where the end-member compositions, A and B, are $MgSi_2O_6H_2$ and $Al_2SiO_6H_2$, respectively

3.3. Phase D–Phase H Phase Boundary

In the MgO-Al_2O_3-SiO_2-H_2O system, we solve for the equilibrium coexistence of solid solution of the phases D and H by identifying the compositions for each in which chemical potentials of the two components in the coexisting phases are equal, thereby minimizing the Gibbs free energy as a function of pressure, temperature [19,20],

$$\left(\frac{dG^{phD}}{dx}\right)_{x=x^{phD}} = \left(\frac{dG^{phH+St}}{dy}\right)_{y=y^{phH+St}} \tag{6}$$

in the system,

$$Mg_{1-x}Al_{2x}Si_{2-x}O_6H_2\,(phD) = Mg_{1-y}Al_{2y}Si_{1-y}O_4H_2\,(phH) + ySiO_2\,(St) \tag{7}$$

where phD, phH, and St refer to the phase D, phase H, and thermodynamically stable SiO_2 structures, respectively. St is in the stishovite structure below 45 GPa and the $CaCl_2$ structure at higher pressures. The energetics for the phase H system along the $MgSiO_4H_2$-δ-AlOOH binary are drawn from [8]. We neglect the further distribution of Al and/or H into the silica phase as the mixing enthalpies are significantly greater than in phase H (e.g., [8,21]).

4. Results

4.1. Phase D Structure

We confirm the structure of phase D as determined through single-crystal x-ray diffraction techniques, with an ensemble average trigonal structure with space group $P\bar{3}1m$. Within the $MgSi_2O_6H_2$ - $Al_2SiO_6H_2$ phase D binary, we find structural consistency across all relaxed structures, in which randomly generated configurations relax towards octahedral coordinations for each Mg, Si, and Al cation with H atoms forming hydrogen bonds across the layer occupied by the MgO_6 octahedra.

Expectation values for the lattice constants and the unit-cell volume are calculated at 1500 K according to Equation (3) and fully reported in Table 1, where 1500 K is chosen to be most relevant to the synthesis conditions of many of the experiments. At zero pressure, we find unit-cell volume decreasing with increasing aluminum content dominated by the decrease in c-axis (Table 1 and Figure 2) consistent with several synthesis experiments, noting significant scatter in the synthesis results. The decrease in unit-cell volume is accommodated almost entirely along the c-axis.

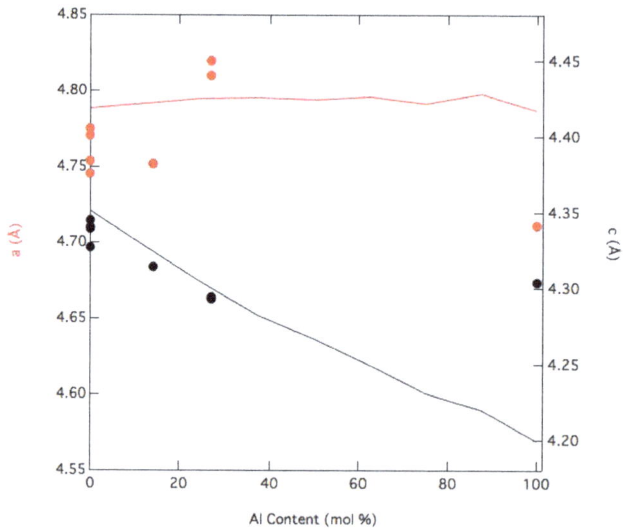

Figure 2. Zero-pressure lattice constants a (red; left) and c (black; right) as a function of aluminum content. Lattice constants as reported in synthesis experiments [1,3,10,12,15,22,23] (circles) are plotted for comparison.

Table 1. Zero-pressure, static, unit-cell volume, and lattice constants as a function of composition for ensemble-averages of each parameter calculated at 1500 K, compared to a selection of synthesis and comparable computational results.

Composition	V_0 Å^3	a Å	c Å
DFT results			
$MgSi_2O_6H_2$	86.41	4.7814	4.3652 [13]
Ensemble averaged DFT results from this work			
$MgSi_2O_6H_2$	85.46	4.788	4.351
$Mg_{0.875}Al_{0.25}Si_{1.875}O_6H_2$	85.06	4.792	4.327
$Mg_{0.75}Al_{0.5}Si_{1.75}O_6H_2$	84.72	4.795	4.304
$Mg_{0.625}Al_{0.75}Si_{1.625}O_6H_2$	84.33	4.795	4.282
$Mg_{0.5}Al_1Si_{1.5}O_6H_2$	83.95	4.794	4.267
$Mg_{0.375}Al_{1.25}Si_{1.375}O_6H_2$	83.70	4.796	4.249
$Mg_{0.25}Al_{1.5}Si_{1.25}O_6H_2$	83.09	4.792	4.230
$Mg_{0.125}Al_{1.75}Si_{1.125}O_6H_2$	82.69	4.798	4.220
$Al_2SiO_6H_2$	82.37	4.787	4.199
Synthesis experiments			
$Mg_{1.1}Si_{1.61}O_6H_{2.66}$	85.46(4)	4.770(1)	4.341(2) [3]
$Mg_{1.11}Si_{1.6}O_6H_{3.2}$	85.66(3)	4.7749(6)	4.3389(7) [22]
$Mg_{1.11}Si_{1.89}O_6H_{2.22}$	84.74(2)	4.7453(4)	4.3450(5) [1]
$Mg_{1.0}Si_{1.7}O_6H_{3.0}$	84.70 (6)	4.754(2)	4.327(3) [23]
$Mg_{0.88}Al_{0.90}Si_{1.33}O_6H_{1.87}$	84.37(6)	4.752(2)	4.314(2) [12]
$Mg_{1.20}Al_{0.90}Si_{1.48}O_6H_{2.13}$	86.41(8)	4.820 (1)	4.294(1) [10]
$Mg_{1.20}Al_{0.90}Si_{1.48}O_6H_{2.13}$	86.03(7)	4.810 (1)	4.293(1) [10]
$Al_{1.54}Si_{0.98}O_6H_{3.5}$	82.74(9)	4.7114(6)	4.3039(7) [15]

4.2. Phase D Disordering and Energetics

A single exchange of neighboring Mg and Si sites within the eight molecular units of $MgSi_2O_6H_2$ leads to an excess formation enthalpy about 10 kJ/mol greater than the fully ordered structure at all pressures between 0 and 60 GPa (Figure 1). As the configurational entropy for the single defect calculation stabilizes the system by 4.7 kJ/mol at 1500 K, we proceed with the assumption that the Mg-end member composition of phase D is minimally disordered up to its melting temperature. We further assume that Mg does not enter the layer composed of edge-sharing sheets for subsequent solid solution calculations.

In contrast, the single exchange of neighboring Al and Si sites within the eight molecular units of $Al_2SiO_6H_2$ in this calculation leads to a change of energy indistinguishable from zero up to 40 GPa, increasing to 2 kJ/mol at 60 GPa. The disordered system, therefore, becomes more stable than the ordered system for all pressures at mantle-relevant temperatures, consistent with previous experiments [15]. As a confirmation, for multiple randomly generated configurations of the aluminous phase D, we find that the ensemble average enthalpy at a given pressure is indistinguishable from the one of the nominally ordered structure.

For each configuration at each composition and pressure, we calculate the probability of each state from Equation (2) and find that the probabilities of configurations across all compositions can be described by a normal distribution, in which 67% of configurations have a probability between 2% and 14% at 1500 K. The configurations with lower probabilities are associated with a greater number of AlO_6-MgO_6–AlO_6 links, whereas higher probability configurations minimize such clusters. We therefore consider the configurational entropy of the system to be broadly ideal for nearly random mixing of Al on Mg-sites with associated random mixing of Al on Si-sites.

We model the binary solution in which the excess enthalpy has the form of a regular solution (Figure 3),

$$H_{excess} = 2Wx_{Al}(1 - x_{Al}) \tag{7}$$

where x_{Al} is the mole fraction aluminum in phase D, and W is the Margules interaction parameter between the end-member compositions. Between 20 and 40 GPa, there is little variation in the interaction parameter of about 23 kJ/mol (Equation (7)). In contrast, the mixing enthalpy is indistinguishable from zero in the same pressure range along the δ-AlOOH-phase H binary.

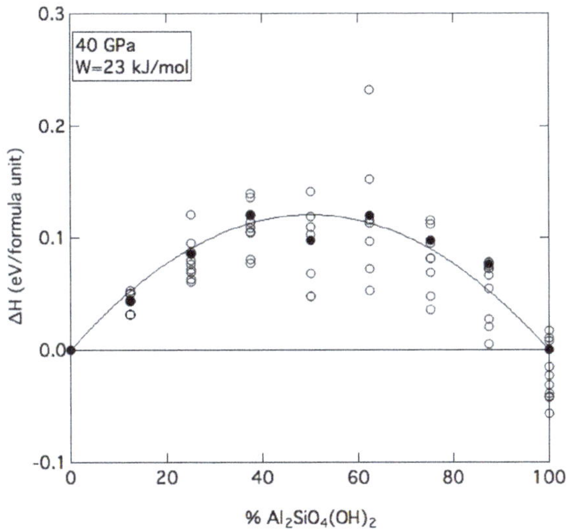

Figure 3. Excess enthalpy of solution for each configuration and composition at 40 GPa (open circles) as a function of aluminum content. The excess enthalpy expectation value (closed circles) at each composition can be well described by a regular solution with an interaction parameter of 23 kJ/mol.

4.3. Hydrogen Bonding

In the fully ordered, $MgSi_2O_6H_2$ structure, we find consistent structural results as [13], with a lengthening of the OH distance and associated shortening of the OH..O distance along a nearly straight OH..O bond, becoming symmetric at about 40 GPa when O–O distances decrease below 2.34 Å. This is consistent with bond symmetrization for phase H and δ-AlOOH (Figure 4). The Al-phase D, however, does not display consistent bond symmetrization: at 60 GPa only about 25% of the OH..O bonds are symmetric.

In the solid solutions, bond distances as a function of pressure exhibit a pattern intermediate to the end-member compositions. In general, with increasing pressure, shortening O–O distances tend to lengthen the OH distance and shorten the OH..O distance; eventually the hydrogen sits in a single-well potential. Symmetrization occurs when the O–O distance reaches about 2.4 Å with H locate equidistant to the oxygen atoms. In contrast to the ordered $MgSi_2O_6H_2$ structure, the solid solutions display a symmetrization that is protracted across more than 20 GPa in pressure (Figure 4). At 30 GPa, 12% of bonds are within 0.06 Å of equidistance between neighboring oxygen atoms; at 40 GPa, the number of symmetric bonds increases to 40%, with 48% of the bonds symmetric at 60 GPa (Figure 4).

These results are in contrast to those of phase H solid solutions, which displays a failure to undergo bond symmetrization. That is due to an asymmetric potential well that develops and deepens under pressure, thus leading to a near-constant OH bond distance with increasing pressure (Figure 4), about 0.2 Å shorter than the hydrogen bond.

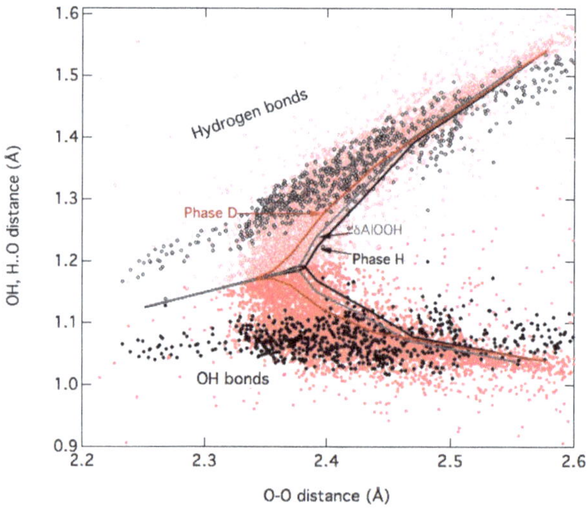

Figure 4. $MgSi_2O_6H_2$ phase D OH and OH..O bond distances as a function of O–O distance (dark red lines), as compared to δ-AlOOH (grey lines) and phase H (black lines). The suite of all phase D OH (pink, filled) and OH..O (pink, open) solid solutions shows an increasing fraction of symmetric bonds with increasing pressure, in contrast with bond distances in phase H solid solutions (black).

4.4. Phase D-Phase H Phase Boundary

Under static conditions where the configurational entropy is neglected, the idealized phase D, $MgSi_2O_6H_2$, breaks down to phase H, $MgSiO_4H_2$ with excess silica at 41 GPa. This is consistent with prior computational work [2] and consistent with synthesis experiments [5] (Figure 5). In contrast, when in the presence of excess brucite, the breakdown occurs at 23.5 GPa, in agreement with other previous computational studies [2] but about 10 GPa lower than identified in synthesis experiments [5]. In the latter, phase D is stable at 1100 °C and 32 GPa but breaks down to phase H at lower temperatures, suggesting a positive Clapeyron slope. This is consistent with a decrease in entropy as hydrogen is incorporated in the denser phase H upon breakdown of brucite with significant proton disorder [24].

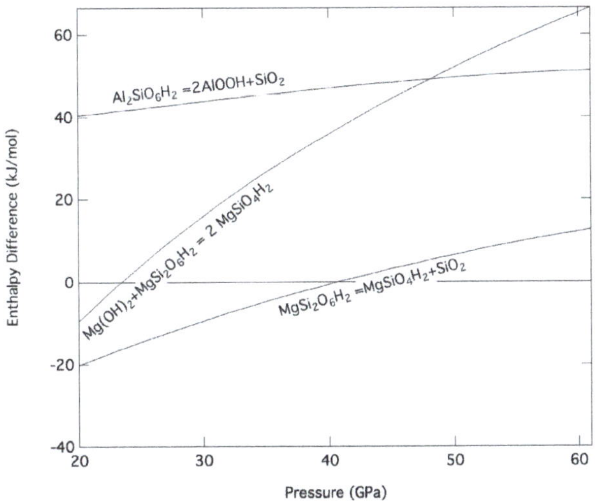

Figure 5. Static (0 K) excess enthalpy difference for the breakdown of phase D to phase H with ideal, end-member compositions.

In contrast to the Mg-rich composition, the ordered, aluminous phase D, $Al_2SiO_6H_2$, is not stable at any pressure relative to 2δ-AlOOH + silica under static conditions (Figure 5).

The aluminous phase D is likely disordered at mantle relevant temperatures, suggesting a thermal stabilization of phase D relative to δ-AlOOH. Assuming a random population of Al and Si in each of the octahedral sites and associated randomization of the OH bonding site, we find that the OH bonds relax to one of four configurations. This possible disordering of the H and the OH bonds could in principle be captured by in situ vibrational observations, like Raman, or by finite-temperature lattice dynamics calculations [25]. While these are both envisageable, they fall outside the purpose of this article. Therefore, for the time being, we approximate the configurational entropy in the aluminous phase D as $S_{config} = R * (ln(4) + ln(2))$ in which the first term represents the number of H bonding configurations, while the second represents the number of configurations for the Al in the Si octahedral layer. If instead there is a random distribution of all Si and Al across all octahedral sites as suggested in [15], the configurational entropy increases to $S_{config} = R * (ln(4) + 1/3ln(2) + ln(3))$. This latter model successfully reproduces the observed, high-temperature stability of aluminous phase D (Figure 6) at 26 GPa, whereas the lower entropy model requires temperatures 200–300 K higher to stabilize phase D relative to δAlOOH and stishovite.

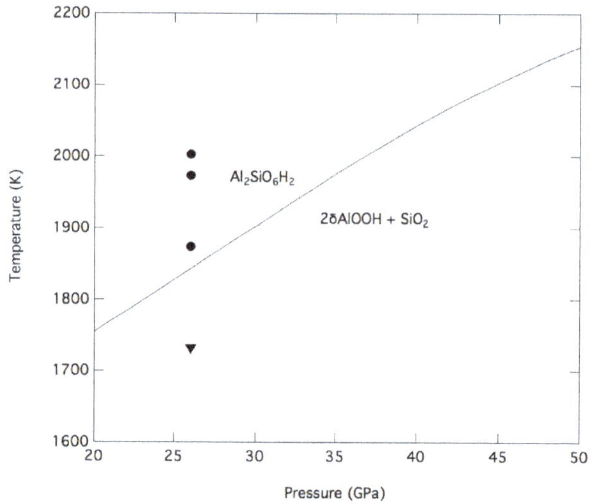

Figure 6. The calculated phase boundary of aluminous phase D and 2δ-AlOOH+SiO_2 as a function of temperature and pressure assuming random distribution of Al and Si across the occupied octahedral sites in $MgSi_2O_6H_2$ (dashed) and a random distribution of Al and Si across all possible octahedral sites (solid). Predicted phase boundaries are compared to synthesis data of phase D (circle) and mixtures of 2δ-AlOOH+SiO_2 (triangle) from [15]. Calculations are in a water-free system that does not capture effects of breakdown to constituent oxides.

For the intermediate compositions along the Mg-Al substitution, we extend the configurational model assuming (i) a composition of $Mg_{1-x}Al_{2x}Si_{2-x}O_6H_2$ for phase D with cations distributed evenly over all octahedral sites:

$$S_{config} = -R(xln(x) + (1-x)ln(1-x)) - R(x/2ln(x/2) + (1-x/2)ln(1-x/2)) \\ - R(x/3ln(x/3) + (1-x/3)ln(1-x/3)) + xRln(4) \quad (8)$$

and (ii) an ideal configurational entropy for $Mg_{1-x}Al_{2x}Si_{1-x}O_4H_2$ phase H as in [8],

$$S_{config} = -R(xln(x) + (1-x)ln(1-x)) \quad (9)$$

In the absence of other phases (e.g., bridgmanite, superhydrous B, phase Egg, and free water), we predict that the Mg-component of phase D is stable to 41 GPa, breaking down to phase H + SiO_2. The addition of aluminum to the system stabilizes an alumina-rich phase H in equilibrium with an alumina-poor phase D, in which the phase H/phase D Al partitioning is 17.5 at 1300 K and 20 GPa, decreasing to about 5 at 41 GPa. At 1800 K and 20 GPa, the phase H/phase D Al partitioning is 2.2, increasing slightly with increasing pressure to 3.5 at 41 GPa Figure 7.

A second phase loop develops in the phase D + phase H mutual stability field, such that at 2100 K phase D is stable for all compositions up to 30.3 GPa, with the aluminous phase D stable to 40 GPa in equilibrium with, the relatively Al-poor phase H + SiO_2, with a phH/phD Al partitioning of 0.95 at 2100 K between 30–39 GPa.

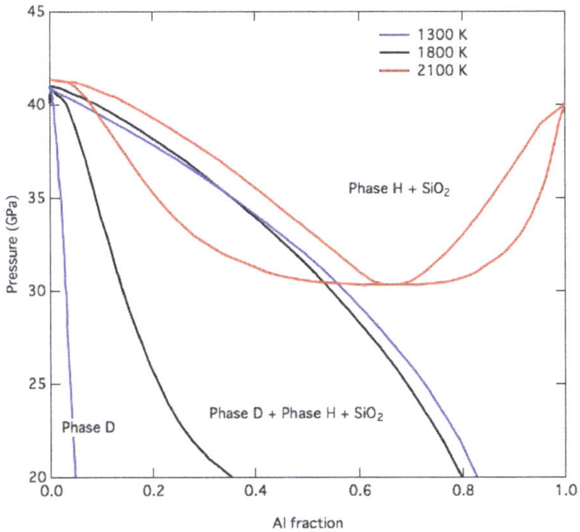

Figure 7. Equilibria between phase D and phase H + SiO_2 as a function of pressure and aluminum content at 1300 K (blue), 1800 K (black), and 2100 K (red).

However, the high-temperature, high-pressure phase stability predicted here of aluminous phase D cannot explain the experimental observation of mixtures of phases D and H in an alumina-rich assemblage at 45–50 GPa and 1273 K [5,12]. At these temperatures the aluminous phase D is not stable relative to δ-AlOOH+SiO_2; at these pressures the stability of phase D occurs only above ~2100 K, suggesting either nonequilibrium process or an unexplored structural component may be active.

5. Discussion

We find that the structure of phase D can be explained by an ordered arrangement of Si and Mg in octahedral layers in the $P\bar{3}1m$ space group, undergoing bond symmetrization at pressures of about 40 GPa. The aluminous phase D, in contrast, exhibits disordering between all octahedral sites at moderate temperatures, consistent with previous XRD refinements [1,15]. Through comparison with the temperature stability field of synthesized aluminous phase D, further disorder of aluminum and silicon must exist across all octahedral sites as suggested by [15].

The broad, mutual stability field of phases D and H in a silica-rich system emphasizes the complexity of these hydrous phases should they persist in the mantle. While broadly consistent with many synthesis experiments in the MSH and MASH system, significant discrepancies persist. Many synthesis experiments exploring phase D report greater hydrogen concentrations than expected from stoichiometry [15] with less than ideal Mg/Si or Al/Si ratio. They imply an additional, unexplored defect mechanism that should include either a hydrogarnet substitution of Si^{4+} = 4 H^+ or a vacancy on the Mg-site as Mg^{2+} = 2 H^+, two of the most common defects stable under transition zone pressures [26,27]. Such additional mechanisms may also help explain the mismatch with multiple synthesis experiments in the a-axis as a function of aluminum content (Figure 2) as well as explain the scatter in structural results.

A few test calculations reveal that the hydrogarnet substitution in the $MgSi_2O_6H_2$ end-member phase D in the presence of brucite is more stable than the hydrogarnet-free phase D system. The difference is 10 kJ/mol decrease in Gibbs free energy at 30 GPa (Figure 5), similar to results for synthesis experiments with excess brucite [6]. This also helps to explain chemical analyses with excess H with a Si deficit [23]. Should this stabilization energy be broadly constant as a function of Al content, this will increase the pressure at which Mg-phase D breaks down but still does not predict the stability of Al-phase D at 50 GPa, while increasing the Mg/Si ratio, counter to what is observed.

If such additional defect mechanisms are active in phase D they may enhance the hydrogen carrying capacity through the Earth's transition zone, while limiting the thermal stability of this phase.

Author Contributions: Conceptualization, W.R.P.; methodology, W.R.P.; software, R.C.; formal analysis, W.R.P.; data curation, W.R.P. and R.C.; writing—original draft preparation, W.R.P.; writing—review and editing, W.R.P and R.C.; visualization, W.R.P. All authors have read and agreed to the published version of the manuscript.

Funding: This research was funded by NSF-EAR grant number 1724693 to Wendy R. Panero. Razvan Caracas was funded by the European Research Council (ERC) under the European Union's Horizon 2020 research and innovation program (grant agreement no. 681818 IMPACT to Razvan Caracas) and by the Research Council of Norway through its Centres of Excellence funding scheme, project number 223272. Razvan Caracas acknowledges access to the GENCI supercomputers (Occigen, Ada, Jean-Zay, and Curie) through the stl2816 series of eDARI computing grants and the TGCC supercomputers (Irene-AMD) through the PRACE grant RA4947.

Acknowledgments: Wendy R. Panero thanks the Wednesday writing group for support.

Conflicts of Interest: The authors declare no conflict of interest.

References

1. Yang, H.; Prewitt, C.T.; Frost, D.J. Crystal structure of the dense hydrous magnesium silicate, phase D. *Am. Mineral.* **1997**, *82*, 651–654. [CrossRef]
2. Tsuchiya, J. First principles prediction of a new high-pressure phase of dense hydrous magnesium silicates in the lower mantle. *Geophys. Res. Lett.* **2013**, *40*, 4570–4573. [CrossRef]
3. Frost, D.J.; Fei, Y. Stability of phase D at high pressure and high temperature. *J. Geophys. Res. Solid Earth* **1998**, *103*, 7463–7474. [CrossRef]
4. Ghosh, S.; Schmidt, M.W. Melting of phase D in the lower mantle and implications for recycling and storage of H2O in the deep mantle. *Geochim. Cosmochim. Acta* **2014**, *145*, 72–88. [CrossRef]
5. Nishi, M.; Irifune, T.; Tsuchiya, J.; Tange, Y.; Nishihara, Y.; Fujino, K.; Higo, Y. Stability of hydrous silicate at high pressures and water transport to the deep lower mantle. *Nat. Geosci.* **2014**, *7*, 224. [CrossRef]
6. Walter, M.; Thomson, A.; Wang, W.; Lord, O.; Ross, J.; McMahon, S.; Baron, M.; Melekhova, E.; Kleppe, A.; Kohn, S. The stability of hydrous silicates in Earth's lower mantle: Experimental constraints from the systems MgO–SiO_2–H_2O and MgO–Al_2O_3–SiO_2–H_2O. *Chem. Geol.* **2015**, *418*, 16–29. [CrossRef]
7. Ohira, I.; Ohtani, E.; Sakai, T.; Miyahara, M.; Hirao, N.; Ohishi, Y.; Nishijima, M. Stability of a hydrous δ-phase, $AlOOH$–$MgSiO_2(OH)_2$, and a mechanism for water transport into the base of lower mantle. *Earth Planet. Sci. Lett.* **2014**, *401*, 12–17. [CrossRef]
8. Panero, W.R.; Caracas, R. Stability of phase H in the $MgSiO_4H_2$–$AlOOH$–SiO_2 system. *Earth Planet. Sci. Lett.* **2017**, *463*, 171–177. [CrossRef]
9. Ballaran, T.B.; Frost, D.J.; Miyajima, N.; Heidelbach, F. The structure of a super-aluminous version of the dense hydrous-magnesium silicate phase D. *Am. Mineral.* **2010**, *95*, 1113–1116. [CrossRef]
10. Xu, C.; Inoue, T. Melting of Al-rich phase D up to the uppermost lower mantle and transportation of H_2O to the deep Earth. *Geochem. Geophys. Geosystems* **2019**. [CrossRef]
11. Liu, X.; Matsukage, K.N.; Nishihara, Y.; Suzuki, T.; Takahashi, E. Stability of the hydrous phases of Al-rich phase D and Al-rich phase H in deep subducted oceanic crust. *Am. Mineral.* **2019**, *104*, 64–72. [CrossRef]
12. Bindi, L.; Nishi, M.; Irifune, T. Partition of Al between Phase D and Phase H at high pressure: Results from a simultaneous structure refinement of the two phases coexisting in a unique grain. *Am. Mineral.* **2015**, *100*, 1637–1640. [CrossRef]
13. Tsuchiya, J.; Tsuchiya, T.; Tsuneyuki, S. First-principles study of hydrogen bond symmetrization of phase D under high pressure. *Am. Mineral.* **2005**, *90*, 44–49. [CrossRef]
14. Mookherjee, M.; Panero, W.R.; Wunder, B.; Jahn, S. Anomalous elastic behavior of phase egg, $AlSiO_3$ (OH), at high pressures. *Am. Mineral. J. Earth Planet. Mater.* **2019**, *104*, 130–139. [CrossRef]
15. Pamato, M.G.; Myhill, R.; Ballaran, T.B.; Frost, D.J.; Heidelbach, F.; Miyajima, N. Lower-mantle water reservoir implied by the extreme stability of a hydrous aluminosilicate. *Nat. Geosci.* **2015**, *8*, 75. [CrossRef]
16. Kresse, G.; Furthmüller, J. Efficient iterative schemes for ab initio total-energy calculations using a plane-wave basis set. *Phys. Rev. B* **1996**, *54*, 11169. [CrossRef]

17. Kresse, G.; Joubert, D. From ultrasoft pseudopotentials to the projector augmented-wave method. *Phys. Rev. B* **1999**, *59*, 1758. [CrossRef]
18. Perdew, J.P.; Burke, K.; Ernzerhof, M. Generalized gradient approximation made simple. *Phys. Rev. Lett.* **1996**, *77*, 3865. [CrossRef]
19. Panero, W.R.; Pigott, J.S.; Reaman, D.M.; Kabbes, J.E.; Liu, Z. Dry (Mg, Fe) SiO_3 perovskite in the Earth's lower mantle. *J. Geophys. Res. Solid Earth* **2015**, *120*, 894–908. [CrossRef]
20. Panero, W.R.; Akber-Knutson, S.; Stixrude, L. Al_2O_3 incorporation in $MgSiO_3$ perovskite and ilmenite. *Earth Planet. Sci. Lett.* **2006**, *252*, 152–161. [CrossRef]
21. Panero, W.R.; Stixrude, L.P. Hydrogen incorporation in stishovite at high pressure and symmetric hydrogen bonding in δ-AlOOH. *Earth Planet. Sci. Lett.* **2004**, *221*, 421–431. [CrossRef]
22. Frost, D.; Fei, Y. Static compression of the hydrous magnesium silicate phase D to 30 GPa at room temperature. *Phys. Chem. Miner.* **1999**, *26*, 415–418. [CrossRef]
23. Hushur, A.; Manghnani, M.H.; Smyth, J.R.; Williams, Q.; Hellebrand, E.; Lonappan, D.; Ye, Y.; Dera, P.; Frost, D.J. Hydrogen bond symmetrization and equation of state of phase D. *J. Geophys. Res. Solid Earth* **2011**, *116*. [CrossRef]
24. Mookherjee, M.; Stixrude, L. High-pressure proton disorder in brucite. *Am. Mineral.* **2006**, *91*, 127–134. [CrossRef]
25. Treviño, P.; Garcia-Castro, A.C.; López-Moreno, S.; Bautista-Hernández, A.; Bobocioiu, E.; Reynard, B.; Caracas, R.; Romero, A. Anharmonic contribution to the stabilization of Mg $(OH)_2$ from first principles. *Phys. Chem. Chem. Phys.* **2018**, *20*, 17799–17808. [CrossRef] [PubMed]
26. Panero, W.R. First principles determination of the structure and elasticity of hydrous ringwoodite. *J. Geophys. Res. Solid Earth* **2010**, *115*. [CrossRef]
27. Pigott, J.S.; Wright, K.; Gale, J.D.; Panero, W.R. Calculation of the energetics of water incorporation in majorite garnet. *Am. Mineral.* **2015**, *100*, 1065–1075. [CrossRef]

minerals

MDPI

Article

Seismic Wave Speeds Derived from Nuclear Resonant Inelastic X-Ray Scattering for Comparison with Seismological Observations

Brent Delbridge * and Miaki Ishii

Department of Earth and Planetary Sciences, Harvard University, Cambridge, MA 02138, USA; ishii@eps.harvard.edu
* Correspondence: bdelbridge@fas.harvard.edu

Received: 15 December 2019; Accepted: 1 April 2020 ; Published: 8 April 2020

Abstract: Nuclear resonant inelastic X-ray scattering (NRIXS) experiments have been applied to Earth materials, and the Debye speed is often related to the material's seismic wave speeds. However, for anisotropic samples, the Debye speed extracted from NRIXS measurements is not equal to the Debye speed obtained using the material's isotropic seismic wave speeds. The latter provides an upper bound for the Debye speed of the material. Consequently, the acoustic wave speeds estimated from the Debye speed extracted from NRIXS (Nuclear Resonant Inelastic X-ray Scattering) measurements are underestimated compared to the material's true seismic wave speeds. To illustrate the differences, the effects of various assumptions used to estimate the Debye speed, as well as seismic wave speeds, are examined with iron alloys at Earth's inner core conditions. For the case of pure iron, the variation of the crystal orientation relative to the incoming X-ray beam causes a 40% variation in the measured Debye speed, and leads to 3% and 31% underestimation in the compressional and shear wave speeds, respectively. Based upon various iron alloys, the error in the inferred seismic shear wave speed strongly depends upon the strength of anisotropy that can be quantified. We can also derive Debye speeds based upon seismological observations such as the PREM (Preliminary Reference Earth Model) and inner core anisotropy model. We show that these seismically derived Debye speeds are upper bounds for Debye speeds obtained from NRIXS experiments and that interpretation of the Debye speeds from the NRIXS measurements in terms of seismic wave speeds should be done with utmost caution.

Keywords: phonon density of states; nuclear resonant scattering; seismic wave speeds; elastic tensor; anisotropy; high pressure; earth materials

1. Introduction

Advances in instrumentation and experimental techniques are allowing for rapid improvement in the use of lattice dynamics to determine the elastic moduli of materials at high pressure (e.g., [1,2]). Inelastic scattering of synchrotron radiation is used to probe the mechanical properties of condensed matter. The importance of this method over other measurements such as ultrasonic methods or Brillouin scattering lies in its ability to be used on small samples under extreme pressure conditions, and has already been successfully applied to Earth materials (e.g., [3]). A review and critical comparison of the variety of techniques and methodologies used to measure sound velocities in metals at megabar pressures can be found in Antonangeli & Ohtani [4], and references therein. Nuclear resonant inelastic X-ray scattering experiments (NRIXS) directly probe the vibrational structure of solids (e.g., [5–8]), and these measurements provide the Fourier-transformed self-intermediate scattering function from which the energy levels of the excited acoustic phonons and the Debye speed can be extracted (e.g., [3]). The Debye speed is a direct link to the material's elastic properties.

One application of NRIXS measurements, which is of particular interest to geophysics, is to understand the alloying effects of nickel and light elements (e.g., S, Si, C, H, etc.) on the elastic properties of iron at inner core conditions. Due to the extreme high pressures and temperatures at the center of the Earth, it has been historically difficult to obtain experimental measurements of the acoustic properties of candidate materials to compare with the seismically observed elastic properties. Recent applications of NRIXS to iron alloy samples in a diamond anvil cell reveal that NRIXS measurements provide an exciting new opportunity to experimentally constrain geophysical models of the Earth's inner core.

This study builds upon the work of Anderson [9,10], who suggested that the Debye speed could be estimated from single crystal elastic constants, and developed techniques which used theories from lattice dynamics to link optical vibrations of a solid to its Debye speed to determine seismic shear wave speeds of materials at high pressures. We show that the Debye speeds calculated from the seismic wave speeds of polycrystalline materials composed of randomly oriented anisotropic crystal with bulk or macroscopically isotropic properties (e.g., powdered samples) yields an upper bound for the material's Debye speed. Furthermore, we show that increasing the strength of anisotropy of a material decreases its Debye speed, even when its bulk isotropic properties remain unchanged.

The study highlights the difficulty of comparing NRIXS derived Debye speeds with seismically observed wave speeds. This is demonstrated by showing the effects of the various approximations to the calculation of the Debye speed for the case of several iron alloys at inner core pressures and temperatures. These results are crucial for using NRIXS measurements of the Debye speed in geophysical applications, and a new methodology is derived that avoids the isotropic assumptions which are currently a large source of error in procedures which extract seismic wave speeds from NRIXS measurements.

2. Theory

Mechanical information can be extracted from phonon density of states $\nu(E)$, since, at low energies, only the acoustic phonon modes are excited by the incident photon and participate in the inelastic absorption. This low energy portion of the density of states exhibits Debye-like behavior in which it is linear with the square of the energy [5]. In the harmonic lattice approximation [11], the density of states can be written as

$$\nu(E) = \frac{M}{\rho}\frac{E^2}{2\pi^2\hbar^3}\frac{1}{v_D^3}, \tag{1}$$

where M is the mass of the resonant nucleus, ρ is the density of the solid, $2\pi\hbar$ is Plank's constant, E is the energy of the photon relative to the resonance energy of the nuclear transition, and v_D is the Debye speed. The Debye speed is defined as an average over all wave speeds such that

$$\frac{1}{v_D^3} = \frac{1}{3}\sum_{m=1}^{3}\int\frac{d\Omega_{\hat{\mathbf{q}}}}{4\pi}\frac{1}{c_m^3(\hat{\mathbf{q}})}, \tag{2}$$

where $c_m(\hat{\mathbf{q}})$ is the wave speed of an acoustic phonon of type m (e.g., longitudinal or horizontally/vertically polarized transverse waves) propagating in the direction of the unit vector $\hat{\mathbf{q}}$, and $d\Omega_{\hat{\mathbf{q}}}$ is the corresponding differential solid angle element. Note that throughout this manuscript, c denotes acoustic wave speeds, and v represents quantities derived from the phonon density of states. For the isotropic elastic materials, there is no directional dependence of the wave speeds, and the expression for the Debye speed v_D can be simplified in terms of its compressional (c_p) and shear (c_s) wave speeds as

$$\frac{1}{v_D^3} = \frac{1}{3}\sum_{m=1}^{3}\frac{1}{c_m^3} = \frac{1}{3}\left(\frac{1}{c_p^3} + \frac{2}{c_s^3}\right). \tag{3}$$

Note that the factor of 2 arises from the fact that the shear modes are degenerate.

The observations from nuclear resonant inelastic scattering provide measurements of the "projected" density of states $g(E, \hat{\mathbf{k}})$, also referred to as the "partial" density of states which is given by (e.g., [8,12])

$$g(E, \hat{\mathbf{k}}) = \frac{M}{\rho} \frac{E^2}{2\pi^2 \hbar^3} \frac{1}{v_{\hat{\mathbf{k}}}^3}, \tag{4}$$

where $v_{\hat{\mathbf{k}}}$ is the "projected mean wave speed", that is, the wave speeds weighted by the projection of the incident photon unit wave vector $\hat{\mathbf{k}}$ (Figure 1) on to the unit phonon polarization vectors of type m, $\hat{\mathbf{p}}_m$. It is defined as

$$\frac{1}{v_{\hat{\mathbf{k}}}^3} = \sum_{m=1}^{3} \int \frac{d\Omega_{\hat{\mathbf{q}}}}{4\pi} \frac{\left| \hat{\mathbf{k}} \cdot \hat{\mathbf{p}}_m(\hat{\mathbf{q}}) \right|^2}{c_m^3(\hat{\mathbf{q}})}, \tag{5}$$

where c_m is the wave speed for acoustic phonon of type m, and $0 \leq \left| \hat{\mathbf{k}} \cdot \hat{\mathbf{p}}_m(\hat{\mathbf{q}}) \right|^2 \leq 1$ for all $\hat{\mathbf{k}}$ and $\hat{\mathbf{q}}$. In practice, the slope of the linear fit to the experimentally observed projected phonon density of states ν versus E^2 is used to estimate $v_{\hat{\mathbf{k}}}$.

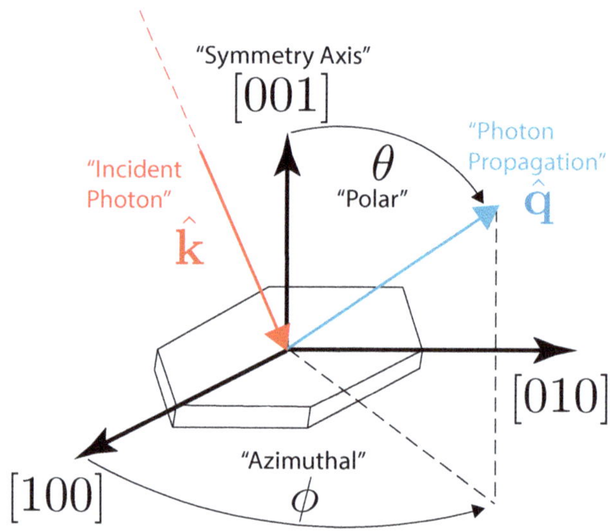

Figure 1. Orientation of incident photons and propagation direction for a hexagonal crystal.

The projected mean wave speed, $v_{\hat{\mathbf{k}}}$, is the link between the nuclear resonant derived measurements of the phonon density of states and the material's elastic constants. In this case, the experimentally observable $g(E, \hat{\mathbf{k}})$ is proportional to $v_{\hat{\mathbf{k}}}^{-3}$, which is related to the inverse cubed acoustic wave speeds c_m^{-3} (Equation (5)). However, the sample orientation with respect to the wave vector is not easily obtained, and the samples are often polycrystalline. For the case of randomly oriented aggregates (e.g., powdered samples), the experimental result is well represented by averaging over the incident wave vector direction $\hat{\mathbf{k}}$ [8,12]. The average over the unit sphere of an arbitrary function $g(\theta, \phi)$ is denoted by $\langle g(\theta, \phi) \rangle = \int g(\theta, \phi) d\Omega / 4\pi = \int g(\theta, \phi) \sin\theta \, d\theta d\phi / 4\pi$ where Ω is the solid angle. The experimentally derived partial density of state is

$$g(E, \hat{\mathbf{k}}) = \frac{M}{\rho} \frac{E^2}{2\pi^2 \hbar^3} \left\langle \frac{1}{v_{\hat{\mathbf{k}}}^3} \right\rangle, \tag{6}$$

and can be directly related to the theoretical Debye speed v_D (see Appendix C) by

$$\left\langle \frac{1}{v_{\hat{\mathbf{k}}}^3} \right\rangle = \int \frac{d\Omega_{\hat{\mathbf{k}}}}{4\pi} \frac{1}{v_{\hat{\mathbf{k}}}^3} = \frac{1}{v_D^3}. \tag{7}$$

while the mean projected sound speed $v_{\hat{\mathbf{k}}}$ has been averaged over the incident wave vector $\hat{\mathbf{k}}$ and the directional dependence of the excitation of the phonon modes has been removed, the acoustic wave speeds $c_m(\hat{\mathbf{q}})$ still depend on their propagation direction $\hat{\mathbf{q}}$.

Alternative Expressions for the Debye Speed under Various Simplifying Assumptions

While randomly oriented anisotropic crystals create a material that appears isotropic in its bulk or macroscopic acoustic properties, the Debye speed measured from this material is not equivalent to the Debye speed that would be obtained from a material with equivalent isotropic properties. This distinction is critical for geophysical applications which seek to derive the seismic P- and S-wave speeds of materials from NRIXS experiments. The discrepancy arises since the procedure commonly used to extract these wave speeds assumes that the material is isotropic (e.g., [3]), while the materials being measured are often anisotropic.

In order to extract isotropic properties such as elastic moduli and seismic wave speeds from anisotropic materials, a mapping of the anisotropic tensor elements to isotropic wave speeds is required. Mathematically, there are several schemes for generating the isotropic elastic moduli from anisotropic fourth-order elasticity tensors under various assumptions such as constant stress condition (e.g., Reuss average). When it comes to constraints for the Earth's interior, the isotropic average information is provided by seismological observations rather than from elasticity tensor with an assumption. Hence, it is critical to understand how the seismological data averages components of the elasticity tensor that ultimately result in isotropic models such as the Preliminary Reference Earth Model [13]. Delbridge & Ishii [14] shows that the relevant averaging scheme for seismic wave propagation is equivalent to the Voigt average, and the elasticity tensor can be written as

$$\Lambda_{ijkl} = \left(\kappa_0^V - \frac{2}{3}\mu_0^V\right)\delta_{ij}\delta_{kl} + \mu_0^V\left(\delta_{ik}\delta_{jl} + \delta_{il}\delta_{jk}\right) + \gamma_{ijkl}, \tag{8}$$

where the subscripts i,j,k,and l are used as Einstein notation and δ_{ij} is the Kronecker delta. The first two terms on the right-hand side describe the isotropic behaviour arising from isotropic contributions while the last term, γ_{ijkl}, describes the anisotropic deviations from isotropy. The superscript V denotes that Voigt averaging schemes are used to obtain the bulk and shear moduli, κ_0^V and μ_0^V. The isotropic seismic waves speeds are related to these elastic moduli as

$$\rho c_p^2 = \kappa_0^V + \frac{4}{3}\mu_0^V \qquad \text{and} \qquad \rho c_s^2 = \mu_0^V, \tag{9}$$

where c_p and c_s are the material's compressional and shear wave speeds, respectively. The seismic wave speeds allow us to estimate the Debye speed under the assumption that it is equivalent to isotropic wave speeds that appear in Equation (3). We denote this Debye speed as v_D^V to indicate the Debye speed of a material based upon seismic wave speeds.

We can write another expression for the Debye speed by assuming that the directionally averaged wave speeds can be used as the isotropic wave speeds, that is,

$$\frac{1}{v_D^3} \approx \frac{1}{\bar{v}_D^3} \equiv \frac{1}{3}\sum_{m=1}^{3}\left\langle c_m^2 \right\rangle^{-3/2} = \frac{1}{3}\left(\frac{1}{\left\langle c_1^2\right\rangle^{3/2}} + \frac{1}{\left\langle c_2^2\right\rangle^{3/2}} + \frac{1}{\left\langle c_3^2\right\rangle^{3/2}}\right). \tag{10}$$

these directionally averaged elastic properties are related to the seismic wave speeds and Voigt moduli as (see Appendix A)

$$\rho\left\langle c_1^2\right\rangle = \rho c_p^2 = \kappa_0^V + \frac{4}{3}\mu_0^V, \quad \text{and} \quad \frac{1}{2}\rho\left(\left\langle c_2^2\right\rangle + \left\langle c_3^2\right\rangle\right) = \rho c_s^2 = \mu_0^V. \tag{11}$$

For isotropic materials $v_D = \bar{v}_D = v_D^V$, but for anisotropic materials, $\bar{v}_D$, and v_D^V may deviate from v_D. We seek a relationship between them in order to understand the implications of various assumptions employed when relating a material's Debye speed to its elastic properties, and hence to associated seismic wave speeds. The Debye speed (v_D) is given by Equation (2), hence

$$\frac{1}{v_D^3} = \frac{1}{3}\sum_{m=1}^{3}\int\frac{d\Omega_{\hat{\mathbf{q}}}}{4\pi}\frac{1}{c_m^3\left(\hat{\mathbf{q}}\right)} = \frac{1}{3}\sum_{m=1}^{3}\int\frac{d\Omega}{4\pi}(f_m)^{-3/2} = \frac{1}{3}\sum_{m=1}^{3}\left\langle f_m^{-3/2}\right\rangle, \tag{12}$$

where $f_m = c_m^2$. On the other hand, the Debye speed based upon the material's seismic wave speeds (Equation (10)) can be written as

$$\frac{1}{\bar{v}_D^3} = \frac{1}{3}\sum_{m=1}^{3}\left\langle f_m\right\rangle^{-3/2}. \tag{13}$$

since f_m is strictly positive and $f_m \to f_m^{-3/2}$ is convex for $f_m > 0$, Jensen's inequality [15,16] tells us that

$$\left\langle f_m^{-3/2}\right\rangle \geq \left\langle f_m\right\rangle^{-3/2}, \tag{14}$$

which implies that

$$\frac{1}{3}\sum_{m=1}^{3}\left\langle f_m^{-3/2}\right\rangle \geq \frac{1}{3}\sum_{m=1}^{3}\left\langle f_m\right\rangle^{-3/2}. \tag{15}$$

Furthermore, we can show that

$$\frac{1}{\left\langle c_1^2\right\rangle^{3/2}} + \frac{1}{\left\langle c_2^2\right\rangle^{3/2}} + \frac{1}{\left\langle c_3^2\right\rangle^{3/2}} \geq \frac{1}{\left\langle c_1^2\right\rangle^{3/2}} + \frac{2}{\left(\frac{\left\langle c_2^2\right\rangle + \left\langle c_3^2\right\rangle}{2}\right)^{3/2}} = \frac{1}{c_p^3} + \frac{2}{c_s^3}, \tag{16}$$

which leads to the conclusion that

$$\frac{1}{3}\sum_{m=1}^{3}\left\langle f_m^{-3/2}\right\rangle \geq \frac{1}{3}\sum_{m=1}^{3}\left\langle f_m\right\rangle^{-3/2} \geq \frac{1}{3}\left(\frac{1}{c_p^3} + \frac{2}{c_s^3}\right). \tag{17}$$

This is equivalent to

$$\frac{1}{v_D^3} \geq \frac{1}{\bar{v}_D^3} \geq \frac{1}{{v_D^V}^3}, \tag{18}$$

from which we can write the desired inequality relating the Debye speed v_D to its approximation using the seismic wave speeds v_D^V

$$v_D \leq v_D^V. \tag{19}$$

Physically, this equation implies that the Debye speed of anisotropic materials (v_D) is not equivalent to the Debye speed estimated from that material's seismic wave speeds (v_D^V), and that the latter is an upper bound to the former. Consequently, since the commonly used procedure to estimate seismic wave speeds from NRIXS experiments assumes that the Debye speed can be related directly to the material's seismic wave speeds, that is, equality in Equation (19), these speeds are often underestimated.

3. Results and Discussion

In order to explore the effect of analyzing anisotropic single crystals and polycrystalline aggregates, elastic tensor elements of five iron alloys with hexagonal-close packed (hcp) [17–20] and one with a body-centered cubic (bcc) [21] structure are considered (Table 1). The five hcp iron materials are—a pure iron at 6000 °K with a density of 13.0 g/cm^3 [17], a Fe-Si alloy with 12.5 at% Si at 360 GPa and 6900 °K [20], a Fe-Ni-Si alloy with 10 at% Ni and 21.25 at% Si with density of 12.5 g/cm^{3}at 360 GPa and 6500 °K [18], a Fe-Si-C alloy with 4.2 at% Si and 0.7 at% C with density of 13.1 g/cm^3 at 360 GPa and 6500 °K [19], and a Fe-S-C alloy with 2.1 at% S and 0.7 at% C with density of 13.1 g/cm^3 at 360 GPa and 6500 °K [19]. The last material considered is a Fe-Si bcc alloy at 360 GPa and 6000 °K with density of 13.6 g/cm^3 [21]. The seismic wave speeds c_p and c_s are calculated from the elastic tensor elements (Table 1) such that they are comparable with the seismically observed wave speeds [14].

Table 1. The elastic constants and seismic wave speeds of iron alloys (column 1; [17–21]). The alloy's density in g/cm^3 (ρ; column 2), seismic compressional wave speed in km/s (c_p; column 3), shear wave speed in km/s (c_s; column 4), the strength of anisotropy (A^L; column 6; [22]), and the elastic tensor elements in GPa (columns 6 through 11) are summarized.

Material	ρ	c_p	c_s	A^L	c_{11}	c_{33}	c_{13}	c_{44}	c_{66}	c_{12}
hcp Fe	13.0	11.8	3.9	1.5	2150	1685	990	140	60	2030
hcp Fe-Si	13.1	11.3	3.9	0.3	1674	1855	1120	176	137	1400
hcp Fe-Si-Ni	12.5	11.8	3.3	1.4	1816	1964	1224	80	49	1718
hcp Fe-Si-C	13.1	11.6	3.7	0.6	1712	2066	1263	164	91	1530
hcp Fe-S-C	13.1	11.8	4.2	0.3	1831	2091	1214	183	173	1485
bcc Fe-Si	13.6	11.5	4.2	1.7	1562	1562	1448	366	366	1448

To investigate how various parameters are influenced by anisotropy, we introduce an absolute measure of anisotropy, A^L , that quantifies the log-Euclidean distance between the Voigt averaged elastic tensor $\left\langle \Lambda_{ijkl} \right\rangle$ and the Reuss averaged elastic tensor $\left\langle \Lambda_{ijkl}^{-1} \right\rangle^{-1}$ [22,23]. The log-Euclidean distance may be written in terms of the Voigt and Reuss averaged isotropic moduli κ_0 and μ_0 as [22]

$$A^L = \sqrt{5 \ln \left(\frac{\mu_0^V}{\mu_0^R} \right)^2 + \ln \left(\frac{\kappa_0^V}{\kappa_0^R} \right)^2}, \quad (20)$$

where the superscripts V and R denote the Voigt and Reuss averages, respectively. The Voigt and Reuss averages coincide for the case of isotropic material [24], and for this case, A^L yields a value of zero.

3.1. Variations in the Debye Speed

We first calculate the mean projected wave speed $v_{\hat{\mathbf{k}}}$ for all possible incident photon wave vectors $\hat{\mathbf{k}}$ (Equation (5)). This requires numerically calculating the wave speeds (eigenvalues) and polarization vectors (eigenvectors) as described in the Appendix A, and then numerically integrating over the phonon propagation directions, $\hat{\mathbf{q}}$ (Equation (5)). The minimum and maximum values represent the spread of possible mean projected wave speeds that may be measured for a given single crystal sample due to changes in orientation of the crystal with respect to the incoming X-ray beam (Figure 1, Table 2). For the transversely isotropic hcp materials considered in this study, the maximum projected mean wave speed corresponds to an incident X-ray beam aligned along the axis of symmetry, and the minimum corresponds to an incident X-ray beam in the plane perpendicular to the axis of symmetry. The range in $v_{\hat{\mathbf{k}}}$ caused by this orientation effect is proportional to the strength of anisotropy, with a 20% increase per unit of anisotropy (Figure 2). For the hcp pure iron, $v_{\hat{\mathbf{k}}}$ varies by about 40%, whereas for the less anisotropic hcp Fe-S-C alloy, it is only 13% (Table 2). As expected from the symmetry in the scattering matrix for cubic materials [11], the projected mean wave speed $v_{\hat{\mathbf{k}}}$ is constant and is identical to the Debye speed for the bcc Fe-Si alloy (Table 2).

Table 2. Exact and approximate Debye speeds calculated from the elastic tensor elements of hcp and bcc iron alloys (column 1). Percent variation (column 2) of the difference of the extreme values of $v_{\hat{\mathbf{k}}}$ (column 3) with respect to the average of the extreme values due to variations in $\hat{\mathbf{k}}$ are given. The true Debye speed v_D (column 4), and the approximate Debye speed obtained using the analytical expressions for c_m given in the Appendix A (column 5) are compared. The last three columns give the values obtained using Equation (3) where the elastic moduli κ and μ are estimated using the Voigt, Reuss, and Voigt-Reuss-Hill averages, respectively. The speeds are all in units of km/s.

Material	±%	$[\min v_{\hat{\mathbf{k}}}, \max v_{\hat{\mathbf{k}}}]$	v_D	v_D^q	v_D^V	v_D^R	v_D^{VRH}
hcp Fe	40%	[3.07, 4.60]	3.35	3.35	4.45	3.18	3.87
hcp Fe-Si	16%	[4.04, 4.76]	4.23	4.22	4.47	4.16	4.32
hcp Fe-Si-Ni	37%	[2.66, 3.85]	2.89	2.96	3.78	2.77	3.32
hcp Fe-Si-C	28%	[3.53, 4.66]	3.79	3.83	4.21	3.67	3.95
hcp Fe-S-C	13%	[4.38, 4.97]	4.54	4.56	4.77	4.47	4.62
bcc Fe-Si	0%	[4.04, 4.04]	4.04	4.04	4.79	3.33	4.13

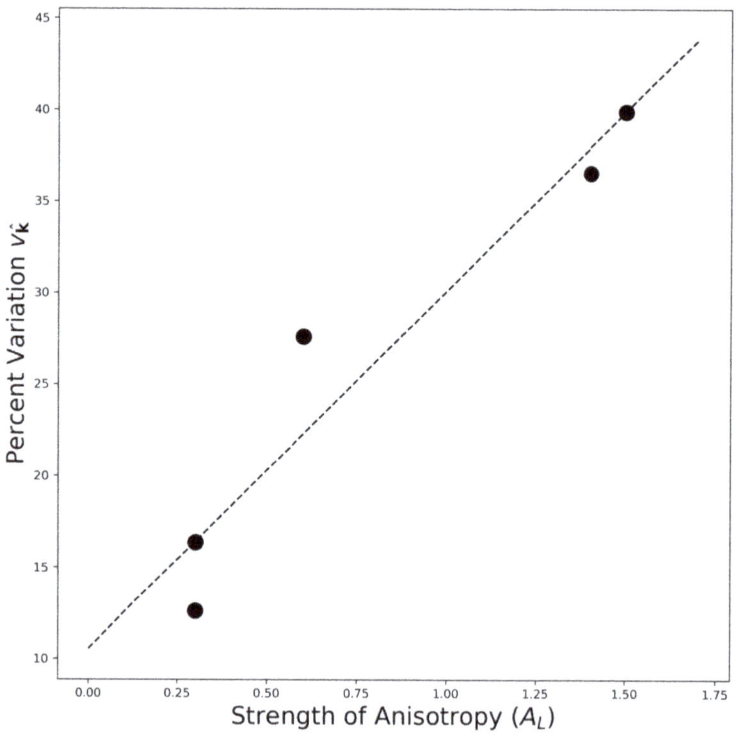

Figure 2. Percent variation of $v_{\hat{\mathbf{k}}}$ as a function of the strength of anisotropy A^L for the five hcp iron alloys (black circles) in Table 1. The dashed line represents a linear best fit to the data.

In order to examine the Debye speed $v_D = \left\langle v_{\hat{\mathbf{k}}}^{-3} \right\rangle^{-1/3}$ (Equation (7)) of powdered samples, an additional integration over the wave vector $\hat{\mathbf{k}}$ is needed. Because of cubic symmetry, $v_{\hat{\mathbf{k}}} = v_D$ for the bcc Fe-Si alloy (Table 1). The Fe-Si-Ni alloy which has the lowest seismic shear wave speed of 3.3 km/s (Table 1) results in the lowest v_D value of 3.0 km/s, and the hcp Fe-S-C alloy which had the highest seismic shear wave speed 4.2 km/s results in the highest v_D value of 4.6 km/s. The Debye speed v_D in Equation (7) can also be calculated approximately using the first-order propagation and polarization vectors (see Appendix A) and obtaining the corresponding wave speeds which we denote by v_D^q. This procedure may either be done by numerical matrix multiplication, or for hexagonal and cubic symmetries, using the analytical expressions as described in the Appendix A. Use of these analytical expressions for wave speeds and associated polarization vectors to calculated Debye speed is only valid for weakly anisotropic materials, however, the approximate values of v_D^q are nearly identical to the exact values v_D (Table 2). Even for the case of the highly anisotropic materials, such as pure iron, the error is less than half a percent. On the other hand, Debye speeds based upon various isotropic averages, v_D^V, v_D^R, and v_D^{VRH} (Equation (10)) from v_D show significant deviations, between 0.5 to 33%, including bcc Fe-Si alloy (Table 2). The largest deviations occur with v_D^V, the averaging scheme that is most relevant for comparison with seismic wave speeds.

The materials in Table 1 only represent a small subset of the possible variations in the elasticity tensor. In order to explore the deviations introduced by the approximations more thoroughly, we generate one million random transversely isotropic elastic tensors whose elements are sampled from a uniform distribution subject to the restriction that the resulting compressive wave speeds are greater than the shear wave speeds. The Debye speed v_D is compared against v_D^V, v_D^R, and v_D^{VRH} (Figure 3). The Debye speeds calculated from the seismically compatible average, v_D^V, are all greater than v_D while the Debye speeds calculated using the Reuss average are all less than v_D, which is consistent with the Voigt and Reuss averages representing upper and lower bounds, respectively, for the elastic moduli. The Debye speeds calculated using the Voigt-Reuss-Hill averages are roughly distributed about zero, and thus are a "best case scenario" that is most consistent with its true value (Figure 3). In fact, Anderson [9] showed that the isotropic values calculated with the Voigt-Reuss-Hill average may be used to accurately estimate a material's Debye speed, however, one must keep in mind that the acoustic wave speeds based upon Voigt-Reuss-Hill averages are incompatible with observed seismic wave speed [14]. Finally, we shows that the seismic wave speeds of Cobalt at pressures of zero to 40 GPa are accurately calculated using the Voigt average and consistent with independent experimental measurements (Appendix B).

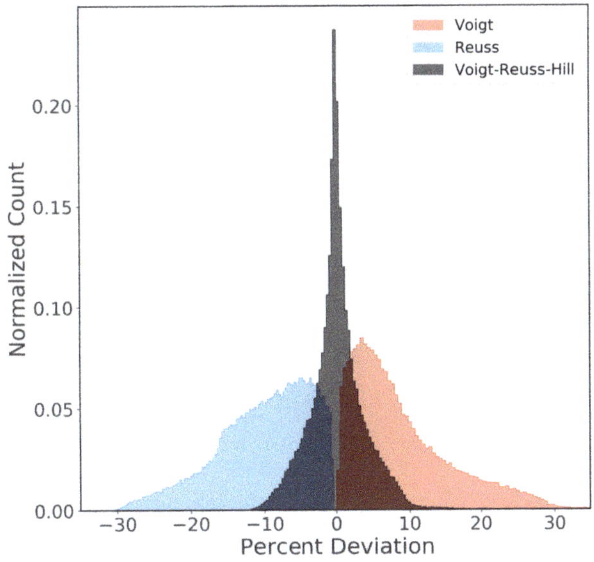

Figure 3. Distribution of the deviation of the Debye speed obtained using acoustic wave speeds estimated using the Voigt (red), Reuss (blue), and Voigt-Reuss-Hill (grey) averaged elasic moduli with respect to the true Debye speed. One million randomly generated tensors of materials with hexagonal symmetry and identical seismic wave speeds are used.

In order to explore the deviation between the Debye speeds estimated using the seismic wave speeds v_D^V and the true Debye speed v_D as a function of the strength of anisotropy A^L, the difference, $v_D - v_D^V$ are examined (Figure 4). The seismic wave speeds for all one million tensors are forced to have identical seismic wave speeds (set to be those of pure iron in Table 1), and yet the difference, despite the scatter in the data, increases with increasing anisotropy roughly at a rate of about 17% per unit of anisotropy. The distribution of the deviations also has some structure (Figure 4). For a given strength of anisotropy, there is an upper and lower limit on the deviation between v_D and v_D^V. These limits, as well as the range in between, result from the partitioning of anisotropy into various elastic tensor elements. Because the elasticity tensors are randomly generated with the only condition being the constant seismic wave speeds, there are tensors for which anisotropy is purely in the bulk modulus but not in shear modulus, and vice versa. The shallowly dipping upper limit corresponds to cases where anisotropy is all in the bulk modulus, only affecting the longitudinal acoustic waves, and the steeper lower limit corresponds to anisotropy that is purely in the shear modulus. The scatter of points between the two limits represent various levels of anisotropy partitioning.

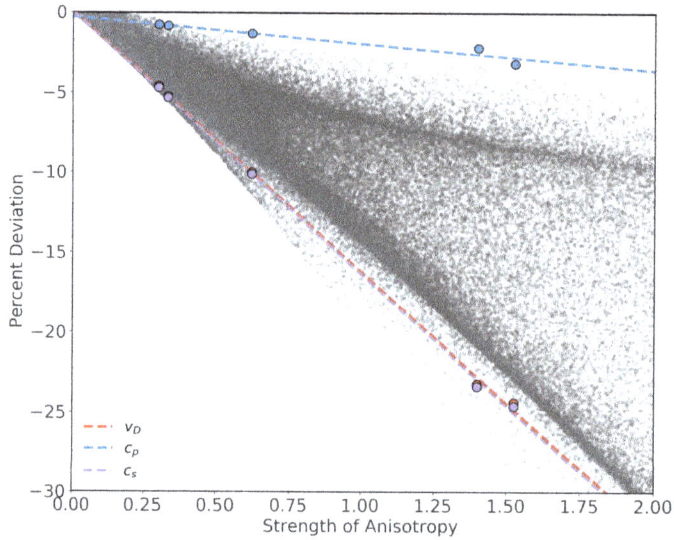

Figure 4. The dependence of the deviation of the Debye speed v_D from that based upon the Voigt average, v_D^V, and the seismic speeds for the hcp iron alloys in Table 1 as a function of the strength of anisotropy A_L. The semi-transparent grey circles represent the difference between v_D (Equation (7)) and v_D^V (Equation (3)) for 10^6 randomly generated transversely isotropic elastic tensors with the same seismic wave speeds as those of pure iron (Table 1). The red circles correspond to the difference between v_D and v_D^V for the five hcp iron alloys (Table 1) and the blue and purple circles denote the corresponding deviation in inferred acoustic longitudinal and transverse wave speeds, respectively (Table 3). The dashed lines represent linear best fits to the acoustic wave speed data.

3.2. Extracting Seismic Wave Speeds

Section 2 and the previous subsection showed that the Debye speed extracted from the partial density of states (Equation (6)), is not equal to the Debye speed estimated from that material's seismically observed wave speeds (Equation (3)). This result stems partly from the fact that the average of the inverse cube of a function (Equation (12)) is not generally equivalent to the inverse cube of the average (Equation (13)) and because seismic waves "see" the material's isotropic properties in a specific manner. Therefore, the Debye speed estimated from seismic wave speeds is an upper bound for the material's Debye speed v_D. This distinction is critical for geophysical applications of

NRIXS, which typically seek to compare the experimentally observed Debye speed $v_{\hat{\mathbf{k}}}$ or v_D with seismically observed acoustic wave speeds c_p and c_s. In this section, we focus on the issues associated with extracting seismic wave speeds based upon v_D.

A common procedure used to extract seismic wave speeds from NRIXS experiments is to use independent estimates of the the bulk modulus κ and density ρ from an equation of state based upon X-ray diffraction (e.g., [25–30]). For isotropic material, manipulating the expressions for the acoustic wave speeds (Equation (9) and noting that $\kappa = \kappa_0^V$ and $\mu = \mu_0^V$ in order to be compatible with the quantities provide by an Earth model such as PREM) combined with Equation (3) results in three equations for the three unknowns μ, c_p, and c_s in terms of the known quantities v_D, κ, and μ such that

$$\begin{aligned} \frac{\kappa}{\rho} &= c_p^2 - \frac{4}{3}c_s^2, \\ \frac{\mu}{\rho} &= c_s^2, \quad \text{and} \\ \frac{1}{v_D^3} &= \frac{1}{3}\left(\frac{1}{c_p^3} + \frac{2}{c_s^3}\right). \end{aligned} \tag{21}$$

these equations are non-linearly dependent on c_p and c_s, and are often solved by first linearizing the equations as is done in Equation (15) of Sturhahn & Jackson [3].

The form of the Debye speed (Equation (3)) used in Equation (21) is only applicable to isotropic materials, and significant errors may result when applied to anisotropic materials. The difference between the seismically relevant Debye speed v_D^V (Equation (3)) and the observed Debye speed v_D (Equation (7)) maps directly into the the estimated acoustic wave speeds, especially into shear wave speed, since the Debye speed is heavily weighted by the shear wave speed.

In order to estimate the magnitude of the errors associated with these approximations, acoustic wave speeds are estimated using the expressions in Equation (21) and the theoretical values of v_D, κ_0^V, and ρ using the five hcp iron alloys in Table 1. Since the full elastic tensors for the materials are available, both the bulk modulus κ_0^V and the Debye speed v_D are calculated exactly. The inferred seismic wave speeds (Table 3) using the linear solution of Sturhahn & Jackson [3] show that the linear approximation results in errors in the compressional wave speed on the order of 10% and errors in the shear wave speed of 10–30%. Therefore, for iron alloys relevant for the Earth's inner core, linearization of Equation (21) results in significant errors due to the terms neglected when linearizing the set of equations. Solving the non-linear system of equations (Equation (21)) captures the compressional wave speed to within several percent, however, significant errors of 25% in shear wave speed remain (Figure 4). The errors associated with these solutions are smaller than those of the linearized solutions, however, they are substantial, especially for the shear wave speed. The source of the error is the assumption that $v_D \approx v_D^V$, that is, relating v_D directly to the seismic speeds even though the material is not isotropic (the third equation in Equation (21)).

Table 3. Comparison of seismic wave speeds obtained from v_D for iron alloys at inner core conditions (Table 1). The material's true compressional wave speed (c_p; column 2) is compared against those obtained using linearized form of Equation (21) (c_p^l; column 3) and solving the full non-linear equations (c_p^{nl}; column 4). Columns 5 through 7 are the same as columns 2 through 4 except for shear waves. The last column gives the shear wave speed estimated from the Debye speed using the relationship of Anderson et al. [31] (c_s^A; Equation (22)).

Material	c_p	c_p^l	c_p^{nl}	c_s	c_s^l	c_s^{nl}	c_s^A
hcp Fe	11.8	10.7 [–10%]	11.5 [–3%]	3.9	2.8 [–28%]	3.0 [–23%]	3.1 [–20%]
hcp Fe-Si	11.2	10.2 [–10%]	11.2 [–1%]	3.9	3.6 [–8%]	3.7 [–5%]	3.9 [–2%]
hcp Fe-Si-Ni	11.8	10.8 [–8%]	11.5 [–2%]	3.3	2.4 [–29%]	2.6 [–22%]	2.7 [–20%]
hcp Fe-Si-C	11.6	10.6 [–9%]	11.4 [–1%]	3.7	3.2 [–13%]	3.4 [–9%]	3.5 [–6%]
hcp Fe-S-C	11.8	10.6 [–10%]	11.7 [–1%]	4.2	3.9 [–7%]	4.0 [–5%]	4.1 [–1%]

Previous estimates of the error due to anisotropy in the seismic speeds obtained from the Debye speed are of the order of a few percent [32], while the errors in this study are found to be considerably larger (Table 3). Bosak et al. [32] significantly underestimates the error due to their choice of averaging scheme. They use an averaging scheme [33] which yields acoustic wave speeds that are similar to those obtained by the Voigt-Ruess-Hill average. However, when comparing with seismic wave speeds from a model such as PREM [13], the wave speeds should be calculated using the Voigt average [14].

Finally, Anderson [10] argued that the ratio c_s/v_D is roughly a constant value of 0.9 ± 0.001, and thus for isotropic materials, the Debye speed may be related to the shear wave speed as [31]

$$c_s \approx 0.9 \times v_D. \tag{22}$$

Surprisingly, this simple estimate of c_s (Table 3) out-performs all other estimates of c_s. Motivated by this result, we perform a simple regression against the strength of anisotropy and find that

$$c_s \approx 1.0 \times v_D + 0.7 \times A^L - 0.7, \tag{23}$$

where c_s and v_D are given in km/s. This relationship is able to provide a good fit to c_s with less than one percent error (Figure 5). Unfortunately, the strength of anisotropy is not typically known, and in that case, the regression

$$c_s \approx 0.45 \times v_D - 2.1, \tag{24}$$

can predict the shear wave speed to within ~5% (Figure 5).

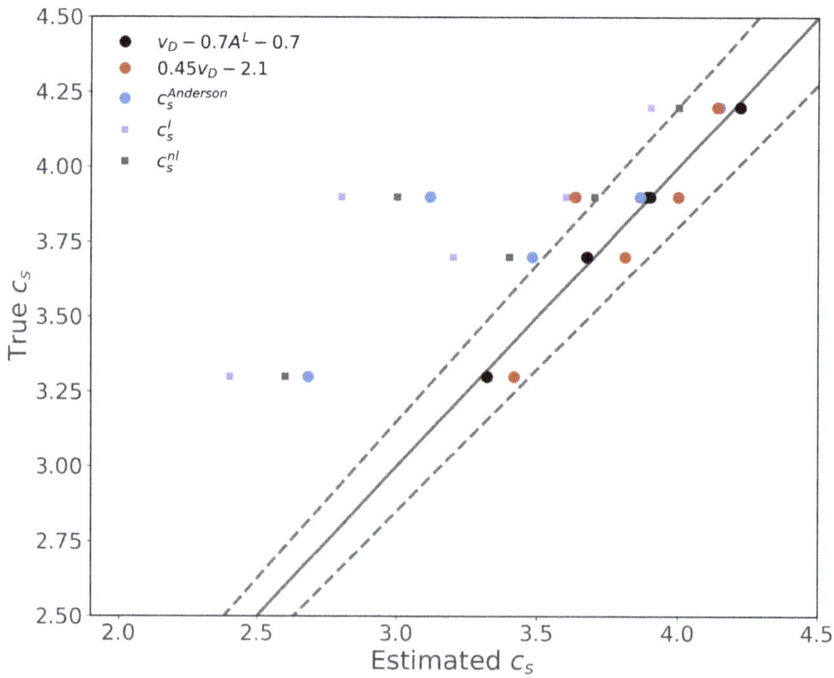

Figure 5. Comparison of c_s against estimates of c_s derived from the Debye speed. The solid grey line represents a one-to-one line. The dashed lines represent ±5% of the shear speed. Points to the left of the grey line underestimate c_s where as points to the right overestimate c_s.

The problem of estimating seismic wave speeds can be turned around to estimate Debye speeds based upon seismically constrained wave speeds. Consider, for illustrative purposes, the seismically observed isotropic wave speeds of the inner core from the PREM model (c_p = 11.1 km/s and c_s = 3.6 km/s; [13]). These isotropic values can be used to calculate Debye speed of v_D = 4.10 km/s if the inner core is isotropic (Equation (3)). Alternatively, using the five elastic constants describing transversely isotropic inner core from seismological observations (c_{11} = 1577 GPa, c_{33} = 1647 GPa, c_{13} = 1259 GPa, c_{44} = 168 GPa, c_{66} = 151 GPa; [14]), the Debye speed is v_D = 4.06 km/s (Equation (2)). This value of the Debye speeds is most similar to that of the bcc iron alloy (4.04 km/s), and least similar to that of the hcp Fe-Si-Ni alloy (2.90 km/s; Table 2). However, comparison of the seismic wave speeds (Table 1) suggests the opposite result, with the hcp Fe-Si-Ni alloy (c_p = 11.8 km/s and c_s = 3.3 km/s) being more consistent with the seismically observed values than those of the bcc Fe-Si alloy (c_p = 11.5 km/s and c_s = 4.2 km/s). This is due to the fact that the higher strength of anisotropy of the bcc Fe-Si crystal lowers its Debye speed, hence the comparison of the Debye speed for anisotropic material ideally should be done with the strength of the anisotropy of the material.

4. Conclusions

The seismic wave speeds of materials at high pressure and high temperature are often estimated from the partial phonon density of states obtained from nuclear resonant inelastic scattering experiments. This manuscript demonstrates that the Debye speeds extracted from these experiments are not equivalent to the Debye speed estimated from that material's seismic wave speeds. Assuming that the experimentally observed Debye speed is equal to that based upon isotropic seismic wave speeds introduces significant error for iron alloys at inner core conditions. The magnitude of these errors depends on the strength of anisotropy, and results in an error of ~20% per unit of anisotropy. A second source of error arises when the approximate linearized solutions (i.e., Equations (15) of Sturhahn & Jackson [3]) are used to estimated compressional and shear wave speeds.

Furthermore, it is shown that the Debye speed estimated from the seismic wave speeds provides an upper bound to the true Debye speed. Consequently, the acoustic wave speeds extracted from the NRIXS experiments are underestimated compared to the true seismic wave speeds as obtained through seismological observations. The compressional wave speed c_p is underestimated by ~10% per unit of anisotropy, and the shear wave speed c_s is underestimated by ~25% per unit of anisotropy. Previous estimates of the error associated with anisotropy are underestimated due to the choice of an averaging scheme that is not the Voigt averaging scheme relevant for models based upon seismological observations. For iron alloys at inner core conditions, new empirical relationships are derived to estimate the seismic shear wave speed from the experimentally derived Debye speed within ~5% error. Additionally, it is shown that if the strength of anisotropy is known, the accuracy of the estimated seismic wave speeds can be dramatically improved. The results in this study are crucial to constraining the inner core's light element composition, and for determining the presence of partial melt.

Author Contributions: B.D. carried out the main investigation of the problem and drafted the paper while M.I. conceived the problem and corrected the manuscript. All authors have read and agreed to the published version of the manuscript.

Funding: This research was funded by the Harvard Faculty of Arts and Sciences Dean's Competitive Fund for Promising Scholarship.

Acknowledgments: We thank Robert Liebermann and three anonymous reviewers for comments that improved the manuscript, Taku Tsuchiya for helpful discussion and providing elastic constants of Fe-Ni-Si alloy, and Joseph Root for pointing us to Jensen's Inequality.

Conflicts of Interest: The authors declare no conflict of interest. The funders had no role in the design of the study; in the collection, analyses, or interpretation of data; in the writing of the manuscript, or in the decision to publish the results.

Abbreviations

The following abbreviations are used in this manuscript:

NRIXS Nuclear Resonant Inelastic X-ray Scattering

Appendix A. Seismic Wave Speeds

The acoustic wave speeds c_m for a given propagation direction $\hat{\mathbf{q}}$ may be determined from the eigenvalues and eigenvectors of the Christoffel Matrix ($M_{ik} \equiv \Lambda_{ijkl}\hat{q}_j\hat{q}_l$, where Λ_{ijkl} are the elements of the fourth-order elasticity tensor) through the Christoffel Equations (e.g., [34–37]),

$$\rho c_m^2(\hat{\mathbf{q}}) = M_{ik}\hat{p}_i^m\hat{p}_k^m = \Lambda_{ijkl}\hat{q}_j\hat{q}_l\hat{p}_i^m\hat{p}_k^m = \hat{\mathbf{p}}_m\hat{\mathbf{q}} : \mathbf{\Lambda} : \hat{\mathbf{p}}_m\hat{\mathbf{q}}, \tag{A1}$$

where $\hat{\mathbf{p}}_m$ is the unit polarization direction with the superscript and subscript m denote the mode of the acoustic phonon. Equation (A1) is solved by using the characteristic equation

$$\det|M_{ik} - \rho c^2\delta_{ik}| = 0, \tag{A2}$$

where δ_{ij} is the Kronecker delta. This equation results in the three eigenvectors ($\hat{\mathbf{p}}_m$) and eigenvalues (c_m). For isotropic materials, the eigenvectors and eigenvalues correspond to a longitudinal wave if the polarization direction is parallel to the propagation direction (i.e., $\hat{\mathbf{p}}_1 \cdot \hat{\mathbf{q}} = 1$) and two transverse waves if the polarization direction is perpendicular to the propagation direction (i.e., $\hat{\mathbf{p}}_2 \cdot \hat{\mathbf{q}} = \hat{\mathbf{p}}_3 \cdot \hat{\mathbf{q}} = 0$). For weakly anisotropic materials, the eigenvectors of M_{ij} are not necessarily orthogonal to $\hat{\mathbf{q}}$.

In order to derive analytical expressions for the wave speeds of hexagonal and cubic symmetries with weak anisotropy, it is useful to consider the problem in the spherical coordinate system with the unit vectors

$$\begin{aligned} \hat{\mathbf{r}} &= \sin\theta\cos\phi\hat{\mathbf{x}} + \sin\theta\sin\phi\hat{\mathbf{y}} + \cos\theta\hat{\mathbf{z}}, \\ \hat{\theta} &= \cos\theta\cos\phi\hat{\mathbf{x}} + \cos\theta\sin\phi\hat{\mathbf{y}} - \sin\theta\hat{\mathbf{z}}, \quad \text{and} \\ \hat{\phi} &= -\sin\phi\hat{\mathbf{x}} + \cos\phi\hat{\mathbf{y}}, \end{aligned} \tag{A3}$$

where $\hat{\mathbf{x}}$, $\hat{\mathbf{y}}$, and $\hat{\mathbf{z}}$ are unit vectors in the Cartesian coordinate system, and θ and ϕ are polar and azimuth angles, respectively, in the spherical coordinate system. These angles are typically defined relative to an axis of crystal symmetry (Figure 1). The propagation unit vector can also be expressed as

$$\hat{\mathbf{q}} = \sin\theta\cos\phi\hat{\mathbf{x}} + \sin\theta\sin\phi\hat{\mathbf{y}} + \cos\theta\hat{\mathbf{z}}. \tag{A4}$$

Appendix A.1. Hexagonal Symmetry

For transversely isotropic materials such as hexagonal close-packed (hcp) iron, the elastic stiffness is described by five independent parameters A, C, F, L, and N [38] that are related to the elasticity tensor Λ as

$$\begin{aligned} &\Lambda_{1111} = \Lambda_{2222} = c_{11} = A \quad &&\Lambda_{1133} = \Lambda_{2233} = c_{13} = F \quad &&\Lambda_{1212} = c_{66} = N \\ &\Lambda_{1313} = \Lambda_{2323} = c_{44} = L \quad &&\Lambda_{1122} = c_{12} = A - 2N \quad &&\Lambda_{3333} = c_{33} = C, \end{aligned} \tag{A5}$$

where the subscripts indicate the indices of the fourth-order tensor. Note that, in the limiting case of an isotropic material with zero anisotropy, $A = C = \kappa + \frac{4}{3}\mu$, $L = N = \mu$, and $F = \kappa - \frac{2}{3}\mu$, where κ and μ are the elastic moduli, the incompressibility and rigidity, respectively.

We assume that the polarization vectors are orthonormal, and use $\hat{\mathbf{p}}_1 = \hat{\mathbf{r}}$, $\hat{\mathbf{p}}_2 = \hat{\theta}$, and $\hat{\mathbf{p}}_3 = \hat{\phi}$. Substituting these expressions for the propagation and polarization vectors (Equation (A3)) and the transversely isotropic elastic tensor (Equation (A5)) into the Christoffel equations (Equation (A1)) gives the first-order perturbation solutions for the longitudinally polarized wave speed c_1 and the two transversely polarized wave speeds c_2, and c_3 (e.g., [39–41]),

$$\begin{aligned} \rho c_1^2 &= A - 2(A - F - 2L)\cos^2\theta + (A + C - 2F - 4L)\cos^4\theta, \\ \rho c_2^2 &= L + (A + C - 2F - 4L)\cos^2\theta - (A + C - 2F - 4L)\cos^4\theta, \quad \text{and} \\ \rho c_3^2 &= N + (L - N)\cos^2\theta. \end{aligned} \tag{A6}$$

These wave speeds are given as functions of the polar angle θ, the angle between the symmetry axis and the propagation direction $\hat{\mathbf{q}}$ (Figure 1). Because c_1 gives speed for a wave propagating in the direction of polarization, this coresponds to seismic P-wave speed while c_2 and c_3 corresopnd to seismic S-wave speeds. Note that the wave speeds is independent of the azimuthal angle ϕ.

Using the the first order perturbation solutions for the wave speeds (Equation (A6)), the Debye speed (Equation (2)) is expressed as

$$\frac{1}{v_D^3} \approx \frac{1}{6}\int_0^\pi \sin\theta \left(\frac{1}{c_1^3(\theta)} + \frac{1}{c_2^3(\theta)} + \frac{1}{c_3^3(\theta)} \right) d\theta. \tag{A7}$$

Note, that this expression with wave speeds defined in Equation (A6) is valid only for weakly anisotropic case where the polarization directions for the three acoustic waves can be assumed to be orthogonal. For strongly anisotropic materials, the polarization vectors and the wave speeds should be calculated numerically via the characteristic equation (Equation (A2)).

Taking the average over the unit sphere, the expression for the directionally dependent seismic wave speeds based upon Equation (A6) become

$$\begin{aligned} \rho \left\langle c_1^2 \right\rangle &= \frac{1}{15}(8A + 3C + 4F + 8L), \\ \rho \left\langle c_2^2 \right\rangle &= \frac{1}{15}(2A + 2C - 4F + 7L), \quad \text{and} \\ \rho \left\langle c_3^2 \right\rangle &= \frac{1}{3}(2N + L). \end{aligned} \tag{A8}$$

These expressions are compatible with the Voigt averaged $\left\langle \Lambda_{ijkl} \right\rangle$ moduli which are given as

$$\begin{aligned} \kappa_0^V &= \frac{1}{9}(4A + C + 4F - 4N), \quad \text{and} \\ \mu_0^V &= \frac{1}{15}(A + C - 2F + 6L + 5N), \end{aligned} \tag{A9}$$

since $\rho\left(\langle c_2^2\rangle + \langle c_3^2\rangle\right)/2 = \mu_0^V$ and $\rho\langle c_1^2\rangle = \kappa_0^V + 4/3\mu_0^V$. However, the Reuss averaged $\left\langle \Lambda_{ijkl}^{-1} \right\rangle^{-1}$ moduli are given as

$$\begin{aligned} \kappa_0^R &= \frac{(A - N)C - F^2}{A - N + C - 2F}, \quad \text{and} \\ \mu_0^R &= \frac{15LN[(A - N)C - F^2]}{6(L + N)[(A - N)C - F^2] + LN(4A + C + 4F - 4N)}, \end{aligned} \tag{A10}$$

which are not compatible with seismological observations. These expressions are consistent with those obtained by other authors such as Watt and Peselnick [42].

Appendix A.2. Cubic Symmetry

For cubic materials such as body-centered cubic (bcc) iron, the elastic stiffness is described by three independent parameters c_{11}, c_{12}, and c_{44} that are related to the elasticity tensor Λ as

$$\begin{aligned} \Lambda_{1111} &= \Lambda_{2222} = \Lambda_{3333} = c_{11}, \\ \Lambda_{1122} &= \Lambda_{2233} = \Lambda_{3311} = c_{12}, \quad \text{and} \\ \Lambda_{3131} &= \Lambda_{2323} = \Lambda_{3131} = c_{44}. \end{aligned} \tag{A11}$$

Note that, in the limiting case of an isotropic material with zero anisotropy, c_{12} corresponds to the Láme parameter $\lambda = \kappa - \frac{2}{3}\mu$, c_{44} corresponds to the shear modulus μ, and $c_{11} = c_{12} + 2c_{44}$. For ease of notation, we will use the parameters λ, μ, and η to describe the cubic material where $\lambda = c_{12}$, $\mu = c_{44}$, and $\eta = c_{11} - c_{12} - 2c_{44}$.

In case of hexagonal symmetry, the unit vectors $\hat{\theta}$ and $\hat{\phi}$ provided the fastest and slowest polarization directions for the two shear waves. This is not the case for cubic material, and even though we can assume $\hat{\mathbf{p}}_1 = \hat{\mathbf{r}}$, the polarization for the shear waves would depend upon an additional angle ξ such that $\hat{\mathbf{p}}_2 = \cos\xi\hat{\theta} + \sin\xi\hat{\phi}$ and $\hat{\mathbf{p}}_3 = \cos(\xi + \pi/2)\hat{\theta} + \sin(\xi + \pi/2)\hat{\phi}$. The angle ξ depends upon θ and ϕ and is obtained through the expression

$$\cot 2\xi = \frac{(1 + \cos^2\theta)\sin^2 2\phi - 4\cos^2\theta}{\sin 4\phi \cos\theta}. \tag{A12}$$

We substitute the expressions for the propagation and polarization vectors and the elasticity tensor into the Christoffel equations (Equation (A1)). The first order perturbation solutions for the longitudinally polarized wave speed c_1 and the two transversely polarized wave speeds c_2 and c_3 are (e.g., [40,43])

$$\begin{aligned} \rho c_1^2 &= \lambda + 2\mu + \eta(\cos^4\phi \sin^4\theta + \sin^4\phi \sin^4\theta + \cos^4\theta), \\ \rho c_2^2 &= \mu + \eta\chi(\theta, \phi, \xi), \quad \text{and} \\ \rho c_3^2 &= \mu + \eta\chi(\theta, \phi, \xi + \pi/2), \end{aligned} \tag{A13}$$

where

$$\begin{aligned} \chi(\theta, \phi, \xi) = &\cos^2\phi \sin^2\theta\,(\cos\xi\cos\phi\cos\theta - \sin\xi\sin\phi)^2 \\ &+ \sin^2\phi\sin^2\theta\,(\cos\xi\sin\phi\cos\theta + \sin\xi\cos\phi)^2 + \cos^2\xi\cos^2\theta\sin^2\theta. \end{aligned} \tag{A14}$$

Using these wave speeds lead to an expression for the Debye speed (Equation (2)),

$$\frac{1}{v_D^3} \approx \frac{1}{12\pi}\int_0^{2\pi}\int_0^{\pi} \sin\theta \left(\frac{1}{c_1^3(\theta,\phi)} + \frac{1}{c_2^3(\theta,\phi)} + \frac{1}{c_3^3(\theta,\phi)}\right) d\theta d\phi, \tag{A15}$$

and the directionally averaged wave speeds

$$\begin{aligned} \rho\left\langle c_1^2 \right\rangle &= \lambda + 2\mu + \frac{3}{5}\eta, \quad \text{and} \\ \rho\left(\left\langle c_2^2\right\rangle + \left\langle c_3^2\right\rangle\right) &= 2\mu + \frac{2}{5}\eta. \end{aligned} \tag{A16}$$

As with the transversely isotropic case, these speeds are consistent with the Voigt averaged moduli given as

$$\kappa_0^V = \lambda + \frac{\eta + 2\mu}{3} \quad \text{and} \quad \mu_0^V = \mu + \frac{\eta}{5}, \tag{A17}$$

since $\rho\left(\left\langle c_2^2\right\rangle + \left\langle c_3^2\right\rangle\right)/2 = \mu_0^V$ and $\rho\left\langle c_1^2\right\rangle = \kappa_0^V + 4/3\mu_0^V$. On the other hand, the Reuss averaged moduli are given as

$$\kappa_0^R = \kappa_0^V \quad \text{and} \quad \mu_0^R = \frac{5\mu\left(\eta + 2\mu\right)}{4\mu + 3\left(\eta + 2\mu\right)}, \tag{A18}$$

and they are not compatible with seismic wave speeds.

Appendix B. Experimental Measurements of Cobalt

All examples in the main text of this paper are based on theoretical calculations of hcp and bcc iron alloys. Here, we briefly report and analyze the results obtain using the elements of the elasticity tensor of cobalt obtained from [44] which have been determined using measurements from inelastic X-ray scattering experiments. Note that, because of weak anisotropy represented by strength of anisotropy (A_L) between 0.03 and 0.09 (Table A1), the wave speeds calculated from the elements of the elasticity tensor following the seismically relevant averaging scheme [14] are consistent with their experimentally measured values (e.g., [45]). At this levels of anisotropy, the NRIXS derived Debye speeds and the expressions in Equation (21) provide accurate measurements of the material's seismic wave speeds.

Table A1. The Debye v_D and seismic wave speeds c_p and c_s of cobalt at pressures of zero to 40 GPa. The alloy's density in g/cm^3 (ρ is given in column 2 [45]), while the strength of anisotropy (A^L; [22]), and the Debye speed calculated using Equation (7) and the elements of the elasticity of cobalt (v_D; [44]) are given in columns 3 and 4. The material's seismic wave speed (c_p; column 5) is compared against longitudinal wave speeds obtained using the full non-linear equations of Equation (21) (c_p^{nl}; column 6), and experimentally measured values determined using ultrasonics (c_p^{US}; column 7; [46]), inelastic X-ray scattering (c_p^{IXS}; column 8; [45]), and impulsive stimulated light scattering (c_p^{ISLS}; column 9; [47]). Columns 10 through 14 are the same as columns 5 through 9 except for shear/transverse wave speeds.

Pressure	ρ	A^L	v_D	c_p	c_p^{nl}	c_p^{US}	c_p^{IXS}	c_p^{ISLS}	c_s	c_s^{nl}	c_s^{US}	c_s^{IXS}	c_p^{ISLS}
0	8.8	0.09	3.44	5.7	5.7	5.8	-	-	3.1	3.1	3.1	-	-
11	9.3	0.06	3.67	6.2	6.2	-	6.3	6.1	3.3	3.3	3.3	3.1	-
40	10.3	0.03	4.18	7.3	7.3	-	7.3	-	3.7	3.7	-	3.9	-

Appendix C. Mean Projected Wave Speed for Randomly Oriented Samples

The partial density of state measured by NRIXS experiments are proportional to $v_{\hat{\mathbf{k}}}^{-3}$ (Equation (5)). However, materials composed of randomly oriented anisotropic crystal with bulk or macroscopically isotropic properties (e.g., powdered samples) are well represented by averaging over the incident wave vector direction $\hat{\mathbf{k}}$. In this section we will show that for all materials,

$$\left\langle v_{\hat{\mathbf{k}}}^{-3}\right\rangle = v_D^{-3}. \tag{A19}$$

The eigenvalues (λ_m) of the Christoffel Matrix $\mathbf{M}$ are equal to the squared seismic the wave speeds ($\lambda_m = c_m^2$; Equation (A1)), such that Equation (5) can be written as

$$\sum_{m=1}^{3} \int \frac{d\Omega_{\hat{\mathbf{q}}}}{4\pi} \frac{\left|\hat{\mathbf{k}} \cdot \hat{\mathbf{p}}_m\right|^2}{c_m^3} = \int \frac{d\Omega_{\hat{\mathbf{q}}}}{4\pi} \sum_{m=1}^{3} \frac{\left|\hat{\mathbf{k}} \cdot \hat{\mathbf{p}}_m\right|^2}{\lambda_m^{3/2}}. \tag{A20}$$

The integrand can be written in matrix form such that

$$\sum_{m=1}^{3} \frac{\left|\hat{\mathbf{k}} \cdot \hat{\mathbf{p}}_m\right|^2}{\lambda_m^{3/2}} = \left(\hat{\mathbf{k}} \cdot \mathbf{P}\right) \cdot \mathbf{D}^{-3/2} \cdot \left(\hat{\mathbf{k}} \cdot \mathbf{P}\right), \tag{A21}$$

where $\mathbf{P}$ is a matrix whose columns are the normalized eigenvectors $\hat{\mathbf{p}}_m$ of $\mathbf{M}$, and $\mathbf{D}$ is a diagonal matrix whose diagonal elements are the eigenvalues (λ_m) of $\mathbf{M}$. Using $\mathbf{P}^T = \mathbf{P}^{-1}$ and $\mathbf{D}^{-3/2} = \mathbf{P}^T \cdot \mathbf{M}^{-3/2} \cdot \mathbf{P}$, the right-hand side of Equation (A21) can be written in terms of the Christoffel Matrix such that

$$\left(\hat{\mathbf{k}} \cdot \mathbf{P}\right) \cdot \mathbf{D}^{-3/2} \cdot \left(\hat{\mathbf{k}} \cdot \mathbf{P}\right) = \left(\hat{\mathbf{k}} \cdot \mathbf{P}\right) \cdot \mathbf{P}^T \cdot \mathbf{M}^{-3/2} \cdot \mathbf{P} \cdot \left(\mathbf{P}^T \cdot \hat{\mathbf{k}}\right) = \hat{\mathbf{k}} \cdot \mathbf{M}^{-3/2} \cdot \hat{\mathbf{k}}. \tag{A22}$$

For ease of notation, let us temporarily denote the elements of $\mathbf{M}^{-3/2}$ by M_{ij}, and represent the incident wave vector by the expression for the unit vector in Equation (A4). The above expression can then be explicitly written out as

$$\begin{aligned}\hat{\mathbf{k}} \cdot \mathbf{M}^{-3/2} \cdot \hat{\mathbf{k}} = &(M_{11} + M_{22}) \sin^2\phi \sin^2\theta + (M_{12} + M_{21}) \sin\phi \sin^2\theta \cos\phi \\ &+ (M_{23} + M_{32}) \sin\phi \sin\theta \cos\theta + (M_{13} + M_{31}) \sin\theta \cos\phi \cos\phi \\ &+ M_{33} \cos^2\theta.\end{aligned} \tag{A23}$$

For randomly oriented sample, the above expression needs to be integrated over all incident wave vector directions, that is, integration over $d\Omega_{\hat{\mathbf{k}}}$, and only the first and last terms of the right-hand side Equation (A23) remain. Thus,

$$\left\langle \hat{\mathbf{k}} \cdot \mathbf{M}^{-3/2} \cdot \hat{\mathbf{k}} \right\rangle = \frac{1}{3}\left(M_{11} + M_{22} + M_{33}\right) = \frac{1}{3}\,\mathrm{Tr}\left(\mathbf{M}^{-3/2}\right), \tag{A24}$$

where Tr denotes the trace. Using the invariance of the trace, $\mathrm{Tr}\left(\mathbf{M}^{-3/2}\right)$ can be directly related to the material's wave speeds as

$$\mathrm{Tr}\left(\mathbf{M}^{-3/2}\right) = \mathrm{Tr}\left(\mathbf{D}^{-3/2}\right) = \sum_{m=1}^{3} \lambda^{-3/2} = \sum_{m=1}^{3} \frac{1}{c_m^3}. \tag{A25}$$

Thus, the mean projected wave speed (Equation (5)), averaged over all incident wave vector directions, is equivalent to the material's Debye speed since

$$\begin{aligned}\left\langle \frac{1}{v_{\hat{\mathbf{k}}}^3} \right\rangle &= \int \frac{d\Omega_{\hat{\mathbf{q}}}}{4\pi} \left\langle \sum_{m=1}^{3} \frac{\left|\hat{\mathbf{k}} \cdot \hat{\mathbf{p}}_m(\hat{\mathbf{q}})\right|^2}{c_m^3(\hat{\mathbf{q}})} \right\rangle = \int \frac{d\Omega_{\hat{\mathbf{q}}}}{4\pi} \left\langle \hat{\mathbf{k}} \cdot \mathbf{M}^{-3/2} \cdot \hat{\mathbf{k}} \right\rangle = \frac{1}{3} \int \frac{d\Omega_{\hat{\mathbf{q}}}}{4\pi}\, \mathrm{Tr}\left(\mathbf{M}^{-3/2}\right) \\ &= \frac{1}{3} \int \frac{d\Omega_{\hat{\mathbf{q}}}}{4\pi} \sum_{m=1}^{3} \frac{1}{c_m^3} = \frac{1}{v_D^3}\end{aligned} \tag{A26}$$

References

1. Sturhahn, W. Nuclear resonant spectroscopy. *Phys. Condens. Matter* **2004**, *416*, S497–S530. [CrossRef]
2. Dauphas, N.; Hu, M.Y.; Baker, E.M.; Hu, J.; Tissot, F.L.; Alp, E.E.; Roskosz, M.; Zhao, J.; Bi, W.; Liu, J.; et al. SciPhon: A data analysis software for nuclear resonant inelastic X-ray scattering with applications to Fe, Kr, Sn, Eu and Dy. *J. Synchrotron Radiat.* **2018**, *25*, 1581–1599. [CrossRef] [PubMed]
3. Sturhahn, W.; Jackson, J.M. *Geophysical Applications of Nuclear Resonant Spectroscopy*; Ohtani, E., Ed.; Advances in High-Pressure Mineralogy; The Geological Society of America: Boulder, CO, USA, 2007; pp. 157–174.
4. Antonangeli, D.; Ohtani, E. Sound velocity of hcp-Fe at high pressure: Experimental constraints, extrapolations and comparison with seismic models. *Prog. Earth Planet. Sci.* **2015**, *2*, 3. [CrossRef]
5. Singwi, K.S.; Sjölander, A. Resonance absorption of nuclear gamma rays and the dynamics of atomic motions. *Phys. Rev.* **1960**, *120*, 1093–1102. [CrossRef]
6. Sturhahn, W.; Toellner, T.S.; Alp, E.E.; Zhang, X.; Ando, M.; Yoda, Y.; Kikuta, S.; Seto, M.; Kimball, C.W.; Dabrowski, B. Phonon density of states measured by inelastic nuclear resonant scattering. *Phys. Rev. Lett.* **1995**, *74*, 3832–3835. [CrossRef] [PubMed]

7. Chumakov, A.I.; Rüffer, R.; Baron, A.Q.R.; Grünsteudel, H.; Grünsteudel, H.F.; Kohn, V.G. Anisotropic inelastic nuclear absorption. *Phys. Rev. B* **1997**, *56*, 10758. [CrossRef]
8. Kohn, V.G.; Chumakov, A.I.; Rüffer, R. Nuclear resonant inelastic absorption of synchrotron radiation in an anisotropic single crystal. *Phys. Rev. B* **1998**, *58*, 8437–8444. [CrossRef]
9. Anderson, O.L. A simplified method for calculating the Debye temperature from elastic constants. *J. Phys. Chem. Solids* **1963**, *24*, 909–917. [CrossRef]
10. Anderson, O.L. An approximate method of estimating shear velocity from specific heat. *J. Geophys. Res. Lett.* **1965**, *70*, 4726–4728. [CrossRef]
11. Sturhahn, W.; Kohn, V. Theoretical aspects of incoherent nuclear resonant scattering. *Hyperfine Interact.* **1999**, *123*, 367–399. [CrossRef]
12. Hu, M.Y.; Sturhahn, W.; Toellner, T.S.; Mannheim, P.D.; Brown, D.E.; Zhao, J.; Alp, E.E. Measuring velocity of sound with nuclear resonant inelastic X-ray scattering. *Phys. Rev. B* **2003**, *67*, 094304. [CrossRef]
13. Dziewoński, A.M.; Anderson, D.L. Preliminary reference earth model. *Phys. Earth Planet. Int.* **1981**, *25*, 297–356.
14. Delbridge, B.G.; Ishii, M.; Reconciling elasticity tensor constraints from mineral physics and seismological observations: Applications to the earth's inner Core. *Geophys. J. Int.* **2019**, submitted.
15. Jensen, J.L.W.V. Sur les fonctions convexes et les inegalites entre les valeurmoyennes, *Acta Math.* **1906**, *30*, 175–193. [CrossRef]
16. Roberts, A.W.; Varberg, D.E. *Convex Functions*; Academic Press: Cambridge, MA, USA, 1973
17. Steinle-Neumann, G.; Stixrude, L.; Cohen, R.E.; Gülseren, O. Elasticity of iron at the temperature of the Earth's inner core. *Nature* **2001**, *413*, 57–60. [CrossRef]
18. Tsuchiya, T.; Kuwayama, Y.; Ishii, M.; Kawai, K. High-P, T elasticity of iron-light element alloys[MR33E-05]. In Proceedins of the 2017 Fall Meeting, AGU, New Orleans, LA, USA, 11–15 December 2017.
19. Li, Y.; Vočadlo, L.; Brodholt, J.P. The elastic properties of hcp-Fe alloys under the conditions of the Earth's inner core. *Earth Planet. Sci. Lett.* **2018**, *493*, 118–127. [CrossRef]
20. Martorell, B.; Wood, I.G.; Brodholt, J.; Vočaldo, L. The elastic properties of hcp-$Fe1\text{-}_xSi_x$ at Earth's inner-core conditions. *Earth Planet. Sci. Lett.* **2016**, *451*, 89–96. [CrossRef]
21. Belonoshko, A.B.; Skorodumova, N.V.; Davis, S.; Osiptsov, A.N.; Rosengren, A.; Johansson, B. Origin of the low rigidity of the Earth's inner core. *Science* **2007**, *316*, 1603–1605. [CrossRef]
22. Kube, C.M. Elastic anisotropy of crystals. *AIP Adv.* **2016**, *6*, 095209. [CrossRef]
23. Ranganathan, S.I.; Ostoja-Starzewski, M. Universal elastic anisotropy index. *Phys.Rev. Lett.* **2008**, *101*, 055504. [CrossRef]
24. Chung, D.H.; Buessem, W.R. The Elastic Anisotropy of Crystals. *J. Appl. Phys.* **1967**, *38*, 2010–2012. [CrossRef]
25. Mao, H.K.; Xu, J.; Struzhkin, V.V.; Shu, J.; Hemley, R.J.; Sturhahn, W.; Hu, M.Y.; Alp, E.E.; Vocadlo, L.; Alfè, D.; et al. Phonon density of states of iron up to 153 gigapascals. *Science* **2001**, *292*, 914–916. [CrossRef] [PubMed]
26. Lin, J.F.; Struzhkin, V.V.; Sturhahn, W.; Huang, E.; Zhao, J.; Hu, M.Y.; Alp, E.E.; Mao, H.K.; Boctor, N.; Hemley, R.J. Sound velocities of iron-nickel and iron-silicon alloys at high pressures. *Geophys. Res. Lett.* **2003**, *30*, 11. [CrossRef]
27. Jackson, J.M.; Hamecher, E.A.; Sturhahn, W. Nuclear resonant X-ray spectroscopy of (Mg, Fe) SiO3 orthoenstatites. *Eur. J. Mineral.* **2009**, *21*, 551–560. [CrossRef]
28. Prescher, C.; Dubrovinsky, L.; Bykova, E.; Kupenko, I.; Glazyrin, K.; Kantor, A.; McCammon, C.; Mookherjee, M.; Nakajima, Y.; Miyajima, N.; et al. High Poisson's ratio of Earth's inner core explained by carbon alloying. *Nat. Geosci.* **2015**, *8*, 220. [CrossRef]
29. Wicks, J.K.; Jackson, J.M.; Sturhahn, W.; Zhang, D. Sound velocity and density of magnesiowüstites: Implications for ultra low-velocity zone topography. *Geophys. Res. Lett.* **2017**, *44*, 2148–2158. [CrossRef]
30. Finkelstein, G.J.; Jackson, J.M.; Said, A.; Alatas, A.; Leu, B.M.; Sturhahn, W.; Toellner, T.S. Strongly anisotropic magnesiowüstite in Earth's lower mantle. *J. Geophys. Res.* **2018**, *123*, 4740–4750. [CrossRef]
31. Anderson, O. L.; Dubrovinsky, L.; Saxena, S. K.; LeBihan, T. Experimental vibrational Grüneisen ratio values for ϵ-iron up to 330 GPa at 300 K. *Geophys. Res. Lett.* **2001**, *28*, 399–402. [CrossRef]
32. Bosak, A.; Krisch, M.; Chumakov, A.; Abrikosov, I.A.; Dubrovinsky, L. Possible artifacts in inferring seismic properties from X-ray data. *Phys. Earth Planet. Int.* **2016**, *260*, 14–19. [CrossRef]
33. Matthies, S.; Humbert, M. On the principle of a geometric mean of even-rank symmetric tensors for textured polycrystals. *J. Appl. Crystallogr.* **1995**, *28*, 254–266. [CrossRef]
34. Fedorov, I. *Theory of Elastic Waves in Crystals*; Plenum Press: New York, NY, USA, 1968.

35. Musgrave, M. *Crystal Acoustics*; Holden-Day Series in Mathematical Physics: San Francisco, CA, USA, 1970; pp. 75–76.
36. Dahlen, F.; Tromp, J. *Theoretical Global Seismology*; Princeton University Press: Princeton, NJ, USA, 1998.
37. Chapman, C. *Fundamentals of Seismic Wave Propagation*; Cambridge University Press: Cambridge, UK, 2004.
38. Takeuchi, H.; Saito, M. Seismic surface waves. *Methods Comput. Phys.* **1972**, *11*, 217–295.
39. Backus, G.E. Possible forms of seismic anisotropy of the uppermost mantle under oceans.*J. Geophys. Res.* **1965**, *70*, 3429–3439. [CrossRef]
40. Every, A.G. General closed-form expressions for acoustic waves in elastically anisotropic solids. *Phys. Rev. B* **1980**, *22*, 1746–1760. [CrossRef]
41. Tromp, J. Normal-mode splitting due to inner-core anisotropy. *Geophys. J. Int.* **1995**, *121*, 963–968. [CrossRef]
42. Watt, J.P.; Peselnick, L. Clarification of the Hashin-Shtrikman bounds on the effective elastic moduli of polycrystals with hexagonal, trigonal, and tetragonal symmetries. *J. Appl. Phys.* **1980**, *51*, 1525–1531. [CrossRef]
43. Every, A.G. General, closed-form expressions for acoustic waves in cubic crystals. *Phys. Rev. Lett.* **1979**, *42*, 1065–1068. [CrossRef]
44. Antonangeli, D.; Krisch, M.; Fiquet, G.; Farber, D.L.; Aracne, C.M.; Badro, J.; Occelli, F.; Requardt, H. Elasticity of cobalt at high pressure studied by inelastic X-ray scattering. *Phys. Rev. Lett.* **2004**, *93*, 215505. [CrossRef]
45. Antonangeli, D.; Krisch, M.; Fiquet, G.; Badro, J.; Farber, D.L.; Bossak, A.; Merkel, S. Aggregate and single-crystalline elasticity of hcp cobalt at high pressure. *Phys. Rev. B* **2005**, *72*, 134303. [CrossRef]
46. Choy, M.M.; Hellwege, K.H.; Hellwege, A.M. *Elastic, Piezoelectric, Pyroelectric, Piezooptic, Electrooptic Constants, and Nonlinear Dielectric Susceptibilities of Crystals: Revised and Expanded Edition of Volumes III/1 and III/2(Vol. 11)*; Springer: New York, NY, USA, 1979.
47. Goncharov, A.F.; Crowhurst, J.; Zaug, J.M. Elastic and vibrational properties of cobalt to 120 GPa. *Phys. Rev. Lett.* **2004**, *92*, 115502. [CrossRef]

minerals

MDPI

Article

Single Crystal Elastic Properties of Hemimorphite, a Novel Hydrous Silicate

Yingzhe Li and Jay D. Bass *

Geology Department, University of Illinois, Urbana, IL 61801, USA; yli200@illinois.edu

* Correspondence: jaybass@illinois.edu; Tel.: +1-217-333-1018

Received: 6 April 2020; Accepted: 7 May 2020; Published: 10 May 2020

Abstract: Hemimorphite, with the chemical formula $Zn_4Si_2O_7(OH)_2{\cdot}H_2O$, contains two different types of structurally bound hydrogen: molecular water and hydroxyl. The elastic properties of single-crystal hemimorphite have been determined by Brillouin spectroscopy at ambient conditions, yielding tight constraints on all nine single-crystal elastic moduli (C_{ij}). The Voigt–Reuss–Hill (VRH) averaged isotropic aggregate elastic moduli are K_S (VRH) = 74(3) GPa and μ (VRH) = 27(2) GPa, for the adiabatic bulk modulus and shear modulus, respectively. The average of the Hashin–Shtrickman (HS) bounds are Ks (HS) = 74.2(7) GPa and and μ (HS) = 26.5(6) GPa. Hemimorphite displays a high degree of velocity anisotropy. As a result, differences between upper and lower bounds on aggregate properties are large and the main source of uncertainty in Ks and μ. The HS average P wave velocity is V_P = 5.61(4) km/s, and the HS S-wave velocity is V_S = 2.77(3) km/s. The high degree of elastic anisotropy among the on-diagonal longitudinal and pure shear moduli of hemimorphite are largely explained by its distinctive crystal structure.

Keywords: elastic properties; Brillouin; hydrogen bonds; sound velocity

1. Introduction

The influence of hydrogen, or "water", on the elastic properties of minerals is a topic of great current interest for possibly identifying hydrous phases in the deep crust and mantle from seismic models for these regions. Adding hydrogen bonds into crystal structures is widely assumed to decrease the density, and elastic moduli of silicate minerals by increasing their volume, at least at ambient conditions. However, the effect of water on the elastic properties of minerals can depend greatly on details of the crystal structure into which hydrogen is incorporated, and on whether hydrogen is present in the form of hydroxyl OH^- or as molecular water. Hydration of wadsleyite, ringwoodite and garnet can change elastic moduli significantly [1,2], in contrast to the olivine-chondrodite group [3], which varies less, with a similar water content. These differences are considered to be caused by different incorporation mechanisms of hydrogen into their structures. To investigate the influence of water on the direction-dependent compressibility, or elastic anisotropy, of minerals, it is necessary to measure the single-crystal elastic moduli matrix, C_{ij}. Hemimorphite, $Zn_4Si_2O_7(OH)_2{\cdot}H_2O$, a hydrated alteration product of willemite, Zn_2SiO_4, is interesting, in that it contains both molecular H_2O and OH^- hydroxyl bonded into its structure. In nature hemimorphite is an economic ore mineral in supergene non-sulfide zinc deposits. Some significant deposits are dominated by hemimorphite, such as Shaimerden supergene deposit in Kazakhstan, the Cho Dien district in Vietnam, and the Skorpion deposit in Namibia [4].

The content of structurally bound H_2O of hemimorphite is 7.18 wt% [5]. It has two phase transitions below room temperature related to ordered versus dynamically disordered H_2O and OH^- in the crystal structure at 20 and 90 K, respectively [6]. At elevated temperatures, hemimorphite dehydrates in two stages. It loses molecular H_2O without breakdown of the structure at 550 °C, then undergoes a phase transition to β-Zn_2SiO_4, involving release of the OH^- group between 725–760 °C. If heated above 960 °C, it transforms into willemite (α-Zn_2SiO_4) [7].

The structure of orthorhombic hemimorphite, with space group *Imm2*, is composed of a three-dimensional framework, in which two $ZnO_3(OH)$ tetrahedra and one SiO_4 tetrahedron form three-membered rings that are corner-shared to form corrugated sheets parallel to (010) [8,9] (Figure 1). SiO_4 tetrahedra in one layer are linked to SiO_4 tetrahedra in an adjacent layer, while the $ZnO_3(OH)$ tetrahedra in adjacent layers also only link to each other, completing the framework. Large open channels run parallel to [001], giving hemimorphite a structural similarity to zeolites. The molecular water, orientated in the (010) plane, is near the center of large cavities of an eight-membered ring and bonded to hydroxyl groups of the $ZnO_3(OH)$ tetrahedra by hydrogen bonds. The hydrogen bonds formed by H of hydroxyl group (H35) are stronger than those formed by H of molecular water (H53) [8]. The presence of both hydroxyl and molecular water, corrugated sheets in a tetrahedral framework, and large zeolitic cavities, make hemimorphite a rather novel silicate mineral. The objective of this project was to measure the elastic properties of hemimorphite, in order to investigate its structure-property relations, and as a first step toward understanding how H_2O and OH^- influence sound velocities and elasticity.

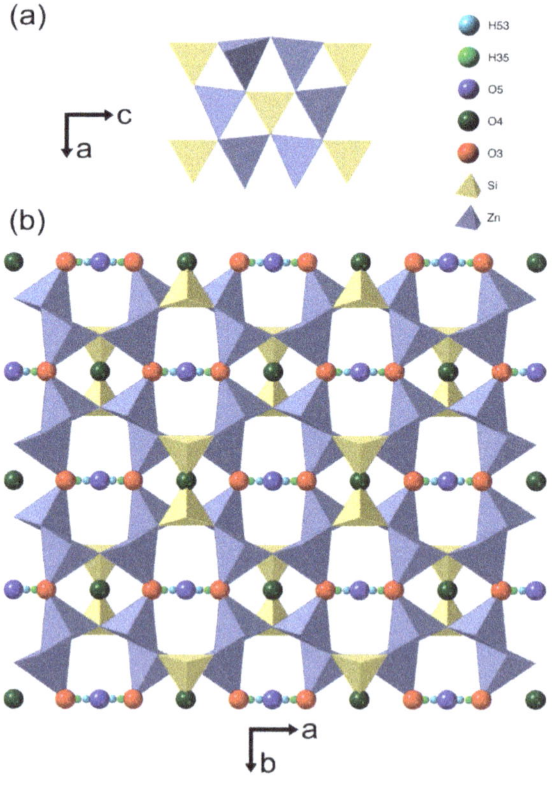

Figure 1. The crystal structure of hemimorphite. (**a**) Here, two zinc and one silicon tetrahedra comprise three-membered ring in (010) plane; (**b**) the structure of hemimorphite projected onto the a–b plane. Six-membered rings and eight-membered rings are in the a–b plane. Large cavities run along the c axis and contain H_2O. The corrugated layers of three-membered rings are also apparent in this projection (modified from Hill, et al., 1977 [8]).

2. Materials and Methods

Our samples of hemimorphite are from the Mapimí mine, Durango, Mexico. The sample contains a "spray" of transparent euhedral crystals with clearly defined faces. Most crystals exhibit tabular planes parallel to (010), and striations parallel to [001]. From single-crystal X-ray diffraction using a four-circle diffractometer, the average lattice parameters obtained on three single crystals are: a = 8.37(2) Å, b = 10.722(4) Å, and c = 5.118(7) Å, with unit cell volume V_0 = 459(1) Å^3. The lattice parameters are consistent with the values reported in earlier crystal structure determinations by Hill et al. [8], McDonald and Cruickshank [9], Takeuchi et al. [10] and Cooper and Gibbs [11].

Scanning electron microscopy-energy dispersive spectroscopy (SEM-EDS) was used to obtain the major element chemical composition of hemimorphite. The results show that our samples are iron-free, nearly pure zinc silicates. The calculated density ρ = 3.48 (3) g/cm^3 was derived from the chemical formula $Zn_4Si_2O_7(OH)_2{\cdot}H_2O$ with Z = 2 and the unit cell volume given by X-ray diffraction.

Three near-principal sections of hemimorphite were polished with parallel faces in the a–b, b–c, and a–c crystallographic planes. The orientations of the polished surfaces were obtained by specular goniometry measurements and the X-ray orientation matrices for each crystal. The accuracy of sample orientation was within 0.5 degrees. For the Brillouin scattering measurements, a single-frequency diode pumped solid-state laser of wavelength 532 nm was used as a light source. All measurements were performed using a 90° platelet symmetric scattering geometry [12] at ambient conditions. The scattered light was analyzed by a piezoelectrically-scanned tandem Fabry–Perot interferometer [13].

For each principal section, acoustic velocities were determined in 26 distinct crystallographic directions by changing the chi angle on the three-circle Eulerian cradle in 15 degrees increments. Measurements were made in a total of 78 distinct crystallographic directions, yielding 156 acoustic mode velocities (78 longitudinal and 78 single shear modes).

3. Results

From the measured Brillouin frequency shifts, $\Delta\nu_B$, velocity v in a given crystallographic direction i is derived from the equation for symmetric platelet geometry [12]:

$$v_i = \frac{\lambda_0 \Delta\upsilon_B}{2\sin\frac{\Theta}{2}} \quad (1)$$

where λ_0 is the wavelength of the incident laser light, and θ is the external angel between the incident and scattered light.

The single-crystal elastic moduli tensor of orthorhombic hemimorphite contains nine independent non-zero elastic moduli C_{ij}. On-diagonal moduli were well constrained by the acoustic velocities in the directions very close to the crystallographic axes. A linearized inversion method of Weidner and Carleton [14], was used to obtain a least-squares best-fit model of the elastic moduli C_{ij} derived from the velocities. Figure 2 and Table 1 show a comparison between model and observed data of phonon velocities. The root-mean-square (RMS) residual in velocity for the final best-fit C_{ij} model is 32.4 m/s. Hemimorphite exhibits considerable longitudinal and shear velocity anisotropy (Figure 2 and Table 2).

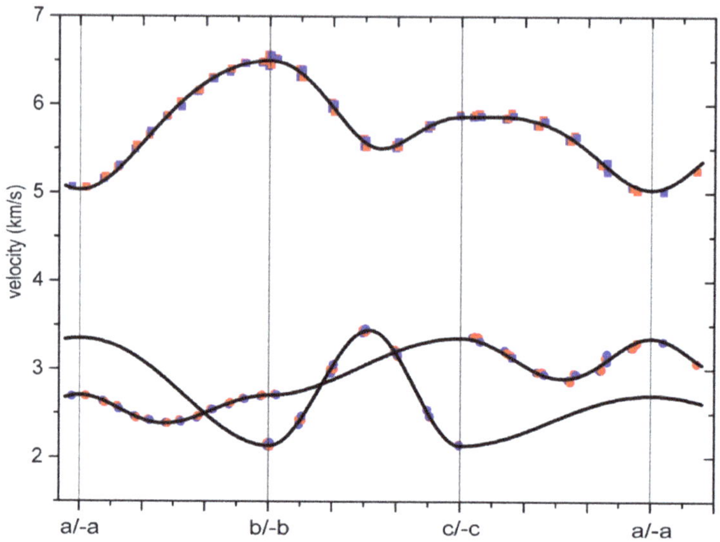

Figure 2. Measured acoustic velocities of hemimorphite (solid symbols), as a function of crystallographic directions projected onto the a–b, b–c, and a–c crystallographic planes. Squares and circles represent V_P and Vs measurements, respectively. The red symbols are measured on (001), (100) and (0$\bar{1}$0); blue symbols are measured on the (00$\bar{1}$), (00) and (010). The best-fit acoustic velocity model is shown by solid lines. Error bars are smaller than the size of symbols.

Table 1. Acoustic measurements of hemimorphite at ambient conditions.

Phonon Direction Cosines			V_P (m/s)			V_S (m/s)		
n_a	n_b	n_c	Obs.	Model	Residual	Obs.	Model	Residual
0.936	−0.005	−0.352	5314	5293	21	3130	3110	20
0.812	−0.009	−0.583	5654	5580	73	2951	2916	35
0.632	−0.013	−0.775	5837	5777	60	2971	2935	36
0.409	−0.016	−0.912	5906	5858	48	3159	3126	33
0.159	−0.016	−0.987	5897	5865	33	3362	3314	48
−0.102	−0.012	−0.995	5881	5863	18	3367	3337	31
−0.355	−0.004	−0.935	5851	5864	−12	3188	3176	12
−0.584	0.006	−0.812	5764	5806	−42	2964	2969	−5
−0.773	0.016	−0.634	5594	5640	−46	2861	2900	−39
−0.91	0.024	−0.414	5318	5371	−53	2998	3048	−50
−0.986	0.028	−0.167	5057	5103	−47	3246	3286	−40
−0.995	0.027	0.092	5027	5057	−31	3299	3331	−31
−0.938	0.021	0.345	5253	5286	−33	3066	3118	−52
−0.942	0.001	0.336	5241	5275	−34	3089	3126	−37
−0.997	0.000	0.083	5011	5051	−40	3312	3335	−23
−0.985	0.003	−0.176	5063	5109	−46	3248	3279	−31
−0.906	0.008	−0.423	5334	5381	−47	2991	3038	−48
−0.767	0.015	−0.641	5596	5647	−51	2878	2898	−20
−0.576	0.022	−0.817	5766	5809	−43	2972	2975	−3
−0.346	0.027	−0.938	5840	5863	−24	3212	3183	28
−0.092	0.027	−0.995	5859	5862	−3	3361	3339	22
0.169	0.023	−0.985	5867	5864	4	3312	3309	4
0.418	0.014	−0.908	5873	5857	16	3133	3118	15
0.64	0.003	−0.769	5811	5773	38	2944	2930	14

Table 1. *Cont.*

Phonon Direction Cosines			V_P (m/s)			V_S (m/s)		
n_a	n_b	n_c	Obs.	Model	Residual	Obs.	Model	Residual
0.817	−0.008	−0.576	5637	5571	65	2943	2920	24
0.939	−0.018	−0.343	5348	5283	64	3175	3120	55
0.004	−1.000	0.003	6446	6497	−51	2126	2136	−10
0.007	−0.965	0.261	6308	6349	−42	2425	2434	−9
0.012	−0.865	0.501	5920	5965	−46	3003	3037	−34
0.018	−0.707	0.708	5521	5570	−49	3422	3438	−16
0.022	−0.500	0.866	5524	5555	−30	3150	3163	−14
0.024	−0.259	0.966	5757	5764	−6	2465	2497	−33
0.022	0.000	1.000	5880	5862	18	2146	2136	10
0.017	0.259	0.966	5772	5763	9	2540	2498	42
0.008	0.500	0.866	5545	5554	−9	3218	3169	49
−0.003	0.708	0.706	5612	5572	41	3451	3437	14
−0.013	0.867	0.499	6020	5971	49	2999	3030	−31
−0.021	0.966	0.256	6400	6354	46	2429	2425	4
−0.025	1.000	−0.003	6563	6497	67	2164	2137	27
−0.008	1.000	0.006	6562	6497	64	2166	2136	29
−0.007	0.964	0.265	6393	6345	49	2471	2443	28
−0.009	0.862	0.506	6004	5954	49	3053	3050	2
−0.012	0.701	0.713	5596	5563	33	3462	3441	21
−0.014	0.492	0.870	5582	5560	22	3163	3148	15
−0.013	0.250	0.968	5774	5770	4	2469	2475	−6
−0.01	−0.010	1.000	5872	5862	10	2140	2137	3
−0.003	−0.268	0.964	5732	5757	−25	2541	2520	20
0.005	−0.508	0.862	5509	5548	−39	3193	3188	5
0.015	−0.713	0.701	5527	5579	−52	3412	3433	−22
0.023	−0.870	0.493	5954	5981	−27	2956	3017	−61
0.028	−0.967	0.252	6309	6359	−50	2372	2416	−44
0.029	-1.000	−0.006	6433	6497	−64	2126	2138	−11
0.029	−0.999	0.034	6524	6494	30	2713	2703	10
−0.228	−0.973	0.041	6470	6445	25	2667	2658	9
−0.469	−0.882	0.047	6306	6276	30	2543	2541	2
−0.68	−0.732	0.049	6028	5992	36	2418	2396	23
−0.845	−0.533	0.048	5651	5623	29	2427	2409	18
−0.954	−0.298	0.045	5288	5250	38	2575	2566	9
−0.998	−0.042	0.040	5055	5042	13	2694	2699	−5
−0.975	0.218	0.033	5179	5155	24	2617	2624	−7
−0.886	0.464	0.025	5534	5502	32	2449	2449	−1
−0.735	0.678	0.015	5871	5891	−20	2387	2387	0
−0.533	0.846	0.005	6161	6211	−50	2467	2481	−14
−0.294	0.956	−0.005	6400	6417	−17	2618	2626	−8
−0.033	0.999	−0.014	6488	6495	−8	2701	2702	−1
−0.043	0.998	−0.036	6471	6492	−21	2692	2703	−11
−0.303	0.953	−0.025	6364	6410	−46	2600	2622	−22
−0.541	0.841	−0.013	6142	6200	−58	2448	2474	−26
−0.742	0.671	−0.001	5860	5878	−18	2389	2387	2
−0.89	0.455	0.011	5493	5488	6	2465	2455	9
−0.978	0.209	0.021	5157	5145	12	2633	2630	3
−0.998	−0.051	0.029	5066	5042	24	2690	2698	−8
−0.951	−0.306	0.034	5309	5261	48	2549	2561	−12
−0.84	−0.541	0.036	5691	5637	54	2419	2407	12
−0.674	−0.738	0.035	5971	6005	−34	2401	2401	0
−0.462	−0.886	0.031	6287	6286	1	2541	2534	7
−0.22	−0.975	0.024	6466	6451	14	2663	2660	3
0.037	−0.999	0.016	6512	6495	17	2708	2702	5

Table 2. Direction of maximum and minimum V_P and V_S.

	Vel (km/s)	Phonon Direction Cosines		
Max V_P	6.497	0	1	0
Min V_P	5.033	1	0	0
Max V_S	3.446	0	−0.682	0.731
Min V_S	2.136	0	1	0

Using the inversion method of Brown [15], we obtained virtually identical results for the C_{ij} and their 1σ uncertainties. Table 3 shows the resulting C_{ij} values for hemimorphite. We note the large differences among the moduli C_{11}, C_{22}, and C_{33}, with C_{11} 40% smaller than $C_{22,}$ are a result of the large acoustic anisotropy of hemimorphite.

Table 3. Single-crystal elastic moduli of hemimorphite (GPa).

C_{11}	C_{22}	C_{33}	C_{44}	C_{55}	C_{66}	C_{12}	C_{13}	C_{23}
88.2(4)	147.0(4)	119.7(4)	15.9(1)	39.1(2)	25.5(2)	73.5(4)	43.9(4)	49.6(5)

From the single-crystal moduli, the Voigt, Reuss, and Hashin–Shtrikman bounds on the adiabatic bulk modulus K_S, and shear modulus μ of an isotropic aggregate were calculated, along with the Hill (Voigt–Reuss–Hill (VRH)) averages (Table 4). The Voigt and Reuss bounds on the bulk modulus and shear modulus differ by factors of 7% and 17%, respectively. The differences are due to the moderately strong anisotropy for both longitudinal waves and shear waves. We note that the Hill average is within the range of the Hashin–Shtrikman bounds.

Table 4. Isotropic aggregate elastic properties of hemimorphite.

Aggregate Properties	Voigt	Reuss	Hill (Voigt–Reuss–Hill (VRH)) Averages	Hashin-Shtrikman Bounds
Bulk modulus, K_S, GPa	76.5(2)	71.3(3)	74(3)	73.6(1)–74.8(1)
Shear modulus, μ, GPa	28.63(6)	24.42(6)	27(2)	26.05 (6)–27.04(6)
V_P, km/s	5.74(1)	5.46(1)	5.6(1)	5.58(1)–5.64(1)
V_S, km/s	2.87(1)	2.65(1)	2.8(1)	2.74(1)–2.79(1)
Density, g/cm^3	3.48(3)			

From the values of Ks (VRH) and μ (VRH), the sound velocities appropriate to an isotropic polycrystalline aggregate were calculated to be $V_{P,aggr} = \sqrt{\left(\frac{K+\frac{4}{3}\mu}{\rho}\right)} = 5.6(1)$ km/s; $V_{S,\,aggr} = \sqrt{\left(\frac{\mu}{\rho}\right)} = 2.8(1)$ km/s (Table 4). The longitudinal and shear velocity anisotropy are defined as (V_P,maximum − V_P,minimum)/V_P,(VRH) = 26% and (V_S,maximum − V_S,minimum)/V_S,(VRH) = 47%), respectively.

Seryotkin et al. [16] studied the structural changes of hemimorphite at pressures up to 4.2 GPa by X-ray diffraction and the diamond-anvil cell. The purpose of their study was to search for a high-pressure phase transition, which they found near 2.5 GPa. We fitted their data on hemimorphite to a 2nd-order Birch Murnaghan equation of state and obtain $K_T = 70(6)$ GPa (K_T' assumed to be 4), in agreement with our result of Ks = 74(3). Their experiments also show that the a-axis of hemimorphite is most compressible, while the b-axis is most rigid, in agreement with our result $C_{22} > C_{33} > C_{11}$. For the purpose of comparing our results and static compression results, we have ignored the difference between the adiabatic and isothermal bulk modulus, which is typically ~1%.

Interestingly, we note that our value of Ks for hemimorphite is in excellent agreement with the value of $K_T = 72(2)$ GPa for bertrandite, $Be_4Si_2O_7(OH)_2$, measured by Hazen and Au [17]. Bertrandite is topologically identical to hemimorphite, but differs chemically and does not contain molecular water in the large open cavities of the structure. The fact that bertrandite and hemimorphite have the same bulk modulus, despite having significant chemical differences, likely indicates a dominant influence of crystal-structural topology on the physical properties of these materials. However, it is also possible that the substitution of Zn for Be, and the presence of H_2O in hemimorphite, have opposite but equal effects on the bulk modulus that nearly cancel. There is some evidence to support this possibility. The Zn–O bond is longer than the Be–O bond and the bulk modulus of the BeO_4 tetrahedron is significantly larger than that of the ZnO_4 tetrahedron [18]. On the other hand, Cooper and Gibbs [11] found that upon dehydrating H_2O from the structure, the large cavities contract. This may suggest that H_2O provides support for the structure and stiffens it. It would be interesting to test this possibility on dehydrated hemimorphite, which would shed light on the effect of molecular water on this structure.

4. Discussion

The compressibility of minerals depends on the compressibility of the constituent cation polyhedra and their linkages to each other via angle-bending forces [17,18]. In their high-pressure X-ray structural study of bertrandite, $Be_4Si_2O_7(OH)_2$, which is structurally identical to hemimorphite Hazen and Au [17], concluded that polyhedral rotation and angle bending were the main compression mechanisms, as opposed to compression of tetrahedra. This is supported by the fact that the bulk modulus of bertrandite, 70(3) GPa, is much smaller than the polyhedral bulk moduli of the BeO_4 and SiO_4 tetrahedra (~200 GPa for both [17]). As noted above, our result for the bulk modulus of hemimorphite is identical to K_T measured by Hazen and Au [17], providing additional support for this interpretation.

The C_{ij} tensor of hemimorphite shows that strong anisotropy is a distinctive property of this mineral. We believe that this high degree of anisotropy is largely due to the topology of the hemimorphite crystal structure. C_{11} and C_{33} are the compressional moduli acting parallel to the corrugated sheets of tetrahedra. C_{11} is a measure of the stiffness in the a direction, normal to the corrugations in the tetrahedral sheets (Figure 3). Strain in this direction can be accommodated by a high degree of angle bending of the corrugations in an accordion-like fashion, and with little or no strain of the tetrahedra themselves. Thus, the value of C_{11} is the lowest of the three compressional moduli.

The strongest elements in the hemimorphite crystal structure are the Si_2O_7 groups that bond across the apical oxygens in adjacent (010) layers. The structure cannot be compressed along [010] without some strain being accommodated by the Si tetrahedra, which are the strongest polyhedra [18]. The Si_2O_7 groups are essentially rigid pylons that stiffen the structure in the [010] direction. Thus, we attribute C_{22} being the largest longitudinal modulus of hemimorphite as being largely due to the Si_2O_7 groups. This situation is reminiscent of the olivine structure, in which SiO_4 tetrahedra are contained in columns along the a direction, yielding C_{11} as the largest longitudinal modulus [19].

C_{55} is the largest of the on-diagonal pure-shear moduli. The C_{55} modulus corresponds to shear within the (010) plane, parallel to the tetrahedral sheets. These corrugated sheets contain only 3-membered rings of Zn and Si tetrahedra. There is virtually no rotational freedom or angle-bending for 3-membered rings in response to a shear stress within their tetrahedral basal planes. Strain within the tetrahedral sheets must be accommodated by either shearing of the tetrahedra themselves, or perhaps the tilting of tetrahedra out of the basal plane. This gives the structure high shear rigidity within the plane of a tetrahedral sheet. In support of this interpretation, we look to mica, which has tetrahedral sheets within the a-b plane, parallel to (001). The shear modulus corresponding to this tetrahedral sheet is C_{66} = 72 GPa [20]. In comparison, the other shear moduli are far smaller, with C_{44} = 16.5 GPa and C_{55} = 19.5 GPa. In the cases of both hemimorphite and muscovite, the polyhedral topology, in particular tetrahedral sheets, dictates which of the pure shear moduli is largest. We suggest that the tetrahedral sheets in hemimorphite are less rigid than those in muscovite, for two reasons. Firstly, in hemimorphite, only 1/3 of the tetrahedra are SiO_4, the strongest tetrahedral unit [18], whereas in muscovite, all the tetrahedra are SiO_4 in composition. Secondly, in muscovite the sheets are planar, whereas in hemimorphite the sheets are corrugated and exhibit some tilting of tetrahedra out of the (010) plane. Both the structural and chemical heterogeneity of hemimorphite make the rigidity of the tetrahedral sheets less than those in muscovite.

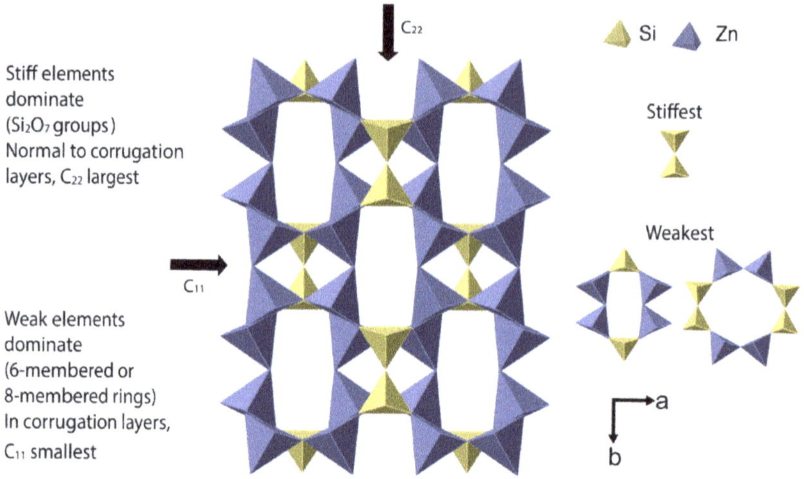

Figure 3. Schematic illustration of a polyhedral model of mineral elasticity. Hydrogens in molecular water and hydroxyls are omitted to emphasize linkages among Si and Zn tetrahedra into corrugated sheets of 3-membered rings parallel to (010), and the bridging between sheets. Stiffness parallel to [010], normal to the sheets, depends on the strongest structural element in this direction, which are Si_2O_7 and $Zn_2O_6(OH)_2$ groups that bridge the sheets along b. Within the plane of the sheets, the moduli depend on the weakest structural elements [19]. C_{11}, measuring stiffness in the a direction, depends, in part, on the stiffness of relatively weak 6-membered and 8-membered rings, which can accommodate strain along [100].

5. Conclusions

The nine single-crystal elastic moduli of hemimorphite have been measured at ambient conditions. The relative magnitudes of the longitudinal elastic moduli, $C_{22} > C_{33} > C_{11}$, can be qualitatively explained by the topology of the hemimorphite crystal structure. Strong Si_2O_7 groups greatly stiffen the structure along [010]. For the pure shear moduli, C_{55} is the largest, due to the lack of rotational freedom of Zn and Si within three-membered rings in sheets parallel to (010). The elastic character of hemimorphite is largely determined by the topology of the crystal structure.

Author Contributions: Conceptualization, J.D.B.; methodology, J.D.B. and Y.L.; software, Y.L.; formal analysis, J.D.B. and Y.L.; investigation, formal analysis, J.D.B. and Y.L.; resources, J.D.B.; data curation, Y.L. and J.D.B.; writing—original draft preparation Y.L.; writing—review and editing, J.D.B.; visualization, Y.L.; supervision, J.D.B.; project administration, J.D.B.; funding acquisition, J.D.B. All authors have read and agreed to the published version of the manuscript.

Funding: This research was supported by the National Science Foundation, under grant EAR 16-20616.

Acknowledgments: The authors thank Jin S. Zhang for helpful suggestions and Paul Ginsberg for help with data analysis. We thank two anonymous reviewers, whose comments and suggestions improved the manuscript.

Conflicts of Interest: The authors declare no conflict of interest.

References

1. Wang, J.; Sinogeikin, S.V.; Inoue, T.; Bass, J.D. Elastic properties of hydrous ringwoodite. *Am. Mineral.* **2003**, *88*, 1608–1611. [CrossRef]
2. O'Neill, B.; Bass, J.D.; Rossman, G.R. Elastic properties of hydrogrossular garnet and implications for water in the upper mantle. *J. Geophys. Res. Solid Earth* **1993**, *98*, 20031–20037. [CrossRef]
3. Sinogeikin, S.V.; Bass, J.D. Single-crystal elastic properties of chondrodite: Implications for water in the upper mantle. *Phys. Chem. Miner.* **1999**, *26*, 297–303. [CrossRef]

4. Hitzman, M.W.; Reynolds, N.A.; Sangster, D.F.; Allen, C.R.; Carman, C.E. Classification, genesis, and exploration guides for non-sulfide zinc deposits. *Econ. Geol. Bull. Soc.* **2003**, *98*, 685–714. [CrossRef]
5. Bissengaliyeva, M.R.; Bekturganov, N.S.; Gogol, D.B. Thermodynamic characteristics of a natural zinc silicate hemimorphite. *J. Therm. Anal. Calorim.* **2010**, *101*, 49–58. [CrossRef]
6. Kolesov, B. Raman investigation of H_2O molecule and hydroxyl groups in the channels of hemimorphite. *Am. Mineral.* **2006**, *91*, 1355–1362. [CrossRef]
7. Taylor, H.F.W. The dehydration of hemimorphite. *Am. Mineral.* **1962**, *47*, 932–944.
8. Hill, R.J.; Gibbs, G.V.; Craig, J.R.; Ross, F.K.; Williams, J.M. A neutron-diffraction study of hemimorphite. *Z. Kristallogr.* **1977**, *146*, 241–259. [CrossRef]
9. McDonald, W.S.; Cruickshank, D.W.J. Refinement of the structure of hemimorphite. *Z. Kristallogr.* **1967**, *124*, 180–191. [CrossRef]
10. Takeuchi, Y.; Sasaki, S.; Joswig, W.; Fuess, H. X-ray and neutron diffraction study of hemimorphite. *Proc. Jpn. Acad.* **1978**, *54*, 577–582. [CrossRef]
11. Cooper, B.J.; Gibbs, G.V. The effect of heating and dehydration on the crystal structure of hemimorphite up to 600 °C. *Z. Kristallogr.* **1981**, *146*, 241–259.
12. Whitfield, C.H.; Brody, E.M.; Bassett, W.A. Elastic moduli of NaCl by Brillouin scattering at high pressure in a diamond anvil cell. *Rev. Sci. Instrum.* **1976**, *47*, 942–947. [CrossRef]
13. Loudon, R.; Sandercock, J.R. Analysis of the light-scattering cross section for surface ripples on solids. *J. Phys. C Solid State Phys.* **1980**, *13*, 2609–2622. [CrossRef]
14. Weidner, D.J.; Carleton, H.R. Elasticity of coesite. *J. Geophys. Res.* **1977**, *82*, 1334–1346. [CrossRef]
15. Brown, J.M. Determination of elastic moduli from measured acoustic velocities. *Ultrasonics* **2018**, *90*, 23–31. [CrossRef] [PubMed]
16. Seryotkin, Y.V.; Bakakin, V.V. Structural evolution of hemimorphite at high pressure up to 4.2 GPa. *Phys. Chem. Miner.* **2011**, *38*, 679–684. [CrossRef]
17. Hazen, R.M.; Au, A.Y. High-pressure crystal-chemistry of phenakite (Be_2SiO_4) and bertrandite ($Be_4Si_2O_7(OH)_2$). *Phys. Chem. Miner.* **1986**, *13*, 69–78. [CrossRef]
18. Hazen, R.M.; Finger, L.W. Bulk modulus-volume relationship for cation-anion polyhedra. *J. Geophys. Res. Solid Earth* **1979**, *84*, 6723–6728. [CrossRef]
19. Bass, J.D.; Weidner, D.J.; Hamaya, N.; Ozima, M.; Akimoto, S. Elasticity of the olivine and spinel polymorphs of Ni_2SiO_4. *Phys. Chem. Miner.* **1984**, *10*, 261–272. [CrossRef]
20. Vaughan, M.T.; Guggenheim, S. Elasticity of muscovite and its relationship to crystal structure. *J. Geophys. Res. Solid Earth* **1986**, *91*, 4657–4664. [CrossRef]

minerals

MDPI

Article

Effects of Composition, Pressure, and Temperature on the Elastic Properties of SiO_2–TiO_2 Glasses: An Integrated Ultrasonic and Brillouin Study

Murli H. Manghnani [1,*], Quentin Williams [2], Teruyuki Matsui [3], Peter C. Schultz [4] and Charles R. Kurkjian [5]

[1] Hawai'i Institute of Geophysics and Planetology, University of Hawai'i, Honolulu, HI 96822, USA
[2] Department of Earth and Planetary Sciences, University of California, Santa Cruz, CA 95064, USA; qwilliam@ucsc.edu
[3] Knowledge Outsourcing Co., Inc., 46-001 Nagoya, Japan; lala-matsui@cocoa.plala.or.jp
[4] Peter Schultz Consulting LLC, 1 Evarts Lane, Madison, CT 06443, USA; pcschultz1@gmail.com
[5] Department of Materials Science and Engineering, Rutgers University, Piscataway, NJ 08855, USA; ckurkjia@scarletmail.rutgers.edu
* Correspondence: murli@soest.hawaii.edu

Received: 2 April 2020; Accepted: 18 May 2020; Published: 25 May 2020

Abstract: We have systematically investigated the elastic properties (ρ, V_P, V_S, K, μ and σ) of eight SiO_2–TiO_2 glasses, varying in composition from 1.3 to 14.7 wt% TiO_2, as a function of pressure up to 0.5 GPa by the pulse superposition (PSP) ultrasonic technique, and two compositions (1.3 and 9.4 wt% TiO_2) up to ~5.7 GPa by Brillouin scattering in a diamond anvil cell. The parameters were also measured after annealing to 1020 °C. Composition–elasticity relationships, except for K and σ, are more or less linear; the annealing simply makes the relationships more uniform (less scatter). There is excellent agreement between the ultrasonic and Brillouin measurements at ambient and high pressure. The pressure-induced anomalous elastic behavior (negative dV_P/dP and dK/dP) becomes more negative (more compressible) with the increasing TiO_2 content. Correspondingly, the acoustic Grüneisen parameters become more negative with increases in the TiO_2 content, reaching a minimum near ~8–10 wt% TiO_2. The comparison of the low- and high-pressure ultrasonic and Brillouin V_P and V_S in two glasses (1.3 and 9.4 wt% TiO_2) shows excellent agreement, defining the reversible elastic behavior at low pressures and irreversible behavior at higher pressures (≥5.7 GPa) well. This result is consistent with our previous high-pressure Raman study showing an irreversible structural change in a similar pressure range.

Keywords: SiO_2–TiO_2 glasses; elastic properties; pressure-temperature dependences; ultrasonic method; Brillouin scattering; equation of state; anomalous compression behavior

1. Introduction

Interest in the SiO_2–TiO_2 glass system has continued to develop in view of its low thermal expansion properties and the ability to tune the coefficient of thermal expansion over wide ranges of temperature by varying compositions (TiO_2 content) and/or annealing. SiO_2–TiO_2 glasses containing up to ~11 wt% TiO_2 have been synthesized as apparently single homogeneous phases [1–3]. At relatively modest TiO_2 concentrations (>~3 wt%), TiO_2 plays the role of a network former just as SiO_2 does; thus, like Si, Ti is tetrahedrally coordinated with oxygen. A SiO_2–TiO_2 glass (ULE®, Corning Code 7971) containing about 7.5 wt% TiO_2 has been demonstrated to have a nearly zero thermal expansion [4,5] in the temperature range of 0 to 300 °C. It has been shown [6–8] that moderate amounts of TiO_2 (up to 10 wt%) added to fused silica cause a systematic and linear decrease in the coefficient of thermal expansion of silica glass. Such composition-dependent variations in the thermal expansion of SiO_2–TiO_2 glasses have been plausibly correlated with changes in the structure and lattice vibrations in the glass network [6,9]. Schultz [7,8] successfully synthesized SiO_2–TiO_2 glasses containing higher TiO_2 contents (up to 16.5 wt%) that were clear, and demonstrated unique annealing methods for developing the low-expansion properties of such glasses, containing even up to 20 wt% TiO_2, across wider ranges of temperature. The synthesized glasses with relatively high TiO_2 contents (16.5–20.0 wt%) were not, however, fully transparent due to phase separation and the presence of crystalline phases of TiO_2 (rutile/anatase).

A number of investigators [10–24] have studied the effects of TiO_2 content on the various physical, thermal, elastic and optical properties of SiO_2–TiO_2 and other related glasses, such as those in the system Na_2O–TiO_2–SiO_2. Some of these studies [10,11] have thrown light on the roles of the structure and coordination of Ti and Si ions in the variations of glass properties. The infrared reflection and Raman scattering studies [23,25,26] on SiO_2–TiO_2 glasses have enabled an elucidation of the structural variations caused by the addition of TiO_2 to SiO_2.

The purpose of this paper is to report more extensive information on the elastic properties of SiO_2–TiO_2 glasses as a function of composition, annealing, pressure and temperature, and to correlate the results, in light of a structural model, with the variations in thermal expansion and other related anharmonic parameters. The pressure dependences of the ultrasonic elastic moduli of the glasses reported here have been described recently [24], but primarily in the context of possible coordination changes in TiO_2 at low titania concentrations. Here, we probe the behavior of the system on annealing, and particularly focus on the acoustic Grüneisen parameters and the minimal anelastic behavior in this system at frequencies exceeding the MHz range via a comparison of new Brillouin data with the ultrasonic measurements.

2. Materials and Methods

The eight SiO_2–TiO_2 glasses used in this study were prepared by the flame hydrolysis boule process [6–8] at the Corning Glass Works laboratory. The TiO_2 content of the glasses ranged from 1.3 to 14.7 wt% (Table 1). Further, a fused silica sample obtained from the Corning Glass Works (Corning Code 7940) is included in this study. The preparation of these glasses has been described in detail previously [6,8]. Briefly, compositions were determined by X-ray fluorescence, with net uncertainties in the compositions of ±0.03 wt% [8]. The abundance of Ti^{3+} in these glasses was minimal; prior to annealing, the abundance was estimated as ~10 ppm [8], and after annealing, the glasses became entirely water-white, implying that the amount of Ti^{3+} was negligible [8].

Table 1. Chemical composition and ultrasonic elastic parameters of seven SiO_2–TiO_2 glasses at an ambient pressure and temperature. The measurements of these glasses after annealing are listed.

Glass Number	Composition in wt%		ρ (g/cm³)	Vp (km/s)	Vs (km/s)	K (GPa)	μ(GPa)	E (GPa)	σ Poisson's Ratio
	SiO_2	TiO_2							
7940A (fused silica)	100	0	2.2007	5.947	3.769	36.16	31.26	728.0	0.165
T4A	98.7	1.3	2.2005	5.911	3.746	35.71	30.88	79.11	0.164
T1A	97.2	2.8	2.2001	5.872	3.717	35.32	30.40	70.88	0.166
T5A	95.4	4.6	2.1995	5.833	3.692	34.87	29.98	69.70	0.166
T6A	94.0	6.0	2.1992	5.803	3.672	34.53	29.65	69.16	0.166
T2A	92.7	7.3	2.1986	5.758	3.635	34.15	29.0.5	67.90	0.169
ULE 7971A	92.5	7.5	2.1993	5.783	3.653	34.42	29.35	62.56	0.168
T3A	90.6	9.4	2.1985	5.719	3.607	33.77	28.60	66.91	0.170

2.1. Density and Ultrasonic Measurements

The density of the glass specimens was measured by the Archimedes method, using distilled water. The density and velocity measurements were made on the glasses both in an "as received, unannealed state," as well as after annealing. The annealing process involved heating the glasses to 1020 °C for about 1–1/2 h, and then cooling them slowly from 1020 to 700 °C at a rate of 5 °C/h, and then allowing them to cool inside the furnace from 700 °C to room temperature.

The ultrasonic velocities were measured by the pulse super-position method at pressures up to 0.5 GPa [19,24]. Samples of the acoustic path length of about 1 cm were employed. The two faces of the sample were lapped flat within about 1 μm and parallel to within about 30 s of the arc. X- and Y-cut 20 MHz quartz transducers, 0.63 cm in diameter, were used for the compressional and shear wave velocity measurements, respectively. The details of the electronics and high-pressure and temperature equipment used, and the procedure to reduce the basic pulse repetition frequency data to obtain the elastic moduli are described elsewhere [19,24]. Briefly, Cook's [27] method (Equation (1)), which solves for the high-pressure density (ρ) at the pressure (P) from the adiabatic bulk modulus (K_S) corrected to isothermal conditions (using Δ, which is equal to $\alpha\gamma T$, the product of the thermal expansion, Grüneisen parameter and temperature), was used to iteratively solve for the density under pressure at progressive intervals of 0.0275 GPa.

$$\ln(\rho/\rho_0) = (1+) \int_{P_0}^{P} \mathrm{d}P/K_S \quad (1)$$

The precision of the frequency (f) measurements is ~ 1 part in 10^5. The phase angle correction, $\gamma/360\,f$, due to the bond between the transducer and the specimen, is negligible (γ does not exceed 2° for a well-prepared bond) and was ignored. The errors in the ultrasonic velocity values reported here are between 0.05% and 0.14%; incorporating the errors in density, our uncertainties on the derived moduli are less than 0.24%. The pressure derivatives of the moduli were calculated from a combination of fits to the velocities and the iteratively-derived densities from Equation (1) [24].

2.2. Brillouin Scattering Measurements

The Brillouin scattering technique was used by deploying a 5-pass Fabry–Pérot interferometer (Sandercock design) coupled with an Ar^+-ion laser (at 100 mW power) for the excitation of the approximately 120 μm diameter sample. Measurements of the acoustic velocities were conducted at an ambient pressure and up to ~7.5 GPa in a diamond anvil cell. The apparatus and methodology are described elsewhere [28]: briefly, the modified platelet geometry approach was deployed with a ~20° external scattering angle, with the angle chosen to optimize the signal. A collection time of 45 min was used for each Brillouin data point. Diamond anvils with 0.8 mm size culets and a T301 stainless steel gasket were used, with a 4:1 methanol/ethanol mixture as the pressure medium. The ruby fluorescence technique was used for the in situ pressure measurements. Due to the relative sharpness and amplitude of the peaks, the compressional velocities were more accurately determined than the shear velocities in these experiments, with estimated accuracies of ±0.4–0.7% for the compressional velocities and ±0.8–1.5% for the shear velocities. The shear velocity error was somewhat larger than the compressional velocity because the lower amplitude of the shear peak generated a smaller signal-to-noise ratio relative to the strong and sharp compressional peak.

3. Results and Discussion

3.1. Elastic Parameters at 1 bar and 25 °C

(A) Effect of composition and annealing. The composition and measured elastic parameters of SiO_2–TiO_2 glasses, including fused silica (Corning Glass Works Code 7940), are listed in Table 1. The data for fused silica and the ULE® SiO_2–TiO_2 glass (Corning Code 7971) containing 7.5 wt% TiO_2 are in good agreement with the previous measurements of McSkimin and Andreatch [29] and the published Corning Glass Works specification [3], but in less good agreement with those of Gerlich et al. [30] The density and bulk moduli values reported by Gerlich et al. for these glasses are appreciably lower.

For end-member fused silica, our results are in good accord with previous ultrasonic and Brillouin results at 300 K [30–32]. Notably, there are variations in the reported values of the velocity in fused silica of up to ~50 m/sec in compressional velocity [33], but velocities are known to vary depending on the thermal history/fictive temperature of the glass [34].

The compositional dependence of the compressional (Vp) and shear (Vs) velocities, density (ρ), Poisson's ratio (σ), and bulk, shear and Young's moduli (K, μ and E, respectively) for these glasses are shown in Figures 1–3. The relations between the TiO_2 content and the elastic parameters, particularly K and σ, are not as uniform and systematic for the unannealed glasses as for the annealed glasses.

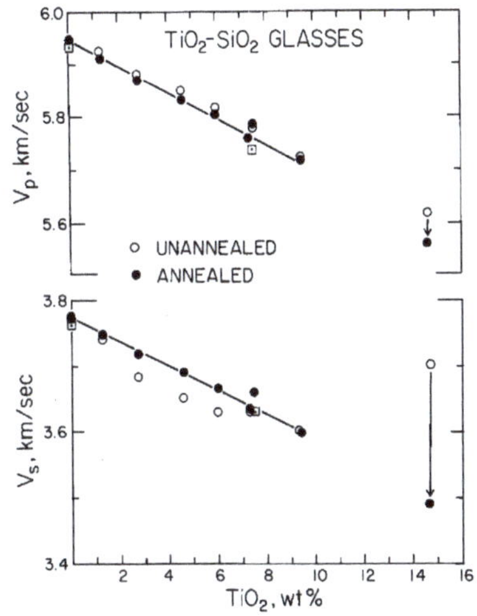

Figure 1. Longitudinal (V_p) and shear wave (V_s) velocities versus TiO_2 content for unannealed (as received) and annealed glasses. For the compositions up to 9.4 wt% TiO_2, annealing causes an increase in V_s and a slight decrease in V_p. Squares are McSkimin's (unpublished) 1972 data for the CorningULE® 7971 glass. For the 14.7 wt% glass, which is partly crystalline, both velocities drastically decrease on annealing (see text for discussion).

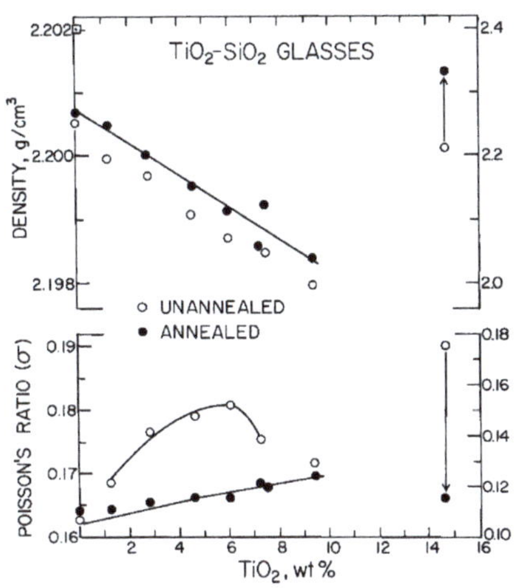

Figure 2. Density (ρ) and Poisson's ratio (σ) versus TiO_2 content. As shown, annealing causes an increase in ρ and decrease in σ for the SiO_2–TiO_2 glasses studied.

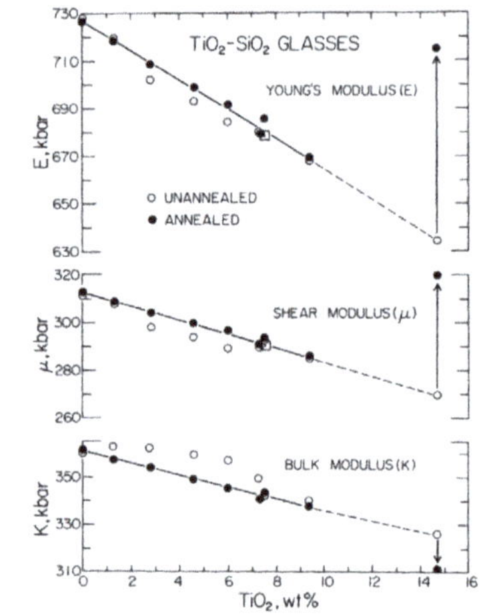

Figure 3. Bulk (*K*), shear (*μ*) and Young's (*E*) moduli versus TiO_2 content. Annealing causes an increase in *K*, and a decrease in *μ* and *E*. Squares are McSkimin's (unpublished) 1972 data. Error bars are smaller than the symbols.

As a result of annealing, the relations between the TiO_2 content and the elastic parameters become uniform and more or less linear for glasses containing up to 9.4% TiO_2 (Figures 1–3). The interesting aspect here is that the effect of annealing, which plausibly equilibrated the glasses closer to their glass transition temperature, T_g, has the effect of increasing the density of the glasses while lowering their compressional wave velocities, while also eliminating the unusually complex behavior of Poisson's ratio, σ, observed in the unannealed samples (Figure 2). Indeed, while the bulk moduli are greater in the unannealed samples, the shear moduli are substantially less (Figure 3), producing the anomalous behavior in the Poisson's ratio of these samples. The origin of the higher bulk moduli of the unannealed samples is not simple to explain: it is possible that there is a complex interplay between the internal strains and compressibility. Moreover, the lower density/higher temperature of the equilibration of the unannealed samples generates a substantial weakening of the shear modulus, implying that the lower density of the unannealed glasses has a particularly strong effect on the shear modulus. An explanation for these observations is that annealing causes the removal of strains and the rearrangement of the glass network such that the structure becomes more compact. This compaction effect of annealing, in particular, is dramatic in SiO_2–TiO_2 glass (T 7) containing 14.7 wt% TiO_2 and, in contrast to the lower Ti-content glasses, results in increases in ρ, μ and E, but decreases in *Vp*, *Vs*, σ and *K*. Optical evidence shows that the annealing of the 14.7 wt% glass generates the unmixing of the glass and crystallization of TiO_2, and hence the elastic results on this phase represent that of a two-phase aggregate. This unmixing is in full accord with the analysis that such high TiO_2 glasses are highly metastable and prone to unmixing to a crystalline TiO_2 phase and a coexisting glass at temperatures beneath the annealing point and as low as 750 °C [8]. The compaction effect in the T7 glasses may be related to an annealing-induced partial coordination change of Ti ions from four to six in the glass structure at high Ti contents. This reasoning is consistent with the infrared reflection [23] and Raman studies [23,25,26]. For the annealed glasses containing up to 9.4% TiO_2, the parameters, V_p, V_s, *K*, μ and *E* decrease linearly with the increase in the TiO_2 content; however, increases linearly with the TiO_2 content.

The decreases in the moduli are certainly caused by the weakening of the silica network. The Ti^{4+} ions have a lower field strength than the Si^{4+} ions, and the average length of the Ti–O bonds is larger than those of the Si–O bonds; hence, the Ti–O and Si–O–Ti bonds in SiO_2–TiO_2 glasses are weaker than the corresponding Si–O and Si–O–Si bonds in endmember silica glass. This view has been supported by infrared absorption studies [9,26].

The density of the glasses is particularly interesting in this respect: despite the larger mass of Ti, the effect of the increased Ti content is to weakly lower the density of these glasses.

Hence, the decrease in with the increasing TiO_2 content (Figure 2) seems, at first, surprising in view of the fact that the substituted Ti^{4+} ions are heavier than the Si^{4+} ions. In crystalline silica, Evans [1] clearly demonstrated that the additions of small amounts of TiO_2 to the SiO_2–TiO_2 solid solution (cristobalite phase) cause the tetragonal a_o and c_o *d*-spacings of this phase to increase. Such an effect is consistent with the observed trend in the density of the SiO_2–TiO_2 glasses. A decrease in density can thus be viewed as being due to the increasing openness of the structure as TiO_2 is added: while Ti may be present five-fold in coordination at low concentrations (below ~3 wt%), and four-fold at higher concentrations [24], it appears that the net effect of both larger Ti ions and potentially weaker polyhedral linkages for more highly coordinated species may each contribute to the unexpected decrease in density. The openness of the structure can also increase if the Si–O–Ti angles increase. This is exactly what is concluded from the optical studies [23,26].

3.2. Densification

The pressure density data for a solid can be used for calculating its bulk modulus, *Ko* and initial pressure derivative $K'o = (\partial K/\partial P)P = 0$ through the Birch–Murnaghan equation of state [35]:

$$P = (3/2\ Ko\ \{(\rho o/\rho)7/3 - (\rho o/\rho)5/3\}\ \{1 - \xi\ [(\rho o/\rho)2/3 - 1]\} \quad (2)$$

where $\xi = 3(4 - K'o)/4$. Correspondingly, if *Ko* and *K'o* are known, $(\rho/\rho o)$ can be evaluated as a function of pressure.

Using Equation (1) and the data for fused silica and the SiO_2–TiO_2 glasses containing 7.3 wt% TiO_2 (Table 2), $(\rho/\rho o)$ was calculated as 6 GPa (Figure 4) using our elasticity data. Here, we deployed our values of the bulk modulus and its pressure derivative to calculate the relative densities: due to the importance of the bulk modulus in determining the compression curve in this pressure range, higher order derivatives do not notably impact the density offset between the two phases. Values of the derivative parameters in Table 2 are derived from the initial (zero-pressure) slopes of the respective elastic parameters (Figures 2 and 3). As shown in Figure 4, the glass containing TiO_2 shows higher densification $(\rho/\rho o)$ under pressure, as anticipated from the lower bulk modulus of this material. At 2 GPa, the difference between the $(\rho/\rho o)$ ratios is expected to be ~7%. Thus, despite the almost identical initial densities of the two glasses (differing by ~0.1% at ambient pressures: Figure 2, top), the effect of pressure is to produce markedly higher densities for the Ti-bearing glass under pressure. The net result that a glass that contains more of a more massive cation is comparable in density to silica at an ambient pressure but becomes denser at a high pressure (as anticipated from its higher mean atomic number) illustrates that the initial structural role of Ti is to expand the overarching Si-dominated network in these glasses, but this Ti-induced expansion of the glass network is reduced under pressure.

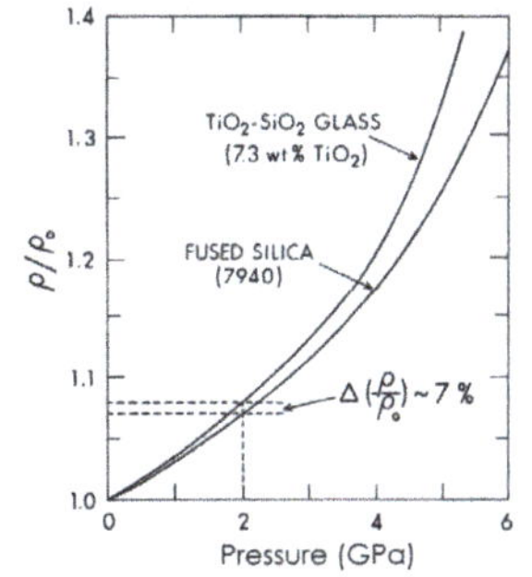

Figure 4. Comparison of the densification of fused silica and SiO_2–TiO_2 glass containing 7.3 wt% TiO_2 up to 6 GPa using the Birch–Murnaghan equation of state.

Table 2. Pressure derivatives of the elastic parameters (moduli in Mbar) of the annealed SiO_2–TiO_2 glasses at ambient conditions.

Glass Number	Composition in wt%		dK/dP	dρ/dP g/cm^3 Mbar^{-1}	dμ/dP	dE/dP	dσ/dP Mbar^{-1}
	SiO_2	TiO_2					
7940A (fused silica)	100	0	−5.37	6.68	−3.53	−8.82	−1.01
T4A	98.7	1.3	−5.68	6.78	−3.56	−9.00	−1.24
T1A	97.2	2.8	−5.42	6.80	−3.56	−8.87	−1.00
T5A	95.4	4.6	−5.97	7.01	−3.55	−9.13	−1.52
T6A	94.0	6.0	−5.85	7.06	−3.54	−9.05	−1.42
T2A	92.	7.3	−5.96	7.20	−3.60	−9.20	−1.47
ULE 7971A	92.5	7.5	−5.88	7.10	−3.60	−9.16	−1.36
T3A	90.6	9.4	−5.77	7.26	−3.59	−9.09	−1.30

3.3. Comparison between Ultrasonic and Brillouin Measurements

A representative Brillouin spectrum from these results, using an apparatus described elsewhere [29], is shown in Figure 5, and Table 3 compares the Brillouin scattering results on two of these glasses within the diamond anvil cell with the lower pressure ultrasonic data. Figure 6 shows the elastic results under pressure from the two techniques. Clearly, on this scale, the two sets of results are in outstanding agreement. The larger pressure range of the Brillouin results reveals the well-known velocity minima occurring under compression in silica-rich glasses, and shows that the minima both shift to slightly higher pressures and the amplitude of the depression increases with higher TiO_2 contents. The pressure at which such minima occur has generally been correlated with the degree of polymerization of the glass [36], and the possible implication here thus might be that progressively more Ti enrichment may induce a greater polymerization of glasses. We speculate that it is more likely that the role of increased Ti is rather to broaden the average T–O–T (tetrahedral cation-oxygen-tetrahedral cation) angles in the glass relative to pure silica, producing a broader pressure range over which these angles may undergo a relatively easy contraction (and softening) of the glass [37].

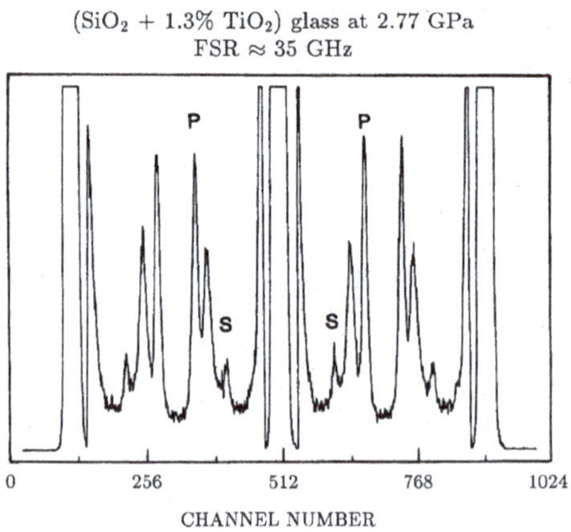

Figure 5. Representative Brillouin spectrum of an SiO_2–TiO_2 glass at 2.77 GPa; P and S represent compressional and shear peaks, respectively. Peaks generated by different orders of the grating are also present, as is a peak produced from the pressure medium between the P and S peaks. The higher signal-to-noise of the compressional peak relative to the shear peak produces a smaller error bar for the compressional velocity determination relative to the shear velocity.

Table 3. Comparison of the ultrasonic and Brillouin scattering measurements of V_p and V_s in SiO_2–TiO_2 glasses at ambient conditions.

Composition TiO_2, wt%	g/cm^3	Brillouin *Vp* (km/s)	Scattering *Vs* (km/s)	Ultrasonic *Vp* (km/s)	Ultrasonic *Vs* (km/s)	ΔVp; %	ΔVs; %
1.3	2.2005	5.933	3.743	5.911	3.746	0.4	−0.08
2.8	2.2001	5.890	3.732	5.872	3.717	0.3	0.4
4.6	2.1995	5.828	3.683	5.833	3.692	−0.09	−0.2
6.0	2.1992	5.800	3.656	5.803	3.672	−0.05	−0.4
7.3	2.1986	5.751	3.622	5.758	3.635	−0.1	−0.3
9.4	2.1985	5.746	3.616	5.719	3.607	0.5	0.2

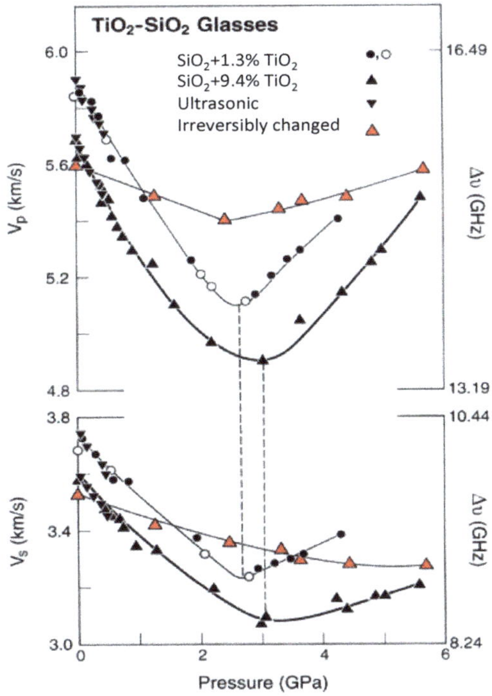

Figure 6. Pressure dependence of the compressional (V_P) and shear (V_S) velocities measured by the ultrasonic method up to 0.5 GPa (inverted filled triangles) and Brillouin scattering (triangles, circles) up to ~5.7 GPa for two SiO_2–TiO_2 glasses. Black filled symbols are compression data and the open are decompression data that have been compressed to pressures below that at which irreversible deformation occurs. Red triangles are decompression points for irreversibly densified glasses compressed to slightly above 5.5 GPa. Error bars for the Brillouin data are in the order of, or smaller than, the size of the symbols; for the ultrasonic data, they are smaller than the symbol sizes.

While Table 3 shows generally excellent agreement between the Brillouin and ultrasonic data sets, there is a small but systematic offset between the two sets of the velocity determinations. Table 3 shows the relative ambient pressure velocities determined within each glass and the offset between the optical and ultrasonic measurements. Here, the Brillouin measurements are observed to yield average compressional velocities higher by ~0.15% relative to the ultrasonic results. Although these velocity differences are small and close to the sums of the respective errors of the two measurements (and, with the less-well determined Brillouin shear velocities, the average velocities are within error), their average sign is in accord with the possible presence of dispersive effects. Notably, the two sets of measurements differ in the frequency of their probes by between a factor of ~450–750, or approaching three orders of magnitude. The average magnitude of difference for compressional velocities is roughly comparable to the ~0.2–0.3% difference across a factor of a ~7 larger frequency range that is observed in silica glass at 300 K [38], and our results thus indicate that weak dispersive effects might be present at ambient temperatures in Brillouin measurements in silica-rich glasses. The origin of these dispersive effects has been proposed to be via anharmonic interactions with localized vibrational states (termed "network viscosity" in [38]), although tunneling between nearly energetically equivalent local structural configurations could also play a significant role [38,39]. If the latter effect predominates, higher temperatures could produce a greater offset between the Brillouin and ultrasonic results in glassy materials, which could produce noticeable discrepancies between highly accurate experiments deploying these respective techniques.

Figure 6 also shows results on decompression from compression to near 6 GPa: this pressure has been previously identified from Raman results as the pressure at which irreversible structural changes within these glasses take place [37]. Such irreversibility is well-documented in silica glass, with the onset of irreversible densification occurring near 9 GPa [40,41]: our results indicate that the onset of irreversible densification occurs at lower pressures within titania-bearing glasses relative to the SiO_2 endmember. The rationale for this lower onset pressure of irreversible densification is almost certainly associated with the weaker Ti–O bonds (and Si–O–Ti linkages) within the glass network: irreversible changes in the ring statistics associated with compaction are likely to be generated more readily due to the presence of titanium in the framework. As an important aside, this irreversible densification will also augment the difference in densities between titania-bearing glasses and silica above 6 GPa (Figure 4).

3.4. Mode Grüneisen Parameters γHT and γLT

The mode Grüneisen parameters γi are evaluated from the pressure dependence of the acoustic mode velocities, Vi:

$$\gamma i = (K_T/Vi)\ (\mathrm{d}Vi/\mathrm{d}P) \quad (3)$$

where K_T is the isothermal bulk modulus. Assuming only contributions from the acoustic modes, the high- and low-temperature limiting values of the Grüneisen parameters, γHT and γLT, can be evaluated from ($\mathrm{d}Vs/\mathrm{d}P$) and ($\mathrm{d}Vp/\mathrm{d}P$) using well-established relations [35,42]. As in fused silica, the γHT values for all the SiO_2–TiO_2 glasses are negative (Figure 7).

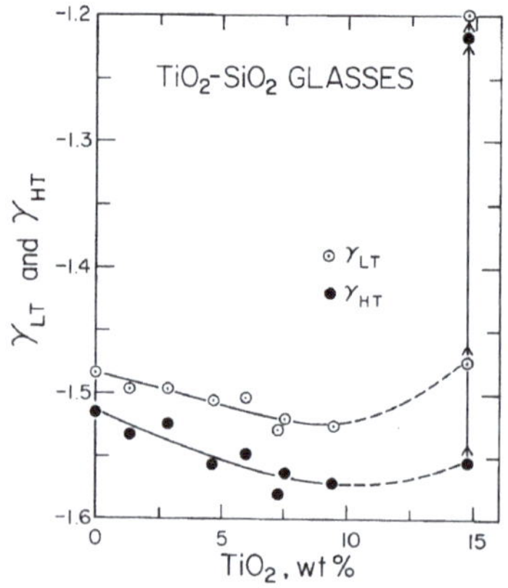

Figure 7. Variation of the low-temperature and high-temperature limits of the acoustic Grüneisen parameter as a function of the TiO_2 content.

Indeed, both γHT and γLT become more negative with the TiO_2 content. Here, the acoustic mode Grüneisen parameters reflect the low-frequency response of the glass, and demonstrate that the near-zero thermal expansion of these glasses reflects a balance between the strongly negative Grüneisen parameters associated with the acoustic modes and the generally positive Grüneisen parameters of the optical modes within these glasses [37]. Hence, the anomalously small thermal expansions of these glasses are generated by the competing and compensating effects of the low-frequency acoustic modes and high-frequency optic modes, and the decrease in thermal expansion with the addition of TiO_2 is likely related to the influence of TiO_2 on the acoustic lattice vibration frequencies.

4. Conclusions

The elastic properties and their pressure derivatives of the SiO_2–TiO_2 glasses studied systematically vary with composition. In general, Vp, Vs, ρ, E, K and μ decrease more or less linearly with the increases in the TiO_2 content. Further, both the acoustic γHT and γLT become more negative with the TiO_2 content. The effect of annealing is to increase ρ, decrease V_P and increase V_S. Nonetheless, the net effect of annealing is complex, and indicates that the annealing of glasses at temperatures closer to their glass transition may generate markedly more systematic elastic behavior relative to higher temperature quenching.

The excellent agreement between the ultrasonic and Brillouin measurements at ambient and high pressure enables a well-defined demonstration of the anomalous, as well as (elastically) the reversible and irreversible, pressure-induced compression behavior, as observed in V_P and V_S vs. the pressure plots for two selected glasses containing 1.3 and 9.4 wt% TiO_2 up to ~5.7 GPa, supporting our previous Raman study regarding the irreversible structural change near ≥5.7 GPa [37].

Further studies of the elastic, thermodynamic and structural properties of the SiO_2_TiO_2 glasses, involving acoustic and spectroscopic, simultaneously under in situ high-P-T environments, would enhance our understanding of their anomalous elastic and irreversible compressional behavior and structural changes under a high pressure and temperature.

Author Contributions: M.H.M. and T.M. conceived the project; P.C.S. and C.R.K. provided the samples and initial guidance in identifying the scientific issues in this glass system; T.M. carried out the ultrasonic measurements and participated in the Brillouin measurements. The plotting, analysis and preliminary interpretation of the data were done by T.M., M.H.M. and Q.W. The manuscript was prepared by joint efforts, fully shared by M.H.M., Q.W., and T.M. P.C.S. and C.K. provided useful comments on the manuscript. All authors have read and agreed to the published version of the manuscript.

Funding: This research project has been supported by an ONR grant, and the NSF-EAR grants 0757137 (to M.H.M.) and 1620423 (to Q.W.).

Acknowledgments: This paper is dedicated to the memory and in honor of the Late Orson L. Anderson, who made significantly important contributions in the physics of the non-crystalline solids and high-pressure mineral physics throughout his career and lifetime. One of us (M.H.M.) remains most grateful to the late Orson Anderson for introducing him in the field of ultrasonic techniques in his High-Pressure Mineral Physics Laboratory at the Columbia University. We are indebted to the late John Balogh for his invaluable support in maintaining the laboratory ultrasonic and high-P-T research facilities at the University of Hawaii, deployed for carrying out this project. We thank three reviewers for helpful comments on the paper.

Conflicts of Interest: The authors declare no conflict of interest in all respects.

References

1. Evans, D.L. Solid solution of TiO_2 in SiO_2. *J. Am. Ceram. Soc.* **1970**, *53*, 418–419. [CrossRef]
2. Nordberg, M.E. Glass Having an Expansion Lower than That of Silica. U.S. Patent 2,326,059, 3 August 1943.
3. Corning Glass Works. *Low Expansion Materials Bulletin*; Corning Glass Works: Corning, NY, USA, 1969.
4. Ricker, R.W.; Hummel, F.A. Reactions in the system TiO_2-SiO_2; Revision of the phase diagram. *J. Am. Ceram. Soc.* **1951**, *34*, 271–279. [CrossRef]
5. DeVries, R.C.; Roy, R.; Osborn, E.F. The system TiO_2-SiO_2. *Trans. Br. Ceram. Soc.* **1954**, *53*, 525–540.
6. Schultz, P.; Smyth, G.T. Ultra-low expansion glasses and their structure in the SiO_2-TiO_2 system. In *Amorphous Materials*; Douglas, R.W., Ellis, B., Eds.; Wiley-Interscience: New York, NY, USA, 1972; pp. 453–461.

7. Schultz, P.C. Method for Producing TiO_2-SiO_2 Glasses. U.S. Patent 3,690,855, 12 September 1972.
8. Schultz, P.C. Binary titania-silica glasses containing 10–20 wt% TiO_2. *J. Am. Ceram. Soc.* **1976**, *58*, 214–219. [CrossRef]
9. Arndt, J. Irreversible compression of glasses of the system TiO_2-SiO_2 by high static pressures. In Proceedings of the 4th International Conference on High Pressure, Kyoto, Japan, 25–29 November 1974; The Physico-Chemical Society of Japan: Tokyo, Japan, 1975; pp. 317–320.
10. Hirao, K.; Tanaka, K.; Furukawa, S.; Soga, N. Anomalous temperature dependence of the sound velocities of SiO_2-TiO_2 glasses. *J. Mater. Sci. Lett.* **1995**, *14*, 697–699. [CrossRef]
11. Kushibiki, J.-I.; Arakawa, M.; Ueda, T.; Fujinoki, A. Homogeneous TiO_2-SiO_2 glass for extreme ultraviolet lithography evaluated by the line-focus-beam ultrasonic material characterization system. *Appl. Phys. Express* **2008**, *1*, 087002. [CrossRef]
12. Carson, D.S.; Maurer, R.D. Optical attenuation in titania-silica glasses. *J. Non-Cryst. Solids* **1973**, *11*, 368–380. [CrossRef]
13. Copley, G.J.; Redmond, A.D.; Yates, B. Influence of titania upon thermal expansion of vitreous silica. *Phys. Chem. Glasses* **1973**, *14*, 73–76.
14. Plummer, W.A.; Hagy, H.E. Precision thermal expansion measurements on low expansion optical materials. *Appl. Opt.* **1968**, *7*, 825–831. [CrossRef]
15. Kurkjian, C.R.; Peterson, G.E. An EPR study of Ti^{3+}-Ti^{4+} in TiO_2-SiO_2 glasses. *Phys. Chem. Glasses* **1974**, *15*, 12–17.
16. Turnbull, R.C.; Lawrence, W.G. The role of titania in silica glasses. *J. Am. Ceram. Soc.* **1952**, *35*, 48–53. [CrossRef]
17. Trap, H.J.L.; Stevels, J.M. Conventional and inverted glasses containing titania, Part I. *Phys. Chem. Glasses* **1960**, *1*, 107–118.
18. Hirayama, C.; Berg, D. Dielectric properties of glasses in the system TiO_2-Na_2O-SiO_2. *Phys. Chem. Glasses* **1961**, *2*, 145–151.
19. Manghnani, M.H. Pressure and temperature dependence of the elastic moduli of Na_2O-TiO_2-SiO_2 glasses. *J. Am. Ceram. Soc.* **1972**, *55*, 360–365. [CrossRef]
20. McSkimin, J.H. Pulse superposition method for measuring ultrasonic wave velocities in solids. *J. Acoust. Soc. Am.* **1961**, *33*, 12–16. [CrossRef]
21. Anderson, O.L.; Dienes, G.J. The anomalous properties of vitreous silica. Chap. 18. In *Non-Crystalline Solids*; Frechette, V.D., Ed.; John Wiley & Sons, Inc.: New York, NY, USA, 1960; pp. 449–490.
22. Barron, T.H.K. On the thermal expansion of solids at low temperatures. *Phil. Mag.* **1955**, *46*, 720–734. [CrossRef]
23. Scannell, G.; Koike, A.; Huang, L. Structure and thermo-mechanical response of TiO_2-SiO_2 glasses to temperature. *J. Non-Cryst. Solids* **2016**, *447*, 238–247. [CrossRef]
24. Williams, Q.; Manghnani, M.H.; Matsui, T. The effect of coordination changes on the bulk moduli of amorphous silicates: The SiO_2-TiO_2 system as a test case. *Am. Miner.* **2019**, *104*, 679–685. [CrossRef]
25. Chmel, A.; Eranosyan, G.M.; Karshak, A.A. Vibrational spectroscopic study of Ti-substituted SiO_2. *J. Non-Cryst. Solids* **1992**, *136*, 213–217. [CrossRef]
26. Chandrasekhar, H.R.; Chandrasekhar, M.; Manghnani, M.H. Phonons in TiO_2-SiO_2 glasses. *J. Non-Cryst. Solids* **1980**, *40*, 567–575. [CrossRef]
27. Cook, R.K. Variation of elastic constants and static strains with hydrostatic pressure: A method for calculation from ultrasonic measurements. *J. Acoust. Soc. Am.* **1957**, *29*, 445–449. [CrossRef]
28. Tkachev, S.N.; Manghnani, M.H.; Williams, Q.; Ming, L.C. Compressibility of hydrated and anhydrous Na_2O-$2SiO_2$ liquid and also glass to 8 GPa using Brillouin scattering. *J. Geophys. Res.* **2005**, *110*, B07201. [CrossRef]
29. Andreatch, P.; McSkimin, H.J. Pressure dependence of ultrasonic wave velocities and elastic stiffness moduli for a TiO_2-SiO_2 glass (Corning 7971). *J. Appl. Phys.* **1976**, *47*, 1299–1301. [CrossRef]
30. Gerlich, D.; Wolf, M.; Yaacov, I.; Nissenson, B. Thermoelastic properties of ULE® titanium silicate glass. *J. Non-Cryst. Solids* **1976**, *21*, 243–249. [CrossRef]
31. Vacher, R.; Pelous, J.; Plicque, F.; Zarembowitch, A. Ultrasonic and Brillouin scattering study of the elastic properties of vitreous silica between 10 and 300 K. *J. Non-Cryst. Solids* **1981**, *45*, 397–410. [CrossRef]

32. Tielburger, D.; Merz, R.; Enhrenfels, R.; Hunklinger, S. Thermally activated relaxation processes in vitreous silica: An investigation by Brillouin scattering at high pressures. *Phys. Rev. B* **1992**, *45*, 2750–2760. [CrossRef] [PubMed]
33. Zhang, J.S.; Bass, J.D.; Taniguchi, T.; Goncharov, A.F.; Chang, Y.-Y.; Jacobsen, S.D. Elasticity of cubic boron nitride under ambient conditions. *J. Appl. Phys.* **2011**, *109*, 063521. [CrossRef]
34. Le Parc, R.; Levelut, C.; Pelous, J.; Martinez, V.; Champagnon, B. Influence of fictive temperature and composition of silica glass on anomalous elastic behavior. *J. Phys. Cond. Matter* **2006**, *18*, 7507–7527. [CrossRef]
35. Birch, F. Elasticity and constitution of the Earth's interior. *J. Geophys. Res.* **1952**, *57*, 227–286. [CrossRef]
36. Sonneville, C.; De Ligny, D.; Mermet, A.; Champagnon, B.; Martinet, C.; Henderson, G.H.; Deschamps, T.; Margueritat, J.; Barthel, E. In situ Brillouin study of sodium alumino silicate glasses under pressure. *J. Chem. Phys.* **2013**, *139*, 074501. [CrossRef]
37. Xu, J.-A.; Manghnani, M.H.; Ming, L.C.; Wang, S.-Y. High-pressure Raman study of TiO_2-SiO_2 glasses: Evidence of the structural change. In *High Pressure Research: Application to Earth and Planetary Sciences*; Syono, Y., Manghnani, M.H., Eds.; Terra Publishing: Tokyo, Japan, 1992; pp. 519–525.
38. Vacher, R.; Courtens, E.; Foret, M. Anharmonic versus relaxational sound damping in glasses. II. Vitreous silica. *Phys. Rev. B* **2005**, *72*, 214205. [CrossRef]
39. Rau, S.; Enss, C.; Hunklinger, S.; Neu, P.; Wurger, A. Acoustic properties of oxide glasses at low temperatures. *Phys. Rev. B* **1995**, *52*, 7179–7194. [CrossRef]
40. Rouxel, T.; Ji, H.; Hammouda, T.; Moreac, A. Poisson's ratio and the densification of glass under high pressure. *Phys. Rev. Lett.* **2008**, *100*, 225501. [CrossRef] [PubMed]
41. Guerette, M.; Ackerson, M.R.; Thomas, J.; Yuan, F.; Watson, E.B.; Walker, D.; Huang, L. Structure and properties of silica glass densified in cold compression and hot compression. *Sci. Rep.* **2015**, *5*, 15343. [CrossRef] [PubMed]
42. Schuele, D.E.; Smith, C.S. Low temperature thermal expansion of RbI. *J. Phys. Chem. Solids* **1964**, *25*, 801–814. [CrossRef]

minerals

MDPI

Article

Limits to the Validity of Thermal-Pressure Equations of State

Ross J. Angel [1,*], Francesca Miozzi [2] and Matteo Alvaro [3]

1 Istituto di Geoscienze e Georisorse, CNR, Via Giovanni Gradenigo, 6, 35131 Padova, Italy
2 Sorbonne Université, UMR CNRS 7590, Muséum National d'Histoire Naturelle, Institut de Minéralogie, de Physique des Matériaux et de Cosmochimie, IMPMC, 75005 Paris, France; francesca.miozzi@upmc.fr
3 Department of Earth and Environmental Sciences, University of Pavia, Via A. Ferrata, 1 I-27100 Pavia, Italy; matteo.alvaro@unipv.it
* Correspondence: rossjohnangel@gmail.com

Received: 27 August 2019; Accepted: 14 September 2019; Published: 17 September 2019

Abstract: Thermal-pressure Equations of State (EoS) such as the Mie-Grüneisen-Debye (MGD) model depend on several assumptions, including the quasi-harmonic approximation (QHA) and a simplified phonon density of states. We show how the QHA is violated by materials exhibiting anisotropic thermal pressure. We also show that at pressures lower than those of the isochor of the reference volume, the static pressure may become sufficiently negative to make the compressional part of the EoS invalid. This limit is sensitive to the combined effects of the EoS parameters K'_0, q and the Grüneisen parameter γ_0. Large values of q, which correspond to a rapid decrease in phonon mode frequencies with increasing volume, can also lead to the bulk modulus becoming zero at high pressures and temperatures that are not particularly extreme for planetary geotherms. The MGD EoS therefore has an extremely limited P and T regime over which it is both valid and has physically-meaningful properties. Outside of this range, additional terms should be included in the thermal pressure that represents the physical properties of the solid. Or, alternatively, 'isothermal' EoS in which the temperature variation of the elastic properties is explicitly modeled without reference to a physical model can be used.

Keywords: equations of state; thermal pressure; Mie-Grüneisen-Debye; planetary materials

1. Introduction

The text book written by Orson Anderson titled "Equations of State of Solids for Geophysics and Ceramic Science" [1] is a masterful summary of the principles, theory, and practical application of Equations of State (EoS). In particular, it is emphasized that *"By physicists' standards, the materials of planets are not well characterized ...... The corresponding physical theories applied to these planetary materials are also necessarily simplified..."* [1] (p. 113). Therefore, Orson Anderson took great care in laying out the assumptions and simplifications underlying each EoS discussed, and in explaining the consequent limitations on the P, T, and V regimes in which they are valid. The presentation of this material was based on the deep knowledge and insights that Orson Anderson and his contemporaries and co-workers developed over their research careers. Unfortunately, as research into planetary interiors moves into new regimes of P and T, and science "frees itself from the tyranny of Earth's geotherm" [2], these limitations on EoS are often forgotten and EoS are blindly used in pressure and temperature regimes where they are unphysical and have no validity. In this contribution, we build on the work of Orson Anderson by first reviewing the properties and behavior of thermal-pressure type EoS, and then indicate further limits to their physical validity, in particular at high-temperature and low-pressure conditions not found on the terrestrial geotherm. These conditions are more likely to occur on the smaller terrestrial planets (i.e., Mars, Mercury) in the solar system. Furthermore, advances in the field of extrasolar planet exploration have led to the discovery of planets with a wide variability in their characteristics from Earth-like planets with different masses and radii, to planets that have no solar system analogues (e.g., [3,4]). As equations of state are the link between astrophysical observations and the interpretation of the interiors of exoplanets [5], knowing the limitations of *PVT* EoS is fundamental for the study of exoplanets.

2. Methods

All numerical calculations were performed with v7.5 of the EosFit7c software [6], which was released in summer 2019. The code has been validated [6] against a wide variety of other software and algebraic solutions, as documented in the help system for the software. Most recently, the calculated properties of the Mie-Grüneisen-Debye thermal pressure EoS have been validated against a Matlab code independently written by Eleanor Berryman.

3. Thermal-Pressure EoS

In thermal-pressure EoS, the pressure at a given volume V and temperature T is treated as the sum of two contributions:

$$P(V,T) = P_{ref}(V,T_0) + \Delta P_{th}(V,T) \quad (1)$$

The pressure P_{ref} is the pressure required to compress the material from its volume V_0 at reference conditions (T_0 and $P = 0$) to the volume V at the same temperature T_0. The second term is the change in thermal pressure ΔP_{th}. This is the pressure change at constant volume due to the temperature difference between T_0 and T. It is thus the pressure change along an isochor of the material (Figure 1). If the volume V of the material at P and T is equal to its volume V_0 at the reference conditions, then no compression at the reference conditions is required to attain the final volume V, so $P_{ref} = 0$. In this case, the total pressure is equal to ΔP_{th} (black line and symbols in Figure 1).

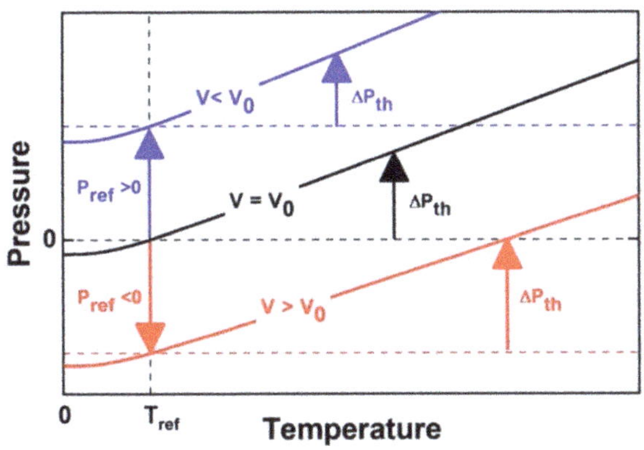

Figure 1. The concepts and definitions of the various pressures and variables used in thermal pressure EoS, illustrated for three different volumes. The solid lines are the isochors for three different volumes.

At pressures above the isochor of V_0 the volume V is smaller than V_0 and this is obtained with a positive pressure P_{ref} (blue in Figure 1). On the other hand, for P, T conditions below the isochor of V_0 the V is greater than V_0, and the material must be expanded at T_0 to obtain the required V. This means that $P_{ref} < 0$ (red in Figure 1). The isochor passing through the reference conditions therefore divides P–T space into two regions: at P above this reference isochor the P_{ref} is positive, but at pressures below the reference isochor the P_{ref} is negative.

3.1. Limitations to Isothermal EoS in Expansion

The volume variation of a material with pressure at T_0 is described by its isothermal equation of state. Such EoS for solids are developed explicitly for describing their behavior under compression, and their validity for negative pressures and volume expansion relative to the reference conditions is limited. This limit should be considered when working with thermal pressure EoS at volumes greater than V_0. The limit is illustrated in Figure 2 which shows that all common isothermal EoS show a divergence of V (and thus V/V_0) to very large values at negative pressures that are a small fraction (typically <25%) of the value of the bulk modulus K_0 at the reference conditions.

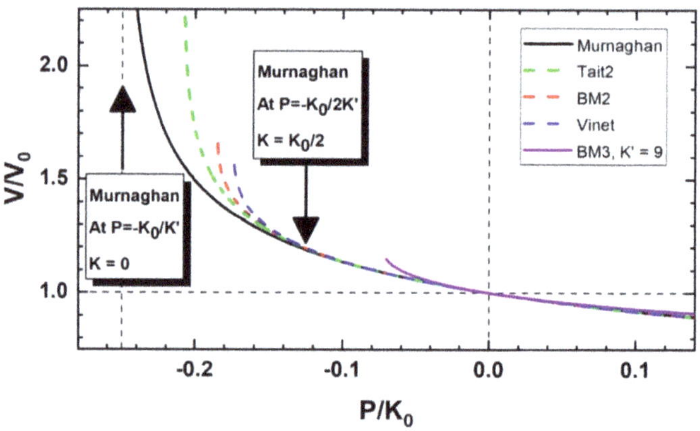

Figure 2. Volume-pressure variation for Murngahan, Vinet, 2nd-order Tait and Birch-Murnaghan isothermal EoS, with $K_0 = 125$ GPa and $K'_0 = 4$, parameters similar to mantle composition olivines and forsterite (e.g., [7]). The 3rd-order Birch-Murnaghan EoS (deep purple line) has the same bulk modulus but $K'_0 = 9$, a value typical of silicate orthopyroxenes, (e.g., [8,9]). At $P \sim -K_0/2K'_0$ the different EoS start to diverge, predicting significantly different volumes. When the curves of V in this graph become vertical the bulk modulus, $K = -V(dP/dV)$, is zero.

This divergence can be most easily understood by reference to the Murnaghan EoS [10] which has the property that the bulk modulus is linear in pressure:

$$K = K_0 + K'_0 P \tag{2}$$

with K'_0 being the pressure derivative of the bulk modulus. It is clear that with $K'_0 > 0$, there is a pressure $P = -K_0/K'_0$ at which the bulk modulus K becomes zero and thus the volume diverges to infinity (Figure 2). The pressure at which this divergence occurs depends on the EoS. It is generally less-negative than for the Murnaghan EoS because other EoS include a variation in $K' = (dK/dP)_T$ which is expressed as $K'' = (dK'/dP)_T$. The value of K'' is zero in the Murnaghan EoS (Equation (2)) but for most materials it is actually negative [1], and the implied value of K'' for the 2nd-order EoS shown in Figure 2 is also negative (e.g., [1,6,11,12]). Therefore, K' becomes more positive at negative pressures in other EoS than in the Murnaghan, and thus the bulk modulus becomes zero at smaller negative pressures. If the value of K'_0 is larger, then the decrease of K is faster, and the limit to the validity of the EoS is at less-negative pressures (deep purple line in Figure 2). This pressure limit becomes even less-negative in 4th-order EoS with K'' more negative than the implied value of the 3rd-order EoS. It follows that there is no simple dependence upon the reference parameters of the EoS of the pressure at which the divergence of volume occurs. Further, before the volume curves show strong divergence, the bulk modulus and its pressure derivatives show values that would be considered anomalous for solid materials. Therefore, rather than just considering the *algebraic* limit (e.g., $K = 0$) to the validity of EoS in expansion, it is necessary to find a practical limit to their *physical reasonableness* or validity. Figure 2 also shows that the difference between the various EoS only becomes significant at pressures more negative than where the bulk modulus of the Murnaghan EoS becomes one-half of K_0, which suggests that $K > K_0/2$ is a reasonable criterium to use for a limit to the validity of EoS. For the Murnaghan EoS this gives a limiting pressure $P_{lim} = -K_0/2K'_0$. For other EoS the expressions for P_{lim} are more complex, but the values of P_{lim} are typically 5–10% less negative than for a Murnaghan EoS with the same values of the parameters K_0 and K'_0. For the example of an olivine-like EoS used in Figure 2, P_{lim} ranges from −14.0 to −15.6 GPa for different types of isothermal EoS. The influence of K'_0 on this limit is significant. For an orthopyroxene-like EoS with $K'_0 = 9$ and the same value of K_0, the EoS is restricted in expansion to only P_{lim} ~ 6–7 GPa (deep purple line, Figure 2).

3.2. Limitations to Thermal-Pressure EoS

Thus, at sufficiently high temperatures, thermal-pressure EoS will have large enough ΔP_{th} that P_{ref} is more negative than P_{lim}, thus making the EoS physically invalid. Given that in planetary science we are only interested in positive total pressures, the largest negative value of P_{ref} will occur for calculations of volume at ambient pressure, for which $P_{ref} = -\Delta P_{th}$. The corresponding temperature limit, T_{lim}, to the validity of the thermal-pressure EoS at room pressure then follows from requiring $-\Delta P_{th} > P_{lim}$. For isochors that are approximately linear and parallel, we can estimate T_{lim} by using the slope of the isochor $(dP/dT)_V = \alpha K \sim \alpha_0 K_0$ and equating it to the slope of the isochor from P_{lim} and T_{ref} to $P = 0$ and T_{lim}, thus:

$$\left(\frac{dP}{dT}\right)_V \approx \alpha_0 K_0 \approx \frac{-P_{lim}}{(T_{lim} - T_0)} \tag{3}$$

If we further take the value of $P_{lim} = -K_0/2K'_0$ from the Murnaghan EoS, one can obtain an estimate of the limiting temperature in terms of the EoS parameters at the reference conditions:

$$T_{lim} \approx T_0 + \frac{1}{(2\alpha_0 K'_0)} \tag{4}$$

For our olivine-like EoS parameters, one obtains a T_{lim} ~ 5000 K, but for orthopyroxene-like EoS parameters, the larger K'_0 reduces T_{lim} to the range 1900–2400 K (depending on the value of the thermal expansion coefficient). If the isochors are not quasi-parallel, a better estimate of T_{lim} can be obtained if one does not equate $\alpha K = \alpha_0 K_0$, thereby obtaining:

$$T_{lim} \approx T_0 + \frac{K_0}{(2K'_0 \alpha_{Plim} K_{Plim})} \tag{5}$$

where the subscript "*Plim*" indicates the parameter values at the pressure P_{lim} and T_0. One can imagine that low-pressure conditions that exceed these temperature limits to the validity of the thermal-pressure EoS may occur in several planetary environments. Certainly, they could occur in the later stages of post-impact recovery from shock, when temperatures remain high but the shock wave-induced pressure has decayed. Alternatively they could occur in the early stages of the evolution of hot small planets.

The thermal pressure in a crystal arises from the excitation of its collective vibrational modes, the phonons. Different thermal-pressure EoS are distinguished by the method used to calculate the thermal pressure and hence the change in thermal pressure, ΔP_{th}, from the reference temperature. It can be modeled either in an ad-hoc thermodynamic approach by fitting parameters such as the thermal expansion coefficient and the bulk modulus and their derivatives to experimental data, or it can be calculated from an analysis of the properties of the phonon modes themselves via statistical mechanics [13]. In the thermodynamic approach there are no further assumptions, and the temperature limit T_{lim} is a direct constraint on the physical meaningfulness of the EoS. On the other hand, for thermal-pressure EoS based on statistical mechanics, T_{lim} may not be of practical relevance if the assumptions behind the derivation of the EoS are already violated at lower temperatures.

Therefore, each EoS formulism of this type needs to be examined individually, as we do for the Mie-Grüneisen-Debye EoS in Section 4. Nonetheless, in order to prevent EoS software from crashing in the large V regime, it is certainly necessary to impose a limit to how negative P_{ref} can become. In the latest release of EosFit software suite [6,14], this limit is now set as the requirement that $K > K_0/2$ for all steps of all EoS calculations.

4. Mie-Grüneisen-Debye EoS

The Mie-Grüneisen-Debye (MGD) EoS is based on the quasi-harmonic approximation (QHA) to the statistical mechanics of the collective vibrational modes (phonons) of crystalline materials [1] (chapter 2). Under the QHA it is assumed that the vibrational modes do not interact and are completely independent of one another, and that their wavenumbers (and thus frequencies) are only dependent upon the molar volume; the effects of P and T on the wavenumbers of the phonon modes are thus *indirect* in the sense that they change the volume, which then changes the wavenumbers. The volume dependence of the phonon frequencies ω_ι is defined by the phonon-mode Grüneisen parameters $\gamma_i = \frac{V}{\omega_i}\frac{\partial\omega}{\partial V}$. In order to derive the Mie-Grüneisen-Debye EoS, further assumptions are necessary. The first is that the volume dependence of all of the phonon frequencies is the same, and can be represented by a single Grüneisen parameter γ. Then the thermal pressure becomes $P_{th} = \frac{\gamma}{V}E_{th}$, where V is explicitly the molar volume and E_{th} is the thermal energy that is the energy of the vibrational modes. E_{th} is further assumed to derive from a simplified phonon density of states, which can be characterized by a single Debye frequency, or alternatively the Debye temperature θ_D, whose energy is represented by the Debye function [1] which is a function of only T and θ_D and is an integral from $T = 0$ to T. Therefore, if a finite temperature is chosen for T_0, the change in thermal pressure from T_0 becomes:

$$\Delta P_{th} = \frac{3nR\gamma}{V}\left[TD\left[\frac{\theta_D}{T}\right] - T_{ref}D\left[\frac{\theta_D}{T_0}\right]\right] \quad (6)$$

In which $D\left(\frac{\theta_D}{T}\right)$ represents the Debye function, R is the gas constant and n is the number of atoms in the formula unit corresponding to the molar volume V.

4.1. Limitations to MGD EoS from Expansion

As demonstrated in [1] (Figure 10.1 therein), the temperature range in which the QHA is valid and the material properties are dominated by harmonic effects is bounded by regions of invalidity at both low and high temperatures. At ambient pressure and low temperatures, below $T \sim 0.8\theta_D$, quantum effects start to become significant, whereas above $T \sim 1.2\theta_D$ there can be significant contributions from anharmonicity. Given that Debye temperatures for minerals are typically in the range 500–800 K, this is a significant restriction, but the range of temperatures in which the QHA remains valid is significantly expanded under high pressures [1] (Figure 10.2 therein) and [15]. If the non-validity of the compressional part of the EoS discussed in Section 3 is to be significant for a MGD EoS, T_{lim} must therefore be of the order of $1 - 2\theta_D$. Figure 3 shows isochors (in black) calculated for a range of values of γ for the BM2 EoS representative of olivine minerals. All of the isochors are drawn from the same pressure close to P_{lim} and therefore all represent the same volume, and the same value of K at T_{ref}. For low values of γ, the thermal expansion coefficients are reasonable ($\alpha_0 \sim 2.6 \times 10^{-5}$ K^{-1}), and the T_{lim} where the isochors cut the temperature axis are far in excess of $T = 2\theta_D$. Extreme values of $\gamma = 5$, which lead to unrealistic values of the volume thermal expansion coefficient at reference conditions (α_0 $\sim 1.3 \times 10^{-4}$ K^{-1}), are required to reduce T_{lim} to the vicinity of $2\theta_D$. In contrast, the orthopyroxene-like EoS has a much larger value of K'_0 and thus a smaller P_{lim}. Thus, although the slopes of the isochors through P_{lim} are very similar to those of the olivine-like EoS, the T_{lim} is in the vicinity of $2\theta_D$ for $\gamma = 2$.

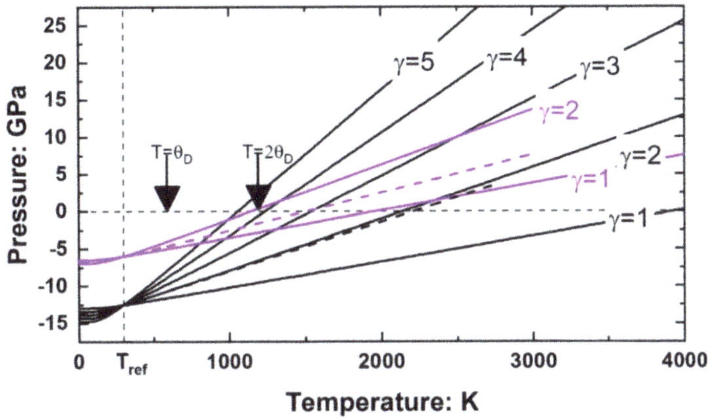

Figure 3. Isochors for two sets of MGD EoS with different values of γ, at volumes corresponding to the respective P_{lim}. The black lines are for the olivine-like compressional BM2 EoS with $K'_0 = 4$, and the purple lines are for the orthopyroxene-like compressional BM3 EoS with $K'_0 = 9$. All the MGD EoS shown have $\theta_D = 600$K. Solid lines have $q = 0$. Dashed lines have $\gamma_0 = 1$ and $q = 5$. The temperatures corresponding to θ_D and $2\theta_D$ are indicated.

The isochors shown as solid lines in Figure 3 are each calculated with a constant value of γ, although it would be expected to increase with increasing volume. The variation in γ with volume is usually modeled in the MGD EoS by $\gamma = \gamma_0\left(\frac{V}{V_0}\right)^q$. The effect of $q > 0$ is to increase the value of γ at high temperatures, increasing the volume thermal expansion coefficient and thus steepening the isochors, as shown by the dashed lines in Figure 3, which both have $\gamma_0 = 1$ and an extreme value of $q = 5$ so as to clearly demonstrate the effects. The influence of large positive values of q is clearly stronger on isochors of the EoS with the smaller K'_0, but they are not expected to bring T_{lim} within the range of temperature for which the QHA will be valid. Therefore, the temperature limit T_{lim} appears to only be relevant for materials with large values of K'_0 for which it is possible for T_{lim} to be in the vicinity of $2\theta_D$ (purple lines in Figure 3).

4.2. The Consequences of q

We have already noted how the value of the γ parameter influences the volume thermal expansion coefficient at room conditions, and that the value of q controls how the value of γ changes, especially at large volumes corresponding to high temperatures. More generally, both γ_0 and q affect the value of both the thermal expansion coefficient and the bulk modulus at all P, T conditions. In this section we take the results of Anderson [1] and show how extreme values of these variables place additional constraints on the validity of the MGD EoS.

Taking the volume differential of Equation (1) and multiplying by the volume V, one obtains an expression for the bulk modulus of a thermal-pressure EoS at any P, T:

$$\begin{aligned} -V\left(\frac{\partial P}{\partial V}\right) &= -V\left(\frac{\partial P_{ref}}{\partial V}\right) - V\left(\frac{\partial P_{th}}{\partial V}\right) \\ K &= K_{Pref} - V\left(\frac{\partial P_{th}}{\partial V}\right) \end{aligned} \quad (7)$$

Thus, the bulk modulus is the sum of two terms. The first is the bulk modulus, K_{Pref}, on the isochor at T_0, which is at the pressure P_{ref}. The second term is the volume derivative of the thermal pressure. At high temperatures the Debye function $D\left(\frac{\theta_D}{T}\right)$ approaches the value of 1, allowing Equation (6) to be used to obtain an approximation for the P_{th} of the MGD EoS at high temperatures (see Equation (2.39) in [1]):

$$P_{th} \approx \frac{3nR\gamma T}{V} \tag{8}$$

By substituting $\gamma_0\left(\frac{V}{V_0}\right)^q$ for γ, taking the derivative of this expression with respect to V and then multiplying by V, one obtains an expression for the last term in Equation (7):

$$-V\left(\frac{\partial P_{th}}{\partial V}\right) \approx \frac{-3nR\gamma_0 T V^{(q-1)}}{V_0^q}(q-1) \tag{9}$$

This term appears in Equation (2.42) in reference [1]. Along an isochor the terms in V remain constant, as do the constants n, R, and γ_0. Therefore, the sign of this contribution to the bulk modulus variation along an isochor depends on $(q - 1)$. For $q = 1$, the derivative (Equation (9)) is zero, and therefore from Equation (7) $K = K_{Pref}$. This means that the bulk modulus remains constant and equal to K_{Pref} along an isochor. The Holland-Powell thermal-pressure EoS has this same property [7,13,16]. In the MGD EoS when $q < 1$, the expression for $-V\left(\frac{\partial P_{th}}{\partial V}\right)$ is positive and the bulk modulus increases with increasing temperature along an isochor. This is also a property of the MGD EoS with $q = 0$, or constant $\gamma = \gamma_0$, that are shown as solid lines in Figure 3. Conversely, for $q > 1$, the bulk modulus decreases as temperature increases along an isochor. It is believed that most minerals and planetary materials exhibit values of q in the range 1 to 2 [1] providing that there are no phase transitions. However, for all values of $q > 1$, there will be a temperature at which the bulk modulus becomes zero, and the EoS again becomes physically invalid. This limit is reached at lower temperatures with larger values of q. For example, along the isochor shown as a purple dashed line in Figure 3 for the orthopyroxene-like EoS with $\gamma_0 = 1$ and $q = 5$, the bulk modulus becomes zero just above 3000 K and 7.6 GPa. At the same time, the volume thermal expansion coefficient becomes infinite, because the product αK_T has to remain equal to the slope $\left(\frac{\partial P}{\partial T}\right)_V$ of the isochor.

4.3. The Consequences of Anisotropic Pth

The basis of the MGD EoS are the concepts of thermal pressure and isochors, with the thermal pressure being the increase in pressure along an isochor. However, the thermal expansion and compressibility which describe the response of a crystal to respectively temperature and hydrostatic pressure are only isotropic for cubic crystals. Therefore, for lower-symmetry crystals, the cell parameters vary by different amounts for a given change in P or T. From the definitions of the thermal expansion tensor α_{ij} and the compressibility tensor β_{ij} [17], the combined effect of a change in P and T is described by the strain tensor ε_{ij}:

$$\varepsilon_{ij} = \alpha_{ij} dT - \beta_{ij} dP \tag{10}$$

It is therefore not possible to maintain all of the cell parameters or the lengths of all directions in the crystal constant when P and T are changed. However, along a line defined by

$$\left(\frac{\partial P}{\partial T}\right) = \frac{\alpha_{ij}}{\beta_{ij}} \tag{11}$$

the strain ε_{ij} will remain constant. The property $\frac{\alpha_{ij}}{\beta_{ij}}$ of a crystal varies with direction in the crystal, but it is not a tensor property. It defines lines in P–T space along which there is no strain in one chosen direction within the crystal. By analogy with the conventional definition of thermal pressure, $\left(\frac{\partial P}{\partial T}\right)_V$, Equation (11) defines the anisotropic thermal pressure of a crystal.

In particular, for uniaxial crystals (in which $a = b$) and orthorhombic crystals, one can calculate from the values of the axial thermal expansion coefficients and axial compressibilities lines in P–T space along which the individual cell parameters a, b and c remain constant. It has long-been recognized [18] that many minerals exhibit the property that directions with large thermal expansion tend to compress more under pressure, which is equivalent to saying that the anisotropic thermal pressure defined by

Equation (11) tends to be close to isotropic in many crystal structures. However, for crystals in which the structural response to P and T appears to be dominated by cation-cation repulsion along one axis, this axis tends to be relatively incompressible (small β) and at the same time exhibits a large thermal expansion coefficient α. These compounds then exhibit significantly anisotropic thermal pressure (Equation (11)). Figure 4 illustrates one way in which the anisotropy of the thermal pressure of a material may be evaluated, directly from the thermal expansion and compressibility tensors measured at room conditions. Figure 4a shows lines of constant a (=b) and c cell parameters calculated from the well-constrained elastic properties of zircon at room conditions [19,20]. These two lines deviate significantly from the isochor, with the consequence that there are significant linear strains along isochors (Figure 4b). Because phonon-mode frequencies change linearly with small strains [21–25], this implies that the phonon-mode frequencies of materials with anisotropic thermal pressure change along the isochors, in violation of the assumptions of the QHA.

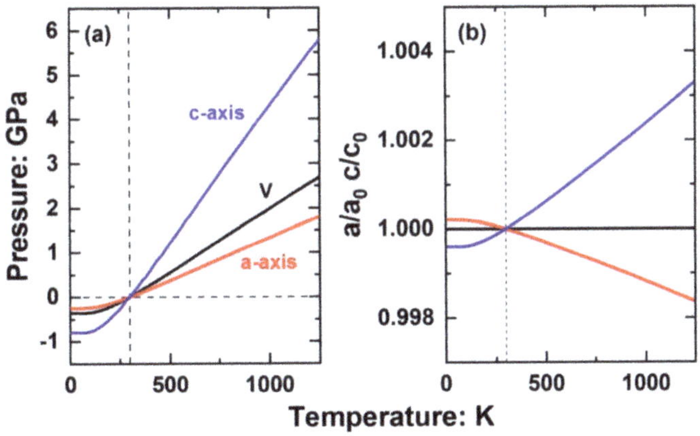

Figure 4. (**a**) The lines of constant a and c cell parameters of zircon passing through room T and P do not coincide with the isochor. (**b**) As a consequence, there are significant changes in the a and c cell parameters along this isochor.

Whether the resulting strains along the isochors are sufficiently large to constitute a significant violation of the QHA and invalidate the MGD EoS depends on the magnitudes of the components of the phonon-mode Grüneisen tensors which relate the phonon frequencies to the strains [19,22,23,25]. One direct test is to compare the measured changes in the phonon-mode frequencies (e.g., the frequencies of Raman bands) with pressure and with temperature. If the isobaric and isothermal change in phonon wavenumbers obey the relationship $\left(\frac{\partial\omega}{\partial P}\right)_T = -\left(\frac{\beta_V}{\alpha_V}\right)\left(\frac{\partial\omega}{\partial T}\right)_P$, then the thermal pressure is effectively isotropic [19]. The isobaric variation of the bulk modulus with temperature is also a sensitive indicator of the validity of the QHA because it is sensitive to the parameter K'_0. If fitting the temperature variation of the bulk modulus requires a value of K'_0 that is inconsistent with that required to fit either the volume variation with pressure or the measured change in the elastic moduli with pressure this may indicate that the anisotropic thermal pressure is significant. EoS such as the MGD which rely on the QHA should not be used for materials in which the thermal pressure is significantly anisotropic.

5. Conclusions

In his book, Orson Anderson [1] took great pains to explain the assumptions behind various EoS formula and to explore the limits of these assumptions. For the MGD EoS, he emphasized the point that the temperature range within which the QHA is valid is extremely limited at ambient pressures, although it expands at higher pressures as both quantum-mechanical and anharmonic effects are suppressed [15]. Further, in his chapter 5 [1] he demonstrated that only 'close-packed' structures have phonon densities of state that are 'Debye-like', meaning that MGD EoS should not be applied to open-framework materials whose structural response to changing pressure and temperature is dominated by tilting of relatively rigid polyhedral structural units such as SiO_4 tetrahedra, (e.g., [18,26]). In this contribution we have identified further limitations on thermal-pressure type EoS. First, that structures that respond by polyhedral tilting [26] and even densely-packed structures such as rutile and zircon can exhibit strong anisotropy in their thermal pressure. This leads to changes of their phonon-mode frequencies along the isochors, and thus invalidates a fundamental assumption behind thermal-pressure EoS including the MGD and the Holland-Powell EoS [16] widely used for petrological and planetary thermodynamics. Fits of a thermal-pressure EoS to volume and elastic data of such materials are often poor, and at the same time can lead to extreme values of EoS parameters [27]. Second, at low pressures below the isochor of the reference volume V_0 (and thus at high temperatures), the static pressure P_{ref} may become sufficiently negative to make the compressional part of the EoS invalid. This limit to the validity of the MGD EoS is sensitive to the combined effects of the EoS parameters K'_0, q, and γ (Figure 3). Third, large values of q which correspond to a rapid decrease in phonon mode frequencies with volume can lead to the bulk modulus becoming zero at pressures and temperatures that are not particularly extreme for planetary geotherms. If one obtains extreme values of especially K'_0, q, and γ through fitting of P–V–T data, it is worthwhile to explore the physical reason for the values and consider whether or not one or more of the assumptions behind the derivation of thermal-pressure EoS are violated by the known structure, properties, or phonon density of states of the material. Possible causes could include the presence of phase transitions, anisotropic thermal pressure, or significant anharmonic contributions (both phonon and electronic) to the thermal pressure. There are at least two possible types of approaches to obtaining more accurate and valid EoS in these cases. One is to include additional terms in the thermal pressure that represent the physical properties of the solid, for example by using additional oscillators to better represent the phonon density of state [28] and explicitly adding terms to describe the anharmonic contributions of the phonons [1,29,30] or electrons, (e.g., [31]), or by determining the anisotropy of the thermal Grüneisen parameter from the variation of the unit-cell parameters with temperature [29,30,32]. The alternative approach is not to use thermal-pressure EoS, but to employ "isothermal" EoS in which the temperature variation of the elastic properties such as K and K' is explicitly modeled [7]. Which approach is tractable and appropriate depends on the data available for the material.

Author Contributions: Conceptualization, methodology, R.J.A., F.M. and M.A.; software, R.J.A., M.A.; validation, R.J.A., F.M. and M.A.; writing, R.J.A., F.M. and M.A.; project administration, M.A.; funding acquisition, M.A.

Funding: This project was funded by the European Research Council (ERC) under the European Union's Horizon 2020 research and innovation programme, grant agreement No 714936 to M. Alvaro. F. M. has received funding from the European Research Council (ERC) under the European Union's Horizon 2020 research and innovation Programme, grant agreement No 670787 to Guillaume Fiquet.

Acknowledgments: We are very grateful to Eleanor Berryman and Tom Duffy (Princeton) for providing calculated properties of the MGD EoS from their own Matlab code that allowed us to validate the EosFit7 code. We thank George Helffrich (Tokyo Institute of Technology) for some observations that provoked our initial investigation into EoS under expansion, and Herbert Kroll and Peter Schmid-Beurmann (Münster) for discussions about thermal-pressure EoS. R.J.A. would like to record here his thanks for the inspiration he received from Orson Anderson during the latter's visits to the Bayerisches Geoinstitut in Bayreuth during the 1990s.

Conflicts of Interest: The authors declare no conflict of interest. The funders had no role in the design of the study; in the calculations reported; in the writing of the manuscript, or in the decision to publish the results.

References

1. Anderson, O.L. *Equations of State of Solids for Geophysics and Ceramic Science*; Oxford University Press: Oxford, UK, 1995; p. 432.
2. Navrotsky, A. Presented at the MSA Centennial Symposium, Washington, DC, USA, 20–21 June 2019.

3. Wagner, F.; Sohl, F.; Hussmann, H.; Grott, M.; Rauer, H. Interior structure models of solid exoplanets using material laws in the infinite pressure limit. *Icarus* **2011**, *214*, 366–376. [CrossRef]
4. Elkins-Tanton, L.; Seager, S. Coreless Terrestrial Exoplanets. *Astrophys. J.* **2008**, *688*, 628–635. [CrossRef]
5. Duffy, T.; Madhusudham, N.; Lee, K.K.M. Mineralogy of super-Earth planets. In *Treatise on Geophysics*, 2nd ed.; Elsevier: Oxford, UK, 2015; Volume 2, pp. 149–178. [CrossRef]
6. Angel, R.J.; Gonzalez-Platas, J.; Alvaro, M. EosFit7c and a Fortran module (library) for equation of state calculations. *Z. für Krist.* **2014**, *229*, 405–419. [CrossRef]
7. Angel, R.J.; Alvaro, M.; Nestola, F. 40 years of mineral elasticity: a critical review and a new parameterisation of Equations of State for mantle olivines and diamond inclusions. *Phys. Chem. Miner.* **2018**, *45*, 95–113. [CrossRef]
8. Angel, R.J.; Jackson, J.M. Elasticity and equation of state of orthoenstatite, $MgSiO_3$. *Am. Mineral.* **2002**, *87*, 558–561. [CrossRef]
9. Webb, S. The elasticity of the upper mantle orthosilicates olivine and garnet to 3 GPa. *Phys. Chem. Miner.* **1989**, *16*, 684–692. [CrossRef]
10. Murnaghan, F. Finite deformations of an elastic solid. *Am. J. Math.* **1937**, *49*, 235–260. [CrossRef]
11. Angel, R.J. Equations of state. In *High-Pressure and High-Temperature Crystal Chemistry*; Hazen, R.M., Downs, R.T., Eds.; MSA: Chantilly, VA, USA, 2000; Volume 41, pp. 35–60.
12. Birch, F. Finite elastic strain of cubic crystals. *Phys. Rev.* **1947**, *71*, 809–824. [CrossRef]
13. Kroll, H.; Kirfel, A.; Heinemann, R.; Barbier, B. Volume thermal expansion and related thermophysical parameters in the Mg,Fe olivine solid-solution series. *Eur. J. Mineral.* **2012**, *24*, 935–956. [CrossRef]
14. Gonzalez-Platas, J.; Alvaro, M.; Nestola, F.; Angel, R.J. EosFit7-GUI: A new GUI tool for equation of state calculations, analyses, and teaching. *J. Appl. Crystallogr.* **2016**, *49*, 1377–1382. [CrossRef]
15. Hardy, R.J. Temperature and pressure dependence of intrinsic anharmonic and quantum corrections to the equation of state. *J. Geophys. Res. B* **1980**, *85*, 7011–7015. [CrossRef]
16. Holland, T.J.B.; Powell, R. An improved and extended internally consistent thermodynamic dataset for phases of petrological interest, involving a new equation of state for solids. *J. Metamorph. Geol.* **2011**, *29*, 333–383. [CrossRef]
17. Nye, J.F. *Physical Properties of Crystals*; Oxford University Press: Oxford, UK, 1957; p. 329.
18. Hazen, R.M.; Finger, L.W. *Comparative Crystal Chemistry*; John Wiley and Sons: New York, NY, USA, 1982; p. 231.
19. Stangarone, C.; Alvaro, M.; Angel, R.; Prencipe, M.; Mihailova, B.D. Determination of the phonon-mode Grüneisen tensors of zircon by DFT simulations. *Eur. J. Mineral.* **2019**, *2019*. [CrossRef]
20. Zaffiro, G. Elastic Geobarometry: In-Situ Single-Crystal X-ray Diffraction Measurements of Inclusions Trapped in Host Minerals to Determine the Entrapment Conditions. Ph.D. Thesis, University of Pavia, Pavia, Italy, 2019.
21. Grüneisen, E. Zustand des festen Körpers. *Handb. der Phys.* **1926**, *1*, 1–52.
22. Key, S.W. Grüneisen tensor for anisotropic materials. *J. Appl. Phys.* **1967**, *38*, 2923–2928. [CrossRef]
23. Cantrell, J.H. Generalized Grüneisen tensor from solid nonlinearity parameters. *Phys. Rev. B* **1980**, *21*, 4191–4195. [CrossRef]
24. Angel, R.J.; Murri, M.; Mihailova, B.; Alvaro, M. Stress, strain and Raman shifts. *Z. für Krist.* **2019**, *234*, 129–140. [CrossRef]
25. Barron, T.H.K.; Collins, J.F.; Smith, T.W.; White, G.K. Thermal expansion, Grüneisen functions and static lattice properties of quartz. *J. Phys. C Solid State Phys.* **1982**, *15*, 4311–4326. [CrossRef]
26. Angel, R.J.; Sochalski-Kolbus, L.M.; Tribaudino, M. Tilts and tetrahedra: The origin of anisotropy of feldspars. *Am. Mineral.* **2012**, *97*, 765–778. [CrossRef]
27. Zaffiro, G.; Angel, R.J.; Alvaro, M. Constraints on the Equations of State of stiff anisotropic minerals: Rutile, and the implications for rutile elastic barometry. *Mineral. Mag.* **2019**, *83*, 339–347. [CrossRef]
28. Blackman, M. The theory of the specific heat of solids. *Rep. Prog. Phys.* **1941**, *8*, 11–30. [CrossRef]
29. Murshed, M.M.; Zhao, P.; Fischer, M.; Huq, A.; Alekseev, E.; Gesing, T.M. Thermal expansion modeling of framework-type $Na[AsW_2O_9]$ and $K[AsW_2O_9]$. *Mater. Res. Bull.* **2016**, *84*, 273–282. [CrossRef]
30. Murshed, M.M.; Mendive, C.B.; Curti, M.; Šehović, M.; Friedrich, A.; Fischer, M.; Gesing, T.M. Thermal expansion of mullite-type $Bi_2Al_4O_9$: A study by X-ray diffraction,vibrational spectroscopy and density functional theory. *J. Solid State Chem.* **2015**, *229*, 87–96. [CrossRef]

31. Fei, Y.; Murphy, C.; Shibazaki, Y.; Shahar, A.; Huang, H. Thermal equation of state of hcp-iron: Constraint on the density deficit of Earth's solid inner core. *Geophys. Res. Lett.* **2016**, *43*, 6837–6843. [CrossRef]
32. Munn, R.W. Role of the elastic constants in negative thermal expansion of axial solids. *J. Phys. C Solid State Phys.* **1972**, *5*, 535–542. [CrossRef]

minerals

MDPI

Article

Equations of State for the Deep Earth: Some Fundamental Considerations

Frank D. Stacey [1,2]

[1] CSIRO Mineral Resources, Pullenvale, Queensland 4069, Australia; Frank.Stacey@csiro.au
[2] Technology Court, Pullenvale, Queensland 4069, Australia

Received: 3 September 2019; Accepted: 15 October 2019; Published: 17 October 2019

Abstract: None of the 40+ equations that have been proposed to describe material properties at the pressures of the Earth's core and mantle have escaped serious criticism. In this paper, some basic algebraic and thermodynamic constraints are reviewed, with the conclusion that the next step should be a re-examination of the relationship between the dependence of the bulk modulus, K, on pressure, P, that is $K' \equiv dK/dP$, and the normalized (dimensionless) pressure, P/K. A linear relationship between $1/K'$ and P/K terminating at the infinite pressure asymptote, at which these quantities become equal, has been used for analysing properties at extreme pressure, but may be inadequate for calculations requiring precise derivatives of an equation of state. A new analysis indicates that $\mathrm{d}(1/K')/\mathrm{d}(P/K)$ increases with compression (or P/K), but there are, at present, no reliable equations representing this. Relationships between higher derivatives of K and the thermodynamic Grüneisen parameter offer the prospect of a resolution of the problem and hence a new generation of fundamentally-based equations of state. Although an earlier conclusion that a completely general 'universal' equation is not possible, in principle, is confirmed in this study, the fundamental relationships present strong constraints for the forms of other proposed equations.

Keywords: equations of state; Grüneisen parameter; algebraic tests; thermodynamic tests; bulk modulus derivatives; core; mantle

1. Introduction

Of the physical properties of geological materials that are central to studies of the lower mantle and core, a basic one is the bulk modulus, $K = \rho \mathrm{d}P/\mathrm{d}\rho = \mathrm{d}P/\mathrm{dln}\rho$, representing the variation of pressure, P, with density, ρ. Seismological observations of $K(P)$ provide data that are required for the equation of state, which is the starting point for any theoretical description. This paper considers the fundamental developments that have been introduced since the pioneering contributions by Birch [1] and Anderson [2]. The regular increase of K with P, $\mathrm{d}K/\mathrm{d}P \equiv K'$, over any depth range of a uniform phase and composition, is a primary equation of state parameter. Here, the fundamental theoretical constraints on its behaviour are explored, in particular, what we can learn from the extrapolation of K' and higher derivatives, $K'' \equiv \mathrm{d}^2K/\mathrm{d}P^2$, K''' and to their infinite pressure asymptotes, K'_∞ and so on. A basic constraint [3] is

$$(K'P/K)_\infty = (\mathrm{dln}K/\mathrm{dln}P)_\infty = 1. \tag{1}$$

This is an algebraic identity that all proposed equations must satisfy. Most of them do so incidentally and not by design, but there is an advantage in making Equation (1) the starting point for the development of an equation of state. The significance of this to an understanding of the deep regions of the Earth is that in the core and deeper parts of the mantle, K' is nearer in value to K'_∞ than it is to K'_0, the zero pressure value, which is commonly used as an equation of state parameter. This means that K'_∞ imposes a useful constraint on any equation in which it appears. As previously pointed out [4], there is a thermodynamic bound, $K'_\infty > 5/3$, and of the 35 proposed equations examined by

Stacey [5], 25 failed this test. There is no corresponding test for K'_0 and, although in some situations, this is accessible to observation, in many cases, it is an extrapolation as extended as K'_∞, requiring an assumed equation of state.

Stacey and Hodgkinson [6] introduced another fundamental constraint using volume derivatives of the thermodynamic Grüneisen parameter, γ, which is, with its thermodynamic definition and an equation used for calculating it from $K(P)$,

$$\gamma = \alpha K_T/\rho C_V = [K'/2 - (f/3)\ (1 - P/3K)]/[1 - (2/3)\ fP/K] \tag{2}$$

The use of this equation has been limited by the fact that f is a parameter that depends on statistical details of atomic thermal vibration and is assigned different values in alternative theories. K_T is the isothermal value of K. For most of the present discussion, it is not distinguished from the adiabatic modulus obtained from seismology. The relative difference diminishes with compression at all pressures and is not noticeable when applied to Equation (2). C_V is the constant volume-specific heat. This equation is the generalized free volume formula, so named because it was first derived from free volume theory with $f = 2$ by Vashchenko and Zubarev [7]. A comparison of the equation with values of γ for common materials under ambient conditions shows that $f < 2$, and there is no requirement that it should be constant, independent of compression, or have the same value for different materials. A derivation of the V-Z formula (Equation (2) with $f = 2$) [8] with graphs of its pressure dependence for various equations of state shows that the formula assumes atomic thermal vibration that is random and uncorrelated with neighboring atoms. This is not valid. Lacking a reliable theory for f, Stacey and Hodgkinson [6] adopted an empirical value, $f = 1.67$, selected to make values of γ for the lower mantle of Equation (2) coincide with the values obtained by an alternative formula, the 'acoustic gamma', which assumes that all elastic moduli are adequately represented by the seismic P- and S-wave speeds. That calculation indicated a negligible pressure dependence. Equation (2) has an advantage for calculations of the kind considered here because, if f can have any value and any compression dependence, then the equation is completely general and, through f, incorporates the effect of the rigidity modulus. This generality extends to the infinite pressure extrapolation, in which f is canceled out by Equations (1) and (2), reducing it to

$$\gamma_\infty = K'_\infty/2 - 1/6. \tag{3}$$

Although f remains finite in this situation, it does not appear in Equation (3), which is, therefore, free of the uncertainty in f, but retains the generality of Equation (2) and is an identity that can be used to bring γ into the quest for fundamental equation of state constraints.

The two relevant derivatives of γ are

$$q = ((\partial ln\gamma)/\partial lnV)_T \tag{4}$$

$$\lambda = (\partial \ln q/\partial \ln V)_T. \tag{5}$$

In the infinite pressure extrapolation, q vanishes, as explained in Section 5 of [6], but λ_∞ is finite positive and is a useful parameter. Using Equation (2) with the assumption that f is constant, Shanker et al. [9] derived the equation

$$\lambda_\infty = [KK''/(1 - K'\ P/K)]_\infty/K'_\infty + K'_\infty. \tag{6}$$

Later derivations [6,10,11] removed the restriction on f and, like Equation (3), Equation (6) is an identity. It can be rewritten as the first of an infinite series of equations for dimensionless derivatives expressed in terms of K'_∞ and λ_∞, all of which are identities:

$$[KK''/(1 - K'\ P/K)]_\infty = -K'_\infty(K'_\infty - \lambda_\infty) \tag{7}$$

$$\left[K^2K'''/(1-K'\,P/K)\right]_\infty = +K'_\infty(K'_\infty-\lambda_\infty)(K'_\infty+\lambda_\infty) \tag{8}$$

$$\left[K^3K^{iv}/(1-K'\,P/K)\right]_\infty = -K'_\infty(K'_\infty-\lambda_\infty)(K'_\infty+\lambda_\infty)(2K'_\infty+\lambda_\infty) \tag{9}$$

and so on, with alternating signs. The quantities on the left-hand sides of these equations are all formally indeterminate, in the sense of being ratios of vanishing quantities. This makes it difficult to use the equations individually, but ratios of them, e.g., $\left(K^2K'''/KK''\right)_\infty = -(K'_\infty+\lambda_\infty)$, can provide useful tests of validity for proposed equations, as in Section 5 of Stacey and Hodgkinson [6]. The fact that the infinite pressure asymptotes of derivatives of K can all be represented in terms of two constants invited an investigation of the possibility of writing a universal equation as a Taylor expansion of K' as a function of P/K, anchored at the infinite pressure limit. All the coefficients in the series can be written in terms of K'_∞ and λ_∞, but are found to be indeterminate for all terms except the first, with denominators involving $(1-K'P/K)_\infty$, which becomes zero. In view of this result, Stacey and Hodgkinson [6] inferred that no universal form of high-pressure equation of state is possible, contrary to some expectations [12,13], and a semi-independent check of this is presented in Section 2.

The difficulty with higher derivatives of K does not affect the first derivative of K' with P/K. This is conveniently written as

$$[\mathrm{d}(1/K')/\mathrm{d}(P/K)]_\infty = 1-\lambda_\infty/K'_\infty \tag{10}$$

because $1/K'$ and P/K become equal at $P\to\infty$, as in Equation (1). The gradient in Equation (10) must be positive because $\lambda_\infty < K'_\infty$, a condition that is basic to a constraint on equations of state, discussed in Section 4, additional to the $K'_\infty > 5/3$ constraint imposed by a thermodynamic argument [3,4]. It needs to be recognised that, although K'_∞ and λ_∞ are referred to here as constants, they are not universal constants, but are properties unique to each material. Numerical values or expressions for them are listed in Table 1 of Stacey and Hodgkinson [6] for a selection of equations that survived the culls prompted by basic tests, such as $K'_\infty > 5/3$. In juggling Equation (10), it is useful to note another identity:

$$[(1-K'P/K)/(1-K'_\infty P/K)]_\infty = \lambda_\infty/K'_\infty. \tag{11}$$

2. A Further Check of the Indeterminacy of Derivative Properties

If we attempt to avoid the problem of indeterminacy arising from denominators involving $(1-K'P/K)_\infty$ in Equations (7) to (9), we can differentiate derivatives with respect to P, yielding the result

$$[\mathrm{dln}(KK'')/\mathrm{dln}P]_\infty = \left[\mathrm{dln}\left(K^2K'''\right)/\mathrm{dln}P\right]_\infty = \left[\mathrm{dln}\left(K^3K^{iv}\right)/\mathrm{dln}P\right]_\infty = \ldots = -\lambda_\infty/K'_\infty. \tag{12}$$

An indeterminacy problem also arises in this case because all the logarithms become infinite (positive or negative), but it is nevertheless possible to manipulate the infinite pressure limits by multiplying or dividing by complete expressions and applying cancellations of common factors within any expression. Therefore, the ratio of the first two expressions is in unity, with the common term, ln$P$, having been canceled, leaving

$$\left[\mathrm{dln}\left(K^2K'''\right)/\mathrm{dln}(KK'')\right]_\infty = \left\{\left[(KK'')/\left(K^2K'''\right)\right]\mathrm{d}\left(K^2K'''\right)/\mathrm{d}(KK'')\right\}_\infty = 1. \tag{13}$$

Since, in the infinite pressure limit, both (KK'') and $\left(K^2K'''\right)$ vanish while their ratio remains finite, by the Bernoulli-l'Hospital rule [14],

$$\left[\mathrm{d}\left(K^2K'''\right)/\mathrm{d}(KK'')\right]_\infty = \left[\left(K^2K'''\right)/(KK'')\right]_\infty. \tag{14}$$

This reduces Equation (13) to 1 = 1, a symptom of a circular argument. It means that the algebraic identities involved in the manipulations cause the indeterminacy of a Taylor expansion using these derivatives, reinforcing the conclusion that there can be no universal form for equations of state.

3. Inner Core Problems

Of the deep regions of the Earth, the one most in need of theoretical input to the understanding of physical properties is the inner core. This is because it is the least accessible observationally. Direct P-wave propagation through it is well-observed and this gives a measure of the combination of elastic moduli $(K + 4\mu/3)$, but requires additional information to separate K from the rigidity modulus, μ. In principle, this would be obtained from the speed of shear waves, but they are not easily observed in the inner core. Conversion from and to compressional waves in the fluid outer core is extremely weak. Models of the Earth, such as Preliminary Earth Model (PREM) [15], rely largely on free oscillation records, which do not effectively resolve the fine structure. The modest variations in K and μ over the limited pressure range in the inner core contribute to the difficulty in discerning their behavior, with the result that models of the inner core structure are seen to be inconsistent with a plausible equation of state.

In the PREM tabulation, the problem is made most immediately obvious by a sharp decrease in K' at the inner core boundary, from an outer core value of 3.75 to 2.34 in the inner core. This could not be caused by the liquid–solid transition, even allowing for a modest compositional difference between the inner and outer cores. At core pressures, both the solid and liquid have close-packed atomic structures and the atoms interact in the same way in both so that, for the same composition, the density difference would be 2% and the differences in K and K' would both be ~1.6% ([3], Section 13). The anomalously low K' in PREM is compensated for by an anomalously high value of $d\mu/dP$, to give an inner core gradient of the observed P-wave modulus, $d(K + 4\mu/3)/dP$, that is not anomalous. This may be only circumstantial evidence of a flaw in the PREM inner core model, but it can be checked by considering the compensating changes in the K and μ profiles, without invoking a comparison with the outer core. For a fixed phase of uniform material, the ratio μ/K is described by the μ-K-P equation ([3], Section 8):

$$\mu/K = A + B(P/K), \tag{15}$$

where B is a negative constant. In the PREM inner core tabulation, μ/K increases strongly with P/K, as in the very short line in Figure 18.7 of [16], giving $B \approx + 0.68$, compared with the expected value, ~−1.2, derived from second-order elasticity theory ([16], Section 18.8). A positive value of B is incompatible with any plausible equation of state for a material in a single stable phase. The fact that iron and the core are normal, in the sense of following Equation (15) with negative B, is evidenced by the very low value of μ/K in the inner core, ~0.12, compared with the value for iron alloys under ambient conditions, ~ 0.48. There is no reason for supposing that the normal behavior prevails down to core pressures, but is reversed in the inner core. As Stacey and Davis [16] argued, the anomalous behavior of μ/K and the implausibly low value of K' in the PREM inner core, are simultaneously avoided by a simple empirical adjustment, leaving $d(K + 4\mu/3)/dP$ unchanged (Appendix F, Table F2 in [16]), because they are two aspects of the same problem However, this adjustment can be questioned because it does not involve any re-examination of the primary (seismological) data. Several authors have addressed the problem by re-assessing the data, but it has not yet been satisfactorily resolved.

Another earth model, ak135 [17], derived with a greater emphasis on body wave travel times, permits an independent check. Its inner core value of $d(\mu/K)/d(P/K)$ is 0.68, coinciding with the PREM value, and the value of K' is at least as low as that of PREM, with the further complication that K' is also low at the bottom of the outer core. Deuss et al. [18] concluded that the inner core gradient of shear wave speed could not be satisfactorily resolved and attributed the average value, 3.6 $k{\cdot}ms^{-1}$, to the whole depth range. If this is assumed, then $d(\mu/K)/d(P/K) \approx -1.25$, coinciding with the theoretical expectation. This may be taken as evidence that a constant shear wave speed is a better approximation than mode variations, but other factors must be considered, especially the implausible gradient of K. Another difficulty is the density. Masters and Gubbins [19] developed a powerful method of choosing free mode periods that selectively emphasized specific depths and concluded that it indicated an inner core boundary (ICB) density increment of about 820 $kg{\cdot}m^{-3}$, 200 $kg{\cdot}m^{-3}$ more than in the standard models. Using normal mode data, de Wit et al. [20] also favored a high ICB density increment, with a range of uncertainty extending to the Masters and Gubbins value. Such a revision reduces, but does not eliminate, the K' and μ/K problems, which still need to be resolved.

Heterogeneity and anisotropy of the inner core introduce both difficulties and uncertainties to equation of state studies, but also offer some insight on its origin and development. The crystal structure of the inner core alloy has generally been considered to be hexagonal close-packed (hcp), but doubts have arisen, with rival suggestions for alternatives and even multiple alternatives [21]. One suggestion was for a subtle modification of the hcp structure (dhcp). This would not cause a dramatic revision of core physics, but another approach, proposed by Belonoshko et al. [22] and subsequently strongly advocated by the same group, was for a return, at the inner core pressure and temperature, to a body-centered cubic (bcc) phase, familiar for iron under ambient conditions. This would necessitate some rethinking of core properties and Stacey and Davis [16] argued that a case against it was that the core equation of state did not satisfactorily extrapolate to the familiar properties of laboratory bcc iron. This bears directly on the anisotropy, which is not easily explained as a property of bcc iron, but has a more natural explanation with the anisotropic hcp structure. Yoshida et al. [23] argued that it is caused by deformation because the inner core grows by the freezing of added material on to an equatorial band and the consequent excess ellipticity relaxes by progressive deformation, causing crystalline alignment. With this interpretation, there is an interesting implication in the irregularity of the anisotropy: polar wander of the inner core. Following Curie's principle of symmetry, which disallows any effect with lower symmetry than the combination of its causes, we see that the symmetry constraints on inner core formation are the superposition of axial rotation on the sphericity. This is a close analogy to the control on the geomagnetic dynamo which, when averaged over the randomizing effect of secular variation, has axial symmetry. Similarly, the driver for equatorial growth of the inner core and consequent deformation has axial symmetry, but its rotational axis moves about in response to the varying electromagnetic torques of outer core dynamo action. Although the average alignment may be axial, the anisotropy of any part depends on its position and orientation at the time of formation or deformation.

The outer core presents a somewhat simpler problem. Internal fluid motion at speeds of tens of km/s ensures homogeneity and a temperature gradient that departs insignificantly from the adiabat. This gives some confidence to the application of equation of state principles, leading to the conclusion that $K'_\infty = 3.05 \pm 0.10$, which, by Equation (3), means $\gamma_\infty = 1.36 \pm 0.05$. It is clear that geophysical data cannot come anywhere near to accommodating the estimate $\gamma_\infty = 0.5$, which has been favored by some shock wave theories (e.g., [24]).

4. Is There a Best Equation of State?

It may appear that this question is answered by the conclusion of Sections 1 and 2 that no universal equation of state is possible, in principle, but that sidesteps the question. We are still interested to know whether one form of equation is most appropriate for all regions of the Earth, and all circumstances, or whether different equations should be applied, for example, to the core and mantle, or to different phases of the same material. Even if a single equation is acceptable for several materials, the parameters and coefficients will differ, but that is another question. The requirement $K'_0 > K'_\infty > 5/3$ is a condition for the plausibility of an equation. However, it does not constrain K'_∞ sufficiently to give reliable values of the derivatives of K that describe thermal properties, as well as compositional differences. Stacey [5] noted that, for both lower mantle and core equation of state fits, $K'_\infty = (3/5)K'_0$ within the uncertainties of the data. If this is used, it is a very strong constraint, but it is empirical, not based on fundamental principles, and may be specific to terrestrial materials or influenced by the equation of state used in the analysis. However, it complies with the general rule proposed by Keane [25]:

$$K'_0/2 < K'_\infty < K'_0 - 1. \tag{16}$$

The search for further tests and constraints continues. Stacey and Hodgkinson [6] showed that λ_∞ must be positive with the restriction

$$\lambda_\infty < K'_\infty - 2f_\infty/3, \tag{17}$$

where f_∞ is the infinite pressure limit of the parameter f in Equation (2). It probably differs very little from the lower mantle value of f, 1.67, but cannot plausibly be outside the range $1.5 < f_\infty < 1.8$, and this imposes a bound on λ_∞:

$$0 < \lambda_\infty < K'_\infty - x,\ 1.0 \le x \le 1.2. \tag{18}$$

By imposing this bound on Equation (10), we have a new equation of state constraint:

$$[\mathrm{d}(1/K')/\mathrm{d}(P/K)]_\infty > x/K'_\infty. \tag{19}$$

The significance of this to the present discussion is that it is quite a tight constraint on the gradient of the approach of a $1/K'$ vs. P/K equation to the infinite pressure asymptote and is more restrictive than the λ_∞ and K'_∞ constraints. In particular, it calls into question the reciprocal K-primed (RKP) equation [4], which was designed to take advantage of the relationship between K' and P/K by building on Equation (1):

$$1/K' = 1/K'_0 + \left(1 - K'_\infty/K'_0\right)(P/K). \tag{20}$$

Table 1 gives values of the quantities that are compared in Equation (19) for a selection of equations. A tick in the final column indicates an equation that passes the test imposed by Equation (19), regardless of its performance in the λ_∞ and K'_∞ tests. Crosses mark equations that clearly fail this test and those that are marginal or doubtful are assigned a ~ symbol.

Table 1. Values of the parameters in Equation (19) for selected equations of state.

Equation	Reference	$[\mathrm{d}(1/K')/\mathrm{d}(P/K)]_\infty$	x/K'_∞	Pass/Fail
Mie	[26]	$K'_\infty/K'_0 - 1 \approx 0.67$	$1/K'_\infty$ to $1.2/K'_\infty$	✓
Murnaghan	[27]	0	$1/K'_0$ to $1.2/K'_0$	×
Birch, 2nd order	[1]	0.71	0.43 to 0.51	✓
3rd order		0.78	0.33 to 0.40	✓
4th order		0.82	0.27 to 0.33	✓
Bardeen	[28]	0.5	0.50 to 0.60	~
Davis-Gordon	[29]	0.5	0.50 to 0.60	~
Keane	[25]	0	0.50 to 0.60	×
Rydberg	[30]	0.5	1.50 to 1.80	×
Holzapfel	[31]	0.8	0.60 to 0.72	✓ *
Logarithmic	[32]	−1	1.0 to 1.2	×
Reciprocal K-primed	[4]	$\left(1 - K'_\infty/K'_0\right) \approx 0.4$	$1/K'_\infty$ to $1.2/K'_\infty$	~

* Holzapfel's theory requires $K'_\infty = 5/3$.

The consistent ticks for the Birch equations in Table 1 make them appear as favorites for selection as the preferred choice, but with some qualifying comments, including the observation that for the highest pressures, the reciprocal K-primed equation (Equation (20)) is more satisfactory, despite the new doubt about it. The Birch equations are generally presented as successive approximations to an infinite series, but should not be considered that way, as they behave quite differently at high pressure. It is misleading to regard them as successive improvements of the same basic equation. The value of K'_∞ increases by 1/3 for each increase in order, 7/3 for the second order, 3 for the third order, and so on, becoming infinite for the complete series. As a result, the third-order Birch equation performs best for the core and the second order for the lower mantle, and the fourth and higher orders cannot be favored at all. Since the second-order equation is a special case of the Mie equation, the general Mie equation comes into contention as a rival. This conclusion is proposing nothing more than a stop gap measure as these equations have frequently been noted to have serious flaws [12,31–33], making them unusable outside the ranges of fitted data (if at all). Both the third- and fourth-order Birch equations, fitted to terrestrial data, lead to negative values of P and K not far from the terrestrial pressure range [16], making their derivatives unsatisfactory for thermal calculations even within that range. There appears to be no short cut to a choice of preferred equation, or modification of an existing one, but Section 5 presents a suggestion for a possible new approach.

5. Concluding Comments

We may be glimpsing a step forward on the erratic path to an analytical form of a high-pressure equation of state, one that will describe thermal, as well as mechanical, properties. Although most of the equations that have been presented so far are supported by logical arguments, and even sophisticated theoretical support, they are empirical, in the sense of relying on assumptions and intuitions that make them vulnerable to questioning. This note pursues a recent development of the questioning process, using only algebraic and thermodynamic principles. One of the outcomes is the conclusion of Stacey and Hodgkinson [6], extended in Section 2, that these principles do not lead to a completely general or 'universal' form of equation. However, they present criteria that all equations need to satisfy, as summarized by Equations (1) and (3), the thermodynamic bound on K'_∞ and the related relationship for λ_∞. Although these principles have sufficed to discount the majority of the 40+ proposed equations of state, their effectiveness is restricted by the appeal to infinite pressure asymptotes. While these introduce some rigorous conclusions by simplifying basic algebraic and thermodynamic relationships, they offer only indirect constraints on properties in the accessible range. Equation (19) presents a further development that addresses this limitation by extending the fundamental approach to finite pressures.

Recognizing that all equations of state must have end points represented by Equation (1), Stacey [4] compared several simple relationships between K' and P/K with terrestrial data and concluded that a reciprocal relationship was the only one that gave a reasonable fit. This was the basis of the RKP equation (Equation (20)), which became a serious candidate for the choice of 'best' equation. Equation (19) and Table 1 now show that it needs improvement and, most importantly, the nature of its shortcoming. Equation (19) gives the gradient of a graph of $1/K'$ vs. P/K for the approach to infinite pressure, showing it to be steeper than the gradient presented by Equation (20), which assumes a uniform gradient from $1/K'_0$ at $P = 0$. This means that $\mathrm{d}(1/K')/\mathrm{d}(P/K)$ increases with P/K and is not constant, as Equation (20) assumes. The graph should be a curve that is concave upwards. An independent indication of that is a graph of $1/K'$ vs. P/K for MgO at 300 K, as calculated by Stacey and Hodgkinson [6] from a pressure-density tabulation in a report of a 'first principles' calculation by Wu et al. [34]. This graph is concave upwards, supporting the conclusion from Equation (19) that Equation (20) needs to be modified in this way.

There appears to be the possibility of a laboratory check of the curvature of a $1/K'$ vs. P/K graph from measurements of derivatives of K at low to moderate pressures. For the RKP equation (Equation (20)), $\mathrm{d}(1/K')/\mathrm{d}(P/K) = (-KK'')/\left[K'^2(1 - K'\,P/K)\right] = \left(1 - K'_\infty/K'_0\right)$ at all pressures and at low pressure, is thus equal to $-K_0K''_0/K_0'^2$. If the suggestion that $K'_\infty/K'_0 \approx 3/5$ is accepted as general, then for a linear $1/K'$ vs. P/K relationship (the RKP equation),

$$\left(-K_0K''_0\right)/K_0'^2 \approx 0.4. \tag{21}$$

A lower value of this quantity is required for a $1/K'$ vs. P/K graph to be concave upwards, as expected by Equation (19), and a higher value would indicate a curve that is convex upwards. This looks like a useful test, but measurements of K''_0 are difficult and available values appear too imprecise to provide a convincing check. This problem needs a more thorough investigation before we can be clear that the RKP equation should be superseded and, until then, it must be considered the best available equation for very high pressures.

It remains to be seen how much further the fundamental approach can go. The Taylor expansion attempt failed, but that was conducted with the aim of finding a 'unique' equation. Perhaps the derivatives in Equations (7) to (9) or (13) could still yield a useful conclusion with a less ambitious target. Now that Equation (19) has qualitatively shown what to expect, further effort on the algebraic-thermodynamic approach will be better directed. Although the $1/K'$ vs. P/K relationship will not necessarily give the needed inspiration, it will indicate whether an idea is on track and appears to be central to further progress. The aim is to establish algebraic relationships that give the reliable derivatives required for the calculation of thermal properties, such as γ, as well as minimizing the use of adjustable constants in data fitting and identifying conflicts between observations and theory.

Funding: This research was supported by CSIRO Mineral Resources.

Acknowledgments: This paper pursues ideas originating from a collaboration with Jane Hodgkinson [6].

Conflicts of Interest: The author declares no conflicts of interest.

References

1. Birch, F. Elasticity and constitution of the earth's interior. *J. Geophys. Res.* **1952**, *57*, 227–286. [CrossRef]
2. Anderson, O.L. *Equations of State of Solids for Geophysics and Ceramic Science*; University Press: New York, NY, USA, 1995.
3. Stacey, F.D.; Davis, P.M. High pressure equations of state with applications to the lower mantle and core. *Phys. Earth Planet. Inter.* **2004**, *142*, 137–184. [CrossRef]
4. Stacey, F.D. The K-Primed approach to high pressure equations of state. *Geophys. J. Int.* **2000**, *143*, 621–628. [CrossRef]
5. Stacey, F.D. High pressure equations of state and planetary interiors. *Repts. Prog. Phys.* **2005**, *68*, 341–383. [CrossRef]
6. Stacey, F.D.; Hodgkinson, J.H. Thermodynamics with the Grüneisen parameter: Fundamentals and applications to high pressure physics and geophysics. *Phys. Earth Planet. Inter.* **2019**, *286*, 42–68. [CrossRef]
7. Vashchenko, V.Y.; Zubarev, V.N. Concerning the Grüneisen constant. *Sov. Phys. Solid State* **1963**, *5*, 653–655.
8. Irvine, R.D.; Stacey, F.D. Pressure dependence of a thermal Grüneisen parameter, with applications to the Earth's lower mantle and outer core. *Phys. Earth Planet. Inter.* **1975**, *11*, 157–165. [CrossRef]
9. Shanker, J.; Dulari, P.; Singh, P.K. Extreme compression behaviour of equations of state. *Phys. B* **2009**, *404*, 4083–4085. [CrossRef]
10. Singh, K.S. Second order and third order Grüneisen parameters at extreme compression. *Phys. B* **2012**, *407*, 668–669. [CrossRef]
11. Shanker, J.; Sunil, K.; Sharma, B.S. Formulation of the third order Grüneisen parameter at extreme compression. *Phys. B* **2012**, *407*, 2082–2083. [CrossRef]
12. Hama, J.; Suito, K. The search for a universal equation of state correct up to very high pressures. *J. Phys. Condens. Matter* **1996**, *8*, 67–81. [CrossRef]
13. Vinet, P. Universal features of the equation of state of solids. *J. Phys. Condensed Matter* **1989**, *1*, 1941–1963. [CrossRef]
14. Adams, P.; Smith, K.; Výborný, R. *Introduction to Mathematics with MAPLE*; World Scientific: Singapore, 2004.
15. Dziewonski, A.M.; Anderson, D.L. Preliminary reference earth model. *Phys. Earth Planet. Inter.* **1981**, *35*, 297–356. [CrossRef]
16. Stacey, F.D.; Davis, P.M. *Physics of the Earth*, 4th ed.; Cambridge University Press: Cambridge, UK, 2008.
17. Kennett, B.L.N.; Engdahl, E.R.; Buland, R. Constraints on seismic velocities in the earth from travel times. *Geophys. J. Int.* **1995**, *122*, 108–124. [CrossRef]
18. Deuss, A. The observation of inner core shear waves. *Geophys. J. Int.* **2000**, *142*, 67–73. [CrossRef]
19. Masters, G.; Gubbins, D. On the resolution of density within the Earth. *Phys. Earth Planet. Inter.* **2003**, *140*, 159–167. [CrossRef]
20. de Wit, R.W.L. Baysian inversion of free oscillations for Earth's radial (an)elastic structure. *Phys. Earth Planet. Inter.* **2014**, *237*, 1–17. [CrossRef]
21. Anderson, O.L. Iron: Beta phase frays. *Science* **1997**, *278*, 821–822. [CrossRef]
22. Belonoshko, A.B.; Ahuja, R.; Johansson, B. Stability of the body-Centred-Cubic phase of iron in the Earth's inner core. *Nature* **2003**, *424*, 1032–1034. [CrossRef]

23. Yoshida, S.; Sumita, I.; Kumazawa, M. Growth model of the inner core coupled with outer core dynamics and the resulting anisotropy. *J. Geophys. Res.* **1996**, *101*, 28085–28103. [CrossRef]
24. Burakovsky, L.; Preston, D.L. Analytical model of the Grüneisen parameter all densities. *J. Phys. Chem. Solids* **2004**, *65*, 1581–1587. [CrossRef]
25. Keane, A. An investigation of finite strain in an isotropic material subjected to hydrostatic pressure and its seismological applications. *Aust. J. Phys.* **1954**, *7*, 322–333. [CrossRef]
26. Mie, G. Zur kinetischen Theorie der einatomigen Körper. *Ann. Physik* **1903**, *11*, 657–697. [CrossRef]
27. Murnaghan, F.D. *Finite Deformation of an Elastic Solid*; Wiley: New York, NY, USA, 1951.
28. Bardeen, J. Compressibilities of the alkali metals. *J. Chem. Phys.* **1938**, *6*, 372–378. [CrossRef]
29. Davis, L.A.; Gordon, R.B. Compression of mercury at high pressure. *J. Chem. Phys.* **1967**, *46*, 2650–2660. [CrossRef]
30. Stacey, F.D.; Brennan, B.G.; Irvine, R.D. Finite strain theories and comparisons with seismological data. *Geophys. Surv.* **1981**, *4*, 189–232. [CrossRef]
31. Holzapfel, W.B. Physics of solids under strong compression. *Rep. Prog. Phys.* **1996**, *48*, 29–90. [CrossRef]
32. Poirier, J.P.; Tarantola, A. A logarithmic equation of state. *Phys. Earth Planet. Inter.* **1998**, *109*, 1–8. [CrossRef]
33. Hofmeister, A.M. Interatomic potentials calculated from equations of state: Limitations of finite strain to moderate K′. *Geophys. Res. Lett.* **1993**, *20*, 635–638. [CrossRef]
34. Wu, Z. Pressure-Volume-Temperature relations in MgO: An ultrahigh pressure-temperature scale for planetary sciences applications. *J. Geophys. Res.* **2008**, *113*, B06204. [CrossRef]

minerals MDPI

Article

A Simple Derivation of the Birch–Murnaghan Equations of State (EOSs) and Comparison with EOSs Derived from Other Definitions of Finite Strain

Tomoo Katsura [1,2,*] and Yoshinori Tange [3]

1 Bayerisches Geoinstitut, University of Bayreuth, 95440 Bayreuth, Germany
2 Center of High Pressure Science and Technology Advanced Research (HPSTAR), Beijing 100094, China
3 Japan Synchrotron Radiation Research Institute (JASRI), Kouto 679-5198, Japan; yoshinori.tange@spring8.or.jp
* Correspondence: tomo.katsura@uni-bayreuth.de; Tel.: +49-921-553791

Received: 19 October 2019; Accepted: 25 November 2019; Published: 30 November 2019

Abstract: Eulerian finite strain of an elastically isotropic body is defined using the expansion of squared length and the post-compression state as reference. The key to deriving second-, third- and fourth-order Birch–Murnaghan equations-of-state (EOSs) is not requiring a differential to describe the dimensions of a body owing to isotropic, uniform, and finite change in length and, therefore, volume. Truncation of higher orders of finite strain to express the Helmholtz free energy is not equal to ignoring higher-order pressure derivatives of the bulk modulus as zero. To better understand the Eulerian scheme, finite strain is defined by taking the pre-compressed state as the reference and EOSs are derived in both the Lagrangian and Eulerian schemes. In the Lagrangian scheme, pressure increases less significantly upon compression than the Eulerian scheme. Different Eulerian strains are defined by expansion of linear and cubed length and the first- and third-power Eulerian EOSs are derived in these schemes. Fitting analysis of pressure-scale-free data using these equations indicates that the Lagrangian scheme is inappropriate to describe P-V-T relations of MgO, whereas three Eulerian EOSs including the Birch–Murnaghan EOS have equivalent significance.

Keywords: high-pressure; compression; equation of state; finite strain; Eulerian scheme; Lagrangian scheme

1. Introduction

Density distributions are one of the most fundamental properties to describe planetary interiors. Materials within planetary interiors are under high pressure and, therefore, have higher densities than under ambient conditions. Material density measurements using high-pressure experimental techniques have a limited pressure range and obtained pressure-density data must often be extrapolated to higher pressures. For this procedure, it is necessary to fit available experimental data to a certain formula, which is referred to as the "equation of state" (EOS).

Various EOS formulas have been proposed. Among them, the third-order Birch–Murnaghan EOS is the most frequently used. This EOS is constructed on the basis of finite elastic strain theory in the Eulerian scheme. When strains are infinitesimal, the dimensions of a body decrease linearly with pressure. However, this situation is not the case when strains are finite because matter becomes increasingly incompressible with pressure. A theory to treat this phenomenon is, therefore, required: the finite elastic strain theory. The Eulerian scheme describes compression using the post-compression state as reference. In contrast, the other scheme, namely, the Lagrangian scheme describes compression using the pre-compressed state as reference.

Historically, finite elastic strain in the Euler scheme was treated by Murnaghan [1] who introduced Eulerian finite strain (see Equation (9)) and expressed pressure as a quadratic function of Eulerian finite strain. Birch [2] extended the theory of Murnaghan [1] to derive a prototype of the EOS, which is referred to as the Birch–Murnaghan EOS. They proposed an EOS using tensors in three dimensions, but other simpler arguments are available.

Poirier [3] plainly derived the Eulerian finite strain and Birch–Murnaghan EOS starting from the infinitesimal length squared with tensors in the three-dimensional space. However, if we ignore elastic anisotropy for matter compressed uniformly, there is no need to use a tensor or any special reason to start from infinitesimal length squared. Anderson [4] also derived the Eulerian finite strain and Birch–Murnaghan EOS. His derivation is, however, not easy to follow, either.

This article presents a more simple derivation of Eulerian finite strain and the Birch–Murnaghan EOS so that its essence can be easily understood. For comparison, other EOSs are constructed based on the Lagrangian scheme and other definitions of finite strain, and the usefulness and applicability of the Birch–Murnaghan EOS is discussed. Because we treat isothermal EOS's in this article, all differentiation is carried out at constant temperature. A partial derivative of a quantity Y with respect to some parameter X at constant temperature is, therefore, simply referred to as a X derivative of Y.

2. Derivations of the Birch–Murnaghan Equation of State (EOS)

2.1. Eulerian Finite Strain

The foundation of the Birch-Murnaghan EOS is Eulerian finite strain, and we therefore begin with its introduction. An important aspect of the Eulerian scheme is that the reference is defined by the post-compression state, whereas the Lagrangian scheme uses the pre-compression state. A second important point is that changes are expanded in squared length before and after compression.

Let us consider a cubic body with edge length X_0. Its volume is accordingly $V_0 = X_0^3$. This cube is uniformly compressed to edge length X and accordingly to a volume of $V = X^3$ (Figure 1). The edge length before compression is expressed by the reference length and length change, or displacement u, as:

$$X_0 = X - u \tag{1}$$

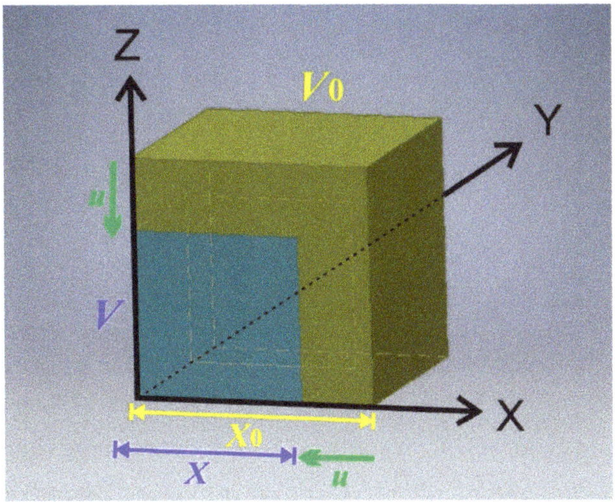

Figure 1. Finite uniform compression of a cube. The initial edge length of the cube X_0 is reduced to the final edge length $X = X_0 + u$. The volume of the cube accordingly decreases from $V_0 = X_0^3$ to $V = X^3$.

Because of compression, $u < 0$.

As mentioned, the change in squared edge length by compression is expanded as:

$$\begin{aligned} X^2 - X_0^2 &= X^2 - (X-u)^2 = 2Xu - u^2 \\ X_0^2 - X^2 &= u^2 - 2Xu \end{aligned} \tag{2}$$

Displacement should be proportional to the reference edge length because of uniform compression and is expressed using a proportional constant at a given compression, c, as:

$$u = cX \tag{3}$$

The quantity c is referred to as the strain in the linear elasticity. Using Equation (3), Equation (2) becomes:

$$X^2 - X_0^2 = 2XcX - c^2 = \left(2c - c^2\right)X^2 \tag{4}$$

Finite strain in the Eulerian scheme, ε_{E2}, is defined as:

$$\varepsilon_{E2} \equiv c - \frac{1}{2}c^2 \tag{5}$$

In this article, finite strain is referred to as the second-power Eulerian strain because it is defined by expanding the second power of length. The change of squared length (Equation (2)) is expressed using the Eulerian finite strain (Equation (5)) as:

$$X^2 - X_0^2 = 2\varepsilon_{E2}X^2 \tag{6}$$

The ratio of the edge length before compression to the reference edge length is expressed by Eulerian finite strain as:

$$\frac{X_0}{X} = (1 - 2\varepsilon_{E2})^{\frac{1}{2}} \tag{7}$$

The ratio of the pre-compression volume to the reference volume is also expressed by Eulerian finite strain as:

$$\frac{V_0}{V} = \left(\frac{X_0}{X}\right)^3 = (1 - 2\varepsilon_{E2})^{\frac{3}{2}} \tag{8}$$

Eulerian finite strain is therefore expressed by the volume ratio as:

$$\varepsilon_{E2} = \frac{1}{2}\left[1 - \left(\frac{V_0}{V}\right)^{\frac{2}{3}}\right] \tag{9}$$

The ε is negative when a body is compressed. To make strain positive by compression, we introduce f_{E2} as:

$$f_{E2} = -\varepsilon_{E2} = \frac{1}{2}\left[\left(\frac{V_0}{V}\right)^{\frac{2}{3}} - 1\right] \tag{10}$$

The quantity f_{E2} instead of ε_{E2} is usually referred to as Eulerian finite strain.

2.2. The Second-Order Birch–Murnaghan EOS

In isothermal EOSs, pressure, P, is expressed as a function of volume, V. From thermodynamics, pressure is the volume derivative of Helmholtz energy, F, as:

$$P = -\left(\frac{\partial F}{\partial V}\right)_T \tag{11}$$

The Helmholtz free energy of matter should increase with compression and may be expressed by a series of the Eulerian finite strain:

$$F = a_0 + a_1 f_{E2} + a_2 {f_{E2}}^2 + a_3 {f_{E2}}^3 + \ldots \tag{12}$$

Because the absolute value of F is arbitrary, the coefficient of the first term in Equation (12) can be $a_0 = 0$. Because pressure should be zero in an uncompressed state, $f_{E2} = 0$, we have:

$$P_{f_{E2}=0} = -\left(\frac{\partial F}{\partial f_{E2}}\right)_{T, f_{E2}=0}\left(\frac{\partial f_{E2}}{\partial V}\right)_{T,\ f_{E2}=0} = -a_1\left(\frac{\partial f_{E2}}{\partial V}\right)_{T,\ f_{E2}=0} = 0 \tag{13}$$

The coefficient of the second term in Equation (12) is therefore $a_1 = 0$.
Here, we truncate Equation (12) to the ${f_{E2}}^2$ term for the first approximation and therefore have:

$$F \cong a_2 {f_E}^2 \tag{14}$$

By substituting Equation (14) into Equation (11), we have

$$P = -\left\{\frac{\partial}{\partial V}\left(a_2 {f_{E2}}^2\right)\right\}_T = -2a_2\left(\frac{\partial f_{E2}}{\partial V}\right)_T f_{E2} \tag{15}$$

From the definition of finite strain (Equation (10)), the volume derivative of Eulerian finite strain is:

$$\left(\frac{\partial f_{E2}}{\partial V}\right)_T = \frac{\partial}{\partial V}\left[\frac{1}{2}\left\{\left(\frac{V_0}{V}\right)^{\frac{2}{3}} - 1\right\}\right] = -\frac{1}{3V_0}\left(\frac{V_0}{V}\right)^{\frac{5}{3}} \tag{16}$$

As shown in Appendix A, the coefficient a_2 is given by:

$$a_2 = \frac{9}{2} K_{T_0} V_0 \tag{17}$$

where K_{T_0} is the isothermal bulk modulus at standard temperature. By substituting Equations (10), (16), and (17) into Equation (15), we have the second-order Birch-Murnaghan EOS:

$$P = -2\frac{9}{2}K_{T_0}V_0\frac{1}{2}\left[\left(\frac{V_0}{V}\right)^{\frac{2}{3}} - 1\right]\left[-\frac{1}{3V_0}\left(\frac{V_0}{V}\right)^{\frac{5}{3}}\right] = \frac{3}{2}K_{T_0}\left[\left(\frac{V_0}{V}\right)^{\frac{7}{3}} - \left(\frac{V_0}{V}\right)^{\frac{5}{3}}\right] \tag{18}$$

This equation contains a subtraction formula between the 7/3 and 5/3 powers of V_0/V. The difference of these two powers, 7/3 − 5/3 = 2/3 is owing to the definition of Eulerian finite strain (Equation (10)), and the power of 5/3 in the second term is because of the volume derivative of Eulerian finite strain (Equation (16)).

2.3. The Third-Order Birch–Murnaghan EOS

The concept of the third-order Birch–Murnaghan EOS is almost identical to that of the second-order equation. The difference is that the Helmholtz free energy expressed by the Eulerian finite strain (Equation (12)) is truncated not up to the second term but to the third term as:

$$F \cong a_2 {f_{E2}}^2 + a_3 {f_{E2}}^3 \tag{19}$$

By substituting Equation (19) into Equation (11) and differentiating it with respect to volume, we have:

$$\begin{aligned} P = -\left\{\frac{\partial}{\partial V}\left(a_2 {f_{E2}}^2 + a_3 {f_{E2}}^3\right)\right\}_T &= -\left(2a_2 f_{E2} + 3a_3 f_{E2}^2\right)\left(\frac{\partial f_{E2}}{\partial V}\right)_T \\ &= -2a_2\left(1 + \xi_1 f_{E2}\right)\left(\frac{\partial f_{E2}}{\partial V}\right)_T f_{E2} \end{aligned} \tag{20}$$

where $\xi_1 = 3a_3/2a_2$. As presented in the Appendix A, the parameter ξ_1 is given by:

$$\xi_1 = \frac{3}{2}\left(K'_{T_0} - 4\right) \tag{21}$$

where K'_{T_0} is the pressure derivative of the isothermal bulk modulus at standard temperature. By substituting the definition of Eulerian finite strain (Equation (10)), its volume derivative (Equation (16)), the parameter a_2 (Equation (17)), and parameter m_1 (Equation (21)) into Equation (20), we have the third-order Birch–Murnaghan EOS as:

$$\begin{aligned} P &= -2\left(\tfrac{9}{2}K_{T_0}V_0\right)\left[1 + \tfrac{3}{2}\left(K'_{T_0} - 4\right)\tfrac{1}{2}\left\{\left(\tfrac{V_0}{V}\right)^{\frac{2}{3}} - 1\right\}\right]\left[-\tfrac{1}{3V_0}\left(\tfrac{V_0}{V}\right)^{\frac{5}{3}}\right]\tfrac{1}{2}\left[\left(\tfrac{V_0}{V}\right)^{\frac{2}{3}} - 1\right] \\ &= \tfrac{3}{2}K_{T_0}\left[\left(\tfrac{V_0}{V}\right)^{\frac{7}{3}} - \left(\tfrac{V_0}{V}\right)^{\frac{5}{3}}\right]\left[1 + \tfrac{3}{4}\left(K'_{T_0} - 4\right)\left\{\left(\tfrac{V_0}{V}\right)^{\frac{2}{3}} - 1\right\}\right] \end{aligned} \tag{22}$$

The second term in the first square bracket appears because of the truncation of the Helmholtz free energy to the higher-order (third) term. The form of the curly bracket is owing to the Eulerian finite strain (Equation (10)).

The third-order equation (Equation (22)) becomes identical to the second-order equation (Equation (18)) when

$$K'_{T_0} = 4 \tag{23}$$

On the other hand, if K'_{T_0} is neglected as $K'_{T_0} = 0$, the third-order equation should differ from the second-order equation.

2.4. The Fourth-Order Birch–Murnaghan EOS

For the fourth-order EOS, Equation (12) is truncated up to the fourth term as:

$$F \cong a_2 {f_{E2}}^2 + a_3 {f_{E2}}^3 + a_4 {f_{E2}}^4 \tag{24}$$

The pressure is then expressed as:

$$P = -\left(2a_2 f_{E2} + 3a_3 f_{E2}^2 + a_4 {f_{E2}}^3\right)\left(\frac{\partial f_{E2}}{\partial V}\right)_T = -2a_2 f_{E2}\left(1 + \xi_1 f_{E2} + \xi_2 f_{E2}^2\right)\left(\frac{\partial f_{E2}}{\partial V}\right)_T \tag{25}$$

where $\xi_2 = 2a_4/a_2$. As given in the Appendix A, the parameter ξ_2 is:

$$\xi_2 = \frac{9{K'_{T_0}}^2 - 63K'_{T_0} + 9K_{T_0}K''_{T_0} + 143}{6} \tag{26}$$

where K''_{T_0} is the second pressure derivative of the isothermal bulk modulus at standard temperature.

Similar to the lower-order EOSs, we have the fourth-order Birch–Murnaghan EOS as:

$$\begin{aligned} P = \tfrac{3}{2}K_{T_0}\left[\left(\tfrac{V_0}{V}\right)^{\frac{7}{3}} - \left(\tfrac{V_0}{V}\right)^{\frac{5}{3}}\right]&\left[1 + \tfrac{3}{4}\left(K'_{T_0} - 4\right)\left\{\left(\tfrac{V_0}{V}\right)^{\frac{2}{3}} - 1\right\}\right. \\ &\left. + \tfrac{1}{24}\left(9{K'_{T_0}}^2 - 63K'_{T_0} + 9K_{T_0}K''_{T_0} + 143\right)\left\{\left(\tfrac{V_0}{V}\right)^{\frac{2}{3}} - 1\right\}^2\right] \end{aligned} \tag{27}$$

This fourth-order equation (Equation (27)) becomes identical to the third-order equation when

$$K''_{T_0} = -\frac{1}{9K_{T_0}}\left(9{K'_{T_0}}^2 - 63K'_{T_0} + 143\right) \tag{28}$$

Again, if K''_{T_0} is neglected as $K''_{T_0} = 0$, the fourth-order equation should differ from the third-order equation.

3. Equations of States from Other Finite Strain Definitions

Through deriving the Birch–Murnaghan EOS, we may have some questions in mind. One is why the Eulerian scheme is more frequently used to describe the compression of bodies instead of the Lagrangian scheme. Another question is why is a change in squared length considered; why not other powers of length? Murnaghan's [1] argument leaves one to reflect that the use of squared length was based on Pythagorean theorem but there is no physical reason. To answer these questions, we derive EOSs from finite strain using the following different definitions:

1. Lagrangian scheme, expansion of squared length, referred to as the second-power Lagrangian EOS;
2. Eulerian scheme, linear expansion of length, referred to as the first-power Eulerian EOS;
3. Eulerian scheme, expansion of cubed length, referred to as the third-power Eulerian EOS.

3.1. The Second-Power Lagrangian EOS

Let us come back to Figure 1. In the Lagrangian scheme, the reference is taken as the pre-compressed state. Hence, the length after compression, X, is expressed using the reference length, X_0, and displacement:

$$X = X_0 + u \tag{29}$$

The change in squared edge length is:

$$X^2 - X_0^2 = 2X_0u + u^2 \tag{30}$$

The displacement is expressed by the reference length as:

$$u = c_0X_0 \tag{31}$$

Note that Equation (31) differs from Equation (3) unless $X \cong X_0$, namely when strain is finite. Using Equation (31), Equation (30) becomes:

$$X^2 - X_0^2 = \left(2c_0 + {c_0}^2\right)X_0^2 \tag{32}$$

We now define the Lagrangian finite strain, ε_{L2}, as:

$$\varepsilon_{L2} \equiv c_0 + \frac{1}{2}{c_0}^2 \tag{33}$$

Like the Eulerian scheme, Lagrangian finite strain is expressed using the volume ratio:

$$\varepsilon_{L2} = \frac{1}{2}\left[\left(\frac{V}{V_0}\right)^{\frac{2}{3}} - 1\right] \tag{34}$$

For convenience, the quantity f_{L2} below is hereafter referred to as the second-power Lagrangian finite strain:

$$f_{L2} = -\varepsilon_{L2} = \frac{1}{2}\left[1 - \left(\frac{V}{V_0}\right)^{\frac{2}{3}}\right] \tag{35}$$

EOSs are constructed based on this strain, which is hereafter referred to as the second-power Lagrangian EOS. The Birch–Murnaghan EOS is hereafter referred to as the second-power Eulerian EOS.

Like the second-order second-power Eulerian EOS, the Helmholtz free energy is expressed with the second-power Lagrangian finite strain squared as:

$$F \cong a_2{f_{L2}}^2 \tag{36}$$

The expression for pressure in the Lagrangian scheme is identical to the Eulerian scheme as:

$$P = -2a_2 f_{L2}\left(\frac{\partial f_{L2}}{\partial V}\right)_T \tag{37}$$

The volume derivative of the second-power Lagrangian finite strain is given by:

$$\left(\frac{\partial f_{L2}}{\partial V}\right)_T = -\frac{1}{3V_0}\left(\frac{V}{V_0}\right)^{-\frac{1}{3}} \tag{38}$$

The formula of parameter a_2 in the Lagrangian scheme is identical to that in the Eulerian scheme (Equation (17)). As a result, the second-order second-power Lagrangian EOS is given by:

$$P = \frac{3}{2}K_{T_0}\left[\left(\frac{V}{V_0}\right)^{-\frac{1}{3}} - \left(\frac{V}{V_0}\right)^{\frac{1}{3}}\right] = \frac{3}{2}K_{T_0}\left[\left(\frac{V_0}{V}\right)^{\frac{1}{3}} - \left(\frac{V_0}{V}\right)^{-\frac{1}{3}}\right] \tag{39}$$

For the third-order EOS, the Helmholtz free energy is expanded to the third term of f_{L2} as:

$$F \cong a_2 {f_{L2}}^2 + a_3 {f_{L2}}^3 \tag{40}$$

By differentiating Equation (33) with respect to volume, pressure is expressed as:

$$P = -2a_2(1 + \xi_1 f_{L2})\left(\frac{\partial f_{L2}}{\partial V}\right)_T f_{L2} \tag{41}$$

The expression of m_1 for the second-power Lagrangian finite strain differs from that for the second-power Eulerian strain and is given by:

$$\xi_1 = \frac{3}{2}{K_{T_0}}' \tag{42}$$

From Equations (17), (28), (34), and (35), we have the third-order second-power Lagrangian EOS as:

$$P = \frac{3}{2}K_{T_0}\left[\left(\frac{V}{V_0}\right)^{-\frac{1}{3}} - \left(\frac{V}{V_0}\right)^{\frac{1}{3}}\right]\left[1 + \frac{3}{4}K'_{T_0}\left\{1 - \left(\frac{V}{V_0}\right)^{\frac{2}{3}}\right\}\right] \tag{43}$$

The third-order equation (Equation (43)) becomes identical to the second-order equation (Equation (39)) when:

$$K'_{T_0} = 0 \tag{44}$$

In contrast to the Eulerian scheme, the neglect of K'_{T_0} as $K'_{T_0} = 0$ is identical to the neglect of a term higher than third order in the Helmholtz free energy (Equation (36)).

3.2. Finite Strains and EOSs from Linear Length

The change of length itself, namely the first power of length, is considered here in the Eulerian scheme and given from Equations (1) and (3) as:

$$X - X_0 = u = cX \tag{45}$$

The first-power Eulerian strain, f_{E1}, is therefore defined as:

$$f_{E1} \equiv -\varepsilon_{E1} = -c \tag{46}$$

Similar to the change in squared length, the first-power Eulerian strain is expressed by the volume change V_0/V as:

$$f_{E1} = \left[\left(\frac{V_0}{V}\right)^{\frac{1}{3}} - 1\right] \quad (47)$$

The volume derivative of the first-power Eulerian finite strain is given by:

$$\left(\frac{\partial f_{E1}}{\partial V}\right)_T = -\frac{1}{3V_0}\left(\frac{V_0}{V}\right)^{\frac{4}{3}} \quad (48)$$

The parameter a_2 for the first-power Eulerian strain is identical to those of the second-power strain in the Eulerian and Lagrangian schemes. The second-order first-power Eulerian EOS is, therefore, given by:

$$P = K_{T_0}\left[\left(\frac{V_0}{V}\right)^{\frac{5}{3}} - \left(\frac{V_0}{V}\right)^{\frac{4}{3}}\right] \quad (49)$$

The parameter ξ_1 is given by:

$$\xi_1 = \frac{3}{2}\left(K'_{T_0} - 3\right) \quad (50)$$

The third-order first-power Eulerian EOS is therefore given by:

$$P = 3K_{T_0}\left[\left(\frac{V_0}{V}\right)^{\frac{5}{3}} - \left(\frac{V_0}{V}\right)^{\frac{4}{3}}\right]\left[1 + \frac{3}{2}\left(K'_{T_0} - 3\right)\left\{\left(\frac{V_0}{V}\right)^{\frac{1}{3}} - 1\right\}\right] \quad (51)$$

3.3. *Finite Strains and EOSs from Cubed Length*

We now consider the change in length cubed, namely the third-power of length in the Eulerian scheme, which is given from Equations (1) and (3) as:

$$X^3 - X_0^3 = \left(3c - 3c^2 + c^3\right)X^3 \quad (52)$$

The third-power Eulerian strain, f_{E3}, is therefore defined as:

$$f_{E3} \equiv -\varepsilon_{E3} = -c + c^2 - \frac{1}{3}c^3 \quad (53)$$

Similar to those with the change in squared length, the first-power Eulerian strain is expressed by volume change V_0/V:

$$f_{E3} = \frac{1}{3}\left[\left(\frac{V_0}{V}\right) - 1\right] \quad (54)$$

The volume derivative of the third-power Eulerian finite strain is given by:

$$\left(\frac{\partial f_{E3}}{\partial V}\right)_T = -\frac{1}{3V_0}\left(\frac{V_0}{V}\right)^2 \quad (55)$$

In the same way as shown previously, the second-order third-power Eulerian EOS is given by:

$$P = K_{T_0}\left[\left(\frac{V_0}{V}\right)^3 - \left(\frac{V_0}{V}\right)^2\right] \quad (56)$$

The parameter ξ_1 is given by:

$$\xi_1 = \frac{3}{2}\left(K'_{T_0} - 5\right) \quad (57)$$

The third-order third-power Eulerian EOS is therefore given by:

$$P = K_{T_0}\left[\left(\frac{V_0}{V}\right)^3 - \left(\frac{V_0}{V}\right)^2\right]\left[1 + \frac{1}{2}\left(K'_{T_0} - 5\right)\left\{\frac{V_0}{V} - 1\right\}\right] \quad (58)$$

4. Discussion

4.1. Comparison of Birch–Murnaghan EOSs of Different Orders

Figure 2 shows pressures obtained by the second-, third- and fourth-order Birch–Murnaghan EOSs of NaCl in the B1-structure, Au, and MgO, which are materials frequently used as pressure standards in high-pressure experiments. The isothermal bulk moduli of these three materials and their first and second pressure derivatives at ambient temperature and zero pressure used for construction of Figure 2 are summarized in Table 1. Note that these EOS parameters were all obtained by the latest studies of sound velocity measurements [5–7].

Table 1. Bulk moduli and their pressure derivatives of frequently used pressure standard materials.

Material	K_{T_0} (GPa)	$K_{T_0}{}'$	$K_{T_0}{}''$ (GPa^{-1})	Reference
NaCl	23.7	5.14	−0.392	[5]
Au	160.44	6.56	0	[6]
MgO	160.64	4.35	0	[7]

The second- and third-order Birch–Murnaghan EOSs give significantly different pressures for NaCl and Au under high compression owing to large deviations of $K_{T_0}{}'$ from 4 (5.14 and 6.56). At $V/V_0 = 0.75$, the second- and third-order EOSs yield 12.1 and 14.3 GPa for NaCl and 85 and 112 GPa for Au, respectively, with differences of 18% and 41% in these cases. Birch [8] justified the validity of the Birch-Murnaghan EOS by identifying the second- and third-order EOSs within experimental uncertainty (e.g., $K_{T_0}{}' \cong 4$). However, modern experimental results have demonstrated that $K_{T_0}{}' \cong K_{S_0}{}'$ and significantly differs from 4 in many materials (see Table 11 in Bass [9]), which therefore compromises the validity of the second-order Birch–Murnaghan EOS.

Figure 2a also shows that the fourth-order equation gives slightly lower pressures than the third-order equation in the case of NaCl. At $V/V_0 = 0.75$, the fourth-order equation gives a pressure of 13.7 GPa, which is about 4% lower than that from the third-order equation. For the third- and fourth-order equations to give identical pressures, the second pressure derivative of the isothermal bulk modulus must not be $K_{T_0}{}'' = -0.392$ GPa^{-1} but rather $K_{T_0}{}'' = -0.267$ GPa^{-1}. Figure 2b,c also show the fourth-order Birch–Murnaghan EOSs of Au and MgO whose second pressure derivatives of the isothermal bulk modulus at zero pressure are set to $K_{T_0}{}'' = 0$. The pressure at the bottom of the Earth's mantle is 136 GPa. To obtain such pressure by the third-order Birch–Murnaghan EOSs of Au and MgO, the compression must be $V/V_0 = 0.736$ and $V/V_0 = 0.678$, respectively. Under these compressions, the fourth-order Birch–Murnaghan EOSs with $K_{T_0}{}'' = 0$ give 160 and 152 GPa. Thus, approximations that ignore higher-order terms of finite strain and equate higher-order derivatives of the isothermal bulk modulus to zero are not identical in the Birch–Murnaghan EOSs.

The differences in pressure given by the two kinds of EOSs above correspond to 470 and 330 km in the case of Au and MgO, respectively. The depth of the mantle phase transition at the base of the Earth's mantle is particularly complicated to argue. Although Hirose et al. [10] determined the post-spinel transition boundary in $MgSiO_3$ using Au and MgO pressure scales to interpret the phase boundary pressure as the D″ layer, their argument contains errors on the order of a few hundred km simply by neglecting K'_{T_0} as $K_{T_0}{}'' = 0$.

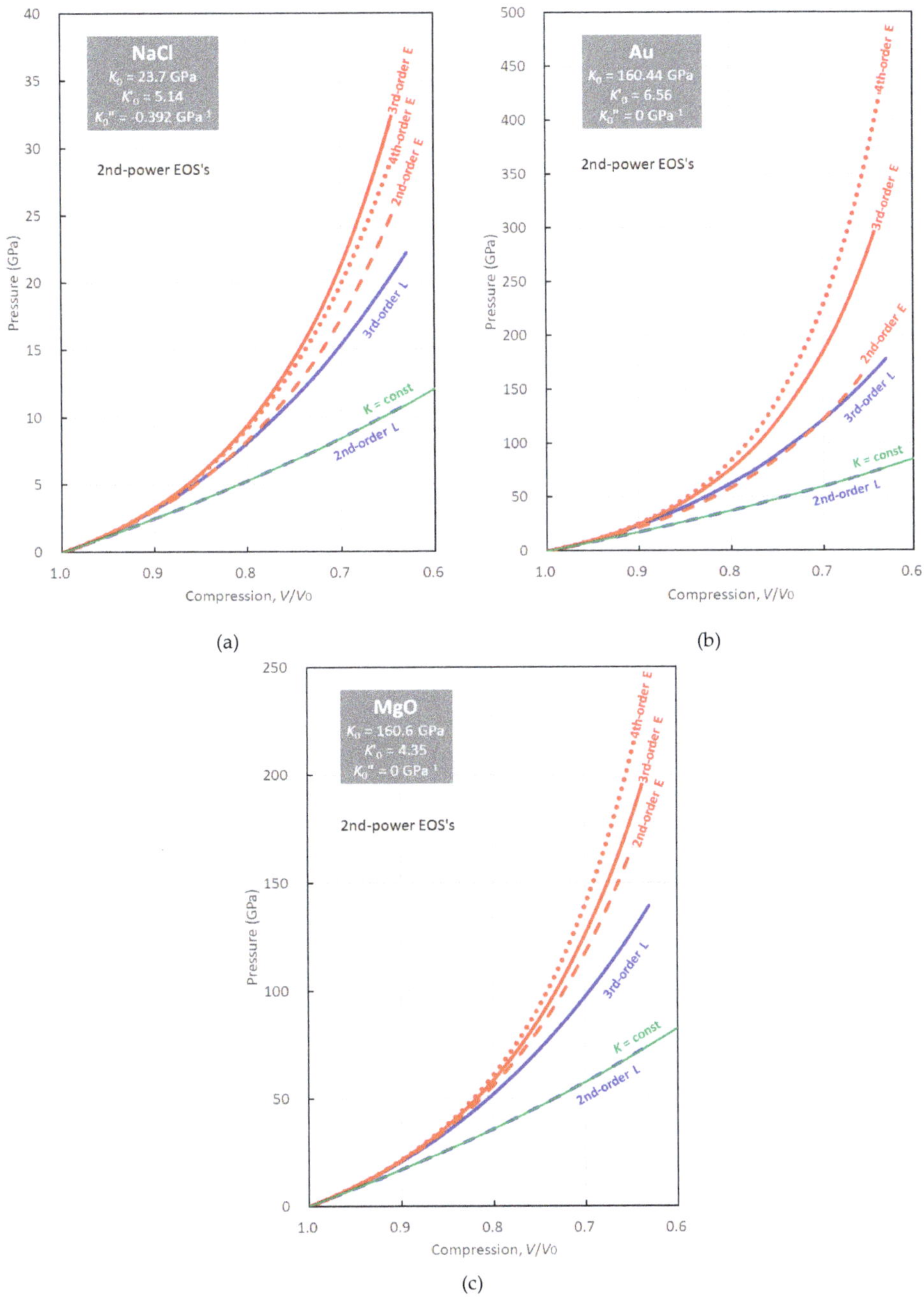

Figure 2. Comparison of the second-, third- and fourth-order Birch–Murnaghan equations of state (EOSs). (**a**) NaCl (B1) data are from Matsui et al. [5]; (**b**) Au data are from Song and Yoneda [6]; (**c**) MgO data are from Kono et al. [7]. Red and blue colors denote the Eulerian and Lagrangian schemes, respectively. The dashed, solid, and dotted curves are of the second-, third- and fourth-order EOSs, respectively. The green solid curve denotes the pressure-volume relation obtained by integration of the definition of the bulk modulus (Equation (59)).

4.2. Eulerian Versus Lagrangian Schemes

An essential difference between the Eulerian and Lagrangian finite strains is that the volume ratio is inversed with respect to one another (Equations (10) and (35)). Because the reference state is before compression in the Lagrangian scheme, the post-compression volume is put in the numerator of the strain formula (Equation (35)). As shown in Figure 3, the Lagrangian finite strain increases nearly linearly upon compression to $V/V_0 = 0.2$ and then approaches a finite value (0.5) with further compression ($V/V_0 \rightarrow 0$). As a result, the Helmholtz free energy and its volume derivative, namely pressure, only moderately increase with compression. However, it is important to note that pressure approaches infinity with compression to zero volume because the volume derivative of the finite strain and one term of the EOS are proportional to the −1/3 power of V/V_0 (Equations (38), (39), and (43)).

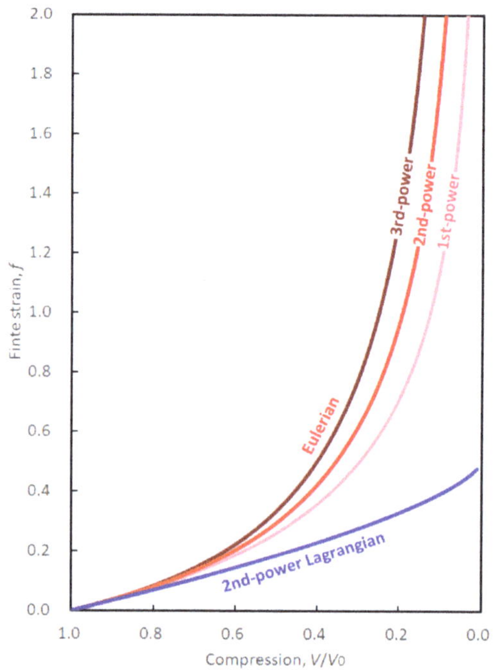

Figure 3. Comparison of the Eulerian and Lagrangian finite strains derived by expansion of the powered length as a function of compression V/V_0. The light red, red, and dark red curves denote the finite strains by expansion of the linear, squared, and cubed lengths in the Eulerian scheme, respectively, and the blue curve denotes those of the squared lengths in the Lagrangian scheme.

In contrast to the Lagrangian scheme, the Eulerian reference state is after compression and the post-compression volume is put in the denominator of the strain formula (Equation (10)). As shown in Figure 3, all Eulerian finite strains rapidly increase and then diverge to infinity as $V/V_0 \rightarrow 0$. As a result, the Helmholtz free energy and pressure also rapidly increase to infinity with $V/V_0 \rightarrow 0$. This behavior in the Eulerian scheme is in better agreement than that in the Lagrangian scheme. It is also important to note that the rate of pressure increase is very similar in both schemes under low compression. In other words, both schemes give the same results if the strain is infinitesimal.

Figure 2a–c show pressures given by the second- and third-order (second-power) Lagrangian EOSs as well as the Eulerian EOS's. For comparison, we also show pressures given by the simplest EOS, which is obtained by integrating the definition of the isothermal bulk modulus as:

$$P = K_{T_0} \ln\left(\frac{V_0}{V}\right) \tag{59}$$

with green curves in Figure 2a–c.

As previously discussed, finite strains given in the Eulerian scheme increase much more rapidly with compression than those of the Lagrangian scheme. The second-order second-power Lagrangian EOS gives almost identical pressures as the integration of a constant bulk modulus (Equation (37)). This means that the second-order Lagrangian EOS gives pressures without considering an increase in bulk modulus with compression. This is reasonable because the third-order Lagrangian EOS becomes identical to the second-order Lagrangian EOS when K'_{T_0} is neglected as $K'_{T_0} = 0$. In contrast, the Eulerian EOSs give substantially higher pressures than when integrating a constant bulk modulus (Equation (59)). As can be seen from the fact that the second- and third-order (second-power) Eulerian EOSs become identical in the case of $K'_{T_0} = 4$, the Eulerian EOSs implicitly contain the effects of an increased bulk modulus with compression.

The EOSs in both schemes converge with increasing order owing to better approximations of the Helmholtz free energy by the finite strains. This is the case for MgO (Figure 2c), however not for Au (Figure 2b). The small difference between the third- and fourth-orders of EOSs of NaCl (Figure 2a) does not imply that expansion to higher orders effectively helps conversion of the two schemes. These observations imply an essential problem with the Lagrangian scheme.

4.3. Equations of State Obtained from Expansions of Different Powers of Length

Figure 4a–c, show the third-order EOSs of NaCl, Au, and MgO, respectively, based on Eulerian finite strain defined by expansion of the first- (linear), second- (squared) and third- (cubed) powers of length given by Equations (22), (51), and (58), respectively. The EOS from the expansion of the second-power of length is the Birch–Murnaghan EOS. Because the finite strains based on higher powers of length increase more rapidly with compression, the EOSs based on higher powers of length are expected to give higher pressures. This is the case for materials with high K" such as Au. As is seen in Figure 4b, the first-, second- and third-power EOSs of Au gives pressures of 73, 76, and 78 GPa at $V/V_0 = 0.8$, respectively. The differences of the first- and third-power EOSs from the Birch–Murnaghan (second-power) EOS are both about −4% and +3%. The discrepancy between different powers of EOSs increases with compression, giving pressures of 170, 184, and 194 GPa, respectively, with differences of about −8% and +5%. The finite strains more rapidly increase under compression with increasing power of V_0/V from 1/3 to 2/3 and then to 1 (Equations (10), (47), and (54)), as shown in Figure 3. Because these three EOSs are obtained in the same way except for the power of length to define the finite strain, we cannot say that the Birch–Murnaghan EOS provides substantially more accurate pressures than the others.

As discussed previously, Birch [8] justified the Birch–Murnaghan EOSs by the identity of the second- and third-order EOSs when $K_{T_0}{}' \cong 4$, which is approximately the case in many kinds of materials. As Equations (51) and (58) show, the second- and third-order Eulerian EOSs based on finite strains defined from the first- and third-powers of length become identical when $K'_{T_0} = 3$ and $K'_{T_0} = 5$, respectively. Again, as summarized by Bass [9], the K'_{T_0} of the majority of solids are larger than 4. In the case of Au, $K'_{T_0} = 6.56$ and the third-power EOS (Equation (58)) should therefore be more appropriate for Au than the Birch–Murnaghan EOS (Equation (22)) if we follow Birch's [8] justification.

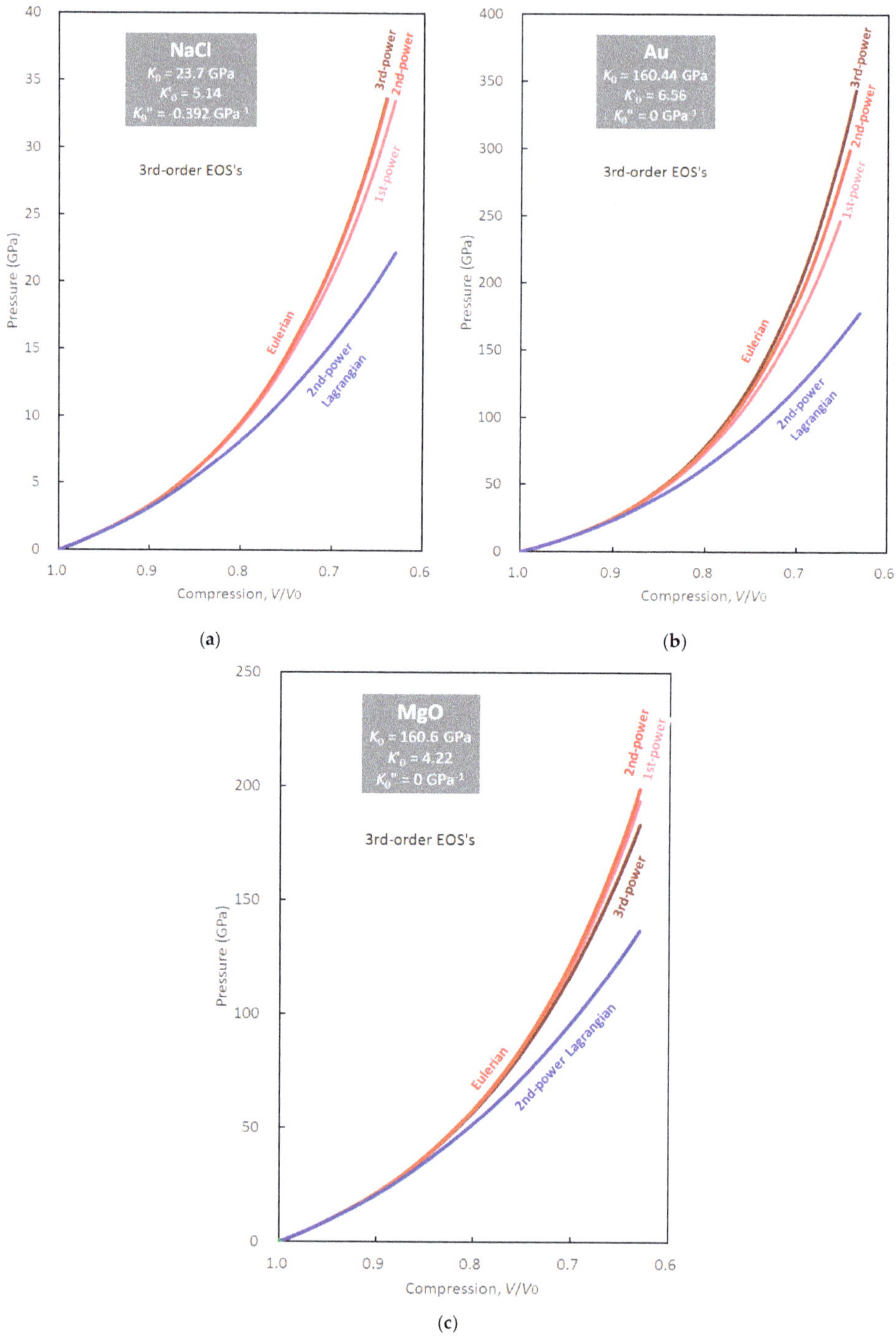

Figure 4. Pressures given by the second- (light red), third- (red) and fourth- (dark red) power Eulerian and the second-power Lagrangian (blue) third-order EOSs. (**a**) NaCl; (**b**) Au; (**c**) MgO.

4.4. Examination of Equations of State Using Pressure Scale-Free Experimental Data

In this section, we examine the validity of the second-power Lagrangian and first- and third-power Eulerian third-order EOS's in comparison with the third-order Birch–Murnaghan (second-power Eulerian) EOS. Specifically, we attempt to obtain parameters of the Mie-Grüneisen-Debye thermal EOS of MgO on the basis of these isothermal EOSs using pressure scale-free experimental data following the method by Tange et al. [11]. "Pressure scale-free" data were data that can be used to build an equation of state, but no other pressure scales such as pressure values obtained using equations of state of other pressure standard materials were used to obtain those data. Use of pressure scale-free data allows us building an equation of state with avoiding any circular argument. The pressure scale-free experimental data sets used here are zero-pressure and high-temperature thermal expansivity, zero-pressure and high-temperature adiabatic bulk moduli, 300-K and compressed adiabatic bulk moduli, and shock compression. Data sources are summarized in Table 2 of Tange et al. [11]. The following parameters at 300 K and 0 GPa are fixed: lattice volume, adiabatic bulk modulus, thermal expansivity, and isobaric heat capacity. The parameters obtained by these fittings are given in Table 3 of Tange et al. [11]. Note that parameters a and b are adjustable to express the volume dependence of the Grüneisen parameter γ, according to the following formula:

$$\gamma(V) = \gamma_0 \left\{ 1 + a \left[\left(\frac{V}{V_0} \right)^b - 1 \right] \right\} \tag{60}$$

Table 2 lists the obtained parameters of the thermal EOSs and Figure 5 shows the reproduction of Hugoniot curves using the obtained parameters.

Table 2. Parameters obtained by fitting pressure-scale-free data to various EOSs.

Parameter	2nd-Power Eulerian EOS (Birch-Murnaghan)	2nd-Power Lagrangian EOS	1st-Power Eulerian EOS	3rd-Power Eulerian EOS
V_0 (Å^3)	74.698 (fixed)	74.698 (fixed)	74.698 (fixed)	74.698 (fixed)
K_{T_0} (GPa)	160.64	160.55	160.64	160.63
K'_{T_0}	4.221	4.909	4.293	4.347
θ_0 (K)	761	761 (fixed)	761 (fixed)	761 (fixed)
γ_0	1.431	1.496	1.436	1.440
a	0.29	0 (fixed)	0.20	0.14
b	3.5		4.4	5.5

V_0: lattice volume at 300 K and 0 GPa; K_{T_0}: isothermal bulk modulus; K'_{T_0}: pressure derivative of the bulk modulus; θ_0: Debye temperature; γ_0: Grüneisen parameter; a, b: parameters to express volume dependence of the Grüneisen parameter given in Equation (59).

Figure 5 indicates that all Eulerian EOSs provide Hugoniot curves in agreement with the experimental data by Marsh [12] and Duffy and Ahrens [13], whereas the Lagrangian EOS does not. The reason for the failure in construction of an EOS in the Lagrangian scheme is that the parameters a and b in Equation (60) must not be less than zero; namely, thermal pressure or thermal expansivity should not increase with increasing pressure. Because the Lagrangian EOS gives much lower pressures at ambient temperature than the Eulerian EOSs, the thermal pressure must be abnormally high in the Lagrangian scheme to reproduce the shock experiment data. As a conclusion, the Eulerian scheme is more appropriate than the Lagrangian scheme to construct an EOS of real materials. On the other hand, linear and cubed length expansions in the Eulerian scheme provide equivalently appropriate EOSs compared with the EOS derived from the squared length expansion to define finite strain.

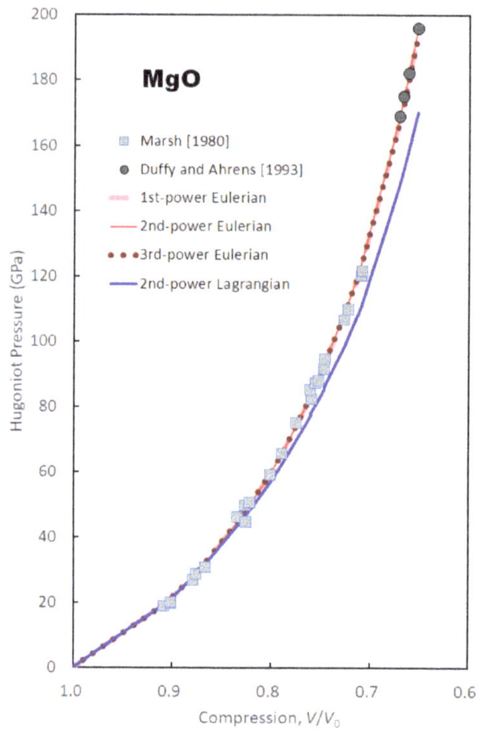

Figure 5. Hugoniot curves of MgO reproduced by the various EOSs using pressure scale-free data following Tange et al. [11]. The dashed light-red, thin-solid red, dotted dark-red, and solid blue curves are obtained from the first-, second- and fourth- Eulerian and second-power Lagrangian EOSs, respectively. The light-gray square and dark-gray circle are experimental data obtained by Marsh [12] and Duffy and Ahrens [13], respectively.

Author Contributions: Conceptualization, T.K.; methodology, T.K. and Y.T.; software, Y.T.; investigation, T.K. and Y.T.; writing—original draft preparation, T.K.; writing—review and editing, Y.T.; visualization, T.K.; funding acquisition, T.K.

Funding: This research was funded by the European Research Council (ERC) under the European Union's Horizon 2020 research and innovation programme (Grant agreement No. 787 527), and also funded by German Research Foundation (DFG), grant numbers KA3434/9-1.

Acknowledgments: T.K. thanks R.C. Liebermann for invitation to the Special Issue, Mineral Physics—In Memory of Orson Anderson, and E. Posner for improvement of English writing. TK also acknowledges A. Yoneda for his useful suggestion.

Conflicts of Interest: The authors declare no conflict of interest.

Appendix A

Here we derive parameters a_2 in Equations (17), ξ_1 in Equations (21), (42), (50) and (57), and ξ_2 in Equation (25), which are related to the isothermal bulk modulus and its first and second derivatives at zero pressures, respectively.

Appendix A.1 Derivation of Parameter a_2

The derivation of parameter a_2 in Equation (17) involves equating two expressions of the first volume derivative of pressure at zero pressure. One expression is based on the isothermal bulk modulus at zero pressure. The other assumes that the change in Helmholtz free energy by compression is proportional to the finite strain squared.

Let us first present the expression from the isothermal bulk modulus, which is defined as:

$$K_T = -V\left(\frac{\partial P}{\partial V}\right)_T \quad \text{(A1)}$$

The partial derivative of pressure with respect to volume at zero pressure is therefore:

$$\left(\frac{\partial P}{\partial V}\right)_{T,P=0} = -\frac{K_{T_0}}{V_0} \quad \text{(A2)}$$

Second, let us present the expression from the Helmholtz free energy by the Eulerian finite strain. By differentiating Equation (15) with respect to volume at constant temperature, we have:

$$\left(\frac{\partial P}{\partial V}\right)_T = \left[\frac{\partial}{\partial V}\left\{-2a_2 f\left(\frac{\partial f}{\partial V}\right)_T\right\}\right]_T = -2a_2\left[\left(\frac{\partial f}{\partial V}\right)_T^2 + f\left(\frac{\partial^2 f}{\partial^2 V}\right)_T\right]_T \quad \text{(A3)}$$

At zero pressure, the strain is zero: $f = 0$. The partial derivative of pressure with respect to volume at constant temperature and zero pressure is therefore:

$$\left(\frac{\partial P}{\partial V}\right)_{T,P=0} = -2a_2\left(\frac{\partial f}{\partial V}\right)_{T,P=0}^2 \quad \text{(A4)}$$

For Equations (16), (38), (48), and (57), the volume derivative of finite strain at zero pressure is identical for all Eulerian and Lagrangian finite strains as:

$$\left(\frac{\partial f}{\partial V}\right)_{T,P=0} = -\frac{1}{3V_0} \quad \text{(A5)}$$

By equating Equations (A2) and (A4) with Equation (A5), we have Equation (17).

Appendix A.2 Derivation of Parameter ξ_1

The essence of this derivation is identical to that of a_2. We present the second volume derivative of pressure, $\left(\partial^2 P/\partial V^2\right)_T$, in this proof.

We first express the volume second derivative of pressure using the isothermal bulk modulus and its pressure derivative. From the definition of the isothermal bulk modulus, we have

$$\left(\frac{\partial P}{\partial V}\right)_T = -\frac{K_T}{V} \quad \text{(A6)}$$

Therefore,

$$\left(\frac{\partial^2 P}{\partial V^2}\right)_T = \frac{K_T}{V^2}\left(K'_T + 1\right) \quad \text{(A7)}$$

At zero pressure, Equation (A7) is:

$$\left(\frac{\partial^2 P}{\partial V^2}\right)_{T,P=0} = \frac{K_{T_0}}{V_0^2}\left(K'_{T_0} + 1\right) \quad \text{(A8)}$$

The second volume derivative of pressure is then expressed using the finite strain. Before obtaining this formula, we obtain the second volume derivative of the second-power Eulerian, the second-power Lagrangian, and the first-and third-power Eulerian finite strains respectively as:

$$\left(\frac{\partial^2 f_{E2}}{\partial V^2}\right)_T = \frac{5}{9V_0^2}\left(\frac{V_0}{V}\right)^{\frac{8}{3}} \tag{A9a}$$

$$\left(\frac{\partial^2 f_{L2}}{\partial V^2}\right)_T = \frac{1}{9V_0^2}\left(\frac{V}{V_0}\right)^{-\frac{4}{3}} \tag{A9b}$$

$$\left(\frac{\partial^2 f_{E1}}{\partial V^2}\right)_T = \frac{2}{9V_0^2}\left(\frac{V_0}{V}\right)^{\frac{7}{3}} \tag{A9c}$$

$$\left(\frac{\partial^2 f_{E3}}{\partial V^2}\right)_T = \frac{2}{3V_0^2}\left(\frac{V_0}{V}\right)^{3} \tag{A9d}$$

At zero pressure, we have:

$$\left(\frac{\partial^2 f_{E2}}{\partial V^2}\right)_{T,P=0} = \frac{5}{9V_0^2} \tag{A10a}$$

$$\left(\frac{\partial^2 f_{L2}}{\partial V^2}\right)_{T,P=0} = \frac{1}{9V_0^2} \tag{A10b}$$

$$\left(\frac{\partial^2 f_{E1}}{\partial V^2}\right)_{T,P=0} = \frac{2}{9V_0^2} \tag{A10c}$$

$$\left(\frac{\partial^2 f_{E3}}{\partial V^2}\right)_{T,P=0} = \frac{2}{3V_0^2} \tag{A10d}$$

The pressure expressed by the volume derivative of the finite-strain polynomial up to the third term, such as Equation (20), is then differentiated with respect to volume as:

$$\begin{aligned}\left(\tfrac{\partial P}{\partial V}\right)_T &= \left[\tfrac{\partial}{\partial V}\left\{-2a_2 f(1+\xi_1 f)\left(\tfrac{\partial f}{\partial V}\right)_T\right\}\right]_T \\ &= -2a_2\left[(1+2\xi_1 f)\left(\tfrac{\partial f}{\partial V}\right)_T^{\,2} + \left(f+\xi_1 f^2\right)\left(\tfrac{\partial^2 f}{\partial V^2}\right)_T\right]\end{aligned} \tag{A11}$$

Equation (A11) is once more differentiated by volume as:

$$\left(\frac{\partial^2 P}{\partial V^2}\right)_T = -2a_2\left[(3+6\xi_1 f)\left(\frac{\partial f}{\partial V}\right)_T\left(\frac{\partial^2 f}{\partial^2 V}\right)_T + 2\xi_1\left(\frac{\partial f}{\partial V}\right)_T^{\,3} + \left(f+\xi_1 f^2\right)\left(\frac{\partial^3 f}{\partial V^3}\right)_T\right] \tag{A12}$$

As is done for Equation (A4), Equation (A12) at zero pressure becomes:

$$\left(\frac{\partial^2 P}{\partial V^2}\right)_{T,0} = -2a_2\left[3\left(\frac{\partial f}{\partial V}\right)_{T,0}\left(\frac{\partial^2 f}{\partial^2 V}\right)_{T,0} + 2\xi_1\left(\frac{\partial f}{\partial V}\right)_{T,0}^{\,3}\right] \tag{A13}$$

By equating Equations (A8) and (A13) with Equations (17), (A5) and (A10a–10d), we obtain Equations (21), (42), (50), and (57) for the second-power Eulerian, second-power Lagrangian, first-power Eulerian, and third-power Eulerian finite strains.

Appendix A.3 Derivation of Parameter ξ_2

To obtain the parameter m_2, we present the third volume derivative of pressure, $\left(\partial^3 P/\partial V^3\right)_T$. By differentiating Equation (A7) with respect to volume, we have:

$$\left(\frac{\partial^3 P}{\partial V^3}\right)_T = -\frac{K_T}{V^3}\left(K_T'^{\,2} + 3K_T' + 2 + K_T K_T''\right) \tag{A14}$$

At zero pressure, Equation (A14) is:

$$\left(\frac{\partial^3 P}{\partial V^3}\right)_{T_0} = -\frac{K_{T_0}}{V_0^3}\left(K_{T_0}'^{\,2} + 3K_{T_0}' + K_{T_0}K_{T_0}'' + 2\right) \tag{A15}$$

The third volume derivative of pressure is then expressed using the finite strain. Before obtaining this formula, we obtain the second volume derivative of the second-power Eulerian as:

$$\left(\frac{\partial^3 f_{E2}}{\partial V^3}\right)_T = -\frac{40}{27V_0^3}\left(\frac{V_0}{V}\right)^{\frac{11}{3}} \tag{A16}$$

At zero pressure, we have:

$$\left(\frac{\partial^3 f_{E2}}{\partial V^3}\right)_{T,P=0} = -\frac{40}{27V_0^3} \tag{A17}$$

The pressure expressed by the volume derivative of the finite-strain polynomial up to the fourth term (Equation (25)) is then differentiated by volume three times as:

$$\begin{aligned}\left(\frac{\partial^3 P}{\partial V^3}\right)_T = -2a_2 \quad & \left\{6\xi_2\left(\frac{\partial f_{E2}}{\partial V}\right)_T^4 + 6(2\xi_1 + 6\xi_2 f_{E2})\left(\frac{\partial f_{E2}}{\partial V}\right)_T^2\left(\frac{\partial^2 f_{E2}}{\partial V^2}\right)_T\right.\\ & +3\left(1 + 2\xi_1 f_{E2} + 3\xi_2 f_{E2}^2\right)\left(\frac{\partial^2 f_{E2}}{\partial V^2}\right)_T^2\\ & +4\left(1 + 2\xi_1 f_{E2} + 3\xi_2 f_{E2}^2\right)\left(\frac{\partial f_{E2}}{\partial V}\right)_T\left(\frac{\partial^3 f_{E2}}{\partial V^3}\right)_T\\ & \left.+\left(f_E + \xi_1 f_{E2}^2 + \xi_2 f_{E2}{}^3\right)\left(\frac{\partial^4 f_{E2}}{\partial V^4}\right)_T\right\}\end{aligned} \tag{A18}$$

At zero pressure, Equation (A18) becomes:

$$\begin{aligned}&\left(\frac{\partial^3 P}{\partial V^3}\right)_{T_0} =\\ &-2a_2\left\{6\xi_2\left(\frac{\partial f_{E2}}{\partial V}\right)_{T_0}^4 + 12\xi_1\left(\frac{\partial f_{E2}}{\partial V}\right)_{T_0}^2\left(\frac{\partial^2 f_{E2}}{\partial V^2}\right)_{T_0} + 3\left(\frac{\partial^2 f_{E2}}{\partial V^2}\right)_{T_0}^2 + 4\left(\frac{\partial f_{E2}}{\partial V}\right)_{T_0}\left(\frac{\partial^3 f_{E2}}{\partial V^3}\right)_{T_0}\right\}\end{aligned} \tag{A19}$$

By equating Equations (A14) and (A19) with Equations (17), (21), (A5), (A9c), and (A17), we have Equation (26) for the second-power Eulerian EOS.

References

1. Murnaghan, F.D. Finite Deformations of an Elastic Solid. *Am. J. Math.* **1937**, *59*, 235–260. [CrossRef]
2. Birch, F. Finite elastic strain of cubic crystals. *Phys. Rev.* **1947**, *71*, 809–824. [CrossRef]
3. Poirier, J.P. *Introduction to the Physics of the Earth's Interior*, 2nd ed.; Cambridge University Press: Cambridge, UK, 2000; p. 312.
4. Anderson, O.L. *Equations of States of Solids for Geophysics and Ceramics Science*; Oxford University Press: New York, NY, USA, 1995; p. 405.
5. Matsui, M.; Higo, Y.; Okamoto, Y.; Irifune, T.; Funakoshi, K. Simultaneous sound velocity and density measurements of NaCl at high temperatures and pressures: Application as a primary pressure standard. *Am. Mineral* **2012**, *97*, 670–1675. [CrossRef]

6. Song, M.; Yoneda, A. Ultrasonic measurements of single-crystal gold under hydrostatic pressures up to 8 GPa in a Kawai-type multi-anvil apparatus. *Chin. Sci. Bull.* **2008**, *52*, 1600–1606. [CrossRef]
7. Kono, Y.; Irifune, T.; Higo, Y.; Inoue, T.; Barnhoorn, A. *P-V-T* relation of MgO derived by simultaneous elastic wave velocity and in situ X-ray measurements: A new pressure scale for the mantle transition region. *Phys. Earth Planet. Inter.* **2010**, *183*, 196–211.
8. Birch, F. Elasticity and constitution of the earth's interior. *J. Geophys. Res.* **1952**, *57*, 227–286. [CrossRef]
9. Bass, J.D. Elasticity of Minerals, Glasses, and Melts. *Miner. Phys. Crystallogr. Handb. Phys. Constants* **1995**, *2*, 45–63.
10. Hirose, K.; Sinmyo, R.; Sata, N.; Ohishi, Y. Determination of post-perovskite phase transition boundary in $MgSiO_3$ using Au and MgO pressure standards. *Geophys. Res. Lett.* **2006**, *33*, L0310. [CrossRef]
11. Tange, Y.; Nishihara, Y.; Tsuchiya, T. Unified analyses for P-V-T equation of state of MgO: A solution for pressure-scale problems in high P-T experiments. *J. Geophys. Res. Solid Earth* **2009**, *114*, B03208. [CrossRef]
12. Marsh, S.P. *LASL Shock Hugoniot Data*; University California Press: Berkeley, CA, USA, 1980; pp. 312–313.
13. Duffy, T.S.; Ahrens, T.J. Thermal expansion of mantle and core materials at very high pressure. *Geophys. Res. Lett.* **1993**, *20*, 1103–1106. [CrossRef]

minerals

MDPI

Article

First Principles Calculation of the Stability of Iron Bearing Carbonates at High Pressure Conditions

Jun Tsuchiya [1,2,*,†], **Risa Nishida** [1,†] and **Taku Tsuchiya** [1,2,†]

1 Geodynamics Research Center (GRC), Ehime University, Ehime 790-8577, Japan; soranotumiki@outlook.jp (R.N.); tsuchiya.taku.mg@ehime-u.ac.jp (T.T.)
2 Earth-Life Science Institute (ELSI), Tokyo Institute of Technology, Tokyo 152-8550, Japan
* Correspondence: junt@ehime-u.ac.jp
† These authors contributed equally to this work.

Received: 16 October 2019; Accepted: 2 January 2020; Published: 8 January 2020

Abstract: Carbonate minerals such as ferromagnesite $(Mg,Fe)CO_3$ are suggested to be a possible major deep-carbon host in the lower mantle, because ferromagnesite is possibly stabilized by Fe spin crossover under pressure. However, the behavior of Fe-bearing carbonates under lower mantle pressure conditions has not been suitably examined thus far. Thus, in this study, we investigate the high-pressure stability of ferromagnesite and possible high-pressure structures with the chemical composition of $(Mg_{0.833}Fe_{0.167})CO_3$ via first principles calculation using internally consistent local density approximation with Hubbard parameter (LDA+U) method, which can more accurately account for the electronic state of Fe than the LDA and generalized gradient approximation (GGA) approaches. The enthalpy values obtained via our calculations suggest that $(Mg_{0.833}Fe_{0.167})CO_3$ undergoes phase transition from the $R\bar{3}c$ structure (high spin) to the $P\bar{1}$ (high spin) at 50 GPa, and to *C2/m* (high-spin) structure above 80 GPa, under static 0 K conditions. Therefore, no spin transitions in these carbonate minerals is expected under the lower mantle pressure conditions.

Keywords: carbonate mineral; high pressure; first principles calculation

1. Introduction

Carbon is one of the important volatile elements in the Earth. Accordingly, it is known to significantly affect the climate as well as dynamics processes on the planet, such as volcanic activity. Carbon is primarily transported into deep Earth in the form of carbonate minerals, such as magnesite ($MgCO_3$), calcite and aragonite ($CaCO_3$), dolomite ($CaMg(CO_3)_2$), and siderite ($FeCO_3$) [1,2]. Among them, magnesite has attracted special attention from researchers because of its high stability under high pressure and temperature conditions in the Earth's interior [3–8]. However, in general, the circulation mechanism of carbon in the deep interior remains largely unknown [9,10].

High pressure and temperature experiments conducted in some previous studies using a diamond-anvil cell (DAC) found that magnesite did not undergo phase transformation at 80 GPa between temperatures of 2000–2500 K [3–5]. In contrast, through an in situ X-ray diffraction experiment conducted in a later work, it was shown that magnesite transformed into an unknown form at ~115 GPa and 2100–2200 K [7]. Recently, in a study wherein in situ X-ray diffraction measurement using a double-sided laser-heated DAC was conducted, a phase transformation from magnesite to magnesite-II (space group *C2/m*) was reported at 80 GPa [8], which had previously been predicted based on ab initio calculations [11,12].

The structural evolution of magnesite under ultrahigh pressure conditions has been explored theoretically [11,12]. Oganov et al. [11] reported that the magnesite structure transformed to space group *C2/m* with a lower enthalpy than that of magnesite above 82.4 GPa; this transformation was predicted based on calculations performed using ab initio evolutionary crystal structure searching algorithm. In addition, they also reported that the *C2/m* crystal structure transformed to one with space group $P2_1$ above 138 GPa. Pickard and Needs [12] also conducted ab initio random structure searching of the $MgCO_3$ system and found that its new crystal structure (space group $P\bar{1}$) has a lower enthalpy than that of the *C2/m* phase in the range of 80–101 GPa.

The presence of Fe in carbonate minerals has been shown to modify their phase stability under high pressure and temperature conditions [13–15]. Ferromagnesite $(Mg,Fe)CO_3$ is a solid solution of magnesite and siderite. Several different results on the high pressure behaviors of ferromagnesite have been reported thus far. In particular, siderite-rich ferromagnesite $(Fe_{0.75}Mg_{0.25})CO_3$ has been reported to transform into a new Fe^{3+}-bearing high pressure carbonate, Fe_3O_4 magnetite, and nano-diamonds above 80 GPa and 1850–2300 K [2]. In contrast, it was found that $(Fe_{0.65}Mg_{0.35})CO_3$ transforms into an orthorhombic phase (phase II with space group *Pmm2*) above ~50 GPa and 1400 K; this was determined via a synchrotron X-ray diffraction experiment using a laser-heated DAC [14]. It should be noted that these previous experiments were strongly affected by the oxygen fugacity in the system at high pressure–temperature conditions[2,14,16].

Moreover, it has been suggested that spin crossover of Fe in high-pressure carbonate phases stabilizes them at relatively lower pressure conditions [14]. In particular, high-spin (HS) to low-spin (LS) transition of ferrous iron (Fe^{2+}) in $(Mg,Fe)CO_3$ ferromagnesite has been indicated via Raman spectroscopy, X-ray emission spectroscopy, and X-ray diffraction measurements at ~50 GPa [14,17–21]. In addition, first principles calculation by Hsu and Huang [22] showed that a spin transition from HS to LS occurs in ferromagnesite at around 45–50 GPa, which is consistent with the previously conducted experiments. Furthermore, Hsu and Huang [22] also found that spin transition pressure does not depend on the Fe concentration in ferromagnesite.

In general, it is believed that the spin transition of Fe significantly affects the structure, elasticity, and thermodynamic properties of minerals owing to a large difference between the ionic radii of Fe in its HS and LS states. However, these effects of Fe spin transitions on the stability of high-pressure phases of ferromagnesite have not been investigated thus far. Thus, in this study, we investigated the structure and stability of ferromagnesite and its possible high-pressure phases via first principles calculation. Because the strongly correlated nature of Fe cannot be properly reproduced using local density approximation (LDA) and generalized gradient approximation (GGA) approaches, we used the internally consistent LDA+U method wherein the Hubbard parameter *U* was determined based on our first principles calculation [23].

2. Results

2.1. Calculation of $MgCO_3$ Polymorphs under Pressure

First, we optimized the structural parameters of $R\bar{3}c$, $P\bar{1}$, and *C2/m* structures of $MgCO_3$ polymorphs and confirmed that our calculations are consistent with previously reported results (Table 1) [11,12]. In particular, the coordination number (CN) of all Mg sites is six in $R\bar{3}c$ structure, whereas $P\bar{1}$ and *C2/m* structures contain three different crystallographic Mg sites (Figure 1). Some of the CNs of Mg in $P\bar{1}$ and *C2/m* structures are larger than six ($P\bar{1}$: $CN_{M1} = 8$, $CN_{M2} = 6$, $CN_{M3} = 8$; C2/m: $CN_{M1} = 8$, $CN_{M2} = 8$, $CN_{M3} = 10$), and thus their polyhedral volumes are significantly larger than that of $R\bar{3}c$ (V_M = 9.0~12.0 Å^3) (Figure 1).

Table 1. Optimized structural parameters of P$\bar{1}$ and C2/m phases of $MgCO_3$ polymorphs.

Atom	Wyckoff position	Atomic coordinates		
P$\bar{1}$ structureat 100 GPa				
a = 5.178 Å, b = 5.206 Å, c = 7.242 Å, $\alpha = 69.85°, \beta = 81.64°, \gamma = 78.04°$				
(Pickard and Needs [12]: a = 5.211 Å, b = 5.238 Å, c = 7.268 Å, $\alpha = 70.030°, \beta = 81.904°, \gamma = 78.272°$ at 100 GPa)				
		x	y	z
Mg1(M1)	2*i*	0.7414	0.2541	0.0022
Mg2(M2)	2*i*	0.5669	0.4731	0.3143
Mg3(M3)	2*i*	0.0636	0.1784	0.6501
C1	2*i*	0.2237	0.2078	0.0295
C2	2*i*	0.0812	0.3351	0.3092
C3	2*i*	0.4398	0.9790	0.3231
O1	2*i*	0.1277	0.1522	0.8965
O2	2*i*	0.0086	0.2930	0.1470
O3	2*i*	0.8794	0.3896	0.4232
O4	2*i*	0.3461	0.9572	0.1548
O5	2*i*	0.3752	0.3872	0.9623
O6	2*i*	0.2030	0.0682	0.4149
O7	2*i*	0.2249	0.5252	0.2622
O8	2*i*	0.5465	0.7397	0.4278
O9	2*i*	0.6110	0.1433	0.2692
C2/m structure at 120 GPa				
a = 8.0417 Å, b = 6.4468 Å, c = 6.8273 Å, $\beta = 103.84°$				
(Oganov et al. [11]: a = 8.0945 Å, b = 6.4881 Å, c = 6.8795 Å, $\beta = 103.98°$ at 120 GPa)				
Atom	**Wyckoff position**	**Atomic coordinates**		
		x	y	z
Mg1(M1)	4*g*	0.0000	0.7536	0.0000
Mg2(M2)	4*i*	0.3229	0.5000	0.6975
Mg3(M3)	4*i*	0.9353	0.5000	0.6555
C1	8*j*	0.3701	0.3214	0.3273
C2	4*i*	0.2315	0.5000	0.0363
O1	4*i*	0.0926	0.5000	0.8961
O2	8*j*	0.3490	0.1665	0.4299
O3	8*j*	0.5081	0.3095	0.2675
O4	4*i*	0.3661	0.5000	0.9729
O5	4*i*	0.3522	0.5000	0.4276
O6	8*j*	0.2253	0.3320	0.1595

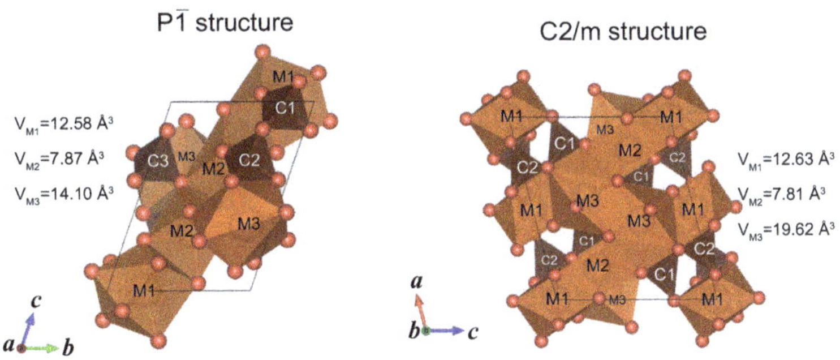

Figure 1. Optimized structures and polyhedral volumes (M1, M2, and M3 sites) of *P*$\bar{1}$ and *C2/m* $MgCO_3$ polymorphs at 120 GPa.

2.2. Structure Modeling of $(Mg_{0.833}Fe_{0.167})CO_3$

For the atomic configuration of Fe in (Mg,Fe)CO_3, we selected the lowest enthalpy configuration of Fe in the cell. In particular, we used the 2 × 2 × 1 supercell of the conventional $R\bar{3}c$ structure, and replaced four Mg with Fe atoms (Figure 2) to avoid the generation of an FeO_6 octahedral layer.

Consequently, there was almost no site preference for Fe configurations in the $R\bar{3}c$-type structure because all Mg sites in the $R\bar{3}c$ structure have a CN of six with a Wyckoff *6b* position. It should be noted that we retain the same crystal structure notations for $(Mg_{0.833}Fe_{0.167})CO_3$ as those of $MgCO_3$ though the symmetry (space group) of $(Mg_{0.833}Fe_{0.167})CO_3$ polymorphs are changed to $P1$ (no symmetry) because of the incorporation of Fe.

In the $P\bar{1}$ structure, there are three independent Mg positions (Figure 1, Table 1); in addition, we confirmed that the structure with Fe in the M3 position shows the lowest enthalpy in the HS state structures, whereas Fe in the M2 position is the most stable in the LS state structures (Figure 3). Owing to the relatively larger ionic radius of HS-Fe, the structure with a large polyhedral volume (V_{M3} = 14.10 Å^3 at 100 GPa) is most stable, whereas, for LS-Fe, the structure with the smallest polyhedral volume (V_{M2} = 7.87 Å^3 at 100 GPa) is most stable (Figure 2).

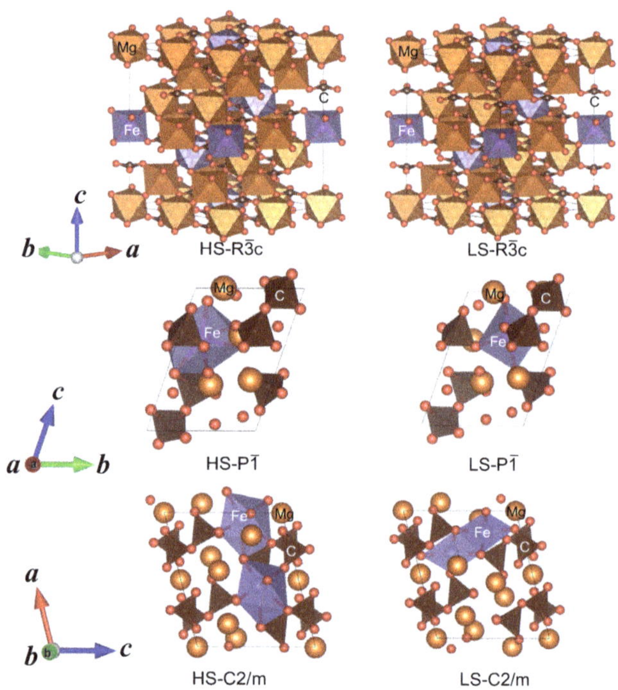

Figure 2. Structural models used for $(Mg_{0.833}Fe_{0.167})CO_3$ polymorphs.

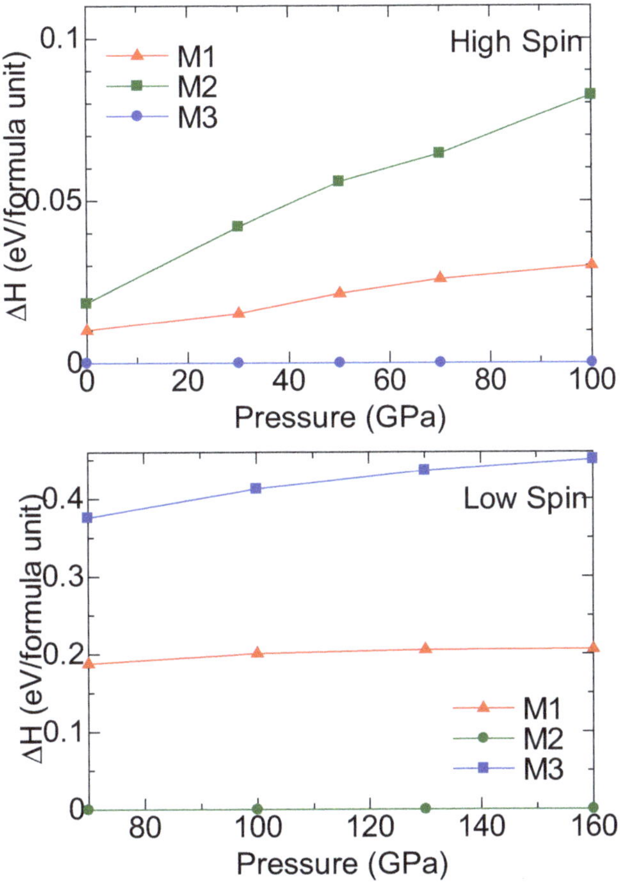

Figure 3. Enthalpies of $P\bar{1}$-($Mg_{0.833}$,$Fe_{0.167}$)CO_3 structures with respect to the lowest enthalpy configuration.

A similar trend was observed for site preference of Fe in the *C2/m* crystal structure, wherein HS-Fe atom is the most stable at the M3 site, which has the largest polyhedral volume, whereas LS-Fe atom is most stable at the M2 site leading to the lowest enthalpy (Figure 4).

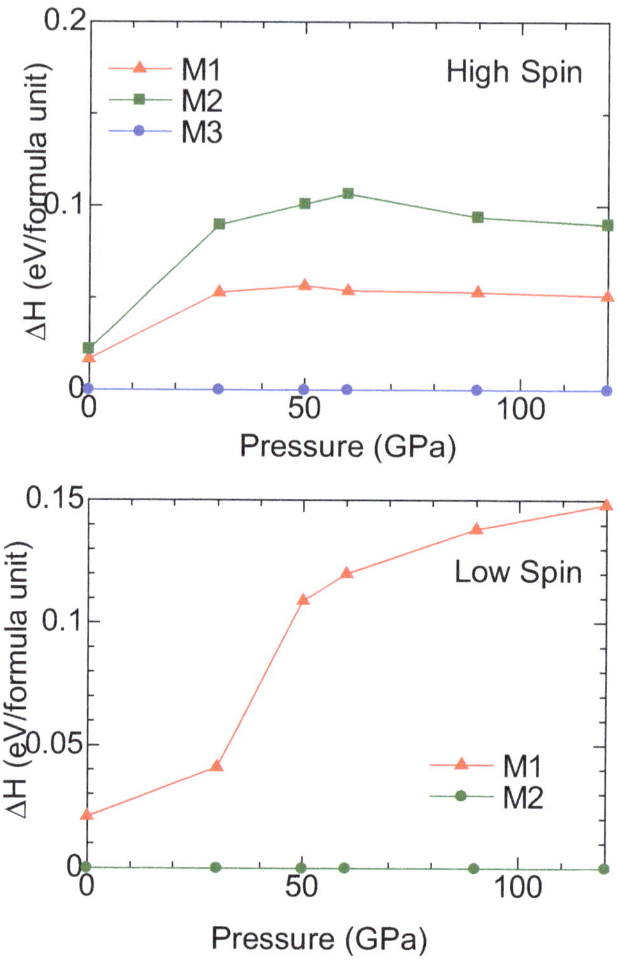

Figure 4. Enthalpies of *C2/m*-($Mg_{0.833}Fe_{0.167}$)CO_3 structures with respect to the lowest enthalpy configurations. Calculations reveal that the structure with LS-Fe atoms at M3 sites was highly unstable.

2.3. High Pressure Transitions of $MgCO_3$ and ($Mg_{0.833}Fe_{0.167}$)CO_3

We have calculated the enthalpy of $MgCO_3$ and ($Mg_{0.833}Fe_{0.167}$)CO_3 with $R\bar{3}c$, $P\bar{1}$, and *C2/m* structures. Figure 5a shows the enthalpy of $MgCO_3$ relative to that of the *C2/m* phase based on LDA. $MgCO_3$-magnesite transforms into $P\bar{1}$ phase at 60 GPa, and the *C2/m* phase is stable above 80 GPa. In contrast, calculations in previous studies indicate the stability of the $P\bar{1}$ phase between 85–101 GPa, and that of the *C2/m* phase between 101–144 GPa based on GGA [12]. It is typically known that LDA/GGA leads to overestimation/underestimation of the bond strength; therefore, the volume calculated using LDA is usually smaller than that using GGA. Our calculated transition pressures for $MgCO_3$ obtained using LDA are 10–15 GPa lower than those obtained using GGA; however, these GGA results are consistent with previously reported GGA results (Table 2) [12]. Hereafter, in our study, we use LDA results to investigate the stability of ferromagnesite. Nevertheless, it should be noted that the difference of about 10–15 GPa in the transition pressures between LDA and GGA results is typical, and thus, it provides lower and upper bounds for the transition pressures, respectively.

Table 2. Comparison between previously reported calculation results for high-pressure transition of $MgCO_3$ and those obtained in this study.

	GGA[12]	GGA (This Study)	LDA (This Study)
$R\bar{3}c \rightarrow P\bar{1}$	85	85	62
$P\bar{1} \rightarrow C2/m$	101	95	80
$C2/m \rightarrow P2_12_12_1$	144	145	130

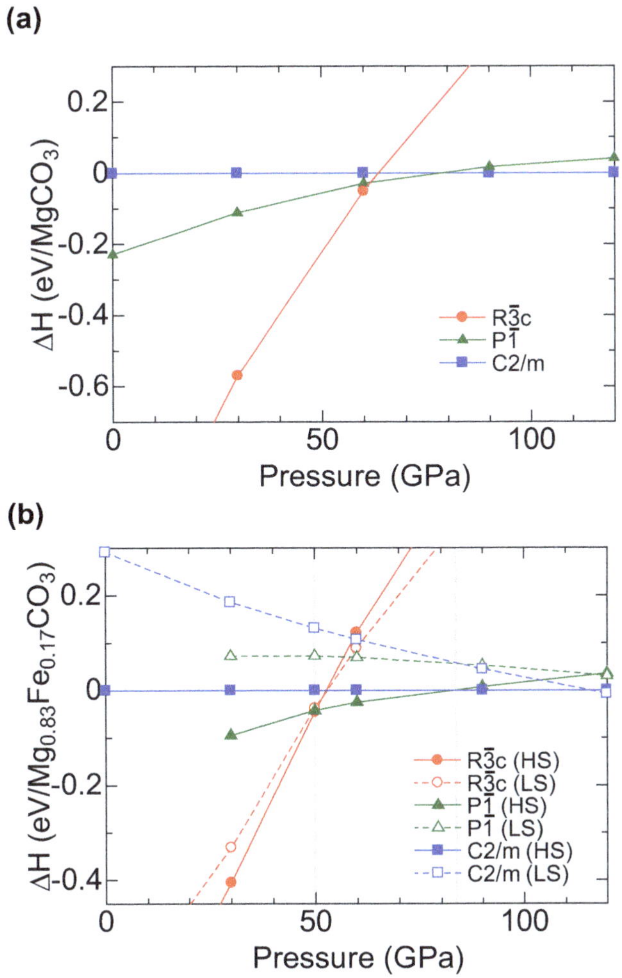

Figure 5. (**a**) Relative enthalpies of $MgCO_3$ polymorphs with respect to that of the *C2/m* phase. These calculations are based on the LDA method. (**b**) Relative enthalpies of $(Mg_{0.833}Fe_{0.167})CO_3$ under pressure with respect to that of the HS-*C2/m* phase. These calculations are based on the LDA+U_{ic} method.

Figure 5b shows the relative enthalpies of ferromagnesite $(Mg_{0.833}Fe_{0.167})CO_3$ and those of its high-pressure candidate phases with respect to the HS-*C2/m* phase obtained using the LDA+U method. The spin crossover of ferromagnesite (from HS-$R\bar{3}c$ to LS-$R\bar{3}c$) occurs at ~50 GPa which is consistent with previously reported calculations [22]. However, the HS-$P\bar{1}$ phase has a lower enthalpy than LS-ferromagnesite above ~50 GPa. In addition, the HS-*C2/m* phase has a slightly lower enthalpy than that of the HS-$P\bar{1}$ phase above about 80 GPa, and the former is stable up to about 120 GPa, suggesting that no spin crossover is expected under lower mantle pressure conditions. This high stability of the HS state in $P\bar{1}$ and $C2/m$ phases can be attributed to the large polyhedral volumes due to Fe^{2+} in these structures (Figures 1 and 2). However, the energy differences between HS and LS states are relatively small (~0.03 eV, equivalent to ~350 K) in the $P\bar{1}$ and $C2/m$ phases above 100 GPa. Therefore, there is still a possibility that the high-pressure transition sequence changes or spin crossover occurs around 100 GPa for above-ambient temperature conditions.

3. Discussion

Carbonate minerals are expected to exist in the subducting cold slabs. Because SiO_2 is one of the most abundant components in a subducting slab, it has been previously indicated that a reaction between $MgCO_3$ and SiO_2 can occur in deeply subducted slabs [8]. We calculated enthalpies of the following chemical reactions to investigate the relative stability of carbonate phases compared with the mantle mineral phases:

$$MgCO_3 + SiO_2 \rightarrow MgSiO_3 + CO_2, \tag{1}$$

wherein the high pressure phase of CO_2 corresponds to the tetragonal $I\bar{4}2d$ phase [11], SiO_2 is stishovite, and the $CaCl_2$-type structure below and above 50 GPa, respectively, and $MgSiO_3$ has bridgmanite-type structure. The relative enthalpy curves shown in Figure 6a indicate that the transformation of $MgCO_3$ and SiO_2 into $MgSiO_3$-bridgmanite along with CO_2 formation is unfavorable, which is consistent with the results reported in previous studies [11,12].

Then, we calculated the reaction between $(Mg_{0.833}Fe_{0.167})CO_3$ and SiO_2,

$$(Mg_{0.833}Fe_{0.167})CO_3 + SiO_2 \rightarrow (Mg_{0.833}Fe_{0.167})SiO_3 + CO_2. \tag{2}$$

For the calculation of enthalpy of $(Mg_{0.833}Fe_{0.167})SiO_3$, we used the supercell of bridgmanite containing 60 atoms with high spin Fe^{2+} at A sites [24]. Although both Fe^{2+} and Fe^{3+} have been observed in bridgmanite, ferric iron Fe^{3+} content in bridgmanite strongly correlates with the concentration of Al in bridgmanite [25]. In addition, the spin crossover of Fe^{2+} from HS to LS at an A site has not been theoretically shown at lower mantle pressure conditions (e.g., [26]). Moreover, the enthalpy of HS-$(Mg_{0.833}Fe_{0.167})SiO_3$-bridgmanite plus CO_2 is higher than that of the Fe-bearing carbonate mineral plus SiO_2 up to ~120 GPa, which also suggests that decomposition of $(Mg_{0.833}Fe_{0.167})CO_3$-carbonate phases is unfavorable under lower mantle pressure conditions (Figure 6a).

Finally, we also investigated the stability of Fe^{2+} between bridgmanite and carbonate minerals by the following Fe exchange reaction,

$$(Mg_{0.833}Fe_{0.167})CO_3 + MgSiO_3 \rightarrow MgCO_3 + (Mg_{0.833}Fe_{0.167})SiO_3. \tag{3}$$

In the case of $R\bar{3}c$ structures, the enthalpy of the mixture of (Mg,Fe)CO_3 and $MgSiO_3$ is higher than that of $MgCO_3$ and (Mg,Fe)SiO_3 (Figure 6b), which indicates that Fe^{2+} is more stable in bridgmanite than in (Mg,Fe)CO_3. HS-Fe^{2+} is more stable at the A site of bridgmanite because the polyhedral volume of the A site of the bridgmanite (16–21 Å^3) is significantly larger than those of ferromagnesite (7–12 Å^3). However, the HS-Fe^{2+} is more stable in $P\bar{1}$ and $C2/m$ structures because HS-Fe^{2+} can be well accommodated at the sites with large polyhedral volumes in these phases.

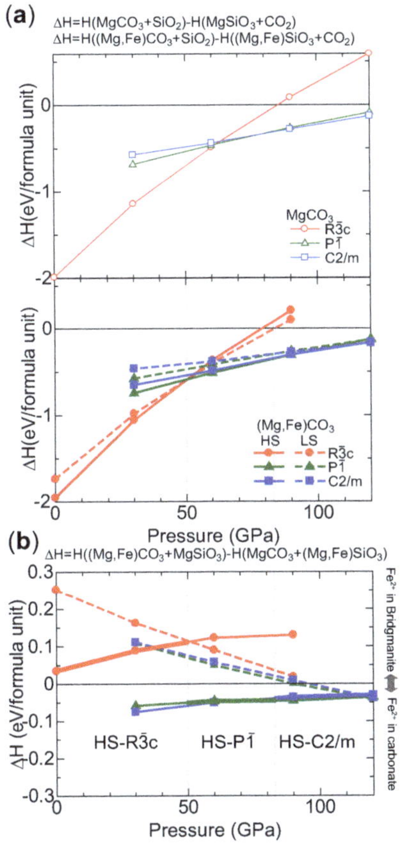

Figure 6. (**a**) The thin lines correspond to the relative enthalpy of $MgCO_3$ + SiO_2 with respect to $MgSiO_3$-bridgmanite + CO_2. The full/dashed lines indicate the relative enthalpy of the HS/LS-$(Mg_{0.833}Fe_{0.167})CO_3$ + SiO_2 with respect to the HS-$(Mg_{0.833}Fe_{0.167})SiO_3$-bridgmanite + CO_2. (**b**) The relative stability of $(Mg_{0.833}Fe_{0.167})CO_3$ + $MgSiO_3$-bridgmanite and HS-$(Mg_{0.833}Fe_{0.167})SiO_3$-bridgmanite + $MgCO_3$ as a function of pressure. The thick lines indicate the enthalpies of the reactions containing the lowest enthalpy phase of $(Mg,Fe)CO_3$ as shown in Figure 5b.

4. Conclusions

In this study, we calculated the enthalpy of ferromagnesite $(Mg_{0.833}Fe_{0.167})CO_3$ to investigate the effect of Fe on the stability of high-pressure phases of magnesite. Our enthalpy calculations suggest that $(Mg_{0.833}Fe_{0.167})CO_3$ undergoes phase transition from the HS-$R\bar{3}c$ structure to the HS-$P\bar{1}$ structure at 50 GPa, and to the HS-*C2/m* structure above 80 GPa at static 0 K conditions. Therefore, no spin transition in $(Mg_{0.833}Fe_{0.167})CO_3$ is expected under the lower mantle pressure conditions. This result also indicates that the effect of Fe on the transition pressure of high-pressure phases of ferromagnesite is not significant.

Furthermore, the calculated enthalpy relationship suggests that the high-pressure phase of ferromagnesite plus SiO_2 does not decompose even up to pressures of ~120 GPa. In addition, the enthalpy of the Fe^{2+} exchange reaction between magnesite and bridgmanite shows a weaker preference in the high-pressure carbonate phases than in bridgmanite under deep lower mantle pressure conditions.

5. Methods

5.1. First Principles Calculation

We performed first principles calculation based on the density functional theory (DFT) [27,28] for the determination of the relative stability of $(Mg_{0.833}Fe_{0.167})CO_3$ ferromagnesite and its possible high-pressure phases. We used the norm-conserving pseudopotentials [29] for Mg and Si, which have also been used extensively in previous studies e.g., [30]. In contrast, we used ultrasoft pseudopotentials for Fe, C, and O [31,32]. The electronic wave functions are expanded into plane-waves with a kinetic cutoff of 50 Ry. The irreducible Brillouin zone of $R\bar{3}c$ (ferromagnesite), $P\bar{1}$, and *C2/m* phases were sampled on 2 × 2 × 2, 4 × 4 × 4, and 4 × 4 × 4 Monkhorst-Pack meshes with shifted grids, respectively [33]. The associated k-point sampling and kinetic energy cutoff are sufficiently converged within 0.001 eV/$(Mg_{0.83}Fe_{0.17})CO_3$. All structure parameters of $MgCO_3$ and $(Mg_{0.833}Fe_{0.167})CO_3$ were fully relaxed at static 0 K and in the range of 0–120 GPa via the damped variable cell-shape molecular dynamics method implemented using a Quantum-espresso code until residual forces were less than 1.0×10^{-4} Ry/a.u.(a.u. = atomic unit) [34].

The number of atoms calculated in the $MgCO_3$ cells are 30 for the $R\bar{3}c$ (Wyckoff positions of Mg, C, and O are *6b*, *6a*, and *18e*, respectively) and $P\bar{1}$ structures, whereas they are 60 for the *C2/m* structure (Wyckoff positions of $P\bar{1}$ and *C2/m* structures are listed in Table 1). For the $(Mg_{0.833}Fe_{0.167})CO_3$ composition, we constructed a supercell containing 120, 60, and 30 atoms for the $R\bar{3}c$, $P\bar{1}$, and *C2/m* structures, respectively (Figure 2).

5.2. LDA+U_{ic} Calculation

Because LDA and GGA typically fail to accurately determine the phase relationship of Fe-bearing mineral oxides owing to the large on-site Coulomb interactions of 3*d* Fe electrons, in this study, we performed calculations based on LDA with Hubbard U correction, where U values were determined using an internally consistent method (U_{ic}, Figure 7) [23]. This method has been successfully applied for mantle mineral stability calculations [24,35,36]. The U_{ic} parameters for HS state of $(Mg,Fe)CO_3$ are obtained in the range of 3.5–4.0 eV, except for ferromagnesite (HS-$R\bar{3}c$), for which they are higher than those reported in a previous study [22]. The large U_{ic} of the HS ferromagnesite could be attributed to the Fe atom occupying the small octahedral site in the $R\bar{3}c$ structure. In contrast, the U_{ic} parameters of the LS state are in the range of 5.0–7.0 eV.

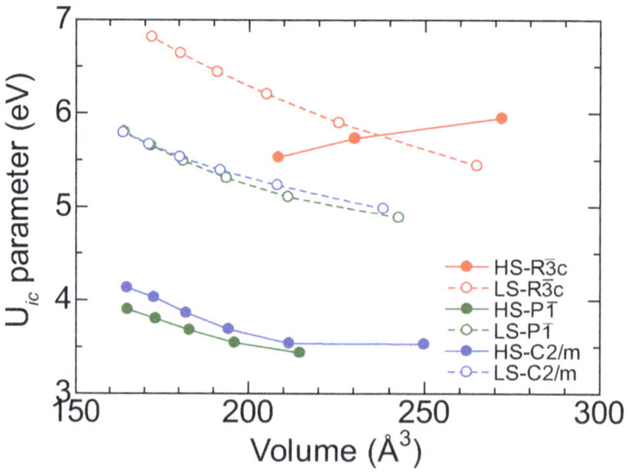

Figure 7. Volume (per 30 atoms) dependence of U_{ic} parameters for $(Mg,Fe)CO_3$.

Author Contributions: J.T. and R.N. conducted the first principles calculation. All authors analyzed the results. J.T. and T.T., J.T. drafted the manuscript. All authors have read and agreed to the published version of the manuscript.

Funding: The research was supported in part by Grants-In-Aid for Scientific Research from the Japan Society for the Promotion of Science (No. JP15H05834 for J.T. and T.T. and No. JP19H01994 for J.T.). In addition, it was also supported by MEXT as "Exploratory Challenge on Post-K computer" (Challenge of Basic Science - Exploring Extremes through Multi-Physics and Multi-Scale Simulations.

Acknowledgments: We are grateful to two anonymous reviewers who helped to improve the manuscript.

Conflicts of Interest: The authors declare no conflict of interest.

References

1. Merlini, M.; Crichton, W.A.; Hanfland, M.; Gemmi, M.; Muller, H.; Kupenko, I.; Dubrovinsky, L. Structures of dolomite at ultrahigh pressure and their influence on the deep carbon cycle. *Proc. Natl. Acad. Sci. USA* **2012**, *109*, 13509–13514. [CrossRef] [PubMed]
2. Boulard, E.; Gloter, A.; Corgne, A.; Antonangeli, D.; Auzende, A.; Perrillat, J.; Guyot, F.; Fiquet, G. New host for carbon in the deep Earth. *Proc. Natl. Acad. Sci. USA* **2011**, *108*, 5184–5187. [CrossRef] [PubMed]
3. Biellmann, C.; Gillet, P.; Guyot, F.; Peyronneau, J.; Reynard, B. Experimental evidence for carbonate stability in Earth's lower mantle. *Earth Planet. Sci. Lett.* **1993**, *118*, 31–41. [CrossRef]
4. Gillet, P. Stability of magnesite ($MgCO_3$) at mantle pressure and temperature conditions: A raman spectroscopic study. *Am. Miner.* **1993**, *78*, 1328–1331.
5. Fiquet, G.; Guyot, F.; Kunz, M.; Matas, J. Andrault, D.; Hanfland, M. Structural refinements of magnesite at very high pressure. *Am. Miner.* **2002**, *87*, 1261–1265.
6. Shatskiy, A.; Litasov, K.D.; Palyanov, Y.N.; Ohtani, E. Phase relations on the K_2CO_3-$CaCO_3$-$MgCO_3$ join at 6 GPa and 900–1400 °C: Implications for incipient melting in carbonated mantle domains. *Am. Miner.* **2016**, *101*, 437–447. [CrossRef]
7. Isshiki, M.; Irifune, T.; Hirose, K.; Ono, S.; Ohishi, Y.; Watanuki, T.; Nishibori, E.; Tanaka, M.; Sakata, M. Stability of magnesite and its high-pressure form in the lowermost mantle. *Nature* **2004**, *427*, 60–63 . [CrossRef]
8. Maeda, F.; Ohtani, E.; Kamada, S.; Sakamaki, T.; Hirao, N.; Ohishi, Y. Diamond formation in the deep lower mantle: A high-pressure reaction of $MgCO_3$ and SiO_2. *Sci. Rep.* **2017**, *7*, 40602. [CrossRef]
9. Palyanov, Y. N. Mantle-slab interaction and redox mechanism of diamond formation. *Proc. Natl. Acad. Sci. USA* **2013**, *110*, 20408–20413. [CrossRef]
10. Kelemen, P.B.; Manning, C.E. Reevaluating carbon fluxes in subduction zones, what goes down, mostly comes up. *Proc. Natl. Acad. Sci. USA* **2015**, *112*, E3997–E4006. [CrossRef]
11. Oganov, A.R.; Ono, S.; Ma, Y.; Glass, C.W.; Garcia, A. Novel high-pressure structures of $MgCO_3$, $CaCO_3$ and CO_2 and their role in Earth's lower mantle. *Earth Planet. Sci. Lett.* **2008**, *273*, 38–47. [CrossRef]
12. Pickard, C.; Needs, R.J. Structures and stability of calcium and magnesium carbonates at mantle pressures. *Phys. Rev. B* **2015**, *91*, 104101. [CrossRef]
13. Franzolin, E.; Schmidt, M.W.; Poli, S. Ternary Ca-Fe-Mg carbonates: subsolidus phase relations at 3.5 GPa and a thermodynamics solid solution model including order/disorder. *Contrib. Miner. Petrol.* **2011**, *161*, 213–227. [CrossRef]
14. Liu, J.; Lin, J.-F.; Prakapenka, V.B. Ferromagnesite as a potential deep-mantle carbon carrier. *Sci. Rep.* **2015**, *5*, 7940. [CrossRef]
15. Solomatova, N.V.; Asimow, P.D. First-principles calculations of high-pressure iron-bearing monoclinic dolomite and single-cation carbonates with internally consistent Hubbard U. *Phys. Chem. Miner.* **2018**, *45*, 293–302. [CrossRef]
16. Boulard, E.; Menguy, N.; Auzende, A.L.; Bureau, H.; Antonangeli, D.; Corgne, A.; Morard, G.; Siebert, J.; Perrillat, J.P.; Guyot, F.; et al. Experimental investigation of the stability of Fe-rich carbonates in the lower mantle. *J. Geophys. Res.* **2012**, *117*, B02208. [CrossRef]
17. Lavina, B.; Dera, P.; Downs, R.T.; Prakapenka, V.; Rivers, M.; Sutton, S.; Nicol, M. Siderite at lower mantle conditions and the effects of the pressure-induced spin-apiring transition. *Geophys. Res. Lett.* **2009**, *36*, L23306. [CrossRef]

18. Lavina, B.; Dera, P.; Downs, R.T.; Yang, W.; Sinogeikin, S. Meng, Y.; Shen, G.; Schiferl, D. Structure of siderite $FeCO_3$ to 56 GPa and hysteresis of its spin-pairing transition. *Phys. Rev. B* **2010**, *82*, 064110. [CrossRef]
19. Mattila, A.; Pylkkanen, T.; Rueff, J.-P.; Huotari, S.; Vanko, G.; Hafland, M.; Lehtinen, M.; Hamalainen, K. Pressure induced magnetic transition in siderite $FeCO_3$ studied by x-ray emission spectroscopy. *J. Phys. Condens. Matter* **2007**, *19*, 386206. [CrossRef]
20. Farfan, G.; Wang, S.; Ma, H.; Caracas, R.; Mao, W.L. Bonding and structural changes in siderite at high pressure. *Am. Miner.* **2012**, *97*, 1421. [CrossRef]
21. Lin, J.-F.; Liu, J.; Jacobs, C.; Prakapenka, V.B. Vibrational and elastic properties of ferromagnesite across the electronic spin-pairing transition of iron. *Am. Miner.* **2012**, *97*, 583. [CrossRef]
22. Hsu, H.; Huang, S.C. Spin crossover and hyperfine interactions of iron in $(Mg,Fe)CO_3$ ferromagnesite. *Phys. Rev. B* **2016**, *94*, 060404. [CrossRef]
23. Cococcioni, M.; de Gironcoli, S. Linear response approach to the calculation of the effective interaction parameters in the LDA+U method. *Phys. Rev. B* **2005**, *71*, 035105. [CrossRef]
24. Wang, X.; Tsuchiya, T.; Hase, A. Computational support for a pyrolitic lower mantle containing ferric iron. *Nat. Geosci.* **2015**, *8*, 556–560. [CrossRef]
25. McCammon, C. Perovskiate as a possible sink for ferric iron in the lower mantle. *Nature* **1997**, *387*, 694–696. [CrossRef]
26. Zhang, F.; Oganov, A.R. Valence state and spin transitions of iron in Earth's mantle silicates. *Earth Planet. Sci. Lett.* **2006**, *249*, 436–443. [CrossRef]
27. Hohenberg, P.; Kohn, W. Inhomogeneous electron gas. *Phys. Rev.* **1964**, *136*, B864–B871. [CrossRef]
28. Kohn, W.; Sham, L.J. Self-consistent equations including exchange and correlation effects. *Phys. Rev.* **1965**, *140*, A1133–A1138. [CrossRef]
29. Troullier, N.; Martins, J.L. Efficient pseudopotential for plane wave calculations. *Phys. Rev. B* **1991**, *43*, 1993–2006. [CrossRef]
30. Tsuchiya, T.; Tsuchiya, J.; Umemoto, K.; Wentzcovitch, R.M. Phase transition in $MgSiO_3$ perovskite in the earth's lower mantle. *Earth Planet. Sci. Lett.* **2004**, *224*, 241–248. [CrossRef]
31. Ichikawa, H.; Tsuchiya, T.; Tange, Y. The P-V-T equation of state and thermodynamic properties of liquid iron. *J. Geophys. Res. Solid Earth* **2013**, *119*, 240–252. [CrossRef]
32. Vanderbilt, D. Soft self-consistent pseudopotentials in a generalized eigenvalue formalism. *Phys. Rev. B* **1990**, *41*, 7892–7895. [CrossRef]
33. Monkhorst, H.J.; Pack, J.D. Special points for Brillouin-zone integrations. *Phys. Rev. B* **1976**, *13*, 5188–5192. [CrossRef] [PubMed]
34. Giannozzi, P.; Baroni, S.; Bonini, N.; Calandra, M.; Car, R.; Cavazzoni, C.; Ceresoli, D.; Chiarotti, G.L.; Cococcioni, M.; Dabo, I.; et al. Quantum ESPRESSO: A modular and open-sourcesoftware project for quantum simulations of materials. *J. Phys. Condens. Matter* **2009**, *21*, 395502. [CrossRef]
35. Tsuchiya, T.; Wentzcovitch, R.M.; da Silva, C.R.S.; de Gironcoli, S. Spin transition in magnesiowustite in earth's lower mantle. *Phys. Rev. Lett.* **2006**, *96*, 198501. [CrossRef]
36. Fukui, H.; Tsuchiya, T.; Baron, A.Q.R. Lattice dynamics calculations for ferropericlase with internally consistent LDA+U method. *J. Geophys. Res.* **2012**, *117*, B12202. [CrossRef]

minerals

MDPI

Article

Local-Basis-Function Equation of State for Ice VII–X to 450 GPa at 300 K

J. Michael Brown * and Baptiste Journaux

Earth and Space Sciences, University of Washington, Seattle, WA 98195, USA; baptiste.journaux@gmail.com
* Correspondence: brown@ess.washington.edu

Received: 7 December 2019; Accepted: 17 January 2020; Published: 21 January 2020

Abstract: Helmholtz energy of ice VII–X is determined in a pressure regime extending to 450 GPa at 300 K using local-basis-functions in the form of b-splines. The new representation for the equation of state is embedded in a physics-based inverse theory framework of parameter estimation. Selected pressures as a function of volume from 14 prior experimental studies and two theoretical studies constrain the behavior of Helmholtz energy. Separately measured bulk moduli, not used to construct the representation, are accurately replicated below about 20 GPa and above 60 GPa. In the intermediate range of pressure, the experimentally determined moduli are larger and have greater scatter than values predicted using the Helmholtz representation. Although systematic error in the determination of elastic moduli is possible and likely, the alternative hypothesis is a slow relaxation time associated with changes in proton mobility or the ice VII to X transition. A correlation is observed between anomalies in the pressure derivative of the predicted bulk modulus and previously suggested higher-order phase transitions. Improved determinations of elastic properties at high pressure would allow refinement of the current equation of state. More generally, the current method of data assimilation is broadly applicable to other materials in high-pressure studies and for investigations of planetary interiors.

Keywords: equation of state; Helmholtz energy; phase transition; ice VII; ice X; NaCl; exoplanets; icy/ocean worlds; local-basis-function; b-spline; Tikhonov inverse

1. Introduction

From icy worlds of our solar system to water-rich super-Earth exoplanets, high-pressure water ices are potential planetary constituents that can exist as distinct mineral species. For example, on Earth, ice VII as a mineral inclusion has been identified in a mantle-derived diamond [1]. Having an accurate equation of state representations for all ice phases is a prerequisite in evaluating questions of origins, evolution, and the potential habitability of ocean worlds [2–9]. A particularly challenging domain for equation of state analysis of ices lies between 2.2 GPa (the pressure of the first order transition between tetragonal ice VI and cubic ice VII) and 450 GPa where three-fold compression is accommodated without first order transitions. First order transitions have been predicted to occur in a regime beyond 300–400 GPa. The experimental and theoretical evidence for higher order transitions at lower pressures is further discussed below.

To provide an improved representation of ice, a new framework for equations of state analysis is described that relies on the thermodynamic representation of Helmholtz energy. The approach follows naturally from long-standing ideas. Birch [10] identified four principle means to obtain equation of state representations: (1) quantum-mechanical treatment, (2) the Thomas–Fermi electron gas approximation, (3) semi-empirical laws for interactions of atoms and ions, and (4) thermodynamic relationships involving elasticity. The latter, a thermodynamic approach using a series of expansions truncated at a low order of Helmholtz energy with parameters expressed in terms of the bulk modulus and its first two pressure derivatives, is the widely practiced approach. This approach can, in addition, be related to special cases of semi-empirical laws for atom/ion interactions [11]. Here, results from quantum mechanical and Thomas–Fermi calculations are combined with measurements within the new framework.

The current work is also guided by the Stacey et al. [11] comments that "Since almost any analytic form may be used to represent a finite data set if sufficient arbitrary or adjustable constants are allowed, we may logically regard as most successful those that use fewest such constants to achieve a particular precision of fit. If this still leaves too much choice, we would prefer a theory in which the adjustable constants are determined most directly from observations." The new framework is responsive to the need to keep the size (number of required parameters) of a model small, while adequately representing measurements, theory, and physical insight garnered from examination of data and calculations. An additional argument developed here is that representational results should not depend on the method of parameterization (i.e., the choice of underlying basis functions) and the size of a model should be flexibly adjusted to meet the physics needs of a particular material.

Our primary findings are straightforward. The traditional physics-blind series expansion to low order of Helmholtz energy, while successful in reproducing measurements and/or theory in restricted regimes of compression, eventually fails over larger regimes. Electronic states and the nature of bonding fundamentally change over a large range of compression. Representations, as developed here, preserve the concept of low-order expansions, using subdomains of compression, while allowing for the necessary changes of the crystal potential (as represented by Helmholtz energy) over larger regimes of pressure and volume compression. Within this framework, concerns that dominate conventional equation of state analysis, including the choice of the expansion order and the definition of a strain metric, become less important. The relevant concerns then appropriately shift to a physics-based discussion: the relative weighting of data from varied sources, the quality of results based on theory, as well as the articulation of appropriate physical insight that should be applied to the representations.

In the following sections we (1) review knowledge concerning the behavior of ice beyond 2.2 GPa, (2) discuss the conceptual foundations of conventional equations of state based on series expansions of Helmholtz energy in powers of strain, (3) describe local-basis-functions that are related to, but more flexible than the polynomial global-basis-functions underlying standard parameterizations, (4) test the new framework against data over an extended regime of pressure for a relatively simple ionic material (NaCl), and finally, (5) use the new framework in analysis of high pressure ices. Mathematical and conceptual details of the framework are given in an exposition on basis functions and inverse theory found in Appendix A.

Phase Behavior of Ice Beyond 2.2 GPa

Although continuous in volume, the complex variations in observed physical properties in the stability field of ice VII and ice X are currently understood as changes in proton dynamics and distortions of the oxygen lattice during compression. Details of transitions remain debated and observations that support the existence of higher-order transitions are summarized in Table 1 and are further discussed here. *International Union of Crystallography* "short name" designations are used to describe the space group of ice phases in the following discussion.

Cubic ice VII (*Pn-3m*) is stable beyond 2.2 GPa until reaching an indistinct transition above about 60 GPa to ice X. The gradual symmetrization of hydrogens between oxygens leads to ionic ice X (which is also *Pn-3m*). The stability of cubic ice X above 150 GPa is experimentally unconstrained, but density-functional calculations of ground state structures [12,13] suggest an eventual transition to the *Pbcm* structure between 300–400 GPa, then to $Pmc2_1$ at 930 GPa and $P2_1$ at 1.3 TPa. Eventually a metallic transition to *C2/m* is predicted at 4.8 TPa. At higher temperatures, ice X transforms into super-ionic ice XVIII above 2000 K [14]. At pressures below 100 GPa, anomalies are reported associated with changes in derivatives or discontinuities in optical spectra, bulk moduli, proton dynamics, and electrical conductivity. The most prominent features include a suggested softening of ice VII volume around 40 GPa [15–17] and the onset of the ice X transition by proton symmetrization above about 65 GPa. The latter is supported by changes in the IR reflectivity trend of $\nu_3{}'$ and translation modes ν_T [18], optical reflectivity [19] and from H-NMR experiments [20].

Table 1. Reported transitions in ice VII and ice X.

Pressure of Transition	Suggested Transition	Type of Measurements	References
		5 GPa	
5 GPa	Tetragonal distortion	Powder X-Ray diffraction (XRD)	Grande et al. [21]
		10–15 GPa	
11 GPa	Lattice distortion	XRD peak splitting	Hirai et al. [22]
14 GPa	Strain in cubic lattice	Powder X-Ray diffraction	Somayazulu et al. [23]
11 GPa	-	Changes in Raman line width trends for the $\nu_1(A_{1g})$ band	Hirai et al. [22], Pruzan et al. [24]
13–15 GPa	-	Raman line pressure trends	Zha et al. [25]
13 GPa	-	Neutron diffraction (220/110 ratio)	Guthrie et al. [26]
10–14 GPa	Lattice distortion	c/a ratio changes in ice VIII	Yoshimura et al. [27]
10–14 GPa	-	Maximum in electrical conductivity	Okada et al. [28]
10–15 GPa	-	Maximum in proton diffusion	Noguchi et al. [29]
20–25 GPa—Possible transition to Ice VII′ with proton dynamic disorder (tunneling and thermal hoping)			
23–25 GPa	-	Bump in the 220/110 ratio from Neutron diffraction	Guthrie et al. [26]
25 GPa	Proton tunneling: ice VII′	IR reflectivity trend of ν_3 and ν_3' trend	Goncharov et al. [18]
20–25 GPa	Proton tunneling: ice VII′	H-NMR	Meier et al. [20]
27 GPa	-	Raman line pressure trends	Zha et al. [25]
		40 GPa	
40 GPa	Softening	Drop in volume reported based on XRD	Hemley et al. [15], Loubeyre et al. [16], Sugimura et al. [17]
44 GPa	-	Raman line pressure trends	Zha et al. [25]
44 GPa	-	Discontinuity in the pressure dependence of Brillouin esound speeds	Noguchi et al. [30]
40 Gpa	-	Changes in trend of reflective index	Zha et al. [19]
40 GPa	-	Drop in Brillouin transverse wave speeds over a narrow P range (<2 GPa) in compression and decompression	Asahara et al. [31]
		>60 GPa transition to ice X	
60 GPa	Proton symmetrization	IR reflectivity trend of ν_3' and translation modes ν_T	Goncharov et al. [18]
62 GPa	-	Raman line pressure trends	Zha et al. [25]
60 Gpa	-	Changes in trend of reflective index	Zha et al. [19]
59 GPa	-	Drop in Brillouin transverse wave speeds in compression	Asahara et al. [31]
70 GPa	Proton symmetrization	H-NMR	Meier et al. [20]
90 GPa	Proton symmetrization	Emergence of the $p20$ Raman mode	Zha et al. [25]

Several other potential higher order transitions have been identified. Tetragonal distortion is possible above 5 GPa [21]. Between 10 GPa and 14 GPa shifts are observed in X-Ray, Neutron, and Raman measurements [22–24,26]. Such behavior is also observed in the proton-ordered analog of ice VII, known as ice VIII [27]. The initiation of proton tunneling above 25 GPa (possible transition to ice VII′) is interpreted on the basis of IR reflectometry [18] and H-NMR data [20].

Equations of state for ice VII and X based on density functional theory, developed by French et al. [32], show variable degrees of agreement with measurements at 295 K depending on the chosen potential and the range of density. They noted that "the fitting of experimental data for such a complex solid may require using a more refined parametric equation of state than the simple Vinet formula, even when the equation of state is split into multiple sections [33]".

As a result of the complex behavior noted here, ice VII and X viewed as a single-phase continuum, cannot be appropriately represented by a single (relatively low order) global polynomial-based equation of state. Previous equations of state, constrained by measurements over limited and differing ranges, present a confusing collection of parameterizations. Extrapolations using any of these equations of states to higher or to lower pressures than the range of fitted data results in significant misfits of other measurements.

2. Materials and Methods

2.1. Helmholtz Energy-Based Equations of State

Here, a brief overview of conventional (global) equation of state parameterizations is provided to differentiate them from the local-basis-function method described later. The derivation of an equation of state based on a series expansion of Helmholtz energy as a function of a strain metric, η, has been extensively discussed [10,11,34–36]. The choice of strain metric is arbitrary, and several have found favor in successful applications of data representation, although the finite strain metric based on an Eulerian representation (as opposed to Lagrangian) is commonly adopted. The Eulerian finite strain metric (as used in the Birch–Murnaghan equation of state) is $\eta = 1/2((V/V_o)^{-2/3} - 1)$, where V is volume and subscript refers to the ambient pressure value. The Vinet strain metric [36] is $\eta = ((V/V_o)^{-1/3} - 1)$. Poirier and Tarantola [35] developed a logarithmic equation of state using $\eta = -1/2\log(V/V_o)$. Other definitions are possible and have been used.

The expansion of the Helmholtz energy about ambient pressure as a function of the strain metric is given as

$$F(\eta) = F_o + F_1\eta^1 + F_2\eta^2 + F_3\eta^3 + F_4\eta^4 + \ldots \tag{1}$$

where the expansion coefficients F_i contain appropriate derivatives of F evaluated at $\eta = 0$. The first term on the right establishes the energy datum. Pressure follows as

$$P(V) = -\frac{dF}{dV} = -\frac{dF}{d\eta}\frac{d\eta}{dV} = \frac{d\eta}{dV}\left(F_1 + 2F_2\,\eta + 3F_3\eta^2 + 4F_4\eta^3 + \ldots\right) \tag{2}$$

The F_1 term is conventionally set to zero, although more formally Equation (2) should predict ambient pressure (0.1 MPa) for $\eta = 0$. The isothermal bulk modulus given as

$$K = -V\frac{dP}{dV} \tag{3}$$

has pressure derivatives indicated by the number of prime superscripts:

$$K' = \frac{dK}{dP} \quad K'' = \frac{d^2K}{dP^2} \quad K''' = \frac{d^3K}{dP^3}.$$

Based on one definition for strain, the Eulerian strain metric, and with algebraic effort, Equation (2), truncated at the fourth order in energy, can be expressed in terms of the ambient pressure bulk modulus and its pressure derivatives (denoted with subscript) as:

$$P = \frac{9K_o}{16}\left(-B_1\left(\frac{V}{V_o}\right)^{-5/3} + B_2\left(\frac{V}{V_o}\right)^{-7/3} - B_3\left(\frac{V}{V_o}\right)^{-9/3} + B_4\left(\frac{V}{V_o}\right)^{-11/3}\right) \tag{4}$$

$$B_1 = K_oK_o'' + (K_o' - 4)(K_o' - 5) + 59/9,$$

$$B_2 = 3K_oK_o'' + (K_o' - 4)(3K_o' - 13) + 129/9,$$

$$B_3 = 3K_oK_o'' + (K_o' - 4)(3K_o' - 11) + 105/9,$$

$$B_4 = K_oK_o'' + (K_o' - 4)(K_o' - 3) + 39/9.$$

Equations (2) and (4) contain the common physical assumption that a low order series expansion of energy is adequate. The focus in Equation (2) is on model parameters that are measures of derivatives of the energy at equilibrium. In Equation (4), with application of a particular strain metric, parameters are cast as elastic properties determined at ambient condition. The use of other strain metrics generates different algebraic relationships but preserves the underlying physics of a truncated series expansion.

While expansion of Equation (1) to arbitrary higher order is straightforward, equation of state applications have been limited to no more than fourth-degree polynomials for several reasons. A general belief has been that an appropriate choice of a strain might lead naturally to a convergent series. Higher order derivatives of the bulk modulus are not well constrained by pressure–volume measurements that may not be sufficiently accurate and/or span a sufficiently large range of compression. The low order (second and third) forms of Equation (4) have provided relatively stable extrapolations to high pressure and have proven adequate to represent many measurements. The fourth-order expansion is often marginally stable or completely unstable in extrapolation. Higher order versions fit to measurements are expected to be less stable. The algebraic and arithmetic effort to relate the energy expansion coefficients to derivatives of the bulk modulus increases with increasing order.

The strength of the strain-expansion-based representation of data is that a small model consisting of three or fewer parameters (F_2, F_3, F_4 or K_o, K_o', K_o'', plus the 1 bar volume) adequately fits many data sets. The weakness lies in the requirement that the underlying energy potential be represented by a physics-blind series expansion arbitrarily truncated to a small order. Only for a sufficiently small interval will a low order polynomial provide a complete (able to replicate all data within uncertainties) representation. Furthermore, high-pressure behavior may be governed by physics that is not represented in the near ambient pressure potential. In the following, a framework is developed that introduces the use of physics-based constraints and allows construction of equations of state using basis functions having greater flexibility.

2.2. Local-Basis-Function Representation of Helmholtz Energy

Numerical methods, developed to meet modern computational speed and accuracy requirements as well as for computer-aided graphics applications, are shown in this section to provide an alternative framework for the representation of Helmholtz energy. Although standard quantities (e.g., trigonometric and logarithmic) can be determined numerically, interpolation of previously calculated values in lookup tables is more efficient. Various interpolation algorithms have been developed and are embedded in both modern hardware and software. Such methods can be applied to representation of Helmholtz energy since any empirically based equation of state is created as an accurate interpolating formula capable of matching all measurements and theory.

The linear interpolation of a list of energies at different volumes is the simplistic approach that can work with sufficiently dense tables. However, accurate smaller models are possible using better interpolating functions and mathematical forms can be chosen to aid construction of representations. Here, piece-wise polynomials on intervals (splines) are used to represent Helmholtz energy over wide ranges of compression. Within each interval, representations are numerically analogous to the conventional series expansion-based equations of state. However, having separate polynomials in different intervals gives greater flexibility in meeting the needs of the underlying physics.

In the following, b-splines [37] are used since they have a number of beneficial numerical properties that are further described below. A b-spline representation, in essence a lookup table, consists of two lists, one containing points defining intervals of the independent variable and a second containing the model parameters. Basis functions then provide the means to interpolate the table. A b-spline of order two (spline order is one greater than the degree of underlying polynomials) is equivalent to linear interpolation of model parameters as a function of the independent variable. In this case, the representation has discontinuous derivatives at interval boundaries. With increasing b-spline order (using higher degree polynomials within each interval) the number of continuous derivatives at interval boundaries increases. The size of a b-spline equation of state is larger than that of a conventional representation. In addition to coefficients required for any global polynomial representation, at least one additional model parameter is needed for each added interval. As shown later, only a few intervals are needed to achieve adequate representations and the model size remains small.

Key properties of b-spline basis functions that make their use advantageous for equation of state representations include:

- B-spline basis function values are available in all computer environments as a call to a function/subroutine. Analogous to the use of exponential or trigonometric functions, no custom (user) programing is necessary for use of b-spline basis functions. The evaluation of equation of state properties then uses universal calling functions that are not material specific.
- The calculation of values and derivatives of a b-spline model are based on linear programing. Interpolation using b-splines is essentially a weighted average of neighboring model parameters with the basis functions providing the normalized weights. This enables efficient computer algorithms for both construction and evaluation of spline models. Arbitrary precision is possible in representing any functional behavior.
- B-spline basis functions are localized. Unlike global polynomial fits of data, spline model parameters pertain to the behavior of the underlying function in a separate restricted regime of the independent variable.
- Details of how intervals are defined allow flexibility in the behavior of function derivatives at interval boundaries. It is possible to allow discontinuities of the function or specified derivatives of the function at a location to meet the needs of a particular equation of state that might involve higher-order transitions.

Additional b-spline details are important in understanding their use for equations of state representations. The full articulation of the underlying mathematics is provided by de Boor [29]. Only a short overview that emphasizes points relevant to the current applications is provided in Appendix A. These include further articulation of spline order, model parameters, and b-spline basis functions. Also included is a description of the interval boundaries, the knots, and how local derivatives are controlled by knot multiplicity. An example representation of Helmholtz energy is given to further illustrate the underlying concepts.

Although spline representations of data can be constructed using standard numerical packages, the concept of local-basis-function (LBF) equation of state representations, as developed here, emphasizes the use of physics-based constraints during their construction. This was introduced in Brown [38], where Gibbs energy was expanded as a function of pressure and temperature. An equation of state for water based on Gibbs energy LBF representations is given in Bollengier et al. [39] and for high pressure ices in Journaux et al. [40]. Here, Helmholtz energy, with volume as the independent variable, is the focus. The representations are extended far beyond the regime of measurement using theory-based constraints for limiting behavior (i.e., a constraint on the pressure derivative of the bulk modulus at infinite pressure based on the Thomas–Fermi electron gas limit).

2.3. *Determination of Helmholtz Energy by Collocation*

Helmholtz energy, a surface in volume and temperature, can be reconstructed through collocation. Collocation is a standard approach for the numerical solution of differential equations in which basis functions are chosen that match differential properties at specified points, the collocation points. In addition, integrating constants must be applied. Here, considering only the volume axis of energy at 300 K, measurements, theory, and physical insight constrain derivatives of the energy surface at specified collocation sites. A sufficiently flexible set of b-spline basis functions then allows representation of any plausible physical behavior for Helmholtz energy and its derivative properties. Here, the arbitrary energy "datum" is set to zero at equilibrium. The challenges lie in the details of data selection, articulation of what constitutes adequate fitting as opposed to overfitting of data, and how to constrain behavior in data-poor regimes.

Such challenges are not unique to equation of state studies and are described within the rubric of parameter estimation using inverse methods [41]. Here, the "damped least square method" or Tikhonov inversion is used to determine model parameters for equation of state representations. The minimization of one or more side constraints (hereafter called regularization) is added to the standard least square solution. A straightforward approach based on physical insight is to minimize a specified derivative of the model. A smooth model would be one with the smallest second derivatives while still adequately fitting data. In the case of equations of state represented by energy potentials, the second derivative of the Helmholtz energy figures into the determination of the bulk modulus (as shown in Equation (3)), a third derivative is needed to determine K' and a fourth derivative is needed to determine K''. Since K' is a frequent quantity of interpretation and since it is typically found to be smooth and trending to an asymptotic value at infinite pressure, minimizing the fourth derivative of the Helmholtz energy emerges as a possible side constraint. Linear inverse methods then allow determination of the model parameters, fitting pressure–volume (and/or bulk moduli) data with a side constraint on the derivative of energy. The explicit minimization of the third derivative of the bulk modulus (to make K' smooth or to require a specific value for K') is undertaken with non-linear analysis. Further details of inverse theory are described in Appendix A.

In summary, a framework is described to find Helmholtz energy by collocation of measurements, theory, and physical insight (through use of regularization) over arbitrarily large regimes of compression. Local-basis-functions (LBF) based on piece-wise polynomials in the form of b-splines are used. A linear problem allows fits of pressure–volume and bulk modulus–volume measurements. Fits involving determinations of the bulk modulus as a function of pressure and its pressure derivatives requires iterative and non-linear analysis. The resulting representations for Helmholtz energy have model parameters in units of energy that can be analytically evaluated for equation of state properties including pressure, the bulk modulus, and derivatives of bulk modulus as a function of volume (or density). The numerical function to evaluate properties based on the LBF representation is universal, no material-specific coding is required.

A number of decisions are required in the creation of an LBF representation for Helmholtz energy. An optimal choice of strain metric may reduce model size. The number of intervals for strain (as well as their distribution and bounds) can be varied. The cost of subdivision is that at least one additional model parameter is required for each added interval of representation. The total parameter count grows slowly since boundary conditions (matching values and derivatives of values at boundaries) reduce the required number of unique parameters. The order of a chosen basis function set is arbitrary and might depend on the physics of a specific material. Smaller intervals allow lower order fits. The derivative of the potential used for regularization is an arbitrary choice. Additional non-linear optimization is required if a specific limiting value for the pressure derivative of the bulk modulus is required.

Using the MATLAB numerical environment, a set of tools based on the concepts given here are provided with the Supplementary Materials. They allow creation and evaluation of LBF equations of state. Scripts using these tools are also provided that reproduce the analyses given in the following sections.

3. Results

3.1. Equations of State for NaCl

3.1.1. Data and Representations

Measured volumes of NaCl provide an internal pressure calibrant in some high-pressure experiments [42]. The widely used pressure scale for NaCl given by Decker (1971) [43] was based on an assumed two-term pair-wise interaction energy potential. Accurate high-pressure data obtained subsequent to Decker's work exhibited systematic deviations from his pressure scale. Brown (1999) [44] reported an energy potential for NaCl that provides a better representation of accurate high-pressure and high-temperature measurements. Here, properties on the zero Kelvin isotherm, tabulated in [44], provide precise quantities to evaluate the current method of representation. The data and an example script illustrating the analysis given in this section are provided in Supplementary Materials. The fitting is constrained by both pressure–volume and bulk modulus data. The relative weight of these data was adjusted to provide a satisfactory balance between simultaneous fits of pressures and the bulk moduli as a function of volume.

In Figure 1 the zero-Kelvin densities (panel a), bulk modulus (panel b), and its pressure derivative (panel c) as tabulated in [44] are plotted as a function of pressure. Lines show equation of state predictions based on the model parameters given in Table 2. Data misfits are listed in Table 2. Values extend from negative pressures (constrained by high-temperature measurements at 1 bar) to the high-pressure limit of the NaCl B1 structure.

Properties are plotted in Figure 1 to pressures beyond the stability regime of the B1 structure to examine the behavior of each equation of state in extrapolation. Both conventional global fits (Birch-Murnaghan third and fourth order) and a non-standard global fit (ninth-order), all based on Equation (2), are illustrated. Local-basis-function (LBF) representations (using both Eulerian and log strain metrics) are also shown.

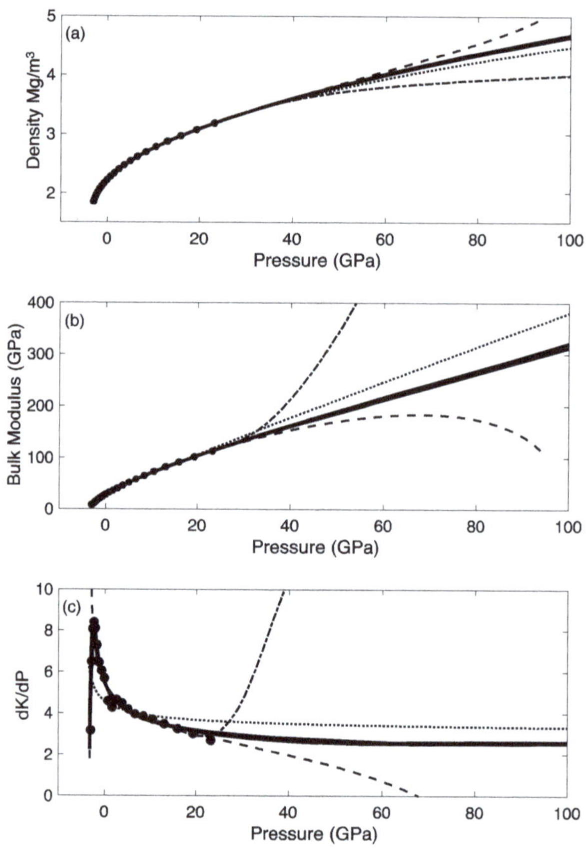

Figure 1. Equation of state properties for NaCl as a function of pressure. (**a**) Density. (**b**) Isothermal bulk modulus. (**c**) Pressure derivative of the isothermal bulk modulus. Solid circles in all panels are the tabulated values given in [44]. Lines are predictions based on differing parameterizations as listed in Table 2. Solid lines: local basis functions using different strain metrics (essentially indistinguishable). Dotted line: global third order finite strain. Dashed line: global fourth-order finite strain. Dash-dot line: global ninth-order finite strain.

Table 2. Equation of state parameters and root-mean-square (rms) misfits of pressure (P_{rms}) and the bulk modulus (K_{rms}) for representations of the NaCl 0 K isotherm. A 1 bar density of 2.226 Mg/m^3 is used in all representations. Units for all parameters are shown.

Type of Function	Parameters			(rms)
	Global-Basis-Function			
Third order/degree Eulerian Finite Strain	K_o = 28.0 GPa	K_o' = 4.5		P_{rms} = 0.15 GPa K_{rms} = 2.3 GPa
Fourth order/degree Eulerian Finite Strain	K_o = 27.4 GPa	K_o' = 5.4	K_o'' = −0.44 GPa^{-1}	P_{rms} = 0.02 GPa K_{rms} = 0.3 GPa
Ninth order/degree Eulerian Finite Strain	K_o = 27.8 GPa	K_o' = 5.4	K_o'' = −0.67 GPa^{-1}	P_{rms} = 0.01 GPa K_{rms} = 0.2 GPa
	(Five more parameters for the ninth order fit are not reported here. See Supplementary Materials)			
	Local-Basis-Function:			
	Knots (strain units):	Coefficients (GPa m^3/Mg)		
Eulerian Finite Strain Order: 6 (degree 5)	[−0.08, −0.035, 0.24, 0.67] (first and last knots are repeated six times)	[0.308, 0.242, −0.155, −0.292, 4.10, 10.9, 19.8, 26.1]		P_{rms} = 0.01 GPa K_{rms} = 0.2 GPa
log Strain Order: 5 (degree 4)	[−0.09, −0.036, 0.2, 0.42] (first and last knots are repeated five times)	[0.326, 0.241, −0.141, −0.326 3.92 16.4 26.0]		P_{rms} = 0.01 GPa K_{rms} = 0.2 GPa

3.1.2. Discussion

The energy potential for NaCl over the extended range of volume compression samples significantly different physical behavior. Expanded state properties (negative pressures) are more strongly influenced by long-range cohesive forces (including Coulomb and inductive). In the compressed state, the overlap of electrons creates large short-range repulsive interactions. The large variation of K' values with a maximum just below ambient pressure provides evidence that the underlying potential is more complex than one that can be represented by a global low order series expansion.

In Figure 1 all equations of state in the regime of measurements provide adequate approximation to values and trends in density and the bulk modulus. Unsurprisingly, as shown in Table 2, misfit is largest for the third-order representation and this lower order fit has an unsatisfactory representation of the pressure dependence of K'. The fourth-order version, while providing a more plausible fit of measurements in the compressed regime, does not have sufficient flexibility to follow the behavior of K' in the expanded state. The fourth-order predicted pressure derivative of the bulk modulus continues to increase as pressure becomes increasingly negative. The data show a maximum in K' slightly below ambient pressure as required by the necessary behavior of the energy of interaction in the expanded state. As a result of a need for significant curvature of the bulk modulus as a function of pressure near equilibrium (requiring a large second derivative of the bulk modulus), the behavior of the bulk modulus for the global fourth-order fit at high pressure is unstable; K' implausibly goes through zero to negative values and the bulk modulus decreases with increasing pressure. A ninth-order fit proved to be the lowest order global representation that could adequately represent all data. As expected, this equation of state cannot be extrapolated.

The LBF equations of state are constructed to have satisfying performance from the expanded state to high compression. They are required to adequately fit data while extrapolating with appropriate physics-based limiting behavior. Two versions (graphically indistinguishable) are shown that fit data equally well. One LBF uses Eulerian finite strains and the other uses log strains. The log strain version allowed use of a lower order spline (order 5) than the Eulerian strains did (order 6). Both LBFs use three intervals (two in the low-pressure regime, and one extending well beyond the regime of measurements). The high-pressure behavior for both was constrained by requiring that the fourth derivative of the potential be small and that the high-pressure limit of K' be greater than 5/3 [45,46]. The high-pressure constraint on K' was enforced through non-linear optimization of an initial linear model. In this case, it is a choice to slightly misfit the pressure derivative of the bulk modulus for the highest-pressure data as a necessary consequence of enforcing the smooth asymptotic behavior for K'. Here, physical insight is given more weight that the accuracy of a single data point at the extreme of the measurements.

The key findings of this section include the following: (1) Standard forms for equations of state based on global representations using series expansions of the Helmholtz energy can neither capture the details of NaCl behavior in regimes of measurement nor do they correctly extrapolate to high pressure. (2) Higher order global fits can be constructed that adequately fit all data, but even modest extrapolation is impossible. (3) In contrast, an LBF representation having three volume intervals (one extending well beyond measurements) can both better represent data and provides plausible (physics-based) behavior well beyond any possible measurement. (4) The choice of strain metric has no impact on the quality of the fit. (5) The behavior of representations in regions of extrapolation can be adjusted to match theory-based constraints. (6) The model size of the LBF representations are larger (a dozen or so parameters) compared to three for a global fourth-order fit. However, this is not a serious impediment to their use.

3.2. *Equations of State for High Pressure Ice (Ice VII–X System)*

3.2.1. Data and Representations

Based on the success of local-basis-function representations in the previous section, attention is focused on the ice VII–ice X system to 450 GPa. The upper pressure limit of this representation is associated with the span of the density-functional (DFT) calculations. Equations of state at 300 K are constructed minimizing the root-mean-square (rms) misfit to 291 selected pressure–volume points from 16 independent datasets. While independently determined values for the bulk modulus have been reported and are discussed below, these data were not directly used as fitting constraints. Data and the numerical analysis are provided as scripts with the Supplementary Materials. Both experimental [15–17,21,33,47–57] and DFT studies [51,52] are considered. Results above 150 GPa are based solely on theory. The results in [15], that used an experimental method for pressure calibration, and measurements reported in [16], that deviate substantially from other work, are excluded in the current fits. Inclusion of these data increases the root-mean-square (rms) misfit but does not change the interpretations given below. No additional quality-of-data assessment is undertaken in the current analysis. Biases associated with systematic and random errors are likely reduced in the current meta-analysis of all data. The sample size may be sufficient to diminish the impact of small disagreements between merged datasets, arising from the use of different pressure gauges, volume measurement techniques, or experimental protocols that may affect biases associated with non-hydrostatic stresses. Pressure–volume data are plotted and compared with published equations of states [47,49,53,58] in Figure 2a.

To adequately explore the information contained in the pressure–volume measurements, an ensemble of representations was created based on differing assumptions for parameterizations and regularization. Examples are plotted in all panels of Figure 2, and parameters for these fits and associated misfits are listed in Table 3. The global fourth-order Eulerian representation, Equation (2), has the largest rms misfit. The LBF versions have reduced rms misfits. Although reduced misfit is expected for models containing more parameters, the global representation has greater systematic misfit and, as shown below, the LBF representations give results that better align with the lowest pressure bulk moduli data and the known complex phase behavior in the ice VII–X system. The LBF representation labeled "low structure" resulted from exploration (based on manual variation of knot placements) in which the goal was to find the smallest number of intervals with the least structure (smallest variations of K'). The impact of discontinuities in higher order derivatives of Helmholtz energy were tested by introducing knot multiplicity at boundaries that are manually aligned to match pressures for proposed higher-order phase transitions. One such fit labeled "transition informed" is reported and is further discussed below. "Agnostic" LBF fits were constructed using a sixth-order spline over 14 internal intervals to allow more than necessary flexibility in representing the equation of state with no preconceptions of where higher-order phase transitions might occur. Two examples demonstrate differing levels of regularization. Additional LBF representations, created to explore hypothesized pressure dependences of the bulk modulus, are discussed and rejected in the following discussion. Regularization, based on minimizing the fourth derivative of Helmholtz energy, was applied to the LBF models. The damping parameter for regularization was varied to find an optimal trade-off between misfit and model smoothness. The tradeoff is illustrated in Figure 3 where the rms average of the Helmholtz energy fourth derivative is plotted against rms misfit of the pressure–volume data. An infinite number of solutions are possible as the damping parameter is continuously changed. As shown in the figure, larger rms values for the fourth derivative allow smaller misfit at the cost of possibly over fitting data. As greater smoothing is enforced (larger damping), misfit increases, eventually becoming unacceptably large. Three points marked on the diagram are associated with the three levels of misfit reported in Table 3. Acceptable solutions for damped solutions are typically chosen near the corner of the trade-off curve, representing a compromise between data misfit and the degree of smoothness.

Table 3. Equation of state parameters and root-mean-square (rms) pressure misfits (P_{rms}) for representations of high-pressure ice VII and X 300 K isotherm. Units for all parameters are shown. V_o for ice VII is taken from Klotz (2017): 12.7218 cm^3/mol or 42.25 $Å^3$ at 300 K.

Type of function	Parameters		(rms)
	Global-Basis-Function		
Fourth order/degree Eulerian Finite Strain	K_o = 19.2 GPa, K_o' = 3.8, K_o'' = −0.09 GPa^{-1}		P_{rms} = 3.0 GPa
	Local-Basis-Function:		
	knots (dimensionless strain):	Coefficients (GPa cm^3/mole)	
"Agnostic" log Strain low damping Order: 6 (degree 5)	[−0.01, 0.02, 0.04, 0.06, 0.08, 0.10, 0.12, 0.15, 0.17, 0.20, 0.24, 0.27, 0.31, 0.36, 0.42] (first and last knots are repeated six times)	[−0.10, −0.07, 0.05, 0.40, 1.37, 3.75, 7.13, 12.5, 20.0, 30.7, 46.7, 72.5, 103, 157, 254, 379, 520, 647, 723]	P_{rms} = 1.7 GPa
"Agnostic" log Strain higher damping Order: 6 (degree 5)	[−0.01, 0.02, 0.04, 0.06, 0.08, 0.10, 0.12, 0.15, 0.17, 0.20, 0.24, 0.27, 0.31, 0.36, 0.42] (first and last knots are repeated six times)	[0.02, −0.03, −0.03, 0.23, 1.16, 3.55, 6.96, 12.3, 19.7, 30.4, 46.9, 71.5, 103, 157, 253, 375, 516, 645, 724]	P_{rms} = 2.0 GPa
"low structure" log Strain (seven intervals) Order: 7 (degree 6)	[−0.01, 0.08, 0.12, 0.16, 0.16, 0.24, 0.3, 0.42] (first and last knots are repeated seven times)	[−0.09, −0.04, 0.43, 2.21, 6.50, 17.5, 46.3, 93.1, 153, 258, 433, 595, 723]	P_{rms} = 2.0 GPa
"transition informed" log Strain (seven intervals) Order: 7 (degree 6)	[−0.01, 0.12, 0.16, 0.16, 0.21, 0.23, 0.26, 0.42] (first and last knots are repeated seven times)	[0.10, −0.27, 0.26, 2.86, 10.6, 25.6, 57.3, 104, 153, 263, 396, 556, 722]	P_{rms} = 1.7 GPa

Fractional deviations from the "low-structure" representation of the pressure volume data and selected literature equations of state are plotted as a function of the logarithm of pressure in Figure 2b. As shown in this panel, the fractional deviations are relatively uniform as a function of pressure with a standard deviation near 5%. The highest-pressure points based on theory have greater precision with unknown accuracy but appear to merge into the more scattered measurements with little apparent offset or change in slope.

In Figure 2c (full pressure range) and d (lower pressure regime), predictions of the isothermal bulk modulus based the current equation of state are compared to reported adiabatic bulk modulus based on Brillouin acoustic measurements [31,59–61]. The difference between adiabatic and isothermal moduli is small in the high-pressure regime and is currently ignored. Although the conversion from sound speeds to elastic moduli requires knowledge of the density, the differences in density between the current representation and those used in the original studies had negligible impact at the scale of these figures. The measurements to 8 GPa reported by Shimizu et al. [60] for single crystals grown in equilibrium with liquid water provide plausible determinations of the bulk modulus. Other studies relied on longitudinal and transverse wave measurements in polycrystalline material with unknown fabric in a non-hydrostatic environment. In spite of the cubic symmetry for the high-pressure ices, such data may suffer from systematic biases.

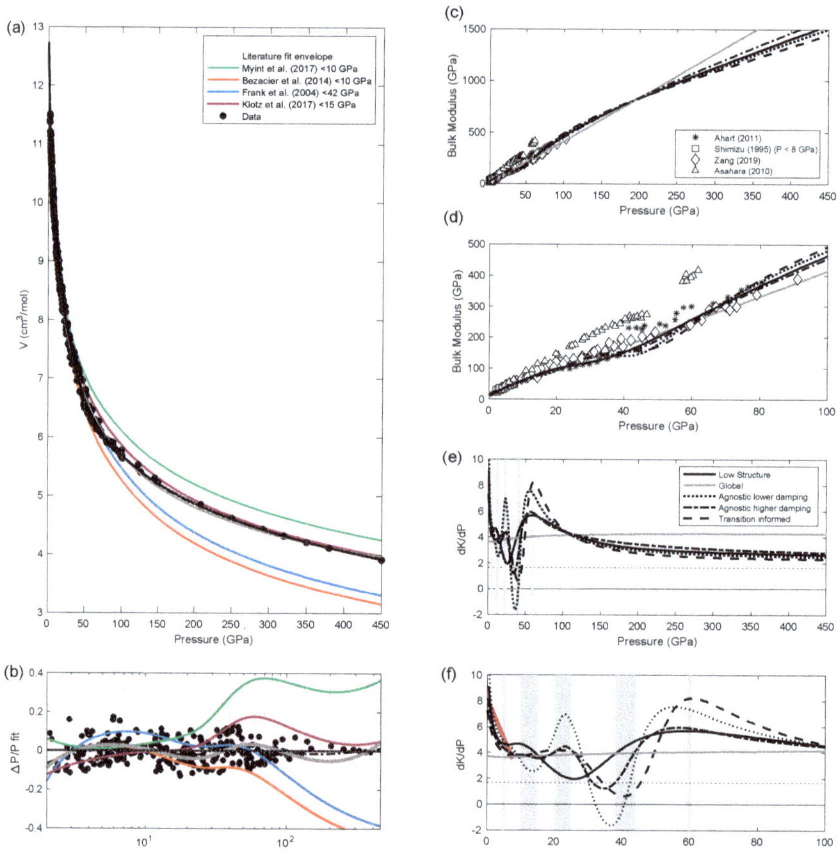

Figure 2. Data and equations of states of ice VII and X at 300K. In all panels, predictions based on the "low structure" local-basis-function (LBF) are shown with thick solid lines, the dotted and dot-dashed lines are the "agnostic" LBFs with lighter or heavier damping, the dashed lines are "transition informed", the gray lines are the global fourth-order fit. (**a**) Specific volume as a function of pressure. All measurements are represented with solid circles. Prior fits are identified in the legend. (**b**) Residuals relative to the preferred LBF representation as a function of the logarithm of pressure. (**c**,**d**) Bulk moduli as a function of pressure. All symbols are identified in the legend. (**e**,**f**) Pressure derivative of the bulk moduli for the four current representations. The 5/3 limit for high-pressure behavior of K′ in represented as a horizontal dotted line. The behavior of K′ based on a quadratic fit to Shimizu et al. [60] is shown for pressures between 2 and 8 GPa a thick line. Vertical bars indicate identified pressure ranges (Table 1) for higher order transitions.

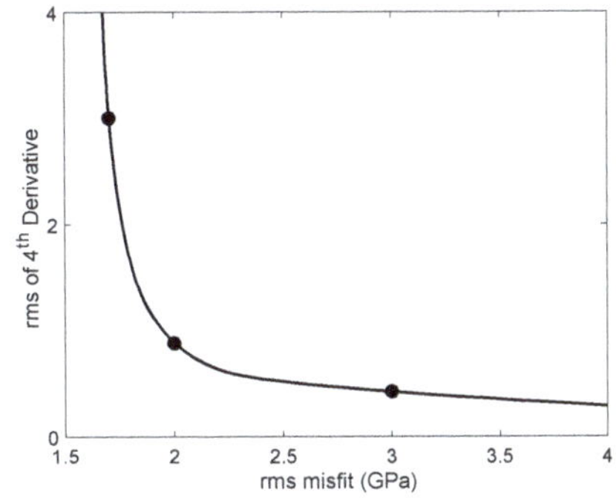

Figure 3. Tradeoff between root-mean-square (rms) data misfit and model smoothness as measured by the root-mean-square average of the fourth derivative of Helmholtz energy. The line defines the locus of solutions obtained by varying the damping factor for regularization. Circles mark points with rms misfits of 1.7 GPa, 2 GPa, and 3 GPa that are associated with fits listed in Table 3.

Pressure derivatives of the bulk modulus based on our representations are shown in Figure 2e (full pressure range) and f (lower pressure regime). Also shown between 2 GPa and 8 GPa is the behavior of K' as constrained by a quadratic fit to the single crystal measurements of Shimizu et al. Vertical bands in these panels indicate the pressure ranges for the suggested higher-order transitions listed in Table 1.

3.2.2. Discussion

The following discussion progresses through considerations of (1) the pressure–volume data, (2) the elastic bulk moduli determinations, and finally, (3) how variations of the pressure derivative of the bulk modulus correlate to pressure ranges where higher-order transitions likely occur. In the absence of transitions and as previously documented in Figure 1 for NaCl, values for the pressure derivative of the bulk modulus, K', can range from more than 6 at low pressure to less than 3 at the highest pressures. The second derivative, K'', is usually small and negative. Possible changes of ice's compressive behavior related to higher order transition as summarized in Table 1 include the following ideas. A possible distortion in the bcc lattice around 15 GPa is suspected to be the expression of the locking of OH-rotational disorder which could change the rate of stiffening under compression [51]. Changing derivatives of the bulk modulus near 25 GPa might reflect the transition from ice VII to ice VII′ associated with the start of dynamic translational proton disorder by tunneling and thermal hoping. The onset of the hydrogen bond symmetrization, with the transition towards H-translational disorder [51], should have an effect on compression beyond 40 GPa. Furthermore, a broad minimum in K' would be the expected behavior for a second order transition between ice VII and a slightly higher density ice X phase.

In prior analyses of individual datasets, equations of state were constrained by data spanning restricted ranges of pressures. The envelopes of such regional fits, extrapolated to cover the entire range of pressure, are illustrated in Figure 2a,b. The various fits are essentially indistinguishable below 50 GPa (Figure 2a). Even at the scale of the residuals (Figure 2b), data scatter is as large as the envelop bounds. Unsurprisingly, predictions in regimes of extrapolation diverge and do not accurately match the highest-pressure DFT results.

The adequacy of representations as measured by the rms misfit to data is not immediately obvious in examination of just the pressure–volume relationships shown in Figure 2a,b. However, the trade-off curve shown in Figure 3 indicates that an rms misfit near 2 GPa is optimal. On this basis, the global fourth-order representation clearly underfits the data, while the low damping agnostic fit and the transition informed fit may overfit the data (as further discussed below).

Greater discrimination between equation of state representations is expected at the level of the derivative properties: the bulk modulus and its pressure derivative (all panels on the right side of

Figure 2). For pressures below 8 GPa, K and K' values determined from sound speed measurements are replicated in the LBF fits, indicating that the underlying pressure–volume data are sufficiently accurate and that the analysis can extract higher order derivatives of Helmholtz energy. In contrast, all global fits, both from prior literature and the current fourth-order, do not match the low-pressure behavior for K'. In order to approximate the volume trends over the entire range of pressures, all global fits require a nearly constant and low (~4) value for K' since such global representations have insufficient flexibility to accommodate both behavior at low pressure and at high pressure. This mirrors the findings in the previous section with the effort to represent the NaCl equation of state. At the highest pressures, where only theory provides constraints, the LBF representations smoothly approach the Thomas–Fermi limit. No global representation provides correct limiting behavior. Furthermore, the current fourth-order global representation shows unphysical behavior with K' increasing from 3.8 near zero pressure to a nearly constant value of approximately 4 at the highest pressure.

In the critical range of higher-order transitions for ice between 15 GPa and 60 GPa, the measured bulk moduli show disagreement between the independent studies and the LBF representations Whether these differences are caused by systematic errors in measurement, differences in non-hydrostatic stresses, or are indicative of material properties that vary with the time-scale of measurement, remains unresolved. A hypothetical relaxation time for a phase transition or for proton mobility that is large compared to the sub-nanosecond periods of acoustic wave measured in Brillouin experiments could rationalize the observation that the static moduli are smaller than the moduli based on sound speeds. Equation of state fits required to match the Brillouin-based bulk moduli in this pressure regime were found to systematically misfit the pressure–volume data. In contrast, the bulk moduli beyond 60 GPa [59,61] converge to a common pressure dependence that is consistent with predictions based on the LBF fits.

Focusing on K' in Figure 2e,f, trends for its pressure dependence separate into three zones. In the low-pressure regime K' decreases from a value near 6.5 at 2.2 GPa (similar to values for ice V and ice VI [40]) to about 4 for pressures between 10 to 15 GPa. As shown in the figure, this normal trend is in agreement with the Shimizu et al. determinations to 8 GPa. Above 60–70 GPa, K' also shows normal behavior, an asymptotic decrease with pressure. In contrast, the intermediate pressure range, the previously identified regime of higher-order transitions, shows the most anomalous behavior for K'. All LBF representations show a minimum between 30 GPa and 40 GPa. The "agnostic" fits have shoulder maxima on the sides of the minimum values. One "agnostic" fit has an optimal misfit of 2 GPa (based on Figure 3). The other "agnostic" fit, with smaller rms misfit (that appears on the overfitting side of the trade-off curve), predicts that K' dips below zero resulting in a local minimum of K as shown in Figure 2d. The "low structure" representation has a shallower minimum and no shoulder maxima around 25 GPa for the same (optimal) 2 GPa misfit. Since K' is negative in the transition region only for the least damped "agnostic" solution, the bulk modulus does not generally "soften" (decrease at higher pressure). Instead, it simply fails to increase normally with pressure. The "transition informed" fit, created to better align features of K' with identified pressures of transitions, has interval boundaries near the transition from dominant rotational disorder to more translational disorder around 15 GPa and the start of hydrogen symmetrization with the increase of translational disorder above 40 GPa, with a full transition to ice X above 60–80. As shown in Table 3, the misfit for the "transition informed" representation is reduced at the cost of a higher rms average for the fourth derivative of Helmholtz energy. Whether this represents over fitting data remains to be determined.

The key findings of this section reinforce those associated with the effort to represent NaCl. Conventional equation of state global representations can provide approximations to all pressure–volume data at the expense of predicted bulk moduli that are not in accord with low pressure measurements and exhibit incorrect high-pressure limiting behavior. Local-basis-function representation for high-pressure ice are able to better satisfy these strong side constraints. However, for the ice VII–X continuum, in the regime of higher-order phase transitions, an infinite number of representations are possible that fit the data within plausible bounds. The ensemble reported here provide examples of trade-off possibilities between misfit to data and allowable structure of the fits. Representations can have features of K' aligned with expected transitions on the basis of a priori constraints although such close alignment is not required by the pressure–volume data alone. The most robust features of all members of successful fits are a decrease in K' from normal values near 15 GPa to a minimum in the 25 to 40 GPa range followed by an increase to 60–70 GPa followed by normal asymptotic decreases at much higher pressures. No fit required to follow measured bulk moduli in the 25 to 60 GPa range provided an adequate representation of the pressure–volume data. Whether this is a result of errors in data or is an indication of a rate-dependent process remains undetermined.

4. Discussion

A representation of Helmholtz energy at 300 K for ice VII–X was developed in order to better understand the complexity of high-pressure ices that exist as mineral phases in this solar system and beyond. Questions of origins and evolution require knowledge of an underlying equation of state. The results of this investigation are summarized in the concluding section. In order to accomplish the principal goal, coverage of a broad set of concepts and analysis was necessary. Topics include a foundational review about the nature of numerical representations for equations of state, an exposition describing a new framework for representations, and articulation of important details associated with the new framework. Both interpolation and inverse problems are widely recognized challenges in data assimilation. Methods, previously found useful in other fields, are adapted and modified to meet the unique requirements of Helmholtz energy representations.

The underlying premise is that equations of state must provide coherent empirical representations of all measurement-constrained and theory-based knowledge of a material. The previous generation of parameterizations under performed in this task as a result of a bias towards using physics-blind models with insufficient flexibility. This lack of flexibility resulted in discussions that shifted to numerical issues such as "What constitutes the best strain metric?" and "Which order expansion is preferred?". The current analysis refocuses attention. The behavior of an equation of state should be physics-based and basis-function independent. The total number of model parameters required is less important than how well the representation matches the constraining information. Finding the best representation involves determining the appropriate relative weighting of data from varied sources, assessing the quality of results based on theory, and applying suitable side-constraints based on the best physical insight. Decisions based on these ideas were specifically expressed during the construction of equations of state for NaCl and high-pressure ice.

The new equation of state analysis is a seminal step beyond the earlier generation of Helmholtz energy-based parameterizations that use polynomial series expansions of an arbitrarily defined strain metric. Here, a single global polynomial is replaced with flexible local piece-wise polynomials in the form of b-splines. The new local-basis-function representations are more than simple spline fits of data since their creation is embedded in an inverse theory framework of parameter estimation by Tikhonov inversion that requires incorporation of physics-based constraints and insight. An important characteristic of inverse theory solutions is the ability to explore features shared by all possible representations of data. A key strength of this approach, demonstrated in the effort to represent ice, is the power to ask both what model features are required by data and conversely what are data requirements to adequately delineate model features.

5. Conclusions

A large body of measurements and theory for ice VII–X were assimilated into an equation of state covering pressures up to 450 GPa. Selected pressure–volume data from 14 experimental studies and two theoretical reports were then accurately represented. In addition, separately measured derivative properties (the bulk modulus and its pressure derivative) were replicated both in low-pressure and high-pressure regimes, indicating that the underlying pressure–volume data are sufficiently accurate and that the current analysis can extract higher order derivatives. Anomalies in the bulk modulus (and its derivative) were found to roughly align with proposed higher-order transitions at intermediate pressures. However, in this range, bulk moduli determined using GHz-frequency Brillouin experiments did not match the static determination. The differences between results from different groups leaves open the possibility that systematic experimental errors remain large. Alternatively, a relaxation time-constant for the underlying high-order phase transition or for proton mobility may be large compared to the frequencies used in acoustic measurements. The pressure–volume data can resolve variations of the second derivative of the bulk modulus with pressure. However, discontinuities in K'' and/or the precise locations of features in the pressure derivative of the bulk modulus are not currently resolvable. Improved high-pressure determinations of derivative properties could provide better constraints for the next-generation representations. More generally, the tools and methodologies developed here can be broadly applied in other high-pressure equation of state studies.

Articulation of the necessary background for b-splines and for parameter estimation is placed in Appendix A. A small toolbox containing five functions and example scripts written in the MATLAB numerical environment are also provided with the Supplementary Materials. MATLAB is convenient for exploration of the ideas presented in this paper. The open source environment OCTAVE is compatible with the provided MATLAB functions and scripts and can be downloaded for no cost. PYTHON and FORTRAN implementations can easily be created.

In the provided MATLAB scripts, the analysis requires four steps: (1) load data, (2) set options, (3) fit data, (4) display results. This provides a straightforward workflow that encourages exploration of how modifications of assumptions provide differences in the resulting equation of state. It is important to try modifications of all adjustable elements in order to explore the parameter sensitivity and quality variations of fits.

Supplementary Materials: The following are available online at http://www.mdpi.com/2075-163X/10/2/92/s1, computer software files.

Author Contributions: Conceptualization, J.M.B. and B.J.; methodology, J.M.B.; software, J.M.B.; validation, J.M.B. and B.J.; formal analysis, J.M.B. and B.J.; writing—original draft preparation, J.M.B. and B.J.; writing—review and editing, J.M.B. and B.J. All authors have read and agreed to the published version of the manuscript.

Funding: This work is partially supported by the NASA Solar System Workings Grant 80NSSC17K0775 and the NASA Astrobiology Institute through the Icy Worlds and the Titan and Beyond; Habitability of Hydrocarbon Worlds (08-NAI5-0021 and 17-NAI8-2-017). B.R.J.'s research was partially supported by an appointment to the NASA Postdoctoral Program at the University of Washington, administered by Universities Space Research Association under contract with NASA.

Acknowledgments: Long running discussions with E. Abramson improved the current presentation. We thank Philip Myint for his encouragement to undertake this work. The authors also thank J-A. Hernandez for fruitful discussions on DFT calculations of ice. J.M.B. expresses gratitude for the influence Orson L. Anderson had on his career. Anderson's research, helping to define the field of mineral physics, provided an impetus for J.M.B., as an undergraduate student, to switch from atomic physics to geophysics. Later, J.M.B. had the privilege to work with O.L.A. in the founding of the Mineral Physics Group within the American Geophysical Union. O.L.A.'s record of scholarship, his enthusiasm, and his support for others are lasting legacies of his career.

Conflicts of Interest: The authors declare no conflict of interest.

Appendix A. B-Spline and Inverse Method Details Related to Equation of State Representations

Appendix A.1. B-Spline Basis Functions

Fundamentals of b-splines are covered in detail in de Boor [29]. Here, a few points are highlighted in association with the use of b-splines in equation of state representations. B-splines are piece-wise polynomials of specified order, on intervals defined by the interval bounding knots. In polynomial representations, x^3 is both a third degree and third order polynomial. However, using accepted convention, a *kth*-order spline is associated with polynomial of degree *k-1*. To avoid confusion in the following discussion, the spline order is parenthetically followed by explicit articulation of the underlying degree of an associated polynomial.

Knots are points along the independent variable (x) axis that define interval boundaries. For a knot sequence ($t_1, t_2, \ldots, t_{p+k}$), where p is the number of coefficients (the model parameters) and k is the order of the b-spline (degree k-1), a recursion relationship (Equation (A1)) defines the spline basis function, $B_{j,k}(x)$, where x lies between the first and last knot and the first order basis function $B_{j,1}$ is equal to 1 for x within the interval between t_j and t_{j+k} and 0 otherwise.

$$B_{j,k}(x) = \frac{x - t_j}{t_{j+k-1} - t_j} B_{j,k-1}(x) + \frac{t_{j+k} - x}{t_{j+k} - t_{j+1}} B_{j+1,k-1}(x) \quad \text{(A1)}$$

Analytic values for basis functions, B_{jk} evaluated at specified locations, are numerically obtained and represented as arrays with a column count equal to the number of spline coefficients (model parameters) and a row count equal to the number of locations of the function evaluation. The underlying algorithms are robust and available as functions/subroutines in all numerical environments. The use of b-spline basis functions requires no more concern on the part of the user than the use of any standard numerical functions (i.e., trigonometric or logarithmic functions). However, in order to develop new equations of state, the user needs to know how to create an acceptable knot series for a specific application. The knots (t_i) used to construct the spline basis functions must meet several requirements. (1) The total number of knots must equal $k + p$. (2) Each repetition of a knot reduces by one the number of continuous derivatives at the interval boundary. (3) The first and last knot are repeated k-fold times in order for arbitrary data to be fit. Further discussion of how knots are chosen is given below.

Appendix A.2. Evaluation of B-Spline Representations

The evaluation of $y = f(x)$, where function f is a b-spline representation, is numerically represented:

$$y_l = \sum_{i=1}^{p} B_{j,k}(x_l) m(i) \quad \text{(A2)}$$

where data are paired values (x_l and y_l), vector $\overline{m}$ contains p spline coefficients (the model parameters), and $B_{j,k}(x_l)$ values are determined using Equation (A1). Using vector and matrix notation Equation (A2) is simply a linear equation of the form:

$$\overline{y} = \overline{\overline{B}}\,\overline{m} \quad \text{(A3)}$$

which leads naturally and advantageously to spline coefficients, $\overline{m}$, being determined through the inverse solution of a linear system (further discussed below). As written in Equation (A2), the determination of a set of modeled $\overline{y}$ values requires that a (possibly large) matrix of basis functions be determined and held in computer memory. The Cox–de Boor algorithm is implemented in all numerical environments and allows Equation (A2) to be more efficiently evaluated (not holding the matrix $\overline{\overline{B}}$ in memory) which reduces computer CPU and memory requirements.

Evaluation of derivative or integral properties of a b-spline representation simply requires analytic determination of the appropriate derivative or integral values of the polynomial basis functions $B_{j,k}(x_l)$:

$$DB_{j,k}(x) = \frac{k-1}{t_{j+k-1} - t_j} B_{j,k-1}(x) - \frac{k-1}{t_{j+k} - t_{j+1}} B_{j+1,k-1}(x) \quad \text{(A4)}$$

followed by a solution using the form shown in Equation (A2).

Appendix A.3. Details of B-Spline Knots and Control Points

As an example of knots for a (k = 4) fourth-order (third degree) b-spline on a domain spanning from 1 to 10, an arbitrary chosen knots sequence containing the nine elements [1, 1, 1, 1, 5, 10, 10, 10,

10] divides the domain into two intervals (1 through 5 and 5 through 10). Bounding knots are repeated k times. Based on the relationship that the number of knots is equal to the sum of the number of spline coefficients (model parameters) and the spline order, five spline coefficients are required to describe the two associated cubic polynomials. While a single third-degree polynomial has four possible coefficients, the additional polynomial associated with a second interval requires one additional coefficient since the matching boundary conditions fixes the other three parameters. The sequence [1, 1, 1, 1, 4, 7, 10, 10, 10, 10] has three intervals using 10 knots that require six spline coefficients. The spacing of intervals is an arbitrary choice to meet needs in a particular application. An alternate sequence with 10 knots having different intervals is [1, 1, 1, 1, 2, 4, 10, 10, 10, 10]. The sequence [1, 1, 1, 1, 2, 4, 4, 10, 10, 10, 10] (with one interior knot repeated) contains 11 knots and requires seven coefficients. The underlying representation is then allowed to have a discontinuous second derivative at location 4.

An important distinction is made between splines as a tool to interpolate precise data and splines as a method to represent, in a least square sense, data with uncertainty. In the historical development of spline methods, data (pairs of dependent and independent variables) were assumed to have high precision and basis functions were constructed to provide accurate interpolation between the data. In this context, the number of knots is chosen such that the number of required model parameters exactly matches the number of constraints (the available data). Knot locations are then chosen such that at least one data point is associated with each interval. For such an exactly constrained problem, the locations of the independent variables are labeled "control points". The resulting model parameters are then directly associated with the specified control points. However, the model, given as a list of the knots (different from the list of control points) and a list of model parameters, does not require that control point locations be specified. In the least square fitting of more data than model parameters, with representations that are constructed through choice of knot locations (not a choice of data locations), the control point locations are ill defined other than that model parameter locations sequentially progress across the span of the knots with a few more model parameters required than intervals. This discussion is provided since the standard presentations of splines tend to emphasize a perspective centered on control points.

Appendix A.4. Local-Basis-Function Equations of State Representations

A simplified equation of state application using b-splines is illustrated in Figure A1. Panel (a) on the left shows Helmholtz energy as a function of volume. The corresponding pressures (the negative derivative of Helmholtz energy) are shown on the right side in panel (b). An energy minimum on the left is associated with zero pressure on the right. Pressures increase as volumes decrease from equilibrium. Pressure becomes negative in the expanded state but tends back to zero for sufficiently large volumes.

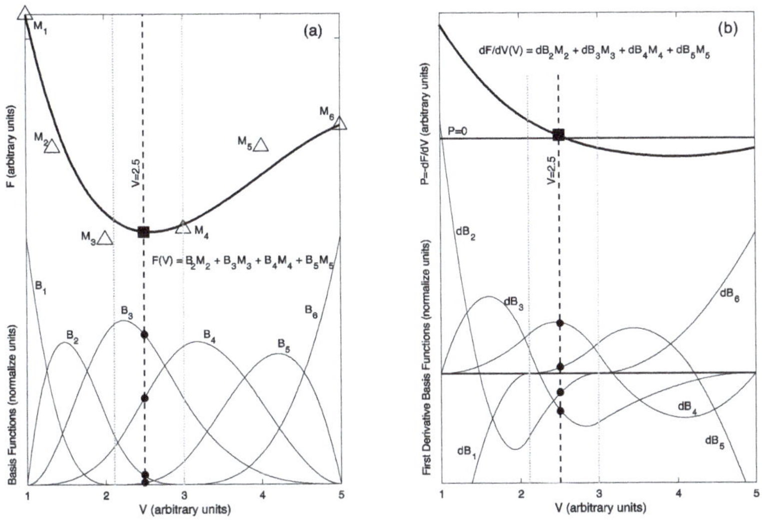

Figure A1. Equation of state example using a cubic b-spline representation on three intervals. Panel (**a**) (left side): Helmholtz energy (F) as a function of volume (V) is shown as a thick line. Panel (**b**) (right side): Pressure, as the negative of the first derivative of Helmholtz energy, is shown. Vertical dotted lines show the internal spline boundaries between three intervals of volume. B-spline basis functions are shown at the bottom of each panel with labels B_i and dB_i. The six model parameters (labeled M_i) required to represent the function are shown in panel (a), plotted at their control point locations. Note that the control point locations are implicit in the distribution of the intervals and are not separately tabulated in the model description. The evaluation of the function at a specified volume (vertical dashed line) is illustrated in each panel where the function is given as a sum of the product of basis function values and model parameters. The evaluation of the functions for any other value of volume is undertaken by the same linear analysis. The model parameters, in units of energy, can be determined through a fit of pressures and volumes based on the linear relationships as shown on the right side.

In this example, three intervals are chosen for a b-spline representation using cubic basis functions (a fourth-order spline for energy). The ranges of the internal intervals are marked by the vertical dotted lines. If three independent cubic polynomials were used for these intervals, 12 separate basis functions and parameters would be required (four degrees of freedom for each cubic polynomial). However, matching the function and its first two derivatives at the two internal boundaries reduces the problem to six degrees of freedom—two additional basis functions are required beyond the four that otherwise would be associated with a single global cubic polynomial. Given specified intervals and spline order, the b-spline basis functions are exactly defined and numerically determined by a subroutine call. The behavior of the six b-spline basis functions is shown at the bottom of each panel.

The b-spline basis functions are non-zero over subdomains of the full model space, extending over k intervals across some interval boundaries. Basis function in the right panel are clearly derivatives of basis functions plotted in the left panel. Model parameters, determined by fitting data on the right side, are plotted in the left panel. The volumes associated with model parameters (the control points as previously described) are not directly aligned with the interval boundaries; instead they are implicit in the distribution of intervals and do not figure into evaluations of the splines. In the limit of small intervals, the model parameters are equal to the function being represented within each interval.

Interpolation using b-splines is a weighted average of neighboring model parameters, the basis functions provide the normalized weights. As shown in Figure A1, both the model and the derivative of the model are evaluated as a linear problem: a sum of basis function values times model parameter values. The same model parameters are used for calculation of the function and for calculation of its derivatives. This provides a path to determine the Helmholtz energy by collocation, using measurements of its derivatives. Implicit is the need to set an integrating constant. The inverse problem, determining model parameters from data, is further discussed below.

Appendix A.5. Defining Equation of State Properties with Derivatives of Helmholtz Energy

In the current framework, equation of state properties are numerically evaluated using analytic representations for the underlying quantities. The necessary relationships associated with derivatives of Helmholtz energy that define pressure, the bulk modulus, and its pressure derivative are summarized. Since Helmholtz energy is parameterized as a function of strain $F(\eta, T = 300\ K)$, the relationship of η as a function of volume figures into the thermodynamic quantities. Derivatives of strain with volume are analytically determined from the variously defined functional forms and derivatives of energy are analytically calculated from the local-basis-function representation. These derivatives are denoted using subscripts for the derivative level:

$$\eta_1 = \frac{d\eta}{dV}, \ \eta_2 = \frac{d^2\eta}{dV^2}\cdots$$

and

$$F_1 = \frac{dF}{d\eta}, \ F_2 = \frac{d^2F}{d\eta^2}\cdots$$

Through application of the chain rule for differentiation, pressure is then calculated as a product of derivatives:

$$P = -\frac{dF}{dV} = -\frac{dF}{d\eta}\frac{d\eta}{dV} = -F_1\eta_1 \tag{A5}$$

In order to calculate the bulk modulus and its pressure derivative, derivatives of pressure with volume are required

$$P_1 = \frac{dP}{dV} = -\ {\eta_1}^2F_2 -\ \eta_2F_1$$

And

$$P_2 = \frac{d^2P}{dV^2} = -\ {\eta_1}^3F_3 - 3\ \eta_1\ \eta_2F_2 -\ \eta_3F_1$$

Using these derivatives, the bulk modulus is

$$K = -VP_1 \tag{A6}$$

and its pressure derivative is

$$K' = \frac{dK}{dP} = -V\left(1 + \frac{P_2}{P_1}\right) \tag{A7}$$

From these relationships it is apparent that pressure and the bulk modulus are determined as linear combinations of strain basis functions while the pressure derivative K' requires a non-linear combination of strain basis functions.

Appendix A.6. Inverse Techniques to Find Model Parameters

As described in Aster et al. [33], parameter estimation based on inverse theory is a foundational problem in diverse fields where data assimilation and interpretation are required. Here a selection of ideas is presented that directly bear on equation of state studies. As shown above, with Helmholtz energy expanded in a strain metric, both pressures and bulk moduli as a function of volume can be expressed using linear relationships. A general linear problem, associated with any equation of state application, is given as:

$$\overline{\overline{B}}\,\overline{m} = \overline{d} \tag{A8}$$

where the array $\bar{\bar{B}}$ on the left-side contains values for the basis functions evaluated at the independent variable locations associated with the dependent variables $\bar{d}$ given in the right-most column vector, and $\overline{m}$ contains the model parameters. For m model parameters and n data, $\bar{\bar{B}}$ is an n by m matrix of values. In the case of conventional equations of state fitting, the basis functions are the polynomial values of strain shown on the right side of Equation (2) (main text) or the fractional powers of (V/V_o) shown on the right side of Equation (4) (main text). For local-basis-function representations, the b-spline basis functions evaluated for each independent variable location, are placed in $\bar{\bar{B}}$.

Following standard numerical analysis, the least square solution for a linear problem (Equation (A8)) for parameters $\overline{m}$ is

$$\overline{m} = \left(\bar{\bar{B}}^t \bar{\bar{B}}\right)^{-1} \bar{\bar{B}}^t \bar{d} \tag{A9}$$

where superscript ^{t} denotes the transpose operation and the negative one power implies determination of a matrix inverse. Equation (A9) can usually be solved if more data than model parameters are available (not rank-deficit). However, even with an adequate amount of data, solutions may be poorly-conditioned when data do not adequately span the parameter space of the model. The typical example of a poorly-conditioned solution is the exercise of trying to fit a linear distribution of data using a higher-order polynomial. Problems can also arise in constraining model parameters for a local-basis-function representation if data are not adequately distributed in regimes containing measurements or in regions of extrapolation.

To overcome problems of poorly-conditioned or rank-deficit inverse problems, additional side constraints are needed and are included through regularization. A common form of regularization is to make the model "smooth" by minimizing its second derivative. More generally, based on physical insight in equation of state representations, a higher-order derivative of Helmholtz energy is minimized, as described in the main text. Basis functions for the specified derivative of energy are added to the array $\bar{\bar{B}}$ and associated zeros are added to the data vector $\bar{d}$:

$$\begin{bmatrix} \bar{\bar{B}}_d \\ \lambda\, \bar{\bar{B}}_r \end{bmatrix} [\overline{m}] = \begin{bmatrix} \bar{d} \\ 0 \end{bmatrix} \tag{A10}$$

where $\bar{\bar{B}}_d$ are the basis functions associated with data, $\bar{\bar{B}}_r$ are the basis functions for regularization (a chosen derivative of Helmholtz energy), and λ is an adjustable (damping) factor to weight the influence of the regularization (small value for no influence and large value to strongly enforce the regularization). The data vector on the right side of Equation (A10) contains both measurements, $\bar{d}$, and as many zeros as rows of regularization in array $\bar{\bar{B}}_r$. The idea is that $\bar{\bar{B}}_r\overline{m}$ should be as close to zero as possible, while simultaneously requiring the model to adequately fit the data. In areas of sparse or no data, regularization controls the nature of the fit. The advantages of a least square solution of Equation (A10) includes an ability to suppress unnecessary structure (overfitting) and to provide constraints in regimes without data. Linear optimization of pressure–volume and bulk modulus–volume data with regularization based on a user-specified derivative of Helmholtz energy is implemented in the numerical toolbox included with the Supplementary Materials.

In order to directly fit measurements that cannot be expressed as a linear combination of model parameters (i.e., derivatives of the bulk modulus with respect to pressure), non-linear optimization of model parameters is required. The form of the non-linear Tikhonov problem is given as:

$$\begin{bmatrix} \bar{\bar{C}} \\ \lambda\bar{\bar{Q}} \end{bmatrix} [\delta\overline{m}] = \begin{bmatrix} \delta\bar{d} \\ -\lambda\bar{\bar{Q}}\overline{m}_o \end{bmatrix} \tag{A11}$$

where $\overline{m}_0$ is an initial guess for the model, $\delta\overline{m}$ are increments to $\overline{m}_0$ that reduce data misfit, and $\delta\overline{d}$ are deviations between data and predictions based on $\overline{m}_0$. Data can be a combination of equation of state measurements including pressures, bulk moduli, or derivatives of the bulk modulus. The array $\overline{\overline{C}}$ contains derivatives of the model predictions with respect to model parameters and the array $\overline{\overline{Q}}$ contains derivatives with respect to model parameters of any regularization quantity to minimize. The least square solution of Equation (A11) gives the increments $\delta\overline{m}$ to $\overline{m}_0$ that provide a better representation of data subject to the side constraint. Using an improved model $\overline{m}_0$ after each step gives results that usually converge after a few iterations. Non-linear optimization that includes constraints on the pressure derivative of the bulk modulus is implemented in the numerical toolbox included with the Supplementary Materials.

References

1. Tschauner, O.; Huang, S.; Greenberg, E.; Prakapenka, V.B.; Ma, C.; Rossman, G.R.; Shen, A.H.; Zhang, D.; Newville, M.; Lanzirotti, A.; et al. Ice-VII inclusions in diamonds: Evidence for aqueous fluid in Earth's deep mantle. *Science* **2018**, *359*, 1136–1139. [CrossRef]
2. Léger, A.; Selsis, F.; Sotin, C.; Guillot, T.; Despois, D.; Mawet, D.; Ollivier, M.; Labèque, A.; Valette, C.; Brachet, F. A new family of planets? *Icarus* **2004**, *169*, 499–504. [CrossRef]
3. Sotin, C.; Grasset, O. Mass-radius curve for extrasolar Earth-like planets and ocean planets. *Icarus* **2007**, *191*, 337–351. [CrossRef]
4. Noack, L.; Höning, D.; Rivoldini, A.; Heistracher, C.; Zimov, N.; Journaux, B.; Lammer, H.; Van Hoolst, T.; Bredehöft, J.H. Water-rich planets: How habitable is a water layer deeper than on Earth? *Icarus* **2016**, *277*, 215–236. [CrossRef]
5. Vance, S.D.; Panning, M.P.; Stähler, S.; Cammarano, F.; Bills, B.G.; Tobie, G.; Kamata, S.; Kedar, S.; Sotin, C.; Pike, W.T.; et al. Geophysical investigations of habitability in ice-covered ocean worlds: Geophysical habitability. *J. Geophys. Res. Planets* **2018**, *123*, 180–205. [CrossRef]
6. Journaux, B.; Daniel, I.; Petitgirard, S.; Cardon, H.; Perrillat, J.-P.; Caracas, R.; Mezouar, M. Salt partitioning between water and high-pressure ices. Implication for the dynamics and habitability of icy moons and water-rich planetary bodies. *Earth Planet. Sci. Lett.* **2017**, *463*, 36–47. [CrossRef]
7. Unterborn, C.T.; Desch, S.J.; Hinkel, N.R.; Lorenzo, A. Inward migration of the TRAPPIST-1 planets as inferred from their water-rich compositions. *Nat. Astron.* **2018**, *2*, 297–302. [CrossRef]
8. Kite, E.S.; Ford, E.B. Habitability of exoplanet waterworlds. *arXiv* **2018**, arXiv:1801.00748 [astro-ph]. [CrossRef]
9. Hendrix, A.R.; Hurford, T.A.; Barge, L.M.; Bland, M.T.; Bowman, J.S.; Brinckerhoff, W.; Buratti, B.J.; Cable, M.L.; Castillo-Rogez, J.; Collins, G.C.; et al. The NASA Roadmap to Ocean Worlds. *Astrobiology* **2019**, *19*, 1–27. [CrossRef]
10. Birch, F. Elasticity and constitution of the Earth's interior. *J. Geophys. Res.* **1952**, *57*, 227–286. [CrossRef]
11. Stacey, F.D.; Brennan, B.J.; Irvine, R.D. Finite strain theories and comparisons with seismological data. *Geophys. Surv.* **1981**, *4*, 189–232. [CrossRef]
12. Benoit, M.; Bernasconi, M.; Focher, P.; Parrinello, M. New High-Pressure Phase of Ice. *Phys. Rev. Lett.* **1996**, *76*, 2934–2936. [CrossRef] [PubMed]
13. Hermann, A.; Ashcroft, N.W.; Hoffmann, R. High pressure ices. *Proc. Natl. Acad. Sci. USA* **2012**, *109*, 745–750. [CrossRef]
14. Millot, M.; Coppari, F.; Rygg, J.R.; Correa Barrios, A.; Hamel, S.; Swift, D.C.; Eggert, J.H. Nanosecond X-ray diffraction of shock-compressed superionic water ice. *Nature* **2019**, *569*, 251–255. [CrossRef] [PubMed]
15. Hemley, R.J.; Jephcoat, A.P.; Mao, H.K.; Zha, C.S.; Finger, L.W.; Cox, D.E. Static compression of H_2O-ice to 128 GPa (1.28 Mbar). *Nature* **1987**, *330*, 737–740. [CrossRef]
16. Loubeyre, P.; Letoullec, R.; Wolanin, E.; Hanfland, M.; Häusermann, D. Modulated phases and proton centering in ice observed by X-ray diffraction up to 170 GPa. *Nature* **1999**, *397*, 503–506. [CrossRef]
17. Sugimura, E.; Iitaka, T.; Hirose, K.; Kawamura, K.; Sata, N.; Ohishi, Y. Compression of H_2O ice to 126 GPa and implications for hydrogen-bond symmetrization: Synchrotron x-ray diffraction measurements and density-functional calculations. *Phys. Rev. B* **2008**, *77*, 214103. [CrossRef]

18. Goncharov, A.F.; Struzhkin, V.V.; Somayazulu, M.S.; Hemley, R.J.; Mao, H.K. Compression of Ice to 210 Gigapascals: Infrared Evidence for a Symmetric Hydrogen-Bonded Phase. *Science* **1996**, *273*, 218–220. [CrossRef]
19. Zha, C.-S.; Hemley, R.J.; Gramsch, S.A.; Mao, H.; Bassett, W.A. Optical study of H_2O ice to 120GPa: Dielectric function, molecular polarizability, and equation of state. *J. Chem. Phys.* **2007**, *126*, 074506. [CrossRef]
20. Meier, T.; Petitgirard, S.; Khandarkhaeva, S.; Dubrovinsky, L. Observation of nuclear quantum effects and hydrogen bond symmetrisation in high pressure ice. *Nat. Commun.* **2018**, *9*, 1–7. [CrossRef]
21. Grande, Z.M.; Huang, C.; Smith, D.; Smith, J.S.; Boisvert, J.H.; Tschauner, O.; Steffen, J.H.; Salamat, A. Bond strengthening in dense H_2O and implications to planetary composition. *arXiv* **2019**, arXiv:1906.11990 [astro-ph, physics:cond-mat].
22. Hirai, H.; Kadobayashi, H.; Matsuoka, T.; Ohishi, Y.; Yamamoto, Y. High pressure X-ray diffraction and Raman spectroscopic studies of the phase change of D_2O ice VII at approximately 11 GPa. *Int. J. High Press. Res.* **2014**, *34*, 289–296. [CrossRef]
23. Somayazulu, M.; Shu, J.; Zha, C. In situ high-pressure x-ray diffraction study of H_2O ice VII. *J. Chem. Phys.* **2008**, *128*, 064510. [CrossRef] [PubMed]
24. Pruzan, P.; Chervin, J.-C.; Gauthier, M. Raman Spectroscopy Investigation of Ice VII and Deuterated Ice VII to 40 GPa. Disorder in Ice VII. *Europhys. Lett.* **1990**, *13*, 81–87. [CrossRef]
25. Zha, C.-S.; Tse, J.S.; Bassett, W.A. New Raman measurements for H_2O ice VII in the range of 300 cm^{-1} to 4000 cm^{-1} at pressures up to 120 GPa. *J. Chem. Phys.* **2016**, *145*, 124315. [CrossRef]
26. Guthrie, M.; Boehler, R.; Tulk, C.A.; Molaison, J.J.; dos Santos, A.M.; Li, K.; Hemley, R.J. Neutron diffraction observations of interstitial protons in dense ice. *Proc. Natl. Acad. Sci. USA* **2013**, *110*, 10552–10556. [CrossRef]
27. Yoshimura, Y.; Stewart, S.T.; Somayazulu, M.; Mao, H.; Hemley, R.J. High-pressure x-ray diffraction and Raman spectroscopy of ice VIII. *J. Chem. Phys.* **2006**, *124*, 024502. [CrossRef]
28. Okada, T.; Iitaka, T.; Yagi, T.; Aoki, K. Electrical conductivity of ice VII. *Sci. Rep.* **2014**, *4*, 5778. [CrossRef]
29. Noguchi, N.; Okuchi, T. Self-diffusion of protons in H_2O ice VII at high pressures: Anomaly around 10 GPa. *J. Chem. Phys.* **2016**, *144*, 234503. [CrossRef]
30. Polian, A.; Grimsditch, M. New High-Pressure Phase of H_2O: Ice X. *Phys. Rev. Lett.* **1984**, *52*, 1312–1314. [CrossRef]
31. Asahara, Y.; Hirose, K.; Ohishi, Y.; Hirao, N.; Murakami, M. Thermoelastic properties of ice VII and its high-pressure polymorphs: Implications for dynamics of cold slab subduction in the lower mantle. *Earth Planet. Sci. Lett.* **2010**, *299*, 474–482. [CrossRef]
32. French, M.; Redmer, R. Construction of a thermodynamic potential for the water ices VII and X. *Phys. Rev. B* **2015**, *91*, 014308. [CrossRef]
33. Wolanin, E.; Pruzan, P.; Chervin, J.; Canny, B.; Gauthier, M.; Häusermann, D.; Hanfland, M. Equation of state of ice VII up to 106 GPa. *Phys. Rev. B* **1997**, *56*, 5781–5785. [CrossRef]
34. Davies, G.F. Quasi-harmonic finite strain equations of state of solids. *J. Phys. Chem. Solids* **1973**, *34*, 1417–1429. [CrossRef]
35. Poirier, J.-P.; Tarantola, A. A logarithmic equation of state. *Phys. Earth Planet. Inter.* **1998**, *109*, 1–8. [CrossRef]
36. Vinet, P.; Rose, J.H.; Ferrante, J.; Smith, J.R. Universal features of the equation of state of solids. *J. Phys. Condens. Matter* **1989**, *1*, 1941. [CrossRef]
37. de Boor, C. *A Practical Guide to Splines*, Rev. ed.; Springer: New York, NY, USA, 2001; ISBN 978-0-387-95366-3.
38. Brown, J.M. Local basis function representations of thermodynamic surfaces: Water at high pressure and temperature as an example. *Fluid Phase Equilibria* **2018**, *463*, 18–31. [CrossRef]
39. Bollengier, O.; Brown, J.M.; Shaw, G.H. Thermodynamics of pure liquid water: Sound speed measurements to 700 MPa down to the freezing point, and an equation of state to 2300 MPa from 240 to 500 K. *J. Chem. Phys.* **2019**, *151*, 054501. [CrossRef]
40. Journaux, B.; Brown, J.M.; Pakhomova, A.; Collings, I.E.; Petitgirard, S.; Espinoza, P.; Ballaran, T.B.; Vance, S.D.; Ott, J.; Cova, F.; et al. Holistic Approach for Studying Planetary Hydrospheres: Gibbs Representation of Ices Thermodynamics, Elasticity, and the Water Phase Diagram to 2300 MPa. *J. Geophys. Res. Planets* **2020**, *125*, e2019JE006176. [CrossRef]
41. Aster, R.C.; Borchers, B.; Thurber, C.H. *Parameter Estimation and Inverse Problems, Waltham*; Academic Press: Cambridge, MA, USA, 2012.

42. Strässle, T.; Klotz, S.; Kunc, K.; Pomjakushin, V.; White, J.S. Equation of state of lead from high-pressure neutron diffraction up to 8.9 GPa and its implication for the NaCl pressure scale. *Phys. Rev. B* **2014**, *90*, 014101. [CrossRef]
43. Decker, D.L. High-pressure equation of state for NaCl, KCl, and CsCl. *J. Appl. Phys.* **1971**, *42*, 3239–3244. [CrossRef]
44. Brown, J.M. The NaCl pressure standard. *J. Appl. Phys.* **1999**, *86*, 5801–5808. [CrossRef]
45. Stacey, F.D.; Davis, P.M. High pressure equations of state with applications to the lower mantle and core. *Phys. Earth Planet. Inter.* **2004**, *142*, 137–184. [CrossRef]
46. Stacey, F.D. The K-primed approach to high-pressure equations of state. *Geophys. J. Int.* **2000**, *143*, 621–628. [CrossRef]
47. Bezacier, L.; Journaux, B.; Perrillat, J.-P.; Cardon, H.; Hanfland, M.; Daniel, I. Equations of state of ice VI and ice VII at high pressure and high temperature. *J. Chem. Phys.* **2014**, *141*, 104505. [CrossRef] [PubMed]
48. Bridgman, P.W. Freezings and Compressions to 50,000 kg/cm^2. *J. Chem. Phys.* **1941**, *9*, 794–797. [CrossRef]
49. Fei, Y.; Mao, H.; Hemley, R.J. Thermal expansivity, bulk modulus, and melting curve of H_2O–ice VII to 20 GPa. *J. Chem. Phys.* **1993**, *99*, 5369–5373. [CrossRef]
50. Frank, M.; Fei, Y.; Hu, J. Constraining the equation of state of fluid H_2O to 80 GPa using the melting curve, bulk modulus, and thermal expansivity of Ice VII1. *Geochim. Cosmochim. Acta* **2004**, *68*, 2781–2790. [CrossRef]
51. Hernandez, J.-A.; Caracas, R. Proton dynamics and the phase diagram of dense water ice. *J. Chem. Phys.* **2018**, *148*, 214501. [CrossRef]
52. Journaux, B.; Caracas, R.; Carrez, P.; Gouriet, K.; Cordier, P.; Daniel, I. Elasticity and dislocations in ice X under pressure. *Phys. Earth Planet. Inter.* **2014**, *236*, 10–15. [CrossRef]
53. Klotz, S.; Komatsu, K.; Kagi, H.; Kunc, K.; Sano-Furukawa, A.; Machida, S.; Hattori, T. Bulk moduli and equations of state of ice VII and ice VIII. *Phys. Rev. B* **2017**, *95*, 174111. [CrossRef]
54. Liu, L. Compression of ice VII to 500 kbar. *Earth Planet. Sci. Lett.* **1982**, *61*, 359–364. [CrossRef]
55. Munro, R.G.; Block, S.; Mauer, F.A.; Piermarini, G. Isothermal equations of state for H_2O-VII and D_2O-VII. *J. Appl. Phys.* **1982**, *53*, 6174–6178. [CrossRef]
56. Olinger, B.; Halleck, P.M. Compression and bonding of ice VII and an empirical linear expression for the isothermal compression of solids. *J. Chem. Phys.* **1975**, *62*, 94. [CrossRef]
57. Sugimura, E.; Komabayashi, T.; Hirose, K.; Sata, N.; Ohishi, Y.; Dubrovinsky, L.S. Simultaneous high-pressure and high-temperature volume measurements of ice VII and its thermal equation of state. *Phys. Rev. B* **2010**, *82*, 134103. [CrossRef]
58. Myint, P.C.; Benedict, L.X.; Belof, J.L. Free energy models for ice VII and liquid water derived from pressure, entropy, and heat capacity relations. *J. Chem. Phys.* **2017**, *147*, 084505. [CrossRef]
59. Ahart, M.; Somayazulu, M.; Gramsch, S.A.; Boehler, R.; Mao, H.; Hemley, R.J. Brillouin scattering of H_2O ice to megabar pressures. *J. Chem. Phys.* **2011**, *134*, 124517. [CrossRef]
60. Shimizu, H.; Nabetani, T.; Nishiba, T.; Sasaki, S. High-pressure elastic properties of the VI and VII phase of ice in dense H_2O and D_2O. *Phys. Rev. B* **1996**, *53*, 6107–6110. [CrossRef]
61. Zhang, J.S.; Hao, M.; Ren, Z.; Chen, B. The extreme acoustic anisotropy and fast sound velocities of cubic high-pressure ice polymorphs at Mbar pressure. *Appl. Phys. Lett.* **2019**, *114*, 191903. [CrossRef]

minerals

MDPI

Article

A Paris-Edinburgh Cell for High-Pressure and High-Temperature Structure Studies on Silicate Liquids Using Monochromatic Synchrotron Radiation

Tony Yu [1,*], Clemens Prescher [1,†], Young Jay Ryu [1], Feng Shi [1,‡], Eran Greenberg [1,§], Vitali Prakapenka [1], Peter Eng [1], Joanne Stubbs [1], Yoshio Kono [2,||], Guoyin Shen [3], Heather Watson [4], Mark L. Rivers [1], Stephen R. Sutton [1] and Yanbin Wang [1]

1 Center for Advanced Radiation Sources, The University of Chicago, Chicago, IL 60637, USA; clemens.prescher@gmail.com (C.P.); ryu@cars.uchicago.edu (Y.J.R.); shifeng@cug.edu.cn (F.S.); erangre@gmail.com (E.G.); prakapenka@cars.uchicago.edu (V.P.); eng@cars.uchicago.edu (P.E.); stubbs@cars.uchicago.edu (J.S.); rivers@cars.uchicago.edu (M.L.R.); sutton@cars.uchicago.edu (S.R.S.); wang@cars.uchicago.edu (Y.W.)

2 Geophysical Laboratory, Carnegie Institution of Washington, Argonne, IL 60439, USA; kono.yoshio.rj@ehime-u.ac.jp

3 HPCAT, X-ray Science Division, Argonne National Laboratory, Argonne, IL 60439, USA; gyshen@anl.gov

4 Physics and Astronomy Department, Union College, Schenectady, NY 12308, USA; watsonh@union.edu

* Correspondence: tyu@cars.uchicago.edu

† Currently at the Institute of Geology and Mineralogy, University of Cologne, Albertus-Magnus-Platz, 50674 Cologne, Germany.

‡ Currently at the State Key Laboratory of Geological Processes and Mineral Resources, China University of Geosciences-Wuhan, Wuhan 430074, China.

§ Currently at the Applied Physics Department, Soreq Nuclear Research Center (NRC), Yavne 81800, Israel.

|| Currently at the Geodynamics Research Center, Ehime University, Matsuyama 790-8577, Japan.

Received: 16 October 2019; Accepted: 14 November 2019; Published: 19 November 2019

Abstract: A Paris-Edinburgh press combined with a multi-channel collimator assembly has been commissioned at the GeoSoilEnviro Center for Advanced Radiation Sources (GSECARS) beamline for monochromatic X-ray scattering, with an emphasis on studying low-Z liquids, especially silicate liquids at high pressure. The Paris-Edinburgh press is mounted on a general-purpose diffractometer, with a pixel array detector mounted on the detector arm. The incident monochromatic undulator beam with energies up to 60 keV is focused both horizontally and vertically to a beam size about 30 × 30 μm. With this setup, background scattering from the surrounding pressure media is completely removed at 2θ angles above 10° for samples larger than 1.05 mm in diameter. Thirty minutes is typically sufficient to collect robust X-ray scattering signals from a 1.6 mm diameter amorphous silicate sample. Cell assemblies for the standard Paris-Edinburgh anvils have been developed and pressures and temperatures up to 7 GPa and 2300 K, respectively, have been maintained steadily over hours. We have also developed a cupped-toroidal Drickamer anvil to further increase pressure and temperature capabilities. The cupped-toroidal Drickamer anvil combines features of a modified Drickamer anvil and the traditional Paris-Edinburgh anvil. Pressures up to 12 GPa have been generated at temperatures up to 2100 K.

Keywords: Paris-Edinburgh cell; multi-channel collimator; X-ray scattering; non-crystalline materials; high pressure; high temperature

1. Introduction

Studying atomic-scale structures of silicate liquids can help us improve our understanding of the properties of magmas under pressure. Such properties play fundamental roles in the evolution of the

Earth and other terrestrial planets. Crystalline silicates require very high temperatures to melt, and the melting points increase significantly with pressure. As a result, studying silicate liquids under high pressure is faced with tremendous technical challenges [1]. Most earlier studies have mainly focused on the structure of silicate glasses in order to gain information of their liquid counterparts. This is based on the general conception that a glass represents a frozen state of a liquid. Many silicate compositions have been reported in spectroscopy studies (see, e.g., review by [2]). However, structures of glasses are shown to depend on their thermal history [3,4]. Many properties of glasses, e.g., density and elastic moduli, also depend on their pressure and temperature history [5–7]. Such differences, as pointed out by Jing and Karato [8], are mainly due to the dominant roles played by entropy in the liquid state. In contrast, the entropic contribution to the Gibbs energy of glasses is negligible. A recent ambient-pressure angle-dispersive X-ray diffraction (ADXD) study on silicate glasses and melts of identical compositions suggests significant differences in their structures [9]. Therefore, developing a method for direct silicate liquid structure measurement under extreme pressure and temperature conditions becomes particularly important.

Funamori et al. [1] reported liquid structures of $MgSiO_3$ and $CaSiO_3$ up to 6 GPa in a cubic-anvil apparatus (DIA [10]) using a multi-angle energy-dispersive X-ray diffraction (MA-EDXD) technique developed by Tsuji et al. [11]. Yamada et al. [12] adapted the MA-EDXD technique to the Paris-Edinburgh press. These developments have enabled liquid silicate structural studies to ~7 GPa. The MA-EDXD technique has the ability to tightly collimate incident X-ray beam, thereby reducing background noise effectively. However, a complete X-ray scattering dataset at a given pressure and temperature condition typically requires several hours to collect, because X-ray scattering signals from liquids are extremely weak. An alternative to MA-EDXD is ADXD. In order to remove background scattering, Mezouar et al. [13] developed a multi-channel collimator system for the Paris-Edinburgh press for liquid structure studies. With the combination of a Paris-Edinburgh press with a multi-channel collimator, the liquid structure and density of iron-rich alloys have been studied up to 17 GPa and 1200 K [14].

For silicate liquids, which have much lower X-ray scattering power than metals and melts at generally much higher temperatures, measuring their structures is challenging. In this paper, we report our recent development at the GeoSoilEnviro Center for Advanced Radiation Sources (GSECARS) beamline 13-ID-C, using a combination of a Paris-Edinburgh press with a multi-channel collimator and an advanced photon-counting area detector for collection of angle-dispersive X-ray scattering data on amorphous materials, especially silicate liquids under high pressure and temperature conditions. The multi-channel collimator, i.e., Soller slits, consists of two arrays of fine slits made of tungsten carbide blades. The slits are aligned so that they rotate about the sample center. By oscillating the slits during data collection, background scattering from the material surrounding the sample can be effectively removed. The excellent spatial selectivity provides an exciting opportunity for glass and liquid structure studies in the Paris-Edinburgh press. Compared with the energy-dispersive X-ray diffraction (EDXD) method, commonly used to minimize unwanted X-ray scattering signal from surrounding pressure media in large-volume multi-anvil experiments, our multi-channel collimator combined with ADXD measurements are performed at fixed energy and undulator gap, eliminating the need for energy dependent absorption and the undulator spectrum corrections to the data as well as allowing us to take full advantage of the brilliance of the Advanced Photon Source synchrotron light source. Our optimized setup results in typical data collection times of 5 to 30 min for a single ADXD pattern, which is 2 to 3 orders of magnitude faster than the EDXD method. The elimination of energy dependent data corrections, the effective use of the brilliance of the source, and the dramatically reduced collection times allow users to collect rich data sets while pushing the sample cell to extreme pressures and temperatures.

2. Experimental Setup

Figure 1a shows the current layout of the GSECARS beamlines at the Advanced Photon Source, Argonne National Laboratory (Argonne, Lemont, IL, USA). The X-rays for the 13-ID-C/D branch are produced by a 3.0-cm-period undulator that delivers high brilliance and energy X-rays to the beamline X-ray optics and experiments performed in either the ID-C or ID-D hutches downstream. Monochromatic X-rays are produced with a cryogenically cooled double crystal silicon monochromator using either a Si(111) or Si(311) crystal set with a maximum energy of 42 keV and 60 keV respectively. These high energies are ideal for large volume high pressure experiments, since the sample is often surrounded by various high X-ray absorbing layers of ceramic cell parts. The beam is focused to 30 μm × 30 μm full width at half maximum (FWHM) using a pair of dynamically bent one-meter-long platinum-coated silicon mirrors in Kirkpatrick-Baez (KB) geometry. A 40 μm diameter clean-up pinhole is located approximately 70 mm upstream of the sample cell assembly. A nitrogen-filled ion chamber mounted immediately upstream of the pinhole is used to monitor the incident X-ray beam intensity and is later used to normalize the intensity of each collected pattern. The experimental setup is shown in the sketch in Figure 1b.

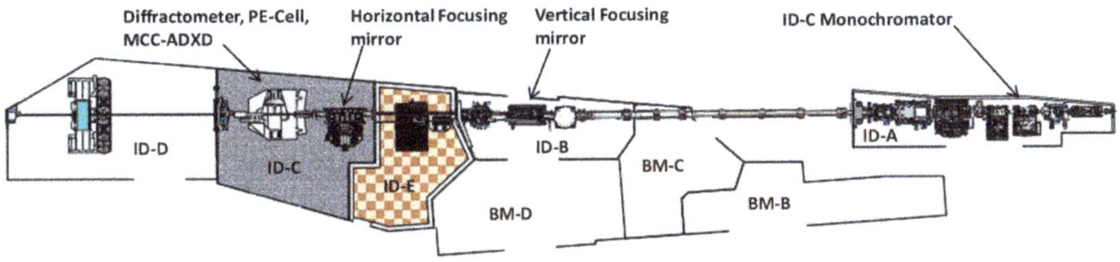

(a)

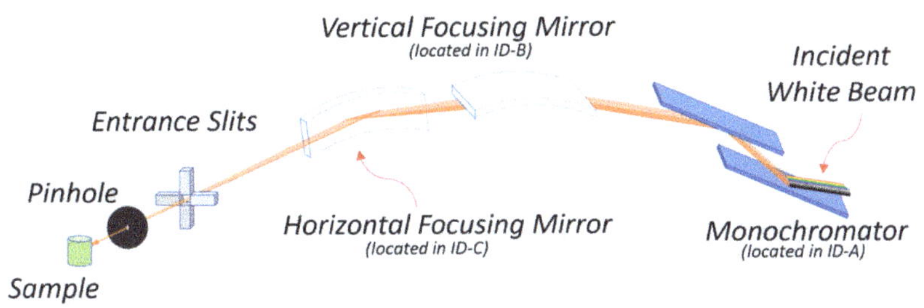

(b)

Figure 1. Current layout of the GeoSoilEnviro Center for Advanced Radiation Sources (GSECARS) bending magnet (BM) and insertion device (ID) beamlines (**a**). X-ray optics of the ID beamlines are mainly installed in ID-A and ID-B hutches. The diffractometer that hosts the Paris-Edinburgh press + multi-channel collimator setup is located in the ID-C hutch (labeled: MCC-ADXD in the gray colored area). (**b**) is a sketch of the experimental setup showing the doubly focused mono beam passing through the entrance slits, pin hole then entering the sample in the Paris-Edinburgh press.

2.1. The Paris-Edinburgh Press

The large-volume Paris-Edinburgh press was originally designed for high pressure neutron scattering experiments due to the large sample volume required to perform meaningful measurements in a reasonable amount of time [15,16]. The VX5 (MG63) Paris-Edinburgh press in our setup with a wide-opening angle of 140° in the horizontal plane for the X-ray scattering signal to exit the sample allows us to easily collect structure data up to 20 A^{-1} in momentum transfer (Q; Q = $\frac{4\pi sin\theta}{\lambda}$; where λ is the wavelength and 2θ is the X-ray scattering angle) at incident X-ray energy of 42 keV. This is beneficial for the ADXD setup. Figure 2 shows the experimental setup in the ID-C hutch from two different viewing angles. The 180-ton VX5 Paris-Edinburgh press, which is mounted on a general-purpose diffractometer [17], has a base of 175 mm in diameter and a height of 212 mm, weighing about 30 kg. The compact form of this "large volume press" makes the apparatus easy to install and manipulate on the diffractometer where space is often limited. XYZ translation stages that allow the press to move along and perpendicular to the beam direction are mounted under the Paris-Edinburgh press. A heavy-duty Huber vertical stage (Series 5103) brings the sample to the beam center. The compression system of the Paris-Edinburgh press is a double-stage setup. The first-stage ENERPAC (Model P142, Menomonee Falls, WI, USA) hydraulic hand pump applies initial compression up to around 30 MPa, while the second-stage syringe pump (High Pressure Equipment Company; Model #37-6-30) is responsible for reaching higher pressure (maximum oil pressure ~190 MPa). By pumping hydraulic oil into the bottom ram, the main cylinder located at the lower part of the Paris-Edinburgh press advances towards the breech that is screwed into the top of the press frame, and we bring the two opposed tungsten carbide anvils towards each other along the uniaxial load direction to generate pressure. The sample pressure inside the cell assembly depends on the size, material, and geometry of the cell assembly. The sample center is determined before each data collection by scanning the Paris-Edinburgh cell in Y (vertical; perpendicular to the beam), X (horizontal; perpendicular to the beam), and Z (horizontal; along the beam) directions to locate the maximum X-ray scattering signal position from the sample.

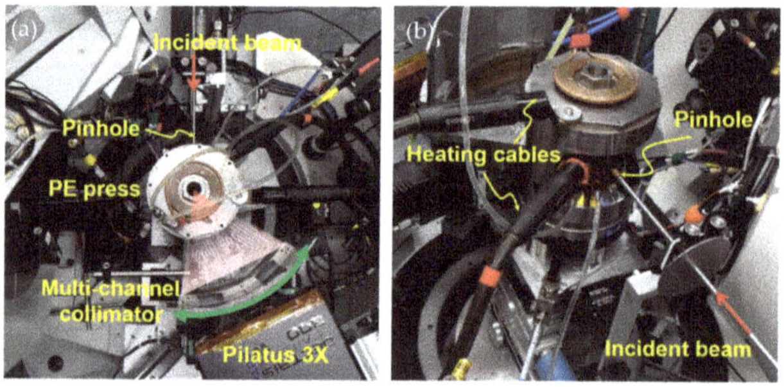

Figure 2. A general-purpose diffractometer is used for glass and liquid structure studies in 13-ID-C. (**a**) is a bird's-eye view of the experimental setup. The incident beam enters the press from the top of the image, the general-purpose diffractometer carries a Pilatus3 X CdTe 300K-W pixel array detector, and the detector arm is capable of rotating with respect to two axes with an angular accuracy of 0.002°. This allows convenient alignment of the setup relative to the focused X-ray beam and large Q coverage. The multi-channel collimator is aligned with respect to the incident X-ray beam, with the Paris-Edinburgh cell located at the rotation center of the general-purpose diffractometer. (**b**) shows the experimental setup with the heating cables and water-cooling hoses attached and the 40 μm wide pin hole projecting into the Paris-Edinburgh press.

Temperature is raised by the conventional resistive heating method. The DC power supply (Hewlett Packard 6671A) provides up to 8 V/200 A, which can bring the cell temperature up to 2100 K routinely. This is sufficient to melt many materials within the accessible pressure range of the Paris-Edinburgh press. Type C (W-5%Re vs. W-26%Re) thermocouples were used to establish heater power-temperature relationships at various loads for both the 3 mm standard and the 2 mm cupped-toroidal Drickamer Paris-Edinburgh cell assemblies. For these off-line temperature calibration runs, the thermocouple was placed at the center of the sample capsule. Such calibrations were conducted at various press loads, to establish a two-dimensional temperature-power grid [18], which was used to estimate temperatures in X-ray scattering experiments without thermocouples. According to the multiple heater power to temperature calibration curves acquired, the error in the estimated temperature is about 20 K at 373 K and about 90 K at 1800 K. The heater power is controlled by a proportional-integral-derivative controller programmed into our operation software. Because of its transparency to the X-ray beam and its good electrical stability under high pressures and temperatures, the most common heater material that we use is high purity graphite for the 3 mm standard cell assembly. A titanium diboride (TiB_2) heater is selected for higher pressure applications, as graphite will transform to diamond at pressures above 10 GPa. TiB_2 is a low density (4.52 g/cm^3) hard material with high compressive strength, which is helpful to keep the sample cell mechanically stable under extreme pressure and temperature conditions. The only drawback for using TiB_2 is the non-linearity of its resistance during heating at high pressure. There is a resistance turnover after reaching the peak value, which is probably due to a phase transition in TiB_2, as the resistance after the turnover follows a smooth curve [19]. To prevent the tungsten carbide anvils from catastrophically failing under high temperature conditions, both the top and bottom anvils are cooled by a constant flow of cold water (~283 K) through copper blocks (heat sinks) that hold the anvils.

2.2. The Multi-Channel Collimator

The multi-channel collimator, a.k.a. Soller slits (custom made by Usinage and Nouvelles Technologies), consists of two arrays of fine slits that are made of tungsten carbide blades (Figure 3). The slits are arranged in two concentric circular arcs. Both arrays consist of 75 slits precisely separated by 0.8° and are located 50 and 200 mm, respectively, from the rotation center. The slit width of the inner array is 0.05 mm and the outer array has a slit width of 0.20 mm [13], therefore covering up to 60° in total angle. During data collection, the combination of a rotation stage and a pair of translation stages (one perpendicular to the beam direction and the other one along the beam direction) that move in the horizontal plane controls the multi-channel collimator oscillation. By oscillating the slits during data collection, background scattering from the material surrounding the sample gets blocked by the tungsten carbide blades and can be effectively removed. Only the sample signal within the diffraction volume defined by the two arrays of slits passes through the slits and reaches the detector. Our analysis shows that with an incident beam collimated at 0.05 mm, a molten sample with a diameter larger than 1.05 mm can be used to collect structural data without being interfered by crystalline peaks from the surrounding pressure media at 2-theta angles greater than 10° (Figure 4). This is a critical component for data collection of weak scattering materials, for example, non-crystalline borates or silicates that are of interest to Earth science related topics. Similar multi-channel collimator assemblies have also been used extensively with the Paris-Edinburgh press at the European Synchrotron Radiation Facility for studying metallic liquids and low-Z materials [13].

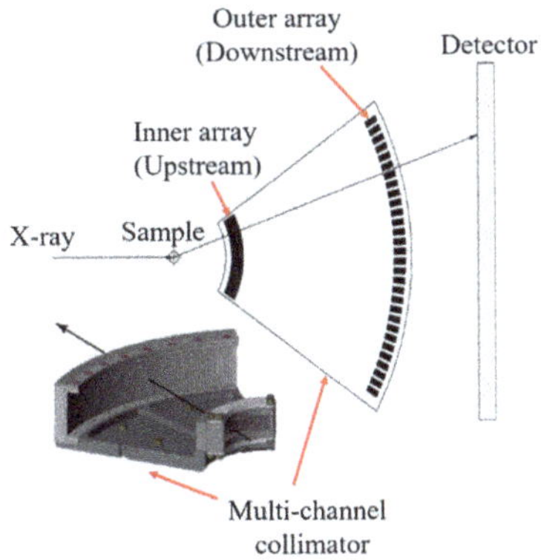

Figure 3. A multi-channel collimator is used for beam collimation on the X-ray scattering side. The inner and outer arrays, with slit opening widths of 0.05 and 0.2 mm, are composed of 75 slits with 0.8° separation. Schematic shows the basic concept of beam collimation (top view) (modified from [13]).

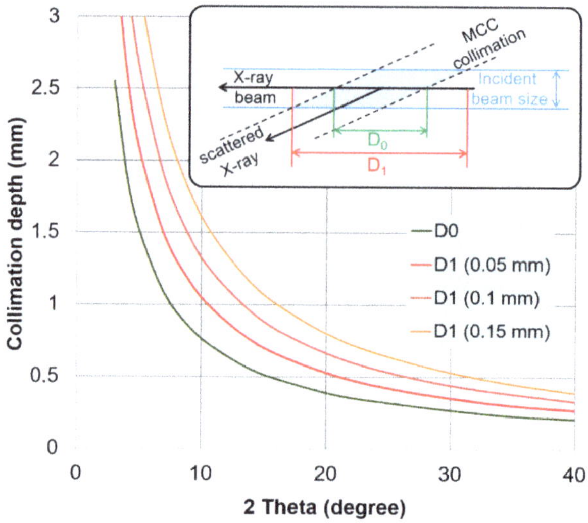

Figure 4. Calculated collimation depth at the sample center (D_0) and the maximum collimation depth (D_1) that takes into account the size of the incident beam. As an example, with an incident beam size of 0.05 mm (red curve), a sample with a diameter >1.05 mm can be used without being interfered by crystalline peaks from the surrounding pressure media at two-theta angles greater than 10°.

The oscillation of the multi-channel collimator, required to produce a continuous X-ray scattering pattern, involves two translation stages and one rotation stage to create a circular path with respect to the multi-channel collimator center. Correction parameters for the rotation are determined visually under a microscope lens. The oscillation motion during data collection is controlled using an in-house Python program. Once the detector is triggered for a data collection, in a fixed time of 15 s, the slits will rotate in one direction by an angle of 3.2° during data collection. This angle is a multiple of 0.8° to mitigate imperfections in the slits and slit spacings. After finishing the first scan, the slits will spend another 15 s moving in the opposite direction by the same 3.2° and return to their original starting position. One oscillation period takes 30 s to complete and it is the minimum data collection time required for a single X-ray scattering pattern in our setup.

2.3. The Detector

The Dectris Pilatus3 X CdTe 300K-W is a hybrid photon counting detector [20]. It is free of both dark current and readout noise, therefore, providing an excellent signal-to-noise ratio (the total intensity of an integrated pattern is less than 1 count for a 5 min dark field exposure). As a result, the visibility of weak signals is greatly improved. This is a great advantage for data collection from amorphous materials, which can have a very weak X-ray scattering signal especially at higher Q. The active area of the detector is 254 mm horizontal by 33.5 mm vertical with a square pixel size of 172 μm (1475 × 195 pixels array size). This active area is tiled horizontally by six large 1 mm thick CdTe crystals (sensors) each with dimensions of 42 mm × 34 mm with a 3-pixel (0.52 mm) gap between crystals and the point-spread function (PSF) of 1 pixel FWHM. These 1 mm sensors have a quantum efficiency greater than 80% between 40 and 75 keV and therefore are well matched to the energy used in our measurements. The combined low PSF, high efficiency and large dynamic range (~1e6/sec/pixel) allows closely spaced signals with huge intensity differences to be accurately resolved and measured. The detector has a fast (ns) electronic gate eliminating the need for a slow mechanical shutter and a readout time of less than 1 ms. The detector is water cooled by a closed loop chiller and requires a low flow of dry N_2 to keep it moisture free. The detector weighs about 7 kg and is mounted on the large detector arm of the general-purpose diffractometer in ID-C with two rotation degrees of freedom allowing the active area of the detector to cover a large scattering angle range.

3. Sample Environment and Preparation

3.1. The Paris-Edinburgh Cell Assembly

Figure 5 shows the two main Paris-Edinburgh cell assemblies that have been developed at the beamline and have been used routinely since our operation began. The "standard" 3 mm Paris-Edinburgh cell assembly (Figure 5a) has been requested the most by users since it can accommodate the largest sample in volume (1.6 mm in diameter; 1.6 mm in height). The cell assembly consists of a boron epoxy ring with an inner MgO ring (pressure marker [21]). A graphite cylindrical heater is used to heat the sample. This cell assembly has been rigorously tested and proven to perform stably up to 7 GPa and 2300 K. The large sample size provides sufficient amount of sample material for good quality X-ray scattering signal. With the large sample volume, one can easily perform additional in-situ measurements of physical properties including density measurement using the absorption method, sound velocity measurement with the ultrasonic interferometry, and viscosity measurement using the falling sphere method, etc. The greatest advantage of our standard Paris-Edinburgh cell assembly is the capability of keeping liquid samples stable in the sample chamber under extreme pressure and temperature conditions, which makes studies of liquid possible.

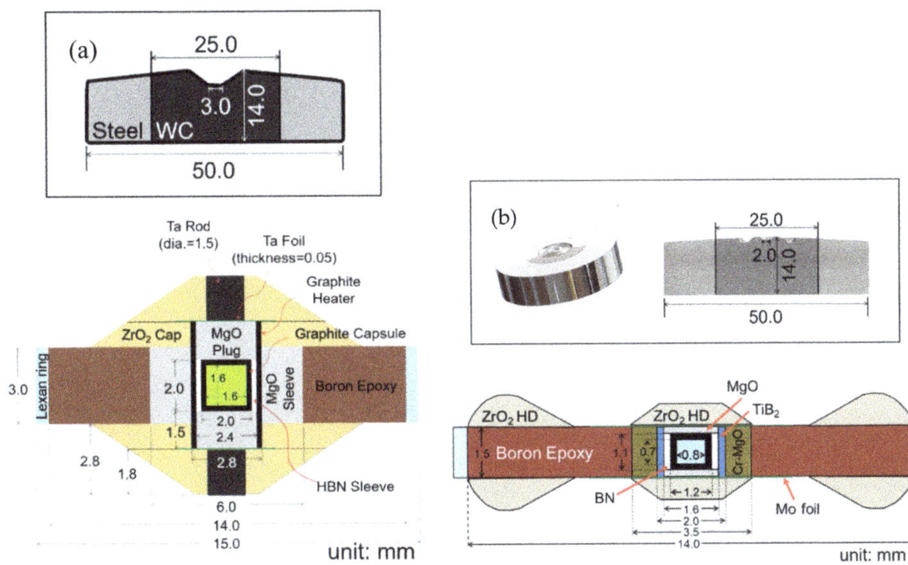

Figure 5. (**a**): The standard 3 mm Paris-Edinburgh cell assembly (bottom) that can reach 7 GPa and 2300 K. The 1.6 mm diameter sample size is the largest that we can accommodate in our setup. The 3 mm Paris-Edinburgh anvil (top) is custom-made by Fujilloy. The center part of the anvil is tungsten carbide with a steel ring surrounding it for lateral support. (**b**) For pressures up to 12 GPa, a 2 mm cell assembly (a TiB_2 heater and a graphite sample capsule) is utilized (bottom). From both the full image and the cross section of the 2 mm cupped-toroidal Drickamer anvil (top), one can clearly see the toroidal-cut groove that helps to prevent the pressure medium from flowing outward under compression. Similar to the 3 mm standard anvils, the cupped-toroidal Dricakmer anvil also consists of a tungsten carbide center part and an outer supporting steel ring.

To increase pressure and temperature range, we have developed and tested a cupped-toroidal Drickamer anvil [18] and a cell assembly for higher pressure experiments. The anvil design, with a central depression, a toroidal groove and a small tapered angle, combines features of a modified Drickamer anvil and the traditional Paris-Edinburgh anvil. By optimizing the parameters for the cupped-toroidal Drickamer anvil design, pressures corresponding to the mantle transition zone can be generated. The 2 mm cupped-toroidal Drickamer cell assembly with thermally insulating materials has been developed (Figure 5b). The cupped-toroidal Drickamer cells can reach >12 GPa in pressure and up to 2100 K in temperature. However, at temperatures above 2100 K, the anvil gap often closes to a point where X-ray scattering signal from the sample is restricted. We are currently working on modifying the cupped-toroidal Drickamer cell design to overcome this gap closing issue.

3.2. Sample Preparation

Sodium-silicate glass samples ($Na_2Si_2O_5$, Na_2SiO_3, etc.) used in the preliminary experiments were prepared by grinding and thoroughly mixing high purity (>99.99%) Na_2CO_3 (Alfa Aesar #10861) and SiO_2 (Alfa Aesar #88777) powders in an aluminum oxide mortar. The oxide mixtures were placed into a furnace and melted in a platinum crucible at temperatures 200 K above each composition's melting point. Glass samples were formed by removing the crucible from the furnace at high temperature and immediately submerging the crucible in room temperature water without contacting the samples. Quenched glasses were crushed, ground into fine powder and re-melted. This process was repeated 3 times to create a homogeneous glass form. The final glass samples were colorless and transparent. The glass samples were examined using both the Raman spectroscopy and X-ray scattering to ensure that they were crystal-free. The glass samples were then machined into cylinders with proper dimensions that fit the sample capsule.

4. Data Collection and Analysis

The collected X-ray scattering data is saved as a 32-bit 2D TIFF image (1475 × 195 pixels) with a size of 1126 kilobytes. The raw image is processed (detector calibration, image integration, etc.) using the software package Dioptas [22]. Limited by the active area of the Dectris Pilatus3 X CdTe 300K-W detector, which is a 254 mm × 33.5 mm rectangular, only fractions of the full scattering rings are recorded. In addition, in order to utilize the entire area of the detector to collect scattering signals at high Q, the detector is setup so that the beam center (direct beam) is positioned at the edge of the detector (Figure 6). One advantage of processing our X-ray scattering patterns using Dioptas is that we can still run the detector calibration with only capturing a fraction of the diffracted powder rings (Figure 6), and the program also works for special setups where the beam center is not present in the X-ray scattering pattern.

To properly analyze X-ray scattering data of amorphous material, various background corrections become critical in order to extract the real X-ray scattering signal. The critical factors that should be considered include: (1) subtracting the background from the surrounding high pressure cell assembly; (2) eliminating multiple scatter of photons inside the experimental station (this may arise from optics components, the press, and other equipment), (3) eliminating collimation effects of the multi-channel collimator. We have developed and upgraded several major hardware in order to routinely collect these correction intensity profiles efficiently. One important improvement worth mentioning is the installation of a manually operated horizontal translation stage for the multi-channel collimator unit to move out of the beam path and to precisely return to its previous position for data collection without having to re-align it. By collecting data of a non-crystalline standard (e.g., a Pyrex glass rod) with and without the multi-channel collimator, one can determine the transfer function and apply the correction to the dataset.

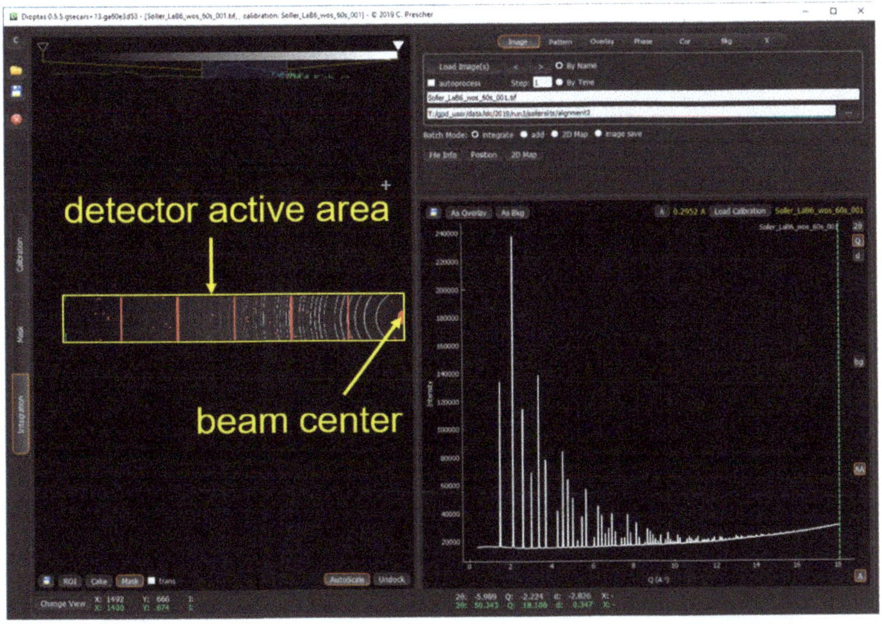

Figure 6. A Dioptas window showing the 2D X-ray diffraction image of the lanthanum hexaboride (LaB_6) powder (NIST 660C) collected using the Pilatus3 X CdTe 300K-W detector (left panel). We can work with image files that only recorded a portion of the diffracted powder rings. The right panel is the integrated diffraction pattern. LaB_6 powder is filled in a 0.5 mm diameter amorphous silica capillary and mounted in the horizontal plane oriented perpendicular to the incident beam. LaB_6 serves as the line position standard for calibrating detector parameters.

A Pyrex glass rod with a diameter identical to the sample diameter is used to establish an X-ray scattering intensity correction for the sample. First, we collect data of the Pyrex glass rod in air with no surrounding cell material, no multi-channel collimator. This is our reference pattern for intensities. The next step is to collect data of the same Pyrex glass rod with the multi-channel collimator in place. We then calculate the transfer function of the multi-channel collimator. The transfer function is essentially the ratio of intensities of the reference pattern to the one with the multi-channel collimator. This is used later to rescale the intensities of XRD measurements under high pressure and temperature. Other patterns that are necessary for the intensity correction are collected subsequently, including the detector dark field, the multiple scattering signal inside the hutch, and the Paris-Edinburgh sample cell background. The latter is based on measurement of an empty (dummy) cell.

The transfer function (Figure 7) provides a correction ratio for the intensity of the sample pattern at each Q value to correct for the intensity differences generated by the air scattering inside the hutch, dark field of the detector, and the multi-channel collimator. Only with a properly derived transfer function can the background be successfully removed from the sample pattern. Since the scattered signal intensity from the sample/cell assembly is sensitive to the beam position, alignment of the pinhole, and the alignment of the multi-channel collimator, etc., each setup for a beamtime period requires its own transfer function for proper intensity correction. Furthermore, for different sample diameters, the geometric correction from the sample will be different.

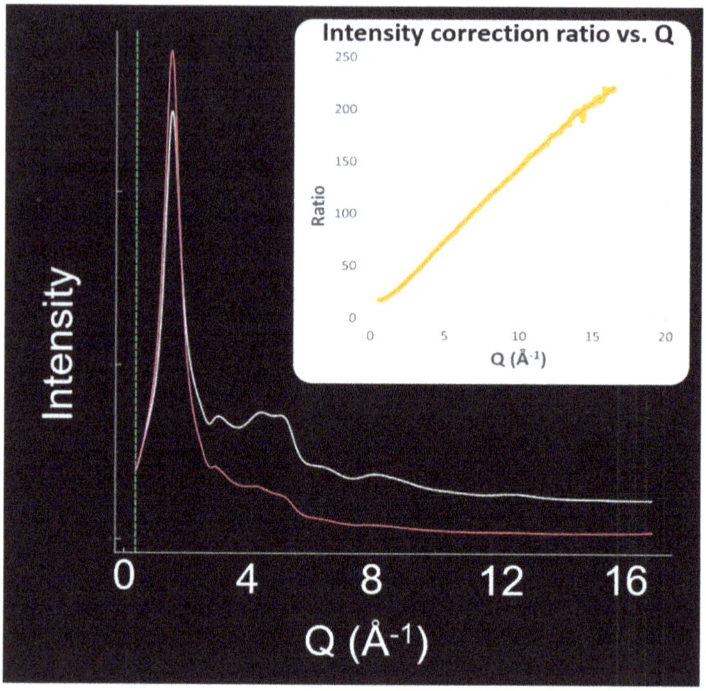

Figure 7. Two sets of X-ray scattering data collected for the same Pyrex glass rod at ambient condition. The red line (scaled in intensity) is the data set collected using the multi-channel collimator while the white one is without the multi-channel collimator. Both were collected for 600 s. The inset shows an example of the transfer function (intensity correction ratio vs. Q) derived from one of the beamtime setups.

The structure factors S(Q) and the pair distribution functions g(r) are calculated from the X-ray scattering patterns using the software package Glassure [23]. Detailed description of the pair distribution function can be found in [24,25]. The Python-based GUI program performs background subtraction, atomic form factor corrections, Fourier transform and optimization of the experimental data (Figure 8). The number density in atoms per cubic Å is used in the calculation. To minimize truncation error ripples in the pair distribution function, the optional Lorch function [26] was applied during the Fourier transforming process. An optimization process following [27] has also been applied to the final g(r).

Here we show results from two different samples: a sodium silicate liquid sample and a sodium disilicate glass sample. Figure 8 shows sodium silicate liquid X-ray scattering data collected at 3 ± 0.1 GPa and 1673 ± 90 K for 10 min. The experiment started with a sodium silicate glass sample with a diameter of 1.6 mm. Using a 42 keV monochromatic X-ray beam, we were able to collect X-ray scattering data of the liquid up to ~17 $Å^{-1}$ in Q. However, the X-ray scattering signal becomes weak above Q = 12 $Å^{-1}$. For better statistics, especially in high Q ranges, a longer data collection time is suggested.

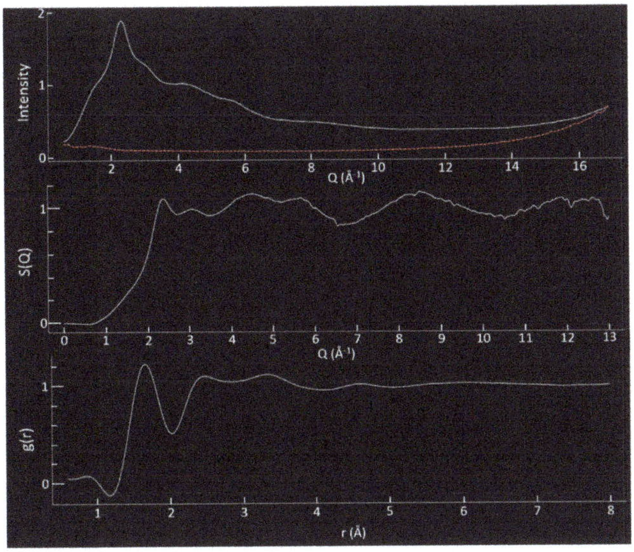

Figure 8. A captured screenshot of the data analysis software "Glassure". The top window shows the collected sodium silicate liquid data at 3 ± 0.1 GPa and 1673 ± 90 K. The red dotted line in the top window is the background collected using an empty cell assembly at ambient condition. The middle window shows the structure factor S(Q) of the liquid sample. The signal becomes noisy at Q > 12 $Å^{-1}$. The bottom window is the pair distribution function of the liquid sample with a Max Q cutoff of the structure factor at 13 $Å^{-1}$. We are currently working on data collection of sodium silicate liquids at higher pressures; detailed analysis will be published in the future.

We collected sodium disilicate (NS2) glass X-ray scattering patterns at ambient condition and compared the bond length results (Figure 9) with previous structure studies of sodium disilicate glass from neutron scattering [28] and molecular dynamics simulations [29,30] (Table 1). Bond lengths of Si–O, Na–O, O–O, and Si–Si were determined by fitting the assigned four peaks of the g(r) with Gaussian functions using the software Origin. Our measured bond lengths are consistent with previously reported values.

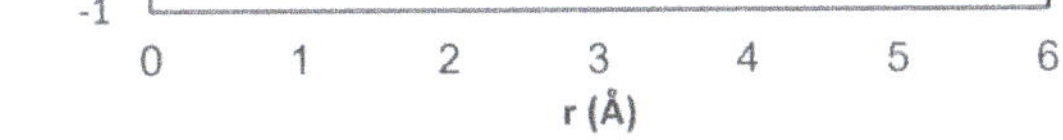

Figure 9. The pair distribution function g(r) of sodium disilicate glass collected at ambient pressure and temperature. Peaks representing atomic distances of Si–O, Na–O, O–O and Si–O are shown.

Table 1. Comparison of the atomic bond lengths of sodium disilicate glass derived from our experiment (Figure 9) with values from previous studies. The results agree well, demonstrating that our setup is capable of collecting reliable structure data of non-crystalline materials (unit: Å).

Atomic Bond Lengths	This Study	Neutron Scattering [28]	MD Simulation [29]	MD Simulation [30]
Si–O	1.61 ± 0.01	1.63	1.7	1.61
Na–O	2.20 ± 0.04	-	2.3	2.42
O–O	2.60 ± 0.02	2.66	2.6	2.62
Si–Si	3.14 ± 0.01	-	3.1	3.16

5. Conclusions

A Paris-Edinburgh press combined with a multi-channel collimator has been successfully commissioned at GSECARS at the Advanced Photon Source for monochromatic X-ray scattering studies on non-crystalline materials. The current Paris-Edinburgh press setup has the capability of reaching pressures and temperatures up to 12 GPa and 2100 K using the newly designed cupped-toroidal Drickamer cell assembly. With doubly focusing KB mirrors, monochromatic X-ray scattering has been conducted up to 60 keV with a beam size below 40 μm. The single-photon counting detector Pilatus3 X CdTe has low noise and yields reliable data. Sodium silicate and borosilicate glasses and melts, as well as Fe_2O_3 and $FeCaSi_2O_6$ liquids have been studied using this new setup. Atomic bond-lengths determined for sodium disilicate glass agree well with values from previous studies, demonstrating that our setup provides reliable structure data. The facility is open to the general mineral physics community for structure and physical property studies on non-crystalline materials at high pressures and high temperatures.

Author Contributions: Conceptualization, Y.W., T.Y.; methodology, Y.W., T.Y., V.P., C.P.; software, C.P. and E.G.; validation, T.Y.; formal analysis, T.Y.; investigation, T.Y.; resources, Y.J.R., F.S., H.W., P.E., J.S., Y.K., G.S.; data curation, T.Y.; writing—original draft preparation, T.Y.; writing—review and editing, all authors; project administration, M.L.R., S.R.S., Y.W.

Funding: This work is supported by the National Science Foundation-Earth Sciences (EAR-1620548). These experiments were performed at GSECARS (Sector 13), Advanced Photon Source (APS), Argonne National Laboratory. The APS is a U.S. Department of Energy (DOE) Office of Science User Facility operated for the DOE Office of Science by Argonne National Laboratory under Contract No. DE-AC02-06CH11357. The Paris-Edinburgh system was developed at GSECARS and HPCAT (Sector 16), APS, Argonne National Laboratory. GSECARS is supported by the National Science Foundation-Earth Sciences (EAR-1634415) and Department of Energy-Geosciences (DE-FG02-94ER14466). HPCAT operations are supported by DOE-NNSA's Office of Experimental Sciences.

Acknowledgments: We are grateful to Mohamed Mezouar for his valuable input and assistance in the design, manufacturing and installation of the multi-channel collimator. We thank Curtis Kenney-Benson, Mike Proskey, Pasquale DiDonna, and Guy Macha for their wonderful help and excellent technical support during the commissioning process.

Conflicts of Interest: The authors declare no conflicts of interest.

References

1. Funamori, N.; Yamamoto, S.; Yagi, T.; Kikegawa, T. Exploratory studies of silicate melt structure at high pressures and temperatures by in situ X-ray diffraction. *J. Geophys. Res.* **2004**, *109*, B03203. [CrossRef]
2. McMillan, P. Structural studies of silicate glasses and melts—Applications and limitations of Raman spectroscopy. *Am. Mineral.* **1984**, *69*, 622–644.
3. Stebbins, J.F.; Ellsworth, S.E. Temperature effects on structure and dynamics in borate and borosilicate liquids: High-resolution and high-temperature NMR results. *J. Am. Ceram. Soc.* **1996**, *79*, 2247–2256. [CrossRef]
4. Angeli, F.; Villian, O.; Schuller, S.; Charpentier, T.; de Ligny, D.; Bressel, L.; Wondraczek, L. Effect of temperature and thermal history on borosilicate glass structure. *Phys. Rev. B* **2012**, *85*, 054110. [CrossRef]

5. Susman, S.; Volin, K.J.; Price, D.L.; Grimsditch, M.; Rino, J.P.; Kalia, R.K.; Vashishta, P.; Gwanmesia, G.; Wang, Y.; Liebermann, R.C. Intermediate-range order in permanently densified vitreous SiO_2: A neuton-diffraction and molecular-dynamics study. *Phys. Rev. B* **1991**, *43*, 1194–1197. [CrossRef]
6. Zha, C.S.; Mao, H.K.; Hemley, R.J.; Duffy, T.S. Elasticity measurement and equation of state of MgO to 60 GPa. *EOS. Trans. AGU* **1997**, *78*, F752.
7. Xu, M.; Jing, Z.; Chantel, J.; Jiang, P.; Yu, T.; Wang, Y. Ultrasonic velocity of diopside liquid at high pressure and temperature: Constraints on velocity reduction in the upper mantle due to partial melts. *J. Geophys. Res. Sol. Earth* **2018**, *123*, 8676–8690. [CrossRef]
8. Jing, Z.; Karato, S. A new approach to the equation of state of silicate melts: An application of the theory of hard sphere mixtures. *Geochim. Cosmochim. Acta* **2011**, *75*, 6780–6802. [CrossRef]
9. Wilding, M.C.; Benmore, C.J.; Weber, J.K.R. In situ diffraction studies of magnesium silicate liquids. *J. Mater. Sci.* **2008**, *43*, 4707–4713. [CrossRef]
10. Inoue, K.; Asada, T. Cubic anvil X-ray diffraction press up to 100 kbar and 1000 °C. *Jpn. J. Appl. Phys.* **1973**, *12*, 1786–1793. [CrossRef]
11. Tsuji, K.; Yaoita, K.; Imai, M.; Shimomura, O.; Kikegawa, T. Measurements of X-ray diffraction for liquid metals under high pressure. *Rev. Sci. Instrum.* **1989**, *60*, 2425–2428. [CrossRef]
12. Yamada, A.; Wang, Y.; Inoue, T.; Yang, W.; Park, C.; Yu, T.; Shen, G. High-pressure X-ray diffraction studies on the structure of liquid silicate using a Paris-Edinburgh type large volume press. *Rev. Sci. Instrum.* **2011**, *82*, 015103–015107. [CrossRef] [PubMed]
13. Mezouar, M.; Faure, P.; Crichton, W.; Rambert, N.; Sitaud, N.; Bauchau, S.; Blattmann, G. Multichannel collimator for structural investigation of liquids and amorphous materials at high pressures and temperautures. *Rev. Sci. Instrum.* **2002**, *73*, 3570–3574. [CrossRef]
14. Morard, G.; Mezouar, M.; Bauchau, S.; Alvarez-Murga, M.; Hodeau, J.L.; Garbarino, G. High efficiency multichannel collimator for structural studies of liquids and low-Z materials at high pressures and temperatures. *Rev. Sci. Instrum.* **2011**, *82*, 023904. [CrossRef]
15. Besson, J.M.; Nelmes, R.J.; Hamel, G.; Loveday, J.S.; Weill, G.; Hull, S. Neutron powder diffraction above 10 GPa. *Phys. B Condens. Matter* **1992**, *180*, 907–910. [CrossRef]
16. Klotz, S.; Besson, J.M.; Hammel, G.; Nelmes, R.J.; Loveday, J.S.; Marshall, W.G. High pressure neutron diffraction using the Paris-Edinburgh cell: Experimental possibilities and future prospects. *High Press. Res.* **1996**, *14*, 249–255. [CrossRef]
17. Rivers, M.L.; Duffy, T.S.; Wang, Y.; Eng, P.J.; Sutton, S.R.; Shen, G. A new facility for high-pressure research at the Advanced Photon Source. In *Properties of Earth and Planetary Materials at High Pressure and Temperature*; Manghnani, M.H., Yagi, T., Eds.; AGU: Washington, DC, USA, 1998; pp. 79–88.
18. Kono, Y.; Park, C.; Kenney-Benson, C.; Shen, G.; Wang, Y. Toward comprehensive studies of liquids at high pressures and high temperatures: Combined structure, elastic wave velocity, and viscosity measurements in the Paris-Edinburgh cell. *Phys. Earth Planet. Int.* **2014**, *228*, 269–280. [CrossRef]
19. Yu, T.; Wang, Y.; Rivers, M.L. Imaging in 3D under pressure: A decade of high-pressure X-ray microtomography development at GSECARS. *Prog. Earth Planet. Sci.* **2016**, *3*, 17. [CrossRef]
20. Brönnimann, C.; Trüb, P. Hybrid pixel photon counting X-ray detectors for synchrotron radiation. In *Synchrotron Light Sources and Free-Electron Lasers*; Jaeschke, E., Khan, S., Schneider, J., Hastings, J., Eds.; Springer: Cham, Switzerland, 2016; pp. 995–1027.
21. Dewaele, A.; Fiquet, G.; Andrault, D.; Hausermann, D. P-V-T equation of state of periclase from synchrotron radiation measurements. *J. Geophys. Res.* **2000**, *105*, 2869–2877. [CrossRef]
22. Prescher, C.; Prakapenka, V.B. DIOPTAS: A program for reduction of two-dimensional X-ray diffraction data and data exploration. *High Press. Res.* **2015**, *35*, 223–230. [CrossRef]
23. Prescher, C. Glassure: An API and GUI program for analyzing angular dispersive total X-ray diffraction data. *Zenodo* **2017**, *2017*. [CrossRef]
24. Wagner, C.N.J. Direct methods for the determination of atomic-scale structure of amorphous solids (X-ray, electron, and neutron scattering). *J. Non Cryst. Solids* **1978**, *31*, 1–40. [CrossRef]
25. Parise, J.B.; Antao, S.M.; Michel, F.M.; Martin, C.D.; Chupas, P.J.; Shastri, S.D.; Lee, P.L. Quantitative high-pressure pair distribution function analysis. *J. Synchrotron Radiat.* **2005**, *12*, 554–559. [CrossRef] [PubMed]

26. Lorch, E. Neutron diffraction by Germania, silica and radiation-damaged silica glasses. *J. Phys. C Solid State Phys.* **1969**, *2*, 229–237. [CrossRef]
27. Eggert, J.H.; Weck, G.; Loubeyre, P.; Mezouar, M. Quantitative structure factor and density measurements of high-pressrure fluids in diamond anvil cells by X-ray diffraction: Argon and water. *Phys. Rev. B* **2002**, *65*, 174105. [CrossRef]
28. Misawa, M.; Price, D.L.; Suzuki, K. Short-range structure of alkali disilicate glasses by pulsed neutron total scattering. *J. Non Cryst. Solids* **1980**, *37*, 85–97. [CrossRef]
29. Smith, W.; Greaves, G.N.; Gillan, M.J. Computer simulation of sodium disilicate glass. *J. Chem. Phys.* **1995**, *103*, 3031. [CrossRef]
30. Huang, C.; Cormack, A.N. The structure of sodium silicate glass. *J. Chem. Phys.* **1990**, *93*, 8180–8186. [CrossRef]

minerals

MDPI

Article

Casting Octahedra for Reproducible Multi-Anvil Experiments by 3D-Printed Molds

Yingxin Liu [1,2,*], Haijian Li [3], Xiaojing Lai [2], Feng Zhu [2], Robert P. Rapp [2] and Bin Chen [2,*]

[1] School of Gemmology, China University of Geosciences, Beijing 100083, China
[2] Hawaii Institute of Geophysics and Planetology, University of Hawaii at Manoa, Honolulu, HI 96822, USA; laixiaoj@hawaii.edu (X.L.); zhufeng@hawaii.edu (F.Z.); rrapp@hawaii.edu (R.P.R.)
[3] China Building Materials Academy, Beijing 100024, China; lhj.1002@163.com
* Correspondence: liuyingxin@cugb.edu.cn (Y.L.); binchen@hawaii.edu (B.C.)

Received: 18 November 2019; Accepted: 14 December 2019; Published: 18 December 2019

Abstract: Making consistent and precise octahedral pressure media is crucial for reproducible high-pressure experiments in the multi-anvil press. Here we report a new approach of casting octahedra using 3D-printed molds, and pressure calibrations for octahedra both with and without pre-existing gaskets ("fins"). The 3D-printed molds for casting octahedra from either Ceramacast 584-OF or 646 cement improve the reproducibility of the octahedra and allow for a pre-existing central hole (for the high-pressure cell assembly) in the final cast product. Pressure and temperature calibrations of the octahedra have been performed based on phase transitions in bismuth (Bi) and silica (SiO_2), respectively, in order to determine the efficiency and reproducibility of pressure generation and thermal insulation for cast octahedra designed for use with 18/12, 14/8, and 10/5 multi-anvil assemblies. The pressure-generating efficiency of the 14/8 and 10/5 octahedra with pre-existing gaskets, cast from the 584-OF cement, is similar to that of the corresponding COMPRES (Consortium for Materials Properties Research in Earth Sciences) octahedra, and more efficient than pre-cast octahedra made from the same material but lacking pre-existing gaskets. The efficiency of pre-gasketed 18/12 octahedra made of the 646 cement is markedly lower than those of the 584 cement. However, the 18/12 large-volume octahedra, cast (with fins) from the ZrO_2-based 646 cement, also provides efficient thermal insulation. Casting octahedral solid pressure media for multi-anvil experiments using 3D-printed "injection" molds is a low-cost and low failure-rate alternative for conducting reproducible experiments at high pressure in the multi-anvil apparatus.

Keywords: mold; octahedron; bismuth; pressure calibration cell assembly; large volume press; 3D printer

1. Introduction

In an effort to replicate in experimental laboratories the often-extreme conditions known to exist within deep interiors of the Earth, other planets, and moons, earth scientists have pioneered many technical developments in the simultaneous generation of high-pressure and high-temperature conditions using various high-pressure devices. These efforts have been directed in general towards studying the behavior and chemical and physical properties of planetary materials under appropriate conditions. In addition to dynamic compression techniques, static high-pressure equipment such as the diamond anvil cell (DAC) and the multi-anvil press (MAP) have been used to compress samples to study variations in material properties and polymorphous phase transitions as a function of pressure and temperature. The MAP technique is typically employed when a relatively large volume of sample is desired. Since the first 6-8 multi-anvil large-volume apparatus was introduced by Kawai and Endo [1], continuous technical modifications and improvements have been made that facilitate broader investigation of geological processes taking place not only in the deep Earth [2] but also in the upper mantle and lithospheric mantle. One of the most important practical developments was made by Walker [3], who designed and built a removable cylindrical containment ring housing all six of the first-stage anvils as well as the eight carbide cubes of the second stage that enclose the octahedron-shaped pressure medium. The cylindrical containment device was small enough that it could be moved in and out of a hydraulic press between experiments, greatly reducing the cost of setting up a multi-anvil laboratory for high pressure experiments. In order to accommodate a larger sample volume, larger carbide cubes (54 and 75 mm) have been used at Bayerisches Geoinstitut [4] and Ehime University [5]. At the same time, higher tonnage [4], and stiffer materials, such as sintered diamond (SD) and nano-polycrystalline diamond (NPD) [5], were also employed to make second-stage cubes capable of achieving much higher pressures. The use of SD cubes has allowed for progressive increases in the maximum achievable pressure, from 41 GPa [6] in 1989, to 63 GPa [7] in 2005, and to 74 GPa [8] by 2010. The highest pressures documented thus far in the multi-anvil apparatus have reached 88 GPa [9] and 125 GPa [10] with the use of NPD cubes.

To consistently and reproducibly achieve well-defined high-pressure conditions in the laboratory, various cell assemblies have been designed for the 6-8 multi-anvil apparatus. Standardized cell assemblies have been developed for in situ MAP experiments [11], in which appropriate "windows" were established through the materials comprising the assembly to improve the passage of X-rays. Over the past decade, Leinenweber et al. [12], supported by the Consortium for Materials Properties Research in Earth Sciences (COMPRES), have designed, tested, and calibrated a series of standardized high-pressure cell assemblies that are available to the high-pressure community for general use and interlaboratory comparisons. However, many laboratories are still using the method introduced by Walker in which pre-cast octahedra are used for multi-anvil experiments, which is both cost-effective and relatively easy to assemble and run [13]. The segmented molds proposed by Walker are composed of different size frames, PVC cubes, and Teflon sheets, allowing the experimentalist to cast their octahedron (effectively, the octahedral "pressure-transmitting" medium) with dimensions, usually expressed in terms of their "octahedral edge-length" (OEL) that is appropriate to the corresponding "truncated edge length" (TEL) on the second-stage cubes ("OEL/TEL") [13]. The Walker approach has been adopted by many multi-anvil laboratories largely because it allows for MAP experiments that are cheaper and more time efficient. The octahedra made by this approach, however, often suffer from pre-formed gaskets or "fins" that are of uneven thickness or that contain cavities on its outer surface from unfilled air bubbles left over from the casting stage. In addition, after drying and sintering in a 1 atm box furnace at ~1000 °C, the octahedra become not only hardened but also more brittle, which makes drilling the central hole in the octahedron, utilized for accommodating the various components of the cell assembly, difficult, time-consuming, and with a high failure-rate.

In this study, we describe a new, low-cost method for casting octahedral pressure-media for multi-anvil experiments from commercially available ceramic cement that makes use of "semi-injection" molds that have been printed by a 3-D printer. Commercially-available 3-D printers are now relatively inexpensive to purchase and easy to operate and are already used in a number of applications in the Earth Sciences, e.g., in crystallography [14,15], and sedimentology [16]. These 3D-printed molds have been fabricated in different sizes appropriate for casting octahedron of various standard OEL to match the standardized TEL of the corresponding cubes. The important difference is that our ceramic octahedra already have a central hole established in the final cast product prior to sintering at 1000 °C.

The new approach presented here can easily be adopted to any MAP laboratory. Our "injection-molded" octahedra were tested in our mineral physics laboratory at the University of Hawaii at Manoa, using both 1000-ton and 2000-ton presses equipped with Walker-design multi-anvil devices through a series of pressure-calibration experiments based upon the room temperature, pressure-dependent phase transitions in bismuth (Bi) [17], and the high-temperature phase transition in SiO_2 from quartz to coesite [18]. We then compare the pressure-generating efficiency and thermal-insulating properties of our cast ceramic octahedra to the calibrations for the corresponding

14/8 and 10/5 COMPRES cell assemblies (where the first number refers to the octahedron's OEL and the second number refers to the cube's TEL).

2. Experimental Methods

2.1. D-Printed Molds for Casting Octahedra

Two types of casting molds were designed and fabricated from PLA filaments using a 3D printer (3D printer model ANYCUBIC i3 MEGA, ShenZhen ANYCUBIC Technology Co., Ltd., Shenzhen, China). The first model is the prototype for ceramic octahedra cast with integrated and continuous gasket fins surrounding all edges of the octahedron and containing a pre-existing, centered, cylindrical hole to accommodate the cell assembly. The mold consists of a two-part casing that encloses the 8 spacing blocks, which impart the overall octahedral shape to the cavity and will be filled with the ceramic slurry; grooves aligned along the edges of the inner faces of these blocks are responsible for creating the pre-formed gaskets. Every two neighboring blocks meet at the $\frac{1}{2}$ height of the gaskets. Thus, the height and width of the pre-formed gaskets are controlled by dimension of the grooves, which were found to be reproducible and consistent when a tight fit of the 8 spacing blocks was achieved by turning the screws to fit together the 2 main casing blocks (Figure 1a). The top and bottom spacing blocks have a central hole of a diameter appropriate to the size of the octahedron being cast, and after the cavity is filled with the ceramic cement (in liquid form), a drill-rod pin is inserted through the top hole, on through the center of the 8 aligned spacing blocks (before the cement can set), and onto the center of the bottom spacing block. When fit together, the 2 main (upper and lower) casing blocks are cylindrical in shape; prior to filling the octahedral cavity with the ceramic slurry, these two blocks can be tightened together with three set screws to prevent leakage of the cement from the mold (additional details are provided in Figure 1a). The insertion of the drill-rod pin also introduces some pressure onto the freshly poured ceramic slurry in the molds, mimicking the "injection"-molding process.

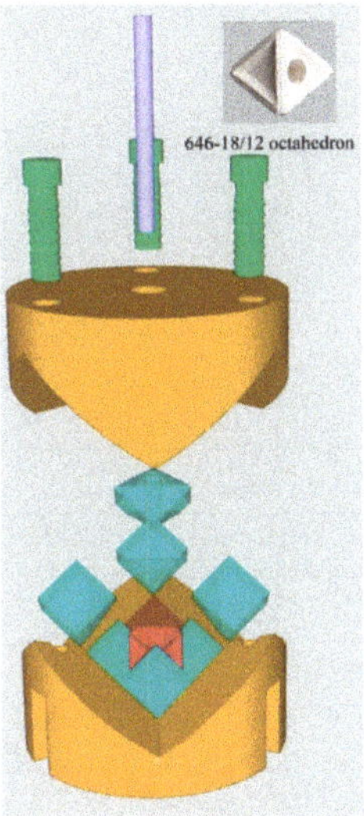

(**a**)

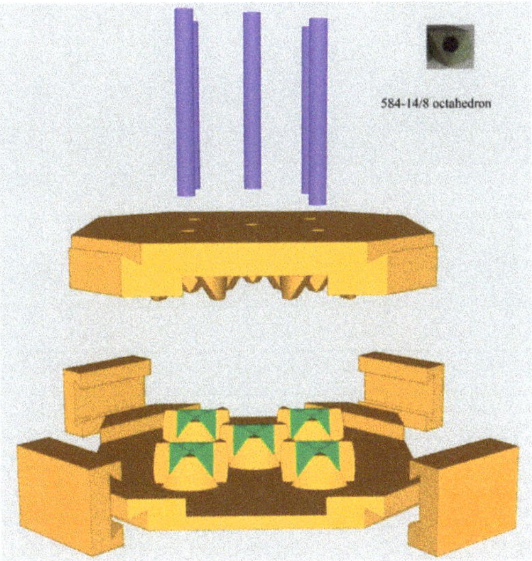

(**b**)

Figure 1. Schematic 3D pictures of the "injection" molds in different sizes with pre-existing hole. (**a**) 3D schematic picture of the mold for casting octahedra with gaskets, the inset is a 18/12 octahedron made by 646 cement; (**b**) 3D schematic diagram of the mold for casting octahedra without gaskets, the inset is a 14/8 octahedron made of 584-OF cement with a 5 wt.% of Cr_2O_3 in it to reduce the radiative heat loss.

The second type of mold is for casting octahedral pressure media without accompanying gasket fins ("finless" octahedra; Figure 1b). Here, the mold casing consists of only two parts; when fit together, the intersecting upper and lower blocks create multiple octahedral cavities that can cast up to (for this mold) 5 octahedra at a time. Four locking brackets are designed to slide into place around the four corners of the two aligned blocks prior to pouring the cement into the mold. Again, five drill-rod pins of appropriate diameter are inserted into holes on the top plate that are aligned with the centers of the five octahedral cavities below.

2.2. Casting Octahedra from Aremco Ceramacast Cement

Castable ceramics from Aremco Ceramacast were first used by Walker [13] for casting octahedral-shaped pressure media for multi-anvil experiments. Ceramacast 584-OF (Old Formula, hereafter referred to as "584") and 646 castable ceramic systems have been used in this study: (1) 584-OF is made from 66% MgO + 33% Al_2O_3 in mass; (2) Ceramacast 646 (hereafter referred to as "646") is comprised predominantly of ZrO_2. The cement powder and activator of different weight and proportions were used in order to completely fill the mold cavity with minimal materials wasted, when casting octahedra (with or without fins) of different sizes (see detailed information for casting 584 octahedra in Table 1. For the 584-OF cement, it was necessary to stir the powder-activator mixture quickly (~10 s) in order to blend evenly and minimize the formation of any air bubbles because the mixture tends to set very quickly. The cement slurry is poured into the mold; the central pin is quickly inserted and forced to the base of the mold; all the cement is allowed to set for 10 min before the drill-rod pin is gently removed. The cast octahedron, still wet, is then allowed to cure in the mold for at least 1 h (preferably overnight) before the two blocks are separated and the octahedron are recovered. The information on the weight and dimension of cast octahedra is listed in Table 1.

Table 1. Dimensions and weights of octahedra from 584 cement [1].

OEL (mm)	Powder (g)	Activator (g)	P/A	W (g)	L (mm)	DH (mm)	WG (mm)	HG (mm)
				with fins				
10	3.5	1.5	2.33	2.25 (0.07)	9.88 (0.21)	4.37	2.4 (0.14)	2.47 (0.04)
14	5.5	2.5	2.2	4.31 (0.16)	14.12 (0.23)	4.80	2.68 (0.06)	2.85 (0.05)
18	8.5	3.9	2.2	9.22 (0.10)	17.98 (0.25)	6.75	2.50 (0.04)	2.86 (0.05)
				without fins or finless [2]				
10	7	3.5	2	0.70 (0.02)	9.97 (0.16)	4.40	4.66	2.22
14	15	7.5	2	1.96 (0.03)	13.67 (0.07)	5.60	4.47	2.82
				COMPRES octahedra without fins [3]				
10			-	1.01 (0.01)	10.02 (0.04)	4.40	4.50	2.36
14			-	2.79 (0.01)	13.57 (0.08)	5.60	4.50	2.83

[1] OEL denotes octahedral edge length; P/A, Powder/Activator ratio of the 584 cement; W, weight after firing; L is the average length of the OEL; DH, diameter of the center hole of the octahedra; WG, width of gasket fins; HG, height of gasket fins. Numbers in parentheses represent the standard deviations. [2] The weight of powder and activator is for casting 5 finless octahedra at a time. WG and HG denote the width and thickness of the pyrophyllite gaskets, respectively. [3] Injection molded octahedra developed by COMPRES [12] for comparison.

The 18 mm-OEL octahedra with fins and pre-exiting center hole were also cast as pressure medium using the ZrO_2-based Ceramacast 646 cement, using the same 18/12 mold and procedure. The activator for the 646 cement is water, with a typical powder/water ratio of 5.61. According to the instructions from the manufacturer, this cement mixes directly with water; in addition to being simple and easy to mix, the 646 cement has the added advantage that, once mixed with its activator (i.e., water), it remains relatively fluid much longer than the 584 cement mixture. After the cement sets, the mold can be opened and a single octahedron with pre-existing fins and central hole retrieved, ready for drying and heat treatment. It should be noted that it does not require controlled humidity to dry the 646-octahedra, and they are typically free of cracks.

2.3. Pressure Calibrations

Room-temperature pressure-load calibrations of 18/12, 14/8 and 10/5 multi-anvil assemblies using octahedra cast with our newly-designed "injection" molds were conducted based on electrical resistance measurements of bismuth (Bi) during compression at room temperature [18,19] and were performed in both the 1000-ton and 2000-ton presses at the University of Hawaii at Manoa, both equipped with Walker-design, cylindrical multi-anvil devices. For these calibrations, Bi powder was packed into a MgO cylinder, and aluminum or copper plugs were placed at both ends of the cylinder, serving as electrical contacts with the exposed corners of two tungsten carbide cubes (Toshiba Tungaloy Grade F tungsten carbide) on either side of the centrally-located, cylindrically shaped high-pressure cell assembly (see dimensions in the insets of Figure 2). Resistance was measured by a high-precision multimeter (model HP 4333a) during the compression of the cell assembly at each run (Figure 2). The changes in the electrical resistance of Bi phase result from phase transitions from the trigonal to monoclinic (Bi I–II) at 2.55 GPa, from monoclinic to tetragonal (Bi II–III) at 2.69 GPa, and from tetragonal to body-centered cubic (Bi III–V) at 7.7 GPa (Figure 2). The Bi II–III and III–V phase transitions are sluggish in solid pressure media, and thus, the minimum in the electrical resistance labeled as "II–III" in Figure 2 was typically placed at 3.1 GPa for the pressure calibration of multi-anvil assemblies [19,20]. The Bi III–V phase transition at 7.7 GPa was indicated by the start of electrical resistivity reduction [17].

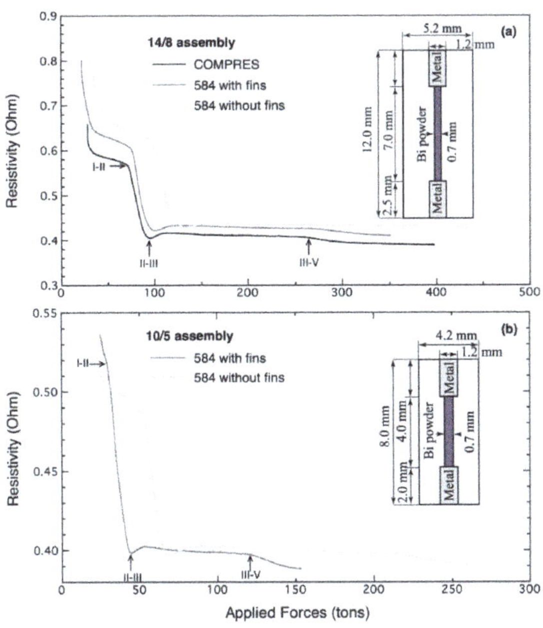

Figure 2. Representative electrical resistivity of bismuth (Bi) across the I–II, II–III, and III–V phase transitions as measured as a function of applied forces in (**a**) 14/8 and (**b**) 10/5 multi-anvil assemblies at room temperature. The insets at the top-right corner are the cross section of the Bi sample assembly with dimensions.

For pressure calibrations at high temperatures using 18 mm octahedra with gaskets cast from 646 ceramic, quench experiments were performed on amorphous or crystalline SiO_2 powders [18] near the pressure-temperature conditions appropriate to the quartz-coesite phase transition (see Figure 3 for the cross section of the high-temperature assembly). After reaching the target pressure/load, the sample was heated at a rate of 1°/s until the target temperature (1000 °C) was achieved by applying

AC voltage to the graphite heater; type-C thermocouples were used to monitor temperature. For the high-temperature pressure calibration, SiO_2 samples were heated at 1000 °C for 2–4 h.

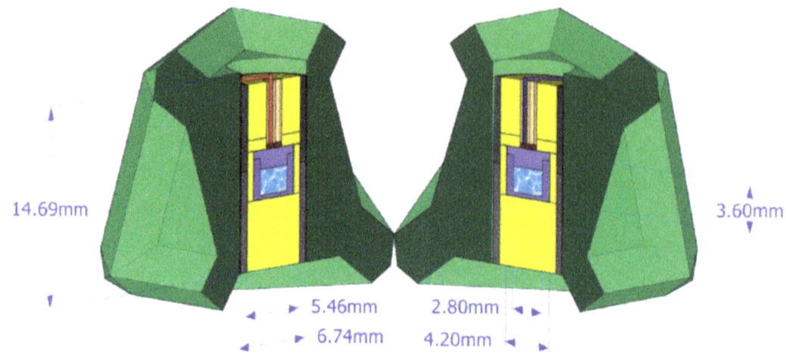

Figure 3. Cross section of the 18/12 assembly with graphite heater (gray), molly capsule (blue), MgO plugs (yellow), and pre-gasketed octahedra cast by ZrO_2-based 646 cement (green). The thermocouples are placed axially into the center hole in an Al_2O_3 4-bore sleeve, with the junction situating on top of the sample chamber.

3. Results and Discussion

3.1. Reproducible Octahedral Pressure Media for Multi-Anvil Experiments

After removal from the mold, all octahedra cast from 584 were allowed to cure for more than 7 days in a suitably humid environment (typically in sealed plastic bags) in order to minimize cracking as the ceramic continued to dry, prior to heating in a muffle furnace from room temperature at a rate of ~100 °C/min up to 1000–1100 °C, where they were held for ~2–10 h. Table 1 provides detailed information on the weight and dimensions of the octahedra cast from the 584 cement using the 3D-printed molds.

After casting a number of these "semi-injection-molded" octahedra, they were found to have the following general features: (1) there are few if any cavities or bubbles on the surface of the octahedra and within the preformed gaskets surrounding the octahedra, which we attribute to the effective pressurization of the ceramic slurry when the central pin is inserted, mimicking the effect of an injection mold; (2) the cast gaskets surrounding the octahedron are of equal thickness around all the edges of the octahedron, which we attribute to the tight seal between the upper and lower blocks that the three alignment screws permit; (3) the central hole is pre-indented in the octahedra and well centered before the cement cures, so that there is no need to drill the octahedron after firing.

Following the outlined procedure, octahedra can be readily and reproducibly made from castable ceramic using our 3D-printed "injection" mold design (see insets in Figure 1 for photos of 584 and 646 octahedra). Pre-gasketed octahedra from 646 cement were cast by following the same procedure for 584 cement. The advantage of the 646 castable cement includes longer pot life (tens of minutes) and longer shelf life (years). The octahedra cast from 646 cement generally demonstrated better surficial quality (insets in Figures 1a and 4) relative to those made from the 584 cement (inset in Figure 1b). The cost of materials for each cast octahedron is estimated to be $1 to $3 US dollars depending on its OEL. The 3D models for the two types of molds, which are available in the Supplementary Materials, can be downloaded and ready for fabrication by a 3D printer.

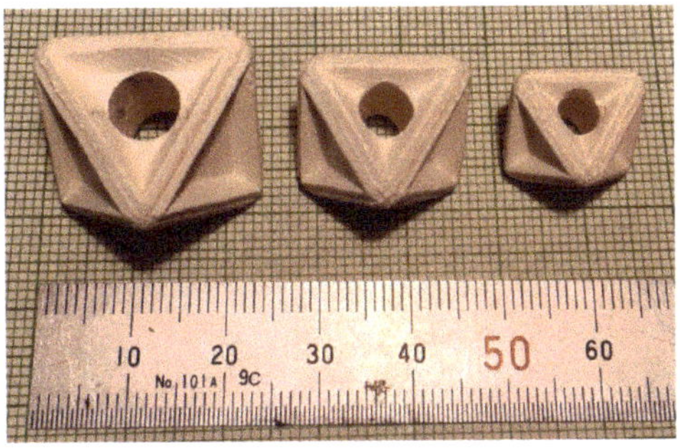

Figure 4. Pictures of octahedra with gaskets and pre-indented center hole cast from Ceramacast 646 cement by 3D-printed molds for the 18/12, 14/8, and 10/5 multi-anvil assemblies.

3.2. Room-Temperature Pressure-Load Calibration of the Cell Assemblies

Figure 5 shows the results of the pressure calibration based on Bi phase transitions for the 10/5, 14/8, and 18/12 assemblies using the cast octahedra and a comparison on calibrations with the COMPRES octahedra [12]. The calibrations of the various assemblies include several experimental runs (six runs for 14/8 and five runs for 10/5 assembly) for octahedra cast by our 3-D printed molds, with or without gaskets, and using both the 584 or 646 cement; all these are shown relative to the calibrations for the corresponding COMPRES assemblies using separately machined and individually glued pyrophyllite gaskets (Figure 5). Our calibrations, based on the Bi I–II phase transition at 2.55 GPa, II–III at 3.1 GPa, and III–V at 7.7 GPa [19], suggest that the pressure-generating efficiency of our mold-cast, pre-gasketed 14/8 and 10/5 octahedra using the Cr-doped 584 cement is equal to or only slightly more efficient than the corresponding COMPRES [12] assemblies calibrated with octahedra surrounded by a cage of pyrophyllite gaskets (Figures 2 and 5). The calibrations in experiments using finless octahedra, cast using our mold from Cr-doped 584 cement, and paired with pyrophyllite gaskets, however, suggest that they are much less efficient at generating pressure; that is, they require 25% more load or tonnage for the 10/5 assembly and 34% or more for the 14/8 assembly in order to reach the I–II and II–III phase transitions of Bi at room temperature. And for the 14/8 and 10/5 assemblies using mold-cast, finless octahedra, it appears that 60% to 70% additional load is needed to reach the Bi III–V phase transition at 7.7 GPa (Figure 5). It should be noted that the finless cast octahedra from the 584 cement is 30% less dense or more porous than those COMPRES octahedra (Table 1), which may be the reason why the cast 584-octahedra without fins are less efficient at generating pressure. Furthermore, the oil loads required for the Bi phase transitions (I–II, II–III, III–V) only varies slightly for replicate experimental runs conducted for the same assemblies (see the data points with error bars in Figure 5).

Room-temperature calibrations based on Bi phase transitions were also carried out on the 18/12 assemblies using pre-gasketed 18 mm octahedra and both the 584 and 646 ceramic cement. For calibration experiments using "finned" 18 mm OEL octahedra cast from the 584 cement, the pressure-load relationship is linear, reaching 7.7 GPa at 500 ton (Figure 5). For calibration experiments using "finned" 18 mm OEL octahedra cast from the 646 ceramic cement, 28% to 32% more load is needed to reach the same pressures; note, however, that we did not attempt to increase our load up to and beyond that needed to reach the Bi III–V phase transition at 7.7 GPa (expected to be >600 ton) in these experiments with the 646 ceramic.

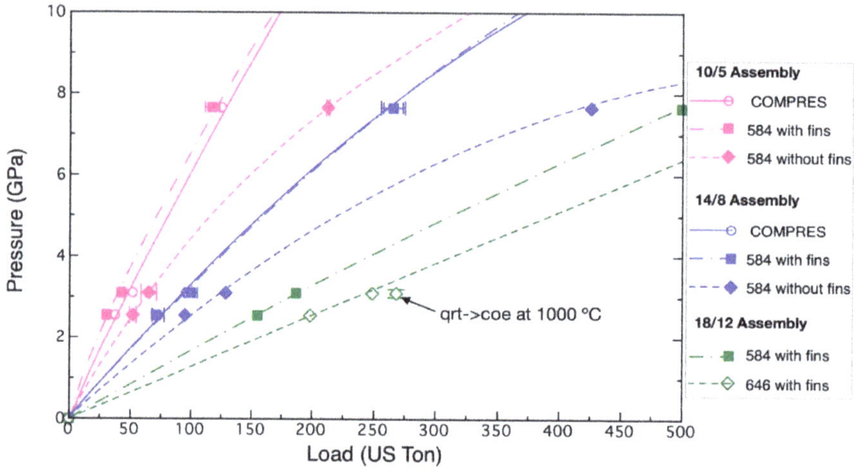

Figure 5. Load-pressure calibrations of the 18/12, 14/8, and 10/5 assemblies with cast octahedra, in comparison with those of COMPRES assemblies. The room-temperature calibrations are based on phase transitions of Bi (I–II at 2.55 GPa, II–III at 3.1 GPa, and III–V at 7.7 GPa) [19,20]. The vertical lines in the 10/5 and 14/8 assembly symbols are the standard deviation ranges of the calibration experiments. High-temperature calibration based on quartz (qrt) to coesite (coe) transition at 3.1 ± 0.1 GPa at 1000 °C [18] was also conducted for the 18/12 assembly with 646 octahedra with gaskets.

3.3. *High-Temperature Calibration of the 646-18/12 Assembly*

Four high-temperature pressure calibration experiments based on the quartz to coesite phase transition were carried out at 1000 °C for 2–4 h. After recovery of the quenched samples, the molybdenum sample capsules were mounted in resin epoxy, sectioned, and polished to expose the sample, and Raman spectroscopy was used to identify which phase or phases of SiO_2 were present (quartz or its high-pressure polymorph coesite) (Table 2). The quartz-coesite phase transition occurs at 3.1 ± 0.1 GPa at 1000 °C [18]. In run PL136, both coesite and minor quartz coexist in the sample capsule; the press load in this experiment was set at 268 ton for the 646-18/12 assembly, corresponding to a sample pressure of 3.1 GPa, which is close to the room-temperature calibration based on Bi phase transitions (Figure 5).

Table 2. Experimental conditions and results for the 18/12 assembly using 646 octahedra with gaskets.

Run Number	Load (ton)	Temperature (°C)	Resultant Phases *
PL133	241	1000	Quartz
PL134	286	1000	Coesite
PL136	268	1000	Coesite + quartz (minor)
PL137	251	1000	Quartz

* Raman spectroscopy was used to identify the phases.

4. Summary and Conclusions

We have conceived, designed, and tested a cost-effective method for quickly and reproducibly casting octahedra for use in high-pressure experiments in the multi-anvil apparatus, using a 3D-printed "injection" mold that produces octahedra with or without preformed gaskets, as well as a pre-existing central hole, eliminating the need for post-sintering drilling or coring of castable ceramic octahedra. Our calibration results indicate that both the 10 mm and the 14 mm OEL "injection"-molded octahedra with gaskets, cast from 584 ceramic cement, have similar pressure-generating efficiency as the corresponding COMPRES octahedra which require individually glued pyrophyllite gaskets. The 10/5 and 14/8 assemblies that use molded, gasketless octahedra cast from the 584 ceramic cement and separate pyrophyllite gaskets require 25% to 30% more load in order to reach the pressures required for the I–II and II–III phase transitions of Bi, respectively, and 60% to 70% more load to reach the Bi III–V phase transition at 7.7 GPa and room temperature. The pressure-load relationship for 18/12 assemblies with pre-gasketed octahedra cast from the 584 ceramic cement is almost linear but requires 28% to 32% less load to reach the same pressure when compared with assemblies run using pre-gasketed octahedra cast from the ZrO_2-based 646 cement. High-temperature pressure calibration of the 18/12 assembly, using ZrO_2-based castable octahedra and based on the phase transition from quartz to coesite at 3.1 ± 0.1 GPa and 1000 °C, show consistent results relative to the room-temperature calibrations. We close by noting that the designs and 3D printer files for the "semi-injection" molds used in casting our ceramic octahedra are available in the public domain, with the intent that they will be utilized by other laboratories conducting high-pressure research in the multi-anvil apparatus.

Supplementary Materials: The 3D models for the two types of molds for casting 18 mm, 14 mm, and 10 mm OLE octahedra are available online at http://www.mdpi.com/2075-163X/10/1/4/s1, and can be downloaded and ready for fabrication by a 3D printer.

Author Contributions: B.C. conceived and designed the research. H.L. and B.C. designed and fabricated the molds and casted the octahedra; Y.L. and B.C. performed the room-temperature calibration experiments for 14/8 and 10/5 assemblies and analyzed the data. X.L., F.Z., R.P.R., and B.C. conducted pressure calibration experiments for the 18/12 assemblies. Y.L., B.C., and R.P.R. wrote the manuscript; All authors have read and agreed to the published version of the manuscript.

Funding: This research was funded by the Fundamental Research Funds for the Central Universities (grant number 2652017090) to Y.L., and the National Science Foundation (grant numbers EAR-1555388 and EAR-1829273) to B.C.

Acknowledgments: We acknowledge the comments and suggestions from the three anonymous reviewers, which improved the manuscript. We thank K. Chao, K. Nichols, and J. Switek for their help with the experiments.

Conflicts of Interest: The authors declare no conflict of interest.

References

1. Kawai, N.; Endo, S. The generation of ultrahigh pressures by a split sphere apparatus. *Rev. Sci. Instrum.* **1970**, *41*, 1178–1181. [CrossRef]
2. Liebermann, R.C. Multi-anvil, high pressure apparatus: A half-century of development and progress. *High Pressure Res.* **2011**, *31*, 493–532. [CrossRef]
3. Walker, D. Some simplifications to multianvil devices for high pressure experiments. *Am. Mineral.* **1990**, *75*, 1020–1028.
4. Frost, D.J.; Poe, B.T.; Trønnes, R.G.; Liebske, C.; Duba, A.; Rubie, D.C. A new large-volume multianvil system. *Phys. Earth Planet. Ineriors* **2004**, *143*, 507–514. [CrossRef]
5. Irifune, T.; Kurio, A.; Sakamoto, S.; Inoue, T.; Sumiya, H. Ultrahard polycrystalline diamond from graphite. *Nature* **2003**, *421*, 599–600. [CrossRef] [PubMed]
6. Ohtani, E.; Kagawa, N.; Shimomura, O.; Togaya, M.; Suito, K.; Onodera, A.; Sawamoto, H.; Yoneda, M.; Tanaka, S.; Utsumi, W.; et al. High-pressure generation by a multiple anvil system with sintered diamond anvils. *Rev. Sci. Instrum.* **1989**, *60*, 922. [CrossRef]
7. Ito, E.; Katsura, T.; Aizawa, Y.; Kawabe, K.; Yokoshi, S.; Nozawa, A.; Funakoshi, K. High-pressure generation in the Kawai-type apparatus equipped with sintered diamond anvils: Application to wurtzite-rocksalt transformation in GaN. In *Advances in High-Pressure Technology for Geophysical Applications*; Elsevier: Amsterdam, The Netherlands, 2005; pp. 451–460.
8. Ito, E.; Yamazaki, D.; Yoshino, T.; Fukui, H.; Zhai, S.M.; Shatzkiy, A.; Katsura, T.; Tange, Y.; Funakoshi, K. Pressure generation and investigation of the post-perovskite transformation in $MgGeO_3$ by squeezing the Kawai-cell equipped with sintered diamond anvils. *Earth Planet. Sci. Lett.* **2010**, *293*, 84–89.
9. Irifune, T.; Kunimoto, T.; Shinmei, T.; Tange, Y. High pressure generation in Kawai-type multianvil apparatus using nano-polycrystalline diamond anvils, C.R. *Geoscience* **2019**, *351*, 260–268. [CrossRef]
10. Kunimoto, T.; Irifune, T. Pressure generation to 125 GPa using a 6-8-2 type multianvil apparatus with nano-polycrystalline diamond anvils. *J. Phys. Conf. Ser.* **2010**, *215*, 012190. [CrossRef]

11. Leinenweber, K.; Mosebfelder, J.; Diedrich, T.; Soignanrd, E.; Sharp, T.G.; Tyburczy, J.A.; Wang, Y. High-pressure cells for in situ mutli-anvil experiments. *High Pressure Res.* **2006**, *26*, 283–292. [CrossRef]
12. Leinenweber, K.D.; Tyburczy, J.A.; Sharp, T.G.; Soignard, E.; Diedrich, T.; Petuskey, W.B.; Wang, Y.; Mosenfelder, J.L. Cell assemblies for reproducible multi-anvil experiments (the COMPRES assemblies). *Am. Mineral.* **2012**, *97*, 353–368. [CrossRef]
13. Walker, D. Lubrication, gasketing, and precision in multianvil experiments. *Am. Mineral.* **1991**, *76*, 1092–1100.
14. Casas, L.; Estop, E. Virtual and printed 3D models for teaching crystal symmetry and point groups. *J. Chem. Educ.* **2015**, *92*, 1338–1343. [CrossRef]
15. Casas, L. 3D-printing aids in visualizing the optical properties of crystals. *J. Appl. Cryst.* **2018**, *51*, 901–908. [CrossRef]
16. Luffman, I.; Nandi, A.; Luffman, B. Comparison of geometric and volumetric methods to a 3D solid model for measurement of Gully erosion and sediment yield. *Geosciences* **2018**, *8*, 86. [CrossRef]
17. Knibble, J.S.; Luginbühl, S.M.; Stoevelarr, R.; van der Plas, W.; van Harlingen, D.M.; Rai, N.; Steenstra, E.S.; van de Geer, R.; van Westrenen, W. Calibration of a multi-anvil high-pressure apparatus to simulate planetary interior conditions. *EPJ Tech. Instrum.* **2018**, *5*, 5. [CrossRef] [PubMed]
18. Bose, K.; Ganguly, J. Quartz-coesite transition revisited: Reversed experimental determination at 500–1200 °C and retrieved thermochemical properties. *Am. Mineral.* **1995**, *80*, 231–238. [CrossRef]
19. Decker, D.L.; Bassett, L.; Merrill, H.T.; Hall, H.T.; Barnett, J.D. High-pressure calibration: A critical review. *J. Phys. Chem. Ref. Data* **1972**, *1*, 773. [CrossRef]
20. Walker, D.; Li, J. Castable solid pressure media for multi-anvil devices. *Matter Radiat. Extrem.* **2019**. [CrossRef]

minerals

MDPI

Article

High-Pressure Sound Velocity Measurements of Liquids Using In Situ Ultrasonic Techniques in a Multianvil Apparatus

Zhicheng Jing [1,2,*], Tony Yu [3], Man Xu [2], Julien Chantel [2,4] and Yanbin Wang [3]

[1] Department of Earth and Space Sciences, Southern University of Science and Technology, Shenzhen 518055, China
[2] Department of Earth, Environmental, and Planetary Sciences, Case Western Reserve University, Cleveland, OH 44106, USA; man.xu2@case.edu (M.X.); julien.chantel@univ-lille.fr (J.C.)
[3] Center for Advanced Radiation Sources, The University of Chicago, Chicago, IL 60637, USA; tyu@cars.uchicago.edu (T.Y.); wang@cars.uchicago.edu (Y.W.)
[4] Unité Matériaux Et Transformations, Bâtiment C6, University of Lille, 59000 Lille, France
* Correspondence: jingzc@sustech.edu.cn

Received: 30 December 2019; Accepted: 29 January 2020; Published: 31 January 2020

Abstract: Sound velocity and equation of state of liquids provide important constraints on the generation, presence, and transport of silicate and metallic melts in the Earth's interior. Unlike their solid counterparts, these properties of liquids pose great technical challenges to high-pressure measurements and are poorly constrained. Here we present the technical developments that have been made at the GSECARS beamline 13-ID-D of the Advanced Photon Source for the past several years for determination of sound velocity of liquids using the ultrasonic techniques in a 1000-ton Kawai-type multianvil apparatus. Temperature of the sound velocity measurements has been extended to ~2400 K at 4 GPa and ~2000 K at 8 GPa to enable studies of liquids with very high melting temperatures, such as the silicate liquids.

Keywords: sound velocity; liquids; high pressure; ultrasonic technique; multianvil; synchrotron X-ray

1. Introduction

Liquids with a range of chemical compositions, including silicates, carbonates, and iron alloys, are present in the Earth's deep mantle. Examples of such liquids include dehydration-induced silicate melts either above [1] or below the mantle transition zone [2], carbonatite melts in the asthenosphere [3] or the mantle transition zone [4], primordial silicate melts [5] or metallic melts [6] in the deep mantle, and slab-derived metallic liquids in the lowermost mantle [7]. The presence of liquids can dramatically modify the density and seismic wave velocities of the mantle, resulting in low-velocity anomalies and high V_P/V_S ratios [8]. Thus, knowledge of sound velocity and equation of state (EOS) of liquids is fundamental in modeling the gravitational stability, migration, and seismic signature of liquids in the mantle.

Unlike solids, for which the X-ray diffraction (XRD) technique can provide easy and accurate measurements to determine the equation of state, the lack of long-range order in liquid structures makes determination of liquid density by XRD virtually impossible. Furthermore, the high mobility and reactivity of liquids at temperatures above the liquidus make it challenging to maintain the pressure and temperature conditions as well as the sample composition during experimental measurements. A number of techniques have been developed for determining the high-pressure density of liquids, including sink/float densitometry [9–11], X-ray absorption/radiography [12,13], X-ray scattering [14], and shock-wave measurements [15]. However, the density data obtained are often sparse and are likely more uncertain than that of their solid counterparts. Due to the well-known strong trade-off between the fitting parameters such as the bulk modulus K and its pressure derivative K' [16], the large errors in density data make determination of the liquid EOS uncertain. The errors would then be propagated to the sound velocities of liquids when these properties are calculated from the EOS [17]. Extrapolating laboratory results to deeper mantle conditions is thus faced with increasingly greater uncertainties.

The sound velocity of liquids can be directly measured in situ using the ultrasonic technique. If combined with a reference density, sound velocity measurements can provide tight constraints on the bulk modulus and the pressure derivative of the bulk modulus, which are critical when extrapolating an EOS to high pressures [18,19]. With recent technical developments, several groups have successfully applied the ultrasonic technique to determine the sound velocity of liquids with relatively low liquidus temperatures, specifically the Fe-light element alloying liquids including Fe-S [18,20–22], Fe-C [23], and Fe-Si [21]. Recently, we have made a series of improvements to enable high-pressure sound velocity measurements for other liquid compositions such as silicates [19] at higher temperatures. In this paper, we present details of the experimental techniques developed at the GSECARS (GeoSoilEnviro-Center for Advanced Radiation Sources) beamline 13-ID-D of the Advanced Photon Source (APS), Argonne National Laboratory, Lemont, IL, USA. We will first describe the overall experimental setup focusing on the improvements of cell assemblies, and then the experimental procedures focusing on the realization of the automated data acquisition system, and finally the data analysis focusing on an interactive data analysis procedure.

2. Experimental Setup for In Situ Sound Velocity Measurements

Figure 1 shows the experimental setup employed in the ultrasonic measurements. This setup was originally developed in [18] for studying Fe-alloying liquids and later modified by [19] for silicate liquids. Here we present some details of the setup and new improvements to the system for better performance and faster data acquisition at very high temperatures (up to ~2400 K).

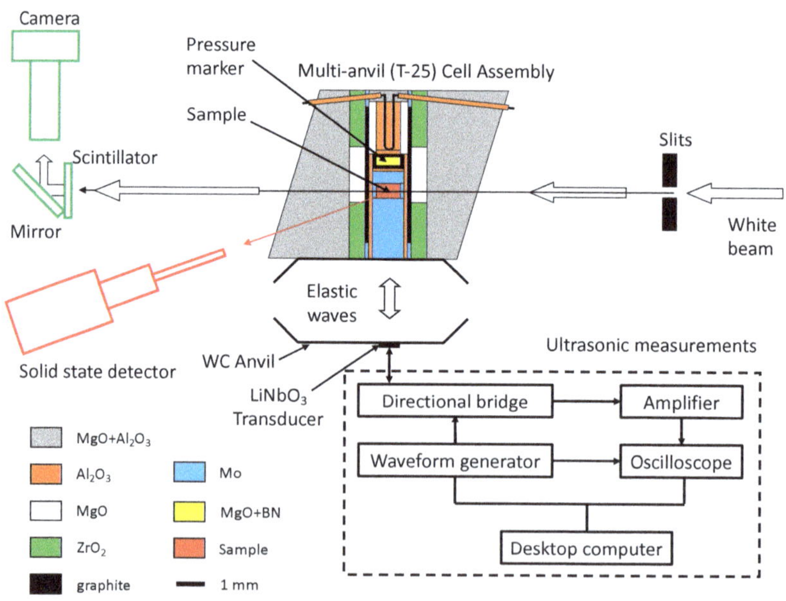

Figure 1. Experimental setup at the GSECARS beamline 13-ID-D of APS for high-pressure ultrasonic sound velocity measurements of liquids in a Kawai-type multianvil apparatus (T-25).

2.1. High-Pressure Multianvil Experiments

We employed the 1000-ton Kawai-type multianvil apparatus (T-25, Rockland Research Corporation, West Nyack, NY, USA) equipped in 13-ID-D [24] to generate high pressures up to ~8 GPa at ~2000 K and ~4 GPa at ~2400 K. Tungsten carbide (WC) second-stage anvils with 25.4 mm edge length and 8 mm truncation edge lengths (TEL) were used to compress the sample cell assemblages. An $MgO-MgAl_2O_4$ octahedron with 14 mm edge length was used as the pressure medium in each experiment [25]. A sleeve furnace made of high-purity graphite was used to provide heating of the sample, with two ZrO_2 sleeves placed outside the furnace for thermal insulation and an MgO sleeve for the passage of X-rays. The chemical inertness of the graphite furnace ensured the stability of heating during the measurements even at 2400 K. However, the pressure of this cell assembly is limited to ~8–10 GPa, beyond which the graphite heaters may transform to diamond at high temperatures and become unstable. The sample temperature was monitored using a C-type W-5%Re-W-26%Re thermocouple, inserted from one end of the cell. A mixture of MgO and h-BN (MgO:h-BN = 3:1 by weight) was used as the pressure standard. Under normal conditions, the pressure was calculated from the cell volume determined by energy dispersive X-ray diffraction, using a P-V-T equation of state of MgO [26]. However, at high temperatures above 1900 K, it was often difficult to identify MgO peaks due to the grain growth of MgO. In this situation, the pressure was estimated by extrapolating the P-T relation obtained at lower temperatures. The determined pressure was cross-checked using the melting curve of diopside and the uncertainty in pressure was estimated to be less than 0.3 GPa when the MgO peaks were identifiable, and about 0.5 GPa at ~2400 K [19].

2.2. Setup for Ultrasonic Measurements

We applied the pulse-echo overlap technique [27] for the ultrasonic measurements. The connection diagram for the ultrasonic measurement system is shown in Figure 1. Electric signals of two to three cycles of sine waves at frequencies of 20 MHz, 25 MHz, 30 MHz, 40 MHz, 50 MHz, and 60 MHz were generated by a programmable function generator (AFG 3252C, bandwidth 250 MHz, Tektronix, Inc., Beaverton, OR, USA) and were sent to a directional bridge (86205A, Keysight Technologies, Santa Rosa, CA, USA) and synched to an oscilloscope (MSO 54, bandwidth 1 GHz, Tektronix, Inc.). The directional bridge guided the input signals from the waveform generator to the piezoelectric transducer rigidly attached to the back of the bottom WC anvil and the output signals from the transducer to an amplifier (Ultrasonic Preamplifier Model 5678, bandwidth 50 kHz–40 MHz, 40 dB, Olympus Corporation, Shinjuku, Japan) and then the oscilloscope. The entire ultrasonic measurement system was physically placed next to the high-pressure device inside the 13-ID-D hutch to reduce cable lengths and hence signal loss. Both the waveform generator and the oscilloscope were connected through the Ethernet and were controlled remotely by a desktop computer outside the hutch.

Both longitudinal (P-) and shear (S-) waves were generated by converting input electric signals using a dual-mode piezoelectric transducer made of an $LiNbO_3$ single crystal (10° Y-cut). The transducer had resonant frequencies of 50 MHz for the P-wave and 30 MHz for the S-wave, but was capable of generating forced oscillations between 20 and 60 MHz with significant amplitudes. Elastic waves propagated through the WC anvil, the buffer rod (BR), the sample, and the backing plate (BP), and were reflected back at the anvil–BR, BR–sample, sample–BP, and other interfaces inside the cell assembly due to the impedance (velocity times density) contrasts between materials across various interfaces (Figure 2a). The reflected elastic waves were converted back to electrical signals by the transducer and were received, displayed, and saved by the oscilloscope. Round-trip travel times of sound waves in the sample were determined by the time difference between the reflected echoes from the BR–sample (R_1 echo in Figure 2a) and the sample–BP interfaces (R_2 echo in Figure 2a).

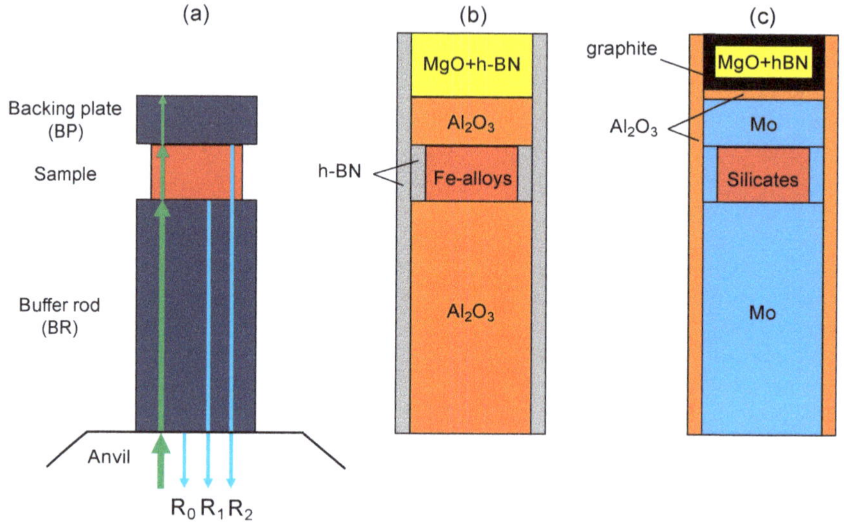

Figure 2. (**a**) Schematic diagram for the sound velocity travel paths in the cell assembly. R_0, R_1, and R_2 represent reflected sound waves from the anvil–BR, BR–sample, and sample–BP interfaces, respectively. (**b**) and (**c**) Schematic drawings of cell assemblies used for sound velocity measurements on Fe-alloying liquids and silicate liquids, respectively.

Sample lengths were determined by white-beam X-ray radiographic imaging. While in the imaging mode, the entrance slits were taken out of the X-ray path. Transmitted X-rays passing through the sample cell were converted to visible light by a YAG scintillator and received by a CMOS camera (Pointgrey Grasshopper 3, FLIR, Richmond, BC, Canada). The pixel length of the camera was calibrated by (1) taking an image of a WC sphere of known size, or (2) taking a series of sample images at different camera vertical positions that were numerically controlled by a motorized stage and then making a plot showing the relation between image position and camera position; the slope of the plot corresponded to the pixel size. Both methods gave consistent calibration results within uncertainties and averaged at 2.425 ± 0.005 μm/pixel if a four-inch-long tubing was used for the camera lens.

3. Improvements for Measurements at High Temperatures

Most of the technical challenges for sound velocity measurements of liquids under high pressures are due to the high mobility and reactivity of the liquids. The liquidus temperatures for Earth and planetary materials are quite high and usually increase with increasing pressure. For example, the melting temperature of diopside ($CaMgSi_2O_6$) is ~1670 K at ambient pressure and raises quickly to ~2177 K at 5 GPa [28]. Such high temperatures would: (1) reduce the viscosity of the liquid and increase mobility, making it difficult to confine the liquid and maintain parallelism of the sample surfaces for sound wave reflection; (2) significantly enhance chemical reactions between the liquid sample and the sample container (including BR and BP) as well as reactions between other cell parts; (3) heat up the transducer crystal, causing thermal stress or phase transitions in the crystal that may weaken the piezoelectric effect. To solve these problems, we made the following technical improvements to enable measurements at high temperatures up to 2400 K.

3.1. Sample Preparations

We are interested in two broad types of liquids: Fe-light element alloying liquids and silicate liquids, relevant to planetary molten cores and mantle melts, respectively. Ultrasonic measurements require that all the interfaces in the cell assembly are perpendicular to the incident waves and hence are parallel to each other, so that reflected waves would be guided directly back to the transducer to maximize the reflected acoustic energy and minimize deviations of the actual lengths of wave path from sample thickness. The shape of a liquid sample confined at high pressures depends highly on the shape and strength of the starting solid sample during the very early stage of compression. If the starting sample is soft or not in perfect shape, the nonsymmetrical deviatoric stress during compression would deform the sample nonuniformly and create tilted surfaces that would significantly deteriorate the signal-to-noise ratio of reflected waves. It is therefore critical to prepare a starting sample that has a relatively high strength and a homogeneous composition that is the same as the target liquid composition. For Fe-alloy compositions, this can be achieved by hot-pressing powder mixtures of Fe and Fe-light element compounds with appropriate ratios in a large-volume device such as a piston-cylinder apparatus at a temperature ~100 K below the solidus temperature of the system. For silicate compositions, depending on the viscosity of the corresponding liquids, it is best to premelt the sample powder mixture at ambient conditions and quench it into a glass. Sometimes multiple cycles of melting and quenching are required to remove bubbles to make a transparent and homogeneous glass. The hot-pressed samples or quenched glasses need to be machined into a cylindrical disk of 1.6–2 mm outer diameter (OD) and ~1–1.2 mm thickness to fit into the cell assembly. Among the several techniques we tested, the best result was achieved by machining the sample piece using a computer numerically controlled (CNC) milling machine (MDX 540S, Roland DGA Corporation, Irvine, CA, USA). Both end surfaces of the sample disks were then carefully polished down to a roughness of 1 μm using diamond paste to achieve optimal mechanical coupling at the BR–sample or sample–BP interfaces.

3.2. Sample Environments for High Temperatures

Two different cell assemblies have been developed to accommodate both Fe-alloy liquids and silicate liquids, separately, aiming at minimizing chemical reactions between the liquid sample and the sample container as well as reactions between other cell parts. The key to a successful cell assembly is to find a suitable BR and BP material that: (1) has a high melting point and does not chemically react with the liquid sample; (2) has a significantly different acoustic impedance and a significantly different X-ray absorption coefficient than the liquid sample, so that the sample can be seen from both ultrasonic measurements and X-ray imaging; (3) is strong enough so that the BR–sample and sample–BP surfaces remain relatively flat and parallel.

Figure 2b shows the cell assembly developed for determining the sound velocity of Fe-alloy liquids. In this case, densified Al_2O_3 is an ideal BR/BP material that satisfies all the requirements listed above. Because of the hardness of Al_2O_3, it is best to use a soft material that has relatively small grain size to help seal the liquid sample. Here we chose an h-BN sleeve sandwiched between BR and BP to confine the sample, and another long h-BN sleeve to protect the graphite furnace from being crushed by the sharp edges of the hard Al_2O_3 parts. This cell assembly has been applied to ultrasonic measurements in Fe-S [18] and Fe-P [29] liquids up to 8 GPa and 2000 K. Chemical analysis using a scanning electron microscope (SEM) from our previous study [29] showed essentially no chemical reactions between these liquids, the h-BN sleeve, and the Al_2O_3 parts. However, such analyses should always be performed when this assembly is applied as potential chemical reactions between h-BN, Al_2O_3, and Fe-alloys are possible depending on experimental conditions and melt compositions [22].

Figure 2c shows the cell assembly developed for silicate liquids. Molybdenum (Mo) BR, BP, and sleeve-shaped sample container were used for their low reactivity and high impedance and absorption contrasts with the sample. We have observed, through many tests, chemical reactions between Mo and many materials including graphite, h-BN, and MgO, but not Al_2O_3, at temperatures higher than ~2000 K. We therefore used a crushable Al_2O_3 sleeve and an Al_2O_3 disk to insulate the Mo parts from the graphite heater and MgO in the pressure standard. An additional layer of graphite was used to prevent the spinel-forming reaction between MgO and Al_2O_3. This cell assembly has been applied to determining sound velocity of diopside liquid up to ~4 GPa and ~2400 K [19]. For this Fe-free system, limited chemical reaction between the melt and Mo was observed in the sample even at 2400 K, based on X-ray imaging and the SEM analysis of three recovered samples (see supporting information in [19]). Among these samples, one contained 1.5 wt % MoO_2 in the quench product and the other two contained essentially no Mo. Reactivity of Mo with Fe-bearing silicate melts and the effect of the Mo-MoO_2 system as an oxygen fugacity buffer on the oxidation state of Fe-components in Fe-bearing melts remain to be tested in the future.

The temperature gradients in both cell assemblies shown in Figure 2b,c were determined at relevant pressure and temperature conditions in offline calibration experiments in a Walker-type multianvil apparatus (Rockland Research Corporation) at Case Western Reserve University, using the spinel layer growth kinetics method described in the studies of [30] and [31]. The calibration experiments and results for the temperature distributions in the assemblies are detailed in the supplementary materials of [29] and [19], respectively, for the cells using Al_2O_3 and Mo BR/BP. For the cell with Al_2O_3 BR/BP (Figure 2b), the temperature of the sample was about 60 K higher than the thermocouple reading, with a temperature variation of about 100 K across the sample. The real temperature gradient in the sample for this assembly, in fact, should be smaller, considering the high thermal conductivity of the metallic sample. For the cell with Mo BR/BP (Figure 2c), on the other hand, the temperature of the sample was about ~160–200 K lower than the thermocouple reading owing to the high thermal conductivity of the Mo BR. There existed a ~100 K temperature difference across the sample in this cell under high pressure and temperature conditions. The highest sample temperature we reached for velocity measurements of liquids was ~2400 K at ~4 GPa using Mo as BR/BP, with a thermocouple reading of 2573 K.

3.3. Protecting the Transducer from High Temperature

The effect of high temperature on the piezoelectric transducer crystal was successfully reduced by the following cooling strategies: (1) A tubing was inserted into the press to guide cooling air directly to the transducer crystal to prevent overheating; (2) a cooling nozzle was used to blow focused cooling air to the cell assembly through gaps between first-stage anvils; and (3) a high-power exhaust fan was used to cool down the entire Kawai module.

3.4. Fast Data Acquisition

Reducing data collection time is also important for measurements at very high temperatures. For each pressure–temperature condition, we need to take ultrasonic measurements at six different frequencies between 20 and 60 MHz to make sure that there is no frequency dependence of the velocity and the sample is relaxed. However, each measurement at a given frequency requires ~60–90 s to complete if the waveform generator and oscilloscope are operated manually. In total, the whole set of measurements at six frequencies plus X-ray diffraction and imaging would take about 12–15 min, and often require two people working simultaneously to operate X-ray diffraction, imaging, and ultrasonic measurements simultaneously. We have recently developed an automated data collection system to minimize human intervention. In this system, both the waveform generator and the oscilloscope are connected through the Ethernet, and are controlled by a desktop computer through TekVISA, the Tektronix implementation of the standard VISA communication protocol. A python procedure (see Supplementary Materials for the python script) using the PyVISA library has been developed to automate the entire ultrasonic measurement process. Figure 3 shows the flow chart for the python procedure. This procedure shortens the operation time of ultrasonic measurements to less than 6–8 min, and reduces human operation errors at the same time. Potentially, the ultrasonic data collection time may be further shortened by using the transfer function method described in [32], in which the response of the system (transfer function) to an input signal of any frequency can be obtained by determining the system response to a single input sinc function ($\sin x/x$). So far, we have not tested the transfer function method at Beamline 13-ID-D of APS as the current function generator (Tektronix AFG 3252C) available cannot be programmed to generate a sinc function, but this could be a future development.

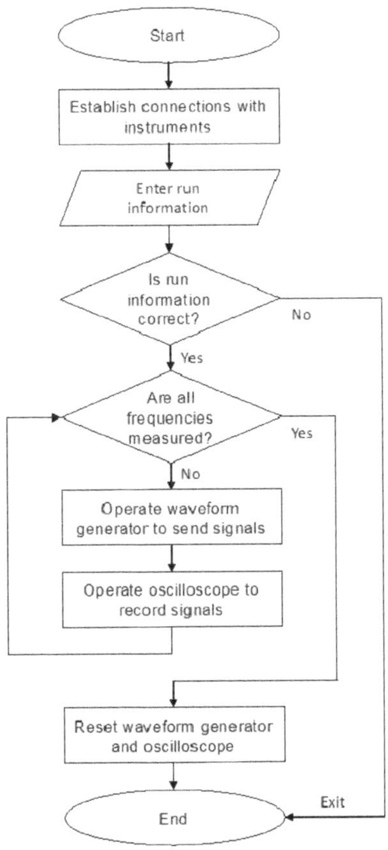

Figure 3. Flowchart for the automated data acquisition program used for ultrasonic sound velocity measurements.

4. Experimental Procedure for Velocity Measurements at High Pressures

In each experimental run, the sample is initially compressed to a target load by increasing the oil pressure in the hydraulic ram. Once the target load is reached, the temperature of the sample is raised by increasing the heating power output at a rate of about 100–200 K/min to a temperature of a few hundred degrees below the solidus temperature of the sample, say 1073 K for Fe-alloys and 1273 K for silicates, at which the sample is annealed for 30–60 min to relax the deviatoric stress in the cell assembly. Manual heating (in contrast to feedback-controlled automated heating) and current control heating mode are used to prevent super heating from potential breaking of the thermocouple or a sudden drop of heater resistance caused by liquid sample leakage. After taking the XRD, X-ray imaging, and ultrasonic measurements, the sample is heated up by 100 K and stopped again for another round of measurements. This heating and measuring cycle continues in temperature intervals of 100 K to a final temperature at a few hundred degrees above the liquidus temperature of the sample, before decreasing to ambient temperature to complete this heating cycle at the first target load.

At each temperature condition, the ultrasonic measurements can be started first by running the automated data acquisition procedure. Then, a sample X-ray image is taken before moving the press and entrance slits to the positions for X-ray diffraction. It takes about 30 s to move all the slits and press motors, 100 s for the XRD of the sample, and 200 s for the XRD of the pressure standard. The press is then moved back and slits out to the imaging position for another image of the sample. The ultrasonic measurements and the X-ray diffraction and imaging can be operated completely independently and in parallel, and therefore the overall time required at each temperature condition is mostly determined by the ultrasonic data collection, which is about 6–8 min.

The complete melting of the sample can be clearly identified from the ultrasonic signals. The P-wave signal from the sample (R_2 echo in Figure 2a) is significantly weakened upon initial melting of the sample due to the scattering/attenuating effect of the partial melts. Immediately after the complete melting of the sample, the amplitude of the R_2 echo increases dramatically to indicate the disappearance of the scattered crystals in the melt. In addition, the S-wave signal of the sample disappears completely once the sample is fully molten. The detailed evolution of ultrasonic signals for an Fe-alloy (Fe-5 wt %P) and a silicate sample ($CaMgSi_2O_6$) across the entire temperature range covering the solidus and liquidus are described in [29] and [19], respectively. For the case of Fe-alloys, the melting of the sample can also be cross-checked by XRD measurements of the sample [29].

Once a heating cycle is completed, if the sample remains in good shape without obvious chemical reactions and leakage, the sample can be compressed more for another one or more heating and measuring cycles. Three or four heating cycles at different loads are typical for an Fe-alloy sample, and two or three heating cycles are possible for silicate samples. The sample is usually quenched by shutting off the heating power at the highest temperature reached in the final heating cycle. The quenched product can then be sectioned and polished for chemical analysis using, for example, SEM.

5. Data Analysis

5.1. Travel Time Analysis

As an example, Figure 4a shows the signals recorded by the oscilloscope for the diopside liquid at ~3.8 GPa and ~2400 K [19]. The echoes reflected at the anvil–BR, BR–sample, and sample–BP interfaces are represented by R_0, R_1, and R_2, respectively. We have developed an interactive program using the Igor Pro package (WaveMetrics, Inc., Portland, OR, USA). The source code of the Igor procedures and the user's manual for the procedures are also provided in the Supplementary Materials. Here we describe the basic steps performed in the program.

(1) The raw signal recording (red curve in Figure 4a) was first corrected to remove the offset in amplitude such that the average amplitude was set to zero.

(2) The raw signals were then Fourier transformed using the Fast Fourier Transform (FFT) algorithm to obtain its frequency characteristics. Because the input signal for this particular measurement was a sine wave of 20 MHz, we see that a large fraction of energy was centered at about 20 MHz. However, there were some low-frequency noises below 10 MHz (and sometimes some high-frequency noises above 70 MHz) caused by the environment as those seen in Figure 2b. Some of these noises can be traced to the high AC currents from the stepper motors that drive the 1000-ton press to the desired position to collect radiographic image or diffraction data. To remove these low- and high-frequency noises, a band-pass filter was applied to remove unwanted information below 10 MHz and above 70 MHz. The resulting signals in the frequency domain (blue curve in Figure 2b) were then back transformed to the time domain. The resulting signals with noises removed are shown as the blue curve in Figure 2a. Further analyses of travel time were based on the corrected signals.

(3) The BR–sample echo (R_1) was selected interactively using cursors from the beginning to the end of the echo. We then took a cross-correlation between the R_1 echo and the R_2 echo to determine the time lag that maximized the absolute value of the correlation. This time lag was the round-trip travel time for the sound wave to propagate through the liquid sample. It should be noted that for the sample cells described in this study, there was a 180° phase shift between the R_1 and R_2 echoes. This was because both BR and BP were made of the same material and depending on the impedance contrast between the sample and BR/BP, either R_1 or R_2 should have a negative reflection coefficient (i.e., a 180° phase shift from the incident signal), resulting in a 180° phase shift between R_1 and R_2. Therefore, the time lag at which the correlation is minimized (a negative value) should be chosen.

Figure 2c shows that the R_1 echo, when corrected for the time lag and flipped for the 180° phase shift, correlated with the R_2 echo very well.95.

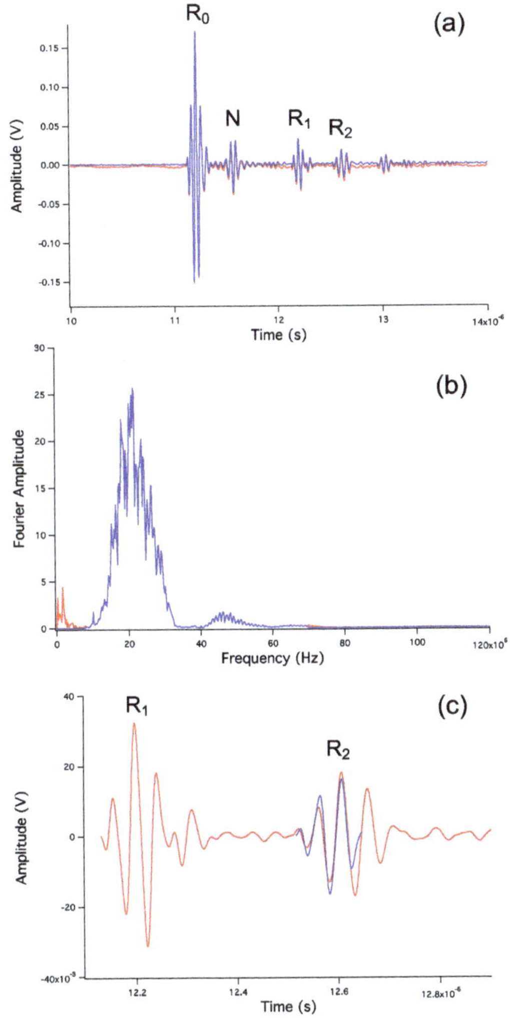

Figure 4. (**a**) P-wave ultrasonic signals of diopside ($CaMgSi_2O_6$) liquid at 3.8 GPa and ~2400 K. R_0, R_1, and R_2 represent reflected sound waves from the anvil–BR, BR–sample, and sample–BP interfaces, respectively. N represents an artificial echo that resulted from the unmatched impedance between the coaxial cable. The time delay between R_0 and N does not depend on the changes that happen in the cell assembly, and is small enough that N does not interfere with R_1 and R_2. The raw data are shown in red. The blue curve shows the data corrected for offset in voltage and band-pass filtered to remove low- and high-frequency noises. (**b**) The Fourier transformed signals in the frequency domain. The red curve shows the entire transformed waveform and the blue curve shows the waveform after removing the low- and high-frequency noises. (**c**) The segment of P-wave signals showing the R_1 and R_2 echoes (red). The blue curve shows the shifted and flipped R_1 echo that results in the maximum correlation between the flipped R_1 and R_2. This time shift is the round-trip travel time for sound waves propagating through the liquid sample.

5.2. Sample Length Analysis

Figure 5a shows a raw sample image of diopside liquid at 3.8 GPa and ~2400 K. The sample was sandwiched by Mo BR and BP, and was enclosed by a sleeve also made of Mo. The wall of the Mo sleeve was thin enough for white X-ray beams to go through, hence creating an absorption contrast between the sample and the BR/BP rods. Because of the plastic deformation of Mo at very high temperatures, the Mo sleeve and the sample–BP interface were somewhat deformed close to the contact area between the sleeve and BP. This deformation is shown as a shaded area right below the sample–BP interface and has a brightness between the sample and the BP. Such a deformation, however, did not affect the sample length measurements much as most of the interface remained flat and identifiable in the image. The image was then processed by taking derivatives to find edges (Figure 5b) using the software package ImageJ [33]. Finally, the intensity of the image was integrated along the horizontal direction to obtain an intensity profile as a function of the vertical pixel position. The details of the sample length determination procedure are provided in the Ultrasonic Data Analysis Manual in the Supplementary Materials. As shown in Figure 5c, the BR–sample and sample–BP interfaces can be identified as the sharp peaks in the intensity profile. The pixel distance of the two interfaces multiplied by the length/pixel ratio from the camera calibration gives the sample lengths. The sound velocity of the liquid can then be calculated from the travel time and sample length. Tilting of the sample/BR and sample/BP interfaces can sometimes be identified from the sample images by observing the orientation and sharpness of the interfaces, but for most cases, it is less than 3°. The uncertainty in sample length measurements caused by such a small tiling angle would be only 0.14% (that is $(1 - \cos 3°)$) and is negligible. Severe tilting (larger than 10°) can be caused by compressing a very soft/porous sample between hard BR and BP. When that happened, the sound velocity data were discarded although the uncertainty in sample length is still only 1.5% at a 10° tilting angle.

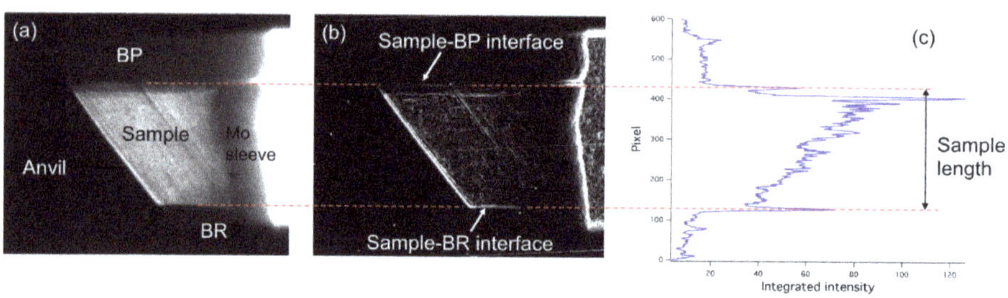

Figure 5. (**a**) Raw image of the liquid diopside sample and cell assembly at 3.8 GPa and ~2400 K. The image is partially saturated on the rightmost part to show the details of the sample–BP interface. (**b**) Processed image to find edges in the image by taking derivatives in the ImagJ software. (**c**) Integrated intensity of the processed image along the horizontal direction as a function of the vertical position. The BR–sample and sample–BP interfaces are identified as sharp peaks in the plot. Red dashed lines indicate the positions of the BR-sample and sample–BP interfaces.

5.3. Sound Velocity at High Pressures

The uncertainties in sound velocities are propagated from the uncertainties in travel time and sample length analyses. Because of the high sampling rate of the oscilloscope (6.25 GS/s), the uncertainty in travel time measurements was less than 0.2 ns, or only about 0.16%. The uncertainties in sample length measurements therefore dominated the overall uncertainty. Depending on how much the BR-sample and sample–BP interfaces have deformed, the uncertainty ranges between 0.5% and 2% [18,19].

The high-pressure velocity data can be combined with any number of density data points (at least one) to fit an EOS (e.g., the Birch–Murnaghan EOS) for a liquid. Thermodynamic properties including the thermal expansion coefficient and specific heat are required to relate the sound velocity and the adiabatic bulk modulus to the isothermal bulk modulus of the liquid. This type of fitting has been performed in several studies [18,19,21] and demonstrated tight constraints on the bulk modulus and the pressure derivative of the bulk modulus of the liquid. For example, Xu et al. [19] constrained the bulk modulus and the pressure derivative of the bulk modulus of diopside liquid to be 23.8 ± 0.4 GPa and 7.5 ± 0.5, respectively.

6. Conclusions

We have made a series of technical improvements to the in situ ultrasonic velocity measurements in the multianvil press at GSECARS beamline 13-ID-D at the Advanced Photon Source. With carefully prepared starting samples, suitable cell assemblies, sufficient cooling, and an automated data acquisition system, the current setup enables velocity measurements of Fe-alloying liquids up to ~2000 K at 8 GPa and silicate liquids up to ~2400 K at 4 GPa, and can be readily extended to measurements on other solid or liquid materials of geological or industrial interest. An interactive program has also been developed for the travel time data analysis. The experimental setup and software developed in this study have been made available at the GSECARS facility to the high-pressure mineral physics and materials science communities.

Supplementary Materials: The following are available online at http://www.mdpi.com/2075-163X/10/2/126/s1. File 1: Ultrasonic_Data_Acquisition.py; File 2: Travel_Time_In_Phase_2.ipf; File 3: Travel_Time_Out_Phase_2.ipf; File 4: Button.ipf; File 5: Ultrasonic_Data_Analysis_Manual.pdf.

Author Contributions: Conceptualization, Z.J., Y.W.; Methodology, Z.J., T.Y., M.X., J.C., Y.W.; Software, Z.J., T.Y.; Validation, Z.J.; Investigation, Z.J., M.X., J.C.; Resources, Z.J., T.Y., Y.W.; Data Curation, Z.J., T.Y., M.X.; Writing—Original Draft Preparation, Z.J.; Writing—Review & Editing, all authors; Visualization, Z.J.; Project Administration, Z.J., Y.W.; Funding Acquisition, Z.J., Y.W. All authors have read and agreed to the published version of the manuscript.

Funding: This work was partially supported by the National Aeronautics and Space Administration (NNX14AN01G), the National Science Foundation (EAR 1619964 and 1620548), and the National Natural Science Foundation of China (41974098). Experiments were performed at GeoSoilEnviroCARS (Sector 13), Advanced Photon Source (APS), Argonne National Laboratory. GeoSoilEnviroCARS is supported by the National Science Foundation - Earth Sciences (EAR-1128799) and Department of Energy-GeoSciences (DE-FG02-94ER14466). This research used resources of the Advanced Photon Source, a U.S. Department of Energy (DOE) Office of Science User Facility operated for the DOE Office of Science by Argonne National Laboratory under Contract No. DE-AC02-06CH11357.

Acknowledgments: We thank Yoshio Kono for his assistance with the ultrasonic setup and data analysis software.

Conflicts of Interest: The authors declare no conflicts of interest.

References

1. Revenaugh, J.; Sipkin, S.A. Seismic evidence for silicate melt atop the 410 km mantle discontinuity. *Nature* **1994**, *369*, 474–476. [CrossRef]
2. Schmandt, B.; Jacobsen, S.D.; Becker, T.W.; Liu, Z.; Dueker, K.G. Dehydration melting at the top of the lower mantle. *Science* **2014**, *344*, 1265–1268. [CrossRef] [PubMed]
3. Gaillard, F.; Malki, M.; Iacono-Marziano, G.; Pichavant, M.; Scaillet, B. Carbonatite Melts and Electrical Conductivity in the Asthenosphere. *Science* **2008**, *322*, 1363–1365. [CrossRef] [PubMed]
4. Rohrbach, A.; Schmidt, M.W. Redox freezing and melting in the Earth's deep mantle resulting from carbon–iron redox coupling. *Nature* **2011**, *472*, 209–212. [CrossRef]
5. Labrosse, S.; Hernlund, J.W.; Coltice, N. A crystallizing dense magma ocean at the base of the Earth's mantle. *Nature* **2007**, *450*, 866–869. [CrossRef]
6. Zhang, Z.; Dorfman, S.M.; Labidi, J.; Zhang, S.; Li, M.; Manga, M.; Stixrude, L.; McDonough, W.F.; Williams, Q. Primordial metallic melt in the deep mantle. *Geophys. Res. Lett.* **2016**, *43*, 3693–3699. [CrossRef]
7. Liu, J.; Li, J.; Hrubiak, R.; Smith, J.S. Origins of ultralow velocity zones through slab-derived metallic melt. *Proc. Natl. Acad. Sci. USA* **2016**, *113*, 5547–5551. [CrossRef]
8. Chantel, J.; Manthilake, G.; Andrault, D.; Novella, D.; Yu, T.; Wang, Y. Experimental evidence supports mantle partial melting in the asthenosphere. *Sci. Adv.* **2016**, *2*. [CrossRef]

9. Agee, C.B.; Walker, D. Olivin flotation in mantle melt. *Earth Planet. Sci. Lett.* **1993**, *114*, 315–324. [CrossRef]
10. Balog, P.S.; Secco, R.A.; Rubie, D.C.; Frost, D.J. Equation of state of liquid Fe-10 wt % S: Implications for the metallic cores of planetary bodies. *J. Geophys. Res.* **2003**, *108*, 2124. [CrossRef]
11. Jing, Z.; Karato, S. Effect of H_2O on the density of silicate melts at high pressures: Static experiments and the application of a modified hard-sphere model of equation of state. *Geochim. Cosmochim. Acta* **2012**, *85*, 357–372. [CrossRef]
12. Sakamaki, T.; Ohtani, E.; Urakawa, S.; Suzuki, A.; Katayama, Y. Density of dry peridotite magma at high pressure using an X-ray absorption method. *Am. Mineral.* **2010**, *95*, 144–147. [CrossRef]
13. Nishida, K.; Ohtani, E.; Urakawa, S.; Suzuki, A.; Sakamaki, T.; Terasaki, H.; Katayama, Y. Density measurement of liquid FeS at high pressures using synchrotron X-ray absorption. *Am. Mineral.* **2011**, *96*, 864–868. [CrossRef]
14. Morard, G.; Siebert, J.; Andrault, D.; Guignot, N.; Garbarino, G.; Guyot, F.; Antonangeli, D. The Earth's core composition from high pressure density measurements of liquid iron alloys. *Earth Planet. Sci. Lett.* **2013**, *373*, 169–178. [CrossRef]
15. Asimow, P.D.; Ahrens, T.J. Shock compression of liquid silicates to 125 GPa: The anorthite-diopside join. *J. Geophys. Res.* **2010**, *115*, B10209. [CrossRef]
16. Jing, Z.; Karato, S. Compositional effect on the pressure derivatives of bulk modulus of silicate melts. *Earth Planet. Sci. Lett.* **2008**, *272*, 429–436. [CrossRef]
17. Weber, R.C.; Lin, P.-Y.; Garnero, E.J.; Williams, Q.; Lognonné, P. Seismic detection of the lunar core. *Science* **2011**, *331*, 309–312. [CrossRef]
18. Jing, Z.; Wang, Y.; Kono, Y.; Yu, T.; Sakamaki, T.; Park, C.; Rivers, M.L.; Sutton, S.R.; Shen, G. Sound velocity of Fe–S liquids at high pressure: Implications for the Moon's molten outer core. *Earth Planet. Sci. Lett.* **2014**, *396*, 78–87. [CrossRef]
19. Xu, M.; Jing, Z.; Chantel, J.; Jiang, P.; Yu, T.; Wang, Y. Ultrasonic Velocity of Diopside Liquid at High Pressure and Temperature: Constraints on Velocity Reduction in the Upper Mantle Due to Partial Melts. *J. Geophys. Res. Solid Earth* **2018**, *123*, 8676–8690. [CrossRef]
20. Nishida, K.; Kono, Y.; Terasaki, H.; Takahashi, S.; Ishii, M.; Shimoyama, Y.; Higo, Y.; Funakoshi, K.-I.; Irifune, T.; Ohtani, E. Sound velocity measurements in liquid Fe-S at high pressure: Implications for Earth's and lunar cores. *Earth Planet. Sci. Lett.* **2013**, *262*, 182–186. [CrossRef]
21. Terasaki, H.; Rivoldini, A.; Shimoyama, Y.; Nishida, K.; Urakawa, S.; Maki, M.; Kurokawa, F.; Takubo, Y.; Shibazaki, Y.; Sakamaki, T.; et al. Pressure and Composition Effects on Sound Velocity and Density of Core-Forming Liquids: Implication to Core Compositions of Terrestrial Planets. *J. Geophys. Res. Planets* **2019**, *124*. [CrossRef]
22. Nishida, K.; Suzuki, A.; Terasaki, H.; Shibazaki, Y.; Higo, Y.; Kuwabara, S.; Shimoyama, Y.; Sakurai, M.; Ushioda, M.; Takahashi, E.; et al. Towards a consensus on the pressure and composition dependence of sound velocity in the liquid Fe–S system. *Phys. Earth Planet. Int.* **2016**, *257*, 230–239. [CrossRef]
23. Kuwabara, S.; Terasaki, H.; Nishida, K.; Shimoyama, Y.; Takubo, Y.; Higo, Y.; Shibazaki, Y.; Urakawa, S.; Uesugi, K.; Takeuchi, A.; et al. Sound velocity and elastic properties of Fe–Ni and Fe–Ni–C liquids at high pressure. *Phys. Chem. Miner.* **2015**, *43*, 229–236. [CrossRef]
24. Wang, Y.; Rivers, M.L.; Sutton, S.R.; Nishiyama, N.; Uchida, T.; Sanehira, T. The large-volume high-pressure facility at GSECARS: A "Swiss-army-knife" approach to synchrotron-based experimental studies. *Phys. Earth Planet. Int.* **2009**, *174*, 270–281. [CrossRef]
25. Leinenweber, K.D.; Tyburczy, J.A.; Sharp, T.G.; Soignard, E.; Diedrich, T.; Petuskey, W.B.; Wang, Y.; Mosenfelder, J.L. Cell assemblies for reproducible multi-anvil experiments (the COMPRES assemblies). *Am. Mineral.* **2012**, *97*, 353–368. [CrossRef]
26. Tange, Y.; Nishihara, Y.; Tsuchiya, T. Unified analyses for P-V-T equation of state of MgO: A solution for pressure-scale problems in high P-T experiments. *J. Geophys. Res.* **2009**, *114*, B03208. [CrossRef]
27. Kono, Y.; Park, C.; Sakamaki, T.; Kenny-Benson, C.; Shen, G.; Wang, Y. Simultaneous structure and elastic wave velocity measurement of SiO_2 glass at high pressures and high temperatures in a Paris-Edinburgh cell. *Rev. Sci. Instrum.* **2012**, *83*, 033905. [CrossRef]
28. Williams, D.W.; Kennedy, G.C. Melting curve of diopside to 50 kilobars. *J. Geophys. Res.* **1969**, *74*, 4359–4366. [CrossRef]

29. Chantel, J.; Jing, Z.; Xu, M.; Yu, T.; Wang, Y. Pressure Dependence of the Liquidus and Solidus Temperatures in the Fe-P Binary System Determined by In Situ Ultrasonics: Implications to the Solidification of Fe-P Liquids in Planetary Cores. *J. Geophys. Res. Planets* **2018**, *123*, 1113–1124. [CrossRef]
30. van Westrenen, W.; Van Orman, J.A.; Watson, H.; Fei, Y.; Watson, E.B. Assessment of temperature gradients in multianvil assemblies using spinel layer growth kinetics. *Geochem. Geophys. Geosyst.* **2003**, *4*, 1036. [CrossRef]
31. Watson, E.B.; Wark, D.A.; Price, J.D.; Van Orman, J.A. Mapping the thermal structure of solid-media pressure assemblies. *Contrib. Mineral. Petrol.* **2002**, *142*, 640–652. [CrossRef]
32. Li, B.; Chen, K.; Kung, J.; Liebermann, R.C.; Weidner, D.J. Sound velocity measurement using transfer function method. *J. Phys. Condens. Matter* **2002**, *14*, 11337–11342. [CrossRef]
33. Schneider, C.A.; Rasband, W.S.; Eliceiri, K.W. NIH Image to ImageJ: 25 years of image analysis. *Nat. Methods* **2012**, *9*, 671–675. [CrossRef] [PubMed]

MDPI
St. Alban-Anlage 66
4052 Basel
Switzerland
Tel. +41 61 683 77 34
Fax +41 61 302 89 18
www.mdpi.com

Minerals Editorial Office
E-mail: minerals@mdpi.com
www.mdpi.com/journal/minerals

www.ingramcontent.com/pod-product-compliance
Lightning Source LLC
Chambersburg PA
CBHW061450220926
51850CB00004B/55
9783039365418